WORLD ORGANISATION FOR ANIMAL HEALTH
Protecting animals, preserving our future

家畜疫病病原生物学国家重点实验室资助出版

# 骆驼传染病
# Camelid Infectious Disorders

Ulrich Wernery，Jörg Kinne
Rolf Karl Schuster 编著
殷 宏 贾万忠 关贵全 储岳峰 独军政 主译

中国农业科学技术出版社

图书在版编目（CIP）数据

骆驼传染病 /（沙特）沃纳瑞，（沙特）凯恩，（沙特）舒斯特著；殷宏等译.
--北京：中国农业科学技术出版社，2016.12

ISBN 978-7-5116-2434-5

Ⅰ. ①骆… Ⅱ. ①沃… ②凯… ③舒… ④殷… Ⅲ.
① 骆驼病－传染病－防治 Ⅳ. ①S858.24

中国版本图书馆 CIP 数据核字（2015）第 317011 号



**责任编辑** 张国锋
**责任校对** 贾海霞
**出 版 者** 中国农业科学技术出版社
北京市中关村南大街12号 邮编：100081
**电 话** （010）8210 6636（编辑室） （010）82109702（发行部）
（010）8210 9709（读者服务部）
**传 真** （010）8210 6631
**网 址** http://www.castp.cn
**经 销 者** 各地新华书店
**印 刷 者** 北京卡乐富印刷有限公司
**开 本** 880 mm × 1230 mm 1/16
**印 张** 34.5
**字 数** 1100 千字
**版 次** 2016年12月第1版 2016年12月第1次印刷
**定 价** 298.00 元

# 骆驼传染病

# Camelid Infectious Disorders

Ulrich Wernery

Jörg Kinne

Rolf Karl Schuster

2014

# 目录

# 序

很荣幸为Ulrich Wernery、Jörg Kinne和Rolf Karl Schuster三位博士编著的《骆驼传染病》一书作序。

骆驼是家养或野生的哺乳动物，与新大陆和旧大陆多个文明有密切的关系。骆驼在驯化家养后用于驮运货物、骑乘，也可用于生产骆驼肉、骆驼奶和纺织品。虽然新旧大陆使它们远隔千里万里，但共有的生理特征使之与其他哺乳动物易于区别。在本书中，对新大陆骆驼和旧大陆骆驼没有区别对待。

一些生理特征（包括适应性体液免疫应答）是骆驼对恶劣环境适应的见证，在本书中多次提到这些不同寻常的特征。从科技角度来看，人们对骆驼越来越感兴趣，有关骆驼的疾病防控知识在近年快速增加，尤其是骆驼传染病学和寄生虫病学的相关知识。有关骆驼传染病和寄生虫病的知识最早见于由Curasson（1947）撰写的非常著名的《骆驼疾病》一书，该书主要描述了单峰驼的疾病，本书中多次引用该著作的内容。

骆驼对多种能感染家养和野生动物的病原也非常敏感，且通常没有临床症状。本书插图丰富，对病原对骆驼的感染和侵染都有详尽的描述，且配有大量的详实资料。

骆驼除保留了自古以来就有的为人们提供基本需求的功能外，还获得了新的功能，如体育功能和娱乐功能。在北美和欧洲有多个与新功能相关的协会。

本书有很高的参考价值，也是一本名副其实的骆驼传染病学和寄生虫病学的字典。

世界动物卫生组织祝贺Ulrich Wernery、Jörg Kinne和Rolf Karl Schuster 3位博士及其同行编著了一本很全面的书籍。我本人也对他们表示衷心的谢忱。

世界动物卫生组织总干事

BERNARD VALLAT 瓦拉

# 引

Ulrich Wernery博士等人编著的有关骆驼传染病学的前两个出版物为全世界临床兽医、诊断人员、骆驼主和养驼业的从业人员提供了非常重要的参考资料。这是两本专门描述骆驼传染病和寄生虫病的专著，而同类其他书籍所涉及的是骆驼所有疾病。

秉承这一传统，他们撰写了《骆驼传染病》这本新书，并由世界动物卫生组织（OIE）在巴黎出版并发行。此举为全世界政府部门的图书馆提供了新的资源，也使得骆驼饲养地区的人们进一步认识到养驼业的重要性。Ulrich Wernery、JörgKinne 和 Rolf Karl Schuster 博士这3位本书作者，是微生物学、病理学和寄生虫病学的知名专家，经验丰富、学术造诣深。

位于阿拉伯联合酋长国迪拜市的中央兽医研究实验室是全球知名的兽医诊断实验室，全体员工致力于骆驼、马、猎鹰和中东地区其他动物疫病诊断技术的改进和提高。中央兽医研究实验室也是 OIE 的马鼻疽和骆驼痘参考实验室的依托单位。除诊断工作外，实验室还研究和生产传染病疫苗和抗蛇毒血清。最著名的成果是研发了控制骆驼痘的高效疫苗，此病曾是中东和亚洲骆驼的瘟疫。他们的工作丰富了相关知识，对养驼业的发展做出了贡献。

新大陆地区驼类的疾病与其他地方骆驼的疾病非常相近，在书中都进行了讨论。书中对新大陆地区的驼类特有的疾病也专门描述。除诊断工作之外，中央兽医研究实验室在骆驼的克隆工作中世界领先。该实验室也支撑着阿拉伯联合酋长国奶驼业的发展，驼奶出口到欧洲国家。

在阿拉伯联合酋长国，能为知名的科学家提供从事骆驼疾病研究的机会，是一件非常了不起的事。尊敬的 Sheikh Mohanmmed bin Rashid Al Maktoum 先生远见卓识，建立了中央兽医研究实验室，他长期支持赛驼业，关注骆驼疫病研究，他应当受到全世界骆驼主的尊敬。骆驼主、骆驼训练师、兽医和多学科的科学家都对其善举表示感激。

本书有多个章节（细菌病、病毒病、真菌病、免疫学疾病和寄生虫病），每一节都含有病原学、流行病学、临床症状、病理变化和防控措施的相关知识。作者希望与全世界的同行共享他们的知识和经验。在此谨向作者的奉献和开阔胸襟表示感谢。

美国加利福利亚大学动物医学系名誉教授

MURRAY E FOWLER博士

# 前　言

阿拉伯联合酋长国中央兽医研究实验室位于迪拜，成立于1985年，该研究所的主要任务之一是开展骆驼传染病的研究。1970年之前，人们对骆驼传染病知之甚少；在之后的40年中，有关骆驼疾病的研究文章大量发表，许多科技刊物也开始刊登有关骆驼疾病研究的成果。更值得一提的是，好几本有关骆驼疾病的书籍，如由美国学者Murry E Fowler等人编撰的《骆驼医学和手术学》，最近出版发行。由印度Bikener公司出版的《骆驼生产与研究杂志》“The Journal of Camel Practice and Research”是一本专门的有关骆驼的杂志。世界动物卫生组织（OIE）专门组织世界各地的专家，对骆驼传染病进行了梳理并发表报告，内容涉及各个方面；2010年在OIE的网站发表了报告的第二版。骆驼研究的里程碑式的大事是骆驼基因组序列的测定，其中，包括由沙特阿拉伯和中国的科学家完成了单峰驼的基因组，由中国科学家完成的野骆驼和双峰驼基因组，美国科学家完成了美洲驼基因组。这些成就对于骆驼的遗传学研究非常重要。另外的一个重大突破是阿拉伯联合酋长国中央兽医研究实验室于2010年克隆了第一个单峰驼。

传染病造成骆驼大量死亡，在新大陆占死亡骆驼总数的50%，在旧大陆高达65%。肺炎、腹膜炎和肠道疾病是新大陆骆驼的主要疾患，而对旧大陆骆驼而言，消化道的传染病是造成死亡的主要原因。

骆驼科的许多个种都能家养并发挥驮运作用，被称为“沙漠之舟”；还为人类提供高品质的纤维、肉品和乳品。在世界上一些环境严酷的地区，一些奶用家畜如奶牛、山羊和绵羊等都无法存活，更不用说产奶了，但旧大陆骆驼能生产出大量营养丰富的优质奶。因此，很难理解如此有价值的动物只在极个别农场规模化饲养。

骆驼具有巨大的使用价值。与其他家养动物不同，它是多功能的，可以用于运输、拖曳、挤奶、产肉、纤维以及护院和当做宠物。有些骆驼价额高昂，也是骆驼主社会地位的标志。一些政府、组织和民众逐渐认识到骆驼是最能适应极端气候的动物。在未来气候变暖，沙漠扩大，水源和饲料减少等情况下，骆驼也许是解决问题的办法之一。在非洲和亚洲一些国家，从可持续发展的角度看，依靠谷物生产是不可行的；目前的共识是，在干旱地区，随时能够迁移的畜牧业才有可能带来稳定长久的收入。骆驼最适宜在高热干旱的地区生活，它们可在长期缺水的条件下存活，并且其泌乳量远远高于其他家养动物。

驼科动物的特殊之处不仅仅是其明显的形态特征和生产性能，其生理学方面也是非常独特的。值得一提的特征是所有骆驼的抗体都缺乏轻链，由于其非常小，被称为纳米抗体。对于骆驼纳米抗体的研究，开创了一项新的“治疗抗体产业”，生产的高免血清可用于治疗毒蛇咬伤和一些致死性的疾病，如肉毒梭菌肠毒素等，也可用于疾病诊断。骆驼的纳米抗体具有广泛的应用前景，如作为治疗性抗体防治一些种类的癌症、热带昏睡病和淀粉样变性。比利时布鲁塞尔大学分子和细胞相互作用学系的研究专家和中央兽医研究实验室在多个项目开展合作，基于此技术，已建立了一个公司。

仍然有很多人认为骆驼的经济价值有限，而且是发展中国家的代名词。

直到最近，人们才逐渐认识到骆驼在很多方面有益于人类。了解和使用好这一给予人类的特殊礼物将有助于饥荒地区驼业的发展、减少人类饥饿和防止沙漠化。

三位欧洲骆驼研究专家撰写此书的目的是为了展现骆驼对于人类的重要性和对我们子孙后代的意义。

Ulrich Wernery博士
Jörg Kinne博士
Rolf Karl Schuster博士
2014年春于迪拜

# 致　谢

我们衷心表达对尊敬的Sheikh Mohammed Bin Rashid Al Maktou将军阁下的感激和谢意，他是迪拜市市长，也是阿拉伯联合酋长国的副总统，在过去20年中，是他支持了骆驼传染病的研究。在他慷慨资助之下，我们与位于迪拜的中央兽医研究实验室合作，建立了被人交口称赞的动物疫病诊断中心，在该中心用单峰驼进行实验，取得了一系列新的研究成果。

我们对中央兽医研究实验室行政所长Ali Ridha博士的合作、帮助和建议表示感谢，他对我们的研究工作非常关注、充满兴趣，也是这么多年来我们在另一种文化方面的导师。

中央兽医研究实验室的以下同行做出了巨大贡献；Renate Wernery，Sunitha Joseph，Anie Riya Thomas，Ginu Syriac，Bobby Johnson，Marina Joseph，Sherry Jose，Nissy Georgy，Shyna Korah Elizabeth，Saritha Sivakumar，John Christopher，Sweena Liddle，Ringu Marita George，Chellappan Viswanathan。我们从内心深处感谢Alamgir Siddique先生，在每一次危机出现时，都是他毫不吝啬地提供无私帮助。没有上述各位的共同努力，我们无法完成本书的撰写。他们工作积极，在实验室引入新技术、疫苗研发和动物实验等方面做了大量的工作。

我们非常感激在“酋长国驼奶及其制品工业公司”工作的3位兽医人员，他们是Jutka Juhasz、Peter Nagy和Abubaker Abbas博士。自2007年成立奶驼场以来，他们一直与中央兽医研究实验室合作，不断地为科研工作提供病料、饲料和奶样。Samer Al Mahamied博士和Ted Wiechowski博士以及在Bin Hamoodah集团公司的 Belane Nanjegowda Kumarb博士给了我们大量无私的帮助。我们对在迪拜市领导小组工作的Ahmed M Billah及其同事Gulraiz Muunawar博士，Mohammed Amir Saeed博士，Mansoor Ali博士，Jahangir Akbar 博士和Ahsan Ul-Haq博士表示感谢，多年来，没有他们源源不断地提供样品，发现骆驼新传染病是不可能的。

此外，还得感谢Rekha Raghavan和Jyothi Thyagarajan在输入本书大部分文稿时的耐心细致；感谢Moideen Jamsheer和Sajith Jalal为文稿排版及参考文献编排所做的工作，感谢David Wernery为本书做图表时的精雕细琢。

感谢Rajan Babu，Maliakkal Dilshad，Nasarullah Chaudhry 和Abdul Latheef为电脑分析数据提供的协助，感谢中央兽医研究实验室员工在动物实验和尸体剖检时提供的帮助。

我们非常感激位于Nakhlee的“骆驼繁殖实验室”的同行，在他们提供的大量生殖系统的样品之中我们首次分离到胎儿毛滴虫和胎儿弯曲杆菌。

我们也感谢比利时的Serge Myldermans博士，与他的团队合作，我们制备了抗多种抗原的纳米抗体。

对很多允许我们使用他们尚未发表的数据的朋友和同事，我们心存感激，尤其是为本书提供非常有价值资料的Murry E Fowler教授和Tarun Kumar Gahlot博士。

非常感谢我们的家人对无法陪他们出席各类社会活动时的理解。最后要感谢的是世界动物卫生组织出版部的同行长期的鼓励和支持，以及他们对本书精彩纷呈的设计。

Ulrich Wernery博士
Jörg Kinne博士
Rolf Karl Schuster博士
2014年春于迪拜

**以下科学家分别审阅了本书的各个章节，世界动物卫生组织（OIE）对他们表示特别的谢忱！**

Daniel Aguirre博士，Martin Beer博士，Mohammed Bengoumi博士，Set Bornstein博士，Daniel De Lamo博士，Marc Desquesnes博士，Bernard Faye博士，Tarun Kumar Gahlot博士，Alberto A. Guglielmone博士，Salah Hammami教授，Mehdi El Harrak博士，Anil Kumar Kataria博士，Abdelmalik Khalafalla博士，K.M.L. Pathak博士，David E. Swayne博士和David F. Twomey博士。

# 常用缩写词

| 缩写词 | 英文 | 中文 |
|---|---|---|
| AFT | Aflatoxin | 黄曲霉毒素 |
| AGID | Agar gel immunodiffusion test | 琼脂免疫扩散试验 |
| AHS | African horse sickness | 非洲马瘟 |
| ALP | Alkaline phosphatase | 碱性磷酸酶 |
| APPD | Avian purified protein derivative | 禽型结核菌素（纯化蛋白衍生物） |
| APZEC | Animalpathogenic and zoonotic *Escherichia coli* | 动物致病性大肠杆菌和人致病性大肠杆菌 |
| BcoV | Bovine coronavirus | 牛冠状病毒 |
| BD | Borna disease | 博尔纳病 |
| BDV | Borna disease virus | 博尔纳病毒 |
| BLV | Bovine leukaemia virus | 牛白血病病毒 |
| BoHV-1 | Bovine herpesvirus 1 | 牛疱疹病毒I型 |
| BPPD | Bovine purified protein derivative | 牛型结核菌素（纯化蛋白衍生物） |
| BPSV | Bovine popularstomatitis virus | 牛丘疹性口炎病毒 |
| BPV | Bovine papilloma virus | 牛乳头瘤病毒 |
| BRSV | Bovine respiratory syncytial virus | 牛呼吸道合胞体病毒 |
| BT | Bluetongue | 蓝舌病 |
| BTV | Bluetongue virus | 蓝舌病病毒 |
| BUN | Blood urea nitrogen | 血液尿素氮 |
| BVD | Bovine viral diarrhoea | 牛病毒性腹泻 |
| BVDV | Bovine viral diarrhoea virus | 牛病毒性腹泻病毒 |
| CATT | Card agglutination test for trypanosomosis | 锥虫卡片凝集试验 |
| CBPP | Contagious bovine pleuropneumonia | 牛传染性胸膜肺炎 |
| CCHF | Crimean-Congo haemarrhagic fever | 克里米亚-刚果出血热 |
| cELISA | Competitive enzyme-linked immunosorbent assay | 竞争性酶联免疫吸附试验 |

| 缩写词 | 英文 | 中文 |
|---|---|---|
| CFT | Compliment fixation test | 补体结合试验 |
| CFU | Colony-forming unit | 克隆形成单位 |
| CLA | Caseous lymphadenitis | 干酪性淋巴腺炎 |
| CMPV | Camelpox virus | 骆驼痘病毒 |
| CMT | California mastitis test | 乳房炎加利福尼亚体细胞检测法 |
| CNS | Central nervous system | 中枢神经系统 |
| CPE | Cytopathic effect | 细胞病变 |
| CSN | Contagious skinnecrosis | 接触传染性皮肤坏死病 |
| CVRL | Central Veterinary Research Laboratory | 中央兽医研究实验室 |
| DGGE | Denaturing gradient gel elctrophoresis | 变性梯度凝胶电泳 |
| DIC | Disseminated intravascular coagulation | 弥散性血管内凝血 |
| EAV | Equine arteritis virus | 马动脉炎病毒 |
| EBL | Enzootic bovine leukosis | 地方流行性牛白血病 |
| EDTA | Ethylenediaminetetraacetic | 乙二胺四乙酸 |
| EEE | Eastern equine encephalitis | 东部马脑炎 |
| EEEV | Eastern equine encephalitis virus | 东部马脑炎病毒 |
| EHV | Equine herpesvirus | 马疱疹病毒 |
| EIAV | Equine infectious anaemia virus | 马传染性贫血病毒 |
| ELISA | Enzyme-linked immunosorbent assay | 酶联免疫吸附试验 |
| EMCV | Encephalomycarditis virus | 脑心肌炎病毒 |
| EPEC | Enteropathogenic *Escherichia coli* | 肠致病性大肠杆菌 |
| ERAV | Equine rhinitis A virus | 马甲型鼻病毒 |
| ERBV | Equine rhinitis B virus | 马乙型鼻病毒 |
| ETEC | Enterotoxogenic *Escherichia coli* | 产肠毒素大肠杆菌 |
| EXPEC | Extraintestinal pathogenic *Escherichia coli* | 肠道外致病性大肠杆菌 |
| FA | Fluorescent antibody | 荧光抗体 |
| FAT | Fluorescent antibody test | 荧光抗体试验 |
| FMD | Foot and mouth disease | 口蹄疫 |
| FMDV | Foot and mouth disease virus | 口蹄疫病毒 |
| FPA | Fluorescent polarization assay | 荧光偏振分析法 |
| FPT | Failure of passive transfer | 被动转移（移植）失败 |
| GGT | Gammaglutamytranferase | 丙种谷酰胺转移酶 |
| HD | Haemorrhagic disease | 出血性疾病 |

| 缩写词 | 英文 | 中文 |
|---|---|---|
| HEY | Herrold’s egg yolk | Herrold卵黄培养基 |
| HIT | Haemagglutination inhibition test | 血凝抑制试验 |
| HIT | Haemagglutination test | 血凝试验 |
| HRP | Horseradish peroxidase | 辣根过氧化物酶 |
| HRV | Human rhinovirus | 人鼻病毒 |
| HS | Haemorrhagic septicaemia | 出血性败血病 |
| IBR | Infectious bovine rhinotracheitis | 牛传染性鼻气管炎 |
| IFAT | Indirect fluorescent antibody test | 间接荧光抗体试验 |
| IFN- $\gamma$ | Interferon gamma | $\gamma$-干扰素 |
| IPB | Infectiouspustular balanoposthitis | 牛传染性脓包性龟头包皮炎 |
| IPV | Infectiouspustular vulvovaginitis | 牛传染性脓包性阴户阴道炎 |
| JLIDS | Juvenile llama immunodeficiency syndrome | 幼龄羊驼免疫缺陷综合征 |
| JSRV | Jaasiekte sheep retrovirus | 南非羊肺炎反转录病毒 |
| LAT | Latex agglutination test | 乳胶凝集试验 |
| LIV | Loop-ill virus | 跳跃病病毒 |
| LJ | Loewenstein-Jensen | 罗氏（Lownstein-Jenson）培养基 |
| LPS | Lipopolysaccharide | 脂多糖 |
| MAP | *Mycobacterium avium* subsp *paratuberculosis* | 禽分枝杆菌副结核亚种 |
| MAPIA | Multiantigen print immunoassay | 多抗原线条免疫分析法 |
| MAT | Microscopic agglutination test | 微量凝集试验 |
| MD | Mucosal disease | 黏膜病 |
| MDBK | Madin-Darby bovine kidney | Madin-Darby牛肾细胞 |
| MERS-CoV | Middle East respiratory syndrome cornonarirus | 中东呼吸综合征冠状病毒 |
| MLV | Modified live virus | 改造的活病毒 |
| MNT | Mouse neutralizationtest | 鼠中和试验 |
| MOT | Old mammalian tuberculin | 旧结核菌素 |
| MRT | Milk ring test | 乳汁环状试验 |
| NCP | Non-cytopathic | 无细胞病变 |
| NPVI | Non-pathogenic viral infection | 无致病性病毒感染 |
| NS | Non-structural protein | 非结构蛋白 |
| NTM | Non-tuberculosis mycobacteria | 非致结核分枝杆菌 |
| NWC | New World camel | 新大陆骆驼 |
| OIE | World Organization for Animal Health | 世界动物卫生组织 |

| 缩写词 | 英文 | 中文 |
|---|---|---|
| OPF | Oesophageal-pharyngeal fluid | 食道咽喉刮取液 |
| ORF | *Parapoxvirus ovis* | 羊副痘病毒 |
| ORFV | *Parapoxvirus ovis* virus | 羊副痘病毒 |
| OTA | Ochatoxin | 棕曲霉毒素 |
| OWC | Old World camel | 旧大陆骆驼 |
| PAGE | Polyacrylamide gel electrophoresis | 聚丙烯酰胺凝胶电泳 |
| PAS | Periodic acid-Schiff | 过碘酸雪夫反应 |
| PCPV | Psuedocowpoxvirus | 伪牛痘病毒 |
| PCR | Polymerase chain reaction | 聚合酶链式反应 |
| PGRP | Peptidoglycan recognition protein | 肽聚糖识别蛋白 |
| PI-3 | Parainluenza virus 3 | 3型副流感病毒 |
| PLD | Phospholipase D | 磷脂酶D |
| PPD | Purified protein derivative | 纯化蛋白衍生物 |
| PPR | Peste des petits ruminants | 小反刍兽疫 |
| PPRV | Peste des petits ruminants virus | 小反刍兽疫病毒 |
| PVI | Pathogenic viral infection | 致病性病毒感染 |
| PVNZ | *Parapoxvirus* of red deer in New Zealand | 新西兰马鹿副痘病毒 |
| QF | Q fever | Q热 |
| R-LPS | Rough lipopolysaccharide | 粗糙型脂多糖 |
| RBC | Red blood cell | 红细胞 |
| RBPT | Rose Bengalplate test | 虎红平板试验 |
| RBT | Rose Bengal test | 虎红试验 |
| RP | Rinderpest | 牛瘟 |
| RT | Rapid test | 快速试验 |
| RT-PCR | Real-time PCR | 实时PCR |
| RVA | Rotavirus | 轮状病毒 |
| RVF | Rift Valley fever | 裂谷热 |
| RVFV | Rift Valley fever virus | 裂谷热病毒 |
| S-LPS | Smooth lipopolysaccharide | 光滑型脂多糖 |
| SAC | South American camelid | 南美骆驼 |
| SBV | Schmallenberg virus | 施马伦贝格病毒 |
| SCC | Somatic cell count | 体细胞计数 |
| SDS | Sudden death syndrome | 猝死综合征 |

| 缩写词 | 英文 | 中文 |
|---|---|---|
| SICTT | Single intradermal comparative tuberculin test | 单次结核菌素皮内比较试验 |
| SNT | Serum neutralisation test | 血清中和试验 |
| STEC | Shiga toxin-producing *Escherichia coli* | 产志贺毒素大肠杆菌 |
| TAT | Tube agglutination test | 试管凝集试验 |
| TEM | Transmission electron microscope | 透射式电子显微镜 |
| TP | Total protein | 总蛋白 |
| TST | Tuberculosis skin test | 结核菌素皮内变态反应 |
| UHT | Ultra heat treatment | 超热治疗 |
| VNT | Virus neutralization test | 病毒中和试验 |
| VS | Vesicular stomatitis | 水疱性口膜炎 |
| VTEC | Verotoxigenic *Escherichia coli* | 产非洲绿猴肾细胞毒素大肠杆菌 |
| WBC | White blood cell | 白细胞 |
| WHO | World Organization for Health | 世界卫生组织 |
| WNF | West Nile fever | 西尼罗河热 |
| WNS | “Wry-neck” syndrome | “曲颈”综合征 |
| WNV | West Nile virus | 西尼罗河热病毒 |
| ZN | Ziehl-Neelsen | 齐尔-尼尔森染色法 |
| ZON | Zearalenone | 玉米赤霉烯酮 |
| ZST | Zinc sulphate turbidity | 硫酸锌浊度 |

# 生物学与进化

几千年来，骆驼为人类提供食物、毛类纤维和油脂，在世界一些地方，骆驼还是驮运工具，保障了干旱和半干旱地区贸易的畅通。对生活在阿拉伯半岛、北非和东非一些环境非常恶劣的地区的贝多因人来说，单峰驼对于他们的生存一直是至关重要的。双峰驼能抵抗非常严寒的气候，主要分布于亚洲的高海拔沙漠地区，在过去的几百年中，将大量的货物沿着丝绸之路由中国驮往欧洲。在中国和蒙古国的戈壁滩上，还生活有一些野骆驼。在南美洲，驼马和原驼仍然为野生状态，而美洲驼和羊驼已被驯化，适应了高海拔地区的生活。

骆驼具有非常良好的生产性能，但是，在过去的几十年中，除个别地方外，全球养驼业陷入危机。在南美洲，骆驼数量在19世纪和20世纪早期大幅减少，但在近几十年中，数量逐渐回升，主要原因是生活在安第斯山脉的人们认识到骆驼产业可以提高他们的收入。这一理念也渗入到欧洲，目前在欧洲许多地方可以看到很多驼群。另一个拉动骆驼数量上升的因素是阿拉伯半岛的赛驼业，这一贝多因人的传统体育项目因引入机器人代替儿童作为赛驼骑士而迅速复兴。

在世界有些地方，骆驼正面临着灭绝的危险或正在被当作有害动物而屠杀。全球人口的爆炸性增长、城市化、自动化以及现代化正在不知不觉地破坏着数千年来骆驼赖以生存的环境。

在澳大利亚，骆驼曾是文化的怪诞之物，在1840年左右被引进，用于开发内陆，在整个社会生活、矿井挖掘、游牧生活以及深度开发和探索等方面发挥了主要作用。然而，时至今日，它们因在寻找水源时破坏房屋、拔断水管，与牛群竞争日渐荒芜的草原，被认为是有害动物，多数人建议将其宰杀，但也有人建议通过发展澳大利亚骆驼产业，以更加人道地对待骆驼。

在印度拉贾斯坦邦，因当地采取屠宰政策，已能明显观察到骆驼数量的下降。自20世纪90年代中期以来，印度的骆驼总数下降一半以上，气候变化，森林、草原和未开发土地被用于工业也起到了推波助澜的作用。全球变暖成为不争的事实，地球上的沙漠正在扩大；在许多国家，沙漠化非常严重。以前适宜放牧牛群的土地变为荒漠，只适合骆驼生存。面对荒漠化问题和生态恶化，急需新的解决办法。位于卡塔尔多哈的联合国教科文组织提出了建立奶骆驼养殖场的新概念，以解决这一难题[3]。骆驼的许多生理特性使得其可以在干旱的地区规模饲养，为当地数百万人造福，同时，在生态危机日趋严重的今天，骆驼也可在解决可持续农业和土地开垦之间的矛盾中发挥作用。

在蒙古国，处于数量大幅下降危险之中的，不仅仅是家养和野生的骆驼，而是包括整个游牧文化，其主要原因为“阻得”（蒙古语，意思是因大雪导致的寒冬），2009—2010年造成家畜数量减少1000万[7]。此外，在中国和蒙古国的濒危野生双峰驼，由于非法金矿或铁矿的开采已处于灭绝的边缘。

虽然有上述缺憾，或者说由于这些缺憾，在过去的几十年，人们对骆驼科动物的兴趣日渐浓厚，骆驼也成为科学研究的焦点。这一点不仅从日益增多的有关骆驼的科技论文和新编骆驼专业期刊（Journal of Camel Practice and Research，Tarun Kumar Gahlot博士主编）看得出来，也反映出欧洲的大学与干旱国家研究机构之间越来越多的合作研究项目，同时也体现于过去几年中在多个国家召开的有关骆驼的国际性学术会议上，其会议论文集大多数已经出版并向公众开放。

骆驼不仅具有良好的生产性能，为人类提供高质量产品和服务，其体液免疫还具有独特性，为高水平和高质量的研究提供源泉。与其他哺乳动物不同，骆驼除正常的抗体外，还拥有非常简单的多样性抗体，它们为一类新的抗体，来源于重链抗体，是单一域的抗原结合片段。这类抗体是刚刚被发现，是一类非常小的能与抗原自然结合的蛋白质结构域，称为单可变域重链（VHS）或纳米抗体[11]。纳米抗体具备一系列的生物物理学特性，在医学和生物技术应用方面具有非常广阔的前景[5]。

骆驼是奶类、肉类和毛类产品的重要来源之一，也是运输和劳动的工具，目前，食物和能源日渐紧缺，应让其发挥更重要的作用[6]。由于特别的生理特征，骆驼比其他家养哺乳动物适应性更强，在极端气候条件下，更能发挥特长[10,27,29,31]。

在相当长的一段时间里，人们都错误地认为双峰驼和单峰驼都来自一个共同的祖先，即野生的双峰驼——野骆驼。究其原因：①不论单峰驼还是双峰驼在胚胎时期都有一个双峰的发育阶段；②双峰驼和单峰驼的杂交后代具有繁殖能力。第一点依据来自130年前Lombardini的解剖学研究结果，他认为单峰驼的胎儿和成年时期有一小的第二驼峰[18]。然而，位于阿拉伯联合酋长国迪拜的中央兽医研究实验室在20多年间解剖分析了大量的单峰驼胎儿和幼畜，没有发现类似的结构[12,13]（图1a，1b，1c）。最近对单峰驼和双峰驼的颅后骨骼发生学研究也证明他们确实来自不同的种[21]。现在，人们逐渐接受以下观点，即在中国和蒙古国一带生活的野生的双峰驼是骆驼的第三个种——野骆驼（*Camelus bactrianus ferus*）。近几年来，对采自野骆驼皮肤的样品进行DNA分析后发现，与双峰驼有3%的差异，据此认为野生双峰驼、驯化双峰驼和单峰驼都源自不同的祖先。这一观点还有待进一步研究确认。

胼足亚目的动物在5 000万～6 000万年（第三纪）源自北美，之后分化为8个不同的科，当时只有野兔般大小[32]。在后来的中新世中期，其中的6个科消亡（表1a）。在500万年前，旧大陆骆驼的祖先穿过白令海峡，迁徙到东北亚地区，现在的旧大陆骆驼都是这些骆驼进化而来，栖息地也向西扩展。在更新世时期（结束于200万年前），生活圈迅速扩大，最远延伸至东欧、北非、东非和东亚[14,15]。但在上述有些地区，最终却没能存活下来。当旧大陆骆驼的祖先穿过白令海峡时，无驼峰的新大陆骆驼也在穿过南北美大陆之间的海峡[24]，它们在南美繁衍，种群扩大，逐渐形成今天的美洲驼（*Lama glama*）、原驼（*Lama guanicoe*）、羊驼（*Lama pacos*）和骆马（*Vicugna vicugna*）（图2a, b, c, d）。美洲驼和羊驼已驯化家养，而原驼和驼马为野生状态。新大陆骆驼和旧大陆骆驼属于胼足亚目的骆驼科（图1）。

在北美，所有的骆驼科动物在1万年前灭绝；最后消亡的是拟驼（*Camellops*），原因很可能是北美印第安人的狩猎活动。在今天的南美，有700万～800万小型骆驼（秘鲁和玻利维亚）（表2）。美洲驼和羊驼在7000年前就被驯化，是最早被驯化的动物之一，也是南美印第安的早期文明之一。

在旧大陆骆驼中，已驯化的有两个种：①为单峰骆驼或单峰驼（*Camelus dromedarius*）；②为双峰驼（*Camelus bactrianus*）（图3a，b）;在中国和蒙古国，还有一小群的野生骆驼（图4）。单峰驼非常适应高温和干旱，在东非、北非和阿拉伯半岛及近东地区，它们主要的用途是驮运货物（图5）和供人骑乘，也为游牧的人群提供驼奶。双峰驼适宜于在寒冷干燥的地区生活，在蒙古国、西西伯利亚、亚洲次大陆、伊朗和阿富汗，它们发挥着与单峰驼同样的作用。双峰驼有很厚的被毛，可以用来制造驼毛制品。双峰驼有很强的驮运能力，曾经将大量的货物沿着丝绸之路由中国内陆驮往地中海地区。图6和表3展示了世界上旧大陆骆驼的分布地区和数量。

令人赞叹的是，旧大陆骆驼要么适宜寒冷气候，要么适应酷热气候，而新大陆骆驼适应高海

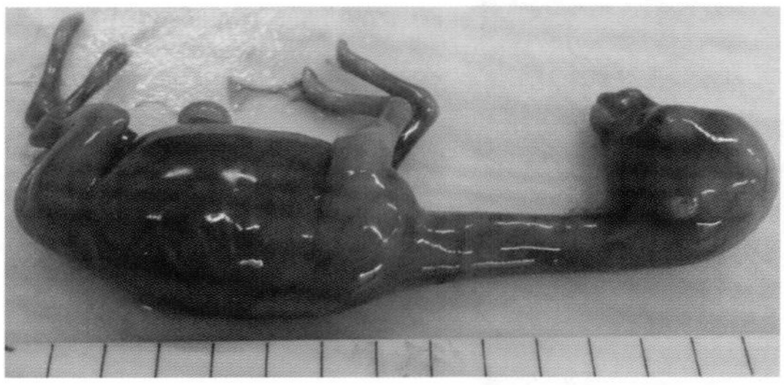

图1a　单峰驼胚胎（顶臀长7.5cm），观察不到驼峰的雏形

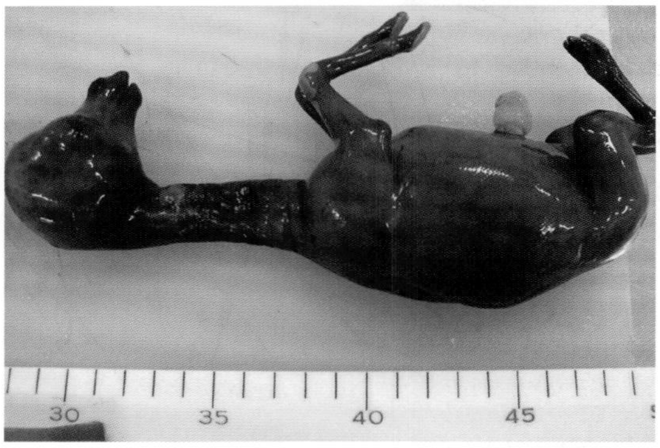


图1b　单峰驼胎儿（顶臀长22cm），没有可见的第二驼峰

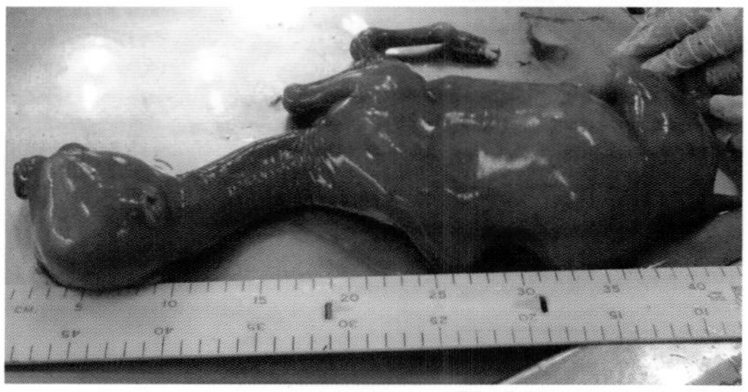

图1c　双峰驼胎儿，有明显可见的前后驼峰

生物学与进化

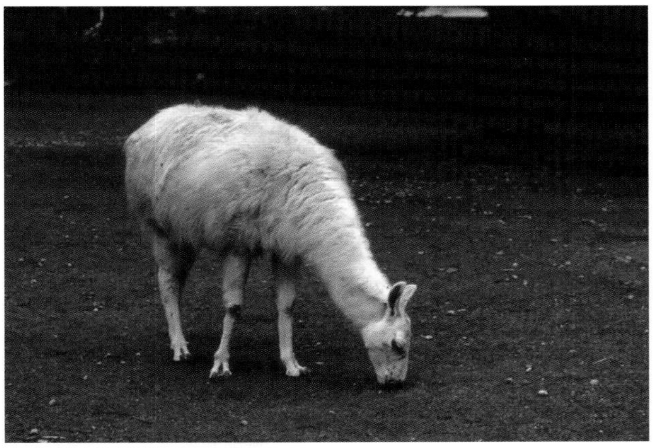

a　美洲驼

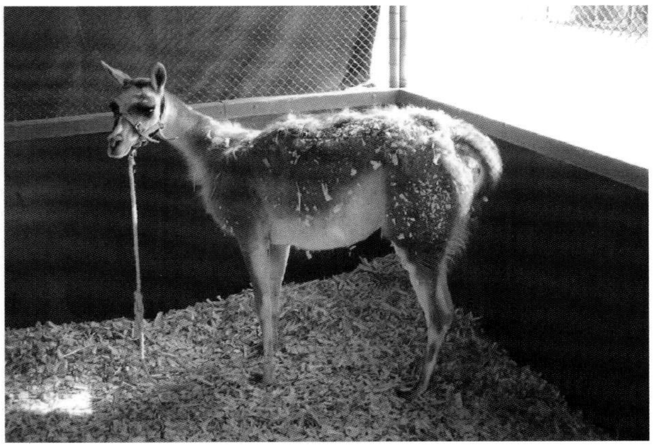

b　原驼

c　厄瓜多尔钦博拉索火山口附近的羊驼群

d　厄瓜多尔阿尔蒂普拉诺高原4 300m高处的驼马

图2　新大陆骆驼的4个种

表1　驼科动物和其他偶蹄目动物的分类[9]

| | | |
|---|---|---|
| 纲 | 哺乳纲 | |
| 目 | 偶蹄目 | |
| 亚目 | 猪形亚目 | 河马、猪、野猪类动物 |
| | 胼足亚目 | 骆驼科 |
| | 旧大陆骆驼 | 单峰驼*Camelus dromedarius* |
| | | 双峰驼*Camelus bactrianus* |
| | | 野骆驼*Camelus bactrianus ferus* |
| | 新大陆骆驼 | 美洲驼*Lama glama* |
| | | 原驼*Lama guanicoe* |
| | | 羊驼*Lama pacos* |
| | | 驼马*Vicugna vicugna* |
| | | 驼马秘鲁亚种*Vicugna vicugna mensalis* |
| | | 驼马阿根廷亚种*Vicugna vicugna vicugna* |
| | 反刍亚目 | 牛、绵羊、山羊、水牛、长颈鹿、鹿、羚羊、野牛 |

表2　新大陆骆驼数量估测[4,25]

| 国家/地区 | 美洲驼 | 羊驼 | 原驼 | 驼马 |
|---|---|---|---|---|
| 阿根廷 | 75 000 | 2 000 | 555 000 | 23 000 |
| 玻利维亚 | 2 500 000 | 300 000 | 未知 | 12 000 |
| 智利 | 85 000 | 5 000 | 20 000 | 28 000 |
| 秘鲁 | 900 000 | 3 020 000 | 1 400 | 98 000 |
| 澳大利亚 | <5 000 | >5 000 | 动物园，数量少 | 0 |
| 加拿大 | >6 000 | >2 000 | 动物园，数量小于100 | >10 |
| 欧洲 | <2 000 | <1 000 | 动物园，数量小于100 | 动物园数量小于100 |
| 美国 | >110 000 | 150 000 | 145，多数在动物园 | 0 |
| ISIS注册的动物园 | 343 | 303 | 397 | 100 |
| 小计 | 3 683 343 | 3 485 303 | 572 142 | 161 210 |
| 总计 | 7 901 998 | | | |

ISIS: **国际物种编目系统**（International Species Inventory System）

a　**单峰驼**

b　**双峰驼**

图3　两种旧大陆骆驼

图4　带着幼驼的野骆驼（照片由野骆驼保护基金会J Hare惠赠）

图5　在印度拉贾斯坦邦的一峰正在拉车的公单峰驼

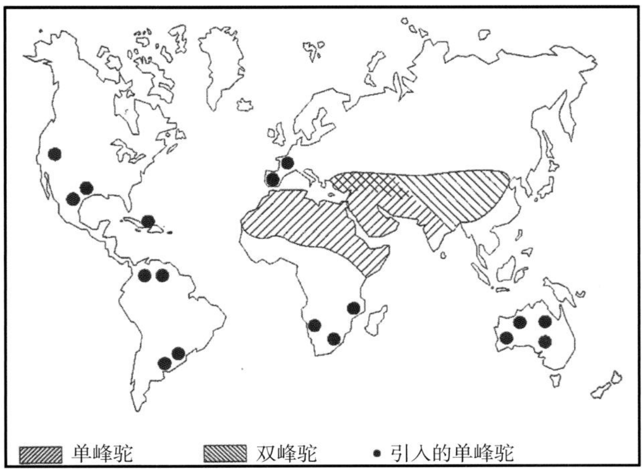

a 旧大陆骆驼：单峰驼和双峰驼

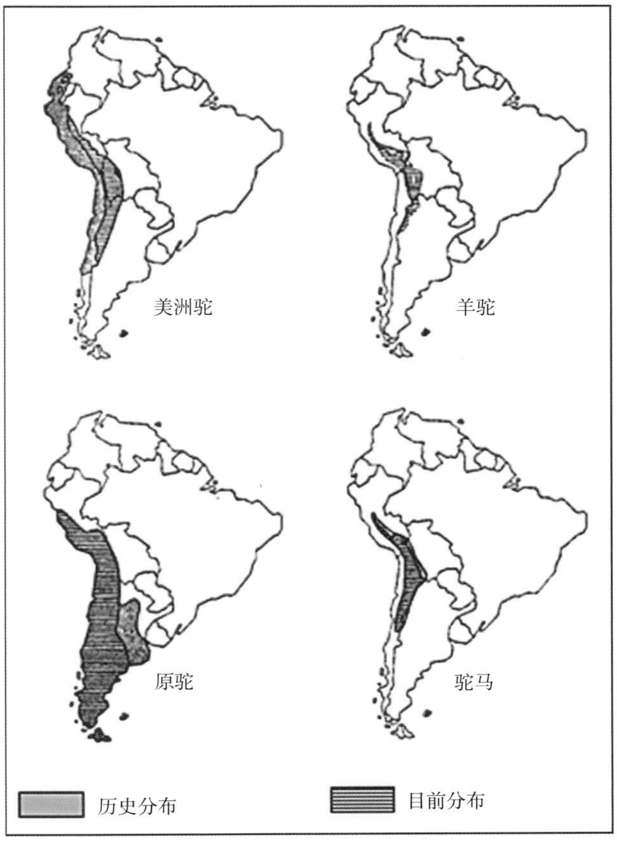

b 新大陆骆驼

图6 旧大陆骆驼和新大陆骆驼的分布

拔地区。这其中肯定有一种复杂的机制，使得骆驼科动物能在极端的气候条件下存活下来。

跟许多的偶蹄目动物一样，骆驼也有反刍现象，将未消化的食物再次咀嚼，然后吞咽，在胃内进行二次消化。不过，确切来讲，骆驼并不属于反刍科动物。骆驼从质量低劣饲料中摄取能量和蛋白质的能力非常强，仅靠一点树叶就可生存。它们的前3个胃称为室（图7），表4列出了驼科动物和反刍动物的差异[28]。

英文的“单峰驼”源自希腊文，意思是“奔跑”，“双峰驼”源自西南亚地区的“Bactria”这一地名[1]。

骆驼不仅可用于犁地和骑乘，还可生产肉品、奶类、皮张和驼毛。比较发现，骆驼肉的脂肪含量远低于牛肉，而蛋白质的含量与牛肉相差无几。骆驼皮张很结实，强度是牛皮的5倍。已经用骆驼皮加工出时尚的服装，柔软的钱包和手提袋。在许多饲养骆驼的国家，驼毛为重要的副产品。驼毛是世界上最昂贵的天然动物纤维。贝多因人用驼毛制造地毯和帐篷。就单峰驼而言，每峰公骆驼可产毛3kg，母骆驼可产毛2kg。双峰驼毛的质量优于单峰驼[2]，每峰公单峰驼可生产10～16kg的优质毛。遗憾的是，驼毛加工业者对这些驼毛不感兴趣，不过，人们对新大陆骆驼产的毛需求量逐渐增加。驼马的毛为最细的动物毛，毛发直径和纺织品与羊绒类似，每年的产毛量只有200g，人们对驼马毛的需求也使得驼马避免了灭绝的命运。

在4种新大陆骆驼中，驼马体格最小，被毛为肉桂色。这种漂亮的动物根据地理分布分为两个亚种，分别生活在秘鲁和阿根廷，现已经出口到厄瓜多尔和南美的其他国家。羊驼有两个培育的品种，分别为“Huacaya”和“Suri”，在南美洲和欧洲，数量较多，不仅产毛，也是宠物。在南美洲，当地印第安人所用的毛类主要来自羊驼，颜色由白色到黑色或者混合色。原驼为新大陆骆驼的另一个种，分布广，有四个地理亚种。所有亚种有共同的毛色特征，身体的上部为深褐色。在出生时捕获可驯化。美洲驼是新大陆骆驼中体格最大者，毛色变化大，体格也有差异。在秘鲁存在两个品种。在几千年前驯化之后，当地人主要用其驮运货物，现在它们和羊驼也都是提供肉品的动物。据估计，目前北美洲羊驼的数量大约为15万峰、美洲驼的数量在10万～12万峰，由于口蹄疫的缘故，无法从南美洲进口，只能从无口蹄疫的国家，如澳大利亚、加拿大、新西兰等国进口，数量很少。近几年来，英国

表3 旧大陆骆驼的数量[3]

| 非洲 | 骆驼数量 | 亚洲、大洋洲和欧洲 | 骆驼数量 |
|---|---|---|---|
| 阿尔及利亚 | 130 000 | 阿富汗 | 270 000 |
| 布基纳法索 | 6 000 | 印度 | 600 000 |
| 乍得 | 540 000 | 伊朗 | 27 000 |
| 吉布提 | 60 000 | 伊拉克 | 250 000 |
| 埃及 | 190 000 | 以色列 | 10 000 |
| 埃塞俄比亚 | 1 000 000 | 约旦 | 15 000 |
| 肯尼亚 | 810 000 | 科威特 | 6 000 |
| 利比亚 | 2 000 000 | 蒙古国 | 580 000 |
| 马里 | 240 000 | 阿曼 | 9 000 |
| 毛里塔尼亚 | 820 000 | 巴基斯坦 | 980 000 |
| 摩洛哥 | 230 000 | 卡塔尔 | 24 000 |
| 尼日尔 | 420 000 | 沙特阿拉伯 | 780 000 |
| 尼日利亚 | 18 000 | 叙利亚 | 5 000 |
| 塞内加尔 | 15 000 | 土耳其 | 12 000 |
| 索马里 | 6 800 000 | 阿联酋 | 250 000 |
| 苏丹 | 2 800 000 | 也门 | 140 000 |
| 突尼斯 | 187 000 | 前苏联 | 200 000 |
| 西撒哈拉 | 92 000 | 中国 | 600 000 |
| | | 澳大利亚 | 1 000 000 |
| | | 加那利群岛 | 2 000 |
| 合计 | 16 358 000 | 合计 | 5 760 000 |
| 总计 | | 22 118 000 | |

表4 骆驼和反刍动物差异比较

| 骆驼 | 反刍动物 |
|---|---|
| **进化** | **进化** |
| 在4 000万年前开始分化 | 在4 000万年前开始分化 |
| **血液** | **血液** |
| 红细胞呈椭圆，较小（6.5 μm），白细胞主要为嗜中性粒细胞 | 红细胞呈圆形，较大（10 μm），白细胞主要为淋巴细胞 |
| **蹄** | **蹄** |
| 具有趾甲和软掌 | 具有分叉和蹄指（趾） |
| 第二、第三趾骨为水平状 | 第二、第三趾骨几乎为垂直状 |

续表

| 骆驼 | 反刍动物 |
| --- | --- |
| **消化系统** | **消化系统** |
| 前消化道发酵，有逆流、再咀嚼和再吞咽动作 | 前消化道发酵，有逆流、再咀嚼和再吞咽动作 |
| 胃：有3个胃室，对臌气有抵抗力 | 胃：有4个胃室，对臌气敏感 |
| 第一胃室有一层鳞片样上皮 | 无腺囊 |
| 第一胃室有两个腺囊，不贮水 | 乳头状上皮 |
| **呼吸系统** | **呼吸系统** |
| 有肺隔 | 无肺隔 |
| 有软腭 | 无软腭 |
| **生殖系统** | **生殖系统** |
| 诱导排卵 | 自然排卵 |
| 无发情周期 | 有发情周期 |
| 有卵泡发育周期 | 无卵泡发育周期 |
| 俯卧交配 | 站姿交配 |
| 胎盘为融合型 | 胎盘为子叶型 |
| 上皮膜包裹胎儿 | 没有上皮膜包裹胎儿 |
| 阴茎末端有软骨凸起 | 阴茎末端无软骨凸起 |
| 射精时间长 | 射精时间短且迅猛 |
| **泌尿系统** | **泌尿系统** |
| 肾脏光滑呈椭圆形 | 肾脏光滑有分叶 |
| 母驼尿道外口有尿道下憩室 | 无尿道下憩室 |
| 背侧尿道凹进 | |
| **寄生虫** | **寄生虫** |
| 特有的虱子和球虫 | 特有的虱子和球虫 |
| 可以感染牛、绵羊和山羊的消化道线虫 | 消化道线虫互相感染 |
| **传染病** | **传染病** |
| 对结核病易感 | 对结核病和口蹄疫非常易感 |
| 双峰驼对口蹄疫易感 | |
| 临床上很少见到牛、羊的其他病毒病 | |

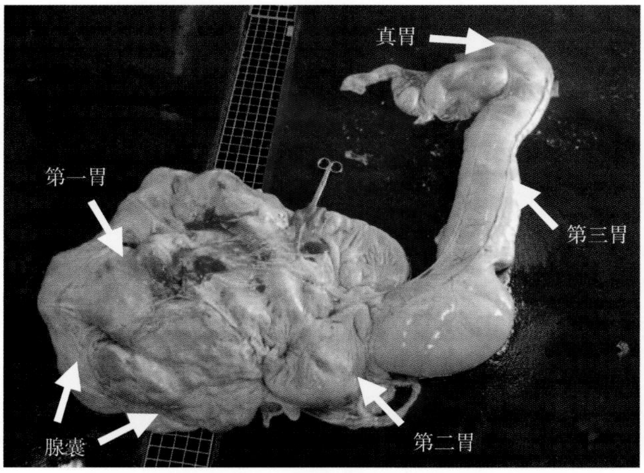


a　没有剖开的骆驼胃

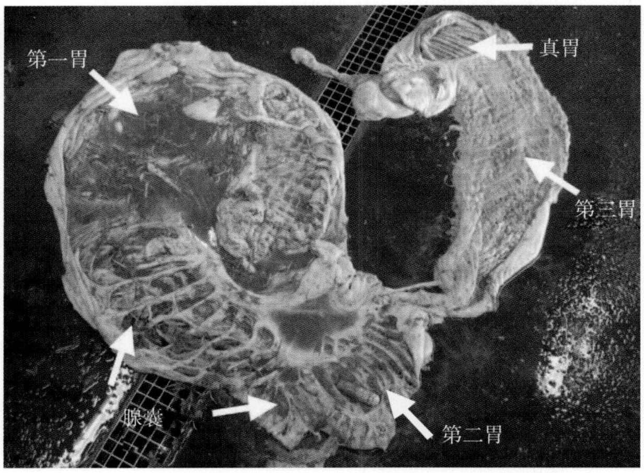


b　已剖开的骆驼胃，可见3个前胃、腺囊和真胃

图7　骆驼胃

图8　双峰驼和单峰驼杂交

图9　公双峰驼和母单峰驼的杂交后代（图鲁，两岁，雌性）

的羊驼和美洲驼也遭受牛结核病之害。

所有驼科动物的染色体数量相同（2×36=72加1对性染色体），在遗传衍化方面相互关联，因此，在所有种之间杂交育种是可行的，已经培育出了几个非常有名的品种。在前苏联的加盟共和国中，单峰驼和双峰驼的杂交非常普遍。杂交产生的具有繁殖能力的后代与父母代比，身体壮，体格大，具有杂交优势（图8）。$F_1$代在哈萨克斯坦、土库曼斯坦和伊朗非常受欢迎[16]，它们的后代被称为“柏图阿（Bertuar）”（♀单峰驼×♂双峰驼）或“勒尔（Ner）”（♀双峰驼×♂单峰驼）。在杂交后代之间再杂交时，产生的$F_2$后代羸弱，而杂交$F_1$代与纯种单峰驼或双峰驼杂交时，后代很强壮。在土耳其西部，单峰驼和双峰驼的杂交后代在斗驼竞赛中表现很出色。与父母代相比，杂交后代体格和体重都大，被称为“图鲁（Tulus）”（图9）。

科学家试图在新大陆骆驼和旧大陆骆驼之间杂交，第一个成功的例子是在阿联酋，一峰公单峰驼和一峰母原驼杂交产生后代（图10），但后代没有繁殖能力。在迪拜的这一繁育中心，还杂交出了其他新大陆骆驼和旧大陆骆驼的后代。

虽然在蒙古国和哈萨克斯坦的史前遗址中发现有双峰驼先祖的遗迹，但对单峰驼先祖了解甚少。一种“巨骆驼”，动物学上称为托氏骆驼（*Camelus thomasi*）(以法国古生物学家Thomas命名)，据认为是单峰驼的先祖[22]。在苏丹、埃及和莱文特西南地区挖掘出的骨头显示它们在形态学上与目前家养的单峰驼一致。考证认为这些骆驼在上一个冰河时期，在北非和内盖夫沙漠地区它们属于野生状态，在1 2000～2 0000年前，因极寒和干旱而灭绝。在撒哈拉地区山脉上发现的有关骆驼的岩画，更进一步为这一理论增添了证据。有关野骆驼的证据只发现过一次，在也门一个称为斯亥的村子，发掘出公元前7 000年前的骆驼遗骸，它源自全新世的早期阶段，这一阶段起始于冰河时期的后期，或者说末次盛冰期。

最近在阿联酋北部靠近阿尔奎瓦德的特尔阿布洛克地区的遗址发掘出骆驼遗骨，考古学和骨发育学研究表明，这些地区一直有野骆驼的存在，大约在公元前2 000年，从考古记录中消失，原因可能是在此部分阿拉伯地区的捕猎导致了灭绝。大约在公元前1 000年，骆驼遗骨再次出现在特尔阿布洛克地区，但骨发育学研究表明，它们比家养骆驼略小。目前，还无法确认最早驯化单峰驼的地区在何处，但很可能是在阿拉伯半岛、巴勒斯坦、莱文特的其他地区或伊朗。在靠近迪拜欧姆阿尔拉的一个非常著名的考古遗迹处发现公元前3000年的骆驼遗骨，曾认为是源自家养骆驼，但后来证实为野生单峰驼的骨头。

在杰利科和沙伊索科塔的贸易遗址发掘出的遗骨表明在公元前1 000年或者之前一个世纪左右，人们开始拥有家养骆驼。圣经故事中也显示在公元前1 100年左右，来自米蒂安的贝多因人部落利用骆驼征服巴勒斯坦。公元前1 000年左右，大群的商运驼队从也门，或者由阿曼经由阿联酋，将香料驮运至地中海地区。沙迦国际机场附近的姆维拉曾在公元前8世纪和9世纪非常兴盛，在这儿的一个考古学遗址中，有证据表明，当时使用骆驼穿越沙漠到达东方。在沙迦考古博物馆中，陈列着一个从该遗址发掘出的制作非常精细的骆驼样陶器。

在道珐的南欧玛尼省的北部，有一个叫石瑟的考古遗址，也许是骆驼商队穿越沙漠的出发点；然后经马里卜（Marib）、迈迪奈（Medina）和佩特拉（Petra）穿过鲁卜哈利沙漠（Empty Quarter），到达地中海。另一条路线是向北直行，直达阿拉伯半岛西侧，途经今沙特阿拉伯达曼附近格哈的一个考古学遗址。在穿越如此复杂坎坷的路途时，骆驼的作用至关重要（图11）。

由于香料贸易的巨额利润，推动家养骆驼育种的飞速发展。在也门东部边境山区的马里卜和约旦南部的佩特拉之间，有一段环境非常恶劣的不毛之地，这些重负的“沙漠之舟”需要50～70天才能穿越。骆驼商队在佩特拉的纳巴特王国（Nabatean）时期达到巅峰，在佩特拉发现的赤土陶器多装饰有单峰驼图案（图12）。然而，随着基督教的兴起，罗马帝国和拜占庭王朝对香料的需求下降，曾经被称为“阿拉伯福地”的阿拉伯半岛西南部失去了主要经济来源，又变回到不毛的鲁卜哈利沙漠。伴随贸易的消失，单峰驼也只有阿拉伯半岛的贝多因部落的人继续使用。

图10 原驼（母亲）和单峰驼（父亲）的杂交后代

图11 香料贸易的路线，出土骆驼骨头的遗址

图 12　在佩特拉出土的赤土陶器

大约在公元前1000年骆驼被驯化之后不久的一段时期，单峰驼很可能通过非洲之角进入东非。从索马里这个人均单峰驼拥有数最多的国家，“沙漠之舟”向西和向北拓展，但直到罗马王朝时代才进入突尼斯和阿特拉斯山脉地区的国家。

美洲驼和羊驼这两种新大陆骆驼在公元前4 000～5 000年被驯化，远早于旧大陆骆驼，因为单峰驼和双峰驼直到公元前1 000～2 000年才被驯化。毫无疑问，这是南美土著印第安人及其先进文化的例证。在印加帝国时代，所有骆驼都是政府的财产，驼马的毛织品只有皇室成员才能使用；人们依赖美洲驼和羊驼制作衣物，在不同地区运输货物、食品和油料。自1532年西班牙入侵之后，在安第斯地区的许多国家都开展了土地的农业化改革，目前，只有农民和游牧民用传统的方式饲养美洲驼和羊驼，这也避免了这些骆驼灭绝的命运。

到了更晚一点的时期，也就是欧洲殖民化时期，单峰驼不仅被引入到气候较为温和的欧洲、南美洲和加勒比地区，还引入到气候炎热的澳大利亚和南非等地。据估计，在1849—1907年，澳大利亚共引入10 000～12 000峰骆驼，供开发干旱内陆的人们骑乘、驮运货物[26]。澳大利亚进口的基本全是单峰驼，这是因为它们更适应当地的炎热气候，随之而来的还有阿富汗的骆驼驯养师，他们也为澳大利亚骆驼养殖的发展做出贡献，但历史几乎没有记载他们。这些人群的另外一个贡献就是他们走到那里，就把枣椰树栽到那里。在当时，一头骆驼可以驮载275kg的货物，只需要少量的草料和饮水就可在不同的地区间长距离运输，而在20世纪20年代，当机动车辆开始在澳大利亚中部地区运行之后，骆驼的使命也就完成，被认为是有害的动物，多数被射杀。在澳大利亚的半干旱沙漠地带，建立了自由活动群，目前的数量大约为100万峰或更多一点，这些野生状态的骆驼散布于干旱的内陆，约有50%在西澳大利亚，25%在北方地区，25%在昆士兰和南澳的北部地区。20世纪60年代，人们对骆驼的兴趣有了新的拓展，在1970年时，澳大利亚已有两个骆驼旅游公司，各地开始赛驼活动[2]。1998年8月，在筹备2000年悉尼奥运会期间，通过阿联酋的帮助，组织了几场赛驼竞赛。

1890年，单峰驼被带入到非洲南部的纳米比亚等地，之后直至第一次世界大战结束，这些骆驼的主要使用者是在纳米比亚的德国镖局社团，其原因有三：①在纳米比亚和卡拉哈里沙漠上，只有单峰驼能够生存；②因牛瘟和口蹄疫，牛只死亡殆尽；③因灾难性的非洲马瘟，马匹受害严重。1906年，Lorenz hagenbeck向纳米比亚斯瓦科普蒙德的一个小前哨站运送了200峰苏丹骆驼。在凡尔赛和约（1919年）签署之后，英国警察掌管了在纳米比亚的剩余骆驼。不过，与澳大利亚的一样，当车辆运输逐渐普遍之后，骆驼被人遗弃；据信是因为狮子捕食和土著布须曼人的猎食，骆驼在20世纪60年代从南部非洲消亡[20]。

在19世纪40年代的墨西哥战争之后，美国也开始利用单峰驼。1857年，从印度运来70峰单峰驼，用于在南部与墨西哥接壤的边境哨所巡逻。遗憾的是，美国国会对购买千峰骆驼成立一个军团的建议没有响应，只能将原有的驼群划归美国邮政局，在新墨西哥州和加利福利亚州之间美国新得的沙漠地区穿越，运送邮件。许多职员不愿与动物打交道，导致许多骆驼逃跑，直到20世纪40年代，还有报告说在美国西南部遥远的地区有“野骆驼”存在。

在欧洲，近20年涌现了很多骆驼社团，骆驼也用于吸引旅游者。1997年8月，在柏林的非常著名的哈普噶凳赛马场，举办了一场有60 000观众参加的赛驼竞赛。实际上，这并不是如宣传所说那样，是欧洲举办的第一场赛驼竞赛。真正的第一次赛驼发生于1969年的科隆草原节，来自摩洛哥的骆驼进行了竞赛[17]。

自1980年以来，骆驼不仅得到了学术界关注，也在发挥着骑乘和运输功能的国家得到广泛关注。驼奶、驼皮和驼肉都得到充分利用，还推动了旅游业。在未来，尽管肯定会遇到现代运输工具和其他家养动物的挑战，但骆驼不会给其他家养动物和野生动物造成威胁。

科学家最近加大了对单峰驼的研究力度，

其中的一个讨论焦点是单峰驼是否有不同的“品种”。截至目前，单峰驼还是按以下的方式分类：以饲养它们的部落冠名，以骑乘或运输功能得名，以颜色、地理背景、生理特征命名，或以产奶、吃肉或竞赛而命名[30]。已界定的单峰驼的品种或品系有50多个，而对双峰驼的品种知之甚少。许多从事骆驼研究的科学家都强调品种特性分析和遗传多样性保护的重要性。最近，中国和沙特科学家在单峰驼、双峰驼和野骆驼的基因组序列分析的成果，将触发这一领域的大规模研究，也将有助于明确野骆驼、双峰驼和单峰驼是否拥有共同祖先。

今天，所有单峰驼的品种和品系都分属以下3个类别。

（1）驮运或役用。

（2）奶用。

（3）骑乘或竞赛用。

骆驼具有丰富的遗传多样性，还有许多迷需要解释。

在圣经故事和早期的伊斯兰征服过程中，骆驼是战士的坐骑，今天依然如此，在中东、亚洲和北非国家，仍然有骆驼分队服务于现代的机械化军队，如在阿联酋的军队中，骆驼部队依然在遥远的沙漠地区巡逻。

公元前557年，希腊历史学家赫洛德最早记录骆驼特种部队，在阿拉伯半岛的伊斯兰征服之前和征服的早期的历史故事中，也有士兵骑在骆驼上作战的描述。在阿联酋的考古发现也证实了以骆驼为坐骑的价值，在梅勒哈和节博布荷，发现有蜷曲埋葬的骆驼，很显然它们是被故意宰杀后作为殉葬品掩埋的。这一习俗一直延续到伊斯兰时代的开端。

虽然罗马人和其他欧洲人使用过骆驼，但真正把骆驼广泛地当做战斗单元的是拿破仑，他充分认识到在沙漠上骆驼优于战马，建立了骆驼营，可驮载700多战士。英国在印度和巴基斯坦也利用过骆驼，1881年，英国警卫队骑乘骆驼前往苏丹解救喀土穆。在第一次世界大战时，英国的Allenby将军利用骆驼，在巴勒斯坦和叙利亚的非常艰险的沙漠地区成功地运送人员和物资，在这些地区，马匹因缺水而踌躇不前；与此同时，作为阿拉伯半岛战士及作家的Lawrence和他的同盟贝多因人，在与土耳其人作战时，也利用骆驼运输货物。

## 赛驼竞赛

在阿拉伯半岛，贝多因人将驯化的骆驼作为他们游牧生活中不可或缺的一部分已有3000年左右的历史。实际上，只有在骆驼驯化以后，在沙漠上生活才有可能，在阿拉伯半岛的真正游牧生活也才成为事实。在半岛的核心地区，也就是鲁卜哈利沙漠，在机动车辆发明之前，也就只有骆驼能以其坚强的耐力、韧性和其他能力，可保证人们在此生活，因此，当地的贝多因人将骆驼称为“天神的礼物”。骆驼也是唯一一种能在如此的严酷条件下存活的家养动物。

除了在没有水源的情况下可长途跋涉的能力之外，单峰驼也可用于产奶、肉和毛。更为重要的是，“沙漠之舟”不仅能在酷热的气候下生存，而且对致命性的一些疾病，如牛瘟、口蹄疫和非洲马瘟有天然的抵抗力，成为养殖者不可或缺的生产资料。

对于古代的贝多因人部落来说，在冬季和春天下雨之后，需要寻找新的放牧草场，很可能需要穿过其他部落控制的地区，这有可能导致双方的摩擦和冲突。很显然在这种情况下，如果骑乘的骆驼速度很快，或者不需要水源就能生活很长时间，就有明显优势，可避免和对手面对面，或者快速通过敌方控制的水井。这种情形一直持续到20世纪中叶，在英国旅行家Wilfred Thesiger爵士的沙漠旅行经典游记——《阿拉伯之沙》中就有记述。对沙漠上生活的人来说，骆驼不仅在双方发生冲突时至关重要，在嫁娶、添丁和竞赛等社会性的活动场合也发挥重要作用。让单峰驼奔跑不太长的距离就可筛选出最快者。

在阿拉伯沙漠及其周围地区，贝多因人部落及其定居的亲戚们饲养和繁育了有重要地位的3种家畜：非常珍贵的阿拉伯马、萨路基犬和单峰驼。在世界上气候最酷热的地方经几百年之久培育出来的阿拉伯单峰驼，其耐力、智力和外形，完全可以与阿拉伯马媲美。这些行动敏捷、体型修长、四肢健美、头部欣长的棕色单峰驼，使得其他任何品种骆驼都逊色不少。多年过去了，昔日的育种记录不复存在，随着时光转换，进入到DNA技术时代，谱系鉴定方法也用于赛驼。

在石油热之后的大约60年之内，赛驼竞赛也发生天翻地覆的变化，如引入在赛马业中使用多年的尿样违禁药物检测。赛驼，这一曾经只是贝多因人部落消磨时光、测试骑士与坐骑耐力的活动，目前已在阿拉伯半岛许多国家成为可与现代赛马一较高低的一项非常受欢迎且科技水平高的体育赛事。

在阿联酋，有约20个赛驼场，每年10月至翌年4月举办竞赛，吸引着阿拉伯海湾许多国家的人带着骆驼参赛。仅在阿联酋。就有100 000峰竞赛用单峰驼，每峰顶级骆驼的价值在几百万美元（表5）。

图13　阿联酋的现代赛驼竞赛，已用机器人代替了年轻的骑士

表5　阿拉伯半岛的单峰驼数量

| | |
|---|---|
| 巴林 | 2 000 |
| 科威特 | 5 000 |
| 阿曼 | 8 000 |
| 卡塔尔 | 24 000 |
| 沙特阿拉伯 | 780 000 |
| 阿联酋 | 250 000 |
| 也门 | 140 000 |
| 合计 | 1 425 000 |

目前在阿联酋和其他海湾国家举办的赛驼与30年前有本质的不同。在1992年成立了骆驼竞赛联合会，推动在所有酋长国骆驼竞赛的标准化。另外一个巨大的变化是用机器人替代了骑士（图13），这使得骆驼竞赛在海湾国家更受欢迎。

2009年，世界动物卫生组织（OIE）邀请旧大陆和新大陆地区的一批专家，首次在其总部所在地巴黎召开骆驼疫病特别会议。在2010年又召开一次会议，撰写了骆驼最重要疫病的报告，可在OIE的网站下载（www.oie.int或者oie@oie.int），本书的附录部分也附有该报告。这一系列令人欢欣鼓舞的事情以及骆驼疫病已收入OIE的2008年版《陆生动物疫苗和诊断手册》，都说明骆驼不再是不合时宜的，也不是昨日黄花。越来越多的人们和著名的组织机构都认识到骆驼科动物在减轻贫困和抗沙漠化方面的无以伦比的作用。在酷热的气候条件下，只需要少量的树叶，就可生产出大量的质量很好的奶、肉和毛，只有骆驼做得到。

能保证骆驼在恶劣环境生存的因素，除了解剖学和生理学上一系列特异之处外[23]（表6），骆驼对多数疫病的易感性很低也是重要因素[8]，对病毒病更是如此，双峰驼的传染病主要是由细菌引起。由于骆驼对多数微生物具有抵抗力[6,19]，再加上以前人们对骆驼的兴趣不大，导致了有关骆驼传染病的研究文章很少的假象。编撰本书的目的是为了填补这一空白，尽可能地搜集整理有关骆驼的病毒病、细菌病、真菌病，以及病理学和寄生虫病学方面发表的文献。书中的大部分内容来自单峰驼疫病研究结果，这是因为双峰驼的内容很少或很难得到，或者根本就不存在。目前，与双峰驼饲养国家的交流逐渐增多，希望能改善这一状况。本书也吸纳了新大陆骆驼的疫病。

表6　单峰驼独特的生理特征

| | |
|---|---|
| 1 | 牛每天经粪便排出20-40L水，而骆驼只有1.3L，饲养骆驼远比饲养牛更能适应沙漠地区缺水的现状。 |
| 2 | 骆驼的热调节能力会因是否有水源而有很大的变化。骆驼对脱水的耐受性非常强，失去体重⅓的水分可以不出现任何疾病的征兆。 |
| 3 | 脱水的骆驼可根据环境温度而改变体温。早晨时，沙漠气温低，骆驼体温也可低至34℃，在下午的晚些时候，体温有可能达到42℃。骆驼随环境温度改变体温的主要目的是防止出汗。体温升高可以节约很多因正常情况下散发热量而需要的水分。血液温度过高会对大脑和眼睛的视网膜细胞造成永久损伤，但骆驼能利用呼出气体中的水分使大脑和眼睛降温。呼出气体中的水蒸气在鼻腔内停留较长的时间，使颈动脉网冷却，进入大脑和眼睛血液供应小网络。 |
| 4 | 如将山羊置于没有阴凉的圈内，在没有水的情况下不出3天即死亡。巴尔基绵羊也在断绝水源3天后死亡，而骆驼可在不饮水的情况下生存20 ~ 30天。 |
| 5 | 一峰600kg的骆驼可在3分钟内补充200L的水，消除脱水状态。骆驼的红细胞对低渗有很强的抵抗力。贝多因山羊在断水4天后，当在8分钟内补充所缺水量的40%时，因溶血而死亡。 |
| 6 | 骆驼在饮水后，水分迅速进入血流中，4小时后，全身大部分组织或器官基本达到平衡状态。其他动物没有如此快的吸收水分进入血液的能力。 |
| 7 | 驼峰在需要能量存储时可以堆积脂肪，也可间接地降低体温。其机理是脂肪在驼峰堆积后，身体的皮下组织脂肪少，有利于热的散发。 |
| 8 | 在脱水的骆驼中，消化道是其水源的唯一供应途径，小肠可不断地吸收水分。骆驼体重的75%为液体，当连续采食时，胃内就会有水分出现。骆驼胃内存储有大量的液体，能在3周甚至更长时间不饮水而继续正常采食。在一个实验中，给骆驼断绝水源51天，只饲喂干草，在实验的后期，骆驼食欲下降，但只损失37%的体重。 |
| 9 | 骆驼第一胃内（瘤胃）有大量的钠和二碳化合物，以及浓度稍低的氯和镁，在唾液和小肠中，这些电解质的浓度也很高，有利于充分利用消化道中的水分。 |
| 10 | 骆驼的肾脏有很长的汗勒氏袢（细尿管袢），当骆驼脱水时，可大幅降低尿液的排量，它对盐也能很好地处理，当骆驼饮用海水时，也没有任何副作用。骆驼排出的汗液中盐的含量可以两倍于海水。 |
| 11 | 骆驼缺水时，可在血液中"储藏"糖分，以防止通过尿液排出水分（糖具有非常强的吸湿能力）。实验表明，血糖浓度高达1300mg/dL时，尿液中检测不出糖。一旦恢复饮水，就会大量排尿，血糖回归正常。 |
| 12 | 脱水的骆驼仍能泌乳。 |
| 13 | 骆驼科动物具有非常特殊的免疫系统，抗体缺乏重链，称为纳米抗体。 |
| 14 | 骆驼的肺脏有肺隔，可防止肺炎。 |
| 15 | 骆驼采用俯卧交配，属于诱导排卵，交配期相对较短，孕期为13个月。 |

## 参考文献


[1] Allen W.R.,Higgins A.J.,Mayhew I.J.,Snow D.H. & Wade J.F. (1992). –In Proceedings of the 1st International Camel Conference,2–6 February 1992,Dubai. R. & W. Publications,Newmarket,UK.

[2] Breulmann M.,Boer B.,Wernery U.,Wernery R.,El Shaer H.,Alhadrami G.,Gallacher D.,Peacock J.,Al-Chaudhary S.,Brown G. & Norton J. (2007). – The camel from tradition to modern times. UNESCO Doha Office,Qatar.

[3] Camels Australia Export. (1995). – The central Australian camel industry. Brochure of the Central Australian Camel Industry Association,Alice Springs,Australia,1–4.

[4] Carpio M. (1991). – Camelidos socio-economia Andina [Camelids and Andean socio-economics]. In Produccion of de Rumiantes Menores: Alpacas,Lima,Peru (C. Novoa & M. Florez,eds). *Rerumen*,3–16.

[5] De Genst E.,Saerens D.,Myldermans S. & Conrad K. (2006). – Antibody repertoire development in camelids. *Dev. Comp. Immunol.*,30 (1–2):187–198.

[6] El-Gayoum S.E.A. (1986). – Study on the mechanism of resistance to camel diseases. Thesis,Tierarztl Institut, Gottingen, Germany,22.

[7] Faye B. & Bonnet P. (2012). – Camel sciences and economy in the world: current situation and perspectives. In Proceedings of the Third Conference of the International Society of Camelid Research and Development,29 January–1 February 2012,Muscat,Oman,Keynote Presentation,2–15.

[8] Fazil M.A. & Hofmann R.R. (1981). – Haltung und Krankheiten des Kamels. *Tierärztl. Praxis*,9,389–402.

[9] Fowler M.E. (2010). – Medicine and surgery of camelids. 3rd Ed. Wiley-Blackwell,Oxford,UK.

[10] George U. (1992). –Überleben. *Geo Spezial. Sahara*,6,47.

[11] Hamers-Casterman C.,Atarhouch T.,Myldermans S.,Robinson G.,Hamers C.,Songa E.B.,Bendahman N. & Hamers R.(1993). – Naturally occurring antibodies devoid of light chains. *Nature*,363 (6428),446–448.

[12] Kinne J.,Wani N.A.,Wernery U.,Peters J. & Knospe C. (2010). – Is there a two-humped stage in the embryonic development of the dromedary? *Anatomica Histologia Embryologica*,39 (5),479–480.

[13] Knospe C.,Kinne J.,Wani N.A.,Wernery U. & Peters J. (2010). – Is there a two-humped stage in the embryonic development of the dromedary? Poster presented at Camels in Asia and North Africa,Workshop,5–6 October 2010, Vienna,Austria.

[14] Koehler J. (1981). – Zur Domestikation des Kamels. Thesis,Hannover Veterinary School,Hannover,Germany.

[15] Koehler-Rollefson I. (1988). – The introduction of the camel into Africa with special reference to Somalia. Working paper,Somali Academy of Sciences and Arts,Mogadishu,Somalia,24.

[16] Kolpakow V.N. (1935). – Tierhaltung,-schutz und –zucht. Ueber Kamelkreuzungen. *Berl. Tieraerztl. Wschr.*,9,617–621.

[17] Leue G. (1969). – Erstmaliges Kamelrennen in Europa 1969 auf der Pferderennbahn in Koln aus veterinarphysiologischer, genetischer and biomechanischer Sicht. *Dtsch. tierärztl. Wschr.*,78 (18),500–502.

[18] Lombardini L. (1879). – Monographia dei cammelli. Tipografia T. Nestri,Pisa. *Ann. Del. Universita Toscane*,259,147–187.

[19] Margan U. (1987). – Vergleichende Untersuchungen zur Bedeutung der alternativen Komplementaktivierung bei Rindern und Kamelen. Thesis,Tierärztl Institut, Gottingen,Germany,33.

[20] Massmann U. (1981). – Kamele in Südwestafrika. *Namib und Meer*,9,31–54.

[21] Peters J. (1997). – Das Dromedar: Herkunft,Domestikationsgeschichte und Krankheitsbehandlung in fruhgeschichtlicher Zeit. *Tierä ztl. Praxis*,25,559–565.

[22] Peters J. (1998). –*Camelus thomasi* Pomel,1893,a possible ancestor of the one-humped camel? *Int. J. Mammalian Biol.*,63,372–376.

[23] Schmidt-Nielsen K. (1964). – Desert animals: physiological problems of heat and water. Clarendon Press,Oxford.

[24] Sielmann H. (1982). – Weltreich der Tiere. Naturalis Verlags- und Vertriebsgesellschaft,mbH,München,Mönchengladbach,Ar bus.

[25] Torres H. (1992). – South American camelids: an action plan for their conservation. South American Camelid Specialist Group,Gland,Switzerland. IUCN/CSE.

[26] Viswanathan L. (1991). – More about camels. The Gazelle,Dubai Natural History Group,6,6.

[27] Wernery U. (1992). – Dromedare,die Rennpferde Arabiens. *Tierärztl. Umschau*,47,801.

[28] Wernery U.,Fowler M.E & Wernery R. (1999). – Color atlas of camelid hematology. Blackwell Wissenschafts-Verlag, Berlin.

[29] Wilson R.T. (1989). – Ecophysiology of the Camelidae and desert ruminants. Springer Verlag,Heidelberg,Germany.

[30] Wilson R.T. (1998). – Camels: The tropical agriculturalist,MacMillan,France,106.

[31] Yagil R. (1985). – The desert camel. Verlag Karger,Basel.

[32] Zeuner F.E. (1963). – A history of domesticated animals. Hutchinson,London.

## 深入阅读材料

Beil C. (1999). – Reproduktion beim weiblichen Kamel (*Camelus dromedarius* und *Camelus bactrianus*). Eine gewichtete Literaturstudie. Thesis,Hannover Veterinary School,Hannover,Germany.

Bhattacharya A.N. (1988). – Camel production research in northern Saudi Arabia: a monograph. Ministry of Agriculture and Water Department of Agricultural Research,UTFN/SAU/008/SAU.

Bitter H. (1986). – Untersuchungen zur Resistenz von Kamelen (*Camelus dromedarius*) unter besonderer Berucksichtigung der Infektion mit Trypanosoma evansi (Steel 1885). Thesis,Hannover Veterinary School,Hannover,Germany.

Doose A. (1990). – Funktionen und Morphologie des Verdauungssystems des einhockrigen Kamels (*Camelus dromedarius*). Thesis,Hannover Veterinary School,Hannover,Germany.

Farid M.F.A. (1981). – Camelids Bibliography. ACSAD-AS,15.

Faye B. (1997). – Guide de l'élevage du dromadaire. Sanofi Sante Nutrition Animale,La Ballastiere – BP126,33501 Libourne, Cedex,France,115 — 116.

Fowler M.E. (1997). – Evolutionary history and differences between camelids and ruminants. *J. Camel Prac. Res.*,4 (2),99 — 105.

Gruendel M. (1988). – Das Blut des einhockrigen Kamels (*Camelus dromedarius*). Eine Literaturubersicht. Thesis,Hannover Veterinary School,Hannover,Germany.

Hare J.N. (1997). – Status and distribution of wild Bactrian camels (*Camelus bactrianus ferus*) in China. *J. Camel Prac. Res.*,4 (2),107 — 110.

Hare J.N. (1998). – The lost camels of Tartary. Little Brown and Company,London.

Mukasa-Mugerwa E. (1981). – The camel (*Camelus dromedarius*): A bibliographical review. International Livestock Center for Africa,ILCA Monograph 5,4 — 119.

Saint-Martin G.,Nitcheman M.F.,Richard D. & Richard M.A. (1990). – Bibliographie sur le dromadaire et le chameau. 2nd Ed., Tome 1,Tome 2: Index.

Skidmore J.A.,Billah M.,Binns M.,Short R.V. & Allen W.R. (1999). – Hybridizing Old and New World camelids: *Camelus dromedarius* ×*Lama guanicoe*. *Proc. R. Soc. Lond. B*,266,649 — 656.

Wensvoort J. (1991). – Camels,camel nutrition and racing camels. The Gazelle,Dubai Natural History Group,6,5.

Wernery U. (1997). – Dromedare in Arabien. Lamas. Haltung and Zucht von Neuweltkameliden,5 (1),34 — 36.

Wernery U. & Kaaden O.-R. (1995). – Infectious diseases of camelids. Blackwell Wissenschaftsverlag,Berlin.

(殷宏译，储岳峰校)

# 1 细菌病

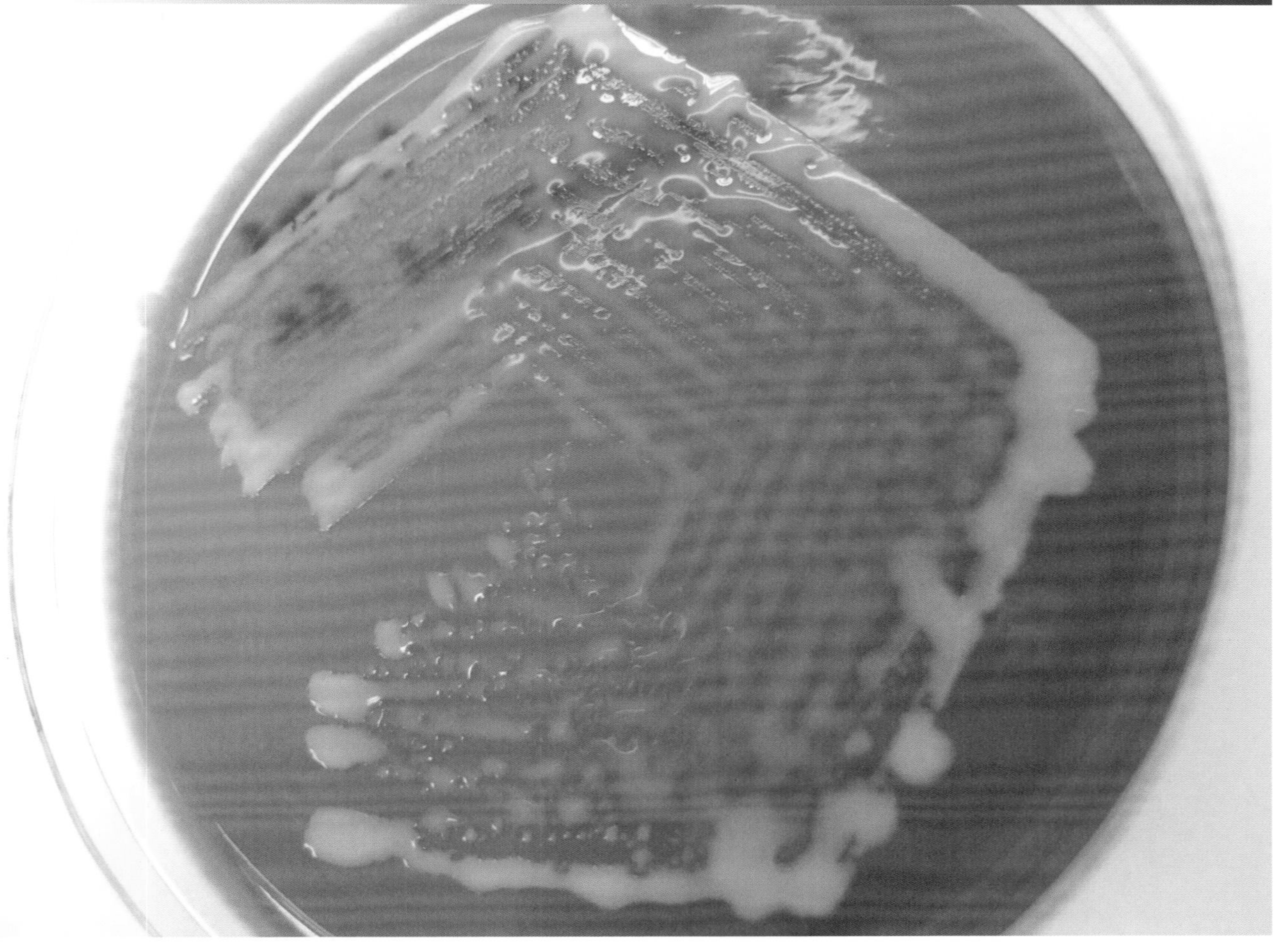

# 1.1 全身性疾病

## 1.1.1 厌氧菌感染

梭菌性疾病一直以来威胁着世界各地的家畜。梭菌都能产生外毒素，这些外毒素决定了其致病性。梭菌一般存在于土壤和动物、人类肠道中，只有在特定条件下才引发疾病。梭菌属细菌的普遍存在导致梭菌病很难被消除，只能通过预防进行控制。新大陆骆驼（NWCs）和旧大陆骆驼（OWCs）都可感染梭菌疾病[14,38,39]，骆驼感染梭菌病的种类和感染部位在表7中显示。

### 公共卫生意义

产气荚膜梭菌产肠毒素，C型产气荚膜梭菌和艰难梭菌都是人和动物致病菌。肠毒素通常可通过污染、处理不当的食物可以将梭菌传染给人类，从而引发食源性传染病。肠毒素是美国第三大食源性疾病，每年引发约250000起病例，该病通常为自限性疾病。然而C型产气荚膜梭菌可能引起坏死性肠炎和节段出血性病理损伤，艰难梭菌是人类和家畜的共同病原，在零售的肉中已经分离出艰难梭菌的基因型[31]。

表7 梭菌病以及在骆驼中的感染位点

| 细菌种类 | 感染部位 | 临床表现 | 动物 | 特征 |
|---|---|---|---|---|
| A型产气荚膜梭菌<br>*C□perfringens* A | 肠 | 突然死亡，没有腹泻，节段性出血性肠炎，坏死性肠炎 | 新大陆骆驼 | 并发 |
| B型产气荚膜梭菌<br>*C. perfringens* B | 肠 | 坏死，出血性肠炎 | 新大陆骆驼<br>旧大陆骆驼 | 罕见<br>罕见 |
| C型产气荚膜梭菌<br>*C. perfringens* C | 肠 | 肠毒血症 | 新大陆骆驼<br>旧大陆骆驼 | 罕见<br>罕见 |
| D型产气荚膜梭菌<br>*C. perfringens* D | 肠 | 小牛心肌变性 | 新大陆骆驼<br>旧大陆骆驼 | 罕见<br>罕见 |
| 腐败梭菌<br>*C. septicum* | 肌肉 | 创口感染，急性全身性疾病 | 新大陆骆驼 | 罕见 |
| 索氏梭菌<br>*C. sordellii* | 肠 | 类似A型产气荚膜梭菌，出血性肠炎 | 旧大陆骆驼 | 罕见 |
| B型诺维氏芽胞梭菌<br>*C. novyi* B | 肌肉 | 黑病，肌坏死 | 新大陆骆驼 | 罕见 |
| A型诺维氏芽胞梭菌<br>*C. novyi* A | — | 突然死亡，没有损伤 | 新大陆骆驼 | 罕见 |
| 溶血梭菌<br>*C. haemolyticum* | 肝脏 | D型诺维氏芽胞梭菌，坏死性肝炎 | 新大陆骆驼 | 罕见 |
| 气肿疽梭菌<br>*C. chauvoei* | 肌肉 | 黑脚病，高热，水肿，四肢肌肉产气，黑腿病 | 旧大陆骆驼 | 新大陆骆驼可实验发病 |
| 破伤风梭菌<br>*C. tetani* | 创口 | 创口，去势污染 | 新大陆骆驼<br>旧大陆骆驼 | 常见<br>常见 |
| 肉毒梭菌<br>*C. botulinum* | 肌肉 | 摄食 | 新大陆骆驼<br>旧大陆骆驼 | 罕见<br>罕见 |

## 病原学

梭菌性疾病由梭菌属细菌引起，这些细菌较大、革兰氏阳性、厌氧、产芽孢，孢子在母细胞上突起。产气荚膜梭菌在动物组织上形成一个非移动的囊。梭菌的氧化酶为阴性，过氧化物酶为阴性，不同种间厌氧需求不同，大多数致病菌能产生一种到多种毒性有差异的外毒素，营养器官能形成孢子从而在土壤中长时间存活，每克被污染的土壤中可含有多达$10^5$个产气荚膜梭菌孢子[29]。

## 流行病学和临床症状

以往的、经典的厌氧菌感染病因分类中，将特殊的临床症状归于一个特定的梭菌，在当今不再被认为是有效的。气相色谱分析和PCR等这些现代鉴别方法的应用使得病原检测更精确。利用这些方法将流行性厌氧综合征分为3类：

（1）气性水肿综合征。

（2）肠毒血症。

（3）毒素性综合征。

图14中总结了这个发展过程（家畜重要传染病和毒素的病原学区分）。

由于梭菌病不同的临床症状和病理学特性，导致该病很难鉴别诊断，A、B、C、D型产气荚膜梭菌、诺维氏芽孢梭菌、气肿疽梭菌和腐败梭菌都已从骆驼中分离出。Akeilla等人检测了埃及屠宰的285头单峰驼的肠系膜淋巴结和肝脏样本，检测出了A、B、D型产气荚膜梭菌，生孢梭菌，索氏梭菌，溶组织梭菌，其中，只有生孢梭菌对豚鼠没有致病性。

## 气性水肿综合征

据Seifert报道[29]，气性水肿综合征的致病病原体能引起以下疾病：

（1）黑腿病；

（2）恶性水肿；

（3）杆状血红蛋白尿；

（4）传染性坏死性肝炎。

这些致病菌很少在骆驼中分离到。很多现有文献已过时，如流行病学研究不完善，传统的分析方法很有可能导致疾病误诊。

现有方法已能将气性水肿综合征的病原体与以下病原体区分开[29]。

（1）气肿疽梭菌和腐败梭菌，野毒株（335份气肿疽梭菌和735份腐败梭菌来自马达加斯加，805份来自墨西哥）；

（2）溶组织梭菌，索氏梭菌，A–C型诺维氏芽孢梭菌，溶血性梭菌；

（3）A–F型产气荚膜梭菌以及野毒株（217份马达加斯加）。

## 黑腿病

已报道的单峰驼气肿疽梭菌感染最早发生在北非、东非、乍得以及印度，但这些有待进一步证实[17]。除了Cross[7]之外，Curasson[8]认为之前很多学者混淆了黑腿病和炭疽杆菌引起的炭疽。这两种病发展趋势相似，开始肩部皮下肿胀，最终导致动物在2～3天内死亡。Hutyra等[19]报道骆驼对气性水肿不易感。然而，Cross[17]对3只单峰驼肌内注射气肿疽梭菌进行人工发病实验，通过肿胀类型可以区分气性水肿和炭疽。据作者所知，到目前为止，还没有骆驼气性水肿的相关出版物。

黑腿病是由气肿疽梭菌毒素在反刍动物体引起四肢充血肿胀和气性产物的疾病，这种杆菌引起的骨骼肌肌肉发炎通常和阉割、剪毛、断尾引起的外伤和创伤有关。

黑腿病已经在羊驼身上做了实验[14]，可以认为该病比起牛，旧大陆骆驼和新大陆骆驼更能抵抗黑腿病感染，诊断有赖于感染组织分离的杆菌或荧光抗体实验，感染早期用高剂量青霉素治疗很有必要。

## 恶性水肿

恶性水肿在多种宿主中引起气性坏疽或产气性的肌坏死。腐败梭菌、气肿疽梭菌、产气荚膜梭菌、索氏梭菌、诺维氏芽孢梭菌都已在恶性水肿动物的典型病变处分离得到[25]，这些梭菌在动物体中普遍存在，也存在于健康骆驼的肠中。这种感染通常源于土壤，伤口往往是感染的入口。深度刺伤和创伤给厌氧菌提供了良好的生长环境，但也会通过肠道感染发生。随着细菌的生长，其会在创伤部位释放神经毒素，然后吸收进入循环系统。同时，局部外毒素引起水肿和黑斑病，引发坏疽，死亡后组织很快发生变化，这些都是梭菌疾病所具有的特点。皮肤坏疽快速恶化，伴随水肿和皮下积液，积液往往呈胶冻状并混有血液。当动物突然死亡后，很难观察到损伤。

恶性水肿容易感染秘鲁的羊驼，具有重大的经济意义，该病同时还和科罗拉多被响尾蛇咬伤的美洲驼有关[14,23]。恶性水肿往往由多种梭菌感染美洲驼[4]和羊驼[14]引起，在新大陆骆驼中该疾病由腐败梭菌引起，典型症状为外伤感染和水肿以及急性全身性疾病，这很有可能使动物迅速毙命，有一例关于母美洲驼自然感染诺维氏芽孢梭菌后很快死亡的报告。

Uzal等[34]报道了阿根廷一只驼马因为右侧腹皮肤擦伤而引起的恶性水肿，尸检结果显示在擦伤部位的皮下组织出现了严重的胶冻状出血性水肿，和梭菌性肌坏死；Tyler[33]也有相同的报道。通过涂片观察到很多革兰氏阳性细菌，腐败梭菌和诺维氏芽孢梭菌通过荧光抗体实验也呈阳性结果。

### 治疗和控制

病畜应尽早治疗，每隔4 ~ 6小时静脉注射高剂量青霉素，手术切口对感染位点的积液排除有一定辅助作用。也可用青霉素预防有感染风险的骆

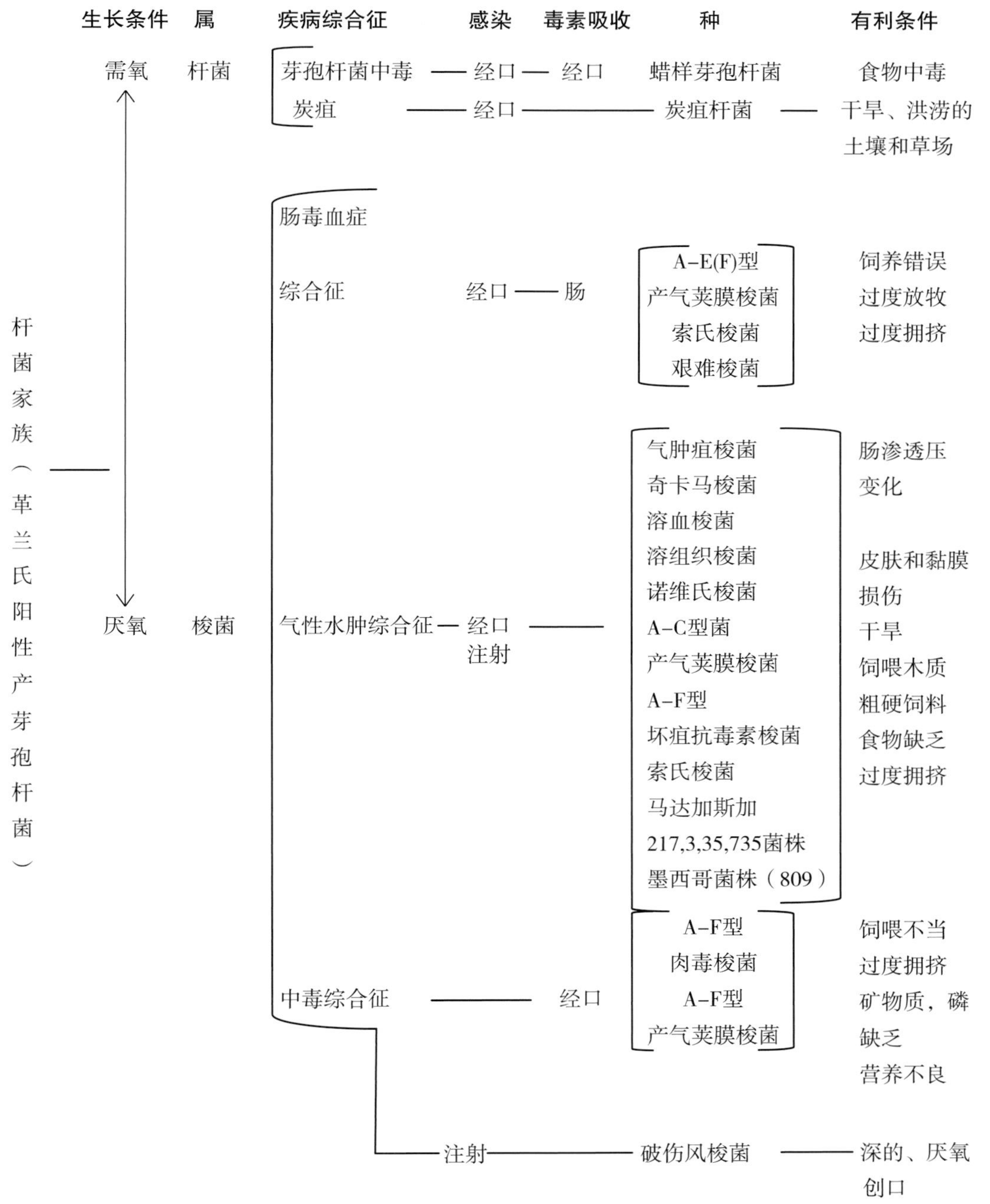


图14 家畜重要传染病和中毒病的病原学分类

驼，注意剪毛、阉割、断尾时的卫生以避免疾病感染。特定细菌类毒素疫苗可以用来保护贵重动物。

### 细菌性血红素尿（黄疸血红素尿）和传染性坏死性肝炎（羊黑疫）

其他两种恶性水肿综合征疾病，细菌性血红素尿和传染性坏死性肝炎，在骆驼科中还未有过报道，两种疾病都由诺维氏芽孢梭菌引起，传染性坏死性肝炎由B型引起，细菌性血红素尿由D型或者溶血梭菌引起。

### 肠毒血症综合征

所有类型的产气荚膜梭菌及索氏梭菌和螺旋形索状芽孢杆菌均能引起肠毒血综合征，产气荚膜梭菌（尤其是A型）在健康动物的肠内也存在，因此很难依据产气荚膜梭菌的存在判定疾病。产气荚膜梭菌引起的肠毒血症在世界范围内均有发生，而且在所有的家畜中均能观察到，据Seifert报道[29]，不合理的饮食，气候的影响，草场的改变，动物的运输和处理都是诱发该病的主要因素。

Moebuu等[22]，Ipatenko[20]，Chauhan 等[6]，Gameel等[15]，Younan和Gluecks等[42]报道了骆驼的急性和亚急性肠毒血症，以及A、C、D型产气荚膜梭菌引起的出血性肠炎，Fowler等报道了A、C、D型产气荚膜梭菌在新大陆骆驼中引起的肠毒血症。

### 产气荚膜梭菌引起的新大陆骆驼肠毒血症

肠毒血症是一种新大陆骆驼重要的细菌疾病，导致新生驼很高的死亡率，也会在成年新大陆骆驼中零星暴发。本病在所有安第斯山饲养骆驼的城市都有发生，在幼崽中周期性感染。该病发生和极端的气候有关，比如暴雨以及出生季天气温暖，身体机能良好的幼崽在2～3周龄时感染。5～6岁最大致死率可达70%，从而打乱了所有计划好的育种工作[24]。

### 病原学

肠毒血症主要由产气荚膜梭菌引起，有A、B、C、D、E五个型。该疾病主要由A型产气荚膜梭菌引起，发生在世界不同地方不同环境的老、幼单峰驼。A型产气荚膜梭菌产生$\alpha$毒素，通过磷脂酶引发溶血、坏死、膜溶解。在成年动物中为并发的疾病，如疥癣、锥虫病、球虫病、第一胃酸中毒都是患病因素。幼年单峰驼中，硒、铜的缺乏以及溶血性埃希氏大肠杆菌和沙门氏菌都能引发该病。

### 新大陆骆驼中的临床症状

A型产气荚膜梭菌在南美羊驼幼崽中是一种非常严重的疾病，动物死亡率在10%～70%[11,13,18,23,26,27,28]，就连管理良好的牧场死亡率也接近50%。羊驼发病的年龄在8～35天，突然死亡或有短时期的卧地，呈现神经系统紊乱症状。

尽管A型产气荚膜梭菌引发的肠毒血症临床症状在所有骆驼品种中都相似，在新大陆骆驼中最普遍的症状描述如下。比较惊人的是腹泻只发生在肠病原中含沙门氏菌或溶血性大肠杆菌的患病幼崽或成年动物中。这也许可以解释感染动物的突然死亡，这意味着骆驼的生理特性不会使腹泻发展。在幼崽中，首先观察到由$\alpha$毒素引起的“突然死亡综合征”，这种毒素很快被肠壁吸收，继而被转运到不同器官。初步调查证实局灶性、对称性脑软化由产气荚膜梭菌毒素引起的酸中毒引发，在小牛也是如此[35]。至于突然死亡综合征，幼崽发展为疝痛、抑郁、腹痛，有时也会有腹部肿胀。尸体剖检结果、诊断、预防和治疗与新大陆骆驼和旧大陆骆驼相同，接下来的部分对此有总结。

### 旧大陆骆驼肠毒血症

在新大陆骆驼中，A型产气荚膜梭菌在很多国家的单峰驼中普遍发生，也有双峰驼的报道，Wernery[41,37]等、Seifert等[30]、Wernery和Kaaden等[38]对阿拉伯联合酋长国（UAE）单峰驼中暴发的A型产气荚膜梭菌做了大量研究，发生在种用和竞赛的单峰驼的恶性和急性肠出血症，还伴有心肌变性和软肾。在3组不同年龄的单峰驼中，诱发病因的相关因素和疾病暴发有关。

临床症状和病理损伤的多样性导致梭菌肠毒血症很难被诊断，临床症状和损伤程度取决于毒素吸收量。很多梭菌有机体产生的毒素同时具有全身和局部效应，包括溶血、局部组织坏死。疾病以急性、亚急性、慢性方式发生于不同的年龄段，在个体动物中零星散发和在群体骆驼中并发。病理损伤因为病因不同而不同，急性和亚急性病例刚开始在不同器官有小的出血点，继而发展为节段性小肠出血，严重出血性肠炎。慢性病例表现为坏死性肠炎、肺水肿、心包积液。病理损伤常伴随气体和液体引起的肠鼓胀。

单纯A型产气荚膜梭菌的急性和亚急性病例，粪便排泄物的颜色变为黑色，粪球变小，非常坚硬，预示便秘。能进一步明确表明该病的指征是白细胞总数下降（WBC）$8\times10^9$/L到$4\times10^9$/L，但分类计数差别不大。可以认为白细胞总数的减少是毒素破坏了白细胞的形成中心，相似的现象也在一头严重内毒素血症的竞赛单峰驼中观察到，在1.1.4中的内因性出血中有相关描述。只有资深实验室开展包括尸检的调查才能区分这两种重要的疾病。内毒素血症中，白细胞总数可降低至$1\times10^9$/L。

在一群90头的育种动物中，71%的单峰驼有急性伊氏锥虫感染[9]。伊氏锥虫可以引起单峰驼的免疫抑制[21]，也是引发此次畜群超急性产气荚膜梭菌暴发的诱发因素，因为营养不良和气候影响已排除。感染骆驼表现为以下症状：疝痛、腹痛；流汗；肌肉震颤；共济失调；攻击性；亢奋；癫痫。被感染动物在上述临床症状发作1小时内死亡，动物尸检病理变化轻微，包括：胸肌组织有出血点；小脑和脑干有出血点；咽黏膜有出血点；胸膜下肺和心外膜有出血点（图15）；瓣胃和真胃黏膜有出血点（图16）；心包积液和纤维蛋白渗出物；黑肾以及实质组织和肾小囊粘连（图17）。

在另一起病例中，竞赛单峰驼中暴发的A型产气荚膜梭菌感染并发沙门氏菌感染[9]，动物发展为顽固性腹泻并在4天内死亡，尸检中在相同器官中发现的病理变化，比上文列出的更严重，包括严重的出血性肠炎、心包积液、纤维蛋白渗出物以及瓣胃和真胃的瘀斑。

在个别竞赛骆驼中，一个很重要的诱发因素就是竞赛前的饮食不当，很多骆驼被饲喂了大量的未压碎的大麦、牛奶、蜂蜜、苜蓿，尸检发现胃系统中有大量未消化的大麦（图18和19），竞赛的应激也是促成极严重肠出血症的重要因素。

Wernary等[40]的调查可以解释为什么只有很少的竞赛单峰驼可以抵抗疾病，他们发现接近85%的竞赛单峰驼体内有很高的先天获得的A型产气荚膜梭菌抗体，因此可以保护免于肠毒血症的发生。据此建议用ELISA检测竞赛单峰驼体内的抗体，对于体内没有或者只有低水平A型产气荚膜梭菌抗体的动物进行免疫。

在年轻的骆驼中，梭菌病发展的重要因素包括血清免疫球蛋白的量变化（见2.1.13部分），并发疾病。硒、铜缺乏也都是诱发因素。骆驼科动物有上皮绒毛膜胎盘，和小牛或马驹一样，出

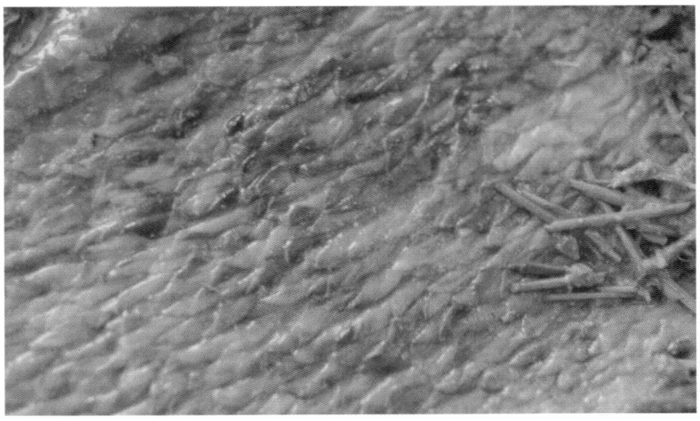

图16　产气荚膜梭菌肠毒血症：瓣胃瘀斑出血

图15　产气荚膜梭菌肠毒血症：胸膜下肺出血

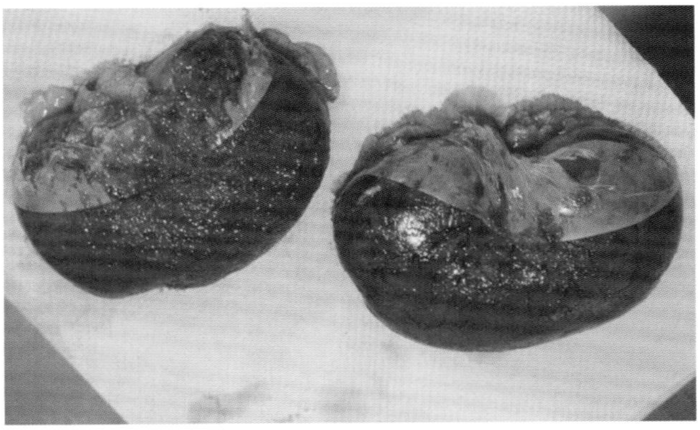

图17　产气荚膜梭菌肠毒血症：肾小囊粘着于肾实质组织

表8 阿拉伯联合酋长国的竞赛单峰驼矿物质和四种维生素的血清参考值

| 参数 | 设备 | 样本量 | 平均标准偏差（mmol/ L） |
|---|---|---|---|
| 铁 | Hitachi 912 | 100 276 | 20.97 ± 5.63 |
| 钙 | Hitachi 912 | 62 390 | 2.67 ± 0.48 |
| 镁 | Hitachi 912 | 613 | 1.04 ± 0.53 |
| 磷 | Hitachi 912 | 58 979 | 2.05 ± 0.34 |
| 钠 | Hitachi 912 | 747 | 149.3 ± 5.03 |
| 钾 | Hitachi 912 | 736 | 4.2 ± 0.40 |
| 氯化物 | Hitachi 912 | 764 | 111.5 ± 6.62 |
| 铜 | FAAS | 14 237 | 11.21 ± 2.62 |
| 锌 | FAAS | 6 190 | 8.22 ± 2.00 |
| 硒 | GFAAS | 14 899 | 1.91 ± 0.61 |
| 维生素$B_1$ | HPLC | 48 551 | 0.142 ± 0.039 |
| 维生素A | HPLC | 132 | 1.14 ± 0.44 |
| 维生素E | HPLC | 1 064 | 4.51 ± 1.63 |
| 维生素C | HPLC | 423 | 22.18 ± 7.80 |

FAAS，火焰式原子吸收光谱仪；GFAAS，石墨炉原子吸收分光光度计；HPLC，高效液相色谱；Hitachi912，德国，曼海姆，柏林格尔

生后通过肠吸收母乳中的免疫球蛋白获得被动免疫。尽管新出生的幼畜有免疫能力，但是在第一个月内在抗体系统不足以产生足够的保护性免疫球蛋白。出生时先天的球蛋白量很低，即使摄入大量初乳中的抗体，球蛋白水平在7天后下降，出生后20 ~ 30天达到最低水平，感染产气荚膜梭菌的幼驼在这段时间里损失最严重。Fowler[14]在新大陆骆驼做了相似的研究，发现新大陆骆驼血清中的球蛋白含量非常低（<52 g/L），食用初乳4 ~ 5天后增加到55 ~ 62 g/L，然而产后3 ~ 4周达到最低水平，幼驼的最高损失发生在这段时间。

除了已经描述的产气荚膜梭菌的临床症状和病理损伤，很多幼驼还表现出心和肾变化，Wernery等[37]报道了4头6周龄大的单峰驼严重的心肌变化（图20和21）和软肾病（图22）。El-Sanousi和Gameel等[12]报道了发生在沙特阿拉伯3 ~ 5周龄骆驼的D型产气荚膜梭菌肠毒血症的心肌萎缩、钙化和坏死，这种疾病的诱发因素似乎是断奶。4 ~ 6周龄的幼驼开始摄入牛奶以外的食物，尸检发现体内有凝固的牛奶，少量粗饲料，胃室中有沙子以及肠出血（图23）。

检测畜群所在的土壤样本，发现每克土壤中含有多达$10^4$个产气荚膜梭菌营生孢子，尽管牧场每天都在打扫，饲养骆驼这几年所在地的土壤中仍有很多梭菌的营生孢子，这给成长中的幼驼带来了持续性的感染威胁。流行病学知识的掌握使得骆驼主越来越频繁的迁移饲养骆驼或者用新鲜沙子替换污染的沙子。为了调查为什么幼驼会咽食沙子，对健康骆驼采集血样，建立矿物和维生素群平衡分析（表8），这些结果都和患病单峰驼在血液水平上进行了比较。

结果表明，血清中硒和铜的水平明显低于A型产气荚膜梭菌肠毒血症的单峰驼，硒低于0.8 mmol/L （参考值1.9 mmol/L），铜低于6.0 mmol/L（参考值11.2 mmol/L），因此认为单峰驼摄食沙子是为补偿营养物质的缺乏，现在认为很多肠毒血症死亡病例的心肌坏死主要是由硒缺乏引起，较小程度上由梭菌毒素引起。

## 诊断

鉴于梭菌在动物死后很快入侵，包括小肠液在内的样本，应从新鲜（<4小时）死亡动物中取

图18 产气荚膜梭菌肠毒血症：竞赛骆驼胃中未消化的牛奶

图19 产气荚膜梭菌肠毒血症：竞赛骆驼胃中未消化的大麦

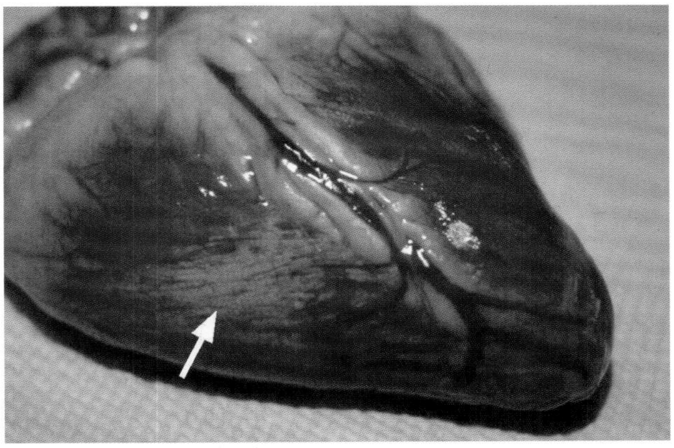

图20 幼单峰驼产气荚膜梭菌：严重的心肌变性是病畜死因

图21 幼单峰驼产气荚膜梭菌：心脏肌肉的玻璃样变性是病畜死因

图22 产气荚膜梭菌肠毒血症：单峰驼的“软肾”

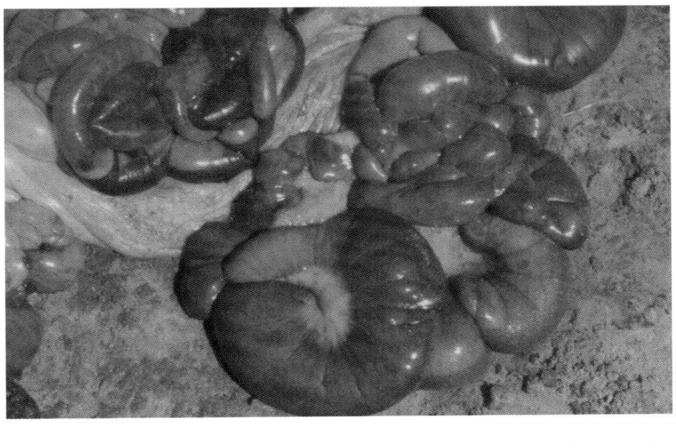

图23 幼单峰驼产气荚膜梭菌肠毒血症：小肠出血产气

样。毒素特别不稳定，因此小肠内容物在处理之前应该尽快冻存。梭菌肠毒素血症的实验室诊断通过检测最近死亡动物十二指肠处的梭菌毒素，肠内容物在动物死后立即除去并深度冷冻，第二天解冻，进行无菌操作过滤后对老鼠进行致病性检测，将1mL的稀释肠内容物腹腔注射实验老鼠，梭菌毒素存在情况下，老鼠在2～8小时内死亡，表现出痉挛和角弓反张。MTT，ELISA，PCR实验已广泛取代了老鼠致死实验并普遍用于肠液中梭菌毒素的检测。MTT，ELISA在液体稀释和建立梯度方面也有优势，梯度越高，肠毒素越多（图24）。

赛普拉斯（比利时，安特卫普）肠毒血症抗原试验，可检测首先在蔡斯勒琼脂中培养，继而在厌氧胰酪胨葡萄糖酵母提取物肉汤培养基中传代培养的可疑产气荚膜梭菌克隆的培养上清[38]，赛普拉斯ELISA可检测$\alpha$-、$\beta$-、$\varepsilon$-产气荚膜梭菌毒素和细菌本身（菌种鉴定）（图25）。

Tansuphasti等[32]和Gkiourtzisia等[16]用多重PCR检测了6个基因$\tau$-、$\varepsilon$-、$\beta$-、$\beta$2-和$\alpha$-毒素和肠毒素。$\alpha$-毒素也叫磷脂酶基因或者plc-$\alpha$，肠毒素ENT或者CPE（图26）。

在阿拉伯联合酋长国单峰驼肠毒素血症的研究过程中，肠毒素基因从未被检测到。A型$\beta$2-毒素只在很少的病例中发现，它不是突变的$\beta$2-毒素，但认为A型株同时拥有$\beta$2-基因和$\alpha$-基因时，致病性更强。在秘鲁肠毒素血症的新大陆骆驼中还发现了$\beta$2-基因。

在少数病例中，用多重PCR除了检测到$\alpha$-毒素基因，还有$\varepsilon$-基因，表明不是所有的肠毒素血症都是由A型毒素引起。发生在新大陆骆驼的A，C，D三种不同类型的产气荚膜梭菌肠毒血症各自都有报道[14]，但是临床症状和病理学是相似的。不管怎样，这些菌株的更多信息对疫苗的生产非常重要。

在可疑肠毒素血症中，新鲜死亡动物小肠黏膜碎屑中出现的大量革兰氏阳性菌，可作为梭菌肠毒素血症的证据（图27）。荧光抗体技术也被普遍用来确诊气肿疽梭菌、腐败梭菌、诺维氏芽孢梭菌和索氏梭菌。

死亡动物器官产气荚膜梭菌在有气体产生装置的厌氧Sahidi-Ferguson琼脂和蔡斯勒琼脂中培养，产气荚膜梭菌容易培养，在双区溶血琼脂中呈标准生长，暴露于氧气中几小时后产生绿色素（图28）。

阿拉伯联合酋长国暴发产气荚膜梭菌期间，用ELISA和PCR检测到一个（939/06）能产生较高$\alpha$-毒素的A型产气荚膜梭菌菌株。该菌株分离自一个死于急性梭菌肠毒血症的两岁赛用母单峰驼，被选用于$\alpha$-毒素细胞苗的生产。

## 治疗和控制

用牛产气荚膜梭菌高免血清治疗患病单峰驼非常有效，一些有价值的竞赛驼通过静脉注射100 mL抗血清被拯救，该过程可重复，无任何副作用。需要提到的是，在牛体内生产的高免抗血清，往往没有足量的抗$\alpha$-毒素，而且很难买到。另外，这些抗血清的生产都来自于世界其他地方的菌株，没法证实在另外的地方是否奏效。因此，中央兽医研究实验室生产了一种同源$\alpha$-毒素高免血清[38]，考虑到所有的骆驼家族都可以产生特殊抗体，称之为纳米抗体（见新生畜腹泻）。这种产物纯度高，包含20 ~ 30 g/L的免疫球蛋白。

肠毒血症同样导致高死亡率，抗生素治疗往往导致预后不良，因此，在急性病例中应该尝试高免血清治疗。此外，预防措施是重点，在重大疫病的预防过程中，卫生状况，饲养状况以及平日的饲养管理都应该优化。注意避免应激，例如突然更换饲料，频繁竞赛，过度拥挤。在濒临灭绝种群中，将金霉素以25 mg/kg的量掺在饲料中饲喂，并配合一些泻药帮助毒素排出。

类毒素疫苗被普遍应用于预防牛、羊肠毒素血症的暴发。因为新生儿的免疫系统尚未成熟，不足以产生足够的抗体，因此疫苗应用于孕畜，在孕畜分娩前2个月予以免疫，产前一月再加强免疫一次。

与其他疫苗一样，梭菌疫苗在骆驼中还未许可。然而，如果应用商业疫苗，则必须包含A型产气荚膜梭菌菌株，而实际情况往往不是这样。牛、羊的C、D型不能对A型感染产生有效的交叉保护，一种包含A型产气荚膜梭菌毒素的疫苗（诺华动物保健）已经在骆驼中应用成功。

梭菌疫苗只有瞄准引发该地区疫病的确切毒素才有效，因此，肠毒素血症的自体疫苗得以产生，例如类毒素疫苗，产自秘鲁，用于美洲驼和羊驼，包含了A型基因和$\beta$2类毒素的混合物，孕畜分娩前应注射两次肠毒素。

939/06菌株的自体疫苗也用于阿拉伯联合酋长国的竞赛单峰驼，它是一种由IDT Dessau（德国，德绍）生产的$\alpha$类毒素细胞疫苗（Bestvac 5890807）。已经证明想要获得至少一年的高水平抗体，3次皮下接种疫苗是非常必要。已经获得高水平先天抗体的新生单峰驼，抗体水平在免疫后不会有明显的上升。因此，正如之前提到的，建议用ELISA对所有竞赛单峰驼的A型产气荚膜梭菌抗体进行检

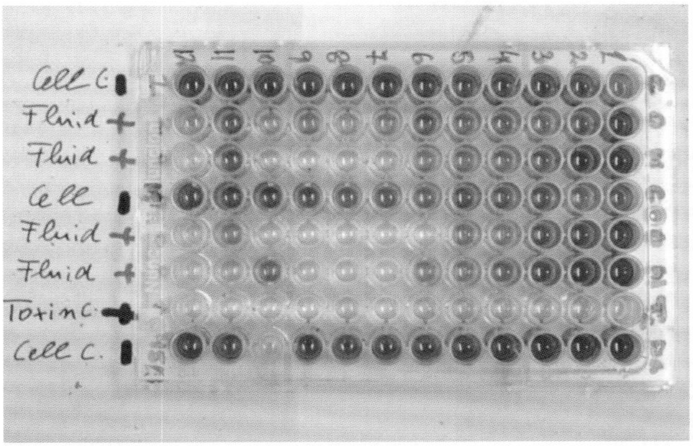


图24 Vero细胞的MTT实验表明梭菌肠毒血症单峰驼的肠液毒素

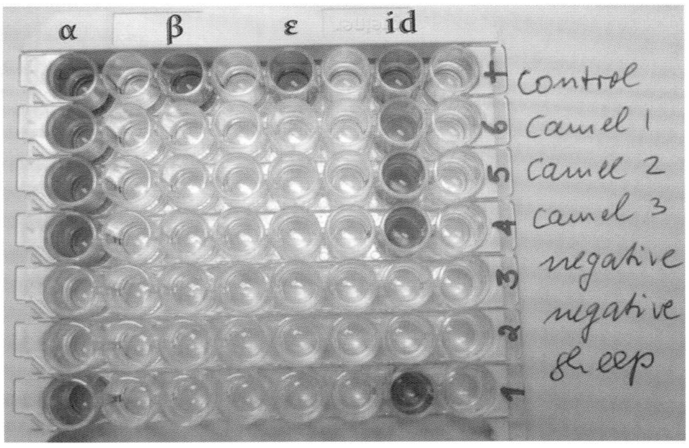


图25 产气荚膜梭菌的ELISA实验结果。ELISA检测到了3个主要毒素（$\alpha$、$\beta$、$\varepsilon$），分离株在对照排中从左往右，1、2、3号骆驼和一只羊只产生$\alpha$毒素

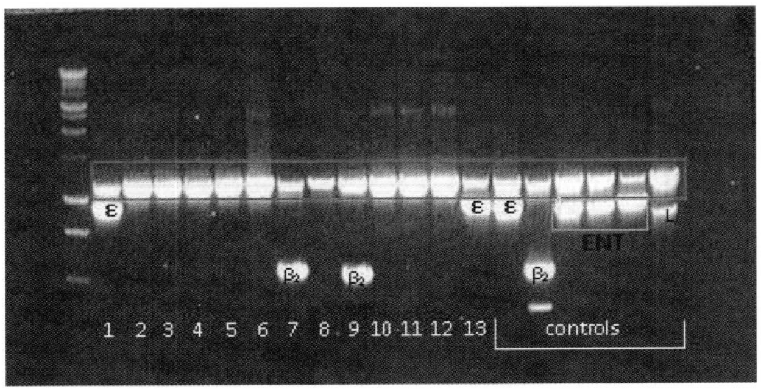


图26 多重PCR反应检测了$\alpha$、$\beta$、$\beta 2$、$\varepsilon$、$\tau$毒素基因和肠毒素基因。所有产气荚膜梭菌菌株上清都能产生$\alpha$毒素，1～13的样本都能产生$\varepsilon$毒素，样本7和9还能产生$\beta 2$毒素。

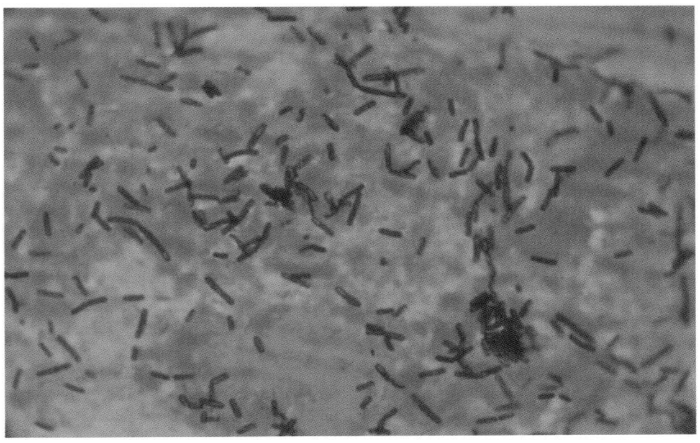

图27 产气荚膜梭菌，来自于梭菌肠毒血症单峰驼小肠黏膜刮的革兰氏阳性菌的数量增多

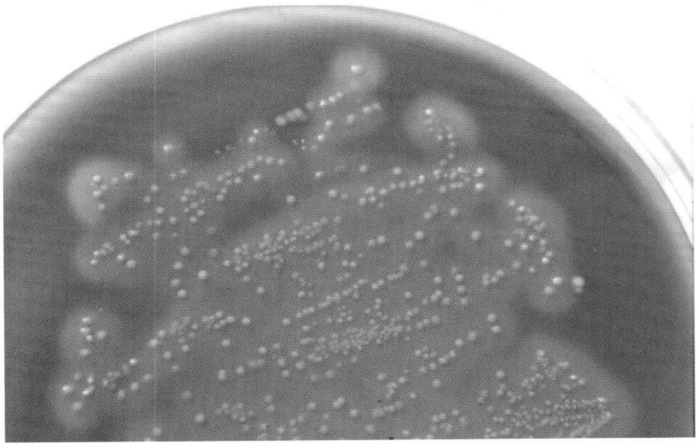

图28 蔡斯勒琼脂培养的产气荚膜梭菌克隆，暴露于氧气后出现溶血和绿色色素

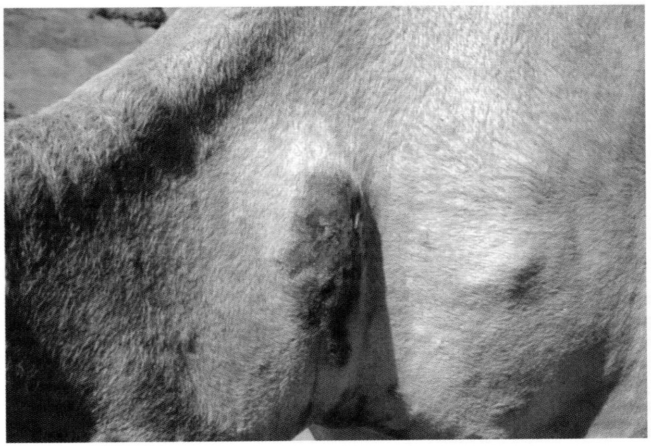

图29 产气荚膜梭菌肠毒血症：皮下接种Monatide佐剂细胞毒素疫苗后的局部反应

测，只对没有或只有很低抗体水平的动物进行免疫。

最近，有报道用商业疫苗分别在0天，21天，42天[3]对美洲驼幼崽进行免疫，用间接ELISA和鼠中和试验（MNT）测量D型产气荚膜梭菌的ε毒素的滴度。ELISA和MNT实验一致显示免疫一次的动物抗体效价不会上升，二免后一周，平均抗体效价明显上升，12只动物中有7只出现了很高的抗体效价，三免后一周平均抗体效价开始下降。

在之前的调查中，皮下接种乳佐剂细胞类毒素疫苗后，30%的免疫单峰驼出现了局部炎症反应（图29）[30]。单峰驼尤其对油佐剂疫苗敏感，因此，氢氧化铝佐剂疫苗开始应用，而且可以同时用于肌肉和皮下组织。在此期间，对没有或者具有很小副反应、能够产生高滴度的抗体单峰驼疫苗佐剂的研究颇受鼓舞。第一个鼓舞人心的发现是由Eckersley等报道的[10]。

## 参考文献

[1] Anon. (1998). – Blackleg. *Vet. Rec.*,143 (12),322.

[2] Akeilla A.M.,Hussein A.S.,Draz A.A. &Mujalli D.M. (2006). – Incidence of certain clostridia in camels. Proceedings of the International Scientific Conference on Camels,10–12 May,Saudi Arabia,465–477.

[3] Bentancor A.B.,Halperin P.,Flores M. &Iribarren F. (2009). – Antibody response to the epsilon toxin of *Clostridium perfringens* following vaccination of *Lama glama crias. J. Infect. Dev. Ctries,*3 (8),624–627.

[4] Birutta G. (1997). –Storey's guide to raising llamas – Care/Showing/Breeding/Packing/ Profiting. 2nd Ed. Pownal,Storey Publishing,105–106.

[5] Bisping W. &Amtsberg G. (1988). –Colour atlas for the diagnosis of bacterial pathogens in animals. Verlag Paul Parey,Berlin and Hamburg.

[6] Chauhan R.S.,Kulshreshtha R.C. & Kaushik R.K. (1985). – A report of enterotoxemia in camels in India. *Indian Vet. J.*,62 (10),825–827.

[7] Cross H.E. (1919). – Are camels susceptible to blackquarter,haemorrhagic septicemia and rinderpest? *Bull. Agric. Res. Inst. Pusa*,80,17.

[8] Curasson G. (1947). – Le chameau et ses maladies. Vigot Freres Editions，Paris，86–88.

[9] Central Veterinary Research Laboratory (CVRL) (1995). – Annual report，7.

[10] Eckersley A.M.，Petrovsky N.，Kinne J.，Wernery R. &Wernery U. (2011). – Improving the dromedary antibody response: the hunt for the perfect adjuvant. *J. Camel Pract. Res.*，18 (1)，30–46.

[11] Ellis R.P.，Todd R.J.，Metelman-Alvis L.A.，Newton A.L.，Thomson T.J.，Johnson L.W. & Ramirez A. (1990). – Response of llamas to *Clostridium perfringens* type C and D vaccines. American Association of Small Ruminants Practitioners Western Region Coordinating Committee Symposium，Corvallis，OR，USA，4–5.

[12] El-Sanousi S.M. &Gameel A.A. (1993). – An outbreak of enterotoxaemia in suckling camels. *J. Vet. Med. A*，40，525–532.

[13] Fowler M.E. (1996). – Husbandry and diseases of camelids. *In* Wildlife husbandry and diseases (M.E. Fowler，ed.). *Rev. sci. tech. Off. int. Epiz.*，15 (1)，155–169.

[14] Fowler M.E. (2010). – Medicine and surgery of camelids. 3rd Ed. Wiley-Blackwell，Oxford，UK，196–201.

[15] Gameel A.A.，El-Sanousi S.M.，Musa B. &. El-Owni E.E. (1986). – Association of some pathogenic bacteria with haemorrhagic enteritis in camels. Camel research paper S.R.C. No. 12，Camel Research Unit，University of Khartoum，Sudan，50–55.

[16] Gkiourtzidis K.，Frey J.，Bourtzi-Hakaopoulou E.，Iliadis N. & Sarris K. (2001). – PCR detection and prevalence of $\alpha$-，$\beta$-，$\beta$2-，$\varepsilon$-，$\iota$- and enterotoxin genes in *Clostridium perfringens* from lambs with clostridial dysentery. *Vet. Microbiol.*，82，39–41.

[17] Gatt Rutter T.E. & Mack R. (1963). – Diseases of camels. Part 1: bacterial and fungal diseases. *Vet. Bull.*，33 (3)，119–124.

[18] Huaman D.，Ramirez A. &Samame H. (1981). –Produccion de alfa toxina de 3 cepas de *Clostridium perfringens* tipo A

aislados de alpacas. *In* Resumenes 5th Congress, Peru. Microbiology and Parasitology (Arequipa), 56.

[19] Hutyra F., Marek J. &Manninger R. (1946). – Special pathology and therapeutics of the diseases of domestic animals. 5th English Ed. Baillière, Tindall and Cox, London.

[20] Ipatenko N.G. (1974). – Infectious enterotoxemia of camels. *Vet. Bull.*, 44 (4), 1481–1484.

[21] Losos G.J. (1986). – Infectious tropical diseases of domestic animals. Avon, The Bath Press.

[22] MoebuuAynurzanaDashdavaIpatenko N.G. (1966). – Infectious enterotoxaemia of camels in Mongolia, caused by *C. perfringens*type *C. Veterinariya. (Moscow)*, 43 (11), 32–35.

[23] Moro Sommo M. (1956). –Contribucion al estudio de las enfermedades de losauguenidos. *Rev. Fac. Med. Vet. (Lima)*, 7 (11), 15–17.

[24] Moro Sommo M. (1963). –Enfermedadesinfecciosas de las alpacas. V. Enterotoxemia odiarrea bacilarpr oducida por *Clostridium welchii,*tipo A. *Rev. Fac. Med. Vet.,(Lima)*, 18 (20), 85–87.

[25] Radostits O.M., Gay C.C., Hinchcliff K.W. & Constable P. (2007). – Veterinary Medicine, 10th Ed. Elsevier Health Sciences, Philadelphia, PA, USA, 830–832.

[26] Ramirez A. &Huaman D. (1980–1981). –Evaluacion de la enterotoxemia en crias de alpacas vacunados. *ResumenesProyectos Invest. Realizados (Lima)*, 1, 48–49.

[27] Ramirez A., Lauerman L., Huaman D. & Vargas A. (1983). –Inducción preliminar de la enterotoxemia A *Clostridium perfringens* tipo A en alpaca. *ResumenesProyectos Invest. Realizados (Lima)* 3, 48–49.

[28] Rath E. (1950). –Sintomas y cuadros anatomo-patologicos de las enfermedades de ganado en el Departamento de Puno. *Rev. Agrop. Perus*, 1 (2), 68–70.

[29] Seifert H.S.H. (1992). – Animal diseases in the tropics. *In* Tropentierhygiene (H.S.H. Seifert, ed.). Gustav Fischer Verlag, Jena, Stuttgart, Germany, 256–266.

[30] Seifert H.S.H., Boehnel H., Heitefuss S., Rengel J., Schaper R., Sukop U. &Wernery U. (1992). – Isolation of *C. perfringens* type A from enterotoxemia in camels and production of a locality-specific vaccine. Proceedings of the 1st International Camel Conference, 2–6 February 1992, Dubai. R. and W. Publications (Newmarket) Ltd, Newmarket, UK, 65–68.

[31] Songer J.G. (2010). – Clostridia as agents of zoonotic disease. *Vet. Microbiol.*, 140, 399–404.

[32] Tansuphasti K., Wongsuvan G. &Eampokalap B. (2002). – PCR detection and prevalence of enterotoxin (cpe) gene in *Clostridium perfringens* isolated from diarrhoea patients. *J. Med. Association Thai*, 85, 624–633.

[33] Tyler J.W., Petersen A., Ginsky J., Parish St., Besser T., Leaters C. & Beyer J. (1996). –Clostridial myonecrosis, hepatitis, and nephritis in a llama with vegetative endocarditis. *J. Vet. Int. Med.*, 10 (2), 94–96.

[34] Uzal F.A., Assis R.A. & Chang Reissig E. (2000). – Malignant oedema in a guanaco (*Lama guanacoe*). *Vet. Rec.*, 147, 336.

[35] Watson P.J. & Scholes S.F.E. (2009). –*Clostridium perfringens* type D epsilon intoxication in one-day-old calves. *Vet. Rec.*, 146, 816–817.

[36] Wernery U., Abraham A.A., Jyothi T., Abubaker Ali & George R.M. (2009c). – Mineral and vitamin contents in the blood of racing dromedaries in the United Arab Emirates. *J. Camel Pract. Res.*, 16 (1), 39–40.

[37] Wernery U., Ali M., Wernery R. & Seifert H.S.H. (1992). – Severe heart muscle degeneration caused by *Clostridium perfringens* type A in camel calves (*Camelusdromedarius*). *Rev. Elev. Méd. Vét. Pays Trop.*, 45 (3–4), 255–259.

[38] Wernery U. &Kaaden O.-R. (2002). – Infectious diseases in camelids. 2nd Ed. Blackwell Science, Berlin, Vienna, Austria, 19–31.

[39] Wernery U., Joseph M., Zachariah R., Jose S., Syriac G. &Raghavan R. (2009b). – New preliminary research in *Clostridium perfringens* in dromedaries. *J. Camel Pract. Res.*, 16 (1), 45–50.

[40] Wernery U., Kinne J., Joseph B., Raghavan R., Syriac G. & Jose S. (2009a). – Natural acquired *Clostridium perfringensα*-toxin antibodies protect dromedaries from clostridial enterotoxaemia. *J. Camel Pract. Res.*, 16 (2), 153–155.

[41] Wernery U., Seifert H.S.H., Billah A.M. & Ali M. (1991). – Predisposing factors in enterotoxemias of camels (*Camelus dromedarius*) caused by *Clostridium perfringens* type A. *Rev. Elev. Méd. Vét. Pays Trop.*, 44 (2), 147–152.

[42] Younan M. &Gluecks I.V. (2007). – *Clostridium perfringens* type B enterotoxaemia in a Kenyan camel. *J. Camel Pract. Res.*, 14(1), 65-67.



## 深入阅读材料

Purdy St. (2008). – Enterotoxaemia in camelids. *In* Proceedings of the International Camelid Health Conference for Veterinarians,Ohio State University,Corvallis,Oregon,148 — 154.

Ramirez A.,Ludena H. & Acosta M. (1983). –Mortalidaden alpacas del centropecuario La Raya – Puno ensieteanos. *ResumenesProyectos Invest. Realizados (Lima)* 3,77 — 78.

（周继章译，殷宏校）

## 1.1.2 肉毒梭菌中毒

肉毒梭菌引起人和动物的肉毒中毒，毒素通过肠道吸收，由血液转运至外周神经细胞引起松弛性瘫痪，因循环障碍和呼吸麻痹引发死亡。骆驼对肉毒梭菌易感[3]，然而，旧大陆骆驼的临床病例报道很少[8]，在新大陆骆驼中未见报道，临床诊断很难进行，因此，很容易被忽略。

### 病原学和临床症状

肉毒梭菌是革兰氏阳性棒状杆菌，在接近或者稍高于中性的pH值时产生近端芽孢，芽孢耐热，在121℃高压15分钟后才能将其杀灭，然而，肉毒梭菌毒素在100℃高压15分钟后可被破坏，这种严格的厌氧菌可产生8种不同的神经毒素，但是微量的氧气就可抑制其生长。

### 流行病学

肉毒中毒易在热带的干旱和半干旱牧场流行，典型特征是麻痹无力。该病最早在牛中发现，这与土壤中缺乏磷有关[6]。如果牧草中缺乏矿物质，动物就会通过消化含磷的动物源饲料以弥补不足，尸体是中毒的来源。1990年，一场毁灭性的肉毒中毒暴发于澳大利亚昆士兰的2个饲养场，造成5 500头公牛死亡[4]，饲料中的鸡骨粉是造成了此次暴发的主要原因。造成巨大损失的肉毒中毒在世界范围内的水禽中均有发现。

肉毒梭菌在土壤和泥土中常见，可存活数年，根据毒素特征已经鉴定出8种不同的血清型和亚型，在表9中列出其分布。很多病例都与摄入青贮饲料、腐败蔬菜或家禽粪便有关。病原在创口（创口型肉毒中毒）和消化道（毒素感染型肉毒中毒）生长并产生毒素是不常见的来源。

表9　肉毒梭菌毒素类型和分布

| 型 | 毒素 | 分布 | 中毒来源 | 易感动物 |
|---|---|---|---|---|
| A | A | 美国西部，乌克兰 | 饲料、肉、鱼、创口 | 人类、水禽、貂 |
| B | B | 美国中、东部，欧洲北部、中部 | 肉和肉产品 | 人、牛、水禽、马 |
| C | $C_\alpha$, $C_1$, $C_2$, C | 美国南、北部，南非、澳大利亚、欧洲 | 丝光绿蝇幼虫、植物、泥 | 水禽 |
| | $C_\alpha$, $C_2$ | 澳大利亚、南非、欧洲 | 变质食物、尸体 | 牛、马、貂 |
| D | D, $C_1$, $C_2$ | 南非、前苏联 | 尸体 | 牛 |
| E | E | 北欧、前苏联、加拿大、阿拉斯加、日本 | 鱼和鱼产品 | 人类 |
| F | F | 苏格兰、美国、丹麦、前苏联 | 肝酱、鱼 | 人类 |
| G | G | 阿根廷—— | | |

肉毒梭菌毒素在对数生长期的最后阶段合成，在细菌细胞溶解时释放，细菌细胞只能产生$C_2$毒素，$C_1$和D毒素只有在噬菌体存在时才可产生[9]。肉毒梭菌和噬菌体之间的联系是了解肉毒中毒的决定性标准。为了介绍噬菌体，将非产毒素性肉毒梭菌菌株转化入毒素性菌株，例如，如果中性的肉毒梭菌菌株感染$C_1$毒性噬菌体，该菌株就会产生$C_1$毒素，变成C型菌株。将D-毒性噬菌体转化到相同的中性菌株，变成D型菌株。甚至有可能将无噬菌体的中性型肉毒梭菌菌株感染与诺维氏芽孢梭菌菌株密切相关的噬菌体，结果菌株变成了诺维氏芽孢梭菌[9]。同时，使人们弄不清不同型、不同肉毒梭菌以及它们独特的细菌噬菌体的菌株之间新的关系。

关于骆驼的肉毒中毒报道很少，Provost等[5]发现了乍得单峰驼C型肉毒中毒的毁灭性暴发。在对150只单峰驼进行观察时，45只已经死亡，40只患病严重。患病动物行走困难，后肢及臀部瘫痪，衰竭，在几小时内死亡。推测毒素来源可能是泉水被含有毒素的尸体污染。所有哺乳动物的肉毒中毒的临床症状相同，最初所有骨骼肌进行性麻痹、震颤，导致斜倚。瞳孔放大、眼睑下垂、舌收缩虚弱，尸检没有损伤。

## 诊断

肉毒中毒通常很难诊断，一般基于病史、临床症状和已经死亡或濒临死亡动物的血清或饲料中的毒素检测做出判断。在可疑饲料中可分离到肉毒梭菌，将患病或最近死亡的动物0.5mL或1mL血清，或无菌第一胃内容物腹腔接种老鼠，如果有毒素，几小时到3天内老鼠可能出现“蜂腰”症状（图30）。

遗憾的是，当检测骆驼之类的大动物时，小鼠实验不够敏感，因为血清或第一胃内毒素浓度太低而检测不到，诊断有赖于病史和临床症状。饲料样本的毒素通过饲喂特异免疫实验室动物或羊来鉴定。Tetracore（美国，罗克韦尔）已有A型、B型、C型、E型、F型的商用肉毒梭菌ELISA试剂盒，但不用于临床样本的诊断，只用于环境样本。其他肉毒毒素的检测方法还包括免疫扩散试验和补体结合试验等，但这些方法还没有商用，除了补体结合试验外，其他试验的敏感性都没有超过小鼠的生物学测定方法。

## 治疗和预防措施

除了注射含有特定毒素型的高免血清，肉毒中毒动物还没有特异的治疗方法。由于要弄明白导致动物疾病的肉毒梭菌血清型想要花费宝贵的时间，一些国家在治疗时采取混合几种抗血清或用多价血清，并在病程的早期注射。抗血清通过静脉注射，价格昂贵，但是，可以挽救非常有价值的骆驼。牛和马每种类型的抗血清注射5 mL，患病的旧大陆骆驼静脉注射5 mL，新大陆骆驼静脉注射3 mL，5 mL包含至少70 000 IU，3 mL至少30 000 IU，24小时之内可能需要重复治疗一次。除治疗以外，患病骆驼治疗期间良好的护理也很重要。在未来，骆驼的重链抗体片段可能被应用于治疗[1,7]。

预防：免疫；磷缺乏的补充；中毒源的移除。商用疫苗有时可作为肉毒中毒和黑腿病的联合疫苗，在骆驼濒临灭绝地区的应该免疫，初免之后应该在5周之后进行二免，一年之后三免。澳大利亚和南非用含有C、D毒素的二价疫苗进行免疫，北美和欧洲应该用B、C毒素的二价苗免疫。

图30 老鼠腹腔注射0.5 mL 含肉毒毒素的血清后出现“蜂腰”

1

细菌病

## 参考文献

[1] Anderson G.P.,Ortiz-Vera Y.A.,Hayhurst A.,Czarnecki J.,Dabbs J.,Vo B.H. & Goldman E.R. (2008). – Evaluation of llama anti-botulinum toxin heavy chain antibody. *Botulinum J.*,1 (1),100－115.

[2] Bisping W. &. Amtsberg G. (1988). –Colour atlas for the diagnosis of bacterial pathogens in animals. Verlag Paul Parey,Berlin and Hamburg,79－100.

[3] Fowler M.E. (2010). – Medicine and surgery of camelids. 3rd Ed. Wiley-Blackwell,Oxford,UK,196.

[4] Jones T. (1991). – Bovine botulism. *In Practice*,13 (3),83－86.

[5] Provost A.,Haas P. &Dembelle M. (1975). – Premiers cas au Tchad de botulisme animal (type C): intoxication des dromadairespar l'eau d'un puits. *Rev. Elev. Méd. Vét. Pays Trop.*,28 (1),9－12.

[6] Seifert H.S.H. (1992). –Tropentierhygiene. Gustav Fischer Verlag,Jena,Stuttgart,263－266.

[7] Thanongsaksrikul J.,Srimanote P.,Maneewatch S.,Choowongkomon K.,Tapchaisri P.,Makino S.,Kurazono H. &ChaicumpaW. (2010). – A VHH that neutralizes the zinc metalloproteinase activity of botulinum neurotoxin type A. *J. Biol. Chem.*,285(13),9657－9666.

[8] Wernery U. &Kaaden O.-R. (2002). – Infectious diseases in camelids. 2nd Ed. Blackwell Science,Berlin and Vienna,31－33.

[9] Westphal U. (1991). –Botulismus bei Vögeln. Aula-Verlag,Wiesbaden.

（周继章译，殷宏校）

## 1.1.3　炭疽

炭疽芽孢杆菌引起人类和动物的炭疽。虽然世界各地分离株之间有一定差异，但炭疽芽孢杆菌只有一个抗原型。在自然条件下，最易感炭疽的动物有牛、绵羊、山羊、水牛、马、驯鹿、大象和水貂。鸟（除了鸵鸟）和爬行动物敏感性低，很少能被感染[2]。猪也能感染炭疽，不过一般仅表现为由咽部原发病灶引起的亚急性或慢性的疾病。炭疽在世界各地均有发生，尤其是在动物密集的地区，如饮水坑、动物贸易市场以及动物舐食岩盐的盐渍地。

炭疽是一种急性、败血性疾病，可感染骆驼。

### 病原学

炭疽芽孢杆菌是一种需氧芽孢杆菌，革兰氏阳性、不运动、呈圆柱状。该菌在宿主体内能形成一个荚膜，可通过特殊染色观察到。在组织脏器涂片中，该菌单个存在或呈短链状，形成所谓的“竹节”。本菌在氧气充足、温度12℃以上的条件下易形成芽孢。炭疽芽孢杆菌在普通固体培养基上生长，血琼脂培养基上不溶血。在低倍镜下观察，菌落呈水母头样或呈卷发状。

### 公共卫生意义

炭疽是一种热带地区最重要的人畜共患病，并且多是通过感染炭疽芽孢杆菌的动物传播。病原可经皮下、肠道或空气传播给人类。Punskii和Zheglova[13]报道了亚洲一例群体性皮肤炭疽，37人因接触感染炭疽的单峰驼肉而发病。在2007年，8个阿富汗人在参加宗教庆祝活动时，吃了被感染的骆驼肉而死于炭疽。Makinde等人[10]运用免疫电泳技术做血清流行病学调查，发现在埃及屠宰场的骆驼炭疽感染率达到62.3%，但他们认为，高感染率主要是由蜡样芽孢杆菌的交叉反应引起的。炭疽芽孢杆菌是一种生物武器，并被美国的疾病控制和预防中心（CDC）列为A类病原。在许多发展中国家，炭疽仍然是牲畜死亡的主要原因，也是人食用感染动物的肉死亡的一个主要原因。人存在消化道炭疽和皮肤炭疽两种炭疽。

### 流行病学及临床症状

炭疽是一种急性烈性传染病，其特点是败血

症和突然死亡。炭疽芽孢杆菌能够在土壤中存活数年。在疾病的最后阶段，从宿主体内排出，并在20～30℃条件下在地面或土壤里形成孢子[15]。土壤一旦被埋葬感染炭疽动物尸体所污染，就在多年内作为传染源，尤其是当放牧动物在饲料短缺时期啃食地面上的牧草时，吸入被污染的尘土导致肺炭疽。Fazil[7]认为，炭疽是肯尼亚骆驼最常见的细菌性疾病，表现为急性、特急性和中风等类型。

骆驼饲养者十分惧怕炭疽。所以，给本病取了提示其危险性的多种名称。Mustafa[12]认为，除了锥虫病和疥癣，炭疽是单峰驼最致命的疾病之一。单峰驼可发生急性炭疽或特急性炭疽，有时没有任何临床症状即突然死亡。炭疽的流行往往与明显的气候和生态变化相关，如，暴雨、洪水或干旱。

叙利亚德国技术公司（The Syrian-German Technical Co.）发行的单峰驼炭疽宣传小册子里，描述了本病的临床症状与病理损伤[16]。叙利亚大草原上的一个100峰骆驼的驼群，饮用了因暴雨形成的洪水淹没过的池塘里的水而感染了本病。受感染的骆驼表现出呼吸困难、颤抖，喉咙、颈部腹面和腹股沟区域明显肿胀。在死亡前，骆驼斜卧，排深黑色便，天然孔流带泡沫的血（图31）。十几只单峰驼死亡，在改用城镇用水作为饮用水后感染得以控制，其余的动物用抗生素进行了治疗，并对驼群进行了疫苗接种。

消化道是本病主要的感染途径，如摄入被污染的饲料或水坑里的水[3]。Curasson[5]曾指出虻可以诱发单峰驼皮肤炭疽，蝇蛆（轻触喉蝇*Cephalopina titillator*）携带的炭疽芽孢杆菌，可通过受损的黏膜进入机体。Barakat等人[1]报道了埃及的一次炭疽暴发，123只单峰驼在4天内死亡，9只中风。Curasson[5]、Gatt Rutter和Mack[9]也曾报道过由炭疽引起突然死亡的类似病例。埃塞俄比亚和尼日利亚也报道过骆驼的炭疽。在哈萨克斯坦也有本病的发生，但发生于骆驼的很少。在1955-1977年间，相对于数百例牛的炭疽，只有8例骆驼炭疽报告。一般来说，由于炭疽的特殊性，炭疽疫情有时不能被如实报道出来。Barakat等人[1]报道了一次特殊的单峰驼群炭疽疫情，并且认为是由于迁徙鸟类引起的。通过采取严格的卫生措施、并对每只骆驼进行普鲁卡因青霉素和50 mL抗炭疽血清治疗5天后，疫情最终得到了控制。

单峰驼炭疽的临床症状类似于牛炭疽[9]：体温升至42℃，焦油状血液从身体天然孔渗出，腹泻、腹痛、腹胀，严重的心肺功能障碍。一些骆驼出现疼痛状，咽喉和颈部水肿。

新大陆骆驼炭疽临床症状类似于旧大陆骆驼炭疽。在没有任何症状的情况下发生猝死以及身体各个部位的皮下水肿。血液从所有天然孔流出，美洲驼可在1～3天后死亡[8]。新大陆骆驼的炭疽在秘鲁及美国已有报道。

### 病理变化

败血性炭疽主要病变是出血、水肿和坏死。单峰驼[16]死后不久就开始腐败，从尸体的鼻腔、口腔和肛门流出带血的液体。尸体发红发黑、凝血不全、全身出现瘀点和瘀斑。脾脏肿大、变软，这是反刍动物尸检中最典型的特征，骆驼也是如此[11]。死后尸僵不全、血液不凝固。Boue[3]和Richard[4]还观察到黑色焦油状巨脾、弥漫性充血和肺水肿等。

### 诊断

来自血液和组织中的炭疽芽孢杆菌易于培养。然而，如果怀疑是炭疽，应避免剖检，以减少排除孢子污染土壤的机会。处理尸体或提交样品时需预防人畜共患的风险。采集少量的血液就足够用于诊断。因为死于炭疽的动物凝血不全，通常采集颈静脉血样，通过涂片或培养以及荧光抗体试验可以确诊（图32）。当尸体高度腐败时，或没有检测到炭疽芽孢杆菌时，可以运用Ascoli氏热沉淀法进行诊断。动物实验首选小白鼠，皮下感染后2～4天内死亡，注射部位可见胶胨状水肿。

### 防控措施

为了防止炭疽芽孢杆菌的芽孢生成，尸体不应剖检，应连同所有受污染的草垫等一起进行焚化，所有接触的器材应严格消毒。

下面是几种可用的消毒液。

（1）10%热苛性钠溶液。

（2）4%甲醛溶液。

（3）7%过氧化氢。

（4）2%戊二醛。

（5）含5%活性氯的漂白粉（次氯酸钙）。

炭疽芽孢杆菌对多种抗生素敏感，包括青霉素和四环素类。在疾病的早期，特别是在其他临床症状出现之前检测到发热时，抗生素治疗可能会有效。使用青霉素每千克体重20 000 IU，2次/天，或者链霉素或土霉素每千克体重5 mg，肌注给药。免疫血清价格比较昂贵，对价值高的动物，注射100 ~ 250 mL/天。抗生素治疗应持续5天。

Pasteur研制的第一个有效的炭疽疫苗，已被Sterne研制的弱毒芽孢杆菌苗所替代，其已经在世界范围内广泛使用，对畜牧业和野生动物健康成长起到了保护作用，产生了巨大的经济效益。单次接种免疫持续期可达9个月，但推荐每年进行一次加强免疫。疫苗包含活的减毒芽孢疫苗，能够在免疫动物体内萌发、增殖从而刺激机体产生免疫力。炭疽对骆驼危害严重，因此，推荐在受威胁地区给骆驼接种疫苗。然而，炭疽疫苗应慎重使用于骆驼，并且剂量应按动物的体重调整，因为曾有小美洲驼因疫苗导致发病的报道[4]。旧大陆骆驼接种剂量与牛相同，而新大陆骆驼推荐使用绵羊剂量。新大陆骆驼刚断奶的幼畜，接种剂量则是绵羊剂量的一半[8]。

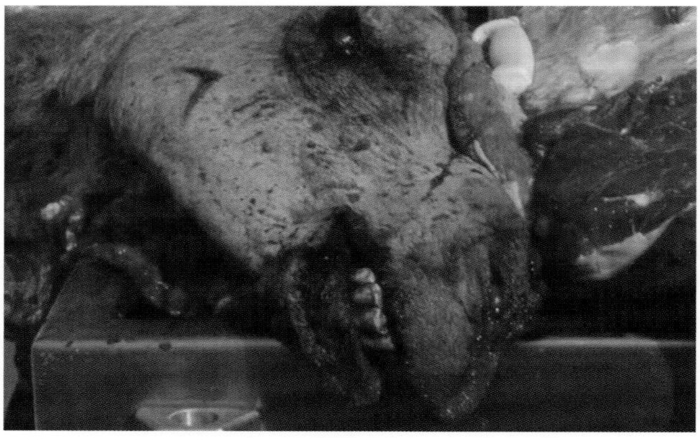

图31 患炭疽的单峰驼鼻孔流出没有凝固的血液

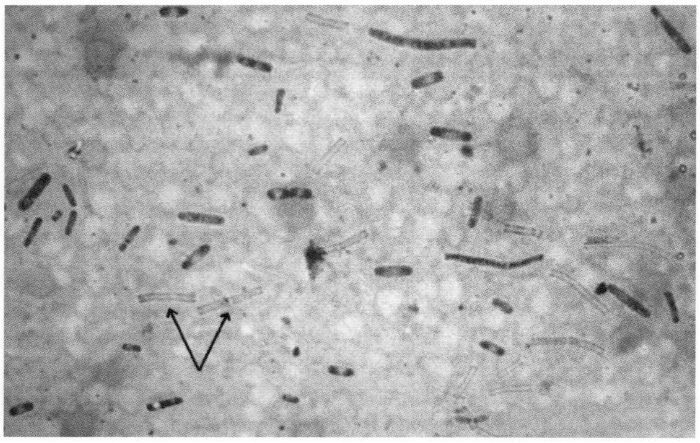

图32 炭疽芽孢杆菌。单峰驼脾脏涂片可见竹节状的炭疽芽孢杆菌

## 参考文献

[1] Barakat A.A.,Sayour E. & Fayed A.A. (1976). – Investigation of an outbreak of anthrax in camels in the western desert. *J. Egypt Vet. Med. Assoc.*,36 (1),183–186.

[2] Bisping W. &. Amtsberg G. (1988). – Colour atlas for the diagnosis of bacterial pathogens in animals. Verlag Paul Parey,Berlin and Hamburg,72–79.

[3] Boue A. (1962). – L'initiation au dromadaire. Service biologique et veterinaire des armees,centre d'instruction du service. Vétérinaire de l'Armee Compiègne,1957 Remis a jour.

[4] Cartwright M.E.,McChesney A.E. & Jones R.L. (1987). – Vaccination-related anthrax in three llamas. *JAVMA*,191 (6),715–716.

[5] Curasson G. (1947). – Le chameau et ses maladies. Vigot Frères Editions,Paris,86–88.

[6] Davis J.W.,Karstad L.H. & Trainer E.D. (1981). – Infectious diseases of wild mammals. 2nd Ed. Iowa State University Press,Ames,IA.

[7] Fazil M.A. (1977). – The Camel. *Bull. Anim. Health Prod. Afr.* 25 (4),435–442.

[8] Fowler M.E. (2010). – Medicine and surgery of camelids. 3rd Ed. Wiley-Blackwell,Oxford,UK,207.

[9] Gatt Rutter T.E. & Mack R. (1963). – Diseases of camels. Part 1: Bacterial and fungal diseases. *Vet. Bull.*,33 (3),119–124.

[10] Makinde A.A.,Majiyagbe K.A.,Chwzey N.,Lombin L.H.,Shamaki D.,Muhammad L.U.,Chima J.C. & Garba,A. (2001). – Serological appraisal of economic diseases of livestock in the one-humped camel (*Camelus dromedarius*) in Nigeria. *Camel Newslett.* 18,62–73.

[11] Manefield,G.W. & Tinson A. (1996). – Camels. A compendium,Sydney,2000. The T.G. Hungerford Vade Mecum Series for Domestic Animals,240–298.

[12] Mustafa I.E. (1987). – Bacterial diseases of the camel and dromedary. OIE 55e Session generale de l'OIE,Office International des Épizooties,Paris,55,18–22.

[13] Punskii E.E. & Zheglova D.V. (1958). – Letter. *J. Microbiol. (Moscow)*,29 (2),78.

[14] Richard B. (1975). – Étude de la pathologie du dromadaire dans la sous-province du Borana (Ethiopie). Thesis,ENV Alfort,Créteil,Paris 75.

[15] Seifert H.S.H. (1992). – Tropentierhygiene. Gustav Fischer Verlag,Jena,Stuttgart,Germany,244–250.

[16] Tabbaa D. (1997). – Anthrax in camels. Pamphlet,Faculty of Veterinary Medicine,Al Baath University,Syria.

[17] Wernery U. & Kaaden O.-R. (2002). – Infectious diseases in camelids. 2nd Ed. Blackwell Science,Berlin and Vienna,33–35.

## 深入阅读材料

Okolo M. (1986). – Anthrax in man and slaughtered animals in Anambra State,*Nigeria. Nigerian Vet. J.*,15 (1–2),39–43.

(储岳峰译，殷宏校)

## 1.1.4 内毒素中毒（内毒素血症）

动物体内许多组成肠道正常菌群的革兰氏阴性菌是内毒素的潜在来源，尤其是当第一胃pH值下降，导致反刍动物和骆驼胃菌群被破坏时，更是如此。饲喂大量的精饲料可引起瘤胃发酵功能障碍，导致食欲不振和乳酸中毒。反刍动物和骆驼急性乳酸中毒，经常可引起内毒素血症或内毒素性休克，这是因为反刍动物肠道内大量革兰氏阴性细菌被破坏导致。由于在酸中毒动物第一胃液中能检测到大量的内毒素，很明显乳酸不是导致中毒的因素。消化道的内毒素不是导致乳酸中毒的原因，而是酸中毒的产物。由于单峰驼的前胃更适合于饲喂粗料，饲喂高度易消化精饲料（过量精饲）常是导致其酸中毒的诱因。阿拉伯半岛地区的沙漠骆驼竞赛日趋流行，使得这种饲喂精料的方式有迅猛发展的势头，但同时却使高能量食物进入了一个实际上并不适合的第一胃（C1）环境。骆驼饲养者和主治兽医应该明白骆驼独特的胃部解剖和生理特征，包括强劲的胃动力、比较长的消化时间和快速的液体通过率等。而且，认识到骆驼只需较低的蛋白质维持量、较低的可代谢能量需要量和较低的基础代谢水平也很重要。骆驼更适合在一个艰苦的、营养贫乏的环境中生存。在所有家畜中，骆驼的胃对发酵纤维素效率最高。可能是因为骆驼的胃黏膜为非角质化上皮，提供了大面积的快速吸收表面的原因。

通过过去10余年的细致研究，已经揭开了一种竞赛用骆驼疾病的神秘面纱，这种疾病曾被称为蜡样芽孢杆菌毒素中毒、出血性素质、出血症（HD）[17,18,19]等。牛只也报道过类似的综合征[1]。

### 病原学

内毒素是革兰氏阴性菌细胞外壁上的脂多糖，当细菌快速增殖或死亡裂解时会被释放。结构上，内毒素包括3个部分。

（1）脂质A，埋藏在细胞壁内，是内毒素的主要毒性组分。

（2）膜外区，具有抗原特异性，是不同菌种特异性抗原组分。

（3）核心区，起连接脂质A和膜外区的桥梁作用。

内毒素毒性极强，浓度为$10^{-9}$ g/mL时即可致死。化学性质稳定，煮沸不能被灭活。其毒性也不能被消化道中的酸或酶类作用降低。胃肠道内规律性产生少量的内毒素，可通过肠黏膜吸收进入血液循环，并在肝脏被解毒。但是，如果肝脏解毒功能受限或内毒素产生过多，即可发生毒血症，并带来严重后果。分布广泛的内皮细胞及其基底组织可能会被破坏，然后内毒素能激活血凝机制，并导致弥漫性血管内凝血（DIC）。DIC的主要临床表现是多系统膜表面点状或斑状出血，静脉穿刺点流血，尤其单峰驼内毒素中毒时，常可观察到上述症状。DIC的另一个特征是广泛的血栓形成、纤维蛋白沉着，造成循环障碍。

在阿联酋竞赛用单峰驼中，有种病已普遍流行多年，由于临床特征和病理特征，曾被称为出血质或出血症[14]。这种病主要发生于竞赛用骆驼，80%是2～4岁或更小的骆驼。该病主要发生于个体，但10头及其以上的驼群也可发病。该病全年均可发生，但夏季发病率最高，因为高温高湿。这种季节性发病不仅与极端气候有关，而且认为跟新赛季前训练的开始和饲喂方式的改变有关，因为此时从饲喂高纤维粗饲料转变为饲喂高蛋白和高能量的精饲料。

## 临床症状与病理变化

发病24～48小时为开始阶段，主要特征性变化是白细胞（WBC）数量的显著下降、发烧至41℃，食欲不振、精神沉郁、反应迟钝。这些症状出现3～4天后，白细胞数量又显著增加，高于正常值（表11）。

部分发病动物可出现咳嗽、咽喉肿胀，伴随一侧或双侧淋巴结肿大（图33）。通常情况下黏膜充血，并出现第一胃迟缓、腹痛和食物倒流等症状。发病的单峰驼直肠检查可观察到被新鲜或褐色血液覆盖的粪球，只有很少的病驼会出现腹泻[10]。

单峰驼发病后3～7天死亡，死前2～3天卧地不起。部分单峰驼出现中枢神经系统失调、流泪、大量流涎等症状。出现中枢神经系统症状是骆驼濒死的信号。本病是竞赛骆驼最严重的疾病，阿拉伯半岛所有开展赛驼活动的国家都有本病的报道。其他养骆驼的国家情况不明。

在过去的20年间，解剖观察超过500只死于内毒素中毒的竞赛用骆驼，最显著的剖检变化是严重的组织脏器和肠道内出血。不同程度的淤血常见于以下脏器：

（1）咽喉和气管（部分单峰驼可形成气管溃疡，图34）。

（2）心外膜和心内膜（图35）。

（3）皱胃（胃黏膜迭层顶部溃疡，部分溃疡面附着血凝块，图36、图37）。

（4）肠道，主要在升结肠（肠道内通常充满新鲜或焦油样血液，图38、图39）。

（5）肾盂（多见出血点，图40）。

所有淋巴结肿大出血，常伴随中心坏死（图41）。肺充血、胸膜下和间质出血（图42）。所有发病动物都表现第一胃酸中毒，pH值为4～6。赛驼病理解剖检时，第一胃液抹片可见革兰氏阳性菌（图43a、b），但找不到原生生物。

## 病理组织学变化

可见脾脏和扁桃体等淋巴组织中，出现中度到严重的淋巴细胞流失。淋巴滤泡中心、集合淋巴结和肠系膜淋巴结的主要组织变化是出血、坏死和细胞核崩裂（图44）。这种病变也提示跟病毒感染有关，但进一步的包括动物试验在内的研究排除了病毒感染的可能。

皱胃、肠道、心内膜下层和心外膜下层常会出血严重。肾直小管和肾曲小管上皮细胞坏死比较常见。大量的肾小球、肾小囊（鲍氏囊）膨大并充满蛋白渗出物。肾小囊壁由于高碘酸希夫氏染色阳性（$PAS^+$）物质的沉积而增厚（图45）。一些肾小球微血管可见到微血栓（休克微粒）。存活时间略长的单峰驼，毛细血管袢出现节段性坏死（纤维素性坏死）。PAS染色阳性的柱状物填充了这些出现小管坏死的扩张的肾小管管腔。剖检发现，单峰驼肝脏出现小叶性脂肪变性以及小叶中心部分坏死（图46）。大脑一般都会充血，常见血管周和脑膜同时水肿。值得注意的是，所有脏器中均不能观察到炎性反应，这可能是由于毒素对淋巴滤泡和外周血白细胞的直接破坏所致。

剖检动物的肝脏、肾脏和第一胃内容物进行化学检验，香豆素及其衍生物、有机磷等外源中毒性因素检查均为阴性。内毒素能对白细胞生成系统

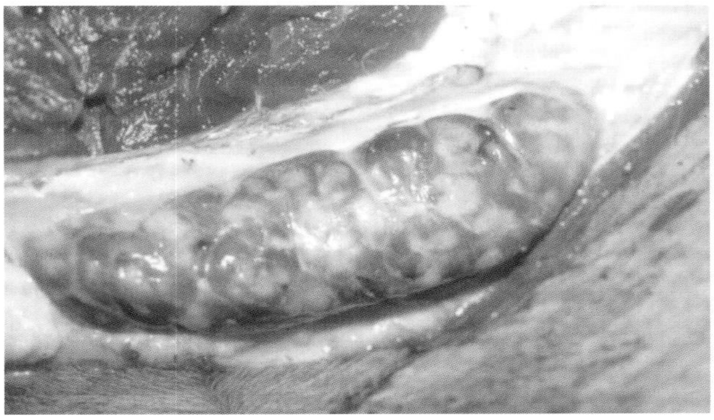

图33　腹股沟淋巴结肿大、出血

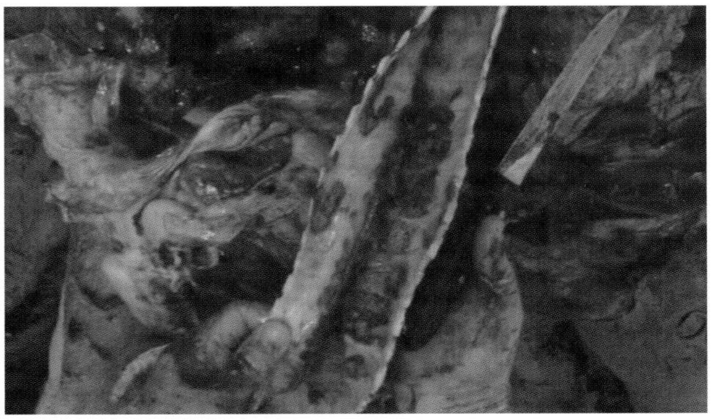

图34　内毒素血症引起的气管溃疡

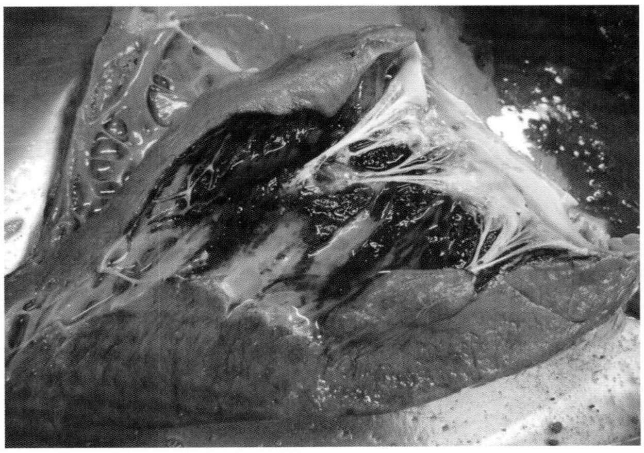

图35　内毒素血症引起的心内膜下出血

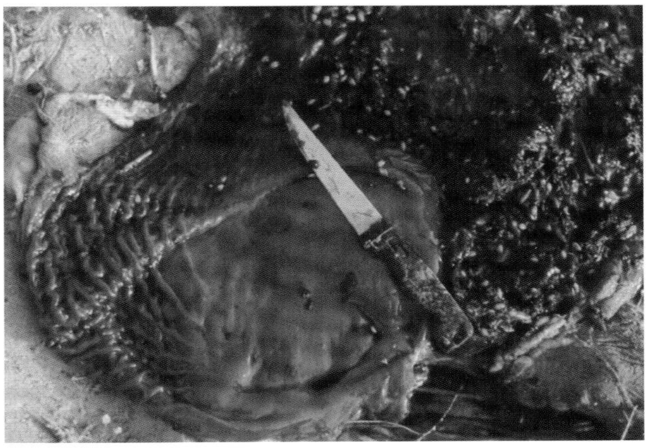

图36　内毒素血症引起的皱胃出血

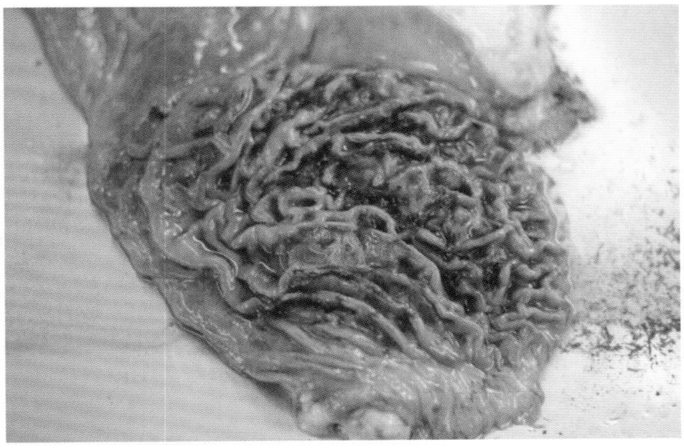

图37　皱胃溃疡，部分表面黏附有血凝块

图38　内毒素血症引起的L结肠皮下血肿

直接产生影响，导致分化停止或者直接坏死，这已经被淋巴结、脾脏、扁桃体和其他淋巴组织的病变观察所证实。内毒素还能对外周血白细胞产生直接的毒性效应，但通常难以确定这种改变就是由于内毒素作用所引起的。由脂多糖引起的粒性白细胞缺乏症可使发病骆驼产生严重的免疫抑制，从而有利于细菌的继发感染。从发病骆驼的不同脏器分离到大量细菌：大肠杆菌、绿脓杆菌、变形杆菌属、肺炎克雷伯氏杆菌、金黄色葡萄球菌、葡萄球菌属和链球菌属等细菌。这些条件致病菌在所有已受损器官中繁殖迅速、在局部产生更多的内毒素。但炭疽芽孢杆菌从未在发病骆驼中分离到过。

由于从死于内毒素中毒的骆驼脏器和饲料中都分离到蜡样芽孢杆菌产毒素菌株，蜡样芽孢杆菌过去被认为是本病的病原，通过静脉注射蜡样芽孢杆菌毒素可复制出本病[12,15,17,20]。但如今已知静脉注射从多种革兰氏阴性细菌纯化的毒素，均可引起系统性的内毒素中毒[8,9,11]。

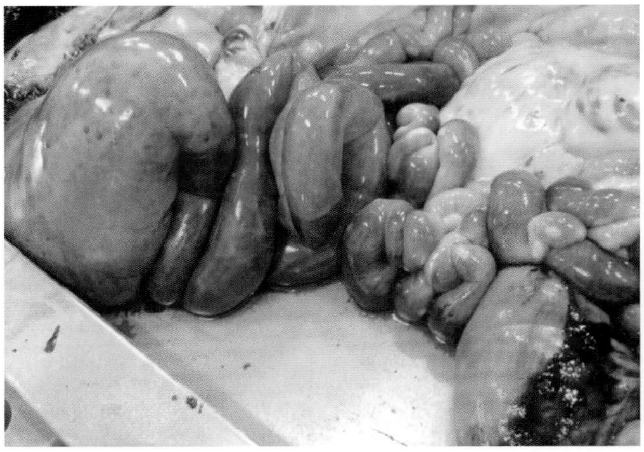

图39　内毒素血症引起的小肠内焦油样血液

图40　内毒素血症引起的肾盂出血

图41　内毒素血症引起的肩胛骨前淋巴结增大、出血和中心坏死

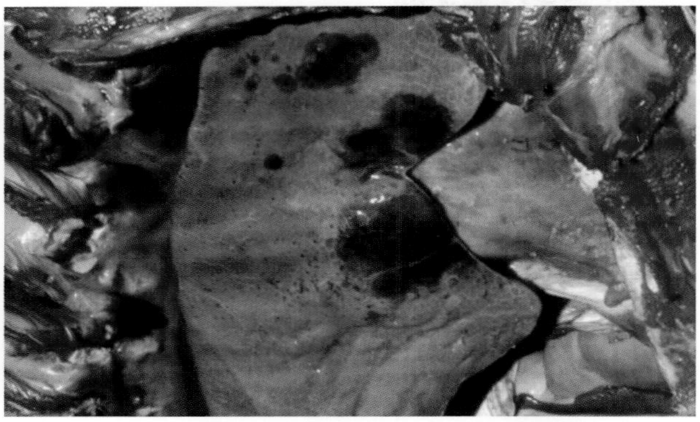

图42　内毒素血症引起的肺间质出血

a. 内毒素血症患驼第一胃的革兰氏阳性菌群

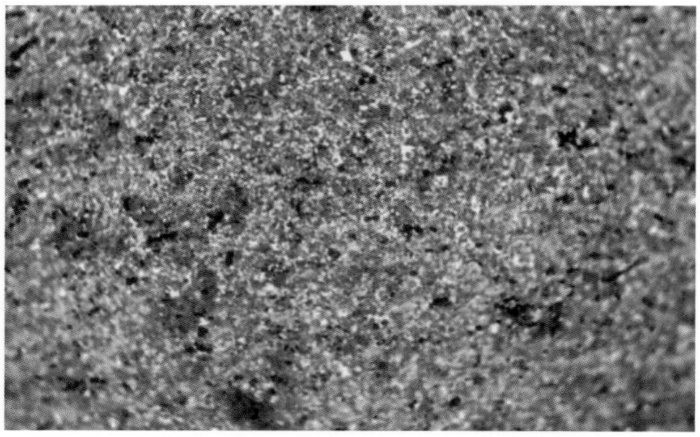

b. 健康骆驼的革兰氏阴性菌群

图43

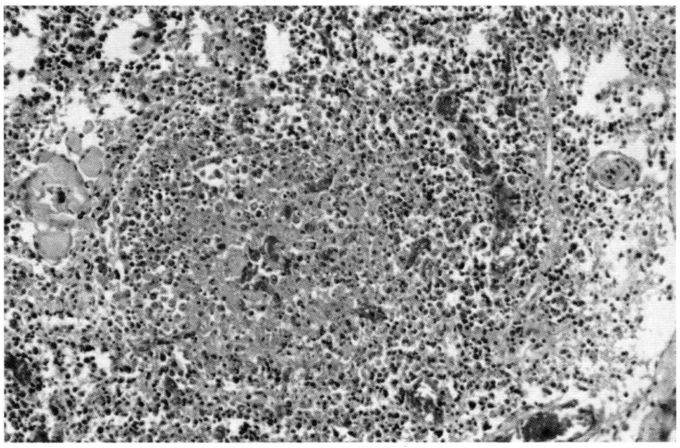

图44　肠系膜淋巴结出血、坏死和细胞核崩解（HE染色）

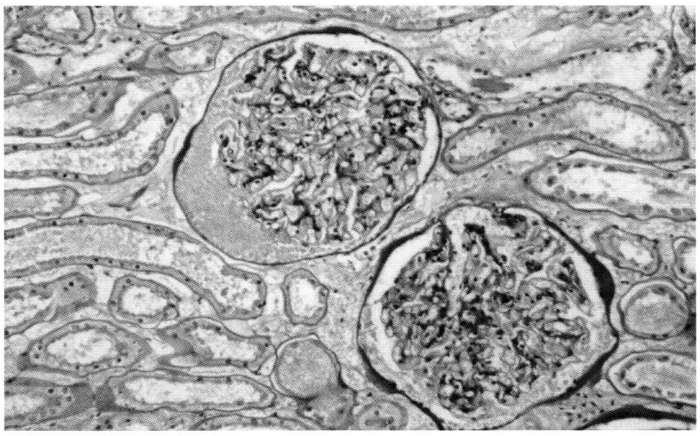

图45　肠毒素血症单峰驼PAS染色阳性物质沉积引起的鲍氏囊增厚

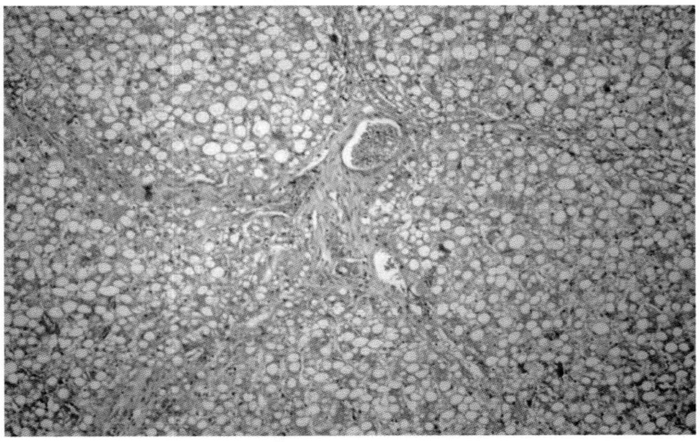

图46　竞赛用骆驼内毒素血症严重的小叶性脂肪变性以及小叶中心部分坏死（HE染色）

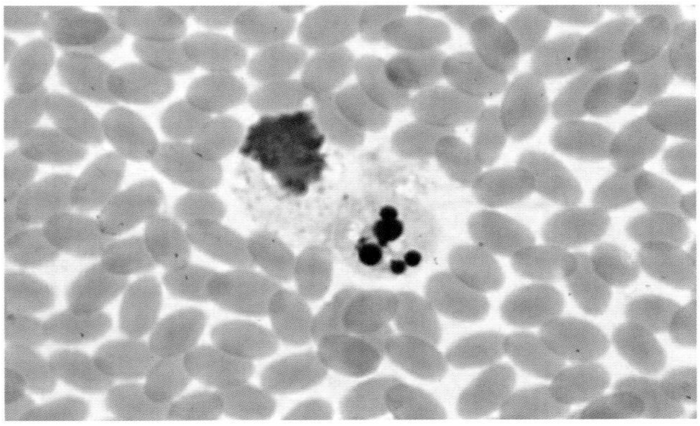

图47　患内毒素血症竞赛用骆驼的白细胞（$1.0\times10^9$个/L），示两个严重核固缩、胞浆形成空泡的无法分类的白细胞

## 临床病理学

白细胞总数和白细胞分类计数的变化是内毒素中毒的典型病理变化。白细胞总数的显著性下降是由于嗜中性粒细胞和淋巴细胞数量的下降导致（表11）。白细胞减少症一般持续1～2天，然后在第3天会出现过量反弹，导致白细胞增多（表10、表11）。

在多数急性病例，当白细胞数量少于$1.0\times10^9$个/L时，由于内毒素对白细胞的毒性作用，已经不能进行白细胞分类计数（图47）[16,19]。白细胞病理变化包括核固缩、胞浆空泡形成和染色不准确。白细胞最低的计数水平通常出现在疾病发生的第1天，并且是本病的特征性诊断标记。在疾病发展过程中，会逐渐朝正常计数回升。由于早期诊断、早期治疗是疾病恢复的关键，所以准确的血液学检查结果显得尤为重要。

发病晚期，血清酶类指标数值急剧上升，提示内脏器官受损。从表12可以看出，一些指标已经超正常值的10倍以上。肌酸酐和尿素氮（BUN）通常会大幅升高，提示肾脏损伤，这与组织学检查结果一致。

病驼表现第一胃迟缓和第一胃酸中毒。骆驼正常的第一胃液pH值高于6.5，但本病的绝大多数病例第一胃液pH值仅为4～6。第一胃（C1）液微生物学检查没有检测到原生生物，细菌染色显示革兰氏阳性微生物菌群占绝对优势。而且，第一胃液泛酸味、发黄并且总是含有少量未消化的大麦残渣。这些病变也同样见于酸中毒的新大陆骆驼[3]。表12列出了内毒素中毒单峰驼和健康单峰驼第一胃（C1）部分参数的差异。

所有发病单峰驼共同的剖检病变表现为不同脏器的出血和严重的肠道内出血，尤其是在结肠（图38）。骆驼的结肠具有极其有效的吸收能力，这就是部分肠道严重出血的原因。据信大部分的毒素已经在第一胃（C1）和第二胃（C2）被吸收，因为在骆驼中，这些胃室内壁为无乳突、光滑的复层鳞状上皮细胞。众所周知，骆驼能快速利用水，胃液和水等，能够非常快的进入血流。这种解剖学上的特征使骆驼对内毒素中毒非常敏感。由于骆驼的强耐受性和肝脏受损，骆驼没有机会对产生于前胃中的内毒素进行脱毒。而且，不仅内毒素在积累，代谢毒素也在积聚，这是由于第一胃积食和肠积食引起的有害代谢活动导致的，第一胃积食和肠积食在竞赛用骆驼上经常出现。由于乳酸中毒，肠道运动基本停止。单峰驼内毒素中毒很少出现腹泻，这归功于它们的结肠和其他部分肠道具有超强的吸收能力。根据肠道出血的位置，新鲜或陈旧焦油状褐色血液可能会通过直肠排出。皱胃内的焦油状血液是由于胃酸作用导致的渗出血液凝集形成的，而结肠的焦油状血液（黑粪症）则是由小肠出血经发酵而成的。

脂多糖可引发弥散性血管内凝血（DIC）。DIC的特征是外周血中纤维蛋白原的下降（表10、表11）、可溶性纤维蛋白和纤维蛋白原变性以及凝血因子的极度缺乏，这些凝血因子可抑制凝血酶活

表10　10头内毒素中毒单峰驼的血液参数和血清酶指标

| 参数 | 单位 | 参考值b | 血液及生化指标参数值 | | | | | | | | | |
|---|---|---|---|---|---|---|---|---|---|---|---|---|
| | | | 1～2天 | | | | | 3～4天 | | | | |
| 白细胞总数 | $10^9$个/L | 6.0～13.5 | 2.5 | 1.6 | 2.6 | 0.8 | 2.9 | 19.3 | 24.8 | 18.0 | 17.3 | 26.6 |
| 中性粒细胞 | % | 50～60 | 70 | a | 66 | a | 65 | 80 | 78 | 82 | 86 | 77 |
| 淋巴细胞 | % | 30～45 | 23 | a | 27 | a | 28 | 12 | 16 | 13 | 12 | 20 |
| 单核细胞 | % | 2～8 | 6 | a | 6 | a | 6 | 8 | 6 | 5 | 3 | 3 |
| 嗜酸性粒细胞 | % | 0～6 | 0 | a | 1 | a | 1 | 0 | 0 | 0 | 0 | 0 |
| 嗜碱性粒细胞 | % | 0～2 | 1 | a | 0 | a | 0 | 0 | 0 | 0 | 0 | 0 |
| 红细胞 | $10^{12}$个/L | 7.5～12.0 | 7.9 | 8.4 | 8.0 | 9.0 | 9.5 | 8.0 | 7.8 | 8.4 | 9.9 | 8.6 |
| 血红蛋白 | g/dL | 12.0～15.0 | 11.1 | 11.3 | 11.1 | 12.2 | 15.0 | 10.4 | 120.0 | 10.9 | 15.1 | 10.8 |
| 血小板 | $10^9$个/L | 350～450 | 168 | 142 | 236 | 116 | 182 | 271 | 372 | 298 | 201 | 291 |
| 肌酸激酶 | U/L | 40～120 | 46 | 81 | 93 | 70 | 32 | 320 | 438 | 594 | 362 | 612 |
| 谷草转氨酶 | U/L | 60～120 | 120 | 104 | 83 | 97 | 110 | 490 | 119 | 257 | 421 | 401 |
| 乳酸脱氢酶 | IU/L | 0～450 | 590 | 390 | 220 | 142 | 350 | 1812 | 675 | 730 | 1557 | 1210 |
| 葡萄糖 | mg/dL | 70～110 | 46 | 70 | 65 | 44 | 48 | 86 | 92 | 99 | 106 | 107 |
| 血尿素氮 | mg/dL | 3～21 | 19 | 21 | 23 | 25 | 21 | 75 | 195 | 60 | 44 | 406 |
| 肌酸酐 | mg/dL | 0～2.2 | 2.0 | 2.2 | 2.0 | 2.0 | 1.8 | 4.5 | 9.3 | 4.2 | 3.7 | 9.6 |
| 纤维蛋白原 | mg/dL | 250～400 | 98 | 102 | 72 | 93 | 106 | 180 | 201 | 305 | 298 | 172 |
| 前凝血酶时间 | S | 17.6 ± 1.6 | 28.2 | 22.4 | 27.0 | 24.8 | 26.3 | 19.2 | 17.4 | 18.9 | 20.2 | 21.6 |
| 部分促凝血酶原激酶时间 | S | 46.9 ± 13 | 82.4 | 60.2 | 62.0 | 54.6 | 70.3 | 50.1 | 48.0 | 47.6 | 53.1 | 60.0 |

注：血液采自表现临床症状后1～2天和3～4天。

a 由于内毒素作用，无法进行分类计数；

b 引自 Wernery等[16]

表11　内毒素中毒存活单峰驼血液参数和血清酶指标

| 参数 | 单位 | 参考值 | 血液及生化指标参数值 | | | | | |
|---|---|---|---|---|---|---|---|---|
| | | | 第1天 | 第2天 | 第3天 | 第6天 | 第11天 | 第25天 |
| 白细胞总数 | $10^9$个/L | 6.0 ~ 13.5 | 1.0 | 1.5 | 5.2 | 22.4 | 20.8 | 9.9 |
| 中性粒细胞 | % | 50 ~ 60 | 82 | 84 | 92 | 84 | 78 | 61 |
| 淋巴细胞 | % | 30 ~ 45 | 12 | 10 | 7 | 10 | 16 | 32 |
| 单核细胞 | % | 2 ~ 8 | 4 | 5 | 1 | 4 | 5 | 4 |
| 嗜酸性粒细胞 | % | 0 ~ 6 | 2 | 0 | 0 | 2 | 1 | 3 |
| 嗜碱性粒细胞 | % | 0 ~ 2 | 0 | 1 | 0 | 0 | 0 | 0 |
| 红细胞 | $10^{12}$个/L | 7.5 ~ 12.0 | 8.3 | 9.1 | 7.6 | 8.3 | 7.2 | 7.6 |
| 血红蛋白 | g/dL | 12.0 ~ 15.0 | 12.2 | 13.5 | 12.2 | 11.5 | 10.1 | 11.1 |
| 血小板 | $10^9$个/L | 350 ~ 450 | 176 | 140 | 193 | 251 | 301 | 483 |
| 肌酸激酶 | U/L | 40 ~ 120 | 67 | 112 | 721 | 882 | 324 | 140 |
| 谷草转氨酶 | U/L | 60 ~ 120 | 196 | 380 | 475 | 680 | 232 | 160 |
| 乳酸脱氢酶 | IU/L | 0 ~ 450 | 338 | 790 | 819 | 1083 | 660 | 379 |
| 葡萄糖 | mg/dL | 70 ~ 110 | 46 | 38 | 44 | 70 | 78 | 108 |
| 血尿素氮 | mg/dL | 3 ~ 21 | 13 | 23 | 51 | 58 | 34 | 22 |
| 肌酸酐 | mg/dL | 0 ~ 2.2 | 1.7 | 3.1 | 3.8 | 4.1 | 2.6 | 1.6 |
| 纤维蛋白原 | mg/dL | 250 ~ 400 | 92 | 103 | 112 | 210 | 370 | 391 |

a 引自 Wernery等[16]

性、纤维蛋白聚合和血小板凝集。部分凝血活酶时间和凝血酶原时间延长（表10）。DIC是内毒素血症的一种严重后果。纤维蛋白不仅可以进一步提高血液黏度，还可以堵塞肾小球，这两种情况在患内毒素血症的单峰驼中经常发生。许多采自发病早期病例的血样离心后仅析出少量血清，离心上清中主要组成是纤维素（图48）。组织学检查可见肾小球内蛋白沉积（图45）。这会导致肾功能衰竭，指标信号是肌酸酐和尿素氮增加。对患病骆驼进行颈静脉抽血即可肉眼观察到凝血因子降低的效应，当抽出采血针头时，采血点会持续出血。这种情况进一步的证据是患驼血小板计数的降低。

图49对骆驼内毒素致病机理进行了汇总说明。

不是每个酸中毒的骆驼都能发展成内毒素中毒。在一个田间试验中，对2只骆驼进行高碳水化合物饲料人工诱导第一胃酸中毒，却没有产生类似于内毒素中毒的临床症状[21]。目前，还不清楚到底是何种机制最终引发了该病。即使饲喂相同的饲料，有些骆驼发病，而有些骆驼不发病，骆驼饲养者对此也感到很困惑。

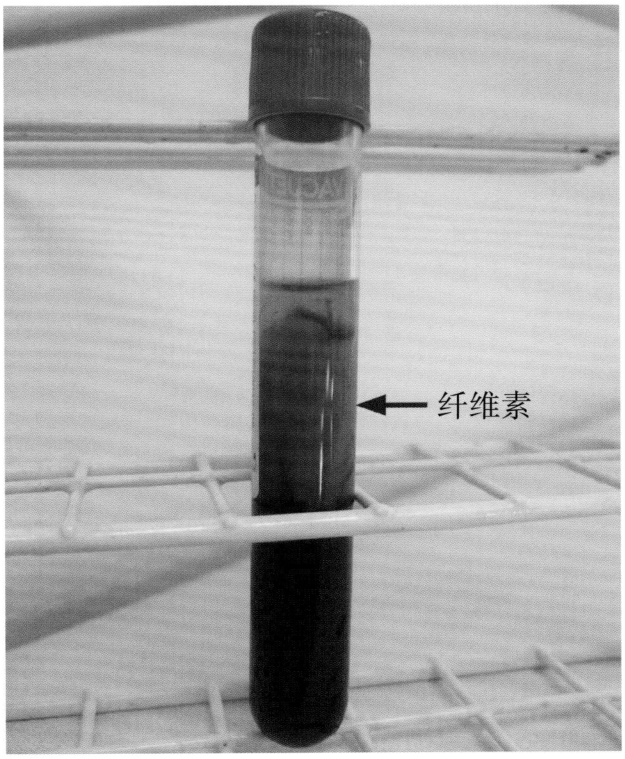


图48　一份患内毒素中毒单峰驼血样，血凝析出的上清主要为纤维素

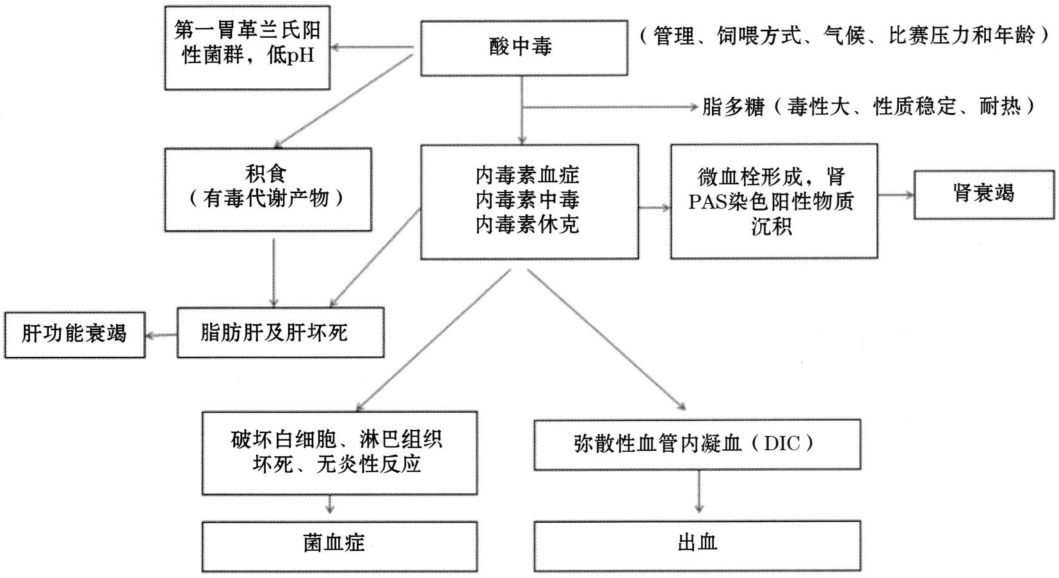


图49　骆驼内毒素血症的致病机制

表12　健康单峰驼和酸中毒单峰驼第一胃（C1）部分属性参数比较

| 属性 | 健康单峰驼 | C1酸中毒单峰驼 |
|---|---|---|
| 颜色 | 橄榄绿-灰绿色 | 发白-黄色 |
| 气味 | 芳香 | 酸味 |
| pH值 | 6.8 | 5.0 |
| 乳酸 | 无 | 0.1% |
| 总酸度mg/mL | 40 | 110 |
| $H_2S$ | +++，变黑色 | 阴性，无颜色反应 |
| $NH_3$(mg/L) | 48 | 24 |
| 吲哚实验 | 红色 | 无颜色反应 |
| TVFA（毫当量） | 10 | 4.5 |
| 原生生物活性 | +++ | +或无 |
| 菌群革兰氏染色 | G- | G+（图43） |

TVFA，总挥发性脂肪酸；+，弱阳性；+++，强阳性

真菌毒素和出血性素质之间的关系已广为人知[2]。阿联酋竞赛用单峰驼饲喂的是质量上乘的饲料，真菌污染的因素可排除。在剖检动物脏器无真菌感染证明出血素质与真菌毒素无关。在单峰驼繁育种群中，其饲养条件标准比竞赛用骆驼略低，每年雨季都会出现真菌毒素导致的死伤[6,7]。真菌毒素中毒可出现与内毒素中毒类似的临床症状，包括粒细胞减少和肠道内出血。但真菌毒素中毒通常伴随灰白色、发臭的腹泻症状。通过对5只单峰驼饲喂真菌高度污染的干草料，成功复制了上述真菌毒素中毒的临床症状。

但在阿布达比酋长国的第二个长期研究单峰驼出血性素质的研究小组，从剖检骆驼的几乎所有脏器上都分离到烟曲霉菌，并在部分血清里检测到了烟曲霉毒素[4]。该报道的作者最后说不能确定这些结果是由于真菌继发感染还是真菌就是出血症的原发病因。利用噻苯哒唑治疗真菌感染，但对疾病的转归没有影响。Pardon等报道欧洲新发现一种新生犊牛的"出血性素质（HD）"[13]，非常类似于竞赛用骆驼的出血症（HD），但只影响一岁以下的犊牛，且不分品种和性别。病例的主要临床表现是各个黏膜面点状出血或瘀血、黑粪症和高热。

血液学检查的典型表现是全血细胞减少症、血小板减少和纤维酶原增加。但在骆驼的出血症（HD）中，还有几个典型特征如淋巴组织损伤、骨髓分化受阻和无炎症反应。在犊牛HD中尝试分离病毒和电镜观察组织中的病毒颗粒均高失败，因此，人们认为该病的发生可能跟某种初乳来源的致病因子有关。

## 防控措施

内毒素血症的治疗完全依赖于早诊断、早治疗。越早诊断，存活概率越高。由于在过去20年间积累了大量关于此病的只是，一旦本病最初的临床症状初现端倪，骆驼饲养者便会立刻通知兽医。内毒素血症是一种非常严重和复杂的疾病，而且很难治愈。当白细胞数少于$1.0\times10^9$个/L时实施早期治疗，是可能被治愈的。但即使是最好的治疗，预后死亡也是可能的。一旦疾病发展到中后期时，通常预后不良。

内毒素血症的治疗原则：

（1）吸附内毒素，并将其清除出机体。

（2）抗酸给药，纠正酸中毒。

（3）输液补液。

（4）对症消炎治疗。

（5）预防胃溃疡。

（6）支持性治疗，以提高肝脏的脱毒能力。

（7）广谱抗生素给药。

（8）激活凝血系统。

（9）预防脑皮质层坏死。

预防内毒素血症，应尤其注意饲喂方式。但在阿联酋，给竞赛用骆驼饲喂牛奶、大枣、过量大麦和新鲜的苜蓿仍比较普遍。实际上，要平衡精饲料和粗饲料。如果碳水化合物过量，应通过胃管给予泻药或通便药，如，石蜡油，以减少食物在胃肠道的停留时间，避免各种积食的发生，防止细菌繁殖和内毒素被机体吸收。如果发生严重的积食情况，要给予效果更好的泻药，并配合使用硫酸镁治疗，每只骆驼500 ~1 000 g 硫酸镁。也有报道通过胃管投喂木炭的方法减少机体对内毒素的吸收，剂量是每天500 g（竞赛用骆驼约重400 kg）。

抗酸是治疗出血症（HD）的主要原则。包括以下几个方面。

（1）益生菌和抗生素。

（2）解酸剂。

（3）膜保护（硫酸铝，20 mg/kg）。

（4）酸抑制剂。

——组胺2型受体阻断剂（H-2R，Ulcer Gard）。

——质子泵抑制剂（PPIs，Gastro Gard）；

——最大限度增加白天的纤维摄入量。

多黏菌素B对内毒素的脂质部分具有很高的亲和力。这种抗菌药对马有严重的副作用，会导致肾损伤，但由于其很强的内毒素结合能力，可以尝试在患内毒素血症的骆驼上使用。硫酸多黏菌素B应用于马的建议剂量是 6 000 IU/kg或2.5 mg/kg，用5 L的生理盐水稀释后，缓慢静脉注射。

有大量报道表明，用抗内毒素核心区的抗血清治疗患马，可以降低死亡率。用于马的推荐剂量是每kg体重1 ~ 2 mL[5]。这种血清有商品化产品，如Veterinary Dynamics 公司生产的Hypermune-J、密苏里州哥伦比亚市的Immvac公司生产的Endoserum。理论上这些血清可以用于严重的急性病例。Stegantox 60是Schering-Plough Animal Health生产的一种冻干的、纯化的特异性抗内毒素IgG，在补体的协同作用下，这种IgG还对多种革兰氏阴性菌具有杀菌活性。目前，该产品已用于骆驼内毒素血症的治疗，1支冻干的制剂为60 mg，200 kg体重的骆驼使用1支。

内毒素血症发生时，氧自由基能引起组织损伤和炎症反应。可通过胃管给予用缓冲盐水或水稀释的10% ~ 20%的二甲基亚砜溶液进行治疗。马属动物推荐剂量是250 ~ 1 000 mg/kg，每12 小时给药一次。因为炎症反应常常导致内毒素性休克，应用非激素类抗炎药对控制炎症反应非常重要。应给予苯基丁氮酮（保泰松）、酮洛芬、氟尼辛等以阻止炎性介质的生成。氟尼辛葡甲胺（Finadyne）具有抗内毒素、止痛、消炎和退热作用，对马属动物非常有效，但对骆驼却有毒性。

由于毛细血管壁的完整性被破坏，大量体液随着内出血流失。纠正外周循环体液不足是避免患内毒素血症骆驼死亡的重要治疗措施。同时，为了纠正酸中毒，也应立即采取精确的体液置换治疗方法。在疾病的早期，每只骆驼静脉注射5 L的5%碳酸氢钠溶液，以后静注60 L的用生理盐水葡萄糖配制的1.3%碳酸氢钠溶液。由于患畜存在低血糖，出现葡萄糖代谢增加、食欲不振，输液时必须加入葡萄糖成分。此外，还需要每天2次通过胃管给予解酸剂治疗，将氢氧化镁溶于温水中，每只骆驼投喂500 g，解酸剂治疗应持续数天。如果是价值极高的骆驼患病，如严重的急性第一胃酸中毒和内毒素血症，应考虑行第一胃切开术。将第一胃（C1）排空并用吸管冲洗，同时，用健康反刍动物，如绵羊、山羊的第一胃营养物移植到患病动物的胃内。

预防胃溃疡，可用苯骈咪唑替代品奥美拉唑口服，剂量为每千克体重0.7 mg。这种药物的作用机制是通过阻断$H^+$–ATP酶和$K^+$ ATP酶活性从而阻止胃酸分泌。内毒素血症也会使肝功能衰竭，由于肝脏在贮存、活化和合成多种维生素以及脱毒方面具有重要作用，因此，应补充多种维生素（包括维生素K）和给予促肝脏活性药物。消耗性凝血紊乱可通过给予肝素治疗，但这种治疗方式还没有在骆驼上试过。为了预防脑灰质软化症，可肌注盐酸硫胺素，剂量为每千克体重2.5 ~ 10 mg。

Endovac-Bov是一种预防大肠杆菌乳腺炎的疫苗，已试用于预防骆驼的内毒素血症，这种疫苗可增强T淋巴细胞和B淋巴细胞反应，并且可与另外一种用于预防其他内毒素相关疾病的灭活苗Re-17联合应用，但目前尚没有资料表明这些疫苗在骆驼上的免疫效率情况。

在阿拉伯半岛，已有数以百计的价值不菲的竞赛用骆驼死于内毒素血症。根据目前对本病发病机理的认识，预防和免疫防控技术应是下一步研究的重点。已有证据表明，即使是饲喂中度的谷物精料也可能引起严重的酸中毒，并导致动物死亡[3]，因此，对于饲喂牛奶、大枣、蜂蜜、过量未研磨大麦和苜蓿的饲喂方式尤其值得注意。应用益生菌、抗内毒素血清和如Baypamun这样的免疫诱导剂是值得考虑的预防性措施。另外，应尽量避免对幼年竞赛用骆驼进行训练，以减少应激诱发疾病。

## 参考文献

[1] Bell C. (2011). – Bleeding disorders in cattle. *In Practice*,33,106–115.

[2] Blood D.C. &. Radostits O.M. (1990). – Veterinary Medicine. 7th Ed. Bailliere Tindall,London,190–191.

[3] Cebra C.K.,Cebra M.L.,Garry F.B. &. Belknap E.B. (1996). – Forestomach acidosis in six New World camelids. *JAVMA*, 208 (6),901–904.

[4] El-Khouly A-B.,Gadir F.A.,Cluer D.D. & Manefield G.W. (1992). – Aspergillosis in camels affected with a specific respiratory and enteric syndrome. *Austr. Vet. J.*,69 (8),182–186.

[5] Gaffin S.L. (1987). – Endotoxins and anti-endotoxin antibodies. *Equine Vet. J.*,32,76.

[6] Gareis M. & Wernery U. (1992). – Determination of mycotoxins in samples associated with cases of intoxications in camels. *In* Proceedings of the 1st International Camel Conference,2–6 February 1992,Dubai. R. and W. Publications (Newmarket) Ltd,Newmarket,UK,403–404.

[7] Gareis M. & Wernery U. (1994). – Determination of Gliotoxin in samples associated with cases of intoxication in camels. *Mycotoxin Res.*,10,2–8.

[8] Huber T.L.,Peed M.C.,Wilson R.C. & Goetsch D.D. (1979). – Endotoxin absorption in hay-fed and lactic acidotic sheep. *Am. J. Vet. Res.*,10 (6),792–794.

[9] Krogh N. (1960). – Studies on alterations in the rumen fluid of sheep,especially concerning the microbial composition,when readily available carbohydrates are added to the food. II. Lactose. *Acta Vet. Scand.*,1,383–410.

[10] Manefield G.W. & Tinson A. (1996). – Camels. A compendium. The T.G. Hungerford Vade Mecum Series for Domestic Animals. University of Sydney Post Graduate Foundation in Veterinary Science,240–298.

[11] Nagaraja T.G. & Bartley E.E. (1979). – Endotoxin shock in calves from intravenous injection of rumen bacterial endotoxin. *J. Anim. Sci.*,49 (2),567–582.

[12] Nothelfer H.B. & Wernery U. (1995). – Hemorrhagic disease in dromedary camels (*Camelus dromedarius*) – etiology and morphology. *In* Proceedings of the International Conference on Livestock Production in Hot Climates,Muscat,Oman, 66,

A57.

[13] Pardon B.,Stenkers L.,Dierick J.,Ducatelle R.,Saey V.,Maes S.,Vercauteren G.,De Clercq K.,Callens J.,De Bleecker K.& Deprez P. (2010). – Haemorrhagic diathesis in neonatal calves: an emerging syndrome in Europe. *Transboundary Emerg. Dis*.,57,135–146.

[14] Walz A. (1993). – Nachweis und Eigenschaften von *Bacillus cereus*-Stammen,isoliert von arabischen Kamelen (*Camelus dromedarius*). Thesis,Ludwig-Maximilians University,Munich,Germany.

[15] Wernery U. (1994). – Neue Ergebnisse zur Diagnose,Prophylaxe und Therapie wichtiger bakterieller und viraler Krankheiten beim Kamel (*Camelus dromedarius*). Habilitationsschrift zur Erlangung der Lehrbefähigung an der Tierärztlichen Fakultät der Ludwig-Maximilians-Universität München.

[16] Wernery U.,Fowler M.E. & Wernery R. (1999). – Colour Atlas of Camelid Haematology. Blackwell Wissenschafts-Verlag,Berlin,36–49.

[17] Wernery U. & Kaaden O.-R. (1995). – Infectious Diseases of Camelids. Blackwell Wissenschafts-Verlag,Berlin,19–27.

[18] Wernery U. & Kaaden O.-R. (2002). – Infectious Diseases in Camelids. 2nd Ed. Blackwell Science Berlin,Vienna,36–49.

[19] Wernery U. & Kinne J. (2001). – Endotoxicosis in racing dromedaries. A review. *J. Camel. Pract. and Res*.,8 (1),65–72.

[20] Wernery U.,Schimmelpfenning H.H.,Seiferk H.S.H. & Pohlenz J.(1992). –*Bacillus cereus* as a possible cause of haemorrhagic disease in dromedary camels (*Camelus dromedarius*). *In* Proceedings of the 1st International Camel Conference (W.R. Allen,A.J. Higgens,I.G. Mayhew,O.H. Snow & J.F. Wade,eds). R. and W. Publications (Newmarket) Ltd, Newmarket,UK, 51–58.

[21] Wernery U. & Wensvoort J. (1992). – Experimentally induced rumen acidosis in a one year old camel bull (*Camelus dromedarius*). A preliminary report. *Brit. Vet. J*.,148,167–170.

(储岳峰译，殷宏校)

## 1.1.5 骆驼马红球菌感染

棒状杆菌科的细菌是一种能够在人和动物中引起多种化脓性疾病的病原菌。马红球菌是马的一种主要的病原菌，主要感染马驹，并在世界范围内广泛分布。马红球菌的感染主要由吸入受污染的灰尘引起并以支气管肺炎和肉芽肿性肺炎为特征[1,6]。它也是一种新发的条件性病原菌，特别在免疫机能不全的动物中常发[14,16,17]。

新大陆骆驼和旧大陆骆驼均对马红球菌易感，尤其是单峰驼，可引起严重的传染性感染[3,4,5,8,9,10,11]。

### 病原学

马红球菌是一种具多形性、需氧、无芽孢、有荚膜的病原菌，革兰氏染色阳性，常呈球状或棒状。该菌在绵羊血平板上生长良好，37℃培养48小时之后可观察到典型的黏液性菌落（图50）。马红球菌可以在细胞内存活，并产生可扩散的“马红球菌因子”，例如，磷脂酶C和胆固醇氧化酶。这些因子以及黏性荚膜、细胞壁酶类，可能在马红球菌的致病机制中发挥着重要的作用。

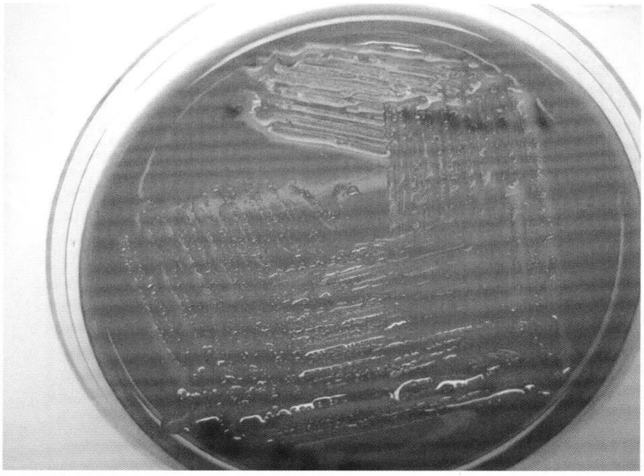

图50　马红球菌在血琼脂平板上37℃培养48小时后的菌落

## 流行病学与临床症状

马红球菌为土壤腐生菌，在有草食动物粪便成分，如醋酸和高温等条件下，生长速度显著增加。在理想的情况下，马红球菌能够在潮湿的土壤中增殖，能在12周龄以下的马驹肠道内大量生长。该菌还能在成年马的粪便中存在，提示骆驼牧场粪便存在污染的可能。因此，在粪便大量露天堆积的牧场，马红球菌会逐步流行。

在马科动物中，马红球菌的感染主要是由于吸入被污染的土壤粉尘引起。马红球菌主要感染2～6月龄的马驹，引起肉芽肿性肺炎。6月龄以上的马若没有免疫缺陷，则具有抵抗力。马红球菌在新大陆骆驼和单峰骆驼中的流行病学情况还不清楚，在双峰驼中还没有此病的报道。有趣的是，在一个数百峰不同年龄的幼畜的奶骆驼农场，此病只发生于泌乳期的成年骆驼。在驼科动物中，马红球菌病的发病部位并不只局限于呼吸道，在分离到马红球菌的美洲驼中，肝脏、肺、脾等部位均可见干酪样脓肿（图51）。也曾有一例马红球菌引起的美洲驼坏死性淋巴结炎的报道[8]。

Kinne等人首次在阿联酋的奶骆驼牧场中诊断马红球菌感染单峰驼的报道，具有重要的意义[9,10]。与马科动物不同的是，马红球菌不会感染骆驼的幼崽，而是只在成年的泌乳骆驼中发生，并引起波及肺、肝脏、脾、纵隔淋巴结和乳房的严重的传染性疾病。从受累器官中分离到了大量的马红球菌，同时，这些器官的VapA质粒（VapA基因和 85 KB的I型质粒）PCR检测均为阳性。正在发病的牧场调查粪便是否在此次流行主要的传染源。高温高湿和持续增长的动物数量可能是此病传播的重要因素。

图51　马红球菌引起的美洲驼脓肿（图片由美国M.E. Fowler教授惠赠）

成年马对马红球菌具有抵抗力[6]，感染病例报道很少[20]。然而，到目前为止，所有的有关单峰驼感染马红球菌的病例中都有成年驼被感染[10]。就严重性和肺部感染程度来看，临床症状温和。奶单峰驼一般不会消瘦，也不会表现出肺炎的症状。然而，大部分感染的单峰驼都表现出乳房炎的症状，乳房可能是这种病感染的入口。在死亡前2～3天时，动物会停止进食，体温L高到40.5℃，在出现临床症状的3天内死亡。

剖检的尸体中，60%～70%的患病骆驼具有严重的肺炎（图52），除了大范围的干酪样区域外，小的棕褐色结节分布在肺部的其他部位。纵隔淋巴结和肝脏肿大。此外，在一个病例中，一些干酪样坏死灶（直径3～5cm）扩散到前胃第三胃室室的胃壁上（图53）。患有此病的单峰驼的乳房分布着小的白色的脓肿，在这些脓肿中可以分离出马红球菌（图54）。组织学检查发现，肺发生多发性的纤维素性炎症，纤维素团块中心坏死，包含了大量的淋巴细胞、浆细胞和嗜中性粒细胞，纵隔淋巴结发生了化脓性炎症，可见大量吞噬了球状病原菌的巨噬细胞。马红球菌肺炎典型的组织学特征，包括吞噬了大量革兰氏阳性球杆菌的巨噬细胞聚集和少量的中性粒细胞[1]，与奶单峰驼的肺动脉炎病变有些相似。

Madarame等[12]和Mariotti等[13]认为，免疫标记技术是诊断马红球菌感染的有用工具。马驹多数自然发生的肺炎是由有毒力的马红球菌引起的。致病性的马红球菌具有大小为15～17 kD的特异性的毒力相关抗原[17]。针对这个15～17 kD抗原（VapA）的特异性的单克隆抗体（单抗10G5）有助于马红球菌肺炎的诊断[15]。这种单克隆抗体同样可用于单峰驼的免疫组化中[10]（图55）。

使用不同的PCR反应可以鉴定马红球菌毒力株。PCR方法有两种，一种依赖于染色体，以编码异柠檬酸裂解酶的异柠檬酸裂解酶基因为靶基因[18]；另一种是根据特异性毒力质粒（VapA基因）而设计[18]。PCR确认从骆驼上分离的菌株[10]含有这种马红球菌毒力株特异性的毒力质粒（VapA基因）[16]。质粒图谱分析表明[12,15]，牧场土壤中存在的有毒力的马红球菌与马红球菌的流行情况没有密切的相关性[15]。

## 诊断

马红球菌感染的最佳诊断方法是从剖检骆驼的病灶直接分离培养出马红球菌。如前所述，马红球菌生长迅速，通过其在血液琼脂平板上的黏液性菌落外观、革兰氏染色特点和生化特征（BBLcrsytal革兰氏阳性细菌鉴定系统，美国BD公司）非常容易鉴定。PCR技术已用于马的临床病例诊断。然而，还没有从骆驼的临床样品（如鼻支气管、支气管肺泡吸取物和粪便样本）中分离培养出这种病原的报道。

血清学检测马红球菌的抗体技术，例如，酶联免疫吸附试验、琼脂糖扩散试验、血清溶血抑制试验在马科动物中已有报道，但作为筛查试验，缺乏准确性。唯一的被用在骆驼上的血清学检测方法，是一种在意大利用于检测羊驼抗体的方法[3]。

## 防控措施

在马科动物上已采取了几种不同的措施控制马红球菌病。例如，定期检测马驹的白细胞总数、纤维蛋白原，超声检查和静脉注射高免血清等。但是，因为这是骆驼一种新的疾病，这些方法还没有用来防控骆驼的马红球菌病。在彻底的研究清楚马红球菌感染的流行病学，弄清传染源之前，控制这种疾病非常困难。现在正在努力从奶骆驼牧场的环境样品中分离病原。

骆驼的马红球菌还没有治疗方法，但是马的

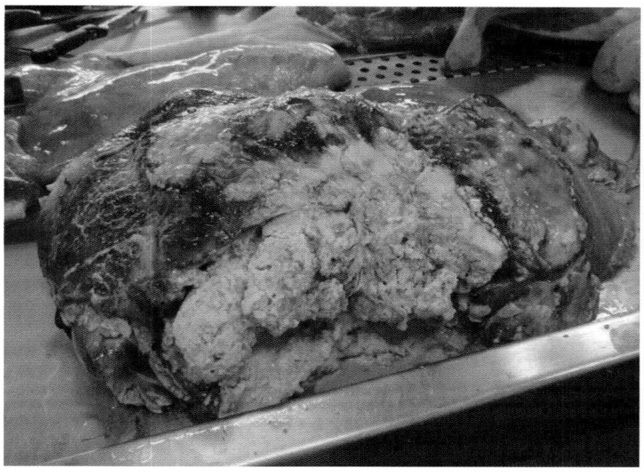

图52　马红球菌引起的单峰驼严重的干酪样肺炎

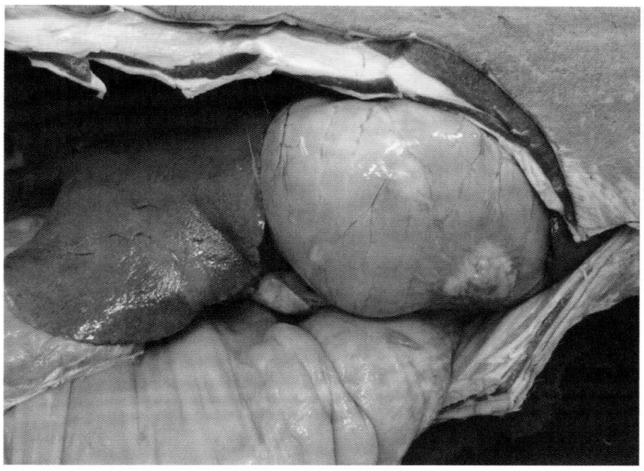

图53　马红球菌感染引起的第三胃室壁多个干酪样坏死和肝肿大

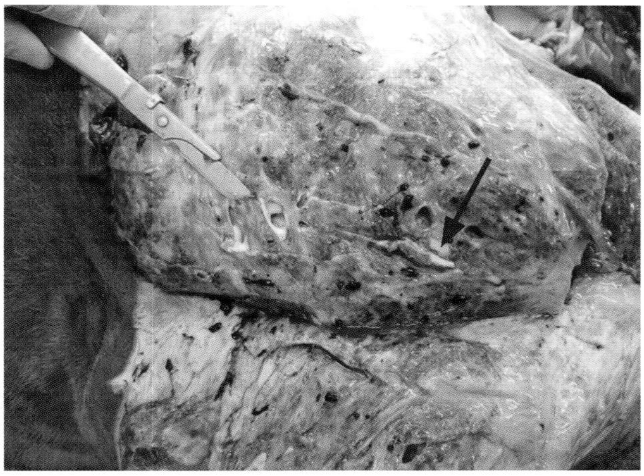

图54　马红球菌引起的奶骆驼乳房上的小脓肿和脓汁

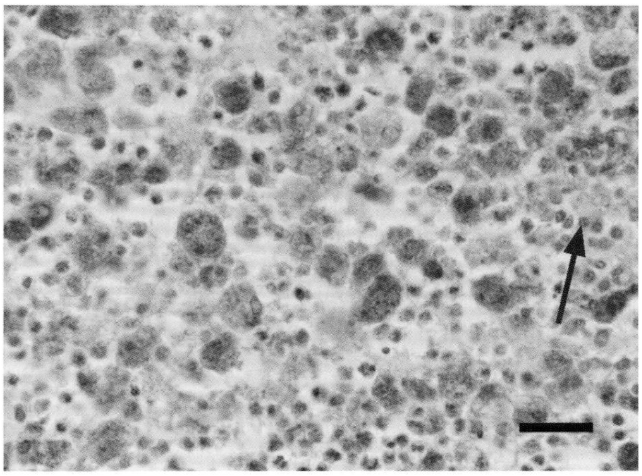

图55　马红球菌感染的免疫组织学：肿胀的巨噬细胞内包含染色阳性的颗粒状细菌，针对致病性马红球菌特异性的15～17 kD抗原染色（图片由日本H. Madarame和S. Taka博士惠赠）

马红球菌病已经有治疗方法。标准的治疗程序是联合运用红霉素和利福平[7]。联合用药的基本原则是使两种药物达到高细胞内浓度。马驹的用药剂量为：红霉素每千克体重25 ~ 37 mg，利福平每千克体重5 ~ 10 mg，每10 ~ 12小时口服一次。近年来，两种大环内脂类药物逐渐被用来治疗马驹的马红球菌肺炎，用量为阿奇霉素每千克体重10 mg，克拉霉素每千克体重7.5 mg，每24小时口服一次[2]。托拉霉素是可注射的大环内脂类药物，已被用于治疗牛和猪的呼吸道疾病。托拉霉素拥有较长的半衰期，并能达到很高的肺组织浓度[19]。

## 参考文献

[1] Ainsworth D.M. (1999). –*Rhodococcus equi* infections in foals. *Equine Vet. Educ.*,11,191–198.

[2] Cohen N. & Giguère S. (2009). –*Rhodococcus equi* foal pneumonia. *In* Infectious diseases of the horse (T.S. Mair &R.E. Hutchinson,eds). Equine Veterinary Journal Ltd,Ely,UK,233–246.

[3] Cuteri V.,Takai S.,Marenzoni M.L.,Morgante M. & Valente C. (2001). – Detection of antibodies against *Rhodococcus equi* in alpaca (*Lama pacos*) in Italy. *Eur. J. Epidemiol.*,17,1043–1045.

[41] Elissalde G.S. & Renshaw H.W. (1980). –*Corynebacterium equi*: an interhost review with emphasis on the foal. *Comp. Immunol. Microbiol. Infect. Dis.*,3,433–435.

[5] Fowler M.E. (2010). – Medicine and surgery of camelids. 3rd Ed. Wiley-Blackwell,Oxford,UK,214–215.

[6] Giguère S. & Prescott J.F. (1997). – Clinical manifestations,diagnosis,treatment and prevention of *Rhodococcus equi* infectionsin foals. *Vet. Microbiol.*,56,313–334.

[7] Hillidge C.J. (1987). – Use of erythromycin – rifampin combination in treatment of *Rhodococcus equi* pneumonia. *Vet. Microbiol.*,14,337–342.

[8] Hong C.B. & Donahue J.M. (1995). –*Rhodococcus equi*-associated necrotising lymphadenitis in a llama. *J. Comp. Pathol.*,113,85–88.

[9] Kinne J.,Jose S. & Wernery U. (2010). –*Rhodococcus equi*-pneumonia in dromedary camels in Dubai. Poster presented at the International Camel Conference,16–20 June,Garissa,Kenya.

[10] Kinne J.,Madarama H.,Takai S.,Jose S. & Wernery U. (2011). – Disseminated *Rhodococcus equi* infection in dromedarycamels (*Camelus dromedarius*). *Vet. Microbiol.*,149,269–272.

[11] Leite R.C.,Negrelli-Filho H. & Langenegger C.H. (1975). – Infection por *Corynebacterium equi* em llama (*Lama glama*). *Pesquisa Agropecuria Brasileira,Servicios Veterinarios*,10,57–59.

[12] Madarame H.,Takai S.,Morisawa N.,Fujii M.,Hidaka D.,Tsubaki S. & Hasegawa Y. (1996). – Immuno-histochemical detection of virulence-associated antigens of *Rhodococcus equi* in pulmonary lesions of foals. *Vet. Pathol.*,33,341–343.

[13] Mariotti F.,Cuteri V.,Takai S.,Renzoni G.,Pascucci L. & Vitellozzi G. (2000). – Immunohistochemical detection of virulenceassociated *Rhodococcus equi* antigens in pulmonary and intestinal lesions in horses. *J. Comp. Pathol.*,123,186–189.

[14] Puthucheary S.D.,Sangkar V.,Feez A.,Karunakaran R.,Raja N.S. & Hassan H.H. (2006). – An emerging human pathogen inimmunocompromised hosts: a report of four cases from Malaysia. *Southeast Asian J. Trop. Med. Public Health.*,37,157–161.

[15] Takai S.,Chaffin M.K.,Cohen N.D.,Hara M.,Nakamura M.,Kakuda T.,Sasaki Y.,Tsubaki S. & Martens R.J. (1991). –Prevalence of virulent *Rhodococcus equi* in soil from five *R. equi*-endemic horse-breeding farms and restriction fragment lengthpolymorphisms of virulence plasmids in isolates from soil and infected foals in Texas. *J. Vet. Diag. Invest.*,13,489–494.

[16] Takai S.,Imai Y.,Fukunaga N.,Uchida Y.,Kamisawa K.,Sasaki Y.,Tsubaki S. & Sekizaki T. (1995). – Identification ofvirulence-associated antigens and plasmids in *Rhodococcus equi* from patients with AIDS. *J. Infect. Dis.*,172,1306–1311.

[17] Takai S.,Sasaki Y.,Ikeda T.,Uchida Y.,Tsubaki S. & Sekizaki T. (1994). – Virulence of *Rhodococcus equi* isolates from

patientswith and without AIDS. *J. Clin. Microbiol.*,32,457–460.

[18] Venner M.,Heyers P.,Strutzberg-Minder K.,Lorenz N.,Verspohl J. & Klug E. (2007a). – Detection of *Rhodococcus equi* by*microbiological* culture and by polymerase chain reaction in samples of tracheo-bronchial secretions of foals. *Berl. Münch. Tierärztl. Wochenschr.*,120,126–133 [in German].

[19] Venner M.,Kerth R. & Klug E. (2007b). – Evaluation of tulathromycin in the treatment of pulmonary abscesses in foals.*Vet. J.*,174 (2),418–421.

[20] Waldrige B.M.,Morresey P.R.,Loynachan A.T.,Reimer J.,Riddle W.T.,Williams N.M. & Bras R. (2008). –*Rhodococcus equip* neumonia in an adult horse. *Equine Vet. Educ.*,20,67–71.

(储岳峰译，殷宏校)



## 1.1.6 鼻疽

鼻疽是可危及马属动物生命的一种传染病。感染鼻疽通常是致命的，已知的鼻疽宿主包括野生猫科动物、熊、狼和犬类等。食肉动物经食用受污染的肉而感染。最近，德黑兰动物园处理了14只感染了鼻疽的狮子[3]。

鼻疽在亚洲、非洲、南美许多国家流行，且2004年在中东地区[11]、2008年在巴基斯坦以及2009年在巴西再次发生。2010年马鼻疽在科威特和巴林首次发生[4]。鼻疽是世界动物卫生组织（OIE）所列疾病之一，编目于《陆生动物卫生法典》的第12章11节[12]。单峰驼对该病易感，Curasson等描述了对双峰驼进行人工感染试验的观察[1]。

### 病原学

鼻疽是由鼻疽伯氏菌引起的疾病，该菌是一种绝对需氧、无芽孢、非运动性的病原菌。革兰氏染色不良，但是两极着色明显（图56）。鼻疽伯氏菌以及类鼻疽伯氏菌（见11.7）都属于实验室最危险的细菌，所有相关试验需在生物安全柜中进行，并且为了避免人员感染要采取所有必要的安全措施。鼻疽伯氏菌生长缓慢，因此需将琼脂平板置37℃孵育至少72小时。在一只感染该病的单峰驼的病例中，孵育48小时后可见奶油色圆形小菌落，继续孵育24小时后菌落变大。在含有甘油的血平板上生长48小时比其他平板上生长的菌落大。用API 20 NE（非肠道菌分析谱）进行生化分析，本菌具有弱氧化酶活性，硝酸脱水酶、精氨酸脱水酶（孵育4天后）、N-乙酰葡萄糖胺以及葡萄糖酸钾反应阳性。依照笔者的经验，从冻存于-20℃的临床样品中培养鼻疽伯氏菌非常困难。因此，冷冻很可能会破坏该病原。

鼻疽损伤部位通常含菌量很少，在显微镜下观察时不容易观察或被忽略（图57）。对雄性成年豚鼠腹腔注射感染性物质诱发其发生斯特劳斯反应，是一种很好的可以富集细菌的方法。在注射12天后这些动物或许会有典型的斯特劳斯反应，并且可以从这些动物的睾丸中获得纯培养的鼻疽伯氏菌（图58）。由于其他细菌的竞争性生长，临床直接分离鼻疽伯氏菌很困难。目前还没有任何一种生化反应试剂盒可以明确鉴别出该菌[9]，只有PCR这种方法比较可靠，甚至可以从福尔马林固定的样品中鉴别出该病原，并且可以区分鼻疽伯氏菌和类鼻疽伯氏菌[2,6]。

### 流行病学及临床症状

骆驼也是鼻疽的易感动物[1]，但骆驼自然感染鼻疽的记录只是刚刚才确认。Curasson[1]提到，两个俄罗斯的研究者Djounkovski和Petrowsky，将来源于马的鼻疽病料通过静脉注射给俄罗斯双峰驼。在注射后11～15天这些双峰驼的鼻腔和许多其他器官出现特征性的结节和溃疡。该病在骆驼、马、长颈鹿之间通过接触传播是可能的。然而Samartsev[5]认为鼻疽对骆驼来说没有致病意义。在1966—1968年期间，对蒙古国的500 000只骆驼进行大规模的鼻疽菌素眼睑皮内注射接种检测，仅有很少的骆驼呈阳性反应，且没有观察到任何临床症状[7]。

最近，报道了一例自然感染鼻疽的单峰驼病例，该单峰驼与两只鼻疽阳性的马有过接触[10]。

患病单峰驼鼻孔流出大量黏脓性的分泌物（图59）、发烧、消瘦、疲劳，但是不表现出皮肤型鼻疽样（Farcy）损伤。

## 病理变化

在单峰驼病例中，可以观察到典型鼻疽损伤的部位是肺、后鼻孔以及鼻中隔。在肺部可以检测到高尔夫球般大小的微红的灰色结节，与带有中央灰色坏死区的结核结节类似。在后鼻孔和鼻中隔，星状疤痕、溃疡以及蜂窝坏死性斑点处覆盖着黄色脓液（图60）。其他器官中没有观察到鼻疽性损伤。对肺部肉芽肿进行组织学检查，可观察到特征性的被脓性肉芽肿样炎性物质包围的中心组织坏死，炎性物质中包含许多嗜中性粒细胞、一些淋巴细胞、巨噬细胞、上皮细胞以及少量的巨细胞。

## 诊断

世界动物卫生组织（OIE）认定的两种诊断马鼻疽的方法，一种是补体结合试验，一种是鼻疽菌素检测。这两种检测方法同样适用于骆驼[10]。补

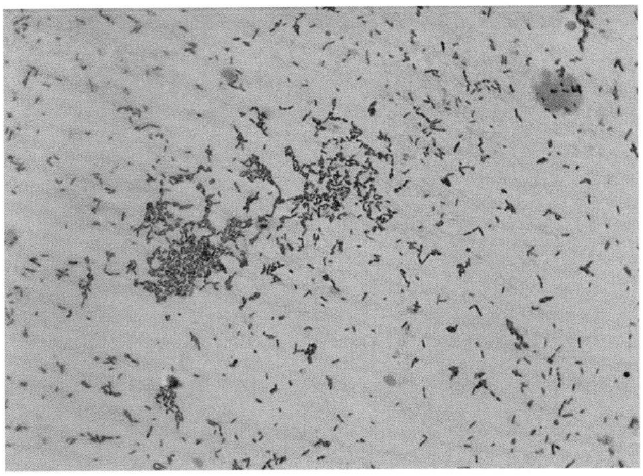

图56　鼻疽伯氏菌革兰氏染色不良

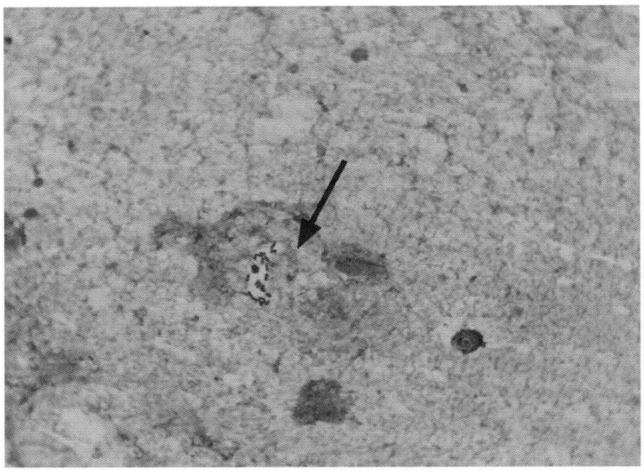

图57　对马鼻疽鼻后孔病变组织革兰氏染色，呈现出少量疑似鼻疽的革兰阴性杆菌

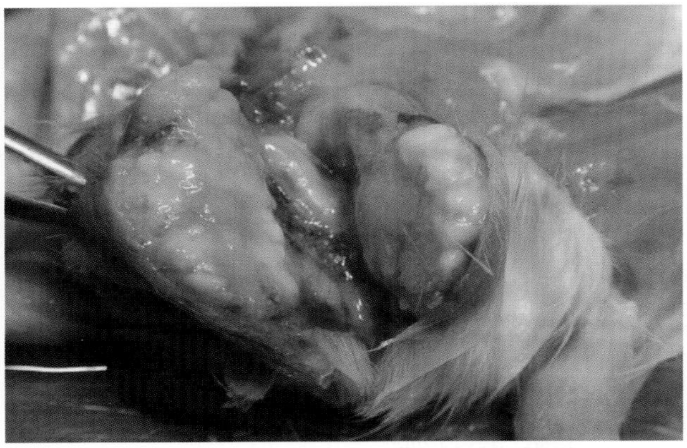

图58　豚鼠腹腔注射鼻疽病料后呈现斯特劳斯反应

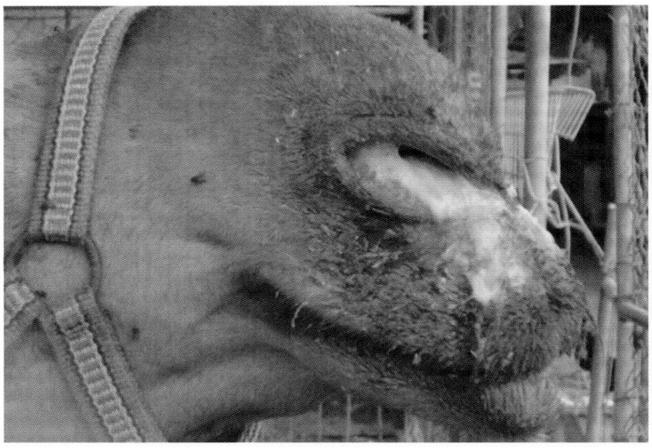

图59　患鼻疽骆驼鼻孔中大量排出黏脓性的分泌物

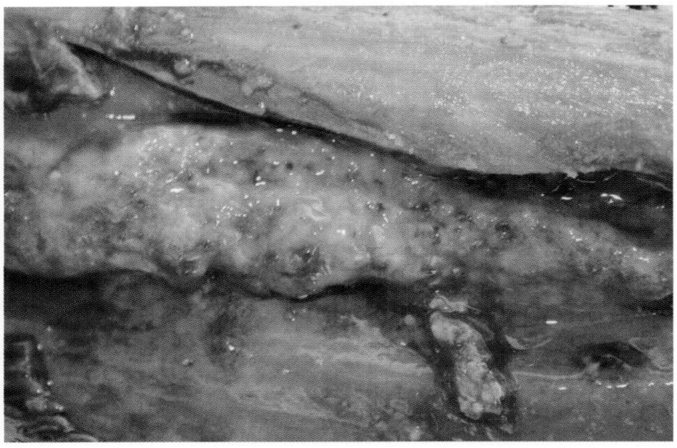

图60　单峰驼鼻后孔鼻疽性损伤

体结合试验的敏感性达90%～95%，但有一些马血清呈现假阳性（交叉反应）。因此开发了一些其他的血清学检测方法。但是目前为止还没有任何一种血清学方法可以区分鼻疽伯氏菌和类鼻疽伯氏菌。即使是最近研发的用于人类类鼻疽血清学诊断的竞争ELISA方法也不行，该方法利用类鼻疽伯氏菌脂多糖上的特异性表位获得的单克隆抗体为基础建立，对人类类鼻疽的检测具有100%的特异性。同样，利用抗鼻疽伯氏菌脂多糖的单克隆抗体（单抗3D11），也建立了一种竞争ELISA方法，已确认可适用于骆驼以及马的鼻疽检测。该方法的优点是，到目前为止还没有观察到交叉反应。前述患鼻疽的单峰驼，利用补体结合试验以及cELISA检测，结果均为阳性。

注射鼻疽菌素的方法可通过检测感染鼻疽的动物的延迟超敏反应来诊断，但目前还没有应用在单峰驼上。因此，无法评论该方法的敏感性。尽管给蒙古的500 000只骆驼注射了鼻疽菌素，但是仅有少数骆驼出现反应，并且没有观察到临床症状[7]。

目前没有抗鼻疽疫苗。尽管鼻疽伯氏菌对抗生素以及磺胺嘧啶药物敏感，但是患该疾病的动物禁止治疗，所以这种疾病只能通过改善卫生措施来预防。患鼻疽的动物必须要处死，并且尸体也必须妥善处理。必须对被感染的动物用具进行消毒处理，对有感染的马厩要进行全面检疫处理。鼻疽是OIE所列马属动物疾病之一，因此如有本病发生，需通报OIE。然而，其他种类动物鼻疽没有列于OIE法定报告传染病列表。所以，鼻疽也应该列于OIE的骆驼传染病列表。

## 参考文献


[1] Curasson G. (1947). – Le chameau et ses maladies. Vigot Frères Editions,Paris,86－88.

[2] 2. Neubauer H.,Sprague L.D.,Joseph M.,Tomaso H.,Al Dahouk S.,Witte A.,Kinne J.,Hensel A.,Wernery R.,Wernery U. & Scholz H.C. (2007). – Development and clinical evaluation of a PCR assay targeting metalloprotease gene (mpr A) of *Burkholderia pseudomallei. Zoonoses Public Health*,54,44－50.

[3] Promed (2011). – Glanders,feline – Iran: (Tehran),tiger,lion. Available at: www.promedmail.org/pls/apex/f?p=2400:1001:260 8383620720133::NO::F2400_P1001_BACK_PAGE,F2400_P1001_PUB_MAIL_ID:1000,86767 (accessed 22 January 2011).

[4] Roberts H.,Lopez M. & Hancock R. (2010). – International disease monitoring,April to June 2012. *Vet. Rec.*,167 (6),194.

[5] Samartsev A.A. (1950). – Infectious pustular dermatitis in camels. *Proc. Kazakh. Res. Vet. Institute*,5,190－197.

[6] Scholz H.C.,Joseph M.,Tomaso H.,Al Dahouk S.,Witte A.,Kinne J.,Hagen R.M.,Wernery R.,Wernery U. & Neubauer H.(2006). – Detection of the re-emerging agent *Burkholderia mallei* in a recent outbreak of glanders in the United Arab Emiratesby a newly developed *flip*-base polymerase chain reaction assay. *Diagn. Microbiol. Infect. Dis.*,54,241－247.

[7] Splisteser H. (2010). – 25 Jahre Tierarzt in der Mongolei. Eike Andreas Seidel,Buchholz,Germany,68-69.

[8] Thepthai C.,Sunithtikam S.,Suksuwan M.,Songsivilai S. & Dharakul T. (2005). – Serodiagnosis of melioidosis by acompetitive enzyme-linked immunosorbent assay using a lipopolysaccharide-specific monoclonal antibody. *Asian Pac.J. Allergy Immunol.*,23,127－132.

[9] Wernery U. (2009). – Glanders. *In* Infectious diseases of the horse (T.S. Mair & R.E. Hutchinson,eds). A peer reviewedpublication,Geerings Print Ltd.,Ashford,UK,253－260.

[10] Wernery U.,Wernery R.,Joseph M.,Al Salloom F.,Johnson B.,Kinne J.,Jose S.,Tappendorf B. & Scholz H.C. (2011). – Firstcase of natural *B. mallei* infection (glanders) in a dromedary camel in Bahrain. *Emerg. Infect. Dis.*,17 (7),1277－1279.

[11] Wittig M.B.,Wohlsein P.,Hagen R.M.,Al Dahouk S.,Tomaso H.,Scholz H.C.,Nikolaou K.,Wernery R.,Wernery U.,Kinne J.,Elschner M. & Neubauer H. (2006). – Ein Übersichtsreferat zur Rotzerkrankung. *DTW*,113,323－330.

[12] World Organisation for Animal Health (OIE) (2008). – Manual of diagnostic tests and vaccines for terrestrial animals. 6th Ed.World Organisation for Animal Health,Paris,France,919－928.


(储岳峰译，殷宏校)

## 1.1.7 类鼻疽

类鼻疽是由类鼻疽伯氏菌引起的细菌性传染病，类鼻疽伯氏菌也是一个潜在的生物武器[8]。类鼻疽症状类似于肺结核，以淋巴结和内脏的化脓性或干酪样病变为特征。类鼻疽与鼻疽的临床症状和病理变化非常相似，而且这两种病原只有通过分子生物学的方法才能区分[11]。类鼻疽可感染很多动物，包括鸟类、海豚、热带鱼和人类[3]。仓鼠、兔子和豚鼠是高度易感动物，是理想的实验动物。本病主要由环境感染到动物，而不是从动物传染到动物。疫病暴发源于土源性感染，在高温高湿的热带地区常在暴雨或洪水之时或之后发生。

Bergin和Torenbeeck[2]、Forbes-Faulkner等[5]、Low Choy等[9]和Wernery等[17]分别在澳大利亚和阿联酋的单峰驼中首次确诊类鼻疽。此外，类鼻疽在新大陆骆驼中也有报道[13]。

### 病原学

类鼻疽伯氏菌是唯一传染源，并且不同种株系在遗传和致病性差异较大。类鼻疽伯氏菌为革兰阴性杆菌，能氧化乳糖，在麦康凯和血琼脂培养基上生长快于鼻疽伯氏菌。通过等位基因图谱分析，本菌与无毒的泰国类鼻疽伯氏菌属于不同分支，但鼻疽伯氏菌却属于本菌的一个克隆分支[8]。

### 流行病学及临床症状

类鼻疽伯氏菌存在于东南亚、澳大利亚北部和墨西哥的水和泥土中。在这些地区，本菌是无处不在的土壤腐生菌，通常存在于水坑，这使得水坑成为类鼻疽的第一疫源地。这也是1997年雨季阿联酋单峰驼感染类鼻疽伯氏菌的最可能的原因[17]。许多营自由生活的变形虫也是本菌的潜在宿主[6]。本菌可通过吸入和食入而感染。在伊朗的山羊、绵羊、两匹马和一匹骡子中分离到本菌，表明进口的家畜可能是另一种传染源。

在疫病流行地区，人类有感染的风险。人类感染类鼻疽主要在东南亚和澳大利亚北部，这种感染与季风和雨季的关系高度一致[4]。研究人员发现，土著居民在雨季接触类鼻疽伯氏菌的方式有所变化：从经皮肤感染转为气溶胶吸入形式，这与阿联酋单峰驼感染病例相似，该单峰驼感染也发生在一个异常的雨季。骆驼类鼻疽主要以弛张热、消瘦、乳腺炎和流产为主要特征。

### 病理变化

在澳大利亚昆士兰两次类鼻疽疫情中，6只单峰驼死亡[2,5]。在所有死亡动物中均观察到严重的坏死性肺炎。Bergin、Torenbeeck和Forbes-Faulkner等人认为生活在潮湿气候中的单峰驼似乎更易感染此病。鉴于类鼻疽在澳大利亚土著居民中流行[4]，也有人感染类鼻疽伯氏菌而死亡的病例发生，在治疗发生肺炎的单峰驼时需要极其小心。Low Choy等报道了几只引入澳大利亚北部地区的单峰驼和羊驼因感染类鼻疽而死亡[9]。阿联酋一只7岁龄雌性单峰驼也被诊断为类鼻疽，死亡前表现为严重消瘦等消耗性疾病症状[17]。剖检病变显示子宫和气管肉芽肿（图61），3/4的肺脏（图62）、纵隔淋巴结（图63）、膈肌（图64）、脾脏、肝脏和肾脏均出现严重的干酪样坏死。

组织病理学调查显示急性坏死性干酪性肺炎（图65）和坏死性淋巴管炎。研究人员推测这一单一的类鼻疽个例是由1997年阿联酋的极端雨季引起的。

类鼻疽不仅在动物园内和自然环境的旧大陆骆驼中发生[13]，而且在新大陆骆驼中也有报道。在澳大利亚北部有一例羊驼类鼻疽病例发生[7,9]，这只羊驼从气候比热带北部温和的南部输入。这只7月龄的雄性羊驼体内外都有多个脓肿，并从脓肿中分离到了类鼻疽伯氏菌。该作者认为骆驼科动物对类鼻疽特别易感（表13）。

表13 澳大利亚北部动物感染类鼻疽的发生率

| 动物种类 | 发生率 | 对类鼻疽的易感性 |
|---|---|---|
| 山羊 | 43 | 高 |
| 绵羊 | 14 | 高 |
| 骆驼 | 3 | 高 |
| 羊驼 | 1 | 高 |
| 猪 | 11 | 中 |
| 鹿 | 2 | 中 |
| 猫 | 5 | 低 |
| 狗 | 2 | 低 |
| 鸟 | 1 | 低 |
| 牛 | 1 | 低 |

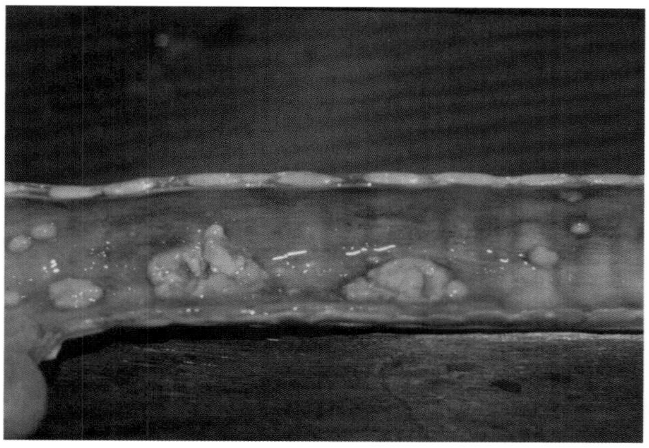

图61 患类鼻疽的单峰驼气管

图62 患类鼻疽的单峰驼肺脏

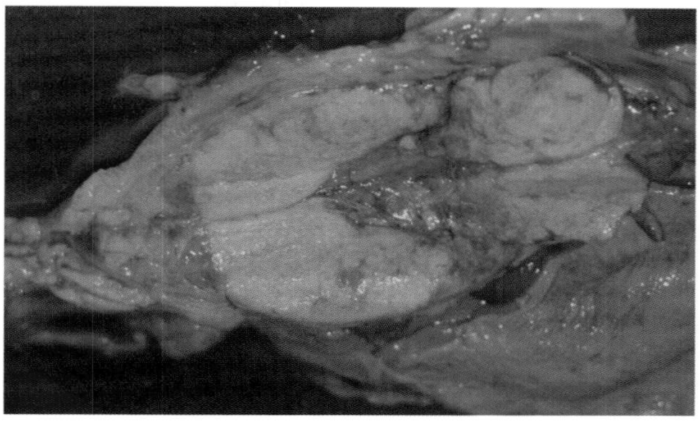

图63 患类鼻疽的单峰驼纵隔淋巴结

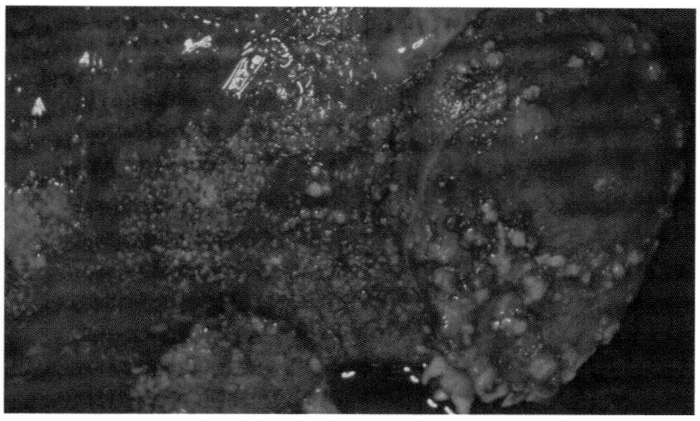

图64 患类鼻疽的单峰驼膈肌结节状病变

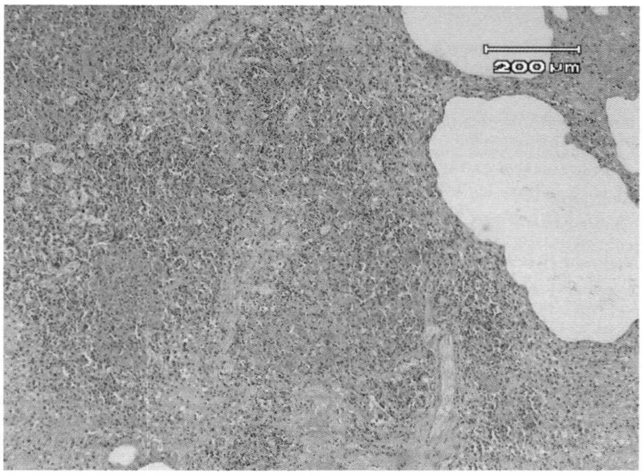


图65 类鼻疽伯氏菌导致的单峰驼小叶性慢性肺炎（苏木精-伊红染色）

## 诊断

类鼻疽伯氏菌在血平板和麦康凯培养基上易于培养，并且易于在不同的选择性培养基，如含甘油和庆大霉素的Ashdown培养基上生长。菌落形态多变，从表面光滑的黏液型到表面粗糙的褶皱波纹型等。在绵羊血平板中完全溶血。通过API 20 NE检测（非肠道菌分析谱）对菌落进行检测（API NO. 1156577），可以准确鉴定为类鼻疽伯氏菌。

以金属蛋白酶基因（mprA）为靶基因的PCR方法，甚至可以在福尔马林固定的石蜡切片中可鉴定出类鼻疽伯氏菌[10,11]。此反应的最低检测限是10 *fg* 纯化的类鼻疽伯氏菌DNA或2个细菌。

已经开发多种血清学试验方法用于人类患者的诊断，例如，细菌凝集试验、间接血凝试验、补体结合试验（CFT）、酶联免疫吸附试验（ELISA）和免疫荧光抗体试验等[12]。在兽医诊断中，血清学方法已用于检测马、山羊和奶牛中的抗体[16]。在泰国最近开展的一项研究中，利用间接血凝试验调查了8153份不同动物血清样品，包括牛、绵羊、山羊、猪以及鹿的血清[14]。结果表明，不同动物中类鼻疽抗体阳性率分别是：山羊（0.33%）、猪（7.23%）、牛（2.56%）、鹿（7.23%）以及绵羊（6.83%）。

补体结合试验（CFT）和鼻疽菌素试验是OIE认定的诊断鼻疽的两种方法。然而这两种方法均不能区分鼻疽伯氏菌与类鼻疽伯氏菌的感染[18]。最新利用针对类鼻疽伯氏菌脂多糖上特异性表位的单克隆抗体，研究出一种诊断人类类鼻疽的特异性达100%的竞争ELISA血清学诊断方法[15]。

**防控措施**

几乎没有收集到治疗患类鼻疽家畜的信息，但有治疗人类该病的建议。在马属动物中，氯霉素显示出有效的治疗效果；对于宠物来说，推荐使用头孢他啶和碳青霉烯类抗生素[9]。使动物远离污染源，如泥水坑是一种有效的预防措施。

## 参考文献

[1] Baharsefat M. & Amjadi A.R. (1970). – Equine melioidosis in Iran. *Arch. Inst. Razi*,22,209–213.

[2] Bergin T.F. & Torenbeeck L.R. (1991). – Melioidosis in camels. *Austr. Vet. J.*,68,309.

[3] Cheng A.C.,Bart J. & Currie B.J. (2005). – Melioidosis: epidemiology,pathophysiology,and management. *Clin. Microbiol. Rev.*,18 (2),383–416.

[4] Cheng A.C.,Jacups S.P.,Ward L. & Currie B.J. (2008). – Melioidosis and aboriginal seasons in northern Australia. *Trans.R. Soc. Trop. Med. Hyg.*,102 (1),26–29.

[5] Forbes-Faulkner J.C.,Townsend W.L. & Thomas A.D. (1992). – *Pseudomonas pseudomallei* infection in camels. *Aust. Vet. J.*,69 (6),148.

[6] Inglis T.J.J.,Rigby P.,Robertson T.A.,Dutton N.S.,Henderson M. & Chang B.J. (2000). – Interaction between *Burkholderia pseudomallei* and *Acanthamoeba* species results in coiling phagocytosis,endamebic bacterial survival,and escape. *Infect. Immun.*,68 (3),1681–1686.

[7] Janmaat A.,Low Choy J. & Currie B.J. (2004). – Melioidosis in an alpaca (*Lama pacos*). *Aust. Vet. J.*,82 (10),622–623.

[8] Leelarasamee A. (2004). – Recent development in melioidosis. *Curr. Opin. Infect. Dis.*,17,131–136.

[9] Low Choy J.,Mayo M.,Janmaat A. & Currie B.J. (2000). – Animal melioidosis in Australia. *Acta Tropica*,74,153–158.

[10] Neubauer H.,Sprague L.D.,Joseph M.,Tomaso H.,Al Dahouk S.,Witte A.,Kinne J.,Hensel A.,Wernery R.,Wernery U.& Scholz H.C. (2007). – Development and clinical evaluation of a PCR assay targeting metalloprotease gene (*mprA*) of *Burkholderia pseudomallei. Zoonoses Public Health*,54,44–50.

[11] Scholz H.C.,Joseph M.,Tomaso H.,Al Dahouk S.,Witte A.,Kinne J.,Hagen R.M.,Wernery R.,Wernery U. & Neubauer H.(2006).– Detection of the re-emerging agent *Burkholderia mallei* in a recent outbreak of glanders in the United Arab Emirates by a newly developed *flip*-base polymerase chain reaction assay. *Diagn. Microbiol. Infect. Dis.*,54,241–247.

[12] Sirisinha S. (1991). – Diagnostic value of serological tests for melioidosis in an endemic area. *Asia Pacific J. Allergy Immunol.*,9,1–3.

[13] Sprague L.D. & Neubauer H. (2004). – Melioidosis in animals: a review on epizootiology,diagnosis and clinical presentation.*J. Vet. Med. B*,51,305–320.

[14] Srikawkheaw N. & Lawhavinit O.-R. (2007). – Detection of antibodies against melioidosis from animal sera in Thailand by indirect haemagglutination test. *Kasetsart J. (Nat. Sci.)*,41,81–85.

[15] Thepthai C.,Sunithtikam S.,Suksuwan M.,Songsivilai S. & Dharakul T. (2005). – Serodiagnosis of melioidosis by a competitive enzyme-linked immunosorbent assay using a lipopolysaccharide-specific monoclonal antibody. *Asian Pac.J. Allergy Immunol.*,23,127–132.

[16] Thomas A.D.,Spinks G.A.,D'Arcy T.L.,Norton J.H. & Trueman F. (1988). – Evaluation of four serological tests for thediagnosis of caprine melioidosis. *Aust. Vet. J.*,65,261–264.

[17] Wernery R.,Kinne J.,Haydn-Evans J. & Ul-Haq A. (1997). – Melioidosis in a seven year old camel,a new disease in the

United Arab Emirates (UAE). *J. Camel Pract. Res.*,4 (2),141－143.

[18] Wernery U.,Wernery R.,Joseph M.,Al Salloom F.,Johnson B.,Kinne J.,Jose S.,Tappendorf B. & Scholz H.C. (2011). – First case of natural *B. mallei* infection (glanders) in a dromedary camel in Bahrain. *Emerg. Infect. Dis.*,17 (7),1277－1279.

(储岳峰译，殷宏校)



## 1.1.8 巴氏杆菌病

巴氏杆菌宿主广泛，呈世界性分布。该属大多数成员是共生菌，存在于动物的上呼吸道和消化道黏膜上。巴氏杆菌的细菌与宿主和环境关系密切。巴氏杆菌可引起肺炎、鼻炎、乳腺炎、败血症、脂膜炎、淋巴管炎和其他疾病（表14）。一些诱因，如并发性病毒、细菌及寄生虫感染，环境条件、气候因素等，在巴氏杆菌的致病性中发挥了重要作用。巴氏杆菌也可引起禽类感染。目前，关于骆驼是否感染出血性败血症以及某一个单独的巴氏杆菌的菌种是否能引起骆驼的疫病仍然存在很大争议。很少对从骆驼体内分离的巴氏杆菌进行了血清型鉴定。目前对于骆驼是否有自己独特的巴氏杆菌

表14　动物的巴氏杆菌病

| 细菌种类 | 动物种类 | 疾病 | 血清型 |
|---|---|---|---|
| 溶血性曼氏杆菌（前称溶血性巴氏杆菌） | 新大陆骆驼 | 脓肿及骨炎（血清型未知） | 12种血清型；1和6最常见；许多没有定型 |
| | 牛、羊 | 肺炎巴氏杆菌 | |
| | | 地方性肺炎 | |
| | | 3月龄以下羔羊败血症 | |
| | | 坏疽性乳房炎 | |
| | | 5~12月龄羔羊败血症 | |
| | 旧大陆骆驼 | 乳腺炎 | |
| 多杀性巴氏杆菌 | 牛 | 船运热及地方性肺炎综合征<br>偶尔严重的乳腺炎 | 血清A型，所有具有荚膜和菌体抗原的血清型 |
| | 羊 | 胸膜肺炎<br>乳腺炎 | |
| | 猪 | 肺炎（往往继发感染） | |
| | 大多数骆驼和野生动物 | 肺炎和其他感染（并发的） | |
| | 兔子 | 胸膜肺炎，脓肿，中耳炎，生殖道感染 | |
| | 家禽 | 禽霍乱 | |
| | 旧大陆骆驼 | 小牛肺炎，小牛败血症（但是血清型未知） | |
| 多杀性巴氏杆菌 | 牛、水牛、野牛、牦牛和其它反刍动物 | 出血性败血症<br>运输动物的咽炎 | 血清B型 |
| | 旧大陆骆驼? | 不清楚 | 血清B型和E型 |
| | 猪，家畜和家禽很少发生 | 有或无支气管败血波氏杆菌引起萎缩性鼻炎<br>继发性肺炎 | 血清D型 |
| | 牛，水牛 | 出血性败血症 | 血清E型 |
| 嗜肺巴氏杆菌 | 啮齿动物，狗和猫 | 肺炎，鼻咽脓肿，不显著 | 血清E型 |

表中“？”表示有疑问或不能确定，全书同。

也不清楚。一些研究者认为，骆驼对于一些牛来源的巴氏杆菌不敏感。

多杀性巴氏杆菌可以从健康新大陆骆驼的呼吸道分离到，然而除了两个病例外，没有巴氏杆菌可以引起新大陆骆驼发病的报道[15]。Fower和Gillespie[16]报道一个美洲驼耳朵发生骨炎，从左外耳的渗出物分离到巴氏杆菌的病例。第二个病例由Dwan等人[12]报道，一个羊驼发生喉炎后分离培养出溶血性曼氏杆菌，作者推测，是由于羊驼出生时未吃足够的初乳而导致对细菌易感。

旧大陆骆驼与其他反刍动物相比，对巴氏杆菌的易感性较低[2]；由巴氏杆菌引起的单峰驼出血性败血症的病例不多见[20]。

## 病原学

巴氏杆菌为革兰氏染色阴性、小杆菌或球杆菌，不形成芽孢，无运动性。本属细菌氧化酶试验阳性，接触酶试验阳性。在富含血液或血清的培养基生长良好。内毒素在巴氏杆菌引起的出血性败血症中起重要作用，然而对于巴氏杆菌的发病机制目前尚不完全清楚。

## 流行病学和临床症状

根据荚膜多糖抗原性的差异，溶血性曼氏杆菌至少有12个血清型。血清1～2，5～9，12～14，16～17是溶血性曼氏杆菌，血清型3，4，10，15是海藻糖曼氏杆菌，血清型11是葡萄糖苷酶曼氏杆菌。此外，还有一些无法定型的菌株。该菌可释放多种毒力因子，包括释放一些白细胞毒素，攻击机体对抗感染的巨噬细胞。毒素结合在白细胞或血小板表面黏附分子CD18，引起坏死和细胞裂解。释放的白细胞毒素常被用来诊断致病性的溶血性曼氏杆菌病。

根据荚膜多糖抗原，多杀性巴氏杆菌可分为A、B、C、D、E和F共6个血清群。根据菌体抗原，可分为1～16个血清型。血清型与疾病密切相关。例如，大多数引起肺炎的巴氏杆菌分离株是荚膜型A型和菌体型3型（多杀性巴氏杆菌A:3）。

5种血清群的分布如下。

（1）A型，普遍存在。

（2）B型，引起出血性败血症，主要分布在亚洲、中东、中非、尼日利亚和喀麦隆。

（3）D型，普遍存在（散发、亚急性型或慢性型）。

（4）E型，引起出血性败血症，主要分布在西非和中非。

（5）F型，在疾病中的作用尚不确切；引起火鸡疾病。

在巴氏杆菌病的诊断中，需要区别不同种的巴氏杆菌引起的疾病。在旧大陆骆驼中，已经报道能引起疾病的巴氏杆菌包括以下几种。

（1）溶血性曼氏杆菌，引起肺炎、败血症、结膜炎、乳房炎（参见乳房炎章节）。

（2）多杀性巴氏杆菌，引起肺炎、败血症。

（3）多杀性巴氏杆菌B型和E型菌株，引起出血性败血症。

在新大陆骆驼中，已经报道能引起疾病的巴氏杆菌有：

溶血性曼氏杆菌，骨炎、喉部脓肿。

应激是引发大动物巴氏杆菌病一个重要因素，例如，奶牛和水牛的出血性败血症、奶牛的运输热。一些病毒，例如，副流感病毒3型、牛疱疹病毒、黏膜病病毒、牛呼吸合胞体病毒，常与巴氏杆菌相伴引起运输热性肺炎。应激常与过度疲劳相关，例如，亚洲水牛在雨季犁地劳役、长途跋涉、车辆运输等易引起肌肉过度紧张。当一个动物发生应激以后，病原微生物的毒力会增加，传播给群体中其他动物，引起同一个圈舍或运输车中的动物感染疾病。

与其他动物一样，巴氏杆菌在骆驼上也为共生菌。该菌在上呼吸道黏膜常在，当机体抵抗力下降，如发生疥藓、锥虫病或热应激时，病菌具有致病性，引发疾病[21]。

尽管有多位研究者报道骆驼的巴氏杆菌感染，然而对于某一种巴氏杆菌引起的疾病的临床表现和致病机制仍有争议。该病与炭疽、沙门氏菌病临床症状相似，临床上会出现误诊，更使得其在骆驼的临床症状难以明确[10,14,31]。根据世界卫生组织（WHO）、世界动物卫生组织（OIE）和联合国粮农组织（FAO）记载[43]，在前苏联时发生野生双峰驼出血性败血症，阿尔及利亚、苏丹和索马里单峰驼也出现出血性败血症，毛里塔尼亚和乍得共和国及撒哈拉共和国也有此类病例。Leese[25]从印度2只发生急性胸膜炎、心包炎和腹膜炎的骆驼渗出物中分离到巴氏杆菌样的病原菌。

出血性败血症是一种传染性强、毒力强的败血性疾病，主要发生在奶牛、斑马、水牛和旧大陆骆驼和一些野生反刍动物。该病是由多杀性巴氏杆菌B型和E型引起的最急性、急性传染病，以发热、呼吸困难、鼻腔有分泌物、口吐白沫为特征。B型引起的出血性败血症主要分布在亚洲、中东、东非、尼日利亚和喀麦隆，而E型在西非和中非流行。

Higgins[21]认为，多杀性巴氏杆菌感染骆驼在临床上有急性型、超急性型和腹痛型3种形式。其中，腹痛型的特征是腹泻，粪便常混有血液。Chauhan等人[5]报道由多杀性巴氏杆菌引起的骆驼出血性败血症是一种高度传染性疾病，该病通过直接接触或间接接触污染饲料和饮水传播。临床上以发热、鼻分泌物增多、流泪、呼吸困难、黏膜充血、喉和颈部肿胀、肺炎为特征。Schwartz和Dioli[39]认为，该病急性型主要表现为出血性败血症，以全身广泛性出血、咽部和肩部疼痛性肿胀、胃肠炎为特征。康复的患病动物在肩前巴结会有脓肿残留。很多学者认为骆驼对出血性败血症有天然抵抗力[6,24]，对牛巴氏杆菌病不易感。

## 出血性败血症

骆驼多杀性巴氏杆菌病的暴发在非洲的许多国家，俄罗斯、印度和伊朗都有报道（表16）。Schwartz和Dioli[39]指出出血性败血症（HS）的相关特征有发热（达到40℃）、脉搏加快、呼吸急促、食欲减退、颈部极度疼痛性肿胀，下颌和颈部淋巴结肿大，几乎所有的病例都伴有出血性肠炎，粪便为焦黑色柏油样，此外，进一步发展为巧克力色尿液。出血性败血症主要发生在雨季和有规律性洪灾的地区。该病在所有年龄骆驼中可见，但主要感染成年骆驼。发病率低，死亡率可达到80%[38]。Momin等[30]报道印度暴发巴氏杆菌病，在发病的14头单峰骆驼中有11头死亡。患病动物以发热、颈部水肿、急性呼吸困难和迅速死亡为特征。血涂片染色后可见两极深染的多杀性巴氏杆菌。上述病例并没有经过实验室确诊，推测可能出现该病与炭疽混淆的情况，因为所描述的临床症状和病变与炭疽类似。Mochabo等[28]等在肯尼亚图尔卡纳的Lapur地区进行调查，以深入了解骆驼的最重要疾病及其临床表现；在肯尼亚北部的这个区域，出血性败血症（7.7%）被确定为继锥虫病（11.4%）、疥癣（10.8%）和蜱感染（7.9%）之后的第四个最常见的疾病。出血性败血症在当地被称为“咯咕瑞”，以水肿和高死亡率为特征。另外，Gliecks等[18]报道称，在肯尼亚牧民中出血性败血症被冠以各种名字，肯尼亚北部和索马里北部游牧民认为，出血性败血症是危害单峰驼健康的排名第1～第4的重要疾病。但是，来自肯尼亚个体动物的392份鼻拭子样品中未分离到B型或E型多杀性巴氏杆菌。

Hassan和Mustafa[20]从苏丹出血性败血症发病死亡的单峰驼的组织和骨髓中分离得到B型多杀性巴氏杆菌。作者也证实该菌能在小牛中引起出血性败血症，且其肉汤培养物引起兔子在24小时内死亡。然而，在单峰驼中复制病例未成功，但是使用牛、羊的细菌苗免疫骆驼，可以控制该病在骆驼中暴发。

在过去的10年中，在阿联酋记录了与溶血性曼氏杆菌和多杀性巴氏杆菌感染相关的单峰驼死亡原因（表15），只有16%的死亡骆驼是因为单独感染了溶血性曼氏杆菌，不管其表现是肺炎还是败血症，84%并发有其他疾病。死于多杀性巴氏杆菌的骆驼中，只有一头50kg的小骆驼死于败血症，有9头有其他并发症。

多位研究者对单峰驼人工感染巴氏杆菌培养物后的临床表现进行报道。Fayed[13]从采自埃及健康单峰驼的100份鼻拭子中分离了6株多杀性巴氏杆菌，所有的分离株对小鼠和兔子均有致病性。然而，两头骆驼经滴鼻感染后经过一个短暂的疾病期后恢复。Cross[6]同样对两头单峰驼接种了牛出血性败血症培养物，发现除轻微的局部红肿外没有全身性的症状。

Awad等[1,2]等报道通过肌内注射和鼻腔感染1型多杀性巴氏杆菌，单峰驼表现为食欲不振、发热、流涎，脉搏、呼吸加快，可以从唾液中重新分离到巴氏杆菌，而从血液中未分离到。

除了败血症外，巴氏杆菌感染还有其他临床表现。Donatien和Larrieu[11]发现患病动物出现肺炎、全身性肌肉腱炎、腹泻以及渗出性心包炎和腹膜炎[9]。Richard[38]认为巴氏杆菌感染时，母畜常常发生流产。但是，这些报道都没有提供有关病原

表15 阿联酋不同年龄段单峰驼死亡原因及与巴氏杆菌感染的关系

| 分离到溶血性曼氏杆菌 | 体重（kg） | 分离到多杀性巴氏杆菌 |
|---|---|---|
| 梭菌肠毒血症 | 55 | — |
| 肺炎 | 35 | — |
| 梭菌肠毒血症 | 350 | — |
| 败血病 | 60 | — |
| 白肌病 | 45 | — |
| — | 胎儿 | 鼻病毒流产 |
| 绞杀 | 100 | — |
| — | 50 | 败血病 |
| 肺炎 | ? | — |
| 吸入性肺炎，脑膜炎 | 300 | — |
| 梭菌肠毒血症 | ? | — |
| 梭菌肠毒血症 | 60 | — |
| 梭菌肠毒血症 | 500 | — |
| 梭菌肠毒血症 | 350 | — |
| 肠破裂 | 350 | — |
| 梭菌肠毒血症 | 250 | — |
| 梭菌肠毒血症 | 550 | — |
| 心脏衰竭 | 180 | — |
| 真菌性皮炎，球虫病 | 130 | — |
| — | 40 | 胸膜肺炎，沙门氏菌感染 |
| 败血病 | 40 | — |
| 梭菌肠毒血症 | 400 | — |
| 白肌病 | 70 | — |
| — | 35 | 脑膜炎 |
| — | 45 | 梭菌肠毒血症 |
| 未知 | 胎儿 | — |
| 白肌病 | 90 | — |
| — | 600 | 子宫破裂 |
| — | 350 | 梭菌肠毒血症 |
| — | 250 | 创伤 |
| — | 500 | 脂肪肝 |
| 皱胃真菌壁脓肿 | 300 | — |
| 吸入性肺炎 | ? | — |
| — | 80 | 白肌病 |
| 白血病 | 450 | — |

—：未分离到巴氏杆菌

的分离、鉴定等信息。

在小规模的田间试验中[42]，两种多杀性巴氏杆菌（血清型B，925株；血清型E，978株）对牛和水牛均有很高的致病力（Smith，1994，个人通信），当用5mL营养肉汤培养物（含有$10^6$CFU/mL菌）通过鼻孔喷入两只健康的9月龄骆驼时，没有表现出任何疾病症状。另外，用同一菌株的5mL营养肉汤（含有$10^6$CFU/mL菌）通过气管接种4只健康的8月龄骆驼。4只骆驼中2只体温L高至39.2℃，白细胞计数有轻微的L高，其中，1只有黏液性脓性鼻液，但是，在鼻液中没有分离到多杀性巴氏杆菌。3天后，骆驼的体温和白细胞计数恢复到正常水平，并且没有检测到脓性鼻液。可以看出，通过气管接种巴氏杆菌的骆驼的临床症状和病变表现轻微，这表明宿主因素对于烈性疾病的发展至关重要。自然发生的疾病往往是在一系列诱因条件下出现，情况则更为复杂。

Tesfaye[41]和Bekele[3]报道在埃塞俄比亚的索马里地区单峰驼发生一种呼吸道疾病，发病率为29.6%，死亡率为6.4%。患驼发热、精神沉郁、食欲减退和严重的脓性鼻液，从发病和死亡动物的肝脏、胸腔液体和全血中分离到溶血性曼氏杆菌。剖检病理变化为胸腔积液、肺炎、肺气肿、心包积液和纤维素性心包炎。早期使用土霉素治疗可以使许多患病骆驼得到康复。作者认为麻疹病毒可能是此次疫情的始作俑者，这次疫情是否与Yigezu等[44]（见1.3.2：肺炎）的报道存在某种联系尚不清楚。这是最近一篇关于单峰驼患败血性巴氏杆菌病的报道，不幸的是，没有从感染样品中分离到病原体。

最近，巴基斯坦骆驼发生一种神秘的疾病，科里斯坦沙漠中有4000头骆驼死亡[36]。当地学者认为患病骆驼因感染多杀性巴氏杆菌而发生呼吸抑制，但是，到目前为止还没有关于此次疫情的科学报道。巴基斯坦此前没有此类大规模疾病的报道，此次疫情很可能与那一时期异常寒冷和潮湿的天气有很大关系。

对骆驼巴氏杆菌病很有必要进行全面科学研究，为了能够与相似临床症状的疾病，如炭疽、沙门氏菌和内毒血症加以区别。虽然在阿联酋巴氏杆菌（多杀性巴氏杆菌和溶血性曼氏杆菌）广泛感染绵羊、山羊和牛，并且骆驼与这些小反刍动物圈舍很近，在过去的20年中，没有证据证明或遇到一例单峰驼出血性败血症病例。如上所述，这很可能因为单峰驼群在阿联酋有良好的管理措施和饲料。因此，巴氏杆菌病成为旧大陆骆驼重要疫病的可能性不大[26]。

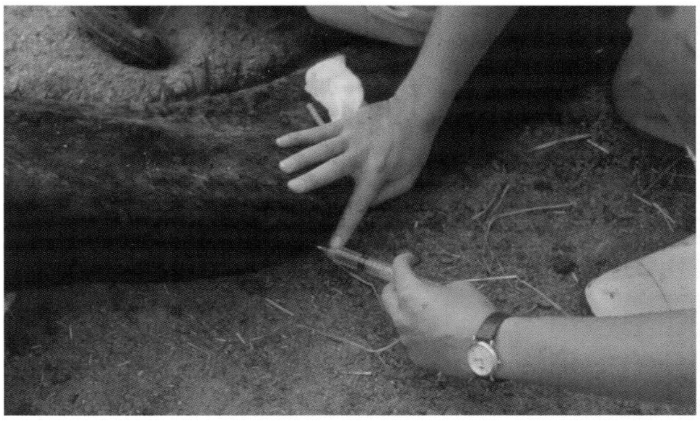

图66　5 mL $10^6$CFU/mL E型多杀性巴氏杆菌通过气管注射到1只8月龄单峰驼

在不同的研究中，许多血清学试验证实，在骆驼体内有多杀性巴氏杆菌血清A型、B型、D型、E型和I型溶血性曼氏杆菌抗体（表16）。这些血清样品均来自健康骆驼，进一步证明了骆驼是巴氏杆菌的宿主，但不发病。

## 诊断

目前大多数关于骆驼巴氏杆菌病的报道是混乱的、且常常是矛盾的。诊断依赖于从流行病学、临床症状、病理变化以及从血液、肝脏、脾脏、肾脏和淋巴结中分离出巴氏杆菌。如果骆驼已死亡一段时间，应提取骨髓样本。有时，为了从含有大量其他细菌的临床样品中分离出巴氏杆菌，进行小鼠腹腔接种病料是必要的。菌株的种和血清型的特异性鉴定对于骆驼是否有特定种巴氏杆菌非常重要。此外，应该在参考实验室对分离株进行血清型鉴定。

## 防控措施

巴氏杆菌病急性发病的特性极大的限制了使用抗生素治疗患病动物的功效。然而，对只有发热表现的骆驼，早期合理使用磺胺类、青霉素或土霉素药物，可以防止该病暴发。

在亚洲和非洲，牛羊通过实施大规模接种疫苗来预防巴氏杆菌病，在一些阿联酋国家，尽管还没有巴氏杆菌病暴发的报道，单峰驼也通过疫苗接种来预防出血性败血症（HS）。在亚洲，水

表16 不同国家旧大陆骆驼巴氏杆菌病汇总表

| 国家 | 作者 | 年份 | 疾病/分离株 |
|---|---|---|---|
| 毛里塔尼亚 | Kane | 1985 | 多杀性巴氏杆菌E型抗体 |
| | Kane | 1987 | |
| 印度 | Leese | 1927 | 颈部水肿 |
| | Ramchandran等 | 1968 | 败血症，多杀性巴氏杆菌 |
| | Dahl | 1987 | |
| | Momin等 | 1987 | |
| 乍得 | Maurice等 | 1967 | 血清学：427份，80%阳性，多杀性巴氏杆菌A,B,E,D和溶血性曼氏杆菌 |
| | Perreau和Maurice | 1968 | |
| | Perreau | 1971 | |
| 埃及 | Awad等 | 1976a,b | 多杀性巴氏杆菌I型（实验感染）食欲不振，发热，流涎 |
| | Fayed | 1973 | 健康骆驼中分离到多杀性巴氏杆菌 |
| 苏丹 | Hassan 和Mustafa | 1985 | 多杀性巴氏杆菌B，HS |
| 法属北非<br>（摩洛哥和阿尔及利亚） | Donatien<br>Donatien和Larrieu | 1921<br>1922 | HS但混淆于炭疽或沙门氏菌病？死亡率50%<br>发热，食欲不振，肌炎，肺炎，腹泻 |
| 伊朗 | Delpy | 1936 | 分离到巴氏杆菌 |
| | Goret | 1969 | |
| | Ono | 1943 | 出血性肠炎 |
| 埃塞俄比亚 | Richard | 1975 | 血清学：161份，65%阳性，多杀性巴氏杆菌A,B,D,E |
| | Bekele | 1999 | 溶血性曼氏杆菌<br>全身性疾病 |
| 突尼斯 | Burgemeister等 | 1975 | 52份血清，没有阳性 |
| 俄罗斯 | Oinakhbaev | 1965 | |
| | Sotnikov | 1973 | 多杀性巴氏杆菌 |
| 阿联酋 | Wernery（未发表） | 2010 | 溶血性曼氏杆菌<br>多杀性巴氏杆菌<br>并发症 |

HS：出血性败血症

牛使用氢氧化铝菌苗或油乳剂疫苗免疫也取得了很大的成功。Hassan和Mustafa [20]、Momin等[30]研究发现，对单峰驼接种菌苗或氢氧化铝疫苗（硫酸铝钾）能够控制出血性败血症暴发。Mohamed和Rahamtalla[29]用间接血凝和小鼠保护试验来评价单峰驼免疫B型出血性败血症菌苗、氢氧化铝疫苗和两者的联合使用后的抗体水平。作者发现用含B型巴氏杆菌抗原的疫苗免疫骆驼产生的血清可以保护小鼠抵抗B型巴氏杆菌的攻毒，但是没有对免疫单峰驼进行攻毒实验。



## 参考文献


[1] Awad F.I.,Salem A.A. & Fayed A.A. (1976a). – Studies on the viability of *Pasteurella multocida* Type 1 under simulated environmental conditions in Egypt. *Egypt J. Vet. Sci.*,13 (1),57–62.

[2] Awad F.I.,Salem A.A. & Fayed A.A. (1976b). – Studies of clinical signs observed on experimentally infected animals with *Pasteurella multocida* type. *Egypt. J. Vet. Sci.*,13 (1),53–56.

[3] Bekele T. (1999). – Studies on the respiratory disease 'sonbole' in camels in eastern lowlands of Ethiopia. *Trop. Anim. Health Prod.*,31,333–345.

[4] Burgemeister R.,Leyk W. & Goessler R. (1975). – Untersuchungen über Vorkommen von Parasitosen,bakteriellen und viralen Infektionskrankheiten bei Dromedaren in Südtunesien. *Dtsch. Tierärtztl. Wschr.*,82,352–354.

[5] Chauhan R.S.,Kaushik R.K.,Gupta S.C.,Satiya K.C. & Kulshreshta R.C. (1986). – Prevalence of different diseases in camels(*Camelus dromedarius*) in India. *Camel Newslett.*,3,10–14.

[6] Cross H.E. (1919). – Are camels susceptible to blackquarter,haemorrhagic septicemia and rinderpest? Bull. *Agric. Res. Inst. Pusa*,80.

[7] Dahl G. (1987). – Seminaire national sur le dromadaire. *In* Seminaire national sur le dromadaire,2–9 décembre 1985,Gao,Mali. *Camel forum*,18,1–111.

[8] Delpy L. (1936). – Sur les maladies contagieuses des animaux domestiques observees en Iran de 1930 à 1935. *Bull. Ac. Vét. Fr.*,9 (4),206–210.

[9] Donatien A. (1921). – El Ghedda,septicemie hémorragique des dromadaires. *Archs Inst. Past. Afr.*,1 (3),242–249.

[10] Donatien A. & Boue A. (1944). – Une epizootie de ghedda dans la region d'Qued Guir (Sahara oranais). *Arch. Inst. PasteurAlger.*,22 (3),171–174.

[11] Donatien A. & Larrieu M. (1922). – Nouvelle épizootie de Ghedda a M' Raier (Sahara) en 1921. Arch. Inst. *Pasteur l'Afrique Nord*,2 (3),316–319.

[12] Dwan L.W.,Thompson H.,Taylor D.J. & Philbey A.W. (2008). – Laryngeal abcessation due to *Mannheimia haemolytica* in an alpaca (*Vicugna pacos*) cria. *Vet. Rec.*,163,124–125.

[13] Fayed A.A. (1973). – Studies on pasteurellosis in buffaloes in Egypt. Thesis,Faculty of Veterinary Medicine,Cairo University,Cairo.

[14] Fazil M.A. & Hofmann R.R. (1981). – Haltung und Krankheiten des Kamels. *Tierätzl. Praxis*,9,389–402.

[15] Fowler M.E. (2010). – Medicine and surgery of camelids. 3rd Ed. Wiley-Blackwell,Berlin,Germany,213.

[16] Fowler M.E. & Gillespie D. (1985). – Middle and inner ear infection in llamas. *J. Zoo An. Med.*,16,9–15.

[17] Gatt Rutter T.E. & Mack R. (1963). – Diseases of camels. Part 1: Bacterial and fungal diseases. *Vet. Bull.*,33 (3),119–124.

[18] Gluecks I.V.,Younan M.,Maloo S.,Ewers C.,Bethe A.,Kehara D. & Kimari J. (2010). – Combination of participatoryapproaches and molecular diagnostics to investigate the epidemiology of haemorrhagic septicaemia in camels (*Camelusdromedarius*). *In* Proceedings of the International Camel Symposium,7–11 June 2010,Garissa,Kenya,5.

[19] Goret P. (1969). – Notes pour servir au cours sur les maladies bacteriennes et virales – la pasteurellose bovo-bubaline. ENS/III–42. Maisons Alfort,IEMVT.

[20] Hassan A.K.M. & Mustafa A.A. (1985). – Isolation of *Pasteurella multocida* type B from an outbreak of haemorrhagic septicemiain camels in Sudan. *Rev. Elev. Méd. Vét. Pays Trop*.,38 (1),31–33.

[21] Higgins A. (1986). – The camel in health and disease. Baillière Tindall,London,103–104.

[22] Kane M. (1985). – Enquête sêrologique sur la pasteurellose des dromadaires dans le cercle de Nara (Mali) et Abdel Bagron(Mauritanie). Forum Working Paper No. 18,Somali Academy of Sciences and Arts,Mogadishu,Somalia,95–96.

[23] Kane M. (1987). – La pasteurellose chez les dromadaires maliens et mauritaniens. *Bull. Liaison ILCA/GRPRC*,9,21–22.

[24] Leese A.S. (1918). –'Tips' on camels for veterinary surgeons on active service. Bailliere Tindall and Cox,London,50.

[25] Leese A.S. (1927). – A treatise on the one-humped camel in health and disease. Vigot Frères Editions,Paris.

[26] Manefield,G.W. & Tinson A. (1996). – Camels. A compendium. The T.G. Hungerford Vade Mecum Series for Domestic Animals,University of Sydney Post Graduate Foundation in Veterinary Science,240–298.

[27] Maurice Y.,Provost A. & Borredon C. (1967). – Presence d'anticorps bovipestiques chez le dromadaire du Tchad. *Rev. Elev. Méd. Vét. Pays Trop*.,20,537.

[28] Mochabo K.O.M.,Kitala P.M.,Gathura P.B.,Oraga W.O.,Catley A.,Eragar E.M. & Kaitho T.D. (2005). – Community perceptions of important camel diseases in Lapur Division of Turkana district,Kenya. *Trop. Anim. Health. Prod*.,37,187–204.

[29] Mohamed G.E. & Rahamtalla M.H. (1998). – Serological response of camel (*Camelus dromedarius*) to haemorrhagic septicaemia(*Pasteurella multocida* infections) vaccines. *J. Camel Prac. Res*.,5 (2),207–212.

[30] Momin R.R.,Petkar D.K.,Jaiswal T.N. & Jhala V.M. (1987). – An outbreak of pasteurellosis in camels. *Indian Vet. J*.,64 (10),896–897.

[31] Mustafa I.E. (1987). – Bacterial diseases of the camel and dromedary. OIE 55e Session générale de l'OIE,Office International des Epizooties,Paris,France,55,18–22.

[32] Oinakhbaev S. (1965). – Study of aetiology of contagious cough in camels. *Veterinariya (Moscow)*,42 (6),105–106.

[33] Ono Y. (1943). – Haemorrhagic enteritis in camels. J. *Vet. Sc*i.,5,113–114.

[34] Perreau P. (1971). –*Pasteurella*. Cours de microbiologie systématique 1971–1972 (Bactériologie) de l'institut Pasteur,Paris.

[35] Perreau P. & Maurice Y. (1968). – Epizootiologie de la pasteurellose des chameaux au Tchad. Enquête sérologique. *Rev. Elev. Méd. Vét. Pays Trop*s.,21 (4),451–454.

[36] Promed (2011). – Undiagnosed lethal disease of camel – Pakistan: haemorrhagic septicaemia suspected. Available at: www.promedmail.org/pls/apex/f?p=2400:1001:2608383620720133::NO::F2400_P1001_BACK_PAGE,F2400_P1001_PUB_MAIL_ID:1000,86780 (accessed 24 January 2011).

[37] Ramachandran P.K.,Ramachandran S. & Joshi T.P. (1968). – An outbreak of haemorrhagic gastroenteritis in camels (*Camelus dromedarius*). *Ann. Parasit. Hum. Comp*.,43 (1),5–14.

[38] Richard B. (1975). – Etude de la pathologie du dromadaire dans la sous-province du Borana (Ethiopie). Thèse Docteur,Ecole Nationale Vétérinaire Alfort,Creteil,Paris,75.

[39] Schwartz H.J. & Dioli M. (1992). – The one-humped camel in Eastern Africa. A pictorial guide to diseases,health care and management. Verlag Josef Margraf,Weikersheim,Germany,159–160.

[40] Sotnikov M.I. (1973). – Camel plague. *In* Orlov,F.M. Maloizvestnye zarazny bolezni zhivotnykh,Izdatel'stvo Kolos,213–222. Résumé in *Vet. Bull*.,44,937.

[41] Tesfaye R. (1996). – Report on the new camel disease (FURROO) in Southern Rangeland Development Project (SORDU),Borena,Ethiopia. *In* Ethiopian Veterinary Association proceedings of the 10th Conference,13–15.

[42] Wernery U. & Kaaden O.-R. (2002). – Infectious diseases of camelids. 2nd Ed. Blackwell Science,Berlin and Vienna,49–54.

[43] World Health Organization (WHO),Food and Agriculture Organization (FAO) and the World Organisation for Animal Health (OIE). (1961). – Animal Health Year Book. Food and Agricultural Organisation of the United Nations,Rome.

[44] Yigezu M.,Roger F.,Kiredjian M. & Tariku S. (1997). – Isolation of *Streptococcus equi* subspecies *equi* (strangles agent) from an Ethiopian camel. *Vet. Rec*.,140,608.

(储岳峰译，殷宏校)

## 1.1.9 鼠疫

在过去的几个世纪，鼠疫耶尔森氏菌（鼠疫杆菌）引起的大流行曾导致数百万人死亡。据称在公元1400年前，“黑死病”造成了大约4000万欧洲人死亡，致使欧洲人口减少了1/3。直到现在，鼠疫仍在许多国家呈地方性流行，例如，非洲国家、前苏联的某些成员国、印度尼西亚、印度、越南以及南美洲北部某些存在自然疫源地的地区。赞比亚人类鼠疫的暴发，与暴雨和洪水引起鼠类迁移到高地地域有关[20]。鼠疫耶尔森氏菌主要通过带菌的啮齿类动物身上的跳蚤传播。猫对兔的鼠疫耶尔森氏菌易感，在本病流行地区会对人的健康形成一定的威胁。鼠疫有两种表现形式，一种是腺鼠疫，细菌侵入局部淋巴结，引起发炎、变软，并致化脓（腹股沟淋巴结炎）。病原通过血液扩散，可导致肺炎和脑膜炎。肺鼠疫是经空气传播的致死性传染病，飞沫和气溶胶可导致人与人之间传染。旧大陆骆驼发生鼠疫已有报道，双峰驼和单峰驼在鼠疫传播给人的过程中起到重要作用[26]。

### 病原学

鼠疫耶尔森氏菌是一种短的、两端钝圆的椭圆形球杆菌，组织或渗出液直接染色时，呈单个或成对存在；在液体培养基中，易于形成链状。鼠疫耶尔森氏菌革兰氏染色阴性、不运动、无芽孢、有荚膜。在甲基蓝染色的组织抹片中，呈现典型的两极着色。该菌在营养琼脂、血琼脂和麦康凯琼脂上均可生长。在剖检疑似感染鼠疫耶尔森氏菌的动物尸体时，须采取严格的防护措施。

### 流行病学和临床症状

数百年来人们已知道骆驼在鼠疫传播中的作用[8,9,11,12]。Wu等[29]和Pollitzer等[23]回顾了之前骆驼鼠疫的报道，认为许多科学家对早期报道的骆驼鼠疫存疑。Fedorov[11]认为鼠疫耶尔森氏菌感染在动物源性和人源性人兽共患病中扮演着同样重要的角色，直到现在还是这样。Sotnikov[26]报道了蒙古、中国、伊拉克、伊朗、非洲和俄罗斯的骆驼鼠疫。双峰驼鼠疫的发生最早可追溯到1911年的俄罗斯，而且多次人类鼠疫的发生是因为接触双峰驼而发生。俄罗斯就发生过一次这样的例子，因食用感染鼠疫的骆驼肉，大量人群被感染[16]。俄罗斯最近一次鼠疫发生在1926年[27]。1907—2001年间，哈萨克斯坦的1951例人鼠疫中，有400例为由骆驼传染[2]，且人致死率超过90%。

近期报道的人类和单峰驼鼠疫发生在毛里塔尼亚和利比亚[28]。从单峰驼的腹股沟淋巴结分离出了鼠疫耶尔森氏菌。据Sacquepee和Gracin报道[25]，发生在法属北非（摩洛哥、阿尔及利亚）单峰驼中的腺鼠疫不仅会影响淋巴结，而且会引起全身性的脓肿。鼠疫耶尔森氏菌不仅可以从这些病灶中分离出来，而且还能从胸膜渗出液中分离出来。此外，皮肤型、败血症型、肺型鼠疫在骆驼中也有发生[18,19]。鼠疫在骆驼中的潜伏期是1～6天，在发病后20天内死亡。Martynchenko[21]、Alonso[1]和Klein等[14]描述了骆驼鼠疫在土库曼斯坦、阿尔及利亚和毛里塔尼亚单峰驼中发生时的临床表现。他们还证明了跳蚤是鼠疫在骆驼中传播的主要媒介。璃眼蜱属和钝缘蜱属的蜱也能机械传播本病[11]。

哈萨克斯坦[24]、约旦[3]、沙特阿拉伯[6]和阿富汗[17]近来都有因食用骆驼肉和生骆驼肝引起的人鼠疫疫情的报道。哈萨克斯坦曼吉斯套地区有5人疑似鼠疫症状被收治入院，他们在3天前食用了骆驼肉。130人与这几个人有过接触的人接受了医疗观察，但最终没有人死亡。

十几年来，几乎每年都有关于腺鼠疫的报道。2005年，骆驼鼠疫在约旦与沙特阿拉伯交界的边境地区沉寂了80余年后再一次出现[3]。12名感染者食用了从骆驼尸体上割下的肉。这些患者表现为咽型鼠疫，之后使用庆大霉素成功治愈。没有从患者身上分离出病原菌，因为最开始认为是土拉菌病。血清学检测发现了抗鼠疫耶尔森氏菌的IgM抗体。对这一地区的流浪狗进行血清学检测，18%为鼠疫阳性。作者怀疑鼠疫在该地区是一个循环性地方性动物疾病，涉及野生啮齿动物、犬、骆驼及其身上的跳蚤等，也包括进入这个循环的被感染的其他草食动物。

沙特阿拉伯发生的咽型鼠疫，4个病人表现为严重的咽炎、下颌下淋巴腺炎、畏寒、精神萎靡和呕吐[6]。所有患者都生食了被屠宰的病骆驼的肝脏。从该患病骆驼的骨髓以及在骆驼的生活圈内捕获的红尾沙鼠（*Meriones libycus*）——一种小型啮齿类动物及印鼠客蚤（*Xenopsylla cheopis*）体内分离出了鼠疫耶尔森氏菌[6]。患者报告，他们饲养的骆驼已有几峰骆驼死亡，极有可能死于鼠疫。

2007年12月，在阿富汗南部尼姆鲁兹省暴发的急性胃肠炎，也是由于食用和处理骆驼肉引起的[17]。83名感染者中17人死亡（死亡率20.5%）。应用聚合酶链式反应和电子喷雾电离质谱对病人临床样品和骆驼的组织样品进行分子生物学检测，检测到鼠疫耶尔森氏菌DNA特征序列。据Pastukhov报道[22]，在哈萨克斯坦，骆驼的主要传染源是大沙鼠（*Rhombomys opimus*），在其他地区则是子午沙鼠（*Meriones meridianus* Pall.）、怪柳沙鼠（*Meriones tamariscinus* Pall.）和红尾沙鼠（*Meriones erythrourus* Gray）。现已知只有当啮齿动物家族发展到住满33%以上用于居住的地洞系统，该病才会发生[10]。现已证明，鼠疫最重要的传播媒介是啮齿类动物携带的跳蚤：粗鬃客蚤（*Xenopsylla hirtipes*）、沙鼠客蚤臀突亚种（*X. gerbilli minax* I）、沙鼠客蚤黑海亚种（*X. g. caspica* J.）、簇鬃客蚤（*X. skrjabini*）、同型客蚤（*X. conformis* W）、黄鼠角叶蚤（*Ceratophyllus tesquorum* W）、毛新蚤（*Neopsylla setosa* W）和谢氏山蚤（*Oropsylla silantiewi*）[13]。

已经证实，骆驼鼠疫是由啮齿动物身上的跳蚤传染，因为骆驼鼠疫只在啮齿动物发病集中的地区发生[22]。从草场回圈舍的骆驼有可能被因被跳蚤叮咬而感染鼠疫，但是，没有明显的临床症状[2]。人有可能在对这些骆驼进行屠宰、剥皮、割肉等处理的时候感染鼠疫。

Lobanov对骆驼进行了实验性感染观察[18]，在接种点形成的脓肿可以快速愈合，全身皮下淋巴结发生急性或亚急性、化脓性或非化脓性炎症。骆驼表现低烧，并在接种2～3周后恢复。有些骆驼浆膜出血。

新大陆骆驼也能感染鼠疫。一只约一岁前腿受伤的雄性美洲驼，在颈部腹侧形成了肿胀，1周之后发展到了胸部，最终治疗无效死亡[4]。剖检可见化脓性和坏死性肺炎，并且在皮肤和肺等不同脏器中分离出了鼠疫耶尔森氏菌。这个发生于新墨西哥州的病例代表了一种致死性败血症型鼠疫。在新墨西哥州，岩松鼠和草原土拨鼠是鼠疫的主要贮存宿主。

### 治疗和控制

预防措施包括避免接触感染的啮齿动物、猫、兔子以及它们身上的跳蚤。

在前苏联，每年调查超过$40\times10^{7}hm^{2}$区域，超过100万只啮齿动物被消灭、大约500万只体表寄生虫接受细菌学检测[22]。系统性消灭啮齿动物有助于减少它们的种群数量，尽管有些种属（如沙鼠）的数量减少是短暂和不稳定的，但这已降低了鼠疫的流行。经验表明，使用黑色氰化盐与六六六（六氯环已烷）混合物，可同时消灭红尾沙鼠和他们身上的跳蚤。结果疾病很快被抑制，效果明显。但这种方法的副作用是鹤和食肉鸟类的大量死亡，就像最近蒙古所报道的一样[5]。

治疗骆驼鼠疫，联合使用链霉素和四环素有效，根据人类治疗方案，应至少连续使用5天[6]。在剖检疑似患有鼠疫的骆驼之前，整个尸体都要使用杀虫剂喷雾来消灭所有的体表寄生虫。

预防骆驼鼠疫，前苏联发明了一种冻干的活鼠疫疫苗[15]，含30亿活菌的疫苗单剂量接种，对成年的骆驼可产生了良好的免疫力。Sotnikov[26]用一种抗鼠疫冻干疫苗对骆驼进行免疫，保护力可以持续6个月。哈萨克斯坦至今仍在使用一种弱毒活疫苗（鼠疫耶尔森氏菌EV株）对人和骆驼进行免疫[2]，每年免疫大量人群（每年26000～67000人次）用于预防鼠疫。但这种疫苗在骆驼中应用没有信息披露。

最近，英国研发了一种针对腺鼠疫的遗传修饰疫苗[28]，主要用于保护驻扎在一些有潜在威胁地区的军队，如鼠疫流行国家或者鼠疫耶尔森氏菌可能被作为生物武器的地区。Blisnick等人[7]使用假结核耶尔森氏菌作为活疫苗来预防鼠疫，因为该菌与鼠疫耶尔森氏菌具有高度的遗传相似性，但毒力却低很多，且遗传稳定性好，而且可以制成口服制剂。初步的结果的确表明，一种减毒的假结核耶尔森氏菌菌株可以成为一种有效、低成本、安全、易生产的抗腺鼠疫口服疫苗。

## 参考文献

[1] Alonso J.M. (1971). – Contribution – L'étude de la peste en Mauritanie. These (Doctorat de médecine),University of Paris,Paris,6,59.

[2] Aikimbajev A.,Meka-Mechenko T.,Temiralieva G.,Bekenov J.,Sagiyev Z.,Kaljan K. & Mukhambetova A.K. (2003). – Plague peculiarities in Kazakhstan at the present time. *Przegl. Epidemiol.*,57,593–598.

[3] Arbaji A.,Kharabsheh S.,Al-Azab S.,Al-Kayed M.,Amr Z.S.,Abu Baker M. & Chu M.C. (2005). – A 12-case outbreak of pharyngeal plague following the consumption of camel meat,in north-eastern Jordan. *Ann. Trop. Med. Parasit.*,99 (8),789–793.

[4] Armed Forces Institute of Pathology (AFIP). (2000). – *Yersinia pestis* pneumonia in a llama (*Lama glama*). *In* The AFIP,Wednesday Slide Conference,2000–2001. Conference 5,Case I – 99–6486 (AFIP 2739540).

[5] Batdelger D. (2002). – Mass mortality of birds in Mongolia. Falco,19,4–5.

[6] Bin Saeed A.A.,Al-Hamdan N.A. & Fontaine R.E. (2005). – Plague from eating raw camel liver. *Emerg. Infect. Dis.*,11 (9), 1456–1457.

[7] Blisnick T.,Ave P.,Huerre M.,Carniel E. & Demeure C.E. (2008). – Oral vaccination against bubonic plague using a live avirulent *Yersinia pseudotuberculosis* strain. *Infect. Immun.*,76 (8),3808–3816.

[8] Christie A.B.,Chen T.H. & Elberg S.S. (1980). – Plague in camels and goats: their role in human epidemics. *J. Infect. Dis.*, 141 (6),724–726.

[9] Curasson G. (1947). – Le chameau et ses maladies. Vigot Frères Editions,Paris,86–88.

[10] Davis S.,Trapman P.,Leirs H.,Begon M. & Heesterbeek J.A.P. (2008). – The abundance threshold for plague as a critical percolation phenomenon. *Nature*,454,634–637.

[11] Fedorov V.N. (1960). – Plague in camels and its prevention in the USSR. *Bull. Org. Mond. Santé*,23 (2–3),275–281.

[12] Gatt Rutter T.E. & Mack R. (1963). – Diseases of camels. Part 1: Bacterial and fungal diseases. *Vet. Bull.*,33 (3),119–124.

[13] Ioff I.G.,Mikulin M.A. & Skalon O.I. (1965). – Key to the fleas of central Asia and Kazakhstan. [in Russian] *Moskva. Izd. Medicina*,370.

[14] Klein J.M.,Alonso J.M.,Baranton G.,Poulet A.R. & Mollaret H.H. (1975). – La peste en Mauritanie. *Med. Mal. Infect.*,5 (4),198–207.

[15] Korobkova E.I. (1946). – *Izv. Irkutsk. Protivo-chum. Inst.*,6,72 (cited by A.K. Strogov,1959).

[16] Kowalesky M.J.M. (1912). – Le chameau et ses maladies d' apres les observations d' auteurs russes. *J. Méd. Vét. Zootechn. (Lyon)* 15,462–466.

[17] Leslie T.,Whitehouse C.A.,Yingst S.,Baldwin C.,Kakar F.,Mofleh J.,Hami A.S.,Mustafa L.,Omar F.,Ayazi E.,Rossi C., Noormal B.,Ziar N. & Kakar R. (2010). – Outbreak of gastroenteritis caused by *Yersinia pestis* in Afghanistan. *Epidemiol. Infect.*,22,1–8.

[18] Lobanov V.N. (1959). – Pathology of experimental plague in camels. *Arkh. Patol.*,21 (7),37–43.

[19] Lobanov V.N. (1967). – La peste chez les chameaux. OMS Seminaire inter-regional de L' OMS pour la lutte coutre la peste,Moscow.

[20] McClean K.L. (1995). – An outbreak of plague in Northwestern Province,Zambia. *Clin. Infect. Dis.*,21,650–652.

[21] Martynchenko V.A. (1967). – Clinical picture of plague in camels infected by means of ectoparasite carriers. *In* Kovalenko Y.R. Maloiznchennye Zabolevaniya Sel' -khoz zhivotnykh,Moscow,Kolos 191–196. *Vet. Bull.*,1968,38 (9),3431.

[22] Pastukhov B.N. (1958). – The epizootic and epidemic situation in the natural foci of plague in the USSR and the prophylactic measures taken. Note submitted to the WHO Expert Committee on Plague,September,401–404.

[23] Pollitzer R. (1954). – Hosts of the infection. Plague. WHO,Geneva,Monogr. Ser. No.,22,305–308.

[24] Promed (2003). – Available at: www.promedmail.org/pls/otn/f?p= 2400:1001:57555::::F2400_P1001_BACK_PAGE, F2400_P1001_ARCHIVE_NUMBER,F2400_P1001_USE_ARCHIVE:1001,20030918.2360,Y (accessed 30 July 2003).

1 细菌病

[25] Sacquepee & Garcin (1913). – N/A. *Arch. Med. Pharm. Abstract: Vet. Bull.*,33 (3),119–124.

[26] Sotnikov M.I. (1973). – Camel plague. *In* Orlov,F.M. Maloizvestnye zarazny bolezni zhivotnykh,Izdatel' stvo Kolos, 213–222. Résumé in *Vet. Bull.*,44,937.

[27] Strogov A.K. (1959). – Plague in camels. Maloizvestnye zaraznye bolezni zhivotnykh Moscow,Sel'-khoz 262–280. *Vet. Bull.*,28,2734.

[28] Titball R.W. & Williamson E.D. (2001). – Vaccination against bubonic and pneumonic plague. *Vaccine*,19 (30),4175–4184.

[29] Wu L.T.,Chu J.W.H.,Pollitzer R. & Wu C.Y. (1936). – Plague: a manual for medical and public health workers. Weishengshu National Quarantine Service,Shanghai,232–235.

（储岳峰译，殷宏校）



## 1.1.10 钩端螺旋体病

钩端螺旋体病呈世界性流行，该病在旧大陆和新大陆骆驼都有相关报道[8,35]。钩端螺旋体在哺乳动物的肾小管中繁殖，并通过尿液排出，可持续数月。河边地带和池塘以及牛舍中的尿液与感染奶牛的奶汁形成的气溶胶是传染源。

钩端螺旋体病被世界卫生组织（World Health Organization，WHO）和世界动物卫生组织（World Organization for Animal Health，OIE）列入必须申报的疾病。钩端螺旋体病在热带国家的发病率最高，有些血清群在某个国家的感染率要高于另外的国家。

除了骆驼（驼科动物），大量的动物物种，包括家养的和野生的，都受到钩端螺旋体病的危害，包括猪、马、牛、大象、狐狸、负鼠、人类以及非人灵长类动物。

### 公共卫生意义

所有感染的物种都可以成为钩端螺旋体的传染源，其感染可能或不可能演变成一种临床症状明显的疾病。钩端螺旋体感染可以通过直接接触或已被该病原体污染的环境因素引起。环境污染由家养和野生动物的尿液所排出的钩端螺旋体病原引起。啮齿类动物，例如，黑鼠和褐家鼠，可以在很多国家传播特定的血清群，例如，出血性黄疸钩端螺旋体（*Leptospira icterohaemorrhagiae*）。人可出现非特异性的流感样症状，表现为发烧、头痛、呕吐。很多病人不用治疗，可在几天内恢复，但有些病人会发展为脑膜炎、呼吸衰竭、急性肝炎及肾功能衰竭。世界卫生组织认为有些国家控制了钩端螺旋体病，例如，中华人民共和国和古巴，对农民使用人用菌苗预防这类致命性的疾病。

### 病原学

螺旋体目包括螺旋体科和钩端螺旋体科，以下列出的为对动物和人类有重要意义的菌属。

（1）螺旋体科，蛇形螺旋体属（*Serpulina*），短螺旋体（*Brachyspira*）、密螺旋体属(*Treponema*)和疏螺旋体属（*Borrelia*）。

（2）钩端螺旋体科，细螺旋体(*Leptospira*)。

钩端螺旋体可根据抗原结构分为多个血清型。在钩端螺旋体属中，只有问号钩端螺旋体（*L. interrogans*）具有医学重要性。所有具有致病性的钩端螺旋体都包括在这一属中。由于具有可变化的抗原结构特征，问号钩端螺旋体由25个血清群组成，并且大约有230个血清型。几种血清型具有相近的抗原性，在微量凝集试验（microscopic agglutination test，MAT）中有交叉反应。研究人员用这一方法确立了不同的血清群。例如，一个由黄疸出血群感染的动物，可以被黄疸出血型、赖型或哥本哈根型等相同的血清型感染。

通过暗视野显微镜观察和荧光抗体实验（FAT），可以在尿液、体液及组织中发现钩端螺旋体。钩端螺旋体需在特殊的培养基中生长，例如，斯氏（Stuart）培养基和柯氏（Korthof）培养基。钩端螺旋体是呈纤细、螺旋状的细菌，运动活泼，易在血液中散播。

### 流行病学

钩端螺旋体在世界范围普遍流行。所有的家养动物，野生动物，尤其是啮齿类，还有人类都很容易感染。对于有些动物，肾损的慢性感染使其成为这类微生物的储藏库[4]。人和动物可能会通过接触已感染的尿液或摄食尿液污染的食物或水源，直接感染或间接感染。一系列的症状都可预测到，从食欲不振到严重的临床症状。根据Seifert的研究[30]，在热带地区的密集养牛区，啮齿类动物和犬是流行病学上最为重要的传染源。人与这些动物紧密接触后可被感染。

Wilson[38]和Higgins[12]认为钩端螺旋体病是旧大陆骆驼非常重要的疾病。在旧大陆骆驼中，还没有人描述过钩端螺旋体病的临床症状，至于骆驼是否对该病易感还存在某些疑问。Rafyi和Maghami[25]，还有Higgins[12]怀疑钩端螺旋体偶尔可能引起血尿症。Wilson[38]也观察到了单峰驼血尿症，但没有找到病因。

在阿联酋，雌雄赛驼都发生过血尿症，但是，其与钩端螺旋体感染无关[37]。采用血清学检查和暗视野显微镜观察了50头患有血尿症的单峰驼，结果均为阴性。血尿症主要在阿联酋的单峰赛驼中观察到，很少在种畜中观察到。微生物学集中检查和血清生物化学检查（肌酸酐和尿素）没有发现因肾脏感染或肾功能不全引起血尿症的证据。另外，尿液的检查并没有看到不断沉积的晶体样或脱落物。感染的单峰驼没有表现出与肾相关的疼痛或系统性疾病。在强化研究单峰驼血尿症期间，一头患有血尿症的骆驼被处以安乐死，同时，检查了它的尿路器官。结果表明，由出血引起的远端肾小管钙化以及局灶性肾小球肾炎（图67和68）。

在印度，5头雄性单峰驼被检查到有慢性血尿症（8～12岁）[33]。这种单峰驼对各种治疗均没有效果，并且血清磷酸钙的比例在正常范围内。在出现血尿症后的8～10个月后会转归死亡。尸检表明，在单峰驼肾髓质有2～3cm大小的钙化灶。组织生理学检查表明，出血的肾小管钙化及局灶性肾小球肾炎。对这种骆驼的日粮分析显示，饲喂钙片过量。因此，肾实质中的钙沉积可能是由于摄入过量的钙而引起。相比于其他的反刍动物，骆驼会吸收更多的钙和磷[33]。

这种分散的肾钙化的病因尚未明确。推测这些沉积物是由于单峰赛驼食用高浓度的矿物质引起。这就解释了为什么在育种的骆驼中很少发生血尿症，它们的饲料主要由干草组成，很少饲喂营养丰富的饲料。血尿症与病原是否相关还有待进一步研究。

Krepkogorskaja[16]是唯一从骆驼的器官中分离到钩端螺旋体的人。俄罗斯Betpak-dal育种站的牛、马和骆驼体内分离并鉴定了以下的种：哈萨克斯坦钩端螺旋体（L.kazachstanica）的Ⅰ型、Ⅱ型及犊牛钩端螺旋体(L. vitulina)，但没有对这些病原引起疾病的临床症状进行描述。其他的研究报道了针对不同钩端螺旋体血清型（表17）特异的凝集性抗体。这些研究来源于非洲的不同国家，以及阿富汗、伊朗、印度、俄罗斯、内蒙古和阿联酋。

Maronpot和Barsoum[21]发现，在埃及，34%被检查的单峰驼体内存在钩端螺旋体抗体。他们认为，单峰驼的亚临床钩端螺旋体病是世界范围性的现象，并且会对人类的健康造成危害。Rafyi和Maghami[25]发现伊朗的5头被检测的骆驼中，1头呈肾脏问号钩端螺旋体黄疸出血性血清型抗体阳性，这只骆驼在检测的1周前出现流产和血尿症。

Wernery和Wernery[37]发现，2.5%的育种驼和5.6%的单峰赛驼对钩端螺旋体有血清学反应。该作者在过去的15年里，没有在临床上观察到单峰驼的钩端螺旋体病。Wernery等[36]最近利用微量凝集试验（MAT）检测164份来自奶单峰驼的血清时发现，其血清阳性率为12%（19份阳性样品）。经过检测，最主要的血清型是：哥本哈根型（Copenhageni）[15]、拜伦型（Ballum）[1]、秋季型（Autumnalis）[1]、塔拉索夫型（Tarassovi）[1]和爪哇型（Javanica）[1]。7头单峰驼的抗体滴度为1:800，这表明了他们是最近感染或活动性感染。然而，在农场中没有观察到钩端螺旋体病的临床症状。因此，有必要做更多的研究来证明钩端螺旋体对骆驼的致病性，并且找到引起感染的传染源。哥本哈根型是最常见的血清型，被视为啮齿动物血清型。最近，Hussin和GarEl Nabi[14]通过平板凝集实验发现，在沙特阿拉伯国家被检的90头骆驼中的6头呈秋季型抗体（6.7%）阳性。然而，通过暗视野显微镜检查已屠宰的36头骆驼的尿液样品，没有观察到钩端螺旋体。

利用微量凝集试验（MAT）检测来自蒙古的9份野生双峰驼血清样品，发现4份含有赛罗型（Sejroe）血清型抗体（44.4%），但是，没有该病发生的报道[2]。从吉布提地区的3头单峰驼血清样品中检测到了钩端螺旋体抗体。微量凝集试验最高的滴度是抗黄疸出血型的抗体，为1:160，随后是致热型（Pyogenes）、萨可斯可宾型（Saxkoebing）、哈勒焦型（Hardjo）和哥本哈根型（Copenhageni）。

钩端螺旋体病已在新大陆骆驼中有所描述[17]，特别是对羊驼[19]和底特律野生动物园中3月龄的原驼[13]。但总体上来说，对羊驼的钩端螺旋体病的研究还不透彻。可以确定的是，钩端螺旋体病的临床症状及病理学变化在这类物种中与其他物种相似。钩端螺旋体通过黏膜或损伤的皮肤进入机体，在血液内扩散后，在实质器官内定居并增殖。在肾脏中，这种微生物会在近曲小管的内腔中散播增殖，并持续长期存在。有些菌株会产生血红蛋白尿（红尿）。大体病变包括黏膜和脂肪黄疸，在组织结构上，可能会出现间质性肾炎和肾小管肾炎。

Marin等[20]研究发现，秘鲁地区的美洲骆驼中钩端螺旋体抗体的检出率为32.4%（30/96），其中，不同的血清型包括：*L.Castellonis* 4.3%（4/93）、巴达维亚型（Bataviae）5.4%（5/93）、哥本哈根型（Copenhageni）6.5%（6/93）、波摩那型（Pomona）9.7%（9/93）、致热型(Pyogenes)1.1%（1/93）和塔拉索夫型(Tarassovi)5.4%（5/93）。另外的一项研究[16]发现，对阿根廷不同地区的健康未接种疫苗的美洲驼，骆马和原驼的494份血清样品检测发现，这些样品呈钩端螺旋体抗体阳性。该项研究利用微量凝集试验检测了钩端螺旋体血清型，结果表明美洲驼的钩端螺旋体阳性率为47.3%～96.2%，原驼为0～13%，骆马中为9%～62.8%。其中，反应活性好的血清型是哥本哈根型（Copenhageni）和*L.Castellonis*，这些血清型已经在阿根廷地区的人和家畜或野生动物中分离出来。Tibary[34]等在阿根廷的新大陆骆驼中发现了抗体，他们检测到的阳性率在美洲驼中为47%～96%，在原驼中为0～13%，在骆马中为9%～63%。其中，最常见的血清型是*L.Castellonis*和哥本哈根型（Copenhageni）。野生新大陆骆驼（原驼和骆马）体内的钩端螺旋体抗体对钩端螺旋体病的流行病学起非常重要的作用。在美国，血清学诊断发现由钩端螺旋体感染导致骆驼流产的就有3例[18]。所有的病例都有很高的针对波摩那型（Pomona）的母源抗体。然而，研究表明，自由放养的新大陆骆驼中似乎并没有钩端螺旋体病的发生，在阿根廷丘布特省的23头自由放养的原驼中也没有检测到钩端螺旋体抗体[15]。

## 诊断

当怀疑有钩端螺旋体病时，尿液样品的暗视野显微镜检测，血涂片的荧光抗体检测技术及组织冰冻切片，还有病原培养和实验动物接种等方法都可用于诊断。然而，仅在疾病的短暂的急性阶段才能成功分离钩端螺旋体。另外，样品必须在收集后的2小时内进行研究，对分离结果的鉴定往往也比较复杂。培养需要持续数周甚至数月，且经常会因污染而毫无收获。微量凝集试验（MAT）检测结果显示血清抗体滴度高于1:100可视为阳性。该试验采用活的钩端螺旋体作为抗原，具有敏感性高，血清型特异性好的特点。由于从单个样品的检测结果很难确诊该病，因此，应收集急

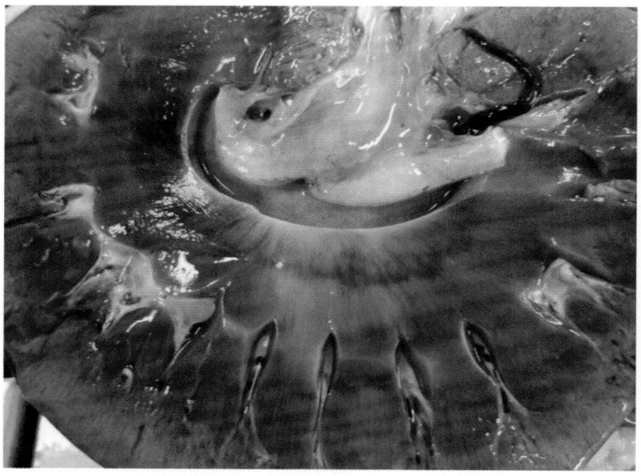

图67　单峰驼血尿症；肾乳头钙化

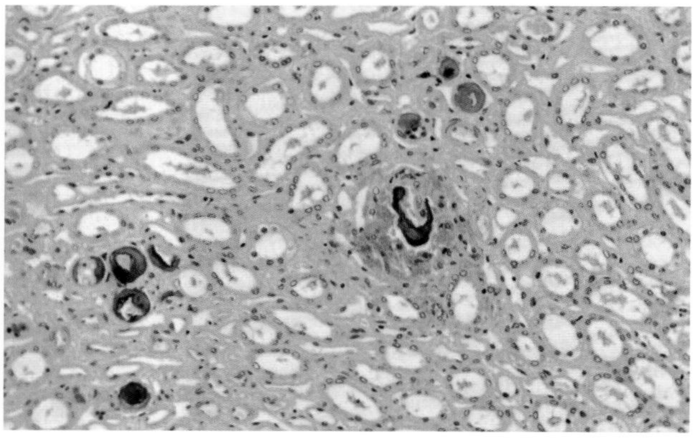

图68　单峰驼血尿肾；伴随出血的斑点状钙化（苏木精-伊红染色）

表17　骆驼钩端螺旋体病：流行和血清型

| 国家 | 作者 | 年份 | 阳性率（%） | 钩端螺旋体血清型 |
|---|---|---|---|---|
| 阿富汗 | Sebek等 | 1972 | | 流感伤寒型L.Gippotyphosa |
| | Sebek | 1974 | 0.8 | |
| | Sebek等 | 1978 | | |
| 苏丹 | Shigidi | 1974 | 0.0 | — |
| 埃塞俄比亚 | Moch等 | 1975 | 15.4 | 流感伤寒型L.Gippotyphosa<br>致热型L.Pyogenes<br>布滕博型L.Butembo<br>L.Borincana |
| 埃及 | Brownlow和Dedeaux | 1964 | | |
| | Maronpot和Barsoum | 1972 | 34.0 | 致热型L.Pyogenes<br>塔拉索夫型L.Tarassovi<br>秋季型L.Autumnalis<br>布滕博型L.Butembo<br>爪哇型L.Javanica |
| | Hatem Ahmed | 1976 | 9.2 | 致热型L.Pyogenes<br>塔拉索夫型L.Tarassovi<br>布滕博型L.Butembo |
| 伊朗 | Rafyi和Maghami | 1959 | 20.0 | 黄疸出血型L.Icterohaemorrhagiae |
| 索马里 | Farina和Sobrero | 1960 | 16.2 | 黄疸出血型L.Icterohaemorrhagiae<br>犬型L.Canicola<br>流感伤寒型L.Gippotyphosa<br>拜伦型L.Ballum |
| | Aresh | 1982 | 0.0 | — |
| | Hayles | 1986 | | |
| 印度 | Mathur等 | 1986 | 51.4 | 犬型L.Canicola<br>黄疸出血型L.Icterohaemorrhagiae<br>拜伦型L.Ballum<br>波摩那型L.pomona<br>沃尔菲型L.Wolfei<br>秋季型L.Autumnalis |
| 突尼斯 | Burgemeister等 | 1975 | 48.0 | 黄疸出血型L.Icterohaemorrhagiae<br>波摩那型L.pomona<br>巴达维亚型L.Bataviae |
| | Gallo等 | 1989 | 0.0 | — |
| 俄罗斯 | Krepokogorskaja | 1956 | | 哈萨克斯坦I型L.Kazachstanica I<br>哈萨克斯坦II型L.Kazachstanica II<br>犊牛型 L.Vitulina（血清学和培养） |
| 蒙古 | Sebek | 1974 | | |
| | Sosa等 | 1988 | | 动物园骆驼 |
| | Anan’ina等 | 2011 | 44.4 | 赛罗型L.Sejroe |
| 阿联酋 | Wernery和Wernery | 1990 | 2.5 | 家养单峰驼 |
| | | | 5.6 | 单峰赛驼 |
| | Afzal和Sakkir | 1994 | 4.1 | 问号钩端螺旋体L.interrogans |
| | Wernery等 | 2008 | 12 | 奶单峰驼<br>根本哈根型L.Copenhageni<br>拜伦型L.Ballum<br>秋季型L.Autumnalis<br>塔拉索夫型L.Tarassovi<br>爪哇型L.Javanica |

性和处在恢复期的血清进行检测。微量凝集试验(MAT)被视为金标法用于血清学病例诊断（图69a和b）。

在兽医学诊断上，PCR是非常有用的方法，即使样品在送达实验室之前保存条件比较差，该检测方法仍然可行，因为核酸物质非常稳定，即使钩端螺旋体已经被损坏。一些钩端螺旋体PCR检测试剂盒已经商业化；但是，这些试剂盒仅适合用于钩端螺旋体菌株的检测，不适用于血清型检测。

## 治疗

钩端螺旋体对抗生素很敏感，目前，没有发现抗药性。治疗方案必须尽可能防止肝脏和肾脏出现病变。临床发病的骆驼（驼科动物）可按照每千克体重25mg，肌内注射二氢链霉素，连续给药5天。

世界动物卫生组织（OIE）也推荐在国际贸易中携带病原的阳性感染动物使用链霉素预防治疗肾脏病变，也可使用阿莫西林进行治疗，因为该药物在肠肝循环系统可很好被吸收。灭活菌苗可适用于农场动物，尤其适用于犬科动物。由于钩端螺旋体病对于骆驼并没有太大的重要性，仅在流行区域推荐适当的菌苗接种。有些地方的美洲驼按照常规的接种疫苗预防钩端螺旋体病。Pugh等[24]对实验室测定的两种疫苗针对5种血清型的血清学反应低且时间短。然而，没有对接种免疫后的动物进行病原攻击试验。

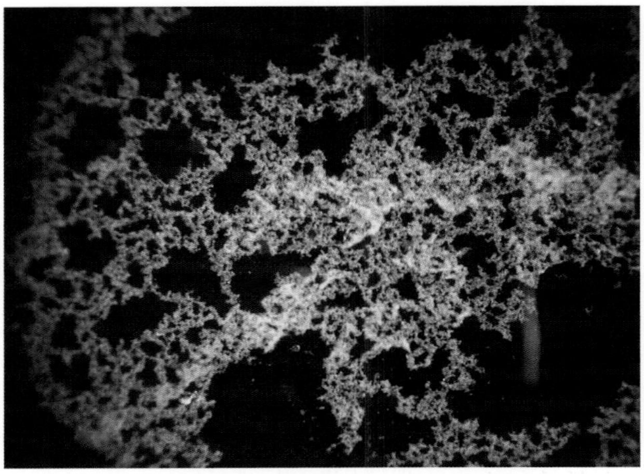

a. 显微镜凝集试验钩端螺旋体阳性反应

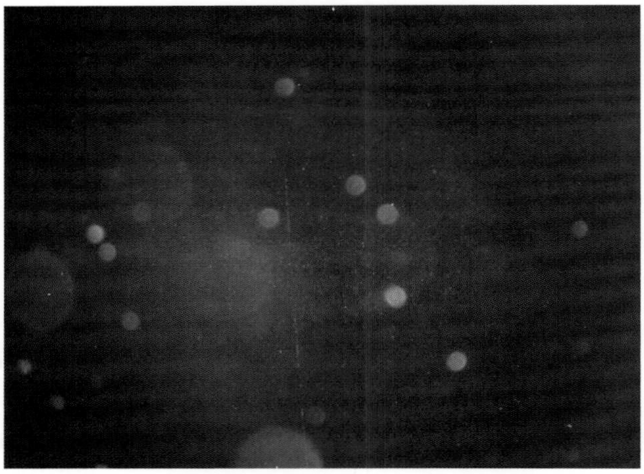

b. 显微镜凝集试验钩端螺旋体阴性反应

图69

# 参考文献


[1] Afzal M. & Sakkir M. (1994). – Survey of antibodies against various infectious disease agents in racing camels in Abu Dhabi,United Arab Emirates. *Rev. sci. tech. Off. int. Epiz.*,13 (3),787−792.

[2] Anan'ina Yu. V,Korenberg E.I.,Tserennorov D.,Savel'eva O.V.,Batjav D.,Otgonbaatar D.,Enkhbold N.,Tsend E. & Erdenechimeg B. (2011). – Detection of leptospirosis infection in certain wild and domestic animals in Mongolia. *Zh. Mikrobiol. Epidemiol. Immunobiol.*,5,36−39.

[3] Arush M.A. (1982). – La situazione sanitaria del dromedario nella Repubblica Democratica Somala. *Bollettino scientifica della facoltà di zootecnia e veterinaria*,3,209−217.

[4] Bisping W. & Amtsberg G. (1988). – Colour atlas for the diagnosis of bacterial pathogens in animals. Verlag Paul Parey,Berlin and Hamburg,277−286.

[5] Brownlow W.J. & Dedeaux J.D. (1964). – Leptospirosis in animals of upper Egypt. *Amer. J. Trop. Med. Hyg.*,13,311−318.

[6] Burgemeister R.,Leyk W. & Goessler R. (1975). – Untersuchungen über Vorkommen von Parasitosen,bakteriellen und

viralen Infektionskrankheiten bei Dromedaren in Sudtunesien. *Dtsch. Tierärtztl. Wschr.*,82,352–354.

[7] Farina R. & Sobrero L. (1960). – Ricerche sierologiche sulla diffusione delle leptospirosi animali in Somalia. *Zooprofilassi*,15 (12),925–936.

[8] Fowler M.E. (2010). – Medicine and surgery of camelids. 3rd Ed. Wiley-Blackwell,Oxford,UK,209.

[9] Gallo C.,Vesco G.,Campo F.,Haddad N. & Abdelmoula H. (1989). – Enquete zoosanitaire chez les chèvres et les dromadaires au Sud de la Tunisie. *Maghreb Vét.*,4 (17),15–17.

[10] Hatem Ahmed M.E. (1976). – Studies on leptospira group of microorganisms with special reference to purification and cultivation. Master of Veterinary Science Thesis,Cairo University,Cairo.

[11] Hayles L.B. (1986). – Leptospirosis. *In* Proceedings of the 1st National Veterinary Symposium,12–15 October 1986,Mogadishu,Somalia. FAO,Rome.

[12] Higgins A. (1986). – The camel in health and disease. Bailliere Tindall,London.

[13] Hodgin C.,Schillhorn T.W.,Fayer R. & Richter N. (1984). – Leptospirosis and coccidial infection in a guanaco. *JAVMA*,185 (11),1 442–1 444.

[14] Hussein M.F. & Gar El Nabi A.R. (2009). – Serological evidence of leptospirosis in camels in Saudi Arabia. *J. Anim. Vet. Adv.*,8,1 010–1 012.

[15] Karesh W.B.,Uhart M.M.,Dierenfeld E.S.,Braselton W.M.,Karesh W.B.,Uhart M.M,Dierenfeld E.S.,Braselton W.M.,Torres A.,House C.,Puche H. & Cook R.A. (1998). – Health evaluation of free-ranging guanaco (*Llama guanocoe*). *J. Zoo. Wildlife Med.*,29 (2),134–141.

[16] Krepkogorskaja T.A. (1956). – La leptospirose des animaux domestiques dans le desert de Betpak-Dal. *Izvest. Akad. Nauouk Kazakh. SSR (Physiol. Méd.)*,7,80–81.

[17] Llorente P.,Leoni L. & Martinez Vivot M. (2002). – Leptospirosis in South American camelids. A study on the serological prevalence in different regions of Argentina. *Arch. Med. Vet.*,34 (1),59–68.

[18] Loehr C.V.,Bildfell R.J.,Heidel J.R.,Valentine B.A. & Schaefer D. (2007). – Retrospective study of camelid abortions in Oregon. *Vet. Pathol.*,44 (5),753.

[19] Ludena J. & Vargus A. (1982). – Leptospirosis en alpacas. *Adv. Vet. Sci. Comp. Med.*,2 (2),27–28.

[20] Marin R.E.,Romero G.,Brihuega B. & Auad G.T. (2007). – Seroprevalencia de endermedades infecciosas en Llamas (*Lama glama*) de la Provincia de Jujuy,Argentina. V° Congreso de Especialistas en Pequenos Rumiantes y Camélidos Sudamericanos,Mendoza,Argentina,1–4.

[21] Maronpot R.R. & Barsoum J.S. (1972). – Leptospiral microscopic agglutinating antibodies in sera of man and domestic animals in Egypt. *Amer. J. Trop. Med. Hyg.*,21 (4),467–472.

[22] Mathur K.N.,Khanna V.K. & Purohit A.K. (1986). – Macroscopic plate agglutination results of serological examination of camels,cattle and goats for leptospirosis. *Indian J. Public Health*,30 (3),170–172.

[23] Moch R.W.,Ebner E.E.,Barsoum J.S. & Botros B.A.M. (1975). – Leptospirosis in Ethiopia: a serological survey in domestic and wild animals. *J. Trop. Med. Hyg.*,78 (2),38–42.

[24] Pugh D.G.,Wright J. & Rowe S. (1995). – Serological response of llamas to a commercially prepared leptospirosis vaccine. *Small Ruminant Res.*,17 (2),193–196.

[25] Rafyi A. & Maghami G. (1959). – Sur la frequence de la leptospirose en Iran: isolement de Leptospira grippotyphosa chez l'homme et chez les bovins. *Bull. Soc. Path. Exot.*,52 (5),592–596.

[26] Raqueplo C.,Davoust B.,Mulot B.,Lafrance B. & Kodjo A. (2011). – Serological study of leptospirosis in equids,camelids and bovids from Djibouti. *Med. Trop.*,71 (5),517–518.

[27] Sebek Z. (1974). – Results of serologic examination of domestic animals for leptospirosis in the Mongolian People's Republic. *Folia Parasit.*,21 (1),21–28.

[28] Sebek Z.,Sery V. &. Saboor A. (1972). – Results of the first leptospirological study carried out in Afghanistan. *J. Hyg. Epidem. Microbiol. Immun. (Prague)*,16 (3),314–324.

[29] Sebek,Z,Blasek K.,Valova M. & Amin A. (1978). – Further results of serological examination of domestic animals for leptospirosis in Afghanistan. *Folia Parasit.*,25 (1),17–22.

[30] Seifert H.S.H. (1992). – Tropentierhygiene. Gustav Fischer Verlag,Jena,Stuttgart,Germany,313–315.

[31] Shigidi M.T.A. (1974). – Animal leptospirosis in the Sudan. *Br. Vet. J.*,130 (3),528–531.

[32] Sosa G.,Santos O.,Duarte C.L.,Hernandez D. & Delgado L. (1988). – Investigación serologica y bacteriologica de leptospirosis realizada en fauna exotica. *Revta Cub. Cienc. Vet.*,19 (3),219–225.

[33] Tanwar R.K. (2006). – Hematuria associated with calcification of renal tubules in camels (*Camelus dromedarius*). *Vet. Path.*,43 (5),826.

[34] Tibary A.,Fite C.,Anouassi A. & Sghiri A. (2006). – Infectious causes of reproductive loss in camelids. *Theriogenology*,66,633–647.

[35] Wernery U. & Kaaden O.-R. (2002). – Infectious diseases in camelids. 2nd Ed. Blackwell Science,Berlin and Vienna,55–58.

[36] Wernery U.,Thomas R.,Raghavan R.,Syriac G.,Joseph S. & Georgy N. (2008). – Seroepidemiological studies for the detection of antibodies against 8 infectious diseases in dairy dromedaries of United Arab Emirates using modern laboratory techniques (Part-II). *J. Camel Pract. Res.*,15 (2),139–145.

[37] Wernery U. & Wernery R. (1990). – Seroepidemiologische Untersuchungen zum Nachweis von Antikorpern gegen Brucellen, Chlamydien,Leptospiren,BVD/MD,IBR/IPV – und Enzootischen Bovinen Leukosevirus (EBL) bei Dromedarstuten (*Camelus dromedarius*). *Dtsch. Tierärztl. Wschr.*,97,134–135.

[38] Wilson R.T. (1984). – The camel. Longman,London and New York.

（牛庆丽译，殷宏校）

## 1.1.11 无浆体病和

## 1.1.12 支原体病

### 立克次体病的流行病学与临床症状

立克次体是一类微小的严格细胞内寄生的细菌，呈革兰氏阴性，缺少鞭毛，基因组很小，由1000～1500kD碱基组成。病原在入侵细胞后，定居于细胞质或细胞核内，是寄生于节肢动物的一类重要寄生虫，在节肢动物的肠道细胞内增殖。立克次体常常被错误的称作大病毒。然而，他们是真正的细菌。他们具有DNA和RNA，通过二分裂繁殖，具有特有的新陈代谢方式，对多种抗生素敏感。立克次体呈杆状和球状，无鞭毛，需氧。不易被碱性的苯胺染料着色，通常被用作革兰氏染色，但瑞特氏染色或姬姆萨染色效果好。大多数的立克次体需要在活细胞内进行分裂增殖。也可在组织培养物或鸡胚卵黄囊里培养。

立克次体是典型的媒介传播性流行病的致病因子，因为哺乳动物必须通过昆虫媒介才可感染立克次体。立克次体属于变形杆菌纲、立克次体科、立克次体目。立克次氏体属包括很多与人类疾病相关的立克次体种，如引起斑点热和斑疹伤寒症的种。根据疾病的临床特征，人类病原的立克次氏体可细分为3个主要的组群。

（a）斑点热组；

（b）斑疹伤寒组；

（c）恙虫病组。

很多研究者已经鉴定了普氏立克次氏体（*Rickettsia prowazekii*），立氏立克次体（*R. rickettsii*），莫氏立克次氏体（*R. mooseri*）和斑疹热立克次（氏）体（*R. conorii*），但未见关于这些微生物对骆驼（驼科动物）引起的相关疾病和损失的报道。Reiss-Gutfreund [30]在埃塞俄比亚地区从单峰驼身上的蜱（麻点璃眼蜱）体内成功分离到普氏立克次体，这些动物并没有表现出立克次体病的临床症状。这种现象被Ormsbee等[27]证实，他们没能从人工感染的幼龄单峰驼的血液中再次分离到普氏立克次体；该作者认为单峰驼在典型的流行性斑疹伤寒证的流行中没有起任何作用。

### 无浆体病与支原体病

在过去，立克次体的分类方法主要是基于形

态学、抗原性、新陈代谢和其他的一些显性特征，但是，最近立克次体目的分类有了一些重要的突破，主要采用系统发育学与分类学方法。这些突破主要是在分子技术发展的基础上，通过比较基因组的核苷酸序列，尤其是通过比较16S rRNA基因序列对立克次体进行分类。然而，分类学不能仅依赖于特定的基因序列，还要综合考虑病原的显性特征。

新的现代生物分类学系统明确指出埃里希氏体科已被废除。立克次体目中除了无浆体科，另外一科是立克次体科，该科包括立克次体属和东方次体属，这些科很可能没有兽医学重要性。血巴尔通体属和附红细胞体属与支原体有关，埃立克体属、考德里氏体属、埃及小体属和新立克次氏体属划归无浆体科，反之，贝氏柯克斯体属于军团杆菌目中的柯克氏体属。表18和19分别列出了无浆体科和支原体科在分类学上的一些变化。

旧大陆骆驼（OWCs）和新大陆骆驼（NWCs）中已发现有无浆体病（埃里希体病）和支原体病（血巴尔通体病，附红细胞体病）。

# 无浆体病

## 病原学

### 旧大陆骆驼的边缘无浆体

无浆体是细胞内的专性寄生虫，可感染不同种类的动物（表18）。边缘无浆体是无浆体属的一个种。目前，无浆体属也包括牛无浆体、扁形无浆体和嗜吞噬细胞无浆体。边缘无浆体，牛无浆体和扁形无浆体居住于红细胞内，而嗜吞噬细胞无浆体寄生于白细胞内。

边缘无浆体是牛和野生反刍动物无浆体病的致病因子，绵羊无浆体感染绵羊和山羊。中央无浆体与边缘无浆体有较近的亲缘关系，引起牛的温和型无浆体病。中央无浆体最早分离于非洲，作为疫苗使用时被引入到澳大利亚、南美和亚洲。边缘无浆体的不同分离株之间存在抗原变体，6个主要表面蛋白抗原之间存在抗原多态性。

早期文献描述单峰驼的无浆体病时，只观察到一些亚临床病例。来自索马里关于健康单峰驼边缘无浆体病例的报道[5,6,25]证明了上述观察。Kornienko-Koneva[19]成功地用感染了无浆体的骆驼血液传播给牛。然而，双峰驼可能对边缘无浆体天然易感。Ristic和Kreier[33]，Ristic[32]和Ajayi等[1]用3种不同的血清学方法检测尼日利亚骆驼的血清，发现10.7%（3/28）存在边缘无浆体抗体。

Alsaad[4]报道了由边缘无浆体引起的伊朗单峰驼病例。在疾病暴发期间，自然感染了边缘无浆体的52头单峰驼表现出黏膜苍白，食欲减退，消瘦，咳嗽，流泪和毛发粗糙的症状。诊断发现红细胞增大，色素性贫血，在感染的红细胞边缘检测到呈球状颗粒的无浆体（图70），同时观察到红细胞、血红蛋白和红细胞压积显著减少，在体表的很多部位发现有蜱寄生。

### 旧大陆骆驼和新大陆骆驼的嗜粒细胞无浆体

嗜粒细胞无浆体寄生在白细胞中，常见于胞浆内的空泡，这些包涵体的大小差别极大，原生小体增殖形成所谓的“桑椹胚”。在发热初期制备的血涂片粒细胞中可见到典型的桑椹胚。无浆体病不仅发生在动物上，还是人畜共患病，人的嗜吞噬细胞无浆体病是由感染动物相同病原的变体所引起的。最常发生于春夏季节，呈现未分化热性疾病。在美国已有死亡病例的报道，但在欧洲未见有死亡病例报道，由硬蜱属传播。

最近一例报道描述了阿曼里亚尔单峰驼感染了嗜粒细胞无浆体[45]。无浆体病的暴发，也称为蜱传热，以精神萎顿，嗜睡，消瘦和卧地不起为特征。感染的骆驼表现为颈下肿胀并延伸至胸骨（图 71）。

尸检显示，单峰驼淋巴结肿大，腹部器官出现纤维蛋白凝集。血涂片，姬姆萨染色显示，在单核细胞和大淋巴细胞的胞质中发现类似嗜粒细胞无浆体的立克次体样微生物（图 72）。

四环素类药物静脉给药5天，肿胀消退，病驼完全康复。

Barlough等[8]报道，从感染粒细胞无浆体病的美洲驼体内鉴定出一株无浆体病原。序列分析显示这株无浆体与嗜粒细胞无浆体的成员亲缘关系较近。从同一个美洲驼饲养牧场收集的太平洋硬蜱中发现同样的病原。呈现非特异性的临床症状，包括

表18　无浆体科在兽医学的重要性及相关疾病

| 种 | 宿主 | 媒介 | 寄生细胞 | 疾病 | 分布 |
|---|---|---|---|---|---|
| 鸡埃及小体（*Aegyptianella pullorum*） | 鸡和其他鸟类 | 锐缘蜱属（波斯类） | 红细胞 | 埃及小体病 | 热带和亚热带地区 |
| 边缘无浆体（*Anaplasma marginale*） | 牛，家养水牛，单峰驼 | 一些属和某些种的蜱，机械性传播 | 红细胞边缘 | 牛无浆体病，骆驼无浆体病 | 热带，亚热带，甚至温带，非洲和中东地区 |
| 中央无浆体（*Anaplasma centrale*） | 牛 | 蜱和机械性传播 | 红细胞中央 | 温和性牛无浆体病 | 南非，活疫苗将该病原散播于其他地方 |
| 绵羊无浆体（*Anaplasma ovis*） | 绵羊和山羊 | 可能由多种蜱和机械性传播 | 红细胞边缘 | 小反刍动物无浆体病 | 热带，亚热带和温带地区 |
| 牛无浆体（埃立克次氏体）（*Anaplasma bovis*） | 牛，家养水牛 | 蜱（花蜱属，璃眼蜱属，扇头蜱属） | 粒细胞 | 牛单核细胞无浆体病 | 非洲，南美洲，亚洲 |
| 嗜粒细胞无浆体（埃立克次氏体）(*Anaplasma phgocytophilum*) | 牛，小反刍动物，马，单峰驼，美洲驼，人 | 蜱（硬蜱属，璃眼蜱属） | 粒细胞 | 蜱传热，放牧热，马无浆体病，嗜吞噬细胞无浆体病，骆驼粒细胞无浆体病 | 欧洲，北美，非洲，中东，美国 |
| 犬无浆体（*Anaplasma canis*） | 狗 | 蜱（扇头蜱属） | 单核细胞 | 犬单核细胞无浆体病，热带犬全血细胞减少症 | 热带，亚热带地区 |
| 翁迪里无浆体（*Anaplasma ondirii*） | 牛 | 未知 | 粒细胞，单核细胞 | 翁迪里病，牛斑点热病 | 东非 |
| 羊无浆体（*Anaplasma ovina*） | 小反刍动物 | 蜱（扇头蜱属） | 粒细胞 | 羊单核细胞无浆体病 | 非洲，亚洲 |
| 反刍兽无浆体（考得里氏体）(*Anaplasma ruminantium*) | 家养反刍动物 | 蜱（钝眼蜱属） | 血管上皮细胞浆 | 心水病 | 撒哈拉以南的非洲，马达加斯加，小安的列斯群岛 |
| 蠕虫新立克次氏体(*Neorickettsia helminthoeca*) | 狗 | 吸虫类 | 白细胞（难） | 鲑鱼肉中毒 | 北美 |
| 立氏新立克次氏体(*Neorickettsia risticii*) | 马 | 吸虫类 | 白细胞 | 波多马克河马热，马单核细胞新立克次氏体病 | 北美，欧洲 |

嗜睡，轻度共济失调和厌食症。美洲驼表现为轻度的淋巴细胞减少，单核细胞和嗜酸性粒细胞增多。在中性粒细胞中检测到无浆体胞质包涵体，据此诊断为嗜粒细胞无浆体病。卧地不起的美洲驼用土霉素治疗后可完全康复。

阿联酋羊驼的单核细胞中也观察到胞质包涵体[44]（图 73）。这种羊驼也嗜睡，厌食，表现为单核细胞和嗜酸性粒细胞增多。

Wernery等[46]采用竞争ELISA（cELISA）方法从奶骆驼的1119份血清中检测出边缘无浆体、绵羊无浆体和中央无浆体抗体（兽医医学研发中心，普尔曼，USA）。过氧化物酶标记的单克隆抗体偶联物直接检测无浆体抗原而不针对宿主种类，因此，可用于骆驼（驼科动物）。尽管骆驼被检测到有嗜驼璃眼蜱感染，但cELISA仅检测到五个阳性（0.5%）。

总之，以上研究表明，驼类可感染边缘无浆体（红细胞内）和嗜粒细胞无浆体（白细胞胞质内）。

## 支原体病

支原体是已知最小的微生物，能够自主复制，与其他细菌不同，它没有细胞壁，基因组较小。支原体中的嗜血支原体，不仅没有细胞壁，而且兼性寄生于红细胞内从而称为嗜血支原体（haemoplasmas），且无法进行体外培养。他们是单独的一类微生物群，属于柔膜体纲，目前已报道有200多种支原体，但这种统计并不够准确，因为新的种类不断的被发现，如嗜血性微生物类。

嗜血支原体，原名血巴尔通体和附红细胞体种，属于立克次体目，在欧洲和南、北美洲的新大陆骆驼中有报道[3]。这些嗜血菌被重新归类支原体属，目前称为嗜血支原体（表19）。现在嗜血支原体在南、北美洲似乎是一类众所周知的细菌，并且

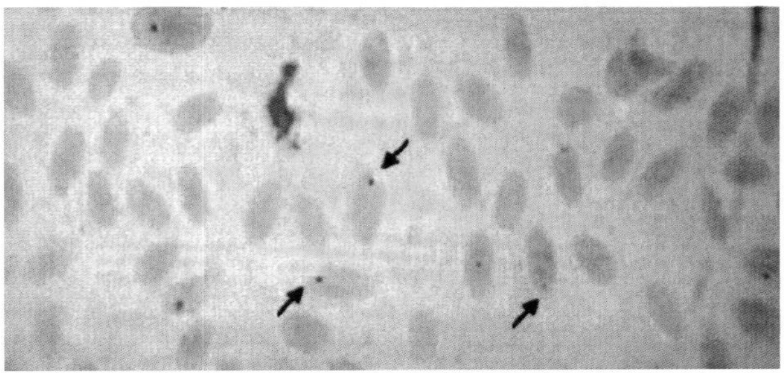

图70　感染的红血细胞内边缘的边缘无浆体（图片由伊朗的K.M.Alssad博士提供）

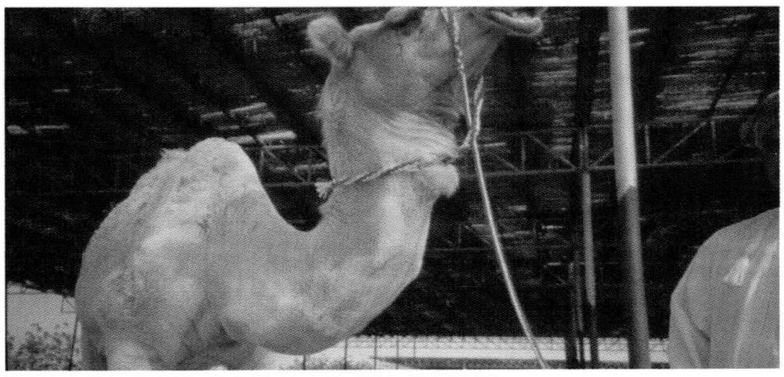

图71　由嗜粒细胞无浆体感染引起的单峰驼颈部腹侧水肿

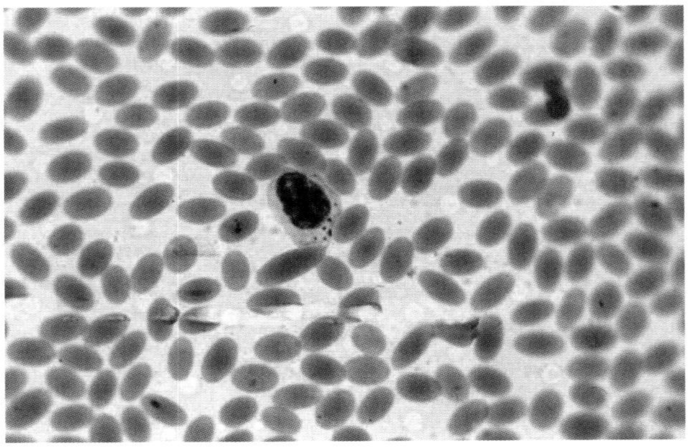

图72　在单核细胞细胞质内观察到的单峰驼嗜粒细胞无浆体（油镜，100×12）

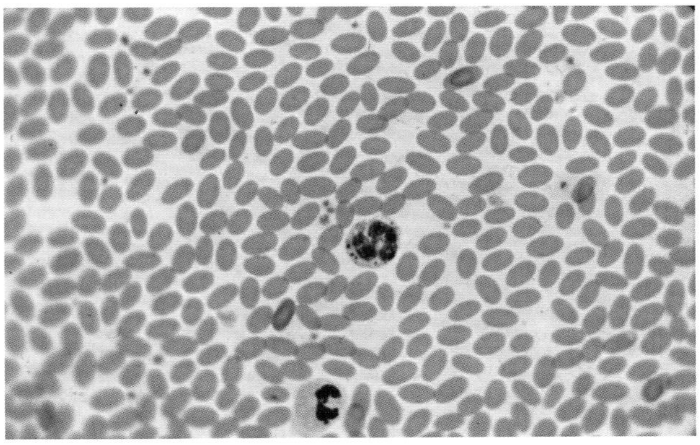

图73　嗜粒细胞无浆体感染羊驼，单核细胞中的胞质包涵体（姬姆萨染色）

表19 兽医学相关的重要支原体

| 宿主 | 属（种类） | 重要种类（更多的种已确定） |
|---|---|---|
| 牛，绵羊，山羊 | 支原体属[12]<br>无胆甾原体属[1]<br>脲原体属[1] | 丝状支原体丝状亚种（2×），无乳支原体<br>丝状支原体山羊亚种，牛支原体<br>山羊支原体山羊肺炎亚种，结膜支原体<br>山羊支原体山羊亚种，绵羊支原体<br>牛支原体属7组 |
| 马 | 支原体属[11]<br>无胆甾原体属[8] | 猫支原体(M. felis)<br>马鼻支原体 |
| 狗和猫 | 支原体属[15]<br>无胆甾原体属[1]<br>脲原体属 | 犬支原体（M. canis），猫血支原体，犬支原体(*M. cynos*)，猫支原体(M. gatae)<br>猫支原体(M. felis) |
| 猪 | 支原体属[13]<br>无胆甾原体属[5]<br>脲原体属[1] | 猪肺炎支原体，猪支原体（猪附红细胞体）<br>猪鼻支原体<br>猪滑液支原体 |
| 骆驼 | 支原体属[2]<br>无胆甾原体属[2]<br>脲原体属（？） | 嗜血支原体（巴尔通体）<br>精氨酸支原体<br>莱德劳（氏）无胆甾原体，眼无胆甾原体 |
| 家禽 | 支原体属[17]<br>无胆甾原体属[2]<br>脲原体属[2] | 鸡毒支原体<br>火鸡支原体<br>滑液支原体 |

在欧洲也有出现，越来越多的南美骆驼（驼科动物）携带该病原。

幼龄美洲驼和羊驼经常确认有支原体感染[9,22,36]。少年美洲驼，从断奶到几岁，发现有明显的免疫缺陷疾病，表现为体重减轻，生长发育不良和急性或复发性感染史。由于预后严重，感染后的美洲驼通常死亡或安乐死。在这些病例中，经常检测到罕见的病原体或机会性微生物感染。尸检常见严重纤维素性浆膜炎累及胸和腹部器官，中度弥漫性非化脓性间质性肺炎，脾肿大，坏死性肠炎，观察肝脏可见广泛血管栓塞形成和缺血性梗死灶。免疫缺陷的美洲驼，经常发现有支原体类微生物感染，有迹象表明，伴随这种疾病，这类微生物经常与贫血有关。这些细菌附着在感染的美洲驼红细胞表面，经常成团出现并朝向细胞边缘（图74）[44]。在英格兰南部，尽管采用聚合酶链反应（PCR）或变性梯度凝胶电泳（DGGE）方法检测38只（29%）羊驼呈阳性，但动物并没有贫血症状。这些亚临床感染动物受到应激反应时，有可能表现出临床症状[9]。此外，已有了南美骆驼的粒细胞无浆体和嗜血支原体的双重感染的报道[21]（图75）。后者很难与无浆体区别，在血涂片检查时，也有一些寄生在红血细胞内。

骆驼嗜血支原体的研究已有很大的进展，在过去的几年，诊断检测技术也有大幅度提高。PCR方法已被俄勒冈州立大学兽医学院兽医诊断实验室用于诊断检测[37,38]。嗜血支原体的特异性检测无法用于其他种类的支原体，比如猪嗜血支原体、猫嗜血支原体和生殖支原体。所有的研究都表明，许多感染临床症状不明显，这些微生物感染后的临床症状表现多样性。临床感染伴随有发热，轻度到重度贫血，精神沉郁，黄疸，不孕，水肿，生长率低下和轻度至重度低血糖症。这些细菌是否可能引起或充当辅助因子造成免疫抑制尚未研究。

Almy等进行了一个有趣的观察，检测了一只4日龄羊驼仔血涂片发现大量嗜血支原体寄生于红细胞内。采集无寄生虫血症母羊驼和驼仔的血液，PCR方法检测时，嗜血支原体感染呈阳性，表明病原可经子宫传播。

Kaufmann等[18]首次在欧洲报道了羊驼的嗜血支原体感染。然而，作者并未阐明这种病原是否来自秘鲁还是瑞士本地，在羊驼全年生活的牧场并没有发现节肢动物媒介的存在。Volokhov等[42]最近从加利福尼亚海狮身上发现了一种新的嗜血支原体。16S rRNA系统进化分析表明，这种新的嗜血支原体与已知羊驼嗜血支原体有很近的亲缘关系。后来甚至有研究者发现幼龄羊驼同时感染了嗜粒细胞无浆体和嗜血支原体。

在伊朗的67只单峰驼也观察到自然感染了支原体病原[26]，姬姆萨染色外周血涂片显示支原体

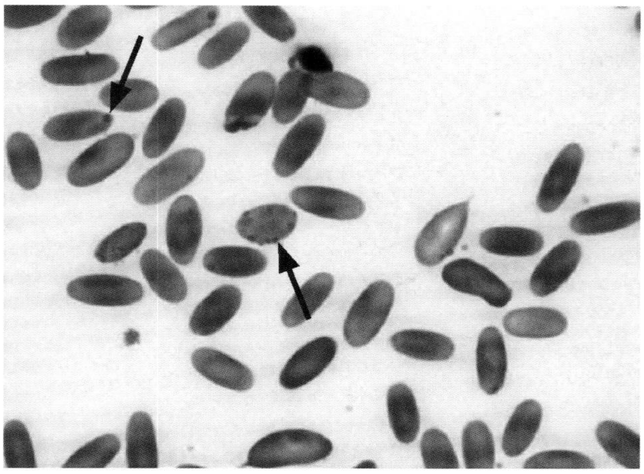

图74 幼龄美洲驼感染嗜血支原体（*M. haemolamae*）后患有免疫缺乏症；细菌贴附红细胞表面（姬姆萨染色）（图片由Tornquist SJ博士提供，USA）

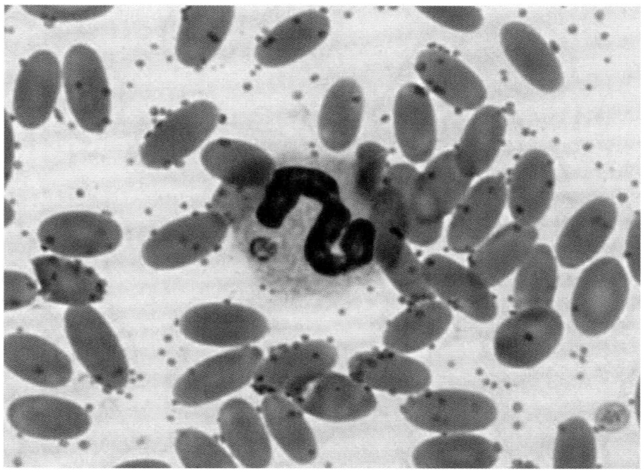

图75 嗜血支原体和粒细胞无浆体感染（中性粒细胞胞质中的桑椹胚）（图片由Lascola K博士提供，USA）



附于红细胞表面。受感染的骆驼，红细胞数量，血红蛋白的浓度和血细胞比容（红细胞压积）明显降低。感染的骆驼出现正常色素性贫血，血糖浓度明显比对照组低。Al-Khalifa等[2]研究发现沙特阿拉伯吉赞地区的20%（N=700）的骆驼感染了附红细胞体病原，该病原现在被称为嗜血支原体。然而，在其他地区没有感染或感染率较低。

表20总结了文献已报道的旧大陆骆驼和新大陆骆驼的立克次体，无浆体和支原体感染情况。

骆驼（驼科动物）是否对引起牛传染性胸膜肺炎（CBPP）的丝状支原体有易感性存在不同的意见。大多数持相反观点的研究者认为，这些支持者混淆了由巴氏杆菌和丝状支原体感染引起的肺部变化。Walker[43]使用皮下“恶性淋巴”未能引发肺部疾病。Samartsev和Arbuzov[34]，Hutyra等[14]，Curasson[10]和Turner[39]认为，骆驼不易患牛传染性胸膜肺炎，虽然还没有提供任何可以支持这一观点的科学证据。然而，认为骆驼易感染牛传染性胸膜肺炎的研究者也无法提供任何证据来支持他们的观点。这些研究者包括Vedernikoff[41]和Kowalesky[20]，他们经常观察到哈萨克斯坦的双峰驼患有呼吸性牛胸膜肺炎，在戴维斯也有类似报道[11]。

Bares[7] 用补体结合试验检测来自乍得地区的单峰驼血清，发现很低（非特异性？）的抗丝状支原体的抗体滴度，认为单峰驼不可能感染牛传染性胸膜肺炎（CBPP），且他们在该病的流行中不起任何作用。早期的文献表明，丝状支原体是引起骆驼肺部变化的一个诱因，因此应保留这种解释。

Paling等[28]发现，在肯尼亚地区检测的49%单峰驼血清有抗山羊传染性胸膜肺炎（支原体菌株F38）抗体。这些研究结果的意义还不清楚，因为没有分离到病原。

虽然还没有从骆驼的肺部病变中分离到丝状支原体病原，但从健康单峰驼的呼吸道已分离培养了其他支原体种。在埃及，Refai[29]从不同部位鉴定出下列菌株。

（1）精氨酸支原体，鼻，肺和纵隔淋巴结。

（2）莱德劳（氏）无胆甾原体，呼吸道。

（3）眼无胆甾原体，鼻腔。

Elfaki等[12]检查了100只单峰驼肺病变组织，从其8.8%样品中分离出精氨酸支原体。分离出支原体的骆驼发展为患有间质性肺炎，然而，是否精氨酸支原体可导致这些病变还不清楚。在秘鲁安第斯地区，利用间接血凝试验检测757只羊驼、美洲驼和骆马是否感染支原体[13]，发现这些动物不显示任何临床症状，但检测出下列支原体抗体：丝状支原体丝状亚种LC，554只羊驼（5%阳性）和141只美洲驼（15.6%阳性）；山羊支原体和F38株，0.9%羊驼和0.2%美洲驼为阳性；骆马仅诊断出一例丝状支原体丝状亚种LC和山羊支原体阳性感染，整群动物未检测到丝状支原体山羊亚种或无乳支原体抗体。

表20 旧大陆骆驼和新大陆骆驼相关的立克次体，无浆体和支原体感染情况

| 种类 | 作者 | 新大陆/旧大陆 | 年份 | 国家 | 血清学检测百分比 | 细菌检测 |
|---|---|---|---|---|---|---|
| 普氏立克次体<br>(*Rickettsia prowazekii*) | Reiss-Gutfreund,<br>Iman 和Labib,<br>Maurice等,<br>Bares Reiss-Gutfreund,<br>Ormsbee 等 | 旧大陆<br>旧大陆<br>旧大陆<br>旧大陆<br>旧大陆<br>旧大陆 | 1955<br>1963<br>1967<br>1968<br>1970<br>1971 | 埃塞俄比亚<br>埃及<br>乍得<br>乍得<br>蒙古<br>埃及 | 蜱<br>44.1<br>1.8<br>11.6<br>实验室感染<br>实验室感染 | 阳性<br>—<br>—<br>—<br>—<br>— |
| 莫氏立克次体<br>(*Rickettsia mooseri*) | Imam 和Labib,<br>Maurice等，Reiss-Gutfreund | 旧大陆<br>旧大陆<br>旧大陆 | 1963<br>1967<br>1970 | 埃及<br>乍得<br>蒙古 | 26.0<br>11.6<br>实验室感染 | —<br>—<br>— |
| 立氏立克次体<br>(*Rickettsia rickettsii*) | Bares,<br>Schmatz等 | 旧大陆<br>旧大陆 | 1968<br>1978 | 乍得<br>埃及 | 1.8<br>3.7 | —<br>— |
| 斑疹热立克次体<br>(*Rickettsia conorii*) | Maurice等 | 旧大陆 | 1967 | 乍得 | 1.0 | — |
| 边缘无浆体<br>(*Anaplasma marginale*) | Monteverde,<br>Anonymous,<br>Anonymous,<br>Ristic 和Kreier,<br>Ristic,<br>基西兽医实验室 | 旧大陆<br>旧大陆<br>旧大陆<br>旧大陆<br>旧大陆<br>旧大陆 | 1937<br>1939<br>1960<br>1974<br>1977<br>1981 | 索马里<br><br><br><br><br>索马里 | —<br>—<br>—<br>—<br>—<br>— | 40%<br>—<br>—<br>—<br>—<br>4.4% |
| 边缘无浆体<br>(*Anaplasma marginale*) | Ajayi等<br>Alsaad | 旧大陆<br>旧大陆 | 1984<br>2009 | 尼日利亚<br>伊朗 | 10.7<br>— | —<br>52份阳性 |
| 嗜吞噬细胞无浆体<br>(*Anaplasma phagocytophilum*) | Wernery 等,<br>Barlough等,<br>Wernery等 | 旧大陆<br>新大陆<br>新大陆 | 2001<br>1997<br>1999 | 阿曼<br>美国<br>阿联酋 | —<br>—<br>— | 3份阳性<br>1个美洲驼<br>1个羊驼 |
| 嗜血支原体<br>(*Mycoplasma haemolamae*) | Crosse等,<br>Kaufmann等,<br>Almy等,<br>McLaughlin,<br>Semrad | 新大陆<br>新大陆<br>新大陆<br>新大陆<br>新大陆 | 2012<br>2007<br>2006<br>1990<br>1994 | 英格兰<br>瑞士<br>美国<br>美国<br>美国 | —<br>—<br>—<br>—<br>— | 29%羊驼（alpacas）（PCR），<br>4份阳性，<br>1羊驼仔，<br>14.1%，<br>1个美洲驼， |
| 反刍兽无浆体<br>(*Anaplasma ruminantium*) | Karrar等,<br>Karrar | 旧大陆<br>旧大陆 | 1963<br>1968 | <br>苏丹 | —<br>— | —<br>— |

注：文献调查

—：未检测

## 诊断

未凝固的血液和感染的组织包括脑（心水）材料送实验室，用于无浆体和支原体的实验室诊断。由于这些微生物难以被革兰氏、姬姆萨、罗曼诺斯基、吉梅内斯、马基阿维韦洛氏法或利什曼染液以及间接荧光抗体（FA）染色法可用于血液和组织涂片。无浆体和嗜血支原体不能在琼脂上生长，因此，必须在鸡胚或组织中培养并添加青霉素和链霉素抑制污染。有些胚胎可能在感染6天后死亡，其余的在12天后检测。通过动物（豚鼠）接种的方法也是可取的，特别是对于一些细菌含量比较低的样本。此外，血清学可用于这些微生物的诊断。

边缘无浆体是专性寄生于红细胞内的细菌，主要感染成熟红细胞，每24～48小时感染的红细胞

表21　无浆体和支原体引起的重要疾病的实验室诊断方法

| 疾病 | 实验室诊断 | 姬姆萨染色涂片病原形态 |
|---|---|---|
| 禽埃及小体病（鸡埃及小体） | 血涂片姬姆萨染色，用感染血通过肠胃途径或皮肤划痕接种易感鸟类 | 紫红色的椭圆形，圆形，环形（直径0.3～3.9μm），红细胞可见大的包涵体 |
| 骆驼的无浆体病和支原体病（边缘无浆体，嗜吞噬细胞无浆体，嗜血支原体） | 姬姆萨，吖啶橙染色，PCR，血清学，CFT，ELISA | 红细胞内可见紫红色球菌或白细胞内可见桑椹胚样物质 |
| 无浆体病（边缘无浆体） | 姬姆萨，吖啶橙染色，FA染色，血清学：间接FA，CFT，ELISA，卡片凝集，PCR | 红细胞内靠近边缘呈紫红色形状多样（直径0.2～0.4μm），超过50%的红细胞被寄生 |
| 羊无浆体病（绵羊无浆体） | 血涂片姬姆萨染色，吖啶橙染色成亮橙色荧光 | 盘状或环形淡紫色微生物（直径0.5～1.0μm），杆状最常见于红细胞边缘 |
| 马无浆体病（嗜吞噬细胞无浆体） | 姬姆萨染色血或白细胞涂片，发病48小时后，可见包涵体。间接FA检测抗体 | 和犬无浆体一样，但细胞或包涵体存在于粒细胞，特别是嗜中性粒细胞 |
| 犬无浆体病（犬无浆体） | 姬姆萨染色血涂片，感染后15天观察最明显，间接FA或ELISA检测抗体 | 单核细胞或淋巴细胞中细胞呈紫色着色（直径0.5μm）或包涵体（桑椹胚）直径4.0μm |
| 反刍动物和人蜱传热（嗜吞噬细胞无浆体） | FA或姬姆萨染色血涂片 | 中性粒细胞、嗜酸性粒细胞、嗜碱性粒细胞和单核细胞中淡紫色包涵体，直径0.7～3.0μm |
| 猫传染性贫血（猫支原体） | 姬姆萨或FA染色血或组织涂片，每天进行血涂片观察，持续一周，病原在红细胞内出现是断续的 | 红细胞表面可见深紫色小球菌或杆状（0.2μm）微生物，偶尔可见少量环形微生物 |
| 猪支原体病（猪支原体） | 姬姆萨或FA染色血涂片<br>血清学：间接FA或CFT | 红细胞表面蓝紫色球菌或环形（最大直径2.5μm），在本属中是最大的菌种 |
| 波托马克马热（新立克次体） | FA或姬姆萨染色血涂片，ELISA或间接FA检测血清中的抗体 | 单核细胞中病原呈紫色着色 |
| 鲑鱼中毒（蠕虫新立克次氏体） | 临床症状和在粪便中可见虫卵（鲑隐孔吸虫）。在呼吸道淋巴结可见病原 | 巨噬细胞的胞浆中可见紫色包涵体，单个球菌（0.3～0.4μm）散落于细胞中 |
| 心水病（反刍动物无浆体） | FA或姬姆萨染色脑组织涂片（大脑皮层）。接种小鼠或易感牛 | 在大脑毛细血管内皮细胞胞浆中可见紫色着色的球菌(0.2～0.5μum）或短杆菌 |

CFT:补体结合试验；ELISA:酶联免疫吸附测定；FA：荧光抗体；PCR：多聚酶链式反应；RBC:红细胞；WBC:白细胞

数量会成倍增加，因此，50%～60%的红细胞可能被感染。血涂片红细胞姬姆萨染色可提供微生物感染的确切诊断依据，用间接荧光抗体（FA）和DNA探针也可确诊。

嗜吞噬细胞无浆体在中性粒细胞内增殖。在发热阶段，在单核细胞和中性粒细胞内可见该细菌，间接荧光抗体(FA)和ELISA也可用于血清学诊断。嗜血支原体感染的红细胞内可看到一些细菌，但对于周期性感染及细菌含量低时，这种方法相对不敏感。一些研究显示，PCR方法对于感染后的检测比血涂片检测更敏感。Meli等[24]建立了一种实时荧光定量PCR(定量PCR)对新大陆骆驼嗜血支原体定量检测。

表21总结了无浆体和支原体实验室检测方法。

### 治疗与控制

四环素和氯霉素可用于治疗无浆体病，急性病例应给药2周。必要时对带菌动物应采取长期治疗方案，例如，边缘无浆体感染。通过控制体外寄生虫加以预防，众所周知，骆驼身上的蜱可传播很多疾病，有些疾病给家畜和人类带来严重威胁。中央无浆体和边缘无浆体的活疫苗主要用于牛，但还没有用于骆驼（驼科动物）。

针对嗜血支原体病的治疗应采取适当的与治疗无浆体感染同样的策略，但是即使重复治疗也无法消除感染。有些研究者建议用四环素治疗50天以上以提高消灭病原的概率[22]。其他抗生素类药物，如氟苯尼考或恩氟沙星也不能清除病原，这两种药物的治疗效果没有明显差异[37]。

## 参考文献


[1] Ajayi S.A.,Onyali I.O.,Oluigbo F.O. & Ajayi S.T. (1984). – Serological evidence of exposure to *Anaplasma marginale* in Nigerian one-humped camels. *Vet. Rec.*,114 (19),478.

[2] Al-Khalifa M.S.,Hussein H.S.,Diab F.M. & Khalil G.M. (2009). – Blood parasites of livestock in certain regions in Saudi Arabia. *Saudi J. Biol. Sci.*,16,63－67.

[3] Almy F.S.,Ladd S.M.,Sponenberg D.P.,Crisman M.V. & Messick J.B. (2006). – *Mycoplasma haemolamae* infection in a 4-day-old cria: support for in utero transmission by use of a polymerase chain reaction assay. *Can. Vet. J.*,47 (3),229－233.

[4] Alsaad K.M. (2009). – Clinical,haematological and biochemical studies of anaplasmosis in Arabian one-humped camels (*Camelus dromedarius*). *J. Anim. Vet. Adv*ances,8 (11),2 106－2 109.

[5] Anon. (1939). – Notes on animal diseases. III. Piroplasmosis and anaplasmosis of animals other than cattle and trypanosomiasis. *E. Afr. Agric. For. J.*,4 (6),463－468.

[6] Anon. (1960). – Notes on animal diseases. III Piroplasmosis and anaplasmosis of animals other than cattle,and trypanosomiasis of domesticated animals. *E. Afr. Agric. For. J.*,25 (3),147－152.

[7] Bares J.F. (1968). – Contribution a l'étude de la pathologie infectieuse du dromadaire au Tchad. These,Ecole,Nationale Vétérinairede de Toulouse,Toulouse,France.

[8] Barlough J.E.,Medigan J.E.,Turoff D.R.,Clover J.R.,Shelly S.M. & Dumler J.S. (1997). – An *Ehrlichia* strain from a llama (*Lama glama*) and llama-associated ticks (*Ixodes pacificus*). *J. Clin. Microbiol.*,35 (4),1 005－1 007.

[9] Crosse P.,Ayling R.,Whitehead C.,Szladovits B.,English K.,Bradley D. & Solano-Gallego L. (2012). – First detection of '*Candidatus Mycoplasma haemolamae*' infection in alpacas in England. *Vet. Rec.*,171,71.

[10] Curasson G. (1947). – Le chameau et ses maladies. Vigot Freres,Editions,Paris,86－88.

[11] Davies G.O. (1946). – Gaiger & Davies veterinary pathology and bacteriology. 3rd Ed. Bailliere,Tindall and Cox,London.

[12] Elfaki M.G.,Abbas B.,Mahmond O.M. & Kleren S.H. (2002). – Isolation and characterization of *Mycoplasma argini* from camels (*Camelus dromedarius*) with pneumonia. *Comp. Imm. Microbiol. Inf. Dis.*,25,49－57.

[13] Hung A.L.,Alvarado A.,Lopez T.,Perales R.,Li O. & Garcia E. (1991). – Detection of antibodies to mycoplasmas in South American camelids. *Res. Vet. Sci.*,51,250－253.

[14] Hutyra F.,Marek J. & Manninger R. (1946). – Special pathology and therapeutics of the diseases of domestic animals. 5th English Ed. Baillière,Tindall and Cox,London.

[15] Imam I.Z.E. & Labib A. (1963). – Complement fixing antibodies against epidemic and murine typhus in domestic animals in U.A.R. *J. Egypt Publ. Health. Ass.*,38,101－109.

[16] Karrar G. (1968). – Letter. *Sudan J. Vet. Sci. and Anim.* Husb.,9 (II),328.

[17] Karrar G.,Kaiser M.N. & Hoogstraal H. (1963). – Ecology and host relationship of ticks (*Ixodoidea*) infesting domestic animals in Kassala province,Sudan,with special reference to *Amblyomma lepidum Doenitz. Bull. Entom. Res.*,54,509–523.

[18] Kaufmann C.,Meli M.L.,Robert N.,Willi B.,Hofmann-Lehmann R.,Wengi N.,Lutz H. & Zanolari P. (2007). – Haemotrophic mycoplasma in South American camelids in Switzerland. *Ver. Ber. Erkrg. Zootiere*,43,13–17.

[19] Kornienko-Koneva Z.P. (1955). – Dissertation,Moscow (cited by A.A. Markov *et al.*,1965,210).

[20] Kowalesky M.J.M. (1912). – Le Chameau et ses maladies d'après les observations d'auteurs russes. *J. Med. Vet. Zootechn. (Lyon)*,15,462–466.

[21] Lascola K.,Vandis M.,Bain P. & Bedenice D. (2009). – Concurrent infection with *Anaplasma phagocytophilium* and *Mycoplasma haemolamae* in a young alpaca. *J. Vet. Intern. Med.*,23,379–382.

[22] McLaughlin B.G.,Evans C.N.,McLaughlin P.S.,Johnson L.W.,Smith A.R. & Zachary J.F. (1990). – An *Eperythrozoon*-like parasite in llamas. *JAVMA*,197 (9),1170–1175.

[23] Maurice Y.,Bares J.F. & Baille M. (1967). – Enquete serologique sur les rickettsioses chez le dromadaire au Tchad. *Rev. Elev. Méd. Vét. Pays Trop.*,20 (4),543–550.

[24] Meli M.L.,Kaufmann C.,Zanolari P.,Robert N.,Willi B.,Lutz H. & Hofmann-Lehmann R. (2010). – Development and application of a real-time Taq Man q PCR assay for detection and quantification of *Candidatus Mycoplasma haemolamae* in South American camelids. *Vet. Microbiol.*,146,290–294.

[25] Monteverde G. (1937). – Anaplasmosi nei cammelli in Cirenaica. *Clin. Vet. (Milano)*,60 (2),73–77.

[26] Nazifi S.,Oryan A. & Bahrami S. (2009). – Evaluation of haematological and serum biochemical parameters in Iranian camels (*Camelus dromedarius*) infected with haemotrophic Mycoplasma (*Eperythrozoon*) spp. *Comp. Clin. Pathol.*,18 (3),329–332.

[27] Ormsbee R.,Burgdorfer W.,Peacock M. & Hildebrandt P. (1971). – Experimental infections of *Rickettsia prowazeki* among domestic livestock and ticks. *Amer. J. Trop. Med. Hyg.*,20 (1),117–124.

[28] Paling R.W.,MacOwan K.J. & Karstad L. (1978). – The prevalence of antibody to contagious caprine pleuropneumonia (*Mycoplasma* strain F38) some wild herbivores and camels in Kenya. *J. Wildl. Dis. Vol.*,14 (7),305–308.

[29] Refai M. (1992). – Bacterial and mycotic diseases of camels in Egypt. *In* Proceedings of the 1st International Camel Conference,Dubai,UAE,2–6 February 1992. R. and W. Publications (Newmarket) Ltd,Newmarket,UK,59–64.

[30] Reiss-Gutfreund R.J. (1955). – Isolement de souches de *Rickettsia prowazeki* a partir du sang des animaux domestiques d'Ethiopie et de leurs tiques. *Bull. Soc. Path. Exot.*,48 (2),602–607.

[31] Reiss-Gutfreund R.J. (1970). – The serological response of Mongolian domestic animals to *Rickettsia prowazeki* and *Rickettsia mooseri* antigens. G. Batt. Virol. Immun.,63 (9–10),455–457.

[32] Ristic M. (1977). – Protozoa. *In* Parasitic protozoa. Vol. 4 (J.P. Kreier ed.),New York Academic Press,New York,235.

[33] Ristic M. & Kreier J.P. (1974). – *In* Bergey's manual of determinative bacteriology. Vol. 8 (D.H. Bergey,R.E. Buchanan & N.E. Gibbons,eds),Williams and Wilkens,Baltimore,MD,907.

[34] Samartsev A.A. & Arbuzov P.N. (1940). – The susceptibility of camels to glanders,rinderpest and bovine contagious pleuropneumonia. *Veterinariya (Moscow)*,4,59–63.

[35] Schmatz H.D.,Krauss H.,Viertel P.,Ismail A.D. & Hussein A.A. (1978). – Seroepidemiologische Untersuchungen zum Nachweis von Antikorpern gegen Rickettsien und Chlamydien bei Hauswiederkauern in Agypten,Somalia und Jordanien. *Acta Tropica*,35,101–111.

[36] Semrad S.D. (1994). – Septicemic listeriosis,thrombocytopenia,blood parasitism and hepatopathy in a Llama. *JAVMA*,204 (2),213–216.

[37] Tornquist S. (2006). – Update on *Mycoplasma haemolamae* in camelids. International Camelid Health Conference for Veterinarians,Columbus,OH,21–25 March 2006,52–54.

[38] Tornquist S. (2008). – Camelid haematology (including *M. haemolamae*) update. Proceedings of the International Camelid Health Conference for Veterinarians,Ohio State University College of Veterinary Medicine,Columbus,OH,18 March 2008,214–217.

[39] Turner A.W. (1959). – Pleuropneumonia group of diseases. *In* Infectious diseases of animals. Volume II (A.W. Stableforth & I.A. Galloway,eds). Butterworths Scientific Publications,London,437.

[40] Veterinary Kisimayo Laboratory (1981). – Annual report of the Veterinary Laboratory,Kismayo. Ministry of Livestock,Forestry and Range,Department of Veterinary Services,Somalia.

[41] Vedernikov V. (1902). – Cited in Curasson (1947) Le chameau et ses maladies. Vigot Frères,Éditions,Paris,86–88.

[42] Volokhov D.V.,Norris T.,Rios C.,Davidson M.K.,Messick J.B.,Gulland F.M. & Chizhikov V.E. (2011). – Novel haemotrophic mycoplasma identified in naturally infected California sea lions *(Zalophus californianus). Vet. Microbiol.*,149,262–268.

[43] Walker J. (1921). – Experiments and observations in connection with pleuropneumonia contagiosa bovum. *Bull. Dept. Agri. (Kenya)*,2.

[44] Wernery U.,Fowler M.E. & Wernery R. (1999). – Color Atlas of Camelid Hematology. Blackwell Wissenschafts-Verlag,Berlin,37–43.

[45] Wernery U.,Musa B. & Kinne J. (2001). – Rickettsia-like disease in dromedaries. *J. Camel Pract. Res.*,8 (1),7–9.

[46] Wernery U.,Thomas R.,Syriac G.,Raghavan R. & Kletzka S. (2007). – Seroepidemiological studies for the detection of antibodies against nine infectious diseases in dairy dromedaries (Part-I). *J. Camel Pract. and Res.*,14 (2),85–90.

（牛庆丽译，殷宏校）

## 1.1.13 Q热

Q热，即Q fever（QF），由贝氏柯克斯体（*Coxilla burnetii*）引起，是一种人兽共患病，但是，人感染多于一半的病例不显示临床症状。除了新西兰，该病在世界范围内分布广泛。贝氏柯克斯立克次体可感染多种动物，在骆驼（驼科动物）中已经检测到Q热抗体。

### 病原学

贝氏柯克斯体是一种专性寄生于细胞内，多形态的球杆菌。一直以来，贝氏柯克斯体都属于立克次体目、立克次体科、立克次体群、贝氏柯克斯体属，同一科的还有立克次体（*Rickettsia*）和东方体属（*Orienta*）。然而，现在的基因序列分析将东方体属划归纳为军团菌目、柯克斯体科。贝氏柯克斯体呈现出两种不同的抗原相变异，即I相和Ⅱ相抗原变异，其中，I相的菌株更具感染性。

### 公共卫生影响

贝氏柯克斯体是引起人兽共患病Q热的病原体。人的感染主要由吸入引起[11]。感染源广泛，例如，被尿液、粪便或动物生产时排出物污染的土壤、空气传播的尘土、羊毛、草垫及其他的材料。很早就有啮齿类动物和驯养的动物作为宿主或贮存宿主感染人的报道。骆驼也不例外，很多的研究人员[28,30]已经证实了通过接触单峰驼而将Q热传播给人的危险性。其中，最大的危险很可能是食用天然的骆驼奶[49]。

Q热已经在世界各地的人群暴发过，症状是高热、严重头痛、心神不安、肌肉痛、咽喉痛、寒颤、出汗、恶心、胃痛和胸痛。贝氏柯克斯体可以通过蜱或吸入污染的尘土而感染。吸入单个的菌体可能会导致疾病。

对于灵长类动物，1.7个剂量的菌体就足以杀死50%的灵长类动物[26]。人的感染通常是通过吸入被污染的尘土发生的，污染源来自空气中散布的已干燥的胎盘物质、分娩时的排出物以及感染动物的排泄物等。最近，在斯洛文尼亚[36]和澳大利亚南部的羊场实习的兽医专业的学生暴发该病，病原体源于山羊屠宰场[37]。另外一例人Q热的疾病发生在苏格兰的屠宰场以及切割车间[50]。在2007年发生于荷兰南部省份暴发一场非常重大的疫情，大概有170人患病。此后，就发生4000多个病例，包括在2010年死亡的11个病例[38]。人感染该病的症状表现为发热，有时伴随肺炎或肝炎。慢性Q热不常见，但可能会引起致命性的心内膜炎。Q热的感染也可以通过食用天然奶而发生。72℃高温15秒的巴氏灭菌法能否消灭奶制品中的贝氏柯克斯体还不确定[39]。

### 流行病学

贝氏柯克斯体对热、干燥及其他常见的消毒剂不敏感。这些特性使得该细菌在环境中可长期存活。贝氏柯克斯体的稳定结构与该细菌的紧凑型小细胞突变株有关，这样的突变会在正常复制时发生，同时，伴随有抵抗性较差的大细胞形态、但它们以代谢静止和孢子样形式存在[33]。到目前为止，在骆驼（驼科动物）中还没有发生过Q热，但是，已经开展了很多的血清学调查（表22）。

最近的血清学检查表明，在埃及、乍得、阿联酋Q热的检出率很高。其中，乍得的阳性检出率高达80%，在以前文献中从未报道过。Afzal和Sakkir[3]及Elamin[13]等人研究表明，骆驼是一种



表22　骆驼Q热抗体检出情况

| 作者 | 年份 | 国家 | 血清阳性率 |
|---|---|---|---|
| Blanc等 | 1948 | 摩洛哥 | — |
| Giroud等 | 1954 | 乍得 | 22.2 |
| Rafyi和Maghani | 1954 | 伊朗 | 2.0 |
| Veeraghavan和Sukumaran | 1954 | 印度 | — |
| Kalra和Taneja | 1954 | 印度 | — |
| Elyan和Dawood | 1955 | 埃及 | 13.9 |
| Brown | 1956 | 肯尼亚 | 20.0 |
| EI-Nasri | 1962 | 苏丹 | 0.0 |
| Imamov | 1964 | 哈萨克斯坦 | 4.8 |
| Maurice等 | 1967 | 乍得 | 13.6 |
| Sabban等 | 1968 | 埃及 | 4.8 |
| Bares | 1968 | 乍得 | — |
| Maurice和Gidel | 1968 | 中非 | — |
| Pathak和Tanwani | 1969 | 印度 | 11.9 |
| Chooudhury等 | 1971 | 印度 | 23.8 ~ 26.9 |
| Harbi和Awad EI Karim | 1972 | 苏丹 | 12.2 ~ 12.8 |
| Kulshreshtha等 | 1974 | 印度 | 17.3 |
| Burgemeister等 | 1975 | 突尼斯 | 15.8 |
| Ghosh等 | 1976 | 印度 | 5.6 |
| Schmatz等 | 1978 | 埃及 | — |
| Mathur和Bhargava | 1979 | 印度 | 6.7 ~ 7.7 |
| Addo | 1980 | 尼日利亚 | 12.0 |
| Harrak | 1986 | 突尼斯 | — |
| Abbas等 | 1987 | 苏丹 | 14.5 |
| Djegham | 1988 | 突尼斯 | 3.06 |
| Gallo等 | 1989 | 突尼斯 | 0.0 |
| Soliman等 | 1992 | 埃及 | 66 |
| Schelling等 | 2003 | 乍得 | 80 |
| Mazyad和Hafez | 2007 | 埃及 | 13.3 |
| Probst等 | 2011 | 德国 | 4.6 |
| Wernery | 2011 | 阿联酋 | 45（普通骆驼） |

注：数据来自文献统计。

—：无资料记载

重要Q热的传染源，尽管在乍得的骆驼有着很高的阳性率，而人的阳性率相对较低(1%)。然而，Schelling[44]等研究表明，在乍得的骆驼饲养员中，Q热的阳性率高于牛饲养员，这有很大的风险性。研究还发现，乍得的骆驼是家畜中Q热阳性率最高的。

研究发现，牛Q热的血清学反应与排泄物没有关系。很多动物的阴道黏膜、粪便及奶汁可排出贝氏柯克斯体，这些动物却表现为血清学阴性。另一方面，Real-time PCR和抗体ELISA检测了12批奶汁样品，其中，PCR检测表明6批奶汁样品（50%）是Q热阳性，ELISA检测表明，10批（83%）样品是阳性[32]。在这项相同的研究中，研究人员针对贝氏柯克斯体，用Real-time PCR检测了45份患有子宫炎的母牛子宫内容物，但只有1份子宫样品呈现强阳性。Guatteo[18]等人研究发现，牛的阳性检出率为20%（动物个体水平）和38%（种群水平）稍高于那些小反刍动物（15%的动物个体水平及25%的种群水平）。他们认为在检测血清抗体时，ELISA和间接免疫荧光比补体结合试验或凝集实验更敏感。

在中东地区，野生动物可能在Q热的流行过程中发挥了重要的作用，因为阿联酋鹿瞪羚（*Gazella dama*）的晚期流产由贝氏柯克斯体感染引起[27]。沙特阿拉伯的利雅得市的227头细角瞪羚中，贝氏柯克斯体抗体的检出率为18.3%；232头山瞪羚中，贝氏柯克斯体抗体的检出率为7.3%；96头阿拉伯大羚羊，贝氏柯克斯体抗体的检出率为46.9%[21]。

在西班牙，贝氏柯克斯体DNA在狍子(*Capreolus capreolus*)中的检出率为5.1%，野猪(*Sus scrofa*)的检出率为4.3%，欧洲野兔(*Lepus europaeus*)的检出率9.1%，在鸟类中，秃鹰(*Gyps fulvus*)的检出率为11%，黑鸢(*Milvus migrans*)的检出率为14%[4]。然而，从这些动物身上收集的340只成年蜱经PCR检测，呈贝氏柯克斯体阴性，这表明在这一地区的蜱不会对贝氏柯克斯体的传播起到重要作用。

研究人员观察到，在单峰驼中没有流产病例的动物相比于有轻微流产的动物有着较高的Q热阳性率[43]。尽管在阿联酋的奶骆驼中Q热的阳性率高达45%[48]，但来自血清学检测为阳性的奶单峰驼的天然奶样品，PCR检测结果显示全部为阴性。

## 临床症状

很多感染的反刍动物是无症状的。在妊娠后期，绵羊和山羊会发生流产，但是，牛很少发生。尽管有着很高的阳性率，但是，在骆驼（驼科动物）中还没有发现由Q热引起的流产[49]。贝氏柯克斯体会引起牛的肺炎，但是，Q热能否引起骆驼（驼科动物）的肺炎尚不清楚。

## 诊断

Q热的临床症状是非特异的，因此，准确地诊断该病需要有适合的实验室检测方法。由贝氏柯克斯体感染引起的流产可以通过在吉梅内斯、斯坦泼或马基阿维韦洛氏法染色的胎盘抹片中发现病原，结合血清学进行检测，如CFT、ELISA或微量凝集试验（表23），得到确诊。最近，用PCR技术在胎盘和阴道黏液检测贝氏柯克斯体的DNA，这从根本上改变了在兽医学领域对Q热的诊断[23]。PCR技术是唯一能够检测隐藏细菌的方法，这对证实牛是否有公共卫生隐患有重要的作用。这一点在奶牛场是非常重要的。利用PCR技术从储存奶制品的储存桶中检测贝氏柯克斯体是一种有用且简单方便的评估奶制品的方法[32]。最近，这种方法已用于骆驼农场[41]。经检测，22个骆驼养殖场的奶汁样品中，只有1.4%（1/70）的样品是贝氏柯克斯体阳性。

表23　Q热的实验室诊断

| 疾病 | 实验室诊断 | 吉姆萨染色涂片的检出率 |
|---|---|---|
| Q热（贝氏柯克斯体*Coxilla burnetii*） | 胎盘物质的荧光抗体或吉姆萨染色涂片鉴定<br>胎盘物质或奶汁的PCR检测<br>血清学检测（CFT、ELISA、微量凝集技术）<br>感染后2～3周抗体检测 | 小的紫红色球菌（0.4～4μm）或细胞中的短杆菌<br>吉姆萨染色中Q热检出率与鹦鹉热衣原体相似 |

PCR技术结合ELISA方法是诊断Q热最有效的方式。

然而到目前为止，没有一种ELISA方法可用于骆驼副结核病的诊断。使用一组标准阳性和阴性血清，尽可能减少假阳性和假阴性，筛选一些抗种特异免疫球G蛋白（IgG）用于骆驼疾病诊断，是一件很重要的事。例如，Soliman[46]等人利用内部竞争ELISA和辣根过氧化物酶（horseradish peroxidase，HRP）蛋白A的ELISA方法检测骆驼血清中的Q热抗体，其结果表明，辣根过氧化物酶蛋白A的ELISA方法比竞争ELISA（competitive ELISA）的敏感性低50%。

**防控措施**

利用四环素进行抗生素治疗通常会降低流产动物的数量及分娩时贝氏柯克斯体的分泌量。然而，这种治疗方法的有效性值得怀疑。在Q热暴发期间，清洁工作要落实到位，以便减少环境的污染，使之不会成为复发感染的持续性传染源。因此，一定要移走并销毁分娩时的排出物及死胎。在生产之后，应将该区域进行消毒，以避免气雾和空气传播感染。粪肥是污染土壤的主要来源，因此，一定要将粪池掩盖好。

目前，已经开发了几种灭活疫苗，但是还没有一种用于骆驼（驼科动物）。



## 参考文献


[1] Abbas B.,Yassin T.T.M. & Elzubir A.E.A. (1987). – Survey for certain zoonotic diseases in camels in the Sudan. *Rev. Elev. Méd. Vét. Pays Trop.*,40 (3),231–233.

[2] Addo P.B. (1980). – A serological survey for evidence of Q fever in camels in Nigeria. *Br. Vet. J.*,136 (5),519–521.

[3] Afzal M. & Sakkir M. (1994). – Survey of antibodies against various infectious disease agents in racing camels in Abu Dhabi,United Arab Emirates. *Rev. sci. tech. Off. int. Epiz.*,13 (3),787–792.

[4] Astobiza I.,Barral M.,Ruiz-Fons F.,Barandika J.F.,Gerrikagoitia X.,Hurtado A. & García-Pérez A.L. (2011). – Molecular investigation of the occurrence of *Coxiella burnetii* in wildlife and ticks in an endemic area. *Vet. Microbiol.*,147 (1 2),190–194.

[5] Bares J.F. (1968). – Contribution à l'etude de la pathologie infectieuse du dromadaire au Tchad. Thèse,Ecole Nationale Veterinaire de Toulouse,Toulouse,France.

[6] Blanc G.R.,Bruneau J.,Martin J.A. & Maurice A. (1948). – Quelques donnes nouvelles sur le virus de la Q fever marocaine. *C.R. Scand. Acad. Sci.*,226 (7),607–608.

[7] Brown R.D. (1956). – La mise en évidence,par tests serologiques,de la fievre Q chez les animaux domestiques au Kenya. *Bull. Epiz. Dis. Afr.*,4,115–119.

[8] Burgemeister R.,Leyk W. & Gossler R. (1975). – Untersuchungen uber Vorkommen von Parasitosen,bakteriellen und viralen Infektionskrankheiten bei Dromedaren in Sudtunesien. *Dtsch. Tierärtztl. Wschr.*,82,352–354.

[9] Choudhury S.,Balaya S. & Mohapatra L.N. (1971). – Serological evidence of *Coxiella burnetii* infection in domestic animals in Delhi and surrounding areas. *Indian J. Med. Res.*,59,1194–1196.

[10] Djegham M. (1988). – À propos de l avortement chez la chamelle en Tunisie. *Maghreb Vét.*,3 (14),60.

[11] European Centre for Disease Prevention and Control (ECDC) (2010). – Technical report: risk assessment on Q fever. European Centre for Disease Prevention and Control,Stockholm.

[12] El-Nasri M. (1962). – A serological survey for the detection of Q fever antibodies in the sera of animals in the Sudan. *Bull. Epiz. Dis. Afr.*,10,55–57.

[13] Elamin E.A.,Elias S.,Dangschies A. & Rommel M. (1992). – Prevalence of *Toxoplasma gondii* antibodies in pastoral camels (*Camelus dromedarius*) in the Butana Palms,mid-eastern Sudan. *Vet. Parasitol.*,43,171–175.

[14] Elyan A. & Dawood M.M. (1955). – A serological survey of Q fever in Egypt. *J. Egypt Public Health. Ass.*,29 (6),185–190.

[15] Gallo C.,Vesco G.,Campo F.,Haddad N. & Abdelmoula H. (1989). – Enquete zoosanitaire chez les chevres et les dromadaires au Sud de la Tunisie. *Maghreb Vét.*,4 (17),15–17.

[16] Giroud P.,Roger F.,Dumas N.,Vouilloux P. & Sacquet E. (1954). – Comportement des animaux domestiques de la region du Tchad vis-à-vis de l'antigène T13. *Bull. Soc. Path. Exot.*,47,644–645.

[17] Ghosh S.S.,Mittal K.R. & Sen G.P. (1976). – Incidence of Q fever in man and animals. *Indian J. Anim. Health.*,15 (1),79–80.

[18] Guatteo R.,Seegers H.,Taurel A.F.,Joly A. & Beaudeau F. (2011). – Prevalence of *Coxiella burnetii* infection in domestic ruminants: a critical review. *Vet Microbiol.*,149 (1 2),1–16.

[19] Harbi M.S.M.A. & El Karim M.H.A. (1972). – Serological investigation into Q fever in Sudanese camels (*Camelus dromedarius*). *Bull. Epizoot. Dis. Afr.*,20,15–17.

[20] Harrak M. (1986). – Contribution à l'etude sérologique de la fievre Q chez le dromadaire en Tunisie. Thèse,Ecole Nationale de Medecine Vétérinaire,Sidi Thabet,Tunisia,283.

[21] Hussein M.F.,Al-Khalifa I.M.,Aljumaah R.S.,Elnabi A.G.,Mohammed O.B.,Omer S.A. & Macasero W.V. (2011). – Serological prevalence of *Coxiella burnetii* in captive wild ruminants in Saudi Arabia. *Comp. Clin. Pathol.*,21,33–38.

[22] Imamov E.D. (1964). – La fièvre Q chez les animaux domestiques de Kirghizie Frunce (cite par P.F. Zdrodowski dans Les rickettsioses en URSS). *Bull. O.M.S.*,31,33–43.

[23] Jones R.M.,Twomey D.F.,Hannon S.,Errington J.,Pritchard G.C. & Sawyer J. (2010). – Detection of *Coxiella burnetii* in placenta and abortion samples from British ruminants using real-time PCR. *Vet. Rec.*,167,965–967.

[24] Kalra S.L. & Taneja B.L. (1954). – Q-fever in India: A serological survey. *Indian J. Med. Res.*,42,315–318.

[25] Kulshreshtha R.C.,Arora R.G. & Kalra D.S. (1974). – Sero-prevalence of Q fever in camels,buffaloes and pigs. *Indian J. Med.Res.*,62,1314–1316.

[26] Lille R.D.,Perrin T.L. & Armstrong C. (1941). – An institutional outbreak of pneumonitis III. Histopathology in man and rhesus monkeys in the pneumonitis due to the virus of 'Q-fever'. *Pub. Health Rep.*,56,1419–1325.

[27] Lloyd C.,Stidworthy M.F. & Wernery U. (2010). – *Coxiella burnetii* abortion in captive dama gazelle (*Gazella dama*) in the United Arab Emirates. *J. Zoo Wildlife Med.*,41 (1),83–89.

[28] Mathur K.N. & Bhargava S.C. (1979). – Sero-prevalence of Q fever and brucellosis in camels of Jorbeer and Bikaner, Rajasthan State. *Indian J. Med. Res.*,70 (11),391–393.

[29] Maurice Y.,Bares J.F. & Baille M. (1967). – Enquete serologique sur les rickettsioses chez le dromadaire au Tchad. *Rev. Elev. Méd. Vét. Pays Trop.*,20 (4),543–550.

[30] Maurice Y. & Gidel R. (1968). – Incidence of Q fever in Central Africa. *Bull. Soc. Path. Exot.*,61 (5),721–736.

[31] Mazyad S.A. & Hafez A.O. (2007). – Q fever (*Coxiella burnetii*) among man and farm animals in North Sinai,Egypt. *J. Egypt Soc. Parasitol.*,37(1),135–142.

[32] Muskens J.,van Maanen C. & Mars M.H. (2011). – Dairy cows with metritis: *Coxiella burnetii* test results in uterine,blood and bulk milk samples. *Vet. Microbiol.*,147,186–189.

[33] Norlander L. (2000). – Q-fever epidemiology pathogenesis. *Microbes Infect.*,2,417–424.

[34] Pathak P.N. & Tanwani S.K. (1969). – Serological investigations in Q-fever. *Indian Vet. J.*,46,551–553.

[35] Probst C.,Speck S. & Hofer H. (2011). – Serosurvey of zoo ungulates in central Europe. *Int. Zoo Yearbook*,1748–1090.

[36] Promed. (2007a). – www.promedmail.org/pls/apex/f?p=2400:1001:7129315385574997::::F2400_P1001_BACK_PAGE,F2400_P1001_ARCHIVE_NUMBER,F2400_P1001_USE_ARCHIVE:1001,20070720.2332,Y

[37] Promed. (2007b). -www.promedmail.org/pls/apex/f?p=2400:1001:140784399314243::::F2400_P1001_BACK_PAGE,F2400_P1001_ARCHIVE_NUMBER,F2400_P1001_USE_ARCHIVE:1001,20070713.2244,Y

[38] Promed. (2011). Q-fever – Australia (02): Background. Available at: www.promedmail.org/pls/apex/f?p=2400:1001:741844215256965::NO::F2400_P1001_BACK_PAGE,F2400_P1001_PUB_MAIL_ID:1000,87070 (accessed 3 April 2011).

[39] Quinn P.J.,Carter M.E.,Markey B.K. & Carter G.R. (1994). – Clinical veterinary microbiology. Wolfe Medical,London,317–319.

[40] Rafyi A. & Maghani C. (1954). – Sur la presence de la fievre Q en Iran. *Bull. Soc. Path. Exot.*,6,766.

[41] Rahimi E.,Ameri M.,Karim G. & Doosti A. (2011). – Prevalence of *Coxiella burnetii* in bulk milk samples from dairy bovine, ovine,caprine,and camel herds in Iran as determined by polymerase chain reaction. *Foodborne Pathog. Dis.*,8 (2),307–310.

[42] Sabban M.S.,Hussein N.,Sadek B. & El Dahabi H. (1968). – Q fever in the United Arab Republic. *Bull. Off. int. Epizoot.*,69 (5 6),745–760.

[43] Schelling E.,Diguimbaye C.,Daoud S.,Nicolet J.,Boerlin P.,Tanner M. & Zinsstag J. (2003). – Brucellosis and Q-fever seroprevalences of nomadic pastoralists and their livestock in Chad. *Prev. Vet. Med.*,61,279–293.

[44] Schelling E.,Diguimbaye C.,Daoud S.,Nicolet J. & Zinsstag J. (2004). – Seroprevalences of zoonotic diseases in nomads and their livestock in Chari-Baguirmi. *Chad. Med. Trop.* (Mars),64 (5),474–477.

[45] Schmatz H.D.,Krauss H.,Viertel P.,Ismail A.S. & Hussein A.A. (1978). – Seroepidemiologische Untersuchungen zum Nachweis von Antikörpern gegen Rickettsien und Chlamydien bei Hauswiederkauern in Agypten,Somalia und Jordanien. *Act. Trop.*,35,101–111.

[46] Soliman A.K.,Botros B.A. & Watts D.M. (1992). – Evaluation of a competitive enzyme immunoassay for detection of *Coxiella burnetii* antibody in animal sera. *J. Clin. Microbiol.*,30 (6),1 595–1 597.

[47] Veeraghavan N. & Sukumaran P.K. (1954). – Q-fever survey in the Nilgiris and Coimbatore districts of Madras state. *Indian J. Med. Res.*,42,5–7.

[48] Wernery U. (2011). – Q fever in camelids with own investigations in dromedaries. *J. Camel. Pract. Res.*,18 (2),213–218.

[49] Wernery U. & Kaaden O.-R. (2002). – Infectious diseases in camelids. 2nd Ed. Blackwell Science,Berlin and Vienna,59–63.

[50] Wilson L.E.,Couper S.,Prempeh H.,Young D.,Pollock K.G.J.,Stewart W.C.,Browning L.M. & M. Donaghy M. (2010). – Investigation of a Q fever outbreak in a Scottish co-located slaughter house and cutting plant. *Zoonoses Public Health*,57, 493–498.

## 深入阅读材料

Bornstein S. (1993). – Camel health and diseases: veterinary projects. The multi-purpose camel: interdisciplinary studies on pastoral production in Somalia by Hjort af Ornaes,Environmental Policy and Society (EPOS),Uppsala,Sweden,189–206.

Davies G.O. (1946). – Gaiger and Davies veterinary pathology and bacteriology. 3rd Ed. Baillière,Tindall and Cox,London.

Mustafa I.E. (1984). A note on the diseases of camels in Saudi Arabia. *In* The Camelid: an all purpose animal (W.R. Cockrill,ed.). Scandinavian Institute of African Studies,Uppsala,Sweden,496–502.

Greiner V. (1985). – Beitrag zum Spektrum der Krankheiten bei Ruminantia und Tylopoda im Munchener Tierpark Hellabrunn. Tierpark Hellabrunn.

Guleed H.A. & Bornstein S. (1987). – Pilot study of the health of Somali camel herds. Camel Forum working paper,Mogadishu and Uppsala,SOMAC/SIAS,23.

Hoste C.,Peyre de Fabregues B. & Richard D. (1985). – Le dromadaire et son elevage. *Elev. Méd. Vét. Pays Trop.*, 145–146.

Leese A.S. (1909). – Two diseases of young camels. *J. Trop. Vet. Sci.*,4,1–7.

McGrane J.J. & Higgins A.J. (1985). – Infectious diseases of the camel; viruses,bacteria and fungi. *Br. Vet. J.*,141,529–547.

Odend'Hal S. (1983). – The geographical distribution of animal viral diseases. Academic Press,New York,99.

Rahamtalla M.H. (1994). – Haemorrhagic septicaemia (*Pasteurella multocida*) in camels: Immunity status,serological and bacteriological studies. Thesis,Faculty of Veterinary Science,University of Khartoum,Sudan.

Refai M. (1992). – Bacterial and mycotic diseases of camels in Egypt. *In* Proceedings of the 1st International Camel Conference,Dubai,UAE,2 6 February 1992. R. and W. Publications (Newmarket) Ltd,Newmarket,UK,59–64.

Richard D. (1986). – Manuel des maladies du dromadaire. Projet de developpement de l'élevage dans le Niger centre-est. IEMVT,Maisons Alfort,Paris.

Shommein A.M. & Osman A.M. (1987). – Diseases of camels in the Sudan. *In* Diseases of camels. *Rev. sci. tech. Off. int. Epiz.*,6 (2),481–486.

Wernery U. (1999). – New aspects on infectious diseases in camelids. *J. Camel Pract. Res.*,6 (1),87–91.

Wernery R.,Ali M.,Kinne J.,Abraham A.A. & Wernery U. (2001). – Mineral deficiency: a predisposing factor for septicemia in dromedary calves. *In* Proceedings of the 2nd International Camelid Conference,Agroeconomics of Camelid Farming,8–12 September 2000,Almaty,Kazakhstan,86.

World Health Organization (WHO),Food and Agriculture Organization (FAO) and World Organisation for Animal Health (OIE). (1990). – Animal health year book. Food and Agriculture Organisation of the United Nations,Rome.

Wilson A.J.,Schwartz H.J.,Dolan R.,Field C.R. & Roettcher D. (1982). – Epidemiologische Aspekte bedeutender Kamelkrankheiten in ausgewählten Gebieten Kenias. *Der praktische Tierarzt*,11,974–987.

（牛庆丽译，般宏校）

# 1.2 消化系统

## 1.2.1 沙门氏菌病

近年来，人沙门氏菌病的发病率增加，而动物被认为是该菌的主要储存宿主。在世界范围内，由于沙门氏菌可以感染所有动物，因而成为众多科学研究的焦点。已知的血清型分布范围进一步扩大，在某种程度上与全球的动物和食品贸易相关。感染沙门氏菌是因为摄取了沙门氏菌污染的食物和饮水，与带菌动物的直接接触也是重要的途径。

### 病原学

目前，已知的沙门氏菌属有2500多个血清型，它们被分为3个种：

（1）肠道沙门氏菌。

（2）邦戈尔沙门氏菌。

（3）地下沙门氏菌。

大部分血清型都属于肠道沙门氏菌，该属的血清型都是以其所发现地区命名的（例如，肠道沙门氏菌肠道亚种慕尼黑血清型、都柏林沙门氏菌）或它们的发病机理（例如，副伤寒沙门氏菌）。每个血清型是以其菌体抗原O和鞭毛抗原H的搭配来定义的。沙门氏菌也有宿主特异性（例如，马流产沙门氏菌）或宿主适应性(都柏林沙门氏菌)血清型。人宿主特异性的血清型伤寒杆菌和副伤寒杆菌都是伤寒的病原。血清型用罗马大写字母表示（例如鼠伤寒沙门氏菌）。沙门氏菌属归属于肠杆菌科，它的成员为革兰阴性球杆菌。除鸡沙门氏菌外，其他所有的沙门氏菌都为周生鞭毛的运动型菌。家畜的沙门氏菌病是由于感染宿主特异性和非宿主特异性的沙门氏菌血清型所引起的。该病以败血症、急性和慢性肠炎其中的一个或多个临床症状为主。健康成年动物感染后可成为无临床症状的带菌者。沙门氏菌定殖于宿主肠道，并且不断地向体外排出。许多动物都可罹患该病，病原最后都寄生于肠系膜的淋巴结上，也存在于脾脏、肝脏和胆囊（骆驼科动物没有胆囊）。这些动物有可能成为长期的病原携带者，成为重要的传染源。

只有绵羊流产性沙门氏菌引起的绵羊或山羊沙门氏菌病是一种须向世界动物卫生组织申报的疾病。

### 公共卫生意义

非伤寒性沙门氏菌病是人的食源性细菌感染的主要原因，它和食品动物的感染密切相关。此病的大暴发和牛奶产品，例如，奶酪、奶粉及冰淇淋都有关系。人类感染的沙门氏菌血清型经常是肠炎沙门氏菌、鼠伤寒沙门氏菌、新港沙门氏菌及海德尔堡沙门氏菌混合感染。肠沙门氏菌主要和禽类有关联，鼠伤寒沙门氏菌的暴发是随着新型噬菌体的出现有关。目前，血清型突变最多的是鼠伤寒沙门氏菌$DT_{104}$。鼠伤寒沙门氏菌$DT_{104}$为多抗药性的菌株，不仅能快速全球性传播，而且抗药性不断增强。

### 流行病学及临床症状

许多学者报道了世界好多地方的骆驼沙门氏菌病或沙门氏菌感染。已报道沙门氏菌感染的国家有苏丹[8]、巴基斯坦[54]、北非法国[10]、美国[6]。目前流行的国家有索马里[7]、埃塞俄比亚[36]、埃及[35、39、54]和阿联酋（UAE）[51]。有关数据参考表24。

沙门氏菌能引起骆驼的肠炎、败血症及流产。慢性沙门氏菌的主要特征是腹泻、体重减少及在几周之后死亡[16]。Pegram和Tareke[36]报道在埃塞俄比亚沙门氏菌病是青年单峰驼最主要的疾病，在该国的部分地区可以导致20%的损失。最近在对小骆驼死亡的原因调查中，许多科学家报道了肠炎沙门氏菌伴随大肠杆菌、骆驼艾美尔球虫、轮状病毒和冠状病毒的严重感染。Faye[15]报道尼日尔单峰驼68.3%的死亡都是由于混合型肠道病原体感染：沙门氏菌、轮状病毒、冠状病毒、大肠杆菌及骆驼艾美尔球虫。同时认为，鼠伤寒沙门氏菌、肠炎型沙门氏菌、肯塔基州沙门氏菌及圣保罗沙门氏菌是骆驼重要的血清型。沙门氏菌病的临床症状为出血性腹泻，伴随脱水及死亡。Berrada[4]等人检查了摩洛哥沙哈拉9个畜群中年龄为1～10周单峰驼仔的27份腹泻粪样，从患病的14.8%幼驼中，分离出5个不同的沙门氏菌菌株。Salih[12,43]等人检查了106头腹泻性幼驼，从14头幼驼的病料培养物中获得沙门氏菌属的细菌。伤寒沙门氏菌是最主要的分离株。沙门氏菌病的高发率在10月份。不仅能从1头病畜病料

表24　骆驼沙门氏菌病和沙门氏菌感染文献总结

| 作者 | 年份 | 国家 | 血清型数量 | 疾病 |
|---|---|---|---|---|
| Kowalevsky | 1912 | 俄罗斯 | 没有分型 | 肠炎 |
| Curasson | 1918 | 苏丹 | 没有分型 | 肠炎 |
| Olitzki | 1942 | 巴勒斯坦 | 1 | 健康 |
| Olitzki and Ellenbogen | 1943 | 巴勒斯坦 | 1 | 肠炎 |
| Donatien and Boue | 1944 | 法国北非 | 没有分型 | 流产，肠炎，败血症 |
| Sandiford | 1944 | 埃及 | 1 | 肠炎 |
| Bruner and Moran | 1949 | 美国 | 2 | 肠炎 |
| Floyd | 1955 | 埃及 | 3 | 健康 |
| Zaki | 1956 | 埃及 | 1 | 健康 |
| Farrag and El-Afify | 1956 | 埃及 | 1 | 健康 |
| Hamada et al. | 1963 | 埃及 | 2 | 健康 |
| Kamel and Lotfi | 1963 | 埃及 | 7 | 健康 |
| Malik et al. | 1967 | 印度 | 6 | 健康 |
| Ramadan and Sadek | 1971 | 埃及 | 8 | 健康 |
| Ambwani and Jaktar | 1973 | 印度 | 5 | 健康 |
| Cheyne et al. | 1977 | 索马里 | 1 | 肠炎 |
| Andreani et al. | 1978 | 索马里 | 1 | 健康 |
| El-Monla | 1978 | 埃及 | 7 | 健康 |
| Sayed | 1979 | 埃及 | 5 | 健康 |
| Pegram and Tareke | 1981 | 埃塞俄比亚 | 2 | 败血症 |
| Elias | 1982 | 埃及 | — | 肠炎 |
| El Nawawi et al. | 1982 | 埃及 | 5 | 健康 |
| Refai et al. | 1984 | 埃及 | 11 | 健康 |
| Yassien | 1985 | 埃及 | 5 | 健康 |
| Selim | 1990 | 埃及 | — | 健康 |
| Pegram | 1992 | 埃塞俄比亚 | 6 | 肠炎 |
| Wernery et al. | 1991 | 阿联酋 | 2 | 并发 |
| Wernery | 1992 | 阿联酋 | 28 | 健康 |
| Anderson et al. | 1995 | 美国 | 美洲驼 | 败血症 |
| Nation et al. | 1996 | 阿联酋 | 3 | 腹泻 |
| Faye | 1997 | 尼日尔 | 4 | 腹泻，败血症 |
| Berrada et al. | 1998 | 摩洛哥 | 5 | 腹泻 |
| Salih et al. | 1998a，b | 苏丹 | 1 | 腹泻 |
| Mohamed et al. | 1998 | 苏丹 | 4 | 腹泻 |
| Moore et al. | 2002 | 阿联酋 | 1 | 腹泻 |
| D’Alterio et al. | 2003 | 英国 | | 脑膜炎 |
| Molla et al. | 2004 | 埃塞俄比亚 | 116 | 健康 |
| Saulez et al. | 2004 | 美国 | 1羊驼 | 腹泻，坏死性肝炎 |
| Whitehead and Anderson | 2006 | 美国 | 羊驼，美洲驼 | 腹泻，败血症 |
| Gluecks | 2007 | 肯尼亚 | 114单峰驼 | 健康 |
| Nour-Mohammadzadeh | 2010 | 伊朗 | 1双峰驼 | 败血症 |
| Muench | 2012 | 阿联酋 | 351单峰驼 | 腹泻，败血症 |

中分离出沙门氏菌[30]，也能从8头健康幼驼的采样中分离出沙门氏菌，也在2头来自阿联酋的青年单峰驼[28]、42只来自苏丹的骆驼中4只（9.5%）中分离到沙门氏菌[26]。一些沙门氏菌的菌株能引起范围较广的临床综合征，包括耳朵、尾巴及四肢的缺血性坏死。Selim[48]在埃及对比了2组单峰驼，发现没有腹泻症状的健康单峰驼携带沙门氏菌的比例为3%，而患有肠炎的单峰驼携带沙门氏菌的比例达17%。

Nothelfer等[31]报道过一例由于沙门氏菌而致使耳朵尖坏死。坏死的部分能被轻易地除去，在耳朵表面发生明显地轻微出血（图76）。这是由于内毒素损伤血管内皮导致的局部血管内凝血而引起的局部贫血。

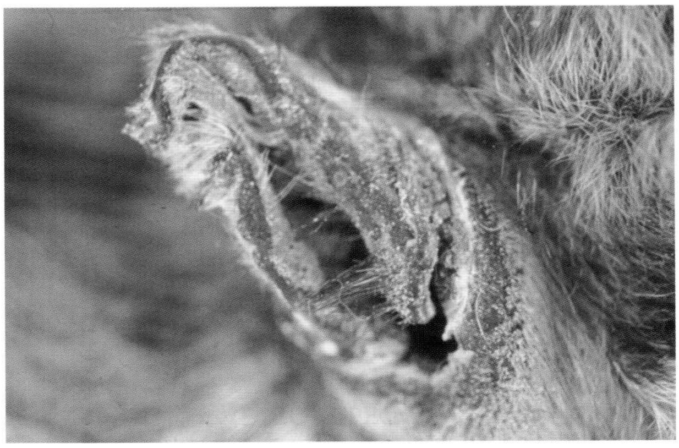

图76　除去耳朵坏死部分后，可见一个清洁、轻度流血的创面

在印度的健康骆驼的粪便中[1,23]以及在阿联酋的屠宰的骆驼的淋巴结和肠道中分离出沙门氏菌[13,19,39,54,55]。Molla[27]等人发现在埃塞俄比亚东部外观健康的屠宰骆驼中的沙门氏菌发生率很高。在714份粪样中，16.2%检测出了沙门氏菌。总共发现有16个不同的血清型，圣保罗沙门氏菌（38.8%）和布伦登芦普沙门氏菌（22.4%）发生最多。

正如表25，从许多国家分离到不同的血清型沙门氏菌。然而，从病畜和健康动物分离到的沙门氏菌是一样的。这种致病力的表现差异与个体抵抗力相关，也与畜种有一定关系。动物对疫病的抵抗力主要是由遗传因素对病原微生物的敏感性决定，但年龄、免疫状态、耐受性和整体状态也有密切相关。此外，病原体的数量和应激因素在年老动物沙门氏菌病的暴发中发挥主要作用。普遍认为沙门氏菌病可以因为动物的营养代谢失调、产仔、圈舍拥挤、外科手术和药物治疗加剧[5]。笔者在过去15年中，在阿联酋竞赛的骆驼上分离出351个沙门氏菌菌株，分属40个血清型（表26）。最常见的血清型是弗林特洛沙门菌和海因德马什沙门氏菌，但只有鼠伤寒沙门氏菌能引起一些幼龄骆驼的腹泻和败血症。鼠伤寒沙门氏菌主要是$DT_{193}$型，在所有的鼠伤寒山门氏菌中，有38%无法定型。

Wernery等[50]报道在阿联酋的竞赛骆驼有产气荚膜梭菌A型暴发。作者认为死亡的骆驼都是由于混合感染圣保罗沙门氏菌和塞罗沙门氏菌。他们在死亡动物所有器官中检查出沙门氏菌，该菌感染为致命肠毒血症产气荚膜梭菌生存和繁殖奠定了基础。在小猪中，感染肠致病性大肠杆菌可以起到同样的作用。Sinkovics[49]发现当小猪的小肠中也感染肠炎型大肠杆菌时，感染的产气荚膜梭菌的活性会升高。

近几年，沙门氏菌病作为人兽共患病的重要性在不断增强。预防措施应该考虑到不恰当的治疗能导致不明显的亚临床症状和动物带菌。长期携带者（慢性的）不仅能威胁同群其他动物，而且由于接触畜产品直接威胁到人类健康。这在非洲的几个食用单峰驼肉的国家非常普遍，例如，埃及、苏丹和索马里。Sandiford等[44]、Sandiford[45]、Ramadan和Sadek[38]及El-Nawawi等人都报道由于食用骆驼肉而引起的食物中毒。在阿联酋，Wernery和Makarem[52]从感染沙门氏菌的人和骆驼的粪便中鉴定出大量血清型相同的沙门氏菌。在埃及，Kamel和Lotfi[20]检查了屠宰的915个骆驼的肠淋巴结和粪样，确认是否有沙门氏菌。从3.1%被检动物中分离出沙门氏菌[鼠伤寒沙门氏菌（15×），圣保罗沙门氏菌（6×）、里丁沙门氏菌（3×）、都柏林沙门氏菌（2×）、伊斯特伯恩沙门氏菌（1×）、肠炎沙门氏菌 （1×）、病牛沙门氏菌（1×）]。这些学者的研究表明，骆驼不仅是沙门氏菌的重要储存宿主，而且威胁人类健康。相反，Rulofson等[41]认为美洲驼不是沙门氏菌的重要来源，因为它们的粪便接触到地表水的可能性很低。Gluecks[18]调查到在未断奶12周龄的幼驼沙门氏菌病的流行及其在幼驼腹泻中的作用。虽然分离到一些不同血清型，但是，仅鼠伤寒沙门氏菌与幼驼腹泻有关。以下是一些分离的沙门氏菌血清型。

（1）病牛沙门氏菌（32.6%）。

表25　不同国家骆驼分离出的沙门氏菌的血清型（除阿联酋）

| 编号 | 沙门氏菌血清型 | 国家 |
|---|---|---|
| 1 | 鼠伤寒沙门氏菌 | 巴勒斯坦、埃及、美国、尼日利亚、美国（美洲驼）、肯尼亚及埃塞俄比亚 |
| 2 | 都柏林沙门氏菌 | 埃及 |
| 3 | 肯塔基州沙门氏菌 | 巴勒斯坦、尼日利亚 |
| 4 | 圣保罗沙门氏菌 | 埃及、埃塞俄比亚及尼日利亚 |
| 5 | 阿哥纳沙门氏菌 | 美国 |
| 6 | 猪霍乱沙门氏菌 | 埃及、索马里、美国（美洲驼、羊驼） |
| 7 | 利密特沙门氏菌 | 印度 |
| 8 | 塞罗沙门氏菌 | 印度 |
| 9 | 鸭沙门氏菌 | 印度、埃及 |
| 10 | 伤寒沙门氏菌副沙门氏菌 | 埃及、印度、苏丹 |
| 11 | 弗林特洛沙门氏菌 | 印度 |
| 12 | 慕尼黑沙门氏菌 | 印度、埃及、埃塞俄比亚及肯尼亚 |
| 13 | 里丁沙门氏菌 | 埃及 |
| 14 | 基夫沙门氏菌 | 印度、埃塞俄比亚 |
| 15 | 伊斯特伯恩沙门氏菌 | 埃塞俄比亚、埃及 |
| 16 | 病牛沙门氏菌 | 埃及、肯尼亚 |
| 17 | 明斯特沙门氏菌 | 埃及 |
| 18 | 布雷得尼沙门氏菌 | 索马里 |
| 19 | 切斯特沙门氏菌 | 埃塞俄比亚、埃及 |
| 20 | 格罗斯特卢浦沙门氏菌 | 埃及 |
| 21 | 肠炎沙门氏菌 | 尼日利亚、摩洛哥、埃塞俄比亚 |
| 22 | 乌干达、新港及科特部斯沙门氏菌 | 埃塞俄比亚 |
| 23 | 海德尔堡沙门氏菌 | 埃塞俄比亚 |
| 24 | 哈瓦那沙门氏菌 | 摩洛哥 |
| 25 | 田纳西沙门氏菌 | 摩洛哥 |
| 26 | 塔拉哈撒沙门氏菌 | 摩洛哥 |
| 27 | 塔那那利佛沙门氏菌 | 美国（美洲驼、羊驼） |
| 28 | 新港沙门氏菌 | 埃塞俄比亚 |
| 29 | 布伦登芦普沙门氏菌 | 美国（美洲驼、羊驼） |
| 30 | 俄亥俄、布旦旦、基安布及阿得雷德沙门氏菌 | 肯尼亚 |

（2）布坦坦沙门氏菌（21.5%）。
（3）鼠伤寒沙门氏菌（11.1%）。
（4）基安布沙门(氏)菌（9.0%）。
（5）慕尼黑沙门氏菌（7.6%）。
（6）阿德莱德沙门氏菌（18.2%）。

Mahzounieh等[22]在伊朗进行了试管凝集试验，在阳性单峰驼体内检测到抗A、抗B、抗C和抗D的菌体抗原。结果表明，外观健康的骆驼可以感染不同血清型的沙门氏菌。尽管在新世界骆驼体内分离到一些沙门氏菌血清型，然而Whitehead和

表26　阿联酋竞赛单峰驼分离的血清型

| 宿主种类 | 沙门氏菌血清型 | 总数 | 宿主种类 | 沙门氏菌血清型 | 总数 |
|---|---|---|---|---|---|
| 成年骆驼 | 阿哥拉沙门氏菌 | 2 | 成年骆驼 | 圣保罗沙门氏菌 | 2 |
| | 阿尔图纳沙门氏菌 | 3 | | 施瓦岑格隆德沙门氏菌 | 1 |
| | 阿姆斯特丹沙门氏菌 | 2 | | S.Subspez I | 6 |
| | 鸭沙门氏菌 | 1 | | S.Subspez I及里丁沙门氏菌 | 1 |
| | 巴赫兰费尔德沙门氏菌 | 2 | | S.SubspezIIIb | 1 |
| | 布洛克兰沙门氏菌 | 1 | | 鼠伤寒沙门氏菌 | 10 |
| | 牛沙门氏菌 | 11 | | 魏尔肖沙门氏菌 | 1 |
| | 布雷得尼沙门氏菌 | 1 | | S.(not typed) | 71 |
| | 塞罗沙门氏菌 | 1 | 小计 | | 261 |
| | 切斯特沙门氏菌 | 1 | 骆驼幼畜 | 阿尔图纳沙门氏菌 | 3 |
| | 吉布提沙门氏菌 | 1 | | 阿姆斯特丹沙门氏菌 | 3 |
| | 肠炎沙门氏菌 | 1 | | 鸭沙门氏菌 | 1 |
| | 弗林特洛沙门氏菌 | 56 | | S.Anti-H42 | 1 |
| | 加明那拉氏沙门氏菌 | 2 | | 病牛沙门氏菌 | 2 |
| | 基夫沙门氏菌 | 1 | | 弗林特洛沙门氏菌 | 6 |
| | 哈达尔沙门氏菌 | 1 | | 基夫沙门氏菌 | 1 |
| | 海因德马什沙门氏菌 | 36 | | 哈达尔沙门氏菌 | 1 |
| | 婴儿沙门氏菌 | 3 | | 海因德马什沙门氏菌 | 12 |
| | 肯塔基州沙门氏菌 | 3 | | 肯塔基州 | 5 |
| | 曼哈顿沙门氏菌 | 1 | | 曼哈顿沙门氏菌 | 2 |
| | 鸡病沙门氏菌 | 5 | | 明斯特沙门氏菌 | 4 |
| | 慕尼黑沙门氏菌 | 2 | | 恩昌加沙门氏菌 | 2 |
| | 明斯特沙门氏菌 | 1 | | 新港沙门氏菌 | 1 |
| | 恩昌加沙门氏菌 | 7 | | S.PolyI 及 II | 1 |
| | 新港沙门氏菌 | 2 | | 斯坦利沙门氏菌 | 2 |
| | 巴拿马沙门氏菌 | 1 | | S.SubspezI | 1 |
| | S.Poly I | 10 | | 鼠伤寒沙门氏菌 | 21 |
| | S.Poly I 及 II | 6 | | 瓦伊勒沙门氏菌 | 1 |
| | S.Poly（乳糖阳性） | 1 | | 沙门氏菌(未分型) | 20 |
| | S.Poly II | 1 | 小计 | | 90 |
| | 里丁沙门氏菌 | 2 | 总计 | | 351 |

Anderson[53]一致认为，沙门氏菌病并不是导致骆驼腹泻的常见病因。一些研究人员调查了新大陆骆驼的幼驼腹泻，从健康羊驼和美洲驼幼驼粪便没有发现沙门氏菌。然而上述作者从不足1月龄的腹泻羊驼和美洲驼分离到多种沙门氏菌血清型。用沙门氏菌血液培养物复制了几例骆驼的败血症病例。这些血清型是俄亥俄沙门氏菌、新港沙门氏菌和猪霍乱沙门氏菌。Saulez等人分离到几种不同的血清型，不仅包括从坏死肝炎羊驼分离的鼠伤寒沙门氏菌，还包括感染肠道羊驼分离出的鼠伤寒沙门氏菌。羊驼

通过抗生素和液体疗法的综合性治疗而使临床症状消除。Anderson等[2]报道两例美洲驼败血症沙门氏菌病。作者认为，将美洲驼与其他动物混合饲养时，产生的应激反应可以增强美洲驼对沙门氏菌的敏感性。

### 病理学

沙门氏菌的特殊毒素是全身性和肠道性的沙门氏菌病的原因。这些毒素包括：

（1）脂多糖。

（2）内毒素。

（3）肠毒素。

（4）细胞毒素。

（5）质粒。

经口感染是沙门氏菌常见的感染途径。细菌穿透进入黏膜固有层，在肠道内分泌细胞毒素和肠毒素引起肠炎。在激发因素如梭菌、球虫或念珠菌存在时，急性肠炎是幼龄骆驼和成年骆驼最普遍的症状。这些动物的粪便有腐败性气味，含有黏液和血液，可能发展成严重的出血性肠炎（图77）。

在成年骆驼感染沙门氏菌的共同特征为慢性肠炎，表现为顽固性的腹泻，伴随间歇热、消瘦和治疗效果不佳等特点。沙门氏菌可以穿过美洲驼肠道的黏膜固有层而进入血液系统，引起败血症。在败血症期间，沙门氏菌可能定殖于头、脑脊膜、怀孕的子宫以及四肢、耳朵、尾巴的末端。也时常定殖于肠系膜淋巴结，幸存的动物不间断排出含细菌的粪便。败血性沙门氏菌在新生幼畜（6个月之内）为并发症状，为急性病例并伴随发热和抑郁，病畜于48小时之内死亡。尸检时每个器官都有出血点，经常伴随肺炎。矿物质缺乏可能是沙门氏菌败血症的诱发因素。Anderson等[2]报道了一起由猪霍乱沙门氏菌引起美洲驼败血性沙门氏菌病。该疾病主要病理变化包括纤维素性心包炎、胸膜炎和腹膜炎。该美洲驼曾经和猪接触过。第二个病例是早产的美洲驼感染了鼠伤寒沙门氏菌。动物表现出心包积液、胸水、肺充血和黏膜渗血。D’Alterio等人[8]记载过1只4日龄羊驼由于新港沙门氏菌引起的脑膜炎。Nour-Mohammadzadeh报道[32]由于鼠伤寒沙门氏菌和无乳链球菌引起的四周龄双峰驼致命败血症，也表现出纤维素性脑膜炎和化脓性脐炎。

图77　鼠伤寒沙门氏菌引起的一头幼龄单峰驼的出血性肠炎

### 诊断

沙门氏菌营养需求简单，可在许多培养基上可以生长。然而，选择性的培养基被用于从混合的菌群样品中分离出沙门氏菌。沙门氏菌的菌落特征可以直接用于该菌的血清型鉴定。血清型以O(菌体)和H(鞭毛)抗原为据。

在临床发病之后或期间，患病动物也可能是假阴性，因为排泄物中排菌是断断续续的。因而，多次连续的取样对确定畜群中传染源是必要的。快速地确定浓缩的粪样是否污染沙门氏菌的方法是采用聚合酶链式反应，甚至是实时定量PCR。然而，进一步研究时，必须进行培养和血清型分型。血清学诊断是在群体层面上确认是否有沙门氏菌感染的一个重要的手段，但是，该方法还没有在骆驼科广泛使用。

### 预防及控制

沙门氏菌主要寄生于恒温动物和冷血动物的肠道，大多数感染的动物就会成为没有临床症状的传染源。沙门氏菌主要通过粪口途径传播，但也可通过结膜黏膜和上呼吸道或者皮肤传播。因此，最为重要预防方法是应设法防止在畜群中引进携带病菌的动物。也应该确保饲料没有被沙门氏菌污染。应遵守骆驼场暴发沙门氏菌时的行动方案：

(a)应该确定带病菌动物，分离病菌，治疗活力强的动物，并反复多次对其检查，直到确定其不携带病菌。

(b)饲料和水都应该远离粪便污染（当心鸽子和啮齿动物）。

(c)限制农场周围动物的活动。

(d)告知所有人与患病骆驼接触后的健康危害。

(e)应该慎重考虑使用疫苗。

支持疗法和良好护理是重要的，尤其患有肠炎性的骆驼幼畜，可口服和注射改善电解质紊乱和平衡酸碱的药物。酸碱异常和代谢性酸中毒在幼畜感染沙门氏菌中常常发生，因此，治疗时应该使用静脉注射或口服含钠液体和碳酸氢盐等高效且离子差异大的制剂或碱性制剂。在群体和后代中避免导致矿物质缺乏因素也是重要的。许多幼畜之后都会遇到内毒素性休克，应使用对动物更加有益的非类固醇的抗炎药，例如，阿司匹林、氟尼辛、葡甲胺和已糖稀释液等。在这些情况下，应该遵循“内毒素中毒的治疗”章节的有关程序。

沙门氏菌抗菌药的使用是有争议的，因为在骆驼上可以使其处于带毒状态及耐药性沙门氏菌菌株的形成。在科威特和阿联酋[40]的人体上分离到对环丙沙星敏感性降低的多重耐药性沙门氏菌。在成年骆驼上，应该使用非肠道型的药物，因为对反刍动物使用口服药能造成反刍动物胃肠菌群严重失衡。对于沙门氏菌的抗菌注射药推荐氨苄青霉素、阿莫西林和甲氧苄氨嘧啶磺酰胺的联合使用。恩诺沙星也是一种非常有效的药物，但是药物的选择应该以培养和药敏性试验为依据（表27）。治疗越早越好，而且应该连续治疗6天。

一些产品化的无毒疫苗已经可以购买到了，它们在牛、猪及羊上是控制沙门氏菌有效的方法，但是，还没有在骆驼上尝试过。不管怎样，沙门氏菌自体疫苗是最有价值的，因为其包含沙门氏菌流行菌株。这些疫苗应该用于有发病的畜群上，对分娩前的动物接种两次以使新生动物免于沙门氏菌的威胁。初乳的免疫作用可以长达8周。自体疫苗必须灭活（麦氏第四号标准，大约$1.20 \times 10^8$cfu/mL）应该按照动物体重，以5 ~ 8mL的剂量肌肉或皮下注射。新驼免疫系统还不成熟，疫苗接种没有作用。在发病畜群中的幼驼，如母驼在分娩前没有进行过疫苗免疫，应在14天内口服总共50mL的畜群流行株特异的疫苗。

表27　骆驼沙门氏菌病抗生素治疗的最低治疗天数

| 抗生素 | 用药途径 | 剂量 |
|---|---|---|
| 羟氨苄青霉素 | 口服（仅幼畜） | 每12小时10mg/kg |
| 克拉维酸-羟氨苄青霉素 | 口服（仅幼畜） | 每12小时12.5mg/kg |
| 黏菌素 | 口服（仅幼畜） | 每12小时50000IU/kg |
| 磺胺增效类药物（甲氧苄氨嘧啶-磺胺嘧啶） | 静脉注射或肌内注射 | 每24小时25mg/kg |
| 氨苄青霉素 | 肌内注射 | 每12小时10mg/kg |
| 头孢噻呋 | 肌内注射或皮下注射<br>肌内注射 | 每12小时2.2mg/kg<br>每24小时5mg/kg |
| 氟苯尼考 | 肌内注射或皮下注射 | 每48小时20mg/kg |
| 氟喹诺酮类 | 口服（幼畜）<br>肌内注射、皮下注射<br>静脉注射 | 根据药物分子(恩氟沙星、达氟沙星、麻保沙星) |

# 参考文献

[1] Ambwani V.R. &. Jaktar P.R. (1973). –*Salmonella* infections of camel in Bikaner. *Indian Vet. J.*,50 (1),100－102.

[2] Anderson N.V.,Anderson D.E.,Leipold H.W.,Kennedy G.A.,Repenning L. &Strathe G.E. (1995). – Septicemic salmonellosis in two llamas. *JAVMA*,206 (1),75－76.

[3] Andreani E.,Prosperi S.,Arush M.A. & Salim A.H. (1978). –Idigane sulla presenza di portatori,di salmonelle trabovini, ovini, caprini e dromedari delle Repubblica Democratica Somala. *Ann. Fac. Med. Vet. Univ. Pisa*,31,65－72.

[4] Berrada J.,Bengouni M.,El Mjyad R.,Touti J. &Fikri A. (1998). –*Salmonella* infection in newborn dromedaries in Moroccan Sahara. *In* Proceedings of the International Meeting on Camel Production and Future Perspectives,2－3 May,Al Ain,UAE.

[5] Blood D.C. &Radostits O.M. (1990). – Veterinary Medicine. 7th Ed. Bailliere Tindall,London.

[6] Bruner D.W. & Moran A.B. (1949). –*Salmonella* infections of domestic animals. *Cornell Vet.*,39 (1),53－63.

[7] Cheyne I.A.,Pegram R.G. & Cartwright C.F. (1977). – An outbreak of salmonellosis in camels in the North-East of the Somali Democratic Republic. *Trop. Anim. Health Prod.*,9 (4),238－240.

[8] Curasson G. (1918). –Une maladie du dromadaire analogue au farcin du boeuf. *Bull. Soc. Cent. Méd. Vét. (Supplement to Rec. Méd. vét.* 94),71,491－496.

[9] D'Alterio G.L.,Bazeley K.J.,Pearson G.R.,Jones J.R.,Jose M. & Woodward M.J. (2003). – Meningitis associated with *Salmonella* Newport in a neonatal alpaca (*Lama pacos*) in the United Kingdom. *Vet. Rec.*,152,56－57.

[10] Donatien A. &Boue A. (1944). –Une épizootie de ghedda dans la region d'Qued Guir (Sahara oranais). *Arch. Inst. Pasteur Alger*,22 (3),171－174.

[11] Elias S.S. (1982). – Preliminary studies on salmonella microorganisms in camels in Egypt. M.V. Sc. Thesis,Faculty of Veterinary Medicine,Cairo University.

[12] El-Monla A. (1978). – Incidence of zoonotic disease (salmonellosis) encountered in animals slaughtered in Egypt. M.V. Sc. Thesis,Faculty of Veterinary Medicine,Cairo University.

[13] El-Nawawi F.,El-Derea H. & Sayed A. (1982). – Salmonellae in slaughter animals. *Arch f. Lebensmittelhygiene*,33,33－36.

[14] Farrag H. & El-Afify A. (1956). –*Salmonella* in apparently normal camels. *J. Egypt. Med. Ass.*,39,698－699.

[15] Faye B. (1997). – Guide de l'élevage du dromadaire. Sanofi Sante Nutrition Animale,La Ballastière,Libourne,France,115－116.

[16] Fazil M.A. & Hofmann R.R. (1981). –Haltung und Krankheiten des Kamels. *Tierärztl. Praxis,*9,389－402.

[17] Floyd T.M. (1955). –*Salmonella* in domestic animals and fowls in Egypt. *J. Egypt. Pub. Health Ass.*,30 (5),177－183.

[18] Gluecks I.V. (2007). – The prevalence of bacterial and protozoal intestinal pathogens in suckling camel calves in Northern Kenia. Thesis,Berlin,Institute of Animal and Environmental Hygiene,Faculty of Veterinary Medicine.

[19] Hamada S.,El-Sawah H.,Sherif I.,Joussef M. &Hidik M. (1963). –*Salmonella* of the mesenteric lymph nodes of slaughtered cattle buffaloes and camels. *J. Arab. Vet Med. Ass.*,23 (4),272－277.

[20] Kamel H. &Lofti Z. (1963). – Types of salmonella prevailing in apparently healthy camels slaughtered for meat. *In* Proceedings of the 4th Arab Annual Veterinary Congress,Cairo,109－114.

[21] Kowalesky M.J.M. (1912). – Le Chameau et ses maladies d'après les observations d'auteurs russes. *J. Méd. Vét. Zootechn.,Lyon*,15,462－466.

[22] Mahzounieh M.,Jafar M.P.,Zahraei Salehi T. &Nazari A. (2006). – Serological survey of antibodies to *Salmonella* groups A,B,C and D in camels of Iran. *In* Proceedings of the International Science Conference on Camels,12－14 May 2006,Qassim,Saudi Arabia,423－428.

[23] Malik P.D.,Datta S.K.,Singh I.P. &Kalra D.S. (1967). –*Salmonella* serotypes from camel in India. *J. Res. Punjab Agric. Univ. Ludhiana*,4,123－126.

[24] Mayr A. (1991). –Neue Erkenntnisse über Entwicklung,Aufbau und Funktion des Immunsystems. *Tierärztl. Praxis*,19,235－240.

[25] Millemann Y.,Evans S.,Cook A.,Sischo B.,Chazel M. &Buret Y. (2010). – Salmonellosis. *In* Infectious and parasitic diseases of livestock (Lefevre P-Ch.,Blancou J.,Chermette R. &Uilenberg G.,eds) Lavoisier,Paris,947－984.

[26] Mohamed M.E.H.,Hart C.A. &Kaaden O.-R. (1998). – Agents associated with camel diarrhea in Eastern Sudan. *In* Proceedings of the International Meeting on Camel Production and Future Perspectives,2–3 May 1998,Al Ain,UAE.

[27] Molla B.,Mohammed A. & Salah W. (2004). –*Salmonella* prevalence and distribution of serotypes in apparently healthy slaughtered camels (*Camelusdromedarius*) in Eastern Ethiopia. *Trop. Anim. Health. Prod.*,36,451–458.

[28] Moore J.E.,McCalmont M.,Jiru Xu,Nation G. Tinson A.H.,Crothers L. &Harron D.W. (2002). – Prevalence of faecal pathogens in calves of racing camels (*Camelusdromedarius*) in the United Arab Emirates. *Trop. Anim. Health. And Prod.*,34,283–287.

[29] Muench S. (2012). – Salmonellosis in animals in the United Arab Emirates. Thesis,Veterinary Faculty,Berlin,Germany.

[30] Nation G.,Moore J.E.,Tinson A.H.,MacAlmont M.,Murphy P.G. &Harron D. (1996). – Treatment of *Salmonella* and Campylobacter-associated diarrhoea in camels (*Camelusdromedarius*) with enrofloxacin. *In* European Congress of Chemotherapy,14–17 May 1996,Glasgow,Scotland.

[31] Nothelfer H.B.,Wernery U. & Akbar J. (1994). –Acral dry gangrene in a camel calf (*Camelusdromedarius*). *J. Camel Prac. and Res.*,1 (2),83–84.

[32] Nour-Mohammadzadeh F.,Sayed Z.B.,Hesaraki S.,Yadegari Z.,NaserAlidadi N. & Saeed SattariTabrizi S.S. (2010). –Septicaemic salmonellosis in a two-humped camel calf (*Camelusbactrianus*). *Trop. Anim. Health. Prod.*,42,1601–1604.

[33] Olitzki L. (1942). – Comparative studies on salmonella strains isolated in Palestine from camels and human beings. *J. Hyg. Camb.*,42,547–548.

[34] Olitzki L. &Ellenbogen V. (1943). – A *Salmonella* strain isolated from camels in Palestine. *J. Comp. Physiol. Ther.*,53 (1),75–79.

[35] Osman A.H.A.E.R. (1995). – Pathological study on intestinal affections in camels. Thesis,Faculty of Veterinary Medicine,Cairo University,1–128.

[36] Pegram R.G. &Tareke F. (1981). – Observation on the health of Afar livestock. *Ethiopian Vet. J.*,5,11–15.

[37] Pegram R.G. (1992). Camel salmonellosis in the horn of Africa. *In* Proceedings of the 1st International Camel Conference,2–6 February 1992,Dubai,UAE (Allen W.R.,Higgins A.J.,Mayhew I.G.,Snow D.H. & Wade J.F.,eds). R & W Publications,Newmarket,UK. 402.

[38] Ramadan F.M. &Sadek I.M. (1971). – Parameters of salmonellosis in Egypt. J. *Egypt. Vet. Med. Ass.*,31 (3–4),193–218.

[39] Refai M.,El-Said W.G.,Osman K.,Lotfi Z.,Safwat E.E. & Elias S. (1984). –*Salmonella* in slaughtered camels in Egypt.*Zagazig Vet. J.*,9,266–267.

[40] Rotimi V.O.,Jamal W.,Pal T.,Sonnerend A. & Albert M.J. (2008). – Emergence of multi-drug resistant *Salmonella* spp. and isolates with reduced susceptibility to ciprofloxacin in Kuwait and the United Arab Emirates. *Diagn. Microbiol. Infect. Dis.*,60,71–77.

[41] Rulofson F.C.,Atuill E.R. & Holmberg C.A. (2001). –Faecal shedding of *Giardia duodenalis,Cryptosporidium parvum,Salmonella*organisms,and *Escherichia coli* 0157:HF from llamas in California. *Am. J. Vet. Res.*,62 (4),637–641.

[42] Salih O.,Shigidi M.T.,Mohamed H.O.,McDoungh M. & Chang Y. (1998a). – The bacterial causes of camel-calf (*Camelus dromedarius*) diarrhoea in Eastern Sudan. *In* Proceedings of the International Meeting on Camel Production and Future Perspectives,2–3 May 1998,Al Ain,UAE,132–137.

[43] Salih O.,Mohamed H.O. &Shigidi M.T. (1998b). – The epidemiological factors associated with camel-calf diarrhoea ineastern Sudan. *In* Proceedings of the International Meeting on Camel Production and Future Perspectives,2–3 May 1998,Al Ain,UAE.

[44] Sandiford B.R.,El Gheriany M.G.,Abdul Ela M. &Keram A.M. (1943). – Food poisoning outbreaks in Egypt associated with *Bacterium aertrycke. Lab. Med. Prog.*,4 (1),14–18.

[45] Sandiford B.R. (1944). – Food poisoning due to *Bacterium typhimurium*(anaerogenes). *J. Path. Bact.*,56,254–255.

[46] Saulez M.N.,Cebra Chr.K. & Valentine B.A. (2004). – Necrotizing hepatitis with enteric salmonellosis in an alpaca. *Can.Vet. J.*,45,321–322.

[47] Sayed A.I.H. (1979). – Studies on *Salmonella* infection in apparently healthy slaughtered animals. M.V. Sc. Thesis,Faculty of Veterinary Medicine,Cairo University.

1

细菌病

[48] Selim A.M. (1990). – Salmonellosis in camel in Egypt. In Proceedings of the International Conference on Camel Production and Improvement,10－13 December,Tobruk,Libya.

[49] Sinkovics G. (1972). – Quantitative changes of clostridia in the intestine of early-weaned pigs diseased with coli-enterotoxemia.*Acta vet. hung*.,22 (21),133－139.

[50] Wernery U.,Seifert H.S.H.,Billah A.M. & Ali M. (1991). – Predisposing factors in enterotoxemias of camels (*Camelus dromedarius*) caused by *Clostridium perfringens*type A. *Rev. Elev. Méd. Vét. Pays Trop*.,44 (2),147－152.

[51] Wernery U. (1992). – The prevalence of *Salmonella* infections in camels (*Camelus dromedarius*) in the United Arab Emirates. *Br. Vet. J*.,148 (5),445－450.

[52] Wernery U. &Makarem E.H. (1996). – Comparative study on *Salmonella* serovars isolated from humans and camels in the United Arab Emirates. *Camel Newslett*er,12 (9),55－59.

[53] Whitehead C.E. & Anderson D.E. (2006) – Neonatal diarrhea in llamas and alpacas. *Small Rumin. Res*.,61,207－215.

[54] Yassien N.A. (1985). –*Salmonella* in slaughtered camels. M.V. Sc. Thesis,Faculty of Veterinary Medicine,Cairo University.

[55] Zaki O.A. (1956). – The incidence of *Salmonella* infections in camels. *J. Egypt. Publ. Health Ass*.,31 (2),75－79.

（李有全译，殷宏校）

## 1.2.2 大肠杆菌病

大肠杆菌是正常肠道菌群的重要组成部分，但也导致了多种疾病，带来了大量的经济损失，尤其是对幼龄动物危害严重。源自动物的有致病性的大肠杆菌又称动物致病性或人畜共患大肠杆菌（APZEC），发病时常见临床症状见表28。

大肠杆菌病和大肠杆菌性败血症在旧大陆骆驼和新大陆骆驼都有报道[3,7,10]。

### 病原学

像沙门氏菌一样，大肠杆菌是革兰氏阴性棒状的直杆菌，属于肠杆菌科。它是普通的肠道菌，在有氧的营养培养基上很容易培养。根据在细胞壁、荚膜、鞭毛上的抗原不同，大肠杆菌被分类成接近200种血清型。仅有一小部分的大肠杆菌血清型具有致病性，被划分为致病性大肠杆菌。

根据大肠杆菌的毒力因子和造成的临床表现，家畜中的重要的大肠杆菌病分类如下。

(a) 肠产毒素性大肠杆菌（ETEC）：具有菌毛黏附素K88/K99以及其他的黏附素，导致新生幼畜大肠杆菌病。

(b) 肠致病性大肠杆菌（EPEC）：不具有肠毒性，但可导致腹泻。

(c) 产志贺氏毒素大肠埃希菌（STEC）或产非洲绿猴肾细胞毒素大肠杆菌（VTEC）。

(d) 肠外致病性大肠杆菌：包括导致败血性大肠杆菌和对鸟类有致病性的大肠杆菌。

(e) 非败血性大肠杆菌。

这五种致病型都具有大肠杆菌的基本特征，能与肠黏膜发生特异交互作用，产生多种毒素，拥有编码毒性因子的质粒。大肠杆菌抗原成分复杂（K：荚膜抗原；O：菌体细胞壁抗原；H：鞭毛抗原；F：菌毛抗原)，用于血清学分类。大肠杆菌中含有一个或多个质粒，编码多个毒力因子以及抗生素抗性。

### 公共卫生意义

大肠杆菌导致的人和动物疫病通常和产志贺样毒素大肠埃希菌（STEC）有关，呈广泛的发生。引起人感染的最常见的产志贺样毒素大肠杆菌为$O_{157}$：$H_7$血清型，但其他的血清型也能引起人的带血或不带血腹泻。一些产志贺样毒素大肠埃希菌属于$O_{157}$：$H_7$血清型驻留在反刍动物的肠道内，在人摄入这些细菌污染的食物或水之后，导致出血性肠炎和溶血性尿毒综合征。世界各地都观察到，分离的菌株对一些抗生素的抗性逐渐增强，因此抗生素的敏感度试验非常重要。

### 流行病学和临床症状

大肠杆菌是所有哺乳动物肠道中某些部位的常在菌，并且从粪便中排泄出来。在饮水和食物存

表28 致病性大肠杆菌引起动物的常见病及菌型

| 动物种类 | 疾病 | 致病型 |
| --- | --- | --- |
| 犊牛 | 犊牛腹泻 | ETEC |
| 羔羊 | 出血性胃肠炎 | STEC |
| 小山羊 | 败血症 | EXPEC |
| 新大陆骆驼幼仔 | 腹泻，脑膜炎 | STEC，EXPEC |
| 旧大陆骆驼犊 | 腹泻，脑膜炎 | ETEC，EPEC |
| 猪 | 仔猪腹泻， | ETEC |
| | 杆菌性断奶仔猪腹泻，断奶仔猪腹泻 | ETEC，EPEC |
| | 猪水肿病 | STEC |
| | 脑膜炎 | EXPEC |
| 牛 | 尿道炎 | 大肠杆菌，APZEC |
| | 膀胱炎 | |
| | 肾盂肾炎 | |
| | 流产 | |
| | 产后子宫内膜炎，子宫内膜炎 | |
| | 耳炎 | |
| | 眼睛感染，角膜结膜炎 | |
| | 关节病 | |
| | 乳房炎（母羊） | |
| | 乳房炎子宫炎无乳综合征（母猪） | |
| 旧大陆骆驼 | 腹泻，肠毒血症 | |

ETEC：肠产毒素大肠杆菌；STEC：产志贺氏毒素大肠杆菌；EXPEC：肠道外致病性大肠杆菌；EPEC：肠道致病性大肠杆菌；APZEC：动物致病性和人畜共患性大肠杆菌

在大肠杆菌中被认为是被粪便污染的的证据。年幼的动物感染大肠杆菌通常与饲养管理不当有关，幼驼的感染通常与沙门氏菌、梭菌、巴氏杆菌、轮状病毒和冠状病毒以及隐形孢子菌的感染有关。临床症状因血清型不同而异。产肠毒素的大肠杆菌菌株能够产生肠毒素，并且能引起肠炎和脱水，而且是农场中动物大肠杆菌腹泻的最常见原因。一些菌株，特别是致病性菌株，能引起溶血[2]。致病性大肠杆菌菌株拥有不同毒力的致病因子，其致病能力因而有差异。这些毒力因子包括内毒素、蛋白黏附因子、$\alpha$和$\beta$溶血素、荚膜膜多糖、热不稳定和热稳定性内毒素和细胞毒素。然而，关于骆驼科中分离大肠杆菌菌株血清型的研究工作开展的很少。最近Cid等[5]用PCR对从秘鲁的安第斯山地区6个养殖场的94头幼驼或羊驼中分离的94株大肠杆菌菌株毒力因子进行了分析。在所有畜群中都检测到了腹泻型大肠杆菌，其中2株（2.13%）属于肠致病性大肠杆菌，10.64%属于非典型肠致病性大肠杆菌，11.7%属于产肠毒素大肠杆菌。

在骆驼饲养场中，大肠杆菌病或大肠杆菌败血症多发于幼龄骆驼，给骆驼饲养场造成巨大的经济损失（可达40%）[14]。Schwartz和Dioli[18]报道在非洲西部的新生单峰驼的发病率为30%。如果治疗不及时，发病驼的死亡率可达到100%。作者认为，畜群饲养环境差、饮用污染的水以及初乳摄入不足均可导致该病的发生。小骆驼的症状为痢疾、腹痛、厌食、脱水，几天之内就死亡。Chauhanetal[3]报道两例单峰驼幼驼的大肠杆菌病，症状为痢疾、发热、全身不适和厌食。作者从发病动物的粪便中分离到了O83血清型大肠杆菌。Rombol[15]也描述了具有严重腹泻的新生单峰驼的大肠杆菌感染。Mohamed 等[14]在苏丹检测了42个

1～3个月龄的单峰驼，发现28%（12/42）感染了致病性大肠杆菌，分离12个菌株中，5个被鉴定为产肠毒素性大肠杆菌，2个被鉴定为肠致病性大肠杆菌，1个被鉴定为产非洲绿猴肾细胞毒素大肠杆菌。Salih等[16,17]从106个腹泻骆驼病料中分离69个具有不同黏附因子的大肠杆菌（K88，F41）。

在肯尼亚北部的幼龄单峰驼（12周龄以内）的腹泻综合征中，大肠杆菌并不是主要病原体，在数百个粪便样品[8]中，未发现O157：H7血清型大肠杆菌。在531个大肠杆菌分离株中，对255株分析了与毒力有关的基因。在75个分离株中检测到毒力相关的基因，不过这些菌株对单峰驼幼驼的致病性没有差异。而且，从健康的骆驼中分离到的大肠杆菌菌株中也发现了同样的毒力因子。

在阿联酋的一些种用单峰驼群里，每年都规律性的发生大肠杆菌感染。在刚出生的骆驼没有临床症状，仅仅发生在2～4周龄的骆驼。对一些种用群造成了巨大的损失。肠杆菌痢疾似乎与最初食用固体食物和沙子有关，这和年轻骆驼的梭菌肠毒血症相似。感染的动物出现淡黄水样腹泻（图78），它们的后腿和尾巴带有干燥的排泄物，由于脱水，眼睛深深的陷入眼眶中。从胃肠道和大部分器官中分离出溶血性大肠杆菌。

在巴林岛的单峰驼中，报道了极少见的由O139血清型溶血性大肠杆菌引起的大肠杆菌肠毒血症的病例[10]。在不同的骆驼饲养中心，成年的种用骆驼中有散发病例，发病率超过50%，死亡率却达到90%。骆驼表现为腹部肿胀，腹腔膨胀，积液达100～150L。眼睑、喉、耳朵和前额水肿。从肠道内容物和腹水中分离出溶血性大肠杆菌。

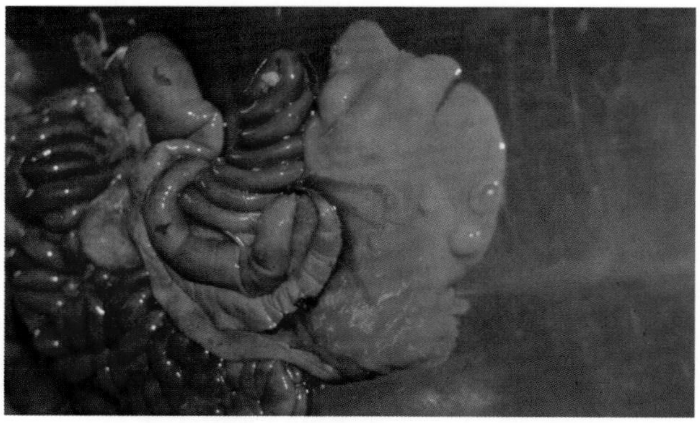

图78　患有大肠杆菌病的3周龄单峰驼，伴有淡黄色的腹泻

Strauss[19]和Fowler[7]指出，幼龄新大陆骆驼的大肠杆菌病是一种重要的疾病，主要发生在营养不良或在出生时没有得到足够母乳的幼驼。在美国，新大陆骆驼的新生幼驼大肠杆菌败血症是一种严重的疾病，重要性位于子宫炎、乳腺炎、脓肿之前。感染的动物症状是急剧腹泻、体重减轻、腹胀和虚弱。Haenichen和Wiesner[9]也描述了动物园中患有严重脑膜炎的羊驼的大肠杆菌败血症。

Mercado 等[13]综述了与美国南部骆驼感染大肠杆菌相关的一些报道，包括一例年轻原驼自然感染大肠杆菌引发的腹泻。在这些动物的采样中，培养出了产生志贺毒素的大肠杆菌。从粪便样品中分离得到的大肠杆菌血清型为O26：H11。表达上皮细胞黏附和抹平（eae）基因和志贺毒素(stx)基因的菌落，体外实验证明能够定植黏附在Hep-2细胞上，并且产生大肠杆菌毒素。

Adams和 Garry[1]、 Whitehead 和Anderson[21]以及 Foster等[6]也报道了在北美和英国的新大陆骆驼中肠型大肠杆菌病引发的败血症。Foster等[6]观察了3例由黏附、入侵型大肠杆菌引起的羊驼的肠损伤。这些大肠杆菌定植到人和动物的红细胞，形成黏附-抹平损伤（AE损伤），在组织切片中清晰可见。在这些病例中，黏附和抹平（eae）基因经常存在。通常情况下，大肠杆菌病常伴发有隐孢子虫病或梭菌感染。

## 病理学

骆驼科的动物患有大肠杆菌病和大肠杆菌败血症时临床出现厌食、虚弱、发热和排淡黄水样便[4]，主要危害6月龄以内的小骆驼。大肠杆菌败血症通常是肠道大肠杆菌病进一步恶化的结果，但也可发生在没有明显肠道症状的动物中。在肠大肠杆菌病和大肠杆菌败血症中，损伤都是非特异性的。大肠型大肠杆菌感染的骆驼科动物会脱水，因为腹泻，后腿及臀部和尾巴粘着有粪便（图79）。尸检可发现小肠充血并伴有卡他性肠炎，胃容物从灰白色到黄绿色，小肠膨胀并充满气体和液体，肠系膜淋巴结水肿。在大肠杆菌败血症中，全身充血，腹膜淤血，脑膜水肿。

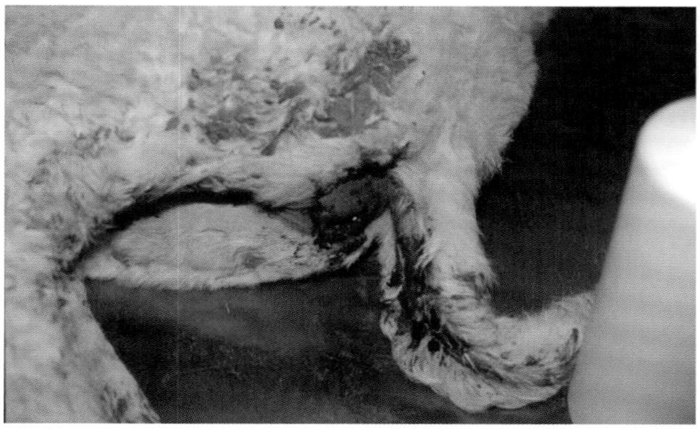

图79 单峰驼大肠杆菌病：后腿和尾巴粘有粪便

年幼的单峰骆驼会发热，体温在40～41℃之间。在2～3天内死亡。尸检时，整个尸体极度苍白（图80），小肠黏膜有炎症反应，通常会在胃内发现不同数量的沙子，特别是在第一胃（图81）。胃肠道内容物都是灰色的，并且混合着刺鼻、污秽的气味。在严重大肠杆菌败血症病例中，腹部器官覆盖有纤维蛋白分泌物（图82）。

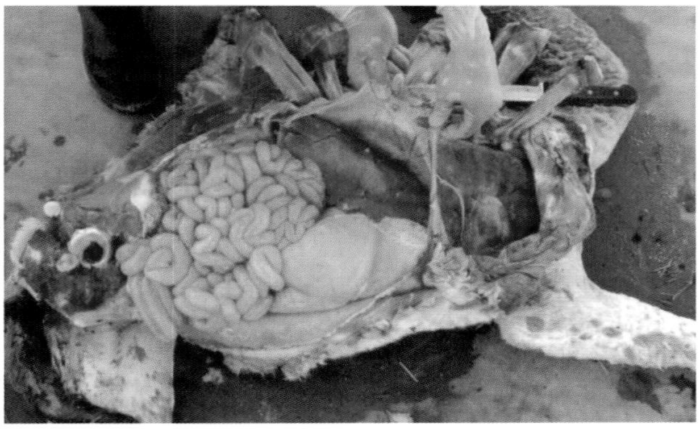

图80 患有大肠杆菌败血症的年轻单峰驼极度苍白

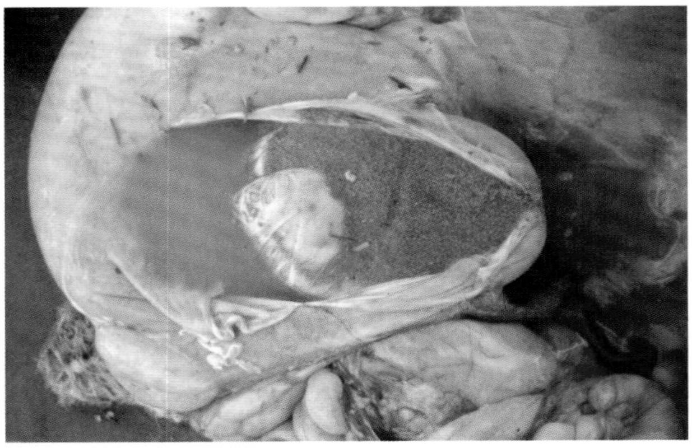

图81 患有大肠杆菌性痢疾的年轻单峰驼的第一胃内有的沙子

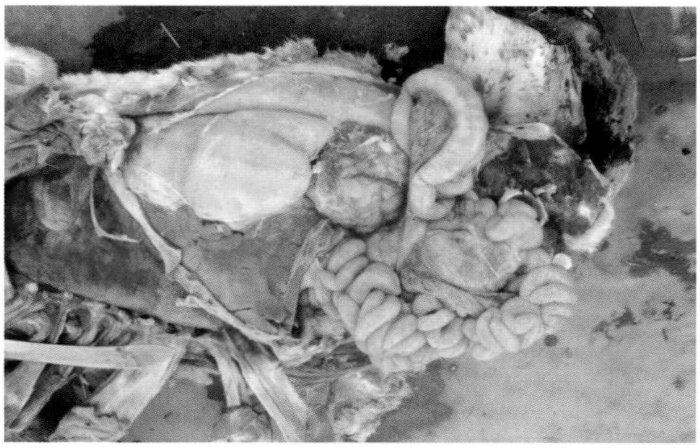

图82 患有大肠杆菌败血症的年轻单峰驼的纤维性渗出覆盖内脏器官

## 诊断

大肠杆菌病的临床症状与轮状病毒、冠状病毒、沙门氏菌、球虫感染的临床症状相似，因此，诊断时要进行微生物学检测。用于细菌学检测样品应包含肠道、肠系膜淋巴结以及不同器官。所有的组织样品应在死后不久无菌采集，大肠杆菌在大部分琼脂培养基上生长良好。需要与肠杆菌科其他的细菌区别时可使用选择性培养基。在一些疾病和个别动物疾病诊断中，某一血清型出现的频率很高。骆驼大肠杆菌的血清型鉴定刚刚起步。在确定大肠杆菌的血清型时，可使用血清定型、毒株定型、肠毒素或细胞毒素的检测以及分子生物学技术，例如PCR检测等方法。

许多实验室都可开展血清定型，以悬浮培养的活菌作为抗原，系列的OK抗血清，采用玻片凝集试验确定。需要提醒的是，并不是所有的大肠杆菌都能在在血平板上产生溶血。$F_4$（K88)的菌落和F18产肠毒素性大肠杆菌几乎都是溶血性的，而F5（K99）型和其他产肠毒性大肠杆菌相关的菌株不溶血。

## 防治

大肠杆菌导致的骆驼腹泻应该采用与沙门氏菌病相同的治疗方案。应口服或肠外补充电解质，以恢复体液平衡，因为死亡通常是由脱水引起的。同时，应限制奶的摄入。但是，在出生的前几个小时，应保证初乳的摄入，以便最大限度地吸收免疫球蛋白。为应对紧急情况，应该建立初乳库。

抗生素治疗：支持疗法通常是一种重要的附

属疗法。在许多病例中，当抗菌治疗效果不佳时，这也是一种可选择的治疗方案。在抗生素治疗之前，要对导致的疫病暴发的大肠杆菌进行药敏测试，而且应该牢记很多大肠杆菌菌株对抗生素有很强的耐药性。大肠杆菌性败血症必须用可注射的抗菌药治疗，如，恩氟沙星、甲氧苄氨嘧啶—磺胺类药、卡那霉素、黏菌素[12]。羊驼和美洲驼的大肠杆菌治疗应遵循以下规则[21]。

(a) 青霉素钾每千克体重22000IU，静脉注射，每6小时一次。

(b) 庆大霉素每千克体重5mg，静脉注射，每天一次，注射5天。

(c) 静脉输液，0.45%氯化钠和2.5%葡萄糖。

为了降低幼龄单峰驼的死亡，应每年给母驼注射畜群特异的自家大肠杆菌疫苗，或者给年幼的骆驼口服疫苗进行免疫。给多价细菌苗注射后的牛补充硒似乎能够增强抗体的特异反应。口服特异的大肠杆菌抗体或口服免疫了致病性大肠杆菌菌株的母鸡产的蛋黄粉，似乎能够阻止大肠杆菌黏附到肠黏液细胞的细胞膜上，而且能够降低临床症状的严重性和死亡率。

由于大肠杆菌的存在，对于集中饲养的动物，例如，在公园或动物园，损失会增加。鉴于这个原因，Strauss[19]应用了一种保护性的母体免疫策略，在预产期前第8周和第4周对怀孕母畜分两次注射疫苗。因此，年幼动物的死亡大大降低。

## 参考文献

[1] Adams R. & Garry F.B. (1992). – Gram-negative bacterial infection in neonatal New World Camelids: six cases (1985—1991). *J. Am. Vet. Med. Ars*.,201,1419－1424.

[2] Bisping W. &Amtsberg G. (1988). –Colour atlas for the diagnosis of bacterial pathogens in animals. Verlag Paul Parey,Berlin und Hamburg,160－169.

[3] Chauhan R.S.,Kaushik R.K.,Gupta S.C.,Satiya K.C. &Kulshreshta R.C. (1986). – Prevalence of different diseases in camels(*Camelusdromedarius*) in India. *Camel Newslett*er,3,10－14.

[4] Chauhan R.S.,Satija K.O.,Tika Ram S.M. & Kaushik R.K. (1987). – Diseases of camels and their control. *Indian Farming*,36,27－31.

[5] Cid D.,Martin-Espada C.,Maturrano L.,Garcia A.,Luna L. &Rosadio R. (2010). –Diarrheagenic *Escherichia coli* strains isolated from neonatal peruvian alpacas (*Vicugna pacos*) with diarrhea. 5th European Symposium on South American Camelids and 1st European Meeting on Fibre Animals,6－8 October,Sevilla,Spain.

[6] Foster A.P.,Otter A.,Barlow A.M.,Pearson G.R.,Woodward M.J. & Higgins R.J. (2008). – Naturally occurring intestinal lesions in three alpacas (*Vicugna pacos*) caused by attaching and effacing *Escherichia coli*. *Vet. Rec*.,162,318－320.

[7] Fowler M.E. (2010). – Medicine and Surgery of Camelids. 3rd Ed. Wiley-Blackwell,Aimes,Iowa,212－213.

[8] Gluecks I.V. (2007). – The prevalence of bacterial and protozoal intestinal pathogens in suckling camel calves in Northern Kenia. Thesis,Institute of Animal and Environmental Hygiene,Faculty of Veterinary Medicine,Berlin.

[9] Haenichen T. &Wiesner H. (1995). –Erkrankungs- und Todesursachen bei Neuweltkameliden. *Tierärtztl. Praxis*,23,515－520.

[10] Ibrahim A.M.,Abdelghaffar A.A.,Fadlalla M.E.,Nayel M.N.,Ibrahim B.A. & Adam A.S. (1998). –Oedema disease in female camels (*Camelus dromedarius*) in Bahrain. *JCPR*,5 (1),167－169.

[11] Jin Y.C. (1985). – Report on the diagnosis and treatment of colibacillosis in camels. *Acta Agricultural College, Yanbian*,2,78－82.

[12] Manefield G.W. &Tinson A. (1996). – Camels. A compendium. *The TG Hungerford Vade Mecum series for domestic animals*,series C,No. 22.

[13] Mercado E.C.,Rodriguez S.M.,Elizondo A.M.,Marcoppido G. &Parreno V. (2004). – Isolation of shiga toxin-producing*Escherichia coli* from a South American camelid (*Lama guanicoe*) with diarrhoea. *J. Clin. Microbiol*.,42 (10),4809－4811.

[14] Mohamed M.E.H.,Hart C.A. &Kaaden O.-R. (1998). – Agents associated with camel diarrhea in Eastern Sudan. *In* Proceedings of the International Meeting on Camel Production and Future Perspectives,2－3 May 1998,Al Ain,UAE.

[15] Rombol B. (1942). – Enzootic bacterium coli infection in new-born camels. *Nuova Vet.*,20,85–93.

[16] Salih O.S.M.,Shigidi M.T.,Mohammed H.O.,McDoungh P. & Chang Y.F. (1997). – Bacteria isolated from camel-calves (*Camelus dromedarius*) with diarrhea. *Camel Newslett*er,13 (9),34–43.

[17] Salih O.,Shigidi M.T.,Mohamed H.O.,McDoungh& Chang Y. (1998). – The bacterial causes of camel-calf (*Camelus dormedarius*) diarrhoea in Eastern Sudan. *Proc. 3rd Annual Meeting Animal Prod. Under arid conditions*,2,*UAE*,132–137.

[18] Schwartz H.J. &Dioli M. (1992). – The one-humped camel in Eastern Africa. A pictorial guide to diseases,health care and management. Verlag Josef Margraf,Weikersheim,Germany.

[19] Strauss G. (1991). – Erkrankungen junger Neuweltkamele imTierpark Berlin –Friedrichsfelde *In* Proc. 11. Arbeitstagung der Zootierärzte im deutschsprachigen Raum. 1–3 November,Stuttgart,Germany,80–83.

[20] Wernery U. &Kaaden O.-R. (2002). – Infectious diseases in camelids. 2nd Ed. Blackwell Science Berlin,Vienna. Austria,78–83.

[21] Whitehead C.E. & Anderson D.E. (2006). – Neonatal diarrhoea in llamas and alpacas. *Small Rumin. Res.*,61,207–215

## 深入阅读材料

Abubakr M.I.,Nayel M.N.,Fadlalla M.E.,Abdelrahman A.O.,Abuobeida S.A. &Algabara Y.M. (1999). – The incidence of bacterial infection in young camels with reference to *Escherichia coli*. *In* International Workshop on the Young Camel,Quarzazate,Morocco,24–26 October,40.

Bornstein S.,Younan M. & Feinstein R. (1999). – A case of neonatal camel colibacillosis. *In* International Workshop on the Young Camel,Quarzazate,Morocco,24–26 October,23.

Dia M.L.,Diop A.,Ahmed O. Mohamed,Diop C. & El Hacen O. Taleb (1999). –Diarrhees du chamelon en Mauritanie: Résultats d’enquête CNERV. *In* International Workshop on the Young Camel,Quarzazate,Morocco,24–26 October,41.

Kane Y. & Diallo B.C. (1999). –Donneessur la pathologie du chamelon en Mauritanie. *In* International Workshop on the Young Camel,Quarzazate,Morocco,24–26 October,42.

Saad M.A.G. & Hussein A.M. (1975). – Isolation of *salmonella*e from camels in Sudan. *In* Sudan Veterinary Association,7th Conference on Meat Industry,Khartoum,Sudan.

Salih O.S.M. (1993). –Aetiological and epidemiological factors associated with camel calf diarrhoea. Thesis,Faculty Veterinary Science,University Khartoum,Sudan.

Younan M. & Bornstein S. (1999). –Colisepticemia in a camel calf. *In* International Workshop on the Young Camel,Quarzazate,Morocco,24–26 October,77.

（李有全译，殷宏校）

## 1.2.3 副结核病

副结核病的临床特征是持续并日益加剧的腹泻、体重减轻、精神萎靡，最终死亡。该病可以致使感染的绵羊、山羊、家养鹿及其他一些家养动物和野生的反刍动物患有慢性的、不可治愈的、并且具有感染性的肠炎。在旧大陆骆驼[7,40]和新大陆骆驼均有该病的报道[3,5,30,33]。

副结核病的发生是世界性的，但该病不是必须申报的疾病，因此，其流行范围被低估了。在热带集约化的奶制品农场，副结核病造成了很严重的经济损失。该病被认为是危害牛羊养殖业最严重的细菌性疾病之一，对于骆驼养殖业也有一定的经济影响。副结核病的病原体是禽分枝杆菌副结核亚种（MAP），它随感染动物的粪便排出体外，可随污染的食物和饮水而被健康动物摄入。病原体定植于肠黏膜和肠系膜淋巴结处并引发慢性炎症。禽分枝杆菌副结核亚种（MAP）可以通过胎盘病传染给胎儿。根据Boschiroli和Thorel的研究，29%的感染动物因临床症状严重被宰杀，70%的感染动物由于继发感染（15%乳腺炎，17%不孕症，17%其他疾病）被宰杀。

### 公共卫生意义

副结核病已在全世界范围内影响到了奶制品产业，该病原在人类克罗恩疾病中的潜在致病作用并未得到解决，巴氏灭菌法并不能使奶制品中的病原彻底灭活，这将会导致人类肠炎症状的疾病发生。有关禽分枝杆菌副结核亚种（MAP）的研究表明，该病原在巴氏灭菌（72℃ 15秒）的商业化奶制品中仍然可以存活。

如果禽分枝杆菌副结核亚种（MAP）和人类的克罗恩疾病之间确实存在联系，那么将会出现公共卫生的问题。最近，从一个患有克罗恩疾病的病人的外周血中分离出来具有活性的禽分枝杆菌副结核亚种的细菌，进一步说明禽分枝杆菌副结核亚种可能是引起人类的克罗恩疾病的病原。虽然动物的副结～核病和人类的克罗恩疾病在病原和临床症状上存在相似性，但它们之间也确实存在不同之处。

### 病原学

禽分枝杆菌副结核亚种（MAP）是一种无鞭毛、无芽孢、需氧的氧化性细菌，因为其细胞壁上有丰富的油脂和分枝杆菌酸，不易被革兰氏染料染色。MAP是抗酸性的，其最适宜染料是萋–尼氏（Zichl-Neelsen）抗酸染色。该病原属于禽分枝杆菌属，在土壤和肥料中可存活一年之久。该疾病可通过聚合酶链反应（PCR）、血清学及致敏反应检测细菌而确诊。Ghosh等[16]发现分离于单峰驼的病原属于绵羊谱系，表明了该病原可能由感染的绵羊传播给单峰驼。

### 流行病学和临床症状

关于骆驼副结核病的流行病学资料少之又少，到目前为止，并没有证据说明骆驼可以将MAP传播给其他动物。MAP随粪便排出，也存在于肠黏膜的巨噬细胞和肠系膜淋巴结。当该病的病原入侵时，细胞介导的免疫应答将会对其作出反应。并不是所有的感染动物都会出现临床症状，但是，它们会持续排泄MAP。MAP经口腔感染后，通过扁桃体和肠黏膜进入淋巴循环。集合淋巴结在肠的内腔收集细菌并通过肠黏膜运送。该病的潜伏期一般为18～24月。

在前苏联，副结核病是最重要和分布最广泛的双峰驼细菌病之一[7,12,20,27]。双峰驼一般比单峰驼更容易感染，这很可能与双峰驼生活环境有关，它们经常与牛圈在一起。内蒙古患有腹泻的双峰驼被证实患有该病，而且自1949年起土库曼斯坦也有该疾病的报道[47]。Strogov[38] 和Buchnev 等[7]曾报道刚断奶的4岁龄的双峰驼更易感染该病。该作者认为患该病的老龄双峰驼可以痊愈。成年感染骆驼痊愈过程较慢，一般需要6个月。从1946—1952年，土库曼斯坦地区双峰驼的年度感染率在0.3%～1.5%。

副结核病也见于单峰驼，但是，发病数量远小于双峰驼，主要与单峰驼所处的环境有关。在印度，Chauhan 等[9]发现105头单峰驼中，其中，只有4头（3.8%）患有副结核疾病。这些单峰驼患有难以治愈的腹泻，并且在一头单峰驼的直肠上皮细胞的活检中发现有抗酸性杆菌存在。这4头骆驼都进行了皮试试验（Johnin），结果都呈现阳性。美国动物园内的一头雌性单峰驼也被报道患有副结核疾病[4]。该单峰驼精神萎靡且瘦弱，并且粪便较稀，有红血丝和黏液；血红蛋白为27 g/L（正常范围为57～75 g/L），明显降低；并检测到有棒状抗酸块存在，而皮试试验和结合菌素检测却显示阴性。Gameel等[15]的研究表明，牛的副结核病可以传染给单峰驼。在伊朗[39]和阿联酋[23]被宰杀的骆驼也被报道患有副结核疾病。在伊朗，单峰驼排泄物被萋–尼氏抗酸染色后检验，阳性率为5.4%。Kinne等[23]的研究称迪拜酋长国地区的具有炎性反应的全身性副结核疾病造成的损伤并不局限于肠道内，还累及脾脏、肝脏及肺部，该病在沙特阿拉伯的单峰驼中也有报道。

动物园的很多动物、自由放养的反刍动物及美洲驼都被报道患有副结核病。前段时间澳大利亚的羊驼被报道患有该病，爱丁堡公园中的美洲驼及美国的其他地区也有该病的报道。在美洲地区，副结核的临床症状多种多样，一些动物表现为日益加剧的腹泻、虚弱和消瘦，6～10天之后死亡。一些羊驼病例发展为虚弱和体重减轻，最终持续3个月的腹泻。另外，有一些美洲驼病例则只表现为虚弱和体重减轻而没有表现出腹泻。所有的感染病例均表现出血液蛋白不足，这也许可以作为一种诊断标准[34]，因为在绵羊上已经按照此法去做。与对照组动物相比，患病的奶牛不仅血液蛋白不足，并且包括血细胞压积、血红蛋白和血细胞计数等值都较低。血清中的铁含量及铁结合能力明显降低。

## 病理学

Strogov[38]、Ivanov和Skaliriskil[20]、Guake等[17]描述了患病的双峰驼的病理损伤；Amanda[4]、Radwan等[29]及Kinner等[23]描述了患病单峰驼的病理损伤；Belknap等[5]、Ridge等[30]和Stehman等[36]描述了患病美洲驼的病理损伤。

俄罗斯一位学者认为，双峰驼的病理变化比牛的病理变化明显。回肠、盲肠和结肠均有病例损伤，并伴有肝、脾和淋巴结炎症。感染动物在腹泻开始后4～6周内死亡。Amanda[4]进行尸体剖检，结果显示肠壁增厚及肠系膜淋巴结增大。Radwan等[29]首次在沙特阿拉伯地区报道了该病，在3000头单峰驼的牧场里发现了60例副结核病例。这些单峰驼大多是2～4周岁，伴有体重下降和慢性间歇性的腹泻，但并没有发热现象。尽管进行了抗生素治疗，但是患病动物还是在出现临床症状后1～4个月内死亡。组织活检时发现回肠、盲肠和结肠壁均有增厚。肠淋巴结增生，肠内和淋巴结及粪便中均发现有抗酸性杆菌。

迪拜的一头患病单峰驼严重消瘦，病理特征明显，肠系膜淋巴结增生（图83）。小肠上有淡灰色的块状突起（图84），结肠黏膜有弥散增生（图85），有轻微的脾肿大，并且有很多灰色结节（直径达10 mm，图86），肝肿大，并有灰色间质增生，且有间质性肺炎病灶（图87和88）。

组织病理学检测显示，弥散性的小肠结节炎增生（图89），黏膜和肠系膜淋巴结上有大量的含有抗酸杆状物的巨噬细胞[23]。肝脏检测有严重的间质性肝炎，含有抗酸杆状物的巨噬细胞群（图90）。肺部表现为间质性肺炎病灶，且支气管周围有含有抗酸杆状物的巨噬细胞群渗入（图91）。自然感染的病例的外肠道损害较少。Hazlett 等[19]研究表明18个月大的患病小型驴子有间质性肝炎，并且胞浆内有大量的抗酸性杆菌。而在一些鹿群中，则倾向于系统感染，Stehman等[37]发现患病的小鹿容易感染包括肠、肝、混合淋巴结在内的组织，肺部也偶尔会有感染。Deutz等[11]从猎获的马鹿的肝部和肺部肉芽肿分离出了MAP。

羊驼的病理学变化与新大陆骆驼相似。患病动物瘦弱，肠内病变明显，肠系膜淋巴结水肿并增生。淋巴结组织切片包含有抗酸性菌落。在动物的空肠、回盲肠的连接处及最接近大肠的位置可见组织增生，分离到MAP病原体的肝、肺及腹膜绒毛膜的淋巴结处有肉芽肿出现。

## 诊断

副结核病可以通过分离培养病原、致敏试验及血清学检测来诊断。从粪便中分离培养微生物，是敏感性和特异性最高MAP病原的检测方法，但该方法需要16周才能得到检测结果。MAP有多种菌株，羊源菌株的培养与骆驼源菌株的培养方法并不相同。骆驼源按照牛源的培养方法得到较好的结果。活检组织肠黏膜及排泄物中分离的菌株经ZN染料染色后镜检可看到大量的MAP病原。然而粪便检测只能检测到症状不明显的排泄物中25%的病原。

从感染器官中分离出来的分枝杆菌，净化和浓缩后需接种至Herrold卵黄斜面培养基和LJ（Loewenstein-Jensen）斜面培养基。Herrold卵黄培养基和LJ培养基接种后要置于37℃培养6周，每周检查生长情况。可疑菌落经ZN染色后检测抗酸性。对于迪拜的那头骆驼病例，从大小肠、肝脏、脾脏和肠内淋巴结分离的分枝杆菌菌落非常小（针尖大小）。培养6周后，Herrold卵黄培养基上有无色湿润菌落出现，但LJ培养基上没有。

禽结核菌素和MAP制备的副结核菌素皮试试验结果不太理想[19]。在俄罗斯，有40%的双峰驼进行禽结核菌素皮试试验后阳性反应。然而，600头皮试反应呈阳性的骆驼的排泄物中并没有分离出抗酸性杆菌。另外一项调查表明，对皮试反应呈强烈阳性的7头双峰驼进行尸体剖检，并没有副结核的典型病理变化。

已建立的血清学方法中，补体结合试验用于检测副结核病获得良好结果[21,22,32]。Burgemeister等[8]

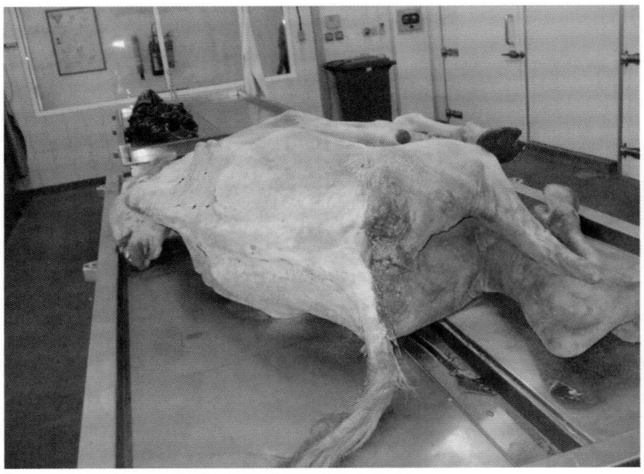

图83 患有副结核病的严重瘦弱单峰骆驼

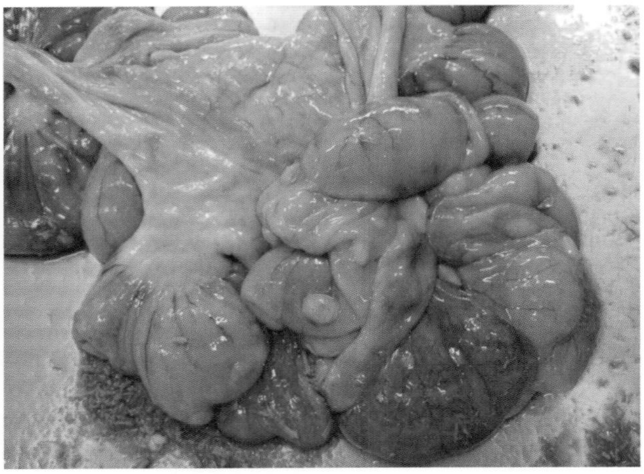

图84 由副结核引起的在整个小肠上明显的浅灰色Peyer结节（派尔结节）

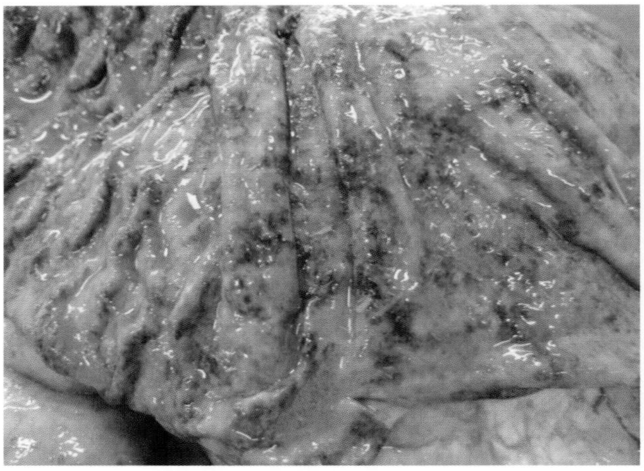

图85 由副结核引起的结肠黏膜的弥漫性增厚

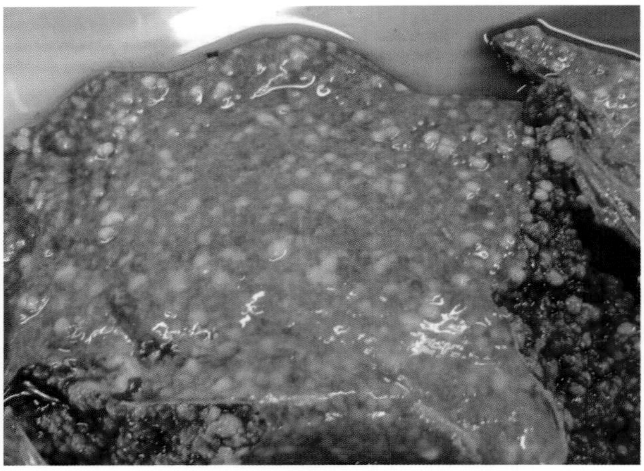

图86 由副结核引起的严重的脾肿大并带有众多直径达10mm的灰色结节

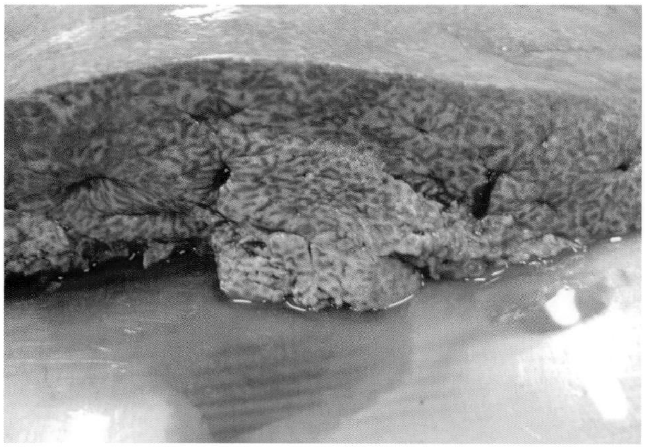

图87 由副结核引起的肝脏肿大，伴有浅灰色间质增厚

图88 由副结核引起的间质性肺炎

对突尼斯的52头单峰驼进行了MAP血清学检测，有11头呈现阳性。Feldmann等[13]对肯尼亚地区进行了该病的血清学检测。然而，血清学检测单一动物获得的结果是不太有说服力的，但这对于群体筛查时，其结果有重要意义。血清学检测包括补体结合试验、琼脂凝胶免疫扩散试验和酶联免疫吸附试验，但是在骆驼上的特异性和敏感性鲜为人知。

用琼脂凝胶免疫扩散试验（AGID）对澳大利亚地区的4头感染羊驼进行检测，只有1头表现为阳性；对科罗拉多的2只发病美洲驼检测，全部为阳性；在对2组表现出副结核临床症状的新大陆骆驼检测时，敏感性很低，只有50%的阳性率。有许多基于不同抗体、结合物和酶作用底物的ELISA方法已建立，但并没有合适的竞争ELISA方法适用于副结核病。在竞争ELISA方法中，结合底物与调查动物物种无关，可以用于多种动物。Probst等[28]用ELISA对来自捷克的1个动物园和德国的10个动物园的75头骆驼进行检测，有1头（1.3%）为阳性。

几乎没有研究涉及MAP病原检测的ELISA方法的特异性。Miller等[25]在间接ELISA方法中利用抗美洲驼结合物对来自美国IDEXX地区的样品进行检测时，其特异性高达98%。Mramsky等[24]改进了的牛的间接MAP-ELISA方法，检测副结核抗原时，其特异性高达99%。一般来讲，物种间多克隆抗体存在交叉反应。单峰驼的IgG与猪的IgG的识别率高达74.3%，与马、牛的IgG的识别率高达73.1%，但与山羊的IgG的识别率仅有61.6%。有些研究者不使用抗骆驼的结合物，而是用辣根过氧化物酶标记的蛋白A或蛋白G来代替。这两种蛋白是葡萄球菌细胞壁的衍生物，与哺乳类的$\gamma$-球蛋白的FcH的重链结构片段结晶化区域发生反应，但与鱼、两栖动物、爬行动物和鸟类不发生反应。IgG结合位点的系统发育树表明在IgA和IgG的结合位点具有高度保守性。为了将假阳性和假阴性结果降到最低，可以用一组阴、阳性血清来筛选可用于骆驼疫病诊断的种特异IgG定。Alhebabi和Alluwaimi[2]用法国Lsivet反刍动物血清副结核筛查试剂盒（结合物使用的是抗反刍动物结合物）检测了沙特阿拉伯的861份单峰驼血清，但并没有评价该盒子对于骆驼的检测效果。他们指出这种ELISA方法的敏感性太低，特别是对于年龄较小的群体。

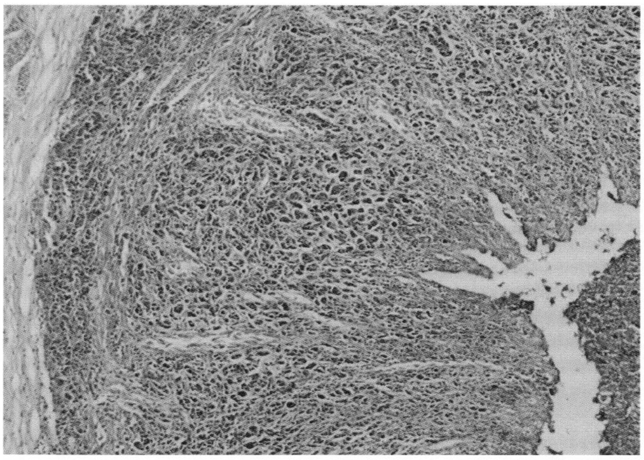

图89　弥漫性增生性肠炎，在黏膜上大量的巨噬细胞含有抗酸的杆菌（Ziehl-Neelsen染色）

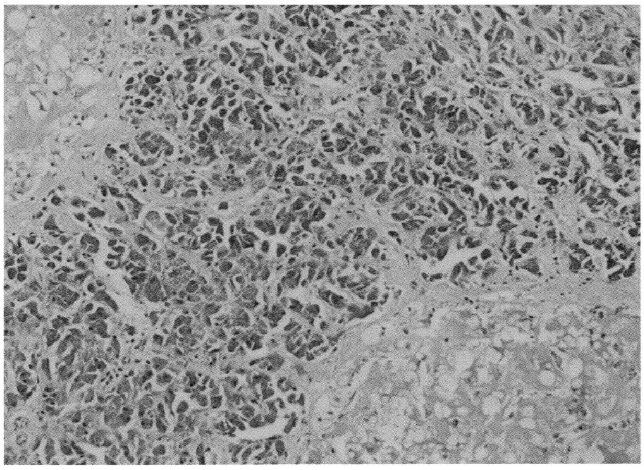

图90　肝脏出现间质性肝炎，巨噬细胞中含有抗酸的杆菌(Ziehl-Neelsen染色)

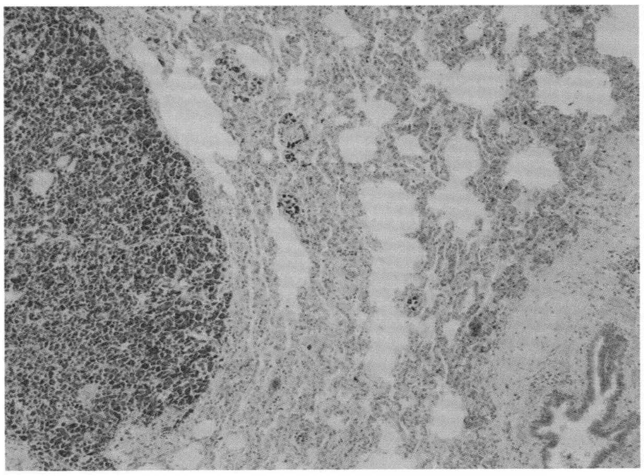

图91　肺脏血管周围和支气管周围巨噬细胞浸润，含有抗酸的杆菌(Ziehl-Neelsen染色)

综上所述，对于检测骆驼的间接ELISA方法的研究是迫切需要的，特别是应用无物种区分的结合物的情况下。Wernery等[41]评估了来自不同国家的不同结合物的副结核病的间接ELISA方法。结果显示，蛋白A结合物与商业化的3个盒子和自制的ELISA都有良好的反应性。骆驼样品与抗反刍动物和抗牛的结合物不反应，所有的检测血清都用枯草分枝杆菌抗体吸附以除去非特异性抗体，提高了检测方法的特异性。

在阿联酋，一些用于竞赛的骆驼也会有零散的病例出现。1985—1993年有5个病例，2010年发现了7个病例。这些动物患有难以治愈的腹泻。在被检排泄物中均有抗酸性菌存在（图92）。在感染动物体内发现补体结合抗体滴度为（1：64）~（1：256），由此可以确诊骆驼患有副结核病。无论是否使用抗生素治疗，所有的患病动物均在一年内死亡。

在病料培养过程中，MAP还可以依据其分枝菌素进行分型。氯化苯甲氢铵经常用来纯化样品。含分枝菌素的HEY培养基用于培养病原体，置于37℃条件下培养，并且每周检测生长情况直至16周。Salgado等[31]用不含分枝菌素的HEY培养基培养来自南美安第斯山脉的501份原驼的排泄物分离物。在这501份样品中，有21份（4.2%）MAP检测呈阳性。相同表观的MAP菌落用基于IS900和F57的实时定量PCR检测，再用引物IS1311进行套式检测，从而鉴别MAP菌株的分型。

聚合酶链式反应和DNA探针也可用于检测副结核病原体。最近，多位点可变数串联重复序列分析（MLVA）也用于MAP菌株的分型研究。MLVA是一种基于PCR原理的一种用微量材料或直接利用病原体进行分型的方法。

### 治疗和控制措施

目前，并没有治疗副结核病有效方法。一旦确诊，为了防止对环境造成污染，应对感染动物实施安乐死，并要有良好的防污染措施。Radwan等[29]曾提出一种根治骆驼副结核病的方法，其包含以下内容。

(1) 需隔离临床疑似病例直到疾病确诊，所有患病骆驼必须屠宰，尸体需妥善处理。

(2) 如果条件允许，幼驼一出生就应和母骆驼分开，并置于无副结核病原的圈里饲养。

(3) 制定适当的卫生清洁条例，防止饲料、水、土壤被副结核病原污染，且池塘和沟渠应用栅栏隔开。

(4) 购新骆驼时，应先用ELISA进行副结核检测。

(5) 应考虑疫苗免疫预防。

很多国家已经有牛、绵羊和山羊的疫苗投入使用。商业化疫苗是通过活菌制备疫苗或灭活的MAP菌株制备的灭活苗，且具有相同效力。疫苗的使用可以降低该病的发病率，但并不能消除该病的发生。一般来说，应在出生1个月内接种疫苗。接种过疫苗的骆驼在其接种部位会出现直径为几厘米的肉芽肿（图93），这使得畜主不愿意使用该疫苗。然而，在补体结合试验中，在注射部位有反应的骆驼血清检测显示为阳性，而没有反应的则呈现为阴性。兽医意外注射给本人时会引起滑膜炎和肌腱炎反应。

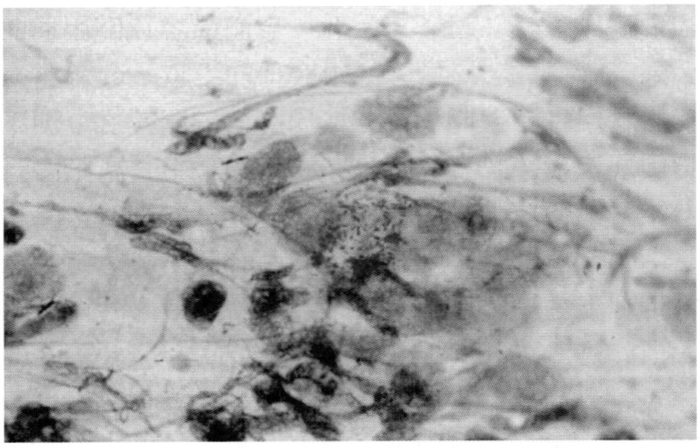

图92　患有副结核病的单峰驼粪便样品中的抗酸杆菌（Ziehl-Neelsen染色）

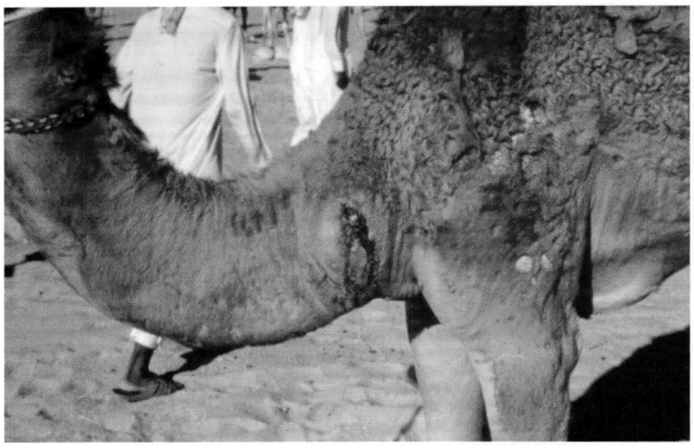

图93　接种疫苗4个月后单峰驼形成的严重脓肿

## 参考文献

[1] Al-Hizab F.A. (2010). – Johne's disease in one humped camels (*Camelus dromedarius*) in Saudi Arabia. *J. Camel Pract. Res.*,17 (1),31–34.

[2] Alhebabi A.M. & Alluwaimi A.M. (2010). Paratuberculosis in camels (*Camelus dromedarius*): The diagnostic efficiency of ELISA and PCR. *Open Vet. Sci*. J.,4,41–44.

[3] Appleby E.C. & Head K.W. (1954). – A case of suspected Johne's disease in a llama (*L. glama*). *J. Comp. Path.*,64,52–53.

[4] Amand W.B. (1974). – Paratuberculosis,*Mycobacterium paratuberculosis* in a dromedary camel. *Proc. Am. Assoc. Zoo Vet.*,150–153.

[5] Belknap E.B.,Getzy D.M.,Johnson L.W.,Ellis R.P.,Thompson G.L. & Shulaw W.P. (1994). – *Mycobacterium paratuberculosis* infection in two llamas. *JAVMA*,204,1805–1808.

[6] Boschiroli M.L. & Thorel M.-F. (2001). – Paratuberculosis. *In* Infectious and parasitic diseases of livestock,Volume 2 (P. Lefèvre,J. Blancou,R. Chemette & G. Uilenberg,eds). Lavoisier,Paris 1097–1108.

[7] Buchnev K.N.,Tulepbaev S.Z. & Sansyzbaev A.R. (1987). – Infectious diseases of camels in the USSR. *In* Diseases of camels. *Rev. sci. tech. Off. int. Epiz.*,6 (2),487–495.

[8] Burgemeister R.,Leyk W. & Gossler R. (1975). – Untersuchungen über Vorkommen von Parasitosen,bakteriellen und viralen Infektionskrankheiten bei Dromedaren in Sudtunesien [in German]. *Dtsch. Tierärtztl. Wschr.*,82,352–354.

[9] Chauhan R.S.,Kaushik R.K.,Gupta S.C.,Satiya K.C. & Kulshreshta R.C. (1986). – Prevalence of different diseases in camels (*Camelus dromedarius*) in India. *Camel Newslett*er,3,10–14.

[10] Cheyne I.A. (1995). – A brief summary of Paratuberculosis (Johne's Disease) in domestic animals with reference to recent reports of the disease in camels. *Pers. Rep.*,1–12.

[11] Deutz A.,Spergser J.,Wagner P.,Rosengarten R. & Kofer J. (2005). – *Mycobacterium avium subsp. paratuberculosis* in wild animal species and cattle in Styria/Austria [in German]. *Berl Münch Tierätzl Wochenschr.*,118 (7–8),314–320.

[12] Fassi-Fehri M.M. (1987). – Les maladies des camelides. *In* Diseases of camels. *Rev. sci. tech. Off. int. Epiz.*,6 (2),315–335.

[13] Feldman B.F.,Keen C.L.,Kaneko J.J. & Parver T.B. (1981). – Husbandry and diseases of camels [in German]. *Tierätzl. Praxis*,9 (3),389–402.

[14] Fowler M.E. (2010). – Medicine and Surgery of Camelids. 3rd Ed. Wiley-Blackwell,Aimes,Iowa,204–207.

[15] Gameel A.A.,Ali A.S.,Razig S.A.,Brown J.,Alhendi S.A. & El-Sanousi S.M. (1994). – A clinico-pathological study on spontaneous paratuberculosis in camels (*Camelus dromedarius*) in Saudi Arabia. *Pakistan Vet. J.*,14 (1),15–19.

[16] Ghosh P.,Hsu C.,Alyamani E.J.,Shehata M.M.,Al-Dubaib M.A.,Al-Naeem A.,Hashad M.,Mahmoud O.M.,Alharbi K.B.J.,Al-Busadah K.,Al-Swailem A.M. & Talaat A.M. (2012). – Genome-wide analysis of the emerging infection with *Mycobacterium avium* subspecies *paratuberculosis* in the Arabian camels (*Camelus dromedarius*). *PLoS One*,7 (2) e31947.

[17] Guake L.K.,Dubba Z.,Tumba K.H. & Abugaliev R.M. (1964). – Letter. *Vet. Moscow*,41,115–116.

[18] Hazlett M.,Henderson R. & Turcotte C. (2004). – *Mycobacterium paratuberculosis* enteritis and hepatitis in a miniature donkey. *AHL Newsletter*,8 (3),7.

[19] Higgins A. (1986). – The camel in health and disease. Bailliere Tindall,London.

[20] Ivanov B.G. & Skalinskii E.I. (1957). – Pathological changes in paratuberculosis in camels [in Russian]. *Trudy Gos. Inst. Eksp. Vet.*,20,186–206.

[21] Khon F.K. (1983a). – Allergical diagnosis of camel paratuberculosis [in Russian]. *Biul. Vses Inst. Eksp. Vet. Moskva*,50,26–28.

[22] Khon F.K. (1983b). – Complement fixation test at camel paratuberculosis [in Russian]. *Biul. Vses Inst. Eksp. Vet. Moskva*,50,30–32.

[23] Kinne J.,Johnson B.,Joseph M. & Wernery U. (2010). – Camel paratuberculosis (Johne's Disease) – a case report. *In* Proceedings of the International Camel Symposium,7–10 June 2010,Garissa,Kenya,59.

[24] Kramsky J.A.,Miller D.S.,Hope A. & Collins M.T. (2000). – Modification of a bovine ELISA to detect camelid antibodies to *Mycobacterium paratuberculosis*. *Vet. Microbiol.*,77,333–337.

[25] Miller D.S.,Collins M.T.,Smith B.B.,Anderson P.R.,Kramsky J.,Wilder G. & Hope A. (2000). – Specificity of 4 serological assays for *Mycobacterium avium* subspecies *paratuberculosis* in llamas and alpacas: a single herd study. *J. Vet. Diagn. Invert.*,12,345–353.

[26] Naser S.A.,Ghobrial G.,Romero C. & Valentine J.F. (2004). – Culture of *Mycobacterium avium* subspecies *paratuberculosis* from the blood of patients with Crohn's disease. *Lancet*,364,1039–1044.

[27] Ovdienko N.P.,Khon F.K.,Sharov V.A. & Yakusheva O.V. (1985). – Diagnosis of paratuberculosis in camels [in Russian]. *Vet. Moscow,USSR*,4,65–68.

[28] Probst C,Speck S. & Hofer H. (2007). – Epidemiology of selected infectious diseases in zoo-ungulates: single species versus mixed species exhibits. *Proc. Inst. Zoo Wildl. Res. Berlin*,7,10–12.

[29] Radwan A.I.,El-Magawry S.,Hawari A.,Al-Bekairi S.J.,Aziz S. & Rebleza R.M. (1991). – Paratuberculosis enteritis (Johne's Disease) in camels in Saudi Arabia. *Biol. Sci.*,1,57–66.

[30] Ridge S.E.,Harkin J.T.,Badman R.T.,Mellor A.M. & Larsen J.W.A. (1995). – Johne's disease in alpacas (*Lama pacos*) in Australia. *Austr. Vet. J.*,72 (4),150–153.

[31] Salgado M.,Herthnek D.,Bölske G.,Leiva S. & Kruze J. (2009). – First isolation of *Mycobacterium avium subsp. Paratuberculosis* from wild guanacos (*Lama guanicoe*) on Tierra del Fuego Island. *J. Wildl. Dis.*,45 (2),295–301.

[32] Schneider J. (1992). – Kamelkrankheiten [in German]. *In* Tropentierhygiene (Seifert H.S.H.,ed.). Gustav Fischer Verlag,Jena ,Stuttgart,Germany,190–199.

[33] Schwarte L.H. (1956). – Johne's Disease suspected in a llama. *Vet. Bull. cited in JAVMA*,128,354.

[34] Scott P.R.,Clarke C.J. & King T.J. (1995). – Serum protein concentrations in clinical cases of ovine paratuberculosis (Johne's disease). *Vet. Rec.*,137,173.

[35] Senturk S.,Metcitoglu Z.,Ulgen M.,Borum E.,Temizel E. & Kasap S. (2009) – Evaluation of serum iron and iron binding capacity in cows with paratuberculosis. *Tierärztl. Prax.*,6,375–378.

[36] Stehman S.M. (1996a). – Paratuberculosis in small ruminants,deer and South American Camelids. *Vet. Clin. N. Am. Food Anim. Pract.*,12 (2),441–454.

[37] Stehman S.M.,Rossiter C.A.,Shin S.J.,Chang Y.F. & Lein D.H. (1996b). – Johne's disease in a deer herd: accuracy of fecal culture. *In* Fifth International Colloquium on Paratuberculosis,29 September–4 October 1996,Madison,WI,20.

[38] Strogov A.K. (1957). – Paratuberculosis in camels. *Trudy Vses. Inst. Eksp. Vet.*,20,120–131.

[39] Taghi T.-B.,Ali S.N. & Taghi Z.S. (2006). – A survey of *Mycobacterium avium paratuberculosis* infection rate in slaughtered camels of Sabzevar (Iran) abattoir and camels of 'Camel Breeding Center' in Shahrood (Iran). *In* Proceedings of the International Science Conference on Camels,10–12 May 2006,Buraydah,Saudi Arabia,416–422.

[40] Wernery U. & Kaaden O.-R. (2002). – Infectious Diseases of Camelids. Second Ed. Blackwell Science,Berlin,Wien,83–87.

[41] Wernery U.,Abraham A.,Joseph S.,Thomas R. Manuel G.,Raghavan R. & Baker T. (2011). – Evaluation of 5 indirect ELISA's for the detection of antibodies to paratuberculosis in dromedaries. *J. Camel Pract. Res.*,18 (1) 47–52.


（李有全译，殷宏校）

## 1.3 呼吸系统

一般来说，骆驼科动物很少发生呼吸系统疾病，这种现象可能是骆驼肺脏基质边缘的胸膜屏障能够清除胸腔液体和附着的巨噬细胞的结果[19]（图94）。

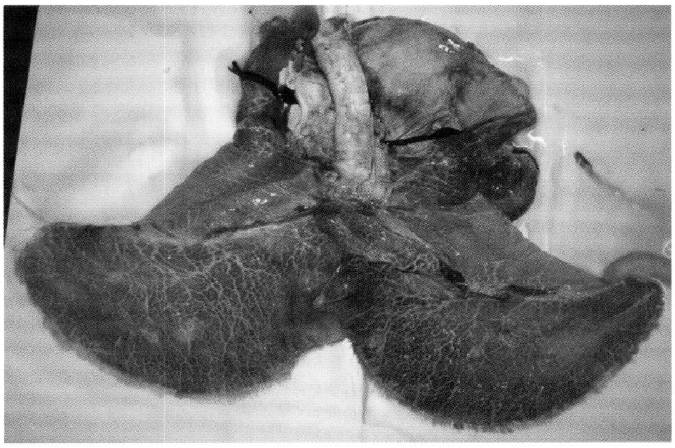

图94 骆驼肺基质边缘的胸膜屏障

骆驼发生肺炎时，一般会有比较明显的诱因，例如，天气突变，卫生条件差[73]，管理不善和一些潜在的可降低机体抵抗力的因素作用等[65,103]。现已发现，毒素中毒、大肠杆菌病、肠毒血症、白血病、慢性皮肤感染和硒/维生素E缺乏均与肺炎的发生有一定的关系。在进行骆驼科动物肺炎诊断时，这些与疾病相关的线索也应予以足够重视。此外，骆驼科动物中一个较为常见奇特的现象是严重的肺脏损伤不会引起呼吸窘迫。

在患有呼吸系统疾病的骆驼中发现了多种细菌，但除了能造成严重肺炎的结核病、马红球菌感染、马鼻疽和类鼻疽病的病原外，并不清楚这些细菌是否一定与疾病的发生有关。上述四种疾病将在本书中分章呈现。

### 1.3.1 结核病

结核病是一种慢性传染性肉芽肿性疾病，它是由结核分枝杆菌引起。结核分枝杆菌可以感染多种脊椎动物，主要病变部位在肺脏和淋巴结，但在其他组织器官也会出现结核病变。病变损伤是被称为结核结节的肉芽肿，但这些结核结节在不同的动物中和不同种结核分枝杆菌感染时有很大的差异。骆驼科动物对结核分枝杆菌不是特别易感，但近些年新大陆骆驼在从原产地国家离开后饲养时，结核病的发生率有所上升，这已引起人们的极大关切。在存有大量骆驼科动物的英国，现已发现一些区域的南美骆驼结核病例明显增多[100]。目前，结核病依然是全球人群中需要上报的主要疾病之一，它引起的人类死亡数量已超过全球战争死亡人数的总和。人类每年大约有900万新增病例报告，每年因其死亡的人数约为170万。结核分枝杆菌广泛感染暴发，目前已引起以下人员及机构的关注，包括公共健康官员、动物保护机构相关人员和负责动物园、野生动物世界以及私人养殖单位、兽医从业人员等。此外值得注意的是，很多结核菌株目前已经具有了临床医疗抗性。引起人们关注的不仅仅是结核分枝杆菌，也包括牛分枝杆菌，人类通过吸入牛分枝杆菌感染性飞沫和饮用感染性生奶引起的人兽共患的人牛分枝杆菌病[95]。结核分枝杆菌和牛分枝杆菌是结核分枝杆菌属中两个最重要的成员，在旧大陆骆驼和新大陆骆驼中都曾分离出相关菌株[46,102,103]。

#### 公共卫生意义

结核分枝杆菌引起的结核病依然是目前需要上报的全球性传染疾病之一，鉴于其病例数量呈上升趋势，世界卫生组织已经宣布全球结核病处于紧急状态。结核分枝杆菌感染人类、其他灵长类动物、犬、金丝雀和鹦鹉等，也有从骆驼科动物分离到本菌的报道[42,76]。

动物群中结核分枝杆菌感染的情况较为少见，在结核病人离开动物生存的环境后，动物群中的结核感染阳性反应就会消失。在发达国家，随着结核扑灭计划的开展和牛奶巴氏灭菌法的推广，与牛分枝杆菌相关的结核病例已经急剧下降，但是人类与该菌感染动物及其产品紧密接触依然存在风险，最近就有羊驼结核通过皮肤感染人的病例报道[101]。罹患皮肤结核后，特别是在感染早期被误诊的情况下，有可能出现极为复杂的临床状况。

牛结核病的高发态势，特别是在免疫功能不全人群中的增长，使本病成为一种重要的人兽共患病。牛分枝杆菌引起的结核病例占人结核病例数的5%～10%，且此比例在儿童群体中更高。结核病在圈养的外部引进动物群体中广泛发生，导致本病从动物群体中持续流向人群。

### 传播方式

结核病主要通过呼吸道和消化道传播，其中，呼吸道感染是主要的传播方式。在草原上，本病也可以通过含有结核菌的粪便污染草场、饮水和食槽造成消化道感染。幼畜饮用污染的奶也可以造成临床感染。

结核病传播途径。

(1) 群内直接接触或间接接触。

(2) 在生活圈内接触感染的家畜（一般是牛）或野生动物（鹿，獾）。

(3) 通过接触污染的设备、饲料和泥土。

(4) 通过接触感染的表演动物或种畜。

### 病原学

结核分枝杆菌属于分枝杆菌属、分枝杆菌科长度各异，为抗酸棒状杆菌，不运动，无芽孢。分枝杆菌属包含多个种（大约50个），各菌种具有不同的致病性，非典型分枝杆菌也已通过鲁尼恩分类法进行聚类。非典型结核分枝杆菌广泛存在于草地、土壤和水中，部分菌种可以感染动物。分枝杆菌菌体外周有一层脂质荚膜，对保护分枝杆菌免受机体破坏具有一定的作用，因而动物在感染中呈现慢性发病过程。此外，这层荚膜也使得人工提取菌体DNA用于聚合酶链式反应（PCR）变得困难。没有一种分枝杆菌菌种是完全的种特异性病原，属内各菌种之间都有相当大的同源性。家畜中最为重要的致病菌种包括：

(1) 牛分枝杆菌:可以感染多种动物和人。

(2) 结核分枝杆菌:主要感染人，偶尔也感染动物。

(3) 禽分枝杆菌复合群中的几个血清型，可以感染禽类、鸟类、猪和马。

(4) 副结核分枝杆菌。

上述四种分枝杆菌感染病例均已经有骆驼科动物中发病的报道，有的是自然感染，也有的是人工感染。

### 流行病学

大约在20世纪初埃及[62]和印度[58,61]就诊断出单峰驼被患有结核病，表29中汇总了骆驼科动物发生结核病的相关文献报道情况。

从表29我们可以看出，均已从旧大陆骆驼和新大陆骆驼中分离到结核分枝杆菌、牛分枝杆菌和非典型分枝杆菌（表30）。非典型分枝杆菌也可以引起类似结核病的临床症状和病理损伤。

游牧状态下的骆驼很少发生结核病，本病主要发生在与其他骆驼一起圈养或者与牛密切接触的骆驼中，这种饲养方式在俄罗斯和埃及较为常见[39,41,42,43,67,68,69]，多数早期出版物都将这些国家作为骆驼结核病的发源地。

Gatt Rutter和Mack报道称，骆驼结核病不会在游牧骆驼中发生，但会发生在与牛一起饲养的骆驼中[48]。目前，该病在骆驼中的传播方式并不明确，猜测其传播方式应与结核在牛群中的传播方式类似，主要为水平传播。一般认为，牛的结核分枝杆菌通过气溶胶的方式感染健康骆驼。例如，采食传播、先天传播、性传播和皮肤传播。这是牛的感染方式，但未见在骆驼结核感染中报道。Kogramanov 等[56]发现亚洲璃眼蜱可以传播结核分枝杆菌到双峰驼中。

由于热带地区的发展中国家对结核病的重视程度较低，该病给畜牧业，尤其是养牛业造成较严重的经济损失。游牧人群因为要消费大量的牛奶和生奶制品，因而本病也对游牧人群健康造成重要威胁[89]，与骆驼饲养相关的人群也面临相似的问题。Donchenko等从俄罗斯712头泌乳骆驼46份混合奶样品中分离到了牛分枝杆菌，在对这些骆驼进行结核菌素检测后发现，9.1%的骆驼呈现阳性反应。除了未加热消毒的骆驼奶之外，马戏团和动物园中处于结核感染活动期的骆驼也对人类健康造成严重的威胁[34,78]。

Zerom等在埃塞俄比亚检测了293峰单峰驼，从61%的被检出有结核病的病理变化的个体中分离出了分枝杆菌，而其中仅68%的分离菌为非结核分枝杆菌（non-tuberculosis mycobacteria，NTM），其余菌株均为结核分枝杆菌，检测结果表明存在极为严峻的人兽共患病感染的风险。此外，从有结核病病理变化的个体中分离出了大量的非结核分枝杆菌，揭示这些病原体对单峰驼有致病力。

在畜牧业生产过程中，结核病的发生有两条不同的途径。

(1) 通过吸入被感染动物污染的含菌气体。被吸入的污染物可能是细菌本身，也可能是含有细菌附着的小尘土颗粒，由此引起的结核病原发病的病

表29　骆驼结核病相关文献报道汇总

| 国别 | 作者 | 年份 | 骆驼品种 | 菌株类型 |
|---|---|---|---|---|
| 埃及 | Littlewood | 1888 | 单峰驼 | — |
| 印度 | Lingard | 1905 | 单峰驼 | — |
| 印度 | Leese | 1908 | 单峰驼 | — |
| 印度 | Leese | 1910 | 单峰驼 | — |
| 苏丹 | Archibald | 1910 | 单峰驼 | — |
| 埃及 | Mason | 1912 | 单峰驼 | |
| 埃及 | Mason | 1917a，b | 单峰驼 | 牛分枝杆菌 |
| | Cross | 1917 | | — |
| 埃及 | Mason | 1918 | 单峰驼 | — |
| 德国 | Andree | 1928 | | 牛分枝杆菌 |
| 索马里 | Pellegrini | 1942 | 单峰驼 | 牛分枝杆菌 |
| 埃及 | El—Afifi等 | 1953 | 单峰驼 | 牛分枝杆菌，结核分枝杆菌 |
| 索马里 | Casati | 1957 | 单峰驼 | 牛分枝杆菌 |
| 马戏团骆驼 | Panebianco | 1957 | | 牛分枝杆菌 |
| 索马里 | Angrisani | 1962 | 单峰驼 | 结核分枝杆菌 |
| 马戏团骆驼 | Dekker和van der Schaaf | 1962 | | 牛分枝杆菌 |
| 俄罗斯 | Abramov | 1963 | | |
| | Abramov | 1964 | | 牛分枝杆菌，结核分枝杆菌 |
| 印度 | Damodaran和Ramakrishnan | 1969 | | — |
| 埃及 | Abd El—Aziz | 1970 | 单峰驼 | 牛分枝杆菌，结核分枝杆菌 |
| 比利时 | Pattyn等 | 1970 | 美洲驼 | 田鼠分枝杆菌 |
| 埃及 | El Mossalami等 | 1971 | 单峰驼 | 牛分枝杆菌，结核分枝杆菌 |
| 俄罗斯 | Fedchenko | 1971a，b | | — |
| 俄罗斯 | Akhundov等 | 1972 | | — |
| 埃及 | Osman | 1974 | 单峰驼 | 牛分枝杆菌，结核分枝杆菌 |
| 俄罗斯 | Donchenko等 | 1975a，b | | 牛分枝杆菌 |
| 俄罗斯 | Kibasov和Donchenko | 1976 | | — |
| 美国 | Thoen等 | 1977 | 美洲驼 | 牛分枝杆菌 |
| 俄罗斯 | Donchenko和Donchenko | 1978 | | — |
| 埃塞俄比亚 | Richard | 1979 | 单峰驼 | — |
| 索马里 | Arush | 1982 | 单峰驼 | — |
| 美国 | Kennedy和Bush | 1978 | 美洲驼 | — |
| 毛里塔尼亚 | Chamoiseau等 | 1985 | 单峰驼 | — |
| 美国 | Bush等 | 1986 | 美洲驼 | — |

续表

| 国别 | 作者 | 年份 | 骆驼品种 | 菌株类型 |
|---|---|---|---|---|
| 索马里 | Hayles | 1986 | 单峰驼 | — |
| 印度 | Chauhan等 | 1986 | 单峰驼 | — |
| 美国 | Thoen | 1988 | 美洲驼 | 牛分枝杆菌 |
| 毛里塔尼亚 | Diatchenko | 1989 | 单峰驼 | — |
| 美国 | Bush等 | 1990 | 双峰驼 | 牛分枝杆菌 |
| 毛里塔尼亚 | Chartier等 | 1991 | 单峰驼 | — |
| 巴基斯坦 | Rana等 | 1993 | 单峰驼 | 牛分枝杆菌 |
| 英国 | Johnson等 | 1993 | 美洲驼 | 堪萨斯分枝杆菌 |
| 阿联酋 | Wernery和Kaaden | 1995 | 单峰驼 | — |
| 英国（实验性） | Stevens等 | 1998 | 美洲驼 | 牛分枝杆菌 |
| 英国 | Barlow等 | 1999 | 美洲驼 | 牛分枝杆菌 |
| 瑞士（动物园） | Oevermann等 | 2003 | 美洲驼 | 田鼠分枝杆菌 |
| 瑞士 | Oevermann等 | 2004 | 美洲驼 | 田鼠分枝杆菌 |
| 印度 | Deen等 | 2004 | 单峰驼 | — |
| 捷克（动物园） | Pavlik等 | 2005 | 双峰驼 | 牛分枝杆菌 |
| 斯洛文尼亚（动物园） | Pavlik等 | 2005 | 单峰驼 | 牛分枝杆菌 |
| 英国 | Anonymous | 2006b | 美洲驼 | 牛分枝杆菌 |
| 阿联酋 | Kinne等 | 2006 | 单峰驼 | 结核分枝杆菌 |
| 印度 | Tanwar | 2006 | 单峰驼 | 牛分枝杆菌 |
| 瑞士 | Robert | 2006 | 美洲驼，羊驼 | 田鼠分枝杆菌 |
| 阿联酋 | Wernery等 | 2007a | 单峰驼 | 牛分枝杆菌 |
| 德国（动物园） | Moser等 | 2008 | 双峰驼 | 鳍脚亚目动物分枝杆菌 |
| 英国 | 匿名 | 2008 | 羊驼 | 田鼠分枝杆菌 |
| 英国 | 匿名 | 2009 | 羊驼 | 田鼠分枝杆菌 |
| 瑞士 | Zanolari等 | 2009 | 美洲驼，羊驼 | 田鼠分枝杆菌 |
| 英国 | Dean等 | 2009 | 美洲驼 | 牛分枝杆菌 |
| 西班牙 | Garcia-Bocanegra等 | 2010 | 羊驼 | 牛分枝杆菌 |
| 尼日利亚 | Kudi等 | 2011 | 单峰驼 | 牛分枝杆菌 |
| 美国、英国、瑞士 | Lyaschenko等 | 2011 | 羊驼，美洲驼 | 牛分枝杆菌 |
| 埃塞俄比亚 | Zerom等 | 2013 | 单峰驼 | 牛分枝杆菌，非结核分枝杆菌 |

变灶在肺脏解剖时可以看到。

(2) 通过采食被病原体污染性粪便、尿液或奶及饲料。由此感染引起的原发病的病变部位在肠道淋巴结。

在过去的几年中，英国牛群中检测出的牛结核病例有增多的趋势。獾和鹿是牛分枝杆菌的储藏宿主，在有獾出没的养牛场也发现结核病例增多。现在，人们开始接受通过消除野生动物感染来降低或消除养牛场结核病的观点。新大陆骆驼现在越来越多地饲养在牛场附近，结核分枝杆菌由野生宿主向牛群和新大陆骆驼传播，或由新大陆骆驼向牛群传播及由牛群向新大陆骆驼传播，都是有可能的。

**表30　骆驼中分离的非典型分枝杆菌**

| 分枝杆菌种类 | 作者 | 年份 | 骆驼品种 | 病例变化 | 来源 | 国别 |
|---|---|---|---|---|---|---|
| 山羊分枝杆菌 | Pate等 | 2006 | 单峰驼 | 弥散性结核 | 欧洲野牛 | 斯洛文尼亚 |
| 田鼠分枝杆菌 | Pattyn等 | 1970 | 美洲驼 | 肺炎 | 啮齿动物 | 荷兰（动物园） |
| 田鼠分枝杆菌 | Oevermann等 | 2003 | 美洲驼 | 普通结核 | 田鼠 | 瑞士 |
| 田鼠分枝杆菌 | 匿名 | 2006a | 美洲驼 | 肺脏，肝脏脾脏结核 | 啮齿类 | 英国 |
| 田鼠分枝杆菌 | Robert | 2006 | 美洲驼，羊驼 | 肺脏，肝脏，脾脏，肠系膜和纵隔淋巴结结核 | 不同来源的啮齿类 | 瑞士 |
| 田鼠分枝杆菌 | Zanolari等 | 2009 | 美洲驼，羊驼 | 肺脏，淋巴结结核 | 6个畜群 | 瑞士 |
| 堪萨斯分枝杆菌，水分枝杆菌 | El Mossalami等 | 1971 | 单峰驼 | 胸膜炎干酪样结节，肺胸膜炎结核 | ? | 埃及 |
| 水分枝杆菌解脲亚种，耻垢分枝杆菌 | Osman | 1974 | | | | |
| 堪萨斯分枝杆菌 | Johnson等 | 1993 | 美洲驼 | 肝肠系膜，淋巴结结核 | 水禽 | 英国 |
| 鳍脚亚目动物分枝杆菌 | Moser等 | 2008 | 双峰驼 | 肺脏，淋巴结结核 | 海狮 | 德国（动物园） |
| 田鼠分枝杆菌 | Lyashchenko等 | 2011 | 羊驼，美洲驼 | 多器官结核 | ? | 美国，英国，瑞士 |

在英国的美洲驼中也分离到了牛分枝杆菌，在当地牛群和因公路交通意外死亡的獾体很容易分离到间接区寡核苷酸类型相似的菌株[98]。也有报道称，感染了结核菌的新大陆骆驼可以很容易地将结核病散播到非感染群内[18]。在西班牙等国家，新大陆骆驼一般饲养在较为偏远的区域，野生动物中的野猪、狐狸或猫鼬取代鹿和獾，成为潜在的病原传播者[28]。携带结核菌的獾会经常出入农场，以及农场的建筑和饲料房，由此可能污染这些场所[35]。通过獾来控制牛结核病的研究已开展数年，第一个用于獾结核病免疫的疫苗（Badge BCG）也已于2010年获得批准使用。用饲养的獾进行的实验研究证实，Badge BCG疫苗可以大大降低獾感染结核分枝杆菌之后的疾病发展速度和严重程度，且能降低细菌的外排[36]。目前，许多国家都制定了消除结核病的防制策略，发达国家期一般期望能在20年左右的时间彻底消除牛结核病。在英国、西班牙和新西兰等国家，野生动物宿主及其贸易是结核病再发以及向畜群散播的主要原因[87]。

美洲驼和羊驼对低剂量的牛分枝杆菌也极易感[91]，在多个国家均有结核病引起高死亡率的报道。这些报道与Fowler在南北美洲的研究结果相反，在上述地区，新大陆骆驼对结核分枝杆菌不是特别易感，在南美洲只有极少数量的骆驼科动物感染结核的自然感染病例，尽管这些区域内生活在一起的牛、绵羊和人感染结核的情况很普遍。在南美洲、北美洲的低感染率反映出这些地方的动物更难接触到结核病原，而不是其自身的抵抗力强。

自然感染和实验室感染的新大陆骆驼结核病例已有报道[20,22,71]。牛分枝杆菌、结核分枝杆菌、禽分枝杆菌和副结核分枝杆菌四种最主要分枝杆菌以及非典型分枝杆菌（堪萨斯分枝杆菌，田鼠分枝杆菌）均有从新大陆骆驼分离到的报道。目前，只有几峰新大陆骆驼发生结核病的报道，因为结核病例少，美洲驼也被认为对结核分枝杆菌不易感。在北美洲的一些地方，鹿和美洲驼常在一起饲养，尽管很多鹿发生了结核病，但只有两峰美洲驼感染结核菌，发展为弥散性肉芽肿结核病。美国兽医服务实验室的研究人员在5年的时间内，从8峰美洲驼的病料中分离培养出牛分枝杆菌[97]，德国的相关人员也从动物园和私人拥有的原驼的病料中分离出结核分枝杆菌和牛分枝杆菌[49]。

Garcia-Boncanegra等在西班牙报道了两峰羊驼的开放性结核病，认为羊驼是潜在的结核病传染源，可以向家畜、野生动物和人类散播结核分枝杆菌。羊驼间通过直接接触传播，最近也有疑似病例报道[99]。上述作者在进行相关研究时，对两种结核发病的病理变化模式进行了描述，其中的两个动物表现为弥散性肺脏和气管结核和气管黏膜溃疡结核，溃疡灶聚集有大量抗酸棒状杆菌和结核结节。那些不形成结节的结核病例可能因为是机体抑制结核损伤和限制病原体扩散失败所造成的。

### 临床症状

结核病是一种慢性消耗性疾病，骆驼科动物在患病时临床症状存在差异，发病动物渐进性消瘦和体重下降为主要症状。一个奇怪的现象是，尽管肺脏有很严重的病变，但无论是新大陆骆驼还是旧大陆骆驼都没有出现明显的呼吸窘迫的表现。结核病是慢性病，骆驼科动物发病后还能够存活几个月甚至几年，而一些发生弥散型粟粒结核的骆驼通常没有临床症状。

### 诊断

生前诊断

患结核病的骆驼科动物的临床表现包括消瘦、精神沉郁、易疲劳、最终死亡，症状均不具有特异性；因此，不论是旧大陆骆驼还是新大陆骆驼，其生前诊断比较困难。除了极个别例外，骆驼科动物患结核病的情况很稀少，但在一些新大陆骆驼进口国，结核病是主要的动物检疫疾病。在2009年，英国共有140个牛结核病例得到确认，其中，68个病例来自羊驼[35]。传统的结核病皮试检测和血清学检测方法等是目前在用的检测方法。它们在用于骆驼科动物结核病检测时，结果均不可靠，也没有任何一种检测方法被批准用于骆驼科动物结核病的检测。这些方法中，结核菌素皮内变态反应（tuberculin skin test，TST）往往出现大量非特异性结果，甚至对病理剖检有典型结核病损伤的个体，也出现阴性结果，而血清学检测方法在20世纪80年代才开始研发和应用。

官方机构在骆驼科动物国际贸易中使用的结核病检测方法是结核菌素皮内变态反应，这也是目前世界范围内唯一能够用于骆驼科动物结核病检测的方法。结核菌素皮内变态反应可以更为准确地称为单侧皮内比较结核菌素检测（single intradermal comparative tuberculin test，SICTT），检测注射部位为腋下，且一些研究人员不支持在诸如颈部、尾部皱褶和耳廓等其他部位进行检测[90]。在人工致敏大羊驼进行的结核菌素皮试研究中发现，腋下注射相对于其他部位，其结果更为理想[52]，相同的部位在单峰驼上也是最优[104]。牛型结核菌素（纯化蛋白衍生物）（bovine purified protein derivative，BPPD）和禽型结核菌素（纯化蛋白衍生物）（avian purified protein derivative，APPD）（各0.1 mL）分别皮内注射于刮毛后的左右腋下选定部位，注射前及注射后72小时分别测定皮肤的厚度。如果BPPD接种部位的皮肤厚度较APPD接种部位的厚度多出1 mm，则认为是阳性反应[8,27]。其他研究人员也可能选择测量双层皮肤皱褶的厚度，高出2 mm作为阳性结果判断标准[100]。在单峰驼的皮试检测中，皮肤厚度的最大差异在结核菌素接种后5天出现[104]（图95）。

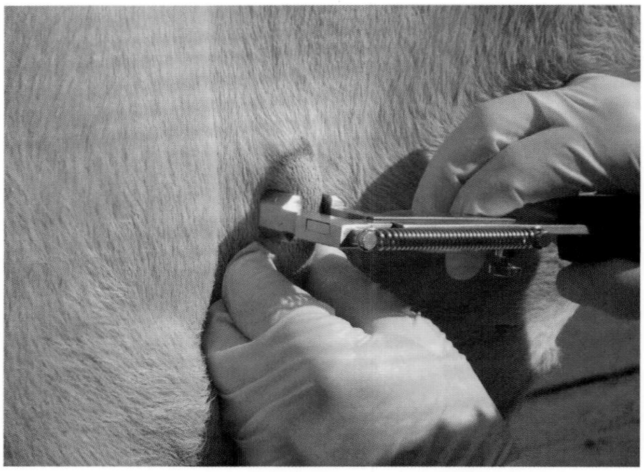

图95　单峰8驼腋下阳性结核菌素皮内检测结果

目前，还有除单侧结核菌素皮内检测之外的其他皮内检测方法，但没有一个变态反应检测方法是被公开认可的。牛结核病是世界动物卫生组织重点关注的动物疾病，骆驼科动物的感染检测也应该遵照世界动物卫生组织给出的指导原则。需要强调的是，目前国际市场上有不同的牛分枝杆菌和禽分枝杆菌蛋白抗原，这些来源于不同生产商的产品差异明显。在结核菌素生产过程中改善制作工艺和加强质量控制，有可能提高结核菌素皮内变态反应的敏感性和特异性。

一些学术文章中报道了在旧大陆骆驼和新大陆骆驼中出现非特异性皮肤反应的情况，文献中还提到单峰驼的皮内结核菌素检测阳性率在印度和肯尼亚分别为1.9%[26]和37%[77]。在俄罗斯874峰双峰驼中有107峰为患有结核病，占总数的12.2%，但其中只有68%的感染骆驼结核菌素检测中出现阳性反应[2]。美国双峰驼的结核菌素检测结果中有大量的假阳性存在，在进行尸检时，这些呈假阳性的骆驼体内既没有结核结节，也没有分离出分枝杆菌，尽管它们在淋巴细胞刺激试验中也呈阳性[53]。澳大利亚单峰驼中禽型和牛型结核菌素试验的阳性率在10%～20%，但动物屠宰后却没有发现特征性的病理损伤[88]。在美国华盛顿一个国家动物园中，19峰双峰驼被诊断出患有牛结核病[18]。鉴于同群中其他骆驼也呈现阳性结核皮试反应，这个骆驼群此后再没有接受旧结核菌素和禽型结核菌素（old mammalian and avian tuberculin，MOT，AOT）以及APPD和BPPD检测，在死后检查中也没有发现临床感染和发病表现。双峰驼接受结核菌素皮内变态反应的部位在近尾部皱褶和颈部位置，有两峰骆驼在数年后出现明显的白细胞增多症，尽管接受了广谱抗生素治疗，白细胞增多症还是持续了3个月之久。这些骆驼在安乐死后，在其肺脏、肠系膜、胰腺、肝脏、脾脏、皮肤、气管和很多外周淋巴结等器官均发现了弥散性的脓性肉芽肿性病变损伤，并且从这些病变部位分离到了牛分枝杆菌。这些骆驼在接受安乐死之前再次进行了颈部MOT、AOT、APPD和BPPD检测，检测结果均为阴性。

调查结果显示，基于单一的皮内结核菌素检测制定的骆驼科动物结核病防控程序有严重缺陷。有一点需要说明的是，牛结核菌素皮内变态反应的检测结果波动很大，与结核菌素的使用剂量、选用纯化蛋白衍生物（purified protein derivative，PPD）、注射位置和检测方案有关[87]，而且检测方法的低敏感性对制定持续有效的结核病的防控措施和清除方案将带来严重风险。此外，通过SICTT方法完成的新调查结果显示，结核皮内检测试验在牛群中有80%～90%的准确度，但在骆驼科动物中只有60%[9]。一些新大陆骆驼饲养人员甚至认为结核菌素皮内变态反应方法只能检出20%的结核病畜[28]，而Twomey等[100]报道称SICTT方法在美洲驼结核病感染检测中的敏感度只有14%。低于60%的特异性以及更低的敏感性使兽医从业人员，特别是骆驼饲养人员认为这些方法极不可靠，甚至有人断定SICTT在骆驼科动物结核病检测中是无意义的[9]。

西班牙的30峰羊驼因为牛结核病被安乐死，但这些羊驼接受的共计300次的皮试检测均没有阳性反应。需要强调的是，在自然感染的结核病牛上频繁进行皮试检测会使机体的敏感性降低[27]，目前还不清楚这种情况在骆驼科动物上会不会出现，但因为存在大量的不确定结果，新大陆骆驼经常接受重复皮试检测。研究发现，频繁的皮内注射却可以急剧增加抗体检测的敏感性[23]，值得一些是，南美的一些结核菌素皮内变态反应检测研究结果刚好与上述情况相反。Fowler[46]报道称，针对大量秘鲁羊驼进行结核菌素检测，没有注意到假阳性和假阴性的反应存在。在阿根廷，对美洲驼进行气管内接种活的牛分枝杆菌后也发现了类似的结果，针对这些动物的皮试敏感性为100%，特异性稍低。造成这些差异的原因目前还不清楚，或许是因为使用了不同的结核菌素抗原。

Ryan等[86]发现肺脏、肾脏、肠道和肠系膜上均出现肉芽肿性损伤，且能够分离培养到牛分枝杆菌的爱尔兰羊驼，在进行皮内检测时结果为阴性。Connolly等据此开展的后续流行病学调查发现，皮试检测耗时且毫无临床价值，而测量白–球蛋白比率和规律性的记录机体状态等其他检测手段更有效。研究还发现，羊驼间直接传播牛分枝杆菌不太可能，羊驼之所以感染是因为生活在当地的牛感染了牛分枝杆菌。

Lyashchenko等[64]在另一调查研究中发现，95%的新大陆骆驼结核病牛皮试检测为阴性，这在新大陆骆驼上再次验证了皮试检测不可靠。这项研究是在英国、瑞士和美国的156峰羊驼和175峰美洲驼上开展的，结果：从35峰羊驼和17峰美洲驼上培养出牛分枝杆菌（44份）或田鼠分枝杆菌（8份）。

这些研究结果促使人们开展了传统的皮试检测，新近建立的血清学检测以及尸检结果之间的比较研究。在沙特阿拉伯开展的一项血清学调查研究发现，来自两个地方的19峰和39峰骆驼只有一峰为血清学阳性反应[70]。Dean等也证实了皮试检测的局限性，并发现多抗原线条免疫分析法（multiantigen print immunoassay，MAPIA）和快速检测（rapid test，RT）在新大陆骆驼结核病体外检测中有应用潜力。MAPIA是被OIE认可的方法，它是通过将一系列抗原结合在硝酸纤维条上，然后与血清样品孵育来完成，它属于多抗原线条免疫分析法，RT（Vet TBSTAT-PAK，Chenbio，New York，USA）是一种选用了3种结核分枝杆菌特异性抗原复合物的便携式横向流色谱检测方法。在Dean等[32]开展的调查研究中，经尸检和细菌分离培养确定的14份美洲驼牛分枝杆菌感染病例中，只要2份为结核菌素皮内变态反应检测阳性，11份RT检测阳性，14份均为MAPIA检测阳性。相似的结果也在Wernery等人[104]的研究中出现，此项研究中的3峰单峰驼被RT和MAPIA检测为牛结核病阳性，并经过了尸检和细菌培养的确认，SICTT检测仅2峰为阳性。研究中还发现，一些没有感染单峰驼接受APPD检测的反应强度比BPPD还强，显示禽型结核分枝杆菌与牛分枝杆菌感染之间有交叉反应。Stutzman[92]在其研究中称血清学检测方法在提高骆驼科动物结核病防控水平中极具潜力，且横向流技术是非常优秀的快速检测手段。

有研究人员提供了新的证据，证实结核菌素皮内变态反应在骆驼科动物结核病检测的结果不可靠。Zanolari等[108]尸检了10峰美洲驼和1峰羊驼，被检动物的肺脏和淋巴结上有典型的结核病的病理变化，经过细菌培养和PCR确定为结核病，但其SICTT检测结果均为阴性。在本研究中，没有进行血清学检测，且鉴定出来有结核分枝杆菌复合群中的田鼠分枝杆菌。Lyashchenko等[64]证实在87份新大陆骆驼样品的田鼠分枝杆菌中，MAPIA和RT两种方法的检出率分别为97%和87%。这些结果显示，对骆驼科动物结核病进行的生前诊断方法中，血清学检测方法比结核菌素皮内变态反应的准确性更高和实用性更强，该结论也得到了Lyashchenko等[64]的证实。研究人员通过来自英国、瑞士和美国（阴性对照）的羊驼和美洲驼对RT和双通道平台检测法进行了对比分析，其在羊驼上的敏感度为89%，在美洲驼上的敏感度为88%，特异性分别为97%和93%。

Younis等[107]用SITT、SCITT和酶联免疫吸附试验（enzyme-linked immunosorbent assay，ELISA）检测了沙特阿拉100峰疑似结核病的消瘦单峰驼，发现30峰为SITT阳性，11峰为疑似；2峰为SICTT阳性，3峰为疑似；3峰单峰驼为ELISA阳性。相关研究人员最后总结认为，SITT不适宜于单峰驼结核病的检测，SCITT和ELISA两种方法的联合检测适宜。

除了MAPIA和RT方法外，另外一种用于检测体液免疫效果的是ELISA方法。$\gamma$-干扰素（gamma interferon，IFN-$\gamma$）和淋巴细胞增殖试验则用于测定细胞免疫效果。上述3种方法均为OIE推荐方法。

Thoen等[96]建立了第一个用于结核病检测的ELISA方法。ELISA方法可以检测出大多数的结核菌素皮内变态反应阳性美洲驼，9个结核菌素皮内变态反应阴性动物8个为ELISA阴性。在另外一个试验中，Stevens等[91]用牛分枝杆菌实验感染了6峰美洲驼，比较了结核菌素皮内变态反应和ELISA两种方法的检测效果，认为只有两种方法的合用才能在结核病诊断中发挥更高的潜力。

因为费时及操作复杂，淋巴细胞增殖试验不能作为例行检测方法用于结核病检测。结核病目前已给新西兰养鹿产业带来严重挑战。新西兰的科学家们已经采用各种诊断方法联合应用于鹿结核病诊断的模式，这些方法包括淋巴细胞刺激试验、ELISA方法和触珠蛋白检测法[46]。

目前，国际贸易被世界动物卫生组织和其他组织认可的结核病检测可选方法IFN-$\gamma$检测试验，是为牛群结核病检测建立的体外检测方法，是通过外周淋巴细胞检测调节细胞的免疫应答检测结核分枝杆菌感染情况。IFN-$\gamma$检测（Bovigam，瑞士）目前已广泛应用于全球牛结核病清除计划，但本方法不适用于骆驼科动物。IFN-$\gamma$检测方法的敏感性和特异性已在大量的国际范围的研究中得到评估，它的牛结核检测中的平均敏感性为88%，特异性为97%。有报道称本方法将要在骆驼科动物结核检测中应用，预计2011年中期可以完成临床应用前的相关准备工作[16]。

此外，曾经就骆驼科动物结核病诊断做过系统综述的Alvarez等表示，尽管数十年来针对骆驼科动物结核病生前诊断方法的研究已经取得很大进展，但这些方法目前还存在严重的局限性[5]。存在的主要问题之一便是这些方法用在实验感染动物检测其敏感性高于自然感染动物。要获得旧大陆骆驼结核病自然感染检测的认可，需要做大量的生前和死后检测，这是不可能或者说极难完成的。

有意思的是，结核病在美国骆驼科动物中的

重要程度要远比在欧洲骆驼科动物中低。在新大陆骆驼中开展的血清学检测工作很少，到目前为止，还没有与血清学诊断相关的文献报道。

结核病在世界许多国家都有发生，也是奶牛业生产中的重要问题之一。本病在发达国家一直处于严格的防控之下，且泌乳奶牛需要定期的接受结核病感染检测，这些防制原则同样应用在阿联酋迪拜的产奶骆驼群体中，所有单峰驼都定期接受RT方法检测以排除结核病动物[105]。

结核病抗体检测方法为血清学调查提供了一个方便、经济、有效的检测手段，且免疫优势蛋白的发现增加了抗体检测的特异性。到目前为止，牛分枝杆菌的MPB 70和MPB 83蛋白是最主要的血清学检测抗原，且适用于检测的蛋白联合应用会提高结核病血清学检测的敏感性。这些在OIE陆生动物操作手册中提到的基于血液的实验室检测方法[106]和其后续改良方法，以及新研发的血清学检测方法后会逐渐淘汰传统的结核菌素皮内变态反应方法。

## 诊断

### 死后诊断

Mason在100年前就描述过埃及单峰驼结核病的病理变化[66,67,68,69]。单峰驼主要的病变器官包括肺脏、支气管、纵隔淋巴结、胸膜和肝脏，气管、肾脏和脾脏也可能受到侵袭，肺脏表面和组织内部有时可见粟粒性结节。Leese[60]、Pellegrini[82]和El Mossalami[42]以及Osman[76]人分别报道了印度、索马里和埃及的单峰驼中受结核菌作用，其器官也有类似病理变化表现。弥散型结核病在旧大陆骆驼中罕有报道，消化道型结核病的病例只在新大陆骆驼中报道过。单峰驼的结核病例很罕见，阿联酋一项持续25年的记录中只有4例结核病[104]。在这些病例中，主要的病理变化部位在纵隔淋巴结和肺脏（图96，97，98）。

Barlow等[15]报道了在英格兰和威尔士交界地域附近的一个小美洲驼群发生牛结核病，一峰状态糟糕的雌性美洲驼在尸检后发现有大量的干酪样坏死，并且从中分离到了牛分枝杆菌。干酪样坏死所在机体部位包括胸膜、肺脏和心包腔，支气管纵隔淋巴结变大且也有干酪样坏死灶（图99）。研究人员同时发现，从美洲驼上分离到的牛分枝杆菌与从当地牛和獾上分离到的菌株同型。

图96 公单峰驼结核病，表现有肉芽肿性胸膜炎，肺黏连和灰变

粟粒样肺脏坏死也是新大陆骆驼结核病的典型病变[10,28,98]，3月龄的美洲驼肺脏和胸淋巴结上也发现了干酪样坏死[101]。坏死主要发生在淋巴结和肺脏，显示出纤维组织和一部分郎罕氏巨细胞参与最初出现并逐渐增强的机体免疫应答。组织病理学检查显示，化脓性肉芽肿内有一个含大量中性粒细胞碎片的高密度核心，外周被上皮样巨噬细胞和一些巨细胞包围（图100），对这些部位的切片进行齐尔-尼尔森染色法染色后，发现有一些抗酸染色菌存在（图101）。Twonmey等[101]称从未在美洲驼的器官切片上发现郎罕氏巨细胞，尽管它们的出现是牛分枝杆菌感染骆驼时的典型特征。

分枝杆菌在培养基上生长缓慢，往往培养2～6周后才能出现肉眼可见的菌落。体外培养方法与动物接种培养一样可靠，目前常用的培养基有罗氏（Lownstein-Jenson，L-J）培养基（图102）和小川培养基，一些分枝杆菌需要营养更为丰富的培养基才能培养成功。结核病也可以用PCR方法进行诊断。

## 治疗和防控措施

结核病是一种在许多国家都存在的疾病，通过大量系统的防控工作，牛结核病在一些国家已基本被消灭。结核病阳性动物必须从畜群中剔除并妥善处理。一些有价值的动物园骆驼有时也允许用异烟肼治疗，其使用剂量为2.4 mg/kg食丸，让双峰驼自由采食[18]。可能是药物超过使用剂量，有几峰骆驼死亡，检查发现有严重的白细胞减少症和血小

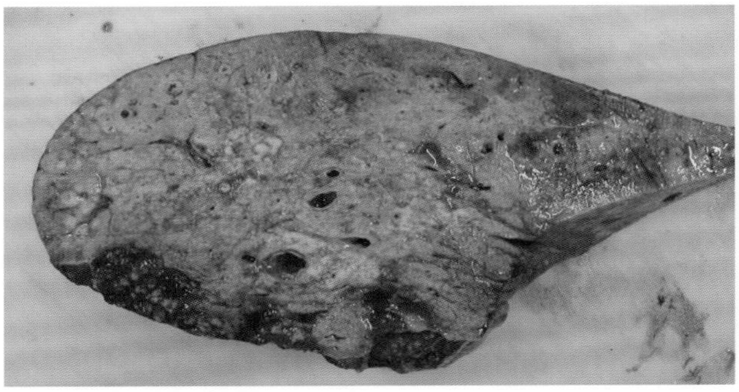

图97　在图96中展示肺脏的切面，因为有大量形状和大小各异的干酪样肉芽肿而呈现灰变

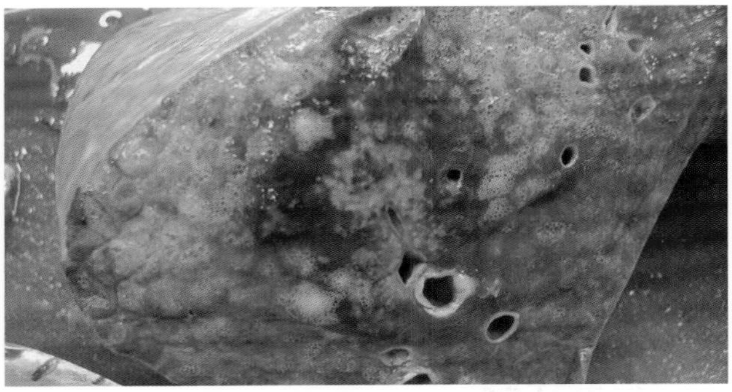

图98　结核病血清学阳性骆驼的肺脏切面，在肺脏中心部位有结核性肉芽肿灰变区

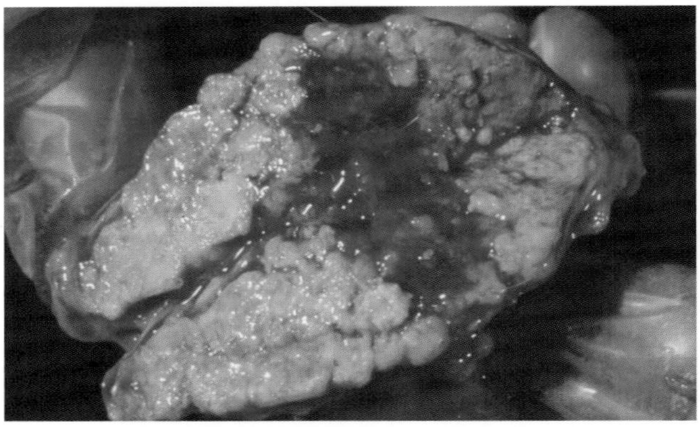

图99　美洲驼气管纵隔淋巴结中的干酪样结节（英国Dr A.M. Barlow赠图）

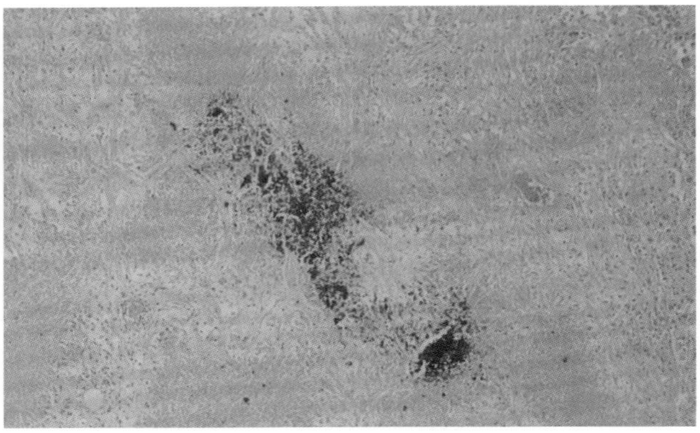

图100　在图96中展示肺脏的组织病理学图片，中心部位坏死肉芽肿部分钙化（苏木素-伊红染色）

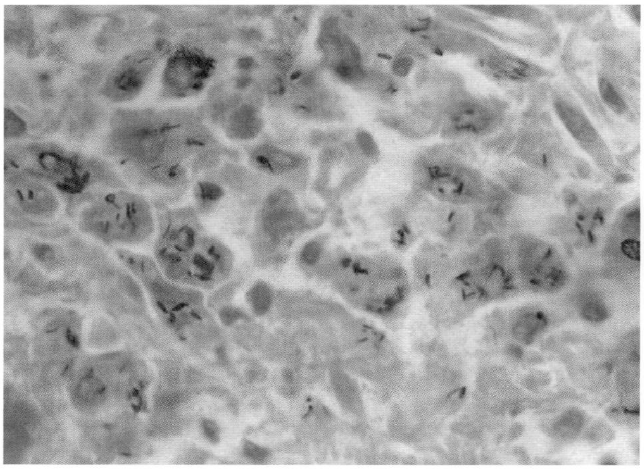

图101　在图96展示肺脏的组织病理学图片，在肺脏肉芽肿的上皮样细胞内可见抗酸染色棒状菌（红染）

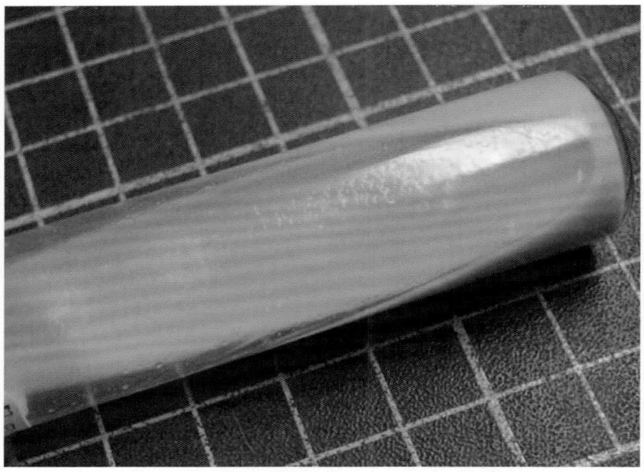

图102　分离自单峰驼，在L-J培养基上培养6周的牛分枝杆菌

板减少症。为了避免感染，骆驼活动场所的表面和器具用3%的福尔马林、2%来苏儿或2.5%的苯酚进行消毒处理。目前尚没有预防骆驼科动物结核病的疫苗可以应用。

基于皮内结核菌素检测结果制订控制骆驼科动物结核病计划还不可能，但多种检测方法联合使用可以明确了解畜群存在本病。清除受感染动物和避免进一步的感染是目前畜群内结核病控制的可靠方式。



## 参考文献

[1] Abd El-Aziz M.A.E. (1970). – Incidence of T.B. in camels and pigs and typing of the isolated organisms. Thesis,Cairo University,Cairo.

[2] Abramov L.P. (1963). – Diagnosis of tuberculosis in camels by the ophthalmic and intradermal tuberculin tests. *Veterinariya (Moscow)*,40,26–29.

[3] Abramov L.P. (1964). – The pathology of tuberculosis in camels. *In* Trudy II Vsesojuzhoj Konferencii Patologiceskoj Anatomii Zhivotnych,Moskovskoj Veterinarnoj Akademii,2,333–337.

[4] Akhundov A.A.,Amanova G.N. & Dubrovskaya V.V. (1972). – Some results of treating pulmonary tuberculosis patients with antibacterial preparations together with 'chal',a product of camel milk. Zdravochranenie,Turkmenistan,16,36–38.

[5] Alvarez J.,Bezos J.,de Juan L.,Vordermeier M.,Rodriguez S.,Fernandez-de-Mera I.G.,Mateos A. & Dominguez L. (2012). –Diagnosis of tuberculosis in camelids: old problems,current solutions and future challenges. *Trans. Emerg. Dis.*,59 (1),1–10.

[6] Andree J. (1928). – Ein Fall von generalisierter Tuberkulose bei einem Kamel. Thesis,Hannover Veterinary School,Hannover,Germany.

[7] Angrisani V. (1962). – Considerazioni sul rapporto tra l'allattamento con latte di bovini e di camelidi,in Somalia,el'eventuada infezione TBC. *Archo. Ital. Sci. Med. Trop. Parasit.*,43,205–210.

[8] Anon. (2003). – Discussion on tuberculosis in camelids. *In* Proceedings of the 10th Meeting of the British Veterinary Camelid Society,24–26 October,Wallingford,UK,20–23.

[9] Anon. (2006a). – Tb update. *In* Proceedings of the 13th Meeting of the British Veterinary Camelid Society,12–15 October,Alfriston,UK,49–53.

[10] Anon. (2006b). Llamas. *Vet Rec.*,158 (21),716.

[11] Anon. (2008). –*Mycobacterium microti. Vet. Rec.*,162 (1),8.

[12] Anon. (2009). – Alpacas. *Vet Rec.*,164 (26),802.

[13] Archibald R.G. (1910). – Acid-fast bacilli in a camel's lung,the gross lesions of which closely simulated miliary tuberculosis. *J. Comp. Path. Ther.*,23,56–57.

[14] Arush M.A. (1982). – La situazione sanitaria del dromedario nella Repubblica Democratica Somala. *Bollettino scientifica dellafacoltà di zootecnia e veterinaria*,3,209–217.

[15] Barlow A.M.,Mitchell K.A. & Visram K.H. (1999). – Bovine tuberculosis in llama (*Lama glama*). *Vet. Rec.*,145,639–640.

[16] Bromage G. (2010). – TB awareness in camelids. Available at: www.britishllamasociety.org/news/files/tb-awareness-info-pack.pdf (accessed 3 October 2010).

[17] Bush M.,Montali R.J. Phillips L.G. & Holobaugh P.A. (1986). – Tuberculosis in Bactrian camel. *In* Proceedings of the Meetingof the American Association of Zoo Veterinarians,Fort Collins,CO,22–23.

[18] Bush M.,Montali R.J. Phillips L.G. & Holobaugh P.A. (1990). – Bovine tuberculosis in a Bactrian camel herd: clinical,therapeutic,and pathologic findings. *J. Zoo Wildl. Med.*,21,171–179.

[19] Buzzel G.R.,Kinne J.,Tariq S. & Wernery U. (2010). – The pleural curtain of the camel (*Camelus dromedarius*). *Anat. Rec.*,293 (10),1776–1786.

[20] Cambre R.,Thoen C.,Lang W.O. & Richards W.L. (1981). – Mycobacteria isolated from exotic animals. *In* Proceedings of the 81st Annual Meeting of the American Veterinary Society of Medical Microbiology,296.

[21] Casati R. (1957). – Osservazione su di un caso di tubercolosi de cammello. *Atti. Soc. Ital. Sci. Vet.*,11,551–554.

[22] Castagnino Rosso D.,Ludena D.H.,Huaman D. & Ramirez A. (1974). – Linea de enfermedades infecciosas. *Bol. Divulg.*,15,145–147.

[23] Chambers M.A. (2009). – Review of the diagnosis and study of tuberculosis in non-bovine wildlife species using immunological methods. *Transbound. Emerg. Dis.*,56,215–277.

[24] Chamoiseau G.,Bah S.O. & Ahmed Vall S.M.O. (1985). – Un cas de tuberculose pulmonaire chez le dromadaire. *Rev. Elev. Méd. Vét. Pays Trop.*,38,28–30.

[25] Chartier F.,Chartier C.,Thorel M.F. & Crespeau F. (1991). – A new case of *Mycobacterium bovis* pulmonary tuberculosis in the dromedary (*Camelus dromedarius*) in Mauritania. *Rev. Elev. Méd. Vét. Pays Trop.*,44 (1),43–47.

[26] Chauhan R.S.,Kaushik R.K.,Gupta S.C.,Satija K.C. & Kulshreshtha R.C. (1986). – Prevalence of different diseases in camels (*Camelus dromedarius*) in India. *Camel Newslett.*,3,10–14.

[27] Coad M.,Clifford D.,Rhodes S.G.,Hewinson R.G.,Vordermeier H.M. & Whelan A.O. (2010). – Repeat tuberculin skin testing leads to desensitisation in naturally infected tuberculous cattle which is associated with elevated interleukin-10 and decreased interleukin-1 beta responses. *Vet. Res.*,41,14–21.

[28] Cobb N. & Cobb G. (2010). – The bio-security imperative – one farm's struggle against a silent killer. *Camelid Q.*,3,1–3.

[29] Connolly D.J.,Dwyer P.J.,Fagan J.,Hayes M.,Rayan E.G.,Costello E.,Kilroy A. & More S.J. (2008). – Tuberculosis in alpaca (*Lama pacos*) on a farm in Ireland. 2. Results of an epidemiological investigation. *Ir. Vet. J.*,61,533–537.

[30] Cross H.E. (1917). – The camel and its diseases. Bailliere,Tindall and Cox,London.

[31] Damodaran S. & Ramakrishnan R. (1969). – Tuberculosis in animals in Madras. *In* Proceedings of the 3rd International Conference on the Global Impacts of Applied Microbiology,7–14 December 1969,Bombay,India,1–86.

[32] Dean G.S.,Cranshaw T.R.,de la Rua-Domenech R.,Farrant L.,Greenwald R.,Higgins R.J.,Lyashchenko K.,Vordermeier H.M. & Twomey D.F. (2009). – Use of serological techniques for diagnosis of *Mycobaterium bovis* infection in a llama herd. *Vet. Rec.*,165,323–324.

[33] Deen A.,Dixit S.K.,Kataria A.K. & Sahani M.S. (2004). – Tuberculosis in camels: case report. *J. Camel Pract. Res.*,11 (1),81–83.

[34] Dekker N.D.M. & van der Schaaf A. (1962). – Een geval van open tuberculose bij een kameel. *Tijdschr. Diergeneesk.*,87,1133–1140.

[35] Department for Environment,Food & Rural Affairs (DEFRA) (2010a). – Tuberculosis in camelids. Available at: www.defra.gov.uk/foodfarm/farmanimal/diseases/atoz/tb/stats/latest.htm (accessed on 20 November 2010).

[36] Department for Environment Food and Rural Affairs (DEFRA) (2010b). – Badger vaccine deployment project. Available at: www.defra.gov.uk/fera/bvdp (accessed on 20 November 2010).

[37] Diatchenko F. (1989). – Contribution à l'étude lésionelle des affections respiratoires du dromadaire. Thèse,École Nationale Vétérinaire d'Alfort,France.

[38] Donchenko A.S. & Donchenko V.N. (1978). – Change of proteins in blood serum of healthy and tuberculosis-diseased camels and cattle. *Vest. Sel'-khoz. Nauki (Alma Ata)*,5,73–76.

[39] Donchenko A.S.,Donchenko V.N.,Fatkeeva E.A. & Kibasov M. (1975a). – Isolation of tuberculosis mycobacteria in camel milk,their survival in 'shubat' and methods of decontamination of these products. *Vest. Sel.'-khoz Nauki (Alma Ata)*,2,119–122.

[40] Donchenko A.S.,Donchenko V.N.,Fatkeeva E.A.,Kibasov M. & Zernova L.A. (1975b). –Destruction of tubercle bacilli in camel's milk and 'shubat',a lactic acid product. *Veterinariya (Moscow)*,2,24–26.

[41] Donchenko A.S.,Donchenko V.N. & Kenzheev S. (1975c). – Effect of tuberculinization on the blood proteins of camels and cows. *Veterinariya (Moscow)*,9,52–53.

[42] El Mossalami E.,Siam M.A. & El Sergany M. (1971). – Studies on tuberculosis-like lesions in slaughtered camels. *Zbl. Vet. Med. B.*,18,253–261.

[43] El-Afifi A.,Zaki R. & Farrag H.F. (1953). – Incidence and typing of tuberculosis in camels in Egypt. *Vet. Med. J.*,1,1–6.

[44] Fedchenko V.A. (1971a). – Tuberculosis in camels. I. Epidemiology in Kazakhstan. *Trudy Kazakh. Nauchno-issled. Vet. Inst.*,14,51–56.

[45] Fedchenko V.A. (1971b). – Tuberculosis in camels. II. Haematological changes. *Trudy Kazakh. Nauchno-issled. Vet. Inst.*,14,57–61.

[46] Fowler M.E. (2010). – Medicine and surgery of camelids. 3rd Ed. Blackwell Publishing,Oxford,UK,201–204.

[47] Garcia-Bocanegra I.,Barranco I.,Rodriguez-Gomez I.M.,Perez B.,Go'mez-Laguna J.,Rodriguez S.,Ruiz-Villamayor E. &Perea A. (2010). – Tuberculosis in alpacas (*Lama pacos*) caused by *Mycobacterium bovis. J. Clin. Microbiol.*,48 (5),1960–1964.

[48] Gatt Rutter T.E. & Mack R. (1963). – Diseases of camels. Part 1: Bacterial and fungal diseases. *Vet. Bull.*,33,119–124.

[49] Hanichen T. & Wiesner H. (1995). – Erkrankungs- und Todesursachen bei Neuweltkameliden. *Tierärztl. Praxis,*23,515–520.

[50] Hayles L.B. (1986). – Proceedings of the 1st National Veterinary Symposium,12–15 October 1986,Food and Agriculture Organization,Mogadishu,Somalia.

[51] Johnson C.T.,Winkler C.E.,Boughton E. & Penfold J.W.F. (1993). –*Mycobacterium kansasii* infection in a llama. *Vet. Rec.*,133,243–244.

[52] Johnson L.W.,Thoen C.O. & Schultheiss P.C. (1989). – Diagnostic procedures for detecting llamas exposed to mycobacteria. *In* 93rd Meeting of the United States Animal Health Association,241–244.

[53] Kennedy S. & Bush M. (1978). – Evaluation of tuberculin testing and lymphocyte transformation in Bactrian camels. *In* Mycobacterial infections in zoo animals (R.J. Montali,ed.). Smithsonian Institution Scholarly Press,Washington DC,139–143.

[54] Kibasov M. & Donchenko A.S. (1976). – Experimental determination of economic losses in camels due to tuberculosis.*Vest. Sel.'-khoz. Nauki (Alma Ata)*,12,5–8.

[55] Kinne J.,Johnson B.,Jahans K.L.,Smith N.H.,Ul-Haq A. & Wernery U. (2006). – Camel tuberculosis – a case report. *Trop. Anim. Health Prod.*,38,207–213.

[56] Kogramanov A.I.,Blagogarny Y.A.,Makarevitch N.M.,Blekhman I.M. & Yakunine M.P. (1971). – Ticks as possible carriers of tubercular infection. *Proceedings Problems of Tuberculosis (Moscow)*,9,60–64.

[57] Kudi A.C.,Bello A. & Ndukum J.A. (2011). – Prevalence of bovine tuberculosis in camels in northern Nigeria. *J. Camel Pract. Res.*,19 (1),81–86.

[58] Leese A.S. (1908). – Camel tuberculosis. Annual report of officer investigating camel disease. India.

[59] Leese A.S. (1910). – Acid-fast bacilli in camel's lung with lesions resembling those of tuberculosis. *J. Comp. Path. Ther.*,23,358–359.

[60] Leese A.S. (1918). –'Tips' on camels for veterinary surgeons on active service. Bailliere,Tindall and Cox,London,50.

[61] Lingard A. (1905). – Camel tuberculosis. Annual report of imperial bacteriologist,Lingard. India. 6.

[62] Littlewood W. (1888). – Camel tuberculosis. Egyptian Official Gazette.

[63] Lyashchenko K.P.,Greenwald R.,Esfandiari J.,Meylan M.,Burri I.H. & Zanolari P. (2007). – Antibody responses in New World camelids with tuberculosis caused by *Mycobacterium microti. Vet. Microbiol.*,125,265–273.

[64] Lyashchenko K.P.,Greenwald R.,Esfandiari J.,Rhodes S.,Dean G.,de la Rua-Domenech R.,Meylan M.,Vordermeier M.H. & Zanolari P. (2011). – Diagnostic value of animal-side antibody assays for rapid detection of *Mycobacterium bovis* or *Mycobacterium microti* in South American camelids. *Clin. Vaccine Immunol.*,18 (12),2143–2147.

[65] Manefield G.W. & Tinson A. (1996). – Camels: a compendium. The T.G. Hungerford Vade Mecum Series for Domestic Animals. University of Sydney Post Graduate Foundation in Veterinary Science,240–298.

[66] Mason F.A. (1912). – Some observations on tuberculosis in camels in Egypt. *J. Comp. Path. Ther.*,25,109–111.

[67] Mason F.A. (1917a). – Tuberculosis in camels. *Agric. J. (Egypt)*,7,2–11.

[68] Mason F.A. (1917b). – Tuberculosis in camels. *J. Comp. Path. Ther.*,30,80–84.

[69] Mason F.A. (1918). – Tuberculosis in the camel. *J. Comp. Path. Ther.*,31,100–102.

[70] Mohammed O.B.,Sandouka M.A. & Abu Elzein E.M.E. (2007). – Disease surveys of livestock in some protected areas where gazelles have been reintroduced in Saudi Arabia. *In* Proceedings of Diseases at the Interface Between Domestic Livestock and Wildlife species,17－18 July,Ames,IA.

[71] Moro Sommo M. (1957). – Investigacion preliminar de la brucelosis en alpacas. *Rev. Fac. Med. Vet. (Lima,Peru)*,12,135－137.

[72] Moser I.,Prodinger W.M.,Hotzel H.,Greenwald R.,Lyashchenko K.P.,Bakker D.,Gomis D.,Seidler T.,Ellenberger C.,Hetzel U.,Wuennemann K. & Moisson P. (2008). –*Mycobacterium pinnpedii:* Transmission from South American sea lion (*Otaria byronia*) to Bactrian camel (*Camelus bactrianus*) and Malayan tapirs (*Tapirus indicus*). *Vet. Microbiol.*,127,399－406.

[73] Mustafa I.E. (1987). – Bacterial diseases of the camel and dromedary. 55th General Session,OIE,Paris,France,55,18－22.

[74] Oevermann A.,Pfyffer G.E.,Zanolari P.,Meylan M. & Robert N. (2004). – Generalised tuberculosis in llamas (*Lama glama*) due to *Mycobacterium microti. J. Clin. Microbiol.*,42 (4),1818－1821.

[75] Oevermann A.,Zanolari P.,Pfyffer G.E.,Meylan M. & Robert N. (2003). –*Mycobacterium microti* infection in two llama (*Lama guanaco F. glama*). *Verh. ber. Erkrg. Zootiere*,41,217－219.

[76] Osman K.M. (1974). – Studies on acid-fast microorganisms in some domesticated animals with special reference to atypical mycobacterium group. PhD Thesis,Faculty of Veterinary Medicine,Cairo University,Cairo.

[77] Paling R.W.,Whaghela S.,Macowan K.J. & Heath B.R. (1988). – The occurrence of infectious diseases in mixed farming of domesticated and wild herbivores and livestock in Kenya. II. Bacterial diseases. *J. Wildl. Dis.*,24,308－316.

[78] Panebianco F. (1957). – Su di caso di tubercolosi del cammello. *Acta Med. Vet.*,3,291－302.

[79] Pate M.,Svara T.,Gombac M.,Paller T.,Zolnir-Dovc M.,Emersic I.,Prodinger W.M.,Bartos M.,Zdovc I.,Krt B.,Pavlik I.,Cvetnić Z.,Pogacnik M. & Ocepek M. (2006). – Outbreak of tuberculosis caused by *Mycobacterium caprae* in a zoological garden. *J. Vet. Med. B Infect. Dis. Vet. Public Health.*,53 (8),387－392.

[80] Pattyn S.R.,Antoine-Portaels F.,Kageruka P. & Gigase P. (1970). –*Mycobacterium microti* infection in a zoo-llama: *Lama vicugna (Molina). Acta Zoolog. Pathologica Antverpiensa*,51,17－24.

[81] Pavlik I.,Trcka I.,Parmova I.,Svobodova J. Melicharek I.,Nagy G.,Cvetnic Z.,Ocepek M.,Pate M. & Lipiec M. (2005). –Detection of bovine and human tuberculosis in cattle and other animals in six Central European Countries during the years 2000—2004. *Vet. Med. Czech.*,50 (7),291－299.

[82] Pellegrini D. (1942). – Tubercoli spontanea del cammello in Somalia. Ricerche diagnostiche esperimentali. *Racc. Stud. Vet. Pat. Somali*,1,33－41.

[83] Rana M.Z.,Ahmed A.,Sindhus. T.A.K. & Mohammed G. (1993). – Bacteriology of camel lungs. *Camel Newslett.*,10,30－32.

[84] Richard D. (1979). – Study of the pathology of the dromedary in Borana Awraja (Ethiopia). Thesis,Debre Zeit,Addis Ababa,Veterinary Faculty,Ethiopia.

[85] Robert N. (2006). – Pathological findings in New World camelids in Switzerland,with emphasis on Dicrocoeliosis and Mycobacteriosis. *In* Proceedings of the American Association of Zoo Veterinarians,56－58.

[86] Ryan E.G.,Dwyer P.J.,Connolly D.J.,Fagan J.,Costello E. & More S.J. (2008). –Tuberculosis in alpaca (*Lama pacos*) on a farm in Ireland. 1. A clinical report. *Ir. Vet. J.*,61 (8),527－531.

[87] Schiller I.,Oesch B.,Vordermeier H.M.,Palmer M.V.,Harris B.N.,Orloski K.A.,Buddle B.M.,Thacker T.C.,Lyashchenko K.P. & Waters WR. (2010). – Bovine tuberculosis: a review of current and emerging techniques in view of their relevance for disease control and eradication. *Transbound. Emerg. Dis.*,57,205－220.

[88] Schillinger D. (1987). – Kamel (*Camelus dromedarius*). Seminar Sonderdruck Veterinaria,Labhard Verlag Konstanz,Germany,9,50－53.

[89] Seifert J. (1992). – Kamelkrankheiten. *In* Tropentierhygiene (H.S.H. Seifert,ed.). Gustav Fischer Verlag,Jena,Stuttgart, Germany, 190－199.

[90] Simmons A.G. (1989). – Alternative site for the single intradermal comparative tuberculin test in the llama (*Lama glama*). *Vet. Rec.*,124,17－18.

[91] Stevens J.B.,Thoen C.O.,Rohonczy E.B.,Tessaro S.,Kelly H.A. & Duncan J.R. (1998). – The immunological response of llamas (*Lama glama*) following experimental infection with *Mycobacterium bovis. Can. J. Vet.* Res.,62,102－109.

[92] Stutzman L.C. (2008). – Serological detection of tuberculosis in camelids. *In* Proceedings of the International Camelid

Health Conference for Veterinarians,19 March 2008,Columbus,OH,84–91.

[93] Tanwar R.K. (2006). – Pulmonary tuberculosis in camels (*Camelus dromedarius*). *In* Proceedings of the International Scientific Conference on Camels,10–12 May,Gassim,Saudi Arabia,411–415.

[94] Thoen C.O. (1988). – Tuberculosis in llamas. *In* Proceedings of the American Association of Zoo Veterinarians (AAZV) Annual Conference,6–10 November 1988,Toronto,Canada,174.

[95] Thoen C.O.,LoBue P. & de Kantor I. (2006). – The importance of *Mycobacterium bovis* as a zoonosis. *Vet. Microbiol.*,112,339–345.

[96] Thoen C.O.,Mills K. & Hopkins M.P. (1980). – Enzyme-linked protein A: an enzyme-linked immunosorbent assay reagent for detecting antibodies in tuberculous exotic animals. *Am. J. Vet. Res.*,41,833–835.

[97] Thoen C.O.,Richards W.D. & Jarnigan J.L. (1977). – Mycobacteria isolated from exotic animals. *JAVMA*,170,987–990.

[98] Twomey D.F.,Crawshaw T.R.,Anscombe J.E.,Farrant L.,Evans L.J.,McElligott W.S.,Higgins R.J.,Dean G.,Vordermeier M.,Jahans K. & de la Rua-Domenech R. (2007). – TB in llamas caused by *Mycobacterium bovis*. *Vet. Rec.*,160 (5),170.

[99] Twomey D.F.,Crawshaw T.R.,Foster A.P.,Higgins R.J.,Smith N.H.,Wilson L.,McDean K.,Adams J.L. & de la Rua-Domenech R. (2009). – Suspected transmission of *Mycobacterium bovis* between alpacas. *Vet. Rec.*,165,121–122.

[100] Twomey D.F.,Gwashaer T.R.,Anscombe J.E.,Barnett J.E.F.,Farrant L.,Evans L.J.,McElligott W.S.,Higgins R.J.,Dean G.S.,Vordermeier H.M. & de la Rua-Domenech R. (2010a). – Assessment of ante mortem test used in the control of an outbreak of tuberculosis in llamas (*Lama glama*). *Vet. Rec.*,167 (13),475–480.

[101] Twomey D.F.,Higgins R.J.,Worth D.R.,Okher M.,Gover K.,Nabb E.J. & Speirs G. (2010b). – Cutaneous TB caused by *M. bovis* in a veterinary surgeon following exposure to a tuberculous alpaca (*Vicugna pacos*). *Vet. Rec.*,166 (6),175–177.

[102] Wernery U. & Kaaden O.-R. (1995). – Infectious diseases of camelids. Blackwell Wissenschafts-Verlag,Berlin,Germany.

[103] Wernery U. & Kaaden O.-R. (2002). – Infectious diseases of camelids. 2nd Ed. Blackwell Science,Berlin and Vienna,91–96.

[104] Wernery U.,Kinne J.,Jahans K.L.,Vordermeier H.M.,Esfandiari J.,Greenwald R.,Johnson B.,Ul-Haq A. & Lyashchenko K.P. (2007a). – Tuberculosis outbreak in a dromedary racing herd and rapid serological detection of infected camels. *Vet. Microbiol.*,122 (1–2),108–115.

[105] Wernery U.,Thomas R.,Syriac G.,Raghavan R. & Kletzka S. (2007b). – Seroepidemological studies for the detection of antibodies against nine infectious diseases in dairy dromedaries (Part 1). *J. Camel Pract. Res.*,14 (2),85–90.

[106] World Organisation for Animal Health (OIE) (2012). – Manual of diagnostic tests and vaccines for terrestrial animals. 7th Ed. Vol. 2. World Organisation for Animal Health,Paris,674–689.

[107] Younis E.E.,Ibrahim M.S.E.,El-Balkemy F.A.,El-Mekrawi M.F. & Newman H.M. (2006). – Comparison between single intradermal and single comparative intradermal tuberculin tests in relation to ELISA in clinically suspected tuberculous camels. *In* Proceedings of the International Scientific Conference on Camels,12–14 May,Buraidah,Saudi Arabia,455–463.

[108] Zanolari P.,Robert N.,Lyashchenko K.P. & Pfyffer G.E. (2009). – Tuberculosis caused by *Mycobacterium microti* in South American Camelids. *J. Vet. Int. Med.*,12,1–7.

[109] Zerom K.,Tessema T.S.,Mamo G.,Bayu Y. & Ameni G. (2013). – Tuberculosis in dromedaries in eastern Ethiopia: Abattoirbased prevalence and molecular typing of its causative agents. *Small Rumi. Res.*,109,188–192.

（娄忠子译，殷宏校）

## 1.3.2 肺炎

最常见的呼吸系统疾病是肺炎，为伴有细支气管炎症和胸膜炎的肺脏实质性炎症。临床上，该病在呼吸系统疾病中的比例呈上升态势，主要症状包括咳嗽、听诊时可听到异常呼吸音和浅呼吸。骆驼发生肺炎时并不表现上述临床症状，因而在骆驼科动物诊断肺炎非常困难。目前，这种现象的成因尚不明确。很多马红球菌感染和结核病导致的肺炎病例都是在剖解时才能诊断，此时，骆驼双肺的大部分已经严重受损（见1.1.5节）。现在有几种鉴别诊断方法用于不同类型肺炎的区分，其中，比较有效的是根据肺炎临床症状和病原体鉴别，它也同样适用于骆驼科动物（表31）。

表31　骆驼科动物肺炎类型（除了结核病，马红球菌病和马鼻疽）

| 肺炎类型 | 微生物分离（比例） | 动物种类 | 病例数 | 作者 | 年份 | 国别 |
|---|---|---|---|---|---|---|
| 肉芽肿性 | 骆驼链丝菌（鼻疽放线菌，鼻疽诺卡菌？） | 单峰驼 | 2 | Mason | 1919 | 埃及，苏丹 |
| | 假结核病 | 单峰驼 | 2 | Leese | 1927 | 埃及 |
| | 荚膜组织胞浆菌 | 单峰驼 | 2 | Chandel和Kher | 1994 | 印度 |
| | 星形诺卡菌 | 美洲驼 | 1 | Ching-Dong等 | 1993 | 美国 |
| | 曲霉菌属真菌和化脓棒状杆菌 | 单峰驼 | 1 | Bhatia等 | 1983 | 印度 |
| 卡他性（急性，亚急性，慢性） | 双球菌 | 双峰驼 | 地方流行 | Semushkin | 1968 | 蒙古 |
| | 双球菌 | 双峰驼 | 地方流行 | Buchnev等 | 1987 | 俄罗斯 |
| | 化脓棒状杆菌（20%） | 单峰驼（屠宰场） | 50 | Farrag等 | 1953 | 埃及 |
| | 溶血性链球菌（12%） | | | | | |
| | 类白喉菌（16%） | | | | | |
| | 草绿色链球菌（20%） | | | | | |
| | 肠道沙门氏菌肠道亚种肠炎血清型（2%） | | | | | |
| | 大肠杆菌群（16%） | | | | | |
| | 粪产碱杆菌（22%） | | | | | |
| | 化脓放线杆菌（2%） | | | | | |
| 脓肿 | 葡萄球菌属 | 单峰驼（屠宰场） | 79 | Moallin和Zessin | 1990 | 索马里 |
| | 绿脓杆菌 | 单峰驼（屠宰场） | 1 | Abdurahman | 1987 | 索马里 |
| | | 单峰驼 | 1 | Gautam等 | 1970 | 印度 |

续表

| 肺炎类型 | 微生物分离（比例） | 动物种类 | 病例数 | 作者 | 年份 | 国别 |
|---|---|---|---|---|---|---|
| 化脓性支气管炎脓肿 | 溶血性链球菌和葡萄球菌 | 单峰驼（屠宰场） | 15 | Vitovec和Vladic | 1983 | 索马里 |
| 坏死性 | 类鼻疽克雷伯氏菌 | 单峰驼 | 4 | Bergin和Torenbeck | 1991 | 澳大利亚 |
| | | | 1 | Wernery等 | 1997 | 阿联酋 |
| 出血性 | 马链球菌马亚种 | 单峰驼 | 动物流行 | Yigezu等 | 1997 | 埃塞俄比亚 |
| 不明确（多类型混合感染） | 表皮葡萄球菌 | 单峰驼（屠宰场） | 20 | El Mossalami和Ghawi | 1981 | 埃及 |
| | 玫瑰色微球菌 | | | | | |
| | 滕黄微球菌 | | | | | |
| | 化脓链球菌 | | | | | |
| | 肺炎链球菌 | | | | | |
| | 金色葡萄球菌 | | | | | |
| | 微球菌属 | | | | | |
| | 浅绿亚种气球菌 | | | | | |
| | 爱德华菌属 | | | | | |
| | 棒状菌属（21%） | 单峰驼（屠宰场） | 63 | Rana等 | 1993 | 巴基斯坦 |
| | 葡萄球菌属（30%） | | | | | |
| | 链球菌属（12%） | | | | | |
| | 大肠杆菌（5%） | | | | | |
| | 假单胞菌属（8%） | | | | | |
| | 变形杆菌属（11%） | | | | | |
| | 克雷白氏杆菌属（6%） | | | | | |
| | 芽孢杆菌属（5%） | | | | | |
| 化脓性 | 肺炎克雷伯氏菌 | 单峰驼 | 6 | Al Darraji，Wajid | 1990 | 伊拉克 |
| | | 单峰驼 | 2 | Arora和Kalra | 1973 | 印度 |
| 慢性非脓性 | | 单峰驼 | 6 | Al Darraji，Wajid | 1990 | 印度 |
| 间质性 | | 单峰驼 | 83 | Al Darraji，Wajid | 1990 | 印度 |
| 淋巴性 | 类梅迪-维斯纳病毒 | 单峰驼 | 6 | Al Darraji，Wajid | 2006 | 阿联酋 |
| | | 单峰驼 | | Kinne，Wernery | 2002 | 瑞士 |
| | | 单峰驼 | 7 | Al Darraji，Wajid | 1990 | 印度 |

续表

| 肺炎类型 | 微生物分离（比例） | 动物种类 | 病例数 | 作者 | 年份 | 国别 |
|---|---|---|---|---|---|---|
| 慢性增生性 | | | | | | |
| 纤维素性（矽肺） | | 单峰驼 | 11 | Abdurahman | 1987 | 索马里 |

## 病原学

肺炎可以由细菌和病毒引起，也可能由真菌、寄生虫和一些物理化学因素共同作用引起。动物的大多数肺炎是吸入引起的支气管肺炎，也有些动物通过血液路径引起，例如，一些幼小动物与败血症相关的肺炎。呼入性肺炎或吸入性肺炎都是比较常见但很严重的肺炎类型。在很多肺炎病例中，鼻腔正常菌群的改变，一种或多种病原体突然大量增加，都是感染肺炎的诱因。细菌被大量吸入肺部，突破机体防御系统后大量增殖，从而引起发病。此外，病毒感染呼吸系统可能是细菌感染性肺炎的诱因，处置、转运、混群和过度拥挤也常常被认为是肺炎发病的前因。

## 流行病学，临床症状和病理学变化

对肺炎的易感性由动物对感染的抵抗力来决定，但一些风险因子，例如，长途转运、畜群过度拥挤、长时间缺乏食物和水等也会是动物发病的原因。在很多国家的屠宰场进行过骆驼科动物肺炎发病情况的调查。Abdel Rahim等[1]在利比亚检查了204峰屠宰的单峰驼，发现一半的肺脏病理解剖学变化是包虫囊和肺炎造成的。Al Darraji和Wajid[4]在伊拉克220峰已屠宰单峰驼中，56%肺脏发现细菌，描述了7种不同类型的肺炎（表31）。Vitovec和Vladik[43]在索马里已屠宰15峰单峰驼中发现了肺脏脓肿，并从中分离到了溶血性链球菌。从病原学上看，化脓性支气管炎易于扩散到肺脏实质，从而引起肺脏病变。Moallin和Zessin[32]在索马里从单峰驼患有化脓性浸润性肺炎病料中分离到了葡萄球菌、铜绿假单胞菌（绿脓杆菌）和弗氏柠檬酸杆菌。Abdurahman[2]从200峰单峰驼的6峰（3%）中分离到假单胞菌、大肠杆菌、双球菌、葡萄球菌和其他一些细菌。Ghawi[21]从埃及肺炎发病骆驼中分离到金色葡萄球菌和克雷伯氏肺炎菌。Farrag等[18]在开罗从屠宰的单峰驼中诊断出大量的肺炎病例，认为一些诱因导致了疾病的发展。在开罗，单峰驼在被屠宰之前要经历没有食物的长时间跋涉才能到屠宰场，然后被安置在肮脏杂乱的畜栏内。按照规则，它们要在两三个月之后才被屠宰。这些应急因素很可能是单峰驼肺炎发病率增高的原因。在50个病例的肺脏的组织学检查中，研究人员从急性肺炎和慢性肺炎样品中鉴定出9个不同菌种。

研究人员从拉合尔和费萨尔巴德的屠宰场采集了100份单峰驼肺脏样品，并对其进行了组织学和细菌学检查。病理学检测发现和分离到的微生物之间的关系见表32[26]。

苏丹也有对丧失功能的单峰驼肺脏进行细菌学和病理学研究的报道[33,34,42]。Tigani等[42]研究了屠宰后的单峰驼肺脏内分离到的细菌与病理变化之间的相关性。他们的研究结果在汇总于表33中。

从各国屠宰场中报道的情况判断，骆驼肺脏出现病理解剖学变化的病例很常见，因而本书早先提到的活骆驼中呼吸系统疾病罕见的事实就比较奇怪。只有几位研究人员报道过活骆驼发生肺炎和支气管肺炎的情况。

Buchnev等[10]报道了一脓毒性肺炎病例，并称之为传染性咳嗽。本病例表现为上呼吸道和肺脏黏膜的急性卡他性炎症，并有高热和其他常见病变。病原为带荚膜的双球菌，对豚鼠有致死能力。根据Semushkin[38]的报道，本病在蒙古也有发现，在当地被称为黑肺病或传染性咳嗽。本病虽然广泛传播，但1920年前未见与之相关的报道[5]。Oinakhbaev[35]报道了一个有5000峰骆驼的驼群由蒙古向哈萨克斯坦转运时暴发咳嗽病的病例，认为是饥饿、苦役和长时间的路途疲劳诱发了呼吸系统疾病，并且降低了骆驼的抵抗力，加重了病情。临床症状出现后，本病持续了1～2个月，症状有发烧、淋巴结肿大、多汗和沉郁。咳嗽症状随着病程延长逐渐变得严重，并出现呼吸困难[28,44]。Diatchenko[16]根据已有文献汇总编写了单峰驼不同的细菌和病毒感染引起的呼吸系统疾病。

Arora和Kalra[7]记录了印度单峰驼慢性支气管肺炎的发病情况，报道称本病仅发生在较冷的月份，基本上只感染成年动物。动物的发病率可达30%，但只有几峰骆驼死亡。这些动物发病后，呈现迁延不愈的病情，这期间均不适合使役，因而带来一定的经济损失。此外，两峰死于支气管肺炎的单峰驼肺脏中分离培养到了克雷伯氏肺炎菌和溶血性双球菌。

许多研究人员都针对单峰驼肺炎进行过报道。Leese[29]报道了一些呼吸系统疾病发展为肺部脓肿病例。Gautam等[20]报道了一例10岁龄单峰驼肺部脓肿病例，骆驼的右肺叶几乎全部被脓肿包



表32　100峰单峰驼肺脏菌群检查和病理变化比较[26]

| 微生物 | 发现个数 | 病理变化 |
|---|---|---|
| 葡萄球菌属 | 7 | 充血 |
| 棒状杆菌属 | 8 | 充血 |
| 克雷伯氏菌属 | 2 | 充血 |
| 假单胞菌属 | 5 | 充血 |
| 葡萄球菌属 | 6 | 肝样变 |
| 链球菌属 | 7 | 肝样变 |
| 克雷伯氏菌属 | 2 | 肝样变 |
| 棒状杆菌属 | 5 | 肝样变 |
| 大肠杆菌属 | 3 | 肝样变 |
| 葡萄球菌属 | 6 | 支气管炎 |
| 芽孢杆菌属 | 2 | 支气管炎 |
| 变形杆菌属 | 4 | 尘肺病 |
| 芽孢杆菌属 | 1 | 尘肺病 |
| 变形杆菌属 | 3 | 包虫囊 |
| 分枝杆菌属 | 2 | 结节瘤 |

表33　分离菌比率与肺脏不同类型病理变化的相关性[42]

| 菌种 | 充血和红色肝变 | 灰色肝变 | 脓肿 | 干酪样结节 | 黏连 |
|---|---|---|---|---|---|
| 葡萄球菌属 | 8（30.7%） | 3（11.5%） | 8（30.7%） | 7（26.9%） | — |
| 链球菌属 | 2（15.4%） | 2（15.4%） | 5（38.5%） | 3（23%） | 1（7.6%） |
| 微球菌属 | 3（43%） | — | 1（14%） | 3（43%） | — |
| 肺炎球菌属 | 8（66.6%） | 1（8.3%） | 1（8.3%） | 2（16.6%） | — |
| 棒状杆菌属 | 7（53.6%） | — | 3（23%） | 2（15.3%） | 1（7.7%） |
| 放线菌属 | — | — | 1（100%） | — | — |
| 嗜血杆菌属 | 2（100%） | — | — | — | — |
| 巴斯德菌属 | 1（100%） | — | — | — | — |
| 肠杆菌属 | 6（66.6%） | 2（22.2%） | 1（11.1%） | — | — |
| 假单胞菌属 | 1（100%） | — | — | — | — |

围，但研究人员没有明确说明发病原因。Mason[31]报道了一例苏丹单峰驼幼仔假结核感染病例，在动物的肺脏可见类似结核分枝杆菌感染的病理解剖学变化。Hansen等[22]报道了一例在索马里发现的单峰驼矽肺病。Kamel[25]在埃及单峰驼中发现了大叶性肺炎病例，Agab等[3]从患有肺炎的骆驼肺中分离到致病性凝结芽孢杆菌。

此外，Khodakaram-Tafti和Mansourian[26]在伊朗调查了屠宰骆驼中肺脏的病理变化，在检查的100份肺脏样品中，58%的样品有病理损伤，这其中27%为双瓣线虫病，18%为间质性肺炎，16%为包虫病，15%为煤肺病，8%为脓性病变，6%为纤维素性及脓性支气管肺炎和4%的胸膜炎。研究结果发现，寄生虫感染是造成伊朗骆驼肺病变的最主要因素。

链球菌是骆驼科动物肺脏感染和其他疾病中的重要病原体之一，但临床上无病变的肺脏中也能分离出链球菌[30,37,39]。目前已有报道称，从戈壁沙漠的双峰驼肺脏中分离到溶血性肺炎球菌[35]，从单峰驼中分离出了绿色链球菌[40]、肺炎双球菌和化脓链球菌[36]。在巴林岛，Ibrahim等[24]从单峰驼鼻腔中培养出了兽疫链球菌2型，该单峰驼因为鼻腔坏死组织完全地阻塞鼻腔通道和额窦，最终窒息而死。一峰雄性澳大利亚单峰驼脓毒性腹膜炎病例中也分离出了兽疫链球菌[23]。

最为重要的一个报道来自Yigezu等人，他们在埃塞尔比亚一场动物流行病暴发期间从一峰发病单峰驼体内分离到马链球菌马亚种[47]。本病传染性极强，高发病率，但低死亡率。其主要的临床症状是高热、流泪、咽喉和眼眶上小窝水肿、食欲下降、咳嗽、呼吸困难和化脓性鼻腔分泌物。剖解后可见肺脏出血和肺小叶间隔膜增厚。这是首次报道引起马属动物发病的病原体会引起骆驼发病，且本次疾病暴发可能与驴有关。

链球菌在新大陆骆驼中的感染也极为普遍，现已从新大陆骆驼脓肿分离出多种链球菌，成年美洲驼中也有败血症肠球菌感染的报道[11]。羊驼高热是一种由兽疫链球菌感染引起的败血病，Thedford和Johnson[41]已有相关报道，应激因素往往是本病发生的诱因。肺炎是新大陆骆驼幼畜中的常见病，现已发现有几种细菌为其病原。在其他种类的动物中（包括新大陆骆驼幼畜），发生败血病的动物常常会出现肺炎症状。大肠杆菌是最为常见的细菌，骆驼放线菌也可能会引起化脓性肺炎。因为患有严重的呼吸困难和苍白病，一峰美洲驼接受安乐死并被剖检，发现其肺脏有坏死性肺炎病变，病变组织样品分离培养出了星形诺卡氏菌。显微镜检后发现，肺脏内弥散有大量小化脓性肉芽肿，内含大量细胞碎片、巨噬细胞和中性粒细胞，外周被一些多核巨细胞包绕。肺胸膜也出现病变，因纤维素浆液性物质聚集而增厚。一般认为，由于该美洲驼接受了超出疗程的抗生素治疗，因而增加了条件性致病诺卡氏菌感染的可能性。

Shigidi[39]检查了64份屠宰后的苏丹单峰驼鼻腔拭子和支气管淋巴结，Chauhan等[14]调查了219份印度健康单峰驼的鼻腔拭子，在表34中进行了两项研究的结果对比。曾有一例4岁龄雌性单峰驼窒息病例，病原最后确定为能引起动物肺炎的马链球菌马亚种，动物在死前表现有厌食、呼吸困难、鼻炎、颌下和咽部区域明显肿大。死后剖检发现，下颌淋巴结有大量化脓灶，切开后有黄色奶油样脓汁流出，组织学检查为化脓性淋巴腺炎[6]。

阿联酋单峰驼肺炎病例很罕见，这很可能是因为赛用驼群和育种驼群受到的管理更为科学。如果有肺炎发生，它很可能与其他全身性疾病相联系而出现，不是独立的一个疾病。肺炎常常与下述几种疾病相联系。

（1）大肠杆菌病。

（2）脐炎。

（3）梭菌性肠毒血症。

（4）硒/维生素E缺乏症。

（5）脓性皮炎。

（6）肺透明膜病。

（7）马红球菌感染。

早产单峰驼幼畜的透明膜病值得关注，本病也发生在其他动物（羔羊、猴子）和人类[12]，可能与动物的机能衰退、肺不张和出生前后窒息有关。单峰驼幼畜活检发现，肺的致密程度较正常动物明显增高。组织学检查发现肺泡壁有透明膜，也可以看到动脉血栓、纤维素性渗出导致的肺泡巨噬细胞脱落和细胞碎片（图103）。透明膜病明显与肺炎有关。

如上所述，阿联酋成年单峰驼中的肺炎发病情况与单峰驼幼畜一样，均与其他疾病相关。单峰驼支气管肺炎往往与白细胞增生症（见2.2.4）和用瓶子经口给药时吸气有关，且因不恰当的给药方

表34 苏丹和印度单峰驼鼻拭子和支气管淋巴结细菌分离结果

| 分离菌 | 苏丹（n=64）（%） | 印度（n=219）（%） |
|---|---|---|
| 需氧菌 | — | — |
| 凝固酶阴性葡萄球菌 | 26.2 | 2.4 |
| 类白喉菌 | 15.9 | 13.7 |
| 曲霉属真菌 | 8.7 | — |
| 化脓放线菌 | 5.4 | 10.9 |
| $\alpha$-溶血性链球菌 | 5.1 | 2.7 |
| 链霉菌 | 4.1 | — |
| 金色葡萄球菌 | 2.6 | 10.5 |
| 大肠杆菌 | 1.0 | 24.7 |
| 肠杆菌属 | 0.5 | — |
| 肺炎克雷伯氏菌 | — | 11.9 |
| 马红球菌 | — | 8.6 |
| $\beta$-溶血性链球菌 | — | 3.7 |
| 溶血性双球菌 | — | 3.7 |
| 溶血隐秘杆菌 | — | 0.9 |
| 奈瑟氏菌属 | — | 0.5 |

式所引起的吸入性肺炎相当常见。除了会引起炎症的化脓性支气管肺炎，大多数病例还会出现胸膜炎（图104）。

## 治疗

目前治疗骆驼科动物肺炎的首选方法是广谱抗生素和抗炎症药物联用，并辅以精心照料和支持疗法。可供选用的抗生素种类有：甲氧苄氨嘧啶/磺胺嘧啶、普鲁卡因青霉素G、庆大霉素和土霉素。抗炎症药包括氟尼辛、葡甲胺和地塞米松。

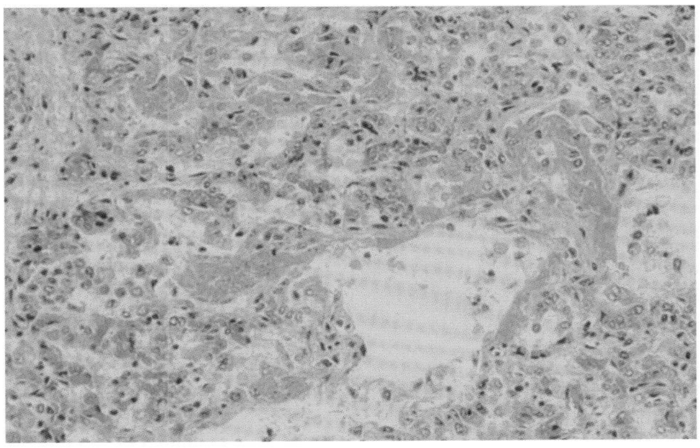

图103 单峰驼幼畜透明膜病：肺泡壁可见透明膜（苏木精-伊红染色）

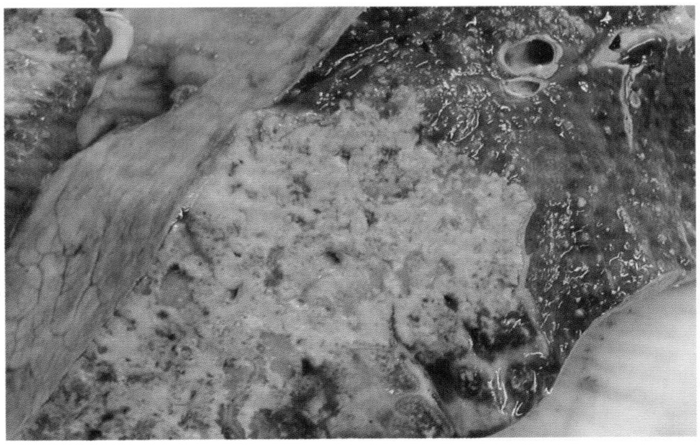

图104 由不恰当经口给药引起的伴有严重胸膜炎病变的吸入性肺炎

## 参考文献

[1] Abdel Rahim A.I.,Benhaj K.M. &Elzurgani M. (1990). – A preliminary study on some Libyan camel affections and the economic losses due to condemnations at slaughter houses. *In* Proceedings of the International Conference on Camel Production and Improvement,10－13 December,Tobruk,Libya,233.

[2] Abdurahman O.A. Sh. (1987). – Pulmonary lesions among slaughtered camels in Mogadishu. *Camel Forum*,20.

[3] Agab H.A.M.,Bakhiet M.R.,El Jack H. &Mamoum I.E. (1993). – The pathogenicity of a *Bacillus coagulans* strain isolated from a camel (*Camelusdromedarius*) in Sudan. *Bull. Anim. Prod. Afr.*,41,269－270.

[4] Al Darraji A.M. &Wajid S.J. (1990). – Etiological and pathological study of camel's lung lesions in Iraq. *Camel Newslett.*,7 (12),77.

[5] Amanzhulov S.A.,Abruzov L.N. &Zhuravlev A.M. (1929). – About one infectious disease of camels with unelucidated aetiology in the Ural province and the experience of experimental challenge. *VeterinarnyiTruzhenic*,1－4.

[6] Anon. (1999). – Strangles-like disease in a she camel. A case report. CARDN/ACSAD/Camel/P 55/1999. Studies on camel disease in Egypt,The Arab Center for the Studies of Arid Zones and Dry Lands,Damascus,Syrian Arab Republic,61.

[7] Arora R.G. &. Kalra D.S. (1973). – A note on isolation of *Klebsiellapneumoniae* and Diplococci from cases of bronchopneumonia in camels. *Ind. J. Anim. Sci.*,43 (12),1 095－1 096.

[8] Bergin T.J. &Torenbeck L.R. (1991). –Melioidosis in camels. *Australian Vet. J.*,68,309.

[9] Bhatia K.C.,Kulshreshtha R.C. & Paul Gupta R.K. (1983). – Pulmonary aspergillosis in camel. *Haryana Vet.*,XXII,118－119.

[10] Buchnev K.N.,Tulepbaev S.Z. &Sansyzbaev A.R. (1987). – Infectious diseases of camels in the USSR. *In* Diseases of camels. *Rev. sci. tech. Off. int. Epiz.*,6 (2),487－495.

[11] Burkhardt J.E.,Janovitz E.B.,Bowerstock T.L. & Higgins R. (1993). – Septicemic enterococcus infection in an adult llama.*J. Vet. Diagn. Invest.*,5,106－109

[12] Caswell J.F & Williams K.J. (2007). – Neonatal respiratory distress syndrome. *In* Jubb,Kennedy,and Palmer's pathology of domestic animals. 5th Ed. Vol. 2 (M.G. Maxie,ed.). Elsevier-Saunders,Philadelphia,PA,571.

[13] Chandel B.S. &Kher H.N. (1994). – Occurrence of histoplasmosis-like disease in camel (*Camelusdromedarius*). *Ind. Vet.J.*,71 (5),521－523.

[14] Chauhan R.S.,Gupta S.C.,Satija K.C.,Kulshreshtha R.C. &. Kaushik R.K. (1987). – Bacterial flora of upper respiratory tract in apparently healthy camels. *Ind. J. Anim. Sci.*,57 (5),424－426.

[15] Ching-Dong C.,Boosinger T.R.,Dowling P.D.,McRae E.E.,Tyler J.W. & Pugh D.G. (1993). –Nocardiosis in a llama. *J. Vet. Diagn. Invest.*,5,631－634.

[16] Diatchenko F. (1989). – Contribution à l'étude lésionelle des affections respiratoires du dromadaire. Thèse,École Nationale Vétérinaire d'Alfort,France,22.

[17] El Mossalami E. &Ghawi A. (1981). – Public health importance of camel lung affections. *Egypt. J. Vet. Sci.*,18 (1－2),109－119.

[18] Farrag H.,Zaki R. & El Hindawi M.R. (1953). – Pneumonia in camels. *Brit. Vet. Rec.*,59,119－122.

[19] Fowler M.E. (2010). – Medicine and surgery of camelids. 3rd Ed. Wiley-Blackwell,Oxford,UK,173－230.

[20] Gautam O.P.,Gulati R.L. & Gera K.L. (1970). – Pulmonary abscess (Malli) in a camel. *Ind. Vet. J.*,47 (4),364－365.

[21] Ghawi A.M. (1978). – Public health importance of camel lung affections. Thesis,Cairo University,Cairo.

[22] Hansen H.J.,Jama F.M. &Abdulkadir O. (1987). – Silicosis in camels. A preliminary report. SIDA Regional Seminar onVeterinary Pathology,Debrezeit,Ethiopia.

[23] Heller M.,Anderson D. &Silveira F. (1998). – Streptococcal peritonitis in a young dromedary camel. *Australian Vet. J.*,76 (4),253－254.

[24] Ibrahim A.M.,Abdelghaffar A.A. &Fadlalla M.E. (1998). –*Streptococcus zooepidemicus* infection in a female camel in Bahrain.*J. Camel Prac. Res.*,5 (1),165－176.

[25] Kamel H. (1939). – Pneumococcus in camels. *Tech. Sci. Service Bull.*,226.

[26] Khodakaram-Tafti A. &Mansourian M. (2010). – Pulmonary lesions of slaughtered camels (*Camelusdromedarius*) in southern Iran. *J. Camel Pract. Res.*,17 (1),21－24.

[27] Kinne J. &Wernery U. (2006). –Lymphosarcoma in dromedaries in the United Arab Emirates (UAE). *In* Proceedings of the International Scientific Conference on Camels,10－12 May 2006,Qassim,Saudi Arabia,714－723.

[28] Kuznetsov S.V. (1962). – Camels with infectious lung inflammation in the Turkmen SSR. *Trudy turkmenskogo,Niskhi,Ashkha bad,*XI.

[29] Leese A.S. (1927). – A treatise on the one-humped camel in health and disease. Vigot Freres,Paris II.

[30] Mahmoud A.Z.,Moustafa S.I. & El-Yas A.H. (1988). – Letter. *Assiut Vet. Med. J.*,20,93.

[31] Mason F.A. (1919). – Pseudo-actinomycosis or Streptotrichosis in the camel. *J. Comp. Path. Ther.*,32 (1),34－42.

[32] Moallin A.S.M. &Zessin K.H. (1990). – Note on diseases of the dromedaries at Beletweyne abattoir of Central Somalia. *Camel Newslett.*,7 (12),69.

[33] Mustafa K.E. (1992). – Study on clinical,etiological and pathological aspects of pneumonia in camels (*Camelusdromedarius*) in Sudan. MVSc Thesis,University of Khartoum Veterinary Medicine Faculty,Sudan.

[34] Nasr N.A. (2003). – Aerobic bacteria associated with respiratory infections of camels: isolation and identification. MVScThesis,University of Khartoum Veterinary Medicine Faculty,Sudan.

[35] Oinakhbaev S. (1965). – Study of aetiology of contagious cough in camels. *Veterinariya (Moscow)*,42 (6),105－106.

[36] Pal M. &Chandel B.S. (1989). – Letter. *Ind. Vet. Med. J.*,13,277.

[37] Rana M.Z.,Ahmed A.,Sindhu S.T.A.K. & Mohammed G. (1993). – Bacteriology of camel lungs. *Camel Newslett.*,10 (6),30－32.

[38] Semushkin N.R. (1968). – Diagnosis of camel diseases. *Sel'khozgiz Moscow*.

[39] Shigidi M.T.A. (1973). – Aerobic microflora of respiratory tract of camels. *Sudan J. Vet. Sci. Anim. Husb.*,14 (1),9－14.

[40] Thabet A. El. R. (1994). – Letter. *Assiut. Vet. Med. J.*,30,188.

[41] Thedford R.R. & Johnson L.W. (1989). – Infectious diseases of New-world camelids (NWC). *Vet. Clin. North Am. Food Anim.Pract.*,5 (3),145－157.

[42] Tigani T.A.,Hassan A.B. &Ababer A.D. (2007). – Bacteriological and pathological studies on condemned lungs of onehumped camel (*Camelusdromedarius*) slaughtered in Tamboul and Nyala abattoirs,Sudan. *In* Proceedings of the International Camel Conference,Khartoum,Sudan,1－13.

[43] Vitovec J. &Vladik P. (1983). – Bronchial disease of camels in Somalia. *Bull. Anim. Health. Prod. Afr.*,31 (3),291－294.

[44] Voikulesku M. (1963). – Streptococcus infection. Infectious diseases. 1. Meridiane,Bucharest,Romania,107－126.

[45] Wenker C.J.,Stuedli A. & Hauser B. (2002). – Malignant lymphoma in 2 vicunas (*Lama vicugna*). *In* Proceedings of the European Association of Zoo and Wildlife Veterinarians,8－12 May 2002,Heidelberg,Germany,65－67.

[46] Wernery R.,Kinne J.,Haydn-Evans J. &Ul-Haq A. (1997). –Melioidosis in a seven year old camel,a new disease in the United Arab Emirates (UAE). *J. Camel Prac. Res.*,4 (2),141－143.

[47] Yigezu M.,Roger F.,Kiredjian M. & Tariku S. (1997). – Isolation of *Streptococcus equi* subspecies equi(strangles agent) from an Ethiopian camel. *Vet. Rec.*,140,608.

## 深入阅读材料

Abu Elgasim K.E.M. (1992). – Study on clinical,aetiological and pathological aspects of pneumonia in camels (*Camelusdromedarius*).Thesis,Faculty Veterinary Science,University of Khartoum,Khartoum,Sudan.

Agab H. & Abbas B. (1999). – Epidemiological studies on camel diseases in the eastern Sudan. *WAR/RMz.*,92,42－51.

Awol N.,Ayelet G.,Jenberie S.,Gelaye E.,Sisay T. &Nigussie H. (2011). – Bacteriological studies on pulmonary lesions of camel(*Camelusdromedarius*) slaughtered at Addis Ababa abattoir,Ethiopia. *African J. Microbiol. Res.*,5 (5),522–527.

El-Metwally A.E.,Yassin M.H. & El-Hallawany H.A. (2010). –Mycoplasmosis in slaughtered she-camels: pathological studies.*Global Veterinaria*,4 (5),474–482.

El Mossalami E. &Ghawi A. (1983). – Public health importance of camel lung affections. *Egypt. J. Vet. Sci.*,18 (1–2),109–119.

Graber M. (1968). – Region of veterinary and zootechnical research of central Africa. Annual report,Farcha Laboratory. 1st research and products 2 pleuropneumonia. Quinquennial report. Fort Lami,Chad. *Vet. Bull.*,38:5 265

Hansen H.,Jama F.M.,Nilsson C.,Norrgren L. &Abdurahman O. Sh. A. (1989). – Silicate pneumoconiosis in camel (*Camelusdromedarius*L.). *ZentralblVeterinarmed A*. 36 (10),789–796.

Kogramanov A.I. (1967). – Microbiology of tuberculosis in the USSR during 60 years,1,28–43.

Sechi L.A.,Roger F.,Diallo A.,Yigezu L.M.,Zanetti S. &Fadda G. (1999). – Molecular characterisation of *Streptococcus equi* subspecies equi isolated from an Ethiopian camel by ribotyping and PCR-ribotyping. *Microbiologica*,22,383–387.

（娄忠子译，殷宏校）

# 1.4 泌尿系统

## 1.4.1 布鲁氏菌病

通过开展强有力的卫生安全防控措施，许多工业化国家已经成功消灭了布鲁氏菌病。然而，在发展中国家，布鲁氏菌病依然在家养及野生动物群内传播，给热带畜牧业造成了巨大的经济损失[108]。在发展中国家，布鲁氏菌病也是一种重要的人兽共患病。在热带奶牛场，牛的感染率可达80%，在萨赫勒地区的规模化畜群中的感染率为25%～30%[108]。Domenech等[46]估计每年因为布鲁氏菌病造成的损失可达6%。布鲁氏菌病发生在哺乳期的牛群会造成20%～25%的经济损失，在拉丁美洲每年的损失可达60亿美元。感染布鲁氏菌病的每个患者的经济损失可达8 000美元。

旧大陆骆驼（OWCs）容易感染布鲁氏菌病，特别是在接触患病的反刍动物的时候[3,19,31,99,112]。新大陆骆驼（NWCs）中布鲁氏菌病很少见，但是具有典型症状的布鲁氏菌病暴发已有报道[54]。人们食用未经巴氏消毒的奶类时，有一定感染布鲁氏菌病的风险[73,76,98,130]。

### 病原学

布鲁氏菌病是一种接触传染性疾病，病原体隶属布鲁氏菌属。分类学上，将布鲁氏菌属分为6个种，种内又分为若干生物型。生物型的划分是依据生化反应和与单一型特异的血清凝集反应的结果。近年来，又从一些海洋哺乳动物体中分离出布鲁氏菌病属的菌株。分子分型方法无法将这些新分离菌株分为已有的6个种的任何一种，所以，将这些菌株命名为：海豚布鲁氏菌(B. cetacea，宿主是海豚)，海豹布鲁氏菌(B. pinnipeda，宿主是海豹、海狗、海象)。布鲁氏菌属的6个种，以及若干生物型的具体分类见表35。

布鲁氏菌是球形或短杆形的革兰氏染色阴性菌，不运动，不形成芽孢，属厌氧菌。某些菌株在培养时需要5%～10% $CO_2$，生长缓慢，但是增菌培养基可以加速其生长。布鲁氏菌的菌落形态由细菌外膜的脂多糖（LPS）决定。光滑型LPS(S-LPS)和粗糙型LPS(R-LPS)的表现型存在差异。大多数布鲁氏菌都是光滑型，只有犬布鲁氏菌和绵羊布鲁氏菌是粗糙型的。布鲁氏菌属某些种和其他细菌的一些蛋白存在血清学交叉反应，交叉反应存在于：

（1）小肠结肠炎耶尔森菌O：9；
（2）赫氏埃希菌；
（3）大肠埃希菌O157：H17；
（4）土拉弗朗西斯菌；
（5）黄单胞菌；
（6）霍乱弧菌O：1；
（7）沙门氏菌血清型N。

除了绵羊布鲁氏菌和流产布鲁氏菌这两种生物型的生长需要加血清或血液外，其他布鲁氏菌在增菌培养基内均可生长，它们还可以在营养琼脂上生长。在进行布鲁氏菌培养时要小心谨慎。

### 公共卫生意义

对人类而言，布鲁氏菌病被称为波状热、马耳他热，是一种严重的公共卫生问题。

人布鲁氏菌病是世界上最常见的人兽共患病之一，每年会有50万新病例发生。布鲁氏菌病在储藏宿主动物中的流行决定了人发病的几率[120]。布

表35 布鲁氏菌分型及生物型分类

| 菌种名称 | 发现者 | 年份 | 生物型 |
|---|---|---|---|
| 马尔他布鲁氏菌（*B. melitensis*） | Bruce | 1886 | 1，2，3 |
| 流产布鲁氏菌（*B. abortus*） | Bang | 1895 | 1，2，3，4，5，6，7，8，9 |
| 猪布鲁氏菌(*B. suis*) | Traum | 1914 | 1，2，3，4，5 |
| 绵羊布鲁氏菌(*B. ovis*) | McFarlane | 1952 | — |
| 沙林鼠布鲁氏菌(*B. neotomae*) | Thomas | 1957 | — |
| 犬布鲁氏菌(*B. canis*) | Carmichael | 1966 | — |

鲁氏菌病属的某些种是潜在的生物威胁，被美国疾病预防与控制中心（CDC）划分为B类动物疫病。马尔他布鲁氏菌和流产布鲁氏菌是患病人体中最常见的两种病原体，马尔他布鲁氏菌能引起人最严重的感染。人布鲁氏菌病主要表现为职业病，在工作中与家畜接触的工作人员易发病。布鲁氏菌病的主要传播方式是接触传播，主要是通过皮肤接触到动物的器官、血液、尿液、阴道分泌物和流产胎儿，尤其是胎盘。另外，饮用生牛奶和一些未经灭菌的奶制品也会感染布病。空气传播感染主要发生在动物圈舍间及实验室[107]和屠宰场内。另外一些病例是由于接种活疫苗引起的。布鲁氏菌病的潜伏期为5～60天，可能时间更长。临床症状不典型，可以急性发病也可以是慢性发病（表36）。

表36　人布鲁氏菌病的症状

| 临床症状 | 病人发生率（%） |
|---|---|
| 发热 | 90～95 |
| 乏力 | 80～95 |
| 身体疼痛 | 40～70 |
| 发汗 | 40～90 |
| 关节痛 | 20～40 |
| 脾肿大 | 10～30 |
| 肝肿大 | 10～70 |

布鲁氏菌病感染怀孕早期的孕妇，可导致高流产率（达40%），感染男性可使其产生睾丸炎和附睾炎。即便经过适当的治疗康复后，马尔他布鲁氏菌病的DNA仍可在人体血液中存在数年[121]。

## 流行病学及临床症状

从某种程度上来说，布鲁氏菌病的流行特点是引起妊娠母畜流产、公畜睾丸炎和附睾炎。布鲁氏菌病在全世界都有分布，可感染牛、猪、绵羊、山羊、骆驼、狗，偶尔可感染马驹。在全球范围内，布鲁氏菌病已经感染多种野生动物，最近海洋哺乳动物也有报道。家畜感染后，可能再传染给野牛、麋鹿、非洲水牛的[103]。一般来说，流产只发生在初次怀孕，而且感染的动物表现为隐性感染。在怀孕动物体内，布鲁氏菌定植于胎盘，并大量存在于流产物，如胎儿的胃、阴道分泌物和初乳等。菌体被单核巨噬细胞吞噬成为胞内寄生菌，仍能够生长繁殖。在未孕动物体内，布鲁氏菌会潜伏在淋巴结内直到怀孕。一些布鲁氏菌感染的母牛分娩的小牛一出生血清反应呈阴性（潜伏感染？），这些小牛长到成年时就会呈现布鲁氏菌病阳性并出现流产现象[79]。这种流行模式在骆驼中也会发生。

感染的发生是通过黏膜系统，包括口鼻、结膜、生殖道黏膜，也可通过皮肤擦伤感染。动物会通过饲料、水源、初乳、污染的牛奶、尤其是舔舐、嗅闻胎盘和流产胎儿感染该病。布鲁氏菌病传播的另一个次要途径是通过性行为。而主要的排毒途径是感染动物子宫阴道分泌物（恶露）和胎盘。依据现有的资料，随着牛的流产，会有$10^{12}$～$10^{13}$个布鲁氏菌排出。布鲁氏菌可以在寒冷和潮湿的环境中生存，但是，在炎热沙漠里，驼群中的2峰单峰驼也感染了马尔他布鲁氏菌，可能是通过距离500 m远的流产骆驼胎儿污染的尘埃颗粒传播引起。母畜流产后，每毫升奶中布鲁氏菌的数量可达$10^3$个，可持续3个月排菌，这具有非常重要的流行病学意义。然而，慢性感染的单峰驼可以产下健康的后代，排出的胎盘和乳汁中都没有分离出布鲁氏菌病；此外，单峰驼幼仔的血液培养和PCR反应都为阴性[120]。有趣的是，有一项研究表明，血清学反应阳性的母驼，它们的幼仔在6个月大的时候血清学反应都为阴性。因此，Ostrovidov[93]，Solonitsynhe和 Pal’gov[111]提议当幼驼在7～8个月大，也就是它们的母源抗体消失时，将其和母驼分开。如果不这样做的话，它们将很有可能在患病母驼下一次分娩时感染布鲁氏菌病。

Von Hieber[120]指出，在两年时间里会有20%的血清反应阳性的母驼转成阴性，5%的母驼存在血清阳性→阴性→阳性的波动变换状态。Gatt Rutter

和Mack[56]以及Ostrovidov[94]已经阐述了布鲁氏菌病患者可自愈的证据。Wernery[125]等进一步研究发现，产下健康幼驼的布鲁氏菌病阳性单峰驼在泌乳阶段，布鲁氏菌隐藏在淋巴结中。布鲁氏菌主要从胸腔内淋巴结中分离出来，这表明布鲁氏菌病通过食入或吸入感染。对这些驼群中的调查清楚地表明，流产单峰驼（急性布鲁氏菌病）和慢性感染不流产的动物，两者存在很重要的流行病学差异。慢性感染是很常见的现象，对牛来说，有75%～90%的母牛只流产一次。

从理论上来说，已知3种布鲁氏菌都可以感染骆驼[68]。然而，据推测马尔他布鲁氏菌病在非洲和中东地区流行，流产布鲁氏菌在前苏联地区流行。Solonitsyn[110]报道了在俄罗斯双峰驼群中流行的是不同布鲁氏菌种的混合感染。表37显示了一些国家不同来源的器官中分离出的布鲁氏菌。

表37 不同国家分离到的旧大陆骆驼的布鲁氏菌情况

| 国家 | 作者 | 年份 | 菌株种类 | 部位（组织器官） |
|---|---|---|---|---|
| 约旦 | Al-Majali | 2006 | 马尔他布鲁氏菌3型 | 流产胎儿、阴道拭子 |
| 俄罗斯 | Solonitsyn | 1949 | 流产布鲁氏菌 | |
| | Pal’gov | 1950 | 流产布鲁氏菌 | 双峰驼的胎儿 |
| 伊朗 | Zowghi and Ebadi | 1988 | 马尔他布鲁氏菌1型 | 淋巴结 |
| 科威特 | Zowghi and Ebadi | 1988 | 马尔他布鲁氏菌3型 | 淋巴结 |
| 利比亚 | Al-khalaf and El-Khaladi | 1989 | 流产布鲁氏菌1型 | 流产胎儿的胃 |
| | Gameel et al. | 1993 | 马尔他布鲁氏菌1型 | 乳 |
| 沙特阿拉伯 | Gameel et al. | 1993 | 马尔他布鲁氏菌1型 | 乳、阴道拭子、流产胎儿 |
| | Radwan et al. | 1992 | 马尔他布鲁氏菌1型、2型 | 乳 |
| | Radwan et al. | 1995 | 马尔他布鲁氏菌1型、2型、3型 | 乳 |
| | Ramadan et al. | 1998 | 马尔他布鲁氏菌 | 腕水囊肿 |
| | Al Dubaib | 2007 | 马尔他布鲁氏菌 | |
| 苏丹 | Agab et al. | 1996 | 流产布鲁氏菌3型 | 乳头、淋巴结、阴道拭子、睾丸 |
| | Omer et al. | 2010 | 流产布鲁氏菌6型 | 淋巴结、睾丸 |
| 秘鲁 | Acosta et al.(羊驼) | 1972 | 马尔他布鲁氏菌 | 阴茎 |
| 阿联酋 | Wernery et al.(来自苏丹的骆驼) | 2007a | 马尔他布鲁氏菌1型、3型 | 乳、淋巴结、胎盘 |
| | Moustafa et al. | 1998 | 马尔他布鲁氏菌 | 乳 |
| 塞内加尔 | Verger et al. | 1979 | 流产布鲁氏菌1型、3型 | |

表38 不同国家的骆驼布鲁氏菌病抗体水平

| 国家 | 作者 | 年份 | 被检骆驼数量 | 患病率（%） | 血清学检测 | | | | |
|---|---|---|---|---|---|---|---|---|---|
| | | | | | ELISA | RBT | CFT | SAT | MRT |
| 埃及 | Abou-Zaid | 1998 | 422 | 10.4～12.3 | | + | + | + | |
| | Ahmed | 1993 | 200 | 3.5 | | | | | |
| | Ayoub et al. | 1978 | 216 | 24.2 | | | | + | |
| | El-Nahas | 1964 | 200 | 4.0 | | + | | | |
| | El-Sawally et al. | 1996 | | 2.3～14.0 | | + | + | + | |
| | Fayed et al. | 1982 | 300 | 6.6 | | + | + | + | |
| | Hamada et al. | 1963 | 175 | 10.3 | | | | + | |
| | Nada | 1984 | 780 | 23.1 | | + | + | | |
| | Nada | 1990 | | 5.3～7.9 | | + | + | + | |
| | Zagloul and Kamel | 1985 | 37 | 8.1 | | | | | |
| | Zaki | 1948 | 200 | 14.0m，26.0f | | | | + | |
| | Ahmed and Nada | 1993 | 360 | 11.6 | | | | + | |

续表

| 国家 | 作者 | 年份 | 被检骆驼数量 | 患病率（%） | 血清学检测 | | | | |
|---|---|---|---|---|---|---|---|---|---|
| | | | | | ELISA | RBT | CFT | SAT | MRT |
| 苏丹 | Abbas et al. | 1987 | 238 | 3.0 | | | | + | |
| | Abu Damir et al. | 1984 | 740 | 4.9 | | + | + | + | |
| | Agab | 1993 | 453 | 30.0(32.9f，15.1m) | | + | | | |
| | Agab | 1998 | | 2.9 | | + | | | |
| | Ali and Ghedi | 1978 | 250 | 10.4 | | | | + | |
| | Bornstein and Musa | 1987 | 102 | 5.9 | | | + | + | |
| | Mustafa and Awad El- | 1971 | 310 | 1.8 ~ 5.8 | | | | + | |
| | Karim | 1971 | 137 | 1.75 ~ 5.75 | | | | | |
| | Mustafa and Hassan | 1987 | 2225 | 8.0 | + | + | | | |
| | Osman and Adlan | 2010 | 153 | 37.5 | | + | + | | |
| | Omer et al. | 1996 | 805 | 10.6 | | | | + | + |
| | Obeid et al. | 2000 | | 14 ~ 44 | | + | | | + |
| | Majid and Goraish | | | | | | | | |
| 索马里 | Ahmed and Ibrahim | 1980 | 802 | 8.0 ~ 11.0 | | + | | | |
| | Andreani et al. | 1982 | 250 | 10.4 | | | | + | |
| | Anonymous | 1981 | | 5.0 ~ 7.8 | | | | + | |
| | Baumann et al. | 1990 | | 3.1 | | | | + | |
| | Baumann and Zessin | 1992 | 1039 | 0.3 ~ 1.9 | | | + | + | |
| | Bishof | 1979 | 47 | 4.0 | | | + | | |
| | Bornstein | 1984 | | 8.5 ~ 11.5 | | + | | | + |
| | Bornstein | 1988 | | 1.3 | | | + | + | |
| | Bornstein et al. | 1988 | 234 | 5.9 | | | + | + | |
| | Elmi | 1982 | 514 | 12.6 | | | | + | |
| | Ghanem et al. | 2009 | 1246 | 3.1 | + | + | | | |
| | | | | 3.9 | | | | | |
| 埃塞俄比亚 | Domenech | 1977 | 977 | 4.4 | | | | + | |
| | Richard | 1980 | 762 | 5.5 | | | | + | |
| | Teshome and Molla | 2002 | 1442 | 5.7 | | | + | + | |
| | | | | 4.2 | | | | | |
| 肯尼亚 | Kagunya and Waiyaki | 1978 | 174 | 4.6 ~ 10.3 | | + | + | + | + |
| | Waghela et al. | 1978 | 172 | 14.0 | | + | + | + | + |
| | Wilson et al. | 1982 | | 6.0 ~ 38.0 | | + | + | + | + |
| 乍得 | Bares | 1968 | 543 | 5.3 | | | | + | |
| | Graber | 1968 | 316 | 3.8 | | | | + | |
| | Schelling et al. | 2003 | 288 | 0.4 | + | + | | | |
| 突尼斯 | Burgemeister et al. | 1975 | 52和150 | 3.9 ~ 5.8 | + | + | | | + |
| | | | 奶样 | 0 | | | | | |
| 尼日利亚 | Okoh | 1979 | 232 | 1.0 | + | + | | + | + |
| | Kudi et al. | 1997 | 480 | 7.5 | | | + | | |
| | Adama et al. | 2007 | 181 | 8.3 | | | | | + |
| 尼日尔 | Bornarel and Akakpo | 1982 | 109 | 8.3 | + | + | + | | + |
| | Saley | 1983 | | 6.2 | | | | | |
| 俄罗斯 | Pal’gov and | 1964 | 500 | 15.0 | | | | + | |
| | Zhulobovski | 1949 | 27 | | | | | + | |
| | Solonitsyn | | | | | | | | |
| 蒙古 | Shumilov | 1974 | 54673 | 1.0 ~ 3.7 | | | + | + | |
| | Mocalov | 1991 | 29300 | 9.7 | | + | + | + | |
| | Erdenebaatar et al. | 2003 | 17 | 23.5 | | + | | | |
| 利比亚 | Ben Faraj et al. | 1990 | 666 | 3.75 | + | + | + | + | + |
| | Gameel et al. | 1993 | 967 | 4.1 | + | + | + | + | + |

续表

| 国家 | 作者 | 年份 | 被检骆驼数量 | 患病率（%） | 血清学检测 | | | | |
|---|---|---|---|---|---|---|---|---|---|
| | | | | | ELISA | RBT | CFT | SAT | MRT |
| 印度 | Kulshrestha et al. | 1975 | 315 | 1.8 | | | | + | |
| | Mathur and Bhargava | 1979 | 210 | 3.8～5.2 | | | | + | |
| 伊朗 | Zowghi and Ebadi | 1988 | 953 | 8.0 | + | + | + | + | + |
| | Ebrahimi et al. | 2007 | 153 | 1.6 | | + | | + | |
| | | | | 3.9 | | | | | |
| | Khorasgani et al. | 2006 | 240 | 11.6 | | + | | + | |
| | | | | 11.2 | | | | | |
| 伊拉克 | Al-Ani et al. | 1998 | 215 | 7.0～17.0 | + | + | | + | + |
| | Jawad | 1984 | 235 | 3.8 | + | + | | + | + |
| 沙特阿拉伯 | Radwan et al. | 1983 | 116 | 2.8～3.5 | | | + | + | |
| | Radwan et al. | 1992 | 2630 | 8.0 | + | + | + | + | |
| | Radwan et al. | 1995 | 2536 | 8.0 | + | + | | | + |
| | Hashim et al. | 1987 | 146 | 1.4 | + | + | + | + | |
| | | | | 3.02 | | | | + | |
| | | | | 1.9 | | | | | |
| | | | | 4.0 | | | | | |
| | Alshaikh et al. | 2007 | 859 | 3.14 | + | + | | + | |
| 科威特 | Al-Khalaf and EL-Khaladi | 1989 | 698和209份奶样 | 14.8 | + | + | + | + | |
| | | | | 8.0 | | | | | + |
| 阿曼 | Harby and Ismaily | 1995 | 550 | 3.6 | | | + | | |
| | Yagoub et al. | 1990 | 1502 | 7.0 | | | + | | |
| | Ismaily et al. | 1988 | 550 | 3.6 | | | + | | |
| 阿联酋 | Afzal and Sakkir | 1994 | 392赛驼 | 0.76 | + | + | | | + |
| | Moustafa et al. | 1998 | 7899 | 0.01 | + | + | + | + | + |
| | Wernery and Wernery | 1990 | 196育种驼 | 2.0 | | | + | + | |
| | | | 348赛驼 | 6.6 | + | + | + | | |
| | Wernery et al. | 2007a | 1119 | 1.9 | + | + | + | + | |
| | Gwida et al.(来自苏丹的骆驼) | 2011 | 895 | 60 | | | + | | |
| | Taha | 2007 | 812 | 0 | + | + | + | | |
| 巴基斯坦 | Ajmal et al. | 1989 | 81 | 2.0 | | + | | + | |
| 秘鲁 | Acosta et al.(羊驼) | 1972 | 1449 | 21 | | + | + | | |
| 约旦 | Al-Majali | 2006 | 590 | 17 | | + | + | | |

注：（1）CFT，补体结合试验；ELISA，酶联免疫吸附试验；f，雌驼;m，公驼；MRT，乳环状试验；RBT，孟加拉玫瑰红试验；SAT，血清凝集试验

（2）来自文献报道。

从表38可以看出，大多数骆驼布鲁氏菌病研究，都采用血清学反应进行确认。Chukwu[44]发表了非洲布鲁氏菌病流行情况的调查报告，发现骆驼布鲁氏菌病的发生率与饲养管理方法有关[101]。前苏联一些地区在大农场中饲养的双峰驼，它们的感染率是15%[96]；在乍得和埃塞俄比亚，骆驼的管理粗放自由，它们的感染率分别是3.8%[62]和5.5%[101]。

沙特阿拉伯的Radwan等[99]和Ghoneim与Amjad[59]已经公开报道了类似的关于血清阳性率的差异现象，他们指出，在农场里精心饲养的骆驼患布鲁氏菌病的概率比自由放牧的骆驼患布鲁氏菌病的概率高。从苏丹的骆驼养殖业来说，布鲁氏菌病在农场（31.5%）比在游牧群（21.4%）中更易流

行[1,9,10]。

Moustafa等[82]针对单峰骆驼做了血清学调查，并计划在5年内根除阿联酋东部地区的布鲁氏菌病。1991年感染率5.8%为最高，1996年感染率0.01%为最低。没有一只骆驼是因为布鲁氏菌病而被屠宰淘汰，所以，人们认为骆驼中布鲁氏菌病的减少是由于绵羊和山羊中布鲁氏菌病的减少。

研究表明，在种用骆驼群中发生过各种形式的布鲁氏菌病，流产是最常见的症状[6,12,53,98,129]。牛羊布鲁氏菌病感染有可能导致死胎、胎衣滞留和产奶量下降。但是，在骆驼科动物布鲁氏菌病中并没有出现胎衣滞留现象，这可能是胎盘附着方式不同的结果[54]。已经从来自不同国家的患病骆驼的乳汁、阴道拭子、流产胎儿、淋巴结和水囊瘤中分离到流产布鲁氏菌病和/或马尔他布鲁氏菌病。

虽然骆驼特别容易感染布鲁氏菌病，但是，从骆驼样本中分离出布鲁氏菌病很不容易。只有最近从乳中分离布鲁氏菌病的尝试才获得成功。从来自塞内加尔的骆驼中分离到流产布鲁氏菌病的1型和3型[119]，Radwan等[99]从来自沙特阿拉伯的100份血清阳性单峰骆驼的奶样中26次分离出马尔他布鲁氏菌1型和2型。这种情况给人类健康造成了严重的威胁，因为骆驼奶很少进行巴氏消毒。同时，Gameel等[55]从利比亚单峰驼中分离出1型马尔他布鲁氏菌，其中，5次从奶样中分离得到，4次在流产胎儿和阴道拭子中分离得到。Zaki[133]将血清阳性单峰驼的奶样注射给豚鼠，同时又在体外培养样本，两者的实验结果均为阴性。Al-Khalaf 和El-Khaladi[21]试验培养了来自科威特单峰驼的209份奶样，这些样本是从流产率递增的牧群中采集的，实验结果均为阴性。然而，他们从5只流产胎儿的胃内容物中成功分离到了流产布鲁氏菌。Pal’gov[95]从来自俄罗斯的双峰驼中分离到流产布鲁氏菌，调查还显示，2%动物在怀孕初期发生流产，15%是布鲁氏菌病阳性。Zowghi 和Ebadi[136]在伊朗从300只屠宰的单峰驼中采集了3 500个淋巴结，用于布鲁氏菌病病原体培养，从1%（3/300）动物的淋巴结中分离到1型和3型马尔他布鲁氏菌。他们认为，这些单峰驼感染布鲁氏菌病的原因在于附近的绵羊和山羊群。

Radwan等[98]检测了沙特阿拉伯一个大的牧群，共2 536只单峰驼，其中，流产率12%，布鲁氏菌病阳性率8%，从流产胎儿中分离到了1型，2型，3型马尔他布鲁氏菌。在他们的调查中，30%的骆驼饲养员和挤奶工患有马尔他热，并且从附近流产绵羊、山羊分离到生物型一样的马尔他布鲁氏菌。

流产布鲁氏菌3型是从苏丹东部自由放养的、有流产、水囊瘤、睾丸损伤病史的单峰驼中分离到的[11,12]，共采集了3种样本，分别是乳腺淋巴结、阴道拭子和腹股沟淋巴结。值得一提的是从塞内加尔和苏丹分离到的3型流产布鲁氏菌是唯一一种氧化酶缺失的生物型。Ramadan等[100]从印度骆驼的水囊瘤中发现了马尔他布鲁氏菌。在阿联酋，研究人员从阳性骆驼的乳汁中两次分离到马尔他布鲁氏菌。

给未孕单峰驼人为注射流产布鲁氏菌（野生株，6×10$^6$个），最后只出现了很微弱的临床症状，如食欲下降、轻微跛行、两眼流泪。在45～65天后，从颅侧和生殖器淋巴结重新分离到布鲁氏菌，可见皮质、副皮质区生发中心活跃，滤泡增生，髓索萎缩和肝窦充血，还会有轻微的间质性肝炎[5]。

布鲁氏菌病并不是新大陆骆驼的主要疾病，但是，在秘鲁，羊驼中暴发了严重的马尔他布病，1449只羊驼中约30%玻板凝集试验呈阳性。超过25%的羊驼饲养员血清布病阳性，有些还患有马耳他热。人们认为，绵羊是这群羊驼感染的罪魁祸首[6]。在美国，人工感染实验发现，美洲驼特别容易感染流产布鲁氏菌，出现的阳性血清效价和病理变化与牛、绵羊和山羊相似[61]。

伦敦动物园的3只美洲驼与从莫斯科新引进的骆驼接触后死亡[54]，诊断发现马尔他布鲁氏菌的血清效价高于1:1000，是属于急性感染死亡。

## 病理学

人们对骆驼感染布病后的病理变化知之甚少。布鲁氏菌偏好的器官有子宫、乳房、睾丸、附属性腺、淋巴结、关节囊和黏液囊，这些部位有可能出现病理变化。Nada和Ahmed[87]在未孕单峰驼体内发现病理损伤，症状包括子宫内膜红肿发炎，上皮组织可见坏死灶，子宫内膜纤维化和子宫角萎缩，还有大量粘连的卵巢囊和滑液囊。粘连发生在卵巢囊和卵巢之间，偶尔也会在卵巢囊和输卵管之间，并造成后者严重硬化。布病阳性单峰驼常见滑液囊炎，导致囊体变大，其中充满琥珀色的液体。到目前为止，流产胎儿中没有发现病理变化，除睾丸炎、附睾炎外，也没有发现阳性公骆驼的病理变化。人工感染流产布鲁氏菌给一头怀孕的美洲驼结膜囊，在接种后43天，怀孕8个月的美洲驼流产，并从胎盘、胎儿样本和母畜的乳腺、淋巴结中分离到流产布鲁氏菌。病理组织学可见轻微的、多病灶的、淋巴细胞和组织细胞及滋养层细胞大量缺失的亚急性胎盘炎。绒毛尿囊膜基质包含大量的坏死和钙化碎片，膨胀的毛细血管中有许多布鲁氏菌[60,61]。Wernery等[125]指出在布病阳性的泌乳期单峰驼很少有病理变化，淋巴结内出现窦状小管水肿，毛囊活跃和组织细胞增生的现象，而生殖道却无损伤。

## 诊断

布病的诊断通常是在实验室进行，方法有血液、乳或脏器采样的细菌培养和抗体检测。布鲁氏菌可从胎盘获得，但是，最便捷的分离方法是胃内容物和流产胎儿肺的纯培养。布鲁氏菌病的首选诊断方法是分离培养，它还可以将分离株分型。然而，现在PCR技术在鉴定与分型中应用广泛。对于骆驼奶样和组织样中病原体的分离培养，推荐使用含有6种抗生素的FM培养基，也可用其他选择性培养基[98]。经过深入调查发现，在沙特阿拉伯的骆驼农场里，有34%的布病阳性的泌乳单峰驼排出布鲁氏菌病。

然而，对骆驼布病的诊断难度较大。造成骆驼的流产和繁殖能力下降的原因通常还有其他疾病，例如，沙门氏菌病、锥虫病或者胎儿弯曲菌、胎儿三毛滴虫感染[123,124,128]，所以，实验室诊断很有必要。只根据血清学试验来诊断布病很容易出现误诊。Sunaga等[113]报告称，有5只引入到日本的单峰驼，血清凝集反应缓慢，但补体结合试验（CFT）结果呈阳性，最终这些单峰驼被直接宰杀。然而，并没有从宰杀的骆驼病料中分离到布鲁氏菌病，而分离到了血清型为0:9的小肠结肠炎耶尔森杆菌。众所周知，其他不同种类的细菌可能会发生假阳性（非特异性）反应[36,51]。

有些人认为，CFT是检测布病最敏感、最特异的诊断方法[56,95,117,122]，对于急性感染和慢性感染来说也的确如此。Shumilov[109]指出，CFT比凝集试验敏感4倍，他用蒙古的布病流行广泛地区的双峰驼做了相应的试验，分两群，试验结果如下：

（1）1号群，3751只，CFT 4.3%，SAT 0.6%；

（2）2号群，54673只，CFT 3.7%，SAT 1.0%。

在血清凝集试验中，研究人员将血清最终凝集价1:20（40 IU）判定为可疑[26,58,95,135]。Fayed等[52]，Salem等[104]，El-Sawally等[50]认为血清凝集试验或试管凝集试验（TAT）比其他试验敏感，检测到更多有反应的布鲁氏菌，是因为它们对IgM的反应比对IgG更敏感。为了消除血清凝集试验中的非特异性反应，Wernery 和Wernery[127]利用5%的苯酚氯化钠溶液。

由于与其他细菌的交叉反应，对布鲁氏菌病的血清学的诊断带来了一定的困难，Zhulobovski和Pal’gov[135]在来自俄罗斯的双峰驼血清试验中研究发现了免疫前带现象，Nada[85]在埃及的单峰驼试验中也有同样的发现。在阿联酋，血清阳性单峰驼中占1.5%的血清在低倍稀释时没有出现可视的阳性反应。在这种情况下，抗免疫球蛋白试验对判定诊断结果很有必要。

有些研究人员利用ELISA方法来检测布病抗体，样本不只是骆驼血样[4,29]，还有骆驼奶样[118]，骆驼奶样ELISA似乎可以作为传统血清学诊断布病的一个替换。一些研究人员对用于布病诊断的不同血清学试验做了评价[3,4,59,88,102]，总体来说，消除布鲁氏菌病的非特异性反应，对于做出正确的诊断是非常关键的。另外，几种试验同时进行对于确诊布病也很重要，但使用5%苯酚氯化钠溶液的TAT在骆驼布病的血清学诊断中是必不可少的。Atwa[27]和Abou-Zaid[4]发现5种不同的血清学试验的一致性在80.6%～95.6%。

Mohammed[81]对虎红平板试验(RBPT)、TAT、CFT在骆驼布病诊断方面的准确性做了评价。他发现，在检测阴阳性血清和前带现象试验中，RBPT和CFT具有同等效果。RBPT的血清抗原比为3：1时最灵敏，然而，CFT要将血清稀释10倍，在54℃灭活30分钟再冷却才可以。TAT则要求使用pH值为3.5的抗原而不是原来中性的抗原才能达到最佳效果。

Radwan等[98]对沙特阿拉伯的一个大农场里

2536只单峰驼做了布病抗体检测，他们用虎红平板试验（RBT）和玻板凝集试验两种实验结合来判定是否为阳性，这个农场原来的流产率为12%，最终他们运用这两种方法淘汰掉阳性动物，成功地消除了该农场的布病。他们之所以选择这两种试验，是因为这两种试验简单、灵敏、实用。

运用间接血清学试验对于控制或消灭布病非常关键，这些试验大多数都可以选择使用，但是对骆驼布病的诊断还没有标准化。OIE建议检测牛布病要用RBT，CFT，ELISA和荧光偏振分析（FPA）。通过不同的血清学试验，根据感染过程中免疫球蛋白的活性可以区分急性感染和慢性感染，IgM和IgG抗体的同时存在表明是一种急性布鲁氏菌病，而慢性布鲁氏菌病的特点是只存在IgG抗体。不同检测方法的敏感性、特异性和可检测的免疫球蛋白的同种型等，列于表39。

表39所列诊断方法都已经应用到骆驼布病检测中，CFT以前经常被用来作为一个验证性方法，现在已经被ELISA 和近年的FPA取代。ELISA具有高敏感性，但是，其特异性很低，对不同种的细菌有交叉反应，特别是与小肠结肠炎耶尔森菌0：9容易发生交叉反应。Alshaikh等[23]明确指出，SAT对于慢性感染骆驼的检测结果的可靠性有限。最近Von Hieber[120]和Gwida等[64]调查发现，从苏丹运送到阿联酋的成百上千的布病阳性骆驼的诊断试验比较结果如下：通过计算K值证实CFT，RBT和SAT的结果具有很高的一致性，但是，与FPA或实时定量PCR相比，三者的敏感性都很低。因此，建议结合实时定量PCR和任意一种血清学试验，例如，RBT，就可以使敏感性提高到100%[63]。

Alshaikh等[23]在沙特阿拉伯已经在使用PCR方法来检测骆驼布病，这种方法的诊断结果可靠，还可以区分马尔他布鲁氏菌和流产布鲁氏菌。

在智利有来自布病阴性牧群的336份羊驼和美洲驼的血清采用FPA和竞争ELISA（cELISA）方法检测，所得结果与RBT、SAT和CFT等传统检测方法相比。用FPA和cELISA检测出两份阳性血清，但这两份阳性血清的效价都很低，传统检测方法没有检出阳性。

与牛奶相比，骆驼奶不能用传统的乳环状试验（MRT）来检测乳汁内的布病抗体，因为骆驼奶缺少可以凝集脂肪球的凝集物质[118]；另外，骆驼奶脂肪球都是微脂粒，不会奶油化产生脂层。表38总结的MRT结果应该谨慎解释，Van Straten等[118]建立的MRT检测方法可以检测骆驼奶中的抗体，他们将这个方法命名为改良MRT，是因为布病阴性牛奶被加入到骆驼奶样中，当有布鲁氏菌病抗体存在时就会产生典型的彩色乳环（图105）。这种方法

图105

表39　敏感性、特异性、免疫球蛋白的检测

| 血清学试验 | 敏感性（%） | 特异性（%） | 检测到的免疫球蛋白 |
|---|---|---|---|
| SAT | 81.5 | 98.9 | IgM |
| CFT | 90 ~ 91.8 | 99.7 ~ 99.9 | IgG，IgM |
| RBT | 87 | 97.8 | IgG，IgM |
| cELISA | 95.2 | 99.7 | IgG，IgM |
| iELISA | 97.2 | 97.1 ~ 99.8 | IgG，IgM |
| FPA | 96.6 | 99.1 | IgG，IgM |
| MRT | 88.5 | 77.4 | |

注：cELISA，竞争性酶联免疫吸附试验；CFT，补体结合试验；FPA，荧光偏振分析；IgG，免疫球蛋白G；IgM，免疫球蛋白M；iELISA，间接ELISA；MRT，乳环状试验；SAT，血清凝集试验

的敏感性很低，但是，它的优点是成本较低，且每月可开展一次重复检测。

检测人工感染流产布鲁氏菌美洲驼抗体的方法中，CFT、标准试管试验、D-tec ELISA的结果没有酸平衡玻板凝集试验、卡片试验、标准平板试验、利凡诺试验的结果可靠[54]。一些研究人员已经利用不同的过敏原进行了布病皮肤试验，特别需要指出的是俄罗斯研究人员在双峰驼上的试验[115]。皮肤试验虽然敏感性偏低，但是具有高度特异性。抗原不能致敏动物的免疫系统，所以不会对该病的诊断造成干扰。

### 治疗及防控

为了消灭动物布病，当在血清学和细菌学方面都证实患病，应立即进行全面检测、扑杀病畜和接种疫苗。扑杀血清阳性动物，并对牧群进行全面监测，直到清除所有的病畜。骆驼和其他动物一样，当2～3种试验结果均为阴性之后，给畜群内所有动物接种疫苗，以免再患病；在此情况下，应充分关注新引入动物的危险性。接种疫苗后而感染的动物仍然是公共卫生的威胁。Abbas和Agab[1]建议，在布病流行不严重的国家全群接种疫苗，在布病流行的国家采取检测和扑杀之后再接种疫苗的措施。布鲁氏菌病是革兰氏阴性球杆菌，对广谱抗生素敏感，但是，在许多国家由于抗生素使用不当、动物产生耐药性而禁用抗生素。

然而，Radwan等[98]用土霉素（每千克体重25mg）/2天，连续用药30天，链霉素（每千克体重25mg）/2天，连续用药16天，两者配合，治疗202只阳性单峰驼。除胃肠外注射治疗外，给泌乳骆驼进行乳房输液，每个乳头10mL土霉素/2天，连续治疗8天，这种治疗方法对于消灭通过乳汁传播的布鲁氏菌病来说很有效果。在经过16个月以后，所有骆驼都转为布病阴性。这就可以使一些有特殊价值，比如用于竞赛的骆驼免于被宰杀的命运，但是，抗生素治疗在群体水平能否成功仍是值得怀疑。

灭活苗和减毒疫苗对旧大陆骆驼都是有效的，可用于骆驼布病免疫的疫苗有流产布鲁氏菌病Buck19株（$B_{19}$或$S_{19}$）[43]和马尔他布鲁氏菌病Rev 1[98]。幼年骆驼使用全剂量，成年骆驼剂量减少，两者在接种疫苗后都会产生布病抗体，幼年骆驼在接种后8个月抗体消退，成年骆驼在接种后3个月抗体消退。接种疫苗后，骆驼就不再发生流产。Agab等[13]给5只单峰驼接种了小剂量（$5\times10^8$U·/2mL）的流产布鲁氏菌病$S_{19}$株疫苗，一周后5只骆驼均产生抗体，在6～7周后抗体水平下降，这些骆驼在14周后诊断均为布病阴性。

## 参考文献


[1] Abbas B. & Agab H. (2002). – A review of camel brucellosis. *Prev. Vet. Med.*,55,46–47.

[2] Abbas B.,Yassin T.T.M. & Elzubir A.E.A. (1987). – Survey for certain zoonotic diseases in camels in the Sudan. *Rev. Elev. Méd. Vét. Pays Trop.*,40 (3),231–233.

[3] Abo El-Hassan D.G.,Mammam H.M.,Youssef R.,Barsoum S.A.,Awad M.M. & Sameh S.M. (1991). – Prevalence of camel brucellosis using different serological tests. *Vet. Med. J. Giza*,39 (3),875–884.

[4] Abou-Zaid A.A. (1998). – Some studies on camel brucellosis. *In* 8th Scientific Congress,15～17 November 1998,Faculty of Veterinary Medicine,Assiut University,Assiut,Egypt,690–707.

[5] Abu Damir H.,Kenyon S.J.,Khalafalla A.E. & Idris O.F. (1984). – *Brucella* antibodies in Sudanese camels. *Trop. Anim. Health Prod.*,16,209–212.

[6] Acosta M.,Ludena H.,Barreto D. & Moro Sommo M. (1972). – Brucellosis en alpacas. *Rev. Invest. Pecu.*,1 (1),37–49.

[7] Adama N.B.,Okoh A.E.J. & Azunku U.J. (2007). – Prevalence of brucellosis in nomadic herds of dromedaries in Borno State,Nigeria. *J. Camel Pract. Res.*,14 (2),135–138.

[8] Afzal M. & Sakkir M. (1994). – Survey of antibodies against various infectious disease agents in racing camels in Abu Dhabi,United Arab Emirates. *Rev. sci. tech. Off. int. Epiz.*,13 (3),787–792.

[9] Agab H. (1993). – Epidemiology of camel diseases in eastern Sudan with emphasis on brucellosis. M.V.Sc. Thesis,University of Khartoum,Khartoum,Sudan.

[10] Agab H. (1998). – Camel pastoralism in the Butana region of eastern Sudan: common diseases with emphasis on brucellosis. *J. Camel Pract. Res.*,5 (1),131–136.

[11] Agab H.,Abbas B.,El Jack Ahmed H. & Mamoun I.E. (1994). – First report on the isolation of *Brucella abortus* biovar 3 from camel (*Camelus dromedarius*) in Sudan. *Rev. Elev. Méd. Vét. Pays Trop.*,47 (4),361–363.

[12] Agab H.,Abbas B.,El Jack Ahmed H. & Mamoun I.E. (1996). – First report on the isolation of *Brucella abortus* biovar 3 from camels (*Camelus dromedarius*) in Sudan. *Camel Newslett.*,12 (9),52–55.

[13] Agab H.R.D,Angus B. & Mamoun I.E. (1995). – Serologic response of camel (*Camelus dromedarius*) to *Brucella abortus* vaccine S19. *J. Camel Prac. Res.*,2 (2),93–95.

[14] Ahmed A. & Ibrahim L. (1980). – Indagine sulla presenza e diffusione delle brucellosi nel dromedario in Somala. Tesi di Laurea,Facultata di Medicina Veterinaria,Mogadishu,Somalia,U.N.

[15] Ahmed M.R. (1993). – The incidence of brucellosis in different domesticated animals in Egypt. *Tech. Bull.*,23,210–231.

[16] Ahmed W.M. & Nada A.R. (1993). – Some pathological affections of testis and epididymis of slaughtered camels (*Camelus dromedarius*). *Int. J. Anim. Sci.*,8,33–36.

[17] Ajmal M.,Ahmad M.D. & Arshad A. (1989). – Sero-surveillance of brucellosis. *Pakistan Vet. J.*,9,115–117.

[18] Al Dubaib M.A. (2007). – Polymerase chain reaction and adapted enzyme linked immunosorbent assay for diagnosis of camel brucellosis. Proceedings of the 9th Scientific Conference,23–26 April 2007,Cairo University,Cairo.

[19] Al-Ani F.K.,Al-Sharrify M. & Khalil F. (1998). – Serological survey on camel brucellosis in camels in Iraq. *Camel Newslett.*,14,32–33.

[20] Ali M. & Ghedi S. (1978). – Indagini siero-epidemiologiche sulla diffusione in Somala della brucellosi degli animali domestici. Tipizzazioni dei primi stipiti isolati nel paese. Tesi di Laurea,Facultata di Medicina Veterinaria,Mogadishu,Somalia,U.N.

[21] Al-Khalaf S. & El-Khaladi A. (1989). – Brucellosis of camels in Kuwait. *Comp. Immunol. Microbiol. Infect. Dis.*,12 (1–2),1–4.

[22] Al-Majali A.M. (2006). – Seroepidemiology of camel brucellosis in Jordan. Proceedings of the 1st Conference of ISOCARD,15–17 April 2006,Al Ain,UAE,79.

[23] Alshaikh M.A.A.,Al Haidary A.,Aljumaah R.S.,Al Korashi M.M.,El Nabi G.R.A. & Hussein M.F. (2007). – Camel brucellosis in Riyadh region,Saudi Arabia. *J. Camel Pract. Res.*,14 (2),113–117.

[24] Andreani E.,Prosperi S.,Salim A.H. & Arush A.M. (1982). – Serological and bacteriological investigation on brucellosis in domestic ruminants of the Somali Democratic Republic. *Rev. Elev. Vét. Méd. Pays Trop.*,35 (4),329–333.

[25] Anon. (1981). – Annual report of the Veterinary Laboratory,Kismayo. Ministry of Livestock,Forestry and Range,Department of Veterinary Services,Somalia.

[26] Arbusov P.N. (1940). – Normal titre of camel serum in relation to brucellosis. Soviet Vet.,5,47–48.

[27] Atwa K.A. (1997). – Brucellosis in camels. M.V.Sc. Thesis,Cairo University,Cairo.

[28] Ayoub N.M.,Shawkat M.A. & Fayed A.A. (1978). – Serological investigation on brucellosis in camels in Egypt. *Assiut Vet. Med. J.*,30,45–50.

[29] Azwai S.M.,Carter S.D.,Woldehiwet Z. & MacMillan A. (2001). – Camel brucellosis: evaluation of field sera by conventional serological tests and ELISA. *J. Camel Pract. Res.*,8 (2),185–193.

[30] Bares J.F. (1968). – Contribution a l etude de la pathologie infectieuse du dromadaire au Tchad. These,Faculte de Médecine de Toulouse,France,30–38.

[31] Barsoum S.A.,El-Sayed M.M. & El-Fayoumy M.M. (1995). – Seroepidemiological study on camel brucellosis. *Bani Suef Vet. Med. Res.*,5 (2),111–117.

[32] Baumann M.P.O.,Nuux H.A. & Zessin K.H. (1990). – Livestock disease survey Central Rangeland of Somalia. Technical report. Vol. III. Herd demographic and disease survey data from herds of camels. CRDP Veterinary Component,Mogadishu,Somalia.

[33] Baumann M.P.O. & Zessin K.H. (1992). – Productivity and health of camels (*Camelus dromedarius*) in Somalia: associations with trypanosomiasis and brucellosis. *Trop. Anim. Health Prod.*,24 (3),145–156.

[34] Ben Faraj S.M.,Azwai S.M.,Gameel S.E.,Shareha A.M.,Benhaj K.M.,Rayes H.M. & Nayil A.A. (1990). – Camel and Libya. *In* Proceedings of the International Conference on Camel Production and Improvement,10–13 December 1990,Tobruk,Libya.

[35] Bishof J. (1979). – Serological examination of blood samples from dromedaries. Serum and Vaccine Institute,Mogadishu,Somalia.

[36] Bisping W. & Amtsberg G. (1988). – Colour atlas for the diagnosis of bacterial pathogens in animals. Verlag Paul Parey,Berlin and Hamburg,246–259.

[37] Bornarel P. & Akakpo A.J. (1982). – Brucelloses animales: Sondages sérologiques dans quatre pays de l'Afrique de l'Ouest (Benin,Cameroun,Haute-Volta,Niger). *Médecine d'Afrique Noire*,29 (12),829–836.

[38] Bornstein S. (1984). – Working paper No. 3,Camel Forum. Somali Academy Science and Art,Mogadishu,Somalia.

[39] Bornstein S. (1988). – A disease survey of the Somali camel. SAREC report,Stockholm,Sweden.

[40] Bornstein S. & Musa B.E. (1987). – Prevalence of antibodies to some viral pathogens,*Brucella abortus* and *Toxoplasma gondii* in serum from camels (*Camelus dromedarius*) in Sudan. *J. Vet. Med. B*,34,364–370.

[41] Bornstein S.,Musa B.E. & Jama F.M. (1988). – Comparison of seroepidemiological findings of antibodies to some infectious pathogens of cattle in camels of Sudan and Somalia with reference to findings in other countries of Africa. *In* Proceedings of the International Symposium of Development of Animal Resources in Sudan,3 ~ 7 January 1988,Khartoum,Sudan,28–34.

[42] Burgemeister R.,Leyk W. & Goessler R. (1975). – Untersuchungen uber Vorkommen von Parasitosen,bakteriellen und viralen Infektionskrankheiten bei Dromedaren in Sudtunesien. *Dtsch. Tierärtztl. Wschr.*,82,352–354.

[43] Chichibabin E.S. (1971). – Results of haemagglutination test with the heat-inactivated sera from camels investigated for brucellosis. *Proc. Kazakh Res. Vet. Inst.*,14,29–30.

[44] Chukwu C.C. (1985). – Brucellosis in Africa. Part I: the prevalence. *Bull. Anim. Health Prod. Afr.*,33,193–198.

[45] Domenech J. (1977). – Enquete serologique sur la brucellose du dromadaire en Chad. *Rev. Elev. Méd. Vét. Pays Trop.*,30 (2),141–142.

[46] Domenech J.,Lucet P.,Vallat B.,Stewart C.,Bonnet J.B. & Hentic A. (1982). – La brucellose bovine en Afrique centrale. III. Resultats statistique des enquetes menees au Tchad et au Cameroun. *Rev. Elev. Méd. Vét. Pays Trop.*,35,15–22.

[47] Ebrahimi A.,Hosseinpour F. & Montazeri B. (2007). – Seroprevalence of brucellosis in dromedaries in Iran. *J. Camel Pract. Res.*,14 (1),43–44.

[48] Elmi,A.M. (1982). – MPVM Thesis,University of California,Davis,CA,USA.

[49] El-Nahas H.M. (1964). – Brucellosis in camels. *In* Proceedings of the Fifth Arab Veterinary Congress,10–13 September 1964,Cairo,Egypt,239–252.

[50] El-Sawally A.A.,Montaser A.M. & Rizk L.G. (1996). – Diagnostic and biochemical evaluations of camel brucellosis. *Vet. Med. J. Giza*,44 (2),323–329.

[51] Erdenebaatar J.,Bayarsaikhan B.,Watarai M.,Makino S.I. & Shirahata T. (2003). – Enzyme-linked immunosorbent assay to differentiate the antibody responses of animals infected with *Brucella* species from those of animals infected with *Yersinia enterocolitica* O9. *Clin. Diagn. Lab. Immunol.*,10 (4),710–714.

[52] Fayed A.A.,Karmy S.A.,Yousef H.I. & Ayoub M.M. (1982). – Serological studies on brucellosis in Aswan Province. *Vet. Med. J.*,30,491–497.

[53] Fazil M.A. & Hofmann R.R. (1981). – Haltung und Krankheiten des Kamels. *Tierärztl. Praxis*,9,389–402.

[54] Fowler M.E. (2010). – Medicine and surgery of camelids. 3rd Ed. Wiley-Blackwell,Ames,IA,207–208.

[55] Gameel S.E.A.,Mohamed S.O.,Mustafa A.A. & Azwai S.M. (1993). – Prevalence of camel brucellosis in Libya. *Trop. Anim. Health Prod.*,25 (2),91–93.

[56] Gatt Rutter T.E. & Mack R. (1963). – Diseases of camels. Part 1: bacterial and fungal diseases. *Vet. Bull.*,33 (3),119–124.

[57] Ghanem Y.M.,El-Khodery S.A.,Saad A.A.,Abdelkader A.H.,Heybe A. & Musse Y.A. (2009). – Seroprevalence of camel brucellosis (*Camelus dromedarius*) in Somaliland. *Trop. Anim. Health. Prod.*,41,1779–1786.

[58] Ghazi Y.A. (1996). – Studies on brucellosis in camels. PhD Thesis,Cairo University,Cairo.

[59] Ghoneim N.A. & Amjad,A.M. (1993). – Brucellosis among sheep,goats and camels in Saudi Arabia in Al Joub region,incidence and comparison between rose bengal test and seroagglutination tube test. *In* Proceedings of 21st Arab Veterinary Medicine Congress,10 14 April 1993,Cairo,273–281.

[60] Gidlewski T.,Cheville N.F.,Rhyan J.C.,Miller L.D. & Gilsdorf M.J. (2000). – Experimental *Brucella abortus* induced abortion in a llama: pathologic effects. *Vet Pathol.*,37,77–82.

[61] Gilsdorf M.J.,Thoen C.O.,Temple R.M.S.,Gidlewski T.,Ewalt D.,Martin B. & Henneger S.B. (2001). – Experimental exposure of llamas (*Lama glama*) to *Brucella abortus*: humoral antibody response. *Vet. Microbiol.*,81,85–91.

[62] Graber M. (1968). – Region of Veterinary and Zootechnical Research of Central Africa. Annual report,Farcha Laboratory,1st Research and Products 2 Pleuropneumonia. Quini quennial Report. Fort Lamy,Chad. *Vet. Bull.*,38,52–65.

[63] Gwida M.M.A.S. (2010). – Isolation,identification and typing of *Brucella* species as zoonotic pathogens by using conventional and molecular biological methods. Thesis,Free University Berlin,Germany.

[64] Gwida M.M.A.S,El-Gohary A.H.,Melzer F.,Tomaso H.,Rosler U.,Wernery U.,Wernery R.,Elschner M.,Khan I. & Heinrich Neubauer H. (2011). – Comparison of diagnostic tests for the detection of *Brucella* spp. in camel sera. *BMC Res. Notes*,4 (1),525–532.

[65] Hamada S.,El-Hidik M.,Sherif I.,El-Sawah H. & Yousef M. (1963). – Serological investigations on brucellosis in cattle,buffaloes and camels. *J. Arab. Vet. Med.*,23,173–178.

[66] Harby H.A.M. & Ismaily S.L.N. (1995). – The prevalence of brucellosis among livestock in the Sultanate of Oman. *In* Proceedings of the International Conference on Livestock Production in Hot Climates,Muscat,Oman,A46.

[67] Hashim N.H.,Galil G.A.,Hulaibi M.A. & Al-Saleem E.M. (1987). – The incidence of brucellosis and species of *Brucella* organisms isolated from animals in Al Hasa,Saudi Arabia. *World Anim. Rev.*,61,32–53.

[68] Higgins A. (1986). – The camel in health and disease. Bailliere Tindall,London.

[69] Ismaily S.I.N.,Harby H.A.M. & Nicoletti P. (1988). – Prevalence of *Brucella* antibodies in four animal species in the Sultanate of Oman. *Trop. Anim. Health. Prod.*,20,269–270.

[70] Jawad A.H. (1984). – Brucellosis in camel in Iraq. *Bull. endem. Dis.*,24–25,45–50.

[71] Kagunya D.K.J. & Waiyaki P.G. (1978). – A serological survey of animal brucellosis in north-eastern province of Kenya. *Kenya Vet.*,2 (2),35–38.

[72] Khorasgani M.R.,Bokaie S.,Moallemzadeh S.A. & Salehi T.Z. (2006). – A note on serologic survey of camel brucellosis in Qum province,Iran. *J. Camel Pract. Res.*,13 (1),51–52.

[73] Kiel F.W. & Khan M.Y. (1987). – Analysis of 506 consecutive positive serological tests for brucellosis in Saudi Arabia. *J. Clin. Microbiol.*,25,1384–1387.

[74] Kudi A.C.,Kalla D.J.U.,Kudi M.C. & Kapio G.I. (1997). – Brucellosis in camels. *J. Arid Environ.*,37,413–417.

[75] Kulshrestha R.C.,Arora R.G. & Kalra D.S. (1975). – Brucellosis in camels and horses. *Indian J. Anim. Sci.*,45 (9),673–675.

[76] Madkow M.M. (1989). – Brucellosis. Butterworths,London.

[77] Majid A.A. and Goraish I.A. (2000). – Seroepidemiological observations of camel brucellosis in eastern and western Sudan. *Camel Newslett.* 2000,17,23–26.

[78] Mathur K.N. & Bhargava S.C. (1979). – Sero-prevalence of Q fever and brucellosis in camels of Jorbeer and Bikaner,Rajasthan State. *Indian J. Med. Res.*,70 (11),391–393.

[79] Millar M. & Stack J. (2012). – Brucellosis – What every practitioners should know. *In Practice*,34,532–539

[80] Mocalov V.I. (1991). – About the prevalence of brucellosis in cattle and camels. *Bull. Sci. Res. Inst. Exp. Vet.-Med. Moscow* [in Russian],41–45.

[81] Mohammed I.M. (1996). – Development,optimization and evaluation of diagnostic immunoassays for camel brucellosis. Thesis,University Khartoum,Khartoum,Sudan.

[82] Moustafa T.,Omar E.A.,Basyouni S.M. & El-Badawi A.S. (1998). – Surveillance of *Brucella* antibodies in camels of the eastern region of the United Arab Emirates. *In* Proceedings of the International Meeting on Camel Production and Future

Perspectives,2–3 May 1998,Al Ain,UAE,160–166.

[83] Mustafa A.A. & Awad El-Karim M.H. (1971). – A preliminary survey for the detection of *Brucella* antibodies in camel sera. *Sudan J. Vet. Sci. Anim Husb.*,12 (1),5–8.

[84] Mustafa A.A. & Hassan A. (1971). – A preliminary survey for the detection of *Brucella* antibodies in camel sera,Sudan. *J. Vet. Sci. Anim. Husb.*,12 (5) 5–8.

[85] Nada A.R. (1984). – Some studies on brucellosis in camels. M.V.Sc.,Faculty of Veterinary Medicine,Cairo University,Cairo.

[86] Nada A.R. (1990). – Further studies on brucellosis in camels. PhD Thesis,Faculty of Veterinary Medicine,Cairo University,Cairo.

[87] Nada A.R. & Ahmed W.M. (1993). – Investigations on brucellosis in some genital abnormalities of she-camels *(C. dromedarius). Int. J. Anim. Sci.*,8 (1),37–40.

[88] Nada A.R.,Ismail E.M.,Shawkat M.E. & Barsoum S.A. (1992). – Evaluation of serotests used in the diagnosis of camel brucellosis. *J. Egypt Vet. Med. Ass.*,52 (4),435–442.

[89] Obeid A.I.,Bagadi H.O. & Mukhtar M.M. (1996). – Mastitis in camels (*Camelus dromedarius*) and the somatic cell content of camel milk. *Res. Vet. Sci.*,61 (1),55–58.

[90] Okoh A.E.J. (1979). – A survey of brucellosis in camels in Kano,Nigeria. *Trop. Anim. Health Prod.*,11 (4),213–214.

[91] Omer M.M.,Musa M.T.,Bakhiet M.R. & Perret L. (2010) – Brucellosis in camels,cattle,humans: association and evaluation of serological tests used for diagnosis of the disease in certain nomadic in Sudan. *Rev. sci. tech. Off. int. Epiz.*,29 (3),663–669.

[92] Osman A.M. & Adlan A.M. (1987). – Sudan. Brucellosis in domestic animals: prevalence,diagnosis and control. *Tech. series,Off. int. Epiz.*,6,67–72.

[93] Ostrovidov P.I. (1954a). – Experiment on rearing healthy camels from dams infected with brucellosis. *Trud. Inst. Vet. Alma-Ata.* 6,62–68.

[94] Ostrovidov P.I. (1954b). – Development of resistance to brucellosis in camels. *Trudy Inst. Vet. Alma Ata*,6,51–56.

[95] Pal'gov A.A. (1950). – Letter. *Trudy Instituta Veterinarnych nauk Kazackskoj SSR*,5,29.

[96] Pal'gov A.A. & Zhulobovski I.Z. (1964). – Diagnosis of brucellosis in camels and methods of eliminating infection from camel herds. *Trudy Instituta Veterinarnych nauk Kazackskoj SSR*,6,43–50.

[97] Radwan A.I.,Asmar J.A.,Frerichs W.M.,Bekairi S.I. & Al-Mukayel A.A. (1983). – Incidence of brucellosis in domestic livestock in Saudi Arabia. *Trop. Anim. Health Prod.*,15,139–143.

[98] Radwan A.I.,Bekairi S.I.,Mukayel A.A.,Albokmy A.M.,Prasad P.V.S,Azar F.N. & Coloyan E.R. (1995). – Control of *Brucella melitensis* infection in a large camel herd in Saudi Arabia using antibiotherapy and vaccination with Rev 1 vaccine. *Bull. Off. int. Epiz.*,14 (3),719–732.

[99] Radwan A.I.,Bekairi S.J. & Prasad P.V.S. (1992). – Serological and bacteriological study of brucellosis in camels in central Saudi Arabia. *Rev. sci. tech. Off. int. Epiz.*,11 (3),837–844.

[100] Ramadan R.O.,Hatem M.E. & Abdin Bey M.R. (1998). – Isolation of *Brucella melitensis* from carpal hygroma in camels. *J. Camel Pract. Res.*,5 (2),239–241.

[101] Richard D. (1980). – Dromedary pathology and productions. Provisional report No. 6. Camels. International Science Foundation (IFS),Khartoum,Sudan,and Stockholm,12 (18–20),409–430.

[102] Rojas X.,Munoz S.,Otto B.,Perez B. & Nielsen K. (2004). The use of polarized fluorescence assay (PF) and competitive ELISA test (c-ELISA) for the diagnosis of brucellosis in South American camelids. *Arch. Med. Vet.*,1 (XXXVI),59–64.

[103] Saegermann C.,Berkvens D.,Godfroid J. & Walravens K. (2010). – Bovine brucellosis. *In* Infectious and parasitic diseases of livestock (P.-C. Lefevre,J. Blancou,R. Chermette & G. Uilenberg,eds),Lavoisier,France,991–1021.

[104] Salem A.A.,El-Gibaly S.M.,Shawkat M.E.,Ibrahim S.I. & Nada A.R. (1990). – Some studies on brucellosis in camels. *Assiut Vet. Med. J.*,23 (45),139–145.

[105] Saley H. (1983). – Contribution a l'etude des brucelloses au Niger: resultats d'une enquete serologique dans 3 departements. These,Doctorat Veterinaire,Veterinary Faculty,Dakar,Senegal,6.

[106] Schelling E.,Diguimbaye C.,Daoud S. & Nicolet J. (2003). – Brucellosis and Q-fever seroprevalences of nomadic pastoralists and their livestock in Chad. *Prev. Vet. Med.*,61,279–293.

[107] Schulze zur Wiesch J.,Wichmann D.,Sobottka I.,Rohde H.,Schmoock G.,Wernery R.,Schmiedel S.,Burchard G.D. & Melzer F. (2010). – Genomic tandem repeat analysis proves laboratory-acquired brucellosis in veterinary (camel) laboratory in the United Arab Emirates. *Zoonoses Public Health*,57 (5),315–317.

[108] Seifert H.S.H. (1992). – Tropentierhygiene. Gustav Fischer Verlag,Jena,Stuttgart,Germany,292–304.

[109] Shumilov K.V. (1974). – Diagnostic value of agglutination and complement fixation test for brucellosis in camels. *Proceedings of All-Union Institute of Experimental Veterinary Medicine*,42,279–282.

[110] Solonitsyn M.O. (1949). – Brucellosis in camels. *Veterinariya (Moscow)*,26 (6),16–20.

[111] Solonitsyn M.O. & Pal'gov A.A. (1950). – Trud. nauchno-issled. *Vet Int. (Alma Ata)*,5,58.

[112] Straten M.O. (1949). – Brucellosis in camels. *Veterinariya (Moscow)*,26 (6),16–20.

[113] Sunaga Y.,Tani F. & Mukai K. (1983). – Detection of *Yersinia enterocolitica* infection in camels serodiagnosed as brucellosis. *Japanese J. Vet. Sci.*,45 (2),247–250.

[114] Taha T.H. (2007). – Pathogens affecting the reproductive system of camels in the United Arab Emirates. Thesis,Swedish University of Agricultural Sciences,Uppsala,Sweden.

[115] Ten V.B. & Cejdachmedova R.D. (1993). – Diagnostica brucelloza verebljudov. *Dositizenie nauki I techniki*,1,31 [in Russian].

[116] Teshome H. & Molla B. (2002). – Brucellosis in camels (*Camelus dromedarius*) in Ethiopia. *J. Camel Pract. Res.*,9 (2),125–128.

[117] Tserendash C. & Shumilov K.V. (1970). – Diagnosis of brucellosis in camels. *Veterinariya*,1,116–117 [in Russian].

[118] Van Straten,M.,Bercovich Z. & Ur-Rahman Z. (1997). – The diagnosis of brucellosis in female camels (*Camelus dromedarius*) using the milk ring test and milk ELISA: A pilot study. *J. Camel Pract. Res.*,4 (2),165–168.

[119] Verger J.M.,Grayon M.,Doutre M.P. & Sagna F. (1979). – *Brucella abortus* d'origine bovine au Sénégal: identification et typage. *Rev. Elev. Méd. Vét. Pays Trop.*,32 (1),25–32.

[120] Von Hieber D. (2010). – Investigation of occurrence and persistence of brucellosis in female camel dams (*Camelus dromedarius*) and their calves. Thesis,Universitat Ulm,Ulm,Germany.

[121] Vrioni G.,Pappas G.,Priavali E.,Gartzonika C. & Levidiotou S. (2008). – An eternal microbe: *Brucella* DNA load persists for years after clinical cure. *Clin. Infect. Dis.*,46 (12),e131–e136.

[122] Waghela S.,Fazil M.A.,Gathuma J.M. & Kagunya D.K. (1978). – A serological survey of brucellosis in camels in north-eastern province of Kenya. *Trop. Anim. Health Prod.* 10 (1),28–29.

[123] Wernery U. (1991). – The barren camel with endometritis – isolation of *Trichomonas fetus* and different bacteria. *J. Vet. Med. B.*,38 (1–10),523–528.

[124] Wernery U. & Ali A. (1989). – Bacterial infertility in camels (*Camelus dromedarius*). Isolation of *Campylobacter fetus. Dtsch.Tierärztl. Wschr.*,96,497–498.

[125] Wernery U.,Kinne J.,Joseph M.,Johnson B. & Nagy P. (2007a). – Where do *Brucella* organisms hide in serologically positive lactating dromedaries? *In* Proceedings of the International Camel Conference,16–17 February 2007,Bikaner,India,68–70.

[126] Wernery U.,Thomas R.,Syriac G.,Raghavan R. & Kletzka S. (2007b). – Seroepidemiological studies for the detection of antibodies against nine infectious diseases in dairy dromedaries (Part- I). *J. Camel Pract. Res.*,14 (2),85–90.

[127] Wernery U. & Wernery R. (1990). – Seroepidemiologische Untersuchungen zum Nachweis von Antikorpern gegen Bruc ellen,Chlamydien,Leptospiren,BVD/MD,IBR/IPV – und Enzootischen Bovinen Leukosevirus (EBL) bei Dromedarstuten (*Camelus dromedarius*). *Dtsch. tierärztl. Wschr.*,97,134–135.

[128] Wernery U. & Wernery R. (1992). – Uterine infections in the dromedary camel. A review. In Proceedings of the 1st International Camel Conference. R. and W. Publications (Newmarket) Ltd,Newmarket,UK,155–158.

[129] Wilson A.J.,Schwartz H.J.,Dolan R.,Field C.R. & Roettcher D. (1982). – Epidemiologische Aspekte bedeutender Kamelkrankheiten in ausgewahlten Gebieten Kenias. *Der praktische Tierarzt.*,11,974–987.

[130] World Health Organization (WHO)/Food and Agriculture Organization (FAO) (1986). – 6th report of the expert committee on brucellosis. *Tech. Rep. Ser.,Geneva*,740,132.

[131] Yagoub I.A.,Mohamed A.A. & Salim M.O. (1990). – Serology survey for Br. *abortus* antibody prevalence in the one humped camel (*Camelus dromedarius*) from eastern Sudan. *Rev. Elev. Méd. Vét. Pays Trop.*,43 (2),167–171.

[132] Zagloul A.H. & Kamel Y. (1985). – Incidence of brucellosis among farm animals in Assiut governorate. *Assiut Vet. Med. J.*,14,117–122.

[133] Zaki R. (1943). – *Br. abortus* infection in buffaloes,ewes and camels. Isolation of the organism from milk. M.V.Sc. Thesis,Faculty of Veterinary Medicine,Cairo University,Cairo.

[134] Zaki R. (1948). – Brucella infection among ewes,camels and pigs in Egypt. *J. Comp. Pathol. Ther.*,58,145–151.

[135] Zhulobovski I.L. & Pal'gov A.A. (1954). – Letter. *Trud. Inst. Vet. Alma-Ata*,6,17 [in Russian].

[136] Zowghi E. & Ebadi A. (1988). – Brucellosis in camels in Iran. *Rev. sci. tech. Off. int. Epiz.*,7 (2),383–386.



## 深入阅读材料

Ahmed M.S.H. (1996). – Some studies on post partum period in the she camels. *Camel Newslett.*,12 (9),27–28.

Chen J.N. (1988). – A serological survey of camel brucellosis. *Gansu J. Anim. Sci. Vet. Med.*,1,8–9.

Khalafalla E.N. & Khan A. (1958). – The occurrence,epidemiology and control of animal brucellosis in the Sudan. *Bull. Epiz. Dis. Afr.*,6,243–247.

Fayza A.O.,El Sheikh O.H.,Zakia A.M.,Halima M.O.,Suliman H.B. & Osman A.Y. (1989 —1990). – Survey of brucellosis among cattle,camels,goats and sheep in the *Sudan. Sudan J. Vet. Res.*,9,36–40.

Hamid A. (1993). – Epidemiology of camel diseases in eastern Sudan,with emphasis on brucellosis. Thesis in Veterinary Science,Faculty of Veterinary Science,University of Khartoum,Khartoum,Sudan,184.

Moro Sommo M. (1957). – Investigacion preliminar de la brucellosis en alpacas. *Rev. Fac. Med. Vet. (Lima)*,12,135–137.

Mousa A.M.,Elhag K.M.,Khogali M. & Marafic A.A. (1988). – The nature of human brucellosis in Kuwait: study of 379 cases. *Rev. Infect. Dis.*,10,211–217.

Richard D.D.,Planchenault D.E. & Giovannetti J.F. (1985). – Production cameline – Rapport final,Project de Développement de l'élevage dans le Niger. Centre – Est,IEMVT,France.

Wang J.L.,Yie H.S.,Zhang Y.B. & Wang Z.X. (1986). – Comparison of four serological tests in diagnosis of camel brucellosis. *Qinghai J. Anim. Sci. Vet. Med.*,Special Issue on Camel,87–89.

（周继章译，殷宏校）

## 1.4.2 传染性繁殖障碍

骆驼科动物的生殖生物学具有一些其他家养动物所没有的重要特性，这些特性是最近20年才发现的，之前一直不被人知晓。

骆驼的卵巢周期很特殊，它们以蹲踞姿势进行交配（图106），诱导排卵，胎儿被一层表皮膜包裹（图107），为左角妊娠。最近20年开展了对骆驼生殖生理学的深入研究，Fowler和Bravo[13]，Gauly[14]，Brown[9]，Bravo[8]，Gauly[15]等深入研究了新大陆骆驼，来自阿联酋的两个团队深入研究了单峰驼[34,35,39]，另一个团队调查子宫感染的原因[46]，Beil[7]对所有科研人员发表的关于旧大陆骆驼生殖系统的文章做了总结。

一般来说，骆驼是生育能力极强的动物，双峰驼和单峰驼结合可得到杂种，新大陆骆驼也可进行杂交繁殖。随着技术的先进，研究已经进入了全新的领域：旧大陆骆驼和新大陆骆驼间的杂交繁育[35]，阿联酋迪拜的科学家们已经对有特殊价值的如专供竞赛和产奶的骆驼进行了克隆[41]。

Wilson[51]认为，尽管普遍认为单峰驼有很强的生育能力，但是，在自然状态下骆驼的出生率很

图106 骆驼以蹲踞姿势交配

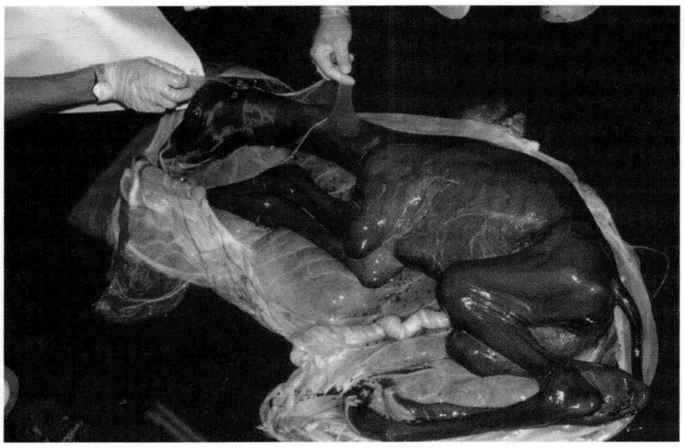
图107 被光滑的羊膜包裹着的新生驼

低。他列举了出生率低的几个可能原因：如生活在亚洲和非洲的单只母驼一生只生产3只幼崽，5岁第一次怀孕，每次孕期长达13个月，生育间隔超过两年；另外，种畜过早屠宰等。

沙特阿拉伯的骆驼生育率为80% ~ 90%，还有1%的骆驼永久不育[5]，Yagil[53]报道了相似的数据，他的实验骆驼达到了100%的生育率。然而，Mukasa-Mugerwa[26]报道了在饲养环境略差的条件下，单峰驼的生育率只有50%，在管理上有了一些改进之后可达65%。营养不足、锥虫病、肺结核、体外寄生虫及体内寄生虫、复发性子宫内膜炎都可以使生育能力下降。新生驼的死亡率可以达到流行病所造成损失，对于羊驼和美洲驼来说，高不孕不育率也是很严重的问题。Tibary[38]等指出，羊驼的年平均生育率只有50%，平均出生率只有45%。在兽医实践中，子宫感染和流产最为常见，多种多样的细菌、真菌、病毒会导致流产、子宫内膜炎和子宫炎的发生。然而，妊娠失败有多种原因，不仅仅和感染有关。

单峰驼的流产率在2% ~ 25%[38]，羊驼和美洲驼的流产率则相对要高一些，双峰驼的数据还未知。

妊娠早期胎儿的损失被称为早期胚胎损失，通常发生在妊娠头40天，在妊娠后期发生的则被称为流产。早期胚胎损失的原因很难确定，且在很多情况下，胎儿会被重吸收或消失而不被发现。流产的原因相对容易确定，但是，不幸的是有很多原因未被确诊。对牛的研究中，超过50%的牛流产原因不明。在美国的俄勒冈州对新大陆骆驼中回顾性流行病学调查中，98例中有68例（69%）流产原因没有确定，只有11%确诊是细菌感染，包括波蒙纳钩端螺旋体、大肠杆菌、单核细胞增生李斯特菌、新孢子虫和假单胞菌感染；15%的流产是由于脐带扭曲导致的[24]。另外，马属动物流产中，有70%原因不明，20%是感染造成的，非感染性原因最常见的就是脐带扭转，可达到15%。在流产病例中，没有症状、病因不明的情况被认为先天性流产，这样的结果并不令人满意，但是，这也至少说明了在胎儿或胎膜中没有发现传染病[50]。

在过去几十年，研究人员展开深入研究以查清牛和马子宫感染的真相以及子宫内分离的细菌和

表40 试验中被屠宰的单峰驼患子宫内膜炎的概率

| 作者 | 年份 | 国家 | 骆驼数量 | 患子宫内膜炎的数量 | 百分比（%） |
|---|---|---|---|---|---|
| Nawito | 1973 | 埃及 | 2075 | 94 | 4.53 |
| Hegazy等 | 1979 | 埃及 | 96 | 24 | 25.0 |
| Laila等 | 1987 | 埃及 | 130 | 97 | 74.6 |
| Fetaih | 1991 | 埃及 | 78 | 67 | 86.0 |
| Al-Ani 等 | 1992 | 伊拉克 | 50 | 2 | 4.0 |

子宫感染之间的关系。从牛、马的子宫内分离出来多种细菌，然而，只有很少数是原发性病原体，其中，大部分细菌是条件致病菌。将细菌学上的结果和生殖道临床症状如子宫炎症、分泌物进行综合判断很重要，还可以通过子宫内膜涂片、子宫活检以获得更多的相关信息[33]。这些方法对骆驼科动物也同样适用。

子宫感染被认为是骆驼繁殖障碍的最常见原因，在不育骆驼的子宫分离出多种细菌，但是，这些细菌在最初的子宫感染中起到什么作用尚不清楚。

## 流行病学和病理学

骆驼科动物和其他家养动物一样，子宫感染被普遍认为是繁殖障碍导致不孕的原因[39]，但是，只有少数科学家对骆驼科的子宫内细菌感染做了检测。最初的研究定位用于屠宰骆驼，通过在不同国家的屠宰场大量的调查而展开，而没有针对处在繁殖状态的试验骆驼[25]。例如，开罗的屠宰场调查中，有55%的单峰驼患有子宫内膜炎，21%的单峰驼患有子宫炎[20]。Al-Delemi[2]调查发现，在伊拉克只有27%的单峰驼患有子宫感染。试验中被屠宰的单峰驼患子宫内膜炎的概率见表40。

最近有大量的关于单峰驼和新大陆骆驼繁殖障碍的文章发表[37,38,40,52]。Wernery和Ali[44]，Wernery[43]，Wernery和Wernery[49]在阿联酋进行了育种骆驼的深入研究，除了双峰驼感染布鲁氏菌病的报道外，很少有关于繁殖障碍的报道。阿联酋科学家们第一次从患有子宫内膜炎的不孕单峰驼的子宫中分离出了胎儿弯曲杆菌和胎儿三毛滴虫，这些发现具有重要的临床意义，因为这些病原可以导致家畜严重的经济损失。最近英国的一个发现也很重要，研究人员从3只流产羊驼的胎盘和胎儿胃内容物中分离出了胎儿弯曲杆菌，还发现了急性纤维坏死性胎盘炎，其中，一只感染母畜伴随出现舌头肿大和白喉假膜的现象[4]。

在阿联酋，对单峰驼的子宫活检时，可见子宫内膜肉芽肿的病例较多（图108），特征为不同大小的淋巴肉芽肿浸润，Nada和Ahmed[27]及Tibary和Ahmed[39]在未孕单峰驼身上也发现了此现象。

Kinne等[22]对内皮下单核细胞浸润现象做了深入研究，他们发现这种现象主要是发生在处女驼和

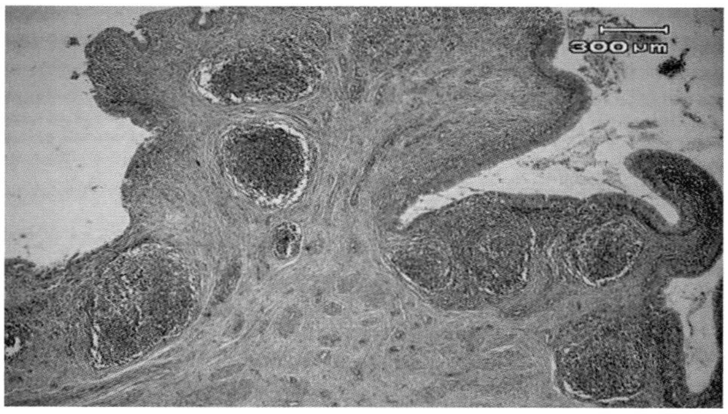


图108 不育单峰驼子宫淋巴肉芽肿（苏木精-伊红染色）

不孕驼。因此，作者假设这种组织学特征可能与生理交配的蹲踞姿势有关，这种姿势极易使细菌污染生殖道，然而单核细胞浸润可能是作为机体抵御子宫感染的屏障。我们仍需要有进一步的研究去证实这个假设，因为牛弯曲杆菌病、肺结核和真菌感染都有相似的特征。

除了还没有从骆驼科动物体分离出马生殖道泰勒菌外，从单峰驼子宫分离的细菌种类同母马和奶牛体内分离的细菌种类一样。为了评价不同种类的微生物在单峰驼子宫感染中起到的作用，迪拜的科学家们建议依据Ricketts[32]和Arthur等[5]对牛马的细菌分类来划分（表41）。可以直接通过子宫拭子、子宫活检、子宫或宫颈分泌物进行细菌学检查，这些拭子应该直接放入到培养基（碳培养基）并送至实验室培养。

除典型的生殖道微生物，如胎儿弯曲菌和胎儿三毛滴虫通过子宫内膜炎导致单峰驼和羊驼不孕外，铜绿假单胞菌、兽疫链球菌和大肠杆菌也起到了很重要的作用[1,6,18,29]。

Pal’gov[30]在哈萨克斯坦研究双峰驼的流产达3年之久，骆驼在怀孕5～6个月的时候流产，流产胎儿出现脐带发炎、心外膜出血、肝脾肿大的现象。结核病、布鲁氏菌病和鼻疽病不是导致流产的原因，而从流产胎儿以及胎儿被膜分离到的化脓放线菌，可能是导致骆驼子宫内膜炎发生的原因[11,23,28]，其他国家也是如此。

Wittek[52]分析了临床上健康经产的母羊驼的阴道拭子样本，描述了阴道菌群的发生和分布（表42），该结果的临床相关性是健康母羊驼的菌群，主要包括非致病性或低致病性的细菌，可能是阴道的生理菌群。而Bravo[8]报道了从普通和育种羊驼

表41 牛马生殖道分离培养的细菌及原生动物的分类方法引用到骆驼生殖道同样发现这些微生物的分类

| 马，牛 | 骆驼 |
|---|---|
| **由细菌和原生动物引起的生殖道感染** | |
| 1.马生殖道泰勒菌 | — |
| 2.肺炎克雷伯菌（荚膜型别1，2，5） | — |
| 3.铜绿假单胞菌 | Nawito[29]，Wernery and Ali[44]，Hassan[16]，Wernery[43]，Nawito[28]，Enany et al.[11] |
| 4.胎儿弯曲菌 | Wernery and Ali[44]，Anonymous[4]［alplaca］ |
| 5.胎儿三毛滴虫病 | Wernery[43] |
| **与子宫内膜炎相关非特异性细菌** | |
| 1.兽疫链球菌 | Nawito[29]，Awad et al.[6] |
| 2.大肠杆菌（溶血） | Nawito[29]，Eidarous et al.[10] |
| 3.金黄色葡萄球菌 | Nawito[29]，Awad et al.[6]，Hegazy et al.[18]，Ali et al.[3]，Wernery and Ali[44]，Hassan[16]，Wernery[43]，Enany et al.[11] |
| 4.变形杆菌属 | Hegazy et al.[18]，Ali et al.[3] |
| 5.肺炎克雷伯菌（荚膜型6，7，21，68） | Awad et al.[6]，Hegazy et al.[18]，Wernery and Ali[44]，Hassan[16] |
| 6.荧光假单胞菌 | — |
| 7.铜绿假单胞菌（非性病株） | — |
| 8.产气肠杆菌 | Hegazy et al.[18]，Eidarous et al.[10] |
| **污染和共生** | |
| 1.粪肠球菌 | Wernery and Ali[44] |
| 2.白色葡萄球菌 | Nawito[29]，Wernery and Ali[44]，Wernery[43] |
| 3.大肠杆菌（不溶血） | Hegazy et al.[18]，Ali et al.[3]，Wernery and Amjad Ali[44]，Hassan[16]，Wernery[43]，Nawito[29]，Enany et al.[11] |
| 4.放线菌属 | Zaki and Mousa[54]，Nawito[29]，Awad et al.[6]，Hegazy et al.[18]，Eidarous et al.[10]，Ali et al.[3]，Hassan[16] |
| 5.奈瑟氏淋球菌属 | — |
| 6.炭疽样微生物 | Zaki and Mousa[54]，Eidarous et al.[10]，Wernery and Ali[44]，Wernery[43] |
| 7.生孢梭菌 | Wernery[43] |
| 8.脆弱拟杆菌 | — |
| 9.梭杆菌属 | — |

表42 临床表现健康羊驼的阴道拭子检查的细菌种类

| 细菌种类 | 数量（占细菌总量的比例%） |
|---|---|
| 大肠杆菌，不溶血 | 35.2 |
| 大肠杆菌，溶血 | 5.6 |
| 大肠杆菌类 | 4.2 |
| 葡萄球菌属 | 16.9 |
| 链球菌属 | 12.7 |
| 化脓隐秘杆菌 | 5.6 |
| 肠球菌属 | 4.2 |
| 棒状杆菌属 | 4.2 |
| 不动杆菌属 | 1.4 |
| 微球菌属 | 1.4 |
| 放线菌属 | 1.4 |
| 嗜血杆菌属 | 1.4 |
| 乳杆菌属 | 1.4 |
| 泡囊短波单胞菌 | 1.4 |
| 恶臭假单胞菌 | 1.4 |
| 洋葱伯克霍尔德菌 | 1.4 |

注：4年检查临床健康羊驼阴道拭子总结

及美洲驼的生殖道分离的细菌的比例。

Bravo[8]培养了相同的细菌以及从难孕种畜的子宫和阴道分离的其他菌，只发现了李斯特菌和棒状杆菌。大量的子宫细菌种类占美国俄勒冈州兽医院的新大陆骆驼中发现菌的35%，并伴随生殖障碍，包括未能受孕、阴道分泌物、流产和死胎；从这些动物子宫内最常分离到的是大肠杆菌和铜绿假单孢菌。在新大陆骆驼检测时，需要着重强调的是要使用完全保护的拭子和细致的技术，因为许多样品可能含有污染物。Hassan和Ahmed[17]检测到埃及单峰驼的生殖器拭子含有支原体和脲原体。研究人员检测了250个拭子，其中，133个为包皮拭子、117个为阴道拭子，还有生殖器官、阴道、子宫颈、子宫及卵巢的采样。从所有被检样本中分离到不同百分比的精氨酸支原体和差异脲原体，但是，没有发现任何由这些微生物造成的临床病理变化。

从患有子宫内膜炎的母驼分离到曲霉菌和毛霉菌[38]，但是由于大量的抗生素治疗它们可能已经进入生殖道，所以，应该先进行活检和适当的真菌染色，再确定结果。

众所周知，其他驯养动物患系统性疾病会导致流产，对骆驼来说也一样。这些系统性疾病包括：

（1）细菌病。

1）结核病；

2）布鲁氏菌病；

3）钩端螺旋体病；

4）Q热（？）；

5）衣原体病；

（2）病毒病。

1）骆驼痘；

2）牛病毒性腹泻；

3）马鼻炎；

4）裂谷热；

5）蓝舌病。

（3）寄生虫病。

1）锥虫病；

2）弓形虫病；

3）肉孢子虫病；

4）新孢子虫病。

Jarvinen 和Kinyon[21]检测了年龄横跨1日龄至16岁的17只美洲驼和13只羊驼的包皮菌群，从这些动物体上采集的拭子按不同的培养基分别进行需氧、微需氧、厌氧微生物和真菌的培养。总共分离了112份菌，属于22个不同的细菌属，从阴茎包皮分离的大部分细菌是共生生物，但是，有些在特定的情况下会成为致病菌。

最近，在阿联酋分离到引起单峰驼流产风暴的一种马病毒[47]，这种病原和马鼻炎病毒（ERAV）一样是小RNA病毒，用其感染两只怀孕单峰驼也出现了流产（见2.1.16）。

单峰驼很容易接触马厩等马的活动范围，从患病的马身上遭受传染。驼科动物的胎盘是弥漫型、有上皮绒膜，与马的相似，这也许就可以从解剖学上部分解释了单峰驼对马病毒的敏感性。

早先从骆驼体分离到两种小RNA病毒，一个来自美国动物收藏中心的单峰驼，另一个来自美国的美洲驼。从成年单峰驼的心脏分离出的第一种病毒被确定为脑心肌炎病毒（心病毒属），病理特征为病灶发白、心肌层和心外膜出血、还有大量的心包液[42]。第二种病毒是从2只流产美洲驼胎儿，根据Stehman[36]等的报道被确定为马鼻炎病毒，报道称美洲驼感染小RNA病毒，在3.5个月内使15只美洲驼出现流产，平均妊娠期为220天（7～8个月），在母畜也会出现并发糖尿病，这与从迪拜的双峰驼分离的一种病毒相似[47]。

条件致病菌在单峰驼子宫感染中起到什么作用还不确定，将分别从有无子宫内膜炎的单峰驼分离到的细菌进行比较（表43），细菌学解释的固有困难就变得明显了。

骆驼和牛马一样，成功的诊断和不孕的治疗关键取决于对诊断评价和结果的解析，这些诊断包括阴道镜检查、子宫内容物培养、子宫细胞学检查、甚至活检。

在过去几年对于旧大陆骆驼的深入研究之后，研究焦点已经开始转向新大陆骆驼。Powers[31]等研究了美洲驼的子宫感染，他们检测了90只有生殖问题的动物，发现了45只有子宫感染（50%）。27只不孕美洲驼中有21只子宫培养物阳性，下面是培养的细菌：

（1）化脓放线菌(7x)；

（2）芽孢杆菌属(6x)；

（3）葡萄球菌属(6x)；

（4）大肠杆菌(6x)；

（5）链球菌属(3x)；

表43　从98只单峰驼分离到的微生物[49]统计

| 有子宫内膜炎 | 无子宫内膜炎 |
| --- | --- |
| 葡萄球菌 | 葡萄球菌属 |
| 金黄色葡萄球菌 | 金黄色葡萄球菌 |
| 链球菌属 | 链球菌属 |
| 需氧杆菌 | 需氧杆菌 |
| 双球菌 | 双球菌 |
| 大肠杆菌 | 大肠杆菌 |
| 生孢梭菌 | 生孢梭菌 |
| 胎儿弯曲杆菌 | |
| 铜绿假单胞杆菌 | |
| 臭鼻克雷伯氏菌 | |
| 沙门氏菌属 | |
| 黏质沙雷氏菌 | |

（6）拟杆菌属(1x)；

（7）坏死梭杆菌(1x)；

（8）混合培养（9x）。

美洲驼样本依据母畜子宫内膜活检的分级系统进行分类，结果见表44，这种分级系统也同样适用于单峰驼（J.Kinne，2004，个人交流）。

**诊断**

评价一只育种母骆驼应该首先要检查它的生育记录，任何一个难孕育种的动物都要考虑是否有子宫感染。对骆驼生殖道彻底检查要求：固定好动物以避免伤害到兽医或该动物，骆驼的固定可以通过使其处在侧卧或前卧姿势，或者是将它固定在直肠触诊溜槽内（图109）。在野外条件下检测时，前卧姿势是唯一可行的固定方法。当骆驼前卧时，可以通过阴道窥镜来检测阴道和宫颈口，样本拭子也可以获得。当有疑似子宫内膜炎发生时，从子宫内膜和宫颈获得的拭子，要做涂片进行姬姆萨染色（图110）。Tibary等[38]和Powers等[31]解释了对不同种的骆驼用什么方法进行子宫冲洗、采集拭子及活检。为了研究目的，Wernery[43]采取了用于分离马生殖道泰勒菌的方法，采集了子宫内膜、阴蒂窝和尿道口三处的拭子。

许多科学家对骆驼科动物展开了子宫的组织学调查和卵泡发育波的研究[7,13,37,38]，因为骆驼不像大多数动物的发情周期，发情期组织学图片变化不能解释什么。

埃及和伊拉克的很多科学家报道了被屠宰骆驼的子宫组织学变化，他们发现子宫内膜炎的概率分别为4.53%（94/2075）、25.0%(24/96)、86%（67/78）、74.6%（97/130）、4.0%（2/50），这些调查的结果专门从无生育史的被屠宰骆驼中

表44　美洲驼的子宫病理学与分离培养的比较

| 子宫级别 | 子宫病理学 | 生长 | 不生长 |
| --- | --- | --- | --- |
| I A | 正常<br>子宫内膜轻微变化 | 2 | 2 |
| I B | 少量淋巴细胞<br>极少的腺体纤维化 | 4 | 1 |
| II A | 子宫内膜炎 | 15 | 3 |
| II B | 子宫内膜炎 | | 0 |
| III A,B | 中度至重度腺体纤维化 | 0 | 0 |

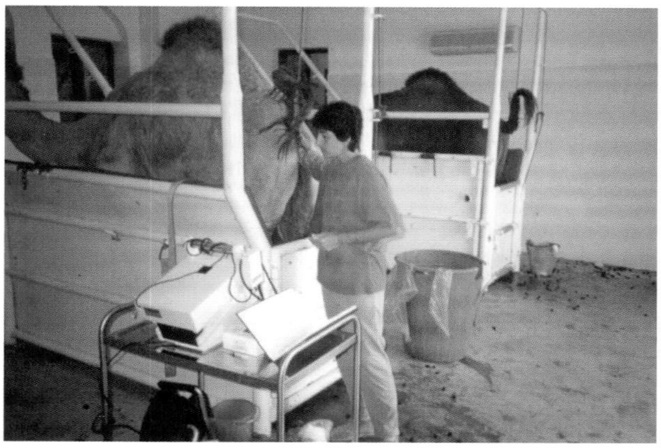

图109　子宫拭子采样法：将单峰驼尾巴固定于上方，使其保持站立姿势，以防下蹲

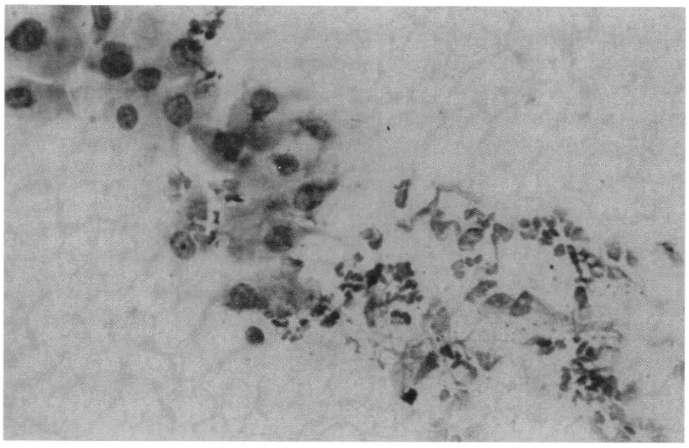

图110　由胎儿弯曲杆菌引起的子宫内膜炎涂片（姬姆萨染色）

获得。

目前，关于旧大陆骆驼的子宫活检、子宫细胞学及子宫培养的报道非常少，尽管运用在旧大陆骆驼和新大陆骆驼的技术与母马的相似。通过母马的子宫颈口将活检穿孔器插入，直接在阴道或直肠操作。根据Tibary等[37]报道，羊驼的子宫活检更困难，需要使用带有一个较小弯曲的活检钳。

这些屠宰骆驼的检查结果显示，它们会罹患不同形式的子宫炎，从中分离出相应的不同种类的细菌（见表45）。

从单峰驼的流产胎儿的不同器官及胎盘分离出了蜡状芽孢杆菌[45]，胎膜显示了严重的出血性坏死性胎盘炎和水肿。

## 治疗及防控措施

治疗骆驼子宫感染与牛马的方法一样，一般可分为子宫局部用药或/和肠胃外注射用药。局部用药包括子宫灌洗、注入温热等渗盐溶液、消毒

表45　在单峰驼子宫炎中细菌分离情况

| 子宫炎 | 作者（地区） | 子宫变化率（%） | 分离的细菌 | 分离出的细菌数 |
|---|---|---|---|---|
| 急性化脓性子宫内膜炎 | Fetaih(1991)(埃及) | | 需氧杆菌 | 4 |
| | | | 金黄色葡萄球菌 | 3 |
| | | | β溶血性链球菌 | 3 |
| | | | 表皮链球菌 | 2 |
| | | | 化脓棒状杆菌 | 2 |
| | | | 大肠杆菌 | 1 |
| | Hegazy et al. (1979) (埃及) | 25 | | |
| | Hussein et al.(2006a) (埃及） | 25(24/96) | 化脓葡萄球菌 | |
| | | | 化脓性放线菌 | |
| 亚急性化脓性子宫内膜炎 | Fetaih(1991)((埃及) | | 化脓棒状杆菌 | 5 |
| | | | 表皮链球菌 | 4 |
| | | | 金黄色葡萄球菌 | 3 |
| | | | 需氧杆菌 | 3 |
| | | | 链球菌属 | 3 |
| | | | 摩根氏变形杆菌 | 2 |
| | | | 非溶血性链球菌 | 2 |
| | | | 大肠杆菌 | 1 |
| | | | β溶血性链球菌 | 1 |
| 卡他性子宫内膜炎 | Nawito(1973) (埃及) | 1.5(31/2075) | 金黄色葡萄球菌 | 2 |
| | | | 白色葡萄球菌 | 2 |
| | | | 链球菌属 | 1 |
| | | | 大肠杆菌 | 4 |
| | Hegazy et al.(1979)( 埃及) | 25(24/96) | | |

续表

| 子宫炎 | 作者（地区） | 子宫变化率（%） | 分离的细菌 | 分离出的细菌数 |
|---|---|---|---|---|
| 慢性卡他性子宫内膜炎 | Fetaih(1991)(埃及) | | 需氧杆菌<br>表皮链球菌<br>金黄色葡萄球菌<br>大肠杆菌<br>摩根氏变形杆菌<br>肺炎克雷伯菌<br>化脓棒状杆菌<br>非溶血性链球菌 | 18<br>16<br>6<br>6<br>4<br>3<br>2<br>1 |
| | Hegazy et al.(1979)( 埃及) | 25(24/96) | | |
| 出血性子宫内膜炎 | Nawito(1973) (埃及) | 0.24(5/2075) | 大肠杆菌<br>白色葡萄球菌<br>大肠杆菌/化脓性微球菌<br>金黄色葡萄球菌<br>$\beta$溶血性链球菌 | 2<br>2<br>1/1<br>2<br>2 |
| 急性化脓性子宫炎 | Fetaih(1991)(埃及) | | 大肠杆菌<br>摩根氏变形杆菌<br>粪产碱菌 | 1<br>1<br>1 |
| 慢性非化脓性子宫炎 | Fetaih(1991)(埃及) | | 芽孢杆菌属<br>表皮链球菌<br>金黄色葡萄球菌 | 3<br>3<br>3 |
| | Al-Ani et al.(1992)（伊朗） | 4.0(2/50) | 化脓棒状杆菌 | 2 |
| 宫腔积脓 | Nawito(1973) (埃及) | 1.9(39/2075) | 铜绿假单胞菌<br>金黄色葡萄球菌<br>链球菌属<br>大肠杆菌<br>白色葡萄球菌<br>$\beta$溶血性链球菌 | 5<br>7<br>10<br>5<br>5<br>6 |
| | Laila et al.(1987)(埃及) | | 表皮链球菌<br>化脓性链球菌<br>变形杆菌 | 83.3%<br>66.6%<br>33.3% |
| 宫腔积脓加浸软胎儿 | Nawito(1973) (埃及) | 0.72(15/2075) | 金黄色葡萄球菌<br>白色葡萄球菌<br>$\beta$溶血性链球菌<br>大肠杆菌 | 5<br>2<br>4<br>3 |
| | Laila et al.(1987)(埃及) | | 表皮链球菌 | |
| 慢性活动性子宫炎 | Fetaih(1991)(埃及) | | 金黄色葡萄球菌<br>表皮链球菌<br>链球菌属<br>非溶血性链球菌<br>大肠杆菌<br>需氧杆菌<br>铜绿假单胞菌<br>摩根氏变形杆菌 | 2<br>2<br>2<br>2<br>1<br>1<br>1<br>1 |
| | Hussein et al.(2006a) (埃及) | | $\alpha$溶血性链球菌<br>大肠杆菌,金黄色葡萄球菌 | 1 |
| 坏死性子宫内膜炎 | Hegazy et al.(1979)( 埃及) | | | |
| 积水性子宫内膜炎 | Laila et al.(1987)(埃及) | | 棒状杆菌<br>大肠杆菌<br>八叠球菌 | <br>75%<br>25% |
| 子宫内膜炎脓肿 | Nawito(1973) (埃及) | 0.05(1/2075) | 金黄色葡萄球菌 | 1 |

剂和/或适量的抗生素溶液。在使用任何抗生素之前，需要做分离的微生物敏感试验。将防腐剂，如露它净（先灵葆雅动物保健）用生理盐水或磷酸缓冲液1：4稀释后通过人工授精管注入，新大陆骆驼注入100mL，旧大陆骆驼注入1000mL。这个过程需要连续进行3～5天，直到子宫培养为阴性。在子宫灌洗并将脓液、碎片清除后，再给注入抗生素。新大陆骆驼和旧大陆骆驼最常用的抗生素包括青霉素K（单峰驼$5\times10^6$ U，南美洲骆驼[SACs]$1.5\times10^6$U），硫酸庆大霉素（单峰驼1000mg，SACs 300mg），替卡西林（单峰驼3g），阿米卡星（单峰驼2g），头孢噻呋（单峰驼1g，新大陆骆驼 400mg）。使用后叶催产素（单峰驼20 IU，新大陆骆驼 10 IU）清洁子宫。治疗的动物需要检测至少14天。为了使药物得到一个最佳的分布，从直肠按摩子宫。严重的宫腔积脓时，在给药之前，子宫需要进行按摩排脓，并通过导管来排出脓液。急性和慢性子宫内膜炎、子宫炎、宫颈炎，可以用利福西亚胺制剂（利福昔明）在子宫产生泡沫来治疗。利福西亚胺是一种新综合生产的抗生素，属于利福霉素类，可以有效抵抗革兰氏阳性和阴性微生物。这种制剂包括抗生素和气体推进剂丁烷/丙烷，可以使子宫角扩大为正常的6倍，这就可以使活性成分到达子宫腔内的所有部位。

严重的子宫感染不仅需要抗生素治疗，而且要进行冲洗。敏感试验具体治疗时并不是总会成功

表46　单峰驼细菌性子宫内膜炎治疗后的生殖情况资料

| 研究序号 | 分离出的细菌 | 治疗 | 是否怀孕 |
|---|---|---|---|
| 1 | 大肠杆菌 | 恩诺沙星[a] | 否 |
| 2 | 双球菌 | 呋喃唑酮[b] | 否 |
| 3 | 链球菌、需氧杆菌 | 新霉素[c] | 否 |
| 4 | 大肠杆菌 | 呋喃唑酮 | 否 |
| 5 | 需氧杆菌 | 氯霉素[d] | 是 |
| 6 | 大肠杆菌 | 呋喃唑酮 | 是 |
| 7 | 大肠杆菌 | 呋喃唑酮 | 否 |
| 8 | 大肠杆菌、葡萄球菌属 | 新霉素 | 否 |
| 9 | $\alpha$溶血性链球菌 | 氨苄西林[e] | 是 |
| 10 | 大肠杆菌、$\alpha$溶血性链球菌 | 呋喃唑酮 | 是 |
| 11 | 大肠杆菌 | 呋喃唑酮 | 是 |
| 12 | 大肠杆菌 | 呋喃唑酮 | 否 |
| 13 | 铜绿假单胞菌 | 新霉素、恩诺沙星 | 否 |
| 14 | 大肠杆菌 | 恩诺沙星 | 是 |
| 15 | 大肠杆菌 | 新霉素 | 否 |
| 16 | $\alpha$溶血性链球菌 | 新霉素 | 否 |
| 17 | 葡萄球菌属 | 氨苄西林 | 否 |
| 18 | 大肠杆菌 | 恩诺沙星 | 否 |
| 19 | 大肠杆菌、$\alpha$溶血性链球菌 | 庆大霉素[f] | 否 |
| 20 | 大肠杆菌、$\alpha$溶血性链球菌 | 呋喃唑酮 | 否 |

a. Bayer提供；

b. SmithKine Beecham提供；

c. Intervet UK提供；

d. Antares Vet. Products提供；

e. Bristol提供；

f. Favet Laboroatoriees，Holland提供；

的。Powers等[31]报道，36只美洲驼中有22只在接受治疗后可以怀孕（67%）。Wernery 和Kumar[48]在单峰驼子宫内膜炎的治疗中很少获得成功，2～5岁不孕的单峰驼在接受抗生素治疗后，可达到30%的生育率。分离到的细菌种类、治疗中使用的抗生素以及成功率总结见表46。其他的研究人员描述到在治疗后怀孕率达60%[31]。

对于严重的子宫感染，未经稀释的卢戈氏碘液可以作为最后的手段。但是，这种治疗手段仍然存在争议。

公畜感染不育症的原因以及导致胎儿和新生儿损失的原因，见表47 a,b,c。

表47　公畜不育以及胎儿和新生儿夭折的原因

表47a　公畜不育

| 疾病 | 影响 |
|---|---|
| 布鲁氏菌病 | 睾丸炎、附睾炎 |
| 兽疫链球菌病 | 睾丸炎、附睾炎、羊驼发热 |
| 锥虫病 | 睾丸变性、精液质量差、垂体功能障碍（不育） |
| 埃氏双瓣线虫病 | 睾丸炎、附睾炎 |
| 牛病毒性腹泻 | 未知 |
| 蓝舌病 | |
| 衣原体病 | |
| Q热 | |
| 支原体病 | |

表47b　胎儿夭折

| 疾病 | 影响 |
|---|---|
| 布鲁氏菌病 | 胎盘炎、流产 |
| 蜡样芽孢杆菌病 | 出血性坏死、胎盘炎水肿 |
| 钩端螺旋体病 | 流产 |
| Q热 | 流产 |
| 锥虫病 | 流产，胎盘炎 |
| 新孢子虫病 | 流产 |
| 肉孢子虫病 | 流产 |
| 蓝舌病 | 流产（？） |
| 裂谷热 | 流产 |
| 牛病毒性腹泻 | 流产 |
| 骆驼痘 | 流产 |

表47c　新生儿夭折（也见2.1.13新生儿腹泻）

| 疾病 | 影响 |
|---|---|
| 肠毒血症 | 全身性疾病 |
| A型、C型产气荚膜梭菌病 | 腹泻 |
| 沙门氏菌病 | 腹泻 |
| 大肠杆菌病 | 腹泻 |
| 轮状病毒病 | 腹泻 |
| 冠状病毒病 | 腹泻 |
| 隐孢子虫病 | 腹泻 |
| 贾第鞭毛虫病 | 腹泻 |
| 艾美球虫病 | 腹泻 |
| 小袋虫病 | 腹泻 |

## 参考文献


[1] Al-Ani F.K.,Zenad K.H. & Al-Shareefi M.R. (1992). – Reproduction failure in female camels during an abattoir survey. Ind. *J. Anim. Sci.* 62 (6),553－555.

[2] Al-Delemi D.H.J. (2007). – Anatomical,physiological,bacteriological and pathological study of the reproductive system of Iraqi she-camels (*Camelus dromedarius*). Thesis,University of Baghdad,Iraq.

[3] Ali L.,Shalaby S.I.,Shalash M.R.,Nawito M.F. & Afiefy M. (1987). – Bacterial status of abnormal genitalia of the camel. *Egypt. J. Vet. Sci.*,24,41－44.

[4] Anon. (2009). – *Campylobacter* species abortion in an alpaca. *Vet. Rec.*,164 (4),101.

[5] Arthur G.H.,Noakes D.E. & Pearson H. (1985). – Veterinary reproduction and obstetrics. Bailliere Tindall,London.

[6] Awad H.H.,El-Hariri M.N. & Omar M.A. (1978). – Bacteriological studies on diseased and healthy reproductive tract

of the she-camel. *Zagazig Vet. J.*,1,57－67.

[7] Beil Ch. (1999). – Reproduktion beim weiblichen Kamel (*Camelus dromedarius* und *Camelus bactrianus*). Eine gewichtete Literaturstudie. Thesis,Hannover Veterinary School,Hannover,Germany.

[8] Bravo P.W. (2002). – The reproductive process of South American camelids. Seagull Printing,Salt Lake City,UT.

[9] Brown B.W. (2000). – A review on reproduction in South American camelids. *Anim. Reprod. Sci.*,58,169－195.

[10] Eidarous A.,Mansour H. & Abdul Rahier A. (1983). – Bacterial flora of the genital system of male and female camel. *Zagazig Vet. J.*,4,24－27.

[11] Enany M.,Hanafi M.S.,El-Ged A.G.F.,El- Seedy F.R. & Khalid A. (1990). – Microbiological studies on endometritis in she-camels in Egypt. *J. Egypt Vet. Med. Assoc.*,50,229－243.

[12] Fetaih A.A.H. (1991). – Some pathological studies on the affections of genital system in she-camel. Thesis,Cairo University,Cairo.

[13] Fowler M.E. & Bravo P.W. (1998). – Reproduction. *In* Medicine and surgery of South American camelids (M.E. Fowler,ed.). 2nd Ed. Iowa State University Press,Ames,IA,381－429.

[14] Gauly M. (1997). – Die Reproduktionsphysiologie von Neuweltkameliden. *Tieraerztl. Praxis*,25,74－79.

[15] Gauly M.,Vaughan J. & Cebra Ch. (2011). – Neuweltkameliden: Haltung,Zucht,Erkrankungen. 3rd Ed. Enke Verlag,Stuttgart,Germany,74－90.

[16] Hassan M.S. (1990). – Some studies on the bacteria of the uterus of the camel. M.V.Sc. Thesis,Cairo University,Cairo.

[17] Hassan N.I. & Ahmed T.M. (1997). – Mycoplasma and ureaplasma of the genital tract of camels in Egypt. *Assiut. Vet. Med. J.*,38 (75),104－118.

[18] Hegazy A.,Youseff H.I. & Selim S.A. (1979). – Bacteriological and histopathological studies on endometritis of the camel. *J. Egypt. Vet. Med. Ass.*,39,81－97.

[19] Hussein F.M.,El-Amrawi G.A.,El-Bawab I.E. & Metwelly K.K. (2006a). – Bacterial and haematological studies in the camel genitalia. *In* Proceedings of the International Camel Conference,12－14 May 2006,Qassim University,Saudi Arabia,492－500.

[20] Hussein F.M.,El-Amrawi G.A.,El-Bawab I.E. & Metwelly K.K. (2006b). – Affections of genital organs of slaughtered shecamels. *In* Proceedings of the International Camel Conference,12－14 May 2006,Qassim University,Saudi Arabia,776－787.

[21] Jarvinen J.A. & Kinyon J.M. (2010). – Preputial microflora of llamas (*Lama glama*) and alpacas (*Vicugna pacos*). *Small Ruminant Res.*,90,156－160.

[22] Kinne J.,Nagy P.,Juhasz J. & Wernery U. (2004). – Uterine histology in dromedaries. In Proceedings of the 15th International Congress on Anim. Reprod. Porto Seguro,8.–12.8.2004,Brazil,Workshop Communications,29－33.

[23] Laila A.M.,Shalaby S.I.A.,Shalash M.R.,Nawito M.F. & Afify M.M. (1987). – Bacterial status of abnormal genitalia of the camels. *Egypt J. Vet. Sci.*,24 (1),41－44.

[24] Lohr C.V.,Bildfell R.J.,Heidel J.R.,Valentine B.A. & Schaefer D. (2007). – Retrospective study of camelid abortions in Oregon. *Vet. Pathol.*,44 (5),753.

[25] Merkt H.,Mousa B.,El-Naggar M.A. & Rath D. (1987). – Reproduction in camels: a review. FAO Animal production health paper,Rome.

[26] Mukasa-Mugerwa E. (1981). – The camel (*Camelus dromedarius*): A bibliographical review. International Livestock Center for Africa. *ILCA Monogr.* 5,4－119.

[27] Nada A.R. & Ahmed W.M. (1993). – Investigations on brucellosis in some genital abnormalities of she-camels (*C. dromedarius*). *Int. J. Anim. Sci.* 8 (1),37－40.

[28] Nawito M.F. (1967). – Some reproductive aspects in the female camel. DVM Thesis,Warsaw Agriculture University,Warsaw,Poland,109.

[29] Nawito M. (1973). – Uterine infections in the camel. *Egypt. J. Vet. Sci.*,10,17－22.

[30] Pal'gov A.A. (1950). – Letter. *Trud. naucho-issled,Vet. Inst. Alma Ata*,5,29.

[31] 31. Powers B.E.,Johnson L.W.,Linton L.B.,Garry F. & Smith J. (1990). – Endometrial biopsy technique and uterine pathologic findings in the llama (*Lama glama*). *J. Am. Vet. Med. Assoc.*,197,1157–1162.

[32] Ricketts S.W. (1981). – Bacterial examination of the mare's cervix: techniques and interpretation of results. *Vet. Rec.*,108,46–51.

[33] Ricketts S.W. (1989). – The barren mare. Diagnosis,prognosis,prophylaxis and treatment for genital abnormality. *In Pract.*,11,119–125.

[34] Skidmore J.A. (1994). – Reproduction in the dromedary camel. Thesis,Cambridge University,Cambridge,UK.

[35] Skidmore J.A.,Billah M.,Binns M.,Short R.V. & Allen W.R. (1999). – Hybridizing Old and New World camelids: *Camelus dromedarius* x *Lama guanicoe*. Proc. R. Soc. Lond. B,266,649–656.

[36] Stehman S.M.,Morris L.I.,Wiesensel L.,Freeman W.,del Piero F.,Zylich N. & Dubovi E.J. (1997). – Case report: picornavirus infection associated with abortion and adult onset diabetes mellitus in a herd of llamas. In Proceedings of American Association of Veterinary Laboratories Diagnosticians Annual Conference,18–24 October 1997,Louisville,KY,43.

[37] Tibary A. & Anouassi A. (1997). – Theriogenology in camelidae: anatomy,physiology,pathology and artificial breeding. Abu Dhabi Printing and Publishing Co.,Mina,Abu Dhabi,UAE.

[38] Tibary A.,Fite C. & Anouassi A. (2006a). – Pregnancy loss and abortion in camelids: Why and what to do? *In* Proceedings of the International Camelid Health Conference,21–25 March 2006,Columbus,OH,356–364.

[39] Tibary A.,Fite C.,Anouassi A. & Sghiri A. (2006b). – Infectious causes of reproductive loss in camelids. *Theriogenology*,66,633–647.

[40] Tillman Ch. B. (2001). – Research updates: Llama and alpaca diseases,project 97LA-08. Morris Animal Foundation (MAF) and Llama Medical Research Group (LMRG),Bend,OR.

[41] Wani N.A.,Wernery U.,Hassan F.A.H.,Wernery R. & Skidmore J.A. (2010). – Production of the first cloned camel by somatic cell nuclear transfer. *Biol. Reprod.*,82,372–379.

[42] Wells S.K.,Gutter A.E.,Soike K.F. & Baskin G.B. (1989). – Encephalomyocarditis virus: epizootic in a zoological collection. *J. Zoo Wildl. Med.*,20,10550–10556.

[43] Wernery U. (1991). – The barren camel with endometritis. Isolation of *Trichomonas fetus* and different bacteria. *J. Vet. Med. B*,38,523–528.

[44] Wernery U. & Ali A. (1989). – Bacterial infertility in camels (*Camelus dromedarius*). Isolation of *Campylobacter fetus. Dtsch. Tierärztl. Wschr.*,96,497–498.

[45] Wernery U.,Ali M. & Cooper J.E. (1996). – *Bacillus cereus* abortion in a nine year old dromedary camel – A case report. *J. Camel Pract. Res.* 3 (2),153.

[46] Wernery U. & Kaaden O.-R. (2002). – Infectious diseases in camelids. Blackwell Wissenschafts-Verlag,Berlin.

[47] Wernery U.,Knowles N.J.,Hamblin Ch.,Wernery R.,Joseph S.,Kinne J. & Nagy P. (2008). – Abortions in dromedaries (*Camelus dromedarius*) caused by equine rhinitis A virus. *J. General Virol.*,89,660–666.

[48] Wernery U. & Kumar B.N. (1994). – Reproductive disorders in dromedary camels due to infectious causes and its treatment. *J. Camel Pract. Res.* 1 (2),85–87.

[49] Wernery U. & Wernery R. (1992). – Uterine infections in the dromedary camel: a review. *In* Proceedings of the 1st International Camel Conference,2–6 February 1992,Dubai,UAE. R. and W. Publications (Newmarket) Ltd,Newmarket,UK,155–158.

[50] Williams N. (2011). – Equine abortion of unknown cause. *Equine Dis. Q.*,20 (1),5–6.

[51] Wilson R.T. (1989). – Reproductive performance of the one humped camel. The empirical base. *Rev. Elev. Méd. vét. Pays Trop.* 42,117–125.

[52] Wittek Th. (2008). – Retrospektive Analyse der bakteriologischen Untersuchungen von Vaginaltupferproben klinisch gesunder Alpakastuten. *Tieraerztl. Praxis*,36 (G),329–332.

[53] Yagil R. (1985)t. – The desert camel. Verlag Karger,Basel.

[54] Zaki K. & Mousa B. (1965). – The bacterial flora of the cervical canal,uterine horn and fallopian tubes in native cows and she-camels. *Fortpfl. Haust.*,1,229–232.

## 深入阅读材料


Abdi-Arush M. (1982). – Diseases of dromedaries in Somalia. Bollettino Scientifica della Facolta di Zootecnia e Veterinaria Somalia,Universita Nazionale,Mogadishu,Somalia,3,209–217.

Ahmed W.M. & Nada A.R. (1993). – Some pathological affections of testis and epididymis of slaughtered camels (*Camelus dromedarius*). *Int. J. of Anim. Sci.*,8,33–36.

Johnson W. (1989). – Llama reproduction. *Vet. Clin. North Aur. Food Anim. Pract.*,5,159–182.

Perle K.M.D.L.,Silveria F.,Anderson D.E. & Blomme E.A.G. (1999). – Dalmeny disease in an alpaca (*Lama pacos*): sarcocystosis,eosinophilic myositis and abortion. *J. Comp. Path*ol.,121,287–293.

Shawki M.M.,El Hariri M.N. & Omar M.A. (1983). – Camel filariasis in Egypt. Incidence,hematological and histopathological studies of testis and epididymis. *J. Egypt. Vet. Med Assoc.*,43,301–308.

Youssif A.H. (1976). – Orchidectomy in camel filariasis. *Egypt. J. Vet. Med. Assoc.*,35,147–157.


（周继章译，殷宏校）



## 1.4.3 衣原体病

家畜的衣原体病是由鹦鹉热衣原体引起、临床上以多种症状为特征的综合征。鹦鹉热衣原体主要感染呼吸道、肠道、生殖系统、关节和眼。沙眼衣原体主要感染人类，而鹦鹉热衣原体不仅感染人类，而且感染各类动物。鹦鹉热衣原体导致羊地方性流产和牛流行性流产[1]。衣原体对旧大陆骆驼的影响还不清楚，然而，对新大陆骆驼的影响是显而易见的[5.6.11]。

### 病原学

衣原体门是由专性细胞内病原体组成的。最近对衣原体科重新分类，共有一个属（衣原体属），由9个种组成（表48）。

从牛和绵羊分离的鹦鹉热衣原体由两个抗原群组成：血清1型和血清2型。属血清1型的衣原体主要导致流产和生殖道、肠道的感染，而属血清2型的衣原体主要导致多发性关节炎、多发性浆膜炎、角膜结膜炎、间质性肺炎和脑膜脑脊髓炎。衣原体是细胞内寄生、革兰氏阴性、非运动性菌，具有独特的发育史。

### 流行病学和病理学

鹦鹉热衣原体无处不在。该菌存在于粪便以及生殖道和呼吸道的分泌物中。也可能通过节肢动物传播。不同的学者从单峰驼中检测到衣原体抗体。Giroud 等[4]从乍得的9只骆驼群中发现2只血清抗体阳性；Burgemeister等[2]发现突尼斯的一群单峰驼有7.7%的个体呈阳性反应；Schmatz等[10]发现埃及的一群单峰驼呈现11%的阳性；Djegham[3]也在突尼斯的一群单峰驼中发现4.4%的阳性率。Wernery 和Wernery[12]对阿拉伯联合酋长国的竞赛单峰驼和育种单峰驼进行衣原体抗体检测，阳性率分别为15%和24%。这些学者认为，衣原体对怀孕的单峰驼没有任何的影响，对畜群的研究发现也没有流产率上升。他们也利用衣原体抗原酶联免疫吸附试验（ELISA）对28份血清反应阳性的单峰驼子宫拭子进行了检测，未发现有任何的反应。

绵羊容易被含有衣原体的生殖道分泌物及与

表48　衣原体科新分类

| 目 | 科 | 属 | 新种 |
|---|---|---|---|
| 衣原体目 | 衣原体科 | 衣原体属 | 沙眼衣原体、猪衣原体、鼠衣原体、鹦鹉热衣原体、流产衣原体、豚鼠衣原体、反刍动物衣原体、猫衣原体、肺炎衣原体 |

流产胎儿接触而被感染。一些感染的绵羊呈现亚临床型的肠道感染，之后大量的病原体通过粪便排出[7]。在阿拉伯联合酋长国，单峰驼的传染源似乎与绵羊和山羊的近距离接触有关，感染率可达50%。

在德国，观察到1只羊驼因衣原体感染而导致流产和1只驼马疑似衣原体性肺炎。在莱比锡动物园的羊驼中发现衣原体病[6]，该病是通过感染的公驼羊而带进动物园的，暴发时呈现典型的临床症状：关节炎、角膜结膜炎、虹膜睫状体炎和葡萄膜炎（图111）。羊驼产下很多死幼崽，几只活幼崽后来发生关节炎。在12年的时间里，出生的53只幼崽32只死于衣原体病。Popovici 等[8]在布加勒斯特的一个动物园幼龄美洲驼死于脑脊髓炎的报道中首先报道了衣原体病的暴发。

对来自捷克的一个动物园和德国的10个动物园的牛科、鹿科和骆驼科共926只偶蹄动物中几个感染性病原体血清抗体的调查发现，这些病原体在种间能够进行传播[9]。用ELISA检测，发现衣原体血清阳性率达32%，表明存在种间传播。

### 诊断

衣原体可在鼠脑内、鸡胚和组织培养中分离。抗原ELISA、免疫荧光和免疫过氧化物酶染色能够快速诊断衣原体病。最近，分子生物学方法也被引入衣原体病的诊断。一般条件下，补体结合试验用来检测鹦鹉热衣原体抗体，但目前被多种抗体ELISA方法所替代。

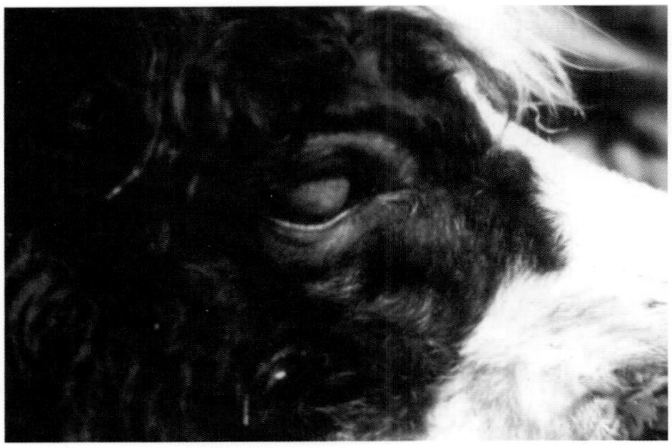

图111　由衣原体病引起的羊驼的角膜结膜炎（照片由德国K. Uilenberger教授惠赠）

### 治疗和控制措施

四环素和氯霉素是治疗衣原体病的最有效药物，因为它们能够抑制衣原体的增殖。然而，Goepner等[6]报道，用抗生素可以控制急性病例，但对慢性和关节炎病例却没有效果。在衣原体暴发期间，将患病动物从健康动物群中隔离出去是非常重要的。灭活疫苗通常用于防控绵羊的衣原体病，健康的羊驼在3周内免疫2次，然后每6个月免疫1次。

重组疫苗和DNA疫苗实验研究表明，不能达到保护性抗体的水平。而且，即使活疫苗和灭活疫苗，在使用衣原体进行攻击时也不能产生完全保护。弱毒疫苗由于存在人兽共患的原因而受到关注，在很多国家不允许使用。

## 参考文献

[1] Beer J. & Wehr J. (1980). – Chlamydien-Infektionen. In Infektionskrankheiten der Haustiere. VEB Gustav Verlag,Jena,Germany. 316–330.

[2] Burgemeister R.,Leyk W. & Goessler R. (1975). – Untersuchungen uber Vorkommen von Parasitosen,bakteriellen undviralen Infektionskrankheiten bei Dromedaren in Sudtunesien. *Dtsch. Tierärtztl. Wschr.*,82,352–354.

[3] Djegham M. (1988). – A propos de l'avortement chez la chamelle en Tunisie. *Maghreb Vet.,3* (14),60.

[4] Giroud P.,Roger F.,Dumas N.,Vouilloux P. & Sacquet E. (1954). – Comportement des animaux domestiques de la regiondu Tchad vis-a-vis de l'antigene T13. *Bull. Soc. Path. Exot.*,47,644–645.

[5] Goepner I. (1999). – Analyse des Krankheitsgeschehens in der Alpakaherde des Zoologischen Gartens Leipzig unter besondererBerucksichtigung der Chlamydiose. Veterinary medicine thesis,Veterinary Faculty of the University of Leipzig,Germany.

[6] Goepner I.,Eulenberger K.,Bernhard A.,Schulz U. & Neubert A. (1999). – Chlamydiose bei Alpakas (*Lama guanacoe F.pacos*). *Verhandlber. Erkrg. Zootiere*,39,199－207.

[7] Morgan K.L.,Wills J.M.,Howard P. & Williams R.C. (1988). – Isolation of *Chlamydia psittaci* from the genital tract of lambs:a possible link with enzootic abortion of ewes. *Vet. Rec.*,123,399－400.

[8] Popovici V.,Hiastru F.,Cociu M.,Mastacan D. & Wagner G. (1970). – Bedsonia (chlamydia) infections in captive ruminantsof the Bucharest Zoological Garden. *Verhandlber. Erkrg. Zootiere*,12,211－213.

[9] Probst C.,Speck S. & Hofer H. (2011). – Serosurvey of zoo ungulates in central Europe. *Int. Zoo Yearbook*,168－182.

[10] Schmatz H.D.,Krauss H.,Viertel P.,Ismail A.S. & Hussein A.A. (1978). – Seroepidemiologische Untersuchungen zumNachweis von Antikorpern gegen Rickettsien und Chlamydien bei Hauswiederkauern in Agypten,Somalia und Jordanien. *Acta Tropica*,35,101－111.

[11] Schroeder H.-D.,Seidel B. & Strauss G. (1998). – Chlamydial infections in ungulates kept in zoological gardens. *In* Proceedingsof the European Association of Zoo and Wildlife Veterinarians,21－24 May 1998,Chester,UK,219－221.

[12] Wernery U. & Wernery R. (1990). – Seroepidemiologische Untersuchungen zum Nachweis von Antikorpern gegen Bruc ellen,Chlamydien,Leptospiren,BVD/MD,IBR/IPV – und Enzootischen Bovinen Leukosevirus (EBL) bei Dromedarstuten (*Camelusdromedarius*). *Dtsch. tierärztl. Wschr.*,97,134－135.

（周继章译，殷宏校）

# 1.5 皮肤

## 引言

骆驼伤口的愈合速度较其他动物慢的认识是不正确的。Purohit 和Chouhan[53]确认骆驼的皮肤中也富含血管，可以使伤口很好的愈合。但是，团蹄亚目的动物易在皮肤上形成脓肿[67]。在骆驼的皮下组织、体表淋巴结和肌肉组织中常见脓肿灶，这可能是因为骆驼喜欢采食刺槐的树叶和小枝。这些长刺（大于5 cm）不仅会刺破皮肤，引起深部感染，也会伤及动物消化道黏膜层（图112）。

图112　单峰驼采食刺槐树

脓肿灶常常位于颅、颈、胸和膝后窝淋巴结，但动物体表一般没有可见的损伤。这些组织损伤通常存在于自由放牧的种用或竞赛用单峰驼上，其远比常年圈养在围栏内的竞赛用骆驼多。在澳大利亚，以采食尖刺和树枝为主的野外生存骆驼中的80%会受到脓肿感染的影响，多数澳大利亚的骆驼主都会观察到骆驼脖子末端有柠檬大小的淋巴结脓肿，这使得骆驼的经济价值大打折扣。

目前，已有多种骆驼皮肤病的相关报道，但其临床症状和病原学都不明确。在一项回顾性调查研究中，Scott等查阅了康奈尔大学在1997～2006年间检测出的68峰羊驼皮肤病的相关资料，这些皮肤病主要为细菌感染（22%）、肿瘤、囊包和血肿（19%）、疑似免疫失调（12%）、和体外寄生虫（10%）。Rosychuk在美洲驼和羊驼上也开展了更为广泛的皮肤病学研究[58,59]。

很多文献报道中并没有说明这些细菌样品是取自封闭性脓肿，还是开放性的脓肿，或伤口。诸如扁盘尾丝虫（Onchocerca fascista）等寄生虫形成的包囊也可能与脓肿混淆[12]。还需要注意的是，很多骆驼在皮下用些药后会出现伴随着机体肿大的严重变态反应[61]，以及前面提到的骆驼对油佐剂疫苗很敏感。

骆驼科动物感染性皮肤病可以由多种不同的病原体引起，包括细菌、病毒、霉菌和寄生虫等。细菌性皮肤病是由化脓棒状杆菌、链球菌、星形诺卡菌、李氏放线杆菌、大肠杆菌和坏死梭菌引起，相关归纳总结详见表49。

本章主要是介绍骆驼科动物中明显影响经济效益的几种皮肤病，包括干酪性淋巴腺炎、金黄色葡萄球菌皮炎和嗜皮菌病。

### 1.5.1　干酪性淋巴腺炎（假结核病）

干酪性淋巴腺炎（caseous lymphadenitis，CLA）或称假结核病在全球绵羊和山羊群中广泛存在，是由假结核棒状杆菌引起的慢性病[11,40,41]，其典型症状为在一个或多个体表淋巴结出现脓肿。假结核棒状杆菌也可以引起动物的肺炎、肝炎、乳腺炎、关节炎、睾丸炎和脑膜炎。本菌也感染马，并能引起牛的溃疡性淋巴管炎。假结核广泛存在于旧大陆骆驼中，是除了兽疥癣之外最常见的皮肤病，可以感染整个单峰驼群。假结核棒状杆菌也曾从新大陆骆驼化脓灶中分离出来[8,9,31]。

在2007年秋天，美国俄勒冈州东部几十匹马中发现了一种名为“鸽子热”的传染病[52]。本病不是由鸟传播，但是由假结核棒状杆菌引起，感染后形成的脓肿，使得马的胸部膨大如鸽子胸。本病在美国加利福尼亚州流行。

#### 病原学

法国兽医Edmond Nocard于1888年最先描述了假结核棒状杆菌，菌体短小、不规则卵圆形、革兰氏染色阳性棒状并接近球菌样，在化脓灶涂片时，本菌呈明显的多形性。在进行病原分离时，取绵羊或公牛血制作平板培养基，并在37℃下孵育至少48小时。假结核棒状杆菌在培养基上生成白色干燥小菌落，周围有窄的溶血带环绕。本菌现在有绵羊/山羊和马/牛两个生物型，绵羊/山羊菌种可以在农

表49　旧大陆骆驼和新大陆骆驼皮肤病的病因

| 临床症状 | 作者 | 年份 | 病原体/疾病 | 国别 | 种类 |
|---|---|---|---|---|---|
| 化脓性皮肤损伤 | Younan和Bornstein | 2007 | B型链球菌和C型链球菌 | 东非 | 单峰驼 |
| 脱毛、结硬皮、痒和鳞状皮 | D’Alterio | 2004 | 足螨 | 英国 | 美洲驼/羊驼 |
| 皮肤坏死、痒、脓肿、扁虱 | El-Allawy等 | 2006 | 疥癣、假结核棒状杆菌、葡萄球菌 | 埃及 | 单峰驼 |
| 下颌肉芽肿 | Kilik和Kirkan | 2004 | 黏性放线菌 | 土耳其 | 单峰驼 |
| 皮肤干燥、鳞状皮、无毛 | Clauss等 | 2004 | 铜和锌缺乏 | 德国（动物园） | 美洲驼/羊驼 |
| 传染性化脓样坏死 | Edelsten和Pegran | 1974 | 无乳链球菌 | 索马里 | 单峰驼 |
| 下颌肉芽性膨大 | Purohit等 | 1988 | 放线菌病 | 印度 | 单峰驼 |
| 头部膨大的肉芽肿性脓肿 | Daneji等 | 1996 | 李氏放线杆菌 | 尼日利亚 | 单峰驼 |
| 脓肿 | Domenech等 | 1977 | B型链球菌、化脓棒状杆菌 | 埃塞俄比亚 | 单峰驼 |
| 皮肤伤口、脓肿 | Qureshi等 | 2002 | 多种细菌感染，主要是金色葡萄球菌 | 印度 | 单峰驼 |
| 脓肿、伤口、鞍疮 | Tadesse和Molla | 2002 | 无乳链球菌、金黄色葡萄球菌、化脓棒状杆菌 | 埃塞俄比亚 | 单峰驼 |

场和屠宰场工人中引起少量的感染病例。感染病人出现慢性纵隔淋巴结肿大，需要手术治疗才能痊愈。

## 流行病学

骆驼假结核病在伊朗[30]、埃及[17,29,42,57]、埃塞俄比亚[25,35]、肯尼亚[12]、澳大利亚[12]、沙特阿拉伯[56]、印度[54]、俄罗斯[60,65]、中国[18]、阿联酋[69,74]和东非[24]均有报道。目前，骆驼中只发现了绵羊/山羊型菌株。从化脓灶中分离假结核棒状杆菌存在困难，且其菌落与链球菌相似，往往被其他伴生菌的菌落覆盖。因而，用感染假结核棒状杆菌的山羊体内分离细菌时有15%的样品为阴性结果[40]。

假结核棒状杆菌的感染是通过消化道、呼吸道或直接的伤口接触引起。本菌也可以通过没有损伤的皮肤侵入机体，被巨噬细胞吞噬后通过淋巴循环进入淋巴结，并可能再次形成脓肿。假结核棒状杆菌为化脓性兼性胞内寄生菌，它可以穿过组织，在内部产生毒素。至少3种毒素：一种毒性胞壁脂质、一种外毒素和破坏哺乳动物细胞膜的磷脂酶D（phospholipase D，PLD），以及一种溶血素在干酪性淋巴腺炎的病程中发挥关键作用。毒性胞壁脂质与细菌毒力有关，溶血素引起出血，从而增加血管的渗透性，增加细菌侵入机体的机会。

与绵羊和山羊的假结核病不同，从骆驼假结核病化脓灶中分离的细菌不仅仅只有假结核棒状杆菌，Domenech等[25]从埃塞俄比亚的假结核病单峰驼中分离到了如下几种细菌。

（1）链球菌（57%）（兰斯菲尔德B型）。

（2）假结核棒状杆菌（37%）。

（3）葡萄球菌（10%）。

（4）化脓棒状杆菌（6.7%）。

除了假结核棒状杆菌外，Radwan等[56]从沙特阿拉伯2500峰单峰驼15%的个体中分离到金色葡萄球菌、牛肾盂炎棒杆菌、马棒状杆菌、志贺氏菌和大肠杆菌。文章还报道了在动物颈、尾和各机体连接处的肌肉和皮下组织内形成化脓灶，部分动物出现不生成化脓灶的淋巴结病变。发病动物同时寄生有璃眼蜱，而且能够从这些蜱体内分离到假结核棒状杆菌。豚鼠腹腔内接种假结核棒状杆菌后，3周后死于多发性脓肿。除了假结核棒状杆菌1型，Hawari[34]从约旦骆驼22份开放性脓肿病例中分离到了金色葡萄球菌（2例）、牛肾盂炎棒状杆菌（1例）和马棒状杆菌（1例）。

Hoste等认为，在假结核致病因素中，化脓放线菌与假结核棒状杆菌有着相似的重要性。Spesivtseva和Noskov[65]以及Dalling等声称，皮疽组织胞浆菌是引起前苏联双峰驼假结核病暴发的病原菌。本次暴发发生在1958年，发病骆驼当时是从中亚向莫斯科周边的农场迁移。发病骆驼的病变集中在肩前淋巴结，淋巴结内可以检测到含有菌丝的细菌和隐球菌样细菌，且在巨噬细胞中也可以观察到隐球菌。

Ismall等[36]报道了1例在埃及村庄内生活的21峰单峰驼暴发假结核病，本次暴发也波及了当地牛群和水牛群。发病动物最初表现为肘部、胸部和外周淋巴结水肿。因为有出血，一些淋巴结出现溃疡病变。从没有发生溃疡的淋巴结中只分离出了假结核棒状杆菌，而从有溃疡的淋巴结中分离出了假结核棒状杆菌和金黄色葡萄球菌。

由刺槐刺、蜱、污染的注射用针头和伯氏肉样瘤破裂造成的皮肤损伤，可能在不经意间成为棒状杆菌侵入机体的入口。

动物的消化道黏膜层可能被刺槐刺和沙漠植物的干硬树枝刺破，假结核棒状杆菌由此进入皮肤和黏膜层，然后被体内白细胞吞噬。细菌体表脂质保护细菌不被白细胞破坏，然后通过机体淋巴循环转移到外周淋巴结。细菌在淋巴结的细胞进行繁殖，细胞死亡后释放出大量细菌，继续侵染周围健康细胞造成大量细胞死亡，形成典型的干酪性淋巴腺炎。在一些病例中，体表淋巴结脓肿会进一步引起病原体通过淋巴循环和血循环扩散，到达机体的肺脏、纵隔淋巴结、肝脏、肾脏和脑，造成潜在的扩散感染风险。也有研究人员认为假结核棒状杆菌不是骆驼科动物淋巴腺炎唯一的病原。通过对一些并不能完全确定取自封闭性化脓灶或开放性化脓灶病例的拭子样品进行分析，Stowe[66]认为，开放性的化脓灶中可以发生球菌引起的二次感染。

Abou-Zaid等[1]检测了339峰埃及单峰驼，其中，10.9%的骆驼发生淋巴腺炎，发病的成年骆驼体表淋巴结出现肿大和形成脓肿灶。病变淋巴结释放出浓稠的、奶样浓汁或钙化物质。在其他的单峰驼中，用纯培养基从62.1%的个体中分离出了假结核棒状杆菌绵羊亚种，其余样品同时有金黄色葡萄球菌和链球菌感染。

Afzal等[2]从阿联酋患有淋巴腺炎的11峰赛用骆驼中分离并培养了假结核棒状杆菌。6个骆驼分离株和1个用作对照的绵羊分离株均可以造成兔的皮肤坏死和红肿。在另一试验中，每种分离株选取一株（引起皮肤坏死、不引起皮肤坏死和绵羊株）用于耳朵根部接种试验用骆驼。接种绵羊株和引起皮肤坏死菌株的骆驼只出现了淋巴结肿大的病变，而不引起皮肤坏死的菌株在接种40天后引起骆驼多发性的脓肿。实验性感染的骆驼在恢复健康后再次接种感染病原菌，感染骆驼机体不出现任何病理变化。研究发现，不含PLD基因的假结核棒状杆菌不能引起淋巴结脓肿[6]。

一些新出版的文献强调了干酪性淋巴腺炎在新大陆骆驼和旧大陆骆驼中的重要性。Muenchau[47]于2003年第一次报道干酪性淋巴腺炎在欧洲骆驼科动物上出现，而早在1991年时就已在英国的绵羊养殖过程中发生过本病。发病后，旧大陆骆驼和新大陆骆驼的死亡率可以分别达到15%和22%，这主要是体内器官损伤的结果。有报道称，患有干酪性淋巴腺炎的骆驼科动物的胃和肠道多见溃疡灶。在单峰驼，作为旅游业工具的加纳利群岛，有数例单峰驼干酪性淋巴腺炎的报道。Tejedor-Junco等报道称[72]，西班牙的大加纳利群岛鲜有假结核棒状杆菌和肠道沙门氏菌感染引起单峰驼胸膜炎、关节炎和关节周炎的病例。他们也报道了几例加纳利岛出现的干酪性淋巴腺炎病例，发现单峰驼的多个部位淋巴结肿大、发凉、柔软和没有痛感。从未破裂的淋巴结脓肿采样后，分离纯培养出16株假结核棒状杆菌，生化鉴定后发现它们分属4个生化类型。在一例年幼单峰驼的病例中，从动物左颈胸部的背侧和腹侧受感染的体表淋巴结中分离出了溃疡棒状杆菌。干酪性淋巴腺炎是单峰驼常见的重症疾病，本病不仅仅是影响了加纳利岛的旅游业，欧洲大陆上因为育种购买的单峰驼和双峰驼也受到假结核病的影响，且本病会因为动物混群而向其他动物扩散[44]。因为动物杂技表演和贸易使得的不同种动物混群，存在严重的干酪性淋巴腺炎扩散风险，尽管疾病造成的死亡率并不高，但由此引起的经济损失异常严重，这其中的一个重要原因是骆驼等易感动物为价值高昂的人类伴侣动物。澳大利亚有几十万峰野养骆驼在大陆腹地游荡，很多人在这些骆驼上见到过外观丑陋的淋巴结脓肿[3]。

干酪样淋巴腺炎不仅影响旧大陆骆驼，在世界很多地方也有新大陆骆驼发生本病的报道，已有研究人员报道过秘鲁发生干酪性淋巴腺炎，且有些病例甚至发生在海拔4000m以上的地区[13,15,73]。安第斯山区羊驼假结核棒状杆菌自然感染的病例出现乳房炎和体表淋巴结脓肿，而且有些病例的脓肿主要出现在肾脏淋巴结。有意思的是，新大陆骆驼上出现的脓肿与在绵羊出现的脓肿不一样，新大陆骆驼出现的脓肿为液化性坏死和严重的肉芽肿增生，周围由厚结缔组织层包围，此类病变在单峰驼上也有发现[74]。此外，研究人员认为病原体可以通过乳房炎传播扩散给哺乳幼畜，而不是经过皮肤伤口传播（剃毛时造成）。

干酪性淋巴腺炎病例在北美也有相关报道，但较为罕见[4]，Anderson等报道过5例羊驼发生干酪性淋巴腺炎的病例。本病的诊断基于微生物培养，治疗手段主要是抗微生物感染和脓肿切除。Beghelli等[10]报道了一例发生在意大利中部的干酪性淋巴腺炎暴发，54峰羊驼由德国引进后与当地一个拥有28头动物的畜群混合。尽管做了各种努力，本病依旧扩散到了进口检疫群和意大利的核心群。本次暴发共有24峰羊驼出现干酪性淋巴腺炎症状，其中，18峰死亡。德国在2003年也有一批羊驼出口到瑞典，此后假结核棒状杆菌在当地的羊驼群中发病流行[49]。

Braga等[16]在成年羊驼上开展了一次实验性感染来复制干酪性淋巴腺炎的临床症状和病理变化，并研究机体针对感染的体液免疫情况。共有9峰骆驼接受了分离自美洲驼（n=4）和羊驼（n=5）的$1.1\times10^6$个菌落形成单位（colony-forming units，CFU）的假结核棒状杆菌，并有4峰羊驼作为阴性对照。观察羊驼的临床状况，并在接种后的16天，58天，93天或128天进行安乐死。用假结核棒状杆菌的胞壁蛋白作为抗原建立间接ELISA方法，用于测定血清样品中的抗体滴度。接种后的羊驼持续发烧，并且有严重的炎症反应和白细胞增多症（$>3.00\times10^{10}$个/L）。体内脓肿主要集中在肾脏淋巴结，没有肺脏损伤。感染初期病变为内部干酪样坏死的典型脓性肉芽肿，后期病变损伤部位含有结缔组织、单核细胞、大量中性粒细胞和液化性坏死。羊驼从感染后第16天能检测到血清抗体滴度，持续到第93天后消失。不同菌株之间造成的病理损伤差别不明显。

在阿联酋，干酪性淋巴腺炎病理也很多见，且主要是影响在沙漠里觅食的单峰驼。Kamat等[37]检查了共计62份从骆驼身上手术取下的化脓性病灶，并从中分离到了化脓放线杆菌（55%）和假结核棒状杆菌（38%）。借助各种实验室检测手段，Wernery和Joseph（未发表）在几年的时间内从迪拜酋长国患有干酪性淋巴腺炎的单峰驼上分离到65株假结核棒状杆菌。研究发现，除了产硝酸盐，这些菌与反刍动物分离株有相似的生化反应特征，产PLD试验均为阳性，均对青霉素敏感，在SfiⅠ和AscⅠ双酶切后凝胶电泳特征相同[20]。这些菌株型除了与从小反刍兽体内分离到的菌株不同外，而且彼此之间也有差异。由此可以判断，源于不同宿主的假结核棒状杆菌可能具有遗传学差异，具有菌株特异性的疫苗可能对其他宿主菌株感染动物不具有保护力。

## 临床症状和病理变化

绵羊和山羊假结核棒状杆菌引起脓肿的潜伏期一般在感染后的25～40天，而骆驼科动物的潜伏期可能更长，甚至超过6个月。有报道称，在最后的干酪性淋巴腺炎阳性动物康复后一年都会使健康动物发病，表明携带病原的动物一直存在或环境一直处于被污染的状态[45]。在用假结核棒状杆菌实验性接种骆驼40天后，Afzal等[2]发现了多发性化脓灶的形成。绵羊和山羊上有广泛的干酪样坏死出现在淋巴结和其他器官（特别是在肺脏）。相比较而言，假结核棒状杆菌在骆驼的内脏器官上造成的病理变化很少见[56]，全身性的皮肤病变也不常见[22,28]。Hawari[34]检查了5峰骆驼尸体，除了在大腿、肩部、肘部、脖颈根部、腋下、下巴以下及各关节处的肌肉和皮下组织中发现脓肿，还在内脏器官，特别是肺脏上发现了多个大的脓肿灶。

本病的特征性病变是在体表淋巴结出现凉的、封闭且无痛的脓肿，大小如柠檬或橘子（图113a,b,c），且主要出现在脖颈根部和肩前淋巴结[62]，也会出现在机体的其他多个部位（图114）。

打开的脓肿灶会流出浓稠的、黄色奶样浓汁，多数脓肿灶都有结缔组织完整的包围着。绵羊和山羊多数病例中的脓肿灶会形成同心片层（洋葱环）结构[11,48]，但在骆驼科动物中没有观察到类似结构。有几个单峰驼病例中出现肋骨脓

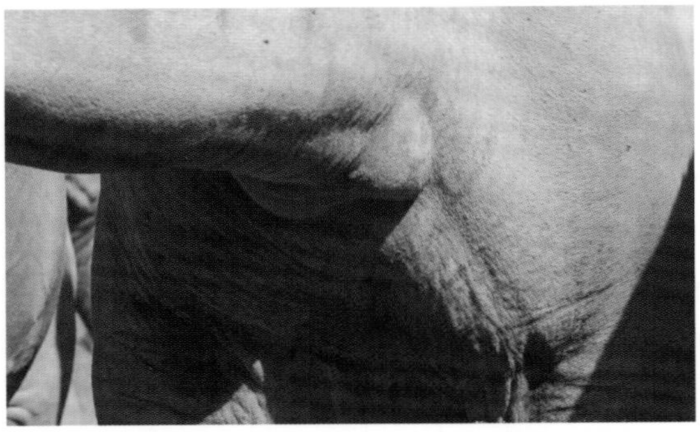

（a）1岁龄单峰驼上柠檬大小的干酪样淋巴结炎

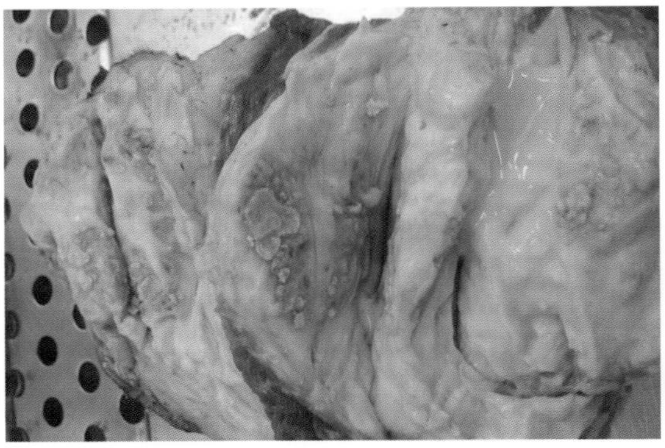

（b）在肯尼亚单峰驼体发现的干酪样淋巴结炎引起的多发性化脓灶

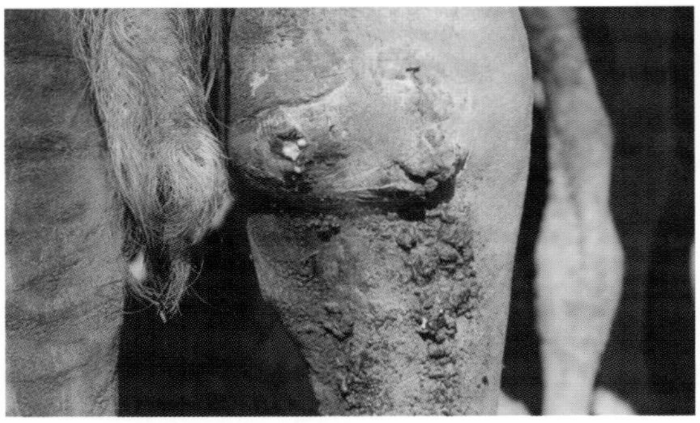

（c）打开的干酪样淋巴结炎脓肿灶

图113　a, b, c

肿破溃，导致脓肿内微生物进入肺脏，引起严重的支气管肺炎及肺部空洞（图115）。

Nashed和Mahmoud[48]描述的显微损伤包括出现淋巴样和上皮状病变的淋巴结干酪样损伤，没有发现巨细胞。Abou-Zaid等[1]进行了受侵害淋巴结的组织病理学检查，发现了急性血浆化脓性和慢性化脓性淋巴结炎（图116）。假结核病主要发生在3岁以上的骆驼[61]。

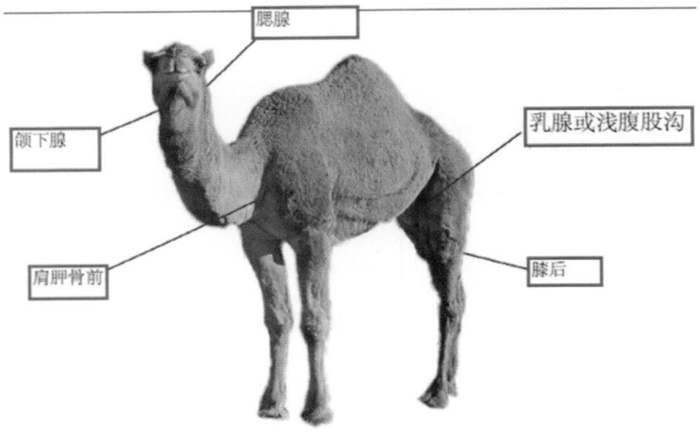


图114　干酪样淋巴结炎脓肿病变的可能出现的部位

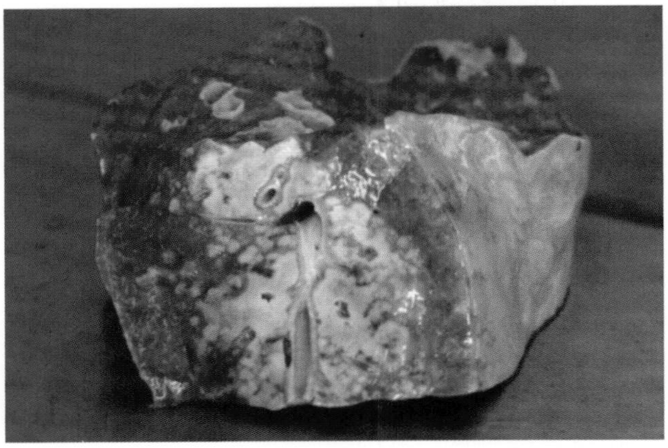

图115　严重的支气管炎及肺空洞病变

## 防控措施

发病动物为干酪性淋巴腺炎的传染源，一旦临床症状出现，应立即采取如下措施。

（1）立刻隔离所有已感染动物。

（2）对畜圈和围栏进行彻底的消毒和清理，清除所有的粪便、垫料和表层土。

（3）体表脓肿灶一旦破裂，应该采取最严格的消毒措施，并彻底销毁或消毒所有被污染的器具和设备。

（4）对畜群进行灭活自家苗免疫接种。

（5）畜舍清理掉金属丝等可能引起皮肤损伤的因素。

（6）杀灭体外寄生虫。

（7）购进的动物必须来自没有化脓病史的

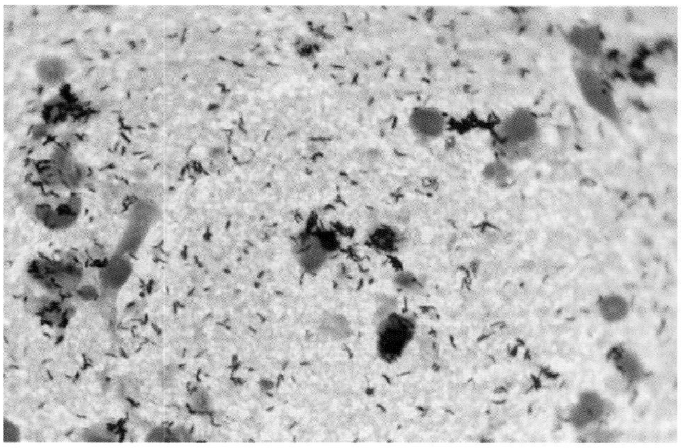

图116　可见假结核棒状杆菌（革兰氏染色）的干酪样淋巴结炎涂片

畜群。

（8）对全群进行血清学筛查，去除或治疗阳性反应个体。

（9）在筛查两次后不应再存在血清学阳性动物。

（10）只能用未感染过的骆驼科动物进行畜群内动物替换。

小反刍动物疾病防控的通用标准是剔除群中所有的淋巴结肿大的动物。这个标准对伴侣动物来说不合实际。病原菌对抗生素敏感，但因为脓肿均被紧密包裹，抗生素治疗不是都有价值，治疗效果通常很差。因此，皮下成熟的脓肿或者是彻底切除，或者通过手术排尽脓肿内容物。为了达到最佳治疗效果，建议采取手术和抗生素两种方法联用进行治疗。

棒状杆菌对青霉素、四环素和头孢菌素极其敏感，但化脓灶中的脓汁会阻止药物到达细菌所在的部位。考虑到红霉素在机体组织中有更强的穿透力，Bergin[12]建议采取青霉素和红霉素联用来治疗骆驼假结核病。另一种治疗成年单峰驼的假结核病的方法是连续12天静脉注射20 mL二甲基亚砜和20 mL恩诺沙星，脓肿最终消退，而且不会复发。

用于绵羊和山羊干酪性淋巴腺炎预防的疫苗目前已商品化，但多数不生产该疫苗的国家没有使用许可，因而无法用于骆驼科动物。这些疫苗均是通过福尔马林灭活含PLD的假结核棒状杆菌培养上清浓缩后制成。澳大利亚的Glanvac 3是一种多组分佐剂疫苗，内含假结核棒状杆菌、D型产气荚膜梭菌和破伤风梭菌超滤抗原。减毒突变疫苗目前也在用。这些疫苗不能完全保护机体在感染后不产生脓肿，但可以大大降低脓肿的数量。对干酪性淋巴腺炎的免疫与机体的抗毒素功能相关，是初级的细胞调节。在经过疫苗免疫和细菌接种之后，免疫绵羊的体内、体外以及总脓肿灶的数量远远比对照绵羊低[51]。假结核棒状杆菌菌体抗原和外毒素的诱导产生抗体的能力进行了检测。在免疫接种后，绵羊为血清学阳性，对照绵羊依旧为血清学阴性。

山羊疫苗免疫接种的保护效果较差，尽管能抵御实验性攻毒，但对自然感染动物具有很少的保护力。母乳免疫也会影响疫苗免疫接种的功效，因此，干酪性淋巴腺炎高发群中的羔羊不应该在12周前接受疫苗免疫接种。

一些研究人员开始研制和生产自家苗用于预防假结核病[5,32]。在意大利羊驼暴发干酪性淋巴腺炎期间，采取动物隔离和治疗无法控制发病后，Beghelli等使用了一种灭活的、含佐剂的自家苗。这种疫苗内含有数种假结核分枝杆菌，它们均是从死于干酪性淋巴腺炎的羊驼脓肿中分离到，并同时用于疫苗生产。该疫苗经3次（第0天、第21天、第150天）皮下接种在64头动物的胸部近肘的皮肤皱褶中。在开始疫苗接种后的4个月，没有新的死亡病例出现，但有几例新发干酪性淋巴腺炎淋巴结病和下颌淋巴结脓肿出现。

Kobera和Poehle[39]也使用了一种群特异性疫苗来防控干酪性淋巴腺炎，这种疫苗使两个羊驼群再没有出现新的死亡，并且降低了脓肿灶的数量。商品化疫苗也有类似的效果。

Braga[14]用20 μg/mL的胞壁酰二肽作为佐剂，评价了羊驼源的假结核分枝杆菌胞壁和毒素成分的免疫保护潜力，12峰成年羊驼在其身体左侧面接受了含低剂量和高剂量细菌粗抗原（250 μg/mL和500 μg/mL胞壁蛋白，133 μg/mL和265 μg/mL 毒素）疫苗的免疫接种。

3周后，在免疫接种过的羊驼和未接种的羊驼在机体右侧皮下接种假结核棒状杆菌，接种量为$1\times10^6$ CFU。未进行疫苗免疫的对照动物出现了持续发热症状，接种部位以及体表和体内淋巴结出现脓肿，只有高剂量接种毒素的羊驼没有出现脓肿病变。与之相对照，那些接种了低剂量毒素的羊驼在接种部位以及外周和肾脏淋巴结脓肿。接种菌体胞壁的羊驼受保护程度要比接种高剂量和低剂量毒素的羊驼都低，出现了体表和体内脓肿病变。此外，所有免疫过的羊驼在攻毒之后都出现迅速而强烈的体液免疫应答，并持续至少3个月。由中央兽医研究实验室研制的高剂量毒素（300 μg/mL）疫苗，配以多种佐剂，近期在肯尼亚进行了实验性应用。

不幸的是，90%的动物在接受商品化疫苗接种后，在接种部位出现了肉芽肿，这使得阿联酋的科研人员终止了在单峰驼上的接种安排。在接种了商品化疫苗之后，大多数接种在脖颈基部的单峰驼都出现了大小不一的肉芽肿（图117）。

Hassan等首次报道了假结核棒状杆菌Cp162株的全基因组测序结果，该菌种分离自一峰英国骆驼的颈部脓肿，但不清楚该菌株分离于双峰驼还是单峰驼。研究人员乐观地认为，测序结果分析会有助于了解该病原的毒性，并可能有助于研发更为有效的疫苗和药物来防控该病原。

新型疫苗或许将来可以提升机体针对干酪性淋巴腺炎的保护力，但之前还有很多工作需要完成。因此，将来也必须重视用于控制和消除本病的其他方法。

### 血清学

目前已有几种不同的血清学检测方法已试用于干酪性淋巴腺炎的血清学诊断，但还没有在骆驼科动物上验证其检测效果。这些方法包括红细胞凝集试验、红细胞凝集抑制试验、琼脂凝胶试验和ELISAs试验，大多数方法的敏感性都很低。新型间接双抗体夹心ELISA方法、γ-干扰素检测方法和免疫印迹试验被认为在山羊群和绵羊群中有较高的特异性和敏感性，所有这些方法在检测PLD外毒素抗体时最为有效。

一些国家的研究人员选取这些血清学方法中的一种或采用多种联用的方法用于干酪性淋巴腺炎检测和的清除[7]。当绵羊出现干酪性淋巴腺炎临床症状，或基于PLD建立的ELISA或蛋白免疫印迹检测为阳性时，应该将其从畜群中清除掉。Piontkowski和Shivvers[51]采用了两种不同的ELISA方法用于检测假结核棒状杆菌的菌体抗原和外毒素。

拥有大量骆驼科动物的几个欧洲国家都制定了自愿对新大陆骆驼定期进行干酪性淋巴腺炎抗体检测的规划，一些农场也已开始按照检查、触诊和血清学方法来清除本病。在进行检测时发现，超过40%的受检新大陆骆驼都有假结核棒状杆菌，但没有任何临床症状。这说明大量的新大陆骆驼都已接触过病原，成为没有临床症状的病原携带者[45,46]。目前，已有用于检测绵羊和山羊血清免疫球蛋白G的直接ELISA试剂盒在进行商品化销售（例如，ELITEST CLA，Hyphen，BioMed，Neuville-sur-Oise，法国），它们借助辣根过氧化物酶（HRP）标记的小鼠抗山羊/绵羊检测抗PLD抗体。正如前文所提及到的，这些ELISA方法均不能用于骆驼科动物。Paule等[50]和Seyffert等[64]介绍了一种通过提取和浓缩的假结核分枝杆菌分泌的免疫反应性蛋白来检测干酪性淋巴腺炎抗体的非商品化ELISA方法。在对阿联酋单峰驼进行干酪性淋巴腺炎血清学诊断时，研究人员同时采用两种ELISA方法（用免疫反应性蛋白建立的未公开ELISA和Hyphen Biomed ELISA）进行了初步的血清学调查研究，试验中用蛋白A替换掉抗绵羊/山羊抗体。研究在21峰病原分离为干酪性淋巴腺炎阳性的单峰驼中取得了血清学也为阳性的好结果，另外1119峰奶用单峰驼为阴性。在一个持续5年的检测研究中，迪拜奶用骆驼农场没有出现任何干酪性淋巴腺炎临床症状，检测结果同时显示出农场中没有干酪性淋巴腺炎携带动物。Braga[14,16]的研究团队通过辣根过氧化物酶（horseradish peroxidase，HRP）标记的蛋白A建立了未公开的间接ELISA方法，并将其用于检测疫苗免疫过的羊驼和自然发生干酪性淋巴腺炎羊驼的血清抗体，且该间接ELISA采用假结核棒状杆菌胞壁抗原作为包被原。结果显示，实验性感染的羊驼在接种后16天产生抗体，持续93天。

假结核病依旧是骆驼科动物最为重要细菌病之一[1,25,29]，畜群中的感染率在10%～60%，肯尼亚东部就有数千单峰驼干酪性淋巴腺炎病例。

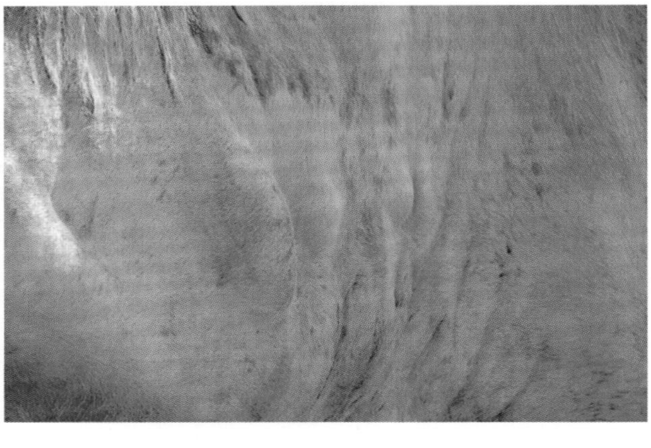

图117　由干酪样淋巴结炎疫苗引起的肉芽肿

## 参考文献


[1] Abou-Zaid A.A.,Selim A.M.,Yousef F.H. & Abd EL-Samea M.M. (1994). – Lymphadenitis in camels. *In* Proceedings of the2nd Veterinary Medicine Congress,11–13 October 1994,Zagazig,Egypt,600–604.

[2] Afzal M.,Sakir M. & Majid Hussain M. (1996). –*Corynebacterium pseudotuberculosis* infection and lymphadenitis (toloa ormala) in the camel. *Trop. Anim. Health. Prod.*,28,158–162.

[3] Anderson D.E. (2000). – Skin abscesses in the dromedary camel. *Austr. Camel News*,5,9–10.

[4] Anderson D.E.,Rings D.M. & Kowalski J. (2006). – Infection with *Corynebacterium pseudotuberculosis* in five alpacas. *JAVMA*,225 (11),1743–1747.

[5] Anon. (1995). – The research work done for the preparation of vaccine against camel abscess. Private report,Camel ResearchCenter,Abu Dhabi,UAE,1–26.

[6] Baird G. (2003). – Current aspects on caseous lymphadenitis. *In Practice*,1 (5),62–68.

[7] Baird G.J. & Malone F.E. (2010). – Control of caseous lymphadenitis in six sheep flocks using clinical examination andregular ELISA testing. *Vet. Rec.*,166,358–362.

[8] Barsallo G.J.A.,Calle E.S. & Samame B.H. (1984a). – Agentes bacterianos en procesos respiratorios que causen mortalidaden alpacas. *Sexto Congr. Peru Microbiol. Parasitol. (Cuzco)*,113,53.

[9] Barsallo G.J.A.,Villena S.C. & Chavera C.A. (1984b). – Abscesos en alpacas. *Sexto Congr. Peru Microbiol. Parasitol. (Cuzco)*,113,53.

[10] Beghelli D.,D'Alterio G.L.,Severi G.,Moscati L.,Pezzotti G.,Foglini A.,Battistacci L.,Cagiola M. & Ayala-Vargas C. (2004).– Evaluation of the immune response to vaccination against *C. pseudotuberculosis* in an alpaca herd in Italy: preliminaryresults. In Proceedings of 4th European Symposium on South American Camelids/DECAMA European Seminar,7–9October 2004,Gottingen,Germany. Wageningen Academic Publisher,Beghelli,Wageningen,133–137.

[11] Behrens H. (1987). – Lehrbuch der Schafkrankheiten. Verlag Paul Parey,Berlin and Hamburg.

[12] Bergin T.J. (1986). –*Corynebacterium pseudotuberculosis* and 'mala' (lymphadenitis) in camels,in FAO the camel: developmentand research. Proceedings of Kuwait seminar,20–23 October 1986,Kuwait.

[13] Braga W.U. (1993). – Aislamiento de *Corynebacterium pseudotuberculosis* en alpacas y llamas adultas. *Rev. Invest. Pec.*,6,128.

[14] Braga W.U. (2007). – Protection in alpacas against *Corynebacterium pseudotuberculosis* using different bacterial components. *Vet. Microbiol.*,119 (2–4),297–303.

[15] Braga W.U.,Chavera A.E. & Gonzales A.E. (2006a). –*Corynebacterium pseudotuberculosis* in highland alpacas (*Lama pacos*) in Peru. *Vet. Rec.*,159,23–24.

[16] Braga W.U.,Chavera A.E. & Gonzalez A.E. (2006b). – Clinical,humoral,and pathologic findings in adult alpacas withexperimentally induced *Corynebacterium pseudotuberculosis* infection. *Am. J. Vet. Res.*,67 (9),1570–1574.

[17] Caprano M. (1934). – Report of Ministry of Agriculture,Technical Science Service. *Vet. Sec. Bull.*,135.

[18] Chen J.J.,Han Z.Y.,Shang Y.Z. & Caimude J. (1984). – Epidemiological survey of corynebacteriosis of Bactrian camel inSubei County,Gansu. *Gansu J. Anim. Sci.* Vet. *Med. Suppl.*,51–54.

[19] Clauss M.,Lendl Chr.,Schramel P. & Streich W.J. (2004). – Skin lesions in alpacas and llamas with low zinc and copperstatus – a preliminary report. *Vet. J.*,167,302–305.

[20] Connor K.M. (2010). – Characterization of United Kingdom isolates of *Corynebacterium pseudotuberculosis* using pulsed-fieldgel electrophoresis. *J. Clinic. Microbiol.*,38,2633–2637.

[21] D'Alterio G.L. (2004). – Skin lesions in UK alpacas (*Lama pacos*): prevalence,aetiology and treatment. In Proceedings of4$^{th}$ European Symposium on South American Camelids/DECAMA European Seminar,07–09 October 2004,Gottingen,Germany. Wageningen Academic Publishers,Wageningen,121–127.

[22] Dalling T.,Robertson A.,Boddie G. & Spruell J. (1966). – Diseases of camels. In The International Encyclopaedia of VeterinaryMedicine (T. Dalling,ed.). W. Green and Son,Edinburgh,1,585.

[23] Daneji A.I.,Djang K.T.F. & Ogunsan E.A. (1996). –*Actinobacillus lignieresi* infection in camels on the Sokoto plains,Nigeria. *Trop. Anim. Health Prod.*,28 (4),315–316.

[24] Dioli M. & Stimmelmayr R. (1992). – Important camel diseases in the one-humped camel in eastern Africa. In A pictorialguide to diseases,health care and management (H.J. Schwartz & M. Dioli,eds). Verlag Joseph Markgraf Scientific Books,Weikersheim,Germany,155–164.

[25] Domenech J.,Guidot T.G. & Richard D. (1977). – Les maladies pyogenes du dromadaire en Ethiopie. Symptomatologie –Etiologie. *Rev. Elev. Méd. Vét. Pays Trop.*,30 (3),251–258.

[26] Edelsten R.M. & Pegram R.G. (1974). – Contagious skin necrosis of Somali camels associated with *Streptococcus agalactiae. Trop. Anim. Health. Prod.*,6,255–256.

[27] El-Allawy T.A.A.,Ahmed L.S. & Hamed Maha I. (2006). – Epidemiological studies on infectious skin affections of camelsin upper Egypt. Proceedings of the Scientific Conference on Camels,10–12 May 2006,Buraydah,Saudi Arabia,819–831.

[28] Eldisougi I. (1984). – A note on the diseases of camels in Saudi Arabia. In The camelid: an 'all purpose' animal (W.R. Cockrill,ed.). Scandinavian Institute of African Studies (The Nordic Africa Institute),Uppsala,Sweden,496–502.

[29] El-Sergamy M.A.,Soufy H.,Lotfi M.M.,Hassanain M.A.,Nasser A.M.,Laila A. & Shash M.S. (1991). – Lymphadenitis inEgyptian camels with special reference to bacteriological and parasitological affections. *Egypt J. Comp. Path. Clinic* Path.,4 (1),25–45.

[30] Esterabadi A.H.,Entessar F.,Hedayati H.,Narimani A.A. & Sadri M. (1975). –*Corynebacterium pseudotuberculosis* infectionin Iranian camels. Archives de l'Institut Razi,27,61–66.

[31] Greenwood A.G. (1991). – Control of pseudotuberculosis in zoos. *Vet. Rec.*,128 (9),215.

[32] Han Z.Y.,Chen J.G. & Shang Y.Z. (1983). – Experiment on immunization against corynebacteriosis of the Bactrian camel. *Acta Agri. Univ.,Gansu*,2,47–58.

[33] Hassan S.S.,Schneider M.P.C,Ramos R.T.J.,Carneiro A.R.,Ranieri A.,Guimaraes L.C.,Ali A.,Bakhtiar S.M.,Pereira U.D.P.,Santos A.R.D.,Soares S.D.C.,Dorella F.,Pinto A.C.,Ribeiro D.,Barbosa M.S.,Almeida S.,Abreu V.,Aburjaile F.,Fiaux K.,Barbosa E.,Diniz C.,Rocha F.S.,Saxena R.,Tiwari S.,Zambare V.,Ghosh P.,Pacheco L.G.,Dowson C.G.,Kumar A.,Barh D.,Miyoshi A.,Azevedo V. & Silva A. (2012). – Whole-genome sequence of *Corynebacterium pseudotuberculosis* strain Cp162,isolated from camel. *J. Bacteriol.*,194 (20),5718–5719.

[34] Hawari A.D. (2008). –*Corynebacterium pseudotuberculosis* infection (caseous lymphadenitis) in camels (*Camelus dromedarius*)in Jordan. *Am. J. Anim. Vet. Sci.*,3 (2),68–72.

[35] Hoste C.,Peyre de Fabregues B. & Richard D. (1985). – Le dromadaire et son elevage. *Etudes et Synthèses Inst. Rev. Elev. Méd.Pays Trop. France,Maisons-Alfort.*,162,145–146.

[36] Ismail M.,Enany M.,El-Seedy F.R. & Shouman M.T. (1985). – Oedematous skin disease of camel in El-Sharkia Governorate. In Proceedings of 1st International Conference on Applied Science,30 March–1 April 1985,Zagazig,Egypt.

[37] Kamat N.K.,Iqbal J. & Yass Al-Juboori A.A. (2006). – Pathogens of internal and external purulent lesions in dromedarycamels. Proceedings of 1st Conference of the International Society of Camelids Research and Development (ISOCARD),15–17 April 2006,Al Ain,UAE,113.

[38] Kilik N. & Kirkan S. (2004). – Actinomycosis in a one-humped camel (*Camelus dromedarius). Vet. Med. A*,51,363–364.

[39] Kobera R. & Poehle D. (2004). – Case reports in South American camelids in Germany. *In* Proceedings of the 4th EuropeanSymposium on South American Camelids/DECAMA European Seminar,7–9 October 2004,Gottingen,Germany. WageningenAcademic Publisher,Wageningen,151.

[40] Lindsay H.J. & Lloyd S. (1991). – Diagnosis of caseous lymphadenitis in goats. *Vet. Rec.*,128,86.

[41] Lloyd S.,Lindsay H.S.,Slater J.D. & Jackson P.G.G. (1990). – Caseous lymphadenitis in goats in England. *Vet. Rec.*,127,478.

[42] McGrane J.J. & Higgins A.J. (1985). – Infectious diseases of the camel: viruses,bacteria and fungi. *Br. Vet. J.*,141,529–547.

[43] Manefield G.W. & Tinson A. (1996). – Camels. A compendium (T.G. Hungerford Vade Mecum Series for Domestic Animals). University of Sydney Post Graduate Foundation in Veterinary Science,240,298.

[44] Muenchau B. (2004a). – Latest information in camelids. *In* Proceedings of the 4th European Symposium on South AmericanCamelids/DECAMA European Seminar,7–9 October 2004,Gottingen,Germany. Wageningen Academic Publisher,Wageningen,299–300.

[45] Muenchau B. (2004b). – Aktuelle Erläuterungen bei Kameliden. 1. Pseudotuberkulose und Abszesse bei Neu-undAltweltkameliden. *Lamas: Haltung und Zucht von Neuweltkameliden* 12 (4),8–12.

[46] Muenchau B. (2005). – Pseudotuberkulose – Erlauterungen zu bisherigen Ergebnissen. *Lamas: Haltung und Zucht vonNeuweltkameliden*,13 (3),13–15.

[47] Muenchau B. (2006). – Pseudotuberkulosis bei Kamelen. *In* Proceedings of Deutsche Gesellschaft fur Zootier Wildtier-undExotenmedizin,61–73.

[48] Nashed S.M. & Mahmoud A.Z. (1987). – Microbiological and histopathological diagnosis of rare cases of Corynebacterium infection in camels. *Assiut. Vet. Med. J.*,18,82–86.

[49] Norgren T. (2008). – *Corynebacterium pseudotuberculosis* hos alpacka: utredning av ett utbrott i en svensk alpackabesattning. Available at: www.uppsatser.se/uppsats/0679675b81/

[50] Paule B.J.A.,Meyer R.,Moura-Costa L.F.,Bahia R.C.,Carminati R.,Regis L.F.,Vale V.L.C.,Freire S.M.,Nascimento I.,Schaer R.& Azevedo V. (2004). – Three-phase partitioning as an efficient method for extraction/concentration of immunoreactiveexcreted-secreted proteins of *Corynebacterium pseudotuberculosis. Protein Expr. Purif.*,34,311–316.

[51] Piontkowski M.D. & Shivvers D.W. (1998). – Evaluation of a commercially available vaccine against *Corynebacteriumpseud otuberculosis* for use in sheep. *JAVMA*,212 (11),1765–1768.

[52] Promed (2007). – Pigeon fever,equine – USA (Oregon). Available at: www.oregonlive.com/newsflash/regional/index.ssf?/base/news-21/119269254119530.xml&storylist=orlocal

[53] Purohit N.R. & Chouhan D.S. (1992). – Wound healing in camels. *In* Proceedings of the 1st International Camel Conference.R. and W. Publications (Newmarket) Ltd,Newmarket,UK,365–370.

[54] Purohit N.R.,Purohit R.K.,Chouhan D.S.,Choudhary R.J.,Mehrotra P.K. & Sharma K.N. (1988). – Suspected cutaneousactinobacillosis in camels. *Austr. Vet. J.*,65,31–32.

[55] Qureshi S.,Kataria A.K. & Gahlot T.K. (2002). – Bacterial microflora associated with wounds and abscesses on camel(*Camelus dromedarius*) skin. *J. Camel Pract. Res.*,9 (2),129–134.

[56] Radwan A.I.,El-Magawry S.,Hawari A.,Al-Bekairi S.I. & Rebleza R.M. (1989). –*Corynebacterium pseudotuberculosis* infectionin camels (*Camelus dromedarius*) in Saudi Arabia. *Trop. Anim. Health. Prod.*,21,229–230.

[57] Refai M. (1992). – Bacterial and mycotic diseases of camels in Egypt. In Proceedings of the 1st International Camel Conference.R. and W. Publications (Newmarket) Ltd,Newmarket,UK,59–64.

[58] Rosychuk R.A.W. (1989). – Llama dermatology. *Vet. Clin. North Am. Food Anim. Pract.*,5,203–215.

[59] Rosychuk R.A.W. (1994). – Llama dermatology. *Vet. Clin. North Am. Food Anim. Pract.*,10 (2),228–239.

[60] Sadykov R.G. & Dadabaev Zh.S. (1976). – On camels with pus lymphangitis (staphylococcosis) in Kazakh,SSR. Infectiousand parasitic diseases of farm animals. *Trudy Instituta Veterinarnych nauk Kazackskoj SSR,Alma Ata (SUN) s.n.*,73–78.

[61] Schwartz H.J. & Dioli M. (1992). – The one-humped camel in eastern Africa. In A pictorial guide to diseases,health careand management (H.J. Schwartz & M. Dioli,eds). Verlag Joseph Markgraf Scientific Books,Weikersheim,Germany.

[62] Schwartz S.,Schwartz H.J. & Wilson A.J. (1982). – Eine fotographische Dokumentation wichtiger Kamelkrankheiten inKenia. *Der prakt. Tierarzt*,11,985–989.

[63] Scott D.W.,Vogel J.W.,Fleis R.I.,Miller W.H. & Smith M.C. (2011). – Skin diseases in the alpaca (*Vicugna pacos*): a literaturereview and retrospective analysis of 68 cases (Cornell University 1997–2006). *Vet Dermatol.*,22 (1),2–16.

[64] Seyffert N.,Guimaraes A.S.,Pacheco L.G.C.,Portela R.W.,Bastos B.L.,Dorella F.A.,Heinemann M.B.,Lage A.P.,Gouveia A.M.,Meyer R.,Miyoshi A. & Azevedo V. (2009). – High seroprevalence of caseous lymphadenitis in Brazilian goat herds revealed by *Corynebacterium pseudotuberculosis* secreted proteins-based ELISA. *Res. Vet. Sci.*,88 (1),50–55.

[65] Spesivtseva N.A. & Noskov A.I. (1959). – Epizootic lymphangitis in camels. *Trudy Vses. Inst. Vet. sanit. Ectoparasit.*,14,86.

[66] Stowe C.M. (1984). – Antimicrobial drug interactions. *JAVMA*,185 (10),1 137–1 141.

[67] Strauss G. (1991). – Erkrankungen junger Neuweltkamele im Tierpark Berlin – Friedrichsfelde 11. In Arbeitstagung derZootierarzte im deutschsprachigen Raum. Nov. 1–3 in Stuttgart,Tagungsbericht,80–83.

[68] Tadesse A. & Molla B. (2002). – Study on skin lesions of camels (*Camelus dromedarius*) associated with bacterial

pathogensin north-eastern Ethiopia. *J. Camel Pract. Res.*,9 (1),37–41.

[69] Tarek M. & Abu-Bakr M. (1990). – Bacteriological studies on dromedaries lymphadenitis in United Arab Emirates. Zag.*Vet. J.*,18 (1),77–90.

[70] Tejedor M.T.,Martin J.L.,Corbera J.A.,Schulz U. & Gutierrez C. (2004). – Pseudotuberculosis in dromedary camels in theCanary Islands. *Trop. Anim. Health. Prod.*,36,459–462.

[71] Tejedor M.T.,Martin J.L.,Lupiola P. & Gutierrez C. (2000). – Caseous lymphadenitis caused by *Corynebacterium ulceranus*in the dromedary camel. *Can. Vet. J.*,41,126–127.

[72] Tejedor-Junco M.T.,Lupiola P.,Caballero M.J.,Corbera J.A. & Gutierrez C. (2009). – Multiple abscesses caused by *Salmonella*enterica and *Corynebacterium pseudotuberculosis* in a dromedary camel. *Trop. Anim. Health. Prod.*,41,711–714.

[73] Villena C. (1985). – Abscesos en alpacas/estudios microbiologicos. DVM thesis,Universidad Nacional Mayor de San Marcos,Lima,Peru,20.

[74] Wernery U. & Kaaden O.-R. (2002). – Infectious diseases in camelids. 2nd Ed. Blackwell Science Berlin,Vienna,134–138.

[75] Younan M. & Bornstein S. (2007). – Lancefield Group B and *C. streptococci* in east African camels (*Camelus dromedarius*). *Vet. Rec.*,160,330–335.

（娄忠子译，殷宏校）

## 1.5.2 金黄色葡萄球菌皮炎

金黄色葡萄球菌是一种人和动物都携带的细菌，该菌主要寄生于皮肤和鼻咽部，在消化道和生殖道也有存在。金黄色葡萄球菌是一种潜在病原，可以引起多种化脓性炎症，在家畜，主要引起牛、绵羊和山羊的乳腺炎。它也会感染不同动物的皮肤，引发以下疾病；

（1）马、山羊、绵羊以及狗的毛囊炎和疖疮。

（2）山羊、仔猪和牛的脓皮病。

（3）绵羊面部或眼睑湿疹。

（4）仔猪脓疱病或角膜下多泡状突起皮炎。

（5）山羊乳房皮炎。

金黄色葡萄球菌还可引发全身性疾病，如马的葡萄状菌病、小羊的脓毒症以及幼年动物的多发性关节炎。脓皮病是旧大陆骆驼最常见的皮肤传染病之一。从一只患皮脓肿的羊驼上，也分离到金黄色葡萄球菌[5]。在此病例中，脓肿被切除，最后被诊断为葡萄状菌病。葡萄状菌病是细菌感染皮肤，有时也会感染内脏所引发的慢性肉芽肿样反应。之所以得此名，是由于脓肿里含有葡萄样颗粒(botryo，希腊语，葡萄的意思)。该病曾被误认为由于真菌感染而引起。1919年，引起葡萄状菌病的细菌性病原体被查明。葡萄状菌病通常由金黄色葡萄球菌引起，但其他细菌如绿脓杆菌也会引起该病。

### 流行病学与病原学

该病难以用药物治疗。骆驼脓皮病主要发生于幼年单峰驼，主要由金黄色葡萄球菌所引起，皮肤发生化脓性、慢性炎症。该病发生时，最初的症状表现为毛囊炎，之后常随着病情发展形成单个的疖疮，或聚集成3～5mm的大肿块。肿块上覆盖可以揭去的小痂，痂下有少量脓汁。当脓汁排除后，就会露出一个坑。肿块也可以变得很大。用小刀切开肿块后，出现白色绿色的脓汁（图118）。在动物的前腿，经常会发现大肿块。

Bornstein[1]在报道发生于不足4月龄小骆驼的淋巴结（腺）炎时，也描述了与脓皮病相似的病变。病灶由一些肿块组成，肿块通常位于脖子下面、前腿之间部位。肿块有温热感，疼痛感，常常有橘子大小。肿块中含有黄色或奶油色的脓汁。受感染动物常常表现不安，健康恶化，有些会死亡。一个群体中常有数个小骆驼感染。从病变部位分离出链球菌似葡萄球菌。

在发生干酪样淋巴结（腺）炎时，位于小骆驼前腿之间的肿块可能在破溃后进入胸腔，引起败血病和/或严重的支气管肺炎，伴发有心包炎和心包积水（图119）。

除了疖疮外，可能还存在感染金黄色葡萄球菌的、含有脓疱的渗出性湿疹。该病可能会慢性发展，由于脓疱含有葡萄球菌及存在其他因素，该病

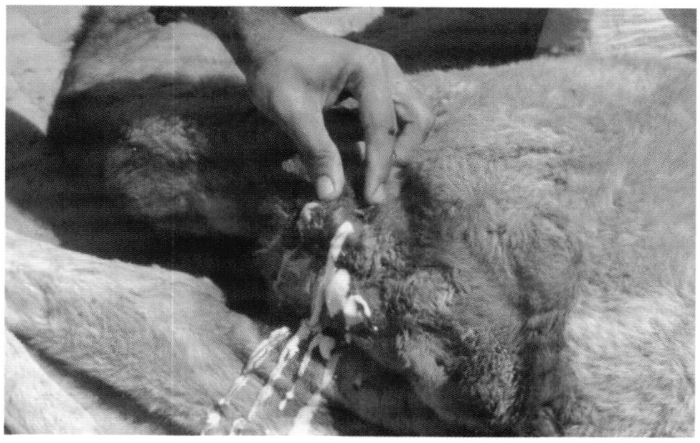

图118 6周龄单峰驼金黄色葡萄球菌脓肿

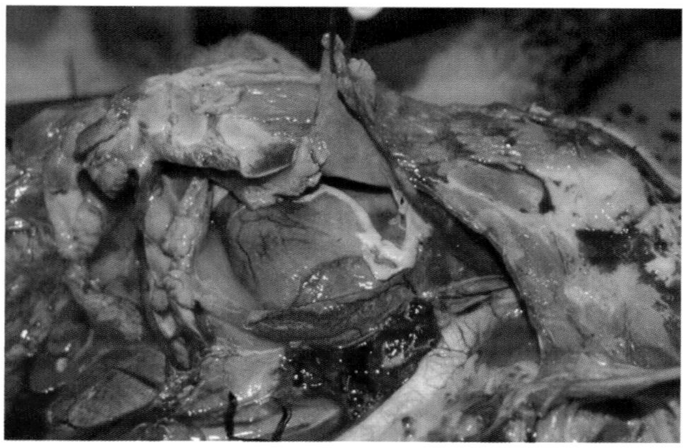

图119 由金黄色葡萄球菌引起的心包炎和心包积液



难于用药物治疗。金黄色葡萄球菌拥有多种毒力因子，毒力因子可以损害宿主，并保护自身免受宿主防御系统的攻击[9]。

在骆驼皮肤脓肿研究中，仅有少数几例进行细菌学研究。Ismail等[6]从还没有排脓的头、肩、胸、腿以及胸腹部脓肿中分离到以下细菌。

（1）金黄色葡萄球菌（*S.aureus*）；

（2）化脓放线杆菌(*Actinomyces pyogenes*)；

（3）伪结核棒状杆菌 (*Corynebaterium pseudotuberculosis*)；

（4）化脓链球菌(*Streptococcus pyogenes*)；

（5）大肠埃希氏杆菌(*Escherichia coli*)；

（6）克雷伯氏菌 (*Klebsiella* spp)；

（7）普通变形杆菌(*Proteus valgaris*)；

（8）奇异变形杆菌(*Proteus mirabilis*)；

（9）绿脓杆菌 (*P.aeruginosa*)；

（10）产气荚膜梭菌 (*Clostridium perfringens*)；

（11）坏死梭杆菌 (*Fusobacterium necrophorum*)。

El-seedy等人[4]从埃及93个驮运骆驼群的干枯管状器官中也分离到类似的细菌。根据Buchnev等人的研究[2]，在中亚双峰驼中，由葡萄球菌感染引起的疾病很普遍。Semushkin[10]将引起5%～20%的双峰驼群体发生皮肤脓肿，导致15%的发病骆驼死亡现象，称为“传染性皮肤脓肿”。直到sadykov和 Dadabaev分离鉴定出导致这种疾病的病原体，其病原学才清楚。双峰驼发生该病时，症状为化脓性淋巴管炎，感染头部、颈部以及肩部的浅表淋巴结。切开脓肿后，里面是浓浓的白色脓汁。一些病例中，脓肿中含有高达500mL的脓汁。其并发症通常是化脓性败血症，许多双峰驼死于并发症。从不同地区的双峰驼病料中，都分离到葡萄球菌。生化试验这些发现从不同地域分离的葡萄球菌，其生化特性一致。这些菌株被命名为骆驼葡萄球菌（*S.cameli* ）。Samartsev在哈萨克斯坦报道了一例由化脓链球菌柠檬色亚种（*S.pyogenes citreus*）引起的传染性脓疮性皮炎。脓疮的直径为0.5～2.0cm，1个月后，在没有采取治疗措施的情况下，脓疮自动消失。

Domenech等人对埃塞俄比亚单峰驼的化脓性感染进行了研究。研究结果显示，有两种定义清晰的皮肤疾病，一种叫马拉（mala），亦即淋巴结（腺）炎，另一种称为马哈（maha），或者杜拉（doula），是由皮肤脓肿在形成溃疡后发生的皮肤坏死病。从这些病灶中分离到金黄色葡萄球菌及B型链球菌。Zaitoun等人对埃及不同地区的单峰驼皮肤疾病进行了调查[11]，对292只单峰驼的检测结果发现，13%的骆驼有皮肤损伤。其中，疥癣是最主要的皮肤疾病，有近60%的骆驼感染。其后是干酪样淋巴结炎（17%），传染性皮肤坏死（14%）以及皮肤脓肿（9%）。在传染性皮肤坏死病料（CNS）中，主要分离出金黄色葡萄球菌，病变主要发生于颈部腹面及胸腹侧面。该病发生率最高的记录是在炎热季节，由于绿饲料缺乏而导致动物缺少营养，以及蜱数量高，所以，该病发生率高。此外，许多骆驼由于感染能抑制机体产生体液免疫及细胞免疫反应的伊氏锥虫而引起皮肤坏死。Zaitoun在其皮肤疾病调查中，将传染性皮肤坏死和皮肤脓肿区分开来。两者都是由金黄色葡萄球菌引起。

在阿联酋的小单峰驼中，化脓性皮炎同样严重。在过去的20年中，研究人员从脓肿、伤口、溃疡以及其他皮肤病灶进行了细菌分离。结果详见表50所示。

### 治疗和控制措施

为了控制该病，应该隔离感染动物治疗。由于一些金黄色葡萄球菌菌株对抗生素具有很强的耐药性，所以，应对所有分离菌株做耐药性试验。皮肤发病部位应每天用5%的碘酒清洗消毒，成熟脓肿应切开引流。在感染严重病例中，应尝试注射抗生素治疗。

如表50所示，70%的脓肿样品中分离出金黄色葡萄球菌。健康犬皮肤中所含的金黄色葡萄球菌数量少[9]，但在病变部位，细菌数量增加了50～100倍。由于脓皮病难于用抗生素治疗，研究人员制作自家疫苗用于单峰驼。自家疫苗适用于群体中单个或小部分动物的免疫预防。

因为金黄色葡萄球菌菌株存在着数量众多的免疫因子及毒力因子，因此，不能商品化生产金黄色葡萄球菌疫苗，所以，必须制作自家疫苗用于防治该病。患金黄色葡萄球菌皮炎的小单峰驼经皮下注射5～8mL用10%福尔马林灭活而制作的自家疫苗后，60%的免疫单峰驼在免疫后数天内出现症状改善，脓肿脱水变干，病变面积缩小。14天后，仅有少数的单峰驼需要再次强化免疫。所有病例均经过这种方式得到成功治疗。对于未发生金黄色葡萄球菌皮炎的动物，也许可以通过接种自家疫苗预防该病，从而阻止该病的传播。自家疫苗的显著成功免疫，整体上是基于通过接种疫苗对机体免疫系统产生非特异性的刺激，以及对引起皮炎的金黄色葡萄球菌所携带的抗原性外毒素及毒力因子所产生的特异性免疫反应共同作用用的结果。致病性葡萄球菌抗吞噬毒力因子被中和后，机体吞噬作用得以恢复。治疗脓皮病的关键在于能够大幅提升动物自身的防御能力。

表50　阿联酋单峰驼皮肤病灶细菌分离情况

| 细菌分离 | 皮肤病变<br>脓肿：开放和闭合 | 伤口/溃疡 | 其他 |
|---|---|---|---|
| 金黄色葡萄球菌 | 162 | 25 | 9 |
| 葡萄球菌 | 17 | 8 | 30 |
| 化脓放线杆菌 | 12 | 3 | 0 |
| 假结核棒状杆菌 | 8 | 0 | 0 |
| 刚果嗜皮菌 | 0 | 0 | 4 |
| 链球菌 | 7 | 7 | 5 |
| 假单胞菌 | 7 | 9 | 5 |
| 变形杆菌 | 4 | 3 | 3 |
| 大肠埃希氏菌 | 6 | 4 | 2 |
| 需氧菌 | 9 | 7 | 8 |
| 总数 | 232 | 66 | 66 |

## 参考文献

[1] Bornstein S. (1995). – Skin diseases of camels. In Camel keeping in Kenya (J.O. Evans,S.P. Simpkin & D.J. Atkins,eds). Range Management Handbook of Kenya. Ministry of Agriculture,Livestock Development,and Marketing,Kenya,3 (8),7 – 13.

[2] Buchnev K.N.,Tulepbaev S.Z. & Sansyzbaev A.R. (1987). – Infectious diseases of camels in the USSR. In Diseases of camels. *Rev. sci. tech. Off. int. Epiz.*,6 (2),487 – 495.

[3] Domenech J.,Guidot T.G. & Richard D. (1977). – Les maladies pyogenes du dromadaire en Ethiopie. Symptomatologie-Etiologie. *Rev. Elev. Méd. Vét. Pays Trop.*,30 (3),251 – 258.

[4] El-Seedy F.R.,Ismail M.,El-Sayed Z.,Enany M.E. & Abdel-Ghani M. (1990). Bacterial species implicated fistulous wither affecting one-humped camels in Egypt. *J. Egypt Med. Ass.*,50,81 – 92.

[5] Fowler M.E. (2010). – Medicine and surgery of camelids. 3rd Ed. Wiley-Blackwell,Oxford,UK,214 – 215.

[6] Ismail M.,Ezzat M.,El-Jakee J.,El-Sayed Z.E. & Abd El-Rahmen M. (1990). Microorganisms associated with closed abscesses of camels in Egypt. *Vet. Med. J. Giza*,38,53 – 62.

[7] Sadykov R.G. & Dadabaev Zh.S. (1976). – On camels with pus lymphangitis (staphylococcosis) in Kazakh,SSR. Infectious and parasitic diseases of farm animals. *Trudy Instituta Veterinarnych nauk Kazackskoj SSR,Alma Ata (SUN) s.n.*,73 – 78.

[8] Samartsev A.A. (1950). – Infectious pustular dermatitis in camels. *Proc. Kazakh Res. Vet. Institute*,5,190 – 197.

[9] Scheels H. (1989). Erfahrungen bei der Behandlung der Pyodermie des Hundes mit Autovaccinen. *Pro Veterinario*,9 (1),3.

[10] Semushkin N.R. (1968). – Diagnosis of camel diseases. Sel'khozgiz,Moscow.

[11] Zaitoun A.M.A. (2007). – Contagious skin necrosis of dromedary camels in south Egypt. *J. Camel. Pract. Res.*,14 (2),125 – 132.

（李学瑞译，殷宏校）

## 1.5.3 嗜皮菌病

由刚果嗜皮菌引起的嗜皮菌病是湿热带地区一种典型的流行病。嗜皮菌病是一种可传播，可感染及接触性传染的疾病，特点为发生渗出性和不发痒的皮炎，随后形成可能覆盖整个皮肤的疤痕和硬壳。该病在非洲、澳大利亚和新几内亚流行广泛。美洲国家中，阿根廷、加拿大和美国也有该病病例报道，而在欧洲，该病则零星发生[35]。气候温和地区的羊和马更易感染该病。嗜皮菌病主要发生于牛、小反刍动物、马、人类和一些野生动物，如斑马、红鹿、食肉动物和爬行动物。澳大利亚圈养鳄鱼最易感染该病。在收到感染病料、开展疾病诊断和研究时，用兔子从病料中“纯化”病原。人可能会通过接触发病动物而感染该病，但鲜有发生。

牛对嗜皮菌病的抵抗力因品种不同而具有本质性、遗传性差异。杂交的欧洲牛极其易感，非洲瘤牛少，西非的恩达玛牛仅有少数感染[35]。该病还有其他名称，如链丝菌病，真菌性皮炎，羊的卷毛病，羊的草莓根腐病及雨斑病。多年来，该病一直受到忽视，或与别的骆驼疾病相混淆。渗出性皮炎可能会覆盖青年单峰驼的全身，导致感染者死亡。新大陆骆驼也感染该病，但鲜有发生[40]。有一篇文献专门谈及美国的新大陆骆驼感染该病的病例[36]，英国也报道了几例北美羊驼感染该病的病例[2]。

### 公共卫生意义

尽管嗜皮菌主要作为一种动物病原，能够引起急性、亚急性及慢性的皮肤疾病，它也可能是人兽共患病原。例如，Burd等人[7]就曾报道过一个在骑马训练营工作的一名15岁女孩，感染脓疱性皮炎的病例。其他不多几个病例主要发生在非洲和澳大利亚，大多数病人都曾与动物接触过。

嗜皮菌病在动物，特别是非洲国家的动物具有高发性，人则只有少量病例发生。比较表明，人对嗜皮菌病具有很强的抵抗力。人嗜皮菌病的特征为：手和前臂有小囊泡和脓疱。病灶渗出物多，随后，是黄或白色的脓水，一旦脓肿破裂，会形成一个粉红的火山口样的溃疡，之后，溃疡上会形成一层红色的痂。病人在不经过特殊治疗的情况下，两周内会康复。到目前为止，所有的病人都与发病动物有过亲密接触。因此，人在接触发病动物时应该特别小心，应该洗手，戴手套。

### 病原学与流行病学

嗜皮菌属于放线菌目。放线菌目有8个科，包括嗜皮菌科，诺卡氏菌科，放线菌科，分枝杆菌科等。嗜皮菌科有两个属，即嗜皮菌属和地嗜皮菌属。地嗜皮菌属只有阴暗地嗜皮菌一个种，该菌生存于土壤中，从动物体内还没有分离到该菌。嗜皮菌属除了刚果嗜皮菌，还有龟嗜皮菌。龟嗜皮菌从龟体分离得到，它是澳大利亚绿海龟的病原菌。刚果嗜皮菌是一种多形态细菌，通常以两种形态存在，即分枝菌丝或球形或卵球形的球菌。分枝菌丝由该菌的球菌形式萌发而形成，这些菌丝由纵向和横向的横隔所分开。菌丝产生可游走的卵球形球菌，它曾被错误地定义为“游走孢子”，在雨季，嗜皮菌主要以卵球形球菌的形态存在，它们通过直接接触传播或媒介（蜱、蝇）传播。刺槐的刺及谷物的麦芒也可能传播卵球形球菌[39]。卵球形球菌每四个聚积成团，或排成链状，似“火车轨”或“钱串”状，链状是刚果嗜皮菌的主要存在方式。从表皮所携带的卵球形球菌萌发而形成的菌丝会攻击发鞘。引发化脓性炎症反应，导致表皮生长缓慢、膨胀，最终与真皮相分离。在原有的表皮与真皮分离后，真皮之上又会形成新的表皮细胞[35]。渗出物干涸后，形成痂，这是该病区分其他病的特征性病变。痂可以被去掉，暴露出一个湿红的部位，分泌脓性、带血性渗出物（渗出性皮炎）[25]。

Khodakaram-Tafti等人认为骆驼携带的璃眼蜱对伊朗境内单峰驼嗜皮菌病的传播起着重要作用。感染嗜皮菌病的单峰驼（14/103，13.6%），其发病部位曾遭蜱严重侵袭。

沙特阿拉伯及阿曼的单峰驼群体中，也有嗜皮菌病发生[4,26]。

刚果嗜皮菌对环境的抵抗力特别强。球菌形式存在的刚果嗜皮菌耐受干燥，即使100℃时仍能存活。在湿润环境中，痂和硬皮中的刚果嗜皮菌42个月后仍具有感染性。令人感兴趣的是，尽管一些嗜皮菌不能培养，它们仍能够感染兔子。

在阿联酋，Wernery和 Ali首先报道了单峰驼嗜皮菌病[38]；在苏丹，Gitao的研究组报道了该病[16,19,20]；在埃塞俄比亚，Samuel等人[33]对该病予以报道；在肯尼亚，由Bornstein报道[5]了该病；在尼日利亚，分别由 Ajogi [3]和Makinde[27]予以报道单峰驼嗜皮菌病。Bornstein[5]研究了不同致病菌株的形态学和生化特性。1976年，在Abu-Samra一篇综述性文献中[1]，没有提及骆驼自然感染嗜皮菌，尽管有不同的研究者已经报道了链丝菌样组织（表51）。

从这些结果或许可以假定传染性皮肤坏死和链丝菌病与嗜皮菌病是同一种病。Abu-Samra[1]通过实验证实单峰驼对刚果嗜皮菌易感。

湿度是促进该病发生的最重要的因素之一。每年第一场雨后，该病开始发生，旱季开始后，该病消失。众所周知，当通过喷雾或浸渍法治疗动物外寄生虫感染时，可能会恶化该病。另外，定期地给动物，特别是马匹洗澡，会引起严重的嗜皮菌病。在雨季末期，病灶干化，病痂脱落，毛发重新长出。然而，即使动物从表面看来健康，它仍是病原携带者，在下一个雨季仍会发病。年复一年，嗜皮菌病不断造成动物体重减轻，消瘦，继发细菌、病毒及真菌感染。伴发有锥虫病等疾病时，可能会导致动物死亡。因此，对动物定期治疗是很重要的。

在阿联酋，近期从单峰驼皮肤病灶分离到一株非溶血性刚果嗜皮菌菌株[22]（图120）。该病皮肤的变化可以与马“泥疥疮”（ gale de boue ）或

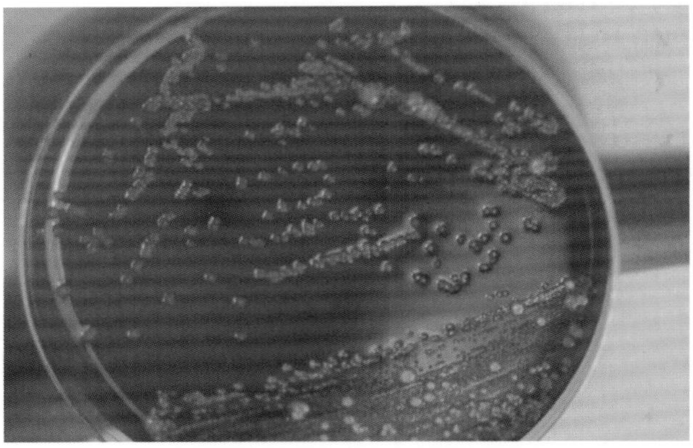

图120　微氧环境下，37℃培养5天后血平皿上生长的分离自单峰驼的非溶血性刚果嗜皮菌

表51　单峰驼传染性皮肤坏死及从病料中分离菌株

| 作者 | 年 | 国家 | 临床症状/菌株 |
|---|---|---|---|
| Cross | 1917 | 印度 | 链球菌 |
| Curasson | 1918 | 非洲 | 皮肤链丝菌病 |
| | 1920 | 非洲 | 骆驼放线菌（诺卡氏） |
| | 1936 | 非洲 | 皮疽诺卡氏菌 |
| | 1947 | 非洲 | 链丝菌病 |
| Mason | 1919 | 印度 | 传染性皮肤坏死 |
| Leese | 1927 | 印度 | 皮肤坏死 |
| Peck | 1938a,b | 索马里 | 传染性皮肤坏死，缺盐 |
| | 1939 | | |
| Edelsten 和Pegram | 1974 | 索马里 | 传染性皮肤坏死，无乳链球菌 |
| Domenech 等 | 1977 | 埃塞俄比亚 | 皮肤坏死，不同种类细菌 |
| Fazi和Hofmann | 1981 | | 皮肤坏死，骆驼放线菌 |
| Schwartz 等 | 1982 | 肯尼亚 | 后腿皮肤坏死，频繁排尿 |
| Wardeh | 1989 | 毛里塔尼亚 | 传染性皮肤坏死 |
| Gitao等 | 1990 | 肯尼亚 | 链丝菌属细菌 |
| Gitao | 1992 | 肯尼亚 | 嗜皮菌病 |
| Gitao | 1993a | 肯尼亚 | 刚果嗜皮菌 |
| Wernery 和Ali | 1990 | 阿联酋 | 嗜皮菌病，刚果嗜皮菌 |
| Joseph等 | 1998 | 阿联酋 | 嗜皮菌病 |
| Makinde等 | 2001 | 尼日利亚 | 嗜皮菌病，血清学方法检测结果 |
| Ajogi等 | 2002 | 尼日利亚 | 嗜皮菌病 |
| 作者不明 | 2007 | 英国（羊驼） | 嗜皮菌病 |

“泥浆兽疥癣”相区别，泥浆兽疥癣在尺骨和球节部位发生。

## 临床症状与病理学

动物发生嗜皮菌病后，临床表现不一。最常见的是慢性过程，症状表现也不一致。很少观察到急性嗜皮菌病。马发生嗜皮菌病时，其临床表现因毛发长度与感染部位而不同。马嗜皮菌病可以分为夏季型和冬季型。其临床表现的差异与Gitao[21]等人对骆驼嗜皮菌病的描述及阿联酋学者对单峰驼嗜皮菌病的描述相一致，分为急性嗜皮菌病和慢性嗜皮菌病。马感染嗜皮菌病时，毛发长的马与毛发短的马，其临床表现截然不同。毛发长的，其渗出液附近变得暗淡无光，特征性表现为“漆刷”效应。可以很容易拔掉暗淡无光的毛发，暴露出粉红色的，充血的伤口表面（图121）；急性感染时，这些区域会被化脓性渗出物所覆盖。高湿度以及雌性单峰驼的频繁的排尿行为，使得后腿及臀部长期湿润，导致皮肤坏死[34]（图122）。

Wernery和 Ali[38]等人描述了发生嗜皮菌病时，毛发短动物其全身几乎全部感染的病例。病灶从小的结节到由厚疤覆盖的较大厚皮区域。将疤移除后，一个同时含有血液和血清渗出液的伤口暴露出来（图123和图124）。嗜皮菌病的特征为结痂，病变部位不发痒，病变可能覆盖整个身体（图125）。

美洲驼感染刚果嗜皮菌会产生严重的毛腐病。由于背部覆盖着厚厚的卷毛，加上高度湿润的气候，使得大羊驼易患此病。病灶通常结痂，特别是在后背部更是如此[36]。

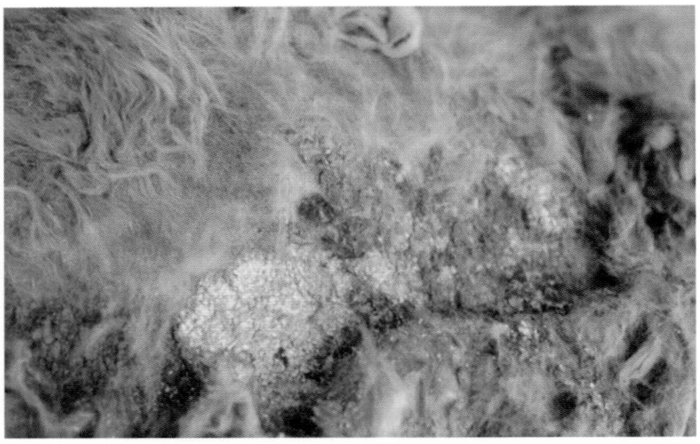

图121　竞赛单峰驼嗜皮菌病。乱蓬蓬的毛发显示“漆刷”效应。毛发移除后，可看到伤口表面

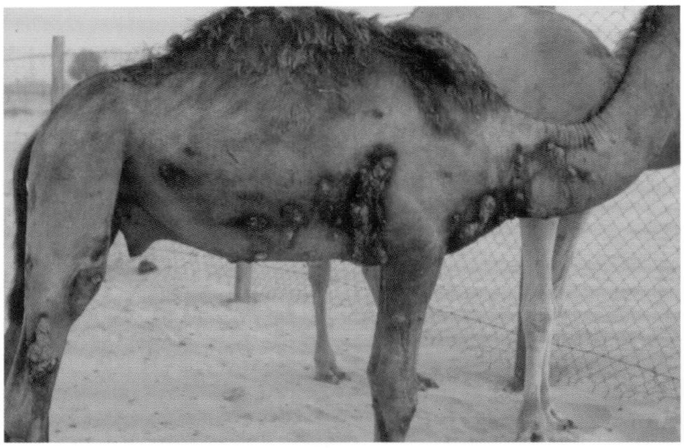

图123　公单峰驼嗜皮菌病例

图125　布基纳法索瘤牛嗜皮菌病例

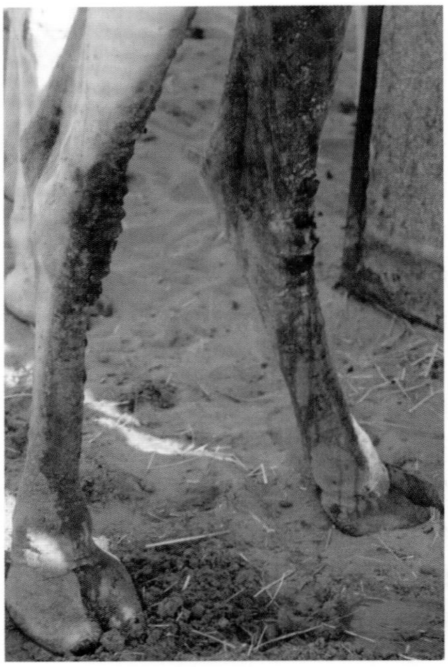

图122　单峰驼后腿发生的皮肤坏死，从坏死病灶分离到刚果嗜皮菌

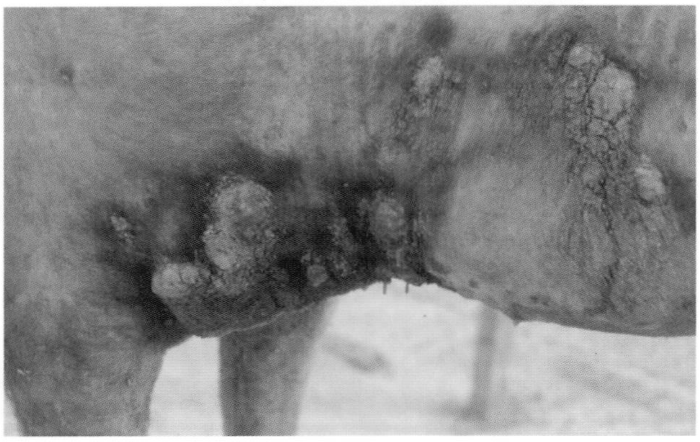

图124　去除硬壳部位流血。这类病灶可见于毛发较短骆驼

Gitao等人[19,20]描述了嗜皮菌病的组织病理变化。其典型的组织病理变化为真皮阻塞、充血及水肿，表皮细胞变质、坏死及过度角化。在皮肤表面，渗出液累积，在表皮及真皮，中性粒细胞浸润。在表皮组织基底层可以发现刚果嗜皮菌分枝、具膈膜的菌丝形态或球菌形态。

## 诊断

嗜皮菌相对容易培养。微量有氧条件下，在绵羊血平皿及牛血平皿上生长良好。培养皿应该在$CO_2$培养箱中培养至5天。该病原在液体培养基中生长不良，当试管上层的培养液还清亮时，在试管的底部就能够观察到片状聚集物。在血平皿上培养不到48小时，小而干的菌落就向琼脂凝胶里面生长。在培养4～5天后，菌落会呈棕色或黄色，表面不规则，起皱褶（粗糙菌落）。刚果嗜皮菌具有强的酶活性，能够发酵多种糖。用疤痕抹片做革兰氏染色，可以发现革兰氏阳性菌排列为钱串状（图126）。直接用显微镜镜检是确认感染的最好方式。可直接检查用磨碎的痂片所做的涂片。当痂片或硬壳与数滴灭菌水混匀后，将其放于载玻片上自然干燥，用革兰氏染液、姬姆萨染液或亚甲蓝染色。也可以用免疫荧光染色。

Gitao[18]等研制了一种酶联免疫吸附方法，通过测定抗体确定骆驼是否感染嗜皮菌。动物在实验室感染刚果嗜皮菌21天后，检测到抗嗜皮菌抗体。他们计划将这套试剂盒用于田间试验。也可以通过琼脂糖凝胶免疫扩散试验和被动血细胞凝集反应去测定抗体。但仅有检测并不能使动物抵御疾病。因此，这些方法仅用于流行病学调查。

### 治疗与控制措施

已经尝试用几种不同方式，或外用（局部），或注射（系统性）治疗嗜皮菌病。按75mg链霉素/千克体重加7.5万单位青霉素/千克体重注射，两次为1个疗程，获得显著疗效。已有人报道用土霉素或普鲁卡因青霉素加链霉素成功治疗单峰驼嗜皮菌病。感染的单峰驼静脉注射盐酸土霉素两次。去掉痂后，暴露部位每日用碘酊消毒，连续7天。病变部位会在4周内完全恢复正常。如果长毛覆盖的严重感染部位，则需要将感染部位的毛剪除，这是减轻病情发展的一个重要措施。将感染动物隔离及控制外寄生虫，以阻断感染循环。

鉴于骆驼属动物中嗜皮菌病呈上升趋势，该病有时与皮肤癣菌病相联系，以及该病属于人兽共患病，所以，应该考虑研制疫苗防控该病。但到目前为止，还没有预防该病的商品化疫苗。商品化疫苗之所以没有成功，其原因在于疫苗效力试验发现，嗜皮菌疫苗对不同刚果嗜皮菌菌株之间的交叉保护力弱，而不同菌株之间的抗原变异或许可以解释免疫结果为何不理想。用抗体作为药物预防的方法也可在雨季来临前使用，但费用仍然昂贵。

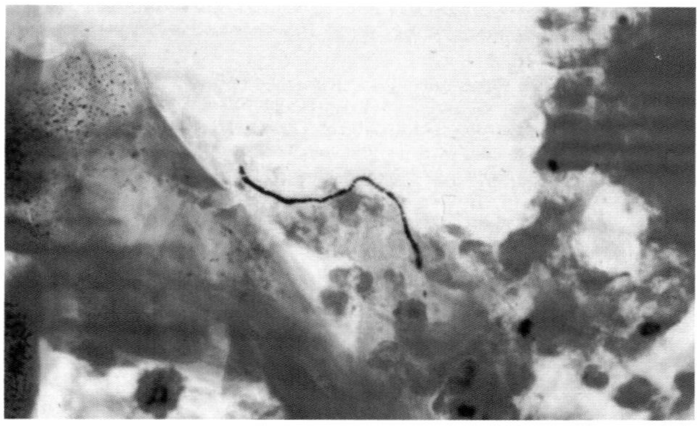

图126　用单峰驼疤痕涂片后染色镜检，刚果嗜皮菌呈钱串状



## 参考文献


[1] Abu-Samra M.T.,Imbabi S.E. & Mahgoub E.S. (1976). – Experimental infection of domesticated animals and the fowl with *Dermatophilus congolensis*. *J. Comp. Path.*,86 (2),157–172.

[2] Anon. (2007). – Alpacas dermatophilosis. *Vet. Rec.*,161 (18),610.

[3] Ajogi I.,Durosilorun A.,Junaidu A. & Tahir F.A. (2002). – Natural (clinical) dermatophilosis in the one-humped camels (*Camelus dromedarius*) slaughtered in Sokoto abattoir,Nigeria. *Bull. Anim Health Prod. Afr.*,50 (1),15–19.

[4] Bakhiet A.O.,Al-Kanzee A.G.,Hassan A.B.,Yagoub S.O. & Mohammed G.E. (2012). – Studies on pathological changes of contagious skin necrosis (CSN) in camels (*Camelus dromedarius*) in Hail region,Kingdom of Saudi Arabia. *In* Proceedings of the 3rd Conference of the International Society of Camelid Research and Development,29 January–1 February 2012,Muscat,Oman,325–327.

[5] Bornstein S. (1995). – Skin diseases of camels. In Camel keeping in Kenya (J.O. Evans,S.P. Simpkin & D.J. Atkins,eds). Range Management Handbook of Kenya. Ministry of Agriculture,Livestock Development,and Marketing,Kenya,3 (8),7–13.

[6] Bucek J.,Pospisil L.,Moster M. & Shalka B. (1992). – Experimental dermatophilosis. *J. Vet. Med. B*,39,495–502.

[7] Burd E.M.,Juzych L.A.,Rudrik J.T. & Habib F. (2007). – Pustular dermatitis caused by *Dermatophilus congolensis*. *J. Clin. Microbiol.*,45 (5),1655–1658.

1 细菌病

[8] Cross H.E. (1917). – The camel and its diseases. Baillière,Tindall and Cox,London.

[9] Curasson G. (1918). – Une maladie du dromadaire analogue au farcin du boeuf. *Bull. Soc. Cent. Méd. Vét. (Supplement to Rec. Med. Vet.* 94),71,491–496.

[10] Curasson G. (1920). – Hygiene et maladies du dromadaire en Afrique occidentale francaise Gorée (SEN),Imprimerie du gouvernement general.

[11] Curasson G. (1936). – Traite de pathologie exotique vétérinaire et comparee. Vigot Freres,Paris,II.

[12] Curasson G. (1947). – Le chameau et ses maladies. Vigot Frères Editions,Paris,86–88.

[13] Domenech J.,Guidot T.G. & Richard D. (1977). – Les maladies pyogenes du dromadaire en Ethiopie. Symptomatologie – Etiologie. *Rev. Elev. Méd. Vét. Pays Trop.*,30 (3),251–258.

[14] Edelsten R.M. & Pegram R.G. (1974). – Contagious skin necrosis of Somali camels associated with *Streptococcus agalactiae. Trop. Anim. Health. Prod.*,6,255–256.

[15] Fazil M.A. & Hofmann R.R. (1981). – Haltung und Krankheiten des Kamels. *Tierärztl. Praxis*,9,389–402.

[16] Gitao C.G. (1992). – Dermatophilosis in camels (*Camelus dromedarius* Linnaeus,1758) in Kenya. *Rev. sci. tech. Off. int. Epiz.*,11 (1–2),309–311.

[17] Gitao C.G. (1993a). – The epidemiology and control of camel dermatophilosis. *Rev. Elev. Méd. Vét. Pays Trop.*,46 (1–2), 309–311.

[18] Gitao C.G. (1993b). – An enzyme-linked immunosorbent assay for the epidemiological survey of *Dermatophilus congolensis* infection in camels (*Camelus dromedarius*). In Biotechnology applied to the diagnosis of animal diseases. *Rev. sci. tech. Off. int. Epiz.*,12 (2),639–645.

[19] Gitao C.G.,Agab H. & Khalafalla A.J. (1998a). – An outbreak of a mixed infection of *Dermatophilus congolensis* and *Microsporum gypseum* in camels (*Camelus dromedarius*) in Saudi Arabia. *Rev. sci. tech. Off. int. Epiz.*,17 (3),749–755.

[20] Gitao C.G.,Agab H. & Khalafalla A.J. (1998b). – Outbreaks of *Dermatophilus congolensis* infection in camels (*Camelus dromedarius*) from the Butana region in eastern Sudan. *Rev. sci. tech. Off. int. Epiz.*,17 (3),743–748.

[21] Gitao C.G.,Evans J.O. & Atkins D.J. (1990). – Natural *Dermatophilosis congolensis* infection in camels (*Camelus dromedarius*) from Kenya. *J. Comp. Path.*,103,307–312.

[22] Joseph S.,Wernery U. & Ali M. (1998). – Dermatophilosis caused by a nonhaemolytic *Dermatophilus congolensis* strain in dromedary camels in the United Arab Emirates. *J. Camel Pract. Res.*,5 (2),247–248.

[23] Khodakaram-Tafti A.,Khordadmehr M. & Ardiyan M. (2012). – Prevalence and pathology of dermatophilosis in camels. *Trop. Anim. Health Prod.*,44,145–148.

[24] Leese A.S. (1927). – A treatise on the one-humped camel in health and disease. Vigot Freres,Paris II.

[25] Losos G.J. (1986). – Infectious tropical diseases of domestic animals. The Bath Press,Avon,UK.

[26] Mahgoub O.,Tageldin M.H.,Nageeb A.,Al-Lawatia S.A.,Al-Bausaidi M.H.,Al-Abri A.S. & Johnson E.H. (2012). – An outbreak of severe dermatophylosis in young Omani camels. In Proceedings of the 3rd Conference of the International Society of Camelid Research and Development,29 January–1 February 2012,Muscat,Oman,325–327.

[27] Makinde A.A.,Majiyagbe K.A.,Chwzey N.,Lombin L.H.,Shamaki D.,Muhammad L.U.,Chima J.C. & Garba,A. (2001).– Serological appraisal of economic diseases of livestock in the one-humped camel (*Camelus dromedarius*) in Nigeria. *Camel Newslett.*,18,62–73.

[28] Mason F.E. (1919). – Pseudo-actinomycosis or streptotrichosis in the camel. *J. Comp. Path. Ther.*,32 (1),34–42.

[29] Pascoe R.R. (1990). – A colour atlas of equine dermatology. Wolfe Publishing Ltd.,London,34–38.

[30] Peck E.F. (1938a). – Notes relating to the camel. *Vet. Rec.*,33 (50),1052–1054.

[31] Peck E.F. (1938b). – The relationship of salt starvation to contagious necrosis and lameness in camels. *Vet. Rec.*,14 (50),409–410.

[32] Peck E.F. (1939). – Salt intake in relation to cutaneous necrosis and arthritis of one-humped camels (*Camelus dromedarius*,L.) in British Somaliland. *Vet. Rec.*,46 (51),1355–1360.

[33] Samuel T.,Tareke F.,Wirtu G. & Kiros T. (1998). – Bacteriological study of Ethiopian isolates of *Dermatophilosis congolensis.*

*Trop. Anim. Health. Prod.*,30,145–147.

[34] Schwartz S.,Schwartz H.J. & Wilson A.J. (1982). – Eine fotographische Dokumentation wichtiger Kamelkrankheiten in Kenia. *Der prakt. Tierarzt*,11,985–989.

[35] Seifert H.S.H. (1992). – Tropentierhygiene. Gustav Fischer Verlag,Jena,Stuttgart,Germany,289–291.

[36] Thedford R.R. & Johnson L.W. (1989). – Infectious diseases of New-world camelids (NWC). *Vet. Clin. North Am. Food Anim. Pract.*,5 (3),145–157.

[37] Wardeh M.F. (1989). – Camel production in the Islamic Republic of Mauritania. *Camel Newslett.*,5,11–17.

[38] Wernery U. & Ali M. (1990). – Dermatophilose in Renndromedaren – Fallbericht. *Tierärztl. Umschau*,45 (3),209–210.

[39] Wilson R.T. (1984). – The camel. Longman,London and New York.

[40] Zanolari P.,Meylan M.,Sager H.,Herrli-Gygi M.,Rufenacht S. & Roosje P. (2008). – Dermatologie bei Neuweltkameliden. Teil 2: Uebersicht der dermatologischen Erkankungen. *Tieräztl. Praxis. Ausgabe G,Grosstiere/Nutztiere*,36 (6),421–427.

（李学瑞译，殷宏校）

# 1.6 乳房

## 1.6.1 引言

在世界上的干旱地区，严重的持续干旱使得牛、绵羊和山羊的数量大幅下降，只有骆驼能够存活下来并不断提供乳品。脱水骆驼不同寻常的特征之一就是它们能够持续不断地泌乳，但是，乳汁是高度稀释的，含水量在90%以上[87,88]。其他反刍动物为了降低体温，用于泌乳的贮备水通过粪便和尿液失去。牛、绵羊和山羊缺乏饮水会导致泌乳停止或分泌出高脂、低水分的奶。骆驼在不饮水的情况下至少能够连续10天，每天分泌20L奶。因此，泌乳骆驼能够为其后代和人类提供丰富的食物和营养来源。然而，必须要通过按摩乳房和幼驼的吮吸刺激才能使骆驼排乳。每次排乳过程很短，因此，挤奶时必须要尽可能的快。牧民都熟悉这种情况，因此，他们挤奶时要两边的乳房同时进行（图 127）。

图127　两名哈萨克斯坦挤奶工站在单峰驼的两侧同时挤奶

据估计，特别优秀的产奶骆驼每天可以产出20～30L的奶，但是每天必须挤奶3～4次，每次挤奶时动作要快（每侧有一个挤奶工），而且，要将幼驼与母驼同群饲养才能达到这一产奶水平。泌乳期可以持续两年。这些数值还有争议，不一定可靠，因为，影响骆驼产奶量的因素还有很多，如畜群的管理水平、营养供给、品种、年龄、季节、分娩次数等。而且，有时候研究人员将幼驼吃掉的奶也计算在内，估计幼驼吃掉的奶有可能达到总产奶量的40%～75%。

在印度、巴基斯坦和中东地区，单峰驼的产奶量可以达到每天20L以上，泌乳期可达8～18个月[15]；与奶牛场的奶牛相比，骆驼的产奶量明显较低。然而，与生活在同一环境的瘤牛（Zebu cattle）相比，单峰驼的产奶量有明显优势。在西方国家，游牧民食用骆驼奶已有数个世纪，由于其营养与医疗价值，现在又重新得到了人们的重视。直到现在，骆驼奶仍然是许多发展中国家人们生存的重要食物来源。例如，在索马里，单峰驼的饲养量达500万头，几乎全是用于产奶。在大部分非洲国家不太可能有可持续的农作物生产，因此，移动式的畜牧业生产就成为土地利用的唯一形式，也是保障人们生计与收入来源的唯一方式。骆驼尤其适合于这种生活方式。它们即使在长期缺水的情况下，仍然能够为人们生产大量的驼奶，在炎热干旱地区，其他家畜是做不到的。

总体来讲，驼奶相比于牛奶更有营养，因为骆奶中的脂肪和胆固醇含量较低，而钾、铁、钠和镁等矿物质元素含量更高；然而，Konuspayeva等[51,54]分析发现，关于驼奶营养成分的研究数据由于地域和年份的不同而变化较大。对驼奶中维生素C含量的研究也有相似结果[52]，但是，总体而言，驼奶中维生素C和烟酸的含量是很高的。驼奶中蛋白质的种类不同于牛奶，这或许可以解释为什么食用驼奶很少出现蛋白质过敏症。尽管驼奶与牛奶中乳糖的含量接近，但是，食用驼奶出现乳糖不耐受症的情况还未见报道。驼奶中胰岛素的含量也高于牛奶，研究发现食用驼奶对糖尿病人也有好的治疗效果。其作用机理还不完全清楚，但驼奶在酸性条件下不发生凝固的事实或许可以部分解释这一现象。驼奶不受酸性环境的影响，可以直接通过人胃的酸性环境到达小肠被吸收。

与牛、羊等其他动物种类相比，驼奶中含有较大量的非饱和脂肪酸，如硬脂酸，油酸和亚油酸，这对于血液中的胆固醇水平有积极的影响。也有一些研究数据显示，驼奶对多种不同的疾病有治疗效果，如自身免疫性疾病（克罗恩氏病、牛皮癣、多发性硬化症）、过敏性疾病（哮喘）、糖尿病、胃溃疡、皮肤癌和肺结核。驼奶似乎对免疫系统功能也具有改善作用。2000年，在哈萨克斯坦阿拉木图举办了一次国际会议[16]研讨这一重要主题，而对于驼奶的免疫增强作用，Shabo等学者也进行了深入的研究[70]。驼奶在食用之前，必须进行巴氏消毒处理，以防止一些危险的动物源性疫病，例如，布鲁氏菌病、结核病、副结核病和Q热等传染给人。巴氏消毒的程序与牛奶的相同。然而，到现在为止，与牛奶类似的驼奶超高温瞬时处理（UHT）技术仍然没有建立[34]。巴氏消毒奶（75℃ 15秒）在冷藏条件下至少能够存贮15天而不变质。Wernery等[79]的一项研究证明驼奶中的许多成分比牛奶中的相应成分具有更好的热稳定性。尤其是维生素C，驼奶中的含量是牛奶中的4～6倍。

检测骆驼乳腺健康状况的标准参数与奶牛的一样，奶牛乳腺健康标准已经应用了几十年。这些参数包括体细胞计数（SSC），加利福尼亚乳腺炎检测（CMT）和其他的一些指标测定。骆驼乳腺健康的微生物学标准与奶牛的一样，包括乳房炎病原分离，碱性磷酸酶（ALP）活性测试。牛奶中的ALP在72℃会被破坏，但不能用这一指标来确认驼奶的热灭活是否恰当。骆奶中已经发现了新的酶标志物。必须强调的是骆奶中已经发现有一种酶可以准确显示巴氏消毒是否恰当，这是未来骆奶及其产品能够出口到国外市场的先决条件。已经有一些科学家研究了驼奶的特性并用于开发产品[9,32,33,35,43,86]。因为驼奶不容易凝固，利用其生产硬奶酪比较困难，但是，位于阿拉伯联合酋长国的中央兽医实验室从幼驼的肠道内膜分离出了一种骆驼凝乳酶[46]，现在用于生产硬奶酪。软奶酪与半硬奶酪也已经研制成功[50,67]。还可以利用驼奶生产高质量的奶粉，利用骆驼奶粉又可以生产出优质的驼奶巧克力[85]。

已经有科学证据表明，单峰驼对口蹄疫有先天的抵抗力，因此，未来全球市场对骆驼奶和其产品的需求会进一步扩大。上述这些特点以及其他的一些性质使得骆驼相对于其他家养产奶反刍动物具有更多优势。然而，在世界上的多数地区，对骆驼科动物仍然存在歧视和误解。

过去十年，位于阿拉伯联合酋长国迪拜的中央兽医实验室已经成为驼奶的研究中心之一，2007年，在这里建成了第一个全自动化操作的驼奶场（图128a,b）。最近，Nagy等[61]首次证实利用单峰驼可以实现驼奶的可持续生产，在传统的粗放式的农业生产中，单峰驼就是专门用于产奶。在大型的驼奶场能够生产出高质量的原料奶。

在过去十年，中央兽医实验室有许多关于驼奶的著述[12,27,41,71,76,82]。然而，不光只有在迪拜的科学家从事驼奶研究，联合国粮农组织（FAO）也出版了一本用单峰驼奶制造奶酪的技术手册[67]，1994年，在毛里塔尼亚首都努瓦克肖特举办一个重要的会议，内容涵盖了驼奶生产的各个方面[24]。数位科学家给位于卡塔尔多哈的联合国教科文组织提出了通过建设现代驼奶场来应对沙漠化的重要建议，这些驼奶场要用当地的一些沙漠植物和盐生植物生产草料[25]。Strasser等[72]的文章则关注驼奶中的抗生素残留。他们的研究结果显示，一些优先用于大宗牛奶样品筛查的商业化试验手段也可以用于驼奶检测。

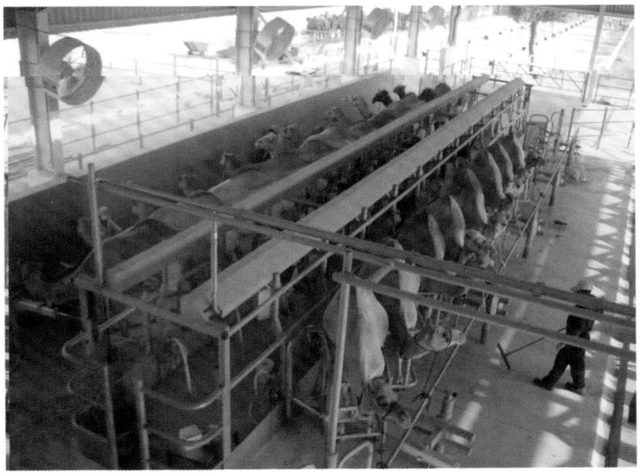

a，带交叉缝式站台的挤奶室

b，正在使用的自动挤奶机

图128　位于阿拉伯联合酋长国的现代化驼奶场

1 细菌病

## 1.6.2 传染性乳房炎

乳房炎是奶牛的一种最重要的疫病，造成严重的经济损失，由于抗生素的使用也导致乳制品中的药物残留，因而受到广泛的研究和关注。筛选出的高产奶性能的奶牛对乳房炎的易感性明显高，这种情况在骆驼还没有发生。这可能是骆驼乳房炎发生率低于其他家养动物的原因之一[6,36,57,66,83]，这或许也是有关骆驼乳房炎的研究报道较少的原因。可能还有其他的一些原因可以解释骆驼比其他乳用家畜较少发生乳房炎。新大陆（OWCs）与旧大陆骆驼（NWCs）的乳房都有四个分区，每个区有一个乳头。奶牛的每个乳头有两个乳导管通入输乳管窦和乳头（图129a,b），但是，对骆驼乳房的结构研究显示乳导管的末端并没有形成像奶牛的窦状结构，而是形成一种海绵状结构，也有人称之为“公共囊”状结构，但是，这些表述都没有正确描述出骆驼乳房的腺体结构（图130）。

在某些情况下，乳区也可能有3个乳导管。每个乳头与2个不相通的腺体相连。乳导管通常很狭窄，直径约1 mm的导管才能通过。这种解剖学的双导管结构和狭窄的乳导管在某种程度上可以防止感染。产奶骆驼经常要装备乳房罩以限制哺乳。乳房罩可以减少对乳头和乳房的损伤，防止乳房污染，并可以杜绝蚊蝇的侵扰。然而，骆驼乳房发生感染的概率较少，更合理的解释可能在于驼奶本身。科学家研究发现驼奶中含有的一些物质可以阻止病原菌的生长[21,28,29,33,46,55]。这些抑菌物质是一些蛋白质成分，例如，溶菌酶、免疫球蛋白、乳铁蛋白和乳过氧化物酶，对这些蛋白的性质已经有了很好的认识。研究证明，驼奶中这些蛋白的浓度或活性均高于牛奶[30,53]。Kappeler[46]发现了一种新的小乳清蛋白，它是一种肽聚糖识别蛋白（PGRP），有助于新生幼驼建立肠道有益微生物种群，尤其能阻止革兰氏阳性菌的生长。科学家对不同国家、不同挤奶方式（手工和机器）的骆驼原料奶细菌学质量进行了评估（表52），也有一些研究人员对反映乳房健康的一些驼奶指标进行了评估[78,81,82,85,90]。

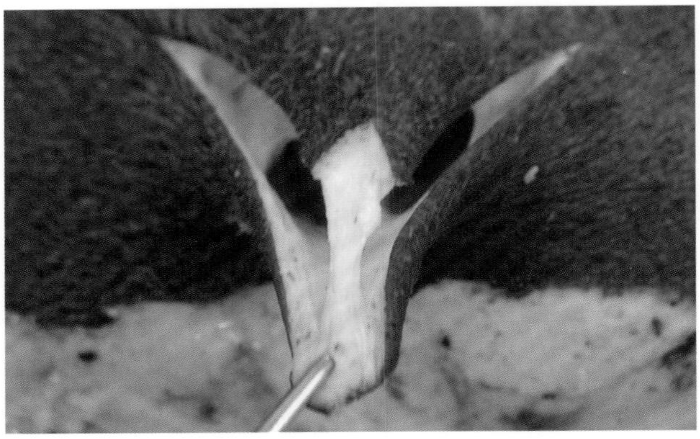

（a）

（b）

图129　每个乳头有2（a）或3（b）个乳导管. a，2个乳导管乳头；b，3个乳导管乳头

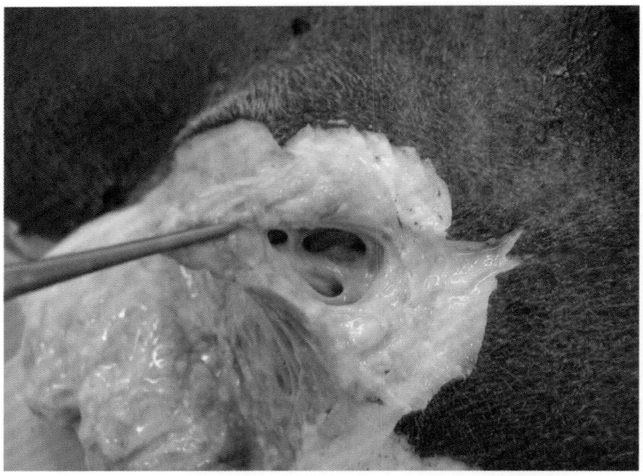

图130　骆驼乳房没有输乳管窦，而是具有一种海绵状结构

### 病原学

据报道，旧大陆和新大陆骆驼均可以发生乳房炎，但在新大陆骆驼发生的情况很少[36]。骆驼乳房炎发病过程与奶牛的相同，特急性的、急性的、慢性的和亚临床型乳房炎均有报道。要依据表53所列的检测指标做出乳房炎的诊断。

表52　单峰驼原料奶的卫生质量检测

| 作者 | 年份 | 国家 | 驼奶样本评估结果 |
|---|---|---|---|
| Semereab，Molla [69] | 2001 | 埃塞俄比亚 | TPC=$4\times10^4\sim4\times10^5$ cfu/mL，57%≤10 cfu/mL大肠菌 |
| Wernery et al [80] | 2002 | 阿拉伯联合酋长国 | TPC 95%≤100000 cfu/mL |
| Aly，Abo-Al-Yazeed [17] | 2003 | 埃及 | 分离到了较多数量的不同霉菌种 |
| Khedid et al [48] | 2003 | 摩洛哥 | $5\times10^4$ cfu/mL TPC，$3\times10^3$ cfu/mL大肠菌 |
| El-Ziney，Al-Turki [31] | 2006 | 沙特阿拉伯 | 24%含有沙门氏菌，46%含有粪大肠菌 |
| Eberlein [27] | 2007 | 阿拉伯联合酋长国 | TPC=$1.1\times10^2$ cfu/mL手挤奶，TPC=$4.2\times10^2$ cfu/mL机挤奶 |
| Hadush et al [39] | 2008 | 埃塞俄比亚 | 质量好的占94% |
| Al Mutery et al [10] | 2008 | 沙特阿拉伯 | $1.6\times10^3$ cfu/mL TPC |
| Akweya et al [8] | 2010 | 肯尼亚 | 23%的样本含有金黄色葡萄球菌和无乳链球菌 |
| Wanjohi et al [75] | 2010 | 肯尼亚 | 75% 不干净，8.9%呈黄色，主要有葡萄球菌和链球菌污染 |
| Dasel et al [26] | 2010 | 肯尼亚 | TPC≤$10^5$ cfu/mL |
| Halbrock [41] | 2010 | 阿拉伯联合酋长国 | TPC=$4.9\times10^3$ cfu/mL机挤奶，TPC=$2.2\times10^3$ cfu/mL手挤奶 |
| Al-Mohizea [14] | 1986 | 沙特阿拉伯 | TPC=$2.09\times10^2$ cfu/mL大肠菌 |
| Wernery et al [83] | 2006 | 阿拉伯联合酋长国 | 86%的机挤奶正常 |
| Tourette et al [74] | 2002 | 毛利塔尼亚 | TPC=$1.65\times10^6$，大肠菌=$3.55\times10^4$ cfu/mL |

TPC，**总平板菌落计数**

表53　乳房炎的诊断方法及判断[18]

| 隐性乳房炎试验（CMT） | 体细胞数（cells/mL） | 乳房炎病原微生物 | |
|---|---|---|---|
| | | - | + |
| -/+ | <100000 | 健康 | 亚临床乳房炎 |
| ++ 或 +++ | >100000 | 无特定病原临床乳房炎 | 特定病原临床乳房炎 |

两种形式的乳房炎有明确的界定。对于临床型乳房炎，奶外观有明显改变，整个乳房或某个乳区有变化，这种变化在亚临床乳房炎是看不到的。很多国家都有关于骆驼乳房炎的文献报道，包括埃及、印度、沙特阿拉伯、索马里、苏旦、阿拉伯联合酋长国、埃塞俄比亚和哈萨克斯坦（表54）。

骆驼的特急性[47]、亚急性[64]和坏疽性乳房炎[45]常见有淋巴结肿大[23]。急性病例的乳房分泌物呈水样，微黄色或带血丝[73]。Kinne和Wernery[49]描述了两例单峰驼急性乳房炎病例，病原是溶血性巴氏德菌（*Pasteurella haemolytica*）和无乳链球菌（*Streptococcus agalactiae*）。病例1的乳房炎是初始病症，而单峰驼最终死于继发性梭菌性肠毒血症。骆驼出现前乳区硬化，两个“乳导管窦”被血液填塞（图131）。组织学检查显示大面积急性出血，导致化脓性乳房炎并伴有小叶间水肿。

病例2急性乳房炎分离到了无乳链球菌（图132）。骆驼的左后乳区疼痛、发热和肿大，乳液呈水样，颜色呈淡红黄色。CMT检测呈阳性（图133）。

Ramadan等[66]报道了3例单峰驼单侧慢性乳房炎病例，因为角蛋白聚集堵塞了乳导管；这种堵塞导致奶产量下降，受影响的乳区膨大，有2

表54 不同国家关于单峰驼乳房炎的研究情况

| 作者 | 年份 | 国家 | 乳房炎类型及症状 | 细菌分离 |
|---|---|---|---|---|
| Mostafa et al [59] | 1987 | 埃及 | 亚临床型 | 产气荚膜梭状芽孢杆菌，金黄色葡萄球菌 |
| Kapur et al [47] | 1982 | 印度 | 特急性炎症 | 肺炎克雷伯氏菌，大肠杆菌 |
| Barbour et al [22] | 1985 | 沙特 | 急性，水肿 | 微球菌，金黄色葡萄球菌，链球菌 |
| Hafez et al [40] | 1987 | 沙特 | 急性，慢性，坏疽性 | 金黄色葡萄球菌，链球菌，巴氏杆菌 |
| Arush et al [19] | 1984 | 索马里 | 临床型 | 链球菌 |
| Abdurahman et al [4] | 1991 | | 临床型 | |
| Obied [62] | 1983 | 苏丹 | 临床型 | 金黄色葡萄球菌，凝固酶阴性；葡萄球菌，无乳链球菌 |
| Quandil，Oudar [64] | 1984 | 阿拉伯联合酋长国 | 亚急性，临床型 | 链球菌，葡萄球菌 |
| Younan [89] | 2002 | 肯尼亚 | 无症状 | 无乳链球菌 |
| Younan et al [90] | 2001 | 肯尼亚 | 多数无症状 | 无乳链球菌，金黄色葡萄球菌 |
| Wernery et al [81] | 2008a | 阿拉伯联合酋长国 | 临床型 | 无乳链球菌，凝固酶阴性；葡萄球菌，金黄色葡萄球菌 |
| Kinne,Wernery [49] | 2002 | 阿拉伯联合酋长国 | 急性 | 溶血性巴氏德菌，无乳链球菌 |
| Almaw,Molla [13] | 2000 | 埃塞俄比亚 | 临床型 | 金黄色葡萄球菌，凝固酶阴性；葡萄球菌，无乳链球菌，链球菌属 |
| Guya [38] | 2006 | 埃塞俄比亚 | 急性 | 金黄色葡萄球菌，溶血性巴氏德菌 |
| Hamouda,Doghaim [42] | 2006 | 埃及 | 急性化脓性和慢性 | 金黄色葡萄球菌，大肠菌，链球菌 |
| Jhirwal et al [45] | 2006 | 印度 | 坏疽性 | 手术 |
| Ahmad et al [7] | 2010 | 巴基斯坦 | 临床和亚临床型 | |
| Bakhiet et al [20] | 1992 | 苏丹 | 亚临床型 | 金黄色葡萄球菌，链球菌 |
| Guliye [37] | 2002 | 以色列 | 亚临床型 | 金黄色葡萄球菌，微球菌，芽孢杆菌，链球菌 |
| Al-Ani,Al-Shareef [11] | 1997 | 伊拉克 | 38%临床型<br>慢性<br>亚临床型 | 金黄色葡萄球菌，化脓棒状杆菌<br>表皮葡萄球菌，链球菌属，溶血性巴氏德菌，大肠杆菌，微球菌 |

注：本表数据来自于文献。

个病例出现了溶血性巴氏杆菌和金黄色葡萄球菌（*Staphylococcus aureus*）的继发感染；第3个病例的奶样是无菌的。Barbour等[22]利用CMT法检测了205份来自沙特阿拉伯的单峰驼奶样，结果显示，大部分发生细菌性乳房炎的单峰驼奶样中的体细胞数升高。与奶牛类似，单峰驼奶样中的体细胞数与乳房炎之间有相关性。Abdurahman等[2,3,5]对东苏丹101头单峰驼的391个乳区采集了391份奶样，分别进行了CMT、体细胞计数和三磷酸腺苷（ATP）试验以检测亚临床型乳房炎，结果也得到了类似的结论。来自致病菌感染乳区的奶样，3种方法检测结果的平均值总体较高，但是，有相当数量的乳区奶样未分离到致病菌，但3种方法检测值也有升高，这一结果说明，亚临床型乳房炎的发生率似乎比人们通常认为的要高。Bakhiet等[20]检测了来自苏丹的49分健康单峰驼奶样，从45%（22/49）的乳样中检出了细菌。其中，葡萄球菌属和链球菌属细菌分离到的频率最高。Guliye[37]研究了内盖夫

图131　骆驼乳房切面，显示急性出血性乳房炎伴有软组织血液凝固

（巴勒斯坦南部地区）沙漠地区单峰驼的亚临床型乳房炎，共检测了临床健康骆驼的奶样86份，有81%的样本细菌检测阳性，其中，有40.7%的样本检出两种以上的细菌，金黄色葡萄球菌、微球菌属（*Micrococcus spp*）、芽孢杆菌属（*Bacillus spp*），停乳链球菌停乳亚种（*S. dysgalactiae*）和大肠杆菌（*Escherichia coli*）是分离到的最重要细菌。体细胞计数值介于$1.0 \times 10^5 \sim 11.8 \times 10^6$细胞/mL。细菌检测阳性的乳区样本平均体细胞计数值显著升高，而且检出金黄色葡萄球菌的乳区样本的平均体细胞计数值最高。Abdurahman[1]的研究结果显示，双峰驼也存在类似的亚临床型乳房炎，共检测了7头健康双峰驼的160份奶样，有22.5%的样本菌检阳性，金黄色葡萄球菌与凝固酶阴性的葡萄球菌是主要的菌种。分离到球菌的乳区体细胞计数与CMT值显著升高。体细胞计数与CMT值都是评估骆驼乳房感染状况的指标。Mody等[58]通过CMT与细菌分离培养的方法研究了146个印度成年单峰驼乳房炎发病程度，结果发现了30个亚临床乳房炎病例，其中，28个病例乳房感染了葡萄球菌属、链球菌属和棒状杆菌等病菌。作者也检测了病原菌的抗药性，发现对庆大霉素和氯霉素高度敏感。最近在沙特阿拉伯进行的一项研究，从30个临床健康的单峰驼采集了120个乳区奶样，采用体细胞计数与CMT进行筛查，阳性样本再进行细菌学检测，结果发现革兰氏阳性球菌是优势乳房病原菌，平均体细胞计数值为$12.5 \times 10^4/mm^3$。四环素对所有种类的细菌仍然有效，但大多对青霉素有抗性，对庆大霉素、硫酸黏杆菌素和磺胺类药的抗性较低[68]。

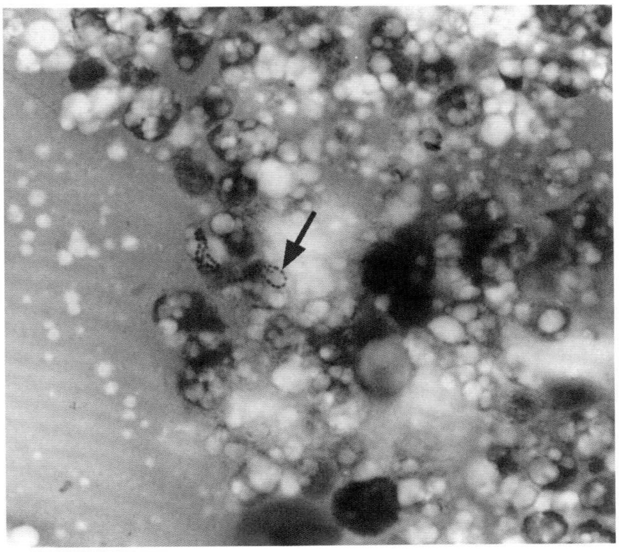

图132　呈链状的无乳链球菌，周围可见巨噬细胞和中性粒细胞

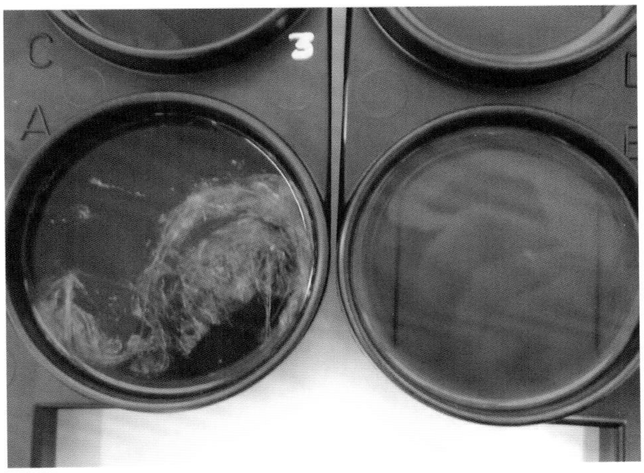


图133　单峰驼乳房炎病例阳性CMT反应（左侧杯）

乳房上可以检出常见的乳房炎病原菌，这些菌可能来自于具有传染性的带菌动物，动物生活的周围环境，如，垫草、粪便和土壤中均有可能存在致病菌。牛的传染性乳房炎病原有金黄色葡萄球菌和无乳链球菌，在挤奶过程中，可以从感染牛传染给未感染牛的乳房，污染的乳罩垫、擦拭用的纸巾、挤奶工的手和蚊蝇等都有可能成为传播媒介。但是，环境中的乳房炎病原主要为除了无乳链球菌之外的其他链球菌，例如，乳房链球菌（*Streptococcus uberis*），停乳链球菌和大肠杆菌、克雷伯氏杆菌（*Klebsiella*）等大肠菌。关于骆驼乳房炎的主要病原菌仍然存在不同的观

点。Barbour等[22]认为微球菌是重要的乳房炎致病菌，而Obied[62]却不认为它是致病菌。Obied等[63]发现链球菌、葡萄球菌、微球菌、产气杆菌（*Aerobacter*）和大肠杆菌是引起乳房炎的主要病原菌；作者没有发现体细胞计数值与乳房感染之间有关系。Al-Ani和Al-Shareefi[11]发现来自伊拉克3个不同群体38%的泌乳期骆驼存在乳房炎。金黄色葡萄球菌和化脓棒状杆菌（*Corynebacterium pyogenes*）是引起慢性乳房炎的主要病原菌，而表皮葡萄球菌（*S. epidermidis*）、链球菌属、溶血性巴氏杆菌、大肠杆菌和微球菌属菌则与亚临床型乳房炎有关。对霉菌性乳房炎仍然知之甚少。Al-Ani和Al-Shareefi[11]从50份奶样中没有分离到任何霉菌，但Quandil和Qudar[64]从亚临床型乳房炎奶样中分离到了白念珠菌（*Candida albicans*），而且Aly和Abo-Al-Alyazeed[17]从来自埃及的感官指标正常的骆驼原料奶样本中分离到了曲霉菌（*Aspergillus*）、青霉菌（*Penicillium*）、支顶孢霉菌（*Acremonium*）、交链孢霉菌（*Alternaria*）和金孢子菌（*Chrysosporum*）。从骆驼乳房炎奶样中分离到频率最多的细菌是金黄色葡萄球菌、溶血性巴氏杆菌和链球菌。很多研究者相信这些细菌是骆驼乳房炎的主要的致病菌[22,40,66]。据认为下列细菌在乳房炎发病过程中居次要地位：

（1）微球菌属（*Micrococcus* spp）；

（2）放线菌属（*Actinomyces* spp）；

（3）大肠杆菌（*E. coli*）；

（4）绿脓假单胞菌（*Pseudomonas-aeruginosa*）；

（5）肺炎克雷伯氏菌（*Klebsiella pneumoniae*）；

（6）拟杆菌属（*Bacteroides* spp）；

（7）产气荚膜梭状芽孢杆菌（*Clostridium perfringens*）。

这些细菌分离自患乳房炎骆驼奶样的单纯和混合培养物[22,44,47,59,64]。最近，Wernery等[81]对阿拉伯联合酋长国机挤单峰驼奶样进行的一项研究，清楚地显示单峰驼乳房炎病原与奶牛乳房培养物分离菌一致，这些菌为：

（1）无乳链球菌（*S. agalactiae*）（到目前为止为主要的乳房炎病原）。

（2）金黄色葡萄球菌（*S. aureus*）。

（3）凝固酶阴性的葡萄球菌（Coagulase-negative *Staphylococcus*，CNS）。

（4）牛链球菌（*Streptococcus bovis*）。

（5）乳房炎链球菌（*S. uberis*）。

（6）停乳链球菌停乳亚种（*S. dysgalactiae*）。

Younan等[89,90]也有相似的研究结果，美丽花蜱（*Amblyomma lepidum*）和嗜驼璃眼蜱（*Hyalomma dromedarii*）、以及防吮乳设备[6]和骆驼痘[90]，都在细菌性乳房炎传播和恶化过程中发挥了作用。

关于双峰驼乳房炎的研究报道很少，可得到的文献也没有描述清楚研究对象是双峰驼还是单峰驼。同样的关于驼奶成分、产奶量、机器挤奶和泌乳反应等的一些研究文献也没有区分骆驼种类。感兴趣的读者可以阅读1.6节结尾列出的继续阅读文献。

重要的是在新鲜骆驼奶中也分离到了布鲁氏菌[65,84]，一些人布鲁氏菌病例是由于食用了骆驼原料奶所致[60]（见布鲁氏菌病部分）。引起新大陆骆驼乳房炎的具体病原菌还没有报道。从特急性的新大陆骆驼乳房炎病例分离到了几种细菌，包括大肠杆菌、肺炎克雷伯氏菌和肠杆菌属某些种。到目前为止，从新大陆骆驼[36]和旧大陆骆驼的乳房样本中都没有分离到支原体。

## 病理学

对于乳房感染过程中发生的病理变化知之甚少。感染乳房通常肿大，变硬、发红，触诊有疼痛反应。在慢性乳房炎病例还可见乳房坏死和脓肿，并见流出绿色的脓汁。Al-Ani和Al-Shareefi[11]描述了一例坏死性乳房炎的组织病理学损伤，病原菌是化脓棒状杆菌。在临床型乳房炎，乳房分泌物呈水样并有出血，含有絮状物和凝块，呈纤维化或变黏稠，颜色和气味也发生改变。在特急性和急性乳房炎，母畜可能会厌食、发热，进而发展为血毒症和败血症，最后死亡。

## 治疗

发生乳房炎后，要采取及时的治疗措施，以防止乳房的损伤和动物的死亡。乳房炎治疗应该依据细菌分离培养和药敏试验的结果进行，而且，治疗人员必须对骆驼乳房的特殊解剖学结构有清醒的认识。应该用直径1mm的导乳管疏通条纹状的乳导管，以防止损伤乳导管。在注入治疗药物之前，应清洁和酒精擦拭消毒乳头，应清空感染的乳区或乳房。对严重的病例，每天要轻柔的按摩乳房3～5次，以清除乳腺中的分泌物。当乳房有痛感时，这种操作可能比较困难，这时候需要围栏保定骆驼，并用绳索固定后腿。乳房炎的注入药品包括青霉素、邻氯青霉素和新霉素，按照制造商的操作说明进行灌注。治疗后一定要严格执行休药期，这一点也很重要。特急性乳房炎或某些急性乳房炎需要进行非肠道给药治疗（静脉注射或肌内注射），同时，进行乳房局部注入给药。南美洲骆驼乳房炎的治疗应遵循Fowler[36]的建议。Younan[89]采用非肠道给药方式治疗了9头肯尼亚单峰驼的无乳链球菌性乳房炎，疗程为3天，每天肌内注射1200万IU（12g）的苄青霉素普鲁卡因双氢链霉素和1000万IU喷沙西林（青霉素G乙胺酯）。首次治疗能够消除了3/4感染乳区的无乳链球菌，二次治疗能够消除2/4感染乳区的无乳链球菌。

乳房卫生护理对于预防乳房炎至关重要。定期观察和触诊乳房可以早期发现乳房炎症，即使是轻微的肿大或颜色变化也容易观察到，可以应用乳房香膏温柔的按摩病区。挤奶前必须对乳房、特别是乳头进行清洁消毒处理（图134）。

为了避免乳房感染，挤奶后对奶头进行药浴是另一个非常必要的措施，操作时要戴保护手套。挤奶后乳管仍然是开放的，药浴则可以预防细菌进入乳导管。

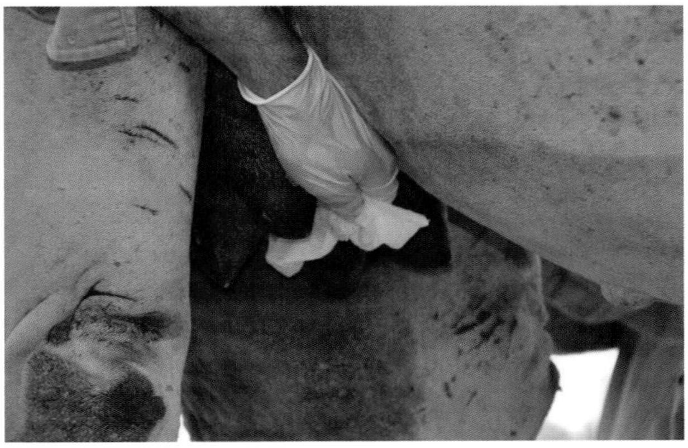

图134 佩戴保护手套并用柔软的纸巾消毒单峰驼的乳头

1 细菌病

## 参考文献


[1] Abdurahman O.A.S. (1996). – The detection of subclinical mastitis in the Bactrian camel (*Camelus bactrianus*) by somatic cell count and California mastitis test. *Vet. Res. Comm.*,20 (1),9－14.

[2] Abdurahman O.A.S.,Agab H. & Abbas B. (1994). – Mastitis in camels (*Camelus dromedarius*) in the Sudan. Relationship between udder infection and inflammatory parameters in Sudanese camels (*Camelus dromedarius*). *Br. Vet. J.*,4,71－76.

[3] Abdurahman O.A.S.,Agab H.,Abbas B. & Astroem G. (1995). – Relations between udder infection and somatic cells in camel (*Camelus dromedarius*) milk. *Acta Vet. Scand.*,36 (4),423－431.

[4] Abdurahman O.A.S.,Bornstein S.,Osman K.S.,Abdi A.M. & Zakrisson G. (1991). – Prevalence of mastitis among camels in Southern Somalia: a Pilot Study. Somali Acad. Arts and Science,Mogadishu. *Camel Forum,Working Paper*,37,1－9.

[5] Abdurahman O.S.,Cooray R. & Bornstein S. (1992). – The ultrastructure of cells and cell fragments in mammary secretions of *Camelus bactrianus*. *J. Vet. Med.*,39,648－655.

[6] Agab H. & Abbas B. (2001). – Epidemiological studies on camel diseases in eastern Sudan. *Camel Newslett*er,18,31－43.

[7] Ahmad S.,Yaqoob M.,Bilal M.Q.,Muhammad G.,Iqbal A. & Khan M.K. (2010). – Genetic and non-genetic factors affecting the prevalence of mastitis in dromedary camels. *In* Proceedings of the International Camel Symposium,7－11 June 2010,Garissa,Kenya,20.

[8] Akweya B.A.,Gitao C.G.,& Okoth M.W. (2010). – The prevalence of common milk pathogens and antibiotic resistance of the organisms in camel milk from North Eastern province. *In* Proceedings of the International Camel Symposium,7－11 June2010,Garissa,Kenya,25.

[9] Al Haj O.A. & Al Kanhal H.A. (2010). – Compositional,technological and nutritional aspects of dromedary camel milk. *Int. Dairy J.*,20,811－821.

[10] Al Mutery A.,Wernery U. & Jose S. (2008). – Quality of raw camel milk from Saudi Arabia. *J. Camel Pract. Res.*,15 (1),61–62.

[11] Al-Ani F.K. & Al-Shareefi M.R. (1997). – Studies on mastitis in lactating one humped camels (*Camelus dromedarius*) in Iraq. *J. Camel Pract. Res.*,4 (1),47–49.

[12] Albrecht C.E.A. (2003). – Production of camel milk. First experiences in machine milking of dromedaries. *Thesis*,Gottingen,Germany,2003.

[13] Almaw G. & Molla B. (2000). – Prevalence and etiology of mastitis in camels (*Camelus dromedarius*) in eastern Ethiopia. *J. Camel Pract. Res.*,7 (1),97–100.

[14] Al-Mohizea I.S. (1986). – Microbial quality of camel's raw milk in Riyadh markets. *Egypt J. Dairy Sci.*,14,173–180.

[15] Al-Sultan S. (1996). – Veterinary care of camels in Saudi Arabia: Mastitis in camels. In Proceedings of the 3rd Meeting of the British Veterinary Camelid Society,14–16 November 1996,UK,48–50.

[16] Alwan A.A. & Tarhuni A.H. (2000). – The effect of camel milk on *Mycobacterium tuberculosis* in man. In Proceedings of the 2nd International Camelid Conference of Agroeconomics of Camelid Farming. 8–12 September 2000,Almaty,Kazakhstan.

[17] Aly S.A. & Abo-Al-Yazeed H. (2003). – Microbiological studies on camel milk in north Sinai,Egypt. *J. Camel Pract. Res.*,10 (2),173–178.

[18] Anon. (2005). – Gesunde Euter-gesunde Milch. Eutergesundheitsmassnahmen zur Mastitisbekaempfung. Intervet Brochure.

[19] Arush M.A.,Valente C.,Compagnucci M. & Hussein H. (1984). – Indagine sulla diffusione delle mastite del dromedario (*Camelus dromedarius*) in Somalia. *Bull. Sci. Fac. Zoot. Vet.*,4,99–104.

[20] Bakhiet M.R.,Agab H. & Mamoun I.E. (1992). – Camel mastitis in Western Sudan. Short communication. *Sud. J. Vet. Sci. Anim. Husb.*,31 (1),58–59.

[21] Barbour E.K., Nabbut N.H., Freriches W.M. & Al-Nakhli H.M. (1984). – Inhibition of pathogenic bacteria by camel milk: relation to whey lysozyme and stage of lactation. *J. Food Prot.*, 47 (11), 838–840.

[22] Barbour E.K., Nabbut N.H., Frerichs W.M., Al Nakhli H.M. & Al Mukayel A.A. (1985). – Mastitis in *Camelus dromedaries* in Saudi Arabia. *Trop. Anim. Health. Prod.*, 17 (3), 173–179.

[23] Bolbol A. (1982). – Mastectomy in she-camel. *Assiut. Vet. Med. J.*, 10, 215.

[24] Bonnet P. (1994). – Dromedaries and camels, milking animals. In Proceedings of the Conference on Camel Milk, 24–26 November 1994, CIRAD, Nouakchott, Mauritania, 107–109.

[25] Breulmann M., Boer B., Wernery U., Wernery R., El Shaer H., Alhadrami G., Gallacher D., Peacock J., Ali Chaudhary S., Brown G. & Norton J. (2006). – The camel: from tradition to modern times. a proposal towards combating desertification via the establishment of camel farms based on fodder production from indigenous plants and halophytes. UNESCO, Doha, Qatar.

[26] Dasel M., Wangoh J., Schelling E., Imungi J., Faraah Z. & Meile L. (2010). – Microbiological quality of raw camel milk within the Kenyan market as well as the effect of aluminum containers and improved container cleaning procedures on milk quality. *In* Proceedings of the International Camel Symposium, 7–11 June 2010, Garissa, Kenya, 80.

[27] Eberlein V. (2007). – Hygienic status of camel milk in Dubai (United Arab Emirates) under two different milking management systems. *Thesis*, Munich, Germany.

[28] El-Agamy E.I. (1998). – Camel' s colostrum. Antimicrobial factors. *In* Dromadaires et chameaux, animaux laitiers. (P. Bonnet, ed). Actes du colloque, 24–26 October 1994, Noaukchott, Mauritanie, 177–179.

[29] El-Agamy S.I., Ruppanner R., Ismail A., Champagne C.P. & Assaf R. (1992). – Antibacterial and antiviral activity of camel milk protective proteins. *J. Dairy Res.*, 59, 169–175.

[30] El-Hatmi H., Levieux A., & Levieux D. (2006). – Camel (*Camelus dromedarius*) immunoglobulin G, α-lactalbumin, serum albumin and lactoferrin in colostrum and milk during the early post partum period. *J. Dairy Res.*, 73 (3), 288–293.

[31] El-Ziney M.G. & Al-Turki A.I. (2006). – Microbiological quality and safety assessment of camel' s milk (*Camelus dromedarius*) in Qassim region (Saudi Arabia). *In* Proceedings of the International Science Conference on Camels, 12–14 May 2006, Buraydah, Saudi Arabia, 2146–2160.

[32] Farah Z. (1993). – Review article: composition and characteristics of camel milk. *J. Dairy Res.*, 60 (4), 603–626.

[33] Farah Z. (1996). – Camel milk. Properties and products. In Laboratory of Dairy Science，Swiss Federal Institute of Technology，ETH-Zentrum，LFO，CH-8092，Zurich.

[34] Farah Z.，Eberhard P.，Meyer J.，Wangho J.，Gallmann P. & Rehberger B. (2010). – Effect of ultra-high temperature treatments on camel milk. *In* Proceedings of the Camel Symposium，7~11 June 2010，Garissa，Kenya，28.

[35] Farah Z. & Ruegg M. (1991). – The creaming properties and size distribution of fat globules in camel milk. *J. Dairy Sci.*，74，2901–2904.

[36] Fowler M.E. (2010). – Medicine and surgery of South American camelids. Iowa State University Press，Ames，306–307.

[37] Guliye A.Y. (1996). – Studies on the compositional and hygienic quality of the milk of Bedouin camels (*Camelus dromedarius*) of the Negev desert. Thesis，University of Aberdeen，Scotland.

[38] Guya M.E. (2006). – Effect of camel (*Camelus dromedarius*) mastitis in eastern Ethiopia. *In* Proceedings of the International Science Conference on Camels，10~12 May 2006，Buraydah，Saudi Arabia，947–958.

[39] Hadush B.，Kebede E. & Kidannu H. (2008). – Assessment of bacterial quality of raw camel’ s milk in Ab-Ala，north eastern Ethiopia. *Livestock Res. Rural Dev.*，20 (9)，36–39.

[40] Hafez A.M.，Fazig S.A.，El-Amrousi S. & Ramadan R.O. (1987). – Studies on mastitis in farm animals in Al-Hasa. First analytical studies. *Assiut Vet. Med. J.*，19，140–145.

[41] Halbrock M.H. (2010). – Zytobakteriologische Untersuchungen von Milch bei maschinen-und handgemolkenen Kamelen (*Camelus dromedarius*) in Dubai. *Thesis*,Munich,Germany.

[42] Hamouda M.A. & Doghaim R. (2006). – Mastitis in she-camels. *In* Proceedings of the International Science Conference on Camels,10–12 May 2006,Buraydah,Saudi Arabia,959–967.

[43] Hashi A.M. (1989). – Preliminary observation on camel milk production and composition. Working paper,30.

[44] Hassanein A.,Soliman A.S. & Ismail M. (1984). – A clinical case of mastitis in she-camel (*Camelus dromedarius*) caused by *Corynebacterium pyogenes. Assiut Vet. Med. J.*,12 (23),239–241.

[45] Jhirwal S.K.,Bishnoi P. & Gahlot T.K. (2006). – Gangrenous mastitis in a camel (*Camelus dromedarius*). *In* Proceedings of the 1st ISOCARD Conference,15~17 April 2006,Al-Ain,UAE,75.

[46] Kappeler S. (1998). – Compositional and structural analysis of camel milk proteins with emphasis on protective proteins. Diss. ETH No. 12947,Zurich,Switzerland.

[47] Kapur M.P.,Khanna B.M. & Singh R.P. (1982). – A peracute case of mastitis in a she-camel associated with *Klebsiella pneumoniae* and *Escherichia coli*. Ind. *Vet. J.*,59 (8),650–651.

[48] Khedid K.,Faid M. & Soulaimani M. (2003). – *Microbiological* characterization of one humped camel milk in Morocco. *J. Camel Pract. Res.*,10 (2),169–172.

[49] Kinne J. & Wernery U. (2002). – Mastitis caused by *Pasteurella haemolytica* and *Streptococcus agalactiae* in dromedary camels. *J. Camel Pract. Res.*,9,121–124.

[50] Konuspayeva G.,Faye B.,Baubekova A. & Loiseau G. (2012). – Camel gruyere cheese making. *In* Proceedings of the 3rd ISOCARD Conference (E.H. Johnson et al.,eds.),29 January–1st February 2012,Mascat (Sultanate of Oman),218–219.

[51] Konuspayeva G.,Faye B. & Loiseau G. (2008b). – The composition of camel milk: a meta-analysis of the literature data. *J. Food Compos. Anal.*,22,95–101.

[52] Konuspayeva G.,Faye B. & Loiseau G. (2011). – Variability of vitamin C content in camel milk from Kazakhstan. *J. Camelid Sci.*,4,63–69.

[53] Konuspayeva G.,Faye B.,Loiseau G. & Levieux D. (2006). – Lactoferrin and immunoglobin content in camel milk from Kazakhstan. *J. Dairy Sci.*,90,38–46.

[54] Konuspayeva G.,Lemarie E.,Faye B.,Loiseau G. & Montet D. (2008a). – Fatty acid and cholesterol composition of camel's (*Camelus bactrianus,Camelus dromedarius* and hybrids) milk in Kazakhstan. *Dairy Sci. Technol.*,88,327–340.

[55] Kosparkov Zh. K. (1975). – Antibacterial properties of camel's milk. *Inst. Veterinarnoi Sanitarri* 51,37–40.

[56] Larska M.,Wernery U.,Kinne J.,Schuster R.,Alexandersen G. & Alexandersen S. (2009). – Differences in the susceptibility of dromedary and Bactrian camels to foot-and-mouth disease virus. *Epidemiol. Infect.*,137 (4),549–554.

[57] Leese A.S. (1927). – A treatise on the one-humped camel in health and disease. *Vigot,Frères,Paris* Ⅱ.

[58] Mody S.K.,Patel P.R. & Prajapati C.B. (1998). – A study on antimicrobial susceptibility of bacteria isolated from mastitic milk of rural camels in India. *In* Proceedings of the International Meeting on Camel Production and Future Perspectives,Al Ain,UAE,2 ~ 3 May1998,138 – 144.

[59] Mostafa A.S.,Ragab A.M.,Safwat E.E.,El-Sayed Z.A.,Abd-El-Rahman M.,El-Danaf N.A. & Shouman M.T. (1987). –Examination of raw she-camel milk for detection of subclinical mastitis. *J. Egypt Vet. Med. Ass.*,47 (1 & 2),117 – 128.

[60] Mousa A.M.,Elhag K.M.,Khogali M. & Marafic A.A. (1988). – The nature of human brucellosis in Kuwait: Study of 379 cases. *Rev. Infect. Dis.*,10,211 – 217.

[61] Nagy P.,Thomas S.,Marko O. & Juhasz J. (2012). – Milk production,raw milk quality and fertility of dromedary camels (*Camelus dromedarius*) under intensive management. *Acta Vet. Hung.*,61,81 – 74

[62] Obied A.I. (1983). – Field investigation,clinical and laboratory findings of camel mastitis. M.V.Sc. *Thesis*,University of Khartoum,Khartoum,Republic of Sudan.

[63] Obied A.I.,Bagadi H.O. & Mukhtar M.M. (1996). – Mastitis in *Camelus dromedarius* and the somatic cell content of camels'milk. *Res. Vet. Sci.*,61 (1),55 – 58.

[64] Quandil S.S. & Oudar J. (1984). – Etude bacteriologique de quelques cas de mammites chez la chamelle (*Camelus dromedarius*) dans les Emirats Arabes Unis. *Rev. Elev. Méd. Vét. Pays Trop.*,135 (11),705 – 707.

[65] Radwan A.I.,Bekairi S.J. & Prasad P.V.S. (1992). – Serological and bacteriological study of brucellosis in camels in central Saudi Arabia. *Rev. sci. tech. Off. int. Epiz.*,11 (3),837 – 844.

[66] Ramadan R.O.,El- Hassan A.M.,El-Abdin Bey R.,Algasnawi Y.A.,Abdalla E.S.M. & Fayed A.A. (1987). – Chronic obstructive mastitis in the camel,a clinical pathological study. *Cornell Vet.*,77 (2),132 – 150.

[67] Ramet J.-P. (2001). – The technology of making cheese from camel milk (*Camelus dromedarius*). FAO Paper,113.

[68] Saleh S.K. & Faye B. (2011). – Detection of subclinical mastitis in dromedary camels (*Camelus dromedarius*) using somatic cell counts,california mastitis test and udder pathogen. *Emir. J. Food Agric.*,23 (1),48 – 58.

[69] Semereab T. & Molla B. (2001). – Bacteriological quality of raw milk of camel (*Camelus dromedarius*) in Afar region (Ethiopia). *J. Camel Pract. Res.*,81 (1),51 – 54.

[70] Shabo Y.,Barzel R.,Margoulis M. & Yagil R. (2005). – Camel milk for food allergies in children. *Immunol. Allerg*,7,796 – 798.

[71] Stahl T. (2005). – Vitamingehalte und Fettsauremuster in Kamelmilch. Thesis,Hannover Veterinary School,Hannover,Germany.

[72] Strasser A.,Zaadhof K.-J.,Eberlein V.,Wernery U. & Martlbauer E. (2006). – Detection of antimicrobial residues in camel milk-suitability of various commercial microbial inhibitor tests as screening tests. *Milchwissenschaft*,61 (1),29 – 32.

[73] Tibary A. & Anouassi A. (1997). – Theriogenology in camelidae. Anatomy,Physiology,Pathology and Artificial Breeding. Abu Dhabi Printing and Publishing Co.,Mina,Abu Dhabi,UAE,51 – 61.

[74] Tourette I.,Messad S. & Faye B. (2002). – Impact des pratiques de traite des eleveurs sur la qualite sanitaire du lait de chamelle en Mauritanie. *Rev. Elev. Méd. Vét. Pays Trop.*,55,229 – 233.

[75] Wanjohi G.M.,Gitao C.G. & Bebora L.C. (2010). – The hygienic quality of camel milk marketed from North Eastern Province Kenya and how it can be improved. In Proceedings of the International Camel Symposium,7~11 June 2010,Garissa,Kenya,23.

[76] Wernery U. (2003). – Novel observations on camel milk. *In* Proceedings of the 9th Kenya Camel Forum,17 – 22 February 2003,Griftu Pastoral Training Centre,Wajir district,Kenya,1 – 15.

[77] Wernery U. (2010). – Evaluation of camel milk parameters in mammary health. *In* Proceedings of the International Camel Symposium held together with the 15th Kenya Camel Forum,7~15 June 2010,Garissa,Kenya,11.

[78] Wernery U.,Fischbach St.,Kletzka S.,Johnson B. & Jose S. (2008b). – Evaluation of some camel milk parameters used in mammary health. *J. Camel Pract. Res.*,15 (1),49 – 53.

[79] Wernery U.,Hanke B.,Braun F. & Johnson B. (2003). – The effect of heat treatment on some camel milk constituents. Preliminary report. *Milchwissenchaft*,58 (5/6),277 – 279.

[80] Wernery U.,Johnson B.,Becker & Maertlbauer E. (2002). – Microbiological status of raw dromedary milk. *J. Camel Pract. Res.*,9 (1),1 – 4.

[81] Wernery U.,Johnson B. & Jose S. (2008a). – The most important dromedary mastitis organisms. *J. Camel Pract. Res.*,15 (2),159 –161.

[82] Wernery U.,Juhasz J. & Nagy P. (2004). – Milk yield performance of dromedaries with an automatic bucket milking machine. *J. Camel Pract. Res.*,11 (1),51–57.

[83] Wernery U.,Juhasz J.,Johnson B. & Nagy P. (2006). – Milk production of dromedaries milked with an automatic milking machine: milk yield,quality control and udder hygiene. *In* Proceedings of the International Science Conference in Camels, 12~14 May 2006,Qassim University,Saudi Arabia,833–844.

[84] Wernery U.,Kinne J.,Joseph M.,Johnson B. & Nagy P. (2007). – Where do Brucella organisms hide in serologically positive lactating dromedaries. *In* Proceedings of the International Camel Conference,16~17 February 2007,Bikaner,India,68–70.

[85] Wernery U. & Wernery R. (2006). – The white gold of the desert (Booklet). Momentum,Dubai,UAE.

[86] Whabi A.A.,Gadir S.E.,Awadelsied A. & Idris O.F. (1987). – Biochemical studies on Sudanese camel milk collected from Butana Area. Sud. *J. Vet. Med.*,34,340–342.

[87] Yagil R. & Etzion Z. (1980a). – Milk yields of camels (*C. dromedarius*) in drought areas. *Comp. Biochem. Physiol.*,67A,207 –209.

[88] Yagil R. & Etzion Z. (1980b). – The effect of drought conditions on the quality of camels' milk. *J. Dairy Res.*,47,159–166.

[89] Younan M. (2002). – Parenteral treatment of *Streptococcus agalactiae* mastitis in Kenyan camels (*Camelus dromedarius*). *Rev. Elev. Méd. Vét. Pays Trop.*,55 (3),177–181.

[90] Younan M.,Ali Z.,Bornstein S. & Mueller W. (2001). – Application of the California mastitis test in intramammary *Streptococcus agalactiae* and *Staphylococcus aureus* infections of camels (*Camelus dromedarius*) in Kenya. *Prev. Vet. Med.*,51,307–316.

## 深入阅读材料

Abera M.,Abdi O.,Abuma F. & Mergersa B. (2010). – Udder health problems and major bacterial causes of camel mastitis in Jijiga,Eastern Ethiopia: implication for impacting food security. *Trop. Anim. Hlth. Prod.*,42 (3),341–347.

Albrecht C.E.A. (2003). – Production of camel milk. First experiences in machine milking of dromedaries. Master thesis,University of Gottingen,Germany.

Albrecht C.E.A. (2006). – Camel milk-production,processing,marketing with special reference to Rajasthan (India) and Dubai (United Arab Emirates). Thesis,University of Göttingen,Germany.

Alexandersen S.,Wernery U.,Nagy P.,Frederiksen T. & Normann P. (2006). – Dromedaries (*Camelus dromedarius*) are of very low susceptibility to experimental,high dose inoculation with FMDV serotype O and do not transmit the infection to direct contact camels or sheep. Abstracts,International Control of Foot-and Mouth Disease: Tools,Trends and Perspectives,Paphos,Cyprus,16–20 August 2010,23.

Almav G. & Molla B. (2000). – Prevalence and etiology of mastitis in camels (*Camelus dromedarius*) in eastern Ethiopia. *J. Camel Pract. Res.*,7 (1),97–100.

Ayadi M.,Hammadi M.,Khorchoni T.,Barmat A.,Atigui & Caja G. (2009). – Effects of milking interval and cisternal udder dairy dromedaries (*Camelus dromedarius*). *J. Dairy Sci.,92* (4):1 452–1 459.

Baimukanov A. (1974). – Machine milking of female camels (Russian). *Borovsk,USSR VNIIFBPSZR*,67–68.

Baimukanov A. (1977). – Stimulation of letting down of milk in lactating female camels (Russian). *Vestnik Sel'skokhozjaistvennoi Nauk Kazachstana*,5,116–117.

Belokobylenko V.T. (1978). – Principles of selecting female camels for machine milking (Russian). *Vestnik Sel'skokhozjaistvennoi Nauk Kazachstana*,Alma Ata,Ministerstvo.,11,65–68.

Dzhumagulov I.K. (1976). – Chemical composition of camel milk and inheritance of its components following interspecies hybridization (Russian). *Izvestija Akademii Nauk Kazachstana*,15,79–81.

Dzhumagulov I.K. (1976). – Milk yield,its fat content and inheritance of these properties in interspecies hybridization of Bactrian and Arabian camels (Russian). *Izvestija Akademii Nauk Kazachskoj. SSR. Seriia Biologicheskaia*,6,69–75.

Dzhumagulov I.K. (1977). – Chemical composition of the milk of camels and the inheritance of its components following interspecies hybridization (Russian). *Izvestija Akademii Nauk Kazachskoj. SSR. Seriia Biologicheskaia*,4,79–81.

Esterabadi A.H.,Entessar F.,Hedayati H.,Narimani A.A. & Sadri M. (1975). – Isolation of *Corynebacterium pseudotuberculosis* from camel in Iran. *Arch. Inst. Razi.*,27,61–66.

Frederiksen T.,Borch J.,Christensen J.,Tjørnehøj K.,Wernery U. & Alexandersen S. (2006). – Comparison of FMDV type O,A and Asia 1 antibody levels after vaccination of cattle,sheep,dromedary camels and horses. Abstracts,International Control of Foot and Mouth Disease: Tools,Trends and Perspectives,Paphos,Cyprus,16–20 August 2006,38.

Han Z.Y.,Chen J.G. & Shang Y.Z. (1983). – Experiment on immunisation against *Corynebacterium pseudotuberculosis* of the Bactrian camel. *Acta Agric. Univers. Gansu.*,2,47–58.

Hassan F.A.,Al-Jaru A.I.,Johnson B.,Wernery U. & Khazanehdari K.A. (2008). – Authentication of camel milk using microsatellite markers. *J. Camel Pract. Res.* 15 (1),39–41.

Ismail M.,Wernery U.,Smales M. & Khazanehdari K. (2010). – Molecular identification and characterization of camel milk insulin. Poster presentation of Abu Dhabi Genomics and Systems Biology Meeting,6–7 January 2010,Abu Dhabi,UAE.

Kheraskov S.G. (1955). – Milk production of camels (Russian). *Sborn. Doklady. Vsesojuznogo Soveschchanija*,174–176.

Kheraskov S.G. (1961). – Composition,properties and nutritive values of camels' milk (Russian). *Voprosy Pitania*,20,69–72.

Kheraskov S.G. (1965). – Camel milk- a valuable food product (Russian). *Konevodsvoi Konny Sport,*35,14–15.

Larska M.,Wernery U.,Kinne J.,Schuster R.,Alexandersen G. & Alexandersen S. (2008). – Differences in the susceptibility of dromedary and Bactrian camels to foot-and-mouth disease virus. *Epidemiol. Infect.*,137,549–554.

Lorenzen P.Chr.,Wernery R.,Johnson B.,Jose Sh. & Wernery U. (2011). – Evaluation of indigenous enzyme activities in raw and pasteurized camel milk. Short communications. *Small Ruminants Res.*,97,79–82.

Mal G.,Suchitra Sena D.,Jain V.K. & Sahani M.S. (1999). – Utility of raw camel milk as nutritional supplement among chronic pulmonary tuberculosis patients. *In* International Workshop on the Young Camel,Quarzazate,Morocco,24–26 October 1999,93.

Restani P.,Gaiascha A.,Plebani A.,Beretta B.,Cavagni G.,Fiocchi A.,Poiesi C.,Velona T.,Ugazio A.G. & Galli C.L. (1999). –Cross-reactivity between milk proteins from different animal species. *Clin. Exp. Allerg.*,29 (7),997–1 004.

Saad,N.M. & A.EI-R. Thabet. (1993). – Bacteriological quality of camel's milk with special reference to mastitis. *Assiut Vet. Med.J.*,28,194–198.

Stahl T.,Sallmann H.P.,Duehlmeier R. & Wernery U. (2006). – Selected vitamins and fatty acid patterns in dromedary milk and colostrum. *J. Camel Pract. Res.*,13 (1),53–57.

Suheir I.A.,Salim M.O. & Yasin T.E. (2005). – Bacteria,mycoplasma and fungi associated with sub-clinical mastitis in camel. Sud. *J. Vet. Res.*,20,23–31.

Wernery U. (2003). – (Booklet) New observations on camels and their milk. Dar Al Fajr，Abu Dhabi，UAE.

Wernery U. (2003). – Novel observations on camel milk. In Proceedings of the 9th Kenya Camel Forum，Griftu Pastoral Training Centre，Wajir district，17–22 February 2003，1–15.

Wernery U. (2004). – FMD and camelids: International relevance of current research. *In* Abstracts of Open Session of the Research Group of the Standing Technical Committee of the European Commission for the Control of Foot-and-Mouth Disease，Chania，Crete，Greece 11–15 October 2004，246–259.

Wernery U. (2006). – Camel milk，the white gold of the desert. *J. Camel Pract. Res.* 13 (1)，15–26.

Wernery U. (2007). – Camel milk-new observations. *In* Proceedings of the International Camel Conference，16–17 February 2007，Bikaner，Rajasthan，India，200–204.

Wernery U. (2007). – Dromedaries have a low susceptibility to Foot-and-Mouth disease-results of 3 infection trials. In Proceedingsof the International Camel Conference，16–17 February 2007，Bikaner，Rajasthan，India，19–22.

Wernery U., Alexandersen S. & Nagy P. (2006). – Foot and Mouth Disease (FMD) in camelids with emphasis on new observationsin dromedaries. *In* Proceedings of the International Science Conference on Camels, Kingdom of Saudi Arabia, 10–12 May 2006, 347–357.

Wernery U., Fischbach St, Johnson B. & Jose Sh. (2008). – Evaluation of alkaline phosphatase (ALP), γ-glutamyl transferase (GGT)and lactoperoxidase (LPO) activities for their suitability as markers of camel milk heat inactivation. *Milchwissenschaft*, 63 (3) 265–267.

Wernery U., Johnson B. & George R.M. (2007). – Gamma-glutamyl transferase (GGT), a potential marker for the evaluation ofheat treatment of dromedary milk. *J. Camel Prac. Res.*, 14 (1), 9.

Wernery U., Johnson B. & Abraham A. (2005). – The effect of short-term heat treatment on vitamin C concentrations in camelmilk. *Milchwissenschaft*, 60 (3), 266–267.

Wernery U., Juhasz J. & Nagy P. (2004). – Milk yield performance of dromedaries with an automatic bucket milking machine.*J. Camel Pract. Res.*, 11 (1), 51–57.

Wernery U. & Kaaden O.-R. (2004). – Foot-and-mouth disease in camelids: a review. *Vet. J.*, 168, 134–142.

Wernery U., Maier U., Johnson B., George R.M. & Braun F. (2006). – Comparative study on different enzymes evaluating heattreatment of dromedary milk. *Milchwissenschaft*, 61 (3), 281–285.

Wernery U., Nagy P., Amaral-Doel C.M., Zhang Z. & Alexandersen S. (2006). – Lack of susceptibility of the dromedary camel(*Camelus dromedarius*) to foot-and-mouth disease virus serotype O. *Vet. Rec.*, 158, 201–203.

Wernery U., Nagy P., Bhai I., Schiele W. & Johnson B. (2006). – The effect of heat treatment, pasteurization and different storagetemperatures on insulin concentration in camel milk. *Milchwissenschaft*, 61 (1), 25–28.

Wernery U. & Wernery R. (2006). – (Booklet) Camel milk, the white gold of the desert. Second edition, Central Veterinary ResearchLaboratory (CVRL), Dubai, UAE.

Wernery D. & Wernery U. (2009). – Camel milk-an udderly healthy product. Poster presentation at the First Camel Workshop, 29~30 September 2009, Debre Zeit, Ethiopia.

Wernery U., Wernery D., Johnson B., Jose Sh. & Wernery R. (2012). – Organoleptic, biochemical and *microbiologica*l investigationsinto the shelf life of camel milk powder-preliminary results. *J. Camel Pract. Res.*, 19 (1), 7–12.

Younan, M. & J.W. Matofari. (1999). –*Streptococcus agalactiae* infection in Kenyan camels (*Camelus dromedarius*). In InternationalWorkshop on the Young Camel, 24–26 October 1999, Quarzazate, Morocco, 73.

（卢曾军译，殷宏校）

1
细菌病

# 1.7 神经系统

## 1.7.1 破伤风（Tetanus）

破伤风的发生范围遍及全世界，在温暖季节的发生频率较高。几乎所有的哺乳动物都可以被感染，但是，不同动物对破伤风毒素的敏感性差异很大。马是除人之外对破伤风最敏感的动物。破伤风以“牙关紧闭”的症状为人们所熟知，是高致死性的神经中毒性疫病，以痉挛性和持续性的肌肉麻痹为特征，当有较大的噪声和皮肤接触等外界刺激时症状加重。该病的致病因子是破伤风梭菌（*Clostridium tetani*）产生的神经毒素。骆驼的破伤风很少见，一般认为，该病不是很重要。旧大陆骆驼和新大陆骆驼对破伤风的易感程度还不清楚。由于骆驼外伤很常见，且在单峰驼体内检测到破伤风抗体时没有观察到临床症状，因而推测骆驼对破伤风具有相当强的抵抗力。

### 病原学与流行病学

破伤风梭菌是一种厌氧性的革兰氏阳性菌，带有末端球状孢子，可见于土壤和粪便中。其繁殖体对热很敏感，不能耐受有氧环境，而孢子体对热和一般消毒剂都有很强的抵抗力。孢子体广泛存在于土壤中，并可以存活许多年。破伤风梭菌可以产生两种毒素，包括具有很强神经毒性的破伤风痉挛毒素和破伤风溶血素。多数情况下，破伤风梭菌主要通过外伤进入机体组织，尤其是较深的刺伤，能够为病菌提供适合生长的厌氧环境。毒血症经常发生于绵羊阉割和断尾之后（特别是伤口用医用胶带包扎后），容易造成较大的经济损失。牛对破伤风不是很易感。

骆驼感染破伤风一般是通过污染的伤口或荆棘的长硬刺造成的刺伤而感染发病。少量破伤风梭菌孢子污染物或许进入刺伤深处。孢子在组织中的增殖也需要有一定的条件，条件之一便是周围组织中的氧分压下降。如果伤口内同时有需氧菌进入，那么马上就可能会产生这种厌氧的环境。破伤风梭菌可以在伤口中存活数月，直到有合适的条件能够让其繁殖引起发病，这种情况或许在初始感染的位置发生二次创伤时才更有可能发生[16]。即使最初的伤口已经痊愈，在周围组织中的氧分压下降时，严格厌氧的破伤风梭菌也能够繁殖。然后，破伤风梭菌的孢子从感染位置扩散到血管和淋巴系统，并进入肝脏、脾脏（突发性破伤风）。高活性的神经毒素随着病原的繁殖和裂解释放出来，并可以通过逆向轴突输送到达中枢神经系统（CNS），产生典型的渐升型破伤风。重度感染情况下，毒素大量产生，可直接攻破血脑屏障，到达中枢神经系统，然后产生渐降型的破伤风[18]。

骆驼破伤风并不是很重要的疫病[5,14,15]。Rabagliati[15]在3年6个月的时间内，从埃及的25000头单峰驼中仅诊断出了4例破伤风，尽管大部分的骆驼有外伤，有一些骆驼的外伤还很严重。Ramon和Lemetayer[17]从没有任何破伤风症状的单峰驼体内检测到了破伤风抗体。

Dioli和Stimmelmayr[6]在他们的书中描述了一种与破伤风很相似的疫病，东非单峰驼的主人称这种病为“硬颈综合征”。该病分为急性和慢性。急性型类似于经典的破伤风，具有肌肉痉挛、颈部僵硬和咀嚼功能紊乱的特征；反射活动增加，噪声和身体接触等轻微的刺激就会使病畜突然发生痉挛。任何年龄的单峰驼都有可能发生该病，但仅有零星动物发生此病。Rabagliati[15]和Morcos[12]也描述了类似的疾病症状。这两个作者都观察到了颈部肌肉僵硬，全身肌肉强直性痉挛，尾巴强直，瞬膜下垂和四肢向侧面伸展（像锯木架样）的临床症状。

有几位研究人员描述了一种被称之为“歪脖”综合征(wry-neck syndrome，WNS)的骆驼疾病[7,8,20]。Agab[3]用1年的时间研究了东苏丹10518头骆驼，其中，有23头（0.22%）出现歪脖症状。由于“歪脖”综合征而致死的骆驼有12头，占总死亡骆驼的比例为2.9%（12/421）。在此之前的一项研究，Agab和Abbas[2]从苏丹的2000头骆驼中发现了8例“歪脖”综合征：其中，7例为布鲁氏菌病血清学阳性。Agab和Ahmed[3]对其中6头“歪脖”综合征病驼进行了更仔细的研究，发现这些骆驼有不同程度的白细胞增多，并伴有淋巴细胞和嗜碱性粒细胞增多，并提到与布鲁氏菌病有一定的关系，因为6头骆驼中有5头是布鲁氏菌病血清学阳性。上述作者都讨论了“歪脖”综合征的病因，例如，蜱、植物毒素和营养缺乏等。到目前为止，还没有对病驼进行综合性的检查，如，超声波诊断、X-线透视颈椎骨等。在山羊、长颈鹿和原驼也发现有类似的疾病。

Wernery等[20]在一篇研究文章中讨论了破伤风

毒素是否可以诱发“歪脖”综合征。作者给原驼颈部注射了少量毒素和菌，24小时后原驼表现出了“歪脖”综合征症状（图135）。原驼由于病情恶化没有被治疗，出于人道主义对其进行了安乐死。

Dioli和Stimmelmayr[7]观察到了一种“硬颈综合征”（stiff-neck syndrome），另外，还有“曲颈综合征”（bent-neck syndrome）和蜱瘫痪症也需要在此提及。目前仍然不清楚这几种病是否有相同的病因。

在秘鲁，有2例羊驼破伤风病例[13]，而在阿根廷[19]和美国分别有1例和几例美洲驼的破伤风病例报道[9,10,11]。

沙特阿拉伯报道了1例雌性成年骆驼的神经系统疾病（当地人称之为Altair）[4]。骆驼表现为腿虚弱无力，特别是受到刺激紧张时不断摇头，出现歪脖等症状。显微镜检查大脑组织显示有软化病灶，并有脑膜充血和水肿现象。检测发现饮水中硫浓度达（13.56 ± 5.8）g/L，这种硫的浓度对家畜而言是过高了，可能是导致骆驼脑灰质炎的一个原因。

## 临床症状

新大陆骆驼与旧大陆骆驼发生破伤风后的临床症状很相似。差不多所有的骆驼破伤风病例都起源于身体不同部位的创伤。Abu Bakr Mekki Ahmed（1992，个人通讯）描述了一个竞赛用单峰驼破伤风病例，由于后足的深割伤而发病。创伤后14天，出现典型的临床症状，下颌痉挛、颈与尾强直、步态僵硬。其他的症状还包括呼吸困难，耳朵直立，双目凝视。治疗期间骆驼躺卧了3周。Wernery等[21]描述了一例单峰驼阉割后发生的破伤风病例，动物食欲不振，步态僵硬、颈伸直，包皮环至腹股沟区域肿大，口角有泡沫状垂涎。动物吞咽困难、牙关紧闭（图136）。到医院就诊后第5天，由于后肢麻痹而不能站立、四肢张开匍匐在地（图137）。

虽然应用抗生素和抗破伤风血清进行了强化治疗，但动物的病情恶化，最后对其实施了安乐死。从阉割的伤口位置，分离到了破伤风梭菌，将经过过滤的巯基乙酸盐肉汤培养物注射两只小鼠，均表现出了后腿痉挛的典型破伤风症状（图138）。

新大陆骆驼的破伤风症状与旧大陆骆驼类似，主要通过外伤和脐带发生感染。临床症状包括

图135　原驼在颈中部注射破伤风毒素后表现出歪脖症状

图136　破伤风公单峰驼口中有泡沫状流涎，牙关紧闭

图137　开放式阉割25天后，骆驼后躯麻痹，四肢张开俯伏在地

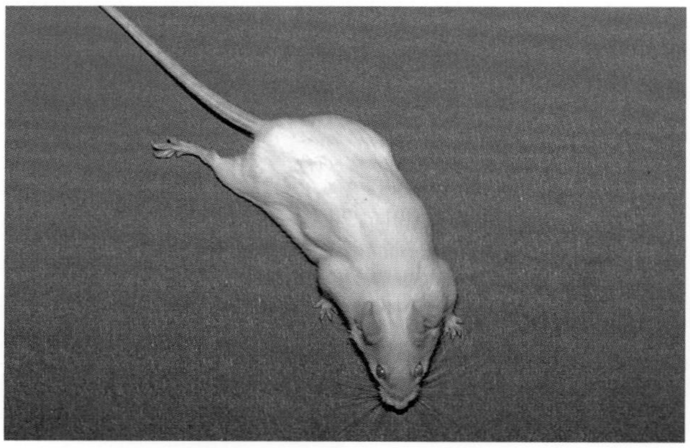

图138 破伤风小鼠后腿痉挛

图139 破伤风美洲驼（图片来自美国的M. E. Fowler教授）

俯伏在地、肌肉僵硬、呼吸困难、强直性肌肉收缩、关节僵硬、双目凝视、耳朵直立、牙关紧闭和发热达41.5℃（107℉）。在阿根廷，有一例美洲驼由于足部的外伤而被诊断为感染破伤风。美洲驼四肢张开，呈“锯木架”样站姿、牙关紧闭变硬、流涎、面部表情僵硬（图139）；同时还表现出呼吸困难、耳朵直立、尾巴抬高僵硬、瞬膜突出等症状。

## 诊断

破伤风的临床症状很特殊，在很多情况下，依据临床症状即可做出诊断。有外伤史并且没有进行免疫预防，也有助于确定临床诊断的方向。破伤风需要与以下几种病进行鉴别，一是狂犬病，患狂犬病动物更具有攻击性；二是番木鳖碱中毒，中毒动物通常只能维持几个小时；三是肠毒血症，患病动物发病更迅速，主要是消化道症状。动物破伤风还要与肉毒中毒（松弛性瘫痪）、高镁血症和脑膜炎进行鉴别。死后检查和临床生化检验可以对上述疫病进行区别。对破伤风几乎都是依据临床症状和病史就可以做出诊断。有时候也有可能从伤口处分离到细菌，通过注射小鼠的动物试验证明产生细菌毒素对诊断很有帮助。也可以尝试用病料涂片进行革兰氏染色，以发现病原菌。

## 控制和预防

骆驼在出现破伤风的早期症状时马上采取治疗措施是可以康复的。通常由于治疗不及时，最后不得不对动物实施安乐死。在阿拉伯联合酋长国报道了两例破伤风病例：其中，一个单峰驼因为治疗及时而存活，而另一个则不得不对其实施了安乐死。在感染后最初的72小时内，采取静脉注射2×100 mL的破伤风抗毒素，并进行清创、抗生素治疗和插胃管人工喂食等治疗措施，单峰驼可以在3周内康复。

Morcos[12]对埃及的两头破伤风病驼进行了治疗，使用的药物为丙酰丙嗪和青霉素。两头骆驼都于12天内康复。只有一头骆驼给予了2×30000IU的破伤风抗毒素。作者认为单峰驼能够快速康复，丙酰丙嗪发挥了主要的治疗作用。

对于贵重动物，应用破伤风抗毒素治疗。对于骆驼用多少剂量还不清楚，但是，300000IU的破伤风抗毒素配合镇静剂或巴比妥酸盐镇静剂对于治疗马破伤风很有效。有一例美洲驼破伤风病例采用牛破伤风的治疗方案：破伤风抗毒素的剂量为每千克体重225IU（一半静脉注射，另一半肌内注射），配合给予抗生素和氯丙嗪每千克体重2.2mg维持96小时。破伤风梭菌对青霉素敏感，应该用全剂量的青霉素连续给药7天进行治疗。在疫病的急性期，也可以同时考虑应用肌肉松驰药。

应将患病动物置于安静、避光的圈舍，在表现痉挛的急性期，良好的饲养护理对于患畜的康复具有重要的作用。如果动物不能吃喝，建议用胃管进行人工喂食。破伤风类毒素疫苗也容易获得，应该在每次手术之前进行免疫预防。美洲驼破伤风类毒素疫苗免疫后抗体应答水平升高。

# 参考文献


[1] Agab H. (1993). – Epidemiology of camel diseases in eastern Sudan with emphasis on brucellosis. M.V.Sc. Thesis,Universityof Khartoum,Khartoum,Sudan,172 pp.

[2] Agab H. & Abbas B. (1999). – Epidemiological studies on camel diseases in the eastern Sudan. *World Anim. Rev.*,92,42－51.

[3] Agab H. & Ahmed K.E. (2003). – Investigation of wry-neck syndrome in camel (*Camelus dromedarius*). *J. Camel Pract. Res.*,10 (2),159－162.

[4] Al-Swailem A.,Al-Dubaib M.A.,Al-Ghamdi G.,Al-Yamani E.,Al-Naeem A.A.,Al-Mejali A.M.,Shehata M. & Mahmoud O.M.(2009). – High sulphur content of water from deep bore wells as a possible cause of polioencephalitis in a camel. *Bulg. J. Vet.Med.*,12 (4),265－270.

[5] Curasson G. (1947). – Le chameau et ses maladies. Vigot Freres Editions,Paris 86－88.

[6] Dioli M. (2010). – Clinical observations on the 'wry-neck' syndrome or torticollis in the camel. *In* Proceedings of theInternational Camel Symposium: Linking Camel Science and Development for Sustainable Livelihoods,7~10 June 2010,Garissa,Kenya,7.

[7] Dioli M. & Stimmelmayr R. (1992). – Important camel diseases. In The one-humped camel in East Africa: A pictorial guideto diseases,health care and management (H.J. Schwatz and M. Dioli,eds). J. Margraf,Weikersheim,Germany,218－219.

[8] Dirie M.F. & Abdurahman O. (2003). – Observations on little known diseases of camels (*Camelus dromedarius*) in the Hornof Africa. *Rev. sci. tech. Off. int. Epiz.*,22 (3),1043－1049.

[9] Fowler M.E. (2010). – Medicine and surgery of camelids. 3rd Ed. Wiley-Blackwell,Oxford,UK,196－198.

[10] Keller D. (1995). – Lockjaw in a llama. *Camel Newslett.*,The Arab Center for the Studies of Arid Zones and Drylands,11,2－4.

[11] Lopez M.J. & Snyder J.R. (1995). – Tetanus in a llama. *Equine Pract.*,17,26－31.

[12] Morcos M.B. (1965). – Treatment of tetanus in the camel. *Vet. Med. Rev.*,2,132－134.

[13] Moro Sommo M. (1961–1962). – Infectious diseases of alpacas. III *Tetanus. Enfermedades infecciosas de las alpacas. IIITetanos. Revista Facultad Medicina Veterinaria (Lima)*,16－17,160－162.

[14] Mustafa I.E. (1987). – Bacterial diseases of the camel and dromedary. OIE 55th General Session of the OIE,Office internationaldes Epizooties,Paris,France,55,18－22.

[15] Rabagliati O.B.E. (1920). – Tetanus in the camel. *J. Comp. Path*ol. *Theriogenology*,33,10－12.

[16] Radostits O.M.,Gay C.C.,Hinchcliff K.W. & Constable P.D. (2007). – Veterinary Medicine. 10th Ed. Bailliere Tindall,London,822－844.

[17] Ramon G. & Lemetayer E. (1934). – Sur l'immunite antitetanique naturellement acquise chez quelques especes de ruminants. *Comm. rel. de la Séa. de la Soci. Biol.*,116 (18),275－277.

[18] Seifert H.S.H. (1992). – Tropentierhygiene. Gustav Fischer Verlag,Jena,Stuttgart,Germany,266－269.

[19] Toucedo G.A. (1965). – Infeccion a *Clostridium tetani* en una llama. *Gaceta Veterinaria (Argentina)*,27 (13),432－436.

[20] Wernery U.,Joseph M.,Johnson B. & Kinne J. (2005). – Wry-neck – a form of tetanus in camelids. *J. Camel Pract. Res.*,12 (2),75－79.

[21] Wernery U.,Ul-Haq A.,Joseph M. & Kinne J. (2004). – Tetanus in a camel (*Camelus dromedarius*) – A case report. *Trop. Anim. Health.*,36,217－224.


（卢曾军译，殷宏校）

## 1.7.2 李斯特菌病（Listeriosis）

李斯特菌（*Listeria*）在自然环境中分布很广，从土壤、农作物、腐烂的植物和pH值大于5.5的青贮饲料中都可以分离到该菌。李斯特菌可以在青贮饲料中繁殖，通常与牛、羊李斯特菌病的发生有关系。李斯特菌病在世界范围内发生，在人和不同的动物表现出不同的临床病症。有趣的是，人和动物的许多李斯特菌病例都是由于食用了污染的食物。单核细胞增生李斯特菌（*Listeria monocytogenes*）感染人主要是通过软奶酪、奶、禽肉和生菜沙拉。由于科学家逐渐认识到该病是一种食物传播性疫病，在该病原菌分类、分离方法、定型和分子诊断方面都取得了较大的进步。然而，需要指出的是许多感染病例都呈散发，而且表现为多种临床症状类型。有记录的李斯特菌感染病例分布于6大洲的40多种家养动物和野生动物。临床症状表现为脑炎、败血症和流产。李斯特菌病在新大陆骆驼和旧大陆骆驼均有报道。

### 病原学

单核细胞增生李斯特菌是一种中等大小，不形成孢子，革兰氏阳性杆菌，直径为0.4～0.5 μm。已经鉴定有5种血清型和很多血清亚型。最初，单核细胞增生李斯特菌是李斯特菌属的唯一的1个种，现在新鉴定的种还包括有伊氏李斯特菌（*L. ivanovii*）、无害李斯特菌（*L. innocua*）、韦氏李斯特菌（*L. welshimeri*）和塞氏李斯特菌（*L. seeligeri*）。单核细胞增生李斯特菌是最主要的致病菌，偶尔可见绵羊李斯特菌引起的临床病例，其他种的李斯特菌极少会成为病原菌。

### 公共卫生意义

人类李斯特菌病很少见，但通常很严重，全身性的李斯特菌病的死亡率20%～40%。引起的病症主要为脑炎，全身性不适，孕妇自发性流产和新生儿感染。而且，畜主和兽医在处理感染幼畜后也容易发生皮肤感染。普遍认为食源性传播是本病最主要的感染方式。单核细胞增生李斯特菌可以从很多食品中分离到，每克面团和软奶酪中可以含有超过1 000个病原菌。污染经常发生在食品加工之后，而且该病原菌可以在冰箱的低温条件下繁殖，对盐有一定的抗性，并能够存活较长的时间。

Ozbey等[16]检测了来自土耳其西南部不同零售市场的100个骆驼香肠样品，从9个样品中分离到了单核细胞增生李斯特菌（9%），从14个样品中分离到了无害李斯特菌（14%），从2份样品中分离到了韦氏李斯特菌（2%）。作者强调在食品加工制造过程中，需要采取适当的卫生预防措施，以避免细菌污染。Rahimi等[17]分析了来自伊朗商业化骆驼屠宰场的样品，于生产线的不同点，从骆驼胴体共采集了94份颈部肉的样品，包括切除内脏之前、切除内脏之后和清洗后3个不同的时间点。上述3个不同时间点的李氏特菌细菌检出率依次分别为7.4%、8.5%与3.2%；单核细胞增生李斯特菌检出率依次为1.1%、2.1%和1.1%；无害李斯特菌检出率依次为3.2%、4.3%与1.1%。作者指出，在清洗后仍然检出单核细胞增生李斯特菌对食品安全有很大的威胁，因为该病原菌在冰箱的低温条件下仍然能够生长繁殖。

### 流行病学

李斯特菌病在世界范围内分布很广。然而，该病更常发生于气候比较寒冷的区域，而且该病的发生往往与饲喂pH值大于5.5的青贮饲料有关系。旧大陆骆驼一般不会饲喂青贮饲料，这或许是旧大陆骆驼很少罹患李斯特菌病的原因。然而，突尼斯、伊朗和尼日利亚的单峰驼体内可以检测到抗体，但骆驼没有表现出临床症状。Salehi等[18]用琼脂扩散试验从380头伊朗单峰驼中仅检测出3头李斯特菌抗体阳性驼，而Oni等[15]利用试管凝集试验检测尼日利亚骆驼，结果李斯特菌抗体阳性率约为4.3%。检测抗体阳性的单核细胞增生李斯特菌血清型为1/2a，1/2b，1/2c和4b。Makinde等[11]报道了尼日利亚单峰驼血清抗体阳性率可达10%，而Burgemeister等[2]发现6头突尼斯单峰驼单核细胞增生性李斯特菌血清型1抗体阳性。

绵羊、山羊和牛发生李斯特菌病的频率更高。本病在新大陆骆驼呈散发[4]。死亡率可达100%。通常只有个别动物感染。Hänichen和Wiesner[6]描述了两例10日龄羊驼发生败血性李斯特菌病，其中1例表现出脑炎症状，调查表明，它们均饲喂质量差的玉米青贮饲料。

## 临床症状和病理学变化

李斯特菌病引起新大陆骆驼的脑膜脑炎，表现为转圈和颤抖、行走盲撞、发热。一些病例发生单侧面神经麻痹，伴有嘴唇、耳朵和眼睑下垂，咽和下颌麻痹导致咀嚼和吞咽功能障碍。病程通常持续3～5天[3,12,13,14,21]。1975年，德国一动物园暴发李斯特菌病，造成6头美洲驼死亡[12]，其中，有3头美洲驼发生脑炎，从大脑组织中分离到了单核细胞增生李斯特菌；另外3头美洲驼则死于败血性李斯特菌病，从机体不同器官均分离到了病原菌。从这次疫情中分离培养到了不同的血清变型，其中3个变型就是通过低温条件下孵育后分离到的。同时提及由于大雨淹没动物园，也造成其他有蹄动物急性发病。当时采用了一种李斯特菌活疫苗进行紧急免疫接种，才阻止疫病的进一步传播扩散[12]。在纽约也有两头成年美洲驼发生了脑炎性李斯特菌病，导致动物流产、共济失调、精神沉郁和面部肌肉麻痹瘫痪，最终死亡[3]；从其中1头美洲驼分离到了单核细胞增生李斯特菌，但两头动物免疫荧光检测均为阳性。Hamir和Moser[7]报道了1头2岁龄雌性美洲驼发生脑膜脑炎，病变局限于脑和脊髓。软脑膜表面粗糙、呈暗红色，有一薄层淡黄色的渗出物。单核细胞增生李斯特菌不仅能引起脑膜脑炎，而且会导致败血症伴发多发性关节炎（羊驼）[24]，中耳炎、内耳炎伴发脑膜脑脊髓炎（美洲驼）[22]，流产（美洲驼）[8,10]，败血症并发血小板减少症和肝病（美洲驼）[20]。在羊驼甚至可见本病引起的壁性心内膜炎和瓣膜性心内膜炎[9]。然而，根据上述作者的研究，反刍动物较少发生左心室心内膜炎。

显微病变主要局限于脑干白质或灰质，尤其是脑桥和延髓部分。在延髓部，可见血管周围单核细胞浸润与小脓肿（图140与图141）。然而，Hamir与Moser[7]认为，新大陆骆驼的李斯特菌脑炎类型不是以脑干部的小脓肿为特征，而是化脓性脑膜炎的表现形式。作者同时也观察到了一种多病灶性的急性坏死性脾炎，通过免疫组织化学检测显示出单核细胞增生李斯特菌免疫过氧化物酶反应为阳性。

美国俄亥俄州立大学的Whitehead等[23]研究了美洲驼与羊驼的神经系统疾病。他们研究了1993—2003年间的185个病例，其中，90例是羊驼，95例为美洲驼。大约有一半的病例最后康复离开了医院，另有一半或病死或最终施行了安乐死。感染性因素包括李斯特菌病和西尼罗河热感染，是继“脑膜蠕虫”感染之后第二个最常见的脑膜脑炎类型。以下是另外两例分别发生于美国和德国的新大陆骆驼李斯特菌病。Frank等[5]从美洲驼雌性胎儿分娩时先露部分采集的血样和7日龄时的尸检组织样本中均分离到了单核细胞增生李斯特菌和大肠杆菌。从脑脊髓液的培养物中也分离到了上述致病菌。该幼驼由于败血症和脑膜脑炎死亡。Seehusen等[19]则报道了1例成年雄性羊驼混合感染李斯特菌和梭菌的病例。动物临床出现食欲不振，消瘦和腹泻。尸检过程中发现有坏死性肝炎、脾炎、结肠炎以及溃疡性和类白喉性的回肠炎（Diphtheroid ileitis）。从羊驼的肠和肝组织中均分离到了单核细胞增生李斯特菌。

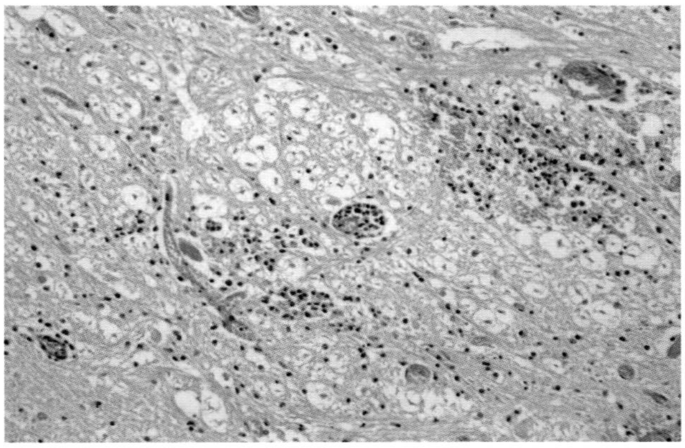

图140　单峰驼李斯特菌脑炎组织病理学观察

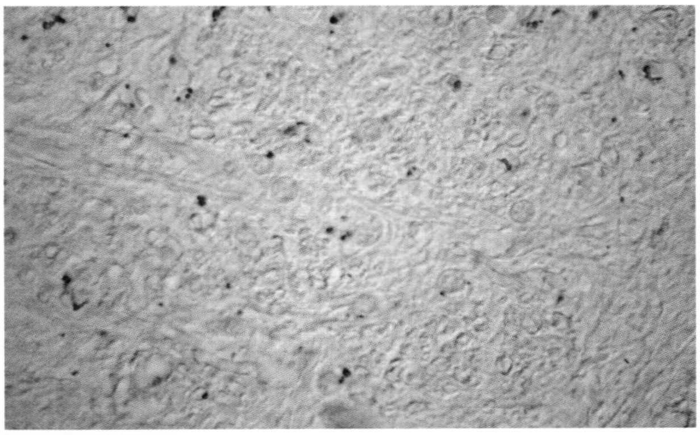

图141　同一单峰驼李斯特菌脑炎的免疫组化检测
（照片由瑞士A Pospischil教授惠赠）

最近才有1例旧大陆骆驼的李斯特菌病例报道[1]。一个6岁龄的雌性骆驼表现出中枢神经系统疾病症状。血液学检测显示，白细胞（总白细胞数为15.3 × $10^9$个/L）和单核细胞（单核细胞占比达到18%）均增多。脑部组织病理学检测显示为急性淋巴细胞性脑膜脑脊髓炎，延髓和脊髓血管周围淋巴细胞浸润并聚集成一层套膜。小脓肿数量较少，仅局限于延脑部。从革兰氏染色的脑组织压片检测到了单核细胞增生李斯特菌样微生物。脑组织培养物分离到的细菌菌落为黑色，与单核细胞增生李斯特菌一致。将菌落涂片进行染色鉴定显示为革兰氏阳性的球杆菌。用单核细胞增生李斯特菌特异性PCR引物，以脑组织中提取的DNA为模板，扩增到了预期大小为453bp的DNA片段，但从肝脏提取DNA中却没有扩增出预期大小的DNA片段。

### 诊断

通过分离和鉴定单核细胞增生李斯特菌可以对本病做出确诊。检测样本可以选择表现有中枢神经症状动物的脑组织，流产胎盘和胎儿组织。在发生败血症时，采集的肝和脾脏组织样本应该先经过培养。如果初次细菌分离失败，样本应该放到4℃条件下数周时间，每周取一些样本进行培养，分离病原菌。免疫荧光和免疫组织化学检测[7]是快速诊断李斯特菌病的方法。要始终注意与狂犬病进行鉴别诊断。

### 治疗和控制

本病一旦出现神经症状，往往预后不良。可以用青霉素按每千克体重44000 IU剂量，2次/天，进行注射，持续1 ~ 2周。在德国的动物园成功地应用了一种活疫苗进行本病的预防。

## 参考文献

[1] Al-Swailem A.A., Al-Dubaib M.A., Al-Ghamdi G., Al-Yamani E., Al-Naeem A.M., Al-Mejali A., Shehata M., Hashad M.E., El-Lithy Aboelhassan D. & Mahmoud O.M. (2010). – Cerebral listeriosis in a she-camel at Qassim Region, central SaudiArabia – a case report. *Veterinarski Arhiv.*, 80 (4), 539−547.

[2] Burgemeister R., Leyk W. & Goessler R. (1975). – Untersuchungen uber Vorkommen von Parasitosen, bakteriellen undviralen Infektionskrankheiten bei Dromedaren in Sudtunesien. *Dtsch. Tierärztl. Wschr.*, 82, 341−354.

[3] Butt M.T., Weldon A., Step D., dela Hunta A. & Huxtable C.R. (1991). – Encephalitic listeriosis in two adult llamas (*Lamaglama*): clinical presentations, lesions and immunofluorescence of *Listeria monocytogenes* in brainstem lesions. *Cornell Vet.*, 81, 251−258.

[4] Fowler M.E. (2010). – Medicine and surgery of camelids. 3rd Ed. Wiley-Blackwell, Oxford, UK, 208−209.

[5] Frank N., Couetil L.L. & Clarke K.-A. (1998). –*Listeria monocytogenes* and *Escherichia coli* septicemia and meningoencephalitisin a 7-day-old llama. *Can. Vet. J.*, 39, 100−102.

[6] Hanichen T. & Wiesner H. (1995). – Erkrankungs- und Todesursachen bei Neuweltkameliden. *Tierätztl. Praxis*, 23, 515−520.

[7] Hamir A.N. & Moser G. (1998). – Immunohistopathological findings in an adult llama with listeriosis. *Vet. Rec.* 143, 477−479.

[8] Loehr C.V., Bildfell R.J., Heidel J.R., Valentine B.A. & Schaefer D. (2007). – Retrospective study of camelid abortions inOregon. *Vet. Pathol.*, 44 (5), 753.

[9] McLane M.J., Schlipf J.W., Margiocco M.L. & Gelberg H. (2008). – Listeria associated mural and valvular endocarditis inan alpaca. *J. Vet. Cardiol.*, 10 (2), 141−145.

[10] McLaughlin B.G., Greer S.C. & Singh S. (1993). – Listerial abortion in a llama. *J. Vet. Diagn. Invest.*, 5, 105−106.

[11] Makinde A.A., Majiyagbe K.A., Chwzey N., Lombin L.H., Shamaki D., Muhammad L.U., Chima J.C. & Garba A. (2001).– Serological appraisal of economic diseases of livestock in the one-humped camel (*Camelus dromedarius*) in Nigeria. *CamelNewslett.*, No.18, CARDN, September 2001, 62−73.

[12] Mayer H. & Gehring H. (1975). – Listeriose bei Lamas. In Verhandlungsbericht 17 Internationalen Symposium

ErkrankungZootiere, Tunis, Tunisia, 17, 307–312.

[13] Moro Sommo M. (1961—1962a). – Infectious diseases of alpacas. Ⅲ Tetanus. Enfermedades infecciosas de las alpacas. Ⅲ Tetanos. *Rev. Fac. Med. Vet. (Lima)*, 16–17, 160–162.

[14] Moro Sommo M. (1961–1962b). – Enfermedades infecciosas de las alpacas. *An. Premier Congr. Nac. Med. Vet. (Lima)*, 129.

[15] Oni O.O., Adesiyun A.A., Adekeye J.O. & Sai' du S.N. (1989). – Sero-prevalence of agglutinins to *Listeria monocytogenes* inNigerian domestic animals. *Rev. Elev. Méd. Vét. Pays Trop.*, 42 (3), 383–388.

[16] Ozbey G., Basri Ertas H. & Kok F. (2006). – Prevalence of *Listeria* species in camel sausages from retail markets in Aydinprovince in Turkey and RAPD analysis of *Listeria monocytogenes* isolates. *Irish Vet. J.*, 59 (6), 342–344.

[17] Rahimi E., Momtaz H. & Nozarpour N. (2010). – Prevalence of *Listeria* spp., *Campylobacter* spp. and *Escherichia coli* O157:H7isolated from camel carcasses during processing. *Bulg. J. Vet. Med.*, 13 (3), 179–185,

[18] Salehi T.Z., Mahzounieh M. & Jafer M.P. (2006). – Seroprevalence of precipitins to *Listeria monocytogenes* in camels in Iran. *In* Proceedings of the International Scientific Conference on Camels, Part 2, 12~14 May 2006, Buraydah, Saudi Arabia, 429–432.

[19] Seehusen F., Lehmbeckera A., Puffa C., Kleinschmidta S., Kleina S. & Baumgartner W. (2009). –*Listeria monocytogenes* septicaemia and concurrent clostridial infection in an adult alpaca (*Lama pacos*). *J. Comp. Path*ol., 139 (2–3), 126–129.

[20] Semrad S.D. (1994). – Septicemic listeriosis, thrombocytopenia, blood parasitism and hepatopathy in a Llama. *JAVMA*, 204 (2), 213–216.

[21] Tapia Cano F. (1965). – Investigacion de *Listeria monocytogenes* en la medula oblongata de alpacas aparentemente normales. B.Sc. Thesis,Universidad Nacional Mayor de San Marcos,Lima,Peru,1–45.

[22] Van Metre D.C.,Barrington G.M.,Parish S.M. & Tumas D.B. (1991). – Otitis media/interna and suppurativemeningoencephal omyelitis associated with *Listeria monocytogenes* infection in a llama. *JAVMA*,199 (2),236–240.

[23] Whitehead C.E.,Anderson D.E. & Saville W.J.A. (2009). – Neurologic diseases in llamas and alpacas. *Vet. Clin. N. Am. FoodAnimal Pract.*,25 (2),385–405.

[24] Wisser J. (1989). – Polyarthritis bei septikaemischer Listeriose eines Alpakas. *In* Proceedings of the International SymposiumErkrankung Wildtiere. Institut fur Zoo- und Wildtierforschung,Berlin,83–88.

（卢曾军译，殷宏校）

# 1.8 其他细菌性疫病

从个别的旧大陆和新大陆骆驼病例还分离到了许多不同的细菌，本节主要描述这些菌所引起的病症。这些细菌感染仅发现于个别骆驼，即使有一些对主人来说非常严重，但几乎没有或仅有轻微的经济影响。这些微生物主要包括以下所述菌：马链球菌兽疫亚种（*Streptococcus equi* subsp. *zooepidemicus*）是秘鲁“羊驼热（alpaca fever）”的病原体，从肯尼亚和索马里[8,20]骆驼及其奶样中均曾分离到该菌。其他链球菌可能会引起幼畜的腹泻，而且可从新大陆骆驼和旧大陆骆驼脓肿病灶中分离到链球菌。“羊驼热”可表现为急性型、亚急性型和慢性型，在南美洲高原地带是一种重要的疫病。传播途径主要为经口摄入病原体或直接接触感染动物。多发性浆膜炎、脑膜炎和内部脓肿、外部脓肿是与此病相关的一些临床病症。当发生全身性感染时，抗生素治疗是唯一有效的方法。Cebra等[5]用兽疫链球菌（*Strep. zooepidemicus*）气管内感染美洲驼，诱发了发热、厌食和精神沉郁等症状。对体液样本进行临床病理学分析证明有炎症反应。病原体快速扩散到身体的其他组织器官。Heller等[11]报道了1头澳大利亚青年单峰驼发生的链球菌性腹膜炎，Hewson和Cebra[12]在1头加拿大7月龄雄性美洲驼也发现了类似病症，病原体均为兽疫链球菌。这头美洲驼的来源牧场还同时饲养有14头不同年龄的美洲驼和几头马，但只有这头美洲驼死亡。此外，Younan等[20]从肯尼亚北部幼驼的关节周围脓肿病灶中分离到了兽疫链球菌，在其他一些病例存在兽疫链球菌与无乳链球菌（*Strep. agalactiae*）的混合感染。Pal'gov[16]报道了7例骆驼（双峰驼）的流产病例，从这些病例中没有证明存在布鲁氏菌（*Brucella*）感染，但是分离到了链球菌属某种（*Streptococcus spp.*）（对小鼠有致病性并造成兔的流产）。从约旦的一些关节严重肿大，出现跛行的单峰驼幼畜病灶中分离到了链球菌、化脓隐秘杆菌（*Arcanobacterium pyogenes*）和金黄色葡萄球菌（*Staphylococcus aureus*）[13]。

幽门螺旋杆菌（*Helicobacter pylori*）是人的一种重要病原菌，引起包括胃癌在内的一些严重疾病。对该病原菌的传播途径仍然不清楚，但是推测这种病原体感染是通过粪→口途径发生的。污染的水和食物可能是人感染本病的重要原因。从一些动物的胃黏膜，包括牛犊、马、猪和羊以及羊奶中均曾分离到该菌，说明这些动物可能是幽门螺旋杆菌的贮藏宿主和传染源。Rahimi和Kheirabadi[17]从55份散装单峰驼奶样中没有分离到幽门螺旋杆菌，但是，从2份样品（3.6%）中扩增到了幽门螺旋杆菌的DNA。

英国有1头表现消瘦的6月龄羊驼被确诊为耶尔森氏菌肠炎（*Yersinia enteritis*），从小肠、肠系膜淋巴结和肝组织样本中分离到了假结核耶尔森氏菌（*Yersinia pseudotuberculosis*）；而从西班牙加那利群岛患有心内膜炎的单峰驼体内同时分离到了假结核耶尔森氏菌和大肠杆菌[10]。

不同种属革兰氏阴性菌感染美国新生新大陆骆驼的病例也有文献报道[1]。从死前的血液样本和尸检样本中分离到的细菌种类包括大肠杆菌（×3），放线杆菌属某种（*Actinobacillus sp.*）（×1）和肺炎克雷伯氏菌（*Klebsiella pneumoniae*）(×1)。在阿联酋腹泻赛驼幼畜体内也分离到了结肠弯曲杆菌（*Campylobacter coli*）[15]。

从旧大陆骆驼和新大陆骆驼内部脓肿、外部脓肿病灶中分离到了许多不同的细菌种类。有1例单峰驼两侧下颌骨的后腹侧长出两个硬的大肉芽肿，造成皮肤损伤，从局部渗出物和坏死组织中分离出黏放线菌（*Actinomyces viscosus*）[14]。在用激光切除病灶后，每天用青霉素与链霉素联合治疗了4周。从2例单峰驼[3]和1例美洲驼[1]的下颌骨肉芽肿组织中也分离到了放线菌（*Actinomyces spp.*）（图142）。

下颌骨肿大的可能原因有牙齿异常和林氏放线杆菌（*Actinobacillus lignieresi*）或坏死梭杆菌（*Fusobacterium necrophorum*）感染。在加利福尼亚发现并鉴定了一个新种命名为羊驼放线杆菌（*Actinomyces lamae*）。Curasson[6]描述了骆驼放线菌（*Actinomyces cameli*）引起的疾病，以全身不同部位脓肿和皮肤坏死性病灶为特征，曾在埃及、苏丹和索马里的雨季大范围的发生，在印度也曾经很常见。已经鉴定化脓放线菌（*Actinomyces pyogenes*）或化脓棒状杆菌（*Corynebacterium pyogenes*）是引起美洲驼体表脓肿以及子宫内膜炎、乳房炎和体内脓肿的病原体，有些病例还伴有坏死梭菌（具有协同作用）、脆弱拟杆菌（*Bacteroides fragilis*）或化脓无色杆菌（*Achromobacter pyogenes*）。Ramosvara等[18]描述了1例羊驼放线菌性脾炎和小肠扭转病例。

从美洲驼蹄皮炎病例曾分离到了几种严格厌氧菌，如，脆弱拟杆菌和坏死梭杆菌，而且还分离到了金黄色葡萄球菌（图143）。

最近，有几位研究人员，如Fahmy等[7]、Tejedor-Junco等[19]、Brightman等[4]和Gionfriddo等[9]研究了新大陆骆驼和旧大陆骆驼的眼病，与其他动物的眼病具有相似性。结膜炎、角膜炎、眼内炎、葡萄膜炎和全眼球炎都有描述，不同的细菌种类与上述眼病有关，包括犬摩拉克氏菌（*Moraxella canis*）[19]、金黄色葡萄球菌和液化摩拉克氏菌（*Moraxella liquefaciens*）[4]。其他的眼部疫病或功能紊乱还有白内障、水肿、结膜囊肿、寄生虫性结膜炎和眼睛创伤。Gionfriddo等[9]有一篇关于美洲驼疾病的详细综述可供大家参考。

蹄部感染治疗起来相当困难，也是因为造成感染的这些厌氧菌对大多数抗生素都有耐药性。由坏死梭形杆菌感染引起的坏死杆菌病不仅发生于蹄叉部位，而且也发生于口唇、腭、舌、咽和喉部。青年骆驼通常有感染的危险。Zanolari等[21]描述了严重溃疡性蹄皮炎病例，从病料中可以分离到葡萄球菌、棒状杆菌属和坏死梭形杆菌。从美洲驼肺和皮肤脓肿病灶中分离培养到了星形诺卡菌（*Nocardia asteroides*）。从骆驼的眼部也可以分离到金黄色葡萄球菌和液化摩拉克氏菌。最近，从加那利群岛单峰驼角膜结膜炎病例中分离到了两株摩拉克氏菌，16S rRNA序列分析显示，分离菌与犬摩拉克氏菌的序列具有高度相似性[19]。

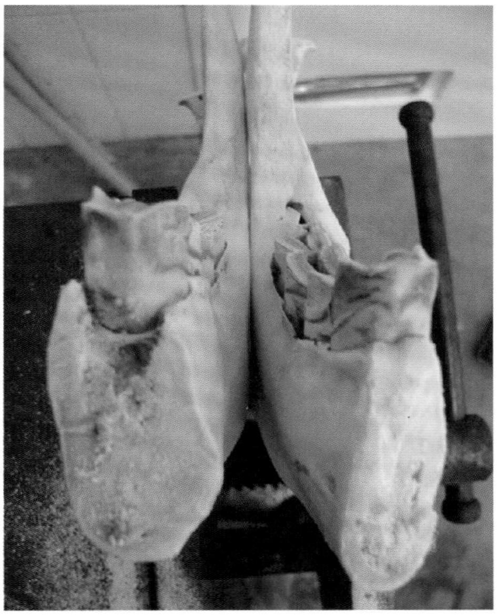

图142　单峰驼下颌骨骨髓炎

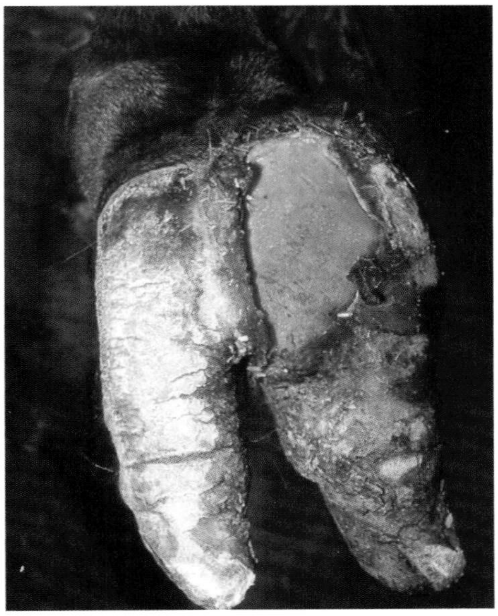

图143　美洲驼传染性蹄皮炎
（图片为美国M.E. Fowler教授惠赠）

## 参考文献


[1] Adams R. & Garry F.B. (1992). – Gram-negative bacterial infection in neonatal New World camelids: six cases (1985—1991). *JAVMA*,201 (9),1419–1424.

[2] Anon. (2007). – Yersiniosis. *Vet. Rec.*,160 (10),320.

[3] Arora R.G.,Kalra D.S. & Gupta R.K.P. (1973). – Granuloma actinomyces in camel (*Camelus dromedarius*). *Archiva Veterinaria*,10 (2),49–52.

[4] Brightman A.H.,McLAughlin S.A. & Brumley V. (1981). – Keratoconjunctivitis in a llama. *Vet. Med.*,1776–1777.

1
细菌病

[5] Cebra C.K.,Heidel J.R.,Cebra M.L.,Tornquist S.J. & Smith B.B. (2000). – Pathogenesis of *Streptococcus zooepidemicus* infection after intradermal inoculation in llamas. JVR,61 (12),1525–1529.

[6] Curasson G. (1947). – Le chameau et ses maladies. Vigot Freres,Paris.

[7] Fahmy L.S.,Hegazy A.A.,Abdelhamid M.A.,Hatem M.E. & Shamaa A.A. (2003). – Studies on eye affections among camelsin Egypt: clinical and bactériological studies. *Sci. J. King Faisal Univ. (Basic and Applied Sciences)*,4 (2),159–175.

[8] Fowler M.E. (2010). – Medicine and surgery of camelids. 3rd Ed. Wiley-Blackwell,Oxford,UK,173–231.

[9] Gionfriddo J.R.,Gionfriddo J.P. & Krohne S.G. (1997). – Ocular diseases of llamas: 194 cases (1980–1993). *JAVMA*,210(12),1784–1786.

[10] 1Gutierrez C.,Schulz U.,Corbera J.A.,Morales I. & Tejedor M.T. (2004). – Vegetative endocarditis associated with *Escherichiacoli* in a dromedary camel. *Vet. Res. Commun.*,28 (6),455–459.

[11] Heller M.,Anderson D. & Silveira F. (1998). – Streptococcal peritonitis in a young dromedary camel. *Aust. Vet. J.*,76 (4),253–254.

[12] Hewson J. & Cebra C.K. (2001). – Peritonitis in a llama caused by *Streptococcus equi* subsp. *zooepidemicus*. *Can. Vet. J.*,42,465–467.

[13] Ismail Z.B.,Al-Rukibat R.,Al-Tarazi Y. & Al-Zghoul B.M. (2007). – Synovial fluid analysis and bacterial findings in arthriticjoints of juvenile male camel (*Camelus dromedarius*) calves. *J. Vet. Med. Physiol. Pathol. Clin. Med.*,54 (2),66–69.

[14] Kilic N. & Kirkan S. (2004). – Actinomycosis in a one-humped camel (*Camelus dromedarius*). *J. Vet. Med. A Physiol. Pathol. Clin. Med.*,51 (7–8),363–364.

[15] Moore J.E.,McCalmont M.,Xu J.,Nation G.,Tinson A.H.,Crothers L. & Harron D.W. (2002). – Prevalence of faecal pathogensin calves of racing camels (*Camelus dromedarius*) in the United Arab Emirates. *Trop. Anim. Health. Prod.*,34,283–287.

[16] Pal'gov A.A. (1954). –*Trud. Inst. Vet. (Alma Ata)*,6,30.

[17] Rahimi E. & Kheirabadi E.K. (2012). – Detection of *Helicobacter pylori* in bovine,buffalo,camel,ovine and caprine milk inIran. *Foodborne Pathog. Dis.*,9,453–456.

[18] Ramosvara J.A.,Kopcha M.,Richter E.,Watson G.L.,Patterson J.S.,Juan-Sallés C. & Yamini B. (1998). – Actinomycoticsplenitis and intestinal volvulus in an alpaca (*Lama pacos*). *J. Zoo Wildlife Med.*,29 (2),228–232.

[19] Tejedor-Junco M.T.,Gutierrez C.,González M.,Fernández A.,Wauters G.,De Baere T.,Deschaght P. & Vaneechoutte M.(2010). – Outbreaks of keratoconjunctivitis in a camel herd caused by a specific biovar of *Moraxella canis*. *J. Clin. Microbiol.*,48 (2),596–598.

[20] Younan M.,Estoepangestie S.,Cengiz M.,Alber J.,El-Sayed A. & Lämmler C. (2005). – Identification and molecularcharacterization of *Streptococcus equi* subsp. *zooepidemicus* isolated from camels (*Camelus dromedarius*) and camel milk inKenya and Somalia. *J. Vet. Med. B*,52,142–146.

[21] Zanolari P.,Meylan M.,Sager H.,Herli-Gygi M.,Rufenacht S. & Roosje P. (2008). – Dermatologie bei Neuweltkameliden.Teil 2: Ubersicht der dermatologischen Erkrankungen. *Tierärztl. Praxis,*36 (G),421–427.

(卢曾军译，殷宏校)

# 1.9 致命污染物

虽然致命污染物与传染病没有太大关系，但还是有必要在本书中提及。这是因为在过去十年中，饲养骆驼的国家因污染物造成的发病比例上升，另外，人们应知晓这一新的问题。塑料污染已是全球性的问题，其残留物可能会进入人的食物链[1]。

因食入塑料而造成骆驼死亡的第一个病例15年前发生于阿联酋，对单峰驼病理解剖才发现病因。大多数骆驼消瘦、停食数天或数周、精神沉郁、虚弱、体温在37℃以下。有些单峰驼突然倒地，数小时之内死亡。在尸体解剖时，在第一胃内可见到数个新鲜的或钙化的塑料块，或者重达50kg的塑料团(图144)。从此之后，已有数百起因食入塑料而造成单峰驼和其他动物死亡的报道，其中包括牛、绵羊、山羊和瞪羚，甚至还有濒临灭绝的阿拉伯大羚羊(Arabian Oryx)。一个反对塑料等“致命污染物”的活动组织，曾试图说服政府禁止使用塑料。经全面调查后发现骆驼食入塑料后死亡的原因有以下几点。

（1）突然死亡：因塑料袋完全阻塞肠道，或虽然肠道没有完全阻塞，但因摄入塑料而继发梭菌性肠毒血症。在后一种情况下，可见到本书1.1.1节所描述的病变，可分离到产毒素厌氧菌。

（2）因器官衰竭而在2～3周内发生的死亡：在这种情况下，食入的塑料释放毒素，进入循环系统，造成肝功能指标[谷氨酸草酰乙酸转氨酶GOT（AST）：γ-谷酰基转移酶（γ-GT），谷氨酸丙酮酸转氨酶-GPT（ALT）]和肾脏功能指标(血尿素氮BUN，肌酸酐)迅速上升，最终器官衰竭。

（3）因饥饿造成的慢性死亡：塑料盖、塑料瓶体、塑料袋、捆扎饲草的塑料绳和其他塑料器具，被食入后长年累月在骆驼胃中积聚，逐渐钙化，形成坚硬的塑料“大石块”，重量可达10～50kg，有时会塞满整个第一胃，形状也如同第一胃。如此大的钙化塑料块会影响骆驼的采食，由于胃经常充盈，使得采食量逐渐减少，最后完全停止。

a 因十二指肠入口阻塞而突然死亡的单峰驼第一胃内的塑料袋

b 在骆驼第一胃内因钙化而形成的坚硬塑料“大石块”，重量有50kg，大约含有5000个塑料袋

图144 a，b

## 参考文献

[1] Weisman A. (2007). – The world without us. Thomas Dunne Books, St. Martin’s Griffin, New York,112 – 128.

# 2 病毒病

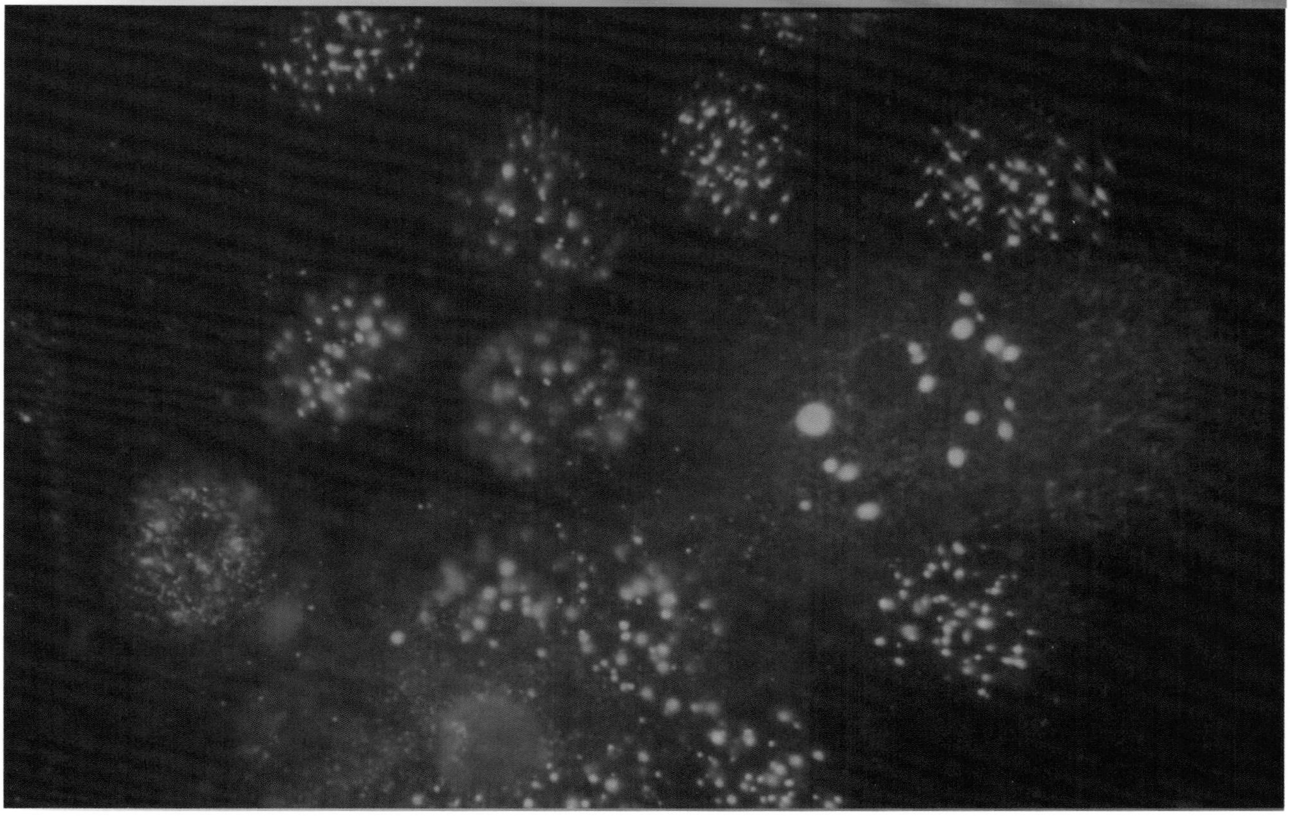

不同属的驼科动物在生理学和解剖学的特征非常相似，以往人们认为它们在对病毒的易感性方面没有差别。然而，在过去10年中一些新的研究证实，不同种的骆驼至少对一种病毒的易感性存在差异。已证实旧大陆骆驼中两个密切相关的两个种，单峰驼和双峰驼，它们对口蹄疫病毒的易感性有明显的差异。双峰驼相对地易被口蹄疫病毒感染，而单峰驼对高剂量的O型和A型口蹄疫病毒有抵抗力。

随着实验室感染实验和田间调查研究的大量开展，有关骆驼疾病的相关知识的积累日趋丰富，骆驼病毒病信息缺乏的局面也不复存在。这一点在新大陆骆驼尤为突出[3]，其原因是在欧洲和北美，饲养新大陆骆驼日益成为人们的爱好或是成为农场饲养家畜，推动人们去探讨有关管理、卫生和疫病等方面的知识，例如，美国最近启动了牛病毒性腹泻抗体检测计划。据估计，美国目前骆驼的数量为300 000 ~ 32 500峰，大约有8 100个羊驼养殖场和26000个美洲驼养殖场[1]。虽然有大量的有关骆驼对病毒感染和其造成的危害的数据，但是，大量的非常重要的危害小反刍动物和牛的病毒病，骆驼是否也易感则知之甚少，尤其是双峰驼。不过，由于人们对驼科动物的兴趣越来越浓，这些空白将在未来几年内迅速被填补。如本书前言所述，世界动物卫生组织（OIE）邀请多个国家的专家，编撰了骆驼传染病名录，包括病毒病、细菌病和寄生虫病，附于本书的最后。名录中有已知的小反刍兽疫、口蹄疫（双峰驼）、鼻疽、嗜血支原体和无浆体，但对于大多数骆驼的疾病，由于缺乏兴趣或经费，没有开展较为细致的调查研究。名录中有些为再发传染病[4]，还有一些疫病，如，裂谷热和西尼罗河热，可能会因为单峰驼在穿越大片沙漠时，因寻找可休息和饮水的绿洲而传播到很远的地方。这些区域因有水而滋生大量的生物媒介，它们可将病毒传播给其他家畜[1]。气候变化而导致的长时间干旱，随后又在短期内大量降水，不仅对骆驼的健康，而且对所有家畜的健康都造成严重的影响[2]。

由病毒引起的驼科动物疾病的数量不断增加，Wernery 和Kaaden[5]编辑的“致病性病毒感染”名录中已加入小反刍兽疫、牛病毒性腹泻、马鼻病毒流产、裂谷热、西尼罗河热、蓝舌病和牛疱疹病毒病等。在表55中列出了骆驼的致病性病毒感染和非致病性病毒感染目录。

虽然骆驼科动物对许多能引起家畜发病的病



表55 骆驼的致病性病毒感染和非致病性病毒感染

| 致病性病毒感染 | 非致病性病毒感染 |
|---|---|
| 狂犬病 | 牛瘟 |
| 博尔纳病 | 非洲马瘟 |
| 骆驼痘 | 呼吸道病毒 |
| 乳头状瘤病 | 反转录病毒 |
| 口蹄疫 | |
| 水泡性口炎 | |
| 小反刍兽疫 | |
| 牛病毒性腹泻 | |
| 裂谷热 | |
| 蓝舌病 | |
| 西尼罗河热 | |
| 东方马脑炎 | |
| 新生畜腹泻（轮状病毒、冠状病毒） | |
| 马和牛疱疹病毒病 | |
| 甲型流感 | |
| 马鼻病毒流产 | |

原是易感的，但是总体来讲，骆驼的传染病比较少，需进一步研究以阐明此现象的本质。

在骆驼上开展了关于病毒的大量血清流行病学调查，表56列出了检测到的抗体和发现的毒株。

血清学检测结果的意义是有限的，它只能证实动物是否接触到病毒并产生抗体，不能说明动物是否发病或发病的程度。血清流行病学调查显示，骆驼能产生针对大量的“致病性”病毒的抗体，但并没有发病。在病毒病这一节的最后，还列出了在热带和亚热带地区广泛存在的“不常见虫媒病毒”（表57），虽然已在骆驼及其寄生蜱的体内分离到这些病毒，其中，有些能引起非常严重的人类疾病，但还不清楚它们是否引起骆驼发病。

另外，在本节最后，还列出了次要的骆驼病毒性疾病名录。

表56　骆驼体内分离的病毒和发现的抗体（骆驼痘除外）

| 疫病/病毒 | 抗原 | 抗体 | 阳性率（%） | 国家/地区 | 作者 | 年份 |
|---|---|---|---|---|---|---|
| 非洲马瘟 | — | x | 5.0 | 埃及 | Awad等 | 1981 |
| | x | x | 23.2 | 苏丹 | Salama等 | 1986 |
| | — | x | 5.6 | 埃及 | Salama等 | 1986 |
| | x | x | 23 | 苏丹 | 美国农业部 | 1988 |
| | — | o | 0.0 | 东非 | Binepal等 | 1992 |
| | — | x | 10.4 | 尼日尼亚 | Baba等 | 1993 |
| 蓝舌病 | — | x | 14.3 | 埃及 | Hafez 和Ozawa | 1973 |
| | — | x | 5.9 | 伊朗 | Afshar和Kayvanfar | 1974 |
| | — | x | 4.9 | 苏丹 | Eisa | 1980 |
| | x | x | 5.6 ~ 16.4 | 苏丹 | Abu Elzein | 1984 |
| | — | x | 16.6 | 苏丹 | Abu Elzein | 1985b |
| | — | x | 13.0 | 也门 | Stanley | 1990 |
| | — | x | 81.0 | 博茨瓦纳 | Simpson | 1979 |
| | — | x | 23.0 | 以色列 | Barzilai | 1982 |
| | — | x | 67.0 | 沙特阿拉伯 | Hafez等 | 1984 |
| | — | x | 21.0（羊驼） | 秘鲁 | Rivera等 | 1987 |
| | — | x | 13.0 | 也门 | Stanley | 1990 |
| | — | x | 1.5（美洲驼） | 美国 | Picton | 1993 |
| | — | x | 5.0 | 阿联酋 | 中央兽医研究实验室年报 | 1998 |
| | — | x | 58 | 沙特阿拉伯 | Abu Elzein等 | 1998 |
| | — | x | 0（原驼） | 阿根廷 | Karesh等 | 1998 |
| | — | o | 0.0（美洲驼） | 阿根廷 | Puntel等 | 1999 |
| | — | x | 58 | 沙特阿拉伯 | Ostrowski | 1999 |
| | — | x | 9,13,7（不同地区） | 印度 | Chandel 和Kher | 1999 |
| | — | x | 18 | 印度 | Chandel等 | 2001 |
| | — | x | 10 | 印度 | Mallik等 | 2002 |
| | — | x | 27（AGID），39（cELISA） | 印度 | Chandel等 | 2004 |
| | — | x | 10.6 | 沙特阿拉伯 | Mohammed等 | 2003 |

续表

| 疫病/病毒 | 抗原 | 抗体 | 阳性率（%） | 国家/地区 | 作者 | 年份 |
|---|---|---|---|---|---|---|
| | — | x | 1.5 | 沙特阿拉伯 | Al-Aflaq等 | 2006 |
| | — | x | 50 | 伊朗 | Mahdavi等 | 2006 |
| | — | x | 30（单峰驼） | 印度 | Chauhan等 | 2007 |
| | — | x | 26（AGID），38（cELISA） | 印度 | Patel等 | 2007 |
| | x | x | 0（南美骆驼） | 德国 | Schulz等 | 2008 |
| | — | x | 21 | 阿联酋 | Wernery等 | 2008 |
| | — | x | 0（双峰驼）0（单峰驼） | 西班牙 | Bellon等 | 2010 |
| | — | x | 0（新大陆骆驼） | 瑞士 | Zanolari | 2010 |
| | — | x | 14（群），47（群） | 德国 | Schulz等 | 2012 |
| | — | x | 11-25 | 摩洛哥 | Touil等 | 2012 |
| 博尔纳病 | x | — | 新大陆骆驼 | 德国（动物园） | Altman | 1975 |
| | x | — | 新大陆骆驼 | 德国（动物园） | Altman等 | 1976 |
| | x | — | 新大陆骆驼 | 德国（动物园） | Schuepel等 | 1994 |
| | — | — | 10,4.8,25 | 肯尼亚、阿联酋、以色列（单峰驼） | Fluess | 2002 |
| | x | — | — | | Kobera和Pohle | 2004 |
| | x | — | — | | Jacobson等 | 2010 |
| 跳跃病 | x | — | — | 赫布里底群岛 | Macaldwie等 | 2005 |
| | x | — | — | 英国 | Cranwell等 | 2008 |
| 东部马脑炎 | x | — | | 美国 | Nolen-Watson等 | 2007 |
| 牛病毒性腹泻 | — | x | 3.9 | 突尼斯 | Burgemeister等 | 1975 |
| | — | x | 6.7 | 阿曼 | Hedger等 | 1980 |
| | — | x | 13 | 美国（单峰驼） | Doyle和Heuschler | 1983 |
| | — | x | 15.7 | 苏丹 | Bornstein等 | 1987/88 |
| | — | x | 11 | 秘鲁（美洲驼） | Rivera等 | 1987 |
| | x | x | 3.4 | 索马里 | Bornstein | 1988 |
| | — | x | 0.0 | 吉布提 | Bohrmann等 | 1988 |
| | — | x | 9.2（种驼）3.6（赛驼） | 阿联酋 | Wernery和Wernery | 1990 |
| | x | — | — | 埃及 | Hegazy等 | 1995/98 |
| | — | x | 11.0 | 埃及 | Hegazy等 | 1993 |
| | — | x | 4.3 | 埃及 | tantawi等 | 1994 |
| | x | — | 6.4（种驼）0.5（赛驼） | 阿联酋 | 中央兽医研究实验室年报 | 1998 |
| | x | — | 无 | 阿根廷（原驼） | Karesh等 | 1998 |
| | x | — | 52.5 | 埃及 | Zaghhana | 1998 |
| | — | — | 2.05（美洲驼） | 阿根廷 | Puntel等 | 1999 |
| | x | x | 无 | 美国（美洲驼) | Belknap等 | 2000 |

续表

| 疫病/病毒 | 抗原 | 抗体 | 阳性率（%） | 国家/地区 | 作者 | 年份 |
|---|---|---|---|---|---|---|
| | — | — | 实验结果 | 美国（美洲驼，羊驼） | Wentz等 | 2003 |
| | x | — | BVDV-1b | 美国（羊驼） | Carmen等 | 2005 |
| | x | — | BVDV-1b | 美国（羊驼） | 匿名 | 2005 |
| | x | — | BVDV-1b | 美国（羊驼） | 匿名 | 2006a |
| | x | — | BVDV-1b | 美国（羊驼） | 匿名 | 2006b |
| | — | x | 1.4 | 欧洲（动物园） | Probst等 | 2007 |
| | — | x | 无 | 阿联酋（单峰驼） | Taha | 2007 |
| | — | x | 25.4（群） | 美国 | Kelling | 2008 |
| | — | x | 1.6 | 阿联酋（奶单峰驼） | Wernery等 | 2008 |
| | — | x | 4.6 | 瑞士（美洲驼，羊驼） | Danuser等 | 2009 |
| | — | x | 20.0 | 美国（羊驼） | Shimeld | 2009 |
| | x | — | 46毒株（BVDV-1b） | 美国（羊驼） | Kim等 | 2009 |
| | — | x | 20 | 美国（羊驼） | Topliff等 | 2009 |
| | — | x | 5.8和3.6(不同年份) | 瑞士（南美骆驼） | Mudry等 | 2010 |
| | x | x | 23（血清学），30.4（抗原） | 中国 | Gao等 | 2013 |
| 疱疹病毒（1型马疱疹病毒，1型牛疱疹病毒） | — | x | 5.8 | 突尼斯 | Burgemester等 | 1975 |
| | — | o | 0.0 | 阿曼 | Hedger等 | 1980 |
| | — | o | 0.0 | 苏丹 | Bornstein和Musa | 1987 |
| | — | x | 5.0（羊驼） | 秘鲁 | Rivera等 | 1987 |
| | — | o | 0.0 | 吉布提 | Borhmann等 | 1988 |
| | | o | 0.0 | 索马里 | Bornstein | 1988 |
| | | o | 0.0 | 阿联酋 | Werner和Wernery | 1990 |
| | x | — | —（美洲驼） | 美国 | Williams等 | 1991 |
| | — | x | 4.4（美洲驼） | 美国 | Picton | 1993 |
| | — | x | 16.7（美洲驼）16.2（羊驼) | 秘鲁 | Rosadio等 | 1993 |
| | — | — | 0.7（美洲驼） | 美国 | Picton | 1993 |
| | x | — | —（美洲驼） | 美国 | Mattson | 1994 |
| | — | x | 11.0（羊驼） | 秘鲁 | Rivera等 | 1997 |
| | — | o | 0.0 | 阿联酋 | 中央兽医研究实验室年报 | 1998 |
| | — | x | 0.77 | 阿根廷 | Puntel等 | 1999 |
| | x | — | —（羊驼） | 美国 | Pursell等 | 1979 |
| | x | — | —（美洲驼） | 美国 | Jenkins | 1985 |
| | x | | —（美洲驼） | 美国 | Rebhun等 | 1988 |
| | x | — | 实验结果（美洲驼） | 美国 | House等 | 1991 |

续表

| 疫病/病毒 | 抗原 | 抗体 | 阳性率（%） | 国家/地区 | 作者 | 年份 |
|---|---|---|---|---|---|---|
| | x | — | —（双峰驼） | 美国 | Bidfell等 | 1996 |
| | — | o | 0.0 | 阿联酋 | 中央兽医研究实验室年报 | 1998 |
| 恶性卡他热 | — | o | 0.0 | 阿根廷 | Puntel等 | 1999 |
| 接触传染性脓疱 | x | o | 0.0（羊驼） | 秘鲁 | Preston Smith | 1940/47 |
| | x | — | 0.0 | 哈萨克斯坦 | Buchnev等 | 1969 |
| | x | — | — | 俄罗斯 | Tulepbaev | 1969 |
| | x | o | 新大陆骆驼 | 南美洲 | Moro Sommo | 1971 |
| | x | — | — | 蒙古国 | Dashatseren等 | 1984 |
| | x | — | — | 肯尼亚 | Munz等 | 1986 |
| | x | — | — | 索马里 | Moallin和Zussin | 1988 |
| | x | — | — | 苏丹 | Ali等 | 1991 |
| | — | x | 37.9（发病群） | 肯尼亚 | Gitao | 1994 |
| | — | x | 0 ~ 68（健康群） | 利比亚 | Azwai等 | 1995 |
| | x | — | — | 阿联酋 | Wernery等 | 1997 |
| | x | — | — | 沙特阿拉伯 | Abu Alzein等 | 1998 |
| | x | — | -新大陆骆驼 | 南美洲 | Fowler | 1998 |
| | x | — | — | 利比亚 | Azwai等 | 1998 |
| | x | — | — | 苏丹 | Khalafalla | 2006 |
| 马鼻病毒甲型 | x | — | 50（单峰驼0） | 阿联酋 | Wernery等 | 2008 |
| 口蹄疫病毒 | — | x | 2.6 | 埃塞俄比亚 | Richard | 1979 |
| O, A, C, As1, SAT1 | — | x | 无 | 肯尼亚 | Paling等 | 1979 |
| O, C, SAT2 | x | — | — | 阿富汗 | Pringle | 1880 |
| O | x | o | 无 | 阿曼 | Hedger等 | 1980 |
| | — | x | 2.6 | 尼尔日 | Richard | 1986 |
| | — | x | 5.4 | 埃及 | Mousa等 | 1986 |
| | x | — | — | 埃及 | Moussa等 | 1987 |
| | — | x | 无（新大陆骆驼） | 南美洲 | Rivera等 | 1987 |
| | — | x | 10.6（ICFT），23.5（ELISA） | 埃及 | Abou Zaid | 1991 |
| | — | x | 无 | 埃及 | Hafez等 | 1993 |
| | — | x | 无 | 沙特阿拉伯 | Hafez等 | 1993 |
| | — | x | 无（新大陆骆驼） | 南美洲 | David等 | 1993 |
| | — | x | 无（新大陆骆驼） | 南美洲 | Viera等 | 1995 |
| | — | x | 实验结果（新大陆骆驼） | 南美洲 | Foundevila等 | 1995 |
| | — | x | 0 | 沙特阿拉伯 | Farag等 | 1998 |

续表

| 疫病/病毒 | 抗原 | 抗体 | 阳性率（%） | 国家/地区 | 作者 | 年份 |
|---|---|---|---|---|---|---|
| | — | x | 24.3 | 埃及 | Moussa和Yousuf | 1998 |
| | — | x | 无（新大陆骆驼） | 南美洲 | Karesh等 | 1998 |
| 组特异 | — | o | 无（美洲驼） | 阿根廷 | Puntel等 | 1999 |
| | — | x | 无（新大陆骆驼） | 南美洲 | Puntel等 | 1999 |
| | — | x | 无 | 阿联酋 | Wernery资料 | 2003 |
| | — | x | 无 | 肯尼亚 | Younan资料 | 2003 |
| | — | x | 无 | 阿联酋 | Wernery等 | 2007 |
| 流感 | x | — | — | 蒙古国 | Anchlan等 | 1996 |
| | — | o | 无 | 阿联酋 | 中央兽医研究实验室年报 | 1998 |
| A | x | — | — | 蒙古国 | Lvov等 | 1983 |
| A | — | x | 4.7 | 苏丹 | El-Amin和Kheir | 1985 |
| A | — | x | 0.6 | 尼日尼亚 | Olaleye等 | 1989 |
| B | — | x | 12.7 | 尼日尼亚 | Olaleye等 | 1989 |
| 疑似流感 | x | — | — | 索马里 | Somac-Sarec | 1982 |
| 新生畜腹泻（轮状病毒、冠状病毒） | — | x | 50（轮状病毒） | 摩洛哥 | Mahin等 | 1983 |
| | — | x | —（轮状病毒，羊驼） | 南美洲 | Rivera等 | 1987 |
| | x | — | —（轮状病毒） | 阿联酋 | Mohammed等 | 1998 |
| | — | x | 87.7（轮状病毒，美洲驼） | 阿根廷 | Puntel等 | 1999 |
| | x | — | —（轮状病毒，原驼） | 阿根廷 | Parreno等 | 2001 |
| | x | | —（冠状病毒，单峰驼） | 美国 | | |
| | x | — | —（单峰驼） | 苏丹 | Ali | 2003 |
| | x | — | —（轮状病毒，原驼） | 南美洲 | Prreno等 | 2004 |
| | x | — | 14（轮状病毒，单峰驼） | 苏丹 | Ali等 | 2005a |
| | — | x | 48（轮状病毒，单峰驼） | 苏丹 | Ali等 | 2005b |
| | x | — | —（冠状病毒，羊驼） | 美国 | Jin等 | 2007 |
| | x | — | 14（轮状病毒，单峰驼） | 苏丹 | Ali等 | 2008 |
| | x | — | —（冠状病毒，羊驼） | 美国 | Genova等 | 2008 |
| | — | x | 100（轮状病毒） | 阿根廷 | Marcoppido等 | 2010 |
| | x | — | 21.4 | 秘鲁 | Lopez等 | 2011 |
| | x | — | 3（A型，轮状病毒) | 阿根廷 | Badaroco等 | 2013 |
| | x | — | 35.7（冠状病毒） | 秘鲁 | Lopez等 | 2011 |

续表

| 疫病/病毒 | 抗原 | 抗体 | 阳性率（%） | 国家/地区 | 作者 | 年份 |
|---|---|---|---|---|---|---|
| 乳头状瘤病 | x | — | | 印度 | Sadan等 | 1980 |
| | x | — | | 索马里 | Munz等 | 1990 |
| | x | — | | 阿联酋 | Kinne和Wernery | 1998 |
| | x | — | | 苏丹 | khlafalla等 | 1998 |
| | x | — | | 阿联酋 | Wernery和Kaaden | 2002 |
| | x | — | | 土耳其（单峰驼） | Kilic等 | 2010 |
| 呼吸道病毒 | — | x | 1.3 | 尼日尼亚 | Olaleye等 | 1989 |
| 腺病毒（牛腺病毒8型，分离株7649） | — | x | 93（美洲驼） | 美国 | Picton | 1993 |
| | x | — | —（美洲驼） | 美国 | Galbreath等 | 1994 |
| | x | x | —（美洲驼、羊驼） | 美国 | Mattson | 1994 |
| | — | x | 5.13（美洲驼） | 阿根廷 | Puntel | 1999 |
| 副流感 | — | x | 22.3 | 尼日尼亚 | Nigeria | 1989 |
| 1 | — | x | 2.5 | 尼日尼亚 | Nigeria | 1989 |
| 2 | — | x | 18.5 | 尼日尼亚 | Nigeria | 1989 |
| 3 | — | x | 3.8 | 埃及 | Singh | 1967 |
| 3 | — | x | 99 | 乍得 | Maurice等 | 1968 |
| 3 | — | x | 80.8 | 突尼斯 | Burgemeister等 | 1975 |
| 3 | — | x | 66 | 索马里 | Frigeri和Arush | 1979 |
| 3 | — | x | 80 | 阿曼 | Hedger等 | 1980 |
| 3 | — | x | 66.7 | 索马里 | Arush | 1982 |
| 3 | — | x | 37 | 尼日尔 | Richard等 | 1985 |
| 3 | — | x | 81.1 | 苏丹 | Bornstein和Musa | 1987 |
| 3 | — | x | 17.3 | 吉布提 | Bohrmann等 | 1988 |
| 3 | — | x | 81.3 | 苏丹 | Bornstein等 | 1988 |
| 3 | — | x | 42.8 | 索马里 | Bornstein | 1988 |
| 呼吸道合胞体病毒 | — | x | 0.6 | 尼日尔 | Olaleye等 | 1989 |
| | x | — | LA7649株（美洲驼实验数据） | 美国 | Tillman | 2001 |
| | x | — | 4（P13株，单峰驼） | 苏丹 | Intisar等 | 2010a |
| | — | x | 82（单峰驼？） | 苏丹？ | Intisar等 | 2010a |
| | — | x | 27.3(牛呼吸道合胞体病毒，单峰驼) | 苏丹 | Intisar等 | 2010b |
| | x | — | 1.4 | | | |
| | — | x | 100（P13，原驼） | 阿根廷 | Marcoppido等 | 2011 |
| | — | x | 9（P13，羊驼） | 南美洲 | Rosadio等 | 2011 |
| 小反刍兽疫 | — | x | 4.2 | 埃及 | Ismail等 | 1992 |
| | x | — | 7.8 | 埃塞俄比亚 | Roger等 | 2001 |
| | — | x | — | 苏丹 | Haroun等 | 2002 |

续表

| 疫病/病毒 | 抗原 | 抗体 | 阳性率（%） | 国家/地区 | 作者 | 年份 |
|---|---|---|---|---|---|---|
| | — | x | — | 埃塞俄比亚 | Abraham等 | 2005 |
| | x | x | — | 苏丹 | Lhalafalla等 | 2005/10 |
| | x | — | 11例 | 沙特阿拉伯 | Abd El-Hakim | 2006 |
| | — | x | 0 | 阿联酋 | Wernery等 | 2007 |
| | — | x | — | 尼日尼亚 | Abubakar等 | 2008 |
| | — | x | — | 土耳其 | Albayrak等 | 2010 |
| 狂犬病 | x | — | —（羊驼） | 南美洲 | Moro Sommon | 1958/59 |
| | x | — | —（单峰驼） | 毛里塔尼亚 | Bar等 | 1981 |
| | x | — | —（单峰驼） | 索马里 | Arush | 1982 |
| | x | — | —（单峰驼） | 阿曼 | Ata等 | 1993 |
| | x | — | —（单峰驼） | 阿联酋 | Wernery和Kumar | 1993 |
| | x | — | —（单峰驼） | 阿联酋 | Afzai等 | 1993 |
| | x | — | —（美洲驼） | 南美洲 | Miller | 1994 |
| | x | — | —（单峰驼） | 尼日尔 | Bloch和Diallo | 1995 |
| | x | — | —（单峰驼） | 以色列 | Perl等 | 1996 |
| | x | — | —（单峰驼） | 印度 | Kumar和Jindal | 1997 |
| | x | — | —（单峰驼） | 苏丹 | El Mardi和Ali | 2001 |
| | x | — | —（单峰驼） | 苏丹 | Ali等 | 2001 |
| 疑似狂犬病 | x | — | —（单峰驼） | 沙特阿拉伯 | Al-Dubiab | 2007 |
| 反录病毒 | — | o | 0.0 | 印度 | Chauhan等 | 1986 |
| 牛白血病 | — | o | 0.0（羊驼） | 秘鲁 | Rivera等 | 1987 |
| | — | o | 0.0 | 阿联酋 | Wernery和Wernery | 1990 |
| | — | o | 0.0（美洲驼） | 美国 | Picton | 1993 |
| | o | o | 0 | 美国 | Cebra等 | 1995 |
| | o | o | 0 | 美国 | Wenker等 | 2002 |
| | o | o | 0 | 阿联酋 | Kinne和Wernery | 2006 |
| | x | x | 1例美洲驼 | 美国 | Lee等 | 2013 |
| 裂谷热 | — | x | 45.0 | 肯尼亚 | Scott等 | 1963 |
| | x | — | — | 埃及 | Imam等 | 1978 |
| | x | — | — | 苏丹 | Eisa | 1981 |
| | — | x | — | 突尼斯 | Slama | 1984 |
| | — | x | 22 | 肯尼亚 | Davies等 | 1985 |
| | — | x | 29 | 毛里塔尼亚 | Saluzzo等 | 1987 |
| | — | x | 33 | 尼日尼亚 | Odaleye等 | 1996 |
| | — | x | — | 毛里塔尼亚 | Nabeth等 | 2001 |
| | — | | — | 沙特阿拉伯 | Shoemaker等 | 2002 |
| | | x | — | — | Paweska等 | 2005 |

续表

| 疫病/病毒 | 抗原 | 抗体 | 阳性率（%） | 国家/地区 | 作者 | 年份 |
|---|---|---|---|---|---|---|
| | — | x | 0.35 | 阿联酋 | Wernery等 | 2008 |
| | x | — | —（羊驼） | 南非 | Gers和Grewar | 2010 |
| | x | x | 33 | 毛里塔尼亚 | Ould等 | 2011 |
| 牛瘟 | x | — | — | 印度 | Haji | 1932-1933 |
| | x | — | — | 俄罗斯 | Samatsev和Arbuzov | 1940 |
| | x | — | — | 印度 | Dhahillon | 1959 |
| | — | o | 0.0 | 肯尼亚 | Soct和MacDonald | 1962 |
| | — | — | 实验结果 | 埃及 | Singh和Ata | 1967 |
| | — | x | 9.7 | 苏丹 | Singh和Ata | 1967 |
| | — | — | 7.7 | 乍得 | Maurice等 | 1967 |
| | — | — | 实验结果 | 肯尼亚 | Taylor | 1968 |
| | — | x | 0.5 | 肯尼亚 | Wilson等 | 1982 |
| | — | o | 0.0 | 印度 | Chauhan等 | 1985 |
| | — | x | 5.2 | 埃及 | Abou Zaid | 1991 |
| | — | x | 11.9 | 埃及 | Ismail等 | 1992 |
| | — | x | 40 | 埃及 | Mankinde等 | 2001 |
| | — | x | 21.3 | 埃塞俄比亚 | Roger等 | 2001 |
| | — | x | 31.2 | 尼日尼亚 | Mankinde等 | 2001 |
| | — | x | 20 | 沙特阿拉伯 | Al-Afaleq等 | 2007 |
| | — | x | 0 | 阿联酋 | Wernery等 | 2007 |
| | — | x | 0 | 阿联酋 | Wernery等 | 2007 |
| 西尼罗河病毒 | x | — | — | 加拿大 | Kutler等 | 2004 |
| | x | — | —（羊驼） | | Dunkel等 | 2004 |
| | x | — | —（羊驼） | 英国 | Yaeger等 | 2004 |
| | x | — | —（羊驼、美洲驼） | 美国 | Whitehead等 | 2006 |
| | — | x | 38 | 阿联酋 | Wernery等 | 2007 |
| **不常见或次要疾病/病原** | | | | | | |
| 脑心肌炎 | x | — | 3000峰死亡（单峰驼） | 土库曼斯坦 | Mischenko等<br>Baborenko | 1994<br>1996 |
| 毒威斯布仑病毒 | | | 60.2（单峰驼） | 尼日尼亚 | Baba等 | 1990 |
| 黄热病 | — | x | 54（单峰驼） | | | |
| 波蒂斯库姆（Potiskum）病毒 | — | x | 66.2（单峰驼） | | | |
| 登革热病毒-1 | — | x | 43（单峰驼） | | | |
| 班齐病毒（Banzi） | — | x | 54（单峰驼） | | | |
| 乌干达病毒 | — | x | 0（单峰驼） | | | |



2 病毒病

续表

| 疫病/病毒 | 抗原 | 抗体 | 阳性率（%） | 国家/地区 | 作者 | 年份 |
|---|---|---|---|---|---|---|
| 托高土病毒（Thogoto） | — | x | —（单峰驼） | 尼日尼亚 | Kemp等 | 1973 |
| 西尼罗河热 | — | x | — | | | |
| 毒威斯布仑病毒(Wesselsbron) | — | x | | | | |
| 马心肌炎病毒 | — | x | PCR(羊驼) | 德国 | Weber等 | 2006 |
| | — | x | <1（羊驼、美洲驼） | 美国 | Hennig等 | 2008 |
| | x | x | | 德国 | Weber等 | 2006 |
| 疑似狂犬病 | x | — | — | 沙特阿拉伯 | Al Dubaib等 | 2008 |
| 呼肠孤病毒 | x | — | 呼肠孤病毒（单峰驼？双峰驼？） | 德国 | Huisinga等 | 2009 |
| 施马伦贝格病毒 | — | x | — | 英国 | Jack等 | 2012 |

**注：—以疾病字母排序（对已有文献汇总）。**
**对于抗原和抗体，“x”为阳性，“—”为没有数据，“o”为阴性**

表57　不常见的虫媒病毒

| 不常见的虫媒病毒 | 来源 | 国家 | 作者 | 年份 |
|---|---|---|---|---|
| 卡达姆病毒（Kadamvirus），披膜病毒科的病毒，黄病毒属的病毒 | 骆驼蜱 | 沙特阿拉伯 | Wood等 | 1982 |
| 卡兰非尔病毒（Quaranfil virus） | 骆驼蜱 | 科威特、伊朗、也门 | Converse和Moussa | 1982 |
| 赤羽病毒，布尼亚病毒科的病毒 | 血清学 | 阿拉伯半岛 | Al-Busaidy 等 | 1988 |
| 多理病毒（Dhori virus） | 骆驼蜱 | 印度 | Anderson和Casals | 1973 |
| 沃诺赖病毒(Wanowrie virus)、托高土病毒、多理病毒 | 骆驼蜱 | 埃及 | Williams 等 | 1973 |
| 刚果出血热病毒 | 骆驼蜱 | 伊朗 | Saidi等 | 1975 |
| | | 俄罗斯 | Hoogstraal | 1979 |
| | | 阿联酋 | Suleiman等 | 1980 |
| | | 伊拉克 | Tantawi等 | 1980 |
| | | 埃及 | Morrill等 | 1990 |
| | | 阿曼 | Scringeour等 | 1996 |

## 参考文献


[1] El-Harrak M., Martin-Folger R., Llorente F., Fernandez-Pacheco P., Brun A., Figuerola J. &Jimenez-Clavero M.A. (2011). – Rift Valley and West Nile virus antibodies in camels, North America. *Emerg. Inf. Dis.*, 17 (12), 2372–2373.

[2] Faye B., Chaibou M. & Vias G. (2012). – Integrated impact of climate change and socioeconomic development on the evolution of camel farming systems. *Brit. J. Environ. Climate Change*, 2 (3), 227–244.

[3] Kapil S., Yeary T. & Evermann J.F. (2009). – Viral diseases of New World camelids. *Vet. Clin. Food Anim.*, 25, 323–337.

[4] Khalafalla A.I. & Bornstein S. (2012). – Emerging infectious diseases in camelids. *In* Conference of the International Society of Camelid Research and Development, 29 January~1 February 2012, keynote presentations, Muscat, Oman, 65–74.

[5] Wernery U. & O.-R. Kaaden (2002). – Infectious diseases in camelids. Blackwell Science. ISBN 3–8263–3304–7, 161–234.


（般宏译，储岳峰校）

# 2.1 致病性病毒感染

## 2.1.1 狂犬病

狂犬病是一种人和所有温血脊椎动物的致死性疾病，通常由带毒动物咬伤引起。一旦暴露，病毒通过外周神经进入中枢神经系统，引起脑炎，器官衰竭，最终导致死亡。由于狂犬病是人兽共患病，一旦出现临床症状，死亡率很高，骆驼容易患狂犬病，因此，该病在新大陆骆驼已进行了广泛研究。动物狂犬病分布于世界各地，但是，一些国家如澳大利亚、新西兰、日本和英国等为无狂犬病国家。一些国家采取免疫预防和疫病防控措施也已逐步消灭了狂犬病。不过，在澳大利亚、北美和北欧一些国家和地区，偶然可通过带毒的哺乳动物或带毒的蝙蝠侵袭而引起狂犬病。

### 病原学

狂犬病病毒属于弹状病毒科。弹状病毒科包括狂犬病病毒属和水疱性口炎病毒属。狂犬病病毒属分为4个血清型和7个基因型。其中，血清Ⅰ型包括所有的狂犬病病毒，其他所有的血清型属于“狂犬相关病毒”。在狂犬病病毒血清型内，狂犬病病毒（血清Ⅰ型为狂犬病病毒本身）和狂犬相关病毒（血清Ⅱ型为拉各斯蝙蝠株、血清Ⅲ型为莫克拉病毒原型株，血清Ⅳ型为杜文海病毒原型株）与分离自非洲、南美和澳大利亚的鸟和嗜血昆虫的其他7个病毒在生物学特性和抗原特性方面存在差异。弹状病毒在形态上呈杆状或子弹状，其基因组是不分节段的单股负链RNA，编码5种结构蛋白。病毒的复制在感染细胞的细胞浆内进行，随着复制，病毒蛋白在细胞浆内聚集，形成组织学可见的狂犬病毒包涵体或内基体。

### 流行病学

狂犬病是由狂犬病毒感染而引起的传染病，以中枢神经系统紊乱、瘫痪和死亡为特征。狂犬病病毒在动物与动物之间以及动物向人类的传播的途径主要通过咬伤。草食动物和人类是终末宿主，通常不会传播病毒。只有肉食动物和吸血蝙蝠维系感染延续。骆驼感染狂犬病毒主要是通过疯狗咬伤引起，也有红狐、臭鼬和蝙蝠引起骆驼感染狂犬病的报道。世界卫生组织估计，每年全球约有75 000人死于狂犬病，但这个数据仅是一个保守的估计，因为大量的病例没有被报道或没有诊断出来。非洲和亚洲是狂犬病的高发区，印度居首位。家养动物、野生肉食动物或蝙蝠是狂犬病病毒寄存的主要宿主。

温血动物和鸟类对狂犬病毒易感，但是，不同动物对狂犬病病毒的易感性有很大差异。狐狸、棉鼠和草原狼对狂犬病毒最为敏感；牛、骆驼、兔子和猫对狂犬病病毒较为敏感；犬、绵羊和山羊对狂犬病病毒不敏感。负鼠可能对狂犬病病毒最不敏感[14]。狂犬病病毒传播媒介生物在不同的国家和地区各不相同。在非洲，家犬仍然是狂犬病病毒传播的主要媒介，但是，黄色猫鼬，侧纹胡狼和蝙蝠耳狐狸具有他们自己的狂犬病循环圈，如纳米比亚的捻角羚一样。在亚洲热带地区，犬是地方性狂犬病传播的唯一动物。犬也是引起美国热带地区狂犬病地方性流行的主要物种；但是，在美国热带地区，在嗜血蝙蝠（吸血蝙蝠，吸血蝠）中存在一个完全独立的狂犬病感染循环圈。吸血蝙蝠引起人和牛感染狂犬病。在其他一些国家，包括澳大利亚，食虫蝙蝠和食果蝙蝠可引起狂犬病[12]。

尽管在非洲和亚洲的许多国家发生过单峰驼的狂犬病，但关于这方面的报道很少[44,45]。骆驼感染狂犬病的报道有摩洛哥[18]，毛里塔尼亚[10]，阿曼[9]，阿联酋[1,51,52]尼日尔[13]，印度[34]，以色列[42]，苏丹[5,20]和伊朗[22]。Arush和Somac/Sarec报道了索马里单峰驼的后肢瘫痪，该病类似狂犬病[48]。

有关骆驼的狂犬病兽疫学和流行病学相互依

图145　患狂犬病的单峰驼：试图不断的打哈欠是典型狂犬病

存关系的资料甚少。依据狂犬病病毒主要宿主和传播媒介动物种类的不同，将狂犬病分为3种类型，即城市型、森林型和蝙蝠型（蝙蝠型狂犬病）。在阿拉伯半岛以森林型狂犬病为主。阿联酋和阿曼的狂犬病主要是红狐引起[9,52]，也门的狂犬病主要是流浪犬（野生犬）引起[49]。由于这方面报道的可用资料很少，推测上述动物可能是阿拉伯半岛狂犬病毒的传播媒介。

在尼日尔，Bloch和Diallo报道患狂犬病的犬引起40头骆驼中7头暴发狂犬病。Al-Dubaib[2]报道了沙特阿拉伯单峰驼中的狂犬病。据管理超过4000多头骆驼的48个牧民的反映，狂犬病在骆驼中的发病率为0.2%。70%的病例是被疯狗咬伤引起，17%的病例是被患狂犬病的狐狸咬伤引起。母骆驼主要是在保护幼驼的时候，其前腿或后腿被咬伤。在理论上，单峰驼不会将狂犬病传染给人类。

在北美洲和南美洲，引起狂犬病流行的传播媒介较多，包括犬、狐狸、浣熊、臭鼬和蝙蝠。但是在秘鲁，羊驼中暴发狂犬病只是由犬咬伤引起[37]。也有患狂犬病的羊驼咬伤其他羊驼传播狂犬病的报道[25]。

新大陆骆驼被犬咬伤后，狂犬病潜伏期在15天～3个月。在临床症状出现6～8天后，患病动物死亡。在美洲驼中成功复制了狂犬病病例[50]，美国报道了多起新大陆骆驼暴发狂犬病[15,16,17,29,31,32,33,36,37,43]。

## 临床症状和病理学

Peck[41]，Mustapha[38]和Bah等[10]都发现单峰驼中有两种类型的狂犬病：“狂躁型”和“静型”。后者在骆驼中很少见到[19,35,38]。但是，据Al-Dubaib[2]报道，在沙特阿拉伯，67%的患狂犬病的单峰驼出现“沉静型”症状，仅有很低的比例出现“狂躁型”症状，且主要发生在雄性单峰驼。当患狂犬病的雄性单峰驼攻击和撕咬附近的东西时，情况非常危险，有时会严重残害自己的身体。经过3周～6个月的潜伏期后，出现“狂躁型”临床症状：焦躁不安、攻击行为、异常兴奋，突然撕咬、奇痒或擦伤并伴随严重的自身伤害、垂涎、肌肉震颤和后肢瘫痪。在单峰驼中，这个兴奋期持续1～3天，之后进入瘫痪期。瘫痪期的患驼侧卧不起，四肢乱蹬，这个阶段可持续1～2天，单峰驼试图不断的打哈欠，之后死亡（图 145）。

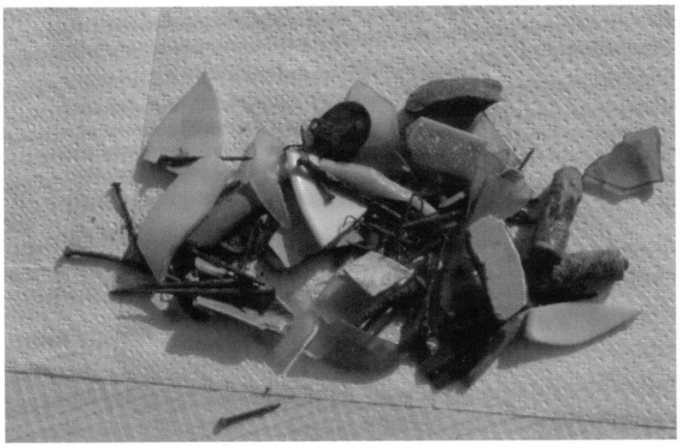

图146　患狂犬病的单峰驼的胃中发现的异物

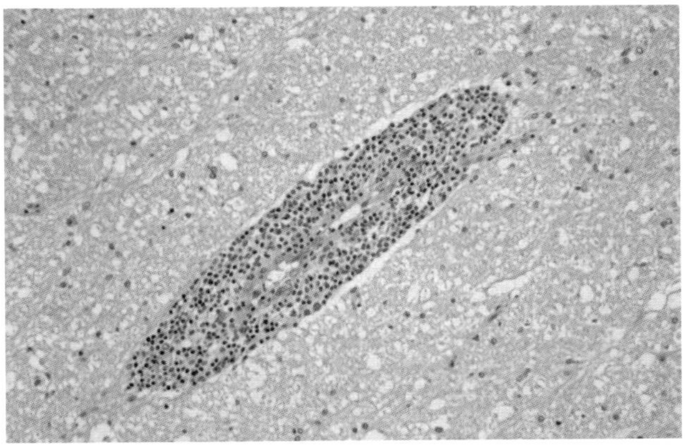

图147　患狂犬病的单峰驼出现非化脓性脑膜炎

在单峰驼中试图不断的打哈欠是狂犬病的典型症状[52]。Blood和Radostits[14]认为这些运动是一种无声的咆哮。Perl等[42]报道以色列一群150头的骆驼中，有一头8岁的单峰驼患有不寻常的狂犬病。患畜表现出狂犬病“沉静型”临床症状，虚弱、颤抖、胸式卧地。尸体检查表明，脊髓周围轻度水肿。用直接免疫荧光试验检测海马、小脑、延髓时，狂犬病病毒呈阴性，但是，小白鼠脑内接种试验呈阳性。小白鼠脑内接种12天后出现瘫痪症状，脑组织荧光试验，狂犬病毒呈阳性。用免疫组化检测骆驼脑组织时，狂犬病毒呈阴性，但是，检测腰部到胸部的脊髓，可检测到狂犬病毒抗原。作者强调指出，在对疑似狂犬病病例的骆驼进行诊断时，必须进行脊髓的检测。在2007年沙特阿拉伯单峰驼中检测到非化脓性脑膜炎，称之为“Dububa”，该名称是沙特阿拉伯东北部的一个地区。患狂犬病的单峰驼以神经紊乱、共济失调、步态蹒跚和偶然的神经兴奋为特征。非化脓性脑膜炎症状出现提示存在病毒感染，但是，包括狂犬病病毒在内的哪种病毒引起了脑膜炎还没有确定[3,4]，引起神经紊乱的原因仍然不清楚。如果有新病例出现，建议检查脊髓，但是，该病似乎已经消失了。

在新大陆骆驼中，狂犬病主要以狂躁型为主，很少出现麻痹综合征（沉静型）。骆驼患狂犬病的主要症状是攻击人、同圈的骆驼、幼驼以及自残等。患狂犬病的动物也会撕咬无生命的物体。新大陆骆驼麻痹型狂犬病的症状以厌食、流涎、转圈、面神经麻痹和咽部痉挛为特征。值得提醒的是，患狂犬病的骆驼由于咽部痉挛不能呕吐[24]。

死于狂犬病的动物其病理变化不一。唯一可见的异常现象是脑部血管栓塞。在整个病程中，动物逐渐消瘦，身体不同部位出现自残的伤口或

图148　用免疫荧光检测到患狂犬病的单峰驼大量的病毒抗原

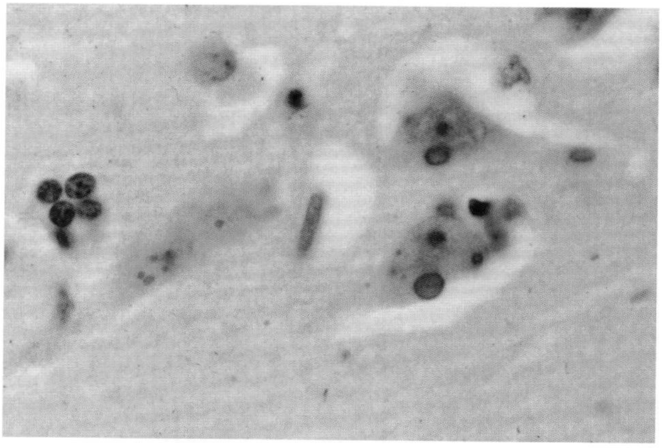

图149　患狂犬病的单峰驼在海马角检测到内基体（苏木素-伊红染色）

损伤。单峰驼第一胃内经常发现异物，如指甲、石头、小电池、玻璃碎片或瓷器碎片等（异食癖）（图 146）。

狂犬病最显著的病理变化在中枢神经系统、颅部和脊髓神经节。狂犬病毒引起血管周围单核细胞聚集的非化脓性脑膜炎（图 147）。在神经元内有局灶性神经胶质增生和转移、神经元变性和胞浆内异物（内基体）。

**诊断**

骆驼出现异常行为应怀疑狂犬病。当可疑动物来自狂犬病疫区时，感染狂犬病的可能性就大大提高。一旦发现疑似狂犬病病例，应向兽医管理机构报告，由兽医人员决定是否对动物进行限制或进行长达14天的观察，经过观察期后，如果出现狂犬病的明显症状，应将动物实施安乐死并进行实验室检查。

动物实施安乐死时避免脑盖骨的损伤。海马角组织常用于狂犬病的诊断。由于不同部位的病灶分布或病毒抗原的含量各不相同，因此，应从大脑和脊髓不同部位采取病料[42]。狂犬病的诊断只能在骆驼死亡后才能进行。采取病料时应谨慎操作，每一个细节必须按照国家和国际制定的操作规程进行，以避免病毒扩散。

狂犬病的标准诊断方法是通过免疫荧光检测新鲜脑组织中的狂犬病毒抗原。免疫荧光是一个标准方法，该技术速度快，特异性强，对实验室工作人员要求不高。作者用免疫荧光法在对所有患狂犬病的单峰驼进行实验室检查时，在大脑，尤其是在海马角发现大小不一的狂犬病毒抗原聚集物（图 148）。

用新鲜甘油保存的脑组织或福尔马林固定的脑组织切片镜检可发现内基体（图 149）。

狂犬病的第3种诊断方法是进行病毒分离，方法是将脑组织悬浮液注射到断奶小白鼠脑部。这种方法非常敏感，但是，比较费时，需要4周或更长时间才能得到结果。通过鼠脑组织病理学检查或免疫荧光检测证实是否分离到病毒。聚合酶链反应（PCR）也是狂犬病诊断的一种方法，对狂犬病毒基因组有关的部分基因组序列进行测序。

正如之前所提到的，在所有的诊断中必须要

表58　用灭活狂犬病氢氧化铝疫苗免疫前后的抗体滴度变化情况

| 骆驼 | 免疫前24小时 | 免疫后14天 | 免疫后7个月 | 免疫后13个月 |
|---|---|---|---|---|
| 1 | < 0.1 | 18.5 | 0.5 | 0.3 |
| 2 | 0.1 | 9.5 | 1.5 | 1.5 |
| 3 | < 0.1 | 3.5 | < 0.1 | < 0.1 |
| 4 | 0.3 | 18.5 | 0.5 | 0.3 |
| 5 | < 0.1 | 2.5 | < 0.1 | < 0.1 |
| 6 | < 0.1 | 1.5 | 0.5 | 0.5 |
| 7 | 0.2 | 7.5 | < 0.1 | < 0.1 |
| 8 | < 0.1 | 2.5 | 0.1 | 0.1 |
| 9 | < 0.1 | 4.5 | 0.3 | 0.3 |
| 10 | < 0.1 | 4.5 | 0.1 | 0.1 |
| 11 | < 0.1 | 7.5 | 0.1 | 0.1 |
| 12 | 0.1 | 5.5 | 0.5 | 0.5 |
| 13 | 0.1 | 28.5 | < 0.1 | < 0.1 |
| 14 | < 0.1 | 4.5 | < 0.1 | < 0.1 |
| 15 | < 0.1 | 1.5 | < 0.1 | < 0.1 |
| 16 | < 0.1 | 2.5 | < 0.1 | < 0.1 |
| 17 | 0.1 | 3.5 | < 0.1 | < 0.1 |
| 18 | 0.1 | 18.5 | 0.5 | 0.5 |
| 19 | < 0.1 | 5.5 | < 0.1 | < 0.1 |
| 20 | 0.1 | 4.5 | < 0.1 | < 0.1 |
| 21 | < 0.1 | < 0.1 | < 0.1 | < 0.1 |
| 22 | < 0.1 | < 0.1 | < 0.1 | < 0.1 |
| 23 | < 0.1 | < 0.1 | < 0.1 | < 0.1 |
| 24 | < 0.1 | < 0.1 | < 0.1 | < 0.1 |
| 25 | < 0.1 | < 0.1 | < 0.1 | < 0.1 |

注：免疫方法：皮下注射；免疫剂量：1mL；免疫抗体检测方法：快速荧光抑制试验。

包括脊髓样品[42]。Fowler[24]强调，新大陆单峰驼狂犬病的准确诊断不能依赖某一项检测。他推荐新大陆单峰驼狂犬病的诊断应包括组织学检测、荧光抗体检测和接种啮齿类动物。在专业实验室，用PCR技术进行狂犬病毒基因组部分序列的测定，确定病毒的基因型。

有几种血清学试验，例如，酶联免疫吸附试验可以鉴定狂犬病病毒的抗体。这些试验主要用来评价疫苗的免疫反应或进行分离病毒的确认。但是，国际贸易中用于狂犬病毒活病毒和组织培养的试验只有两种，即快速荧光灶抑制试验和荧光抗体病毒中和试验。所有相关技术在世界动物卫生组织的《陆生动物诊断试验和疫苗手册》的2.1.13章中有详细描述[53]。

## 治疗和控制

目前，动物狂犬病无药可治。狂犬病是一种病毒性疾病，可通过疫苗免疫进行有效预防控制。用活疫苗、弱毒疫苗以及灭活疫苗对狐狸进行主动预防免疫。目前，已有病毒在原代细胞或传代细胞培养而制备的一系列安全、有效的动物用狂犬病灭活疫苗。免疫后产生的中和抗体是保护动物免受狂犬病病毒感染的主要成分。中和抗体滴度≥0.5IU/mL时，可保护动物免受狂犬病病毒攻击[11,21,47]。抵抗狂犬病病毒攻击的有效免疫持续期在1～3年。由于骆驼对狂犬病病毒敏感，应该用疫苗进行预防免疫。如果母畜在怀孕期间进行了免疫，草食动物幼畜应在4月龄或9月龄时进行免疫，每年加强免疫1次。旧大陆骆驼和新大陆骆驼用狂犬病灭活疫苗免疫具有良好的免疫效果。下面是应用氢氧化铝为佐剂的灭活疫苗（皮下注射1mL）免疫阿联酋的小骆驼获得的一些血清学结果。在整个13个月的试验期间，血清抗体检测了4次，结果汇总在表58。

该试验由法国里昂和德国图宾根动物病毒病联邦研究所的Rhone Merieux完成。在牛上抗体滴度高于0.5IU/mL可抵抗狂犬病毒的攻击[11]。抗体滴度用国际单位表示（IU/mL）。21～25号动物为对照。

用灭活狂犬病疫苗免疫单峰驼，14天后产生令人满意的免疫应答。但是，在免疫7个月后，狂犬病抗体滴度下降到较低水平，甚至消失。Sihvonen等人[47]在驯鹿中获得同样的结果。数据表明，单峰驼注射一个剂量（1mL）灭活狂犬病疫苗可诱导抗体转阳，但持续时间很短。因此，在免疫后6～8个月应加强免疫1次，以保证产生保护性抗体水平。印度的免疫数据表明，疫苗在单峰驼中的免疫持续期有很大的不同。这些研究人员发现组织培养的狂犬病灭活疫苗可诱导产生较长的免疫持续期。Kalanidhi 等[28]认为，抗体持续期差异的原因主要在于试验中所用的骆驼品种不同，或动物个体之间对疫苗存在差异反应。

Fowler[24]推荐单峰驼只能使用狂犬病灭活疫苗进行免疫（对新大陆骆驼也是如此），其原因是有人用改造的活疫苗免疫290只羊驼后14～30天，10%的羊驼暴发狂犬病。用灭活狂犬病疫苗免疫美洲驼可诱导产生抗体，其滴度可达到在其他种的动物中具有保护性的程度。在南美洲不同地区的美洲驼曾经感染狂犬病，因此，应该每年用狂犬病灭活疫苗进行免疫。

在狂犬病病毒相关病毒中，杜文海格病毒与血清1型的狂犬病病毒在抗原性上最相近，免疫狂犬病疫苗可抵抗杜文海格病毒的攻击，但对莫科拉病毒的攻击保护性很低。很难确定从阿联酋的骆驼中分离的狂犬病毒属于狂犬病毒属哪一种血清型。由阿曼和阿联酋分离的狂犬病病毒株所建立的系统进化树显示，这些狂犬病病毒群聚在欧洲/中东谱系[39,40]，并分隔成两个不同的分支。对于不同国家骆驼中的狂犬病毒是否具有相同的抗原结构，尤其是能引起“沉静型”临床症状的狂犬病病毒是否具有相同的抗原结构将是研究的热点。

## 2.1.2 博尔纳病

博尔纳病（BD）是一种渐进性、病毒性、麻痹性脑脊髓炎，主要感染马、驴（其他马科动物中很少）、绵羊和其他动物，该病主要局限在欧洲中部，在德国新大陆骆驼中确诊了该病[6,7,27,30,46]。

### 病原学

从1927年以来，已经明确了博尔纳病的病原学。博尔纳病病毒（BDV）是有囊膜的病毒，内含单股负链RNA。病毒在感染细胞的核内复制。尽管该病病毒与单股负链病毒目的其他病毒具有相同的理化特性，最近，国际病毒分类学委员仍将该病毒划分到新建立的博尔纳病毒科博尔纳病毒属。所有分离的病毒似乎在抗原性上完全相同，但是毒力具有明显的差异。在自然条件下，病毒的宿主范围包括马、骆驼、绵羊、牛、犬、猫，同时，很有可能还有人。

### 流行病学

实验条件下，博尔纳病病毒能够实验感染很多动物物种和各种不同的细胞，但是，其传播模式仍尚未探明。由于在鼻分泌物、唾液和尿液中可检测到病毒，病毒很可能是通过直接和间接接触感染。最近从患神经障碍病人的血清和脑脊液中检测到博尔纳病毒的特异性抗体。

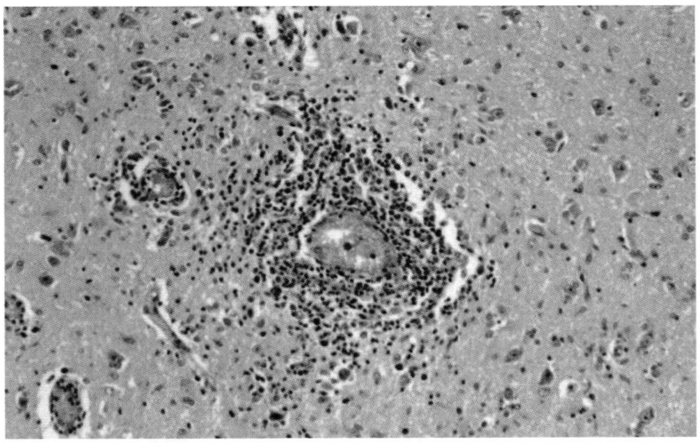

图150　羊驼典型的博尔纳病脑膜脑炎，出现血管周围白细胞聚集现象（苏木精-伊红染色）

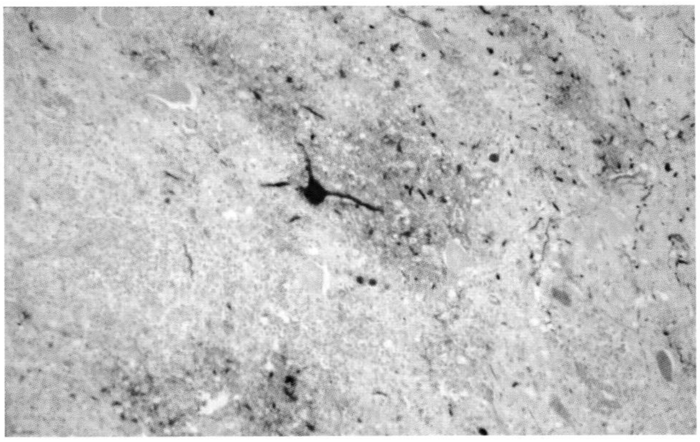

图151　羊驼海马角博尔纳病病毒阳性标记



## 临床症状和病理学

在德国的两个动物园，美洲驼和羊驼被博尔纳病毒感染，在暴发初期出现厌食和体重迅速下降，最终导致死亡。羊驼感染该病不出现任何神经症状。但在德国发现，一峰成年羊驼从巴伐利亚借给德国黑森北部的地区时，表现慢性抑制性性欲和急性惊厥。在惊厥发作几次后，尽管采取多种治疗措施，动物在很短时间内死亡[27]。在另一个病例中，由于后肢共济失调症状越来越严重，羊驼被实施安乐死[30]。

## 诊断

博尔纳病的诊断通过组织病理学研究确诊。所有检测的羊驼都表现出非化脓性脑膜脑炎症状。Schueppel等通过免疫组化技术确诊了4头羊驼中2头患博尔纳病。在海马区神经细胞的细胞核以及在炎症附近齿状回、纹状体可观察到博尔纳病病毒的阳性标记（图150和图151）。Jacobsen等分离的博尔纳病病毒毒株序列分析提示，该毒株序列与巴伐利亚I型地方流行株的同源性几乎一样，表明种畜在巴伐利亚已经感染了该病毒，然后再传染到黑塞羊驼群中。这个发现进一步表明博尔纳病病毒遗传聚类与动物种类无关，并与原始地域相对应。

另外，在海马角检测到核内包涵体（Joest-Degen小体），这是典型的博尔纳病病毒感染特征。

## 诊断

用博尔纳病病毒感染兔胚胎或大鼠脑细胞，或者将病毒脑部注射兔子，可在感染脑的脑悬液或脑脊液中分离到该病毒，病毒抗原也可通过免疫组化方法检测到。如果在神经细胞中出现核内Joest-Degen小体，对博尔纳病的诊断也是有帮助的。

博尔纳病的诊断也可通过感染细胞的间接免疫荧光血清学方法确诊（IFA）。Fluess[23]在她的论文中用间接免疫荧光从不同国家单峰骆驼血清中检测博尔纳病抗体。在肯尼亚，10%（5/50）的单峰骆驼有博尔纳病抗体，阿联酋为4.8%（5/1030），以色列为25%（1/4）。对这现象如何解释还比较困难，因为目前仅知道该病在德国、瑞士和奥地利发生，将来应在这方面投入更多的研究。

## 治疗和预防

博尔纳病是一种必须报告的疾病，通过扑杀政策进行控制。但是在德国动物园，美洲驼和羊驼每年用改造的活病毒疫苗免疫取得了很好的效果。

## 参考文献

[1] Afzal M., Khan I.A. & Salman R. (1993). – Clinical signs and clinical pathology of rabies in the camel. *Vet. Rec.*, 133, 220.

[2] Al-Dubaib M.A. (2007). – Rabies in camels at Qassim region of central Saudi Arabia. *J. Camel Pract. Res.*, 14 (20), 101–103.

[3] Al-Ghamdi G.M., Al-Naeem A.A.,Wuenschmann A., Wernery U., Al-Dubaib M., Al-Swailem A.M., Hamouda M., Al-Yamani E., Shehata M., El-Lithy D.A., Mahmoud O.M. & Al-Mujali A.M. (2009). – Dubdubasyndrome: non-supparative meningoencephalomyelitis in dromedary camels in Saudi Arabia. *J. Camel Pract. Res.*, 16 (1), 19–23.

[4] Al-Hizab F.A. & Abdel Salam E.B. (2007). – Non-suppurative meningoencephalitis in dromedary camel in Saudi Arabia. *J. Camel Pract. Res.*, 14 (2), 105–108.

[5] Ali Y.H., Saeed I.K. & Zakia A.M. (2004). – Camel rabies in Sudan. *Sudan J. Vet. Sci. Anim. Husb.*, 43, 231–234.

[6] Altmann D. (1975). – Die wichtigsten Erkrankungen der Alt- und Neuweltkamele. *In* Verhandlungsbericht 17 Internationalen Symposium Erkrankung Zootiere (Tunis), 17, 53–60.

[7] Altmann D., Kronberger H., Schueppel K.-F., Lippmann R. & Altmann I. (1976). – Bornasche Krankheit (meningoencephalomyelitis simplex enzootica equorum) bei Neuwelttylopoden und Equiden. *In* Verhandlungsbericht 18 Internationalen Symposium Erkrankung Zootiere, Innsbruck, 127–132.

[8] Arush M.A. (1982). – La situazione sanitaria del dromedario nella Republica Democratica Somalia. *Bollettino scientifica della facoltà di zootecnia e veterinaria*, 3, 209–217.

[9] Ata F.A., Tageldin M.H., Al Sumry H.S. & Al-Ismaily S.I. (1993). – Rabies in the Sultanate of Oman. *Vet. Rec.*, 132, 68–69.

[10] Bah S.O., Chamoiseau G., Biha M.L.O. & Fall S.M.O.A. (1981). – Un foyer de rage Cameline en Mauritanie. *Rev. Elev. Méd. Vét. Pays Trop.*, 34 (3), 263–265.

[11] Barrat J., Guillemin F., Brun A., Lacoste F., & Precausta P. (1992). – Cattle vaccination against rabies. Immunity duration and challenge three years after vaccination. Paper read at the Pan-American Health Organization.

[12] Blancou J. & Fooks A.R. (2010). – Rabies. *In* Infectious and parasitic diseases of livestock (P.-C. Lefevre, J. Blancou, R. Chermette & G. Uilenberg, eds). Lavoisier, 342–357.

[13] Bloch N. & Diallo I. (1995). – A probable outbreak of rabies in a group of camels in Niger. *Vet. Microbiol.*, 46 (1–3), 281–283.

[14] Blood D.C. & Radostits O.M. (1990). – Veterinary medicine. 7th Ed. Bailliere Tindall, London.

[15] Centers for Disease Control (1990a). – Rabies in a llama  Oklahoma. *MMWR Morb. Mortal. Wkly. Rep.*, 39 (12), 203–204.

[16] Centers for Disease Control (1990b). – Rabies in a llama – Oklahoma. *J. Am. Vet. Med. Assoc.*, 263 (16), 2766.

[17] Centers for Disease Control (1991). – Rabies in a llama. *Wkly. Epidemiol. Rec.*, 65 (38), 294.

[18] Chevrier L. (1959). – Epidémiologie de la rage au Maroc. *Rev. Elev. Méd. Vét. Pays Trop.* 12 (2), 115–120.

[19] Curasson G. (1947). – Le chameau et ses maladies. Vigot Freres Editions, Paris, 86–88.

[20] El Mardi O.I. & Ali Y.H. (2001). – An outbreak of rabies in camels (*Camelus dromedarius*) in North Kordofan State. Sudan J. Vet. Res., 17, 125–127.

[21] El-Ahwal A.M. (1969). – Rabies problem and eradication in U.A.R. *J. Egypt Vet. Med. Ass.*, 29 (3–4), 121–129.

[22] Esmaeli H., Ghasemi E. & Ebrahimzadeh H. (2012). – An outbreak of camel rabies in Iran. *J. Camel Pract. Res.*, 19 (1), 19–20.

[23] Fluess M. (2002). – Ein Beitrag zur epidemiologic des Bornaschen Krankheit. Doctoral Thesis, University of Giessen, Giessen, Germany.

[24] Fowler M.E. (2010). – Medicine and surgery of camelids. 3rd Ed. Wiley-Blackwell, Oxford, 173–176.

[25] 2Franco E. (1968). – Brote de rabia en alpacas de una hacienda del Departamento de Puno. Bol. Extraordinario, 3, 59–60.

[26] Higgins A. (1986). – The camel in health and disease. Balliere Tindall, London.

[27] Jacobsen B., Algermissen D., Schandien D., Venner M., Herzog S., Wentz E., Hewicker-Trautwein M., Baumgartner W. & Herden C. (2010). – Borna disease in an adult alpaca stallion *(Lama pacos). J. Comp. Path.*, 143, 203–208.

[28] Kalanidhi A.P., Bissa U.K & Srinivasan V.A. (1998). – Seroconversion and duration of immunity in camels vaccinated with tissue culture inactivated rabies vaccine. *Veterinarski Arhiv.*, 68 (3), 81–84.

[29] Kapil S., Yeary T. & Evermann J.F. (2009). – Viral diseases of New World Camelids. *Vet. Clin. Food Anim.*, 25, 323–337.

[30] Kobera R. & Pohle D. (2004). – Case reports in South American camelids in Germany (M. Gerken & C. Renieri, eds). *In* Proceedings of the 4th European Symposium on South American Camelids and DECAMA European Seminar, 7–9 October 2004, Gottingen, Germany, 151.

[31] Krebs J.W., Holman R.C., Hines U., Strine T.W., Mandel E.J. & Childs J.E. (1992). – Rabies surveillance in the United States during 1991. *J. Am. Vet. Med. Assoc.*, 201 (12), 1839.

[32] Krebs J.W., Strine T.W. & Childs J.E. (1993). – Rabies surveillance in the United States during 1992. *J. Am. Vet. Med. Assoc.*, 203 (12), 1721.

[33] Krebs J.W., Strine T.W., Smith J.S., Rupprecht C.E. & Childs J.E. (1995). – Rabies surveillance in the United States during 1994. *J. Am. Vet. Med. Assoc.*, 207 (12), 1562–1575.

[34] Kumar A. & Jindal N. (1997). – Rabies in a camel a case report. *Trop. Anim. Health Prod.*, 29 (1), 34.

[35] Leese A.S. (1927). – A treatise on the one-humped camel in health and disease. Vigot Frères Editions, Paris.

[36] Miller P. (1994). – Rabies on rise. *Llamas*, 8 (3), 49–55.

[37] Moro Sommo M. (1958–1959). – Sobre un brote de rabia en alpacas. *Rev. Fac. Med. Vet. (Lima)*, 13–14, 35–40.

[38] Mustapha I.E. (1980). – IFS provisional report No. 6 on camels, 399. International Foundation for Science, Stockholm, Sweden.

[39] Nowotny N., Zilahi E., Al-Hadj M.A., Wernery U., Kolodziejek J., Lussy H., Al Dhahry S., Al-Kobaisi M.F.M., Al Rawahi A. & Al Ismaily S. (2005). Molecular epidemiology of camelpox and rabies isolated in the United Arab Emirates and in Oman. In 6th Annual U.A.E. University Research Conference, 24–26 April 2005, Al Ain, UAE, 164–170.

[40] Nowotny N., Zilahi E., Al-Hadj M.A., Wernery U., Kolodziejek J., Lussy H., Al-Dhahry S., Al-Kobaisi M.F., Al-Rawahi A. & Al-Ismaily S. (2006). Molecular epidemiology of rabies viruses isolated in the United Arab Emirates and in Oman. In Proceedings of the Panamerican Congress of Zoonosis, 10–12 May 2006, La Plata, Argentina, Supplemento, 3, 272.

[41] Peck E.F. (1966). – Rabies. In Encyclopaedia of Veterinary Medicine (T. Dalling, A. Robertson, G.E. Boddie & J.S. Spruell, eds). 1st Ed. W. Green and Son, Edinburgh, 577.

[42] Perl S., van Straten M., Jakobson B., Samina I., Sheikhab N. & Orgad U. (1996). – Hind limb paralysis associated with rabies in a camel (*Camelus dromedarius*). In 8th International Symposium of Veterinary Laboratory Diagnosticians, 4–8 August 1996, Jerusalem, Israel, 14.

[43] Reid-Sanden F.L., Dobbins J.G., Smith J.S. & Fishbein D.B. (1990). – Rabies surveillance in the United States during 1989. *J. Am. Vet. Med. Assoc.*, 197 (12), 1576.

[44] Richard D. (1980). – Dromedary pathology and productions. Provisional report No. 6 camels. International Science Foundation (IFS), Khartoum, Sudan, and Stockholm, Sweden, 12 (18–20), 409–430.

[45] Richard D. (1986). – Manuel des maladies du dromadaire, Projet de developpement de l elevage dans le Niger centre-est. Maisons Alfort, IEMVT, France.

[46] Schueppel K.-F., Kinne J. & Reinacher M. (1994). – Bornavirus – Antigennachweis bei Alpakas (*Lama pacos*) sowie bei einem Faultier (*Choloepus didactylus*) und einem Zwergflußpferd (*Choeropsis liberiensis*). *Verh. ber. Erkrg. Zootiere*, 36, 189–193.

[47] Sihvonen L., Kulonen K., Soveri T. & Nieminen M. (1993). – Rabies antibody titres in vaccinated reindeer. *Acta Vet. Scand.*, 34, 199–202.

[48] Somac/Sarec. (1982). – Camel research project report by a Somali/Swedish Mission, 10–26 March, 18–23.

[49] Stanley M.J. (1990). – Rabies in Yemen Arab Republic, 1982 to 1986. *Trop. Anim. Health Prod.*, 22, 273–274.

[50] Tamayo M.D. (1905). – La rabia experimental en la llama. *Cron. Med.*, 22, 269–272.

[51] Wernery U. & Kaaden O.-R. (1995). – Infectious diseases of camelids. Blackwell Wissenschafts-Verlag, Berlin, 75–81.

[52] Wernery U. & Kumar B.N. (1993). – Rabies in the U.A.E. *Tribulus, Bulletin of the Emirates Natural History Group*, 3 (1), 5–21.

[53] World Organisation for Animal Health (OIE) (2012). – Rabies, Chapter 2.1.13. *In* OIE manual of diagnostic tests and vaccines for terrestrial animals. OIE, Paris, 263–282.

**深入阅读材料**

*Sidya Ould Bah, Chamoiseau G., Mohamed Lemine Ould Biha & Sidi Mohamed Ould Ahmed Fall (1981). – Un foyer de rage cameline en Mauritanie. Rev. Elév. Méd. Vét. Pays Trop., 34 (3), 263–265.*

（殷相平译，殷宏校）



## 2.1.3 骆驼痘

单峰驼和双峰驼均可发生骆驼痘，新大陆骆驼（NWCs）可实验性地感染骆驼痘[61,113]。骆驼痘病毒（camelpox virus，CMPV）首先感染幼龄动物引起一种增生性皮肤病[33,69,87,88,90,94]。副痘病毒、乳头状瘤病毒可引起骆驼科动物痘样的病理变化，最近的研究表明，新大陆骆驼还可以感染牛痘[26]。

### 病原学

痘病毒为痘病毒科家族成员。痘病毒科可分为两个群，即脊椎动物痘病毒亚科（感染脊椎动物）和昆虫痘病毒亚科（感染昆虫）。痘病毒是体积最大、结构最复杂的病毒，具有砖块形的外貌。目前，已知痘病毒科正痘病毒属中有11个种，骆驼正痘病毒则是其中之一，它是一种大的，有囊膜的，双股DNA病毒，是引起骆驼痘的病原。

对骆驼痘病毒核酸序列测定发现，它与已在全世界消灭的人类的天花病毒很接近[4,44,78]。

### 流行病学

病毒可通过皮肤微小创口感染，通过气溶胶经呼吸道感染，也可通过节肢动物的叮咬机械性传播。在雨季，骆驼痘发病率高且病症严重[73,112,114]，而在旱季，病程相对温和[82]。雨季里大量节肢动物滋生，导致驼群中病毒毒力和病毒量都加大，从被称为骆驼蜱的嗜驼璃眼蜱中都分离到了骆驼痘病毒。不同的骆驼痘病毒毒株具有不同的毒力[73,76,80,81,83,115]，这也解释了为什么一些毒株可引起全身性的感染，而另一些毒株则仅引起局部性感染[110]。DNA限制性酶切分析表明，从不同非洲国家分离的骆驼痘病毒毒株的基因组不同，因此，不同的毒株具有不同的毒力[73]，这也是在疫苗生产和攻毒实验时要考虑的一个重要因素[17]。

感染动物可获得终生免疫。疫病的流行周期与雨季、昆虫的密度、驼群中具有免疫力的骆驼数量有关。已经证实，该病为人畜共患病。在多篇文章中都报道了经临床观察骆驼正痘病毒可以传播给人[92]，最近的一篇相关报道来自印度[19]。此前有报道称，饲养骆驼的牧民会定期出现皮肤丘疹，但目前的实验室诊断方法尚不能确定骆驼痘病毒就是引起这种皮肤丘疹的病原[51,63,111]。最近，位于印度哈里亚纳邦（Haryana）希萨尔市（Hisar）的兽医菌种保藏中心的科学家报道骆驼痘病毒确能感染人[19]。

2009年，印度西北部地区暴发骆驼痘期间，发现了3例人感染骆驼痘的病例，证实骆驼痘病毒可在人畜间传播。虽然只从采集的骆驼样品中分离到了病毒，但从3个训驼师采集的样品，包括血液样品，经分子生物学和血清学检测呈明显的骆驼痘病毒阳性。训驼师还有皮肤丘疹、水泡、溃疡，以及后来在手指及手上形成结痂等临床症状。

痘病是骆驼最常见的病毒病，各地都有发生（表59）。迄今为止，只有澳大利亚的单峰驼群中从未发现过骆驼痘。

表59 骆驼痘暴发报告（按国家和作者排列）

| 国家 | 作者 | 年份 | 国家 | 作者 | 年份 |
|---|---|---|---|---|---|
| 阿富汗 | Odend'Hal | 1983 | 巴基斯坦 | Al-Hendi et al. | 1994 |
| 巴林 | Higgins et al. | 1992 | 俄罗斯 | Vedernokov | 1893 |
| 埃及 | Tantawi et al. | 1974 | | Vedernokov | 1902 |
| | Tantawi | 1974 | | Amanzhulov et al. | 1930 |
| | Tantawi et al. | 1978 | | Bauman | 1930 |
| 埃塞俄比亚 | Shommein and Osman | 1987 | | Ivanov | 1934 |
| | Tefera and Gebreah | 2001 | | Samartsev and Praksein | 1950 |
| 印度 | Leese | 1909 | | Vyshelesskii | 1954 |
| | Cross | 1917 | | Likhachev | 1963 |
| | Chauhan et al. | 1985 | | Borisovichand Orekhov | 1966 |
| | Chauhan et al | 1986 | | Buchnev and Sadykov | 1967 |
| | Chauhan and Kaushik | 1987 | | Semushkin | 1968 |
| | Khanna et al | 1996 | | Vedernikov | 1969 |
| | Bera et al. | 2011 | | Borisovich | 1973 |
| 伊朗 | Baxby | 1972 | | Marennikova et al. | 1974 |
| | Ramyar and Hessami | 1972 | | Buchnev et al. | 1987 |
| 伊拉克 | Al-Falluji et al. | 1979 | 沙特阿拉伯 | Hafez et al. | 1986 |
| 肯尼亚 | Davies et al. | 1975 | | Hussein et al. | 1987 |
| | Schwartz et al. | 1982 | | Hafez et al. | 1992 |
| | Wilson et al. | 1982 | | Abdel Baky et al. | 2006b |
| | Kropp | 1985 | | Saad and Alrobaish | 2006 |
| | Gitao | 1997 | | Abdelhamid Elfadi and Ommer Dafalla | 2007 |
| 利比亚 | Carter and Azwai | 1996 | 索马里 | Kriz | 1982 |
| 毛里塔尼亚 | Wardeh | 1989 | | Arush | 1982 |
| 摩洛哥 | Fassi-Fehri | 1987 | 苏丹 | Shommein and Osman | 2007 |
| | El-Harrak et al. | 1991 | | Khalafallaand Mohamed | 1996 |
| | Touil et al. | 2012 | | Khalafalla et al. | 1998 |
| 尼日尔 | Richard | 1986 | 阿拉伯联合酋长国 | Kaaden et al. | 1992 |
| | Ba-Vy et al. | 1989 | | Wernery et al. | 1997a,b |
| 阿曼 | Shommein and Osman | 1987 | 也门 | Odend' Hal | 1983 |
| 巴基斯坦 | Odend'Hal | 1983 | 叙利亚 | AL-Zi'abi et al. | 2007 |
| | Ghulam et al. | 1998 | | | |

不管是局部感染还是全身型感染，病毒都首先在最初的侵入位点增殖。在全身型感染中，病毒在组织和器官中增殖，形成原发性病毒血症，接着病毒在引流淋巴结中增殖，导致继发性病毒血症和皮肤继发感染。

血清学研究揭示了骆驼痘病毒在不同国家广泛流行。Davies等[35]用血清中和实验证明，无论是放牧的驼群还是农场中的驼群，都有较高的骆驼痘抗体。在肯尼亚，用血清中和实验对6个临床健康的驼群进行检测，在5个驼群中检测到了抗体。Munz等[74]报告苏丹有95%的阳性病例，Khalafalla等[57]确证的比率为72.5%。Pfeffer 等[84]用ELISA对阿拉伯联合酋长国的1000峰单峰驼进行检测，发现流行率为88%～100%。在利比亚，Azwai等[10]调查了6个驼群中的520峰单峰驼，发现仅有10%的动物呈阳性。在摩洛哥，对骆驼痘的血清调查发现，在7年里[104]流行率从23%增长到37%，一些科学家认为这与全球变暖的影响有关。血清学调查对于评估驼群的免疫状态意义不大，众所周知，在正痘病毒感染时，细胞介导的免疫对动物的保护比循环抗体好[41]。

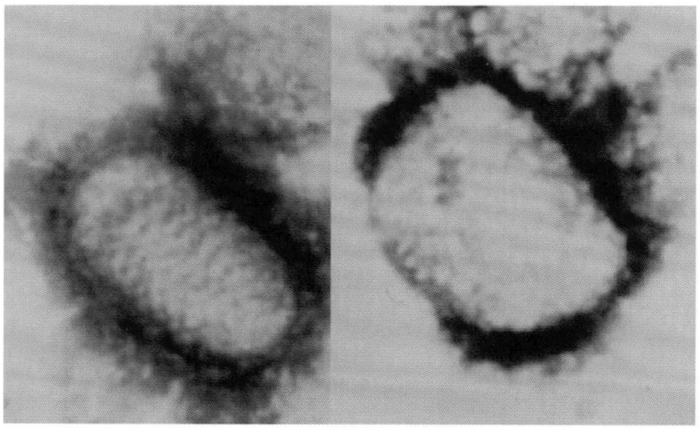

图152　来自同一单峰驼的骆驼痘病毒（左）和副痘病毒电镜照片

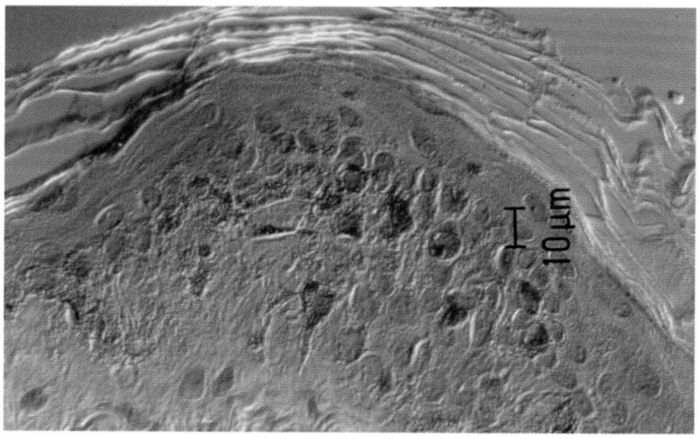


图153　巨噬细胞、纤维细胞和内皮细胞上的骆驼痘病毒抗原（金黄色颗粒表示骆驼痘病毒抗原阳性）

## 诊断

一些学者致力于骆驼痘病毒的特征和种系研究[12,20,28,45,67,68,73,86,91]。基于形态学、化学、物理学和生物学特征，骆驼痘病毒被列为痘病毒科正痘病毒属的代表种。它与这个群中的其他种，如痘病毒中的牛痘/天花病毒亚群免疫学关系非常接近。种属鉴定和实验室鉴定对于划分正痘病毒和副痘病毒非常重要，这两个属的病毒可同时出现在同一骆驼身上[111]（图152）。

利用相对简单的方法，在实验室可将骆驼痘病毒与其他正痘病毒属的病毒区别开[16,67]。区别不同种属病毒的经典方法有鸡胚培养，观察细胞培养中的细胞病变[18]、兔体皮内试验、鸡羽囊试验等。新的技术方法有单抗ELISA技术，DNA限制性酶切分析[74,75]和地高辛标记DNA探针的斑点检验[72]。Czerny等[34]、Johann和Czerny[53]和Pfeffer等[84]叙述了多个实验室诊断骆驼痘的方法，包括电子显微镜观察、ELISA、免疫组化和PCR。其中，PCR是诊断骆驼痘快速可行的方法[98]。免疫组化是检测骆驼痘病毒抗原的一项新技术，该方法不必使用电子显微镜，在实验室很容易进行（图153）。免疫组化在组织病理学研究中也有一定的意义，可以了解痘病毒感染后的形态学变化。免疫组化技术还有一个优点，用这种方法能够检测形成数年的嵌入组织块，因此，该方法非常适用于回顾性研究[60]。

## 临床症状和病理变化

在轻度感染的病例中，经过9～13天的潜伏期，在鼻孔、眼睑以及口腔和鼻黏膜上产生脓疱（图154）。

严重的病例即呈现骆驼痘的全身性临床症状，如发热、精神沉郁、腹泻、厌食，全身布满丘疹（图155）。

实验感染原驼可出现全身性骆驼痘（图156）。

Buchnev和Sadykov[25]描述了骆驼正痘病毒感染骆驼科动物导致流产，从流产胎儿体内分离到骆驼痘病毒。AL-Zi'abi等[7]报道，叙利亚妊娠母驼的流

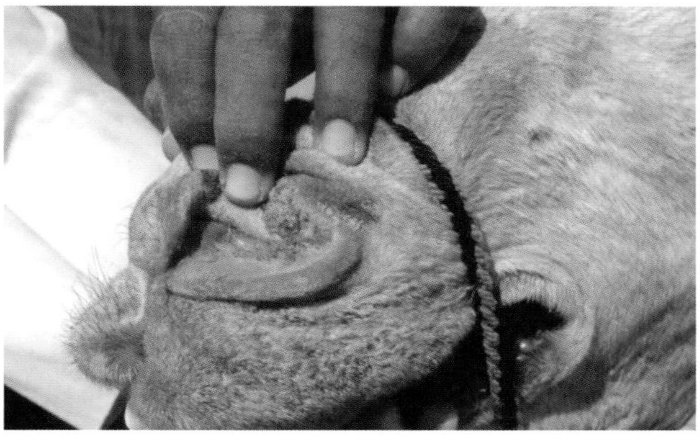

图154　鼻黏膜上的骆驼痘

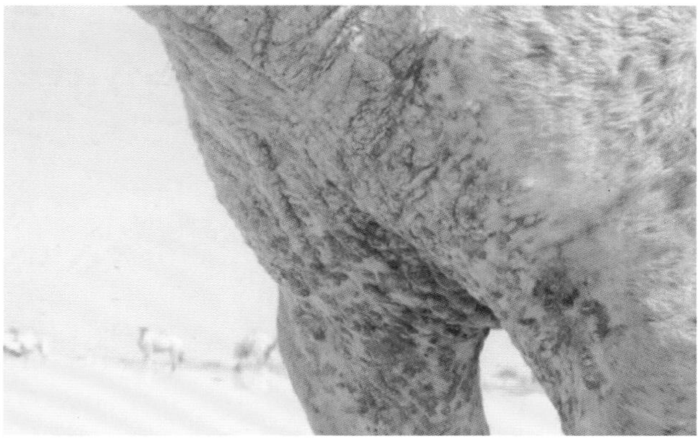

图155　单峰驼全身性骆驼痘

图156　实验感染原驼的全身性骆驼痘

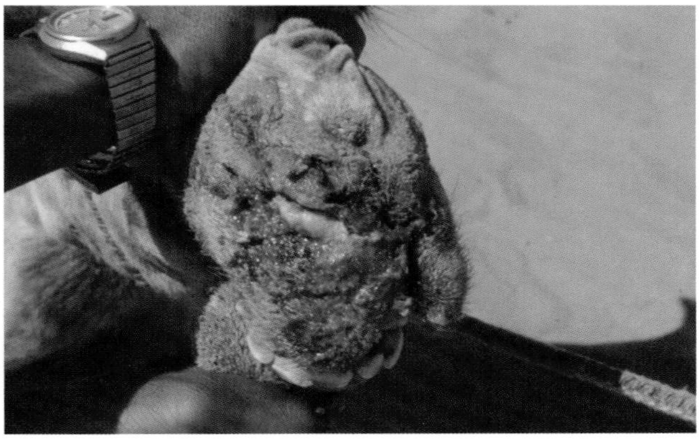

图157　骆驼痘继发金黄色葡萄球菌感染

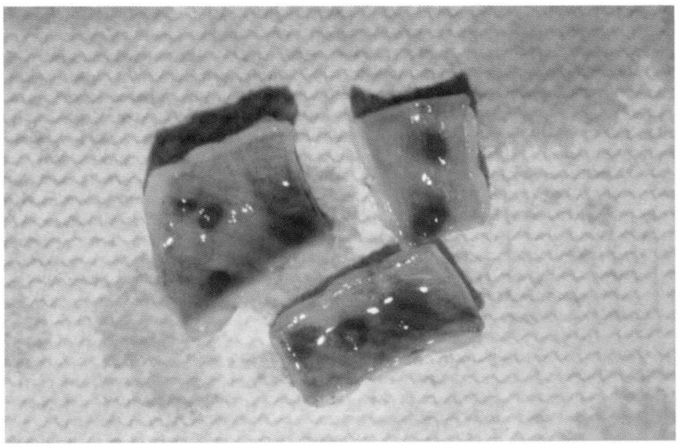

a 气管病变

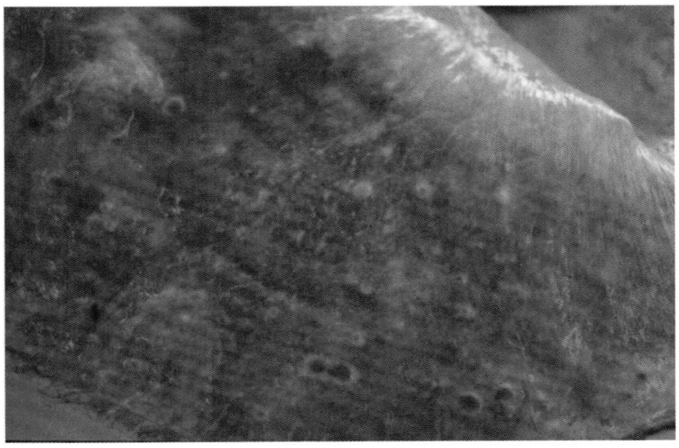

b 肺部病变

图158　9月龄单峰驼的骆驼痘病变

产率高达90%。已报道的发病率和死亡率详见表60。

发病率和死亡率差异较大，这通常取决于是全身性骆驼痘还是局部性骆驼痘。细菌和霉菌继发感染，使骆驼痘的病程复杂化（图157）。

痘斑也可见于幼驼的气管和肺中（图158a，b）[60,111]。

典型的痘斑起初是红色的斑痕，接着形成丘疹和水泡。水泡最终发展成中央扁平，周围有红晕的脓疱，称之为痘疮。脓疱破裂后，形成硬的结痂。脓疱持续4～6周后平复，有的留下斑痕，有的则不留斑痕。痘病毒一般嗜上皮，发生病变的皮肤角质细胞肿胀，空泡化，细胞破裂形成水泡。脓

表60 骆驼痘发病率和死亡率

| 国家 | 作者 | 年份 | 死亡率(%) | 发病率(%) |
|---|---|---|---|---|
| 索马里 | Jezek et al. | 1983 | 28 | — |
| 沙特阿拉伯 | Abdel Baky et al. | 2006b | 13 | 50 |
| | Saad and Alrobaish | 2006 | 8 | 38 |
| | Abdelhamid Elfadil and Ommer Dafalla | 2007 | 3.6 | 41 |
| 叙利亚 | Al-Zi'abi et al. | 2007 | 1 ~ 15 | 30 ~ 90 |



疱周围上皮细胞的增生又逐渐使脓疱增大。血管周边单核细胞浸润，真皮肿胀，常见中性粒白细胞和嗜酸性白细胞。Kinne等[60]描述了骆驼痘致上呼吸系统的病理变化。疫病引起气管、食管和肺部散落着病灶。肺部的病理变化表现为肺组织实变，散落着直径1 ~ 10mm的病灶。肺切片用苏木精-伊红染色，可见增生性肺泡炎和细支气管炎，正常的肺组织结构部分或全部坏死，被纤维化的组织替代（图159）。

免疫组化技术检测病灶发现，支气管上皮细胞中含大量痘病毒抗原阳性细胞（图160）。免疫组化技术也用于皮肤中痘斑的检测[77,84]。

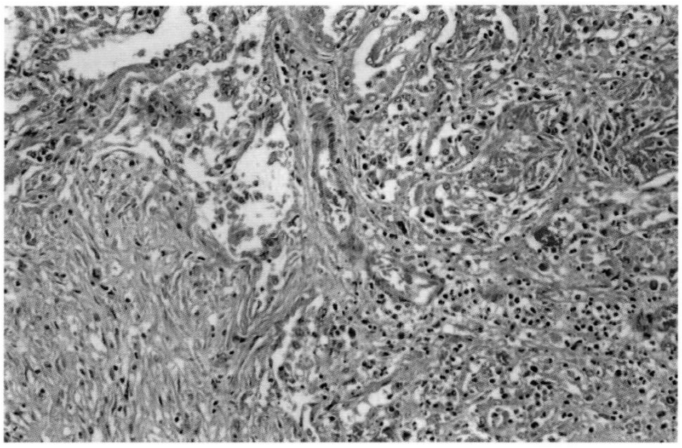

图159 单峰驼骆驼痘肺组织由残余的肺泡，纤维组织和成熟的胶原蛋白形成的病灶

注：嗜酸性单核细胞的细胞质和细胞核残骸（苏木精-伊红染色，x220）

## 治疗和控制

目前，尚无治疗骆驼痘感染的方法。对不严重的病例可局部注射或全身注射广谱抗生素和维生素，以减少继发感染。尽管骆驼痘具有重要的经济意义，但疫苗研发人员并不多。畜主也知道骆驼痘的重要性，他们给这个病创造了很多名字。直到今天，畜主将感染动物痂皮溶解在奶里，再刷在幼驼嘴唇的划痕处，来保护幼驼[48,65]。

关于骆驼痘疫苗的最早报道来自前苏联[21,25,93,96]，但是疫苗毒株、安全性和效力等详细信息不全。Buchnev和Sadykov[25]用氢氧化铝胶疫苗免疫骆驼，未能保护骆驼对骆驼痘的感染。Mary[71]撰文灭活的痘疫苗不能对任何痘病毒感染产生保护。由于灭活疫苗中的病毒在宿主体内不能增殖，不能产生足够的特异性痘病毒抗体。只有当病毒滴度大于$10^{7.0}TCID_{50}$，且在3 ~ 5周后重复免疫的动物，才

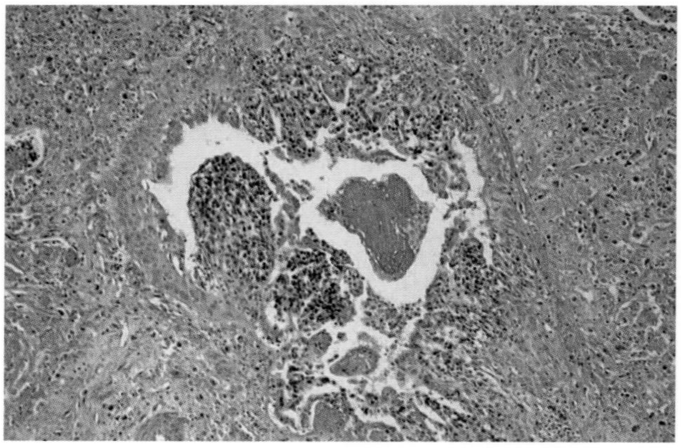

图160 大量痘病毒抗原结合的支气管上皮细胞增生和细胞脱粒

注：腔内坏死（生物素-亲和素标记，x120）

能得到保护。摩洛哥[38,39]，沙特阿拉伯[46]和阿联酋[54,111,113,114,115]等都进行了疫苗试制，成功生产了骆驼痘疫苗。摩洛哥研发的是一种灭活疫苗，从1991年起应用于疫病预防，该疫苗须每年进行注射，此后又研究筛选了数个克隆株，其中，现在应用的A28毒株安全性高免疫力好[38,39]。沙特阿拉伯和阿联酋研发的是弱毒疫苗，埃及也研发了一种骆驼痘弱毒活疫苗[13]。

Khalafalla和El-Dirdiri对摩洛哥灭活疫苗和Ducapox弱毒疫苗进行了田间实验[56]，两个疫苗都安全有效，免疫原性好，产生了体液免疫和细胞免疫应答，免疫的单峰驼能抵抗田间骆驼痘病毒的攻击，1个剂量的Ducapox疫苗有效保护单峰驼的免疫持续期至少为1年。在沙特阿拉伯，用骆驼痘病毒 Jouf-78株生产的弱毒疫苗，注射兔和豚鼠体产生了抗体，显示该疫苗在这些动物中具有可靠的免疫力[1]。阿联酋的一个研究小组构建了一种永生化的幼驼肾细胞系（Dubca），用于分离骆驼痘病毒[54,55,62]。阿联酋从1994年开始应用弱毒驼痘疫苗（称为Ducapox，迪拜骆驼痘疫苗）[110]，具有好的功效，也能对新大陆骆驼的骆驼痘产生保护[113]。Wernery和Zachariah[115]做了一个实验，用一个剂量的Ducapox疫苗，对12月龄的单峰驼进行免疫，可保护单峰驼免受骆驼痘感染6年甚至更长的时间。作者也强调说明，用于这种长期的动物实验的骆驼数量较少。疫苗生产者建议对6～9月龄的骆驼应加强免疫，以避免母源抗体引起的免疫失败。由于骆驼痘病毒和牛痘苗病毒之间的抗原关系，因此，可用已知的牛痘苗病毒株免疫骆驼。Higgins等[49]报道，在巴林岛使用 Lister株牛痘苗，使骆驼痘达到已控制，但还是有一次暴发，Baxby等[17]的报道则显示，在伊朗用EA8牛痘疫苗毒株免疫的单峰驼，能够抵抗骆驼痘病毒。应该提及的是，针对人的天花，总部在瑞士日内瓦的世界卫生组织用痘苗病毒进行计划免疫，已在全世界范围内消灭了天花。

骆驼痘是世界动物卫生组织（OIE）规定的必须报告的动物疫病，在最近的2个报告中[23,37]，有学者计划根除骆驼痘。牛瘟的消灭，以及正在进行的对脊髓灰质炎等病毒病消灭的战斗，激发了人们根除骆驼痘的想法。实际上，骆驼是骆驼痘病毒的唯一宿主，再加上有效的弱毒疫苗，将来一定能实现根除骆驼痘。



## 参考文献


[1] Abdel Baky M.H., Al-Sukayran A., Mazloum K.S., Al-Boky A.M. & Al Mujalli D.M. (2006a). – Immunogenicity of camelpox virus, Jouf-78 strain in Baskat rabbits and guinea pigs. In Proceedings of the International Scientific Conference on Camels, 12–14 May 2006, Buraydah, Saudi Arabia, Part 2, 358–369.

[2] Abdel Baky M.H., Al-Sukayran A., Mazloum K.S., Al-Boky A.M. & Al Mujalli D.M. (2006b). – Isolation and standardization of camel pox virus from naturally infected cases in the central region of Saudi Arabia during 2004. *In* Proceedings of the International Scientific Conference on Camels, 12–14 May 2006, Buraydah, Saudi Arabia, Part 2, 370–380.

[3] Abdelhamid Elfadil A.M. & Ommer Dafalla M.A. (2007). – Epidemiologic and clinical features of camelpox in Jazan Region, Saudi Arabia. *Vet. Res.*, 1 (3), 65–67.

[4] Afonso C.L., Tulman E.R., Lu Z., Zsak L., Sandybaev N.T., Kerembekova U.Z., Zaitsev V.L., Kutish G.F. & Rock D.L. (2002). – The genome of camelpox virus. *Virology*, 295, 1–9.

[5] Al-Falluji M.M., Tantawi H.H. & Shony M.O. (1979). – Isolation, identification and characterization of camelpox virus in Iraq. *J. Hyg. Camb.*, 83, 267–272.

[6] Al-Hendi A.B., Abuelzein E.M.E., Gameel A.A. & Hassanien M.M. (1994). – A slow-spreading mild form of camelpox infection. *J. Vet. Med. B.*, 41, 71–73.

[7] Al-Zi'abi O., Nishikawa H. & Meyer H. (2007). – The first outbreak of camelpox in Syria. *J. Vet. Med. Sci.*, 69 (5), 541–543.

[8] Amanzhulov S.A., Samarzev A.A. & Arbuzov L.N. (1930). – Sur la variole du chameau de la region d'Oural. *Bull. Inst. Pasteur*, 29, 96, abstract.

[9] Arush M.A. (1982). – La situazione sanitaria del dromedario nella Repubblica Democratica Somala. *Bollettino scientifica della facoltà di zootecnia e veterinaria*, 3, 209–217.

[10] Azwai S.M, Carter S.D., Woldehiwet Z. & Wernery U. (1996). – Serology of *Orthopoxvirus cameli* infection in dromedary camels: analysis by ELISA and western blotting. *Comp. Immunol. Microbiol. Infect. Dis.*, 19 (1), 65–78

[11] Ba-Vy N., Richard D. & Gillet J.P. (1989). – Proprietes d'une souche d'orthopoxvirus isolée des dromadaires du Niger. *Rev. Elev. Méd. Vét. Pays Trop.*, 42 (1), 19–25.

[12] Bartenbach G. (1973). – Charakterisierung und Systematisierung eines Kamelpockenvirus. Veterinary Medicine Dissertation, Munich, Germany.

[13] Bassiouny A.I., Aboulsoud A., Hussein H.A., Soliman S.M. & El-Sanousi A.A. (2007). – Preparation of an enhanced live attenuated camelpox virus vaccine. *Egyptian J. Virol.*, 4, 31–39.

[14] Bauman V. (1930). – The camel. Sel'khozgiz, Moscow and Leningrad.

[15] Baxby D. (1972). –.Smallpox-like viruses from camels in Iran. *Lancet*, 2, 1063–1065.

[16] Baxby D. (1974). – Differentiation of smallpox and camelpox viruses in cultures of human and monkey cells. *J. Hyg. Camb.*, 72, 251–254.

[17] Baxby D., Ramyar H., Hessami M. & Ghaboosi B. (1975). – Response of camels to intradermal inoculation with smallpox and camel pox viruses. *Infect. Immun.*, 11 (4), 617–621.

[18] Bedson H.S. (1972). – Camelpox and smallpox. *Lancet*, 9, 1253.

[19] Bera B.C., Shanmugasundaram K., Barua S., Venkatesan G., Virmani N., Riyesh T., Gulati B.R., Bhanuprakash V., Vaid R.K., Kakker N.K., Malik P., Bansal M., Gadvi S., Singh R.V., Yadav V., Sardarilal, Nagarajan G., Balamurugan V., Hosamani M., Pathak K.M. & Singh R.K. (2011). – Zoonotic cases of camelpox infection in India. *Vet. Microbiol.*, 152, 29–38.

[20] Binns M., Mumford J. & Wernery U. (1992). – Analysis of the camel pox virus thymidine kinase gene. *Br. Vet. J.*, 148, 541–546.

[21] Borisovich Y.F. (1973). – Little-known infectious diseases in animals. Kolos, Moscow, 32–42.

[22] Borisovich, Y.F. & Orekhov M.D. (1966). – Camel pox. *Veterinaryia, Moscow.* Dated in *Vet. Bull.*, 36 (3), 50–52.

[23] Bray M. & Babiuk S. (2011). – Camelpox: target for eradication? *Antiviral Res.*, 92, 164–166.

[24] Buchnev K.N., Tulepbaev S.Z. & Sansyzbaev A.R. (1987). – Infectious diseases of camels in the USSR. *In* Diseases of camels. *Rev. sci. tech. Off. int. Epiz.*, 6 (2), 487–495.

[25] Buchnev K.N. & Sadykov R.G. (1967). – Contribution to the study of camelpox. *In* Proceedings of the 3rd All-UnionConference on Virology, Moscow, Russia, Part II, 152–153.

[26] Cardeti G., Brozzi A., Eleni C., Polici N., D'Alterio G., Carletti F., Scicluna M.T., Castilletti C., Di Caro A., Autorino G.L. & Amaddeo D. (2011). – Cowpox virus infection in farmed llamas in Italy. *In* Proceedings of the International Meeting on Emerging Diseases and Surveillance (IMED), 4–7 February 2011, Vienna, Austria, 91.

[27] Carter S.D. & Azwai S.M. (1996). – Immunity and infectious diseases in the dromedary camel. *In* Proceedings of the British Veterinary Camelid Society, 14–16 November, Burford, UK, 23–36.

[28] Chandra R., Chauhan R.S. & Garg S.K. (1998). – Camel pox: A review. *Camel Newslett.*, 14, 34–45.

[29] Chauhan R.S. & Kaushik R.K. (1987). – Isolation of camelpox virus in India. *Br. Vet. J.*, 143, 581–582.

[30] Chauhan R.S., Kaushik R.K., Gupta S.C., Satiya K.C. & Kulshreshta R.C. (1986). – Prevalence of different diseases in camels (*Camelus dromedarius*) in India. *Camel Newslett.*, 3, 10–14.

[31] Chauhan R.S., Kulshreshtha R.C. & Kaushik R.K. (1985). – Epidemiological studies of viral diseases of livestock in Haryana State. *Ind. J. Virol.*, 1 (1), 10–16.

[32] Cross H.E. (1917). – The camel and its diseases. Bailliere, Tindall and Cox, London.

[33] Curasson G. (1947). – Le chameau et ses maladies. Vigot Freres Editions, Paris.

[34] Czerny C.-P., Meyer H. & Mahnel H. (1989). – Establishment of an ELISA for the detection of orthopox viruses based on neutralizing monoclonal and polyclonal antibodies. *J. Vet. Med. B.*, 36, 537–546.

[35] Davies F.G., Mbugna H., Atema C. & Wilson A. (1985). – The prevalence of antibody to camelpox virus in six different herds in Kenya. *J. Comp. Path.*, 95, 633–635.

[36] Davies F.G., Mungai J.N. & Shaw T. (1975). – Characteristics of a Kenyan camelpox virus. *J. Hyg. Camb.*, 75, 381–385.

[37] Duraffour S. Meyer H., Andrei G. & Snoeck R. (2011). – Camelpox virus. *Antiviral Res.*, 92, 167–186.

[38] El-Harrak M. (1998). – Isolation of camelpox virus, development of an inactivated vaccine and prophylactics application in Morocco. In International Meeting on Camel Production and Future Perspectives, 2–3 May 1998, Al Ain, UAE, 736.

[39] El-Harrak M., Loutfi C. & Bertin F. (1991). – Isolement et identification du virus de la variole du dromadaire au Maroc. *Ann. Rech. Vet.*, 22, 95–98.

[40] Fassi-Fehri M.M. (1987). – Les maladies des camelides. In Diseases of camels. *Rev. sci. tech. Off. int. Epiz.*, 6 (2), 315–335.

[41] Fenner F., Wittek R. & Dumbell K.R. (1988). – The orthopox viruses. Academic press, New York, 100–133.

[42] Ghulam M., Khan M.Z. & Athar M. (1998). – An outbreak of generalised pox among draught camels in Faisalabad city. *J. Camel Pract. Res.*, 5 (1), 127–129.

[43] Gitao C.G. (1997). – An investigation of camelpox outbreak in 2 principal camel (*Camelus-dromedarius*) rearing areas of Kenya. *Rev. sci. tech. Off. int. Epiz.*, 16 (3), 841–847.

[44] Gubser C. & Smith G.L. (2002). – The sequence of camelpox virus shows it is most closely related to variola virus, the cause of smallpox. *J. Gen. Virol.*, 83, 855–872.

[45] Gunther G. (1990). – Isolierung und Charakterisierung des Kamelpockengenoms. Institute of Virology, Hannover Veterinary School, Hannover, Germany.

[46] Hafez S.M., Al-Sukayran A., dela Cruz D., Mazloum K.S., Al-Bokmy A.M., Al-Mukayel A. & Amjad A.M. (1992). –Development of a live cell culture camelpox vaccine. *Vaccine*, 10 (8), 533–537.

[47] Hafez S.M., Eissa Y.M., Amjad A.M., Al-Sharif A.K. & Al-Sukayran A. (1986). – Preliminary studies on camel pox in Saudi Arabia. In Proceedings of the 9th Symposium on the Biological Aspects of Saudi Arabia, Riyadh, Saudi Arabia, 24–27 March 1986, 78.

[48] Higgins A. (1986). – The camel in health and disease. Bailliere Tindall, London, 92–95.

[49] Higgins A.J., Silvey R.E., Abdelghafir A.E. & Kitching R.P. (1992). – The epidemiology and control of an outbreak of camelpox in Bahrain. *In* Proceedings of the 1st International Camel Conference. R. and W. Publications (Newmarket) Ltd, Newmarket, UK, 100–104.

[50] Hussein M.F., Hafez S.M. & Gar El-Nabi M. (1987). – A clinico-pathological study of camelpox in Saudi Arabia. In Proceedings of the 10th Symposium on the Biological Aspects of Saudi Arabia, Riyadh, Saudi Arabia, 20–24 April 1987, 8–14.

[51] Ivanov P.V. (1934). – Camel breeding. Kazakhskoe kraevoe izdatel'stvo, Alma-Ata, Kazakhstan.

[52] Jezek Z., Kriz B. & Rothbauer V. (1983). – Camelpox and its risk to the human population. *J. Hyg. Epidem. Microbiol. Immun.*, 27 (1), 29–42.

[53] Johann S. & Czerny C.-P. (1993). – A rapid antigen capture ELISA for the detection of orthopox viruses. *J. Vet. Med. B.*, 40, 569–581.

[54] Kaaden O.-R., Walz A., Czerny C.-P. & Wernery U. (1992). – Progress in the development of a camel pox vaccine. In Proceedings of the 1st International Camel Conference. R. and W. Publications (Newmarket) Ltd, Newmarket, UK, 47–49.

[55] Kaaden O.-R., Wernery U. & Klopries M. (1995). – Camel fibroblast cell line DUBCA and its use for diagnosis and prophylaxis of camel diseases. In Proceedings of the International Conference on Livestock Production in Hot Climates, 8–10 January 1995, Muscat, Oman, A49.

[56] Khalafalla A.I. & El Dirdiri G.A. (2003). – Laboratory and field investigation of live attenuated and inactivated camelpox vaccines. *J. Camel Pract. Res.*, 10 (2), 191–200.

[57] Khalafalla A.I., Mohamed M.E.M. & Ali B.H. (1998). – Camelpox in the Sudan. Part I and Part II. *J. Camel Pract. Res.*, 5 (2), 229–238.

[58] Khalafalla A.I. & Mohamed M.E.M. (1996). – Clinical and epizootiological features of camelpox in Eastern Sudan. *J. Camel Prac. Res.*, 3 (2), 99–102.

[59] Khanna N.D., Uppal P.K., Sharma N. & Tripathi B.N. (1996). – Occurrence of pox infections in camels. *Ind. Vet. J.*, 73 (8), 813–817.

[60] Kinne J., Cooper J.E. & Wernery U. (1998). – Pathological studies on camelpox lesions of the respiratory system in the United Arab Emirates (UAE). *J. Comp. Path.*, 118, 257–266.

[61] Kinne J. & Wernery U. (1999). – Experimental camelpox infection. In Abstracts of the 17th Meeting of the European Society of Veterinary Pathology (ESVP), 14–17 September 1999, Nantes, France, 129.

[62] Klopries M. (1993). – Etablierung und Charakterisierung einer Kamelhautzellinie (Dubca). Veterinary Medicine Dissertation, Ludwig-Maximilian University, Munich, Germany.

[63] Kriz B. (1982). – A study of camelpox in Somalia. *J. Comp. Path.*, 92, 1–8.

[64] Kropp E.M. (1985). – Kamelpocken – eine synoptische Darstellung sowie der Nachweis von Antikörpern in ostafrikanischen Dromedarseren mit einem ELISA. Veterinary Medicine Dissertation, Ludwig-Maximilian University, Munich, Germany.

[65] Leese A.S. (1909). – Two diseases of young camels. *J. Trop. Vet. Sci.*, 4, 1–7.

[66] Likhachev N.V. (1963). – Goats and sheep pox virus. *In* Guidance on veterinary virology (V.N. Sjurin, ed.). Kolos, Moscow, 622–625.

[67] Mahnel H. (1974). – Labordifferenzierung der Orthopockenviren. *Zbl. Vet. Med. B*, 21, 242–258.

[68] Mahnel H. & Bartenbach G. (1973). – Systematisierung des Kamelpockenvirus. *Zbl. Vet. Med. B.*, 20, 572–576.

[69] Mahnel H. & Munz E. (1987). – Zur derzeitigen epizootologischen Lage bei den Tierpocken. *Tierärztl. Umschau*, 42 (1), 5–14.

[70] Marennikova S.S., Shenkman L.S., Shelukhina E.L. & Maltseva N.N. (1974). – Isolation of camelpox virus and investigation of its properties. *Acta Virol.*, 18, 423–428.

[71] Mayr A. (1999). – Geschichtlicher Überblick uber die Menschenpocken (Variola), die Eradikation von Variola und den attenuierten Pockenstamm MVA. *Berl. Münch. Tierärztl. Wschr.*, 112, 322–328.

[72] Meyer H., Osterrieder N. & Pfeffer M. (1993). – Differentiation of species of genus *Orthopoxvirus* in a dot blot assay using digoxigenin-labeled DNA-probes. *Vet. Microbiol.*, 34, 333–334.

[73] Munz E. (1992). – Pox and pox-like diseases in camels. *In* Proceedings of the 1st International Camel Conference. R. and W. Publications (Newmarket) Ltd, Newmarket, UK, 43–46.

[74] Munz E., Kropp E.-M. & Reimann M. (1986). – Der Nachweis von Antikorpern gegen *Orthopoxvirus cameli* in ostafrikanischen Dromedarseren mit einem ELISA. *J. Vet. Med. B.*, 33, 221–230.

[75] Munz E., Linckh S. & Renner-Mueller I.C.E. (1992). – Infektionen mit originarem Kuhpockenvirus und kuhpockenahnlichen Erregern bei Mensch und Tier: Eine Literaturubersicht. *J. Vet. Med. B.*, 39, 209–225.

[76] Munz E., Otterbein C.K., Meyer H. & Renner-Mueller I. (1997). – Laboratory investigations to demonstrate a decreased virulence of two cell adapted African camelpox virus isolates as possible vaccine candidates. *J. Camel Pract. Res.*, 4 (2), 169–175.

[77] Nothelfer H.B., Wernery U. & Czerny C.P. (1995). – Camel Pox: antigen detection within skin lesions – immunocytochemistry as a simple method of etiological diagnosis. *J. Camel Pract. Res.*, 2 (2), 119–121.

[78] Nowotny N., Zilahi E., Al Hadj M.A., Wernery U., Kolodziejek J., Lussy H., Al Dhahry S., Al-Kobaisi M.F.M., Al Rawahi A. & Al Ismaily S. (2005). – Molecular epidemiology of camelpox and rabies viruses isolated in the United Arab Emirates and in Oman. *Emirates Med. J.*, 23 (1), 164–170.

[79] Odend'Hal S. (1983). – The geographical distribution of animal viral diseases. Academic Press, New York, 99.

[80] Otterbein C.K. (1994). – Phaeno- und genotypische Untersuchung zweier Kamelpockenvirusisolate vor und nach Attenuierung durch Zellkulturpassagen. Dissertation in Veterinary Medicine. Ludwig-Maximilians-Universitat Munchen, Munich, Germany.

[81] Otterbein C.K., Meyer H., Renner-Mueller I. & Munz E. (1995). – Charakterisierung zweier afrikanischer Kamelpockenvirus- Isolate. *Mitt. Oesterr. Ges. Tropenmed. Parasitol.*, 17, 7–16.

[82] Pfahler W.H.E. & Munz E. (1989). – Camelpox. *Int. J. Anim. Sci.*, 4, 109–114.

[83] Pfeffer M., Meyer H., Wernery U. & Kaaden O.-R. (1996). – Comparison of camelpox viruses isolated in Dubai. *Vet. Microbiol.*, 49, 135–146.

[84] Pfeffer M., Wernery U., Kaaden O.-R. & Meyer H. (1998). – Diagnostic procedures for poxvirus infections in camelids. *J. Camel Pract. Res.*, 5 (2), 189–195.

[85] Ramyar H. & Hessami M. (1972). – Isolation, cultivation and characterization of camelpox virus. *Zbl. Vet. Med. B.*, 19, 182–189.

[86] Renner-Mueller I.C.E., Meyer H. & Munz E. (1995). – Characterization of camelpox virus isolates from Africa and Asia. *Vet. Microbiol.*, 45 (4), 371–381.

[87] Richard D. (1979). – Study of the pathology of the dromedary in Borana Awraja (Ethiopia). These, Universite Paris-Est Creteil, Paris.

[88] Richard D. (1980). – Dromedary pathology and productions. Provisional report No. 6. Camels. International Science Foundation (IFS), Khartoum, Sudan, and Stockholm, Sweden, 12 (18–20), 409–430.

[89] Richard D. (1986). – Manuel des maladies du dromadaire. Projet de developpement de l'elevage dans le Niger centre-est. Maisons Alfort, IEMVT, France.

[90] Rohrer H. (1970). – Traitè des maladies a virus des animaux. Vigot, Paris.

[91] Roslyakov A.A. (1972). – Comparison of the ultrastructure of camel pox virus, the virus of pox-like disease of camels and contagious ecthyma virus. *Voprosy Virusologii*, 17, cited in *Vet. Bull.*, 42 (1), 26–30.

[92] Saad A.M. & Alrobaish T. (2006). – A herd outbreak of camelpox in Qassim region. *In* Proceedings of the International Scientific Conference on Camels, 12–14 May 2006, Buraydah, Saudi Arabia, Part 2, 381–386.

[93] Samartsev A.A. & Praksein S.T. (1950). – Camel pox study. Proceedings of the Kazakh Research Veterinary Institute, 5, 198–200.

[94] Schwartz H.J. & Dioli M. (1992). – The one-humped camel in Eastern Africa: A pictorial guide to diseases, health care and management. Verlag Josef Margraf, Weikersheim, Germany.

[95] Schwartz S., Schwartz H.J. & Wilson A.J. (1982). – Eine fotographische Dokumentation wichtiger Kamelkrankheiten in Kenia. *Der prakt. Tierarzt*, 11, 985–989.

[96] Sedov V.A. (1973). – Official communication: measures for the prevention and eradication of camelpox. *Veterinariya, Moscow*, (cited in *Vet. Bull.* 1974, 44, 295) 12, 63–64.

[97] Semushkin N.R. (1968). – Diagnosis of camel diseases. Sel'khozgiz, Moscow.

[98] Sheikh Ali H.M., Khalafalla A.I. & Nimir A.H. (2009). – Detection of camel pox and vaccinia viruses by polymerase chain reaction. *Trop. Anim. Health Prod.*, 41 (8), 1637–1641.

[99] Shommein A.M. & Osman A.M. (1987). – Diseases of camels in the Sudan. In Diseases of camels. *Rev. sci. tech. Off. int. Epiz.*, 6 (2), 481.

[100] Tantawi H.H. (1974). – Comparative studies on camelpox, sheeppox and vaccinia viruses. *Acta Virol.*, 18, 347–351.

[101] Tantawi H.H., El-Dahaby H. & Fahmy L.S. (1978). – Comparative studies on poxvirus strains isolated from camels. *Acta Virol.*, 22, 451–457.

[102] Tantawi H.H., Saban M.S., Reda I.M. & El-Dahaby H. (1974). – Camelpox virus in Egypt. I. Isolation and characterization. *Bull. Epiz. Dis. Afr.*, 22, 315.

[103] Tefera M. & Gebreah F. (2001). – A study on the productivity and diseases of camels in eastern Ethiopia. *Trop. Anim. Health Prod.*, 33, 265–274.

[104] Touil N., Cherkaoui Z., Lmrabih Z., Loutfi C., Harif B. & El Harrak (2012). – Emerging viral diseases in dromedary camels in Southern Morocco. *Transbound. Emerg. Dis.*, 59 (2), 177–182.

[105] Vedernikov V. (1893). – Camel diseases. *Arch. Vet. Med. (Saint Petersburg) I*, V, 149.

[106] Vedernikov V. (1902). – Cited from Curasson (1947) Le chameau et ses maladies. Vigot Freres Editions, Paris.

[107] Vedernikov V.A. (1969). – Pox. In Epizootiology (R.F Sosov, ed.). Kolos, Moscow, 158–164.

[108] Vyshelesskii S.N. (1954). – Pox. Particular epizootiology. *Sel'khozgiz*, 195–212.

[109] Wardeh M.F. (1989). – Camel production in the Islamic Republic of Mauritania. *Camel Newslett.*, 5, 11–17.

[110] Wernery U. (1994). – Neue Ergebnisse zur Diagnose, Prophylaxe und Therapie wichtiger bakterieller und viraler Krankheiten beim Kamel (*Camelus dromedarius*). Habilitationsschrift zur Erlangung der Lehrbefahigung an der Tierarztlichen Fakultat der Ludwig-Maximilians-Universitat Munchen.

[111] Wernery U. & Kaaden O.-R. (1995). – Infectious diseases of camelids. Blackwell Wissenschafts-Verlag, Berlin, 176–186.

[112] Wernery U., Kaaden O.-R. & Ali M. (1997b). – Orthopox virus infections in dromedary camels in United Arab Emirates (UAE) during winter season. *J. Camel Pract. Res.*, 4 (1), 51–55.

[113] Wernery U., Kinne J. & Zachariah R. (2000). – Experimental camelpox infection in vaccinated and unvaccinated guanacos. *J. Camel Pract. Res.*, 7 (2), 149–152.

[114] Wernery U., Meyer H. & Pfeffer M. (1997a). – Camel pox in the United Arab Emirates and its prevention. *J. Camel Prac. Res.*, 4 (2), 135–139.

[115] Wernery U. & Zachariah R. (1999). – Experimental camelpox infection in vaccinated and unvaccinated dromedaries. *J. Vet. Med. B.*, 46, 131–135.

[116] Wilson A.J., Schwartz H.J., Dolan R., Field C.R. & Roettcher D. (1982). – Epidemiologische Aspekte bedeutender Kamelkrankheiten in ausgewahlten Gebieten Kenias. *Der praktischeTierarzt*, 11, 974–987.



## 深入阅读材料

*Binns M., Mumford J. & Wernery U. (1992). – Development of camelpox virus as a vaccine vector. Proceedings of the 1st International Camel Conference. R. and W. Publications (Newmarket) Ltd, Newmarket, UK, 97–99.*

*El-Harrak M. & Loutfi C. (1999). – La variole du dromadaire chez le jeune. International Workshop on the Young Camel, 24–26 October 1999, Quarzazate, Morocco, 39.*

*El-Harrak M., Loutfi C. & Harif B. (1994). – Isolation and characterisation of camelpox virus in Morocco. 2nd International Conference on Vaccines, New Technologies and Applications, 21–33 March 1994, Virginia, 2–8.*

*Gitao C.G., Nyaga P.N. & Evans J.O. (1996). – Pathogenicity of sheep skin-cell-propagated camel pox virus in camels (Camelus dromedarius). Ind. J. Anim. Sci., 66 (6), 535–538.*

*Guake L.K., Dubba Z., Tumba K.H. & Abugaliev R.M. (1964). – Letter. Vet. Moscow, 41, 115–116.*

*Nowotny N., Zilahi E., Al Hadj M.A., Wernery U., Kolodziejek J., Lussy H., Al Dhahry S., Al-Kobaisi M.F.M., Al Rawahi A. &Al Ismaily S. (2005). – Molecular epidemiology of camelpox and rabies viruses isolated in the United Arab Emirates and Oman.6th Annual UAE University Research Conference, 24–26 April 2005, Al Ain, UAE, 164–170.*

*Meyer H. & Rziha H.-J. (1993). – Characterization of the gene encoding the A-type inclusion protein of camelpox virus andsequence comparison with other Orthopoxviruses. J. Gen. Virol., 74, 1679–1684.*

*Munz E., Kropp E.M. & Reimann M. (1986). – Serological and etiological investigations on camelpox in African dromedaries. Proceedings of the 5th Conference of the Institute of Tropical Veterinary Medicine, Kuala Lumpur, Malaysia, 75–76.*

*Wernery U. (1999). – New aspects on infectious diseases of camelids. J. Camel Pract. Res., 6 (1), 87–91.*

（张强译，殷宏校）

## 2.1.4 接触传染性脓疱

副痘病毒和正痘病毒无亲缘关系，两种病毒之间无交互免疫力。接触传染性脓疱病毒（或者叫羊口疮病毒，ORFV）可引起局部的水疱脓疱性疹，该病在全球广泛分布。该病可感染绵羊、山羊和野生反刍动物，如马鹿、驯鹿、塔尔羊、北山羊、野大白羊、欧洲盘羊等[11]。它是绵羊最严重的病毒病之一。形成的丘疹、水疱脓疱、结痂不仅见于嘴唇、舌，也常见于乳头、蹄和生殖器官。该病在幼龄动物上表现比较严重，能够引起死亡，特别是当有继发感染时。该病即为大家所知的“口疮”、疮痂嘴、羊口疮、传染性化脓性皮炎。

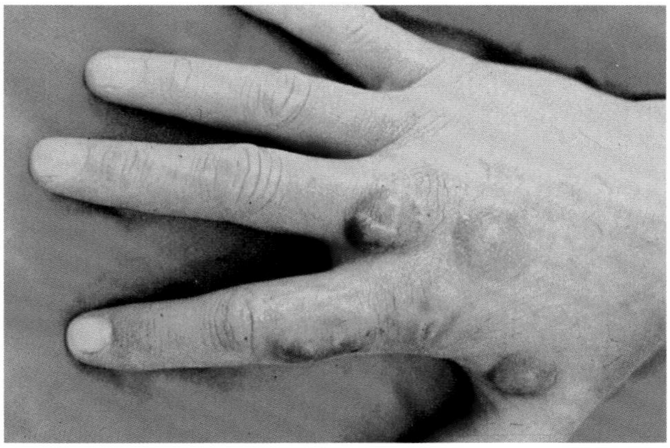

图161 人手上的接触传染性脓疱（图片由美国M.E.Fowler教授惠赠）

### 病原学

接触传染性脓疱病毒属于副痘病毒属，该属中有以下几个成员。

（1）羊口疮病毒（ORFV）

（2）牛丘疹性口炎病毒（BPSV）

（3）伪牛痘病毒（PCPV）

（4）新西兰马鹿副痘病毒（PVNZ）

根据自然宿主的范围、病理学特征以及限制性核酸酶切分析、DNA/DNA杂交等技术分析，将副痘病毒分为4个群。研究表明，副痘病毒基因组中心区域高度保守，末端区域同源性较低[27]。

### 公共卫生意义

上述的3个副痘病毒，都是人畜共患病病原，都能传染给人[16,17,25,26,27]。最近报告新西兰马鹿副痘病毒也能感染人。与患病动物直接接触，或与污染的器具间接接触都能感染。人的潜伏期为3～9天。给遭母羊遗弃的感染小羊人工哺乳，或通过伤口、擦痕和切口沾染是最常见的病毒扩散的方式。水疱脓疱病灶常见于手上（图161）、胳膊、颈部、头和腿部。当有金黄色葡萄球菌属和链球菌属及其他细菌继发感染时，体温升高，有时还可引起严重的眼炎。

### 流行病学

副痘病毒对旧大陆骆驼和新大陆骆驼都感染[5,14,15,40]。在临床上很难区分骆驼的接触传染性脓疱病和骆驼痘。许多国家报道了旧大陆骆驼的接触传染性脓疱病，该病也称为传染性脓疱性皮炎或者疮痂嘴。Tulepbaev[38,39]对其进行了深入研究，在哈萨克斯坦将西亚骆驼的接触传染性脓疱病称为“auzdyk”。长期以来骆驼饲养者一直认为此病不是传染病，只是由带刺的植物引起[8]。现在人们认识到荆棘类植物只是损伤了动物的嘴，使得副痘病毒能够侵入[10]。Roslyakov[36]用电子显微镜对骆驼的这种副痘病毒超显微结构进行了研究，因其与接触传染性脓疱病毒非常相似，命名为骆驼皮肤病病毒，称骆驼的这种病为“骆驼脓疱性皮炎”。Khokhoo[22]也研究了这种病毒的生物学特征。

俄罗斯[9]、蒙古[12]、肯尼亚[13,15,30]、索马里[23,28]、苏丹[5,18]、利比亚[6,7]、阿拉伯联合酋长国（阿联酋）[41]、沙特阿拉伯[2]、巴林[1]和印度[31]也纷纷报道了在旧大陆骆驼上发生的接触传染性脓疱。

骆驼接触传染性脓疱主要发生在幼龄，最大至3岁的动物。不同国家的学者记录的发病率和死亡率具有很大差异（表61）。

对副痘病毒是如何传播的知之甚少。与痂皮直接接触，或感染动物水疱脓疱液通过抓痕渗透入皮肤都可引起感染，与污染的环境或污染物间接接触也可发生自然传播。1g的痂皮中有$10^9$～$10^{10}$个病毒感染粒子。病毒在恢复期动物皮肤内能存活1个多月。最新发现，带毒绵羊在疫病传播中起着非常重要的作用。有证据显示，在多年没有放牧的荆棘类植物非常丰富的牧场，将易感羊群在此再放牧时，又感染了该病[24]。Azwai等[7]用ELISA检测利比亚驼群，具有临床症状的单峰驼接触传染性脓疱

表61　单峰驼接触传染性脓疱的发病率和死亡率

| 学者 | 年份 | 国家 | 检测骆驼数量 | 月龄 | 发病率(%) | 死亡率(%) |
|---|---|---|---|---|---|---|
| Dashtseren et al. | 1984 | 蒙古 | 478 | 成年 | 10 ~ 80 | 0 |
| Munz et al.(两个研究) | 1986 | 索马里 | — | 成年 | 100<br>10 ~ 20 | 0 |
| Ali et al. | 1991 | 苏丹 | 700 | 14 | 6 | 0 |
| Gitao | 1994 | 肯尼亚 | 600 | 8 | 100 | 0 |
| Wernery et al. | 1997 | 阿拉伯联合酋长国 | 30 | 16 | 20 | 0 |
| Abu-Elzein et al. | 1998 | 沙特阿拉伯 | 700 | 幼畜和成年 | 24 | 0 |
| Khalafalla | 1998 | 苏丹 | — | 12 | 60.2 | 8.8 |



的血清阳性率是38%，临床健康的单峰驼血清阳性率是0 ~ 7%。Gitao[15]认为，夜晚让幼驼挤在一个场地里扩大了病毒接触传播的概率，他也证明，当附近的小山羊发生了副痘病毒感染时，幼驼暴发疫病。Munz等[30]和Robertson[35]也有类似的发现，他们检测到了美洲驼感染“羊口疮”。但Abu-Elzein等[2,3]的发现与上述结果不同。他们在易感的单峰驼上复制了接触传染性脓疱病，但实验感染的绵羊对骆驼的病毒具有抵抗力。同样，单峰驼也抵抗源自绵羊的羊口疮病毒的感染。感染羊源羊口疮病毒的绵羊表现出典型的临床病理症状，并产生了相应的抗体，但是，感染了骆驼副痘病毒的单峰驼尽管表现出典型的临床病理症状，却没有产生相应的抗体。Wernery U.（个人交流材料，2008）用羊口疮病毒实验性地感染绵羊却未获成功。因此说单峰驼与绵羊羊口疮的发生无关，绵羊和山羊也不在骆驼发生副痘病毒感染时发病[4]。Abu-Elzein等[3]进一步研究证明，在骆驼痘和骆驼“羊口疮”之间没有交叉保护。他们在沙特阿拉伯对4只3 ~ 4岁的单峰驼注射感染骆驼“羊口疮”病毒，这些骆驼在一个月前已患骆驼痘，正处于恢复期，又产生了典型的骆驼“羊口疮”的症状。

Wernery和Kaaden[40]报告，在阿联酋一项骆驼痘免疫实验中，3只8月龄的单峰驼死于真性骆驼痘和副痘的混合感染。作为对照的骆驼，在用骆驼痘病毒人工感染时，可能已经感染了接触传染性脓疱病毒或者是隐性带毒动物。在电子显微镜下，可见

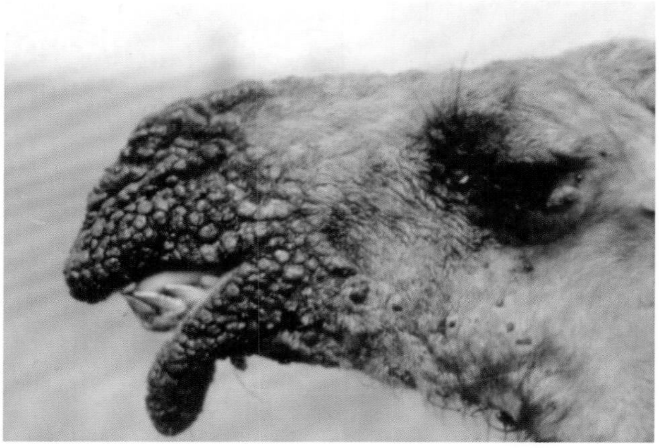

a. 一只苏丹单峰驼的接触传染性脓疱病（A.Khalafalla博士惠赠，苏丹）

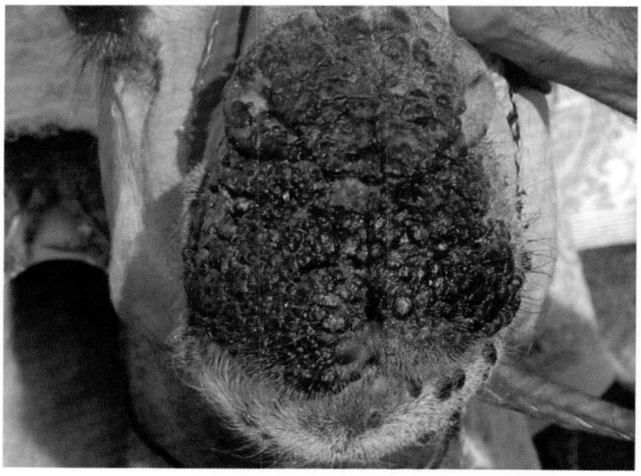

b. 一只阿联酋单峰驼赛驼的接触传染性脓疱病，喷雾可以预防继发细菌感染和蝇蛆病

图162　骆驼接触传染性脓疱病

两种病毒相邻存在（图152）。

前面已提到新大陆骆驼也易感接触传染性脓疱病毒[14,29,32,33,34,37]。感染的新大陆骆驼在嘴角上皮中出现典型的增生病灶，病变也会扩大到脸部和会阴部。若母畜乳头处有病灶，当幼畜从母畜哪里吮乳时也会感染。绵羊的羊口疮病毒可致人的手指、四肢及面部产生严重的溃疡灶（图161）。

## 临床症状和病理变化

感染后2～6天，在病毒侵入机体位点出现原发性病灶，此时主要是局部的皮肤病变，病灶位置、大小、严重性不同。单个或多个原发副痘病毒灶出现在嘴唇和口鼻处的皮肤上，也常出现在眼睑、头部其他部位、口腔里上腭，切齿下的牙床等处。病灶起先是微红色的丘疹，几天后变成淡黄色的水泡。在结痂前，出现溃疡和出血。继发细菌和真菌感染以及蝇蛆病可引起嘴唇和口腔病灶的恶化。浅表淋巴结明显增大。

镜检可见感染的皮肤棘层细胞角化不全，角蛋白细胞气球样变性，真皮炎性肿胀。病灶常伴随溃疡、嗜中性粒细胞和嗜酸性粒细胞浸润，表面可见细菌和真菌菌落。显微观察可用于早期、急性期诊断，在陈旧病变中（6天或更多）可见肿胀表皮细胞里的胞质包涵体。

患接触传染性脓疱病的骆驼通常在面部出现痘样病变（图162a，b）。据报道，东非单峰驼严重全身性副痘病毒感染与骆驼痘无法区分[26]。

Munz等[30]描述了肯尼亚一个450峰单峰驼群的副痘暴发过程。最初，在动物唇部可见增生性病变，病变偶尔扩展到鼻子和口腔黏膜，年轻单峰驼具有全身性感染的特点和趋势。开始时是丘疹，接着丘疹发展成脓疱，继而结痂。结痂最终变为深棕色，在6～10周后跌落。严重感染的单峰驼将形成数个圆形的、表皮轻度变厚的黑色无毛区域，长达6个月。一些骆驼的眼睑、嘴唇和鼻翼，甚至整个头部出现浮肿。Moallin和Zessin[28]报告了索马里，Khalafalla报告了苏丹[19]驼群的类似临床症状。Gitao[15]观察到许多肯尼亚单峰驼幼驼中出现的颈部和下颌淋巴结肿大和水肿。在大多数的皮肤病变中，在痂皮下都有浓稠的、淡黄色脓液。作者未检测母畜乳房上的所有病变和成年骆驼皮肤上的所有病变。

肯尼亚单峰驼群中发病率可达100%。Dashtseren等[12]描述了蒙古西亚驼群中的接触传染性脓疱病，在嘴的周围，4～12天里病变会发展成为直径4mm的大水泡。2～5个月后，皮肤病变融合形成5～15mm厚的厚痂，有时可见厚痂被皮肤褶皱分开，未见到死亡病例。10%～80%的蒙古成年驼、10%～20%的肯尼亚成年驼会罹患这种病。

## 诊断

骆驼接触传染性脓疱病与骆驼痘、乳头瘤、疥癣或嗜皮菌病难以区分，因此兽医诊断时实验室对新鲜病变组织的活组织检查非常重要。由于病毒不能用鸡胚绒毛尿囊膜分离，须用电子显微镜进行活组织或皮肤检查。病毒可在绵羊或小牛肾、睾丸细胞、人或猴羊膜、青年兔肾细胞上增殖。有时需要传几代才能获得细胞病变，CPE须用免疫荧光或电子显微镜鉴别。在病变处含有大量的感染粒子（每克$10^9$～$10^9$病毒粒子，注：原文疑错误，应为每克$10^9$～$10^{10}$病毒粒子）。根据病毒形状、大小特征，在电子显微镜下很容易观察到负染的病毒粒子。近期发展的PCR技术可用于检测副痘病毒感染，尤其是当电子显微镜观察为阴性结果时[11]。Khalafalla等[21]研究了骆驼接触传染性脓疱病毒的分子特征。分离的病毒在胎牛食管细胞和牛肾细胞（BK-KL3A，Bayer AG，德国）上生长。通过不同的方法证明5个分离株是副痘病毒，所有的5个分离株图谱彼此相同，一个与牛丘疹性口炎病毒、牛副痘病毒1型图谱类似。最近的基因分析显示，分离自巴林和沙特阿拉伯[1]、印度[31]的骆驼副痘病毒与伪牛痘病毒关系密切。Roslyakov[36]比较了骆驼痘和副痘病毒的超微结构。

间接免疫荧光、ELISA和western blot技术可用于检测骆驼接触传染性脓疱病毒抗体[6]，但由于该病以细胞免疫为主，用抗体检测结果来分析免疫状态并不可靠。Khalafalla等[21]研究，在PCR中使用区别副痘病毒和正痘病毒的引物，可快速鉴别和区分骆驼接触传染性脓疱和骆驼痘。

## 治疗和控制

本病为病毒病，不值得治疗。对已感染动物的控制非常重要。有时“羊口疮”是慢性病，从公共卫生角度考虑，由于疑似感染动物可能隐性带毒，应从畜群中隔离直至完全康复。病毒在外界环境中能存活多年，在已污染的牧场上禁止放牧，尤其是牧场中有荆棘类植物存在时。应尽量降低家畜的饲养密度，减少继发感染。用大剂量合成青霉素进行全身治疗是防止葡萄球菌继发感染的最佳途径。检查和治疗患“羊口疮”的骆驼时应始终戴手套。

免疫动物和自然感染动物都不能产生持久的免疫，如感染8个月后已康复的绵羊仍可再次感染。弱毒疫苗用于绵羊和山羊的免疫，但在骆驼科动物上没有任何效果。Dashtseren等[12]研究表明，痘苗病毒和羊口疮病毒疫苗都不能保护骆驼感染副痘病毒，但该作者用一个骆驼副痘病毒株适应鸡胚研制的疫苗对这种病能起保护作用，免疫骆驼科动物可保护至少6个月。

## 参考文献


[1] Abubakr M.I., Abu-Elzein E.M., Housawi F.M., Abdelrahman A.O., Fadlallah M.E., Nayel M.N., Adam A.S., Moss S.,Forrester N.L., Coloyan E. & Gameel A. (2007). – Pseudo cowpox virus: the etiological agent of contagious ecthyma (Auzdyk)in camels (*Camelus dromedarius*) in the Arabian peninsula. *Vector Borne Zoonotic Dis.*, 7 (2), 257–260.

[2] Abu-Elzein E.M.E., Coloyan E.R., Gameel A.A., Ramadan R.O. & Al-Afaleq A.I. (1998). – Camel contagious ecthyma inSaudi Arabia. *J. Camel Pract. Res.*, 5 (2), 225–228.

[3] Abu-Elzein E.M.E., Housawi F.M.T., Al-Afaliq A.L., Gameel A.A., & Ramadan R.O. (2004b). – A note on experimentallyinduced severe camel Orf (Auzdyk disease) in dromedary camel. *J. Camel Pract. Res.*, 11 (2), 101–102.

[4] Abu-Elzein E.M.E., Housawi F.M.T., Al-Afaleq A.I., Ramadan R.O., Gameel A.A & Al-Gundi O. (2004a). – Clinico-pathologicalresponse of dromedary camels and sheep to cross-experimental infection with two virulent orf viruses originating fromcamels and sheep. *J. Camel Pract. Res.*, 11 (1), 15–19.

[5] Ali O.A., Kheir S.A.M., Abu Damir H. & Barri M.E.S. (1991). – Camel (*Camelus dromedarius*) contagious ecthyma in theSudan. A case report. *Rev. Elev. Méd. Vét. Pays Trop.*, 44 (2), 143–145.

[6] Azwai S.M., Carter S.D. & Woldehiwet Z. (1995). – Immune responses of the camel (*Camelus dromedarius*) to contagiousecthyma (Orf) virus infection. *Vet. Microbiol.*, 47 (1–2), 119–131.

[7] Azwai S.M., Carter S.D. & Woldehiwet Z. (1998). – An immunological study of contagious pustular dermatitis in camels. International Meeting on Camel Production and Future Perspectives, 2–3 May 1998, Al Ain, UAE, 108.

[8] Borisovich Y.F. &. Orekhov M.D. (1966). – Camel pox. *Veterinaryia, Moscow.* (Cited in *Vet. Bull.*, 3, 50–52).

[9] Buchnev K.N., Sadykov R.G., Tulepbayev S.Z. & Roslyakov A.A. (1969). – Smallpox-like disease of camels Auzdyk. *TrudyAlma-Ata Tinskogo Zootekhnicheskogo Instituta*, 16, 36–47.

[10] Buchnev K.N., Tulepbaev S.Z. & Sansyzbaev A.R. (1987). – Infectious diseases of camels in the USSR. *In* Diseases of camels.*Rev. sci. tech. Off. int. Epiz.*, 6 (2), 487–495.

[11] Buttner M., von Einem C., Mclnnes C. & Oksanen A. (1995). – Klinik und Diagnostik einer schweren Parapocken-Epidemiebeim Rentier in Finnland. *Tierärztl. Praxis*, 23 (6), 614–618.

[12] Dashtseren T., Solovyev B.V., Varejka F. & Khokhoo A. (1984). – Camel contagious ecthyma (*pustular dermatitis*). *Acta Virol.*,28, 122–127.

[13] Dioli M. & Stimmelmayr R. (1992). – Important camel diseases. *In* The one-humped camel in eastern Africa: a pictorialguide to diseases, health care and management (H.J. Schwartz & M. Dioli, eds). Verlag Joseph Markgraf Scientific Books,Weikersheim, Germany, 155–164.

[14] Fowler M.E. (2010). – Medicine and surgery of camelids. Iowa State University Press, Ames, IA, 178–179.

[15] Gitao C.G. (1994). – Outbreaks of contagious ecthyma in camels (*Camelus dromedarius*) in the Turkana District of Kenya. *Rev. sci. tech. Off. int. Epiz.*, 13 (3), 939–945.

[16] Hartmann A.A., Buettner M., Stanka F. & Elner P. (1985). – Sero- und Immunodiagnostik bei Parapoxvirus – Infektionendes Menschen. *Der Hautarzt*, 36, 663–669.

[17] Hartung J. (1980). – Lippengrind des Schafes. *Tierärztl. Prax.*, 8, 435–438.

[18] Khalafalla A.I. (1998). – Epizootiology of camel pox, camel contagious ecthyma and camel papillomatosis in the Sudan. Proceedings of the International Meeting on Camel Production and Future Perspectives, 2–3 May 1998, Al Ain, UAE, 105.

[19] Khalafalla A.I. (2000). – Camel contagious ecthyma: Risks in young calves. *Rev. Elev. Vet. Pays Trop.*, 53 (2), 173–176.

[20] Khalafalla A.I., Buttner M. & Rziha H.-J. (2005). – Polymerase chain reaction (PCR) for rapid diagnosis and differentiation ofpara- and orthopox virus infections in camels. In Applications of gene-based technologies for improving animal productionand health in developing countries. Springer, Netherlands, 335–342.

[21] Khalafalla A.I., Rziha H.J. & Buttner M. (2006). – Isolation and molecular characterization of the camel contagious ecthymavirus. Proceedings of the International Scientific Conference on Camels, 12–14 May 2006, Buraydah, Saudi Arabia, 387–396.

[22] Khokhoo A. (1982). – Biological properties of camel contagious ecthyma virus. Thesis, Veterinary Institute, Brno, CzechRepublic.

[23] Kriz B. (1982). – A study of camelpox in Somalia. *J. Comp. Path.*, 92, 1–8.

[24] Lewis C. (1996). – Update on orf. *In Practice*, 18 (8), 376–381.

[25] Liess H. (1962). – Lippengrind (*Ecthyma contagiosum*) der Schafe als Zooanthroponose. *Zbl. Bakteriol. Microbiol. Hyg.*, A,183, 1969–1983.

[26] Mahnel H. & Munz E. (1987). – Zur derzeitigen epizootologischen Lage bei den Tierpocken. *Tierärztl. Umschau*, 42 (1), 5–14.

[27] Mercer A., Fleming S., Robinson A., Nettleton P. & Reid H. (1997). – Molecular genetic analyses of parapoxviruses pathogenicfor humans. *Arch. Virol.*, 13 (Suppl. 13), 25–34.

[28] Moallin A.S.M. & Zessin K.H. (1988). – Outbreak of camel contagious ecthyma in central Somalia. *Trop. Anim. Health Prod.*,20, 185–186.

[29] Moro Sommo M. (1971). – Ectima. In La Alpaca. Enfermedades Infecciosas y Parasitarias. *Boletin del InstitutoVeteranario deInvestigaciones Tropicales y de Altura (IVITA)*, Universidad Nacional Mayor de San Marcos, Lima, Peru, 30.

[30] Munz E., Schillinger D., Reimann M. & Mahnel H. (1986). – Electronmicroscopical diagnosis of *Ecthymacontagiosum* incamels (*Camelus dromedarius*): first report of the disease in Kenya. *J. Vet. Med.* B., 33, 73–77.

[31] Nagarajan G., Ghorui S.K., Kumar S. & Pathak K.M. (2010). – Complete nucleotide sequence of the envelope gene ofpseudocowpox virus isolates from Indian dromedaries (*Camelus dromedarius*). *Arch. Virol.*, 155 (10), 1725–1728.

[32] Preston Smith H. (1940). – The camelids of Peru, ecthyma of alpacas [Los camello Peruanos, anguenidos, alpacas ectima]. Ministry of Agriculture Bolivia.

[33] Preston Smith H. (1947). – Ectima de los animales del Peru, dermatitis pustular contagiosa. *Ganaderia (Peru)*, 1 (1), 27–32.

[34] Ramirez A. (1980). – Ectima contagioso en alpaca. En aspectos sanitarios en la alpaca. Curso sistema de producción pecuariaen los altos Andes. Peruvian Association of Animal Production, Lima, Peru, 94.

[35] Robertson A. (1976). – Handbook on animal diseases in the tropics. 3rd Ed. British Veterinary Association, London, 9–11.

[36] Roslyakov A.A. (1972). – Comparison of the ultrastructure of camelpox virus, the virus of pox-like disease of camels andcontagious ecthyma virus. *Voprosy Virusologii*, 17 (Abstract cited in Vet. Bull., 1, 26–30).

[37] Thedford R.R. & Johnson L.W. (1989). – Infectious diseases of New-world camelids (NWC). *Vet. Clin. North Am. Food Anim.Pract.*, 5 (3), 145–157.

[38] Tulepbaev S.Z. (1969). – Sensitivity of domestic and laboratory animals to the virus of smallpox-like disease of camels('Auzdyk'). *Trudy Almata-atinskogo Zootekhnicheskogo Instituta*, 16, 41–42.

[39] Tulepbaev S.Z. (1971). – Pox-like disease ('Auzdyk') of the camels in Kazakhstan. Thesis, Veterinary Faculty, Alma-Ata,Kazakhstan.

[40] Wernery U. & Kaaden O.-R. (2002). – Infectious diseases of camelids. Blackwell Wissenschafts-Verlag, Berlin and Vienna,187–192.

[41] Wernery U., Kaaden O.-R. & Ali M. (1997). – Orthopox virus infections in dromedary camels in United Arab Emirates(UAE) during winter season. *J. Camel Pract. Res.*, 4 (1), 51–55.

## 深入阅读材料

*Guo S.Z. (1988). – Serological comparison of the pathogens of aphthosis in camel, sheep and goat. Chinese J. Vet. Med. Tech., 5, 35–37.*

*Khalafalla A.I. (1999). – Camel contagious ecthyma and its risk to young calves. International Workshop on the Young Camel,24–26 October 1999, Quarzazate, Morocco, 45.*

*Khalafalla A.I., Agab H. & Abbas B. (1994). – An outbreak of contagious ecthyma in camels (Camelus dromedarius) in easternSudan. Trop. Anim. Health Prod., 26, 253–254.*

*Khalafalla A.I. & Mohamed M.E.M. (1997). – Epizootiology of camel contagious ecthyma in eastern Sudan. Rev. Elev. Méd. Vét.Pays Trop., 50 (2), 99–103.*

*Mustapha I.E. (1980). – IFS Provisional report No.6 on camels, 399. International Science Foundation (IFS), Khartoum, Sudan,and Stockholm, Sweden.*



（张强译，殷宏校）

## 2.1.5 乳头状瘤

乳头状瘤（疣）是生长在皮肤和黏膜上的良性肿瘤，在世界范围内广泛分布，人和大多数动物均可发生。乳头状瘤是由种特异性乳头瘤病毒引起的，与鳞状细胞癌的发生有一定关系。

牛比其他家养动物更易受到乳头状瘤病毒感染，已鉴定出6个型的牛乳头状瘤病毒。在人和牛上发现了70个以上的乳头状瘤病毒型，但在其他种类的动物，每种动物到目前为止仅鉴定出一个病毒型。

乳头状瘤病毒也能感染骆驼，引起典型的皮肤损伤[2,3,5,6,7,10]。已报道的乳头状瘤病仅发生于旧大陆骆驼。

### 病原学

乳头状瘤病毒属于乳头状瘤病毒属（*Papilloma virus*），乳多空病毒科（*Papovaviridae*）成员。病毒粒子直径约为50nm，呈球形，具有二十面体对称结构，72个壳粒组成一个病毒粒子的衣壳结构，每个壳粒至少由3个蛋白组成。乳头状瘤病毒的鉴别依赖于对病灶的组织学特征观察，以及通过核酸杂交或PCR进行DNA鉴定。对马、小反刍动物或者骆驼乳头瘤病毒的研究很少。在牛上已经鉴定的6个型，包括牛乳头瘤病毒BPV-1、BPV-2、BPV-3、BPV-4、BPV-5、BPV-6。BPV-1、BPV-2、BPV-5引起纤维乳头状瘤，BPV-3、BPV-4、BPV-6引起真上皮细胞性乳头状瘤（true epithelial papillomas）。可感染骡和驴的马肉样瘤似乎与BPV-1和BPV-2有关。

### 流行病学

乳头状瘤病仅在旧大陆骆驼有报道，但很少发生，几乎不会造成经济损失。该病通常发生在2岁以下的骆驼，疣病变与痘病变有很大区别，疣病变主要在口唇和颌下区域，不会影响感染动物的健康[3]。然而，Munz等[7]报道了在索马里中部暴发的一起乳头状瘤病，主要感染6月龄到2岁的动物。病变很难与痘病毒和副痘病毒感染区分，观察到全身型的乳头状瘤。只有通过电子显微镜观察等实验室方法，才能弄清楚致病因子。Sadana等[8]报道了发生于印度单峰驼的一例极为罕见的乳头状瘤病例。一头15岁龄单峰驼的球节上生长了一颗重约2 kg的疣，手术切除后没有并发症。一般认为这种生长物不是真乳头状瘤，而是一种纤维乳头状瘤。

在阿拉伯联合酋长国也有幼龄单峰驼乳头状瘤病例报道[10]。没有观察到全身型的疣，仅在口唇和鼻孔有个别病变，呈茎状疣，很容易与骆驼痘或者骆驼触染性痘疮等病区分（图163）。

Kinne和Wernery[5]等描述了阿拉伯联合酋长国一小群单峰驼发生的乳头状瘤病，10头单峰驼中有3头在口内外出现了增生性茎状疣（图164）。用电

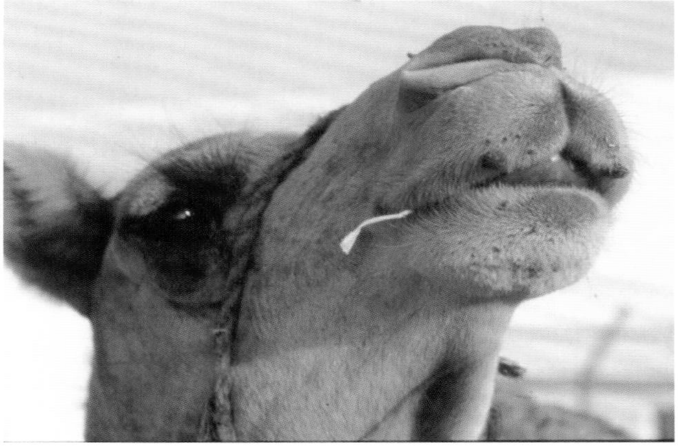

图163 单峰驼乳头状瘤病

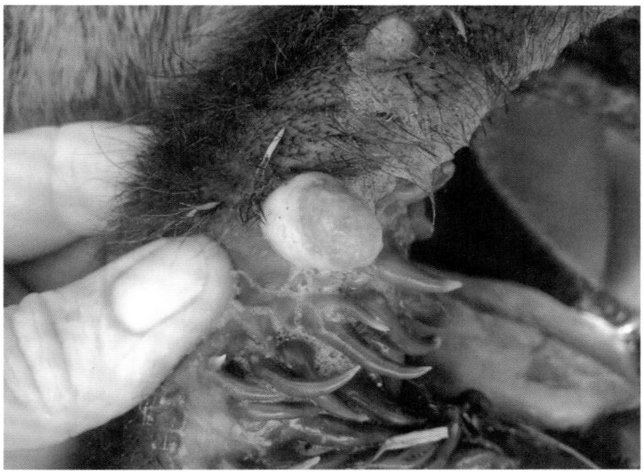

图164 单峰驼口部的疣（图片由阿曼的L. Donders夫人惠赠）

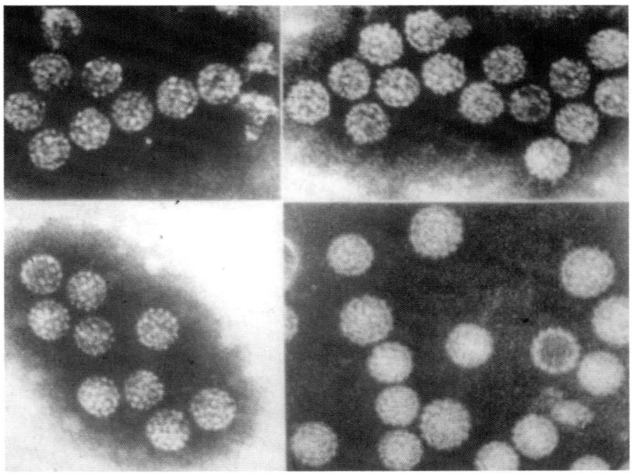

图165 从单峰驼口部疣组织中分离到的乳头状瘤病毒样粒子（电子显微镜，×125000）

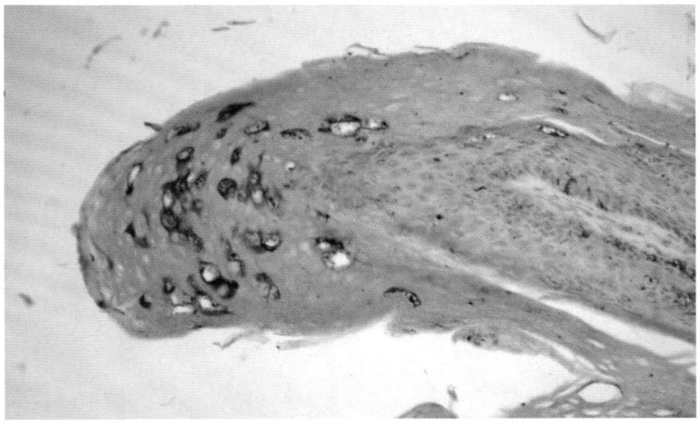

图166 免多抗免疫组化检测显示疣组织上皮细胞乳头状瘤病毒抗原阳性，少量病毒抗原位于上棘层细胞核内，颗粒层多数细胞呈棕色



子显微镜观察，可见病灶样本中含有乳头状瘤病毒样粒子（图165）。

在土耳其单峰驼发生了一例罕见的由乳头状瘤病毒引起的单峰驼角膜乳头状瘤病例[4]。动物表现为严重的慢性角膜结膜炎和左眼角膜堆积。组织学观察可见嗜碱性核内包涵体，免疫组织化学检测显示有BPV-1存在，在进行了角膜切除术后单峰驼康复。据Ure等[9]报道，2009年8月在苏丹一家单峰驼农场暴发了乳头状瘤病，该农场共有55头动物，只有3~7月龄的青年动物发病，发病动物中有44%（11/25）出现了病灶，病灶主要在口唇和下颌。表现出两种类型的皮肤病变：只有一头动物出现了菜花样结节，大多数动物出现了带裂缝的圆形或椭圆形结节。从这两种不同的病灶中检测到单峰驼乳头状瘤病毒1和2两个基因型的病毒（*Camelus dromedarius* papilloma virus 1 and 2），暂时被归入δ-乳头状瘤病毒属（*Deltapapillomavirus*）。

乳头状瘤病毒通常通过皮肤擦伤和微损伤在动物间传播。梳理设备、绳索、污染的工具都可能传播该病毒。该病毒也有可能通过节肢动物的叮咬传播。Khalafalla等[3]认为乳头状瘤病和骆驼触染性痘疮之间有密切的关系。作者发现大多数骆驼乳头状瘤病发生在多雨季节，和触染性痘疮的暴发时间相一致。Dioli和Stimmelmayr[1]发现肯尼亚的骆驼痘和乳头状瘤病之间有联系。

## 病理学

许多研究者描述了骆驼乳头状瘤病的病理形态[1,3,5,7,10]。病灶呈圆形或菜花样乳头状瘤，直径0.3～4cm，通常有茎，不影响骆驼的健康。其临床症状与骆驼痘和副痘区别很大，后二者所引起的皮肤病变通常经历小水泡和结痂过程。在乳头状瘤病的早期阶段，病灶部皮肤为玫瑰色、呈充血性隆起。Munz等[7]描述了索马里暴发的一起乳头状瘤病，许多单峰驼嘴唇和鼻孔内出现脓疱和结痂，以及全身性增生性大小结节和肿瘤样病变。有些骆驼在耳朵、眼睑、腹股沟、生殖器区、腿部出现病灶。患病率较高，但死亡率为零。显微镜观察，受感染上皮过度增生皱折，形成增生性赘生物。上皮细胞增生以棘层显著增厚、角化不全和角化过度为特征，并伴有表皮突伸长。表皮突延伸进入下层的皮肤结缔组织，并可能转化为增生。在颗粒层的单个和/或细胞群可能出现肿大、透明胞浆、多形性大透明角质样颗粒（中空化细胞）。

### 诊断

对乳头状瘤病不难做出临床诊断，可以通过乳头状瘤典型的显微特征来进行鉴别诊断。Kinne和Wernery[5]用兔抗牛乳头状瘤病毒多抗血清建立了一种免疫组织化学检测方法（图166），可用于该病诊断。

### 治疗和控制

乳头状瘤病一般情况下是一种较轻微的、自限性的疾病，通常不需要预防和治疗。而且，骆驼疣通常在3～6个月会自行脱落。然而，在阿拉伯联合酋长国针对两起乳头状瘤病暴发，用切除的疣组织经甲醛灭活处理后制成自体疫苗进行治疗。根据单峰驼体重，皮下注射3～7 mL疣组织疫苗。免疫后8～10天疣体消退。由于乳头状瘤病毒存在抗原变异株，建议根据不同群体内流行毒株，研发特异性的自体疫苗进行治疗。在疣形成的早期阶段不建议用外科手术和冷冻外科手术切除，因为残留的疣体仍会生长增大反而延长了治疗过程。

## 参考文献

[1] Dioli M. & Stimmelmayr R. (1992). – Important camel diseases. *In* The one-humped camel in Eastern Africa: a pictorial guide to diseases, health care and management (H.J. Schwartz & M. Dioli, eds). Verlag Joseph Markgraf Scientific Books, Weikersheim, Germany, 155–164.

[2] Khalafalla A.I. (1998). – Epizootiology of camel pox, camel contagious ecthyma and camel papillomatosis in the Sudan. International Meeting on Camel Production and Future Perspectives, 2–3 May 1998, Al Ain, UAE, 105.

[3] Khalafalla A.I., Abbas Z. & Mohamed M.E.H. (1998). – Camel papillomatosis in the Sudan. *J. Camel Pract. Res.*, 5 (1), 157–159.

[4] Kilic N., Toplu N., Aydogan A., Yaygingül R. & Ozsoy S.Y. (2010). – Corneal papilloma associated with papillomavirus in a one-humped camel (*Camelus dromedarius*). *Vet. Ophthalmol.*, 13 (1), 100–102.

[5] Kinne J. & Wernery U. (1998). – Papillomatosis in camels in the United Arab Emirates. *J. Camel Pract. Res.*, 5 (2), 201–205.

[6] Munz E. (1992). – Pox and pox-like diseases in camels. Proceedings of the 1st International Camel Conference. Wade: R. and W. Publications (Newmarket) Ltd, Newmarket, UK, 43–46.

[7] Munz E., Moallin A.S.M., Mahnel H. & Reimann M. (1990). – Camel papillomatosis in Somalia. *J. Vet. Med. B.*, 37, 191–196.

[8] Sadana J.R., Mahajan S.K. & Satija K.C. (1980). Note on papilloma in a camel. *Indian J. Anim. Sci.*, 50 (9), 793–794.

[9] Ure A.E., Elfadl A.K., Khalafalla A.I., Gameel A.A.R., Dillner J. & Forslund O. (2011). – Characterization of complete genomes of *Camelus dromedarius* papillomavirus 1 and 2. *J. Gen. Virol.*, 92 (8), 1769–1777.

[10] Wernery U. & Kaaden O.-R. (2002). – Infectious diseases of camelids. Blackwell Wissenschafts-Verlag, Berlin and Vienna, 192–195.

（曹轶梅译，卢曾军、殷宏校）

## 2.1.6 口蹄疫

口蹄疫（FMD）仍然是最重要的动物传染病。一些畜牧业高度发达的国家极为害怕发生该病。口蹄疫是高度传染性病毒病，几乎牛、绵羊、山羊和猪等家养和野生偶蹄动物都能感染。该病以水泡性病变为特征，但在小牛也可观察到坏死性心肌变性。也有人感染口蹄疫的报道，但发病率较低，一般认为该病对人类公共卫生没有威胁。

旧大陆骆驼和新大陆骆驼生活于北非和东非、中东和远东以及南美的一些国家，在这些国家口蹄疫呈地方性流行。在过去数年，通过田间观察，尤其是试验感染，提高了人们对骆驼在口蹄疫病毒（FMDV）易感性方面的认识[11,23,46]。

### 病原学

口蹄疫是由隶属于小RNA病毒科（Picornaviridae）口蹄疫病毒属（*Aphthovirus*）的口蹄疫病毒引起的。已经确定口蹄疫病毒有7个不同的血清型，其中欧洲发生过的口蹄疫病毒型主要是A型、O型和C型。O型是世界范围内分布最广的，而C型分布范围最小。各血清型间无交叉免疫原性。在7个血清型内已确定有60个以上的血清亚型。

### 流行病学

口蹄疫在欧洲部分国家、中东、印度、远东和南美的部分地区呈地方性流行。北美、澳大利亚、新西兰和西欧的许多地区和国家没有口蹄疫的流行。这些地区和国家有防止口蹄疫病毒传入的严格法规和措施。口蹄疫与骆驼养殖户的利益密切相关，有骆驼生活的地区口蹄疫也多呈地方性流行。例如，沙特阿拉伯骆驼饲养量有800000只，从非洲、亚洲和澳大拉西亚（泛指大洋洲和太平洋岛屿国家）进口活畜约650万只，大部分为绵羊和山羊。来自非洲和亚洲的动物携带有这些国家特有的口蹄疫病毒毒株，造成在沙特阿拉伯和相邻国家的游牧畜群中传播。了解骆驼在口蹄疫病毒传播中的作用具有重要意义。口蹄疫病毒的自然宿主是偶蹄动物，包括牛、绵羊和山羊，但许多不同种类的野生动物也有可能是其宿主。许多有蹄野生动物种也有可能携带病毒，但不表现临床症状。例如，非洲水牛（*Syncerus caffer*）可在咽部携带病毒达2年之久而不表现水泡性病变（带毒状态）。多数情况下口蹄疫病毒通过气溶胶的方式传播，当动物紧密接触时通常可发生传播，在某些环境条件下病毒也可随风传播到很远的地方。口蹄疫流行病学研究方法已经彻底改变，现在可以利用分子技术来研究每个病毒株的特征，从而可以追踪毒株在不同国家之间的传播路线[18]。

### 新大陆骆驼口蹄疫

南美洲和北美洲已有许多关于新大陆骆驼对口蹄疫病毒易感性的研究，这些研究都是以家养新大陆骆驼为对象进行的，包括美洲驼（*Lama glama*）和羊驼（*Lama pacos*）。也有一些田间新大陆骆驼对口蹄疫易感性的报道和血清学调查。但是，没有关于原驼（*Lama guanacoe*）和骆马（*Vicugna vicugna*）等野生新大陆骆驼对口蹄疫易感性的报道。

### 新大陆骆驼实验感染和自然感染口蹄疫

在自然条件下和实验条件下都证明了新大陆骆驼对口蹄疫病毒有易感性。在秘鲁普诺发生了一起牛口蹄疫疫情，与之接触的羊驼也表现出轻微的临床症状；从发病牛病料中分离到了A24 口蹄疫病毒[26]。但是，对其他的田间疫情调查不能证实新大陆骆驼对口蹄疫有易感性。1981年在印度阿萨姆地区暴发口蹄疫疫情，该区有包括骆驼科在内的大量有蹄动物，美洲驼和单峰驼都没有被感染[36]。在阿根廷曾发生过牛口蹄疫地区的天然牧场，采集了460份不同性别与年龄的美洲驼咽-食道部分泌物（OPF）样本，从中并没有检测或分离到口蹄疫病毒[5,43]。

在秘鲁[25]，美国外来动物疫病诊断实验室[25]和阿根廷布宜诺斯艾利斯的国家农业技术研究所[10]，3个不同实验室都进行了新大陆骆驼感染和传播口蹄疫病毒研究：所有试验都表明，新大陆骆驼可以通过不同的感染途径感染不同血清型的口蹄疫病毒。美洲驼和羊驼表现出了或轻微或严重的临床症状。然而，只有两组研究数据，虽然所使用的动物数量很少，但证明能够将病毒传播给美洲驼和其他易感动物[24,25]。

Fondevila等[10]精心设计了一项试验，以评估美洲驼在直接或间接接触已感染家畜后对口蹄疫病毒的易感性。通过不同途径给6头猪接种了用$A_{79}$、$O_3$和$O_1$亚型口蹄疫病毒，然后将30头美洲驼和受感染猪放在一起，随后再将接触了感染猪的30头美洲驼与另外30头美洲驼一起饲养。然后再将40头易感家畜（包括猪、犊牛、绵羊和山羊）加入上述60头美洲驼中以检测美洲驼是否可以传播口蹄疫病毒。结果，30头直接接触了口蹄疫感染猪的美洲驼中仅有2头出现了轻微症状，产生了口蹄疫病毒抗体，血液和OPF中均检测到了病毒。还有1头美洲驼也检测为口蹄疫病毒血清抗体阳性，但没有出现症状，也没有扩散病毒。其他间接接触动物都没有表现出临床症状，也没有产生口蹄疫病毒抗体，更没有检测到有病毒感染。

所有对新大陆骆驼进行的口蹄疫感染试验结果都清楚地显示，新大陆骆驼在与感染畜同群接触14天后血液或OPF中就没有口蹄疫病毒存在了，这一点至关重要。牛可以携带口蹄疫病毒达3年以上[19]，而美洲驼不会成为病毒携带者[5,24]。口蹄疫病毒在感染动物咽-食道部位持续存在28天以上则该动物被认为是病毒携带者。持续性感染不一定具有传染性。需要指出的是，并不是所有接触了口蹄疫病毒的动物都会形成持续性感染。在试验感染条件下，大约有50%的牛成为病毒携带者。动物的种类和病毒株都是形成持续性感染的决定因素。除了非洲水牛（*Syncerus caffer*）和牛以外还没有确信的证据能够证实其他持续感染动物能够传播口蹄疫[39]。猪不会成为病毒携带者[6]，而是最终清除了病毒[19]。

## 血清学

David等[5]，Rivera等[34]，Viera等[43]，Karesh等[17]和Puntel等[32]对南美的美洲驼和羊驼进行了田间血清流行病学调查，结果显示这些动物没有任何血清型的口蹄疫抗体。这些美洲驼与羊驼的来源农场还混养了牛、绵羊、山羊以及其他野生有蹄动物；在有些农场曾发生过口蹄疫，牛的颊、舌和唇部还表现出了临床症状[5]。

然而，通过不同的感染途径和用不同的口蹄疫血清型进行的试验感染都证明新大陆骆驼产生了口蹄疫抗体[10,24,25]。在这些研究中应用的常规血清学检测方法与结果也证明这些方法能够用于检测口蹄疫抗体[24]。

## 双峰驼口蹄疫

前苏联的许多旧资料描述了真驼属（*Camelus*）动物发生口蹄疫，只是没有区分双峰驼和单峰驼两个品种。这引起了学界对骆驼口蹄疫的一些争议，直到Larska[23]用大量感染试验证明口蹄疫只能感染双峰驼，而不能感染单峰驼。

在哈萨克斯坦和俄罗斯，有些研究者如Vedernikov[42]、Kowalewsky[21]、Krasovskij[22]、Orlov[28]和Terentieva[40]等都观察到了双峰驼发生口蹄疫后表现出了明显的临床症状。他们的研究结果在Skomorchov[37]、Bojko和Suljak[3]编著的兽医学书中有总结。Vedernikov[42]是最早报道双峰驼口蹄疫的研究人员之一，他描述了骆驼口蹄疫的两种表现形式：口部病灶和蹄部病灶型。Kowalewsky[21]描述了双峰驼口蹄疫的临床症状与牛的很相似，首先在病毒进入位点发生早期口疮，体温升高，大量流涎，全身无力和严重的蹄部病灶。另一位俄罗斯研究者Krasovskij[22]观察到了发生于1921—1927年间的几次双峰驼口蹄疫，出于兴趣他开始了双峰驼感染试验。在他的试验中，他将一只奶牛的病毒材料画线刺种到一只双峰驼唇部，将猪源口蹄疫病毒材料注射于一只骆驼的蹄叉部皮肤，将发生口蹄疫奶牛的淋巴液静脉注射给一只双峰驼。只有通过静脉感染的双峰驼发生了口蹄疫。因此，他得出的结论是自然条件下双峰驼很少发生口蹄疫，人工感染也只有通过静脉注射高剂量的病毒才能成功。在他的书中也写到1958/1959年莫斯科动物园暴发的一次口蹄疫，涉及欧洲盘羊、野山羊、鹿、驯鹿、麝鹿（Mochus deer）、美洲驼和骆驼在内的55种有蹄动物。在这次口蹄疫疫情过程中，美洲驼和骆驼（单峰驼？）没有感染，而其他有蹄类动物遭受了严重的损失。Bojko和Suljak[4]在他们的书中提到，1958年在阿斯特拉罕附近另外一次口蹄疫暴发。在一家骆驼养殖场，32头双峰驼出现了严重的口蹄疫临床症状，31头只出现轻微的症状。骆驼食欲不振、体温升高、蹄匣脱落，在腕关节和跗关节以及胸部和膝垫处部分皮肤坏死脱落，在坏死皮肤下能看到明显的渗出液。Kobec[20]全面地描述了一起发生于中亚地区的双峰驼自然感染口蹄疫疫情，具体如下。

第一天：乏力，卧地不起，反刍停止，食欲不振，体温40.2℃，嘴唇和牙龈充血。

第二天：体温40.5℃，唇内、牙齿黏膜以及舌尖小水疱，里面可见黄色液体，食欲不振，卧地不起，流浊涕。

第三天：体温39.4℃，有些水疱破裂而发生溃疡，流涎。

第四天：体温37℃，溃疡开始愈合，骆驼开始采食、反刍，6天后骆驼恢复正常。

作者没有提到蹄部的任何损伤，但骆驼卧地不起。

Krasovskij[22]描述了人工感染双峰驼的口蹄疫症状。感染48小时后在接种部位首先出现水疱，随后在蹄部出现水疱，所有感染动物在7天后康复。Bojko[3]也描写了骆驼的口蹄疫临床症状，与牛相似，但是腿部症状较少。哺乳犊牛吃了感染母牛的奶后会发生病毒血症，进而发生胃肠炎而死亡。他也描写了在乌兹别克斯坦发生的特急性猝死综合征病例，可能与大家熟知的犊牛虎斑心类似。骆驼口蹄疫的传染源是绵羊和山羊，它们一般情况下都饲养在一起。

甚至在今天，东亚地区仍然经常发生口蹄疫疫情。从1997–2000年口蹄疫侵入东亚国家。所有的疫情都是由O型口蹄疫引起的。2000年3月，口蹄疫传至俄罗斯东部和蒙古[35]。在蒙古，O/MNG/2000毒株感染牛、绵羊、山羊和双峰驼（*Camelus bactrianus*），表现出典型的口蹄疫临床症状[46]。2010年5月，蒙古暴发另一起O型口蹄疫疫情，25000头以上活畜受影响，20000头以上被扑杀[30]。在此次流行中，双峰驼也被感染。Hohoo等[16]描述了2000年蒙古双峰驼发生口蹄疫时的临床症状，与牛表现的症状很相似。遗憾的是，在这些口蹄疫流行过程中没有一个骆驼样本被送往研究中心进行检测；有些口蹄疫病毒毒株被送到位于俄罗斯弗拉基米尔的全俄动物卫生研究所，但这些病毒材料都来源于牛而无一例来源于双峰驼。另外，位于英国珀布赖特（Pirbright）的世界口蹄疫参考实验室也曾收到不同时间采集自骆驼的用来确认口蹄疫的样品，包括来自于中东的样品，但是从这些样品中都没有分离到口蹄疫病毒。

由于这些不确定性，来自迪拜中央兽医研究所（CVRL）和丹麦的科学家进行了单峰驼和双峰驼的感染试验[23]。将含有A型口蹄疫病毒的牛水疱皮病毒悬液通过舌皮下注射2头双峰驼（图167）。两头双峰驼都产生了中等到严重的口蹄疫临床症状，后蹄跛行、发热、卧地不起（图168a-d）。接种14天后，2头骆驼后腿蹄匣脱落。接种21天后，损伤愈合，动物恢复健康。在舌头接种部位也没有观察到损伤。接种7~10天，2头骆驼都产生了较高滴度的口蹄疫病毒抗体，但仅在口腔拭子和探杯样品中检测到少量的口蹄疫病毒。这些结果表明，虽然双峰驼可急性感染口蹄疫病毒，但它们长期携带病毒的可能性很小。

在2010年蒙古口蹄疫暴发期间，许多双峰驼都出现了典型的口蹄疫症状，最主要的特征性症状之一是蹄匣脱落（图169）。从1头表现典型口蹄疫临床症状的双峰驼分离到了口蹄疫病毒，这也是首次从自然感染的双峰驼分离到口蹄疫病毒。

目前还没有关于骆驼口蹄疫病毒受体的研究资料，Du等[7]认为，细胞表达的整联蛋白受体可能是影响偶蹄动物对口蹄疫病毒易感性的重要分子。他们研究表明，双峰驼整联蛋白基因与猪、牛的整联蛋白基因关系密切。他们因此推断，口蹄疫病毒宿主嗜性可能与不同种动物整联蛋白亚基的差异相关。

## 单峰驼口蹄疫

最近Wernery和Kaaden[46]总结了骆驼口蹄疫的一些知识，自然感染和试验感染的所有观察结果都未能证实单峰驼具有与牛或猪一样的口蹄疫易感性。

到现在为止，只有2例从田间单峰驼分离到口蹄疫病毒的报道，而且，仅有一个国家利用一个血清型的口蹄疫病毒对少数单峰驼进行了人工感染试

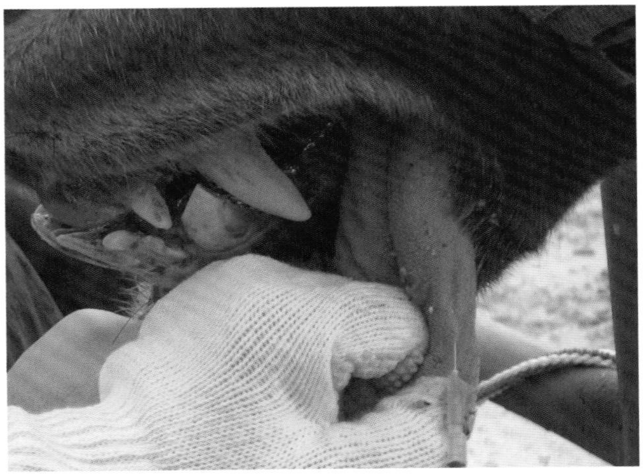

图167 给双峰驼舌皮下注射A型口蹄疫病毒

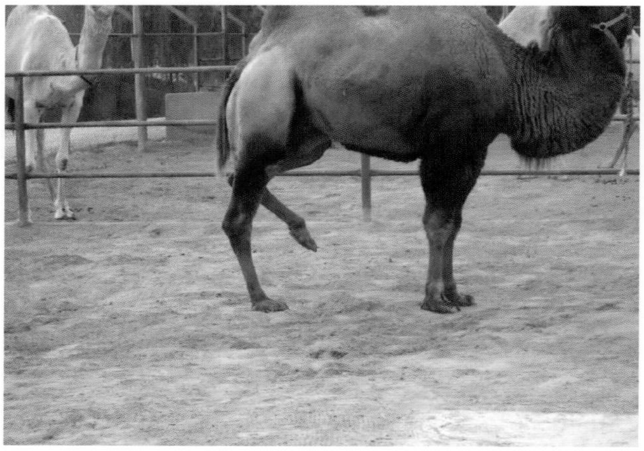

a. 后蹄跛行，前蹄正常

b. 后蹄跛行，前蹄正常

c. 卧地不起，精神沉郁

d. 后腿蹄掌的严重损伤

图168　双峰骆驼口蹄疫临床症状

图169　双峰驼自然感染病例显示蹄匣脱落

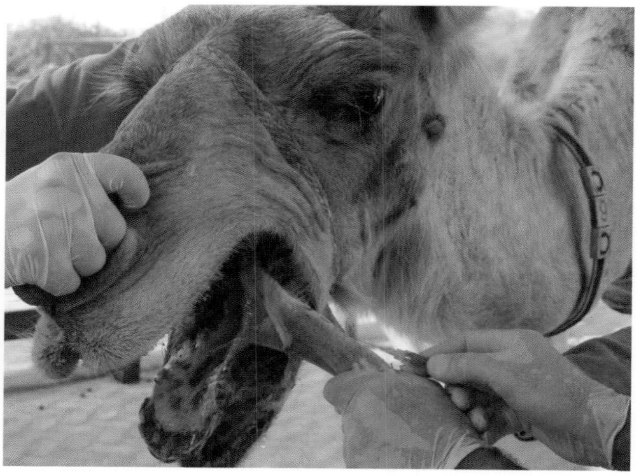

图170　给骆驼皮下注射口蹄疫病毒

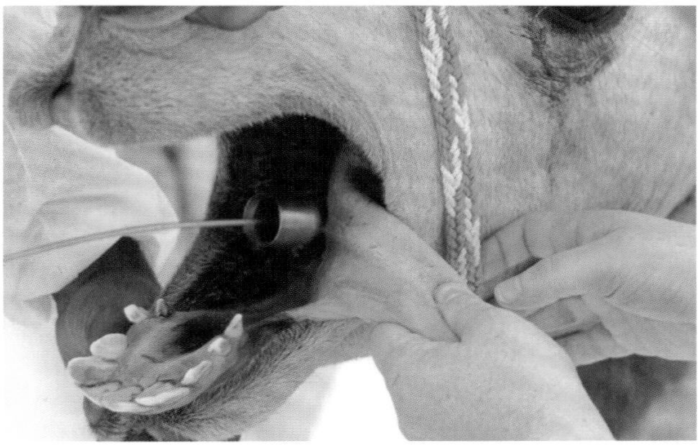

图171 用食道探杯在单峰驼体采样

验，试验的结果也仅发表于不知名的杂志，大多试验设计和操作过时，因此，该研究结论也存有疑问。

因此，来自迪拜中央兽医研究实验室和丹麦林霍尔姆的科学家在迪拜进行了单峰驼感染口蹄疫病毒试验，分别用O型和A型口蹄疫病毒感染了4组单峰驼[2,45,47]。将高剂量的口蹄疫病毒舌皮下接种23头单峰驼（图170）。将几个单峰驼与绵羊饲养在同一个圈舍，使其可以与骆驼直接接触。2头小母牛和2只绵羊分别接种O型和A型口蹄疫病毒作为阳性对照。采用现代实验室技术，如病毒分离、实时荧光RT-PCR、酶联免疫吸附试验（ELISA）、病毒中和试验来检测是否感染。对所有接种和接触动物进行观察并定期采样（包括咽部探杯样品）进行检测（图171）。

接种的单峰驼和直接接触的动物都没有表现出任何口蹄疫临床症状，而阳性对照绵羊和小母牛都出现了典型的口蹄疫临床症状。发病动物都检测到了口蹄疫病毒抗体。虽然接种了高剂量口蹄疫病毒，没有一头单峰驼出现病毒血症，也没有产生特异性抗体。单峰驼对高剂量口蹄疫病毒具有抵抗力，不会向易感动物传播口蹄疫病毒。

### 单峰驼血清学

有一些关于单峰驼在野外和人工感染条件产生口蹄疫抗体的科技论文，但结果是相互矛盾的。有些血清学调查检测到了口蹄疫抗体，而有一些则没有检测到。在大多数调查中，采血的单峰驼与牛、绵羊、山羊、自由放养的野生食草动物一起放牧。许多单峰驼整日和感染的反刍动物直接接触[8,15]，也没有观察到任何的口蹄疫临床症状。而且，几位科学家已经检测了数千份来自非洲和阿拉伯联合酋长国的骆驼血清，结果都为口蹄疫抗体阴性。Wernery等[48]采用竞争ELISA法检测了1119份产奶单峰驼的血清，结果口蹄疫病毒非结构蛋白3ABC抗体均为阴性。Habiela等[14]检测了176份苏丹骆驼，也没有检测到任何口蹄疫抗体。

在埃塞俄比亚和埃及有人报道单峰驼血清口蹄疫抗体呈阳性。但是Moussa[27]认为Richard[33]鉴定的抗体是骆驼血清中常见的非特异性抑制物。而Abou-Zaid[1]发现单峰驼口蹄疫血清学检测的不同结果可归因于实验室方法不同所致。在检测口蹄疫抗体方面，ELISA比其他方法更敏感。在单峰驼上进行口蹄疫感染试验时发现动物发生了血清抗体转阳。这个结果也得到了Frederiksen[12]的证实，他们给5头骆驼免疫了A型、O型、Asia 1型口蹄疫疫苗。骆驼对首免的免疫应答有限，在二免以后有相当好的免疫应答。但是，在加强免疫后仅120天，血清抗体就降到很低水平。

### 诊断

依据临床症状很难区分口蹄疫与水疱性口炎、猪水疱性疹（杯状病毒）和猪水泡病（小RNA病毒科肠病毒属），因此要进行实验室诊断。实验室诊断方法主要包括补体结合试验、ELISA、病毒中和试验、琼脂扩散试验。有一些商品化的ELISA试剂盒可以用于区分自然感染抗体和疫苗免疫抗体，可以利用包括骆驼胚胎肾细胞在内的不同细胞系进行病毒分离[9]。

### 防控

该病不可治疗。最有效的预防措施是禁止从有疫情国家向无疫情国家引入动物或者动物产品。许多欧洲国家禁止进行常规的口蹄疫疫苗免疫，因为大多数疫情的暴发与疫苗抗原灭活不当或者疫苗生产厂的病毒泄露有关。而且，反刍动物尤其是牛在接触感染后病毒可在其咽部持续存活。发生疫情时，即使是疫苗免疫动物，再次接触田间病毒后仍可成为病毒携带者。牛的带毒时间可长达3年。口蹄疫疫苗是一种灭活疫苗；利用分子生物学新技术研制口蹄疫疫苗的尝试也都没有取得成功。口蹄疫疫苗免疫持续期很少能超过6个月[18]。在使用疫苗的国家，发生口蹄疫疫情时必须分离流行毒株并确定其病毒型是否与疫苗毒同源。2010年蒙古发生口蹄疫期间，给双峰驼免疫了口蹄疫疫苗。在蒙古的苏赫巴托尔（Sukhbaatar）和肯特（Khentii）两省区有超过10000头骆驼免疫了O型口蹄疫疫苗[31,49]。但是尚不清楚免疫后双峰驼是否有血清抗体应答和产生了保护性免疫力。

### 结论

对口蹄疫的研究已经进行了很多年，这反映出口蹄疫对贸易和家畜健康有重要的影响。现在口蹄疫诊断方法和疫情信息交换方面已取得了很大的进步，也建立了全球性的口蹄疫防控协作网络。骆驼科动物口蹄疫防控也是其中的内容之一，通过认真的田间流行病学调查与感染试验使人们对骆驼科动物的口蹄疫易感性也有了更清晰的认识。而且，世界动物卫生组织对骆驼科动物疾病也给予了很大关注。

依据双峰驼口蹄疫病毒感染试验结果，以及长达一个世纪的田间临床症状观察，均证实双峰驼和单峰驼这两个密切相关的驼种，其口蹄疫易感性有显著的差异。

现已确认双峰驼能感染口蹄疫。通过临床症状观察，人工感染试验和从田间病例中分离到口蹄疫病毒，都支持这一论断。然而，双峰驼似乎与新大陆骆驼一样，在其咽部藏匿病毒不会超过14天。

双峰驼能感染口蹄疫，而且试验感染也相对容易。双峰驼与单峰驼对口蹄疫易感性的差异，这种现象也见于大象。如非洲象（*Loxodonta africana*）对口蹄疫病毒有抵抗力，而亚洲象（*Elephas maximus*）则较易感。然而，双峰驼不太可能成为口蹄疫病毒长期携带者。今后对双峰驼口蹄疫研究应多关注自然病例，从可疑临床病例收集样品并送往口蹄疫参考实验室，以便对双峰驼口蹄疫流行病学有更好的理解。

田间感染调查与试验感染结果均表明单峰驼不易感染口蹄疫。单峰驼即便是与易感动物密切接触，也不会传播口蹄疫。田间调查和试验感染结果已表明，新大陆骆驼对口蹄疫有易感性。但是，南美驼种对口蹄疫易感性不高，病毒在咽黏膜藏匿不会超过14天，在口蹄疫传播方面不是一个重要的风险因素。

### 其他小RNA病毒

Stehman等[38]报道一种小RNA病毒感染引起15头美洲驼流产。流产一般发生于孕期的220天左右，约3.5个月期间。随着小RNA病毒的感染，成年美洲驼观察到糖尿病。从2个流产胎儿中可分离到病毒，从流产胎儿的胎液和2个有糖尿病症状的美洲驼群中都可以检测到该小RNA病毒的血清中和抗体。

脑心肌炎病毒（EMCV）也是一种小RNA病毒，在一次美国动物样本采集活动中，从一头2岁大的单峰驼样本中分离到了脑心肌炎病毒。大体的病理变化包括心包积液、心外膜出血、心肌内白色点状病灶，可以从心脏分离到脑心肌炎病毒。啮齿动物可能散播脑心肌炎病毒。

## 2.1.7 水疱性口炎

水疱性口炎（VS）是与口蹄疫相似的另一种水泡性疫病，是由弹状病毒（rhabdovirus）引起，该病毒有新泽西和印第安纳两个主要型。很少有关于新大陆骆驼对水疱性口炎病毒（VSV）易感性的研究。一般认为自然条件下，新大陆骆驼很少感染水疱性口炎病毒。因为，美洲驼与水疱性口炎发病牛直接接触也没有发生感染[41]；美洲驼甚至与病牛同饮同食，也没有检测到有水疱性口炎病毒血清抗体。在俄勒冈州对270头美洲驼进行水疱性口炎病毒血清学检测，两种型的水疱性口炎病毒抗体均为阴性[29]。然而，有一例羊驼族成员（Lamoids）自然感染水疱性口炎病毒的病例报道[11]。羊驼和美洲驼都可试验感染水疱性口炎病毒；在接种部位、舌面出现水疱，动物发烧、厌食[13]；将采集的水疱液注射牛能引起发病。目前尚没有旧大陆骆驼发生水疱性口炎的报道。

# 参与文献


[1] Abou-Zaid A.A. (1991). – Studies on some diseases of camels. PhD Thesis, Zagazig University, Zagazig, Egypt.

[2] Alexandersen S., Wernery U., Nagy P., Frederiksen T. & Normann P. (2006). – Dromedaries (*Camelus dromedarius*) are of very low susceptibility to experimental high dose inoculation with FMDV serotype O and do not transmit the infection to direct contact camels or sheep. *In* Report of the Session of the Research Group of the Standing Technical Committee of the European Commission for the Control of Foot and Mouth Disease, 16–20 October 2006, Paphos, Cyprus, 158–164.

[3] Bojko A.A. (1964). – FMD and its eradication. Kolos Izdatel'stvo, Moskva, 55–56.

[4] Bojko A.A. & Suljak F.S. (1971). – The role of wild ungulates in distribution of foot and mouth disease. *In* Foot and mouth disease: biological and ecological aspects of the problem (A.A. Bojko & F.S. Suljak, eds), Moscow, USSR, 117–161.

[5] David M., Torres A., Mebus C., Carrillo B.J., Schudel A., Fondevilla N., Blanco Viera J. & Marcovecchio F.E. (1993). – Further studies on foot-and-mouth virus in the llama. In Proceedings of the Annual Meeting of the United States Animal Health Association (USA HA), Las Vegas, Nevada, and Richmond, Virginia, 25 October 1993, 97, 280–285.

[6] Donaldson A.I. (1987). – Foot and mouth disease: the principal features. *Irish Vet. J.*, 41, 325–327.

[7] Du J., Gao S., Chang H., Cong G., Lin T., Shao J., Liu Z., Liu X. & Cai X. (2009). – Bactrian camel (*Camelus bactrianus*) integrins αv$\beta$1 and αv$\beta$6 as FMDV receptors: molecular cloning, sequence analysis and comparison with other species. *Vet. Immunol. Immunopathol.*, 131 (3–4), 190–199

[8] Farag M.A., Al-Sukayran A., Mazloum K.S. & Al-Bukomy A.M. (1998). – The susceptibility of camels to natural infection with foot and mouth disease virus. *Assuit Vet. Med. J.*, 40 (79) 201–211.

[9] Farid F., Tantawi H.H., Abd El Galil G.A., Saber M.S. & Shalaby M.A. (1974). – Multiplication and titration of foot and mouth disease virus in the foetal camel kidney tissue culture. *J. Egypt Vet. Med. Ass.*, 34 (3–4), 384–392.

[10] Fondevila N.A., Marcovechio F.J., Blanco Viera J., O'Donnell V.K., Carrillo B.J., Schudel A.A., David M., Torres A. & Mebus C.A. (1995). – Susceptibility of Llamas (*Lama glama*) to infection with foot-and-mouth-disease virus. *J. Vet. Med. B.*, 42, 595–599.

[11] Fowler M.E. (2010). – Medicine and surgery of camelids. 3rd Ed. Blackwell Publishing, Oxford, 179–181.

[12] Frederiksen T., Borch J., Christensen J., Tjornehoj K., Wernery U. & Alexandersen S. (2006). – Comparison of FMDV type O, A and Asia 1 antibody levels after vaccination of cattle, sheep, dromedary camels and horses. *In* Report of the Session of the Research Group of the Standing Technical Committee of the European Commission for the Control of Foot and Mouth Disease, 16–20 October 2006, Paphos, Cyprus, 243–247.

[13] Gomez D. (1964). – Tests on the sensitivity of camelids to vesicular stomatitis. *In* Anales Congresso Nacionale Medicina Veterinaria Zoo, Lima, Peru, 403–406.

[14] Habiela M., Alamin M.A.G., Raouf Y.A. & Ali Y.H. (2010). – Epizootiological study of foot and mouth disease in Sudan: the situation of 2 decades. *Veterinarski Arhiv*, 80 (1), 11–26.

[15] Hedger R.S., Barnett T.R. & Gray D.F. (1980). – Some virus diseases of domestic animals in the Sultanate of Oman. *Trop. Anim. Health*, 12, 107–114.

[16] Hohoo A., Zerendordsh S., Zerendagava S. & Zohmonbaatar H. (2001). – Einige Besonderheiten der klinischen Symptome der in der Mongolei aufgetretenen Maul- und Klauenseuche. Translation from *Mongolian Vet. Med. J.*, 37 (2), 17–22.

[17] Karesh W.B., Uhart M.M., Dierenfeld E.S., Braselton W.E., Torres A., House C., Puche H. & Cook R.A. (1998). – Health evaluation of free-ranging guanaco (*Lama guanicoe*). *J. Zoo Wildl. Med.*, 29 (2), 134–141.

[18] Kitching R.P. (1998). – A recent history of foot and mouth disease. *J. Comp. Path.*, 118, 89–108.

[19] Kitching R.P. (2002). – Identification of foot and mouth disease virus carrier and subclinically infected animals and differentiation from vaccinated animals. In Foot and mouth disease: facing the new dilemmas (G.R. Thomson, ed.). *Rev. sci. tech. Off. int. Epiz.*, 21 (3), 531–538.

[20] Kobec A.I. (1936) – Cited from: Skomorokhov A.L. (1952). – Foot and mouth disease. State edition of agricultural literature, Moscow and Leningrad, USSR, 123.

[21] Kowaleskij M.J.M. (1912). – Le Chameau et ses maladies d'apres les observations d'auteurs russes. *J. Méd. Vét. Zootechn.*

*(Lyon)*, 15, 462–466.

[22] Krasovskij V.V. (1929). – Zur Frage der Empfaenglichkeit von Kamelen gegenueber MKS. *Zeitschrift der modernen Veterinaermedizin*, 24.

[23] Larska M., Wernery U., Kinne J., Schuster R., Alexandersen G. & Alexandersen S. (2008). – Differences in the susceptibility of dromedary and Bactrian camels to Foot and mouth disease virus. *Epidemiol. Infect.*, 137 (8), 549–554.

[24] Lubroth J., Yedloutschnig R.J., Culhane U.K. & Mikiciu P.E. (1990). – Foot and mouth disease virus in the llama (*Lama glama*): diagnosis, transmission, and susceptibility. *J. Vet. Diagn. Invest.*, 2 (3), 197–203.

[25] Mancini A. (1952). – Tests on susceptibility of South American camelids to Foot and mouth disease [Ensayos sobre la receptividad de los anguenidos a la fiebre aftosa]. *Bol. Inst. Nac. Antiaftosa (Lima)*, 1 (2), 127–145.

[26] Moro M. (1971). – Ectima: En La Alpaca. Enfermadades Infecciosas y Parasitarias. Boletin del InstitutoVeteranario de Investigaciones Tropicales y de Altura (IVITA), Universidad Nacional Mayor de San Marcos, Lima, Peru, 30.

[27] Moussa A.A. (1988). – The role of camels in the epizootiology of FMD (foot and mouth disease) in Egypt. In FAO. The camel: development research. Proceedings of Kuwait Seminar, 20–23 October 1986, Kuwait, 162–173.

[28] Orlov M. (1963) – Animal virus disease. Agricultural Literatures and Journalists, Moscow, 97–98.

[29] Picton R. (1993). – Serologic survey of llamas in Oregon for antibodies to viral diseases of livestock. MS Thesis, Oregon State University, Corvallis, OR.

[30] Promed@promed.isid.harvad.edu – Foot and mouth disease in Mongolia (accessed 13 November 2010).

[31] Promed-ahead-edr@promed.isid.havard.edu – Foot and mouth disease in Mongolia (accessed on 16 November 2010).

[32] Puntel M., Fondevila N.A., Blanco Viera J., O'Donnell V.K., Marcovecchio J.F., Carrillo B.J. & Schudel A.A. (1999). – Serological survey of viral antibodies in llamas (*Lama glama*) in Argentina. *Zentralbl. Veterinarmed. B.*, 46 (3), 157–161.

[33] Richard D. (1979). – Study of the pathology of the dromedary in Borana Awraja (Ethiopia). Thèse 75, Université de Créteil, Paris, France.

[34] Rivera H., Madwell B.R. & Ameghino E. (1987). – Serological survey of viral antibodies in the Peruvian alpaca *(Llama pacos). Am. J. Vet. Res.*, 48, 189–191.

[35] Sakamoto K. & Yoshida K. (2002). – Recent outbreaks of foot and mouth disease in countries of East Asia. In Foot and mouth disease: facing the new dilemmas (G.R. Thomson, ed.). *Rev. sci. tech. Off. int. Epiz.*, 21 (3), 459–463.

[36] Sarma G., Das S.K. & Dutta P.K. (1983) – Outbreak of foot and mouth disease in deer in the Assam state zoo. *Vet. Rec.*, 113 (18), 420–421.

[37] Skomorochov A.L. (1952). – Foot and mouth disease. State Edition of Agricultural Literature, Moscow and Leningrad, USSR, 123.

[38] Stehman S.M., Morris L.I., Wiesensel L., Freeman W., Del Piero F., Zylich N. & Dubovi E.J. (1997). – Case report: picornavirus infection associated with abortion and adult onset diabetes mellitus in a herd of llamas. In Proceedings of the American Association of Veterinary Laboratory Diagnosticians Annual Conference, 18–24 October 1997, Louisville, USA, 43.

[39] Sutmoller P. & Casas Olascoaga R. (2002). – Unapparent foot and mouth disease infection (sub-clinical infections and carriers): implications for control. *In* Foot and mouth disease: facing the new dilemmas (G.R. Thomson, ed.). *Rev. sci. tech. Off. int. Epiz.*, 21 (3), 519–529.

[40] Terentieva S.M. (1975). – The breeding of animals. Kolos Izdatel'stvo, Moscow, 208–209.

[41] Thedford R.R. & Johnson L.W. (1989). – Infectious diseases of New-world camelids (NWC). *Vet. Clin. North Am. Food Anim. Pract.*, 5 (3), 145–157.

[42] Vedernikov W. (1893). – Das Kamel als landwirtschaftliches Nutztier. Archiv fur Veterinarwissenschaften Bd I, H.4.

[43] Viera B.J., Marcovecchio F., Fondevilla N., Carillo B., Schudel A., David M., Torres A. & Mebus C. (1995). – Epidemiology of foot and mouth disease in the llama (*Lama glama*). *Veterinaria Argentina*, 12 (119), 620–627.

[44] Wells S.K., Gulter A.E., Soike K.F. & Baskin G.B. (1989). – Encephalomycocarditis virus: epizootic in a zoological collection. *J. Zoo Wildl. Med.*, 20 (3), 291–296.

[45] Wernery U. (2010). – Dromedaries are resistant to Foot and mouth disease (FMD), Bactrians are not. A review. In Proc. Int. Camel Symposium, 7–11 July 2010, Garissa, Kenya, 3.

[46] Wernery U. & Kaaden O.-R. (2004). – Foot and mouth disease in camelids: a review. *Vet. J.*, 168 (2), 134–142.

[47] Wernery U., Nagy P., Amaral-Doel C.M., Zhang Z. & Alexandersen S. (2006). – Lack of susceptibility of the dromedary camel (*Camelus dromedarius*) to foot and mouth disease virus serotype O. *Vet. Rec.*, 158, 201–203.

[48] Wernery U., Thomas R., Syriac G., Raghavan R. & Kletzka S. (2007). – Seroepidemological studies for the detection of antibodies against nine infectious diseases in dairy dromedaries (Part 1). *J. Camel Pract. Res.*, 14 (2), 85–90.

[49] World Organisation for Animal Health (OIE) (2010). – Event summary: foot and mouth disease, Mongolia. Available at: www.oie.int/wahis/public.php?page=event_summary&reportid=9653 (accessed on 13 November 2010).

## 深入阅读材料

*Abu Elzein E.M.E., Neumann B.J., Omer E.A. & Haroon B. (1984–1985). – Prevalence of serum antibodies to the foot and mouth disease virus infection associated antigen (via) in camels, sheep and goats of the Sudan. Sudan J. Vet. Res., 6, 58–60.*

*Guo S.Z. (1988). – Serological comparison of the pathogens of aphthosis in camel, sheep, and goat. Chinese J. Vet. Med. Tech., 5, 35–37.*

*Sutmoller P. (1999). – Risk of disease transmission by llama embryos. Rev. sci. tech. Off. int. Epiz., 18 (3), 719–728.*

（曹轶梅译，卢曾军、殷宏校）



## 2.1.8 小反刍兽疫

小反刍兽疫（Pest des petits ruminants, PPR）是家养和野生小反刍动物的一种高度接触性病毒病，在流行区该病被认为是一种严重的经济病。在以放牧为生的地区绵羊和山羊比牛具有更大的经济价值，因此小反刍兽疫对这些地区的食品供应影响巨大。自1942年首次报道该病以来，该病一直在不断扩大蔓延，目前在非洲、中东和亚洲的大部分地区都有发现（图172）。小反刍兽疫可轻易跨越边境传播，在撒哈拉沙漠以南的非洲地区及印度次大陆目前将该病作为一种严格控制的疾病。

全球消除牛瘟使得该病毒病的地位上升。最新研究发现小反刍兽疫可感染骆驼科动物，并可引起单峰驼发病[13]。虽然目前尚未发现新大陆骆驼（New World camels，NWCs）也可感染小反刍兽疫，但饲养者和兽医应该警惕该病，因为小反刍兽疫发生国家拥有越来越多的新大陆骆驼。小反刍兽疫是世界动物卫生组织（OIE）法定报告的疾病。

### 病原学

由于牛瘟（RP）和小反刍兽疫有许多相同的临床症状，因此之前认为小反刍兽疫病毒（PPRV）是牛瘟病毒（RPV）发生变异适应小反刍兽而来，同时病毒对牛的致病性降低。但20世纪70年代之后，通过血清学鉴别试验和交叉免疫保护试验均发现牛瘟病毒和小反刍兽疫病毒之间有明显的区别。小反刍兽疫病毒与牛瘟病毒、犬瘟热病毒以及麻疹病毒都属于副黏病毒科麻疹病毒属，此后这一病毒科中还发现了海豹瘟病毒和海豚、鼠海豚等鲸类动物的麻疹病毒。目前根据基因序列分析结果来看，小反刍兽疫病毒显然不是牛瘟病毒的突变株，它与牛瘟病毒之间的亲缘关系甚至低于与麻疹病毒之间的亲缘关系。已知小反刍兽疫病毒有四个基因分支，基因Ⅰ和Ⅱ型病毒仅在西非发现，而基因Ⅲ型发生在东非、印度南部和阿拉伯，基因Ⅳ型则近来在亚洲和中东出现。虽然这些基因型与非洲的基因Ⅰ型亲缘关系很近（图172），但目前对这些毒株的起源无从得知。Kinne等[15]报道，在阿联酋的半游牧区几种野生反刍兽暴发小反刍兽疫，

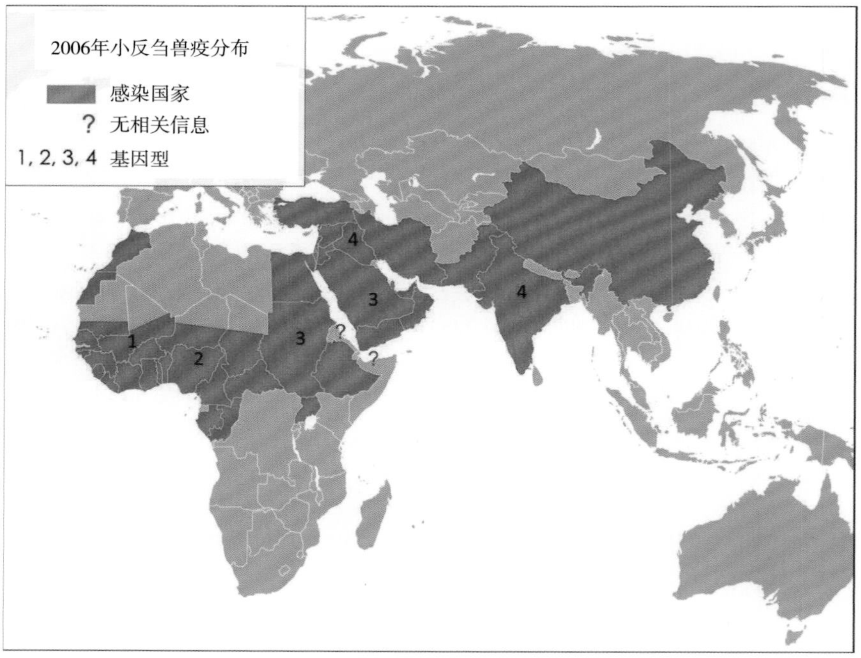


图172　小反刍兽疫地理分布图以及已鉴定的基因型

野生反刍兽出现了腹泻症状并造成100多只死亡，其毒株类型属于小反刍兽疫病毒基因Ⅳ型。该作者认为野生反刍兽在病毒感染家养小反刍兽过程中起到了重要的流行病学作用，目前已有充分的证据表明，病毒跨种传播会使新毒株出现的概率增大。

### 流行病学

小反刍兽疫主要是通过易感动物口鼻接触感染动物的分泌物传播，几乎所有的暴发都是由家畜活动导致。小反刍兽疫临床症状包括发热、肠胃炎、黏膜组织糜烂和支气管炎症引起的呼吸困难，但牛瘟感染的家畜不出现支气管炎症状。小反刍兽疫引起的死亡率差异很大，对小反刍兽死亡率可高达60%～70%。但是，经常有感染动物表现亚临床症状。在部分小反刍兽疫疫情暴发过程中，仅山羊发病而绵羊不发病。许多野生小反刍动物小反刍兽疫也表现得比较温和。在不同国家研究表明，用同一小反刍兽疫病毒毒株感染实验动物引起的临床症状和病理变化在很大程度上不同。Hamdy和Dardiri[10]用小反刍兽疫病毒感染白尾鹿，结果实验动物的表现不同，从没有明显症状的亚临床到死亡。Al-Neaem等[6]用从羚羊组织中分离的小反刍兽疫病毒强毒株感染沙特阿拉伯当地的绵羊和山羊，结果发现人工感染试验的动物症状不一致；此外还有研究发现在不同地理环境和不同季节自然感染动物的发病症状也不相同[4]；虽然所有的人工感染试验羊均表现出典型的临床病理变化，但其病变没有自然感染发病羚羊的病变严重。目前尚不清楚引起病毒致病力发生如此巨大差异的原因。基于这些研究发现，目前很有必要开展人工感染试验以明确水牛、奶牛、骆驼在小反刍兽疫流行病学中的作用。

1995–1996年，埃塞俄比亚发生了一种以发热和高度传染性呼吸道综合征为主的新型动物传染病，成千上万只骆驼被感染，发病率在90%以上，死亡率在5%～70%。主要临床症状为流脓性鼻液、流泪、咳嗽、呼吸困难、呈腹式呼吸，其中一些动物出现颌下肿大和腹泻现象。同时，在一些发病的骆驼体内分离出了马链球菌和溶血性曼氏杆菌，但在此次疫情流行病学调查中并没有对这两种病原所起作用进行调查，研究者推测一种类似于牛瘟或PPR的麻疹病毒属的病毒可能与此次的骆驼感染和死亡有关，在一年内导致埃塞俄比亚全国200万只骆驼感染。在此次疫情中收集的病理样本中检测到小反刍兽疫抗原和核酸，但是没有分离出病毒[16,17,18]；用实时定量PCR鉴定存在有两个麻疹病毒属的毒株，分别命名为埃塞俄比亚96 CAMEL 1和埃塞俄比亚96 CAMEL 2，这两株毒株与小反刍兽疫病毒具有较近的亲缘关系，且分别属于小反刍兽疫病毒基因 II型和基因III型；随后用竞争ELISA检测分析来自疫区和非疫区单峰驼血清中的牛瘟和小反刍兽疫病毒抗体，结果显示疫区单峰驼小反刍兽疫病毒血清阳性率为7.8%，牛瘟病毒血清阳性率为21.3%，而非疫区单峰驼血清两者均为阴性。研究者认为，该疫病可能是由牛源或羊源的小反刍兽疫病毒或牛瘟病毒传播给骆驼导致，也可能是由一种与小反刍兽疫病毒或牛瘟有较近亲缘关系的新型麻疹属病毒引起。

用竞争ELISA检测牛瘟和小反刍兽疫病毒抗体时会出现交叉反应。虽然已宣布在全球范围消灭了牛瘟，但在宣布后的数年时间里依然可以在动物体内检测到循环抗体（circulating antibody），特别是在免疫动物体内，因此设计了一种可检测小反刍兽疫病毒和牛瘟病毒抗原、且两者之间无交叉反应的ELISA方法。但Ismail等人[12]报道，用病毒中和试验（VNT）检测埃及单峰驼血清时发现牛瘟病毒 Kabete O毒株和小反刍兽疫病毒之间存在交叉反应。虽然阿拉伯半岛的家养小反刍动物和羚羊经常暴发小反刍兽疫，但Wernery等人[21]使用英国BDSL公司的竞争ELISA试剂盒对来自奶骆驼（dairy camel）的1119份血清进行血清流行病学调查，结果没有检测到牛瘟和小反刍兽疫的抗体，这表明阿拉伯单峰驼中可能不存在小反刍兽疫。

在一个未发表的小试验中，研究者将小反刍兽疫病毒阳性山羊肺脏（用BDSL公司的抗原捕获ELISA诊断）处理后的上清液5mL皮下注射给一只12岁公骆驼，结果该骆驼没有出现任何症状，血清也没有任何变化，该小反刍兽疫病毒毒株经鉴定属于基因Ⅳ型。用血清学调查的方法进一步对小反刍兽疫感染单峰驼的可能性进行评价，结果显示骆驼可能对小反刍兽疫有易感性，虽然目前尚未观察到任何临床症状[2,3,5,9,11,12,19]。

Ismail等人[12]用病毒中和试验检测埃及单峰驼血清中小反刍兽疫病毒和牛瘟病毒抗体，结果阳性率分别为4.2%和11.9%。近年来已有在沙特阿拉伯[1]和苏丹[13,14]单峰驼暴发小反刍兽疫的报道，这些事实清楚地表明，单峰驼可感染小反刍兽疫病毒并且在小反刍兽疫病毒传播山羊的过程中起重要作用。但上述两个国家所观察到的临床表现不同：在沙特阿拉伯，50只单峰驼中只有4只出现高热、流鼻涕和轻微咳嗽；而苏丹，该病表现为健康单峰驼突然死亡、出血性腹泻和流产等特点。在这两起疫情中，可用Vero细胞、Madin-Darby牛肾细胞、牛肾细胞及鸡胚尿囊腔分离病毒；在毒株基因型方面，苏丹的毒株属于基因Ⅳ型，而沙特阿拉伯的毒株属于基因Ⅱ型和Ⅳ型。

## 临床症状和病理变化

已知小反刍兽疫是一种烈性传染病，但经常也可观察到不同的表现形式。这些形式包括：最急性、急性、亚急性、隐性。

小反刍兽疫在单峰驼的临床病理变化差异很大。迄今为止，在单峰驼上暴发小反刍兽疫有记录和描述只有3起，分别在埃塞俄比亚、苏丹和沙特阿拉伯，因此很难对小反刍兽疫及其导致的病理变化进行全面的描述。此外，似乎成年单峰驼比幼年单峰驼更容易感染该病。以下是目前已报道的有关该病的临床症状和病理变化/病理组织学变化。

（1）无症状突然死亡。

（2）发热。

（3）呼吸困难，严重的腹式呼吸、流鼻涕和咳嗽。

（4）淡黄至血性腹泻。

（5）流产。

（6）支气管周边单核细胞浸润和支气管表面变性。

（7）肺泡隔充血、单核细胞浸润。

（8）肺水肿和肺气肿。

（9）淋巴结中的淋巴滤泡萎缩。

（10）消化道上皮溃烂。

对7种麻疹病毒属病毒（海豚麻疹病毒、鼠海豚麻疹病毒、麻疹病毒、牛瘟病毒、小反刍兽疫病毒、海豹瘟病毒和犬瘟热病毒）最新研究表明，其中一些病毒可使感染动物出现脱髓鞘症状[20]。目前科学家们正在探讨麻疹病毒在人类多发性硬化症中的致病机制。

## 诊断

单峰驼发生小反刍兽疫后临床症状多样，很容易与其他疾病混淆，因此，小反刍兽疫只能通过准确的实验室检测进行确诊。存活动物可采集凝血、EDTA抗凝血及鼻眼分泌物拭子、粪便拭子送到实验室；理想的病料是采集3～4只动物的肺脏、淋巴结、脾脏和肠道黏膜组织，然后在冷藏条件下送到实验室。

琼脂糖凝胶免疫扩散试验、对流免疫电泳、免疫捕获ELISA和PCR已广泛应用于检测含牛瘟病毒和小反刍兽疫病毒抗原的样品。其中，免疫捕获ELISA基于单克隆抗体进行抗原检测，其敏感性高，易于操作，并且牛瘟和小反刍兽疫抗原间无交叉反应。小反刍兽疫诊断的金标准是病毒分离，但无论采用多次传代法或利用多种不同细胞系包括原代单层羊睾丸细胞，病毒粒子还是难以复制繁殖[15]。因此，有研究人员使用其他病毒分离技术，如使用同源滋养层支持的淋巴母细胞在旋转培养瓶中培养至少14天。

免疫组化是另外一种可用来快速诊断麻疹病毒感染的方法，可在小肠上皮细胞、支气管和细支气管上皮细胞、融合细胞、胆管上皮细胞、肝细胞以及肠淋巴结细胞的胞质和胞核中发现麻疹病毒抗原[15]。

此外，也可以通过血清学方法如病毒中和试验或抗体竞争ELISA方法来确定小反刍兽疫病毒感染。竞争ELISA的优点是直接结合针对病毒蛋白的抗体，因此可用于任何动物种类，OIE推荐竞争ELISA作为国际贸易中的首选检测方法。但值得注意的是，此方法不能区分免疫动物和非免疫动物，并且在疫病流行期间，小反刍兽疫竞争ELISA中检测阳性的血清用牛瘟竞争ELISA检测也为阳性，这表明这两种检测方法之间存在交叉反应。

## 防控

在小反刍兽疫暴发期间，应严格控制受感染地区动物的活动，骆驼不应与小反刍动物混合饲养。阿联酋不存在混合养殖的情况，这可能是在阿联酋骆驼没有暴发小反刍兽疫和没有血清学阳性的原因。小反刍兽疫可通过免疫接种进行控制，但在1989年小反刍兽疫疫苗研制成功之前，牛瘟弱毒疫苗被用于保护动物抵抗小反刍兽疫感染。由于牛瘟疫苗在全球范围内被禁止使用，小反刍兽疫弱毒疫苗成为用于保护小反刍动物抵抗小反刍兽疫感染的主要疫苗，该弱毒疫苗可为免疫动物提供终生的免疫保护力。近年来，通过利用海藻糖超速脱水法将小反刍兽疫基因重组到羊痘病毒基因组上的方法，显著地提高了小反刍兽疫疫苗的有效性。具有热稳定性的重组羊痘病毒疫苗可双重保护小反刍动物抵抗两种重要的经济性疫病[7]，但目前这种疫苗尚未应用于单峰驼。

## 参考文献

[1] Abd El-Hakim U. (2006). – The role of camels in dissemination of peste des petits ruminants virus among sheep and goats in Saudi Arabia. *Assiut. Vet. Med. J.*, 52 (110), 132 – 145.

[2] Abraham G., Sintayehu A., Libeau G., Albina E., Roger F., Laekemariam Y., Abayneh D. & Awoke K.M. (2005). – Antibody seroprevalence against peste des petits ruminants (PPR) in camels, cattle, goats and sheep in Ethiopia. *Prev. Vet. Med.*, 70, 51 – 57.

[3] Abubaker A.B., Sanda A.B., El-Yuguda A.D. & Baba S.S. (2008). – Seroprevalence of *Morbillivirus* antibody and abattoir survey of one humped slaughtered camels (*Camelus dromedarius*) in Maiduguri municipality abattoir Maiduguri, Nigeria. *Asian J. Sci. Res.*, 1 (1), 85 – 89.

[4] Abubaker M., Jamal S.M., Arshed M.J., Hussain M. & Ali Q. (2009). – Peste des petits ruminants virus (PPRV) infection; its association with species, seasonal variations and geography. *Trop. Anim. Health Prod.*, 41, 1197 – 1202.

[5] Albayrak H. & Gur S. (2010). – A serological investigation for peste des petits ruminants infection in sheep, cattle and camels (*Camelus dromedarius*) in Aydin province, West Anatolia. *Trop. Anim. Health Prod.*, 42 (2), 151 – 153.

[6] Al-Naeem A., Abu-Elzein E., Abdelsalam E.B., Al-Hizab F., Housawi F. & Al-Afaleg B. (2009). – Clinico-pathological response of indigenous sheep and goats to a virulent gazelle peste des petits ruminants virus (PPRV). *Sci. J. King Faisal Univ.*, 10 (1), 181 – 185.

[7] Berhe G., Minet C., Le Goff C., Barrett T., Ngangnou A., Grillet C., Libeau G., Fleming M., Black D.N. & Diallo A. (2003). – Development of a dual recombinant vaccine to protect small ruminants against peste-des-petits ruminants virus and apripoxvirus infections. *J. Virol.*, 77 (2), 1571 – 1577.

[8] Diallo A. (2010). – Peste des petits ruminants. *In* Infectious and parasitic disease of livestock (P.-C. Lefevre, J. Blancou, R. Chermette & G. Uilenberg, eds). *Lavoisier*, 245–261.

[9] El Amin M.A.G. & Hassan A.M. (1998). – The seromonitoring of rinderpest throughout Africa, phase III results for 1998. International Atomic Energy Agency and Food and Agriculture Organization, Vienna.

[10] Hamdy F.M. & Dardiri A.H. (1976). – Response of white-tailed deer to infection with peste des petits ruminants virus. *J. Wildlife Dis.*, 12, 516–518.

[11] Haroun M., Hajer I., Mukhtar M. & Ali B.E. (2002). – Detection of antibodies against peste des petits ruminants virus in sera of cattle, sheep and goats in Sudan. *Vet. Res. Commun.*, 26 (7), 537–541.

[12] Ismail T.M., Hassan H.B., Nawal M.A., Rakha G.M., Abd El-Halim M.M. & Fatebia M.M. (1992). – Studies on the prevalence of antibodies against rinderpest (RP) and peste des petits ruminants (PPR) viruses in camel in Egypt. *Vet. Med. J. (Giza)*, 10 (2), 49–53.

[13] Khalafalla A.I., Saeed I.K., Ali Y.H., Abdurrahman M.B., Kwiatek O., Libeau G., Obeida A.A. & Abbas Z. (2010). – An outbreak of peste des petits ruminants (PPR) in camels in Sudan. *Acta Tropica*, 116, 161–165.

[14] Khalafalla A.I., Saeed I.K., Ali Y.H., El Hassan A.M. & Obeida A.A. (2005). – Poster: Morbillivirus infection of camels. New emerging fatal and contagious disease. Proceedings of the International Conference on Emerging Infectious Diseases, 26 February–1 March 2005, Al Ain, UAE, 9.

[15] Kinne J., Kreutzer R., Kreutzer M., Wernery U. & Wohlsein P. (2010). – Peste des petits ruminants in Arabian wildlife. *Epidemiol. Infect.*, 138 (8), 1211–1214.

[16] Roger F., Yesus M.G., Libeau G., Diallo A., Yigezu L.M. & Yilma T. (2001). – Detection of antibodies of rinderpest and peste des petits ruminants viruses (*Paramyxoviridae, Morbillivirus*) during new epizootic disease in Ethiopian camels (*Camelus dromedarius*). *Revue. Med. Vet.*, 152 (3), 265–268.

[17] Roger F., Yigezu L.M., Hurard C., Libeau G., Mebratu G.Y., Diallo A. & Faye B. (1998). – Investigation on a new pathology of camels in Ethiopia. Proceedings of the 3rd Annual Meeting for Animal Production under Arid Conditions. Camel production and future perspectives, 2–3 May 1998, Al Ain, UAE.

[18] Roger F., Yigezu L.M., Hurard C., Libeau G., Mebratu G.Y., Diallo A. & Faye B. (2000). – Investigations on new pathological condition of camels in Ethiopia. *J. Camel. Pract. Res.*, 7 (2), 163–165.

[19] Saeed I.K., Ali Y.A., Khalafalla A.I. & Rahman-Mahasin E.A. (2010). – Current situation of peste des petits ruminants (PPR) in the Sudan. *Trop. Anim. Health Prod.*, 42, 89–93.

[20] Sips G.J., Chesik D., Glazenburg L., Wilschut J., De Keyser J. & Wilczak N. (2007). – Involvement of morbilliviruses in the pathogenesis of demylinating disease. *Rev. Med. Virol.*, 17 (4), 223–244.

[21] Wernery U., Thomas R., Syriac G., Raghavan R. & Kletzka S. (2007). – Seroepidemiological studies for the detection of antibodies against nine infectious diseases in dairy dromedaries (Part I). *J. Camel Pract. Res.*, 14 (2), 85–90.

（窦永喜译，殷宏校）

## 2.1.9 牛病毒性腹泻

牛病毒性腹泻病毒（BVDV）感染牛可导致两种不同的症状，包括牛病毒性腹泻和黏膜病。前者发病率较高死亡率较低，后者通常为散发，但多以患病动物死亡为结局。BVDV具有BVDV-1和BVDV-2两个基因型，各基因型的病毒均含有非致细胞病变型（ncp）和致细胞病变型（cp）两种生物型。不同生物型BVDV在牛、驼科动物感染后疾病流行、致病机制中发挥着重要作用。只有非致细胞病变生物型BVDV感染可导致动物持续感染，而致细胞病变生物型毒株是由非致细胞病生物型毒株突变而来，其感染不会导致动物出现病毒血症。只有非致细胞病变型毒株感染后可以排毒并感染同群的其他动物。同时，非致细胞病变型毒株还可通过胎盘屏障，导致妊娠早期胎儿发生感染。

虽然牛病毒性腹泻和黏膜病两种疾病的病原均为瘟病毒，但二者的流行规律、发病机理具有较大差异。动物出生后在不同年龄阶段均可感染瘟病毒，出现亚临床或多种临床症状，包括急性腹泻、急性出血综合征、急性死亡等。瘟病毒在世界范围牛群中广泛流行，该病毒也被分离自新大陆骆驼和旧大陆骆驼[25,34,42]。

近年来，研究人员对驼科动物尤其是新大陆骆驼的瘟病毒感染进行了大量研究，并发表了一系列的研究报告（表62）。BVDV感染新大陆骆驼可导致腹泻、繁殖障碍、呼吸道疾病以及其他播散性疾病[38]。BVDV感染新大陆骆驼的研究见相关综述[57]。

## 病原学

BVDV是一种小的RNA病毒，属于黄病毒科瘟病毒属，同属病毒还包括边界病毒、猪瘟病毒，其抗原性密切相关。从新生牛犊和持续感染牛分离的毒株通常为ncp型BVDV，而从发生黏膜病的牛组织中分离的毒株为cp型BVDV。在BVDV两种血清型（由于BVDV毒株无血清型的差别，因此用“基因型”比较确切，译者注）中，BVDV-1是世界广泛流行的基因型，而BVDV-2主要流行于美国、加拿大，另外还报道于意大利、荷兰和英国[50]。每一种血清型（“基因型”，译者注）都有两种生物型，即ncp和cp型。

## 流行病学

动物产后主要通过呼吸道和消化道感染BVDV，并产生中和抗体，但这一感染过程通常不易被察觉。另外，BVDV感染非免疫的怀孕母畜时，病毒可通过胎盘屏障侵害胎儿（遗传性传播）。虽然母畜可发生血清转阳、不出现疾病症状，但可导致妊娠早期胎儿发生免疫耐受。这种先天感染可导致胎儿异常，包括胎儿死亡、先天性缺陷或终生持续感染，主要决定于病毒感染时胎儿的发育程度。BVDV还可通过精液传播，主要发生于持续感染的公牛，可数年通过精液排毒。

历经数年，研究人员才得以阐述瘟病毒感染的复杂性，尤其是病毒性腹泻和黏膜病之间的关联。

目前，人们认识到瘟病毒感染是导致新大陆骆驼和旧大陆骆驼发生严重疾病的重要原因[24]。驼科动物对瘟病毒具有较高的易感性，并且可发生血清转阳。驼科动物，尤其新大陆骆驼发病的报道主要见于北美、英国、瑞士等国家和地区[17,41,44,22]。与牛感染BVDV类似，羊驼和美洲驼可发生严重的疾病，包括腹泻、繁殖障碍、病畜体弱和死亡[8]。BVDV在驼科动物造成的感染，引起了北美、欧洲国家尤其英国、瑞士和德国这些新大陆骆驼数量多的国家的高度重视。与牛感染瘟病毒类似，瘟病毒感染南美驼后可导致全身性疾病出现，还可在妊娠早期感染无免疫力的胎儿，导致其持续感染。田间数据表明，高达80%的初生动物由于其母体妊娠早期感染瘟病毒而发生持续感染[7]。在群体水平，牛BVDV的持续感染为1%。至今为止，除一例美洲驼胎儿感染cp型BVDV[6]，包

表62　有关新大陆骆驼和旧大陆骆驼动物BVDV的近期文献

| 作者 | 年份 | 国家/地区 | 物种 | 结果 |
|---|---|---|---|---|
| **新大陆骆驼** | | | | |
| Rivera等 | 1987 | 秘鲁 | 羊驼 | 血清学：11% |
| Karesh等 | 1998 | 阿根廷 | 骆马 | 血清学：全部阴性 |
| Belknap等 | 2000 | 美国 | 美洲驼 | 分离到病毒、免疫荧光试验阴性、免疫酶试验阳性、抗体中和试验阴性 |
| Wentz等 | 2003 | 美国 | 美洲驼、羊驼 | 实验感染、未发病、血清学阳性率较低、无致死性感染 |
| Carmen等 | 2005 | 美国 | 羊驼 | 全身性疾病、流产、分离获得病毒、持续感染、BVDV-1b基因亚型 |

续表

| 作者 | 年份 | 国家/地区 | 物种 | 结果 |
|---|---|---|---|---|
| Foster等 | 2005 | 英国 | 羊驼 | 全身性疾病、流产、获得无CPE型毒株、BVDV-1b基因亚型 |
| 匿名 | 2005b | 英国 | 羊驼 | 腹泻、BVDV-1b基因亚型 |
| Evermann | 2006 | 美国 | 美洲驼、羊驼 | 关于美洲驼和羊驼的重要文献 |
| Henningson | 2006 | 美国 | 羊驼 | 持续感染、病理损伤等 |
| Bromage | 2006 | 英国 | 羊驼 | 重要文献 |
| 匿名 | 2006a | 英国 | 羊驼 | BVDV-1 PCR阳性、流产 |
| 匿名 | 2006b | 英国 | 羊驼 | 流产、新生畜孱弱、BVDV-1PCR阳性 |
| 匿名 | 2006c | 美国 | 驼科动物 | 重要文献 |
| Bedenice | 2006 | 美国 | 南美驼 | 重要文献 |
| Evemann等 | 2006 | 美国 | 驼科动物 | 重要文献 |
| Mueller等 | 2007 | 美国、英国 | 驼科动物 | 重要文献 |
| Probst等 | 2007 | 欧洲（动物园） | 驼科动物 | 血清学：1.4% |
| Kelling | 2008 | 美国 | 羊驼 | 血清学：25.4%的动物群阳性 |
| Byers | 2008 | 美国 | 驼科动物 | 重要文献 |
| Bedenice | 2008 | 美国 | 羊驼 | BVDV-1病毒持续感染 |
| Danuser等 | 2009 | 瑞士 | 美洲驼、羊驼 | 血清学阳性率为4.6% |
| Shimeld | 2009 | 美国 | 羊驼 | 血清学阳性率为20%、动物持续感染 |
| Kim等 | 2009 | 美国 | 羊驼 | PCR检测出46株BVDV、均为BVDV-1b |
| Mudry等 | 2010 | 瑞士 | 南美驼 | 血清学5.8%和3.6%、未发现病毒RNA |
| Topliff等 | 2009 | 美国 | 新大陆骆驼 | 血清学阳性率为20%、6%幼畜持续感染 |
| **旧大陆骆驼** | | | | |
| Doyle等 | 1983 | 美国（动物园） | 单峰驼 | 血清学阳性率为13% |
| Fahmy | 1999a, b, c | 埃及 | 单峰驼 | 试验感染 |
| Yousif等 | 2004 | 埃及 | 单峰驼 | 分离获得BVDV-1和BVDV-2毒株 |
| Al-Afaleq等 | 2006 | 沙特阿拉伯 | 单峰驼 | 血清学阳性率为18% |
| Taha | 2007 | 阿拉伯联合酋长国 | 单峰驼 | 抗体阴性 |
| Wenery等 | 2008 | 阿拉伯联合酋长国 | 单峰驼 | 血清学阳性率为1.6% |
| Intisar等 | 2010 | 苏丹 | 单峰驼 | 血清学阳性率为84.6%、RT-PCR 7%阳性 |
| Gao等 | 2013 | 中国 | 双峰驼 | 抗原ELISA30.4%、抗体阳性率23%、检出ncp BVDV-1病毒6个基因亚型 |

括初生动物持续感染，几乎所有南美驼感染为ncp BVDV所致。Shimeld以NADL（BVDV-1型）c125（BVDV-2型）毒株对美国426份羊驼血清进行中和试验，证实20%的羊驼为血清学阳性，为BVDV单一基因型感染或两个基因型感染所致[52]。

持续感染动物不仅可以向外散播体内的ncp型BVDV，而且还可散播变异病毒产生cp型BVDV，因此持续感染动物在该病的流行中发挥着重要作用，所以应及时从群体中剔除持续感染的动物。美国2004—2007年间对12 000以上的羊驼进行荧光定量RT-PCR方法进行病毒筛查，以发现持续感染的动物，对获得的46株病毒进行基因组序列分析发现BVDV-1b是羊驼中流行的主导基因亚型。虽然羊驼对其他基因亚型BVDV易感，但还不清楚羊驼为何主要感染1b BVDV基因亚型，也未阐明牛源BVDV因何进化为可感染羊驼[41]。目前认为，羊驼BVDV感染主要来源于牛[24]。在美国，一些研究人员分离获得了BVDV-1b毒株，在羊驼中存在同样的基因亚型也报道于英国[2,29,30]和瑞士[22]。Mueller等报道，持续感染羊驼可存活30个月，这是迄今发现的最老的持续感染羊驼[44]，然而持续感染牛一岁龄前的死亡率高于50%。大多持续感染牛临床健康，但可大量向外排毒，其抗原性与动物体内持续感染毒株一致。BVDV感染驼科动物具有类似致病机制，这一点目前只在新大陆骆驼得到证实。Carmen等证实，BVDV感染胎儿可导致初生幼畜免疫耐受[19]。

美洲驼和羊驼对BVDV易感，感染后表现腹泻、精神委顿、呼吸道症状以及流产。牛病毒性腹泻可由于牛结核病而加剧并导致动物死亡[20]。与公牛类似，雄性驼科动物在引种时，即使血清转阳，也可能为持续感染动物。抗体检测ELISA方法不能用于持续感染并产生免疫耐受的动物的检测。

一些研究人员认为，虽然南美羊驼群的血清阳性率高达25%[40]，并且羊驼的抗体阳性率为20%[52]，这并不能说明BVDV在南美驼的普遍存在。美国羊驼研究基金会认为羊驼BVD的发生率只有0.05%[24,52]。Topliff等发现在63个羊驼群中，25%的幼畜呈血清学阳性，BVDV的持续感染率为6%。

Belknap等发现在新大陆骆驼BVDV的血清学阳性率为4.4%[8]，而Everman证实可高达53%[24]。Mudry等对2008年和2000年瑞士南美驼和羊驼的BVDV感染率进行了比较分析，发现抗体阳性率和病毒感染率分别为5.8%和0%（2008年）以及3.6%和0%（2000年），在BVDV的根除过程中可以忽略瘟病毒的影响。

1995年，Hegazy等从埃及成年发生腹泻的单峰驼和新生先天缺陷的单峰驼分离获得两株BVDV[33]。后续Yousif对该两株病毒进行基因亚型分析，发现Giza4和Giza7分属BVDV-1和BVDV-2[62]。

## 临床症状和病理变化

血清学研究表明，新大陆骆驼和旧大陆骆驼均对BVDV易感。研究表明，单峰驼的抗体阳性率为2%～50%。单峰驼BVDV的抗体阳性率在突尼斯为3.9%[16]，阿曼6.7%[32]，苏丹15.5%～15.7%[10,11]，索马里3.4%[12]。Bohrmann在吉布提单峰驼中未发现BVDV抗体阳性动物存在[9]。Wernery等证实在阿拉伯联合酋长国舍饲单峰驼的BVDV抗体阳性率为9.2%，高于放牧单峰驼的BVDV抗体阳性率3.6%，是因为前者群体密度较大并与牛群密切接触[61]。后续研究发现在552峰单峰驼中，赛驼群的BVDV抗体阳性率为0.5%，而育种群的BVDV抗体阳性率为6.4%，进一步证实了Wernery的研究结果[21]。在埃及单峰驼BVDV的抗体阳性率为11%，个别地区可达23%[35]。Tantawi发现埃及单峰驼的BVDV阳性率为4.3%[54]，Zaghana发现埃及单峰驼BVDV中和抗体阳性率可达52.5%[63]。

Doyle等利用Singer毒株进行抗体中和试验检测美国动物园单峰驼血液样品24份，发现抗体阳性率为13%[23]。最近血清学调查证实沙特阿拉伯和阿拉伯联合酋长国的奶单峰驼抗体阳性率分别为18%[1]和1.6%[60]，然而在另外的研究中Taha在812份单峰驼血液样品中未发现BVDV抗体[53]。

研究人员在中国西部双峰驼中检出了BVDV病毒抗原和抗体[31]，他们利用抗体中和试验发现该病原的抗体阳性率为23%。与之类似，血清学调查也用于新大陆骆驼BVD的研究（表62）。在秘鲁117峰与牛羊共同放牧的羊驼中，BVDV抗体阳性率为11%[51]。Picton报道，美国俄勒冈州270峰美洲驼的抗体阳性率为4.4%[47]。Puntel报道，阿根廷3个不同省份的9个农场中美洲驼BVDV的感染率为

2.05%（8/390）。

牛罹患BVD和黏膜病后会出现消化道损伤，而且后者的病理变化较前者更为严重。黏膜病的损伤多见于上消化道。在两者中，病理变化都包含不同程度的糜烂、溃疡。然而这些病理变化未出现在新大陆骆驼，只有持续感染的羊驼出现与牛类似的多发性肝坏死、支气管肺炎和严重的胸腺萎缩[36]。在羊驼，BVD病毒抗原主要分布于神经组织、血管外膜和中膜以及胃肠道黏膜下层巨噬细胞、肺组织、大脑组织。作者推断，持续感染羊驼胸腺萎缩可能与持续感染牛类似，为宿主防御机能不全的表现。

埃及单峰驼感染BVDV后可表现为死胎、死产、新生畜先天性缺陷、初生驼呼吸应激综合征、成年驼急性出血性肠炎[34]。可从发病驼淋巴组织、脾脏、脑、肾组织分离获得BVDV，在牛肾细胞培养物中可产生CPE。免疫荧光显示病毒存在于多种组织中。后续Yousif对该两株病毒进行基因亚型分析，发现Giza4和Giza7分属BVDV-1和BVDV-2[62]。目前无相关文献证实ncp型BVDV可在单峰驼分离或持续感染的报道。Fahmy在怀孕山羊进行了埃及BVDV分离株和其余牛源分离株的感染研究，包括对动物繁殖、胎儿和新生儿的影响、病理学变化等。他们发现利用NADL毒株感染妊娠65天的母畜可导致60%母畜早期流产，而cp型驼源分离株导致的流产率为25%，流产胎儿较小、身体自溶严重，表明已在胎内死亡时间较长。母畜感染后7天白细胞严重减少、淋巴细胞减少。淋巴细胞减少可持续至第28天。虽然有些感染母畜可生出健康幼畜，但幼畜的体重明显低于对照组[27,28,29]。

Hegazy等认为BVDV感染是导致单峰驼流产的主要病因，在某些群体50%流产是由BVDV导致[33]。这一观点受到质疑，因为在其余国家单峰驼并未发现如此高的流产率。例如，在阿拉伯联合酋长国，免疫荧光实验常用于对死亡的成年单峰驼和牛进行BVDV筛查，但结果常为BVDV阴性[59]。

研究人员首次在中国西部与牛和牦牛密切接触的双峰驼的血液样品56份中检测到了BVDV抗原和抗体。对17份抗原阳性的样品进行特异的荧光定量RT-PCR检测发现所有毒株均为BVDV-1。利用牛肾细胞MDBK分离获得了12株ncp型BVDV，分属于BVDV-1a、BVCDV-1b、BVDV-1c、BVDV-1m、BVDV-1o、BVDV-1p和潜在的新亚型BVDV-1q。另外，13份（23%）样品中检测到可中和cp型BVDV C24V毒株的中和抗体。由于在牛也有类似病毒亚型的报道，因此双峰驼的BVDV感染可能来源于牛源病毒的传播[31]。

近年来，关于驼科动物BVD的信息与日俱增，都表明新大陆骆驼、旧大陆骆驼对BVDV均易感。然而，还需要进行深入的田间观察和实验室研究以便获得BVDV在驼科动物的致病方式。目前，牛的BVD流行方式、致病机制已经得到较好的阐明，使得人们对BVDV和MD的复杂的流行病学和致病机理有所了解，在驼科动物也有望取得可喜进展。

从驼科动物分离获得的BVDV毒株见表63。

## 诊断

牛病毒性腹泻和黏膜病的诊断需要依赖于病毒分离、血清抗体测定等实验室诊断方法。皮肤活

表63 从驼科动物分离获得的BVDV毒株

| 物种 | 病毒 |
|---|---|
| 新大陆骆驼 | |
| 美洲驼 | cp型BVDV-1（少见） |
| 羊驼、美洲驼 | ncp型BVDV-1和BVDV-2 |
| | BVDV-1b基因亚型 |
| 旧大陆骆驼 | |
| 单峰驼 | BVDV-1、BVDV-2（均为cp型） |
| 双峰驼 | BVDV-1 |
| | ncp型 |
| | 6个基因亚型和一个潜在的新亚型 |

组织检查利用免疫组织化学方法检测BVDV，在持续感染动物表现为阳性[13]。这种方法适用于驼科动物BVD的检测。

牛BVD的诊断有两层次。首先应利用中和试验或ELISA明确群体的血清学状态。目前，竞争ELISA方法已经问世，是适用于驼科动物BVDV血清学诊断的理想方法。如果血清学诊断阴性，表明该群体未感染BVDV。其次，若在动物中检测出BVDV抗体，还需进一步筛查找到持续感染的动物。持续感染动物可以导致整个群体受到病毒的威胁，因此鉴定、剔除持续感染动物是该病防控的重中之重。捕获ELISA、PCR方法可用于持续感染的诊断，后者还可用于BVDV-1和BVDV-2的鉴别诊断。

### 防控

BVD的防控技术和信息比较容易获得，并且欧洲一些国家已经开展了国家BVD防控计划[15]。

该病的经济损失主要来源于产前感染。因此，剔除持续感染动物并对出产母畜前期进行疫苗免疫是该病的重要防控措施[55]。引入血清学阳性并带有持续感染胎儿的母畜是导致畜群疾病再发的最大威胁。因为该病可导致驼科动物发生流产，因此有必要在驼科动物采用与牛BVD类似的防控方法。在一些国家，已经广泛采用活疫苗和灭活疫苗用于防控该病。由于活疫苗具有较大的副反应，在驼科动物应慎用该种疫苗。灭活疫苗较为安全并可提供良好的保护效力。科技的发展使利用ncp生物型毒株制备疫苗成为现实。研究表明，利用灭活疫苗进行3次免疫可使新大陆骆驼获得免疫力。

## 参考文献


[1] Al-Afaleq A., Abu-Elzein E.M.E. & Hegazy A.A. (2006). – Serosurveillance for antibodies against some viral diseases of livestock in camels (*Camelus dromedarius*) in Saudi Arabia. Proceedings of the International Scientific Conference on Camels, 10–12 May 2006, Buraydah, Saudi Arabia, 338–346.

[2] Anon. (2005). – Miscellaneous mammals. *Vet. Rec.*, 156 (23), 728.

[3] Anon. (2006a). – Bovine viral diarrhea virus in camelids. In International Camelid Health Conference for Veterinarians, 18 March 2008, Columbus, OH, 275–278.

[4] Anon. (2006b). – Miscellaneous mammals. *Vet. Rec.*, 159 (16), 510.

[5] Anon. (2006c). – Miscellaneous mammals. *Vet. Rec.*, 159 (20), 654.

[6] Bedenice D. (2006). – Bovine viral diarrhea virus (BVDV) in South American camelids. *In* International Camelid Health Conference for Veterinarians, 18 March 2008, Columbus, OH, 256–262.

[7] Bedenice D. (2008) – Immunological responses in BVDV persistently infected alpacas. *In* Proceedings of the International Camelid Health Conference for Veterinarians, 18 March 2008, Columbus, OH, 167–169.

[8] Belknap E.B., Collins J.K., Larsen R.S. & Conrad K.P. (2000). – Bovine viral diarrhea virus in New World camelids. *J. Vet. Diag. Invest.*, 12, 568–570.

[9] Bohrmann R., Frey H.R. & Liess B. (1988). – Survey on the prevalence of neutralizing antibodies to bovine viral diarrhea (BVD) virus, bovine herpes virus type 1 (BHV-1) and parainfluenza virus type 3 (PI-3) in ruminants in the Djibouti Republic. *Dtsch. Tierärztl. Wschr.*, 95, 99–102.

[10] Bornstein S. (1988). – A disease survey of the Somali camel. SAREC report, Stockholm, Sweden.

[11] Bornstein S. & Musa B.E. (1987). – Prevalence of antibodies to some viral pathogens, *Brucella abortus* and *Toxoplasma gondii* in serum from camels (*Camelus dromedarius*) in Sudan. *J. Vet. Med. B.*, 34, 364–370.

[12] Bornstein S., Musa B.E. & Jama F.M. (1989). – Comparison of seroepidemiological findings of antibodies to some infectious pathogens of cattle in camels of Sudan and Somalia with reference to findings in other countries of Africa. In Proceedings of the International Symposium of Development of Animal Resources in Sudan, 3–7 January 1989, Khartoum, Sudan, 28–34.

[13] Braun U., Schoenmann M., Ehrensberger F., Hilbe M. & Strasser M. (1999). – Intrauterine infection with bovine virus diarrhoea virus on alpine communal pastures in Switzerland. *J. Vet. Med.*, 46, 13–17.

[14] Bromage G. (2006). – BVDV in alpacas. In Proceedings of the British Veterinary Camelid Society, 12–15 October 2006, Alfriston, UK, 45–47.

[15] Brownlie J., Thompson I. & Curwen A. (2000). – Bovine virus diarrhoea virus-strategic decisions for diagnosis and control. *In Practice*, 22 (4), 176–187.

[16] Burgemeister R., Leyk W. & Goessler R. (1975). – Untersuchungen uber Vorkommen von Parasitosen, bakteriellen und viralen Infektionskrankheiten bei Dromedaren in Sudtunesien. *Dtsch. Tierärztl. Wschr.*, 82, 352–354.

[17] Byers S. (2008). – Update on BVDB, BTV & WNV in camelids. *In* Proceedings of the International Camelid Health Conference for Veterinarians, 18 March 2008, Columbus, OH, 171–174.

[18] Byers S.R., Snekvik K.R. & Righter D.J. (2009). – Disseminated bovine viral diarrhoea virus in persistently infected alpacas (*Vicugna pacos*). *J. Vet. Diagn. Invest.*, 21, 145–148.

[19] Carman S., Carr N., DeLay J., Baxi M., Deregt D. & Hazlett M. (2005). – Bovine viral diarrhea virus in alpaca: abortion and persistent infection. *J. Vet. Diag. Invest.*, 17, 589–593.

[20] Cobb N. & Cobb G. (2010). – The bio-security imperative – one farm's struggle against a silent killer. *Camelid Quarterly*, March 2010, 1–3.

[21] Central Veterinary Research Laboratory (1998). – Annual Report. Central Veterinary Research Laboratory, Dubai, UAE, 19.

[22] Danuser R., Vogt H.R., Kaufmann T., Peterhans E. & Zanoni R. (2009). – Seroprevalence and characterization of pestivirus infections in small ruminants and new world camelids in Switzerland. *Schweiz. Arch. Tierheilkd.*, 151 (3), 109–117.

[23] Doyle L.G. & Heuschele W.P. (1983). – Bovine viral diarrhea virus infection in captive exotic ruminants. *JAVMA*, 183 (11), 1257–1259.

[24] Evermann J.F. (2006). – Pestiviral infection of llamas and alpacas. *Small Ruminant Res.*, 61, 201–206.

[25] Evermann J.F., Berry E.S., Baszler T.V., Lewis T.L., Byington T.C. & Dilbeck P.M. (1993). – Diagnostic approaches for the detection of bovine virus diarrhea (BVD) virus and related pesti-viruses. *J. Vet. Diagn. Invest.*, 5, 265–269.

[26] Evermann J.F., Byers S., Parish S.M., Tibary A., Bradway D.S., Rurangirwa F., Ridpath J.F. & McFarland D.W. (2006). – Bovine viral diarrhea virus in camelids: An emerging pathogen and ways to monitor herd infection. Proceedings of the International Camelid Health Conference for Veterinarians, 18 March 2008, Columbus, OH, 266–274.

[27] Fahmy L.S. (1999a). – I – Studies on pregnant goats experimentally infected with bovine viral diarrhea virus isolated from camels. In Studies on camel diseases in Egypt (L.S. Fahmy, ed.). Arab Center for Studies of Arid Zones and Drylands, Damascus, 7–27.

[28] Fahmy L.S. (1999b). – II – Studies on the effect on the foeti and newly born kids. *In* Studies on camel diseases in Egypt (L.S. Fahmy, ed.) Arab Center for Studies of Arid Zones and Drylands, Damascus, 7–60.

[29] Fahmy L.S. (1999c). – III – Clinicopathological studies on pregnant goats infected with bovine viral diarrhoea virus isolated from camels. Brochure CARDN/ACSAD. Arab Center for Studies of Arid Zones and Drylands, Damascus, Camel/P55/1999, 41–60.

[30] Foster A.P., Houlihan M., Higgins R.J., Errington J., Ibata G. & Wakeley P.R. (2005). – BVD virus in a British alpaca. *Vet. Rec.*, 156 (22), 718.

[31] Gao S., Luo J., Du J., Lang Y., Cong G., Shao J., Lin T., Zhao F., Belak S., Liu L., Chang H. & Yin H. (2013). – Serological and molecular evidence for natural infection of Bactrian camels with multiple subgenotypes of bovine viral diarrhea virus in western China. *Vet. Microbiol.*, 163, 172–176.

[32] Hedger R.S., Barnett T.R. & Gray D.F. (1980). – Some virus diseases of domestic animals in the Sultanate of Oman. *Trop. Anim. Health.*, 12, 107–114.

[33] Hegazy A.A., El Sanousi A.A., Lotfy M.M. & Aboellail T.A. (1995). – Pathological and virological studies on calf mortality: B- mortalities associated with bovine virus diarrhea virus infection. In Proceedings of the 22nd Arab Veterinary Medical Congress, 19–23 March 1995, Cairo, Egypt. *J. Egypt Med. Assoc.*, 55 (Nos. 1 & 2), 493–503.

[34] Hegazy A.A., Fahmy L.S., Saber M.S., Aboellail T.A., Yousif A.A. & Chase C.C.L. (1998). – Bovine virus diarrhea infection causes reproductive failure and neonatal mortality in the dromedary camel. *In* International Meeting on Camel Production and Future Perspectives, 2–3 May 1998, Al Ain, UAE.

[35] Hegazy A.A., Lotfia S.F. & Saber M.S. (1993). – Prevalence of antibodies common in viral diseases of domestic animals

among camels in Egypt. In Project (91-H-2-4) NARP (Annual report). Epidemiological, clinical and pathological studies on some diseases in camel in Egypt.

[36] Henningson J.N., Topliff C.L., Steffen D.J., Brodersen B.W., Smith D.R. & Kelling C.L. (2006). – Viral antigen distribution in alpacas persistently infected with bovine viral diarrhea virus. *Vet. Pathol.*, 43 (5), 827.

[37] Intisar K.S., Ali Y.H., Khalafalla A.I., Rahman-Mahasin E.A., Amin A.S. & Taha K.M. (2010). – The first report on prevalence of pestivirus infection in camels in Sudan. *Trop. Anim. Health Prod.*, 42, 1203–1207.

[38] Kapil S., Yeary T. & Evermann J.F. (2009). – Viral diseases of New World camelids. *Vet. Clin. Food Anim.*, 25, 323–337.

[39] Karesh B.W., Uhart M.M., Dierenfeld E.S., Braselton W.E., Torres A., House C., Puche H. & Cook R.A. (1998). – Health evaluation of free-ranging guanaco (*Lama guanicoe*). *J. Zoo Wildlife Med.*, 29 (2), 134–141.

[40] Kelling C. (2008). – Prevalence of BVDV infected US alpaca herds and factors associated with BVDV seropositive herd status. In Proceedings of the International Camelid Health Conference for Veterinarians, 18 March 2008, Columbus, OH, 176–181.

[41] Kim S.G., Anderson R.R., Yu J.Z., Zylich N.C., Kinde H., Carman S., Bedenice D. & Dubovi E.J. (2009). – Genotyping and phylogenetic analysis of bovine viral diarrhea virus isolates from BVDV infected alpacas in North America. *Vet. Microbiol.*, 136, 209–216.

[42] Mattson D.E. (1994). – Update on llama medicine. Viral diseases. *Vet. Clin. North Am. Food Anim. Pract.*, 10 (2), 345–351.

[43] Mudry M., Meylan M., Regula G., Steiner A., Zanoni R. & Zanolari P. (2010). – Epidemiological study of pestiviruses in South American camelids in Switzerland. *J. Vet. Intern. Med.*, 24, 1 218–1 223.

[44] Mueller K. & Broadbent R. (2007). – General discussion on BVD and bluetongue. Proceedings of the British Veterinary Camelid Society, Rugby, UK, 44–46.

[45] Pastoret P.-P. (2010). – Bovine virus diarrhoea and mucosal disease of cattle. *In* Infectious and parasitic diseases of livestock (P.-C. Lefevre, J. Blancou, R. Chermette & G. Uilenberg, eds). Lavoisier, Paris, France, 505–520.

[46] Peterhans E., Bachofen C., Stalder H. & Schweizer M. (2010). – Cytopathic bovine viral diarrhea viruses (BVDV): Emerging pestiviruses doomed to extinction. *Vet. Res.*, 41, 44, 1–14.

[47] Picton R. (1993). – Serologic survey of llamas in Oregon for antibodies to viral diseases of livestock (MS thesis). Oregon State University, Corvallis, Oregon, USA.

[48] Probst C., Speck S. & Hofer H. (2007). – Epidemiology of selected infectious diseases in zoo-ungulates: single species versus mixed species exhibits. *Verh. ber. Erkrg. Zootiere.*, 43, 10–12.

[49] Puntel M., Fondevila N.A., Blanco Viera J., O'Donnell V.K., Marcovechio J.F., Carillo B.J. & Schudel A.A. (1999). – Serological survey of viral antibodies in llamas (*Lama glama*) in Argentina. *J. Vet. Med. B.*, 46, 157–161.

[50] Reed C. (2010). – Import risk analysis: llamas (*Lama glama*) and alpacas (*Vicugna pacos*) from specified countries. Policy and Risk MAF Biosecurity, New Zealand, 24–30.

[51] Rivera H., Madewell B.R. & Ameghino E. (1987). – Serologic survey of viral antibodies in the Peruvian alpaca (*Lama pacos*). *Am. J. Vet. Res.*, 48 (2), 189–191.

[52] Shimeld L.A. (2009). – Serological survey of alpacas living or breeding in Southern California. In Proceedings of the 113th Annual Meeting of the USA Health Association, 8–14 October 2009, San Diego, CA.

[53] Taha H.T. (2007) – Pathogens affecting the reproductive system of camels in the United Arab Emirates. Thesis, Swedish University of Agricultural Sciences, Uppsala, Sweden.

[54] Tantawi H.W., Youssef R.R., Arab R.M., Marzouk M.S. & Itman R.H. (1994). – Some studies on bovine viral diarrhea disease in camel. *Vet. Med. J.*, 32 (3), 9–15.

[55] Thiel H.-J., Becker P., Baroth M., Koenig M. & Orlich M. (1999). – Auftreten von MD nach Impfung. 3rd Berlin Meeting on Cattle Diseases, 10/1998.

[56] Topliff C.L., Smith D.R., Clowser S.L., Steffen D.J., Henningson J.N., Brodersen B.W., Bedenice D., Callan R.J., Reggiardo C., Kurth K.L. & Kelling C.L. (2009). – Prevalence of bovine viral diarrhoea virus infections in alpacas in the United States. *JAVMA*, 234, 519–529.

[57] Van Amstel S. & Kennedy M. (2010). – Bovine viral diarrhea infections in new world camelids – a review. *Small Ruminant*

*Res.*, 21, 121–126.

[58] Wentz A.P., Belknap E.B., Brock K.V., Collins J.K. & Pugh D.G. (2003). – Evaluation of bovine viral diarrhea virus in New World camelids. *JAVMA*, 223 (2), 223–228.

[59] Wernery U., Schimmelpfennig H.H., Seifert H.S.H. & Pohlenz J. (1992). –*Bacillus cereus* as a possible cause of haemorrhagic disease in dromedary camels (*Camelus dromedarius*). In Proceedings of the 1st International Camel Conference, 2–6 February 1992, Dubai, UAE, 51–58.

[60] Wernery U., Thomas R., Raghavan R., Syriac G., Joseph S. & Georgy N. (2008). – Seroepidemiological studies for the detection of antibodies against 8 infectious diseases in dairy dromedaries of the United Arab Emirates using modern laboratory techniques – Part II. *J. Camel Pract. Res.*, 15 (2), 139–145.

[61] Wernery U. & Wernery R. (1990). – Seroepidemiologische Untersuchungen zum Nachweis von Antikorpern gegen Brucellen, Chlamydien, Leptospiren, BVD/MD, IBR/IPV- und Enzootischen Bovinen Leukosevirus (EBL) bei Dromedarstuten (*Camelus dromedarius*). *Dtsch. Tierärztl. Wschr.*, 97, 134–135.

[62] Yousif A.A., Braun L.J., Saber M.S., Aboelleil T. & Chase C.C.L. (2004). – Cytopathic genotype 2 bovine viral diarrhea virus in dromedary camels. *Arab. J. Biotech.*, 7 (1), 123–140.

[63] Zaghana A. (1998). – Prevalence of antibodies to bovine viral diarrhoea virus and/or border disease virus in domesticruminants. *J. Vet. Med. B.*, 45, 345–351.

## 深入阅读材料

*Abou-Zaid A.A. (1991). – Studies on some diseases of camels. PhD Thesis (Infectious Diseases). Zagazig University, Zagazig, Egypt. Eisa M.I. (1998). – Serological survey of some viral diseases in camels in Sharkia Governorate, Egypt. Proceedings of the International Meeting on Camel Production and Future Perspectives, 2–3 May 1998, Al Ain, UAE.*

*Hegazy A.A. & Fahmy L.S. (1997). – Epidemiological, clinical and pathological studies on some diseases of camel. Camel Newslett., 13 (9), 21–22.*

*Thedford R.R. & Johnson L.W. (1989). – Infectious diseases of New-world camelids (NWCs). Vet. Clin. North Am. Food Anim. Pract., 5 (3), 145–157.*

*Wernery U. (1999). – New aspects on infectious diseases of camelids. J. Camel Pract. Res., 6 (1), 87–91.*

(高闪电译，殷宏校)

## 2.1.10 裂谷热

裂谷热（Rift Valley fever，RVF）是一种节肢动物传播的人畜共患的病毒病，大部分病例发现于反刍动物。人的感染主要是由于接触感染尸体及病料而引起[10]。除了对人类健康的危害之外，裂谷热的流行经常对动物饲养者造成严重经济损失，包括动物产量下降和死亡，以及怀孕的任何阶段都高达100%的流产率。尤其重要的是，到目前为止，裂谷热的流行主要发生于严重的雨季之后，这提示大的昆虫群可能是该病流行的先决条件[11]。在干燥的地区不会发生裂谷热。近些年，已经明确新大陆骆驼（NWCs）和旧大陆骆驼（OWCs）能感染该病。在新大陆骆驼，裂谷热病毒（RVFV）不但可引起流产，而且可引起全身性疾病。然而，在单峰驼，除了出血热和肺炎，任何怀孕阶段的流产似乎是主要的临床症状。裂谷热在非洲是一种从急性到特急性的家养反刍动物的传染病。洪水后易出现的伊蚊属的蚊子以及叮咬蝇类被认为是该病的传播媒介。

### 病原学

裂谷热病毒是布尼亚病毒科白蛉病毒属的成员。布尼亚病毒呈球形，直径为80～120微米（μm），有来自宿主细胞的脂质双层囊膜，囊膜中嵌有病毒编码的糖蛋白纤突。裂谷热分离株之间没有明显的抗原差异，然而已发现不同毒株的毒力存在差异。裂谷热病毒是一个单血清媒介传播的RNA病毒。非结构蛋白（NS）是多功能的蛋白，能使病毒入侵后逃避宿主的抗病毒应答。在布尼亚病毒该蛋白是独特的，它能在细胞核内形成丝状体结构，然而病毒在胞浆中复制。总之，已证明该非结构蛋白拥有多种生物学功能去对抗宿主的细胞干扰素防御机制[22]。

### 公共卫生意义

在人类，裂谷热95%的病例显示流感样的症状，该病易与疟疾混淆，温和的症状可能持续一周。然而，在埃及与毛里塔尼亚的裂谷热暴发期间，一些严重的病例出现眼部、脑膜的出血症状。出血症状是最严重的表现，因为这经常导致病人死亡。该病的预防取决于动物流行病学监测预警系统的实施。目前有人用的疫苗，但是并未商业生产，仅用于高风险易感的人群。

### 流行病学

在过去的70多年，裂谷热以延长的间隔流行于非洲的东部和南部。人们认为该病毒流行于原始森林，在蚊子和脊椎动物之间传播，大雨可造成蚊子的滋生范围扩大并将疫病扩散至饲养家畜的地区。白蛉病毒属的病毒可被包括按蚊(疟蚊)、库蚊、伊蚊、曼蚊等属的23种蚊子传播。反刍动物感染裂谷热后在非怀孕成年动物是不明显的。但在大暴发时，可引起流产和新生畜死亡，特别在欧洲品系和新大陆骆驼。该病毒1931年首次发现于肯尼亚裂谷的一个农场，现在流行于非洲亚撒哈拉的大部分地区，也发生于西非[25]。该病已经传播到了埃及、苏丹、肯尼亚和毛里塔尼亚，扩散至其他地区的可能性很大。在南非，裂谷热的发生有固定的时间间隔，第一次已知的暴发报道于1950年左右，1974–1976年有一次大暴发，1981年和1999年有过小范围暴发，2008年再次发生，2009年有零星暴发，之后2010年有过一次影响南非大部分地区的大流行。2000年，该病首次感染非洲以外的沙特阿拉伯的南部和也门的人和家畜被确诊[13,27]。贸易的全球化和气候变化是未来裂谷热传播的关键，这是因为裂谷热病毒作为致病原能够利用多种蚊子媒介进行传播。毛里塔尼亚的最后一次暴发是在2010年9月底和10月初，当时前所未有的降雨在阿德拉尔和毛里塔尼亚北部撒哈拉地区的绿洲造成了大量的水塘。过去数十年未见如此降雨，1956年的情况与此类似（当地人将1956年称为“热病年”）（图173）。

这次气候变化会导致一些植物茂盛生长，吸引着遥远的地方牧羊人和农场主前往，包括毛里塔尼亚的南部和东南部地区。这也有助于蚊子的高密度繁殖，主要为库蚊和按蚊[致倦库蚊、法伦按蚊、普罗托里按蚊（A. prototiensis）、斑点库蚊（C. poicilipes）、冈比亚按蚊、刺扰伊蚊、触角库蚊、麻点按蚊（A. rufipes）、常型曼蚊、佐曼按蚊(A. ziemani)]，包括大部分虫媒病毒的传播媒介。

雨季之后的几周，在阿德拉尔地区的几个绿洲暴发了严重的疟疾和裂谷热。有趣的是，家畜中第一个被报道的病例是一头单峰驼，它于2010年10月的最后一周在Aoujeft地区发病[20]。在这个地区的单峰驼裂谷热免疫球蛋白M（IgM）的血清阳性率为33%，临床症状表现有两种：一种为特急性，在24小时内突然死亡；一种为急性，表现为发热、共济失调、鼻分泌物带血、失明、牙龈出血以及中枢神经症状，也发生流产。从4头单峰驼分离到了裂谷热病毒。这提示单峰驼在裂谷热暴发的流行病学中扮演重要角色。

在裂谷热流行期间或者发生之后总是进行流行病学研究，在苏丹、肯尼亚和埃及的裂谷热流行中也是如此，有几个研究也涉及了单峰驼。Scott等[26]报道，肯尼亚大量降雨之后在牛群中暴发了裂谷热，同样也发现单峰驼的流产率明显上升。此次暴发期间45%的单峰驼具有裂谷热抗体。作者认为裂谷热病毒引发了流产率的升高。然而，没有进行病毒学的研究以支持这个推测。Meegan等[15]在埃及裂谷热流行期间也发现了单峰驼流产率的上升。这次埃及的裂谷热流行被认为由苏丹的单峰驼带入[1,10]，因为那时苏丹北部正在流行裂谷热[7]。在此期间，Hoogstraal等[10]观察了31头感染裂谷热的单峰驼，除了流产率上升之外，没有其他临床症状[5]。在埃及，Aly等[2]利用血凝素抑制试验测试发现15.6%的单峰驼抗体阳性，同时，在裂谷热暴发期间Walker等[28]报道了青年单峰驼的流产和死亡。然而，Peters和Meegan[23]仅仅观察到了裂谷热的亚临床感染。Olaleye等[19]在尼日利亚用血凝素抑制试验和血清中和试验检测了180头单峰驼，发现3.3%的阳性率。作者认为骆驼传播了裂谷热病毒。Paweska等[21]用裂谷热抑制ELISA在迪拜1119份奶用骆驼的血清样品中发现4份（0.35%）是阳性的，该方法对骆驼科有100%的特异性和敏感性。这4份样品来自成年奶用单峰驼，驼犊没有血清转阳的。由于阿拉伯半岛降雨较少，不适合蚊子的大量繁殖，所以裂谷热的发生率比较低。然而，应该注意的是，应从经常发生裂谷热国家进口的单峰驼进行检测[29]。

Imam等[12]和Eisa等[6]从一头健康的、自然感染的单峰驼分离到了裂谷热病毒。用裂谷热病毒试验感染未怀孕的单峰驼没有引发临床症状。尽管单峰驼有比较高的裂谷热抗体效价，作者并不能确定感染单峰驼流产率是否升高。

近年在东非已经发生了严重的裂谷热流行[3]。从1997年12月至1998年1月，肯尼亚、南苏丹和北坦桑尼亚的大部分地区的家养动物和人深受其害。据报道，1998年肯尼亚裂谷热的暴发几乎使驼犊群全部死亡，死亡数量也许达到150 000头[16]。1998年毛里塔尼亚疫情期间，在单峰驼检测到了IgG抗体，但未分离到病毒[18]。由于该病是人畜共患的，因此完成相关研究，阐明单峰驼在裂谷热流行中的作用是非常重要的。Munyua等[17]在2006–2007年肯尼亚裂谷热暴发期间进行了深入研究，当时大量的牛、绵羊、山羊和单峰驼均受影响。单峰驼的主要症状是流产，患病率为38%。然而，研究人员并未尝试着从流产的胎儿和胎盘分离病毒。2011年报道称，2010年和2011年南非许多的美洲驼和羊驼死于裂谷热。在南美，新大陆骆驼经常用于保护羊群，驱赶豺和豹等捕食者。在一些农场也用于生产驼毛，其毛具有羊绒的质量。2010年1月西开普敦暴发了裂谷热。当羊驼感染裂谷热病毒后，并不是全部死亡，一些羊驼出现流感样的症状并很快痊愈。这次暴发的发病率为12.5%，死亡率为0.13%。有趣的是，尽管发生死亡的是一个包括各种年龄段羊驼的农场，但是仅是一两岁的幼驼受到了影响。该病也见于幼年美洲驼。2009年、2010年和2011年末，裂谷热在南非暴发。尽管南非有大约3000只羊驼，但仅有22只于2010年和2011年死于裂谷热。

## 临床症状

在最近东非裂谷热发生期间，世界卫生组织收到了受影响地区骆驼高死亡率的许多报告。发病率和死亡率的描述提示发生了骆驼痘和触染性脓疱（痘疮），头部和颈上部肿胀、眼部水肿、嘴部溃疡且伴有黏膜脱落。

然而该病的常见症状主要是发热、流产，有时也见新生胎儿死亡和黄疸。由于在暴发期间从骆驼上并未分离到裂谷热病毒，所以该病是否由裂谷热引起并不清楚。然而，2000年沙特阿拉伯和2006年肯尼亚裂谷热暴发期间许多单峰驼出现死亡和流产[24]，但是没有对其进行深入的研究。在南非裂谷热暴发期间，观察到了新大陆骆驼的症状，该病突然发生，病程短，持续4～6小时。4只感染的羊驼幼仔出现厌食、精神委顿、发热和腹部不适，同时也伴有呼吸困难，尽管采取对症治疗，但最终死亡。剖检发现，口腔黏膜、皮下组织、浆膜、心内膜、心外膜、肝脏、肺脏和淋巴结等有广泛的瘀血和瘀斑。组织病理学显示，肝脏出现凝固性至溶解性坏死。在两只动物检测到了核内包涵体，实时PCR和免疫组化染色确诊为裂谷热[8]。

裂谷热通常侵袭多种动物，但是临床症状随动物年龄变化有所不同，在绵羊发现了最急性、急性、亚急性和不明显等多种症状。10日龄以内的牛犊经常表现为急性，高温、恶臭性腹泻、精神委顿并伴有呼吸困难，在48小时内迅速死亡，死亡率可能超过50%。山羊对裂谷热的抵抗力较绵羊强，但是山羊羔易受影响，症状与绵羊羔相同。

在单峰驼，经历短时病毒血症后，流产似乎是仅有可被观察到的症状，无单峰驼幼犊感染该病后的相关报道。目前没有双峰驼裂谷热的报道，因为在双峰驼生活的地区没有发生裂谷热的相关报告。

## 诊断

由于其他节肢动物传播的病毒病与裂谷热发生于相同的气候条件下，所以裂谷热的确诊需要病毒学和血清学诊断。尤其是韦瑟尔斯布朗病

（Wesselsbron），该病同样可引起羔羊和犊牛死亡、母羊流产。然而，裂谷热可引起较高的致死率和流产率。裂谷热和韦瑟尔斯布朗病感染幼年动物后肝脏的病变有所不同。裂谷热的肝脏病变较韦瑟尔斯布朗病范围小，实验室诊断的标本应该包括来自流产胎儿的肝素抗凝血、肝脏、脾脏、肾脏、淋巴结和脑组织，用于在Vero和BHK-21细胞上或者乳鼠和断奶小鼠分离病毒。裂谷热的抗体能用补体凝集试验、琼脂免疫扩散、血凝抑制试验和ELISA等方法来检测。Paweska等[21]研究发现抑制ELISA是一种准确的裂谷热诊断方法，可用于该病的监测、防控以及疫苗免疫后免疫效果的评价，其优点是可用于不同种动物的检测。Martin-Folgar等[14]在2006/2007年肯尼亚裂谷热暴发后用竞争ELISA测定单峰驼血清验证了抑制ELISA方法。用感染组织进行涂片后通过免疫荧光方法也可测定病毒抗原。需要注意的是，商业IgM ELISA试剂盒不能用于骆驼。

## 防控

传播媒介的化学控制、向高海拔地区转移家畜或者将家畜饲养在防蚊舍等防控措施经常不太实用或者太迟。免疫预防是保护家畜的唯一有效途径。

尽管不能确定单峰驼是否感染裂谷热，但Guillaud和Lancelot[9]已经关注疫苗的生产，他们发现弱毒疫苗株MVP-22已在单峰驼上取的满意效果。单剂量皮下接种后，22只单峰驼中18只产生了中和抗体，但未进行裂谷热病毒的攻毒试验。正如其他病毒病中描述的，骆驼对裂谷热易感，然而为了进一步理解该病毒对骆驼的致病性，对其深入研究是很有必要的。

没有专门用于骆驼裂谷热的疫苗。在南非，绵羊用疫苗已被用于羊驼，但是否具有很好的效果目前不得而知，这需要详细的注苗跟踪才能确定。

图173　Lefrass绿洲，毛里塔尼亚北部，洪水，此地曾暴发了一次大的裂谷热（该图片由F.Claes博士惠赠）

在迪拜，两只单峰驼用减毒活疫苗（纳米比亚：NSR0580）和灭活疫苗（纳米比亚：NSR0966）接种，在这两种疫苗均产自Onderstepoort，且为牛、绵羊和山羊而研制。用牛的注射剂量，5次注射后，用抑制ELISA未测到抗体。

3种弱毒活疫苗市场有售，分别为分离于乌干达的Smithburn株；来自埃及Zagazig548株的MVP12株，以及重配病毒R566株。第三种是一个非常好的、无副反应的疫苗候选株。灭活疫苗仅诱导产生短期免疫力，且成本较高，但是能用于任何年龄段动物的免疫，包括怀孕动物。利用山羊痘病毒和痘苗病毒产生的重组疫苗已经问世，但目前未用于临床免疫。

## 参考文献


[1] Abd El-Rahim I.H.A., Abd El-Hakim U. & Hussein M. (1999). – An epizootic Rift Valley fever in Egypt in 1997. *Rev. sci. tech. Off. int. Epiz.*,18 (3), 741–748.

[2] Aly R.R. (1979). – Study of Rift Valley fever in camels in Egypt. M.V.Sc. Thesis, Cairo University, Cairo.

[3] Anon. (1998). – Rift Valley fever in Africa. *Vet. Rec.*, 143 (2), 34.

[4] Anon. (2007). – An update on Rift Valley fever. *OIE Bulletin*, 2, 39.

[5] Davies F.G., Koros J. &Mbugua H. (1985). – Rift Valley fever in Kenya: the presence of antibody to the virus in camels(*Camelusdromedarius*). *J. Hyg. Camb.*,94, 241–244.

[6] Eisa M. (1984). – Preliminary survey of domestic animals of Sudan for precipitating antibodies to Rift Valley fever virus.*J. Hyg.*, 93, 692–637.

[7] Eisa M., Abeid H.M.A. & El Sawi A.S.A. (1977). – Rift Valley fever in the Sudan. 1. Results of field investigations of theepizooty in Kosti district, 1973. *Bull. Santé Prod. Anim. Afr.*, 25 (4), 356–367.

[8] Gers S. &Grewar J. (2010). – Rift Valley fever infection in alpacas. *Vet. Pathol.*,47 (6), 3S-59S.

[9] Guillaud M. & Lancelot R. (1989). –Essais de vaccination des ruminants domestiques (bovins, ovins, caprins, camelides) contre la fievre de la vallee du Rift avec la souche MVP-12 au Senegal et enMauritanie. Rapport d'execution, IEMVT-CIRAD,Paris, France.

[10] Hoogstraal H., Meegan J.M., Khalil G.M. &Adham F.K. (1979). – The Rift Valley fever epizootic in Egypt 1977,1978. 2. Ecological and entomological studies. *Trans. R. Soc. Trop. Med. Hyg.*, 73 (6), 624–629.

[11] Huebschle O.J.B. (1983). –ExotischeVirusseuchen der Wiederkauer II. Rift-Tal-Fieber. *Tierärztl. Umschau*, 38, 268–273.

[12] Imam I.Z.E., Karamany R. &Darwish M.A. (1978). – Epidemic of RVF in Egypt. Isolation of RVF virus from animals. *J. EgyptPublic Health Assoc.*, 23, 265–269.

[13] Khan A.S., Rollin P.E., Swanepoel R., Ksiazek T.G. & Nichol S.T. (2002). – Genetic analysis of viruses associated withemergence of Rift Valley fever in Saudi Arabia and Yemen, 2000–2001. *Emerg. Inf. Dis.*, 12, 1 415–1 420.

[14] Martin-Folgar R. Bishop R.P. Younan M. Gluecks I. &Brun A. (2010). – Detection of anti-RVFV antibodies in camels(*Camelusdromedarius*) using a novel monoclonal antibody based-competitive ELISA. Proceedings of the International CamelSymposium, 7–11 June 2010, Garissa, Kenya, 51.

[15] Meegan J.M., Hoogstraal H. & Moussa M.I. (1979). – An epizootic of Rift Valley fever in Egypt in 1977. *Vet. Rec.*, 105, 124–125.

[16] Mungere J.G. (2000). – Camels in the new millennium. Proceedings of the 6th Kenya Camel Forum, 7–11 March 2000,Maralal, Samburu District, Kenya, 8–12.

[17] Munyua P., Murithi R.M., Wainwright S., Githinji J., Hightower A., Mutonga D., Macharia J., Ithondeka P.M., Musaa J.,Breiman R.F., Bloland P. &Njenga M.K. (2005). – Rift Valley fever outbreak in livestock in Kenya, 2006–2007. *Anim. J. Trop.Med. Hyg.*, 83 (Suppl. 2), 58–64.

[18] Nabeth P., Kane Y., Abdalahi M.O., Diallo M., Ndiaye K., Ba K., Schneegans F., Sall A.A. &Mathiot C. (2001). – Rift Valleyfever outbreak. Mauritania, 1998: seroepidemiologic, virologic, entomologic, and zoologic investigations. *Emerg. Inf. Dis.*,7 (6), 1052–1054.

[19] Olaleye O.D., Tomori O. & Schmitz H. (1996). – Rift Valley fever in Nigeria: infections in domestic animals. *Rev. sci. tech. Off. int. Epiz.*,15 (3), 937–946.

[20] Ould El Mamy A.B., Ould Baba M., Barry Y., Isselmou K., Dia M.L., Hampate B., Diallo M.Y., OuldBrahim M., Diop M.,Moustapha L.M., Thiongane Y., Bengoumi M., Puech L., Plee L., Claes F., de La Rocque S. &Doumbia B. (2011). – UnexpectedRift Valley fever outbreak, northern Mauritania. *Emerg. Inf. Dis.*, 17 (10), 1894–1896.

[21] Paweska J.T., Mortimer E., Leman P.A. &Swanepoel R. (2005). – An inhibition enzyme-linked immunosorbent assay forthe detection of antibody to Rift Valley fever virus in humans, domestic and wild ruminants. *J. Virol. Methods*, 127, 10–18.

[22] Pepin M., Bouloy M., Bird B.H., Kemp A. &Paweska J. (2010). – Rift Valley fever virus (*Bunyaviridae: Phlebovirus*): an updateon pathogenesis, molecular epidemiology, vectors, diagnostics and prevention. Vet. Res., 41, 61–95.

[23] Peters C.J. &Meegan J.M. (1981). – RVF. In CRC handbook series in zoonoses (G. Beran, ed.). CRC Press, Boca Raton, FL, 403.

[24] World Organisation for Animal Health (2007). – Rift Valley fever, ruminants-Kenya: OIE. WAHID Interface, AnimalHealth Information (OIE Disease Alert Message 10 Jan 2007 available at www.oie.int/wahid-prod/public.php?page=event_summary&this_country_code=KEN&reportid=4487) (In ProMED www.promedmail.org/direct.php?id=20070111.0118).

[25] Saluzzo J.F., Chartier C., Bada R., Martinez D. &Digoutte J.P. (1987). – La fievre de la vallee du Rift enAfrique de l'Ouest. *Rev. Elev. Méd. Vét. Pays Trop.*, 40 (3), 215–223.

[26] Scott G.R., Coakley W., Roach R.W. &Cowdy N.R. (1963). – Rift Valley fever in camels. *J. Path. Bact.*, 86, 229–231.

[27] Shoemaker T., Boulianne C., Vincent M.J., Pezzanite L., Al-Qahtani M.M., Al-Mazrou Y., Khan A.S., Rollin P.E., SwanepoelR., Ksiazek T.G. & Nichol S.T. (2002). – Genetic analysis of viruses associated with emergence of Rift Valley Fever in SaudiArabia and Yemen, 2000–2001. *Emerg. Inf. Dis.*, 8, 1 415–1 420.

[28] Walker J.S. (1975). – RVF, foreign animal diseases, their prevention, diagnosis and control committee on foreign animaldiseases of United States. *Anim. Health Assoc.*, 6, 209–221.

[29] Wernery U., Thomas R., Raghavan R., Syriac G., Joseph S. &Georgy N. (2008). –Seroepidemiological studies for thedetection of antibodies against 8 infectious diseases in dairy dromedaries of the United Arab Emirates using modern laboratorytechniques – Part II. *J. Camel Pract. Res.*, 15 (2), 139–145.

（独军政译，殷宏校）

## 2.1.11 蓝舌病

属于不同分类学家族和种属的虫媒病毒主要通过吸血昆虫或者蜱来传播，虫媒病毒可以在人类和不同种属动物中引发严重的疾病。表64是对虫媒病毒感染的总结。

蓝舌病（Bluetongue，BT）是一种急性的、虫媒传播疾病，主要感染羊、牛和野生反刍动物。蓝舌病病毒（Bluetongue virus， BTV）经库蠓进行传播，目前已经发现BTV有26个血清型。该病的主要症状为口鼻腔黏膜发绀、蹄叶炎、冠状垫炎、头部和颈部的水肿、口腔炎症和溃疡。此病最初仅限于非洲，且只感染绵羊，但在过去的几十年里已经扩散到欧洲、北美和南美及澳大利亚。可以说BTV能感染分布于热带和亚热带地区所有国家的反刍家畜。该病主要感染绵羊，在2006年已经扩散到北欧，包括英格兰，严重影响了当地畜牧业的发展。在欧洲西北部的暴发主要是由血清型8型（BTV-8）引起，也引起了牛的蓝舌病。BTV-8的扩散基本上不受地理或者地形边界的影响。蓝舌病和另外一种由库蠓传播的鹿流行性出血热也在北美发生。蓝舌病为一种必须申报的疾病，但不感染人。因为BTV-8在欧洲的牛和其他反刍动物造成严重疾病是出乎意料的，同时也因为新的传播媒介出现且病毒可以垂直传播，所以目前在欧洲肆虐的BTV-8将在不久的未来成为特别的挑战，并在国际范围内造成广泛的威胁。在欧洲，另外6种BTV血清型已经被诊断，然而非洲已诊断出的血清型有21种、北美5种、印度11种和澳大利亚8种[26]。在过去的几年中，新型疫苗及比较寒冷的冬季已经在很大程度上降低了欧洲地区蓝舌病的发生。

表64 感染人类和动物的虫媒病毒

| 病毒 | 疾病（血清型或者血清组的种类） | 易感物种 | 储存宿主 | 媒介 |
|---|---|---|---|---|
| RNA病毒 | | | | |
| 呼肠孤病毒科 | 蓝舌病病毒（26个血清型） | 绵羊，骆驼 | 牛（其他野生反刍动物） | 库蠓属 |
| 环状病毒 | 非洲马瘟（9个血清型） | 家养马及其他马属动物 | 野生马（其他？） | 库蠓属 |
| | 鹿流行性出血热（9个血清型）。血清型7：茨城马脑炎（7个血清型） | 野生反刍动物、牛、骆驼 | 驴（斑马？） | 库蠓属（库蠓属？）库蠓属，蚊子 |
| | 巴尼亚姆血清群感染（在6种抗原复合物有15种病毒） | 马<br>牛、羊 | | |
| 弹状病毒科 | 一日热 | 牛、家养水牛 | | |
| | 水疱性口炎（2个血清型） | 马、牛、猪、小型反刍动物（野生动物？）、新大陆骆驼 | 家畜和野生反刍动物<br>野生猪（其他？） | 蚊子（库蠓属？）<br>蚋（蚋属），蚊子（其它？） |
| 布尼亚病毒科 | | | | |
| 内罗病毒属 | 内罗毕病 | 绵羊、山羊 | 小型反刍动物、蜱、（啮齿动物？蚊子？） | 蜱（硬蜱科，扇头蜱属） |
| 白蛉病毒属 | 裂谷热 | 人类、反刍动物、骆驼科动物 | | 蚊子 |
| 布尼亚病毒属 | 赤羽病 | 牛 | 家畜和野生反刍动物 | 库蠓属 |
| 披膜病毒科 | | | | |
| 甲病毒属 | 西方马脑炎 | 马、人类 | 鸟类 | 蚊子 |
| | 东方马脑炎 | 马、人类、新大陆骆驼 | 鸟类 | 蚊子 |
| | 委内瑞拉马脑炎 | 马、人类 | 鸟类、马 | 蚊子 |
| | | | 啮齿动物 | |
| 黄病毒科 | | | | |
| 黄病毒属 | 日本脑炎 | 野猪、马、人类 | 鸟类 | 蚊子 |
| | 西尼罗河热 | 马、羊、人类 | 鸟类 | 蚊子 |
| | | 骆驼科动物 | | |
| | 维塞尔斯布朗病 | 羊、家养反刍动物 | 牛(啮齿动物？) | 蚊子 |
| | 羊脑脊髓炎（跳跃病） | 羊、牛、马、新大陆骆驼 | 绵羊 | 蜱（硬蜱科） |
| | 中欧蜱媒脑炎 | 羊、牛、山羊 | 啮齿动物，蜱 | 蜱（硬蜱科） |

## 病原学

蓝舌病病毒属于呼肠孤病毒科（*Reoviridae*）环状病毒属（Orbivirus）。环状病毒能引发家畜的一些重要虫媒病毒病，包括蓝舌病病毒、非洲马瘟病毒、马脑炎病毒和流行性出血热病毒，它们都是由吸血库蠓传播的。最近多种血清型蓝舌病病毒在欧洲许多地区包括斯堪的那维亚半岛以及美国的部分地区的传播，使得人们不禁要问，为什么在原先没有感染的地区突然出现了外来BTV毒株？[28]。

BTV是第一个发现具有双链RNA基因组的家畜病毒。BTV的外壳由VP2和VP5两种主要蛋白组成，这两种蛋白成分占所有蛋白成分的43%，与型特异中和抗体的产生相关。核衣壳由VP3和VP7两种主要蛋白以及VP1、VP4和VP6 3种次要蛋白组成。病毒在宿主细胞中复制时产生3种非结构蛋白（NS1、NS2、NS3和NS4），这些非结构蛋白并没有组合进病毒粒子中[26]。

## 流行病学

蓝舌病分布主要局限于热带和亚热带地区以及一些有大量牛羊的高降雨量地区。在这种条件下，许多库蠓可以将BTV传播到最易感的绵羊体内。已经从世界不同地区的多种库蠓体内分离出来BTV，其中最重要的是拟蚊库蠓。雌性昆虫吸取血液后，病毒在其唾液腺内进行复制。感染病毒的蠓将终生带毒。在感染BTV 10天后，蠓可通过叮咬将病毒传给动物。蠓可以存活30天，每隔3～5天叮咬一次。

虽然现在已经有关于在新大陆骆驼和旧大陆骆驼中BTV血清阳性的许多报道，但仅有几篇科技文章中提及骆驼的蓝舌病，并提及尝试从临床样品中分离病毒或者是提取BTV RNA。Bellon等报道了在西班牙北部3个动物园中BTV-1的暴发[9]。一个患病的单峰驼显示BTV PCR阳性，ELISA阳性，但是遗憾的是没有关于该病的详细信息。用实时定量PCR对德国的92个南美骆驼群进行检测没有发现病毒RNA，但是1743份血清样品中阳性率为14.3%[41]。不同地区血清学上的差异性与BTV在羊与牛上由西到东的分布梯度相一致。Chandel等[12]尝试从血清阳性的印度单峰驼血液中分离BTV 的RNA，但是用RT-PCR在205份血样中没有发现病毒或BTV基因组。Gerken等[20]和Meyer等[31]报道了美洲驼致死性的BTV-1感染。在2008年法国南部BTV-1严重流行期间，美洲驼也被感染。在首次暴发期间，已呈现血清阳性的9只美洲驼均未出现临床症状，其中一只用RT-PCR检测时持续了42天病毒血症的美洲驼也未出现临床症状。然而，在法国南部的第二次暴发期间，一只雌性驼流产了10个月大的胎儿，并且一头雄性和一头雌性美洲驼在呼吸症状出现后24小时内死亡。从两头死亡美洲驼的肺和脾脏样品在BHK21细胞上分离到了传染性BTV，并经PCR鉴定确认。Henrich等[23]也报道了用细胞放射免疫测定方法检测出5岁龄怀有胎驼的雌性羊驼感染了致死性的蓝舌病，同时美国最近也有羊驼感染急性、致死性蓝舌病的报道[33]。一头15岁龄的雌性羊驼在经历呼吸抑制和卧地不起后死亡。在尸检中，发现胸膜积水、心包积水、肺部水肿和左心室有瘀斑。用RT-PCR可以检测出BTV的 RNA，但是用cELISA没有检测出抗体，这表明BTV感染是急性的。

与此同时，用BTV人工感染了羊驼、美洲驼和单峰驼。

Batten等[8]用BTV-1实验感染3头单峰驼，并对它们连续观察75天。每头骆驼都注射10mL的病毒，皮下注射4mL、肌内注射4mL和颈静脉注射2mL。病毒的滴度为6.5$\log_{10}$/mL。在75天观察期内均未观察到临床症状。然而，在3头单峰驼接种病毒后11天和16天，用血清中和试验发现血清转阳，产生了BTV-1抗体。而且，这3头骆驼从第7天开始都有病毒血症出现，至少28天内在它们的血液中检测到了BTV RNA。该实验证明单峰驼可作为BTV的来源，且有可能具有感染蠓的能力。

在另外一个实验中，Schulz等[40]在德国用分离自绵羊的BTV以$10^{5.6}$$TCID_{50}$的量对3头羊驼和3头美洲驼进行皮下注射。所有的动物在接种6～8天后均血清转阳，在20～25天时可以用RT-PCR方法检测出病毒的基因组，但是病毒分离未能证明病毒具有感染性。这个研究能进一步证明新大陆骆驼对BTV具有敏感性，但是这个病毒在BTV-8流行病学上的作用可以忽略不计。

Zanolari 等[50]也从他们的研究中得出相似的结论。他们检测了瑞士新大陆骆驼总数的10%，共包括354头免疫过的南美驼和27头未免疫过的南美驼（共170头美洲驼和201头羊驼）。用cELISA对其进行监测均未发现BTV-8抗体。对于这些现象，作者的观点是南美小型驼对于BTV-8只有一定程度的易感性或者病毒在宿主体内不能充分地进行复制。2008年和2009年在德国实施了大规模血清学和病毒学调查，用ELISA对新大陆骆驼进行检测，其中249头新大陆骆驼（比例为14.3%）发现血清阳性，91个畜群中的43个（比例为47.3%）群，每群至少有一个血清阳性动物。然而，在这些血清阳性样品中均未检测出BTV RNA。虽然在德国的几个地区BTV的血清阳性率较高，但病毒学结果提示，新大陆骆驼在BTV感染的流行病学中并不重要[42]。

表65　蓝舌病的血清学

| 作者 | 年份 | 国家 | 物种 | 试验 | 阳性率（%） |
|---|---|---|---|---|---|
| Hafez和Ozawa | 1973 | 埃及 | 单峰驼 | AGID | 14.3 |
| Afsh和Kayvanfar | 1974 | 伊朗 | 单峰驼 | AGID | 6 |
| Eisa等 | 1979 | 苏丹 | 单峰驼 | AGID | 5 |
| Simpson | 1979 | 博茨瓦纳 | 单峰驼 | AGID | 81 |
| Barzilia | 1982 | 以色列 | 单峰驼 | AGID | 23 |
| Hafez等 | 1984 | 沙特阿拉伯 | 单峰驼 | AGID | 61 |
| Abu Elzein | 1984 | 苏丹 | 单峰驼 | AGID | 15 |
| Abu Elzein | 1985a | 苏丹 | 单峰驼 | AGID | 6 |
| Abu Elzein | 1985b | 苏丹 | 单峰驼 | AGID | 17 |
| Rivera 等 | 1987 | 秘鲁 | 羊驼 | AGID | 21 |
| Stanley | 1990 | 也门 | 单峰驼 | AGID | 13 |
| Picton | 1993 | 美国 | 羊驼 | AGID | 1.5 |
| Karesh 等 | 1998 | 阿根廷 | 驼马 | AGID,cELISA | 0 |
| Abu Elzein等 | 1998 | 沙特阿拉伯 | 单峰驼 | AGID | 58 |
| Chandel和Kher | 1999 | 沙特阿拉伯 | 单峰驼 | AGID | 7,9,13(不同品种） |
| Ostrowaki | 1999 | 沙特阿拉伯 | 单峰驼 | cELISA | 50 |
| Puntel 等 | 1999 | 阿根廷 | 美洲驼 | AGID | 0 |
| Chandel 等 | 2001 | 印度 | 单峰驼 | AGID | 18 |
| Mallik 等 | 2002 | 印度 | 单峰驼 | AGID | 10 |
| Wernery 和Kaaden | 2002 | 阿拉伯联合酋长国 | 单峰驼 | AGID,cELISA | 1.0,5 |
| Mohammed 等 | | 沙特阿拉伯 | | | |
| Chandel 等 | 2003 | 印度 | 单峰驼 | cELISA | 10,6 |
| Chauhan 等 | 2003 | 印度 | 单峰驼 | AGID,cELISA | 22,34 |
| Mahdavi 等 | 2004 | 伊朗 | 单峰驼 | AGID,cELISA | 27,39 |
| Al-Afaleq 等 | 2006 | 沙特阿拉伯 | 单峰驼 | cELISA,AGID | 50 |
| Patel 等 | 2006 | 印度 | 单峰驼 | SNT | 1.5 |
| Chandel 等 | 2007 | 印度 | 单峰驼 | AGID,cELISA | 26,38 |
| Wernery 等 | 2007 | 阿联酋 | 单峰驼 | cELISA | 30 |
| Schulz 等 | 2008 | 德国 | 单峰驼 | cELISA | 21 |
| Chauhan 等 | 2008 | 印度 | 南美骆驼 | cELISA | 14 |
| Zanolari 等 | 2009 | 瑞士 | 单峰驼 | | 不同阳性率 |
| Bellon 等 | 2010b | 西班牙 | 新大陆骆驼 | cELISA | 0 |
| Touil 等 | 2010 | 摩洛哥 | 双峰驼、单峰骆 | cELISA | 0 |
| Schulz 等 | 2012 | 德国 | 单峰骆驼 | cELISA | 11~25 |
| | 2012 | | 新大陆骆驼 | cELISA | 47.3 |

注：AGID，琼脂免疫扩散试验；cELISA,竞争性酶联免疫吸附试验；NWC，新大陆骆驼；SAC，南美骆驼；SNT，血清中和试验

2 病毒病

许多国家有关于BTV血清阳性骆驼的报道。在单峰驼，蓝舌病抗体的阳性率在1%～80%，但在新大陆骆驼的流行率是0～21%[19]（表65）。

在蓝舌病呈地方性流行的苏丹，Eisa团队报道单峰驼感染蓝舌病的阳性率为4.9%[17,18]，Abu Elzein报道14.6%[1]，Abu Elzein [2]报道16.6%。根据Abu Elzein[3]，这些阳性率显示该病在这个国家存在小范围的流行，而在当地BTV对牛的感染率为93%、对山羊和绵羊的感染率分别为86%和73%。比较低的感染率说明单峰驼可能对BTV易感性较低。同样在BTV呈地方性流行的埃及，Hafez和Ozawa[21]报道感染率仅只有14.3%。然而根据Hafez等[22]的报道，在检测的3头沙特阿拉伯单峰驼中，BTV感染率达到67%。在博茨瓦纳，Simpson[43]发现81%的单峰驼群有BTV抗体。在阿拉伯半岛，也发现有血清阳性的单峰驼。在也门，Stanley[44]发现13%单峰驼有BTV抗体。在阿联酋的一项血清学调查中，绵羊对BTV的感染率达35%，但用琼脂扩散试验（agar gel immunodiffusion test, AGID)发现同地区的1023头单峰驼中，BTV阳性率低于1%，用cELISA发现211头骆驼中5%为BTV阳性[10]。但是在这个地区的羊和骆驼体内均未分离出BTV。值得一提的是，Ostrowski[34]在沙特阿拉伯（一个同阿联酋一样干旱的国家）的骆驼中发现58%的阳性率。造成蓝舌病在这些国家中流行存在差异性的原因还不清楚。在伊朗，Afshar 和Kayvanfar [5]发现骆驼的感染率为5.9%。在以色列的单峰驼中也发现了BTV抗体[7]，调查的单峰驼中23%显示BTV-4抗体阳性。

Abu Elzein [1]用免疫扩散试验第一次从苏丹单峰驼中确定了BTV抗原，89头动物中5.6%带有BTV抗原。基于以上结果，作者认为，单峰驼可能在蓝舌病的传播过程中发挥一定的作用。

在新大陆骆驼也做了几个血清学调查[17]。在秘鲁，Rivera等[38]血清学调查发现，114头羊驼中21%有BTV抗体，而在美国俄勒冈州，270头美洲驼中只有1.5%BTV抗体阳性[36]。然而，在对俄勒冈州美洲驼的研究中，大部分的羊驼来自家畜动物中没有蓝舌病流行的地区。Puntel等[37]对阿根廷3个不同省份的9个牧场的390份美洲驼的血清样品进行检测，均未检测到BTV抗体。

## 临床症状和致病机理

关于骆驼科动物感染BTV后的临床症状或者病理变化已有报道。蓝舌病的致病机理涉及靶组织中小血管的损伤，因此病变特征为口腔和上消化道有出血、溃疡、心肌和骨骼肌坏死、蹄叶炎、肺部水肿、心包积水、胸腔和腹腔积液等。在反刍动物中BTV可以经胎盘传播引起怀孕母畜流产[27]。BTV主要在单核细胞、内皮细胞、淋巴细胞以及其他细胞中进行复制。感染BTV的反刍动物和骆驼科动物通常有一个长期的但不持续感染的病毒血症，因此BTV对吸血库蠓的感染也是不连续的。

在法国，死于BTV-1的美洲驼主要症状有厌食、嗜睡，并伴有呼吸急促、呼吸困难及打嗝式呼吸等急性呼吸窘迫。两头动物均在出现呼吸症状后24小时内死亡。类似的临床症状也出现在一头5岁龄的雌性羊驼，出现了咳嗽、打嗝式呼吸等呼吸道症状[23,24]。对南美骆驼的尸检发现舌头、腭和颊黏膜出现糜烂和溃疡、间质性肺炎和纤维性心包炎。

蓝舌病的临床症状不但在不同反刍动物是变化的，而且在不同品种的绵羊表现也不同。蓝舌病在美国绵羊的症状比在非洲绵羊的症状更温和一些。在阿联酋30%的绵羊有BTV抗体，但是作者未观察到任何的临床症状。绵羊发生蓝舌病的症状包括发热、呼吸困难，口鼻、唇以及耳部充血。其他的症状有浮肿、舌头发绀、跛行和肌肉坏死。口腔黏膜和齿龈可能出现糜烂和溃疡。蹄部冠状带可能出现炎症和水肿。跛行是感染畜群的早期症状，这可能会在诊断上与口蹄疫混淆。牛感染蓝舌病后一般呈隐性感染，但也有一些会出现与羊感染蓝舌病后相似的症状。

## 诊断

在许多国家，蓝舌病是一种必须申报的疫病，无论什么时候有蓝舌病的可疑疫情，将样品送到实验室进行检测都是强制的。在尸检中，可以收集脾脏、肝脏、淋巴结和肺脏用于分离病毒，但是为了防止破坏病毒，这些组织样品不能冷冻。

蓝舌病经常被误诊为光过敏症、口蹄疫、牛病毒性腹泻/黏膜病、牛鼻气管炎、恶性卡他热、流行性出血热和传染性脓疱等，因此该病需要通过病毒分离或者血清学方法进行确诊。可以用鸡胚、培养细胞和易感绵羊分离BTV。BTV也可以通过脑内接种在乳鼠进行增殖，然后可以用血清中和试验和荧光免疫试验进行病毒鉴定，血清学检测包括ELISA、补体结合试验、琼脂免疫扩散试验（AGID）和血清中和试验（SNT）。Chandel 等[11]对176头单峰驼用琼脂免疫扩散试验（AGID)和竞争性酶联免疫吸附试验（cELISA）进行检测，发现cELISA的敏感性略高于AGID。

另外一种敏感、特异的检测方法是聚合酶链式反应（PCR）。从编码VP3、VP7或者NS1蛋白的基因设计的引物用于群特异PCR，可检测BTV的26种血清型。从编码VP2蛋白的基因设计的引物也可以用于BTV 26个血清型的分型鉴定。多重PCR方法也已经建立。

## 防控

在有风险地区减少库蠓的数量能有效地控制蓝舌病，这可以用杀虫剂和通过辐射雄性库蠓绝育来实现。然而，预防蓝舌病最有效切实的方法为预防接种。虽然减毒弱活苗具有高效性，但是随之而来的问题是怀孕母畜接种此疫苗时可能会出现流产或者胎儿先天畸形。一些品种的羊接种此疫苗后已出现上述不良反应。另外需要关注的是，疫苗毒株和野生毒株之间可能发生重组并产生新的病毒株。

Zanolari 等在瑞士和英格兰已经对不同的蓝舌病疫苗进行了尝试[45,49]。42头新大陆骆驼（25头羊驼和17头美洲驼），包括几头怀孕母畜，均接种了来自英特威（Inter-Vet）的两种灭活疫苗（Bovilis BTV 8 和 BTV Pur AISAP8）。两种疫苗均为灭活疫苗且由BTV-8毒株组成。用cELISA和RT-PCR检测这些动物均未感染BTV。所有的南美小型驼在相隔3周分两次皮下免疫接种1mL量的疫苗。在怀孕母畜或者未怀孕动物均未观察到副反应，且怀孕母畜能产出健康胎儿。在加强免疫后所有的羊驼和美洲驼血清转阳，接种后用RT-PCR检测发现所有动物为BTV阴性。在英格兰，用同一种疫苗接种羊驼得到了相似的结果。必须强调的是，BTV有26个血清型，疫苗必须包括在一个国家或者特定地区出现的同种血清型，例如在南非，用5种血清型的疫苗接种羊只。

在2005年，Sanderson等[39]发现在欧洲的一个动物园中不同种的动物对BTV有敏感性差异。BTV-8在感染一些反刍动物时能引起较高的死亡率和发病率，尤其是欧洲和亚洲本土的一些牛科家族的动物，但对动物园内的涉及20个物种的200头非洲反刍动物没有影响。



## 参考文献


[1] Abu Elzein E.M.E. (1984). – Rapid detection of bluetongue virus antigen in the sera and plasma of camels, sheep and cattlein the Sudan, using the gel immunodiffusion test. *Arch. Virol.*, 79, 131–134.

[2] Abu Elzein E.M.E. (1985a). – Bluetongue in camels: a serological survey of the one-humped camel (*Camelus dromedarius*)in the Sudan. *Rev. Elev. Méd. Vét. Pays Trop.*, 38 (4), 438–442.

[3] Abu Elzein E.M.E. (1985b). – Bluetongue in the Sudan. *Rev. sci. tech. Off. int. Epiz.*, 4 (4), 795–801.

[4] AbuElzein E.M.E., Sandouka M.A., Al-Afalea A.I., Mohammed O.B. & Flamand J.R.B. (1998). – Arbovirus infections ofruminants in Al-Rub Al-Khali desert. *Vet. Rec.*, 142, 196–197.

[5] Afshar A. & Kayvanfar H. (1974). – Occurrence of precipitating antibodies to bluetongue virus in sera of farm animals in Iran. *Vet. Rec.*, 94, 233–235.

[6] Al-Afaleq A., Abu-Elzein E.M.E. & Hegazy A.A. (2006). – Serosurveillance for antibodies against some viral diseases oflivestock in camels (*Camelus dromedarius*) in Saudi Arabia. *In* Proceedings of the International Scientific Conference on Camels, 10–12 May 2006, Buraydah, Saudi Arabia, 338–346.

[7] Barzilai E. (1982). – Bluetongue antibodies in camels' sera in Israel. *Refuah Vet.*, 39 (3), 90–93.

[8] Batten C.A., Harif B., Henstock M.R., Ghizlane S., Edwards L., Loutfi C., Oura C.A.L. & El Harrak M. (2011). – Experimental infection of camels with bluetongue virus. *Res. Vet. Sci.*, 90 (3), 533–535.

[9] Bellon F.H., Borragan S., Gonzalez A. & Romeu P.J. (2010). – Infection by bluetongue virus serotype 1 (BTV-1) in zooartiodactyls. Proceedings of the International Conference on Diseases of Zoo and Wild Animal, 6–10 March 2010, 188–189.

[10] Central Veterinary Research Laboratory (CVRL) (1998). – Annual Report. Central Veterinary Research Laboratory, Dubai,UAE, 19.

[11] Chandel B.S., Chauhan H.C. & Kher H.N. (2003). – Comparison of the standard AGID test and competitive ELISA fordetecting bluetongue virus antibodies in camels in Gujarat, India. *Trop. Anim. Health Prod.*, 35, 99–104.

[12]Chandel B.S., Chauhan H.C., Kher H.N., Bhalodia S.D., Shah N.M. & Bulbule N.R. (2007). – Seroprevalence of bluetongue indromedary camels in Gujarat. Proceedings of the International Camel Conference, 16–17 February 2007, Bikaner, India, 11.

[13] Chandel B.S., Chauhan H.C., Kher H.N. & Shah N.M. (2001). – Detection of precipitating antibodies to bluetongue virusin aborted and clinically healthy ruminants in North Gujarat. *Indian J. Anim. Sci.*, 71, 25–26.

[14] Chandel B.S & Kher H.N. (1999). – Seroprevalence of bluetongue in dromedary camels in Gujarat. *J. Camel Pract. Res.*, 6,83–85.

[15]Chauhan H.C., Chandel B.S., Gerdes T., Vasava K.A., Patel A.R., Kher H.N., Singh V. & Dongre R.A. (2004). – Seroepidemiology of bluetongue in camels in Gujarat. *J. Camel Pract. Res.*, 11, 141–145.

[16] Chauhan H.C., Chandel B.S., Kher H.N., Dadawala A.I. & Parsani H.R. (2009). – An overview of bluetongue in camels.*J. Camel Pract. Res.*, 16 (1), 89–92.

[17] Eisa M. (1980). – Considerations on bluetongue in the Sudan. *Bull. Off. int.* Epiz. 92 (7–8), 491–500. XVLIII Session Gènèrale,Rapport No. 2.4. Bull. Off. Int. (OIE).

[18] Eisa M., Karrer A.E. & AbdElrahim A.H. (1979). – Incidence of bluetongue virus precipitating antibodies in sera of somedomestic animals in the Sudan. *J. Hyg. Camb.*, 83 (3), 539–545.

[19] Fowler M.E. (2010). – Medicine and Surgery of Camelids. 3rd Ed. Wiley-Blackwell, 177–178.

[20] Gerken M. (2010). – Animal Health. Funftes Europaisches Symposium uber Neuweltkameliden in Sevilla, Spanien. Europaisches Symposium, 18 (4), 21–26.

[21] Hafez S.M. & Ozawa Y. (1973). – Serological survey of bluetongue in Egypt. *Bull. Epizoot. Dis. Afr.*, 21 (3), 297–303.

[22] Hafez S.M., Radwan A.I., Beharairi S.I. & Al-Mukayel A.A. (1984). – Serological evidence for the occurrence and prevalenceof bluetongue among ruminants in Saudi Arabia. *Arab Gulf J. Sci. Res.*, 2 (1), 289–295.

[23] Henrich M., Reinacher M. & Hamann H.P. (2007). – Lethal bluetongue virus infection in an alpaca. *Vet. Rec.*, 161, 764.

[24] Kapil S., Yeary T. & Evermann J.F. (2009). – Viral diseases of New World camelids. *Vet. Clin. Food Anim.*, 25, 323–327.

[25] Karesh W.B., Uhart M.M., Dierenfeld E.S., Braselton E.W., Torres A., House C., Puche H. & Cook A.R. (1998). – Healthevaluation of free-ranging guanaco (*Lama guanicoe*). *J. Zoo Wildlife Med.*, 29 (2), 134–141.

[26] Lefevre P.-C., Mellor P.S. & Saergerman C. (2010). – Bluetongue. In Infectious and parasitic diseases of livestock (P.-C. Lefevre,J. Blancou, R. Chermette & G. Uilenberg, eds). Lavoisier, Paris, France, 663–687.

[27] MacLachlan N.J., Drew C.P., Darpel K.E. & Worwa G. (2009). – The pathology and pathogenesis of bluetongue. *J. Comp. Path.*, 141, 1–16.

[28] MacLachlan N.J. & Guthrie A.J. (2010). – Re-emergence of bluetongue, African horse sickness, and other Orbivirus diseases. *Vet. Res.*, 41, 35.

[29] Mahdavi S., Khedmati K. & Sabet P.L. (2006). – Serologic evidence of bluetongue infection in one-humped camels (*Camelusdromedarius*) in Kerman province, *Iran. Iranian J. Vet. Res.*, 7 (3), 85–87.

[30] Mallik T.M., Dahiya S., Ramesh K., Pawan K. & Prasad G. (2002). – Bluetongue virus antibodies in domestic camels (*Camelusdromedarius*) in northern regions of Rajasthan, India. *Indian J. Anim. Sci.*, 72, 551–552.

[31] Meyer G., Lacroux C., Leger S., Top S., Goyeau K., Deplance M. & Lemaire M. (2009). – Lethal bluetongue virus serotype1 infection in Llamas. *Emerg. Infect. Dis.*, 15 (4), 608–609.

[32] Mohammed O.B., Sandouka M.A. & Abu Elzein E.M.E. (2003). – Disease surveys of livestock in some protected areas wheregazelles have been reintroduced in Saudi Arabia. In Proceedings of Diseases at the interface between domestic livestock

andwildlife species, 17–18 July 2003, Ames, IA.

[33] Ortega J., Crossley B., Dechant J.E., Drew C.P. & MacLachlan N.J. (2010). – Fatal bluetongue virus infection in an alpaca(*Vicugna pacos*) in California. *J. Vet. Diagn. Invest.*, 22(1), 134–136.

[34] Ostrowski S. (1999) – Health management of the Arabian oryx (*Oryx lencoryx*) reintroduction. *In* Proceedings of the 1st Abu Dhabi International Arabian Oryx Conference, 27 February–1 March 1999, Abu Dhabi, UAE.

[35] Patel A.R., Chauhan H.C., Chandel B.S., Dadawala A.I., Patel N.P., Smital-Patel, Agrawal S.M. & Kher H.N. (2007). –Seroprevalence of Bluetongue in camels in organized farms in Gujarat. *J. Camel Pract. Res.*, 14, 97–100.

[36] Picton R. (1993). – Serologic survey of llamas in Oregon for antibodies to viral diseases of livestock. MS Thesis, Oregon State University, Corvallis, OR.

[37] Puntel M., Fondevila N.A., Blanco Viera J., O'Donnell V.K., Marcovechio J.F., Carillo B.J. &Schudel A.A. (1999). – Serological survey of viral antibodies in llamas (*Lama glama*) in Argentina. *J. Vet. Med. B.*, 46, 157–161.

[38] Rivera H., Madewell B.R. & Ameghino E. (1987). – Serologic survey of viral antibodies in the Peruvian alpaca (*Llama pacos*).*Am. J. Vet. Res.*, 48 (2), 189–191.

[39] Sanderson S., Garn K. & Kaandorp J. (2008). – Species susceptibility to bluetongue in European zoos during the bluetongue virus subtype 8 (BTV8) epizootic. Meeting of the European Association of Zoo and Wildlife Veterinarians, 30 April–3 May2008, Leipzig Germany, 225–227.

[40] Schulz C., Eschbaumer M., Rudolf M., Konig P., Gauly M., Grevelding C.G., Beer M. & Hoffmann B. (2011). – Experimentalinfection of South American camelids with bluetongue virus serotype 8. *Vet. Microbiol.*, 154, 257–265.

[41] Schulz C., Eschbaumer M., Wackerlin R., Gauly M., Grevelding C.G., Beer M., Hoffmann B. & Bauer C. (2008). – Serologicaland virological survey of bluetongue virus infection in South American Camelids in Germany. 5th European Symposium onSouth American Camelids and First European Meeting on Fibre Animals, Gottengen, Germany.

[42] Schulz C., Eschbaumer M., Ziller M., Wäckerlin R., Beer M., Gauly M., Grevelding C.G., Hoffmann B. & Bauer C. (2012).– Cross-sectional study of bluetongue virus serotype 8 infection in South American camelids in Germany (2008/2009). *Vet. Microbiol.*, 160, 35–42.

[43] Simpson V.R. (1979). – Bluetongue antibody in Botswana's domestic and game animals. *Trop. Anim. Health Prod.*, 11 (1), 43–49.

[44] Stanley M.J. (1990). – Prevalence of bluetongue precipitating antibodies in domesticated animals in Yemen Arab Republic. *Trop. Anim. Health Prod.*, 22, 163–164.

[45] Tellez S. (2009). – Report on the United States Animal Health Association. 2009 Annual Meeting. 113th Annual Meeting,8–14 October 2009, San Diego, CA.

[46] Touil N., Cherkaoui Z., Lmrabih Z., Loutfi C., Harif B. & El Harrak (2012). – Emerging viral diseases in dromedary camels in Southern Morocco. *Trans. Emerg. Dis.*, 59 (2), 177–182.

[47] Wernery U. & Kaaden O.-R. (2002). – Infectious diseases of camelids. Blackwell Wissenschafts-Verlag, Berlin and Vienna,214–216.

[48] Wernery U., Thomas R., Raghavan R., Syriac G., Joseph S. & Georgy N. (2008). – Seroepidemiological studies for the detection of antibodies against 8 infectious diseases in dairy dromedaries of the United Arab Emirates using modern laboratory techniques-Part II. *J. Camel Pract. Res.*, 15(2), 139–145.

[49] Zanolari P., Bruckner L., Fricker R., Kaufmann C., Mudry M., Griot C. & Meylan M. (2010a). – Humoral response to 2 inactivated bluetongue virus serotype-8 vaccines in South American camelids. *J. Vet. Intern. Med.*, 24, 956–959.

[50] Zanolari P., Chaignat V., Kaufmann C., Mudry M., Griot C., Thuer B. & Meylan M. (2010b). – Serological survey of bluetongue virus Serotype-8 infection in South American camelids in Switzerland (2007–2008). *J. Vet. Intern. Med.*, 24, 426–430.

（独军政译，殷宏校）

## 2.1.12 西尼罗河病毒脑炎和其他病毒脑炎

1937年，研究人员从乌干达西尼罗河地域一名轻微发热的女性血液中分离获得了西尼罗河病毒（WNV）。从此，关于该病发生的报道遍布世界各地。1999年，该病在美国发生，至今已蔓延至整个美国以及中美、南美和加勒比海。在驼科动物，只在新大陆骆驼见到关于该病毒感染的报道[6,16]，而在单峰驼只检出病毒抗体[15]。

### 病原学

西尼罗河病毒脑炎是一种由蚊媒传播的黄病毒引起的传染病。其他黄病毒感染可分别导致日本乙型脑炎、跳跃病、墨累谷脑炎等疾病。黄病毒科共包含70余种病毒，大多为重要的医学病毒，例如黄热病毒、登革热病毒等。目前，通过基因型分析可将西尼罗河病毒分为2个谱系和一些基因亚型。其中谱系Ⅰ主要包含非洲、南北美洲、澳大利亚、欧洲、印度和中东地区的毒株。该谱系包含多数强毒株和美洲的神经侵袭性毒株。谱系Ⅱ主要包含亚撒哈拉以南地区和马达加斯加地区的毒株，该谱系并不导致人或马的脑炎。

### 流行病学

西尼罗河病毒只可通过蚊虫传播。蚊虫在带毒西尼罗河病毒鸟类吸血时感染病毒，成为该病毒传播的媒介，进而可感染人、马、驼和其他哺乳动物、爬行动物以及鸟类。病毒在这些动物体内增殖不足以导致再次吸血的蚊虫感染，因此只会形成终末感染[4]。

上述终末宿主并不能传播病毒[1]。迄今为止，西尼罗河病毒已在美国造成12000例脑膜炎和脑炎以及1000人死亡。该病毒已经造成候鸟尤其是乌鸦的大量死亡，另外数以千计的马也曾发病。据报道，120余种鸟类可以携带该病毒。

动物感染该病毒可无临床症状，或发生不同程度的疾病，包括轻微发热、致死性脑炎等。Fowler报道，驼类感染该病毒后通常不表现临床症状，而一旦发病，其死亡率很高[5]。

### 临床症状和病理变化

Whitehead等于1993–2003年对北美185峰美洲驼、羊驼进行检查，发现其中一些动物出现神经症状。他们在这些动物血清和脑脊液中检测到了病毒G蛋白的抗体。南美驼感染西尼罗河病毒后通常不表现明显的西尼罗河热症状，但是一旦出现神经症状，即使精心治疗，也难以避免动物死亡。感染后，南美驼也可表现为发热、精神沉郁、食欲不良、昏睡、步态僵硬、痉挛、斜颈、角弓反张，其转归为横卧不起以至3～4天内死亡[3,8,17]。显微组织检查可见发病动物脑部弥散性淋巴浆细胞性脑膜脑炎。

### 诊断

该病的诊断需要经过不同的诊断方法进行实验室确诊，主要包括竞争ELISA以及用于检测脑、肾组织中的病毒核酸的实时荧光定量RT-PCR。也可利用多种细胞和乳鼠进行病毒分离。免疫组织化学用于检测西尼罗河病毒抗原。组织病理学变化为大脑和脊髓的非化脓性脑脊髓灰质炎。

研究人员利用竞争ELISA对阿拉伯联合酋长国的1119峰单峰驼进行西尼罗河病毒抗体筛查，发现该群体的阳性率为38%（425/1119）。但是动物并未表现出临床症状，所以西尼罗河热在单峰驼中的致病作用还有待进一步研究[15]。在7年时间里，摩洛哥驼类西尼罗河热的阳性率由10%增至13%[14]，研究人员推测这与气候变暖有密切关系。目前并无双峰驼感染西尼罗河病毒的相关报道。

### 防控

目前该病无特异的治疗方法。患病的骆驼应隔离至安静的环境中。采用灭蚊和加强动物免疫的方法可以获得较好的防控效果。西尼罗河病毒在北美和南美洲的迅速蔓延促进了灭活疫苗、减毒活疫苗以及重组DNA疫苗的研发。一种全病毒灭活佐剂疫苗以及重组金丝雀痘苗病毒可用于马的免疫，具有高效和安全性。

利用福尔马林灭活的马用西尼罗河病毒疫苗对28峰羊驼和56峰美洲驼进行接种试验，接种3次后动物血清抗体转阳，而且该疫苗对接种动物安全[7]。对所有的南美驼利用1mL疫苗肌肉接种3次，并对其中的55峰第4次加强接种，所有的动物抗体转阳，但由于无攻毒试验，并不能说明疫苗的效力。

研究人员利用超免的马血浆对羊驼进行被动免疫。在疾病的早期进行被动免疫可以取得理想的结果，但在疾病后期并无效果。且被动免疫利用的是非同源动物的血浆，因此输血反应的危险也未知[7]。

### 跳跃病

跳跃病病毒也是黄病毒成员，主要在英国引起一种蜱传疾病。多种蜱传脑炎病毒相互之间密切相关，包括跳跃病病毒，中欧蜱传脑炎病毒，分布于日本、美国、俄罗斯、马来西亚的远东蜱传脑炎病毒以及亚洲的一些其他在医学和兽医学中并不重要的病毒。

跳跃病主要发生于英国绵羊，但也可能波及一些其他动物。该病在赫布里底群岛中的哈里斯岛美洲驼[9]、英国美洲驼[2]都有报道。该病毒主要通过蓖子硬蜱传播，但感染病毒的动物并不能传播该病毒。该病在南美驼比较罕见。该病的特征性症状为精神沉郁、食欲废绝、共济失调以及由于脑炎导致的进行性瘫痪。

该病的诊断主要通过血清学诊断、病毒分离和脑切片免疫组织化学染色进行确诊。疫苗免疫、药浴可用于羊跳跃病的预防，但对于新大陆骆驼目前无切实可行的方法。建议利用羊用灭活疫苗进行南美驼的接种研究。

### 东部马脑炎病毒

东部马脑炎病毒是一种节肢动物传播病毒，由蚊子传播。东部马脑炎主要流行于美国西南以及南美洲中北部，病毒主要通过蚊虫/鸟循环存在。东部马脑炎病毒属于披膜病毒科甲病毒属，与西尼罗河病毒类似，人、马、南美驼是东部马脑炎病毒的终末宿主。该病毒的主要媒介是黑尾脉毛蚊。除上述区域外，目前该病还未在其他地区报道，说明该病毒只在上述特定区域内完成病毒循环。但在美国东海岸的一些州还有少量驼类病例的报道，2009年南美驼的感染也传播至以前从未报道该病的州[13]。目前还不清楚该病的传播是否与全球气温变暖导致的当地气候变化有关。南美驼发病症状与其他病毒引起的脑炎相似。引入该病流行区域的其他动物感染该病毒后可能发生特征性的致死性疾病[11]。疫苗可用于马，但未批准用于南美驼。驼科动物可能对其他脑炎病毒更为易感，例如西方马脑炎病毒和委内瑞拉马脑炎病毒[6]。



## 参考文献

[1] Castillo-Olivares J. & Wood J. (2004). – West Nile virus infection in horses. *Vet. Res.*, 35, 467–483.

[2] Cranwell M.P., Josphson M., Willoughby K. & Marriott L. (2008). – Louping-ill in a llama. *Vet. Rec.*, 162 (1), 28.

[3] Dunkel B.J., Del Piero F., Wotman K.L., Johns I.C., Beech J. & Wilkins P.A. (2004a). – Encephalomyelitis from West Nile flavivirus in 3 alpacas. *J. Vet. Int. Med.*, 18 (3), 365–367.

[4] Espada M., Jimenez P.C., Evelyn C., Vasquez C. & Dolores M. (2010). – South American camelids: health status of their cria. *Rev. Comp. Cien. Vet.*, 4 (1), 64–65.

[5] Fowler M.E. (2010). – Medicine and surgery of camelids. 3rd Ed. Blackwell Publishing, Ames, IA,185–186.

[6] Kapil S., Yeary T. & Evermann J.F. (2009). – Viral diseases of New World camelids. *Vet. Clin. Food Anim.*, 25, 323–337.

[7] Kutzler M.A., Baker R.J. & Mattson D.E. (2004b). – Humoral response to West Nile virus vaccination in alpacas and llamas. *JAVMA*, 225, 414–416.

[8] Kutzler M.A., Bildfell R.J., Gardner-Graft K.K., Baker R.J., Delay J.P. & Mattson D.E. (2004a). – West Nile virus infection intwo alpacas. *JAVMA*, 255 (6), 921–924.

[9] Macaldowie C., Patterson I.A., Nettleton P.F., Low H. & Buxton D. (2005). – Louping-ill in llamas (*Lama glama*) in the Hebrides. *Vet. Rec.*, 156 (13), 420–421.

[10] ]Murray K.O., Mertens E. & Despres P. (2010). – West Nile virus and its emergence in the United States of America. *Vet.Res.*, 41 (6), 67.

[11] ]Nolen-Watson R., Bedenice D., Rodriguez C., Rushton S., Bright A., Fecteau M.E., Short D., Majdalany R., Tewari D.,Pedersen D., Kiuipel M., Maes R. & Del Piero F. (2007). – Eastern equine encephalitis in 9 South American camelids. *J. Vet.Int. Med.*, 21 (4), 846–852.

[12] ]Reed C. (2010). – Import risk analysis: Llamas (*Lama glama*) and alpacas (*Vicugna pacos*) from specified countries. MAFBiosecurity New Zealand, Wellington, New Zealand. Available at: www.biosecurity.bovt.nz/regs/imports/ihs/risk.

[13] Tellez S. (2009). – Report on the United States Animal Health Association. 2009 Annual Meeting. 113th Annual Meeting,8–14 October 2009, San Diego, CA.

[14] Touil N., Cherkaoui Z., Lmrabih Z., Loutfi C., Harif B. & El Harrak (2012). – Emerging viral diseases in dromedary camelsin southern Morocco. *Trans. Emerg. Dis.*, 59 (2), 177–182.

[15] Wernery U., Thomas R., Syriac G., Raghavan R. & Kletzka S. (2007). – Seroepidemiological studies for the detection ofantibodies against nine infectious diseases in dairy dromedaries (Part I). *J. Camel Pract. Res.*, 14 (2), 85–90.

[16] Whitehead C.E., Anderson D.E. & Saville W.J.A. (2006). – Neurological diseases in llamas and alpacas: a retrospective study of 185 cases (1993–2003). *In* Proceedings of the 1st Conference of the International Society of Camelids Research and Development (ISOCARD), 15–17 April 2006, Al Ain, UAE, 23.

[17] Yaeger M., Yoon K.J., Schwartz K. & Berkland L. (2004). – West Nile virus meningoencephalitis in a suri alpaca and a Suffolkewe. *J. Vet. Diagn. Invest.*, 16, 64–66.



(高闪电译，般宏校)

## 2.1.13 新生畜腹泻

与小反刍兽类似，驼科动物特别能适应在干燥环境中生活，主要通过其肠道大量吸收水分，使粪便为硬球。螺旋形的结肠具有很大的表面积，是水分吸收的主要场所。其他种动物可能由于上消化道的损伤导致早期腹泻，而驼类动物由于肠道大量水分的吸收，可以掩盖消化道损伤。因此，驼类发生腹泻的情况较少，但是一旦发生就应特别重视。

新生牛犊腹泻是给养牛业造成巨大损失的主要病因之一。田间和实验室研究表明，腹泻是由多种病原造成的，比较复杂，往往是多种细菌、病毒和寄生虫共同作用的结果。常见的新生畜腹泻病原包括产肠毒素大肠杆菌、轮状病毒、冠状病毒、隐孢子虫和沙门氏菌，临床上难以区分。轮状病毒、冠状病毒分别特异地侵蚀回肠黏膜上皮或肠黏膜上皮，二者均可侵蚀空肠黏膜上皮。粪便中病毒存在并不一定表明疾病的存在，只有病毒复制导致上皮绒毛功能异常时才导致临床症状的出现[27]。

### 病原学

轮状病毒属于呼肠孤病毒科、轮状病毒属。轮状病毒通常根据其感染的宿主进行命名。该病毒为无囊膜病毒，其形态如轮状，核酸由11个节段的双链RNA组成。所有的轮状病毒具有共同的群抗原。目前共发现有7个不同的血清群（A-G），牛和羊的轮状病毒属于A群，而且已有报道从新大陆骆驼以及单峰驼分离出该血清型轮状病毒。

冠状病毒属于巢状病毒目、冠状病毒科。该病毒含有3个组，其中牛冠状病毒属于第2分组，禽冠状病毒属于第3分组。曾造成人大量死亡、严重急性呼吸道综合征（SARS）的冠状病毒属于第2分组。冠状病毒为囊膜病毒，其基因组为单股RNA。

### 流行病学

#### 轮状病毒

A群轮状病毒在世界范围内广泛存在，所以很难找到血清学阴性的动物。不同家畜轮状病毒可以发生基因片段的交换，即基因重排。该病毒主要通过粪-口途径传播，发生腹泻的动物可以大量排出病毒。该病的潜伏期通常为18～48小时，发病动物排出白色水样、乳脂样，有时绒球样粪便，称为“牛奶泻”。轮状病毒可引起牛犊、幼驼、猪仔、羔羊和马驹等幼畜的传染性腹泻。许多轮状病毒感染为亚临床感染，该病毒单独感染通常不致死。人轮状病毒感染可导致婴儿的急性肠道疾病。

### 冠状病毒

牛冠状病毒可特异地引起感染牛发生疾病，但该病毒还可感染羊。从发生呼吸道疾病的新大陆骆驼分离的冠状病毒属于第1组，而从发生腹泻的新大陆骆驼和单峰驼分离的冠状病毒属于第2组。冠状病毒主要通过直接接触和气溶胶的方式传播。该病在幼畜的潜伏期与轮状病毒相同，为24～48小时。发病动物常排出带有黏膜或血液的黄色稀便。该病毒与船运热密切相关。

冠状病毒不仅与大肠杆菌、轮状病毒、隐孢子虫和沙门氏菌共同导致新生畜的腹泻，还可导致犊牛呼吸道疾病以及成年牛的冬季腹泻。在驼科动物也有冠状病毒感染的报道。

### 新生驼腹泻

目前关于新生驼腹泻的信息较少，但比较明确的是导致6月龄内单峰驼的致死性疾病通常为腹泻[47]。新生驼的死亡率在不同地区差别较大。Agab等报道，苏丹单峰驼幼畜的死亡率可达30%以上[1]。Gluecks报道，肯尼亚单峰驼幼畜的死亡率远低于东非其他地区，发生疾病幼畜通常只有23%发生腹泻，腹泻病例死亡率只有2%[35]。在南美，腹泻也是导致新大陆骆驼幼畜死亡的主要疾病[49]。

有关单峰驼的冠状和轮状病毒研究很少，在双峰驼更是没有任何报道。在单峰驼的粪样中发现过冠状和轮状病毒，血清中检测到抗体。

Wunschmann等报道了在美国明尼苏达州6周龄单峰驼幼畜发生的肠道冠状病毒感染[83]。该病驼持续腹泻5天，虽经精心的对症治疗，该畜最终死亡，可能由于继发性梭菌肠毒血症。研究人员通过电子显微镜术在其粪便中发现了冠状病毒粒子，利用第二组冠状病毒抗体进行组织免疫化学实验证实在结肠上皮细胞存在冠状病毒抗原。

研究人员利用电子显微镜在阿拉伯联合酋长国发生腹泻的单峰驼粪便中发现了轮状病毒和冠状病毒[47,56]。另外，Alis研究小组对苏丹单峰驼的轮状病毒感染进行了深入研究[3,5,6]。他们利用ELISA、乳胶凝集实验、免疫层析、聚丙烯酰胺凝胶电泳等方法对几百份的样品中的轮状病毒抗原进行检测，结果在发生腹泻的幼驼粪便样品中检测到了轮状病毒抗原。血清学调查发现轮状病毒抗体阳性率为48%，与动物临床腹泻症状相符合。在临床健康的幼驼中存在较高的抗体阳性率。这些结果说明，轮状病毒在单峰驼腹泻的流行中发挥着重要作用。

据报道，轮状病毒在骆驼中存在季节性感染，多发生于初冬（10月份）。研究人员成功利用猴胎儿肾细胞分离了粪便中的轮状病毒，并利用ELISA和电镜对获得的病毒进行了鉴定。利用胰酶处理样品后进行病毒传代，在初次和再次传代后，该病毒在3～5天可产生细胞病变。利用聚丙烯酰胺凝胶电泳分析显示为轮状病毒特征性基因组。但是，由于从粪便中提取基因组存在较大难度，该研究只获得了部分轮状病毒的核酸电泳分析结果。

Mahin等发现，摩洛哥单峰驼中50%（27/55）为轮状病毒阳性，说明单峰驼对轮状病毒比较易感。

Ali等利用轮状病毒抗体检测ELISA对苏丹530份单峰驼血清进行轮状病毒抗体检测，结果抗体阳性率为48%，而且临床健康的双峰驼的抗体阳性率很高，最高者可达70%[4]。他们已经证实苏丹单峰驼存在较高的轮状病毒感染[2]。另外，El-sayed等证实单峰驼奶样中的轮状病毒抗体效价可高达1：256[24]。

血清学和病毒分离试验证实，在美国和南美也存在轮状病毒和冠状病毒感染新大陆骆驼动物。目前，不仅在四种新大陆骆驼发现轮状病毒抗体[52,53,63,68,70]，研究人员还从智利、秘鲁和阿根廷的美洲驼、骆马以及羊驼的粪便中检测到了轮状病毒[12,17,18,49,63,64,82]。在美国也有冠状病毒的报道，而南美主要以轮状病毒流行较多。目前，这些病毒是否在欧洲新大陆骆驼中流行还不清楚，需要进一步研究。

美国俄勒冈州兽医诊断实验室对未断奶的7月龄的美洲驼和羊驼的共45份粪便样品进行检测，发现冠状病毒是电子显微镜最常检出的病原，可占病例的42%，波及研究64%的动物群。此外，冠状病毒也在发病的成年新大陆骆驼动物检出。与此形成鲜明对照的是，轮状病毒的检出率只有2%（1/45）[17,18]。

冠状病毒感染导致美国西北部大多数农场中所有年龄的新大陆骆驼发生腹泻。这些疾病的暴发与外出表演的动物返回密切相关，他们两天后发生腹泻，并迅速传播给与这些动物密切接触的骆驼[17]。Crosley报道了与此类似的传播方式，即骆驼由于1型冠状病毒感染导致的急性呼吸道综合征（见下文节肢动物传播病毒和其他病毒感染）。

Genova等在表现出体重减轻、营养应激、水样腹泻等症状的4岁龄羊驼检测出2型冠状病毒[30]。通过荧光定量RT-PCR、基因组测序证实该2型冠状病毒与牛冠状病毒密切相关。研究人员还利用电镜从12日龄、9月龄的美洲驼粪便中检测到冠状病毒，但未能分离获得该病毒[54]。

Parreno等在阿根廷的两个骆马场动物中检测到轮状病毒和冠状病毒，并利用分子生物学对病毒进行了鉴定[63]。两个农场动物都发生严重腹泻，发病动物为7～40日龄幼畜，发病率可达100%，死亡率为83%。该病呈急性经过，一旦发生腹泻，动物在2～6日内死亡。研究人员从两峰新生驼粪便中分离获得两株轮状病毒，分别为RV/Arg/RioNegro/98和GRV/Arg/Chubut/99，均属于AG8RV组病毒。进化分析表明二者与美国、日本和瑞士的牛轮状病毒存在较高的亲缘关系。血清学调查结果表明，轮状病毒已经遍及两个农场，其阳性率高达95%。ELISA、粪便中病毒抗原检测结果表明两个农场中冠状病毒均为阴性。

A群轮状病毒在世界广泛分布，非常难以找到抗体阴性的动物。在新大陆骆驼也是如此，表明该病毒广泛存在于南美驼的各物种。Rivera等在阿根廷的羊驼和美洲驼中检测到了轮状病毒抗体[68]。近期的血清学调查表明，阿根廷骆马、美洲驼和驼马的轮状病毒抗体阳性率分别高达77%、98%和90%[63]。

Puntel等对包括牛轮状病毒在内的多种病毒进行了血清学调查，发现阿根廷3个省390峰美洲驼中轮状病毒的抗体阳性率为88%（342/390），但所有农场的动物均未出现腹泻症状。表明美洲驼对轮状病毒易感，而且通常发生亚临床感染。Chang-Say等也证实羊驼对轮状病毒易感[19]。

野生南美驼包括原驼和骆马，曾广泛分布于安第斯山脉区域，但20世纪后期数量骤减，驼马几度濒临灭绝，但由于近20年的保护措施，其数量呈现渐增之势。而且，一些试验站，例如国家农业技术研究所的Abra Pampa试验站，已经进行繁养并返送至当地牧民。该试验站目前驼马的数量为1400峰，其剪毛周期为半年。该试验站对导致胃肠炎的重要病原微生物例行监测，其结果可参见不同研究人员的报道[18,49,64]。Marcoppdio利用A群轮状病毒ELISA检测阿根廷安第斯高原128份野生驼马的血清样品，发现病毒阳性率为100%，但在来自同群44份粪便样品中未检出轮状病毒抗原[52]。Badaracco等首次发现并利用分子生物学方法鉴定了驼马感染的轮状病毒。他们对来源于59峰健康新生和青年小羊驼粪便样品进行ELISA用于检测轮状病毒抗原。两份样品（3%）为G8轮状病毒阳性，其中一株病毒为RVA/vicuna-wt/ARG/C75/2010/G8P，与来源于牛、羊、羚羊、骆马和人的轮状病毒具有相似性。

## 诊断

发生腹泻病畜的粪便样品可利用透射电镜鉴定轮状病毒和冠状病毒的病毒粒子，该方法是鉴定这两种病毒的“金标准”，但用于例行检测并不现实。ELISA检测结果可靠，敏感性较好，已经广泛用于新生畜病毒性腹泻的检测，并且可用于大量样品的检测。利用细胞培养进行病毒分离难度较大，成功率较低。用于检测轮状病毒和冠状病毒抗体的方法已经建立，包括中和试验、血凝抑制试验和ELISA等，但用于驼类动物腹泻诊断实用性较差。

## 轮状病毒

捕获ELISA可用来检测粪便样品中的病毒粒子。捕获ELISA通过检测轮状病毒保守抗原VP6来用于人轮状病毒的检测，但由于所有的轮状病毒具有相同的群抗原，该方法也可用于所有牲畜的轮状病毒。利用该方法在10～15分钟可获得结果，因此可用于现场检测（图174）。但是，轮状病毒致病的最直接证据是小肠的病变和利用小肠切片免疫染色检出的小肠上皮细胞中的轮状病毒。另外，还可利用RT-PCR检测轮状病毒感染。

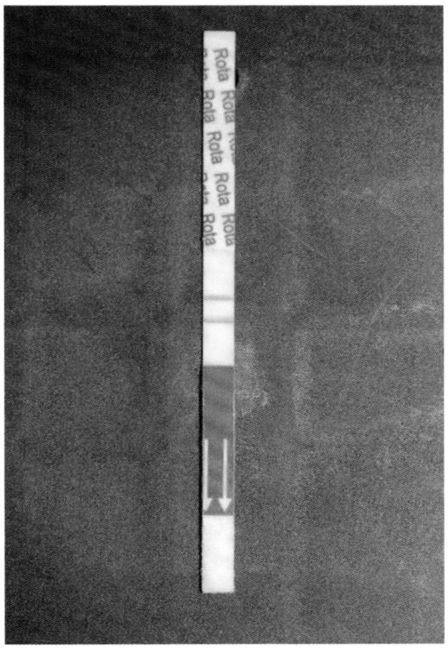


图174　轮状病毒捕获ELISA检测试剂盒，阳性样品（红色）和对照（蓝色）

## 冠状病毒

RT-PCR、捕获ELISA可用于检测粪便和呼吸道中的冠状病毒粒子。感染后3周以内，还可用RT-PCR检测粪便、呼吸道拭子的病毒粒子，该方法已经替代了病毒分离和电镜检测方法。冠状病毒可在牛肾细胞、非洲绿猴肾细胞生长，胰酶、培养维持液有利于病毒的生长。

大多驼幼畜初生并无冠状病毒感染，可以安全度过新生期，但应了解新生期的管理措施，这些相关文献见表66。

初生幼畜在前2天可出现以下情况。

（1）新生畜应在出生后30～60分钟站起，在2～4小时应进行护理。

（2）胎衣应在2～4小时排出，母畜不自食胎衣。

（3）在24～36小时应排出胎便。

免疫球蛋白与被动免疫密切相关，新生畜腹泻通常由继发免疫球蛋白缺失引起。免疫球蛋白具有多种同种型（IgG、IgM、IgA、IgE、IgD），并且还分为多个亚类。大多数研究主要基于IgG、IgM 或IgA（表67）。

所有驼类动物尤其是单峰驼在血液存在特

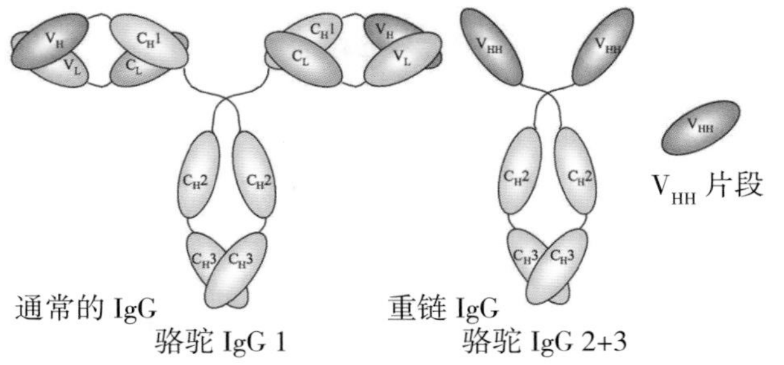


图175　骆驼IgG分子的结构

殊的抗体，这些抗体只由重链二聚体组成而缺乏轻链。从动物血液中可以纯化获得重链抗体，这些抗体与抗原性较差的抗原的反应性高于普通抗体[60]。这与抗体的多样性理论相抵触。目前，共鉴定了3个亚类的驼科动物抗体（IgG1、IgG2、IgG3），其中IgG2和IgG3缺乏轻链（图175）[7,20,37]：

（1）IgG1组成常规抗体，占驼类IgG的25%，与蛋白A和蛋白G反应较强；

（2）IgG2和IgG3由短的重链组成，只有Fc区而无CH1区，占血清抗体的75%。

Ghahroudi等发现，美洲驼具有IgG1a和IgG1b常规抗体以及由IgG2a、IgG2b和IgG33个重链组成的抗体。另外，Woolven等报道IgG2c这一第四类抗体的存在[84]。

在单峰驼至少存在5类IgG同种型，即IgG1a、IgG1b（常规抗体）以及IgG2a、IgG2c和IgG3（重链抗体，对应美洲驼抗体同种型）[61]。Van der Linden等研究了细胞裂解物、氟苯尼考与载体蛋白偶联物等不同抗原与驼类动物不同抗体同种型之间的关系[75]。

研究表明，驼科动物血清蛋白中75%以上为缺乏重链的IgG分子[37]。IgG2、IgG3只有重链，分子量约为100 kD。这些抗体和其抗原结合区由于分子较小，在组织分布、组织渗透方面都优于常规抗体，所以被称为纳米抗体，而且其第三互补决定区形成的环可以深入酶活性中心，可以充分中和酶活性[39,48,59,69]（图176）。

总体来看，驼科动物抗体是一类特殊的抗体，在中和酶活性、分子大小、稳定性方面都优于常规抗体[61]。Van der Linden等证实，美洲驼特异性抗体的抗原结合区对温度的耐受性很高，具有很

表66　驼类新生畜管理措施的相关文献

| 作者 | 年份 | 物种 |
|---|---|---|
| Ryan | 1999 | 羊驼 |
| Bravo | 2002 | 南美驼 |
| Millard | 2004 | 羊驼 |
| Mueller | 2004 | 新大陆骆驼 |
| Whitehead | 2006 | 南美驼 |
| Gerspach | 2006 | 美洲驼、羊驼 |
| Gerspach等 | 2006 | 美洲驼、羊驼 |
| Whitehead | 2006 | 南美驼 |
| Gerspach | 2008 | 南美驼 |

表67　不同家畜的IgG、IgM和IgA亚类

| 物种 | IgG亚类 | IgM | IgA | 来源 |
|---|---|---|---|---|
| 马 | Ga、Gb、Gc、G(B)、G(T) | M | A | |
| 牛 | G1、G2 (G2a、G2b) | M | A | Tizard [73] |
| 绵羊 | G1(G1a)、G2、G3 | M | A1、A2 | |
| 猪 | G1、G2、G3、G4 | M | A1、A2 | |
| 羊驼 | G | M | —[a] | Gramendia [28] |
| 美洲驼 | G1a、G1b（常见） | | | Ghahroudi [34] |
| | G2a、G2b、G3（重链） | | | Woolwen等[84] |
| | G1a、G1b | | | |
| | G2a、G2b、G2c、G3 | | | |
| 骆驼 | G1、G2 | M | A | Grover等[36] |
| | 3个亚类 | M | —[a] | Azwai等[8]，carter[16] |
| | G+相关蛋白 | —[a] | —[a] | Ungar-Waron等[74] |
| | G1、G2[b]、G3[b] | —[a] | —[a] | Hamers-Casterman等[37] |
| | G1a、G1b（常见）、G2a、G2b、G3（重链） | | | Nguyen等[61] |

注：a 未确定；b 重链抗体；lg 免疫球蛋白

好的稳定性，其中2个抗原结合区在温度高达90℃仍可结合抗原[76]。

从母源抗体获得被动免疫是新生畜生存的重要机制。免疫球蛋白，尤其是IgG可在出生后通过初乳从母体获得。血清中和抗体衰减率较快，可使幼畜获得保护。IgG的半衰期为9～21天，IgM为3～5天。不能从母源抗体获得被动免疫是动物免疫缺陷的重要成因之一，多种出生后感染与之密切相关。母源抗体由循环血液转移至初乳、经新生畜肠道最终进入循环系统是一复杂的过程，可在多个环节被破坏。多种因素都可导致被动免疫的失败，包括：

（1）初乳缺乏；
（2）初生重；
（3）发育不良；
（4）初次分娩；
（5）难产；
（6）母畜营养不良等。

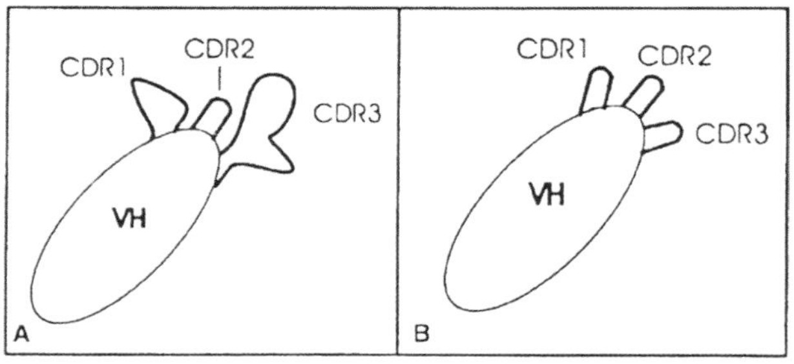


图176　骆驼IgG重链可变区（A）和常规IgG的重链可变区（B）分子结构

CDR：互补决定区；VH：重链可变区

驼科动物具有较厚的上皮绒毛膜胎盘，限制了IgG经胎盘转移。因此幼畜需要经肠道吸收初乳中的IgG获得被动免疫。图177表明，单峰驼幼畜在获得初乳前几乎没有血清中和抗体（无丙种球蛋白血症）。

虽然新生驼在出生时具有针对外源微生物的免疫潜质，但此时还未获得自身的免疫力，需要获得被动体液免疫。初生驼被动免疫获得失败增加了腹泻、肠炎、败血病、关节炎、脐炎和肺炎的风险。被动免疫成功获得的标志是新生畜在48小时龄的血清IgG的水平达90 g/L[10,55,82]。研究表明，驼科动物初乳中的IgG浓度约为220g/L[14]。10kg的幼畜需要吸收20g IgG才足以保证其血清IgG达90 g/L，需要获得初乳的量为100mL。

对68峰幼驼的研究表明，在采食初乳以前，其IgG水平较低[（0.26 ± 0.23）g/L][40]，这与其他家畜类似。驼类动物出生时为无丙种球蛋白血症，初乳为其球蛋白的主要来源。因此，使初生畜及早获得初乳是非常重要的。美洲驼在妊娠后期乳腺中的IgG浓度可达血清IgG浓度10倍以上[15]。分娩前

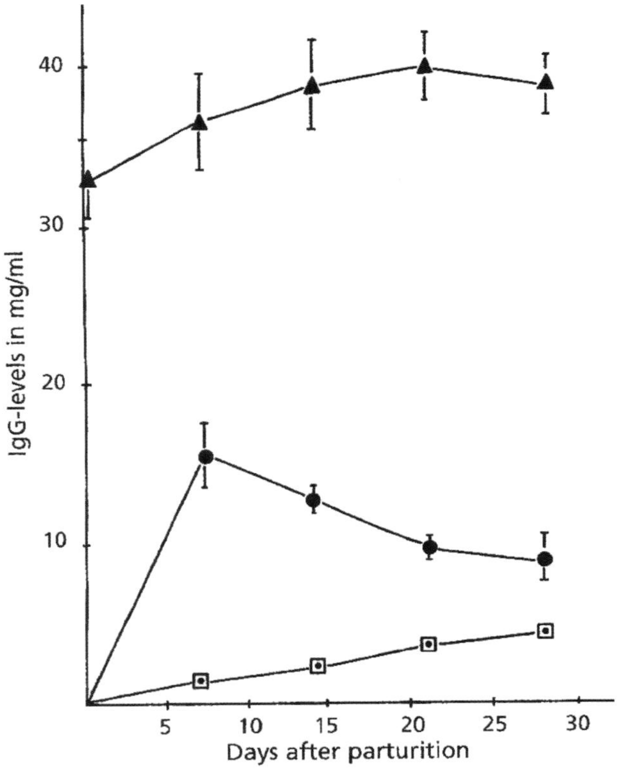


图177　单峰驼母畜（实心三角形）、初乳喂养的单峰驼幼畜（实心园）、未经初乳喂养的单峰驼幼畜（带中心点的方框）的血清免疫球蛋白值（图来源于Uugor-Waron［74］）

初乳中的IgG浓度约为240g/L，之后1～2天降低至一半（120g/L），至第6天时初乳中不再含有IgG。驼类在出生后24小时内应获得其体重10%～12%的初乳，以12小时内获得最好[57]。

虽然初生畜在出生时具有免疫潜质，但其在出生后1个月内，内源抗体的产生量不足以获得保护性免疫。幼畜出生后18～30小时内其IgG浓度最大可达（21.1 ± 11.7）g/L，而母畜分娩当日血清IgG浓度为（23.9 ± 7.5）g/L。驼类动物初乳IgG浓

表68　驼科动物初乳IgG浓度[40]

| 物种 | IgG（g/L） | 作者 |
| --- | --- | --- |
| 羊驼 | 10～280 | Garmendia等（1987） |
| 骆驼 | 70～220 | Ungar –Waron等（1987） |
| | 58.6 ± 15.4 | Kamber（1996）、Kamber等（1996） |
| | 25.56～84.15（lg1） | EI-Agmy（1998） |
| | 1.81～6.02（lg2） | EI-Agmy（1998） |
| | 15.8 ± 1.2 | Sena等（2006） |

度见表68。

与其他新生畜类似，驼类动物在初生24小时后对免疫球蛋白的吸收为线性衰减。牛免疫球蛋白的吸收与活化的IgG1特异受体有关，由于驼类初乳中同样存在IgG（IgG与IgM为7：1），人们推测，其IgG选择性吸收机制与牛类似。

新生畜体内IgG达到高峰之后会在两周之内迅速降低。由于在两周之后幼畜自身才可产生抗体，在出生后1～2个月体内的抗体才可达10g/L。在出生后2～5星期是幼畜感染的危险期。在出生后4个月，血清IgG可达到较高的水平，表明其免疫系统已经成熟。

Ungar-Waron等[74]和Hannant等[38]最早进行了单峰驼IgG的研究。Fowler等研究了新大陆骆驼血清蛋白[26]，发现如果血清总蛋白低于50g/L表明动物获得被动免疫失败；在50～60g/L时可疑；而只有血清总蛋白达60g/L以上时才说明被动免疫成功获得。新大陆骆驼和旧大陆骆驼的幼驼出生后3～5周血清球蛋白达到最低水平。Wernery等证实IgG在新生驼中的关键作用[79]。

对新生驼免疫力的评估是至关重要的，有利于IgG的合理用药。目前一些方法已经用于IgG的测定，包括硫酸锌浊度试验、硫酸钠沉淀法（一些商业化的方法包括Llama-STM、VMRD Inc、Pullman、WA等），但这些方法都不能特异地对血清IgG浓度进行测定。单向放射免疫扩散法是可以特异测定血清IgG的唯一方法（图178）。新大陆骆驼IgG测定的商品化试剂盒有Llama IgG测定试剂盒和Llama Vet-RID。Hutchision利用528份美洲驼血液样品对两种方法进行了比较，发现二者存在较大差异[42]。

Bourke[13]、Wernery[79]、Millard[55]、Whitehead[81]Wernery[80]比较了硫酸锌测定法、总蛋白测定法、免疫球蛋白测定法，发现这三种方法均可以用来评估驼类新生畜的IgG状态（表69）。

除了上述3种方法，γ-谷氨酰转移酶测定也可用于评估被动免疫[55,79]。当血清总蛋白小于50g/L，IgG小于8～10g/L，球蛋白小于2.0g/L，硫酸锌测定小于30，γ-谷氨酰转移酶小于45IU/L，说明被动免疫获得失败。

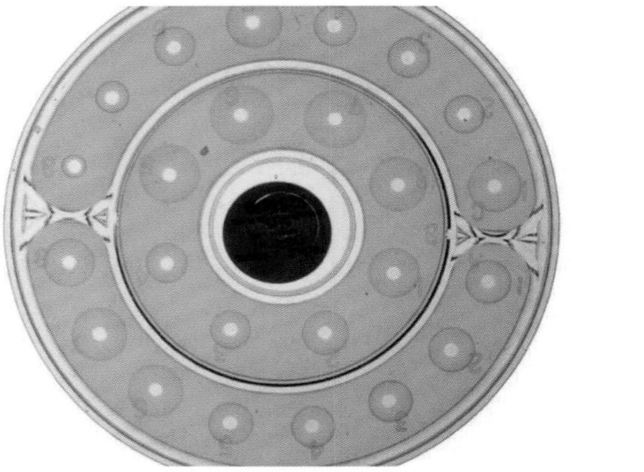

图178　驼类动物IgG单向放射免疫扩散检测法。通过沉淀区沉淀环直径判定IgG水平。通过标准曲线判定IgG浓度。较小的沉淀环表明血清中的IgG水平较低[77]

最近，用于检测驼类血清IgG的ELISA方法已建立，成为鉴定被动免疫失败的有效工具[25,40]。该方法利用96孔间接ELISA模式进行检测，所用抗驼IgG为利用骆驼IgG免疫母鸡之后从卵黄提取获得[62]。

初乳的获得对新生畜的生存至关重要。被动免疫获得失败是败血症发生的重要原因，也与新生畜早期肠道和呼吸道疾病的严重程度、死亡率密切相关。多种因素可以影响新生畜的血清IgG水平。从初乳中获得IgG的水平主要取决于初乳中的IgG总量以及新生畜的吸收效率这两个关键因素。

关于新生驼IgG缺乏的相关报道较少。个别文献表明，被动免疫获得失败是新生羊驼死亡的主要原因[11,28,29,41,46,58]，但是目前关于旧大陆骆驼被动免疫获得失败没有报道。Wernery等报道，在阿拉伯联合酋长国单峰驼由于继发性IgG缺乏症而发生败血症死亡[78]。这种综合征是由于铜缺乏导致。患病幼畜并不吃奶，而是食用大量沙土以弥补铜缺乏。

表69　硫酸锌浊度、总蛋白实验、球蛋白测定法测定的驼类免疫球蛋白状态

| IgG转移状态 | 硫酸锌浊度 | 总蛋白实验 | 球蛋白测定法 |
|---|---|---|---|
| 无或低 | <30 | <50 | <2.5 |
| 中等 | 30～40 | 50～55 | 3～12 |
| 足量 | >40 | >55 | >12 |

美国华盛顿州立大学报道了在青年美洲驼发生的免疫缺陷综合征。由于遗传因素导致了该动物B淋巴细胞发生缺陷。研究人员因此建立了流式细胞术用于检测新生畜的免疫缺陷综合征[22]。研究表明，羊驼也可发生免疫缺陷综合征。发生免疫缺陷综合征时，B淋巴细胞数量降低至1%～5%，其原因为B淋巴细胞发生过程中存在染色体隐性遗传缺陷。

## 防控

骆驼发生腹泻时通常由于脱水导致死亡，因此对发病驼的治疗应首先进行补液和纠正电解质失衡。可以根据脱水的严重程度，选择经口或其他非肠道途径进行补液。抗生素用于防止细菌的继发感染。及时隔离发生腹泻的动物。

免疫球蛋白的成功转移可以使幼畜的感染率降低，成活率提高[50]。治疗性施以IgG可以提供针对多种病原的抵抗力，在兽医临床上具有较广泛的应用。另一常用的方法是进行初乳的储备，在动物发生初生后腹泻时可投喂含有10%初乳的奶。这种方法可以提供保护能力，可抵御2～3周风险。如果无骆驼初乳，可给美洲驼喂以20%体重的羊初乳[66,67]，也可选择牛初乳。新鲜初乳或冻存初乳均可用于投喂。初乳在冰箱中可冷冻至一年还可使用。对新生畜应在出生当日每隔2小时投喂一次，用管喂奶较好。对被动免疫获得失败的羊驼和单峰驼，可分别给予每5kg体重200mL和每40kg体重1.61mL的血浆。血浆可以采自同一养殖场的3岁龄以上的成年动物。年龄较大的动物含有较高的IgG水平[71]。驼类动物血型一致，因此不会导致输血反应[57]。输血前应将血浆加温至37℃，输入时间为30～60分钟。美洲驼的高免血清可从西雅图雷德蒙德的Triple J农场获得。

由于病原复杂多样、病程较快，所以新生畜病毒性腹泻很难防控。母体免疫是防控该病的另一途径。在产犊前1～3个月对母体进行疫苗接种可以降低轮状病毒和冠状病毒的发生率。在母畜妊娠最后3个月接种灭活冠状病毒、轮状病毒和大肠杆菌F5（以前称K99）联苗，可以激发较高的抗体效价，从而使初生幼畜得到保护。在发生新生畜腹泻的疫区，应开展疫苗防控计划。可抵抗病毒的多种单克隆抗体疫苗可用于幼畜和羔羊的口服免疫，这些疫苗也可安全用于幼驼。由于幼驼的免疫系统并不成熟，不需直接对其进行疫苗接种。



## 参考文献


[1] Agab H. & Abbas B. (1998). – Epidemiological studies on camel diseases in eastern Sudan: Ⅱ. Incidence and causes ofmorality in pastoral camels. *Camel Newslett.*, 14 (4), 53–58.

[2] Ali Y.H. (2003). – Camel calf diarrhoea with emphasis on rotavirus infection. PhD Thesis, University of Khartoum, Khartoum,Sudan.

[3] Ali Y.H., Khalafalla A.I. & El Amin M.A. (2005a). – Epidemiology of camel calf diarrhoea in Sudan: seroprevalence of camelrotavirus infection. *JAVMA*, 4 (3), 393–397.

[4] Ali Y.H., Khalafalla A.I. & El Amin M.A. (2006). – Epidemiology of camel calf diarrhoea in Sudan: seroprevalence of camelrotavirus infection. *In* Proceedings of the International Science Conference on Camels, 12–14 May 2006, Buraydah, SaudiArabia, 397–410.

[5] Ali Y.H., Khalafalla A.I., Gaffar M.E., Peenze I. & Steele A.D. (2005b). – Rotavirus-associated camel calf diarrhoea in Sudan. *JAVMA*, 4 (3), 401–406.

[6] Ali Y.H., Khalafalla A.I., Gaffar M.E., Peenze I. & Steele A.D. (2008). – Detection and isolation of group A rotavirus fromcamel calves in Sudan. *Veterinarski Archiv.*, 78 (6), 477–485.

[7] Azwai S.M. & Carter S.D. (1995). – Monoclonal-antibodies against camel (Camelus dromedarius) IgG, IgM and light chains. *Vet. Immunol. Immunopath.*, 45 (1–2), 175–184.

[8] Azwai S.M., Carter S.D. & Woldehiwet Z. (1993). – The isolation and characterization of camel (*Camelus dromedarius*)

immunoglobulin classes and subclasses. *J. Comp. Path.*, 109, 187–195.

[9] Badaracco A., Matthijnssens J., Romero S., Heylen E., Zeller M., Garaicoechea L., Van Ranst M. & Parreno V. (2013). –Discovery and molecular characterization of a group A rotavirus strain detected in an Argentinean vicuna (*Vicugna vicugna*). *Vet. Microbiol.*, 161, 247–254.

[10] Barrington G.M., Parish S.M. & Garry F.B. (1999). – Immunodeficiency in South American camelids. *J. Camel Pract. Res.*,6 (2), 185–190.

[11] Barrington G.M., Parish S.M., Tyler J.W., Pugh D.G. & Anderson D.E. (1997). – Chronic weight loss in an immunodeficient adult llama. *J. Am. Vet. Med. Assoc.*, 211 (3), 295–298.

[12] Berrios P.E. (1988). – Rotavirus en pequenos ruminantes del zoologicos nacional de Santiago. *Monografias de MedicinaVeterinaria*, 10.

[13] Bourke D.A. (1996). – Determination of passive immune status in llama neonates. *In* Proceedings of the 3rd Meeting of the British Veterinary Camelid Society, 14–16 November 1996, Burford, UK, 39–45.

[14] Bravo P.W., Garnica J. & Fowler. M.E. (1997). – Immunoglobulin G concentrations in periparturient llamas, alpacas and their crias. *Small Ruminant Res.*, 26, 145–149.

[15] Bravo W. (2002). – The reproductive process of South American camelids. Seagull Printing, Salt Lake City, UT, 36–100.

[16] Carter S.D. & Azwai S.M. (1996). – Immunity and infectious diseases in the dromedary camel. *In* Proceedings of the 3rdMeeting of the British Veterinary Camelid Society, 14–16 November 1996, Burford, UK, 1996, 23–36.

[17] Cebra C. (2004). – Diarrhoea in llamas and alpacas. In Proceedings of the 11th Meeting of the British Veterinary Camelid Society, 12–14 November 2004, Shaftesbury, UK, 12–18.

[18] Cebra C.K., Mattson D.E., Baker R.J., Sonn R.J. & Dearing P.L. (2003). – Potential pathogens in feces from unweaned llamasand alpacas with diarrhea. *J. Am. Vet. Med. Assoc.*, 223 (12), 1806–1808.

[19] Chang-Say F., Rivera H. & Samame H. (1985). – Reporte preliminar sobre prevalencia de virus influenza tipo A y rotavirus en alpacas. *Res Convencion Int. Camelidos Sudamericanos*, Cuzco, Peru, 37.

[20] Conrath K.E., Wernery U., Muyldermans S. & Nguyen V.K. (2003). – Emergence and evolution of functional heavy-chain antibodies in Camelidae. *Dev. Comp. Immunol.*, 27, 87–103.

[21] Crosley B., Mock R. & Barr B. (2008). – Diagnostic investigation of acute respiratory syndrome in alpacas. *In* Proceedings of the American Association of Veterinary Laboratory Diagnosticians 51st Annual Conference, 22–27 October 2008,Greensboro, NC, 71.

[22] Davis W.C., Heirman L.R., Hamilton M.J., Parish S.M., Barrington G.M., Loftis A. & Rogers M. (2000). – Flow cytometric analysis of an immunodeficiency disorder affecting juvenile llamas. *Vet. Immunol. Immunopathol.*, 74, 103–120.

[23] El-Agamy E.I. (1998). – Camel's colostrum: antimicrobial factors. *In* Dromadaires et chameaux, animaux laitiers (P. Bonnet,ed.). Actes du colloque, 24–26 October 1994, Noaukchott, Mauritania, 177–179.

[24] El-Sayed E.I., Ruppaner R., Ismail A., Champagene C.D. & Assaf R. (1992). – Antibacterial and antiviral activity of camel milk protective proteins. *J. Dairy Res.*, 59, 169–175.

[25] Erhard M.H., Kouider S.A., Dabbag M.N., Schickel F. & Stangassinger M. (1999). – Determination of serum IgG levels in camels by a bovine specific sandwich ELISA. *J. Camel Prac. Res.*, 6 (1), 15–18.

[26] Fowler M.E. (2010). – Medicine and surgery of Camelids. 3rd Ed. Blackwell Publishing, Oxford, UK, 514–518.

[27] Freitag H., Wetzel H. & Espenkötter E. (1984). – Aus der Praxis. Zur Prophylaxe der Rota-Corona-Virus-bedingten Kälberdiarrhoe. *Tierärztl. Umschau,* 39 (10), 731–736.

[28] Garmendia A.E. & McGuire T.C. (1987). – Mechanism and isotypes involved in passive immunoglobulin transfer to newborn alpaca (*Lama pacos*). *Am. J. Vet. Res.*, 48, 1465–1471.

[29] Garmendia A.E. Palmer G.H., DeMartini J.C. & McGuire T.C. (1987). – Failure of passive immunoglobulin transfer: a major determinant of mortality in newborn alpacas (*Lama pacos*). *Am. J. Vet. Res.*, 48, 1472–1476.

[30] Genova S.G., Streeter R.N., Simpson K.N. & Kapil S. (2008). – Detection of an antigenic group 2 coronavirus in an adult alpaca with enteritis. *Clin. Vaccine Immunol.*, 15, 1629–1632.

[31] Gerspach C. (2006). – The premature cria. *In* Proceedings of the International Camelid Health Conference for

Veterinarians,21 – 25 March 2006, Corvallis, OR, 90 – 92.

[32] Gerspach C. (2008). – How much plasma does my cria need? Comparative aspects of passive transfer. In Proceedings of the 11th International Camelid Conference Vet., 18 March 2008, Corvallis, OR, 66 – 76.

[33] Gerspach C., Anderson D.E., Saville W.J.A. & Whitehead C.E. (2006). – Premature/dysmature syndrome in crias: aretrospective study of 42 cases (1999-2005). In Proceedings of the 1st Conference of the International Society of Camelids Research and Development (ISOCARD), 15 – 17 April 2006, Al-Ain, UAE.

[34] Ghahroudi K.B., Desmyter A., Wyns L., Hamers R. & Muyldermans S. (1997). – Comparison of llama VH sequences fromconventional and heavy-chain antibodies. *EMBO J.*, 17 (13), 3512 – 3520.

[35] Gluecks I.V. (2007). – Prevalence of bacterial and protozoal intestinal pathogens in suckling camel calves in northern Kenya. Thesis, Free University of Berlin, Berlin, Germany.

[36] Grover Y.P., Kaura Y.K., Prasad S. & Srivastava S.N. (1983). – Preliminary studies on camel serum immunoglobulins. *Ind.J. Biochem. Biophys.*, 20, 238 – 240.

[37] Hamers-Casterman C., Atarhouch T., Muyldermans S., Robinson G., Hamers C., BajyanaSonga E., Bendahman N. &Hamers R. (1993). – Naturally occurring antibodies devoid of light chains. Nature, 363, 446 – 448.

[38] Hannant D., Mumford J.A., Wernery U. & Bowen J.M. (1992). – ELISA for camel IgG and measurement of colostral transfer. Proceedings of the 1st International Camel Conference. R. and W. Publications (Newmarket) Ltd, Newmarket, UK, 93 – 95.

[39] Hoelzer W., Muyldermans S. & Wernery U. (1998). – A note on camel IgG antibodies. *J. Camel Pract. Res.*, 5 (2), 187 – 188.

[40] Huelsebusch C. (1999). – Immunoglobulin G status of camels during 6 months post natum. Hohenheim Tropical Agricultural Series, Magraf Verlag, Hohenheim, Germany.

[41] Hutchison J.M., Garry F.B., Belknap E.B., Geky D.M., Johnson L.M., Ellis R.P., Quackenbusch S.L., Ravnak J., Hoover E.A. & Cockerell G.L. (1995b). – Prospective characterisation of the clinicopathologic and immunologic features of an immunodeficiency syndrome affecting juvenile llamas. *Vet. Immunol. Immunopathol.*, 49, 209 – 227.

[42] Hutchison J.M., Salman M.D., Garry F.B., Johnson L.W., Collins J.K. & Keefe T.Y. (1995a). – Comparison of two commercially available single radial immunodiffusion kits for quantitation of llama immunoglobulin G. *J. Vet. Diagn. Invest.*, 7, 515 – 519.

[43] Jin L., Cebra C.K., Baker R.J., Mattson D.E., Cohen S.A., Alvarado D.E. & Rohrmann G.F. (2007). – Analysis of the genomesequence of an alpaca coronavirus. *Virology*, 365 (1), 198 – 203.

[44] Kamber R. (1996). – Untersuchungen uber die Versorgung von neugeborenen Kamelfohlen (*Camelus dromedarius*) mitImmunglobulin-G. Thesis, Zurich University, Zurich, Switzerland.

[45] Kamber R., Farah Z., Rusch P. & Hassig M. (2001). – Studies on the supply of immunoglobulin G to newborn camel calves(*Camelus dromedaries*). *J. Dairy Res.*, 68, 1 – 7.

[46] Kennel A.J. & Wilkens J. (1992). – Causes of mortality in farmed llamas in North America. Necropsy findings from thedatabase of a major insurer. International Llama Association and Research Committee, Rochester, MN.

[47] Khanna N.D., Tandon S.N. & Sahani M.S. (1992). – Calf mortality in Indian camels. *In* Proceedings of the 1st International Camel Conference. R. and W. Publications (Newmarket) Ltd, Newmarket, UK, 89 – 92.

[48] Lauwereys M., Ghahroudi M.A., Desmyter A., Kinne J., Hoelzer W., de Genst E., Wyns L. & Muyldermans S. (1998). – Potent enzyme inhibitors derived from dromedary heavy-chain antibodies. *EMBO J.*, 17 (13), 3 512 – 3 520.

[49] Lopez W., Chamorro M., Antonio E. & Garmendia B. (2011). – Deteccion rapida de rotavirus y coronavirus en crias de alpaca (*Vicugna pacos*) con diarrea en la region del Cusco, Peru. *Rev. Inv. Vet. Peru*, 22, 407 – 411.

[50] McGuire T.C., Pfeiffer N.E. & Weikel J.M. (1976). – Failure of colostral immunoglobulin transfer in calves dying from infectious diseases. *J. Am. Vet. Med. Assoc.*, 169, 713 – 718.

[51] Mahin L., Schwers A., Chadli M., Maenhoudt M. & Pastoret P.-P. (1983). – Receptivite du dromadaire (*Camelus dromedarius*)à l'infection par rotavirus. *Rev. Elev. Méd. Vét. Pays Trop.*, 36 (3), 251 – 252.

[52] Marcoppido G., Parreno V. & Vila B. (2010). – Antibodies to pathogenic livestock viruses in a wild vicuna (*Vicugna vicugna*) population in the Argentinean Andean altiplano. *J. Wildlif. Dis.*, 46, 608 – 614.

[53] Marin R., Rodriguez D. & Parreno V. (2009). – Prevalenc ia Sanitaria en Llamas (Lama glama) de la Provincia de

Jujuy,Argentina. Revista Veterinaria Argentina. Direccion Provincia de Desarrollo Ganadero. Gobierno de la Provincia de Jujuy.Ministerio de Produccion. Ley Ovina N 25.422. Proyecto FAO N 2552/07.

[54] Mattson D.E. (1994). – Update on llama medicine. Viral diseases. *Vet. Clin. North Am. Food Anim. Pract.*, 10 (2), 345–351.

[55] Millard C. (2004). – Failure of passive transfer in neonatal alpacas – a comparison of diagnostic tests. *In* Proceedings of the 11th Meeting of the British Veterinary Camel Society, 12–14 November 2004, Shaftesbury, UK, 36–39.

[56] Mohamed M.E.H., Hart C.A. & Kaaden O.-R. (1998). – Agents associated with camel diarrhea in eastern Sudan. *In* Proceedingsof the International Meeting on Camel Production and Future Perspectives, 2–3 May 1998, Al Ain, UAE.

[57] Mueller K. (2004). – Passive transfer in camelids. Proceedings of the 11th Meeting of the British Veterinary Camel Society, 12–14 November 2004, Shaftesbury, UK, 39–47.

[58] Murphy P.J. (1998). – Obstetrics, neonatal care and congenital conditions. *Vet. Clin. North Am. Food Anim. Pract.*, 5, 1983–202.

[59] Muyldermans S., Atarhouch T., Saldanha J., Barbosa J.A.R.G. & Hamers R. (1994). – Sequence and structure of VH domainfrom naturally occurring camel heavy immunoglobulins lacking light chains. *Protein Eng.*,7, 1129–1131.

[60] Muyldermans S., Saerens D., Stijlemans B., Vincke C., Pellis M., Kinne J., Wernery U., De Genst E., Pardon E., Conrath K.& Revets H. (2006). – Biotechnological opportunities of dromedary antibodies. *In* Proceedings of the International ScientificConference on Camels, 12–14 May 2006, Buraydah, Saudi Arabia, 509–517.

[61] Nguyen V.K., Hamers R., Wyns L. & Muyldermans S. (2000). – Camel heavy-chain antibodies: diverse germline VHH andspecific mechanisms enlarge the antigen binding repertoire. *EMBO J.*, 19 (5), 921–930.

[62] Nikbakht Brujeni Gh.,Tabatabaei., S., Khormali M. & Ashrafi I. (2009). – Characterization of IgG antibodies, developed inhens, directed against camel immunoglobulins. *Int. J. Vet. Res.*, 3 (1), 37–41.

[63] Parreno V., Bok K., Fernandez F. & Gomez J. (2004). – Molecular characterization of the first isolation of rotavirus in guanacos(*Lama guanicoe*). *Arch Virol.*, 149, 2465–2471.

[64] Parreno V., Constantini V., Cheetham S., Blanco V.J., Saif L.J., Fernandez F., Leoni L. & Schudel A. (2001). – First isolationof rotavirus associated in the neonatal diarrhea in guanacos (*Lama guanicoe*) in the Argentinean Patagonia region. *J. Vet. Med.B Infect. Dis. Vet. Public Health*, 48 (9), 713–720.

[65] Parreno V. & Marcoppido G. (2006). – Estudio de la sanidad en camelidos: Avances a partir de la obtencion de muestrasde camelidos silvestres. In Investigacion, conservacion y manejo de vicunas (B. Vila, ed.). Proyecto MACS – Universidadde Lujan, Argentina, 147–164.

[66] Pugh G.H. (1992). – Prepartum care of the pregnant llama and neonatal care of the cria. Proceedings of the Annual Meeting of the Society for Theriogenology, Hastings, NE, 198–201.

[67] Pugh G.H. & Belknap E.B. (1997). – Perinatal and neonatal care of South American camelids. *Vet. Med.* 92 (3), 291–295.

[68] Puntel M., Fondevila N.A., Viera B.J., O'Donnell V.K, Marcovechio J.F., Carillo B.J. & Schudel A.A. (1999). – Serologicalsurvey of viral antibodies in llamas (*Lama glama*) in Argentina. *J. Vet. Med. B.*, 46, 157–161.

[69] Riechmann L. & Muyldermans S. (1999). – Single domain antibodies: comparison of camel VH and camelised human VH domains. *J. Immunol. Methods*, 231 (1–2), 25–38.

[70] R ivera H., Madwell B.R. & Ameghina E. (1987). – Serological survey of viral antibodies in the Peruvian alpaca (*Llama pacos*).*Am. J. Vet. Res.*, 48, 189–191.

[71] Ryan D. (1999). – Neonatal care in alpacas. *In* Proceedings of the 6th Meeting of the British Veterinary Camelid Society,1–3 October 1999, Penrith, UK, 2–7.

[72] Sena D.S., Tuteja F.C., Dixit S.K. & Sahani M.S. (2006). – Immunoglobulin in neonatal camel calves and their clams. Proceedings of the International Scientific Conference on Camels, 12–14 May 2006, Buraydah, Saudi Arabia, 518–525.

[73] Tizard I. (1992). – Veterinary immunology – an introduction. W.B. Saunders Co., Philadelphia, PA, 498.

[74] Ungar-Waron H., Elias E., Gluckman A. & Trainin Z. (1987). – Dromedary IgG: purification, characterization and quantitationin sera of dams and newborns. *Isr. J. Vet. Med.*, 43 (3) 198–203.

[75] van der Linden R.H.J., de Geus B., Stok W., Bos W., van Wassenaar D., Verrips T. & Frenken I. (2000). – Induction of immuneresponses and molecular cloning of the heavy chain antibody repertoire of *Lama glama*. *J. Immunol. Methods*, 240,

185–195.

[76] van der Linden R.H.J., Frenken L.G.J., de Geus B., Harmsen M.M., Ruuls R.C., Stok W., de Ron L., Wilson S., Davis P. &Verrips C.T. (1999). – Comparison of physical chemical properties of llama V-HH antibody fragments and mouse monoclonal antibodies. *Biochem. Biophys. Acta*, 1431 (1), 37–46.

[77] Wernery U. (2001). – Camelid immunoglobulins and their importance for the new-born – a review. *J. Vet. Med. B.*, 48, 561–568.

[78] Wernery U., Ali M., Kinne J., Abraham A.A. & Wernery R. (2001b). – Copper deficiency: a predisposing factor to septicemia in dromedary calves. *In* Proceedings of the 2nd Camelid Conference, Agroeconomics of Camelid Farming, 8–12 September2000, Almaty, Kazakhstan, 86.

[79] Wernery U., Anand P. & Kinne J. (2001a). – Camelid immunoglobulins and their importance for the newborn camel. *In* Sixth Annual Conference for Animal Production under Arid Conditions, 11–13 November 2001, Al Ain, UAE.

[80] Wernery U. & Wernery R. (2006). – Camel milk – the white gold of the desert. Momentum Brochure, Zuded, Dubai, UAE.

[81] Whitehead C. (2006). – Camelid neonatology. *In* Proceedings of the British Veterinary Camelid Society, 21–25 March 2006,Rugby, UK, 1–8.

[82] Whitehead C.E. & Anderson D.E. (2006). – Neonatal diarrhea in llamas and alpacas. *Small Ruminant Res.*, 61, 207–215.

[83] Wünschmann A., Frank R., Pomeroy K. & Kapil S. (2002). – Enteric coronavirus infection in a juvenile dromedary (*Camelus dromedarius*). *J. Vet Dign. Invest.*, 14, 441–444.

[84] Woolven B.P., Frenken L., van der Logt P. & Nicholls P.J. (1999). – The structure of llama heavy-chain constant genes reveals a mechanism for heavy-chain antibody formation. *Immunogenetics*, 50, 98–101.

## 深入阅读材料

*Berrada J., Bengoumi M. & Hidane K.(1999). – Diarrhees neonatales du chamelon dans les provinces Sahariennes du sud du Maroc:Etude bacteriologique. 1999. In International Workshop on the Young Camel, 24–26 October 1999, Ouarzazate, Morocco, 53.*

*Broadbent R. (1994). – The neonate. In Proceedings of the British Veterinary Camelid Society, Bristol, UK, 50–64.*

*Dioli M. (1999). – Diseases and pathological conditions of camel calf in eastern Africa. In International Workshop on the YoungCamel, 24–26 October 1999, Ouarzazate, Morocco, 43.*

*Gandega B.E., Bengoumi M., Fikri A., El Abrak A., Aissa M. & Faye B. (1999). –In International Workshop on the Young Camel,24–26 October 1999, Ouarzazate, Morocco, 32.*

*Kataria A.K. (1999). – Molecular characterisation and immunobiology of camel serum and lacteal immunoglobulin classes and subclasses and their role in immunity. Thesis, Rajastan University of Veterinary and Animal Sciences, Bikaner, India.*

*Kaufmann B. (1999). – Camel calf losses in pastoral herds of northern Kenya – a systems comparison. In International Workshop on the Young Camel, 24–26 October 1999, Ouarzazate, Morocco, 33.*

*Leipold H.W., Hiraga T. & Johnson L.W. (1994). – Congenital defects in the llama. Vet. Clin. North Am. Food Anim. Pract., 10 (2),401–402.*

*Nagpal G.K. & Purohit G.N. (1999). – Disease prevalence and calf mortality in camel rearing areas in Bikaner. In International Workshop on the Young Camel, 24–26 October 1999, Ouarzazate, Morocco, 34.*

*Sheriff S. & Constantine K.L. (1996). – Redefining the minimal antigen-binding fragment. Nat. Struct. Biol., 3(9), 733–736.*

*Spinelli S., Frenken L., Bourgeois D., de Ron L., Bos W., Verrips T., Anguille C., Cambillau C. & Tegoni M. (1996). – The crystalstructure of a llama heavy chain variable domain. Nat. Struct Biol., 3 (9), 752–757.*

*Tandon S.N. & Sahani M.S. (1999). – Morbidity and mortality rate in dromedary calves. In International Workshop on the YoungCamel, 24–26 October 1999, Ouarzazate, Morocco, 35.*

*Wernery U. (1999). – New aspects on infectious diseases of camels. J. Camel Pract. Res., 6 (1), 87–91.*

(高闪电译，殷宏校)

## 2.1.14 疱疹病毒

### 马疱疹病毒

比较有趣的是，目前并未发现专门感染驼科动物的特异的疱疹病毒。1型马疱疹病毒（EHV-1）可导致马鼻肺炎和流产，这一疾病也在旧大陆骆驼和新大陆骆驼有相关报道[3,14,27,32,33]。

### 病原学

目前在马体已经分离到至少有9个型的马疱疹病毒，其中最重要的为EHV-1和EHV-4。前者是导致马流产和神经异常的病原，后者与马鼻肺炎有关。二者均为疱疹病毒科水痘病毒属成员。

研究人员已经从马体分离获得了许多EHV-1型的病毒株，对其中一些进行了较全面的分析。近期研究主要集中于与临床疾病特征密切关联的毒株变异。EHV-1属于疱疹病毒甲亚科(Alpha herpesvirinae)，具有双链DNA基因组。其他疱疹病毒甲亚科病毒还有与EHV-1密切相关的EHV-4，但二者基因相似性存在较大差异。EHV-3为马交合疹病毒，引起生殖道疾病。EHV-2、EHV-5属于疱疹病毒丙亚科（Gamma herpesvirinae），主要引起角膜结膜炎和呼吸道疾病[7]。

大多数疱疹病毒具有宿主特异性，但其感染其他宿主也可引起严重的疾病，甚至死亡。EHV-1在双峰驼、单峰驼、羊驼和美洲驼引起的此类疾病已有报道。

### 流行病学和病理学

EHV-1在世界范围内广泛流行，虽然被认为是导致马属动物疾病的病原，但也在其他动物分离获得该病毒，例如牛、斑马和羚羊等。最近相关报道证明EHV-1和EHV-9自然重组株可感染北极熊[11]。EHV-1可导致一两岁龄马驹发生呼吸道疾病，也可导致母马妊娠后期发生流产。同时，还可导致由于脑膜脑炎引起的神经紊乱。近年来，由EHV-1导致的马神经疾病频频报道，以美国多见。研究表明，病毒DNA聚合酶的单碱基的突变与其致神经疾病有关，可以通过DNA测序发现这一突变点。新大陆骆驼和旧大陆骆驼也可由于感染发生严重的疾病。在美国，一个外来动物养殖场曾报道了羊驼和美洲驼EHV-1感染[16,26,27]，从智利引进的100峰羊驼，与美洲驼、骆驼、牛羚和羚羊曾密切接触后发病。从死亡的羊驼和美洲驼分离获得了疱疹病毒，与EHV-1高度一致，这些动物表现失明、眼球震颤、斜颈或麻痹等症状。研究人员认为，发病动物失明与脉络膜和视网膜感染或视神经炎有关[27]。

EHV-1感染羊驼和美洲驼后的致病机制不同。研究表明，利用从发生严重神经症状的羊驼脑组织分离获得的EHV-1经鼻腔试验感染3峰美洲驼，其中2峰表现严重的神经症状，之后1峰死亡1峰安乐死，第三只美洲驼只表现轻微的神经症状。从发病死亡美洲驼丘脑中再次分离获得了病毒。EHV-1感染美洲驼和感染马的表现不同。感染马后的病毒初始复制后会导致病毒血症，而感染新大陆骆驼后，病毒首先在鼻腔黏膜进行复制，主要影响嗅觉神经和视觉神经，进而影响中枢神经。目前并无EHV-1导致驼科动物流产的报道。EHV-1不仅可感染新大陆骆驼，还可感染旧大陆骆驼。Bildfell等从濒死的发生严重神经症状的双峰驼脑组织中分离获得了EHV-1病毒[3]。

组织病理学检查可见发病双峰驼脑组织存在非化脓性脑膜脑炎，伴有脉管炎、坏死和水肿。这些特征与感染EHV-1发生神经症状的羊驼或马的病理特征相似，但并未发现眼部病变。这些数据表明新大陆骆驼和旧大陆骆驼在感染EHV-1后可出现相关的神经症状、失明甚至死亡。EHV-1可导致新大陆骆驼抗体转阳，但并无旧大陆骆驼EHV-1抗体阳性的报道。研究人员在感染EHV-1的羊驼和美洲驼21份血清中发现病毒抗体[27]。但是来自俄勒冈州的270份美洲驼血清中只有1份EHV-1抗体阳性。Puntel等报道，在阿根廷3个不同省份的9个养殖场中美洲驼的抗体阳性率为0.77%（3/390）。不同的是，利用夹心ELISA对阿拉伯联合酋长国的500份单峰驼样品中未发现EHV-1抗体阳性者[10]，利用抗体中和试验对20份放牧的阿根廷骆马进行检测，也未发现抗体阳性者存在[18]。

驼科动物对EHV-1易感。一些国家新大陆骆驼动物养殖业的兴起、新大陆骆驼和旧大陆骆驼与马接触的机会增多，因此需要进行EHV-1以及其他马病毒病的鉴别诊断（见2.1.16部分）。关于EHV-1在驼科动物流产和新生畜疾病中的作用，以及驼科动物是否有物种特异的疱疹病毒，还需进一步研究。

包括EHV-1在内的许多疱疹病毒初次侵染宿主后会形成潜伏感染，以致在发病动物和临床健康动物均可检出这些病毒。马在感染康复后，可能成为长期潜伏的病毒携带者，驼科动物也可能如此。当这些动物发生应激或注射类固醇类药物后，其体内的病毒可能活化，之后向外界散布病毒。病毒主要通过鼻分泌物散布[28]。

### 诊断

EHV-1感染不能单通过临床症状进行判定。可以利用多种细胞培养物，例如兔肾细胞（PK13）、非洲绿猴肾细胞、马皮肤细胞等进行病毒分离，进而以免疫荧光抗体、免疫组织化学法检测细胞或神经组织中的病毒抗原，来进行EHV-1的确诊。对发生急性感染和抗体转阳的驼科动物进行血清学检查也是重要的检测方法。目前已有ELISA方法用于EHV-1和EHV-4的鉴别诊断。另外，多种PCR、荧光定量PCR方法的建立，使EHV-1和EHV-4的鉴别诊断和神经致病型的EHV-1鉴定成为可能[7]。

### 防控

目前并无EHV-1感染的特异疗法。虽然常用类固醇类药物和抗生素对EHV-1感染的羊驼进行注射，但并无疗效。抗病毒治疗（阿昔洛韦）常用于发生神经症状的马的治疗。

对美洲驼接种EHV-1灭活疫苗，可产生抗体[20]。但目前并无对疫苗接种的动物进行攻毒试验来评价疫苗的效力，而且疫苗免疫并不能清除EHV-1潜伏感染。在旧大陆骆驼无EHV-1疫苗应用的先例。在驼类不推荐使用EHV-1弱毒活疫苗。

## 牛疱疹病毒1型（甲型）

反刍动物疱疹病毒较多，分属于3个疱疹病毒亚科中的2个，即疱疹病毒甲亚科和疱疹病毒丙亚科。疱疹病毒甲亚科的成员牛疱疹病毒1型和2型，分别与牛传染性鼻气管炎传染性脓疱样外阴道炎和传染性龟头包皮炎以及假疙瘩皮肤病/牛乳腺炎相关。第三种疱疹病毒甲亚科成员为牛疱疹病毒5型，与牛疱疹病毒1型遗传关系密切，可以导致犊牛的神经症状。而疱疹病毒丙亚科中的牛疱疹病毒4型通常不引起临床症状。疱疹病毒丙亚科中的恶性卡他热病毒，可导致牛发生严重的疾病。

目前已经从羊、马鹿、驯鹿、非洲水牛、长颈鹿、野牛、黑斑羚等动物分离获得与牛疱疹病毒1型相关的众多病毒[11]。牛疱疹病毒还可进一步分为1.1和1.2 a/b不同的亚型，分别可以引起不同的临床症状[28]（表70）。

虽然传染性鼻气管炎/传染性脓疱样外阴道炎在世界范围内广泛分布，但也有些国家和地区根除了牛疱疹病毒，如丹麦、瑞士、奥地利和德国巴伐利亚自由州。虽然大多感染家畜不表现临床症状，但感染动物在其三叉神经和骶骨神经可潜伏感染病毒，动物在应激时可向外排毒。

目前已经在驼科动物发现牛疱疹病毒抗体阳性病例，但是目前还不清楚牛疱疹甲型病毒在新大陆骆驼和旧大陆骆驼动物疾病中的作用[8,30]。

目前认为单峰驼对牛疱疹病毒甲型并不易感。Hedger、Bornstein、Bohrmann、Wernery等并未在单峰驼发现该病毒抗体[4,5,6,12,35]。在阿拉伯联合酋长国，Taha等利用阻断ELISA对81峰单峰驼进行抗体筛查，结果为阴性[31]。Burgemister发现突尼斯5.8%的单峰驼含有较低的抗体效价（1：5），Moussa发现埃及单峰驼的抗体阳性率为13.5%[21]。Al-Afalaq利用血清中和试验检测沙特阿拉伯2968峰单峰驼，发现抗体阳性率为13%。但并未出现临床症状，且所有抗体阳性者来源于沙特阿拉伯东部，该地区同时有牛传染性鼻气管炎的发生[1]。研究人员对沙特阿拉伯其他地区496峰单峰驼进行抗体筛

表70　牛疱疹病毒1型及其引起的临床症状

| 型 | 传染性鼻气管炎 | 传染性脓疱样外阴道炎和传染性龟头包皮炎 | 流产 |
|---|---|---|---|
| 1.1型 | + | − | + |
| 1.2a型 | + | + | + |
| 1.2b型 | + | + | − |

查，阳性率为13%[2]。

研究人员利用夹心ELISA对阿拉伯联合酋长国804峰单峰驼（717峰赛驼、77峰育种驼、10峰幼畜）进行抗体检测，未发现牛疱疹病毒甲型抗体[10]。迪拜研究人员利用$10^5$ $TCID_{50}$/mL病毒鼻内接种两峰单峰驼，结果动物均未表现临床症状，也未产生抗体。

Nawal等首次报道了在埃及单峰驼检出牛疱疹病毒甲型抗体[23]。最近，Intisar等利用夹心ELISA检测苏丹186份肺炎单峰驼肺组织样品的病毒抗原，发现3份（1.6%）阳性样品。利用荧光抗体染色、PCR证实了牛疱疹病毒甲型的存在。利用牛肾细胞在阳性的肺组织中分离获得了牛疱疹病毒甲型。他们进一步利用间接ELISA检测发现，来自周边地区的260份单峰驼血液中76.9%呈抗体阳性[15]。

新大陆骆驼比旧大陆骆驼对牛疱疹病毒甲型更具易感性。曾有报道，在感染溶血性巴斯德菌并发生支气管肺炎的3岁龄美洲驼分离获得该病毒[36]。但并不清楚牛疱疹病毒甲型是导致该动物死亡的原因。组织学检查发现急性轻微的多病灶的中性粒细胞性支气管肺炎，与细菌早期感染一致。另外，研究人员在发生支气管肺炎的3峰美洲驼分离获得了牛疱疹病毒甲型，并利用间接免疫荧光试验验对病毒进行了鉴定。这些病例的临床症状为渐进性咳嗽[20]。他们还从1.5岁龄的非化脓性脑炎并产生神经症状的美洲驼分离获得了牛疱疹病毒甲型。

Leoni等利用牛疱疹病毒甲型接种7日龄胚胎研究病毒经胚胎传播的机制。利用培养基对接种病毒的胚胎进行冲洗并不能清除牛疱疹病毒甲型，利用PCR还可检出特异的病毒DNA[19]。Rosadio等在秘鲁羊驼和南美驼中检测到了牛疱疹病毒甲型抗体。在与牛、绵羊和山羊共同放牧的美洲驼和羊驼中病毒抗体阳性率最高，分别高达16.7%和16.2%，而单独放牧的羊驼病毒抗体阳性率只有5.1%[30]。Rivera等发现秘鲁其他地区羊驼中病毒抗体阳性率为5%[29]，而研究人员对美国俄勒冈州270峰美洲驼进行检测，发现抗体阳性率只有0.7%[24]。

在阿根廷，对423峰美洲驼病毒抗体筛查结果显示抗体阳性率为31.4%（133/423）[25]。Karesh在23峰自由放牧的骆马中未发现抗体阳性者[18]。Celedon等利用血清中和试验（洛杉矶毒株）对智利74峰羊驼、43峰美洲驼、48峰骆马以及34峰驼马进行抗体检测，也未发现阳性者[9]。

疱疹病毒丙亚科的恶性卡他热病毒可感染反刍类150余种动物。Hong等利用竞争抑制ELISA对美国41份美洲驼血清进行检测，发现样品均为阴性[13]。

以上研究表明，新大陆骆驼对牛疱疹病毒甲型易感并可发病，但发病率不高，而旧大陆骆驼对该病毒的易感性较差。

### 防控

牛疱疹病毒甲型在驼科动物病原中并不占有重要地位。目前，有两种疫苗可以应用：减毒活疫苗和灭活疫苗。减毒活疫苗可以经鼻内或肌肉接种。疫苗接种可以减轻疾病的症状，但不能阻止病毒感染[22]。减毒活疫苗用于预防牛感染牛疱疹病毒甲型具有较高的效力[34]。在有牛疱疹病毒甲型暴发或病毒感染导致临床疾病时，可对驼科动物进行疫苗接种，但是目前牛疱疹病毒甲型疫苗只被批准用于牛的接种。

## 参考文献

[1] Al-Afaleq A., Abu-Elzein E.M.E. & Hegazy A.A. (2006). – Serosurveillance for antibodies against some viral diseases oflivestock in camels (*Camelus dromedarius*) in Saudi Arabia. In Proceedings of the International Scientific Conference on Camels, 10–12 May 2006, Buraydah, Saudi Arabia, 338–346.

[2] Al-Afaleq A.I., Abu Elzein E.M.E., Hegazy A. & Elnaeem A. (2007). – Serosurveillance of camels (*Camelus dromedarius*) to detect antibodies against viral diseases in Saudi Arabia. *J. Camel Pract. Res.*, 14 (2), 91–96.

[3] Bildfell R., Yason C., Haines D. & McGowan M. (1996). – Herpesvirus encephalitis in a camel (*Camelus bactrianus*). *J. Zoo Wildlife Med.*, 27 (3), 409–415.

[4] Bohrmann R., Frey H.R. & Liess B. (1988). – Survey on the prevalence of neutralizing antibodies to bovine viral diarrhea(BVD) virus, bovine herpes virus type 1 (BHV-1) and parainfluenza virus type 3 (PI-3) in ruminants in the Djibouti Republic.*Dtsch. tierärztl. Wschr.*, 95, 99–102.

[5] Bornstein S. & Musa B.E. (1987). – Prevalence of antibodies to some viral pathogens, *Brucella abortus* and *Toxoplasma gondii*,in serum from camels (*Camelus dromedarius*) in Sudan. *J. Vet. Med.* B., 34, 364–370.

[6] Bornstein S., Musa B.E. & Jama F.M. (1988). – Comparison of seroepidemiological findings of antibodies to some infectiouspathogens of cattle in camels of Sudan and Somalia with reference to findings in other countries of Africa. Proceedings ofthe International Symposium of Development of Animal Resources in Sudan, 3–9 January 1988, Khartoum, Sudan, 28–34.

[7] Brosnahan M.M. & Osterrieder N. (2010). – Equine herpes virus-1: A review and update. In Infectious diseases of the horse(T.S. Mair & R.E. Hutchinson, eds). Equine Veterinary Journal, Maidstone, UK, 41–50.

[8] Burgemeister R., Leyk W. & Goessler R. (1975). – Untersuchungen uber Vorkommen von Parasitosen, bakteriellen und viralen Infektionskrankheiten bei Dromedaren in Sudtunesien. *Dtsch. Tierärztl. Wschr.*, 82, 352–354.

[9] Celedon N.M.O., Osorio J. & Pizarro J. (2006). – Aislamiento *e* identificacion de pestivirus obtenidos de alpacas (Lama pacos) y llamas (*Lama glama*) de la Region Metropolitana, Chile. *Archivos de Medicina Veterinaria*, 38, 247–252.

[10] Central Veterinary Research Laboratory (CVRL) (1998). – Annual Report. Central Veterinary Research Laboratory, Dubai,UAE, 19.

[11] Greenwood A.D., Tsangaras K., Ho S.Y.W., Szentiks C.A., Nikolin V.M., Ma G., Damiani A., East M.L., Lawrenz A., Hofer H.& Osterrieder N. (2012). – A potentially fatal mix of herpes in zoos. *Curr. Biol.*, 22 (18), 1727–1731.

[12] Hedger R.S., Barnett T.R. & Gray D.F. (1980). – Some virus diseases of domestic animals in the Sultanate of Oman. *Trop. Anim. Health*, 12, 107–114.

[13] Hong L., Shen D.T., Jessup D.A., Knowles D.P., Gorham J.R., Thorne T., O'Toole D. & Crawford T.B. (1996). – Prevalence of antibody to malignant catarrhal fever in wild and domestic ruminants by competitive-inhibition ELISA. *J. Wildl. Dis.*,32 (3), 437–443.

[14] House J.A., Gregg D.A., Lubroth J., Dubovi E.J. & Torres A. (1991). – Experimental equine herpesvirus-1 infection in llamas(*Lama glama*). *J. Vet. Diagn. Invest.*, 3, 101–112.

[15] Intisar K.S., Ali Y.H., Khalafalla A.I., Mahasin E.A. & Amin A.S. (2009) – Natural exposure of dromedary camels in Sudanto infectious bovine rhinotracheitis virus (bovine herpes virus-1). Acta Trop., 111 (3), 243–246.

[16] Jenkins D. (1985). – Alpacas and llamas are susceptible to an equine disease. *Llama Magazine* Nov./Dec., 15–16.

[17] Kapil S., Yeary T. & Evermann J.F. (2009). – Viral diseases of New World camelids. *Vet. Clin. Food Anim.*, 25, 323–337.

[18] Karesh B.W., Uhart M.M., Dierenfeld E.S., Braselton W.E., Torres A., House C., Puche H. & Cook R.A. (1998). – Healthevaluation of free-ranging guanaco (*Lama guanicoe*). *J. Zoo Wildlife Med.*, 29 (2), 134–141.

[19] Leoni L., Cheetham S., Lager I., Parreno V., Fondevila N., Rutter B., Martinez Vivot M., Fernandez F. & Schudel A. (2001).– Prevalencia de anticuerpos contra enfermedades virales del ganado, en llama (*Lama glama*), guanaco (*Lama guanicoe*) y vicuña (*Vicugna vicugna*) en Argentina. *In* Ⅱ Congreso Latinoamericano de Especialidad en Pequenos Rumiantes y CamelidosSudamericanos y XI Congreso Nacional de Ovinocultura, Merida, Yucatan, Mexico, Proceedings CD 2001.

[20] Mattson D.E. (1994). – Viral diseases. *Vet. Clin. North Am. Food Anim. Pract.*, 10 (2), 345–351.

[21] Moussa A.A., Saber M.S., Nafie E., Shalaby M.A., Ayoub W.N., El Nakashly S., Mohsen A.Y., Madbouly M.M., El SanousiA. A., Fathia M.M., Sami A., Allam I. & Reda I.M. (1990). – Serological survey on the prevalence of BHV1 in domestic animals in Egypt. *Vet. Med. J. Giza*, 38, 87–94.

[22] Muylkens B., Thiry J. & Thiry E. (2010). – Infectious bovine rhinotracheitis. *In* Infectious and parasitic diseases of livestock(P.-C. Lefevre, J. Blancou, R. Chermette & G. Uilenberg, eds). Lavoiser, Paris, 449–460.

[23] Nawal M.A.Y., Gabry G.H., Hussein M. & Omayma A.A.S. (2003). – Occurrence of Parainfluenza type 3 and bovine herpesvirus type 1 (BHV-1) viruses (mixed infection) among camels. *Egypt. J. Agric. Res.*, 81 (2), 781–791.

[24] Picton R. (1993). – Serologic survey of llamas in Oregon for antibodies to viral diseases of livestock. MS thesis, Oregon State University, Corvallis, OR.

[25] Puntel M., Fondevila N.A., Blanco Viera J., O'Donnell V.K., Marcovechio J.F., Carillo B.J. & Schudel A.A. (1999). –

2 病毒病

Serological survey of viral antibodies in llamas *(Lama glama)* in Argentina. *J. Vet. Med. B.*, 46, 157–161.

[26] Pursell A.R., Sangster L.T., Byars T.D., Divers T.J. & Cole J.R. (1979). – Neurological disease induced by equine herpesvirus1. *J. Am. Vet. Med. Assoc.*, 175, 473–474.

[27] Rebhun W.C., Jenkins D.H., Riis R.C., Dill S.G., Dubovi E.J. & Torres A. (1988). – An epizootic of blindness and encephalitis associated with a herpes virus indistinguishable from equine herpes virus I in a herd of alpacas and llamas. *JAVMA*, 192 (4),953–956.

[28] Reed C. (2010). – Important risk analysis: llamas (*Lama glama*) and alpacas (*Vicugna pacos*) from specified countries. Policyand risk, MAF Biosecurity New Zealand, 41–45. Available at: www.biosecurity.bovt.nz/regs/imports/ihs/risk

[29] Rivera H., Madwell B.R. & Ameghina E. (1987). – Serological survey of viral antibodies in the Peruvian alpaca (*Llama pacos*).*Am. J. Vet. Res.*, 48, 189–191.

[30] Rosadio R.H., Rivera H. & Manchego A. (1993). – Prevalence of neutralising antibodies to bovine herpesvirus-1 in Peruvianlivestock. *Vet. Rec.*, 132, 611–612.

[31] Taha H.T. (2007) – Pathogens affecting the reproductive system of camels in the United Arab Emirates. Thesis, Swedish University of Agricultural Sciences, Uppsala, Sweden.

[32] Thedford R.R. & Johnson L.W. (1989). – Infectious diseases of New-world camelids (NWC). *Vet. Clin. North Am. Food Anim.Pract.*, 5 (3), 145–157.

[33] Torres A., Dubovi E.J., Rebhun W.D. & King J.M. (1985). – Isolation of a herpesvirus associated with an outbreak of blindnessand encephalitis in a herd of alpacas and llamas. Abstract in the 66th Conference of Research Workers in Animal Diseases.

[34] Van der Poel W.H., Kramps J.A., Quak J., Brand A. & Van Oirschot J.T.(1995). – Persistence of bovine herpesvirus-1-specific antibodies in cattle after intranasal vaccination with a live virus vaccine. *Vet. Rec.*, 137, 347–348

[35] Wernery U. & Wernery R. (1990). – Seroepidemiologische Untersuchungen zum Nachweis von Antikörpern gegen Brucellen,Chlamydien, Leptospiren, BVD/MD, IBR/IPV- und Enzootischen Bovinen Leukosevirus (EBL) bei Dromedarstuten (*Camelusdromedarius*). *Dtsch. tierärztl. Wschr.*, 97, 134–135.

[36] Williams J.R., Evermann J.F., Beede R.F., Scott E.S., Dilbeck P.M., Whetstone C.A. & Stone D.M. (1991). – Association ofbovine herpesvirus type 1 in a llama with bronchopneumonia. *J. Vet. Diagn. Invest.*, 3, 258–260.

## 深入阅读材料

*Fowler M.E. (1996). – Husbandry and diseases of camelids. In Wildlife husbandry and diseases (M.E. Fowler, ed.). Rev. sci. tech.Off. int. Epiz., 15 (1), 155–169.*

*Plowright W. (1981). – Herpesvirus of wild ungulates, including malignant catarrhal fever. In Infectious diseases of wild mammals(J.W. Davis, L.H. Karstad and D.O. Trainer, eds). Iowa State University Press, Ames, IA,126–146.*

*Richard D., Planchenault D. & Giovannetti J.F. (1985). – Production cameline – Rapport final. Projet de Développement de l'élevagedans le Niger. Centre-Est, IEMVT, France.*

*Scott S., Dilbeck P.M., Whetstone C.A. & Stone D.M. (1991). – Association of bovine herpesvirus type 1 in a llama withbronchopneumonia. J. Vet. Diagn. Invest., 3, 258–260.*

*Wernery U. (1999). – New aspects on infectious diseases of camelids. J. Camel Pract. Res., 6 (1), 87–91.*

（高闪电译，殷宏校）

2

病毒病

## 2.1.15 流感

正黏病毒科（流感病毒）包括A、B、C和D四个属，其中D属包含蜱传病毒，如多理病毒和托高土病毒（见下文稀有虫媒病毒和其他次要病毒传染病）。D属病毒在生物学特性方面与A、B、C属病毒有所不同，后三者通常经过气溶胶方式直接传播。尽管反刍动物骆驼作为中间媒介并不是流感病毒的宿主，但是在19世纪70年代末到80年代期间，在蒙古的双峰驼群中暴发过几次流感疫情[1,8,14]。

### 病原学

流感病毒属于正黏病毒科。A、B、C、D四种流感病毒均可感染人类，但仅有A型流感病毒能造成疫病流行。在动物上同样如此，A型流感病毒危害最大[10]。流感疫情主要流行于禽类、猪群和马群，但也感染狗、貂、海豹和鲸类[7]。仅一株H1N1流感病毒分离于蒙古的双峰驼。

流感病毒粒子直径在80～100nm，其外膜为源自宿主的两个囊膜纤突组成，即按照4∶1或5∶1的比例构成血凝素和神经氨酸酶糖蛋白。病毒复制过程中，病毒蛋白在宿主细胞胞质中进行合成。当在同一个体中流感病毒混合感染可能会发生基因重组，从而出现抗原漂移，而且自然界中也确实分离到了重组病毒。

### 流行病学

尽管目前在新大陆骆驼还没有A型流感暴发的报道，但在双峰驼上发现了与流感病毒有关的疫病。在1978–1988年间，蒙古国不同地区61个农场的骆驼发生了19起严重呼吸道疾病[8,14]。自1979年，疫情与H1N1流感病毒有了关联。通过鸡胚培养法，从92份鼻拭子样品中分离出13株病毒，血凝实验检测证实为H1N1流感病毒。H1N1流感病毒在骆驼群中开始流行的同时，在蒙古病人体内也分离了4株H1N1流感病毒，测序结果表明4株病毒的所有基因与骆驼分离株具有高度同源性。一种假说认为，骆驼群中分离的流感病毒来源于1978年在列宁格勒研制出来的一株紫外灭活的重配疫苗株（PR8XUSSR/77），该疫苗同时在蒙古人群中接种使用[1]。但是仍有许多问题存在：该病毒是如何从人群中传播到驼群中的，并且在这种通常不易感动物体内获得了致病力。1978–1988年的疫情发生于蒙古的61个农场，在其中的一次疫情中，4 000只骆驼出现了严重的呼吸道疾病症状，并有部分骆驼死亡。骆驼感染后的主要临床症状如下：致死率为9.1%，流产率为2.6%，精神委顿占6.7%。急性期的临床症状还包括干咳、支气管炎、肺炎和发热。临床症状还有黏液眼和流鼻涕。临床表现可持续一周。

34只3～4岁健康双峰驼用于H1N1骆驼分离株的感染实验，所选骆驼均呈流感病毒抗体阴性。每组3只骆驼经鼻腔、气管或者肌肉途径进行病毒感染。1985-1986年所做的3个独立实验结果显示，尽管感染后在骆驼体内可以再次分离到病毒，并且血清抗体阳转，血凝抑制效价在1∶（16～128），但是实验感染骆驼仅出现了和自然感染骆驼相似的轻微临床症状，表现为发热、咳嗽、支气管炎和鼻眼流涕等症状（图179）。感染动物均恢复健康。自此以后就没有报道过骆驼群出现流感疫情。然而，实验研究中无法复制和田间一样的严重病例，说明田间细菌继发感染发挥了重要作用。

引起蒙古骆驼群流感暴发的病毒是两株人源流感病毒的重配毒株，其在非流感病毒自然宿主的骆驼群中引起小范围的流行。因此，冷适应重配毒株的安全性是否有待提高或者列宁格勒病毒是否仅在骆驼上发生重配和感染个例仍然有待于进一步的研究[12]。

Auguadra[2]和Somac/Sarec[13]报道了在索马里的骆驼上出现了流感样的流行，但是并没有进一步分离病毒，他们报道了骆驼出现呼吸道症状并伴随有鼻炎和结膜炎，这说明该症状有可能是由多种病毒或细菌感染引起的。其他研究报道了多个非洲国家在骆驼群中进行了流感病毒抗体的血清学检测。Olaleye等人[9]在尼日利亚东北的单峰驼屠宰场进行采样，他们发现0.6%的样品呈现A型流感病毒抗体阳性，12.7%的样品呈现B型流感病毒抗体阳性。El-Amin和Kheir[5]报道了7.8%的苏丹的骆驼呈现流感病毒抗体阳性。

在A型流感病毒的16种不同血凝素（H）和9种神经氨酸酶（N）中，只有H1N1组合仅在1978–1988年感染过蒙古的双峰驼。因为在中东和亚洲的很多国家，马和骆驼可以密切接触，一些人认为马流感病毒有可能会从马传播到骆驼。很多的病原体与骆驼的呼吸道症状有关，但是关于流感病毒的仅有一个报道。而且，在对阿联酋的500只骆驼采用A/equine-1/Prague/1956(H7N7)和A/equine-2/Miami/1963(H3N8)两株病毒进行血凝抑制检测（HIT），并未检测到血清阳性的样品[4]。

许多国家和地区时常会出现一些无法确诊的骆驼疾病。从2010年的11月到2011年初，一股冷空气严重影响了印度（旁遮普省和信德省）、阿富汗、巴基斯坦、伊朗、伊拉克的许多完全依靠骆驼和其他家畜为生的家庭。这种未确诊的疾病影响了数百头单峰驼，同时出现大批死亡，骆驼群主要症状表现为发热、厌食、眼有黏性渗出、双侧带血，黏性鼻涕阻塞鼻腔导致呼吸困难[11]。妊娠晚期流产和早产也是此次疫情的主要特征。几种疾病如巴氏杆菌病、流感、小反刍兽疫及鼻疽都被认为这些疫情的可能原因，但是与病原学诊断最为一致的是巴氏杆菌和溶血性曼氏杆菌[8]。然而，为了确定病原，复制此类未确诊疫病病原的实验没有成功。最近，另外一项研究在埃塞俄比亚的骆驼中鉴定了很多与肺脏病变相关的细菌感染病例[3]。



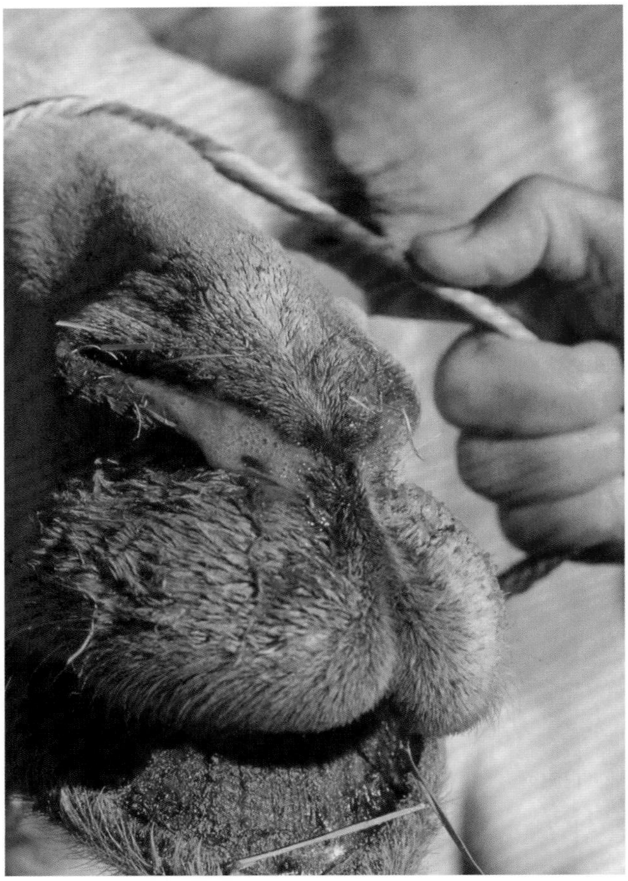
图179　双峰驼流鼻涕症状，其有可能由包括流感在内的多种病毒或细菌引起

## 诊断

骆驼的呼吸道疾病与副流感病毒3型[6]、呼吸道合胞体病毒[7]、A型流感病毒[12]以及一系列的引起肺部损伤的细菌感染有关[3]，包括多杀性巴氏杆菌、溶血性曼氏杆菌、马红球菌、葡萄球菌、链球菌、大肠杆菌、土拉热弗朗西斯菌、黄杆菌、支气管败血波氏菌、嗜水气单胞菌、奈瑟氏菌、无乳链球菌、金黄色葡萄球菌、海藻糖巴氏杆菌、鸭疫巴氏杆菌、绿脓杆菌、微球菌和分枝杆菌。

由于释放病毒的时间很短，因此在发热和咳嗽的开始就应该采集样本并进行病毒分离。鼻咽拭子采集后应置于病毒培养基中，并尽快低温送往实验室。流感病毒可在鸡胚或者MDBK细胞上进行病毒培养。为了成功分离病毒，有时还需要进行几次连续的传代。血凝实验用于病毒鉴定，而用特异性血清进行HIT试验可对流感病毒定型。HIT试验也用来检测流感病毒抗体。用于检测马流感的FLU-A检测试剂盒（Becton Dickinson，New Jersey）已经研制成功，该试剂盒可以用来尝试检测患有流感样的骆驼样品。该试剂盒是一种流感病毒抗原捕获ELISA（酶联免疫吸附试验），用于检测人临床样本中A型和B型流感病毒的核蛋白，也同样适用于检测其他物种，如禽类和马的流感抗原（图180）。巢氏反转录PCR（聚合酶链式反应）也同样可以快速检测临床的流感病毒样本，和病毒分离方法有着同样的灵敏性。配对血清样品的抗体滴度的升高（4倍）必须采用血清学方法检测。

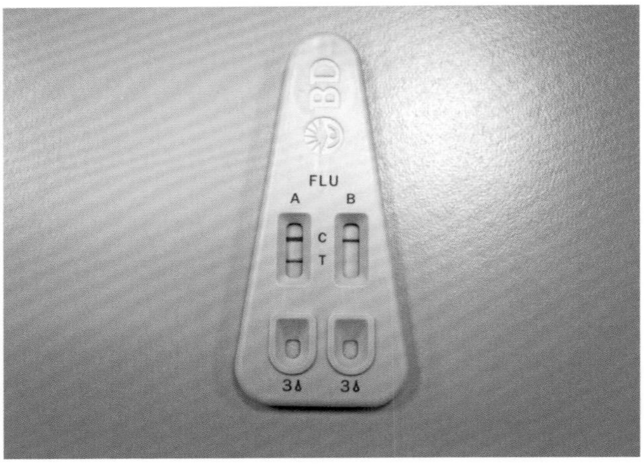

图180　采用FLU-A检测试剂盒检测的阳性结果（Becton Dickinson，New Jersey）

### 防控

控制流感疫情的最有效的方式就是进行疫苗免疫及限制动物的运输或运动。目前在骆驼上尚未使用过流感疫苗，但若在骆驼群中发生流感疫情时，有必要制定疫苗免疫的程序。

## 参考文献

[1] Anchlan D., Ludwig S., Nymadawa P., Mendsaikhan J. & Scholtissek C. (1996). – Previous H1N1 influenza A viruses circulating in the Mongolian population. *Arch. Virol.*, 141 (8), 1553–1569.

[2] Auguadra P. (1958). – Grippe O influenza del dromedario Somalo. *Arch. ital. Sci. med. Trop. Parasit.*, 34, 215–222.

[3] Awol N., Ayelet G., Jenberie S., Gelaye E., Sisay T. & Nigussie H. (2011) Bacteriological studies on pulmonary lesions ofcamel (*Camelus dromedarius*) slaughtered at Addis Ababa abattoir, Ethiopia. *Afr. J. Microbiol. Res.*, 5 (5), 522–527.

[4] Central Veterinary Research Laboratory (CVRL) (1998). – Annual Report. Central Veterinary Research Laboratory, Dubai,UAE, 19.

[5] El-Amin M.A.C. & Kheir S.A. (1985). – Detection of influenza antibody in animal sera from Kassala region, Sudan, byagargel diffusion test. *Rev. Elev. Méd. Vét. Pays Trop.*, 38 (2), 127–129.

[6] Intisar K.S., Ali Y.H., Khalafalla A.I., Rahman M.E. & Amin A.S. (2010a). – Respiratory infection of camels associated withparainfluenza virus 3 in Sudan. *J. Virol. Methods*, 163 (1), 82–86.

[7] Intisar K.S., Ali Y.H., Khalafalla A.I., Rahman M.E. & Amin A.S. (2010b). – Respiratory syncytial virus infection of camels(*Camelus dromedarius*). *Acta Trop.*, 113 (2), 129–133.

[8] Lvov D.K., Yamnikova S.S., Shemyakin I.G., Agafonova L.V., Miyasnikova I.A., Vladimirtseva E.A., Nymadava P., Dachtzeren P.,Bel-Ochir Z.H. & Zhadanov V.M. (1982). – Persistence of genes of epidemic influenza viruses. *Voprosi. Virusol.*, 27 (4),401–405.

[9] Olaleye O.D., Baba S.S. & Omolabu S.A. (1989). – Preliminary survey for antibodies against respiratory viruses amongslaughter camels (*Camelus dromedarius*) in north-eastern Nigeria. *Rev. sci. tech. Off. int. Epiz.*, 8 (3), 779–783.

[10] Plateau E. (2010). – Equine Influenza. In Infectious and parasitic disease of livestock (P.-C. Lefevre, J. Blancou, R. Chermette& G. Uilenberg, eds). Lavoisier, Paris, 287–298.

[11] ProMed-mail<promed@promed.isid.harvard.edu – Undiagnosed lethal disease of dromedary camel – Pakistan,14.1.2011,22.1.2011 and 17.2.2011. Archive numbers: 20110114.0160, 20110122.0274 and 20110217.0524.

[12] Scholtissek C. (1995). – Potential hazards associated with influenza virus vaccines. *Dev. Biol. Stand.*, 84, 55–58.

[13] Somac/Sarec (1982). – Camel research project report by a Somali/Swedish Mission, 10–26 March, 18–23.

[14] Yamnikova S.S., Mandler J., Bel-Ochir Z.H., Dachtzeren P., Ludwig S., Lvov D.K. & Scholtissek C. (1993). – A reassortantH1N1 influenza A virus caused fatal epizootics among camels in Mongolia. *Virology*, 197 (2), 558–563.

## 深入阅读材料

Wernery U. (1999). – New aspects on infectious diseases of camelids. *J. Camel Prac. Res.*, 6 (1), 87–91.

（朱启运译，殷宏校）

## 2.1.16 甲型马鼻病毒

小RNA病毒科拥有多个不同的属、种和血清型，包括许多人类和其他脊椎动物病毒。这些病毒分别属于肠病毒属、心病毒属、口蹄疫病毒属、肝炎病毒属、马鼻病毒属以及捷申病毒属等。肠道病毒包括含有多个血清型的甲型、乙型和丙型人鼻病毒、人肠道病毒、牛肠道病毒和猪肠道病毒。另外还从马体分离获得了一些小RNA病毒，与鼻炎相关，称为鼻炎病毒，主要从血液、粪便、消化道和呼吸道分离而来[4]。马鼻病毒起初被分为3类，但最近分为口蹄疫病毒属和马鼻病毒属。马鼻病毒甲型以前称为1型马鼻病毒，属于口蹄疫病毒属。马鼻病毒乙型属于马鼻病毒属，包括3个型，为酸稳定的马小RNA病毒，以前分别称为2型马鼻病毒和3型马鼻病毒。马鼻病毒甲型是口蹄疫病毒属的四个种之一，其余3个种分别为牛鼻病毒甲型和乙型以及口蹄疫病毒。1962年，英国首次报道了马鼻病毒甲型[1]，随后在世界各地马群均发现了该病毒。该病毒的感染比较普遍，大多数马均存在抗体。在人也有分离出鼻病毒的报道。在迪拜、阿拉伯联合酋长国流产的单峰驼也分离出了鼻病毒[3]。人鼻病毒是世界分布的常见呼吸道病原，可导致人的普通感冒，具有季节性。据估计，人一般每年可发生2～5次的普通感冒[2]。

### 临床症状

鼻病毒感染通常导致发热、鼻炎、咽炎、咽淋巴结炎，3～8天后出现大量鼻分泌物，伴有咳嗽，可持续2～3周。该病比较轻微，可自愈。在马群发生感染时传播速度很快。

研究人员从疑似马鼻病毒感染的单峰驼流产胎儿器官匀浆中分离获得了可致细胞产生病变的病毒。在沙漠自由放牧的258峰育种用单峰驼中，约有10%在妊娠6～8月时发生流产。该单峰驼群与其活动区域外围的57匹退役马发生了直接接触，该马群中至少有80%为马鼻病毒抗体阳性。单峰驼感染很可能由马群中的病毒传播导致，在该驼群并无呼吸道疾病症状。利用非洲绿猴肾细胞培养的$10^4$TCID50的马鼻病毒接种两峰妊娠7月龄的母驼，均发生流产，而且从发病动物重新分离出了病毒。

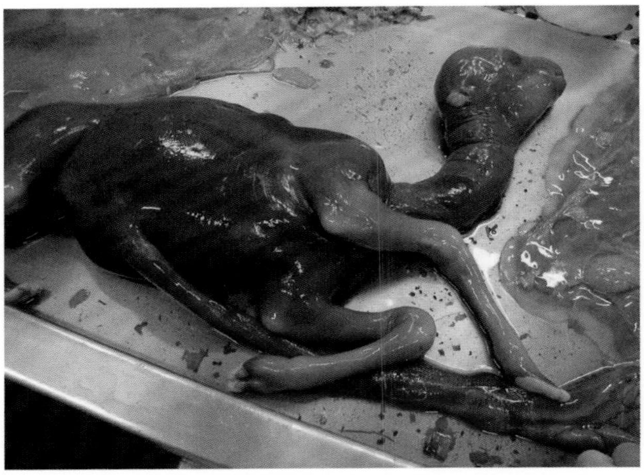

a：感染马鼻病毒的流产胎儿发生严重皮下水肿，以背部最为明显

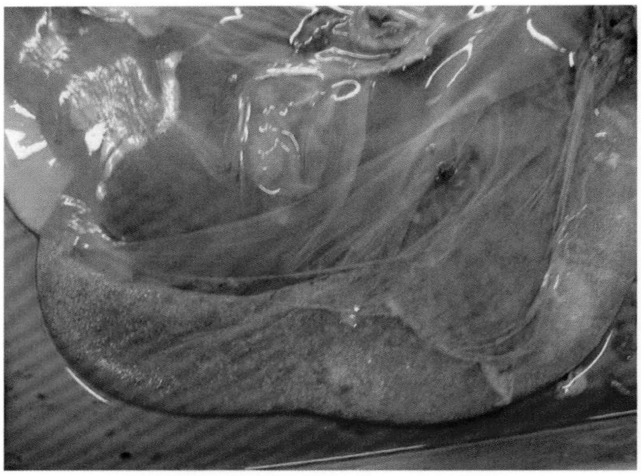

b：马鼻病毒引起胎盘普遍水肿

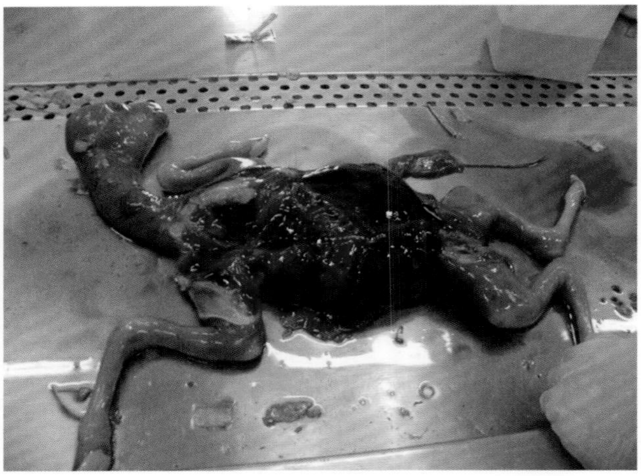

c：流产胎儿的大体病理变化

图181a-c

### 病理变化

田间自然流产和实验流产的胎儿的病理变化类似，主要为胎盘和脐带水肿（图181a）。流产胎儿普遍皮下水肿，尤其是头部至驼峰背部，腿部也有少许水肿（图181b）。所有病例中，流产胎儿腹腔积有黑色但未凝固的血水（图181c）。值得注意的是，驼类的胎盘与马属动物类似，均为扩散的上皮绒毛膜胎盘，其生理结构的相似性是驼科动物对马鼻病毒具有易感性的原因之一。从驼科动物分离获得2株小RNA病毒，其中之一来源于美国动物园的单峰驼，另一株来源于美国的美洲驼。

### 诊断

目前该病的诊断可以利用血清中和试验进行血清学方法筛查，也可利用非洲绿猴肾细胞从胎盘或胎儿器官进行病毒分离。在单峰驼发生流行性流产时，可以进行病毒分离，进而还可利用PCR检测病毒分离物中的病毒RNA。

### 防控

避免将驼类与马混饲。目前还没有疫苗可用于预防鼻病毒引起的流产。



## 参考文献

[1] Plummer G. (1962). – An equine respiratory virus with enterovirus properties. *Nature*, 195, 519–520.

[2] Simmonds P., McIntyre C., Savolainen-Kopra C., Tapparel C., Mackay I.M. & Hovi T. (2010). – Proposals for the classification of human rhinovirus species C into genotypically assigned types. *J. Gen. Virol.*, 91, 2409–2419.

[3] Wernery U., Knowles N.J., Hamblin C., Wernery R., Joseph S., Kinne J. & Nagy P. (2008). –Abortions in dromedaries(*Camelus dromedarius*) caused by equine rhinitis A virus. *J. Gen. Virol.*, 89, 660–666.

[4] Wood J.L.N., Newton J.R. & Smith K. (2009). – Rhinitis and adenovirus infection of horses. *In* Infectious diseases of the horse (T.S. Mair & R.E. Hutchinson, eds). Geerings Print Ltd., Ashford, Kent, UK, 52–55.

(高闪电译，殷宏校)

## 2.2 非致病性病毒感染

### 2.2.1 前言

自1970年以来，发表的有关驼科动物的文献数量急剧增长，占到了该领域总文献的70%以上，显示出人们对这些动物的兴趣日益浓厚。目前关于旧大陆骆驼的文献有8000余篇，关于新大陆骆驼的文献有3000篇[5,23,24]。在这些文献中，有1000余篇是关于病原微生物的，大多数已在本书中引用[6]。

Wernery和Kaaden主编的《驼科动物疾病》中讲到的病毒性感染疾病在本书进行了删减，在本章节只保留一部分。这些病毒感染只能在试验感染中导致轻微的症状，例如非洲马瘟病毒和腺病毒，可以从驼科动物血液中分离出来，但感染动物无任何临床症状。随着将来关于驼科动物研究的深入，会发现更多的疾病，不但有病毒病，还包括细菌病、真菌病和寄生虫病。

如表56所示，最近30年关于驼科动物的血清学阳性数据表明，多种病毒在驼科动物造成感染并引起机体的免疫反应。但是一些病毒病只检测到了抗体，而且大多数报道是关于待宰的旧大陆骆驼的感染情况，并无这些动物的来源等背景。

关于驼科动物感染虫媒病毒的信息较少，目前只有血清学数据以及从这些动物寄生的蜱体分离病毒的数据。这些信息见2.2.5。

### 2.2.2 呼吸道病毒

这里主要讲述分属于不同病毒科的3种病毒，它们可以感染反刍动物呼吸道，这些病毒在驼科动物疾病中的作用不是太重要[19]，主要包括：

（1）3型副流感病毒(PI-3)；

（2）牛呼吸道合胞体病毒(BRSV)；

（3）腺病毒。

在世界范围内的驼科动物均检出了这些病毒的抗体。

3型副流感病毒属于副黏病毒科，该病毒科成员可以导致严重的动物疾病。3型副流感病毒和牛呼吸道合胞体病毒是导致牛“船运热”肺炎综合征的主要病原。“船运热”主要发生于动物运输或动物受到其他应激时。多年来，对3型副流感病毒是否为呼吸道细菌继发感染的诱因一直有争议。3型副流感病毒可以感染人和多种动物，包括犬、马、猴等。研究人员利用凝集抑制试验从旧大陆骆驼和新大陆骆驼检出了病毒抗体，但没有其导致发病的报道。

比较有趣的是在干旱的沙漠地带，3型副流感病毒具有较高的流行，分别为突尼斯81%[4]、乍得99%[13]、苏丹81%[3]以及索马里亚42.8%[2]。其在不同国家流行的差异可能与气候状况、管理措施密切相关[1]。

Intisar等研究了3型副流感病毒在苏丹单峰驼呼吸道感染中的作用。该研究从苏丹4个地区屠宰场以及呼吸道疾病暴发区域共采集281份肺组织样品，利用抗原捕获ELISA进行检测，发现6份样品（2.1%）病毒抗原阳性，并利用间接荧光抗体检测法证实了这一结果。取病毒抗原阳性的4份样品进行RT-PCR检测，发现有2份为核酸阳性。利用牛肾细胞进行病毒分离，获得了1株致细胞病变毒株。由于没有进行动物接种试验，所以并不清楚该分离株是否为导致动物发生肺炎的病原。作者还利用间接ELISA对495份单峰驼血液进行抗体测定，发现阳性率为82.2%，与其他国家的阳性率相一致。

3型副流感病毒和呼吸道合胞体病毒也可感染新大陆骆驼，但目前并无动物感染后出现临床症状的报道[16,20]。

Karesh对阿根廷20峰放牧的骆马进行3型副流感病毒抗体检测，并未发现阳性者[10]。Rivera等利用血清中和试验测定秘鲁100份羊驼血清，发现抗体阳性率为16.5%[20]。研究表明，羊驼对多种病毒具有易感性，包括3型副流感病毒、呼吸道合胞体病毒、A型流感病毒、轮状病毒、水泡性口炎病毒、口蹄疫病毒和传染性脓疱皮炎病毒等。在秘鲁南部山区，肺炎是导致成年羊驼和幼驼死亡的重要病因，但目前并无从病驼成功分离病毒的报道，可能是由于血样中存在较高效价的中和抗体，影响了病毒在培养细胞中的复制。因此，需要进行盲传以获得细胞病变，这一过程通常需要11～21天。鉴于这些困难，间接免疫荧光染色成为替代方法。尸体剖检，肠道气肿和水肿是该病最明显病变。另外剖检还可见支气管炎、支气管壁出现多核合胞体。

Marcoppido等对从阿根廷黑河迁移至布宜诺斯艾利斯地区养殖场的11峰骆马幼畜的病毒感染情况进行监控[11]。在起初并未检测到3型副流感病毒、1型牛疱疹病毒、牛病毒性腹泻或蓝舌病病毒抗体，但是所有动物在数月后出现了3型副流感病毒抗体但没有任何临床症状。该研究表明，野生骆马进行舍饲时对家畜的病毒易感[11]。

Rosadio等[21]对24峰因急性肺炎死亡的羊驼幼畜进行了研究。这些幼畜为7～39日龄，剖检可见中等至非常严重的肺淤血和肺水肿，以及中等程度的化脓性、坏死性、纤维素性支气管肺炎。在死亡的24病例中，22病例可以分离获得3型副流感病毒（9例）和（或）多杀性巴氏杆菌及溶血曼海姆菌[21]。这一结果表明，新生羊驼幼畜感染病毒和细菌的方式可能与其他反刍动物的感染方式类似，例如"船运热"。

牛呼吸道合胞体病毒通常可导致6月龄以内的犊牛发生肺炎，但成年牛也会受到该病毒的影响。该病毒通过气溶胶在易感动物间迅速传播[18]，发病率为30%～50%，但死亡率很低，还可伴随腺病毒的感染[18]。

Intisar等[7]发现苏丹单峰驼的牛呼吸道合胞体病毒抗体阳性率为27.3%（135/495）。抗体最高的区域为苏丹西部（33.5%），其次为中部（31.6%）和东部（23.5%）。他们还从苏丹4个地区屠宰场以及呼吸道疾病暴发区域共采集281份肺组织样品，利用抗原捕获ELISA进行牛呼吸道合胞体病毒检测，发现4份为阳性。这些样品来源于苏丹中部的Tambool屠宰场。ELISA结果与间接免疫荧光抗体检测结果一致，利用RT-PCR检测病毒基因组，结果1份样品检测出病毒核酸[8]。

以上研究结果并不能说明病毒感染是导致动物发生肺炎的原因。

腺病毒科包含5个属，即富腺胸病毒属、禽腺病毒属、鱼腺病毒属、哺乳动物腺病毒属和唾液酸酶腺病毒属。哺乳动物腺病毒属可感染哺乳动物，而禽腺病毒属可感染鸟类。哺乳动物腺病毒可感染犬、马、猪、绵羊、山羊、鹿、家兔和牛。在牛至少存在10个以上血清型的腺病毒，感染后动物不发病或只表现轻微的呼吸道症状。该病毒基因组为双链DNA。

在旧大陆骆驼和新大陆骆驼均检出了腺病毒抗体，但目前只在新大陆骆驼分离获得了病毒。

Olaleye等[15]发现尼日利亚东北部单峰驼的腺病毒抗体阳性率为1.3%。他们还发现这些动物感染1型、2型和3型副流感病毒的阳性率分别为22.3%、2.5%和18.5%，呼吸道合胞体病毒的抗体阳性率为0.6%，但是关于这些病毒感染的流行病学意义还需进一步研究。

研究人员对美国俄勒冈地区21个牧场的270峰美洲驼进行了血清学调查，发现腺病毒（7649分离株）的感染率为93%[16]。虽然感染率较高，但大都分为亚临床感染。Galbreath等报道了4峰美洲驼感染腺病毒发病，从发生肺炎和肝炎的5月龄幼驼肺组织分离获得腺病毒[6]。他们还检测到了肺组织和肝组织腺病毒特异的细胞核内包涵体。两峰美洲驼带有黏液脓性的鼻分泌物，并伴有纤维素性胸膜肺炎和血清样渗出物。1峰美洲驼呈慢性消耗病。利用猪腺病毒结合物进行荧光抗体试验从4峰美洲驼肺部均检出了腺病毒。

大多数腺病毒感染通常会导致上呼吸道疾病，但也可导致肠炎和脑膜脑炎[14]。腺病毒LA7649在美洲驼较为普遍，但该病毒还可导致腹泻、肺炎和胎儿感染甚至流产[12]。在美国，研究人员从发生腹泻的羊驼和美洲驼分离获得了腺病毒[12]。其中1峰死亡的美洲驼患有坏死性肠炎和结肠炎，但是其IgG水平很低，所以其发病原因为免疫缺陷综合征与继发腺病毒感染。美国研究人员对体重持续减轻的患肺炎和其他症状的5周龄羊驼进行尸检，发现其病理变化为小管间质性肾炎，在嗜碱性粒细胞和嗜酸性粒细胞含有细胞核内包涵体。电镜检查为腺病毒特异的包涵体[9]。

腺病毒在新大陆骆驼感染非常普遍。Tillman通过实验感染的方法研究腺病毒LA7649和其他腺病毒在美洲驼疾病中的作用。美洲驼源的腺病毒5530可在眼结膜、鼻腔、咽喉以及肠道复制并致病。该病毒在肺组织复制并不导致任何病理性变化。人工感染的美洲驼只出现轻微的症状，这一结果可能受实验动物良好的饲养环境和营养条件所影响。在其他条件下，该病毒可能导致严重的疾病。

Puntel等发现阿根廷某美洲驼养殖场的抗体阳性率（对牛腺病毒BadⅧ毒株）为5.13%（20/390）。

## 防控

腺病毒、副流感病毒和牛呼吸道合胞体病毒在驼科动物致病中不太重要。但利用减毒活疫苗进行鼻内接种防控牛的由1型疱疹病毒和3型副流感病毒导致的呼吸道疾病，是行之有效的。因此推荐当类似疾病在驼科动物发生时，使用这两种疫苗进行接种。

一些减毒活疫苗和灭活疫苗可用于控制牛呼吸道合胞体病毒感染，但这些疫苗的效力评价研究较少。这些疫苗还未应用于驼科动物接种。腺病毒可能主要通过口、鼻、咽途径进行传播，另外该病毒还具有自限性，控制其感染难度较大。

## 参考文献

[1] Afzal M. & Sakkir M. (1994). – Survey of antibodies against various infectious disease agents in racing camels in Abu Dhabi,United Arab Emirates. *Rev. sci. tech. Off. int. Epiz.*, 13 (3), 787–792.

[2] Bornstein S. (1988). – A disease survey of the Somali camel. Report to Sarec, Sweden.

[3] Bornstein S. & Musa B.E. (1987). – Prevalence of antibodies to some viral pathogens, *Brucella abortus* and *Toxoplasma gondii* in serum from camels (*Camelus dromedarius*) in Sudan. *J. Vet. Med. B.*, 34, 364–370.

[4] Burgemeister R., Leyk W. & Goessler R. (1975). – Untersuchungen über Vorkommen von Parasitosen, bakteriellen und viralen Infektionskrankheiten bei Dromedaren in Sudtunesien. *Dtsch. Tierärztl. Wschr.*, 82, 352–354.

[5] Gahlot K., Tibary A., Wernery U. & Zhao X.X. (2002). – Selected bibliography on camelids 1991–2000. The CamelidPublishers, Bikaner, India.

[6] Galbreath E.J., Holland R.E., Trapp A.L., Baker-Belknap E., Maes R.K., Yamini B., Kennedy F.A., Gilardy A.K. & Taylor D.(1994). – Adenovirus-associated pneumonia and hepatitis in four llamas. *Am. Vet. Med. Ass.*, 204 (3), 424–426.

[7] Intisar K.S., Ali Y.H., Khalafalla A.I., Rahman M.E.A. & Amin A.S. (2010a). – Respiratory infection of camels associated with parainfluenza virus 3 in Sudan. *J. virol. Methods*, 163 (1), 82–86.

[8] Intisar K.S., Ali Y.H., Khalafalla A.I., Rahman M.E.A. & Amin A.S. (2010b). – Respiratory syncytial virus infection of camels(*Camelus dromedaries*). *Acta Tropica*, 113 (2), 129–133.

[9] Jakowski R., Keating J., Boucher G.G. & Bedenice D. (2008). – Renal adenoviral infection in an alpaca. *Vet. Pathol.*, 45 (5), 746.

[10] Karesh W.B., Uhart M.M., Dierenfeld E.S., Braselton W.E. Torres A., House C., Puche H. & Cook R.A. (1998). – Health. evaluation of free-ranging Guanaco (*Lama guanicoe*). *J. Zoo Wildlife Med.*, 29 (2), 134–141.

[11] Marcoppido G., Olivera V., Bok K., Parreno V. (2011). – Study of the kinetics of antibodies titres against viral pathogens and detection of rotavirus and Parainfluenza 3 infections in captive crias of guanacos (*Lama guanicoe*). *Transbound. Emerg.Dis.*, 58 (1), 37–43.

[12] Mattson D.E. (1994). – Viral diseases. *Vet. Clin. N. Am. Food Anim. Pract.*, 10 (2), 345–351.

[13] Maurice Y., Quéval R. & Bares J.F. (1968). – Enquete sur l'infection a virus parainfluenza 3 chez le dromadaire tchadien.*Rev. Elev. Méd. Vét. Pays Trop.*, 21 (4), 443–449.

[14] Murphy F.A., Gibbs E.P.J., Horzinek M.C. & Studdert M.J. (1999). – Adenoviridae. *In* Veterinary virology (F.A. Murphy,E.P.J. Gibbs, M.C. Horzinek & M.J. Studdert, eds). Academic press, San Jose, CA, 327–334.

[15] Olaleye O.D., Baba S.S. & Omolabu S.A. (1989). – Preliminary survey for antibodies against respiratory viruses amongslaughter camels (*Camelus dromedarius*) in north-eastern Nigeria. *Rev. sci. tech. Off. int. Epiz.*, 8 (3), 779–783.

[16] Picton R. (1993). –*Serologic survey of llamas in Oregon for antibodies to viral diseases of livestock.* Master's thesis, Oregon StateUniversity, Corallis, Oregon, USA.

[17] Puntel M., Fondevila N.A., Viera B.J., O'Donnell V.K., Marcovechio J.F., Carillo B.J. & Schudel A.A. (1999). – Serological survey of viral antibodies in llamas (*Lama glama*) in Argentina. *J. Vet. Med. B.*, 46, 157–161.

[18] Radostits O.M., Gay C.C., Hinchcliff K.W. & Constable P.D. (2007). – Veterinary medicine, 10th Ed. Saunders Elsevier,USA, 1343–1348.

[19] Reed Ch. (2010). – Import risk analysis: Llamas (*Lama glama*) and alpacas (*Vicugna pacos*) from specified countries. MAFBiosecurity New Zealand, 24–30.

[20] Rivera H., Madwell B.R. & Ameghino E. (1987). – Serological survey of viral antibodies in the Peruvian alpaca (*Llama*

*pacos).Am. J. Vet. Res.*, 48, 189–191.

[21] Rosadio R., Cirilo E., Manchego A. & Rivera H. (2011). – Respiratory syncytial and Parainfluenza type 3 viruses coexisting with Pasteurella multocida and Mannheimia hemolytica in acute pneumonias of neonatal alpacas. *Small Ruminant Res.*,97 (1), 110–116.

[22] Tillman Ch.B. (2001). – Research Updates: Llama and alpaca diseases. Tillman Llamas and Suri Alpacas, Bend, OR.

[23] Wernery U., Fowler M.E. & Wernery R. (1999). – Color atlas of camelid hematology. Blackwell Wissenschaffsverlag Berlin,Wien, 1–5.

[24] Wernery U. & Kaaden O.-R. (2002). – Infectious diseases in camelids. Blackwell Science Berlin, Vienna, 161–167.

(高闪电译，殷宏校)



## 2.2.3 牛瘟

在新发病不断出现，同时几乎被遗忘的疫病再次出现的时候，高度致死性的牛瘟已经在我们的星球上灭绝。这是历史上人类第一次扑灭一种动物疾病，也是继1980年消灭天花（smallpox）之后第二次通过人类的努力消灭的疾病。这个疾病曾经导致广泛的饥荒，并使数百万的家养和野生动物死亡。当19世纪牛瘟入侵非洲时，导致数百万的家畜和野生动物死亡。据估计，仅该病流行时埃塞俄比亚1/3的人口死于饥饿。已知的最后一次暴发是2001年在肯尼亚。1992年联合国粮农组织（FAO）启动了全球范围根除牛瘟的计划，号召在2010年全面根除该病。1994年在巴基斯坦北部的牛瘟暴发使50000头黄牛和水牛死亡。牛瘟从未在北美出现，仅1921年曾经在巴西暴发。几个世纪以来，人们已熟知了牛瘟的临床表现和危害及对养牛业造成的灾难。为了控制该病，欧洲建立了第一个兽医学院并且颁布了控制传染病的法规。牛瘟是一种偶蹄动物感染的急性的或者亚急性的、高度接触传染的热性病毒病。该病的特征是高死亡率、出血性/败血病变、消化道的糜烂。在大家畜中，黄牛、水牛及牦牛对该病易感。不同品种的黄牛对该病的抵抗力不同。在小家畜中，绵羊、山羊及猪在自然条件下对该病是易感的。野生动物具有不同程度的易感性，在该病的流行病学中起重要作用。有证据表明该病感染的主要方式是通过含有病毒的气溶胶传播[20]。当两头动物距离在2m以内时最容易传播，这也解释了为什么牛瘟发生在水坑或者市场附近。牛瘟病毒在环境中不易存活，排毒期少于15天。

### 病原学

牛瘟病毒是属于副黏病毒科麻疹病毒属的负链RNA病毒。它与犬瘟热病毒、麻疹病毒和小反刍兽疫病毒关系较近。牛瘟病毒在不同动物的致病性不同，病毒的毒力变化很大，从特强毒（沙特阿拉伯株）到温和毒（埃及株）。在牛瘟被确诊前，弱毒株可能已在未检测的动物中流行了很久。

### 流行病学

牛瘟的自然宿主是偶蹄目的所有动物。在新大陆骆驼从未报道过牛瘟，然而实验研究发现羊驼有轻微的发热症状，病程3～5天[11]。如口蹄疫一样，旧大陆骆驼是否感染牛瘟存在争议。

### 临床症状和病理变化

20世纪初的报告描述了牛瘟在骆驼中的暴发。Curasson[8]报道了1892年尼日尔牛瘟的大流行，单峰驼发生了严重腹泻和血尿症。Vedernikov（1902，引自8）和Tschegis（1902，引自8）于1898年在阿塞拜疆巴库周边的地区看到了双峰驼患牛瘟的病例。1899年，Tartakowsky用4峰双峰驼进行人工感染实验，观察到了牛瘟，其中有1峰死亡；在用两峰单峰驼实验时，只产生了轻微的临床反应。Lingard[14]用感染牛瘟病毒的黄牛血接种5峰印度单峰驼，单峰驼出现了发热、水泡、嘴部溃疡、发疹性皮肤病变，有1峰出现腹泻，1头公牛接种1头单峰驼的血后出现了牛瘟症状。Cross等[7]将牛瘟病毒接种3峰印度单峰驼，其中1峰后来死亡。Conti[6]在厄立特里亚也在单峰驼发现了牛瘟症状。

在印度，Haji[12]发现黄牛和水牛牛瘟流行期间，单峰驼也被感染，症状为发热、第一胃无力、眼部有分泌物、精神沉郁、偶尔带血的严重腹泻，唇部和硬腭有水泡并发展成溃疡，死亡率在20%～40%。Dhillon[9]在印度报道了相似的发现，在1948年和1958年，他在单峰驼观察到了15次牛瘟暴发，有的死亡率达到100%。单峰驼的牛瘟症状与黄牛相似。Srinivasan[26]利用感染牛瘟的山羊血接种单峰驼成功地控制了一次单峰驼的牛瘟暴发。

与上述报告相反，多个团队报道骆驼对牛瘟不是易感的。Littlewood在埃及[15]、Pecaud在乍得[21]、Samartsev和Arbuzov[23]在俄罗斯亚细亚地区都曾报道骆驼对牛瘟是不易感的。Leese[13]在印度没有看见或听说单峰驼发生牛瘟，作者也不排除因症状轻微而被忽略的可能。

直到20世纪中期，当与牛瘟相似临床症状的疫病流行时，其诊断是基于临床症状以及与单峰驼临近的偶蹄动物之间扩散的趋势。牛瘟的实验室诊断方法是后来才被应用的。Scott和MacDonald[24]于1960年在肯尼亚北部的野生动物中确诊了一次严重的牛瘟暴发。在该地区的单峰驼没有发生牛瘟。用兔化致弱的牛瘟病毒抗原检测60峰单峰驼的血清是阴性的。Chauhan等[3]检测了来自印度的283峰单峰驼的血清，没有检测到牛瘟抗体。然而，Maurice等[17]发现来自乍得的单峰驼血清中，7.7%为阳性。Singh和Ata[25]从苏丹和埃及的单峰驼测得抗体阳性率为10%，且Abou-Zaid[1]在埃及的536峰单峰驼发现了5.2%的牛瘟中和抗体阳性。

用牛瘟病毒实验感染单峰驼产生了更多关于单峰驼对牛瘟易感性的资料。用含有牛瘟病毒的气溶胶感染10峰单峰驼，只有一峰出现了亚临床的、非接触性的感染，观察到了白细胞减少，且在血清中测到了牛瘟抗体。与骆驼接触的瘤牛没有发病也没有牛瘟抗体产生[22]。Singh和Ata[25]在实验中用了两株强毒和两株致弱的疫苗株，皮下感染单峰驼后未出现牛瘟。接种强毒株后体温轻微上升，接种弱毒株的单峰驼仅有低效价的抗体，然而接种强毒的单峰驼在第28天出现较高的中和抗体。该实验也表明感染的单峰驼没有将病毒传播到易感牛。

Taylor[28]通过进一步接种证实了以上结果，且完成了更多的实验，结果如下。

（1）用牛瘟强毒株（Kabete O）人工静脉感染后，第3天和第8天从感染单峰驼的血液中再次分离到了病毒，该动物也产生了中和抗体。

（2）用强毒RGK/1感染的两峰单峰驼均产生了中和抗体，其中的1峰有轻微的病毒血症，持续了6天。

（3）3峰与感染牛瘟的公牛接触了的单峰驼中，两峰出现了轻微的病毒血症。1峰单峰驼出现轻微发热，这也是实验期间唯一可被观察的症状。

（4）尽管牛瘟感染来源于牛，但是感染的单峰驼并不能将牛瘟病毒传播至牛或者其他单峰驼。

为了确定骆驼对实验性牛瘟感染的易感性，Chauhan等[4]完成了进一步实验。他们将从感染牛瘟的水牛牛犊采集的脾脏制成10%的悬液，用10mL接种两峰8～12月大的健康骆驼。一峰骆驼皮下注射，另一峰骆驼静脉注射。除了有一点可视黏膜出血和轻微腹泻之外，感染牛瘟的一峰骆驼剖检并未发现明显病变。Chauhan等[4]进一步发现，在发热反应高点时收集骆驼的血液注射到两头易感的水牛牛犊，在注射后6天内牛犊表现出典型的牛瘟临床症状和病变。

这些实验显示旧大陆骆驼对牛瘟是易感的，能表现轻微的临床症状，特别是在和感染牛接触的情况下。这些骆驼作为牛瘟病毒媒介的可能性不大，它们在牛瘟流行病学中的作用不大。有个别研究者利用不同的测试方法测定了牛瘟抗体，完成了血清流行病学研究。Mankinde等[16]用免疫电泳法测定了埃及488峰单峰驼的血清，发现40%左右的血清呈阳性。作者认为高水平的牛瘟抗体提示骆驼与牛瘟病毒有密切接触。Al-Afaleq等[2]用血清中和试验发现沙特阿拉伯20%的骆驼抗体阳性。作者认为在骆驼的血清中出现牛瘟病毒抗体可能是由于接种了小反刍兽疫疫苗，该疫苗在沙特阿拉伯广泛应用于绵羊和山羊，或者是由于温和牛瘟病毒株的流行。然而，Wernery等[29]用竞争性酶联免疫反应（c-ELISA）测定牛瘟病毒抗体，从1119峰阿联酋的单峰驼并未测到抗体。C-ELISA是OIE推荐的在国际贸易中使用的方法。与在阿联酋的结果想似，在沙特阿拉伯两个地区的60峰单峰驼未检测到牛瘟病毒抗体[19]。

1998年，尼日利亚宣布消灭了牛瘟，并停止了接种牛瘟疫苗。然而，用c-ELISA测定220峰骆驼血清，发现9.3%的血清牛瘟抗体阳性（尽管骆驼没有接种疫苗），这是一个重要的发现，应该通过病毒分离进一步深入研究[10]。

### 诊断

有几个操作手册和科技论文详细描述了牛瘟的诊断。基于临床症状和大体病例变化在牛上能做出推测性诊断，但在骆驼是比较困难的。尽可能快地获得实验室确诊结果是非常重要的，这对预防控制措施的实施至关重要。Mirchamsy等[18]报道牛瘟病毒在骆驼肾细胞上容易生长。可用的诊断方法包括以下几种：

（1）免疫荧光素和免疫过氧化物酶；

（2）反向免疫电泳；

（3）ELISA；

（4）聚合酶链反应；

（5）病毒分离和细胞微量中和反应。

许多血清学的方法已用于牛瘟诊断，包括：

（1）补体结合试验；

（2）血凝抑制试验；

（3）病毒中和试验；

（4）c-ELISA。

### 防控

牛瘟暴发确诊后通过扑杀来控制，包括全部感染和接触的动物，也包括动物移动控制和严格检疫。

在流行地区，牛瘟的预防要求两岁以内的牛犊每年接种弱毒kabete O 疫苗。该疫苗是一种便宜的、冻干的高效疫苗，但未用于骆驼[5]。用痘病毒载体表达牛瘟病毒蛋白的重组疫苗也已研制成功，这种疫苗可诱导保护性的、长期的免疫应答。



## 参考文献


[1] Abou-Zaid A.A. (1991). – Studies on some diseases of camels. PhD thesis, Faculty of Veterinary Medicine, Zagazig University,Zagazig, Egypt.

[2] Al-Afaleq A.I., Abu-Elzein E.M.E., Hegazy A.A. & Al-Naeem A. (2007). –Serosurveillance of camels (*Camelus dromedarius*) to detect antibodies against viral diseases in Saudi Arabia. *J. Camel. Pract. Res.*, 14 (2), 91–96.

[3] Chauhan R.S., Kaushik R.K., Gupta S.C., Satiya K.C. &Kulshreshta R.C. (1986). – Prevalence of different diseases in camels(*Camelus dromedarius*) in India. *Camel Newsletter*, 3, 10–14.

[4] Chauhan R.S., Kulshreshtha R.C. & Kaushik R.K. (1985). – Epidemiological studies of viral diseases of livestock in HaryanaState. *Ind. J. Virol.*, 1 (1), 10–16.

[5] Coetzer J.A.W., Thomson G.R. & Tustin R.C. (1994). – Infectious diseases of livestock with special reference to Southern Africa, vol. 2. Oxford: Oxford University Press, 1518–1535.

[6] Conti G. (1913). – A serious prophylactic problem: Rinderpest in camel. *ModernaZoolatro Torino*, 24, 1–12.

[7] Cross H.E. (1919). – Are camels susceptible to blackquarter, haemorrhagic septicaemia and rinderpest? *Bull. Agric. Res.Inst. Pusa.*

[8] Curasson G. (1947). – Le chameau et ses maladies. Vigot Freres, Paris, 86–88.

[9] Dhillon S.S. (1959). – Incidence of Rinderpest in camels in Hissar district. *Indian Vet. J.*, 36, 603–607.

[10] El-Yuguda A.D., Abubakar M.B., Baba S.S. &Ngangnou A. (2010). – Competitive ELISA Rinderpest virus antibody in slaughtered camels (*Camelusdromedarius*): Implication for Rinderpest virus elimination from Nigeria. *Afr. F. Biomed. Res.*,13, 83–85.

[11] Fowler M.E. (2010). – Medicine and surgery of Camelids. 3rd Ed. Blackwell Publishing, Ames, IA, 181–182.

[12] Haji C.S.G. (1932–1933). – Rinderpest in camels. *Ind. Vet. J.*, 9, 13–14.

[13] Leese A.S. (1927). – A treatise on the one-humped camel in health and disease. Vigot Freres, Paris, II.

[14] Lingard A. (1905). – Report on the preparation of Rinderpest Serum, Calcutta. Abst. *Bull. Inst. Pasteur*, 4, 235.

[15] Littlewood W. (1905). – Camels are not susceptible to Rinderpest. *J. Comp. Path.*, 18, 312.

[16] Mankinde A.A., Majiyagbe K.A., Chwzet N., Lombin L.H. *et al.* (2001). – Serological appraisal of economical diseases of livestock in the one-humped camel *(Camelusdromedarius)* in Nigeria. *Camel Newslett.*,18, *Camel Applied Research and Development Network (CARDN)*, 62–72.

[17] Maurice Y., Provost A. &Borredon C. (1967). – Presence d'anticorps antibovipestiques chez le dromadaire du Tchad. *Rev. Elev. Méd. Vét. Pays Trop.*, 20 (4), 537–542.

[18] Mirchamsy H., Bahrami B., Amighi M. & Shafyi A. (1971). – Development of a camel kidney cell strain and its use in virology. *Arch. Inst. Razi*, 23, 15–18.

[19] Mohammed D.B., Sandonka M.A. & Mon Elzai E.M.E. (2003). – Disease surveys of livestock in some protected areas where gazelles have been reintroduced in Saudi Arabia. In Proceedings of diseases at the interface between domestic livestock and wildlife species, 17–18 July, Ames, Iowa, USA.

[20] Munz E. (1983). – Die heutige Situation auf demGebiet der tropischenTierseuchen, ihre derzeitige Gefahr fur die Landwirtschaft der Bundesrepublik Deutschland und ihre Bedeutung fur den Tierarzt. *Der prakt. Tierarzt.*,11, 993–1006.

[21] Pecaud G. (1924). – Contribution a l'etude de la pathologie vétérinaire de la colonie du Tchad. *Bull. Soc. Path. Exot.*,17 (3),196–207.

[22] Provost A., Maurice Y. &Borredon C. (1968). – Note sur la peste bovine expérimentale du dromadaire. *Rev. Elev. Méd. Vét. Pays Trop.*, 21 (3), 293–296.

[23] Samartsev A.A. &Arbuzov P.N. (1940). – The susceptibility of camels to glanders, rinderpest and bovine contagious pleuropneumonia.*Veterinariya Moscow*, 4, 59–63.

[24] Scott G.R. & MacDonald J. (1962). – Kenya camels and Rinderpest. *Bull. epizoot. Dis. Afr.*, 10 (4), 495–497.

[25] Singh K.V. & Ata F. (1967). – Experimental Rinderpest in camels. A preliminary report. *Bull. epizoot. Dis. Afr.*, 15, 19–23.

[26] Srinivasan V. (1940). – Active immunisation of camels against Rinderpest with goat blood virus. *Ind. Vet. J.*, 16, 259–260.

[27] Tartakowsky M.M. (1899). – Letter. *Arch. Sci. biol. St. Petersburg*, 8, 11.

[28] Taylor W.P. (1968). – The susceptibility of the one-humped camel (*Camelus dromedarius*) to infection with Rinderpest virus. *Bull. epizoot. Dis Afr.*, 16, 405–410.

[29] Wernery U., Thomas R., Syriac G., Raghavan R. &Kletzka S. (2007). –Seroepidemiologicalstudies for the detection ofantibodies against nine infectious diseases in dairy dromedaries (Part I). *J. Camel Pract. Res.*, 14 (2), 85–90.

（独军政译，殷宏校）

## 2.2.4 非洲马瘟

非洲马瘟（African horse sickness, AHS）是一种感染马、骡、驴等马属动物的高度致死的虫媒病毒病。已知该病在非洲存在数百年。非洲马瘟是一种水泡性血管内皮病，具有多种不同的表现。病毒毒力的不同在马表现的临床症状各异[7]。

### 病原学

非洲马瘟病毒（AHSV）属于呼肠孤病毒科环状病毒属，它与另外的环状病毒如蓝舌病病毒和马器质性脑病病毒特性相似。非洲马瘟病毒粒子由双层衣壳包裹的10个双链RNA基因片段组成。

与蓝舌病病毒相似，病毒粒子的外壳由VP2和VP5两种蛋白组成，而内壳由两种主要蛋白VP3和VP7以及3种次要蛋白VP1、VP4、VP6组成。在非洲马瘟病毒感染的细胞，发现了NS1、NS2、NS3和NS3A四种非结构蛋白。外层衣壳的VP2蛋白负责病毒的血清型、抗原性以及中和抗体的产生，9种血清型的非洲马瘟病毒除了6型和9型外，无交叉

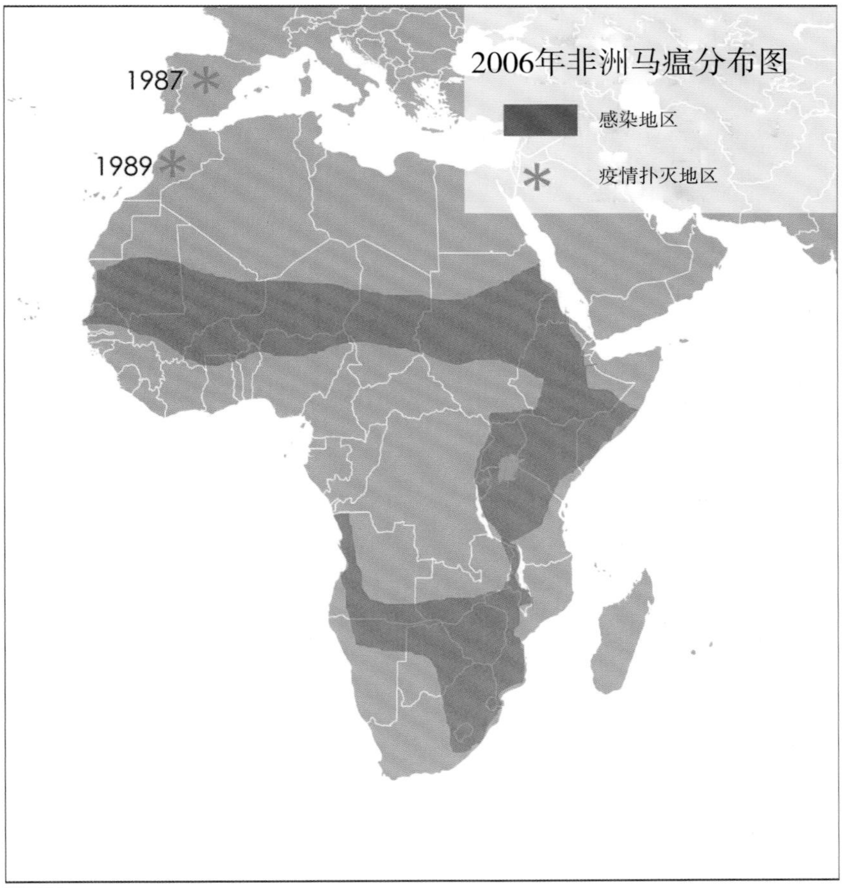


图182 非洲马瘟的地理分布[16]

反应性。

在培养细胞上出现的病毒噬斑的大小代表病毒的毒力，小噬斑的毒株比大噬斑的毒株毒力更强。非洲马瘟病毒可在BHK-21、Vero和猴stable（MS）细胞上生长并出现细胞病变。

## 流行病学

目前，非洲马瘟的大暴发依旧经常出现。非洲马瘟首次于1719年在南非的开普敦暴发。此后，严重的暴发陆续报道，包括1943–1944年在埃及和巴勒斯坦，以及1959–1960年在中东和东南亚300 000只马匹死亡。1965年，该病传播到摩洛哥、阿尔及利亚和突尼斯，1966年传至欧洲大陆。然而，通过疫苗免疫和扑杀政策，血清型9很快在西班牙和北非扑灭。1987年，血清型4在马德里、托雷多和阿维拉等省暴发，2000多匹马在1989年死亡，该病在1991年扑灭[16]。该病由从纳米比亚进口到西班牙的斑马带入。这次暴发从1987年持续到1990年，并于1989年传播到葡萄牙和摩洛哥。非洲马瘟在西班牙持续了4年可能是由于拟蚊库蠓能在南西班牙过冬。由血清型9引发的非洲马瘟1989年报道于沙特阿拉伯，1997年报道于也门和沙特阿拉伯[12]。

该病流行于撒哈拉南部非洲的广大热带和亚热带地区，西到塞内加尔，东到埃塞俄比亚，南至南非（图182）。这些地区的大部分国家饲养单峰驼。

该病的传播受气候条件的影响很大，气候影响非洲马瘟传播媒介库蠓的繁殖。目前已经确定在南非非洲马瘟的流行时间与厄尔尼诺现象带来的气候变化有直接的联系[4]。随着厄尔尼诺现象带来的全球气温变化，重要虫媒传播疾病已受人们关注。血清型1～8型对马具有高毒力，致死率可达95%，血清型9的致死率可达70%。非洲马瘟病毒能感染所有的马属动物和犬科动物。马是最易感的，其次是骡。驴和斑马具有一定的抵抗力，经常表现为亚临床症状。该病毒由库蠓传播，对拟库蠓和bolitinos库蠓是最重要的媒介。这些库蠓能随气流移动数十千米。蜱在该病毒传播中作用不大，但非洲马瘟病毒能够在嗜驼璃眼蜱体内复制，该蜱主要寄生于骆驼[2]。非洲马瘟病毒的储藏宿主目前仍不清楚。

### 临床症状和病理变化

在非洲单峰驼血清学测定发现，非洲马瘟病毒抗体阳性率在埃及为5%[1]，苏丹是23%[10]。Salama等[13]检测了134只苏丹骆驼和266只埃及骆驼，抗体阳性率分别为23.2%和5.6%。在尼日利亚，Baba等[3]用血凝抑制试验发现96只单峰驼中10.4%的抗体阳性。作者认为骆驼和犬是非洲马瘟的重要储藏宿主。Binepal等[5]在24只东非单峰驼中没有检测到抗体，同样，Wernery（1992，未发表资料）对阿联酋的500只单峰驼用琼脂免疫扩散试验（AGID）也未测到抗体。2003年和2009年用竞争ELISA分别在南摩洛哥的556和836份单峰驼血清中测到了抗体[15]，阳性率总体比较低（1%）。然而，2003年的血清学调查发现，20世纪90年代马属动物暴发过非洲马瘟的地区，骆驼的抗体阳性率为10%[14]。在新大陆骆驼没有非洲马瘟血清学的相关调查。

Salama等[13]从两只健康的骆驼用小鼠分离了两株9型非洲马瘟病毒。该病毒的第五代被注射到两只易感马后，马出现了典型的非洲马瘟症状。

在埃及，从单峰驼的蜱（嗜驼璃眼蜱）中也分离到了非洲马瘟病毒。用带毒的蜱感染这些动物并未出现症状。在2089只蜱中，通过小鼠接种测试发现17%的蜱带有非洲马瘟9型病毒[14]。感染病毒的嗜驼璃眼蜱的幼虫和若虫能够传播病原至易感动物并出现非洲马瘟症状。感染的幼虫也能在成虫期传播该病[10]。Dardiri和Salama[9]也从埃及南部的单峰驼血液中分离到了非洲马瘟病毒。嗜驼璃眼蜱能够传播病毒，但是骆驼并未表现临床症状。Boinas等[6]在他们的科技论文中写道，骆驼对非洲马瘟的感染是不明显的，且很罕见。感染的程度和病毒血症尚不清楚，目前认为骆驼在非洲马瘟的流行病学中并不重要。

所有以上研究表明，单峰驼能够作为非洲马瘟病毒的储藏宿主，但到目前为止，并未见在旧大陆骆驼或新大陆骆驼发生该病的报道。已报道旧大陆骆驼可感染非洲马瘟病毒，但是其在非洲马瘟流行病学中的作用至今尚不清楚[11]。

该病可表现为4种形式。肺型在3～5天的潜伏期后发生，表现为发热和严重的呼吸困难，病理表现为明显的肺小叶广泛水肿。心脏型一般在比较长的潜伏期后发生，以间歇热和心力衰竭为特征，剖检可见躯体前部广泛皮下水肿，在器官表面有淤血和淤斑，且伴有心包积水。第三种形式为肺和心的混合型，比较罕见。最后一种形式为温和型，这可见于部分免疫动物，出现流感样症状后痊愈。

### 诊断

在马属动物，通过非洲马瘟的临床症状和肉眼病变可做出初步诊断。然而，要确诊本病，需要完成病毒学实验。应该从发热期的血液和剖检马的肺脏、脾脏以及淋巴结分离到病毒。血液是动物死亡前用于病毒分离的最佳样品。应该在血液中加入乙二胺四乙酸（EDTA）抗凝剂，置于4℃环境运送至实验室。样品不应于-20℃冻存。应该用BHK-21、Vero或MS细胞以及乳鼠的脑内接种来分离病毒。因为临床症状出现之前发热期间病毒的滴度最高，所以病毒分离应该尽可能快地完成。病毒分离物应该用补体结合试验（CFT）、琼脂免疫扩散试验（AGID）、间接荧光抗体（IFA）和酶联免疫吸附试验（ELISA）进行鉴定。利用型特异的抗血清通过病毒中和进行非洲马瘟血清型的鉴定。ELISA是检测抗体最好的血清学方法。用ELISA方法检测发现注射了疫苗的马9年后抗体依然呈阳性[8]。AGID、CFT、HIT已被用于骆驼非洲马瘟抗体的测定。迪拜中央兽医研究实验室用一种夹心ELISA检测了293只阿联酋单峰驼，未检测到抗体[8]。更快的、敏感的、特异的新型诊断技术已用于非洲马瘟的诊断，如聚合酶链反应或基因组探针法。

目前已能从血液、肺脏和脾脏中检测非洲马瘟病毒的基因组，能区分9种血清型非洲马瘟病毒的反转录-聚合酶链反应（RT-PCR）方法已经建立。

### 防控

没有用于治疗非洲马瘟的具体方法，感染非洲马瘟的马匹应该安乐死并严格处理尸体。

由于非洲马瘟有多个血清型，在该病流行地区预防非洲马瘟应使用多价疫苗。多价疫苗是适应细胞培养的弱毒疫苗，非洲有多个国家生产，但是主要的生产商家是南非的Onderstepoort生物制品公司（Onderstepoort Biological Products，OBP）。OBP的非洲马瘟疫苗分装在两个疫苗瓶，包含7个毒株，一个瓶内含有血清型1、3、4，另外一个瓶含有血清型2、6、7、8。该疫苗不包含血清型5和9。血清型5由于严重的副反应于1993年被去掉，血清型9被去掉是因为它与血清型6有交叉反应。尽管接种程序比较严格，但是已经在接种过的马匹发现过非洲马瘟病例。灭活疫苗不含有活的、具有潜在危险的病毒，也可用于该病的预防。如弱毒疫苗一样，为了保持高水平的保护，多次接种是必要的。然而，目前市场上没有灭活疫苗，新型亚单位疫苗正在研制。

该病可通过在日落之后使马匹推迟进圈舍和日出之前提前出圈舍进行预防。库蠓在夜间不活动，这样它们将不能进入圈舍。杀虫剂对控制非洲马瘟也有积极作用。赛马一般不注射非洲马瘟疫苗，因为在国际贸易或竞赛时可能检测不过关。由于骆驼对非洲马瘟病毒不易感，不必要对其注射疫苗。



## 参考文献


[1] Awad F.I., Amin M.M., Salama S.A. & Aly M.M. (1981). – The incidence of African Horse Sickness antibodies in animals of various species in Egypt. *Bull. Anim. Health Prod. Afr.*, 29, 285–287.

[2] Awad F.I., Amin M.M., Salama S.A. & Knide S. (1981). – The role played by *Hyalommadromedarii* in the transmission of African horse sickness virus in Egypt. *Bull. Anim. Health Prod. Afr.*, 29, 337–340.

[3] Baba S.S., Olaleye O.D. & Ayanbadejo O.A. (1993). – Haemagglutination-inhibiting antibodies against African Horse Sicknessvirus in domestic animals in Nigeria. *Veterinary Research*, 24 (6), 483–487.

[4] Baylis M., Mellor P.S. & Meiswinkel R. (1999). – Horse sickness and ENSO in South Africa. *Nature*, 397 (2), 574.

[5] Binepal V.S., Wariru B.N., Davies F.G., Soi R. & Olubayo R. (1992). – An attempt to define the host range for African horsesickness virus (Orbivirus, Reoviridae) in East Africa, by a serological survey in some Equidae, Camelidae, Loxodontidae and Carnivore. *Vet. Microbiol.*, 31 (1), 19–23.

[6] Boinas F., Calistri P., Domingo M., Aviles M.M., López M.M., Sónchez B.R. & Sanchez-Vizcaino J.M. (2009). – Scientific review on African Horse Sickness. Scientific report published by the European Food Safety Authority, 1–61.

[7] Coetzer J.A.W., Thomson G.R. & Tustin R.C. (1994). – Infectious diseases of livestock with special reference to Southern Africa, vol. 2. Oxford University Press, Oxford, 1518–1535.

[8] CVRL (1998). – Annual Report. Central Veterinary Research Laboratory, Dubai, United Arab Emirates, 19.

[9] Dardiri A.H. & Salama S.A. (1988). – African horse sickness: An overview. *J. Equine Vet. Sci.*, 8 (1), 46–49.

[10] United States Department of Agriculture (1988). – Foreign animal disease report. Animal and Plant, Health Inspection Service. Veterinary Services, Washington DC, 16 (4), 26–31.

[11] Guthrie A.J. (2008). – African horse sickness. In Foreign animal diseases (The Gray Book), 7th United States Anim., Health Assoc., St. Joseph, 103–109.

[12] MacLachlan N.J. & Guthrie A. (2010). – Re-emergence of bluetongue African horse sickness and other orbivirus diseases.*Vet. Res.*, 41, 35.

[13] Salama S.A., Abdallah S.K., El-Bakry M. & Hassanein M.M. (1986). – Serological studies on African Horse Sickness virusin camels. *Assiut Vet. Med. J.*, 16 (31), 379–390.

[14] Salama S.A., El-Husseini M.M. & Abdalla S.K. (1979 and 1980). – 3rd & 4th Ann. Rep. US AHS Project 169, Cairo, 55–69,91–98.

[15] Touil N., Cherkaoui Z., Lmrabih Z., Loutfi C., Harif B. & El Harrak (2012). – Emerging viral diseases in dromedary camels in Southern Morocco. *Trans. Emerg. Dis.*, 59 (2), 177–182.

[16] Zientara St. (2010). – African horse sickness. *In* Infectious and parasitic diseases of livestock (P.-C. Lefevre, J. Blancou,R. Chermette & G. Uilenberg, eds). Lavoisier, 689 – 704.

（独军政译，殷宏校）



## 2.2.5 逆转录病毒感染

### 引言

逆转录病毒是一类RNA病毒，可感染大部分脊椎动物和一些无脊椎动物。马传染性贫血病毒（EIAV）是早在1904年第一个报道的逆转录病毒，确认马传染性贫血的致病原是病毒性病原。逆转录病毒科由许多RNA病毒组成，包括7个属，其中有家畜疾病相关的有3个属。

（1）$\beta$逆转录病毒属——绵羊肺腺瘤病毒（JSRV）、山羊肺腺瘤病毒、地方性鼻内肿瘤病毒。

（2）$\delta$逆转录病毒属——牛白血病病毒（BLV）。

（3）慢病毒属——牛免疫缺陷病毒、EIAV、小反刍兽慢病毒。

人免疫缺陷病毒（HIV）也属于慢病毒属，自从1980年左右第一次报道HIV以来，2 500万人们已经死于该病，全世界大约有4 000万人被感染。大动物感染逆转录病毒的病例已在多个大洲出现。本病既能水平传播也能垂直传播。水平传播主要通过初乳或牛奶、直接接触和昆虫叮咬。医源性的传播也有发生。出生后胎儿与母亲立即分离可避免被感染[6]。

由牛白血病病毒引起的地方性牛白血病是牛的一种恶性肿瘤病，主要在淋巴结出现肿瘤，它是一种恶性的淋巴瘤或淋巴肉瘤。该病可能起源于欧洲，19世纪末，通过东普鲁士出口的黄牛传入新大陆。地方性牛白血病容易被传到驼峰牛、水牛和小反刍兽[9]，特别在绵羊被感染时容易发生淋巴肉瘤，且感染能被实验性地传播到不同种类的动物，该病没有传给人的风险。

值得注意的是，从散发形式的地方性牛白血病病例中没有分离到牛白血病病毒或其他传染性病原，且在这些病例也未检测到抗体。

牛白血病病毒是一种RNA病毒，属于逆转录病毒科白血癌病毒亚科，该病毒1963年在牛的淋巴细胞培养物中被发现[8]。

### 流行病学

地方性牛白血病能很容易地传播到许多种动物，但是骆驼似乎对该病不易感，到目前为止，Lee等[5]在幼年羊驼报道了唯一的病例。

在新大陆骆驼和旧大陆骆驼，几个关于地方性牛白血病抗体检测的血清学研究已经报道。Chauhan等[3]在283只单峰驼没有发现地方性牛白血病抗体。Wernery等[18]用琼脂免疫扩散试验测定了986只阿联酋单峰驼的血清，所有血清为阴性，然而，他们已在阿联酋单峰驼诊断出了淋巴性白血病。在20年间，所有患淋巴性白血病的单峰驼在血清或器官匀浆中没有出现牛白血病病毒的抗体[4]。患病的单峰驼全部在8岁以上，全部出现白细胞增多症（表71），主要由淋巴母细胞组成（图 183）。

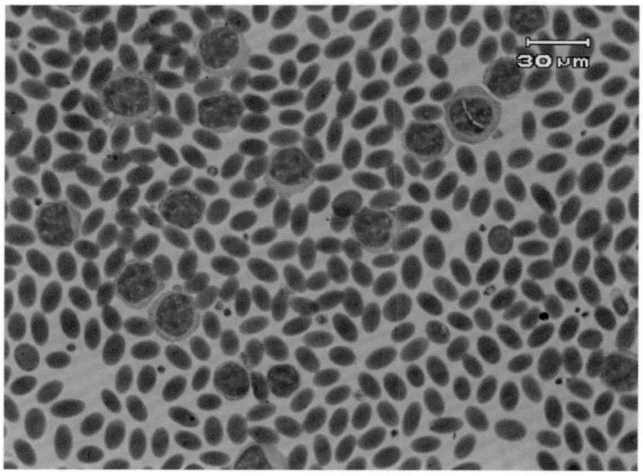


图183　单峰驼的淋巴母细胞白血病（苏丹黑染色）

表71 阿联酋单峰驼淋巴母细胞白血病病例

| 病例 | 白细胞数 ×$10^9$个/L | 红细胞数 ×$10^{12}$个/L | 血红蛋白（g/dL） | 细胞分类数 | | | |
|---|---|---|---|---|---|---|---|
| | | | | 淋巴细胞 | 中性粒细胞 | 嗜酸性细胞 | 单核细胞 |
| 1 | 818.5 | 8.5 | 12.0 | 99 | 1 | 0 | 0 |
| 2 | 126.4 | 9.2 | 12.6 | 99 | 1 | 0 | 0 |
| 3 | 157.4 | 9.6 | 13.7 | 100 | 0 | 0 | 0 |
| 4 | 142 | 9.6 | 14.0 | 99 | 1 | 0 | 0 |
| 5 | 44.7 | 6.8 | 11.0 | 98 | 2 | 0 | 0 |
| 6 | 949.3 | 7.5 | 12.0 | 94 | 6 | 0 | 0 |
| 7 | 45.1 | 7.3 | 11.6 | 98 | 2 | 0 | 0 |
| 8 | 204.8 | 5.3 | 8.9 | 92 | 8 | 0 | 0 |
| 9 | 226.0 | 7.9 | 10.3 | 98 | 2 | 0 | 0 |
| 10 | 217.0 | 3.3 | 7.0 | 98 | 2 | 0 | 0 |

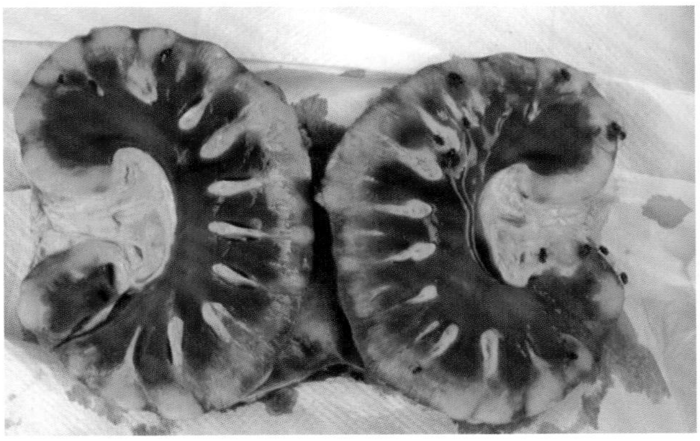

a. 肾脏白细胞增生病灶

b. 纵隔淋巴结白细胞增生病灶

图184

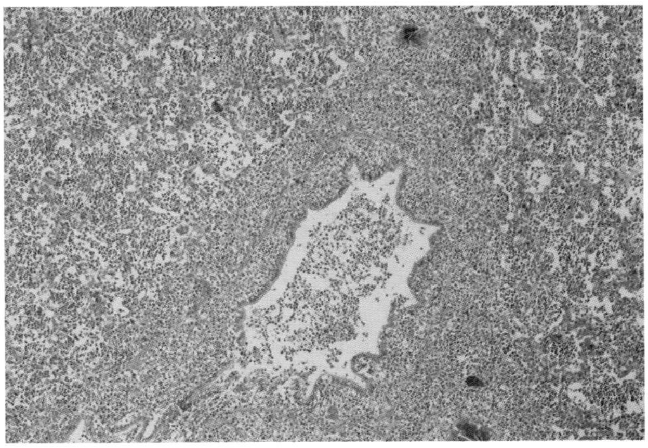

图185 患有淋巴母细胞白血病的竞赛骆驼肺脏新生淋巴样细胞（苏木精-伊红染色）

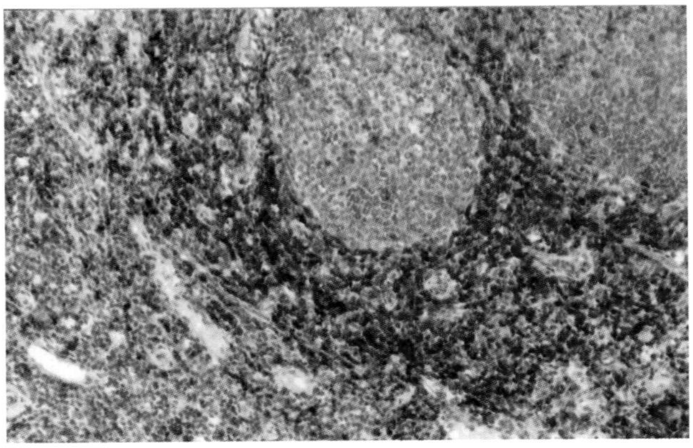

图186 单峰驼肺脏白细胞增生的免疫组化染色（抗人CD3抗体）

### 致病机理

已知的该病流行形式有两种：地方性流行和散发。地方性牛白血病主要见于成年牛（成年淋巴肉瘤）。诊断为白血病的全部单峰驼在6个月内死亡。剖检可见，肺部和纵隔淋巴结广泛出现白细胞浸润。Afzal和hussein[1]及Wernery和Kumar[17]观察到淋巴结肿大、继发性肾盂肾炎、支气管肺炎、子宫内膜炎。组织病理检查显示，在肺脏、脾脏和淋巴结有肿瘤性淋巴样细胞浸润（图 184），新生淋巴样细胞可见于不同器官的组织切片（图 185）。

用抗人的抗体（DAKOA0452，CD3-标记）测得80%肿瘤细胞染色阳性，进一步肯定了T细胞白细胞增生（图186）。

利用透射电镜对血淋巴细胞和福尔马林固定的肺组织检测未发现病毒粒子。诊断为白血病的两只单峰驼产出了健康的后代，其血细胞计数无异常。

从两只患有白血病的单峰驼采取10mL肝素抗凝血，通过静脉注射至两只健康骆驼，跟踪一年，血液未见异常，说明该病无传染性。

在秘鲁，与绵羊一起饲养的羊驼未测到绵羊肺腺瘤病毒抗体[12]。尽管在美国俄勒冈州绵羊感染了羊肺腺瘤病毒，但是该地区的270只美洲驼血清并未转阳[10]。在秘鲁的一项血清学研究没有发现牛白血病病毒抗体阳性的证据[12]。Puntel等在阿根廷的3个省9个农场的390只美洲驼没有发现牛白血病病毒抗体[11]。然而，如单峰驼一样，有几个关于美洲驼发生类似牛白血病病毒诱发淋巴肉瘤的报道[7]。Cebra等[2]在北美的10只新大陆骆驼诊断出了淋巴肉瘤，且Wenker等[16]在瑞士的两只骆马也诊断出了该病。在北美，被感染的是4个月至15岁的美洲驼和羊驼。5只动物的肿瘤组织用电镜检查，与在单峰驼的结果相似，没有检测到逆转录病毒粒子。在13岁和10岁大的雌骆马诊断出了T细胞起源的淋巴肉瘤，她们的父母亲相同，均来自苏黎世动物园。该病被怀疑来自遗传因素。在新大陆骆驼，淋巴瘤是最为普遍报道的肿瘤[2,14]，但是仅报道了一例幼年羊驼的牛白血病淋巴肉瘤[5]。血清学检查（酶联免疫吸附试验和琼脂免疫扩散试验）和原位聚合酶链反应（PCR）检测到了牛白血病病毒。该羊驼产于美国，下颌出现的淋巴瘤使其衰弱。病毒的来源是不清楚的。从有免疫缺陷症状的美洲驼分离到了一株逆转录病毒[13]，但是并未能证实该病毒可引起免疫缺陷症状[15]。

在旧大陆骆驼和新大陆骆驼，淋巴性白细胞增生的罕见发生、大部分病例的老龄化以及逆转录病毒的缺乏已使我们推测这是零星的白细胞增生症。然而，马的淋巴瘤和牛淋巴瘤的散发形式与骆驼淋巴瘤存在不同，尤其是器官分布（肺）和患病动物的年龄（大部分在8岁以上）。马淋巴瘤和牛淋巴瘤主要影响年轻动物，且很少出现肺部病变，所以我们建议将该病命名为散发性成年骆驼白细胞增生症（SACL）。

## 参考文献


[1] Afzal M. & Hussein M.M. (1995). – Acute prolymphocytic leukemia in the camel. *Camel Newsletter*, 11 (9), 22–24.

[2] CebraCh.K., Garry F.B., Powers B.E. & Johnson La Rue W. (1995). –Lymphosarcoma in 10 New World Camelids. *J. Vet.Int. Med.*, 9 (6), 381–385.

[3] Chauhan R.S., Kaushik R.K., Gupta S.C., Satiya K.C. &Kulshreshta R.C. (1986). – Prevalence of different diseases in camels(*Camelus dromedarius*) in India. *Camel Newsletter*, 3, 10–14.

[4] Kinne J. &Wernery U. (2006). –Lymphosarcoma in dromedaries in the United Arab Emirates (UAE). *In* Proceedings of the International Scientific Conference on Camels, 12–14 May, Qassim University, Saudi Arabia, 714–723.

[5] Lee L.C., Scaratt W.K., Buehring G.C. & Saunders G.K. (2012). – Bovine leukemia virus infection in a juvenile alpaca with multicentric lymphoma. *Can. Vet. J.*, 53 (3), 283–286.

[6] Leroux C., Mornex J.-F. &Cadore J.-L. (2010). – An introduction to animal retroviruses. *In* Infectious and parasitic diseases of livestock (P.-L. Lefevre, J. Blancou& R. Chermette, eds). Lavoisier, 561–583.

[7] Mattson D.E. (1994). – Viral diseases. *Vet. Clin. N. Am. Food Anim. Pract.*,10 (2), 345–351.

[8] Miller J.M., Miller L.D., Olson C. &Gilette K.G. (1969). – Virus-like particles in phytohemmagglutinin-stimulated lymphocyte cultures with reference to bovine lymphosarcoma. *J. Natl Cancer Inst.*, 43, 1297–1305.

[9] Parodi A.-L. (2010). – Enzootic bovine leukosis. In Infectious and parasitic diseases of livestock (P.-L. Lefevre, J. Blancou&R. Chermette, eds). Lavoisier, 561–582.

[10] Picton R. (1993). –*Serologic survey of llamas in Oregon for antibodies to viral diseases of livestock.* Master's thesis, Oregon State University, Corvallis, Oregon, USA.

[11] Puntel M., Fondevila N.A., Blanco Viera J., O'Donnell V.K., Marcovechio J.F., Carillo B.J. &Schudel A.A. (1999). – Serological survey of viral antibodies in llamas (*Lama glama*) in Argentina. *J. Vet. Med. B.*, 46, 157–161.

[12] Rivera H., Madwell B.R. &Ameghina E. (1987). – Serological survey of viral antibodies in the Peruvian alpaca *(Llama pacos).Am. J. Vet. Res.*, 48, 189–191.

[13] Underwood W.J., Morin D.E., Mersky M.L., Haschek W.M., Zuckermann F.A., Petersen G.C. &Scherba G. (1992). – Apparent retrovirus-induced immunosuppression in a yearling llama. *J. Am. Vet. Med. Assoc.*, 200, 358–362.

[14] Valentine B.A. & Martin J.M (2007). Prevalence of neoplasia in llamas and alpacas (Oregon State University, 2001–2006).*J. Vet. Diagn. Invest.*,19, 202–204.

[15] Vogel P. (1992). – Retroviral basis for immunosuppression remains to be proven. *J. Am. Vet. Med. Assoc.*, 201, 1318.

[16] Wenker C.J., Stuedli A. & Hauser B. (2002). – Malignant lymphoma in two vicunas (*Lama vicugna*). *In* Proceedings of the European Association of Zoo and Wildlife Veterinarians (EAZWV) and European Wildlife Disease Association (EWDA),8–12 May 2002, Heidelberg, Germany, 65–67.

[17] Wernery U. & Kumar B.N. (1996). – Pulmonary lymphosarcoma in a 16 year old dromedary camel – a case report. *J. CamelPract. Res.*, 3 (1), 49–50.

[18] Wernery U. &Wernery R. (1990). –SeroepidemiologischeUntersuchungen zum Nachweis von Antikoerpern gegen Brucellen,Chlamydien, Leptospiren, BVD/MD, IBR/IPV – und Enzootischen Bovinen Leukosevirus (EBL) bei Dromedarstuten (*Camelus dromedarius*). *Dtsch. tierärztl. Wschr.*,97, 134–135.

## 深入阅读材料

*Duncan R.B., Scarratt W.K. &Buehring G.C. (2005). – Detection of bovine leukaemia virus by in situ polymerase chain reactionin tissues from a heifer diagnosed with sporadic thymiclymphosarcoma. J. Vet. Diagn. Invest.,17, 190–194.*

*Martin J.M., Valentine B.A. &Cebra C.K. (2010). – Clinical, ultrasonographic, and laboratory findings in 12 llamas and 12 alpacaswith malignant round cell tumours. Can. Vet. J., 51, 1379–1382.*

*Martin J.M., Valentine B.A., Cebra C.K., Bildfell R.J., Lohr C.V. & Fischer K.A. (2009). – Malignant round cell neoplasia in llamasand alpacas. Vet. Pathol.,46, 288–298.*

*Pusterla N., Colegrove K.M., Moore P.F., Magdesian K.G. &Vernau W. (2006). –Multicentric T-cell lymphosarcoma in an alpaca.Vet. J., 171, 181–185.*

*Tageldin M.H., Al-Sumry H.S., Zakia A.M. &Fayza A.O. (1994). – Suspicion of a case of lymphocytic leukaemia in a camel (Camelus dromedarius) in the Sultanate of Oman. Rev. Elev. Méd. Vét. Pays Trop., 47 (2), 157–158.*

*Underwood W.J. & Bell T.G. (1993). –Multicentriclymphosarcoma in a llama. J. Vet. Diagn. Invest.,5, 117–121.*

（独军政译，殷宏校）

### 2.2.6 不常见的虫媒病毒和其他次要病毒感染

虫媒病毒（节肢动物传播的病毒）是主要的媒介病毒，它们在吸血昆虫或者蜱体内繁殖，并通过昆虫叮咬和毒刺传播到脊椎动物。它们广泛分布于热带或亚热带地区，但是其在骆驼上的重要性目前还不太清楚。一个属于布尼亚病毒科新的正布尼亚病毒被命名为施马伦贝格病毒，该病毒近来已在欧洲的成年反刍动物的血液和新生胎儿的中枢神经系统中被检测到，并引起胎儿死亡和神经症状[21]。用一种反刍动物酶联免疫吸附试验在10只羊驼中发现2只有施马伦贝格病毒抗体，但是并未观察到临床症状[15]。

表72总结了当前已知的虫媒病毒以及其他次要的病毒感染。

RT-PCR，反转录聚合酶链反应

在沙特阿拉伯，Wood等[29]从一只已死亡单峰驼附近收集的嗜驼璃眼蜱中分离到了卡达姆山（Kadam）病毒。是否这只单峰驼死于卡达姆山病毒感染不得而知。卡达姆山病毒对单峰驼、牛和人类的致病性还不清楚。

Converse和Moussa[7]从科威特、伊拉克和也门采集的嗜驼璃眼蜱分离到了5株夸兰菲尔（Quaranfil）病毒，这些发现的重要性还未进一步证实。

赤羽病毒能在牛、绵羊和山羊引发疫病流行、自然流产、早产和先天性畸形。该病毒广泛分布于非洲和亚洲，且在阿拉伯半岛呈地方性流行。Ai Busaidy等[1]发现在阿曼50%的单峰驼有赤羽病毒的中和抗体，是否赤羽病毒能引起单峰驼流产还不确定。

Anderson和Casals[3]从印度单峰驼收集的蜱（嗜驼璃眼蜱）体内分离到了4株托里（Dhori）病毒，而且他们在50份单峰驼血清中发现48份含有托里病毒的中和抗体，但是没有观察到临床症状，这

表72 骆驼虫媒病毒和其他次要病毒感染

| 病毒 | 骆驼种类 | 样品 | 参考 | 国家 | 年份 |
|---|---|---|---|---|---|
| 卡达姆山病毒 | 单峰驼 | 骆驼璃眼蜱 | 28 | 沙特阿拉伯 | 1982 |
| 夸兰菲尔病毒 | 单峰驼 | 骆驼璃眼蜱 | 7 | 科威特、伊拉克和也门 | 1982 |
| 赤羽病毒 | 单峰驼 | 血清抗体50% | 1 | 阿曼 | 1988 |
| 托里病毒 | 单峰驼 | 骆驼璃眼蜱 | 3 | 印度、亚洲和非洲 | 1973 |
| 未分类病毒、Wanowire病毒、索戈托病毒、托里病毒 | 单峰驼 | 骆驼璃眼蜱 | 28 | 埃及 | 1973 |
| 韦瑟尔斯布朗病毒、黄热病毒、波蒂斯库姆病毒、1型登革热病毒、Banzi和乌干达S病毒 | 单峰驼 | 屠宰的<br>血清学 | 4 | 尼日利亚 | 1990 |
| Thogoto病毒、西尼罗河病毒和韦瑟尔斯布朗病毒 | 单峰驼 | 血液病毒分离物 | 16 | 尼日利亚 | 1973 |
| 克里米亚-刚果出血热病毒 | 单峰驼 | 血清学14% | 18 | 埃及 | 1990 |
| 施马伦贝格病毒 | 羊驼 | 血清 | 15 | 英国 | 2012 |
| 伪狂犬病或脑心肌炎病毒 | 单峰驼 | 不同器官 | 17<br>5 | 土库曼斯坦 | 1994<br>1996 |
| 马动脉炎病毒 | 羊驼 | RT-PCR胎儿组织 | 26 | 德国 | 2006 |
| 马动脉炎病毒 | 羊驼、美洲驼 | 血清学<br>低于1% | 12 | 美国 | 2008 |
| 未诊断 | 羊驼、美洲驼 | 呼吸道症状 | 10 | 美国 | 2010 |
| 类狂犬病毒 | 单峰驼 | 脑炎（Dubduba征） | 2 | 沙特阿拉伯 | 2008 |
| 呼肠孤病毒 | 单峰驼？<br>双峰驼？ | 器官 | 14 | 德国 | 2009 |

个病毒也在西南亚和非洲的单峰驼蜱中被分离。

William等[28]在埃及单峰驼璃眼蜱中发现了3种未分类的虫媒病毒：Wanowire病毒、索戈托病毒和托里病毒。这些病毒随动物的迁徙而被扩散，包括骆驼旅行队。这些病毒在兽医方面的重要性还不清楚。

在尼日利亚对269只屠宰的骆驼进行了黄病毒属病毒抗体调查[4]。针对虫媒病毒的抗体阳性率分别为：韦斯布伦（Wesselsbron）病毒 60.2%、黄热病病毒 54.0%、波蒂斯库姆（Potiskum）病毒 66.2%、登革热1型4.5%、阪兹Banzi病毒 5.4%和乌干达S（Uganda S）病毒 0%。虽然并没评估一些高比例的黄病毒抗体的重要性，但是作者认为感染的骆驼在这些病毒传给人和家畜的过程中可能起重要作用。Kemp等[16]的发现与此类似，他们从注射到幼年小鼠的骆驼血中分离到了3种病毒：索戈托病毒、西尼罗河病毒和韦瑟尔斯布朗病毒。

克里米亚-刚果出血热（CCHF）广泛分布于非洲、中东、南欧、东欧以及亚洲[13]。该病主要呈地方性流行，但在许多动物如牛、绵羊、山羊、骆驼和野兔并无症状[22]。30种蜱能成为储藏宿主和媒介，尤其是璃眼蜱。人类能通过蜱虫叮咬或与感染动物或人类的密切接触被感染。几个报告称从不同动物检测到了CCHF抗体，且从动物分离到了病毒。Causey等[6]在尼日利亚从牛血、刺猬肝脏和脾脏、四种蜱虫以及库蠓（*Culicoides* spp）分离到了35株病毒。用琼脂免疫扩散试验在伊朗的男人（13%）、绵羊（38%）、山羊（36%）、牛（18%）和小型哺乳动物（3%）发现了CCHF抗体，但是在骆驼未发现抗体阳性[20]。然而，用琼脂扩散和间接荧光抗体技术从进口的埃及的骆驼发现14%（600/4301）CCHF抗体阳性[17]。Hassanein等[11]用反向间接血凝抑制试验对沙特阿拉伯的人和家畜进行了血清学调查，在人（0.8%）、绵羊（4.1%）、山羊（3.2%）和牛（0.6%）发现了抗体，但在马和骆驼未检测到抗体。在伊拉克[25]、阿联酋[24]和阿曼[23]人间流行与璃眼蜱密切相关。然而，尽管骆驼对蜱有高的感染率，但在沙特阿拉伯进行的一项蜱流行病学调查中并未从骆驼分离到CCHFV[9]。CCHFV对骆驼的危害还不确定。阿联酋一小群单峰驼出现了出血症，但血清学检测是阴性的[27]（见1.1.4 内毒素血症）。

1992年，Mischenko等[17]和Baborenko等[5]报道在土库曼斯坦的两群骆驼中暴发了严重的伪狂犬病。患病的单峰驼饲喂于猪圈附近，表现典型的伪狂犬病症状，包括严重的瘙痒症、食欲不良、精神沉郁、前胃迟缓、肌肉颤抖、后肢麻痹，约有300只单峰驼被感染，几乎全部死亡，患病动物被安乐死。病理变化主要是喉部出现血斑、心肌外膜和内膜炎及心肌失调、肺脏充血、肝脏出血。剩余单峰驼用灭活的伪狂犬病疫苗接种后，疾病的传播被控制。

用病毒中和试验监测了美国驼群中的650只美洲驼和羊驼，发现了马动脉炎病毒感染，然而，只有不到1%的新大陆骆驼有马动脉炎病毒抗体。尽管感染率比较低，但是Hennig等[12]认为兽医、新大陆骆驼的所有者和饲养者应该意识到马动脉炎病毒能够感染美洲驼和羊驼。德国的5只羊驼也被发现马动脉炎中和抗体。用反转录-聚合酶链反应（RT-PCR）在胎儿组织中检测到了马动脉炎病毒的基因组片段。尽管羊驼和马的直接接触被排除，但有可能通过间接接触（访问者和看管人）传播[26]。除了马的A型鼻炎病毒，马动脉炎病毒是可能诱发骆驼流产的另一病毒。

2007年的夏天和秋天，Fowler[10]在美洲驼和羊驼报道了一种新的呼吸道疾病，使其交易和展出的合同取消。该病是高度接触性的，且传播很快，在部分动物更加严重的表现为呼吸困难、肺部水肿和肺炎，以及流产、死胎和死亡[8]。这也可能由继发感染引起。Huisinga等[14]报道了在两头旧大陆骆驼发生未分类呼肠孤病毒感染的疑似病例。在濒死前两头骆驼均出现了严重的水样腹泻。尸检诊断为急性弥漫性卡他肠炎，以及溃疡性食道炎和胃炎。组织学显示多发性干细胞坏死，以及不同器官的血管及周围出现成淋巴细胞浸润。从不同器官分离到了呼肠孤病毒。

近来中东一种命名为中东呼吸综合征冠状病毒（MERS-CoV）在患有严重呼吸道疾病的病人被分离到。这个新的冠状病毒来源还不清楚。Rusken等[19]在阿曼单峰驼检测到了MERS-CoV中和抗体，是否单峰驼携带该病毒需要深入研究。

## 参考文献

[1] Al Busaidy S.M., Mellor P.S. & Tayler W.P. (1988). – Prevalence of neutralizing antibodies to Akabane virus in the Arabian Peninsula. *Vet. Microbiol.*, 17 (2), 141 – 149.

[2] Al-Dubaib M.A., Al-Swailem A., Al-Ghamdi G. Al-Yamani E., Al-Naeem A.A., Al-Mejali A.M., Shehata M., Hashad M., El-Lithy D.A. & Mahmoud O.M. (2008). – Dubduba syndrome: An emerging neurological disease of camels with a possible viral etiologic agent. *J. Camel Pract. Res.*, 15 (2), 147 – 152.

[3] Anderson C.R. & Casals J. (1973). – Dhori virus, a new agent isolated from *Hyalomma dromedarii* in India. Ind. *J. Med. Res.*, 61 (10), 1416 – 1420.

[4] Baba S.S., Omilabu S.A., Fagbami A.H. & Olaleye O.D. (1990). – Survey for antibodies against flaviviruses in slaughter camels (*Camelus dromedarius*) imported to Nigeria. *Prev. Vet. Med.*, 10, 97 – 103.

[5] Baborenko E.P. (1996) – Immunological properties of the Aujeszky-virus [in Russian]. Thesis, Russian State Control Institute for the Standardisation and Certification of Veterinary Products, Moscow.

[6] Causey O.R., Kemp G.E., Madbouly M.H. & David-West T.S. (1970). – Congo virus from domestic livestock, African hedgehog, and arthropods in Nigeria. *Am. J. Trop. Med. Hyg.*, 19 (5), 846 – 850.

[7] Converse J.D. & Moussa M.I. (1982). – Quaranfil virus from *Hyalomma dromedarii* (Acari: Ixodoidea) collected in Kuwait, Iraq and Yemen. *J. Med. Entomol.*, 19 (2), 209 – 210.

[8] Crossley B.M., Barr B.C., Magdesian K.G., Ing M., Mora D., Jensen D., Loretti A.P., McConnell T. & Mock R. (2010). – Identification of a novel coronavirus possibly associated with acute respiratory syndrome in alpacas (*Vicugna pacos*) in California, 2007. *J. Vet. Diagn. Invest.*, 22, 94 – 97.

[9] El-Azazy O.M.E. & Scrimgeour E.M. (1997). – Crimean-Congo haemorrhagic fever virus infection in the Western Province of Saudi Arabia. *Trans. R. Soc. Trop. Med. Hyg.*, 91, 275 – 278.

[10] Fowler M.E. (2010). – Medicine and surgery of camelids. 3rd Ed. Wiley-Blackwell, Oxford, UK, 186.

[11] Hassanein K.M., Elazazy O.M.E. & Yousef H.M. (1997). – Detection of Crimean Congo Haemorrhagic Fever virus antibodies in humans and imported livestock in Saudi Arabia. *Trans. R. Soc. Trop. Med. Hyg.*, 91 (5), 536 – 537.

[12] Hennig J., Timoney P., Shuck K., Balasuriya U., Baker R., Williamson L. & Niehaus A. (2008). – Serosurveillance of US camelid population for evidence of equine arteritis virus infection. In Proceedings of the Second International Workshop on Equine Viral Arteritis, 13 – 15 October, Lexington, Kentucky, USA, 43.

[13] Hoogstraal H. (1979). – The epidemiology of tick-borne Crimean Congo Hemorrhagic fever in Asia, Europe and Africa. *J. med. Entomol.*, 15 (4), 307 – 417.

[14] Huisinga M., Köhler K., Förster C. & Reinacher M. (2009). – Reovirus infection a possible cause for disease in Old World Camelids [in German]. In Proceedings of the Meeting of the European Society of Veterinary Pathology, Fulda, Germany, 21.

[15] Jack C., Anstaett O., Adams J., Noad R., Brownlie J. & Mertens P. (2012). – Evidence of seroconversion to SBV in camelids. *Vet. Rec.*, 170, 603.

[16] Kemp G.E., Causey O.R., Moore D.L. & O Connor E.H. (1973). – Viral isolates from livestock in Northern Nigeria, 1966 1970. *Am. J. Vet. Res.*, 34 (5), 707 – 710.

[17] Mischenko V.A, Baborenko E.P., Nikitni V.A., Chuchorov V.M., Zacharov V.M., Kornicenko Y.N., Puzankova O.S., Bajbikov T.Z., Mamajeva L.I., Smirnov A.B. & Metlikina N.A. (1994). – Diagnosis of the Aujeszky disease in camels [in Russian]. *Veterinarija*, 4, 23 – 25

[18] Morrill J.C., Soliman A.K., Imam I.Z., Botros B-A.M., Moussa M.I. & Watts D.M. (1990). – Serological evidence of Crimean– Congo haemorrhagic fever viral infection among camels imported into Egypt. *J. Trop. Med. Hyg.*, 93, 201 – 204.

[19] Reusken C.B., Haagmans B.L., Müller M.A., Gutierrez C., Godeke G.J., Meyer B., Muth D., Raj V.S., Vries L.S., Corman V.M., Drexler J.F., Smits S.L., El Tahir Y.E., De Sousa R., van Beek J., Nowotny N., van Maanen K., Hidalgo-Hermoso E., Bosch B.J., Rottier P., Osterhaus A., Gortázar-Schmidt C., Drosten C. & Koopmans M.P. (2013). – Middle East respiratory syndrome coronavirus neutralising serum antibodies in dromedary camels: comparative serological

study. *Lancet Infect. Dis.* [Epubahead of print].

[20] Saidi S., Casals J. & Faghih M.A. (1975). – Crimean Hemorrhagic Fever–Congo (CHF-C) virus antibodies in man and in domestic and small mammals in Iran. *Am. J. Trop. Med. Hyg.*, 24 (2), 353–357.

[21] Schulz C., Beer M. & Hoffmann B. (2012). – Schmallenberg – Virus: A risk for camelids? *Llamas*, 20 (2), 6–10.

[22] Schwarz T.F., Nsanze H., Longson M., Nitschko H., Gilch S., Shurie H., Ameen A., Zahir A.R.M., Acharaya U.G. & Jager G. (1996). – Polymerase chain reaction for diagnosis and identification of distinct variants of Crimean–Congo hemorrhagic fever virus in the United Arab Emirates. *Am. J. Trop. Med. Hyg.*, 55 (2), 190–196.

[23] Scrimgeour E.M., Zaki A., Mehta F.R., Abraham A.K., Al-Busaidy S., El-Khatim H., Al-Rawas S.F.S., Kamal A.M. & Mohammed A.J. (1996). – Crimean–Congo hemorrhagic fever in Oman. *Trans. R. Soc. Trop. Med. Hyg.*, 90, 290–291.

[24] Suleiman M.N.H., Muscat-Baron J.M., Harries J.R., Satti A.G.O., Platt G.S., Bowen E.T.W. & Simpson D.I.H. (1980). – Congo/Crimean Haemorrhagic Fever in Dubai. An outbreak at the Rashid Hospital. *The Lancet* II, 939–941.

[25] Tantawi H.H., Al-Moslih M.I., Al-Janabi N.Y., Al-Bana A.S., Mahmud M.I.A., Jurji F., Yonan M.S., Al-Ani F. & Al-Tikriti S.K. (1980). – Crimean Congo Hemorrhagic Fever virus in Iraq: isolation, identification and electron microscopy. *Acta virologica*, 24, 464–467.

[26] Weber H., Becknann K. & Haas L. (2006). – Equine arteritis virus (EVA) as the cause of abortion in alpacas? Case report. *Dtsch. Tierrztl. Wschr.*, 113 (4), 162–163.

[27] Wernery U., Schimmelpfennig H.H., Seifert H.S.H. & Pohlenz J. (1992). – *Bacillus cereus* as a possible cause of haemorrhagic disease in dromedary camels *(Camelus dromedarius)*. *In* Proceedings of the First International Camel Conference, 2–6 February 1992, Dubai, UAE, 51–58.

[28] Williams R.E., Hoogstraal H., Casals J., Kaiser M.N. & Moussa M.I. (1973). – Isolation of Wanowrie, Thogoto and Dhori viruses from Hyalomma ticks infesting camels. *Egypt. J. Med. Entomol.*, 10 (2), 143–146.

[29] Wood O.L., Moussa M.I., Hoogstraal H. & Buettiker W. (1982). – Kadam virus (Togaviridae, Flavivirus) infecting camelparasitizing *Hyalomma dromedarii* ticks (Acari: Ixodoidea) in Saudi Arabia. *J. Med Entomol.*, 19 (2), 207–208.

（独军政译，殷宏校）

# 3 真菌病

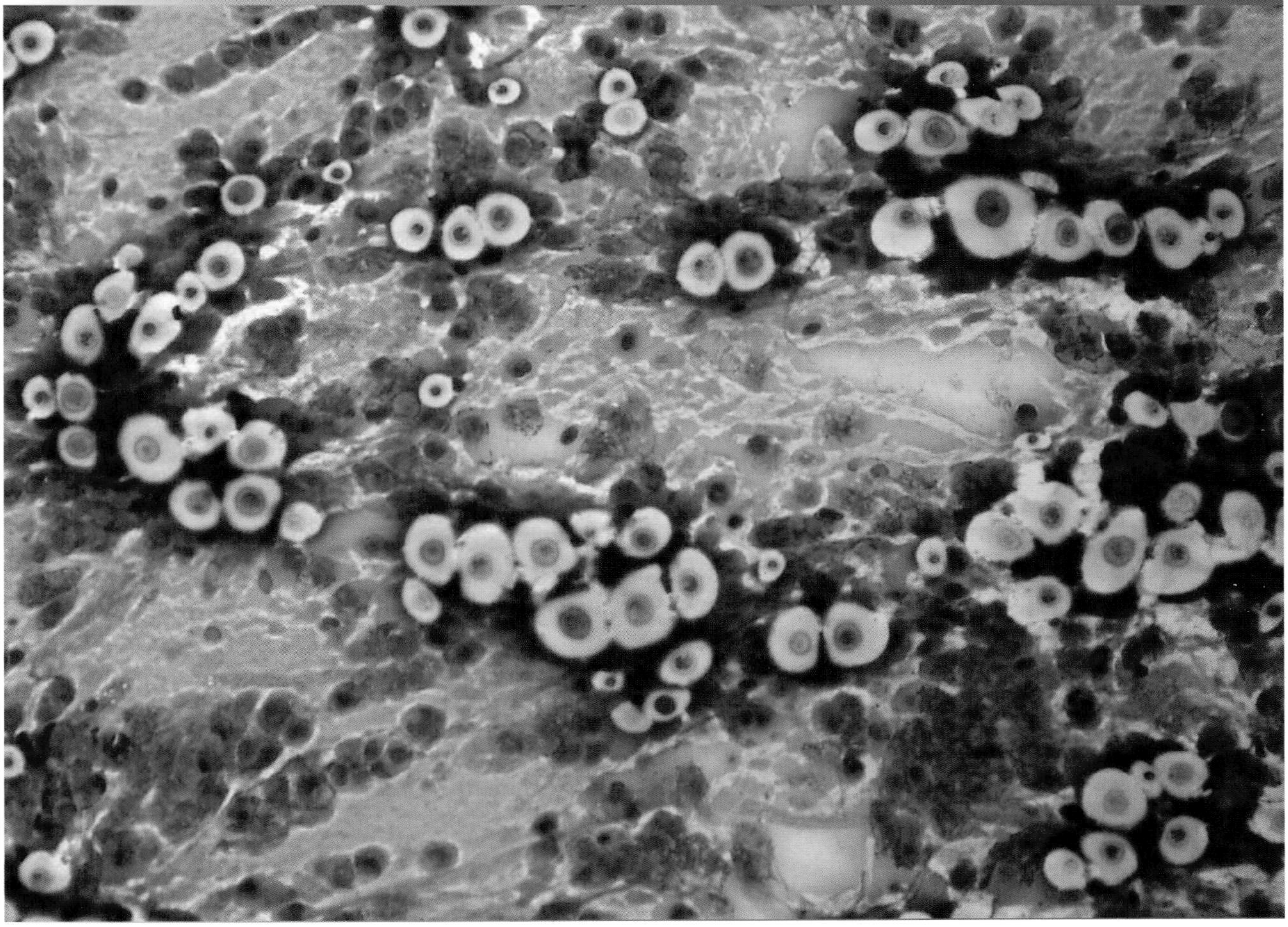

# 前言

大多数真菌病的病原以腐生菌的形式存在于土壤、腐烂的蔬菜和粪便中。大部分真菌感染主要来源于土壤，通过吸入、摄入土壤或与感染的个体或设备相接触而引起。在看似健康的个体上，病原真菌感染引起了如网状内皮细胞真菌病、球孢子菌病和酵母菌病等疾病，这些通常被认为是主要的系统性真菌病。机会致病菌常常在因应激、代谢性酸中毒、营养不良或肿瘤而引起的抵抗力下降时感染宿主。长时间使用抗菌药物或免疫抑制物质可以增加机会致病菌如曲霉菌、毛霉菌、隐球菌、念珠菌感染的可能性。

脚癣（癣菌病）是由叫做皮肤癣菌的几种真菌引起的一种角质化组织（皮肤、毛发、指甲）感染。所有家畜易感，病原菌在全世界均被发现。正常栖居于土壤（嗜土性）的几种皮肤癣菌（如石膏样小孢子菌）可使动物和人发病。一些皮肤癣菌（如小孢子菌）主要感染人，却很少感染动物（亲人性的）。其他的主要是动物致病菌（如犬小孢子菌、毛癣菌），但也可引起人类发病（嗜兽性）。

过去10年，我们了解了更多关于骆驼科动物真菌疾病的知识，新的科学文章已发表于该学科的国际期刊。本文尝试对当前这些微生物的知识作一概括。

# 3.1 霉菌性皮炎

### 病原学

各种真菌均可引起外皮感染，其中引起的癣菌病最常见于骆驼科动物。皮肤癣菌与一组利用角蛋白生长的真菌密切相关（表73）。

已知的38种皮肤真菌中，引起动物感染的是其中的两个属，小孢子菌属和毛癣菌属。

在3岁以下的旧大陆骆驼中，皮肤真菌病是一种常见的皮肤病，3~12月龄为一个发病高峰。然而新大陆骆驼却很少发病。从不同国家的南美骆驼中分离到疣状毛癣菌、须癣毛癣菌、石膏样小孢子菌和矮小孢子菌。实验室培养的小孢子菌和毛癣菌大小分生孢子的差异如图187所示。

骆驼科皮肤癣菌肉眼观察（培养）和显微特征如图188a-e所示。呈现的所有真菌培养物均在沙氏葡萄糖琼脂27℃培养14天。

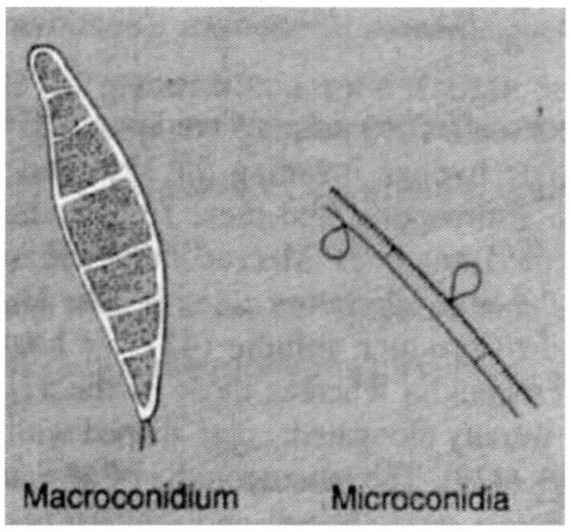


a.小孢子菌属

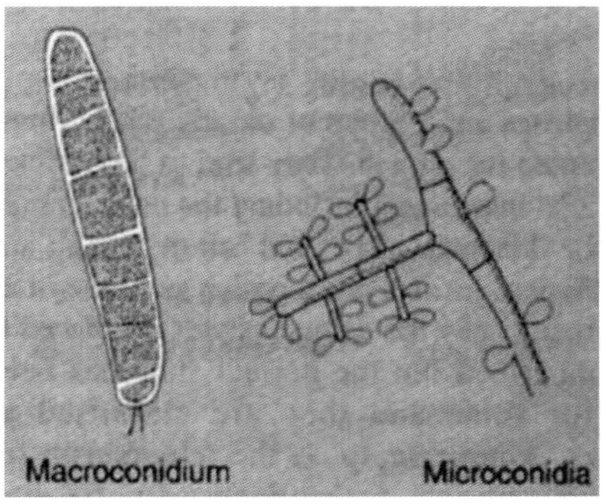


b.毛癣菌属

图187　小孢子菌属和毛癣菌属的差异

### 流行病学

大多数情况下，皮肤癣菌仅在凋亡的角质化组织中生长。活细胞或发炎组织可阻止感染进一步发展。感染开始于生长的毛发或皮肤角质层，分生孢子在此发育成线状菌丝，菌丝穿透并侵入发干造成损伤，在头发向上生长的同时菌丝向下生长。癣菌主要沿着毛发的外表面（毛外癣菌），而不是在内表面（毛内癣菌）产生簇状的分节孢子。骆驼科动物癣病的流行病学尚未有人调查，但人们认为，与感染动物和污染物直接和间接接触是皮肤癣菌的传播方式。高湿、过于拥挤和营养失衡（最可能是维生素A缺乏）等因素有助于本病的暴发。在受感染的骆驼群中，多达80%的幼驼表现出临床症状。Khamiev检查了200峰患有皮肤病变的骆驼，其中90峰呈疣状毛癣菌(*T. verrucosum*)阳性，并命名为骆驼毛癣菌(*T. camelius*)。这90峰感染骆驼90%不

表73　从旧大陆和新大陆骆驼霉菌病的皮损中分离的真菌

| 病原 | 作者 | 物种 |
|---|---|---|
| **皮肤癣菌** | | |
| 疣状毛癣菌 (*Trichophyton verrucosum*) | Curasson(1947),Nasser(1969),Torky and Hammad(1981),Khamiev(1981,1982,1983),El-Kader(1985),El-Kader(1985),El-Tamavy et al.(1988),Mahmoud(1993),Fadlelmula et al.(1994),Abou-Zaid(1995),Al-Ani and Al-Rawahi(2006),Ebrahimi et al.(2007),Khosravi et al.(2007) | 单峰驼 |
| 脱毛毛癣菌(*Trichophyton tonurans*) | Ebrahimi et al.(2007) | 单峰驼 |
| 须癣毛癣菌(*Trichophyton mentagrophytes*) | Refai and Miligy(1968),Kuttin et al.(1986),Mahmoud(1993,Al-Ani and Al-Rawahi(2006),Ebrahimi et al.(2007) | 单峰驼 |
| 许兰毛癣菌(*Trichophyton schoenleinii*) | Kamel et al.(1997),Chatterjee et al.(1978),Al-Ani et al.(1995),Al-Ani and Al-Rawahi(2006),Ebrahimi et al.(2007) | 单峰驼 |
| 萨氏毛癣菌(*Trichophyton sarkisovii*) | lvanova and Polyakov(1983),El-Jaouhari et al.(2004) | 双峰驼 |
| 丹卡利毛癣菌（*Trichophyton dankaliense*） | Dalling et al.(1996) | 单峰驼 |
| 石膏样小孢子菌(*Microsporum gypseum*) | Boever and Rush (1975),Kamel et al.(1997),Fischman et al.(1987),Mancianti et al.(1998),Gitao et al.(1998) | 单峰驼 |
| 猪小孢子菌(*Microsporum nanum*) | Ebrahimi et al.(2007) | 单峰驼 |
| 石膏样小孢子菌(*Microsporum gipseum*)* | Al-Ani and Al-Rawahi(2006) | 单峰驼 |
| 犬小孢子菌（*Microsporum canis*） | El-Kader(1985),El-Tamavy et al.(1998),Abou-Zaid(1995) | 单峰驼 |
| **其他** | | |
| 申克孢子丝菌 (*Sporothrix schenckii*) | Curasson(1947) | 单峰驼 |
| 白色念珠菌 (*Candida albicans*) | Wernery et al.(2007),Pal et al.(2007),Tuteja et al.(2010),Khosravi et al.(2008,2009) | 单峰驼 |
| 马拉色菌属 (*Malassexia* sp.) | Zanolari et al.(2008) | 南美骆驼科 |
| 葡萄酒色青霉（*Penicillium vinaceum*），Pseudorotium spp. 假蛛网霉属（*Pseudoarachniotus* spp.）霉样真菌属（*Allescheria* spp.），无孢菌类(Mycelia sterile) | Singh and Singh(1969) | 单峰驼 |
| 新型隐球菌（*Cryptococcus neoformans*） | Goodchild et al.(1996)<br>Bildfell et al.(2002) | 羊驼<br>美洲驼 |
| 金孢子菌属(*Chrysosporium*) | Mahmoud(1993) | 单峰驼 |
| 诡谲腐霉（*Pythium insidiosum*） | Wellehan et al.(2004) | 单峰驼 |

*注：应与*M. gypseum*为同一种，译者注。

到两岁。疣状毛癣菌和须毛癣菌的厚壁孢子可在毛发、刮擦的动物细胞碎片和留下的黏附污染物中，保持活力长达4.5年。

Driot等[16]描述摩洛哥的单峰驼中癣菌病的流行病学和组织病理学。这次调查作者发现，癣菌病主要见于年轻骆驼，而疥癣主要发生在成年骆驼。在摩洛哥的南部，癣菌病的平均发病率为16%，成为一种频发的皮肤病。尽管作者没有鉴定这些癣菌病的病原，但前人在这一地区鉴定出了萨氏毛癣菌。

Khosravi等人在伊朗报道了一起在单峰驼上由疣状毛癣菌和星形诺卡氏菌引起的混合感染。100峰骆驼中，总计有74峰受到感染，表现易怒，在肩、颈、体侧和四肢上部出现无毛斑、溃疡性结节和微脓肿。Al-Ani 和 Al-Rawahi[5]报道了在约旦和阿曼的幼驼中，不同皮肤癣菌引起的混合感染发病率为13%。幼驼和犊牛混合时感染率更高。在美国，Wellehan 等[77]报道了一例在单峰驼上由诡谲腐霉引起的罕见的藻菌病。患病骆驼右侧面部因软组织肿胀被切除，分离到诡谲腐霉并成功培养。骆驼短期恢复后，6个月后死于第三胃胃炎，病灶内含有该病原菌丝。

## 临床症状

尽管骆驼主人熟悉癣病，能够将皮炎与其他皮肤感染相区别，但皮癣的临床症状却表现为多种形式。骆驼的癣病有两种临床类型：第一种表现为典型的灰白色病变，特征是范围小，呈圆秃状，主要发生于幼龄动物的腿、颈和头部。第二种相对来讲是全身性感染，发生于头、颈、四肢和体侧，这些病变最初可能与兽疥癣相混淆。（图189、190）

本病是人畜共患病，操作者常被感染，表现在手和胳膊上有典型的癣病变。

骆驼皮肤癣菌的大体特征（培养）和微观特性如图191所示。所有的真菌均在沙氏葡萄糖琼脂27℃培养14天。

## 病理学

表皮因钉突向下延伸而增厚，痂皮由组织碎片、炎性细胞、干燥血清和真菌成分构成。在与微脓肿、毛囊炎和毛肉芽肿相关的毛囊内，常检测到真菌成分。组织学检查揭示了角质层发生的角化过度、角化不全和棘皮症等病征。苏木精-伊红染色很难观察到特征菌丝的细丝；用高碘酸 - 希夫（PAS）和罗科特甲胺银染色时最清楚。

## 诊断

直接镜检毛发或皮屑，可发现特征菌丝和/或分节孢子。然而最有效和最特定的诊断手段还是真菌培养，尽管通常需要孵育10 ~ 14天。取可疑部位外缘的毛或刮屑在载玻片上进行真菌成分检查，将其湿润后（20%的氢氧化钾）加热，盖玻片压扁，室温孵育10 ~ 20分钟，方便镜检。通过Hollaender等建立的一种用吖啶橙的荧光染色法，使在组织切片与临床标本中真菌的鉴定更可靠。

小孢子菌和毛癣菌及其他真菌应在沙氏葡萄糖琼脂和真菌检测琼脂（欧罗维特）中培养，27℃孵育10 ~ 14天（图192）。确诊和鉴定种属，需要用醋酸胶带去除菌丝和在菌落表面的大分生孢子，并用乳酚棉兰染色，镜检。当预期有腐生物污染时，在真菌检测琼脂上培养鉴定就特别有效。

许多角质增生性皮炎已见于骆驼科动物，但并不是所有的都由皮肤癣菌引起，因此有必要对任何皮肤损伤都认真调查，对多发性深部皮肤刮下的（带血的）碎屑应送往实验室诊断。

## 防治

通过早期诊断和将感染骆驼与未感染骆驼隔离，可有效控制癣菌病的传播。为避免重复感染，有必要对马厩和设备进行适当消毒。病变部位应首先用温的干净肥皂水擦洗，并祛除所有结痂。各种常见的杀真菌剂和抑真菌剂，如碘、5%的硫芝麻油、5%的水杨酸、煤焦油酚（3.25%）、铜醋酸（0.58%）和羟基喹啉，可以作为癣药膏在感染部位外敷。

克菌丹是观赏植物的杀真菌剂。使用克菌丹时需注意[2]，对感染动物进行喷涂的溶液要作1:200稀释。该混合物混合后一周也很稳定，在病变部位及外缘使用，连续两周。

灰黄霉素在牛的皮肤癣菌病的治疗上非常有效[13]，但在骆驼上会引起副作用，如恶心和腹泻，因此不推荐使用[67]。星状诺卡氏菌和疣状毛癣菌引起的混合感染，用甲氧苄氨嘧啶和磺胺甲基异恶唑治疗，剂量为每日每千克体重5毫克，并用

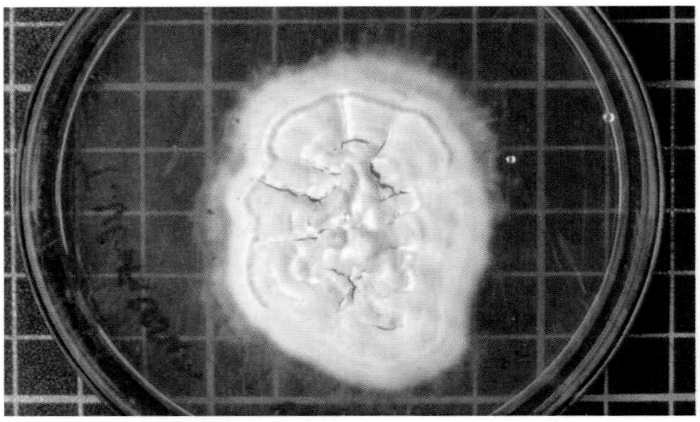

a 疣状毛癣菌

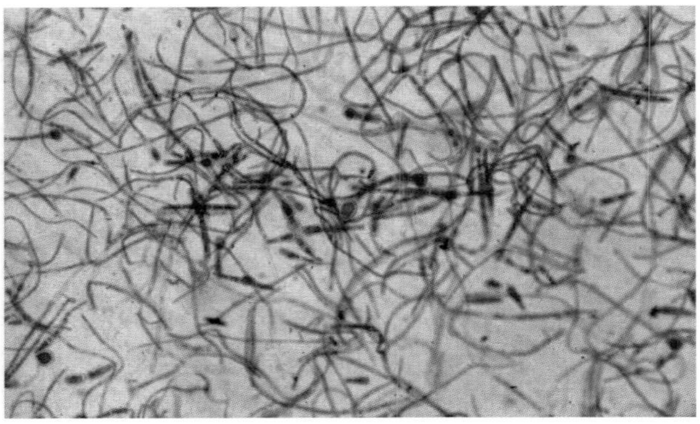

b 疣状毛癣菌

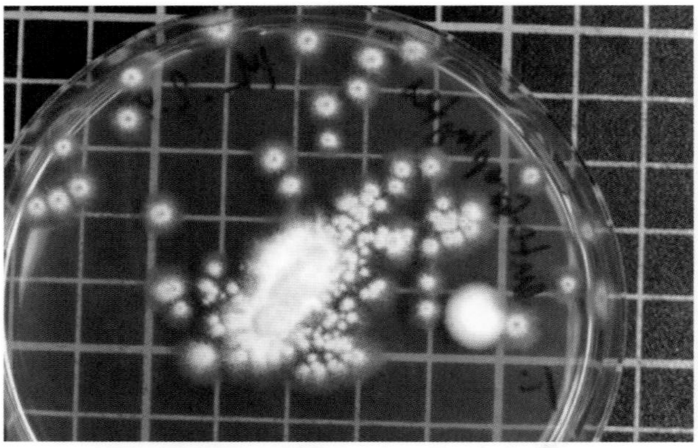

c 须癣毛癣菌

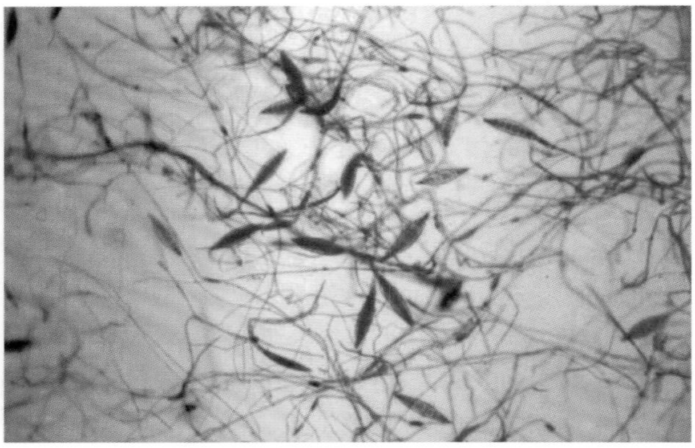

d 须癣毛癣菌

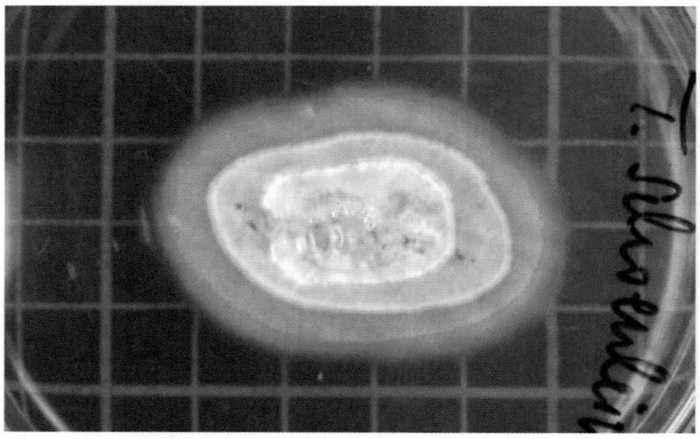

e 许兰毛癣菌

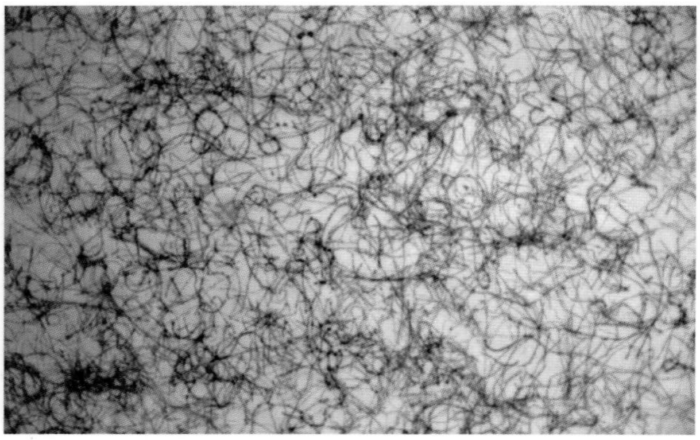

f 许兰毛癣菌

g 石膏样小孢子菌

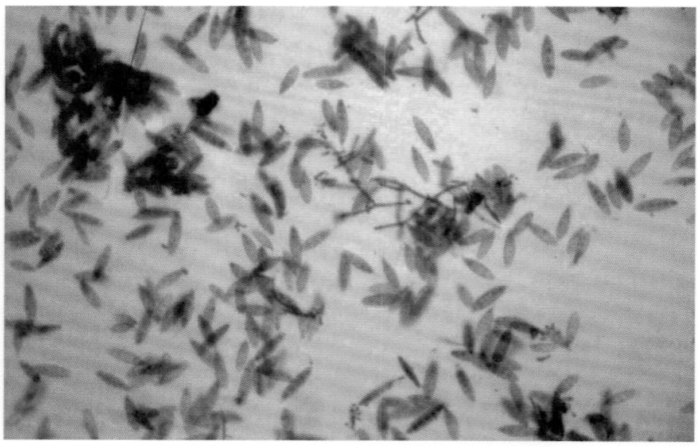

h 石膏样小孢子菌

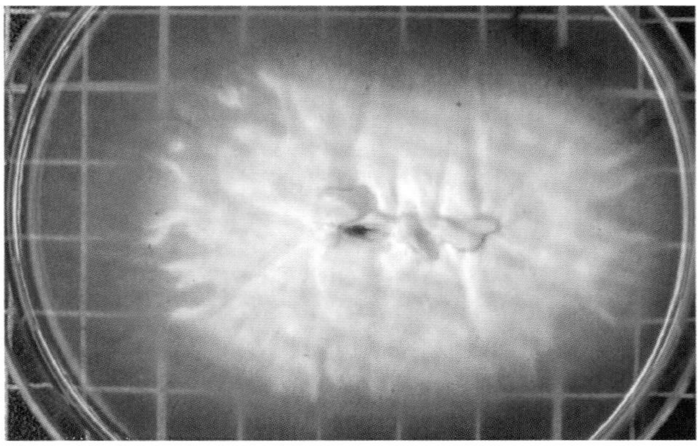

i 犬小孢子菌

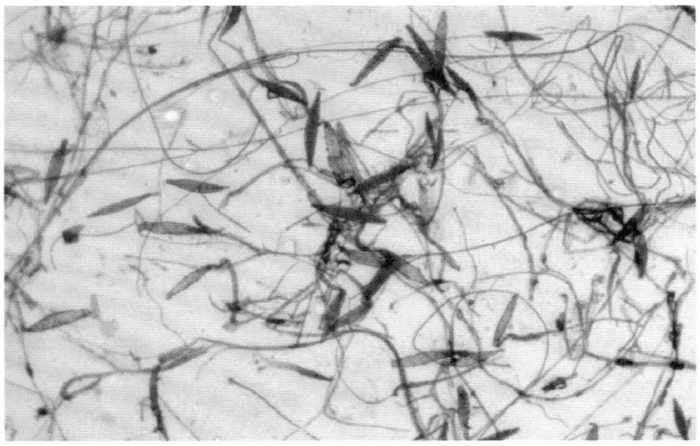

j 犬小孢子菌

图188

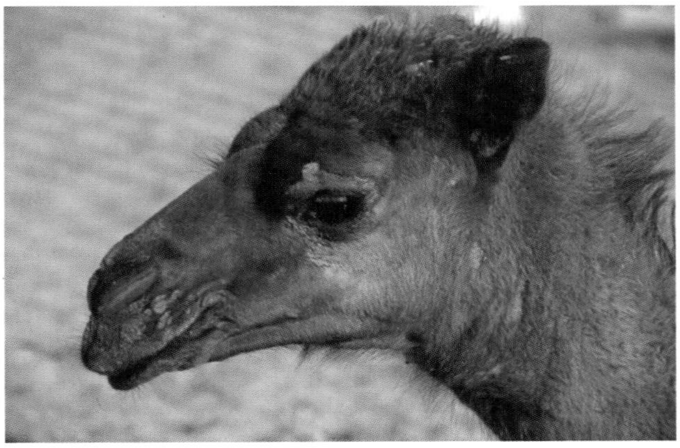

图189 单峰驼幼驼体上的典型的皮肤癣菌病

图190 单峰驼左后肢上大面积皮肤癣菌病

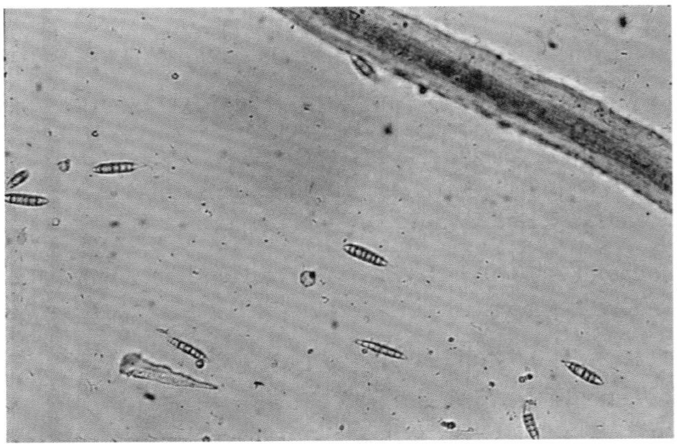

a.毛癣菌属

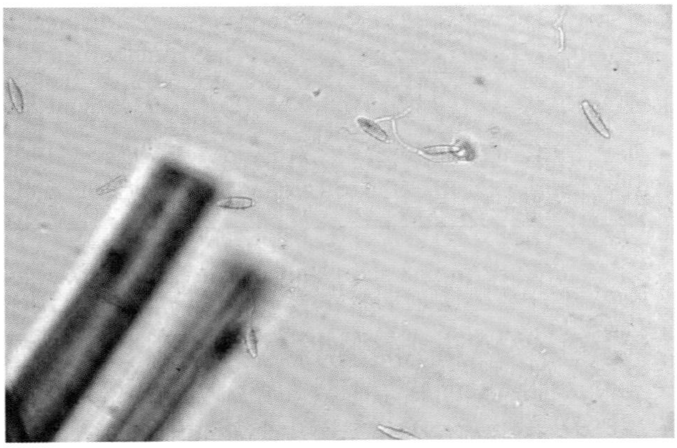

b.小孢子菌属

图191 患皮癣骆驼体表的毛癣菌属和小孢子菌属（用氢氧化钾制备湿玻片）

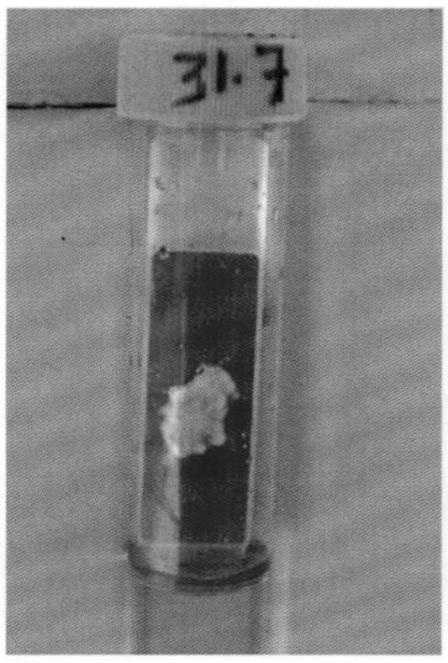


a.疣状毛癣菌

b.犬小孢子菌

图192　疣状毛癣菌真菌玻片培养（新版）和犬小孢子菌的真菌检测培养（室温下培养12天）

那他霉素清洗皮肤病变部位，每日两次。两个月后，感染的单峰驼全部痊愈。在哈萨克斯坦，使用疫苗接种骆驼成功地预防了毛癣菌属和小孢子菌属感染。Camelvac Tricho（IDT Dessau-Tornau，德国）已应用于哈萨克斯坦，并对来自12个农场的34 302 峰双峰驼开展了调查。表74显示癣菌病在这些畜群中的发生率。

表74　癣菌病在哈萨克斯坦双峰驼中的发生率

| 年龄 | 发生率(%) |
| --- | --- |
| 5日～4月龄 | 21.5 |
| 5～12 月龄 | 60.1 |
| 13月龄～3岁 | 17.1 |
| 4 岁及以上 | 1.3 |

这批骆驼中，3 300峰接种了 Camelvac Tricho 的幼驼，数年内无癣病复发。该疫苗的使用很成功，不仅用于预防，也可用于治疗。患病的骆驼注射Camelvac Tricho一两次后治愈。作者在阿联酋的几个骆驼群中成功使用了Camelvac Tricho，患癣病的幼龄单峰驼接种一次后，病灶在14天内减弱，4周后消失。非常遗憾的是，该疫苗不再生产。

（郑福英译，殷宏校）

# 3.2 曲霉病

曲霉，尤其是烟曲霉，与家畜的胎盘和呼吸系统的感染有关，可引起乳腺炎和瘤胃炎。发霉的垫料和饲料常被怀疑为曲霉菌病暴发的来源。曲霉病作为一种机会致病性真菌的感染，在羊驼和单峰驼上已有报道[7,20,28,59,68]。Quist等人[61]从一峰治疗肺圆线虫病的美洲驼上分离到曲霉，治疗期间，该驼继发成胃肠溃疡和肺曲霉病。

### 病原学

已知的曲霉属有数百种，据估计，在动物的曲霉感染中，烟曲霉占到了90%～95%。其他曲霉属，包括黑曲霉、黄曲霉、土曲霉和构巢曲霉，偶尔也会引起感染。黄曲霉与黄曲霉毒素中毒有关，遍布于世界的黄曲霉几乎感染所有的家养动物，以及鸟类和许多野生物种。带有分隔菌丝的曲霉增殖迅速。很多曲霉属由于有着色孢子（分生孢子）（图193），产生有颜色的菌落（黑色、绿色或黄色）。曲霉属可感染人，引起人类中毒和过敏反应。

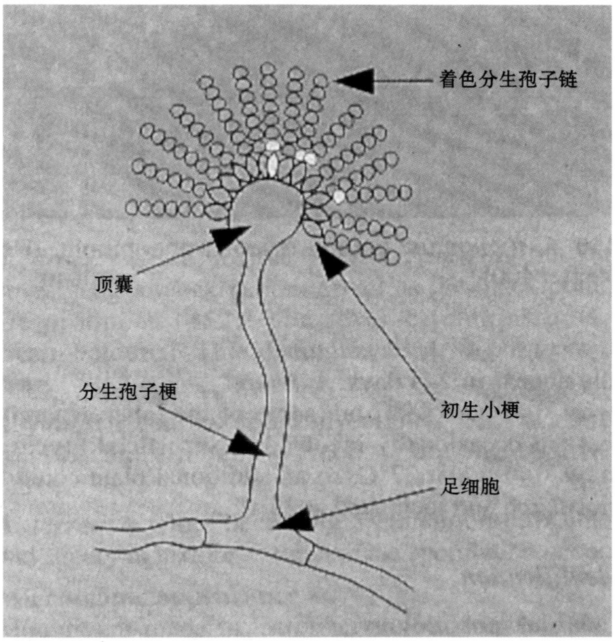

图193　曲霉头端示意图(仿 Quinn et al.)

### 流行病学

普遍存在的烟曲霉菌，并不常感染哺乳动物。因过大应激而疲惫不堪、代谢性酸中毒、营养不良或患有肿瘤的患者易得曲霉病。过量使用或接触抗菌剂或免疫抑制物质，在真菌感染的发展中起着重要作用。曲霉病通过吸入或食入真菌孢子进行传播。

### 诊断

直接镜检用氢氧化钾处理的组织刮屑或其他材料，组织病理切片用PAS 染色。分离的曲霉菌用沙氏葡萄糖琼脂培养，将组织碎片轻轻推入琼脂，培养物在37℃孵育5天。菌落通常在孵育2～5天时出现，通过菌落和子实体微观形态进行鉴定。荧光免疫技术可用来鉴定组织切片中的菌丝，也可应用聚合酶链反应进行鉴定。

琼脂凝胶免疫扩散试验是一种可靠的血清真菌抗体诊断技术，使用酶联免疫吸附测定技术可提高该技术灵敏度。

### 临床症状和损害

EL-Khouly等报道了在阿联酋赛骆驼上的一种特别的呼吸道和肠道综合征。患病骆驼食欲减退，精神委顿，有些出现轻微的干咳。大多数情况下，骆驼喉咙肿胀并伴有颌下淋巴结肿大，最坏时，一些骆驼出现出血性腹泻。发病骆驼体温略有增加，出现临床症状后5～7天死亡。40峰骆驼的尸检结果一致表明，肠道和内脏呈广泛性出血。从解剖的骆驼的许多器官中培养出烟曲霉，病灶直接涂片证实含有真菌菌丝及分生孢子。在部分病例中，从组织和血清中也能检测到黄曲霉毒素。然而作者称，这并不能够证实该真菌是引起此综合征的主要原因。

Wernery等描述了一个非常类似的疾病，出血性素质（参见1.1.4章节：内毒素中毒）。在Gareis和Wernery描述的霉菌毒素中毒病例中，单峰驼表现以严重的水样腹泻、出血、低白细胞数的典型症状，最后死亡。由于大雨和保管不当，使喂养骆驼的干草变质发霉，一些干草中含有大量的曲霉、青霉、交链孢霉、镰刀菌及帚霉的一些种（CFU/g）。细胞培养生物测定试验证明，提取的干草样品和来自剖检骆驼的体液及肠内容物，是具有高度细胞毒性（MIT（3（4-，5-二甲基-2-基）-2，5-二苯基溴化））的。后续分析表明存在环二巯二氧哌嗪霉菌毒素，并第一次证明了饲料中天然存在这种霉菌毒素。在阿联酋3～6

岁的赛骆驼中，Osman等[56]描述了一起极有可能由黄曲霉毒素引起的疾病暴发。发病的单峰驼嗜睡，出现热病，精神委顿，下颌水肿并伴有淋巴结肿大。100峰骆驼中有31峰出现临床症状，其中有20峰一周内死亡。与对照组相比（$B_1$=为3.0ng/mL，$M_1$=15.0ng/mL），所有患病骆驼的血清测试中，黄曲霉毒素含量均很高（$B_1$=10.0ng/mL时，$M_1$=40.0ng/mL）。骆驼尸检显示，肝脏肿大、黄染、易碎并充血。组织病理学检查显示肝小叶中心坏死，肝细胞呈空泡状。

Saad等[64]，Osman 和Abdel-Gadir[57]及Motawee等[52]从来自阿联酋和埃及的单峰驼大量的奶样中，发现了黄曲霉毒素M1和未分类的黄曲霉毒素。作者担心骆驼奶对人类健康构成威胁，强调需对骆驼奶进行连续测试，以确保人类摄入的黄曲霉毒素保持在最低水平。Elmaraghy[21]在利比亚报道了黄曲霉毒素污染骆驼饲料的情况。Bhatia 等[7]报道了一例印度的9岁骆驼的肺曲霉病。病驼肺部的几个结节发现被发暗变硬的肺组织包绕，后者含有半固体的干酪样坏死物。许多脓肿扩散到肺实质，确诊为坏死性化脓性肺炎。观察到与曲霉菌相类似的带有分支隔膜的真菌，并从肺中分离到化脓放线菌。

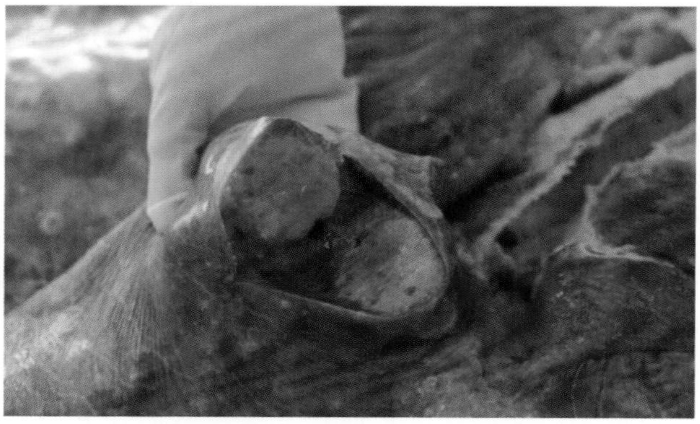

a.肺部感染

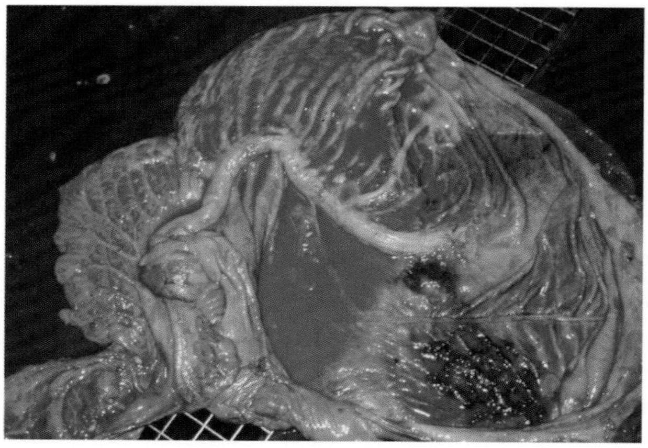

b. 胃嵌塞

图194 单峰驼肺部烟曲霉感染（a）和原驼瘤胃患有胃嵌塞（黑色区域）

在阿联酋饲养的一峰种用单峰驼体内发现的曲霉病肉芽肿，有些直径达5cm（图194a）。该驼患全身性骆驼痘数周，经四环素治疗了一段时间。Pickett 等[59]和Severo 等[68]报道了两峰羊驼的侵袭性曲霉病。该病在传播后引起小的脓肿，在骆驼肺部、心脏、脾和肾出现多病灶的坏死区。其中一峰病驼，从一只眼睛坏死的视网膜、睫状体和后晶状体囊检测到大量的分支状有隔真菌菌丝。病畜失明，伴有歪头和间歇盘旋。这两个病例中，组织学切片中观察到的菌丝形态与曲霉菌相吻合，但未尝试培养真菌。

一峰阿联酋的羊驼由于憩室发炎阻塞了十二指肠，导致胃受到挤压，被诊断患有烟曲霉瘤胃炎（图194b）。

在羊驼上，Muntz[54]报道了一个伴有支气管扩张的黑曲霉脓肺炎病例。由于预后不良，羊驼被实行安乐死。剖检发现，羊驼肺的气道里有脓性物质，从中分离出黑曲霉，同时还分离到大量相关的草酸结晶。推测晶体可能是真菌产生的。

## 防治

曲霉病的治疗效果不令人满意。已使用的药物包括噻菌灵，氟胞嘧啶和两性霉素B等，但对它们在骆驼科动物的效果知之甚少。噻菌灵作为抗真菌剂，在治疗赛骆驼曲霉病时无效。根据El-Khouly等[20]、Manefield和Tinson[48]的经验，曲霉病是一个与应激相关的疾病，尽量减少导致应激的因素，是实现预防的最佳途径。

（郑福英译，殷宏校）

## 3.3 念珠菌病

念珠菌病是由酵母类念珠菌（最常见的是白色念珠菌）引起的一种常见的消化道偶发病。在家禽、狗、猫、马、猪和野生动物上引起的感染，世界范围内均有报道[50]。念珠菌感染可以引起牛的乳腺炎和流产，口腔黏膜真菌病（口疮），幼畜舌炎，皮肤感染和阴道炎。它可能通过肠道感染其他器官。感染常见于幼龄动物，通常与一些诱发因素有关。已报道的一例欧洲新生幼驼的胃念珠菌病和多例阿联酋幼龄单峰驼上的胃念珠菌病，均采用抗生素延长治疗[32]。Wernery等[79]和 Tuteja等[75]描述了皮肤的念珠菌感染。此外，美国的Smith和Reel报道了脑和肾的念珠菌病[71]。系统性白色念珠菌在两峰羊驼[44]和一峰原驼上引起的感染[36]分别由德国的Kramer和法国的Keckr所报道。此外，马拉色菌属似乎可引起南美洲骆驼的皮肤感染。

### 病原学

白色念珠菌是常见的病原，已鉴定的还有其他类似酵母菌种。白色念珠菌在人类和许多动物的肠道和生殖道黏膜所共生。因此，确定这种真菌感染与某一种疾病是否相关，有时很困难。

### 流行病学

从黏膜或组织切片中分离出白色念珠菌，并不能错误地诊断为念珠菌病。许多情况下，白色念珠菌属于消化道的正常菌群。众所周知，白色念珠菌致病性并不强，细胞壁糖蛋白似乎具有类似内毒素的活性。一些如营养不良、长期的免疫抑制或抗菌治疗等诱因常引起念珠菌病。摄入被污染的食物或水可能造成该菌的传播。

### 临床症状和危害

Hajsig[32]报道了一峰新生美洲驼出现淡黄色腹泻的病例，尽管采用了抗生素和电解质进行治疗，幼驼5天后死亡。 尸检显示，瘤胃和网胃的壁增厚、水肿，观察到一个灰白色的厚达几毫米的伪膜。显微镜下，黏膜上皮坏死并感染有大量的假菌丝和芽殖酵母细胞。Wernery 等[78]在阿联酋报道了类似的临床表现和病变，出生8～48小时的幼驼发生念珠菌病，出现了淡黄色腹泻。尸检发现，幼驼小肠患有黄色的伪膜性结肠炎（图195）并伴有卡他性胃肠炎，肠淋巴结肿胀和明显的类白喉结肠炎。幼驼的消化系统中没有乳状物，皱胃中发现数量不等的沙粒。

尸检中，肠黏膜涂片检查发现有白色念珠菌和产气荚膜念珠菌（图196）。显微观察表明，因酵母菌侵入造成的黏膜坏死仅限于上皮组织（图197）。 这些幼单峰驼出现了大肠杆菌性败血症，部分幼驼发生了由产气荚膜念珠菌造成的肠毒血症。作者已证实，幼驼体内的铜含量非常低，因此摄入沙子来阻止梭菌孢子生长。受大量白色念珠菌侵袭（图199），长期用抗生素治疗的成年骆驼，皱胃呈多发性溃疡（图198）。

同一作者还诊断了由白色念珠菌引起的皮肤病变（图200），与嗜皮菌引起的感染相类似（见第1.5节）。几峰6周大的幼驼的驼峰附近形成厚痂，用PAS染色证实有菌丝存在其中（图201）。

Tuteja等[75]在印度的单峰驼确诊了一个类似病例。病变最初出现在背部，靠近驼峰，后来延伸到腹部并覆盖全身。这种念珠菌病在拉贾斯坦邦普遍存在，无需治疗，于翌年换毛时消失。正在泌乳的母畜没有受到该病的影响，而患有皮肤念珠菌病的幼驼情况却很糟糕，体重低于同龄的健康幼驼。据认为，幼驼的不良免疫状态与矿物质缺乏有关。

其他研究人员，如Pal等[58]报道了一例白色念珠菌引起的中耳炎。Khosravi等[42]在伊朗发现，18.6%的单峰驼外耳道含有白色念珠菌，他们还从5.8%的没有受感染骆驼的眼睛和10.9%健康骆驼的鼻子中分离到该菌（图202）。

Keck等[36]从法国报道了原驼上的系统性念珠菌病。该病感染全身所有器官，引起淋巴结肿大，出现肾炎、肝炎、肠出血性坏死并伴有伪膜性结肠炎和系统性肉芽肿。许多真菌菌丝侵入多个器官的微循环，导致血栓形成和坏死。该原驼患有不明疾病，长期使用抗生素治疗。德国的Kramer等[44]报道了两峰羊驼的全身白色念珠菌感染，这两只动物可能因淋巴组织衰竭引起潜在的免疫抑制，在肺、心肌、肾脏、胰腺和脑有许多的化脓灶，瓣胃出现溃疡，肾脏发生局灶性化脓性肾炎。所有病灶均分离出白色念珠菌。此外作者警告称，长期用糖皮质激素和抗生素治疗，最有可能造成这些异常情况。

在美国，Smith和Reel[71]报道了一峰美洲驼上的大脑和肾的念珠菌病。由于治疗效果不佳，骆驼被实行安乐死。尸体剖检时，在肾皮质的实质部分发现多个光滑坚实的浅黄色肉芽肿，直径为1～6毫米。瓣胃的黏膜受到侵蚀，出现很多病灶。显微镜检查发现，大脑和小脑内含有大量的髓内脓肿，从这些病灶中分离到白色念珠菌。作者认为，最初使

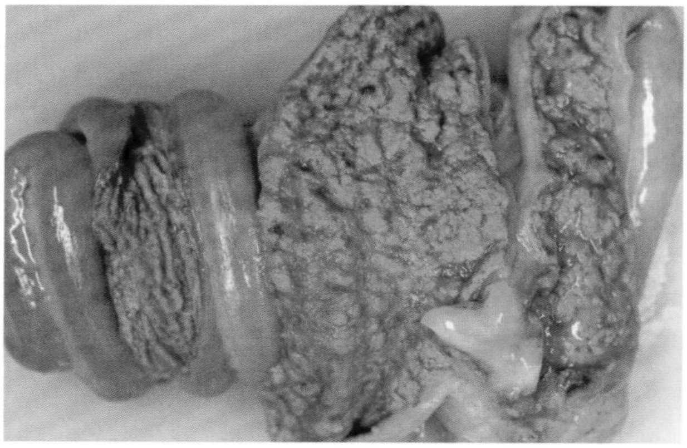

图195　白色念珠菌引起新生幼单峰驼小肠中的伪膜性结肠炎

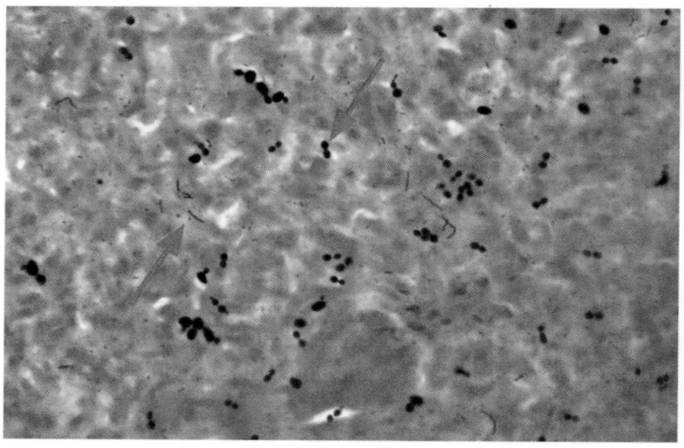

图196　白色念珠菌（红色箭头）和产气荚膜梭菌（绿色箭头）（姬姆萨染色）

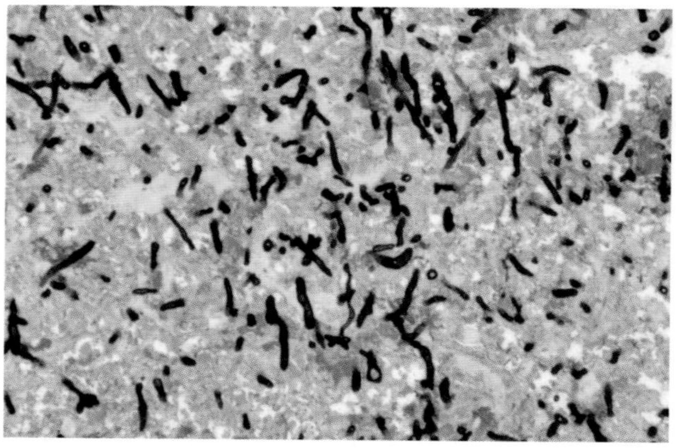

图197　酵母菌侵入引起的黏膜坏死（PAS染色）

图198　一峰成年骆驼皱胃观察到的多发性溃疡

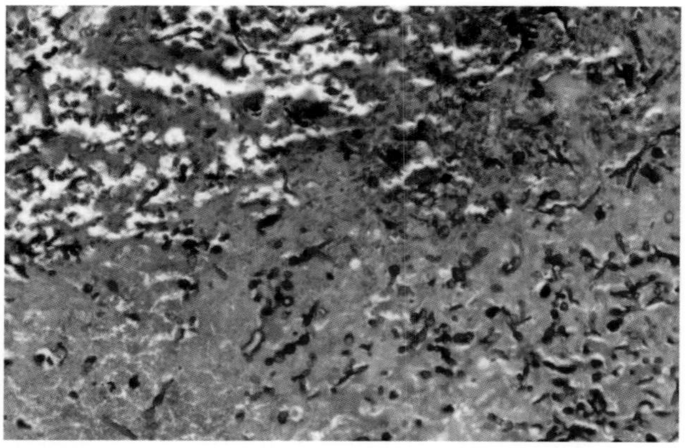

图199　有大量白色念珠菌侵入的溃疡（PAS染色）

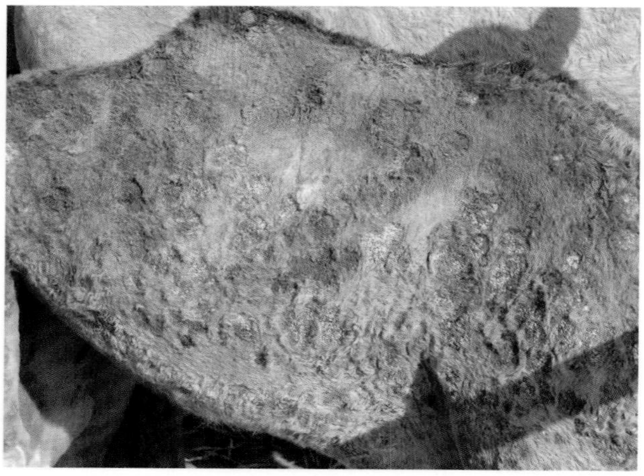

图200　白色念珠菌在单峰幼驼引起的皮肤病变

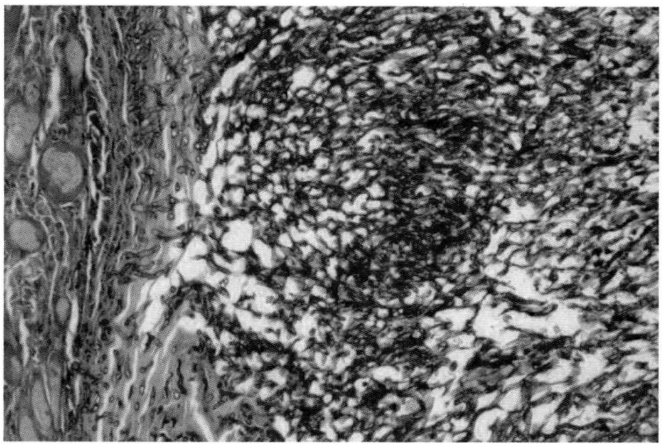

图201 PAS染色法证明驼峰附近的厚痂含有白色念珠菌菌丝

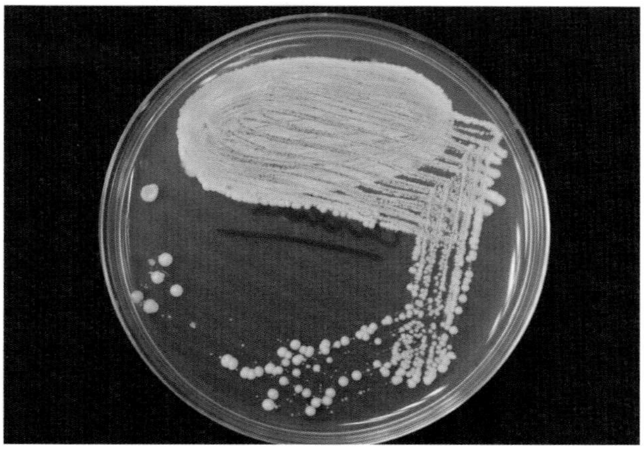

图202 白色念珠菌在沙氏琼脂37℃孵化3天后

用皮质激素治疗高热，诱发了念珠菌病。

**诊断**

真菌生物大量存在于增生的组织，可以通过培养或检查黏膜刮取物或制备组织切片进行诊断。白色念珠菌呈卵圆形，芽殖酵母细胞（芽生孢子，直径为3～6μm）或以链状形式出现，产生假菌丝。真菌用乳酚棉蓝、姬姆萨或革兰染色时着色良好，可见丝状的，规则的真菌菌丝。白色念珠菌可以在沙氏葡萄糖琼脂或普通琼脂，如血液和营养琼脂上培养，条件均为室温或37℃。24～72小时内生长为白色的，有光泽的，凸起的菌落。来自法国生物（梅里埃）的用于酵母识别Api（Api 20 CAUX）可用于酵母的鉴定，它是精确识别最常见酵母菌种类的诊断工具。

**防治**

制霉菌素，咪康唑和酮康唑已被推荐用于猪和牛的白色念珠菌的肠道感染，在感染的骆驼科动物上的应用，尚未报道。在作者治疗的病例中，母驼补充铜和硒，幼驼经口免疫10mL埃希氏菌属自家疫苗，静脉注射20mL产气荚膜梭菌抗血清（中央兽医研究实验室），24小时内分两次静脉注射10mg Stegantox（先灵葆雅动物保健）。骆驼未使用任何抗生素治疗。

通过最大限度地减少诱发因素，可很好地预防念珠菌病。因此，有必要检查并去除这些诱因。优化畜群管理，包括免疫接种（见4疫苗接种计划）以及适当的矿物质补充，对幼龄骆驼科动物的生存至关重要。

在皮肤念珠菌病的病例中，一个用硫芥子油或单独用芥末油进行治疗的传统兽医方法，取得很好的治愈效果。然而治疗开始前，动物须先剪毛，用含有Lotagen 溶液的温水小心地去掉所有的痂皮。皮肤干燥后，用一种由硫（6g）、水杨酸（3g）、芥子油（10mL）组成的药膏每日涂敷于患部，至少7天。所有受感染的幼驼补充10天的矿物质[75]。

Fowler[25]建议，患有皮炎的成年骆驼在感染部位局部使用制霉菌素和醋酸氯已定软膏，但治愈需要较长时间（超过60天）。

（郑福英译，殷宏校）

## 3.4 球孢子菌病

球孢子菌病（Coccidioidomycosis）是人类和动物呼吸道的一种真菌感染，它以一种散播性形式或以皮炎形式出现。新大陆骆驼对这种真菌似乎很易感，旧大陆骆驼的球孢子菌病尚无报道。

### 流行病学

引起球孢子菌病的粗球孢子菌（*Coccidioides immitis*）是一种二相性真菌，不能在动物与动物间传播。球孢子菌病通过从环境中吸入分生孢子而发生。美国人口每年估计有10万人感染该病，其中有70人死亡[65]。许多动物都确诊了该病，狗是最常见的感染动物[82]。粗球孢子菌在感染期以分节孢子的形式存在，在动物组织中转换成小球体，Fowler[25]描述了它的生活史。粗球孢子菌引起的感染，在欧洲和亚洲未被证实，似乎仅在北美和南美的一些地区流行。感染仅发生在特定的地区，这些地区炎热的气候条件、干旱的天气有利于土壤中真菌的生存。土壤的破坏使菌体暴露，形成气溶胶，随风进行长距离传播。气溶胶使得分节孢子易于吸入。

Muir和Pappagianis[53]，Fowler等[26]报道了美洲驼上的球孢子菌病。经胎盘感染的球孢子菌病也有报道。

### 临床症状和损害

已报道的有呼吸型和皮肤型，呼吸型表现为呼吸困难和咳嗽，皮肤型表现为大部分身体表面出现结节性病变。Muir和Pappagianis[53]报道了一峰患有后肢麻痹美洲驼，因为预后不良被实行安乐死。尸检发现，T-10脊髓受到散播性内脏肉芽肿和硬膜外脓肉芽肿严重挤压，并从这些病灶中分离到了粗球孢子菌（图203）。

在散播型中，身体的每个器官都可能受到感染。肉芽肿直径在1～5cm，或聚结成大的团块。镜检发现，灰色坚实的结节含有大量的小球体（图204）。整个身体表面包括四肢、面部及会阴部在内，可以观察到病变。

### 诊断

血清学试验可以对球孢子菌病进行诊断，如补体结合试验、琼脂凝胶扩散、荧光抗体或胶乳凝集法，也可通过活组织显微镜检查或尸检时得到确诊。琼脂凝胶扩散是迄今使用的最敏感和最特异的血清学试验，但因为基准数据尚未公布，往往难以对检测结果进行判定。真菌可以在选择培养基，如放线菌酮——氯霉素琼脂糖上培养，但这应仅限于那些有能力处理危险感染性培养物的实验室。

### 防治

两性霉素B是首选药物，但在驼类上效果不佳。连续用药超过6周，未成功治疗美洲驼疾病，也没能阻止病原经胎盘通道感染胎儿。新大陆骆驼喜在干土里打滚，产生灰尘。避免灰尘是防止感染的唯一方式，但这极难实现。

尽管在人类和灵长类动物已有相关疫苗，但针对驼类的疫苗尚未研发出来。

（郑福英译，殷宏校）

图203　粗球孢子菌引起的肺肉芽肿（图片由M. E. Fowler教授提供，美国）

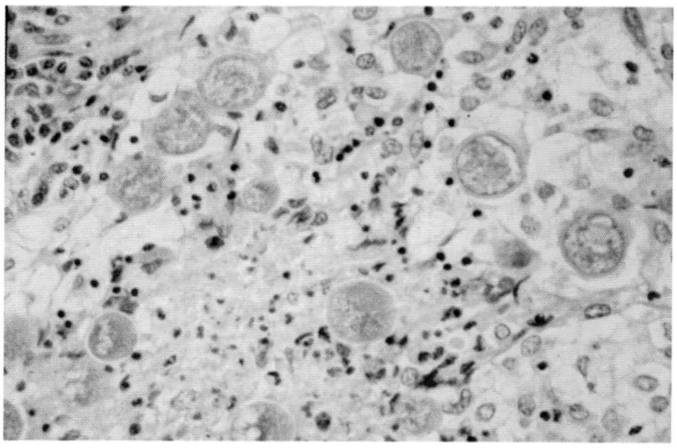

图204　厚壁小球充满内生孢子（图片由M .E. Fowler教授提供，美国）

# 3.5 毛霉菌病

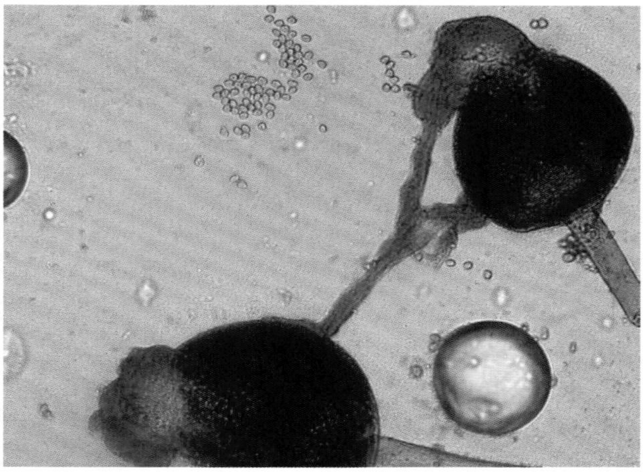

图205　带有分生孢子梗和分生孢子的毛霉菌（×400），透明胶带制备

## 病原学

毛霉目中11个属的22个种与本病有关。最重要的属是毛霉属、犁头霉属、根霉属和被孢霉属。毛霉属中致病性最强且耐热的菌，现在归入一个新属：根毛霉属。该属的任何一个种引起的疾病都被称为毛霉菌病。这些无处不在的真菌常见于土壤、粪便和腐烂的植物中。毛霉菌感染常继发于其他病症，可导致许多动物部分器官的肉芽肿病变。毛霉菌病作为牛胎盘炎和流产的一个原因，显得尤为重要。至今仅从一峰美洲驼上分离到毛霉目中的一个属——根霉属[25]。

## 临床症状

该美洲驼患有与第七脑神经面瘫相关的散播性的多系统感染。患病期间，骆驼吞咽困难，体重减轻。用内窥镜检查发现，鼻腔有一个带有白斑的黑膜。

## 诊断

镜检发现有宽阔的、分支状、无隔膜和不规则菌丝，可以确诊为毛霉菌病（图205）。借助毛霉目各属特异的荧光素抗球蛋白，使用荧光抗体技术可鉴别诊断组织切片中的真菌。在报道的美洲驼的病例中，发生坏死性鼻炎区域的表面，霉菌呈丝状生长，脑腹侧的脑脊膜发炎，脑神经处有肉芽肿。

（郑福英译，殷宏校）

## 3.6 其他真菌感染

在旧大陆和新大陆骆驼上的一些其他真菌感染病例也有报道，但很罕见，可见表75。

在中央兽医研究实验室观察到的一个藻菌病例中，一峰种用单峰驼尸检显示，瓣胃大范围增厚的胃壁组织中，可见大量真菌菌丝（图206）。

发现有趣的是，骆驼的尿液对真菌毒素和真菌的生长具有抑制作用[3]。此外，Al-Zahrani[6]证明了不同浓度的尿液样本对真菌生长有影响，培养的最初几天阻止了孢子的形成。

Bokhari [10]在沙特阿拉伯调查单峰驼的饲料中真菌和毒性代谢物（毒素）的存在情况。沙特阿拉伯，一个拥有约90万峰单峰驼的王国，在2008年至少有5 000峰单峰驼死亡，数千峰发病。在这之后，Bokhari发起了这项调查。这些灾祸被归咎为，一些含霉菌毒素的饲料或因全球气候变暖引起的其他传染性疾病。然而我们现在知道，有意在饲料中加入盐霉素是造成这次损失的原因。利用现代实验技术对7个地区收集的不同饲料的20个样本，检测真菌、黄曲霉毒素（AFT）、赭曲霉素（OTA）和玉米赤霉烯酮（ZON）的含量。从天然饲料中分离出10个属38种真菌。此外，在配合饲料样品中发现16个属32种真菌，包括曲霉菌和镰刀菌。所检样品菌落数都未超过1.0 × 10$^6$（cfu）/g。不少样品含有真菌毒素，其中仅有4种高于每份20/1 × 10$^9$（μg/kg）的国际上限。从这四种饲料中发现了黄曲霉毒素、赭曲毒素和玉米赤霉烯酮，赭曲霉素最高浓度是44μg/kg。

在阿联酋需定期检测不同骆驼饲料中霉菌毒

表75　骆驼科的其他真菌感染

| 疾病 | 病原 | 物种 | 临床症状 | 作者 |
|---|---|---|---|---|
| 结合菌病 (Entomophthormycosis) | 冠耳霉 *Conidiobolus coronatus* | 美洲驼 | 慢性,嗜酸性皮炎 鼻皮肤炎 | French and Ashworth(1994) |
| | | 美洲驼 | 鼻孔外部的结节性皮肤病 | Moll et al.(1992) |
| 藻菌病 | 未提及 | 单峰驼 | 胃溃疡 | Satir et al.(1993) |
| 组织胞浆菌病 | 荚膜组织胞浆菌 *Histoplasma capsulatum* | 单峰驼 | 常见的肺黑斑病 | Chandel and Kher(1994) |
| | | 美洲驼 | 肺炎 | Woolums et al.(1995) |
| 隐球菌病 | 新型隐球菌 *Cryptococcus neoformans* | 小羊驼 | 脑膜炎和肺炎 | Griner(1983) |
| | | 美洲驼 | 系统性疾病 | Bildfell et al.(2002) |
| | | 羊驼 | 脑膜炎 | Goodchild et al.(1996) |
| | 兔脑炎原虫 *Encephalitozoon cuniculi* | 羊驼 | 胎盘炎，围产期死亡 | Webster et al.(2008) |
| | 念珠菌属,*Candida*, 毛孢子霉属*Trichosporon*, 地霉菌属*Geotrichum* 克鲁维酵母属 *Kluyveromyces* 红酵母属*Rhodotorula*, 短梗霉属*Aureobasidium* 隐球菌*Cryptococcus* 原壁菌属*Prototheca* | 单峰驼 | 生殖道健康 | Shokri et al.(2010) |
| 互隔交链孢霉病 | 交链孢霉 *Alternaria alternata* | 单峰驼 | 几乎全身病变 | Tuteja et al.（2010） |

表76　骆驼饲料中微生物标准

| 饲料种类 | 正常菌群 | | 增加菌群 | | 异常菌群 | |
|---|---|---|---|---|---|---|
| | 细菌（百万每克) | 真菌(千每克) | 细菌（百万每克) | 真菌(千每克) | 细菌（百万每克) | 真菌(千每克) |
| 血粉<br>骨粉 | <1 | <10 | 1 ~ 2 | 10 ~ 40 | >2 | >40 |
| 鱼粉 | <1 | <20 | 1 ~ 2 | 20 ~ 50 | >2 | >50 |
| 除玉米外的谷物 | <2 | <40 | 2 ~ 3 | 40 ~ 100 | >3 | >100 |
| 玉米 | <1 | <10 | 1 ~ 2 | 10 ~ 40 | >2 | >40 |
| 面粉 | <2 | <40 | 2 ~ 3 | 40 ~ 80 | >3 | >80 |
| 木薯粉 | <3 | <50 | 3 ~ 4 | 50 ~ 100 | >4 | >100 |
| 大豆 | <1 | <20 | 1 ~ 2 | 20 ~ 50 | >2 | >50 |
| 颗粒料 | <1 | <20 | 1 ~ 2 | 20 ~ 40 | >2 | <40 |
| 干草 | <1 | <20 | 1 ~ 2 | 20 ~ 40 | >2 | <40 |
| 紫花苜蓿（草） | <1 | <20 | 2 ~ 3 | 20 ~ 40 | >3 | <40 |

**卫生状况正常; II，卫生状况差;III，卫生状况很差（不适合食用）**

**霉菌毒素：黄曲霉毒素≥20μg/kg是有毒的;伏马菌素≥2μg/kg是有毒的**

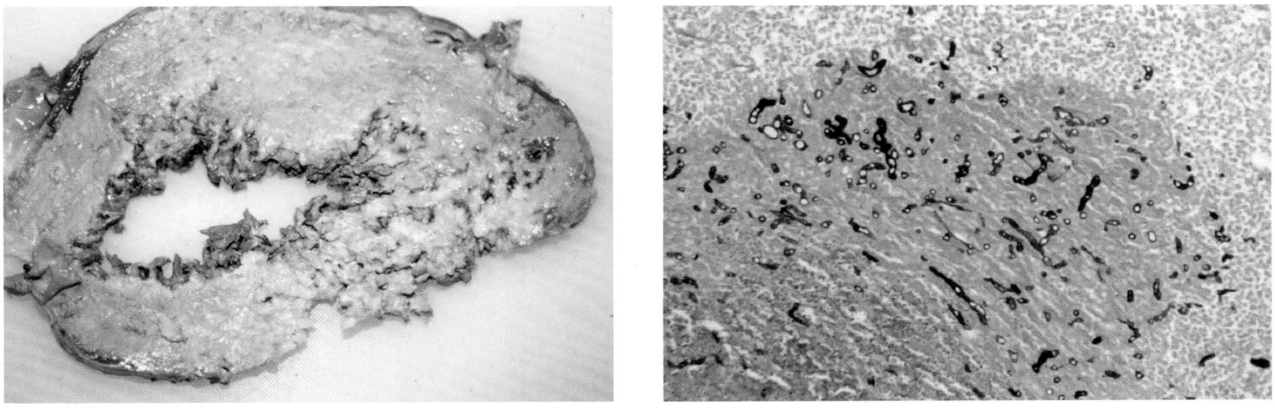

**a.患藻菌病的育种单峰驼瓣胃壁增厚**

**b.同一病例中用Grocott染色的真菌菌丝**

图206　患藻菌病的育种单峰驼

素及每克饲料中真菌菌落总数。多年来，兽医和骆驼业主已按照表76列出的质量指南进行操作。

在发展中国家，饲料中真菌和其毒性代谢物的存在几乎不可避。一些人已经开始通过蒸气处理产品，以减少真菌和细菌的污染量。这种措施具有非常好的效果。

Tuteja等[74]描述了在印度拉贾斯坦邦的单峰驼上的由交链孢霉引起的皮肤链格孢病。在病畜包括臀部和乳房在内所有部位均发现病变，不易与经典癣区别。重复检查皮肤刮屑仅能发现交链孢霉。去除患部所有的皮肤刮屑，然后用含有硫磺、水杨酸和芥子油的软膏涂敷，可治愈病变。

## 参考文献


[1] Abou-Zaid A.A. (1995). – Studies on ringworm in camels. *In* 3rd Scientific Congress, Egyptian Society for Cattle Diseases,3 ~ 5 December, Assiut, Egypt, 158–163.

[2] Ainsworth G.C. & Austwick P.K.C. (1973). – Fungal diseases of animals. 2nd Ed. Commonwealth Agricultural Bureau, Slough.

[3] Al-Abdalall A.H.A. (2010). – The inhibitory effect of camel s urine on mycotoxins and fungal growth. *Afr. J. agric. Res.*, 5 (11), 1331–1337.

[4] Al-Ani F.K., Al-Bassam L.S. & Al-Salahi K.A. (1995). – Epidemiological study of dermatomycosis due to *Trichophyton schoenleinii* in camels in Iraq. *Bull. Anim. Health Prod. Afr.*, 43, 87–92.

[5] Al-Ani F.K. & Al-Rawahi A. (2006). – Epidemiology of fungal infection in camels in relevance to human dermatophytosis in Jordan and Sultanate of Oman. *In* Proceedings of the International Scientific Conference on Camels, 10 ~ 12 May, Kingdom of Saudi Arabia, 501–508.

[6] Al-Zahrani S.M. (2002). – Study on the effect of female camels urine (virgin and fertilized) on the A*spergillus niger* fungus. *Arab Gulf J. Sci. Res.*, 20 (2), 115–122.

[7] Bhatia K.C., Kulshreshtha R.C. & Paul Gupta R.K. (1983). – Pulmonary aspergillosis in camel. *Haryana Vet.*, XXII, 118–119.

[8] Bildfell R.J., Long P. & Sonn R. (2002). – Cryptococcosis in a llama *(Lama glama). J. Vet. Diag. Invest.*, 14, 337–339.

[9] Boever W.J. & Rush D.M. (1975). – Microsporum gypseum infection in a dromedary camel. *Vet. Med. Small Anim. Clin.*, 70 (10), 1190–1192.

[10] Bokhari F.M. (2010). – Implications of fungal infections and Mycotoxins in camel diseases in Saudi Arabia. *Saudi J. biol. Sci.*, 17, 73–81.

[11] Chandel B.S. & Kher H.N. (1994). – Occurrence of histoplasmosis-like disease in camel (*Camelus dromedarius*). *Ind. Vet. J.*, 71 (5), 521–523.

[12] Chatterjee A., Chakraborty P., Chattopadhyay D. & Sengupta D.N. (1978). – Isolation of *Trichophyton schoenleinii* from a camel. *Ind. J. Anim. Health*, 17 (1), 79–81.

[13] Coetzer J.A.W., Thomson G.R &. Tustin R.C. (2004). – Mycoses. *In* Infectious diseases of livestock. Vol. 3. 2nd Ed. (J.A.W. Coetzer and R.C. Tustin), Oxford University Press, Oxford, UK, and Cape Town, South Africa, 2095–2104.

[14] Curasson G. (1947). – Le chameau et ses maladies. Vigot Freres, Editeurs, Paris, 86v88.

[15] Dalling T. (1966). – International encyclopaedia of veterinary medicine, Vol. I. Edinburgh, Green and Son, London, Sweet and Maxwell Ltd., 586.

[16] Driot C., Kamili A., Bengoumi M., Faye B., Delverdier M. & Tligui N. (2011). – Study on the epidemiology and histopathology of sarcoptic mange and ringworm in the one-humped camel in south of Morocco. *J. Camel Pract. Res.*, 18 (1), 107–114.

[17] Ebrahimi A., Montazeri B. & Lotfalian Sh. (2007). – Dermatophytes isolated from the hair coat/skin scrapping of healthy

dromedaries in Iran. *J. Camel Pract. Res.*, 14 (2), 133–134.

[18] El-Jaouhari S., Ouhelli H. & Yaddine M. (2004). – A propos de cas de teignes du dromedaire au Maroc. *J. de myco. Medic.* 14 (2), 83–87.

[19] El-Kader A. (1985). – Studies on skin diseases of camels with special reference to mycotic causes and treatment in Assiut Province. MVSc. Thesis, Assiut University, Egypt.

[20] El-Khouly A-Ba., Gadir F.A., Cluer D.D. & Manefield G.W. (1992). – Aspergillosis in camels affected with a specific respiratory and enteric syndrome. *Austr. Vet. J.*, 69 (8), 182–186.

[21] Elmaraghy S.S.M. (1996). – Fungal flora and aflatoxin contamination of feedstuff samples in Beida Governorate, Libya. *Folia Microbiologica*, 41 (1), 53–60.

[22] El-Tamavy M.A., Seddik I. & Atia M. (1988). – Camel ringworm in Upper Egypt. *Assiut Vet. Med. J.*, 20 (39), 54–59.

[23] Fadlelmula A., Agab H., Le Horgue J.M., Abbas B. &. Abdalla A.E. (1994). – First isolation of *Trichophyton verrucosum* as the aetiology of ringworm in the Sudanese camel *(Camelus dromedarius). Rev. Elev. Méd. Vét. Pays Trop.*, 47 (2), 184–187.

[24] Fischman O., Siguera P.A. & Baptista G. (1987). – *Microsporum gypseum* infection in a grey wolf (*Canis lupus*) and a camel (*Camelus bactrianus*) in a zoological garden. *Mykosen*, 30 (7), 295–297.

[25] Fowler M.E. (2010). – Medicine and surgery of camelids. 3rd Ed. Oxford, Wiley-Blackwell, 187–196.

[26] Fowler M.E., Pappagianis D. &. Ingram I. (1992). – Coccidioidomycosis in llamas in the United States, 19 cases (1981 1989). *JAVMA*, 201 (10), 1609–1614.

[27] French R.A. & Ashworth C.D. (1994). – Zygomycosis caused by *Conidiobolus coronatus* in a llama (*Lama glama*). *Vet. Pathol.*, 31, 120–122.

[28] Gareis M. & Wernery U. (1994). – Determination of Gliotoxin in samples associated with cases of intoxication in camels. *Mycotoxin Res.*, 10, 2–8.

[29] Gitao C.G., Agab H. & Khalafalla A.J. (1998). – An outbreak of a mixed infection of *Dermatophilus congolensis* and *Microsporum gypseum* in camels (*Camelus dromedarius*) in Saudi Arabia. *Rev. sci. tech. Off. int. Epiz.*, 17 (3), 749–755.

[30] Goodchild L.M., Dart A.J., Collins M.B., Dart C.M., Hodgson J.L. & Hodgson D.R. (1996). – Cryptococcal meningitis in an alpaca. *Austr. Vet. J.*, 74 (6), 428–430.

[31] Griner L.A. (1983). – Camelidae. In Pathology of 200 animals (L.A. Griner, ed.). San Diego Zoological Society, San Diego, California, USA, 501–505.

[32] Hajsig M., Naglio T., Hajsig D. & Herceg M. (1985). – Systemic mycoses in domestic and wild ruminants. I. Candidiasis of forestomachs in the lamb, calf, kid and newborn llama. *Vet. Arch.*, 55 (2), 53–58.

[33] Hollaender H., Keilig W., Bauer J. & Rothemund E. (1984). – A reliable fluorescent stain for fungi in tissue sections and clinical specimens. *Mycopathologia*, 88, 131–134.

[34] Ivanova L.G. & Polyakov I.D. (1983). – *Trichophyton sarkisovii Ivanova* and *Polyakov* sp. *nov.*, a new species of the pathogenic fungus inducing dermatomycosis in camels. *Mikol. Fitopatologiya,* 17 (5), 363–366.

[35] Kamel Y.Y., Ahmed M.A. & Ismail A.A. (1977). – Dermatophytes in animals, birds and man. Animals a potential reservoir of dermatophytes to man. *Assiut Vet. Med.*, 4 (7), 149–159.

[36] Keck N., Libert C., Rispail P. & Albaric O. (2009). – Systemic candidosis in a guanaco (*Lama guanicoe*). The *Vet. Rec.*, 164 (8), 245.

[37] Khamiev S.K. (1981). – Camel ringworm. *Buyll. Vses. Inst. Eksp. Vet.*, 42, 14–17.

[38] Khamiev S.K. (1982). – Epidemiology of ringworm (*Trichophyton* infection) among camels in Kazakhstan [in Russian]. *Veterinariya*, 9–42.

[39] Khamiev S.K. (1983). – Clinical symptoms of trichophytosis in camels. *Rev. Med. Vet. Mycology*, 17 (1), 147 (Abstract).

[40] Khosravi A.R., Shokri H. & Niasari-Naslaji A. (2007). – An outbreak of a mixed infection of *Trichophyton verrucosum* and *Nocardia asteroides* in dromedary camel in Iran. *J. Camel Pract. Res.*, 14 (2), 109–112.

[41] Khosravi A.R., Shokri H. & Sharifzadeh A. (2009). – Fungal flora of the eye and nose of healthy dromedary camels (*Camelus dromedarius*) in Iran. *J. Camel Pract. Res.*, 16, 63–67.

3

真菌病

[42] Khosravi A.R., Shokri H., Ziglari T. & Niasari-Naslaji A. (2008). – A study of mycoflora of the external ear canals in dromedary camels in Iran. *J. Camel Pract. Res.*, 15, 153–157.

[43] Kozlowska E.A. & Nuber D. (1995). – Leitfaden der praktischen Mykologie. Einfuhrung in die mykologische Diagnostik. Blackwell Wissenschaftsverlag, Berlin, Wien, 44–57.

[44] Kramer K., Haist V., Roth C., Schroder C., Siesenhop U., Baumgartner W. & Wohlsein P. (2008). – Systemic *Candida albicans* infection in two alpacas (*Lama pacos*). *J. Comp. Path.*, 139, 141–145.

[45] Kuttin E.S., Al-Hanaty E., Feldman M., Chaimovits M. & Muller J. (1986). – Dermatophytosis of camels. *Rev. Med. Mycol.*, 24, 341–344.

[46] Mahmoud A.L.E. (1993). – Dermatophytes and other associated fungi isolated from ringworm lesions of camels. *Folia Microbiologica,* 38 (6), 505–508.

[47] Mancianti F., Papini R. & Cavicchio P. (1988). – Dermatofizia da *Microsporum gypseum* in un Cammello (*Camelus dromedarius*). Ann. Fac. Med. Vet. Univ. Pisa, 4, 233–237.

[48] Manefield G.W. & Tinson A. (1996). – Camels. A compendium. The T.G. Hungerford Vade Mecum Series for Domestic Animals, 240, 298.

[49] Marks R., Knight A. & Laidler P. (1986). – Atlas of skin pathology. Vol. 36. MTP Press Limited, Lancaster, 39.

[50] Merck (1991). The Merck veterinary manual. Merck and Co. Inc., Rahway, New Jersey, USA, 342–343.

[51] Moll H.D., Schumacher J. & Hoover T.R. (1992). – Entomophthormycosis conidiobolae in a llama. JAVMA, 200 (7), 969–970.

[52] Motawee M.M., Bauer J. & McMahon D. (2009). – Survey of aflatoxin M1 in cow, goat, buffalo and camel milks in Ismailia-Egypt. Bull Environ. Contam. Toxicol., 83, 766–769.

[53] Muir S. & Pappagianis D. (1982). – Coccidioidomycosis in the llama: case report and epidemiologic survey. *JAVMA*, 181 (11), 1334–1337.

[54] Muntz F.H.A. (1999). – Oxalate-producing pulmonary aspergillosis in an alpaca. *Vet. Path.*, 36 (6), 631–632.

[55] Nasser M. (1969). – The zoonotic importance of dermatophytes in U.A.R. PhD Thesis, Cairo University, Egypt.

[56] Osman N., El-Sabban F.F., Al Khawlis A. & Mensah-Brown E.P.K. (2004). – Effect of foodstuff contamination by aflatoxin on the one-humped camel (*Camelus dromedarius*) in Al Ain, United Arab Emirates. *Australian Vet. J.*, 82 (12), 759–761.

[57] Osman N.A. & Abdel-Gadir F. (1991). – Survey of total aflatoxins in camel sera by enzyme-linked immunosorbent assay (ELISA). *Mycotoxin Res.*, 7, 35–38.

[58] Pal M., Patil D.B., Kelawala N.H., Parikh P.V., Barvalia D.R. & Patel D.M. (2007). – Occurrence of fungal otitis in camels. *J. Camel Pract. Res.*, 14, 73–74.

[59] Pickett J.P., Moore C.P., Beehler B.A., Gendron-Fitzpatrick A. &. Dubielzig R.R. (1985). – Bilateral chorioretinitis secondary to disseminated aspergillosis in an alpaca. *JAVMA*, 187 (11), 1241–1243.

[60] Quinn P.J., Carter M.E., Markey B.K. & Carter G.R. (1994). – Mycology: introduction to the pathogenic fungi. *In* Clinical veterinary microbiology (P.J. Quinn, M.E. Carter, B.K. Markey & G.R. Carter, eds). Wolfe Publishing, Prescott, Arizona, USA, 381–421.

[61] Quist C.F., Dutton D.M., Schneider D.A. & Prestwood A.K. (1998). – Gastrointestinal ulceration and pulmonary aspergillosis in a llama treated for parelaphostrongylosis. *JAVMA*, 212 (9), 1438–1441.

[62] Ramadan R.O., Fayed A.A. &. El-Hassan A.M. (1989). – Textbook of dermatology. Vol. 2. 4th Ed. Blackwell Scientific Publications, Oxford, 2 (4), 911–915.

[63] Refai M. & Miligy M. (1968). – Soil as a reservoir of *Trichophyton mentagrophytes*. *J. Egypt. Vet. Med. Ass.*, 28, 47–52.

[64] Saad A.M., Abdelgadir A.M. & Moss M.O. (1989). – Aflatoxin in human and camel milk in Abu Dhabi, United Arab Emirates. *Mycotoxin Res.*, 5, 57–60.

[65] Salfelder K. (1990). – Atlas of fungal pathology. Dordrecht, Kluwer Academic Publishers, 101.

[66] Satir A.A., Abu Bakr M.I., Abalkheil A., Abdel Ghaffar A.E. & Babiker A.E. (1993). – Phycomycosis of the abomasum in *Camelus dromedarius*. *J. Vet. Med. Ass.*, 40, 672–675.

[67] Schwartz H.J. & Dioli M. (1992). – The one-humped camel in Eastern Africa. A pictorial guide to diseases, health care and management. Weikersheim, Verlag Josef Margraf, 205.

[68] Severo L.C., Bohrer J.C., Geyer G.R. & Ferreiro L. (1989). – Invasive aspergillosis in an alpaca (*Lama pacos*). *J. Med. Vet. Mycol.*, 27, 193–195.

[69] Shokri H., Khosravi A., Sharifzadeh A. & Tootian Z. (2010). – Isolation and identification of yeast flora from genital tract in healthy female camels (*Camelus dromedarius*). *Vet. Microbiol.*, 144, 183–186.

[70] Singh M.P. & Singh C.M. (1969). – Mycotic dermatitis in camels. *Ind. Vet. J.*, 46 (10), 855.

[71] Smith S.H. & Reel D.R. (2008). – Cerebral and renal candidosis in a llama (*Lama glama*). *Vet. Rec.*, 162 (15), 485–486.

[72] Toleutajewa S.T. (1994). – Widerstandsfahigkeit des Erregers der Trichophytie der Kamele in der Umwelt und vergleichende Aktivitat von Vakzinen bei dieser Erkrankung. Thesis, Russische Akademie der Landwirtschaftswissenschaften, Moskau.

[73] Torky H.A. & Hammad H.A.S. (1981). – Trichophytosis in farm animals and trials for treatment. *Bull. Anim. Health Prod. Afr.*, 29 (2), 143–147.

[74] Tuteja F.C., Ghorui S.K. & Narnaware S.D. (2010). – Cutaneous alternariosis in dromedary camel. *J. Camel Pract. Res.*, 17 (2), 225–228.

[75] Tuteja F.C., Pathak K.M.L., Ghorui S.K., Chirania B.L. & Kumar S. (2010). – Skin candidiasis in dromedary camel calves. *J. Camel Pract. Res.*, 17 (1), 59–61.

[76] Webster J.D., Miller M.A. & Vemulapalli R. (2008). – *Encephalitozoon cuniculi*-associated placentitis and perinatal death in an alpaca (*Lama pacos*). *Vet. Pathol.*, 45, 255–258.

[77] Wellehan J.F.X., Farina L.L., Keoughan C.G., Lafortune M., Grooters A.M., Mendoza L., Brown M., Terrell S.P., Jacobson E.R. & Heard D.J. (2004). – Pythiosis in a dromedary camel (*Camelus dromedarius*). *J. Zoo Wildlife Med.*, 354 (4), 564–568.

[78] Wernery R., Ali M., Kinne J., Abraham A.A. & Wernery U. (2000). – Mineral deficiency: a predisposing factor for septicemia in dromedary calves. *In* Proceedings of the 2nd Camelid Conference on Agroeconomics of Camelid Farming, Almaty, Kazakhstan, 8–12 September, 86.

[79] Wernery U., Kinne J. & Nagy P. (2007). – Candida dermatitis in camel calves – a case report. *In* Proceedings of the International Camel Conference, 16–17 February, Rajasthan Agricultural University, India, 226.

[80] Wernery U., Schimmelpfennig H.H., Seifert H.S.H. & Pohlenz J. (1992). – *Bacillus cereus* as a possible cause of haemorrhagic disease in dromedary camels (*Camelus dromedarius*). *In* Proceedings of the 1st International Camel Conference, Dubai, UAE, 51–58.

[81] Wilson R.T. (1998). – Camels. The tropical agriculturalist, London, MacMillan, 106.

[82] Wolf A.M. & Pappagianis D. (1981). – Canine coccidioidomyiosis – treatment with new agent. *Calif. Vet.*, 5, 25–27.

[83] Woolums A.R., DeNicola D.B., Rhyan J.C., Murphy D.A., Kazacos K.R., Jenkins S.J., Kaufman L. & Thornburg M. (1995). – Pulmonary histoplasmosis in a llama. *J. Vet. Diagn. Invest.*, 7 (4), 567–569.

[84] Zanolari P., Meylan M., Sager H., Herrli-Gygi M., Rufenacht S. & Roosje P. (2008). – Dermatologie bei Neuweltkameliden. Teil 2: Ubersicht der dermatologischen Erkrankungen. Tierarztl. Praxis, 36 (G), 421–427.

## 深入阅读材料

*Abdel Samed G.H. (1983). – Yeast flora in the digestive tract of the one-humped camel. Thesis, University Khartoum, Sudan.*

*Connole M.D. (1990). – Review of animal mycosis in Australia. Mycopathologica, III (3), 133–164.*

*Glawischnig W. & Khaschabi D. (1999). – Generalisierte Aspergillose bei einem juvenilen Alpaca (Lama pacos). Wien. Tieäztl. Mschr., 86, 317–319.*

*Holmes L.A., Frame N.W., Frame R.K., Duff J.P. & Lewis G.C. (1999). – Suspected tremorgenic mycotoxicosis (ryegrass staggers) in alpacas (Llama pacos) in the UK. Vet. Rec., 145, 462–463.*

（郑福英译，殷宏校）

3 真菌病

# 4 疫苗接种计划

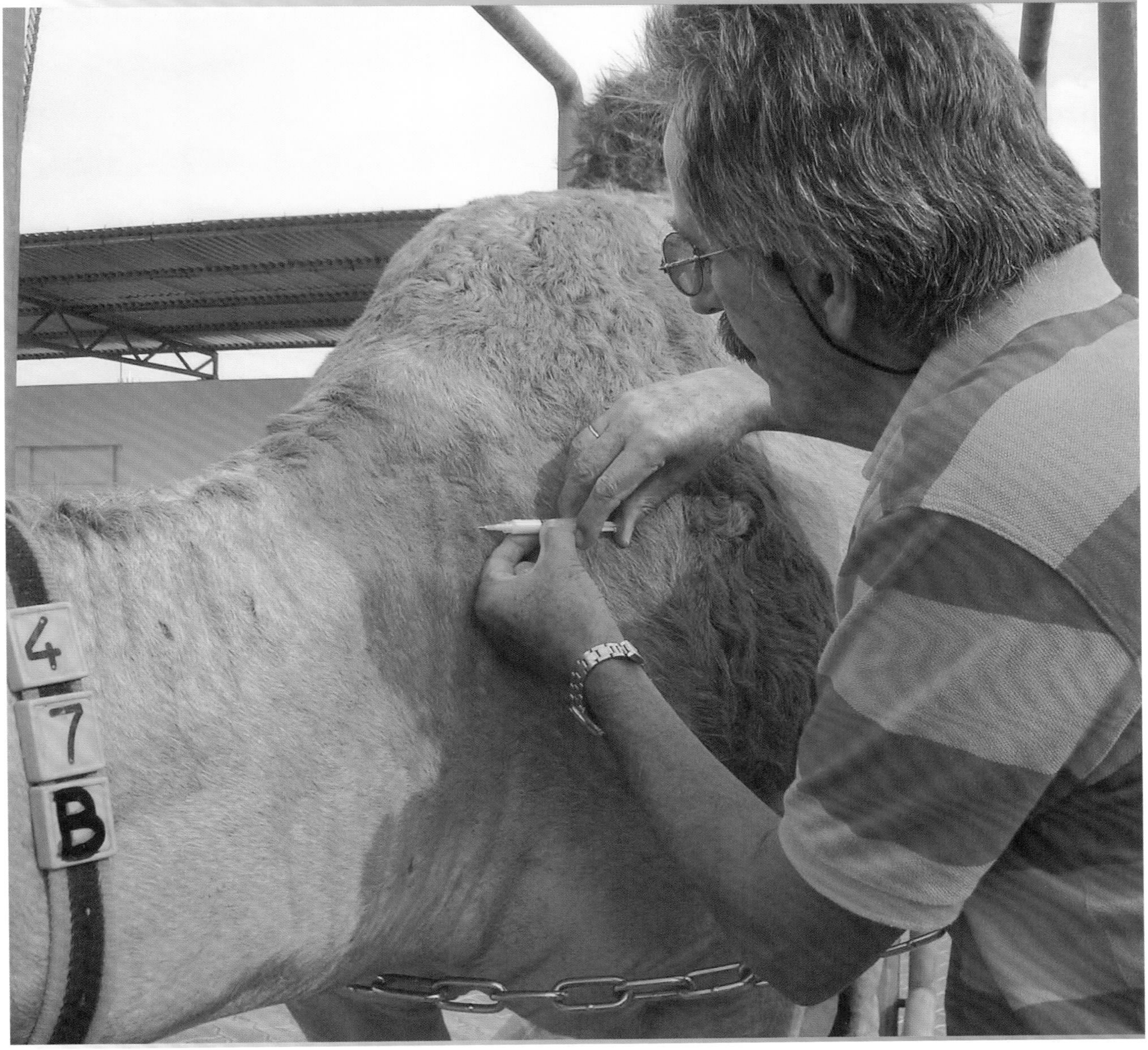

对于具有经济价值的动物，在生产实践中有多种方法可以降低感染的风险，如检测与淘汰、加强卫生管理、免疫等。20世纪兽医学上最突出的成就是用免疫的方法预防和控制了大量的传染病。

没有免疫预防，就无法进行规模化畜牧业生产。目前，市场上的疫苗种类已超过数百种，每种疫苗在许可使用前，必须由权威机构进行有效性和安全性检测。

在此无法详述每一种疫苗的使用方法，只能介绍在疫苗免疫中应遵守的基本原则。由于母源抗体能被动地保护新生幼畜，过早免疫反而不一定取得预期效果。如希望新生幼畜获得免疫力，应对母畜在怀孕后期进行免疫。应掌握好疫苗接种时机，以便幼畜在抗体水平最高时采食初乳。幼畜的被动免疫消退后进行疫苗接种才能获得最佳免疫效果。由于无法预测幼畜母源抗体消失的确切时间，为确保幼畜获得切实的免疫力，应至少接种两次疫苗。

对于各种疫苗在骆驼上的效果如何知之甚少。在美国，没有任何疫苗被准许用于骆驼[1]。Fowler[1]和Mayr[2]推荐了一些用于新大陆骆驼的疫苗，根据他们的意见和我们的经验，在表77和表78中列出了预防病毒、细菌和真菌病的免疫程序。

在骆驼中使用灭活疫苗时非常重要的一点是所使用的佐剂能提供坚实的免疫力但副反应应尽可能小。因此，Eckersley等[3]和Kinne等[4]开展了一项研究，以确认哪一种佐剂的效果最好。在7个不同公司生产的7种佐剂中，Advax HCXL的效果最佳，能有效提高针对病毒和细菌抗原的特异性抗体的效价。在未来，还有必要进一步开展更加安全、动物耐受性更高的佐剂。

表77 驼科动物细菌病的免疫程序

| 疫病 | 疫苗 | 首免时间 | 加强免疫 | 重复免疫 | 注意事项 |
|---|---|---|---|---|---|
| 破伤风（破伤风梭菌） | 类毒素疫苗 | 2～3月龄 | 4周之后 | 1～3年 | 疑似感染时可使用高免血清 |
| 内毒素血症（A型、B型、C型、D型产气荚膜梭菌） | 类毒素和细菌苗 | 1～2月龄 | 4周之后 | 每年一次 | 可用当地菌株生产自家苗，发病时静脉注射100mL单峰驼体制备的高免血清 |
| 气性水肿综合征（气肿疽梭菌、腐败梭菌和诺维氏芽孢梭菌） | 类毒素和细菌苗 | 1～2月龄 | 4周之后 | 每年一次 | 受威胁区使用 |
| 炭疽（炭疽芽孢杆菌） | 弱毒苗 | 2～3月龄 | — | 每年一次 | 流行区使用 |
| 大肠杆菌腹泻 | 灭活疫苗（每头剂＞$10^{10}$CFU） | 口服（20～50mL） | 连用10天 | | |
| | | — | 幼畜使用自家苗 | | |
| | 灭活疫苗 | 分娩前6周 | 分娩前2周 | 每年一次 | 怀孕骆驼使用自家苗 |
| 沙门氏菌病 | 灭活疫苗（每头剂＞$10^{10}$CFU） | 口服（20～50mL） | 连用10天 | | |
| | | — | 幼畜使用自家苗 | | |
| | 灭活疫苗 | 分娩前6周 | 分娩前2周 | 每年一次 | 怀孕骆驼使用自家苗 |
| 钩端螺旋体病 | 适宜血清型灭活疫苗 | 2月龄 | 4周之后 | 4月 | 流行区使用，适宜血清型菌株 |

表78　驼科动物病毒病和真菌病的免疫程序

| 疫病 | 疫苗 | 首免时间 | 加强免疫 | 重复免疫 | 注意事项 |
|---|---|---|---|---|---|
| 狂犬病（弹状病毒） | 灭活细胞培养苗 | 3月龄 | 3周之后 | 每年一次 | 受威胁区使用（Rabinsin） |
| 骆驼痘（正痘病毒） | 弱毒细胞培养苗 Ducapox | 6～9月龄 | 4周之后 | 终生免疫? | 商业化疫苗 (Ducapox，中央兽医研究所) |
|  | 灭活佐剂苗 | 6～9月龄 | 4周之后 | 每年一次 | 摩洛哥，ROBAT 生物和兽医药品公司生产 |
| 接触传染性脓疱（副痘病毒） | 弱毒细胞培养苗 | 1～2月龄 | 6周之后 | 6～8月 | 土耳其生产 |
| 乳头状瘤病（乳头状瘤病毒） | 灭活的乳头瘤组织 | 治疗用（皮下） | 每5天一次，剂量递增，连用3次 | — | 自家苗 |
| 牛病毒性腹泻/黏膜病(黄病毒科瘟病毒) | 灭活疫苗 | 2～4周龄 | 2月之后 | 每年一次 | 在牛病毒性腹泻引起流产的地区使用 |
| 新生畜腹泻（轮状病毒、冠状病毒） | 灭活疫苗 |  | 分娩前第4周和第2周 | 每年一次 | 需要时 |
| 马疱疹 | 灭活疫苗 | 8～12周龄 | 3～4周之后 | 每年一次 | 需要时 |
| 皮肤真菌病 | 弱毒脱毛毛癣菌 | 4月龄（也可用于治疗） | 14天之后 | ? | 有商业化的牛用苗 |

？表示不确定

## 参考文献

[1] Fowler M.E. (2010). – Medicine and surgery of camelids. 3rd Ed. Wiley-Blackwell, Oxford, UK, 183–185.

[2] Mayr A. (1998). – Nutzung des Immunsystems fur die Schutzimpfung und Paraimmunisierung von Neuweltkameliden. Lamas, Haltung und Zucht, 6 (2), 14–23.

[3] Eckersley A.M., Petrovsky N., Kinne J., Wernery R. & Wernery U. (2011). – Improving the dromedary response: the hunt for the ideal camel adjuvant. J. Camel Pract. Res., 18 (1), 35–46.

[4] Kinne J., Eckersley A.M. & Wernery U. (2012). – Search for the best adjuvants for use in dromedaries. In Proceedings of the 3rd Conference of the International Society of Camelid Research and Development, Muscat, Oman, 29 January–1 February 2012, 62–63.

（殷宏译，储岳峰校）

# 5 寄生虫病

# 引言

骆驼的寄生虫包括宿主特异的种（如艾美耳球虫、肉孢子虫、虱和鼻喉蝇）和宿主范围广的种（如伊氏锥虫、刚地弓形虫和新孢子虫/哈芒球虫等成囊球虫、吸虫、肠道绦虫、蜱、吸血蝇，以及麻蝇科和丽蝇科的能引起蝇蛆病的蝇类）。

骆驼寄生虫的种群取决于环境因素和饲养管理水平，旧大陆骆驼和新大陆骆驼最早都生活在非常恶劣环境条件下，使得寄生虫和宿主的相互作用比较简单。然而，有一些寄生虫，却适应了这种严酷的环境，如长干血矛线虫（*Haemonchus longistipes*），流行于干旱地区，潜伏期明显延长，在雨季产卵；一种寄生于骆驼的蜱，嗜驼璃眼蜱（*Hyalomma dromedarii*），可根据外界环境情况，以三宿主蜱、二宿主蜱甚至一宿主蜱的方式完成其生活史。

主动寻找宿主（蜱、蝇及可引起蝇蛆病的蝇类、虱蝇），媒介传播一些寄生虫（伊氏锥虫）或生活史中有中间宿主（斯氏副柔吸虫Parabronema skrjabini、骆驼泡首线虫Physocephalus dromedarii）以及旧大陆骆驼的食粪癖等，都增加了寄生虫感染的可能性。

在温和的气候条件下，寄生虫种群及其数量会发生变化。当新大陆骆驼引入到北美洲和欧洲后，与反刍动物合群放牧时，这一现象更加明显，会感染一些在原来生活地所不存在的寄生虫（矛形歧腔吸虫Dicrocoelium dendriticum、丝状网尾线虫Dictyocaulus filaria）。

与其他农场动物一样，宿主密度高时有利于寄生虫种群的增值，也进一步影响着寄生虫病的临床症状和病程。

下图显示了寄生虫在旧大陆骆驼（左）和新大陆骆驼（右）的寄生部位：(a) 头；(b) 皮肤；(c) 肺；(d) 肌 (e)肝；(f) 胃；(g) 小肠；(h) 大肠；(i) 循环系统；(j)睾丸。

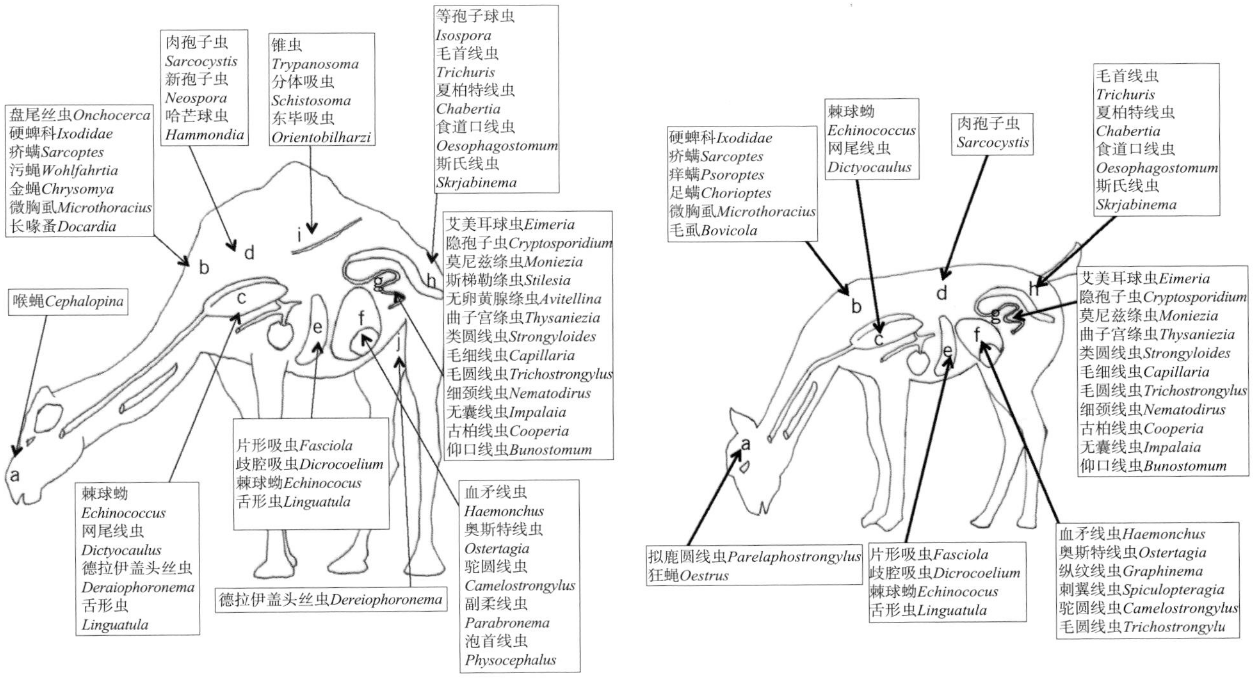

## 5.1.1 锥虫病

锥虫病是由锥虫属的鞭毛体引起的一类人畜共患疾病，其传播形式分为粪便型和唾液型两种。克氏锥虫（*Trypanosoma cruzi*）通过猎蝽科昆虫（reduviid bugs）的粪便进行传播，其主要分布于美国南部热带地区，除此以外，其他所有粪便型锥虫都无致病性。感染骆驼的锥虫都是唾液型的，其通过吸血蝇类的叮咬进行病原传播。伊氏锥虫[*Trypanosome evansi*，(同物异名 *T. ninaekohlyakimovi*❶)]是造成单峰驼伊氏锥虫病（苏拉病）的一种重要寄生虫病原，该病原于1880年由Griffith Evans首次分离于印度旁遮普省的德拉伊斯梅尔汗地区的单峰驼和马属动物中[33]。“苏拉”为印第语，意思是精神沉郁或者机体组织糜烂，在非洲及亚洲大多数国家被用于命名由伊氏锥虫引起的疾病。然而，在拉丁美洲该病原又被命名为马锥虫病（如Mal de caderas、Murrina和Derrengadera）。而在前苏联的中亚地区的共和国及蒙古国双峰驼中的伊氏锥虫病又被称为Su-Auru，该名字在哈萨克族中的意思“水边的疫病”，原因是其传播媒介为马蝇，而马蝇需要在沼泽地中繁殖[60]。

在采采蝇活动的地带，一些锥虫如布氏锥虫（*T. brucei*）、刚果锥虫（*T. congolense*）、活跃锥虫（*T. vivax*）和猴锥虫（*T. simiae*）对单峰驼这一重要宿主有很高的致死率，主要是因为采采蝇反复叮咬的结果。尽管锥虫对单峰驼有较强的治病作用，但单峰驼主要生活在没有采采蝇的地区，其影响范围非常有限。

尽管在南美洲存在伊氏锥虫（*T. evansi*）和活跃锥虫（*T. vivax*），但由于生态的原因该地区并没有新大陆骆驼锥虫病的相关报道，然而在从智利引入美国的美洲驼中已经发现了锥虫的存在[23]。实验数据显示，分离自迪拜骆驼场的伊氏锥虫能导致原驼出现寄生虫血症和锥虫病的临床特征[30]。

### 形态学

伊氏锥虫是锥虫科锥虫亚属的模式种。哺乳动物中，虫体形态单一，呈细长单体，长为15～36μm，宽1.5～2.2μm（图207和图208）。虫体有一明显的波动膜，一根自由活动的鞭毛和位于后端的小的动基体。形态学上伊氏锥虫与处于生长过渡期的布氏锥虫在长度上没有区别，自然条件下伊氏锥虫动基体中DNA高度浓缩使其减少的现象非常普遍。这种形式是锥虫亚属机械性传播的特征，这种传播形式的病原无需在采采蝇体内发育。

有学者认为伊氏锥虫可能起源于布氏锥虫。被感染的宿主离开采采蝇活动区后，虫体逐渐适应非循环传播模式，可通过其他嗜血蝇类以机械性方式传播。

### 流行病学

在毗邻采采蝇分布带及其以北的所有非洲国家都有锥虫病的流行，骆驼为最主要的宿主。在亚

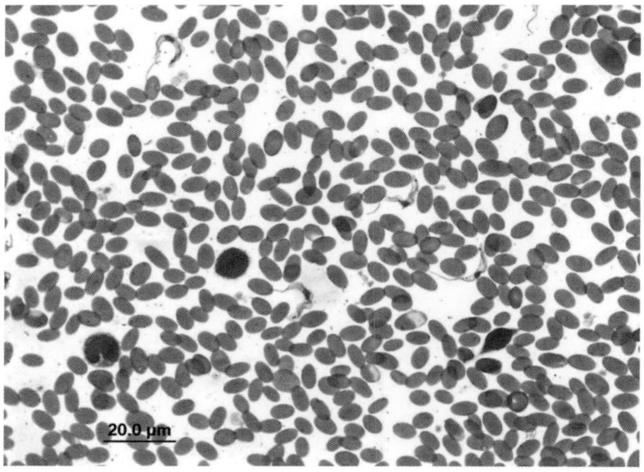


图207　姬姆萨染色血涂片中的伊氏锥虫

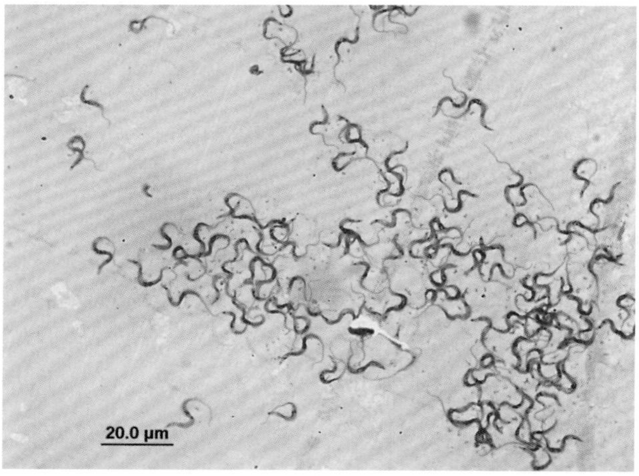


图208　用离子交换法分离出的伊氏锥虫姬姆萨染色照片

❶ 在俄语文献中，*T. ninaekohlyakimovi* 是比 *T. evansi.* 更晚的同物异名

表 79 不同诊断方法调查非洲国家单峰驼伊氏锥虫病流行特征

| 国家 | 检测数据 | 百分率(%) | 方法 | 参考文献 |
|---|---|---|---|---|
| 摩洛哥 | 1460 | 14.1 | CATT | 5 |
| | | 18.2 | Ab- ELISA | |
| 毛里塔尼亚 | 2073 | 1.3 | 血涂片 | 16 |
| | | 16.2 | CATT | |
| | | 25.2 | IFAT | |
| 马里 | 1093 | 30.6 | CATT | 17 |
| 布基纳法索 | 67 | 18 | MHCT | 15 |
| | | 46 | CATT | |
| 尼日利亚 | 110 | 9.1 | 血涂片 | 24 |
| 尼日尔 | 184 | 49.2 | Ab- ELISA (1)[a] | 59 |
| | | 47.1 | Ab- ELISA (2) | |
| | | 49.7 | CATT | |
| | | 49.7 | TL | |
| 乍得 | 2831 | 5.3 | MHCT | 11 |
| | | 30.5 | CATT | |
| 埃及 | 193[b] | 43.5 | CATT | 1 |
| | | 56.9 | PCR | |
| 苏丹 | 1738 | 5.4 | MHCT | 20 |
| | | 31.3 | Ag-ELISA | |
| 埃塞俄比亚 | 179 | 12.8 | 血涂片r | 65 |
| 索马里 | 3000 | 5.33 | MHCT | 18 |
| 肯尼亚 | 549 | 5.3 | MHCT | 43 |
| | | 26.6 | PCR | |
| | | 45.9 | CATT | |

a样品采于1995年，−70℃保存于比利时热带医学研究所（Institute of Tropical Medicine in Belgium.），使用了两种ELISA 方法；

b发病动物；

Ab-ELISA，抗体酶联免疫吸附试验 antibody enzyme-linked immunosorbent assay; Ag-ELISA,抗原酶联免疫吸附试验antigen enzyme-linked immunosorbent assay; CATT，锥虫病的卡片凝集试验card agglutination test for trypanosomosis; IFAT，简介荧光抗体试验 indirect fluorescence antibody test; MHCT，微红细胞积压法 micro haematocrit; PCR,聚合酶链式反应 polymerase chain reaction; TL，溶锥虫试验 trypanolysis test

洲，伊氏锥虫的宿主极其广泛，它们主要包括旧大陆骆驼中的单峰驼和双峰驼、黄牛、水牛、马和猪。据推测，伊氏锥虫病因为骆驼商队的贸易而传入北非、中东以及南亚。

对于锥虫病在非洲骆驼中的流行情况，已有很多文献进行了详细的描述（表79）。

据认为，南非好望角和德兰士瓦地区的锥虫病也是由感染的骆驼带入的。19和20世纪由于牛瘟的原因，引进大批的骆驼被用于战争。因此伊氏锥虫病也随之从西南部非洲（纳米比亚）被带入到德国人控制的非洲西南部、好望角（南非）和罗得西亚（津巴布韦）等地区。

此外，位于中东地区约旦、科威特、沙特阿拉伯、阿拉伯联合酋长国、阿曼、伊朗以及以色列等国家的骆驼中均有检测到伊氏锥虫的报道。

以往的资料报道显示，19世纪末亚洲的印度及周边国家的马和骆驼中有几次伊氏锥虫的大暴发[33]。伊氏锥虫病也被带入中国，但仅在新疆（中国西部）有感染骆驼的报道，而在中国南方的水牛、马以及驴中更常见[32]。

伊氏锥虫曾经在蒙古的科布多省、乌布苏省以及周边国家的双峰驼中均有发现(G. Battsetseg, 乌兰巴托兽医研究所，2012， 私人通讯)。最近报道表明，在印度以及巴基斯坦，伊氏锥虫病是一种非常重要的寄生虫病，该病原在当地骆驼、水牛、马属动物以及犬均有发现[47,25,53]。有少量的报道显示，在印度地区的人体也有伊氏锥虫病的发生[54]。

伊氏锥虫病在很早以前可能由于丝绸之路的骆驼商队而传播到中亚国家，但直到1913年才由Yakimoff报道[33]。在土库曼斯坦和哈萨克斯坦广泛传播的这种疾病被冠名为Su-Auru ，但是相关的流行病学数据在目前却很难获得。Ibragimov和Sulejmenov报道[28]，19世纪20~30年代伊氏锥虫病在哈萨克斯坦境内的骆驼中广泛传播，但在19世纪40年代几乎消失，直至70年代又复发。

曾经在加那利群岛几乎所有的骆驼中均发现抗伊氏锥虫病的抗体[40]，粗略估算阳性率约为9%，但只有在1.3%的样品中检出伊氏锥虫。这种感染性病原可能是从邻近的摩洛哥及毛里塔尼亚被带入的。尽管采取了严格的控制措施，但仍然有伊氏锥虫病的再次感染甚至新病例出现，这可能是因为啮齿类动物也是该病原的贮藏宿主的原因。然而，通过对捕获的田鼠及家鼠进行锥虫病原的PCR检测，发现这些病原是路氏锥虫（*Trypanosoma lewis*）而不是伊氏锥虫（*Trypanosoma evansi*）[50]。

2006年加那利群岛感染了锥虫病的骆驼将病原带入法国，但这些病原却成功得到了消灭[14]。通过水运将单峰驼从加那利岛运往西班牙阿利坎特省的一个农场里之后，伊氏锥虫病便在当地暴发。12/21的单峰驼和2/67的马检测为阳性，疫情成功得到了处治[25]。2004年德国汉诺威市的单峰驼也被诊断出感染伊氏锥虫，而这些动物则是从加那利群岛的凡吐拉市被引进的（U. Wernery, 迪拜，中央兽医研究实验室，2004，个人通讯）。

1907年，澳大利亚的黑德兰港仔进口的骆驼中检测到锥虫病，这些骆驼全部被捕杀。之后再没有证据表明这个病原在澳大利亚的骆驼或者其他动物体中存在[49]。

大量的虻（马蝇）（见本书5.3.6节）都可作为伊氏锥虫的媒介。在非洲，以下这些虻可作为传播媒介：大型牛虻（*Tabanus bigyttanus*）、二代原虻（*Tabanus gratus*）、陆氏牛虻（*Tabanus leucastomus*）、莫氏牛虻（*Tabanus mordax*）、帕氏牛虻（*Tabanus par*）、萨菲氏牛虻（*Tabanus sufis*）、小袋牛虻（*Tabanus taeniola*）、柯氏麻虻（*Haematopota coranata*）、细小麻虻（*Haematopota tenuis*）、磁性矩虻（*Pangonia magnettil*）、晚育安虻（*Ancala latipes*）、吸引安虻（*Ancala fasciata*）、非洲安虻（*Ancala Africana*）、田地黄虻（*Atylotus agrestis*）、迪氏黄虻（*Atylotus diurnis*）、伊氏黄虻（*Atylotus ilinea*）[18,44,46]。能传播伊氏锥虫的亚洲种有安氏牛虻（*Tabanus amaenus*）、牛虻（*Tabanus bovines*）、道氏牛虻（*T. dorsilinea*）、线状牛虻（*T. lineola*）、曼氏牛虻（*T. mandarinus*）、红色牛虻（*T. rubidus*）、中华牛虻（*T. sinensis*）、纹状牛虻（*T. striatus*）几个种[29, 37,59]。

宿主罹患高虫血症、媒介的高密度、宿主的易感性及对病原的敏感性，感染动物与健康动物之间密切接触，都是蝇虫作为传播媒介机械性传播伊氏锥虫的重要条件[13]。

那些牛虻属的成员是特别高效的机械传播媒介，其主要因素为[34]：

（1）叮咬宿主引起的疼痛促使其对这些牛虻进行驱赶，造成吸血暂停。

（2）尽管有宿主的驱赶但牛虻仍然会叮咬并吸血。

（3）在叮咬吸血过程中获得病原并在短暂的吸血间歇之后传播病原。

（4）在被称为唇瓣的口器部有抗凝血的功能。

推测一些双翅目的吸血类昆虫像螫蝇、角蝇以及羊蜱蝇（keds）等都可能成为伊氏锥虫的传播媒介，但随后的传播实验均没有成功[41,45]。但一些实验室进行的厩螯蝇（*Stomoxys calcitrans*）及尼日尔刺蝇（*Stomoxys niger*）传播实验是成功的[31,56]。

很少有人关注可在骆驼喉蝇（*Cephalopina titillator*）的体液中发现伊氏锥虫这一现象，而这些骆驼喉蝇则来自埃及的感染了伊氏锥虫病的单峰驼[27]；将骆驼喉蝇的幼虫研磨物注射到实验室条件下饲养的啮齿动物体内，在超过两个月的时间内并没有发现寄生虫血症的出现，这意味着骆驼喉蝇在骆驼中传播伊氏锥虫仍然是未知的。

在索马里有报道称，骆驼可因周期性叮咬传播方式感染刚果锥虫（*T. congolense*）和布氏锥虫（*T. brucei*）[18]。刚果锥虫是牛的那加那病的病原体之一，猿猴锥虫通常寄生于肯尼亚猪体内，而发现两者在骆驼体内也存在[39]。

## 临床症状及病理变化

在该病流行时间长的地区，伊氏锥虫在骆驼体内产生较轻微的体征变化，譬如在南部撒哈拉沙漠的边缘地带，寄生虫与宿主之间有相对理想的平衡关系，且导致宿主死亡的例子很少。在这种病原流行特征相对稳定的情况下，骆驼可带虫很多年且一些带虫宿主最终可以自行康复并消除寄生虫（自愈）。然而，当一些非本地伊氏锥虫病原被带入时，可侵袭一些易感动物，并可能会对当地的经济造成一定的影响。

大多数情况下，环境及宿主因素的影响是促使锥虫病流行的主要原因；例如营养状况、年龄、孕育、接触或者通过其他疾病刺激造成的免疫抑制等。伊氏锥虫病在单峰驼中的临床症状表现形式多样且略有差异。典型的临床症状包括，动物体重下降、发育受阻，行动迟缓、无法长途跋涉，脚部、胸肋、下腹部和眼睑水肿，被毛粗糙；发热，颤抖，食欲不良以及轻微腹泻;由于贫血的原因引起的黏膜苍白并且细胞体积下降至25%以下等[51]。

一项单峰驼受伊氏锥虫感染的血液学的参数表明，血红素、血细胞数、红细胞数以及铁离子等都有明显下降，受感染宿主的白细胞值明显高于正常值[61]。

伊氏锥虫感染怀孕期宿主使其流产是一种普遍现象。感染母驼产出的幼驼虚弱，可出现寄生虫血症，并在两周之内死亡；而泌乳期的动物产奶量会明显下降。

总之，伊氏锥虫造成的病理变化没有明显的特征。但可能会有一定程度的贫血，骨骼肌和心肌苍白，淋巴结、脾脏肿大；在急性和亚急性病例，肝脏和肾脏实质、浆膜表面出现淤血斑等病理变化，在中枢神经系统有水肿及中度非化脓性脑炎并有脑膜性脑炎病灶出现（图209）。

一般情况下，在大脑组织切片中，灰质中的特征性病变为血管套的形成（图210），因为脑膜中的阳性颗粒结构通过嗜酸性过碘酸品红（Eosinophilic periodic acid-Schiff，PAS）染色很容易被发现（图211）。这些结构被称为“拉塞尔或罗素小体（Russel or Mott bodies）”，是非洲人感染锥虫的锥虫病的特点。

## 诊断

尽管伊氏锥虫病的临床症状不具特征性，但精神沉郁、淋巴结肿大以及贫血等均可以作为伊氏锥虫病的诊断依据。之前已经有相关动物锥虫病诊断方法的报道[19,35]。OIE[64]推荐的诊断伊氏锥虫感染的方法如下。

1. 病原识别

（1）传统的田间方法。

(a) 湿血膜片法；

(b) 厚涂片染色；

(c) 薄涂片染色；

(d) 淋巴结活组织检验法。

（2）浓缩技术。

(a) 血细胞压积离心；

(b) 暗视野/血沉棕黄层技术；

(c) 小型阴离子交换离心技术。

（3）动物接种法。

（4）DNA检测法。

(a) DNA探针；

(b) 聚合酶链反应（PCR）。

（5）抗体检测。

2. 血清学检测

（1）间接免疫荧光抗体检测（IFAT）；

（2）酶联免疫吸附试验（ELISA）；

（3）锥虫病卡片凝集实验（CATT）；

（4）免疫溶锥虫检测（Immune trypanolysis test）。

薄血涂片仅适用于每毫升血液中有500000个病原的高寄生虫血症病原检测，使用该方法时可直接证实锥虫存在，应采集周边血管而非颈静脉血液用于检测。

浓缩技术，像微血细胞比容离心技术

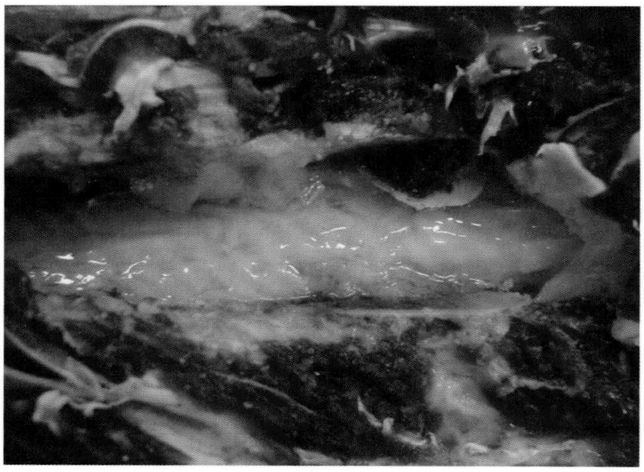

图209 锥虫感染骆驼脊髓形成的脑膜水肿

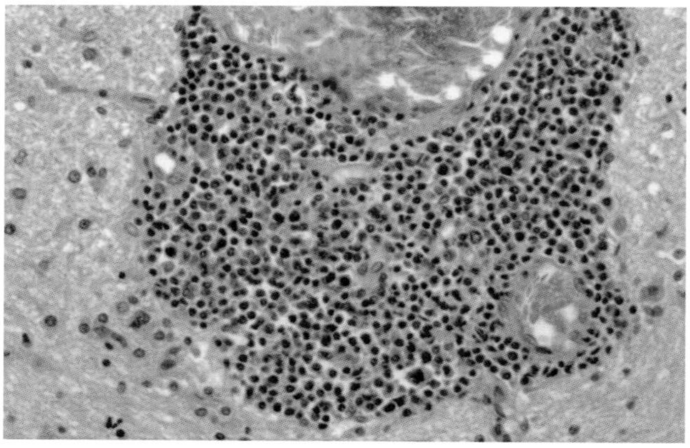

图210 大脑中的血管套

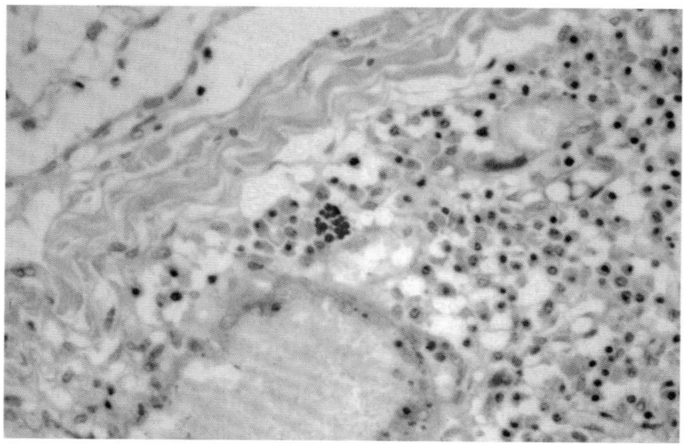

图211 脑膜中的嗜酸性过碘酸品红微粒结构

（mHCT）或者暗视野/相位对比血沉棕黄层技术均是一种较为敏感的方法。mHCT技术是将锥虫浓缩在一个毛细管的白细胞层，并可直接在显微镜下观察。一种将mHCT的改进的方法称为利用白细胞涂层技术（BCT），将白细胞层推挤到载玻片上，直接在暗视野检查虫体。

曾经有报道描述小型阴离子交换离心技术（MAECT）在检测低寄生虫血症感染的布氏锥虫上有过应用[36]。该项技术经改进后主要用于非洲人群锥虫病的检测，该方法可用于检测每毫升的血液量中低于50个锥虫的样本[3]。在MAECT技术中磷酸盐葡萄糖缓冲液（PSG）使宿主血细胞带有负电荷，这些宿主细胞随后吸附至阳离子交换柱，在不影响活力的情况下锥虫得以洗脱。

将疑似感染伊氏锥虫的骆驼血液接种至小白鼠，这种方法被认为是最敏感的诊断方法。采自疑似感染伊氏锥虫的骆驼血液经过EDTA处理后，用0.3 ~ 0.5mL腹腔接种实验室饲养的小白鼠。理论上讲，当接种0.5mL血液时，该方法相当于可以检测到每毫升宿主血液中有2个虫体的水平。

PCR广泛用于哺乳动物和媒介昆虫感染锥虫DNA的检测，该技术比普通检测寄生虫的方法更具敏感性，核糖体DNA转录间隔区是检测锥虫较为适合的靶标分子之一。特异性的PCR可以区分刚果锥虫（*T. congolense*），活动锥虫（*T. vivax*）和猴锥虫（*T. simiae*），但是不能区别伊氏锥虫和布氏锥虫[10,24]。

锥虫DNA检测和感染活动期是一致的。在对感染动物治疗后寄生虫血症消失后的1 ~ 4天锥虫的核酸也随之消失。然而，常规PCR检测仍然是一种复杂、昂贵方法并且需要特殊的实验设备。

利用间接检测方法对骆驼进行大规模筛选已经有过相关的描述，然而基于酶联免疫法则需要昂贵的实验设备。抗体ELISAs(Ab-ELISA)技术是利用离心法对寄生虫处理后获取其粗的可溶性抗原的一种检测方法[14,64]。近年来，研发出了用伊氏锥虫重组抗原的ELISA技术[57]，Ab-ELISA的不足之处是在锥虫经治疗消除后其抗体的持续时间长，然而它具有良好的敏感性，常被用于评估伊氏锥虫感染骆驼种群之间的关系。

抗原检测ELISAs (Ag-ELISA) 能用于检测处于活跃期的锥虫感染，但是当锥虫从血液中清除后其检测结果为阴性。

表80　可治疗动物锥虫病的抗锥虫药

| 类别名称 | 商品名 | 溶液浓度 | 剂量 | 注射途径 | 备注 |
|---|---|---|---|---|---|
| 苏拉明(Suramin) | Naganol | 10% | 10mg/kg (1mL/10 kg) | i.v. | 主要用于抗骆驼伊氏锥虫 |
| 重氮氨苯脒乙酰甘氨酸盐(Diminazene aceturate) | Berenil, Ganaseg, Trypazen, Veriben | 7% | 7mg/kg (2mL/20 kg) | i.m. | 主要用于牛和小反刍兽 |
| 溴化乙锭(Homidium bromide) | Ethidium bromide | 2.5% | 1mg/kg (1mL/25 kg) | i.m. | 主要用于牛和小反刍兽。应溶解于热水中，可能致癌 |
| 胡米氯铵(Homidium chloride) | Ethidium c, Novidium | 2.5% | 1mg/kg (1mL/25 kg) | i.m. | 见上，但溶解于冷水中 |
| 喹匹拉明硫酸甲酯(Quinapyramine methyl sulphate) | Antrycide, Trypacide, Noroquin, Quintrycide | 10% | 5mg/kg (1mL/20 kg) | s.c. | 目前主要用于防治牛和马的伊氏锥虫及布氏锥虫 |
| 美拉索明(Melarsomine) | Cymelarsan | 0.5% | 0.25~0.5mg/kg (1~2mL/20 kg) | i.m.或 s.c. | 仅用于骆驼伊氏锥虫 |
| 氮氨菲啶氯化物(Isometamidium chloride) | Samorin, Trypamidium | 1%, 2% | 0.25~0.5mg/kg (1.25~2.5mL/50kg) 1.0mg/kg (2.5mL/50 kg) | i.m. | 主要用于牛，作治疗用效率较低，做预防用则效率较高；含有胡铵，因此被认为有潜在的致癌作用 |

注：i.m.，肌内注射；i.v.，静脉注射；s.c.，皮下注射

目前用于检测野外骆驼锥虫最广泛的方法是CATT法，最初该方法主要应用于人的冈比亚锥虫（*Trypanosoma gambiense*）的检测，后来将这种方法非常成功地改进，用于各类动物的锥虫病检测，因此该方法又被称作CATT/伊氏锥虫检测法，这是一个直接卡凝集试验检测方法。直接卡凝集试验用于特定抗体的检测，该方法是基于将伊氏锥虫主要抗原类型（VAT Ro Tat 1.2）的克隆化虫体进行纯化、固定以及染色后，制成冻干悬浮抗原。CATT/伊氏锥虫检测方法是用于骆驼锥虫的标准检测方法。

乳胶凝集试验（LAT）是基于冻干的乳胶悬液表面包被伊氏锥虫抗原（VAT Ro Tat 1.2）的一种检测方法，虽然该方法研发成功较早，但用于伊氏锥虫的检测依然没有得到充分的评估且有待进一步完善。

### 治疗及防控

在表80中列出了用于治疗家畜锥虫病的7种复合药物，其中4种药物对单峰驼的伊氏锥虫病有效，分别为苏拉明、喹匹拉明、氮氨菲啶和美拉索明。其他用于治疗牛的伊氏锥虫病药物在骆驼上使用时没有疗效，例如溴化乙锭；或者对单峰驼有很强的毒性，如重氮氨苯脒乙酰甘氨酸盐，此药即使使用治疗剂量，也可引起单峰驼肝脏中毒和神经毒性；然而，此药能用于治疗感染伊氏锥虫的双峰驼。

苏拉明应静脉注射，漏出静脉时可引起静脉炎，同时，这类药物代谢缓慢，具有预防效用。该药物用于防控伊氏锥虫已有75年的历史，但有些虫株有抗药性。

喹匹拉明比苏拉明易于溶解，可以用于皮下注射。喹匹拉明硫酸甲酯具有治疗作用，而喹匹拉明硫酸甲酯和喹匹拉明氯化物的混合物可以用于疾病预防，但剂量过大可能导致出现马钱子样（curare-like）中毒症状。有多个伊氏锥虫虫株对这类药物已产生耐药性。

氮氨菲啶氯化物对伊氏锥虫病的治疗只是温和效用，只在出现对苏拉明和喹匹拉明具有双重抗药性的虫株感染等特殊情况下使用。这3种药物只作用于心血管系统中的锥虫，而不能透过血脑屏障，只有含砷的化合物才能通过血脑屏障。

美拉索明是一类有机砷并且在治疗骆驼伊氏锥虫病方面广泛应用。然而，其穿过血脑屏障的效果并未证实。有必要强调的是美拉索明在治疗锥虫亚属的几个种时效果显著，而该类药物对刚果锥虫（*T. congolense*）和活跃锥虫（*T. vivax*）病的治疗效果一般。骆驼一旦感染伊氏锥虫就会形成循环传播，此时就不得不使用其他药物进行治疗。

美拉胂醇（Melarsopol）是一种治疗人昏睡症的药物。研究发现，以体重3.5mg/kg的剂量通过静脉注射对于骆驼伊氏锥虫病的治疗有一定的效果。但是由于其治疗指数较低并没有在兽用药品中广泛应用。在巴基斯坦苏莱曼地区，伊氏锥虫病极为常见，牧民们把此药贴于动物耳根以提醒昏睡的骆驼。为了阻止骆驼的渐进性消瘦，他们驱使骆驼喝冰冷的从植物中提取的苦涩收敛[49]。在亚洲，其他植物如蓝桉（*Eucalyptus globules*）、珊瑚油桐（*Jatropha podagrica*）、锯棕榈（*Vitris repens*）作为传统药物用于治疗疟疾、痢疾、癌症和肺部疾病，对锥虫也有良好的治疗效果[7]。

除了药物控制伊氏锥虫病外，还可以寻找其他方法，其中，控制传播媒介可能会是另外一种解决途径[26]。

虻科物种的多样性和实验室条件下很难培养的缘故，使得采采蝇雄性不育技术不适用于虻科动物。

污水及沼泽地有益于虻科类的繁殖，因此从污水治理及保护环境的角度看，对于虻的防控是一种可取的办法。然而，在特定条件下，一些排水系统实际上为虻的幼虫提供了栖息地，因此向虻的筑巢环境中使用杀虫剂会有短期的效果，但这种方法也能杀灭其他的昆虫。使用杀虫剂对虻筑巢区进行处理有利于减少这些昆虫的滋扰[6,38]，这种方法在大量的实验支持下将会是一种很有价值的尝试。

当使用杀虫剂预防虻对宿主的侵袭时，不宜使用耳签给药，喷雾或喷洒的药剂应该用于适宜虻虫生长的土壤或叮咬部位。大环内酯类药物（伊佛霉素、多拉菌素、依普菌素等）对虻类的全身性袭扰方面的作用不理想。不同类型的诱捕方法（Manitoba, canopy, malaise）已被用于捕捉虻类，尽管大量的虻被捕获，但仅局限于检测蝇的种类和密集程度，而不能进行有计划的防控。大量的拟寄生物、天敌尤其是黄蜂影响着虻的生存，在成年黄蜂活动的季节，其从家畜体表捕食马蝇，尽管黄蜂不能持久地影响虻的数量。据估计，从虫卵至成虫的整个发育过程中，仅需2%的存活率就可以维持虻的种群数量[22]。

## 参考文献

[1] Abdel-Rady A. (2008). – Epidemiological studies (parasitological, serological and molecular techniques) of *Trypanosoma evansi* infection in camels (*Camelus dromedarius*) in Egypt. Vet. World, 1 (11), 325 – 328.

[2] Abo-Shehada M.N., Anshassi H., Mustafa G. & Amr Z. (1999). – Prevalence of Surra among camels and horses in Jordan. *Prev. Vet Med.*, 38, 289 293.

[3] Al-Khalifa A.S., Hussein H.S., Diab F.M. & Khalil G.M. (2009). – Blood parasites in certain regions in Saudi Arabia. *Saudi J. Biol. Sci.*, 16, 63 – 67.

[4] Al-Taqi M.M. (1989). – Characterization of *Trypanosoma* (*Trypanozoon*) *evansi* from dromedary camels in Kuwait by isoenzyme electrophoresis. *Vet. Parasitol.*, 32, 247 – 253.

[5] Atarhouch T., Rami M., Bendahman M.N. & Dakkak A. (2003). – Camel trypanosomosis in Morocco 1: results of a first epidemiological survey. *Vet. Parasitol.*, 111, 277 – 286.

[6] Bauer B., Blank J., Heile C., Schein E. & Clausen P.-H. (2006). – Pilotstudie zur Bewertung der Effizienz insektizid-behandelter Netzzaune zum Schutz von Pferden gegen Weidefliegen im nordlichen Brandenburg. *Berl. Münch. Tierärztl. Wochenschr.*, 119 (7/8), 2 – 5.

[7] Bawn S. (2010). – Studies of antitrypanosomal activities of medical plants. PhD thesis, Hokkaido University.

[8] Berlin D., Nasereddin A., Amzir K., Ereqat S., Abdeen Z. & Baneth G. (2010). – Longitudinal study of an outbreak of *Trypanosoma evansi* infection in equids and dromedary camels in Israel. *Vet. Parasitol.*, 174 (3 – 4), 317 – 322.

[9] Buscher P., Mumba Ngoyi D., Kabore J., Lejon V. & Robays J. (2009). – Improved models of mini anion exchange centrifugation technique (mAECT) and modified single centrifugation technique (MSC) for sleeping sickness diagnosis and staging. *PLoS Negl. Trop. Dis.*, 3 (11), e471.

[10] Cox A., Tilley A., McOdimba F., Fyfe J., Eisler M., Hide G. & Welburn S. (2005). – A PCR based assay for detection and differentiation of African trypanosome species in blood. *Exp. Parasitol.*, 111, 24 – 29.

[11] Delafosse A. & Doutoum A.A. (2004). – Prevalence of *Trypanosoma evansi* infection and associated risk factors in camels in Eastern Chad. *Vet. Parasitol.*, 119, 155 – 164.

[12] Derakhshanfar A., Mozaffari A.A. & Zadeh A.M. (2010). – An outbreak of trypanosomiasis (surra) in camels in the Southern Fars Province of Iran: clinical, hematological and pathological findings. *Res. J. Parasitol.*, 5 (1), 23 – 26.

[13] Desquesnes M., Biteau-Coroller F., Bouyer J., Dia M.L. & Foil L. (2009). – Development of a mathematical model for mechanical transmission of trypanosomes and other pathogens of cattle transmitted by tabanids. *Int. J. Parasitol.*, 39, 333 – 346.

[14] Desquesnes M., Bossard G., Patrel D., Herder S., Pataut O., Lepetitcolin E., Thevenon S., Berthier D., Pavlovic D., Brugidou R., Jaquiet P., Schelcher F., Faye B., Touratier L. & Cuny G. (2010). – First outbreak of *Trypanosoma evansi* in camels in metropolitan France. *Vet. Rec.*, 162, 750 – 752.

[15] Dia M.L. (2006). – Parasites of the camel in Burkina Faso. *Trop. Animal Health Prod.*, 38 (1), 17 – 21.

[16] Dia M.L., Diop C., Aminetou M., Jazquiet P. & Thiam A. (1997). – Some factors affecting the prevalence of *Trypanosoma evansi* in camels in Mauritania. *Vet. Parasitol.*, 72 (2), 111 – 120.

[17] Diall O., Bajyane S.E., Magnus E., Kouyate B., Diallo B., Van Meirvenne N. & Hamers R. (1994). – Evaluation of a direct serologic card agglutination test for the diagnosis of camel trypanosomiasis caused by *Trypanosoma evansi. Rev. sci. tech. Off. int. Epiz.*, 13 (3), 793 – 800.

[18] Dirie M.F., Wallbanks K.R., Aden A.A., Bornstein S. & Ibrahim M.D. (1989). – Camel trypanosmiasis and its vectors in Somalia. *Vet. Parasitol.*, 32, 285 – 291.

[19] Eisler M.C., Dwinger H., Majiwa P. & Picozzi K. (2003). – Diagnosis and epidemiology of African animal trypanosomiasis. *In* The Trypanosomiasis (I. Maudlin, P.I. Holmes & M.A. Miles, eds). Cromwell Press, Trowbridge, 253 – 267.

[20] Elamin E.A., El Bashir M.O.A. & Saeed E.M.A. (1998). – Prevalence and infection pattern of *Trypanosoma evansi* in camels in Mid-Sudan. *Trop. Animal Health Prod.*, 30, 107 – 114.

[21] Erdenebileg O. (2001). – Camel diseases [in Mongolian]. Dalanzadgad, Ulaanbaatar, Mongolia.

[22] Foil L.D. & Hogsette J.A. (1994). – Biology and control of tabanis, stable flies and horn flies. *Rev. sci. tech. Off. int. Epiz.*, 13 (3), 1 125 – 1 158.

[23] Fowler M.E. (2010). – Medicine and surgery of camelids. 3rd Ed. Wiley-Blackwell, Oxford, UK.

[24] Garba H.S. & Alabi N.T. (2004). – Preliminary studies on tick and tickborne parasites of the one humped camel (*Camelus dromedarius*) in Sokoto, Nigeria. *ACSAD Camel Newsletter*, 20, 57 – 60.

[25] Gutierrez C., Desquesnes M., Touratier L. & Buescher P. (2010). – *Trypanosoma evansi*, recent outbreaks in Europe. *Vet. Parasitol.*, 174, 26 – 29.

[26] Hall M.J.R. & Wall R. (2003). – Biting flies: their role in the mechanical transmission of trypanosomes to livestock and methods for their control. *In* The Trypanosomiasis (I. Maudlin, P.I. Holmes & M.A. Miles, eds). Cromwell press, Trowbridge, 583 – 594.

[27] Hilali M. & Fahmy M.M. (1993). – Trypanozoon-like epimastigotes in the larvae of *Cephalopina titillator* (Diptera: Oestridae) infesting camels (*Camelus dromedarius*) infected with *Trypanosoma evansi. Vet. Parasitol.*, 45, 327 – 329.

[28] Ibragimov B.S. & Sulejmenov M.Z. (1997). – Prophylaxis and therapy of Su Auru in camels [in Russian]. *Vestn. Vet.*, 7, 45 – 48.

[29] Shekogar R., Powar R.M., Joshi P.P., Bhargagava A., Dani V.S., Katti R., Zare V.R., Khanande V., Jannin J. & Truc P. (2006). Human trypanosomiasis caused by *Trypanosoma evansi* in a village in India: preliminary serological survey of the local population. *Am. J. Trop. Med. Hyg.*, 75 (5), 869 – 870.

[30] Kinne J., Wernery U. & Zachariah R. (2001). – Surra in a guanaco (*Lama guanicoe*). *J. Camel Pract. Res.*, 8, 93 – 98.

[31] Latif B., Kadim F. & Ali S. (1992). – The role of stable fly (*Stomoxys calcitrans*) in the transmission of *Trypanosoma evansi*. *In* Proceedings premier seminaire international sur les Trypanosomoses Animales non Transmises par les Glossines, 14 – 16 October, Annecy, France, 124.

[32] Liao D. & Shen J. (2010). – Studies of quinapyramine-resistance of *Trypanosoma evansi* in China. *Acta Trop.*, 116, 173 – 177.

[33] Luckins A.G. (1988). – *Trypanosoma evansi* in Asia. *Parasitol. Today*, 4 (5), 137 – 142.

[34] Luckins A.G. (1998). – Epidemiology of Surra: unanswered questions. *J. Protozool. Res.*, 8, 106 – 119.

[35] Luckins A.G. & Dwinger R.H. (2003). Non-tsetse transmitted animal trypanosomiasis. *In* The Trypanosomiasis (I. Maudlin, P.I. Holmes & M.A. Miles, eds). Cromwell Press, Trowbridge, 269 – 281.

[36] Lumsden W.H.R., Kimber D.D., Evans D.A. & Doig S.J. (1979). – *Trypanosoma brucei*: Miniature anion-exchange centrifugation technique for the detection of low parasitaemias: adaption for field use. *Trans. Roy. Soc. Trop. Med. Hyg.* 73, 312 – 317.

[37] Lun Z-R., Fang Y., Wang C-J. & Brun R. (1993). – Trypanosomiasis of domestic animals in China. *Parasitol. Today*, 9 (2), 41 – 45.

[38] Maia M., Clausen P.-H., Mehlitz D., Garms R. & Bauer B. (2010). – Protection of confined cattle against biting and nuisance flies (Muscidae: Diptera) with insecticide-treated nets in the Ghanaian forest zone at Kumasi. *Parasitol. Res.*, 106, 1307 – 1 313.

[39] Mihok S., Zweygarth E., Munyoki E.N., Wambuac J. & Kock R. (1994). – *Trypanosoma simiae* in the white rhinoceros (*Ceratotherium simum*) and the dromedary camel (*Camelus dromedarius*). *Vet. Parasitol.*, 53 191 – 196.

[40] Molina J.M., Ruiz A., Juste M.C., Corbera J.A., Amador R. & Gutierrez C. (2000). – Seroprevalence of *Trypanosoma evansi* in dromedaries (*Camelus dromedarius*) from the Canary Islands (Spain) using an antibody Ab-ELISA. *Prevent. Vet. Med.*, 47, 53 – 59.

[41] Ngeranwa J.J.N. & Kialo D.C. (1994). – The ability of *Stomoxys calcitrans* and mechanical means to transmit *Trypanosoma (brucei) evansi* from goats to camels in Kenya. *Vet. Res. Com.*, 18, 307–312.

[42] Njiru Z.K., Constantine C.C., Guya S., Crowther J., Kiragu J.M., Thompson R.C. & Davila A.M. (2005). – The use of ITS1 rDNA PCR in detecting pathogenic African trypanosomes. *Parasitol. Res.*, 95, 186 – 192.

[43] Njiru Z.K., Constantine C.C., Ndung'u J.M., Robertson I., Okaye S., Thompson R.C. & Reid S.A. (2004). – Detection of *Trypanosoma evansi* in camels using PCR and CATT/*T. evansi* tests in Kenya. *Vet. Parasitol.*, 124, 187 – 199.

[44] Ouhelli H. & Dakkak A. (1987). – Protozoal diseases of dromedaries. *In* Diseases of camels. *Rev. sci. tech. Off. int. Epiz.*, 6 (2), 417 – 422.

[45] Oyieke F.A. & Reid G. (2003). – The mechanical transmission of *Trypanosoma evansi* by *Haematobia minuta* (Diptera: Muscidae) and *Hippobosca camelina* (Diptera: Hippoboscidae) from infected camel to a mouse and the survival of trypanosomes in fly mouthparts and gut. *Folia Vet.*, 47, 38 – 41.

[46] Rahman A.H.A. (2005). – Observation on the trypanosomosis problem outside tsetse belts of Sudan. *Rev. sci. tech. Off. int. Epiz.*, 24 (3), 962–972.

[47] Ravindran R., Rao J.R., Mishra A.K., Pathak, K.M.L., Babu N., Satheesh C.C. & Rahul S. (2008). – *Trypanosoma evans*i in camels, donkeys and dogs in India: comparison of PCR and light microscopy for detection – short communication. *Veterinarski Arhiv.*, 78 (1), 89 – 94.

[48] Raziq A., de Verdier K. & Younas M. (2010). – Ethnoveterinary treatments by dromedary camel herders in the Suleiman Mountainous Region in Pakistan: an observation and questionary study. *J. Ethnobiol. Ethnomed.*, 6, 1 – 12.

[49] Reid S.A. (2002). – *Trypanosoma evansi* control and containment in Australia. *Trends in Parasitol.*, 18 (5), 219 – 224.

[50] Rodriguez N.F., Tejedor-Junco M.T., Hernandez-Trujillo Y., Gonzalez M. & Gutierrez C. (2010). – The role of wild rodents in the transmission of *Trypanosoma evansi* infection in an endemic area of the Canary Islands (Spain). *Vet. Parasitol.* 174, 323 – 327.

[51] Rottcher D. & Zweygart E. (1995). – Trypanosomiasis in the camel. *In* Camel keeping in Kenya (J.O. Evans, S.P. Simpkin & D.J. Aitkins, eds). English Oress Limited, Nairobi, Kenya, 7:1 – 7:6.

[52] Shah S.R., Phulan M.S., Memon A.A., Rind R. & Bhatti W.M. (2004). – Trypanosomes infection in camels. *Pakistan Vet. J.*, 24 (4), 209 – 210.

[53] Shahzad W., Munir R., Khan M.S., Ahmad M.D., Ijaz M., Ahmad A. & Iqbal M. (2010). – Prevalence and molecular diagnosis of *Trypanosoma evansi* in Nili-Ravi buffalo (*Bubalus bubalus*) in different districts of Punjab (Pakistan). *Trop Anim. Health Prod.*, 42, 1 597 – 1 599.

[54] Shegokar V.R., Powar R.M., Joshi P.P., Bhargava A., Dani V., Katti R., Zare V.R., Khanande V.D., Jannin J. & Truc P. (2006). – Human trypanosomiasis caused by Trypanosoma evansi in a village in India: preliminary serologic survey of the local population. *Am. J. Trop. Med. Hyg.*, 75 (5), 869 – 870.

[55] Srivastava V.K., Obeid H.M. & Elbosadi S.M. (1984). – Trypanosomiasis in camels in the Sultanate of Oman. *Trop. Anim. Health Prod.*, 16, 148.

[56] Sumba A.L., Mihok S. & Oyieke F.A. (1998). – Mechanical transmission of *Trypanosoma evansi* and *T. congolense* by *Stomoxys niger* and *S. taeniatus* in a laboratory mouse model. *Med. Vet. Entomol.*, 12, 417–422.

[57] Tran T., Claes F., Verloo D., De Greeve H. & Buscher P. (2009). – Towards a new reference test for surra in camels. *Clin. Vaccine Immunol.*, 16 (7), 999 – 1002.

[58] Uilenberg G. (1998). – A field guide for the diagnosis, treatment and prevention of African animal trypanosomosis. Food and Agriculture Organization, Rome.

[59] Veer V., Parashar B.D. & Rao K.M. (1999). – Notes on Tabanidae (Diptera) that are surra vectors or pestiferous with description of a new species of *Tabanus* in India. *J. Oriental Insects*, 33, 247 – 266.

[60] Wells E.A. (1984). – Animal trypanosomiasis in South America. *Prev. Vet. Med.*, 2, 31 – 41.

[61] Wernery U. (1995). – Blutparameter und Enzymwerte von gesunden und kranken Rennkamelen (*Camelus dromedarius*). *Tierärztl. Praxis*, 23, 187 – 191.

[62] Wernery U., Thomas R., Syriac G., Raghavan R. & Kletzka S. (2007). – Seroepidemiological studies for the detection of antibodies against nine infectious diseases in dairy dromedaries (Part I). *J. Camel Pract. Res.*, 14 (2), 85–90.

[63] Wilson R.T. (2008). – Perceptions and problems of disease in the one-humped camel in southern Africa in the late 19th and early 20th centuries. *J. S. Afr. Vet. Assoc.*, 79 (2), 58 – 61.

[64] World Organisation for Animal Health (OIE) (2012). – *Trypanosoma evansi* infection (surra). *In* Manual of diagnostic tests and vaccines for terrestrial animals (mammals, birds and bees), 1 – 15. Available at: www.oie.int/fileadmin/Home/eng/Health_standards/tahm/2.01.17_TRYPANO_ SURRA.pdf (accessed on 12.08.2013).

[65] Zeleke M. & Bekele T. (2001). – Effect of season on the productivity of camels (*Camelus dromedarius*) and the prevalence of their major parasites in eastern Ethiopia. *Trop. Animal Health Prod.*, 33 (4), 321 – 329.

（罗金译，刘光远、殷宏校）

## 5.1.2 贾地鞭毛虫病（梨形鞭毛虫属）

贾地鞭毛虫病是由鞭毛虫属引起的原生动物肠溶性疾病，包括脊椎动物宿主、人类等。目前资料显示有50多种鞭毛虫被记载，但仍有大部分还没有得到确认；基于形态学标准可对贾地鞭毛虫属的6个种进行区分，例如寄生于人和哺乳动物的十二指肠贾第虫（*G. duodenalis*）、寄生于啮齿动物鼠贾第虫（*G. muris*）和微小贾第虫（*G .micron*），寄生于苍鹭的苍鹭贾第虫（*G. ardeae*），寄生于鹦鹉的鹦鹉贾第虫（*G. psittacci*）以及寄生于两栖类的多动贾第虫（*G. agilis*）。十二指肠贾第虫（同物异名：蓝氏贾第虫*G. lamblia*，肠贾第虫G. *intestinalis*）复合群能够引起人类、家畜和伴侣动物的疾病，其形态单一，但其7种基因型表现出遗传基因多样性[7]。

贾地虫滋养体背部隆起，腹部扁平，大小为(9~2)μm×(15~12)μm，双细胞核，四对鞭毛，分别为前侧鞭毛、后侧鞭毛、腹鞭毛和尾鞭毛各1对（图212）。滋养体在肠道内营纵二分裂法繁殖。

如果滋养体落入肠腔而随食物到达回肠下段或结肠腔后形成包囊，随粪便排出。包囊呈椭圆形或梨形，大小为(8~15)μm×(7~10)μm（图213）。贾第虫有复杂的生活循环史，宿主之间通过食入了包囊污染的食物或水而传播。土壤表层的包囊在适宜的条件下仍有感染力，可存活数周甚至在冷水中存活达3个月之久。

1987年首先在美国威斯康星州的美洲驼中发现了贾第虫包囊[5]。研究报告显示，在肠内有贾第虫存在的情况下，宿主并未表现出任何临床症状。对加利福尼亚州的33个农场中选取345只美洲驼进行调查，仅有12只（3%）美洲驼中发现贾第虫包囊，大多数幼崽在1～4月龄间无临床症状[6]。选取美国俄勒冈州45只患有腹泻的美洲驼和羊驼粪便样品进行检查，仅有8只（18%）发现贾第虫包囊[2]。同样，在俄亥俄州58例腹泻病例中33%有包囊，感染年龄主要在7～120日龄[9]。在最新的调查中，来自秘鲁的274例羊驼粪便样品检查中发现了137例（50%）贾第虫基因阳性[4]，最小的动物为一周龄。感染率随着年龄的增长而升高，8周左右感染率高达80%。扩增贾地鞭毛虫基因可获得至少两种片段的序列。尽管幼崽更容易排出贾第虫包囊，这并不局限于年轻的动物:在12月龄大的羊驼也能发现贾第虫包囊[8]。旧大陆骆驼中也有贾第虫包囊的报道，但与新大陆骆驼一样，还没有确认贾第虫包囊引起有关临床症状的报道[1,3]。

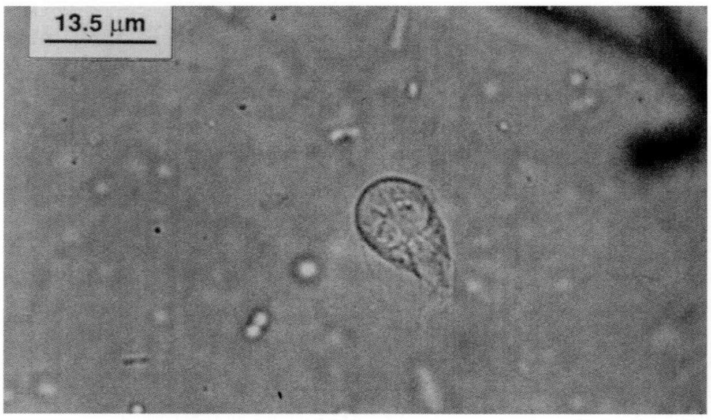


图212　贾第虫滋养体

（照片来自 Sloss, Kemp和 Zajac教授编著的《兽医临床寄生虫学》，第六版，1994，美国衣阿华州立大学出版社）

贾第虫可通过粪便进行检测，也可用饱和氯化锌或硫酸锌溶液悬浮法进行检测。虽然盐溶液能使贾第虫包囊变形，但仍能进行分辨。直接免疫荧光实验能直接用于粪便样品中贾第虫包囊的检测，这种方法是基于单克隆抗体的制备并能检测隐孢子虫卵母细胞在哺乳类动物中的感染情况（图 214 ）。

芬苯达唑以每千克体重50mg的剂量每天口服给药，连续5天，对治疗新大陆骆驼的贾第虫效果良好[9]。

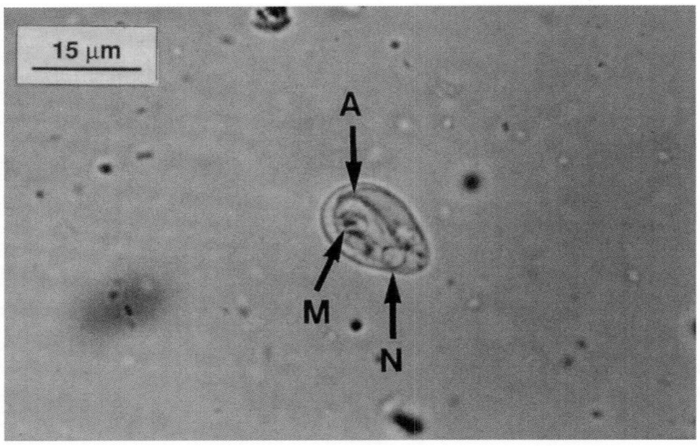


图213　贾第虫包囊

（照片来自 Sloss, Kemp和 Zajac教授编著的《兽医临床寄生虫学》，第六版，1994，美国衣阿华州立大学出版社）

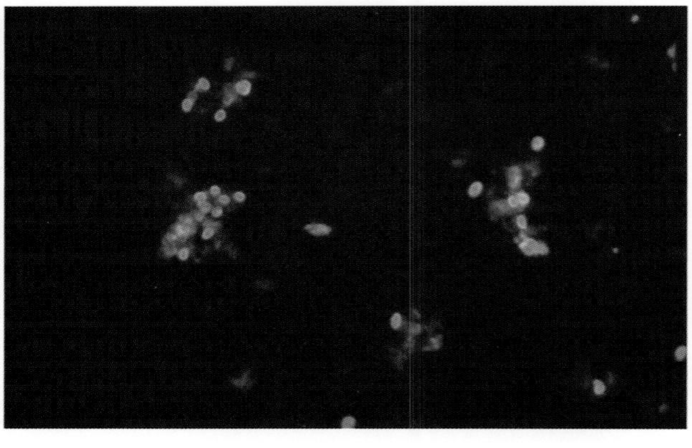

图214　直接免疫荧光法发现的贾第虫包囊



## 参考文献

[1]　Al-Jabr O.A., Mohammed G.E. & Al–Hamdan B.A. (2005). – Giardiosis in camels (*Camelus dromedarius*). *Vet. Rec.*, 157,

350–352.

[2] Cebra C.K., Mattson D.E., Baker R.J., Sonn R.J. & Dearing P.L. (2003). – Potential pathogens in feces from unweaned llamas and alpacas. *J. Am. Vet. Med. Assoc.*, 223 (12), 1 806 – 1 808.

[3] De Verdier E., Erlingsson B., Erlingsson S., Reineck H. & Bornstein S. (2010). – Endoparasites in young, suckling Bactrian camel calves in Sweden. *In* International Camel Symposium: Linking Camel Science and Development for Sustainable Livelihoods, 7 – 10 June, Garissa, Kenya, 56.

[4] Gomez-Couso H., Ortega-Mora L.M., Aguado-Martinez A., Rosadio-Alcantara R., Maturano-Hernandez L., Luna-Espinosa L., Zanabria-Huisa V. & Pedraza-Diaz S. (2012). – Presence and molecular characterisation of *Giardia* and *Cryptosporidium* in alpacas (*Vicugna pacos*) from Peru. *Vet. Parasitol.* 187, 414 – 420.

[5] Kiorpes A.L., Kirkpatrick C.E. & Bowman D.D. (1987). – Isolation of *Giardia* from llama and from sheep. *Can. J. Vet. Res.*, 51, 277 – 280.

[6] Rulofson C., Atwill E.R. & Holmberg C.A. (2001). – Fecal shedding of *Giardia duodenalis, Cryptosporidium parvum, Salmonella* organisms and *Escherichia coli* O157:H7 from llamas in California. *Am. J. Vet. Res.*, 62 (4), 673 – 642.

[7] Thompson A.R.C. & Monis P.T. (2011). – Taxonomy of *Giardia* species. *In* Giardia a model organism (H.D. Lujan & S. Svard, eds). Springer, Wien, 3 – 16.

[8] Trout J., Satin M. & Fayer R. (2008). – Detection of assemblage A, *Giardia duodenalis* and *Eimeria* spp. in alpacas on twoMaryland farms. *Vet. Parasitol.*, 153, 203 – 208.

[9] Whitehead C.E. & Anderson D.E. (2006). – Neonatal diarrhea in llamas and alpacas. *Sm. Rumin. Res.*, 61, 207 – 215.

（罗金译，刘光远、殷宏校）

## 5.1.3 三毛滴虫病

三毛滴虫病是由胎儿三毛滴虫(*Tritrichomonas foetus*)引起的，可导致家畜不育或流产的一种性病。有关单峰驼种畜群患三毛滴虫病的报道仅有一例，在48头患有子宫内膜炎动物中有24头分离出这种鞭毛虫（图215）[2]。由于毛滴虫是具有严格的宿主特异性的寄生虫，因此所分离的鞭毛虫似乎不大可能是胎儿三毛滴虫。

在沙特阿拉伯的 9 头腹泻骆驼的排泄物中被确诊存在三毛滴虫，但对种类没有深入的研究[1]。

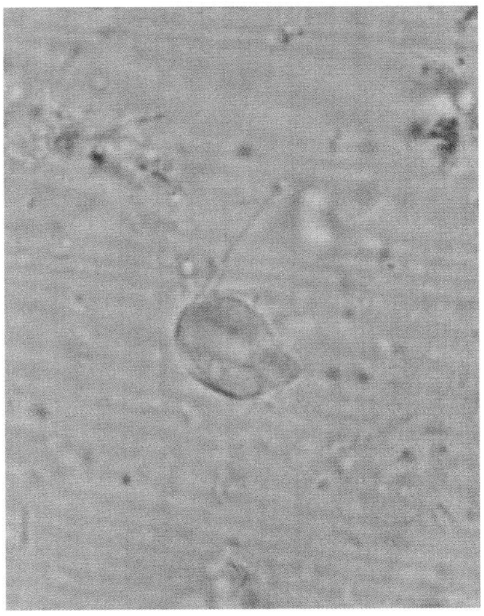

图215 患子宫内膜炎单峰驼的子宫内膜组织涂片中的三毛滴虫

### 参考文献

[1] Baksh A. & El Jalii I. (2010). – A clinical report on some internal protozoa in camels (*Camelus dromedarius*) in Saudi Arabia. *In* Proceedings of the 26th World Buiatrics Congress, 14 – 18 November, Santiago de Chile, Chile.

[2] Wernery U. (1991). – The barren camel with endometritis. Isolation of *Trichomonas* foetus and different bacteria. *J. Vet. Med.,B.*, 38, 523 – 528.

（谢俊仁译，刘光远、殷宏校）

## 5.1.4 纤毛虫病（Infundibuloriosis）

由纤毛虫感染引起的肠功能紊乱通常称为小袋虫病，这是由结肠小袋纤毛虫（*Balantidium coli*）引起的一类疾病。该病通常在猪体上发生，而其他单胃哺乳动物和人很少感染。反刍动物中也有发生该病的报道，但该病例可能是与牛的腹泻相关的有沟布克斯顿纤毛虫（*Buxtonella sulcata*）相混淆[3,10]。

Bozhenko记述的骆驼纤毛虫（*Infundibulorium cameli*）是一种骆驼消化道内的纤毛虫[4]。纤毛虫属（*Infundibulorium*）与包括布克斯顿纤毛虫属（*Buxtonella*）在内的其他10个属均被分在海珠科（Pycnotrichidae）中[7]。它们都是共生体，其特征是口腔均为一个长的前庭沟，至少占体长的一半以上。然而，人们几乎不了解骆驼纤毛虫，仅仅在印度的刊物上有所报道[5]，有人怀疑，文献所描述的牛盲肠分离的有沟布克斯顿纤毛虫（*B. sulcata*），是骆驼纤毛虫（*I. cameli*）的一个较新同物异名[6]。

骆驼纤毛虫的滋养体（即寄生阶段）的大小变化不一，含有一个吞咽食物的细胞口；滋养体被覆纤毛，并含有两个核（图216）；它们是通过二分裂法进行增殖的，而不是配子生殖方式；当其随排泄物排出时，滋养体变成一个淡灰色无纤毛的类球形的包囊，直径为40 ~ 60μm。

此前已对猪的结肠小袋纤毛虫感染的发病机制和临床进行过详细的研究。发现结肠小袋纤毛虫能够侵入肠道黏膜，引起出血和多点组织坏死，之后可演变成溃疡。在组织学切片中，带有特征性大核的滋养体不仅发现于溃疡中，而且在黏膜和黏膜下层的周边组织中也存在。

在骆驼中与腹泻有关的纤毛虫曾在苏丹[2,9]和伊朗的巴赫拉姆[1]有所报道。然而在反刍动物和骆驼的前胃内也存在许多非致病性的共生纤毛虫[6]。它们属于消化道正常的微生物区系，其包囊可在显微镜下观察到；在摄入富含碳水化合物和蛋白的饲料后，前胃内容物中的纤毛虫的密度能达到$10^7$/mL；而从埃及的11只健康的单峰驼的胃室中总计也发现了24种纤毛虫[8]。

在新鲜的排泄物或尸体剖解时用前胃和肠道内液体制作的涂片中能观察到纤毛虫滋养体（图217），可用于确诊感染。若用液体漂浮则会将滋养体破坏，虽然用沉淀技术能找到纤毛虫包囊，但不能用于种的鉴定。

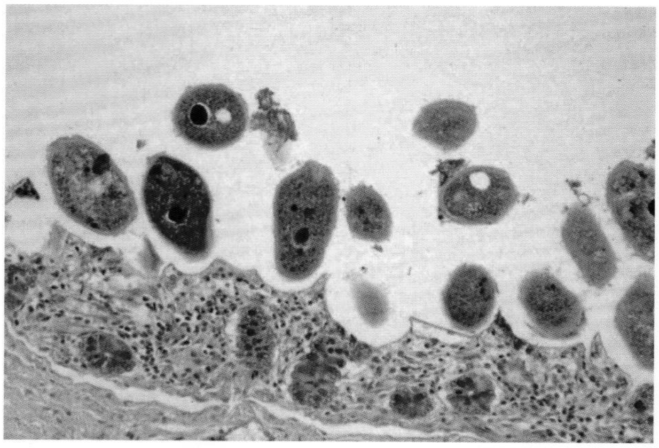

图216 骆驼结肠中的纤毛虫，组织切片

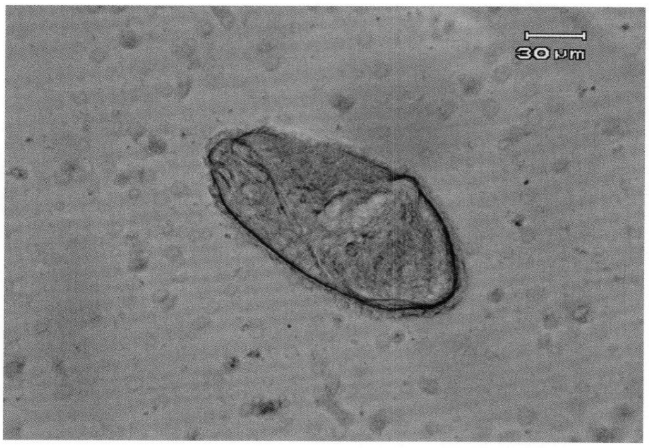


图217 从骆驼的结肠内容物中分离的纤毛虫滋养体

## 参考文献


[1] Abubakr M.I., Nayel N.M., Fadlalla M.E., Abdelrahman A.O., Abuobeida S.A. & Elgabara Y.M. (2000). Prevalence of gastrointestinal parasites in young camels in Bahrain. *Rev. Elev. Méd. Vét. Pays Trop.*, 53, 267 – 273.

[2] Ali B.H. & Abdelaziz M. (1982). – Balantidiasis in a camel. *Vet. Rec.*, 110, 506.

[3] Al-Saffar T.M., Suliman E.G. & Al-Bakri H.S. (2010). – Prevalence of intestinal ciliate *Buxtonella sulcata* in cattle in Monsul. *Iraqi J. Vet. Sci.*, 24, 27 – 30.

[4] Bozenko V.P. (1925). – Ciliates of the intestinal tract of the camel [in Russian]. *Vestn. Mikrobiol. Epidemiol. i Parazitol.*, 4, 56 – 57.

[5] Gill H.S. (1976). – Incidence of *Eimeria* and Infundibulorium in camel. *Ind. Vet. J.*, 53, 897 – 898.

[6] Levine N.D. (1985). – Veterinary protozoology. Iowa State University Press, Ames, Iowa.

[7] Lynn D.H. (2008). – The ciliate protozoa. 3rd Ed. Springer Dordrecht, Heidelberg.

[8] Selim H.M., Imai S., Yamato O., Miyagava E. & Maede Y. (1996). – Ciliate protozoa in the fore stomach of the dromedary camel (*Camelus dromedarius*), in Egypt, with description of new species. *J. Vet. Med. Sci.* 58 (9), 833 – 837.

[9] Shommain A.M. & Osman A.M (1987). Diseases of camels in the Sudan. *In* Diseases of camels. *Rev. sci. tech. Off. int. Epiz.*, 6 (2), 481 – 486.

[10] Tomczuk K., Kurek L., Stec A., Studcinska M. & Mochol J. (2005). – Incidence and clinical aspects of colon ciliate *Buxtonella sulcata* infection in cattle. *Bull. Vet. Inst. Pulawi*, 49, 29 – 33.


（谢俊仁译，刘光远、殷宏校）

## 5.1.5 艾美尔球虫病

球虫亚纲的代表是属于复顶亚门（syn. 孢子虫纲）的隐孢子虫科、艾美尔科和肉孢子虫科，这些胞内寄生虫的共同特征是在侵袭性阶段的细胞（子孢子、裂殖子）的前端拥有一个顶复合体。复合体与圆锥体和极环一样，也是由分泌性的细胞器组成。这些顶复合体的细胞器参与吸附和穿透宿主细胞。

从严格意义上来讲，球虫病是由球虫纲的胞内寄生虫引起的一类疾病。然而，这一术语常被动物饲养者、动物保护者和兽医从业者用来描述艾美尔属和等孢子属的寄生虫感染，实际上艾美尔球虫病（Eimeriosis）这个术语并不通用。艾美尔球虫是在肠上皮细胞中寄居的胞内寄生虫，它们在肠道进行无性繁殖和有性繁殖，而孢子生殖却发生在宿主体外。

### 形态学

在新大陆骆驼（NWCs）中发现了6种艾美尔球虫（图218），而在单峰驼和双峰驼中发现了5种以上的艾美尔球虫（图219，图220）。其形态学特征列于表81中。

骆驼球虫的卵囊有小于30μm的（普诺艾美尔球虫 *Eimeria punoensis*、派氏艾美尔球虫 *E. pellerdyi* 和羊驼艾美尔球虫 *E. alpacae*），有在30～40μm中等大小的（单峰驼艾美尔球虫 *E. dromedarii*、双峰驼艾美尔球虫 *E. bactriani*、秘鲁艾美尔球虫 *E. peruviana*、拉氏艾美尔球虫 E. rajasthani和无峰驼艾美尔球虫 *E. lamae*），也有与感染马属动物鲁氏艾美尔球虫 （*E. leuckarti*）大小相当的大型卵囊（伊维塔艾美尔球虫 *E. ivitaensis*、骆驼艾美尔球虫 *E. cameli* 和马库沙里艾美尔球虫 *E. macusaniensis*）。它们的形状由球形、椭圆形到卵圆—梨形变化不一，除了秘鲁艾美尔球虫❶外，所有的球虫卵囊均拥有一个卵膜孔。此外，在6种球虫中，卵膜孔上均覆盖着一个极帽。新鲜的卵囊中包含着一个圆形的母孢子，而经过孢子化后，每个艾美尔球虫卵囊包含四个孢子囊，每个孢子囊中含两个子孢子。

在新鲜的粪便样品中，能发现骆驼艾美尔球虫卵囊的表面包被着前文提及的透明的双层膜[9]，这个膜形成了朝向卵囊狭长卵孔的一个漏斗形的凹陷（图221）。

普诺艾美尔球虫 *Eimeria punoensis*
羊驼艾美尔球虫 *Eimeria alpacae*
无峰驼艾美尔球虫 *Eimeria lamae*
马库沙里艾美尔球虫 *Eimeria macusaniensis*
伊维塔艾美尔球虫 *Eimeria ivitaensis*

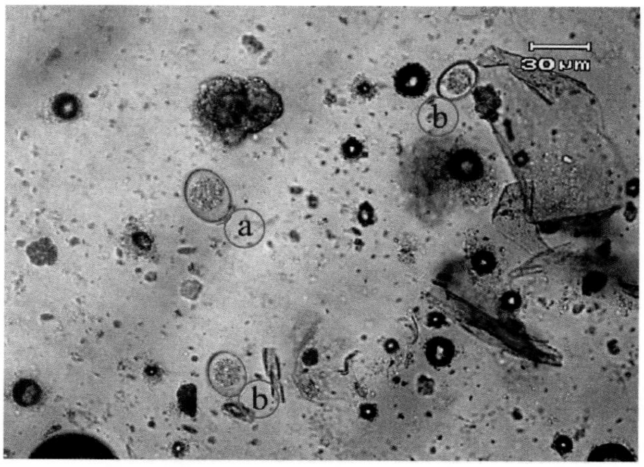


图219　旧大陆骆驼艾美尔球虫的卵囊：拉氏艾美尔球虫（*E. rajasthani*）（a）和单峰驼艾美尔球虫（*E. dromedarii*）（b）

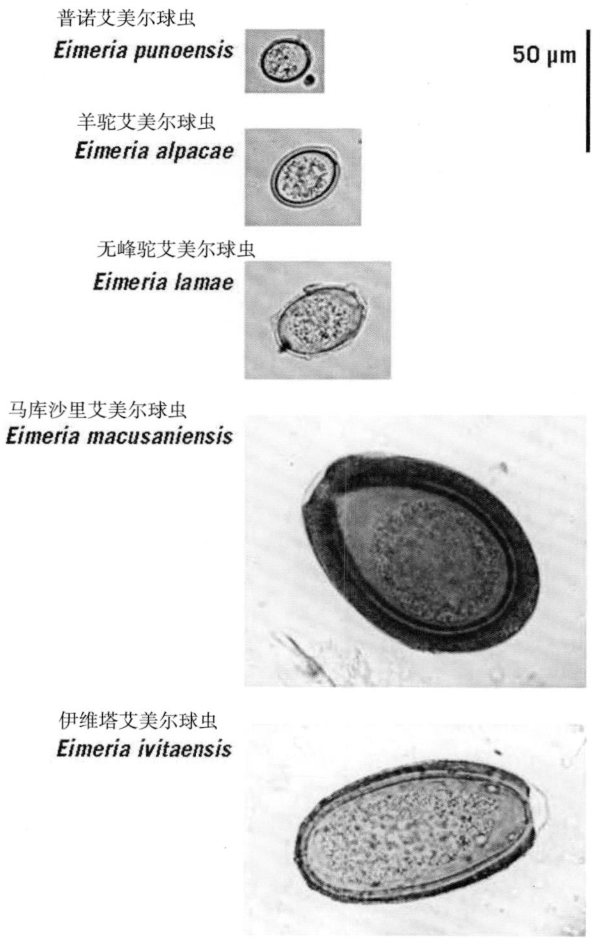


图218　新大陆骆驼艾美尔球虫的卵囊（图片由Ch. Bauer教授提供，Institute of Parasitology, Justus Liebig Universität, GieBen）

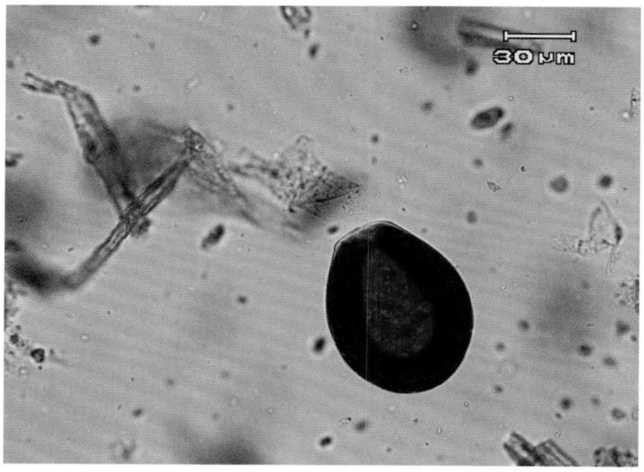


图220　骆驼艾美尔球虫的卵囊，一个旧大陆骆驼引人注目的艾美尔球虫虫种

## 生活史

在骆驼中还没有进行过详细的研究对艾美尔球虫的寄生生活史（图222），但认为它们也遵循着众所周知的在其他宿主体内寄生的球虫的相同规律。因而，总的来说球虫的生活史是以世代交替进行无性增殖（分裂生殖和孢子增殖）和有性增殖（配子生殖）为特征的。在其感染阶段，孢子化的

❶　由于只有两篇文章描述了秘鲁艾美尔球虫（*E. peruviana*），因此对其是否为独立种尚存疑问。

表81　骆驼科动物中艾美尔球虫的形态特征（参考 Rohbeck[21]和Gerlach[9]）

| 虫种 | 宿主 | 大小 (长 × 宽 μ m) | 形态 | 卵膜孔 | 极帽 |
|---|---|---|---|---|---|
| 马库沙里艾美尔球虫 *E. macusaniensis* | NWC | (81 ~ 110) × (61 ~ 84) | 卵形—梨形 | + | + |
| 骆驼艾美尔球虫 *E. cameli* | OWC | (80 ~ 100) × (55 ~ 94) | 梨形 | + | - |
| 伊维塔艾美尔球虫 *E. ivitaensis* | NWC | (88 ~ 98) × (49 ~ 59) | 椭圆形 | + | - |
| 无峰驼艾美尔球虫 *E. lamae* | NWC | (30 ~ 40) × (26 ~ 30) | 椭圆形—卵形 | + | + |
| 拉氏艾美尔球虫 *E. rajasthani* | OWC | (34 ~ 39) × (23 ~ 29) | 椭圆形 | + | + |
| 秘鲁艾美尔球虫 *E. peruviana* | NWC | (28 ~ 38) × (18 ~ 23) | 卵形 | - | - |
| 双峰驼艾美尔球虫 *E. bactriani* | OWC | (21 ~ 34) × (20 ~ 28) | 圆形—椭圆形 | + | - |
| 单峰驼艾美尔球虫 *E. dromedarii* | OWC | (23 ~ 33) × (19 ~ 25) | 卵圆形—圆形 | + | + |
| 羊驼艾美尔球虫 *E. alpacae* | NWC | (22 ~ 27) × (18 ~ 24) | 椭圆形—卵形 | + | + |
| 佩氏艾美尔球虫*E. pellerdyi* | OWC | (22 ~ 24) × (20 ~ 28) | 圆形—椭圆形 | + | - |
| 普诺艾美尔球虫 *E. punoensis* | NWC | (17 ~ 22) × (14 ~ 18) | 椭圆形—卵形 | + | + |

注：NWC，新大陆骆驼；OWC，旧大陆骆驼

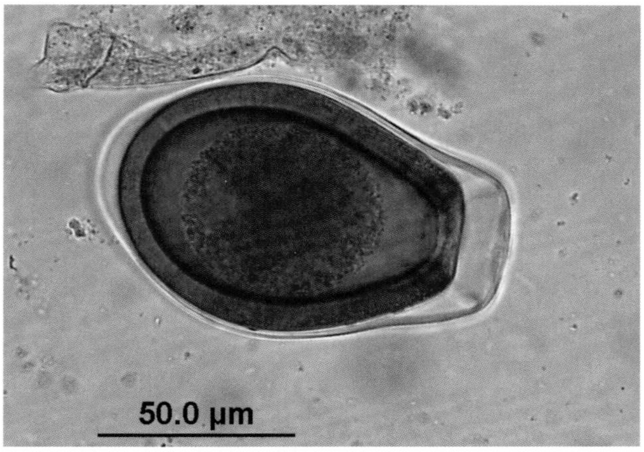


图221　新鲜粪便样品中的骆驼艾美尔球虫卵囊包被着一层膜

卵囊被宿主吞食后脱囊，子孢子被释放出来，并侵入肠黏膜上皮样细胞，紧接着进行所谓的分裂生殖（卵块发育）的无性发育阶段。子孢子发育成一个可刺激宿主细胞进行巨型生长的滋养细胞。首先，寄生虫的核分裂多次，成为一个分裂体。这个阶段的细胞包含几百个核，充满整个宿主细胞，将宿主细胞的核挤到外缘。其后，在每个寄生虫核的周围分化出细胞质和细胞器。因此，在卵块发育期间出现大量的纺锤形裂殖子。裂殖子是在所谓的裂殖体的发育阶段形成的，不断生长的寄生虫引起宿主细胞的破裂，释放出来的裂殖子又入侵其他细胞。这个过程可以重复2~3次。在最后一代裂殖体中形成的裂殖子侵入新的宿主细胞后，或者发育成为雄配子体（即小配子母体），其能发育出数个雄配子；或者发育成为雌配子体（即大配子母体），其只能发育成一个单独的雌配子。在接下来的配子生殖阶段，大配子和小配子融合后形成一个可发育为卵囊

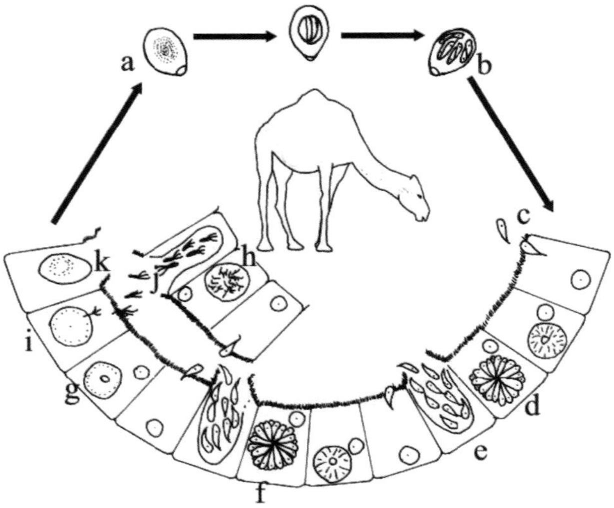


图222　球虫生活史

（a）未孢子化的卵囊；（b）孢子化卵囊；（c）子孢子；（d）Ⅰ型裂殖体；（e）裂殖子；（f）Ⅱ型裂殖体；（g）大配子体（雌配子体）；（h）小配子体（雄配子体）；（i）大配子（雌配子）；（j）小配子（雄配子）；（k）合子

5 寄生虫病

的合子。成熟的卵囊随粪便排出，在体外进行下一个阶段的发育。卵囊经过孢子化后形成四个孢囊，每个孢囊中各含两个子孢子，这个过程是在外界环境中进行的（图223）。依据外界牧场的条件，孢子化的卵囊能保持感染力达几个月之久[5]。骆驼球虫孢子化所需时间、潜伏期和显露期（patent period）的数据概括于表82中。

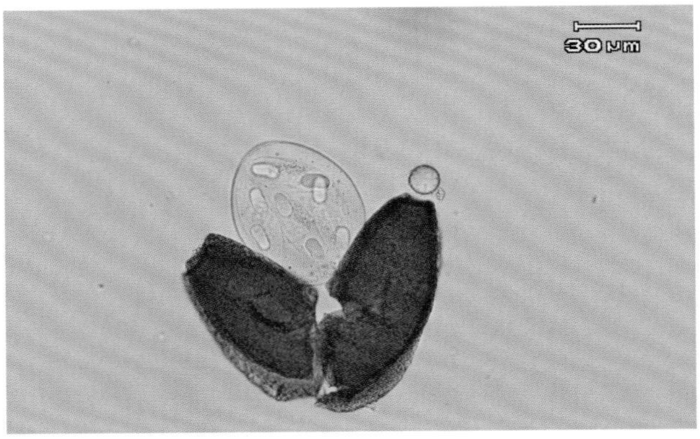


图223　孢子化的骆驼艾美尔球虫（*Eimeria cameli*）卵囊

棕色的外壳被机械性撑开后可看到四个孢子囊，每个孢子囊中都含有两个子孢子。

## 流行病学

### 经口感染

不同种的艾美尔球虫一般具有宿主特异性，但能感染同属不同种的宿主动物，例如骆驼和大羊驼。艾美尔球虫病是骆驼最常见的肠道寄生虫感染的疾病之一[8,20]，骆驼因吞食了孢子化卵囊而被感染。旧大陆骆驼的食粪癖行为增加了感染风险，但新大陆骆驼在固定地方（公厕）排便的行为降低了患球虫病的风险。对南美洲的新大陆骆驼的研究表明，球虫病的感染率在30%～100%[17,19]。据Cafrune等的报告，伊维塔艾美尔球虫（*E. ivitaensis*）在秘鲁和阿根廷的流行率非常低，分别只感染羊驼和美洲驼，而缺乏驼马和原驼慧球虫病的统计数据。

除了秘鲁艾美尔球虫（*E. peruviana*）外，感染新大陆骆驼的其他艾美尔球虫在南美洲以外的地方也存在。发现羊驼艾美尔球虫（*E. alpacae*）、无峰驼艾美尔球虫（*E. lamae*）、曼库塞尼艾美尔球虫（*E. macusaniensis*）和普诺艾美尔球虫（*E. punoensis*）在北美洲[26]和澳大利亚[4]也存在。此外，在欧洲的新大陆骆驼中也检测到伊维塔艾美尔球虫[24]。在奥地利和德国南部的研究表明，曼库塞尼艾美尔球虫、羊驼艾美尔球虫和普诺艾美尔球虫已感染率比较高，但其OPG值（每克粪便中卵囊数）却较低[2, 21, 22]。

旧大陆骆驼中艾美尔球虫感染似乎并不普遍。唯一可查的中国双峰驼球虫病的数据显示，49.53%的双峰驼感染了众所周知的旧大陆骆驼所有的5种艾美尔球虫[25]。Partani在印度拉贾斯坦邦收集单峰骆驼的数据[18]表明，总感染率为25%，其鉴定出的病原为单峰驼艾美尔球虫（*E. dromedarii*）、骆驼艾美尔球虫（*E. cameli*）、派氏艾美尔球虫（*E. pellerdyi*）和拉氏艾美尔球虫（*E. rajasthani*），球虫感染率最高的数据是来源于不足一岁的幼年骆驼。Mahmoud等[16]在沙特阿拉伯盖西姆地区单峰骆驼中发现球虫感染率仅为

表82　骆驼球虫孢子化所需时间，潜伏期和显露期的数据

| 种 | 孢子化所需时间(天) | 潜伏期(天) | 显露期(天) | 参考文献 |
|---|---|---|---|---|
| 羊驼艾美尔球虫 *E. alpacae* | 4 | 16~18 | 9 | 15 |
| 伊维塔艾美尔球虫 *E. ivitaensis* | 未知 | <15 | 未知 | 14 |
| 无峰驼艾美尔球虫 *E. lamae* | 5 | 16 | 10 | 10 |
| 马库沙里艾美尔球虫 *E. macusaniensis* | 9~21 | 32~36 | 39~43 | 21 |
| 普诺艾美尔球虫 *E. punoensis* | 4 | 10 | 22~26 | 8 |
| 骆驼艾美尔球虫 *E. cameli* | 9~15 | 16~40 | 12~48 | 9 |
| 单峰驼艾美尔球虫 *E. dromedarii* | 6~8 | 未知 | 未知 | 6 |
| 拉氏艾美尔球虫 *E. rajasthani* | 7~8 | 未知 | 未知 | 7 |

13%，并鉴定出骆驼艾美尔球虫（*E. cameli*）、单峰驼艾美尔球虫（*E. dromedarii*）和拉氏艾美尔球虫（*E. rajasthani*）为致病病原。在阿拉伯联合酋长国内也存在同样的球虫。在迪拜中央兽医研究实验室中检测了11474份竞赛骆驼粪便样品，其中14.4%的样品为骆驼艾美尔球虫阳性，而单峰驼艾美尔球虫和拉氏艾美尔球虫的感染率分别是5.6%和3.4%。所有这三种球虫的最高感染率是源自当年第二季度的流行病学的调查数据。

艾美尔球虫卵囊在宿主体外进行孢子化的过程需要适当的温度和湿度条件，在干燥、高温条件下，薄壁型卵囊很快失去了感染力，而厚壁型卵囊则能耐受较高温度。试验研究表明，单峰驼粪便中的骆驼艾美尔球虫的卵囊在实验室条件下能进行孢子化过程，甚至暴露在60℃以上的温度条件下几个小时后也依然可以孢子化[9]。

马库沙里艾美尔球虫（*E. macusaniensis*）的孢子化卵囊能保持感染力达几个月，但这与外界牧场的条件有一定的关系[5]，而在实验室条件下，它们能存活达8年之久[13]。

由于没有持久地在组织内寄生的阶段，如果不发生再次感染，骆驼的艾美尔球虫病具有自限性。而再次感染时潜伏期延长、显露期变短、卵囊密度降低，表明球虫病能引起宿主产生一定的免疫力。

## 临床症状与病理变化

在球虫分裂生殖和孢子生殖阶段，动物肠黏膜上皮样细胞遭到破坏，引起卡他性－出血性肠炎（图 224），并导致动物出现食欲不振、体重降低、疝痛、便秘和腹泻，尤其是幼年动物。若继发细菌感染可能会急剧加重病情，引起死亡。尽管成年旧大陆骆驼的每克粪便中卵囊数（OPG值）很高，但由球虫感染引起腹泻的症状却并不明显[8]。

根据Foreyt与Lagerquist[8]和Trout 等[23]的报道，普诺艾美尔球虫（*E. punoensis*）和羊驼艾美尔球虫（*E. alpacae*）没有致病性，即使对幼年动物也不致病；而新大陆骆驼中较大的艾美尔球虫，即马库沙里艾美尔球虫（*E. macusaniensis*）和伊维塔艾美尔球虫（*E. ivitaensis*），则能导致动物死亡。在空肠和回肠中可观察到大体的病理变化，包括黏膜出现水肿、增厚、脱落和发红并伴有轻微凸起的白色结节病灶。派伊尔结节（Payer's patches）的直径达到4cm以上。

发育阶段的裂殖体、配子体、受精卵和卵囊均在肠黏膜上进行（图 225）。伊维塔艾美尔球虫的裂殖体仅限于空肠和回肠，而曼库塞尼艾美尔球虫的裂殖体也可在盲肠和结肠上段检测到。这两个球虫的大配子体也可在大肠上皮样细胞中找到。组织学检查主要的病理变化是黏膜遭到破坏、杂乱无序，并带有出血和炎性细胞（嗜酸性粒细胞和巨噬细胞）浸润。

## 诊断

临床特征提示可能为球虫病时，确诊的依据是使用漂浮法（饱和食盐水或饱和蔗糖溶液）在排泄物中检测到球虫卵囊，或在肠黏膜涂片或者尸检时组织切片中发现组织中处于发育阶段的球虫。

依据孢子化卵囊的形态学特征可将球虫鉴定到种。球虫卵囊孢子化需在2.5%的重铬酸钾溶液中25℃条件下持续大约10天时间才能完成。

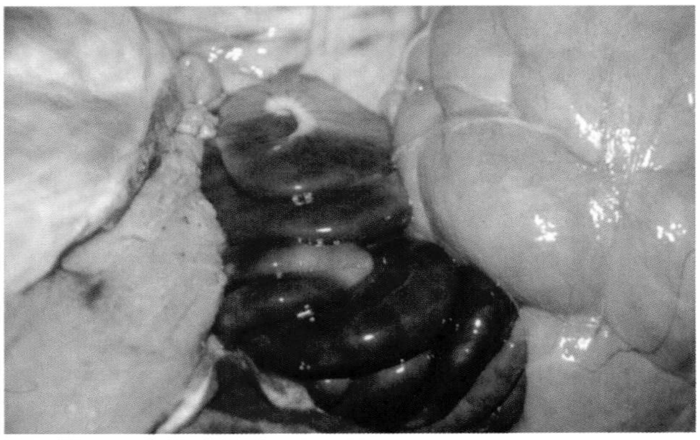

图224　由骆驼艾美尔球虫引起的出血性肠炎

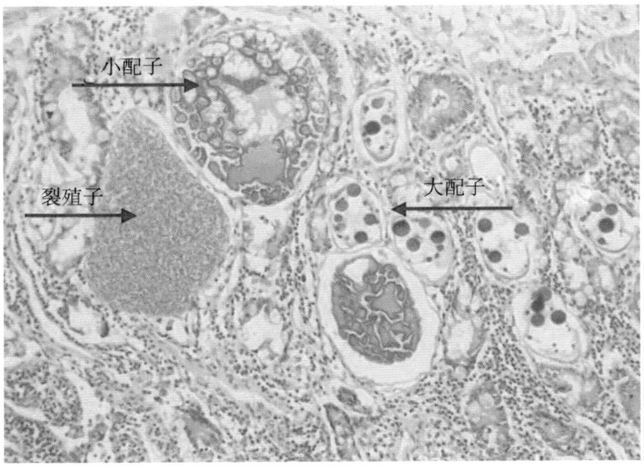


图225　在组织切片中的骆驼艾美尔球虫

### 防控

在大规模的养禽业、养羊业和养牛业中，主要以化学药物预防的方式，采用常见抗球虫药来控制家畜的球虫病，但几乎不知道这类药物用于预防骆驼球虫病的效果如何。治疗已形成卵囊的动物个体已经太迟，这是因为抗球虫药只针对裂殖体和配子体有效，而不能杀死已形成的卵囊。在文献中可以看到使用磺胺类（Sulfonamides）药物治疗骆驼球虫病的报道，如磺胺二甲嘧啶（Sulfadimidine）以水悬浮液的形式每千克体重30mg的剂量通过口服10天，通常用来治疗单峰驼幼畜[12]。其他作者推荐用甲醛磺胺噻唑（Formosulfathiazole）膏以每千克体重100～200mg的剂量治疗3～5天[11]。另外一种磺胺类化合物——磺胺地托辛（Sulfadimethoxine）常以每千克体重15mg的剂量用于治疗新大陆骆驼的球虫病。

托曲珠利（Toltrazuril）是一种用在禽、猪和反刍动物中的很有潜力的抗球虫药，在骆驼的药代动力学研究中表明，血清含量水平可达到很高，但无法阻断骆驼艾美尔球虫（*E. cameli*）在单峰驼体内完成其生活史[9]。而通常广泛用于控制禽类球虫病的离子运载类抗生素（莫能菌素、拉沙里菌素、沙利霉素）对骆驼是有毒性的[1]，可引起骨骼肌的退化（图 226）。

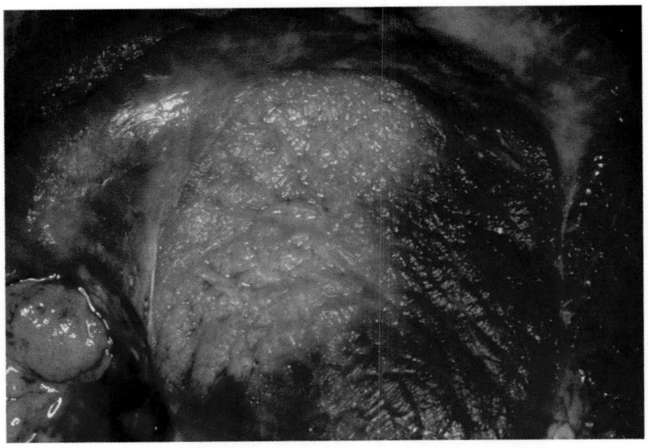

图226　沙利霉素中毒后单峰驼骨骼肌退化

## 参考文献

[1] Al-Nazawi M.H. & Homeida A.M. (2009). – Kinetics and tolerance of salinomycin in camels. *Res. J. Pharmacol.*, 3, 48 – 51.

[2] Burian E. (2010). – Endoparasitenstatus von Neuweltkameliden in Oberosterreich und im Burgenland. Thesis, Veterinarmedizinische Universitat Vienna.

[3] Cafrune M.M., Martin R.E., Rigalt F.A., Romero S.R. & Aguirre D.H. (2009). – Prevalence of *Eimeria macusaniensis* and Eimeria ivitaensis in South American camelids of Northwest Argentina. *Vet. Parasitol.*, 162, 338 – 341.

[4] Carmichael I.H., Judson G.J., Ponzoni R.W., Hubbard D.J., Howse A., McGregor B.A. (1999). – Diagnosis and control of parasites of alpacas in winter rainfall areas of Australia. *In* Australian Alpaca Conference Proceedings, Australian Alpaca Association, Mitcham North, VIC, Australia, 69 – 73.

[5] Cebra C., Valentine B., Schlipf J., Bidfell R., McKenzie E., Heidel J., Cooper B., Lohr C., Barid K. & Saulez M. (2007). *Eimeria macusaniensis* infection in 15 llamas and 34 alpacas. *J. Am. Vet. Med. Ass.*, 230, 94 – 100.

[6] Dariush A.I & Golemansky V.G. (1993). – Coccidia (Apicomplexa, Eucoccidia) in camels (*Camelus dromedarius* L.) from Syria. *Acta Zool. Bulg.*, 46, 10 – 15.

[7] Dubey J.P. & Pande B.P. (1963). – On eimerian oocysts recovered from Indian Camel (*Camelus dromedarius*). *Ind. J. Vet.* Sci. *Anim. Husbandry*, 34, 28 – 34.

[8] Foreyt W.J. & Lagerquist J. (1992). – Experimental infections of *Eimeria alpacae* and *Eimeria punoensis* in llamas (*Lama glama*). *J. Parasitol.*, 78, 906 – 909.

[9] Gerlach F. (2008). – Kokzidiose beim Dromedar (*Camelus dromedarius*). Thesis, Freie Universitat, Berlin.

[10] Guerrero C.H., Bazalar H. & Tabachini L (1970). – Infeccion experimental de alpacas con *Eimeria lamae. Bol. Ext. IVITA*, 4, 79 – 83.

[11] Hanichen T., Wiesner H. & Gobel E. (1994). – Zur Pathologie, Diagnostik und Therapie der Kokzidiose bei Wiederkauernim Zoo. *Verhandlungsber. Erkr. Zootiere*, 36, 375 – 380.

[12] Hussein H.S., Kasim A.A. & Shawa J.R. (1987). – The prevalence and pathology of *Eimeria* infections in camels in Saudi Arabia. *J. Comp. Pathol.*, 97, 293 – 297.

[13] Jarvinen J.A. (2008). – Infection of llamas with stored *Eimeria macusaniensis* oocysts obtained from guanaco and alpaca faeces. *J. Parasitol.*, 94 (4), 969 – 972.

[14] Leguia P.G. & Casas A.E. (1996). – *Eimeria ivitaensis n.* sp. (Protozoa: Eimeriidae) en alpacas (*Lama pacos*). *MV Rev. Cienc. Vet.*, 12, 113 – 114.

[15] Leguia P.G. & Casas A.E. (1999). – Enfermedades parasitarias y atlas parasitologico de camelidos Sudamericanos. Editorial de Mar, Lima, Peru.

[16] Mahmoud O.M., Haroum E.M., Magzoub M., Omer O.H. & Sulman A. (1998). – Coccideal infection in camels of Gassim Region, Central Saudi Arabia. *J. Camel Pract. Res.*, 5 (2), 257 – 260.

[17] Palacios C.A., Perales R.A., Chavera A.E., Lopez M.T, Braga W.U. & Moro M. (2006). – *Eimeria macusaniensis and Eimeria ivitaensis* co-infection in fatal cases of diarrhoea in young alpacas (*Lama pacos*) in Peru. *Vet Rec.*, 158, 344 – 345.

[18] Partani A.K., Kumar D. & Manohar G.S. (1999). – Prevalence of *Eimeria* infections in camels (*Camelus dromedarius*) at Bikaner (Rajasthan). *J. Camel Pract. Res.*, 6 (1), 69 – 71.

[19] Pelayo P.A.R. (1973). – Prevalencia de coccidias (Protozoa: Eimeriidae) en llamas (*Lama glama*). Bachelor's thesis, National University of San Marcos, Lima.

[20] Rickard L.G. & Bishop J.K. (1988). – Prevalence of *Eimeria* sp. (Apicomplexa: Eimeriidae) in Oregon llamas. J. *Protozool.*, 35, 335 – 336.

[21] Rohbeck S. (2006). – Parasitosen des Verdauungstraktes und der Atemwege bei Neuweltkameliden: Untersuchungen zu ihrer Epidemiologie und Bekampfung in einer sudhessischen Herde sowie zur Biologie von *Eimeria macusaniensis*. Thesis, Justus- Liebig-Universitat Giessen, 132 pp.

[22] Schlogl C. (2010). – Erhebungen zum Vorkommen von Endo- sowie Ektoparasiten bei Neuweltkameliden. Thesis, Ludwig- Maximilians-Universitat, Munich, 94 pp.

[23] Trout J.M., Satin M. & Fayer R. (2008). – Detection of assemblage A, *Giardia duodenalis* and *Eimeria* spp. in alpacas on two Maryland farms. *Vet. Parasitol.*, 153, 203 – 208.

[24] Twomey D.F., Allen K., Bell S., Evans C. & Thomas S. (2010). – *Eimeria ivitaensis* in British alpacas. *Vet. Rec.*, 167, 797 – 798.

[25] Wei J.G. & Wang Z. (1990). – A survey of the *Eimeria* species in the double-humped camels of the Inner Mongolian Autonomous region, China. *Chin. J. Vet. Med.*, 16, 23 – 24.

[26] Whitehead C.E. (2006). – Coccidiosis. *In* Current veterinary care and management of lamas and alpacas. *In* International Camelid Health Conference for Veterinarians, Ohio State University, 21 – 25 March 2006, Abstracts, 31.

（谢俊仁译，刘光远、殷宏校）

## 5.1.6 隐孢子虫病

隐孢子虫病是由隐孢子属球虫（*Cryptosporidium*）引起的一种寄生虫性的人兽共患病。在哺乳动物中，病原主要感染胃肠道上皮样细胞，引起急性水样腹泻。该病在全世界分布，所有的脊椎动物类宿主都罹患此病。在放牧动物群中，它是新生幼畜腹泻综合征的最主要的致病病原之一。

### 形态学

由于卵囊体积很小，几乎没有可用于种鉴定的形态学特征，因而其种的归属是依据基因和生物学的差异。到目前为止，文献所记载的不同毒力的隐孢子虫大约有20个种和超过40个基因型[3,4,10,15]。

隐孢子虫卵囊略呈卵圆形，其大小在5.0 μm × 4.5 μm（小隐孢子虫，*C. parvum*）到7.4 μm × 5.6 μm（小鼠隐孢子虫，*C. muris* 和安氏隐孢子虫*C. andersoni*）之间，并含有4个无孢子囊的子孢子。

## 生活史

已进行过详细的个体发育描述的主要是小隐孢子虫（*C. parvum*）❶[8]。隐孢子虫是胞内寄生虫，但其后的发育阶段位于细胞外，卵囊是体外发育的唯一阶段。卵囊被适宜的宿主吞食后，释放出4个能运动的子孢子，主要入侵胃肠道上皮样细胞，通常能在上皮样细胞的微绒毛表面找到其组织发育阶段。I 型裂殖体形成8个裂殖子，每个裂殖子侵入其他新的宿主细胞，变成下一代的I型裂殖体，或者发育成含 4 个裂殖子的Ⅱ型裂殖体。后者在宿主的其他肠细胞中发育成大配子体和小配子体，在宿主细胞中形成两种类型的卵囊：厚壁型卵囊和薄壁型卵囊。厚壁型卵囊随粪便排出体外，而薄壁型卵囊在宿主体内释放出子孢子并引起自体感染。

## 流行病学

### 经口感染

宿主吞食了被隐孢子虫卵囊污染的食物和饮用水而遭受感染。隐孢子虫病是一种全世界分布的重要的水源性的人兽共患的寄生虫病，大部分已知的隐孢子虫无宿主特异性，但对骆驼科动物中隐孢子虫病的情况知之甚少。据记载，美国动物园双峰驼曾患由小鼠隐孢子虫引起的隐孢子虫病[7]。小隐孢子虫也被列为双峰驼的一种寄生虫[6]，而经鉴定安氏隐孢子虫可感染中国的双峰驼[19]。小鼠隐孢子虫卵囊在埃及单峰驼中的检出率为3.37%，这些卵囊略大于文献所记载的尺寸。从表面上看，被感染的动物呈健康状态，但血清中胃蛋白酶原的含量水平呈现升高的趋势[16]。

在埃及[1]和伊朗[2,12,14]的单峰驼中发现过隐孢子虫卵囊，但没有进一步的描述。

在秘鲁对5 163份羊驼粪便样品进行了隐孢子虫感染的大型流行病学调查[11]揭示，排放隐孢子虫卵囊的羊驼仔的平均检出率为13%，其中新生幼仔（1～3日龄）检出率最低（2%），而13～15日龄幼仔检出率最高（20%）。作者认为牧场利用率低和管理水平低下是其主要的风险因素。分子生物学方法鉴定的结果揭示，感染秘鲁羊驼的隐孢子虫是小隐孢子虫和泛在隐孢子虫（*C. ubiquitum*）[9]。

在南美洲以外国家也有许多报道有关新大陆骆驼中存在隐孢子虫的病例，但在病原学上鉴定出的致病病原只有小隐孢子虫[18]。

## 临床症状与病理变化

水样腹泻（通常粪便为黄色）是感染小隐孢子虫的幼年动物主要症状，可能出现的其他症状包括：食欲不振、脱水、发热、精神抑郁和体重减轻。隐孢子虫病是一种新生动物疫病，动物主要在1～3周龄时罹患该病，腹泻待续4～6天，而后在1～2周内恢复。然而，严重的感染可能是致死性的，特别是免疫缺陷性的宿主。在病理剖解时，小肠和大肠膨胀，肠内有黏液样黄色内容物（图227）。在组织切片中可看到，处于发育时期的隐孢子虫在肠细胞表面下呈小圆点状（图228）。

除了两例动物园双峰驼患慢性隐孢子虫病时出现腹泻、被毛粗乱、体重减少和慢性白细胞增多的报告之外[5,7]，目前还没有其他描述与双峰驼有关的临床症状的文献。

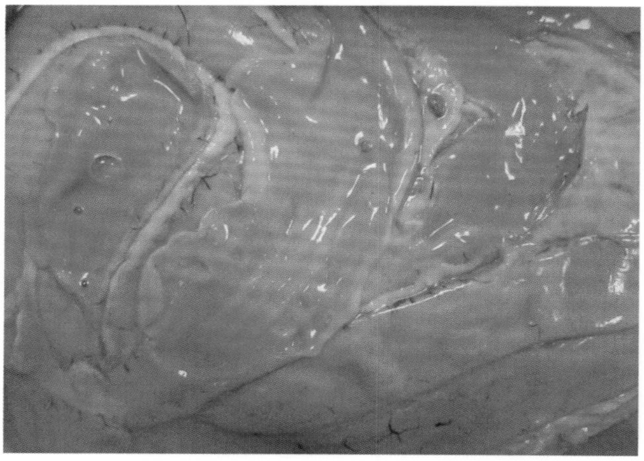

图227　骆驼幼仔的隐孢子虫病。膨胀的结肠中充满黏液样黄色内容物

❶ 在严格意义上，目前应将小隐孢子虫定义为“小鼠”基因型，而鼠型隐孢子虫（*C. pestis*）在形态学上与前者无法区分，被正式地确认为人兽共患的“牛”基因型[17]

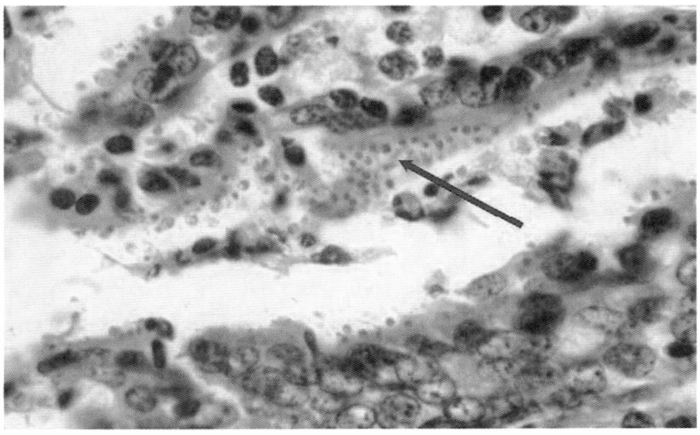

图228 在肠黏膜细胞表面下处于发育阶段的隐孢子虫（箭头所示）。用苏木精和伊红染色的小肠切片

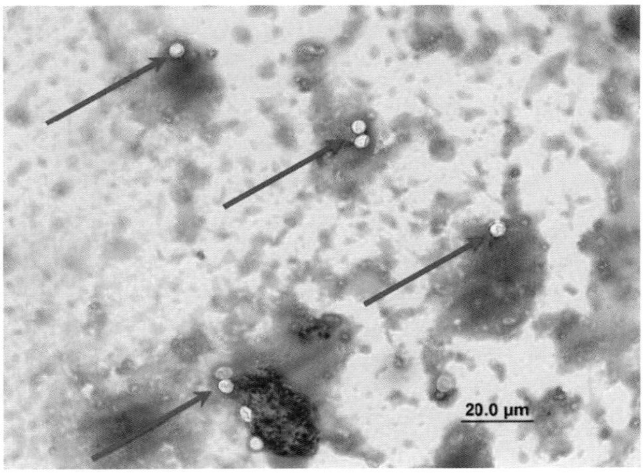


图229 在石炭酸—碱性品红染色的粪便涂片中的隐孢子虫卵囊。新鲜制作的涂片中染料并未渗入卵囊壁，在红色的背景下能清楚地看到带有折射光层的无色的卵囊（箭头所示）

## 诊断

动物粪便涂片可用改进的抗酸染色法或石炭酸负染法染色后鉴定隐孢子虫卵囊[13]。

后一种技术包括：将一滴腹泻的粪便样品与同量的未稀释的石炭酸混合于载玻片上，待载玻片风干后，加一滴香柏油，并盖上盖玻片，然后在400 × 显微镜下检查。在新鲜制作的涂片中背景呈红色，隐孢子虫卵囊在形态上略呈椭圆形，以没有着色的结构形式在红色背景中出现（图229）。也可用饱和蔗糖溶液漂浮粪便中的隐孢子虫卵囊进行诊断。要准确鉴定到种，可应用分子生物学方法[19,20]。还有一种用于检测哺乳动物粪便样品中的隐孢子虫卵囊的直接免疫荧光检测技术（Merifluor *Cryptosporidium*/*Ciardia*, Meridian Bioscience, Inc., Cincinnati. OH）。这个方法是以单克隆抗体技术建立的，也能检测贾第虫包囊。

## 治疗

对于放牧动物的隐孢子虫病还没有特别的治疗方法。用电解质溶液进行的对症治疗是必需的，以避免渐进性的脱水。阴冷、潮湿的天气，牧场卫生条件差和高密度饲养均会增加患隐孢子虫病的风险。

# 参考文献


[1] Abu-Eisha A.M. (1994). – Cryptosporidial infection in man and farm animals in Ismailia Governorate. *Vet. Med. J. Giza*,42, 107 – 111.

[2] Borji H., Razmi G., Movassaghi A.R., Naghibi A.G. & Maleki M. (2009). – Prevalence of *Cryptosporidium* and *Eimeria* infections in dromedary (*Camelus dromedarius*) in abattoir of Mashhad, Iran. *J. Camel Pract. Res.*, 16 (2), 167 – 170.

[3] Caccio S.M. (2005). – Molecular epidemiology of human cryptosporidiosis. *Parassitologia*, 47, 185 – 192.

[4] de Graaf D.C., Vanopdenbosch E., Ortega-Mora L.M., Abbassi H. & Peeters J.E. (1999). – A review of the importance of cryptosporidiosis in farm animals. *Int. J. Parasitol.*, 29, 1 269 – 1 287.

[5] Fagasinski A. & Zuchowska E. (1992). – Cryptosporidiose bei einem Trampeltier (*Camelus bactrianus*). In Erkrankungen der Zootiere (R. Ippen & H.D. Schroder, eds). Verhandlungsbericht des 34. *In* Internationalen

Symposiums uber die Erkrankungen der Zoo- und Wildtiere, 27 – 31 May 1992, Santander, Spain, 307 – 308.

[6] Fayer R. (2004). – *Cryptosporidium*: a water borne zoonotic parasite. *Vet. Parasitol.*, 126, 37 – 56.

[7] Fayer R., Phillips L., Anderson B.C. & Bush M. (1991). – Chronic cryptosporidiosis in a Bactrian camel (*Camelus bactrianus*). *J. Zoo Wildlife Med.*, 22 (2), 228 – 232.

[8] Fayer R., Speer C.A. & Dubey J.P. (1997). – The general biology of *Cryptosporidium. In Cryptosporidium* and cryptosporidiosis (R. Fayer & L. Xiao, eds). CRC Press, Boca Raton, Florida, 1 – 41.

[9] Gomez-Couso H., Ortega-Mora L.M., Aguado-Martinez A., Rosadio-Alcantara R., Maturrano-Hernandez L., Luna-Espinoza V., Zanabria-Huisa V. & Pedraza-Diaz S. (2012). – Presence and molecular characterisation of *Giardia* and *Cryptosporidium* in alpacas (*Vicugna pacos*) from Peru. *Vet. Parasitol.*, 187, 414 – 420.

[10] Joachim A. (2004). – Human cryptosporidiosis: an update with special emphasis on the situation in Europe. *J. Vet. Med. B*, 51, 251 – 259.

[11] Lopez-Urbina M.T., Gonzales A.E., Gomez-Puerta L.A., Romero-Arbizu M.A., Prerales-Camacho R.A., Rojo-Vazquez F.A., Xiao L. & Cama V. (2009). – Prevalence of neonatal cryptosporidiosis in Andean Alpacas (*Vicugna pacos*) in Peru. *Open Parasitol. J.*, 3, 9 – 13.

[12] Nazifi S., Behzadi M.A., Raayat Jahromi A., Mehrshad S. & Tamadon A. (2010). – Prevalence of Cryptosporidium isolated from dromedary camels (*Camelus dromedarius*) in Qeshm Island, Southern Iran. *Comp. Clin. Pathol.*, 19, 311 – 314.

[13] Potters I. & van Esbroeck M (2010). – Negative staining technique of Heine for the detection of *Cryptosporidium* spp.: a fast and simple screening technique. *Open Parasitol. J.*, 4, 1 – 4.

[14] Razawi S.M., Oryan A., Bahrami S., Mohammadalipour A. & Gowhari M. (2009). – Prevalence of *Cryptosporidium* infection in camels (*Camelus dromedarius*) in a slaughterhouse in Iran. *Trop. Biomed.*, 26 (3), 267 – 273.

[15] Ryan U. (2010). – *Cryptosporidium* in birds, fish and amphibians. *Exp. Parasitol.*, 124, 113 – 120.

[16] Saleh M.A. & Mahran O.M. (2007). – A preliminary study on cryptosporidiosis in dromedary camels at Shalatin area, Egypt. *Assuit Vet. Med. J.*, 53, 195 – 208.

[17] Šlapeta J. (2011). – Naming of *Cryptosporidium pestis* is in accordance with the ICZN Code and the name is available for thistaxon previously recognized as *C. parvum* 'bovine genotype'. *Vet. Parasitol.*, 177, 1 – 5.

[18] Twomey D.F., Barlow A.M., Bell S., Chalmers R.M., Elwin K., Giles M., Higgins R.J., Robinson G. & Stringer R.M. (2008). – Cryptosporidiosis in two alpaca (*Lama pacos*) holdings in the South-West of England. *Vet. J.*, 175, 419 – 422.

[19] Wang R., Zhang L., Ning C., Feng Y., Jian F., Xiao L., Lu B., Ai W. & Dong H. (2008). – Multilocus phylogenetic analysis of *Cryptosporidium andersoni* (Apicomplexa) isolated from a Bactrian camel (*Camelus bactrianus*) in China. *Parasitol. Res.*, 102, 915 – 920.

[20] Xiao L. & Ryan U.M. (2008). – Molecular epidemiology. In *Cryptosporidium* and cryptosporidiosis (R. Fayer & L. Xiao, eds). CRC Press, Boca Raton, Florida, 119 – 172.

（谢俊仁译，刘光远、殷宏校）



### 5.1.7 等孢球虫病

等孢球虫病是由奥氏等孢球虫（*Isospora orlovi*）感染引起的旧大陆骆驼的一种球虫病。迄今为止，该病仍然未能在南美洲骆驼体内得到确诊。奥氏等孢球虫感染最初报道于哈萨克斯坦骆驼幼崽体内[10]，由于描述不全面，有很长一段时间认为是假性寄生❶[5, 7]。

等孢球虫属和艾美耳球虫属的原虫相关，两者的生活史很相近，但卵囊结构不同。然而，有学者建立进化树进行了比较分析之后发现，哺乳动物

❶ 假寄生虫（False parasite），等孢球虫卵囊常见于燕雀类，当其他并不是真正的宿主摄食了含有这些卵囊的粪便后，对其不造成损伤。因奥氏等孢球虫没有经实验证实，因此Pellerdi[7] 和 Levine[5]都对其是否为独立的骆驼寄生虫虫种提出质疑

等孢球虫属的种和一些能形成包囊的球虫种，如贝诺孢子虫、新孢子虫、哈芒球虫、刚地弓形虫病和肉孢子虫的亲缘关系很近[6]。

## 形态学

没有孢子化的卵囊呈椭圆形，中心有一个母孢子，大小为(27 ~ 33)μm × (15 ~ 20)μm。孢子化后的形态大小多变，取决于孢子囊的排列顺序，孢子化卵囊含有两个孢子囊，每个孢子囊含有四个子孢子。孢子囊呈椭圆形，大小为20 μm × 15 μm，而子孢子呈香蕉形，大小为(2 ~ 14)μm × (3 ~ 4)μm。

## 生活史

等孢球虫属的生活史目前尚未通过实验证实，但与其相关的猪等孢球虫（*I. suis*）的生活史已详细研究[2,9]。

早期感染阶段，3种无性繁殖过程主要发生于宿主的空肠末端和回肠的肠上皮细胞。第一次增殖是在感染后第二天，以孢内生殖的形式形成两代裂殖体，有性繁殖阶段发生在感染后5 ~ 6天；在感染后8 ~ 9天，第二代裂殖体形成，这就是在感染后5 ~ 9天和14 ~ 11天分别出现两个显露期的原因。

有人将实验小猪在无菌状态下经腹膜注射肝脏、脾脏和淋巴组织进行感染试验，显露期在注射后10 ~ 12天出现。这一无菌实验证实了肠外发育阶段的存在。

## 流行病学

口腔感染：感染源尚不清楚。奥氏等孢球虫病最初报道于哈萨克斯坦的小骆驼体内[10]。在印度[8]、阿拉伯联合酋长国[3,4]和肯尼亚[1,11]，小的单峰驼体内也有报道。在阿联酋，2 ~ 4周龄的骆驼最易感染，该病主要暴发于冬季，从12月份持续到翌年3月份，而在肯尼亚主要发生于6月和7月母畜产仔的高峰期。在阿拉伯联合酋长国，等孢球虫病只见于单个小骆驼体。而在肯尼亚的调查中，该病在放牧骆驼中的流行率11.2%，而舍饲骆驼为6.3%[1]。一年当中，幼畜的繁殖高峰期恰好是该病流行的高发期，其余的时间则是在围产期母畜体内进入肠外的发育阶段，该病的发病率随之降低。等孢球虫病泌乳传播的可能性高于经胎盘传播。

在阿拉伯联合酋长国发现等孢球虫对成年骆驼造成的病灶，在尸体剖解中发现，该病与马红球菌属的感染相关。这一现象进一步表明应激反应可以刺激肠外的发育阶段转入肠道内发育[4]。

## 临床症状和病理变化

幼畜感染后常常表现为腹泻，腹泻物呈水样状，腹泻持续时间达10余天，颜色由苍白变化到黄色（图231），黄色排泄物在显露期结束后又恢复到正常的黏度。对大量的因病致死的动物剖检发现，病理变化表现为严重的类白喉样结肠炎和出血性结肠炎[3]（图232）。组织切片表明，病变部位的黏膜组织紊乱和破坏，并伴有出血和炎症细胞的浸润，浸润细胞主要是嗜酸性细胞和巨噬细胞。球虫的大多数发育阶段（包括小配子母细胞、大配子母细胞、合子、卵囊和不同阶段孢子的形成）可见于结肠黏膜（图234）。

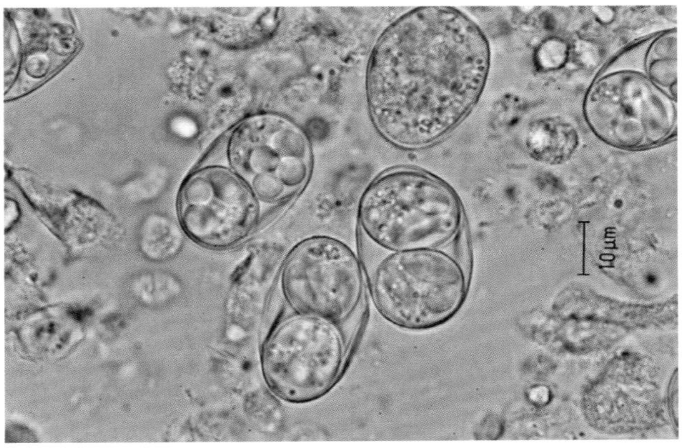


图230　奥氏等孢球虫（*Isospora orlovi*）卵囊

图231　等孢球虫病：幼驼结肠中的黏性的黄色内容物

## 诊断

2～4周龄患病的小骆驼表现为严重的腹泻。等孢球虫感染后，通常用浮选法或染色法检查排泄物中典型的卵囊数目（图235）。甚至在新鲜的排泄物样品中，正在孢子化的卵囊也占一定的比例。

在尸体剖检中，通过检查严重的类白喉样出血性结肠炎，可以确诊小骆驼是否患有等孢球虫病。可以在遭到损坏的大肠黏膜的组织碎片中发现有大量卵囊。

## 治疗

目前尚无文献报道有治疗等孢球虫的特效药物，常用托曲珠利治疗猪等孢球虫。

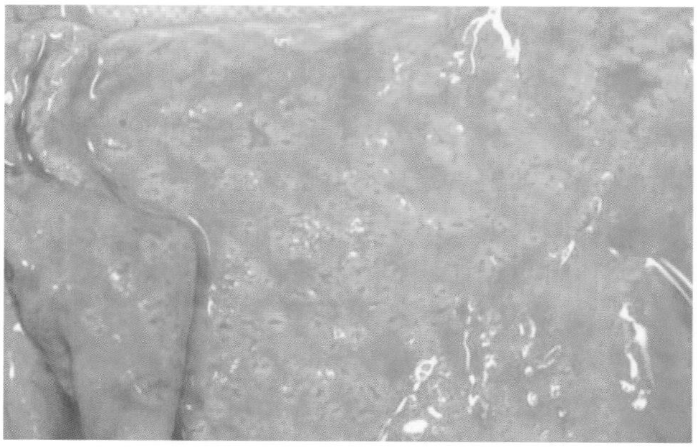

图232　奥氏等孢球虫引起的幼驼类白喉样结肠炎

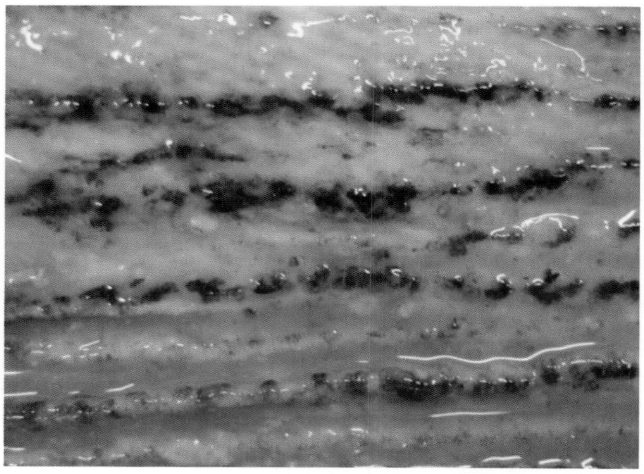

图233　奥氏等孢球虫引起的幼驼类出血性结肠炎

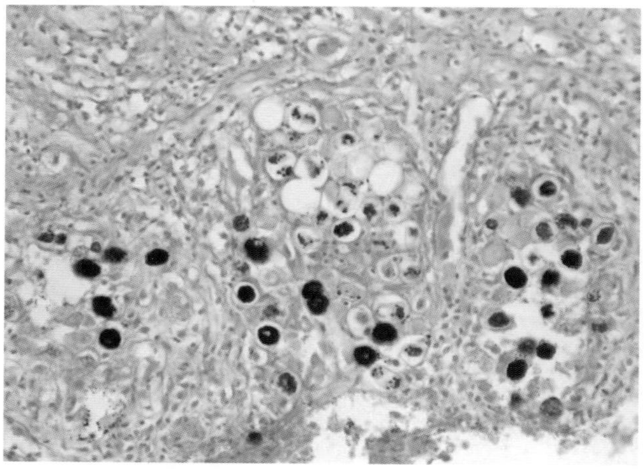

图234　等孢球虫病：过碘酸希夫反应染色的组织切片中的奥氏等孢球虫的合子和卵囊

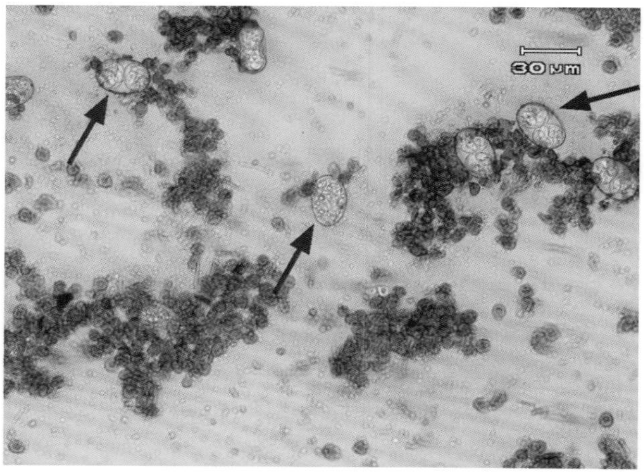


图235　石炭酸—品红染色的粪便涂片中的等孢球虫卵囊



## 参考文献

[1]　1. Bornstein S., Gluecks I.V., Younan M., Thebo P. & Mattsson J.G. (2008). – *Isospora orlovi* infection in suckling dromedary camel calves (*Camelus dromedarius*) in Kenya. *Vet. Parasitol.*, 152, 194–201.

[2]　Harlemann J.H. & Meyer R.C. (1984). – Life cycle of *Isospora suis* in gnotobiotic and conventionalized piglets. *Vet. Parasitol.*, 17, 27–39.

[3] Kinne J., Ali M. & Wernery U. (2001). – Camel coccidiosis caused by *Isospora orlovi* in the United Arab Emirates. *Emir. J. Agric. Sci.*, 13, 62–65.

[4] Kinne J., Ali M., Wernery U. & Dubey J.P. (2002). – Clinical large intestinal coccidiosis in camels (*Camelus dromedarius*) in the United Arab Emirates: description of lesions, endogenous stages and redescription of *Isospora orlovi* Tsygankov, 1950 oocysts. *J. Parasitol.*, 88 (3), 548–552.

[5] Levine N.D. (1985). – Veterinary protozoology. Iowa State University Press, Ames, Iowa.

[6] Morrison D.A., Bornstein S., Thebo P., Wernery U., Kinne J. & Mattsson J.G. (2004). – The current status of the small subunit rRNA phylogeny of the coccidian (Sporozoa). *Int. J. Parasitol.*, 34, 501–514.

[7] Pellerdy L. (1965). – Coccidia and coccidiosis. Akademiai Kiado, Budapest.

[8] Raisanghani P.M., Monahar G.S. & Yadav J.S. (1987). – *Isospora* infection in the Indian camel *Camelus dromedarius*. *Ind. J. Parasitol.*, 11 (1), 93–94.

[9] Sayd S.M.O. & Kawazoe U. (1998). – Experimental infection of swine by *Isospora suis* Biester 1934 for species confirmation. *Mem. Inst. Oswaldo Cruz*, 93 (6), 851–854.

[10] Tsygankov A.A. (1950). – Contribution on the revision of coccidian fauna of camels [in Russian]. *Izv. AN Kaz. SSR.*, 8, 174–185.

[11] Younan M., McDonough S.P., Herbert D., Saez J. & Kibor A. (2002). – *Isospora* excretion in scouring camel calves (*Camelus dromedarius*). *Vet. Rec.*, 151, 548–549.

## 5.1.8 弓形虫病

弓形虫病是由刚地弓形虫（*Toxoplasma gondii*）引起的一种人兽共患病。该病病原体于1908年在摩洛哥的刚地（gondi）发现，20世纪后50年，对猫在弓形虫生活史中的作用才有了全面的了解[7,17]。

### 形态学

弓形虫的发育经历3种具有诊断意义的阶段：在其中间宿主中有速殖子和缓殖子两种形态，在终末宿主猫体内主要的形态是卵囊。速殖子在细胞内快速繁殖，典型形态是半月形，大小为(4～7)μm×(2～4)μm，速殖子在宿主体外迅速死亡，经胃传代不能存活。缓殖子与速殖子形态相似，但比速殖子小，在各种宿主组织细胞的包囊内进行胞内生殖；这些薄壳型的组织包囊大小为150μm，不能继续在组织内进行分裂；未孢子化的卵囊形状呈圆形，直径为9～14μm，随猫的粪便排出后在宿主体外形成孢子（图236）。等孢球虫样的孢子化卵囊含有两个孢子囊，每个孢子囊含有四个子孢子。

### 生活史

刚地弓形虫病的生活史比较复杂，家猫和其他猫科动物均是弓形虫的终末宿主，因捕食中间宿主含缓殖子或包囊的组织而感染；感染的缓殖子在小肠绒毛的上皮细胞内分裂成五种不同类型的裂殖体，在感染后第3天，进行两性生殖，这一过程比较短，仅持续3~10天；其中的一些缓殖子进入终末宿主的其他组织器官，且依照孢内生殖的方式进行细胞内的无性繁殖。

假若猫感染了大量的孢子化卵囊，其潜伏期较长，持续18～36天。因为卵囊最初在肠道外进行发育，然后在小肠内进行两性繁殖。

当中间宿主感染了含有弓形虫的包囊或包囊

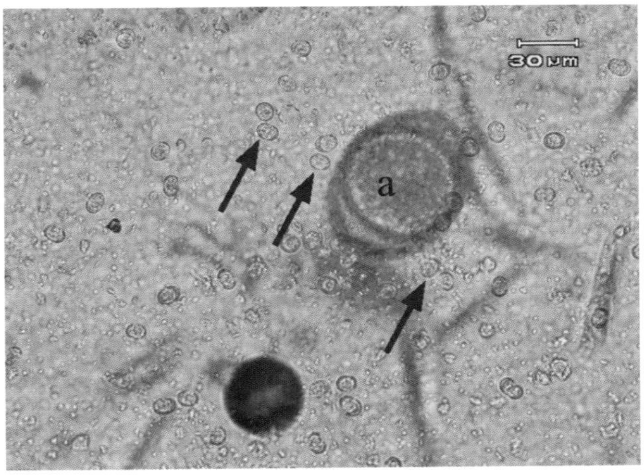


图236　猫的排泄物中的大量弓形虫卵囊（箭头所示）和一个上唇弓首蛔虫虫卵(a)，刚地弓形虫和哈氏哈芒球虫的卵囊在形态上无法区分

组织，子孢子或裂殖子进入肠系膜淋巴结、肝脏、肺脏和骨骼肌组织后，在有核细胞内以二分裂方式增殖；孢内生殖产生的速殖子进入宿主细胞，形成假包囊；当假包囊破裂后，速殖子便释放出来，通过血液再入侵其他组织器官，在组织器官中以缓殖子的形式进行分裂增殖，最终形成包囊。

当免疫力低的中间宿主或终末宿主在怀孕期间感染了弓形虫后，可经胎盘传给胎儿，使其后代发生先天性感染。

## 流行病学

自然条件下，弓形虫病的感染方式是中间宿主食入含有孢子化卵囊或组织包囊，通过垂直传播的方式经母体胎盘传播给胎儿。有一些宿主，像单峰驼，可以通过泌乳的过程感染弓形虫病[18]。

弓形虫病呈世界性分布，猫科动物（主要是家猫）是唯一的终末宿主。以作者的经验，即使猫的血清学阳性率很高，但只有个别猫排出卵囊[33]。感染的猫在很短的潜伏期可排出数亿卵囊，污染环境，在阴暗潮湿的环境中这些卵囊可以存活数月。猫可产生一定的抵抗力，但在应激情况下肠外的发育阶段被激活，随后进行肠内发育，产生卵囊。

弓形虫对中间宿主的选择没有特异性，所有的温血动物宿主都可能感染。啮齿类动物是一种常见的感染源。

骆驼食入含有弓形虫的卵囊后感染弓形虫。雌性骆驼通过哺乳将弓形虫传给后代[18]。

已在埃及[9,15,23,24,29,30,36]、苏丹[1,4,10,11,20,22]、尼日利亚[26]、沙特阿拉伯[16]、阿拉伯联合酋长国[2,37]、伊朗[31]、土库曼斯坦[3]和阿富汗[21]等国使用不同的血清学检测方法，调查单峰驼弓形虫病的流行情况，其阳性率在0～85%。目前尚无弓形虫在双峰驼体内的报道。在秘鲁[5,6,12,28,31,32,38]、智利[13,27,38]、阿根廷[23]、美国[7]和德国[38]等使用血清学检测方法，调查新大陆骆驼弓形虫病的流行情况，结果表明，这些国家弓形虫的流行率较低。在其他大多数中间宿主体，血清学调查发现，弓形虫病的感染率随着年龄的增加而增加[5,12,27,28,32]。

## 临床症状和病理变化

弓形虫病可引起绵羊和山羊的流产和消瘦，驼科动物感染弓形虫病没有明显的临床症状，然而已证实血清学阳性母驼所产的幼崽更易表现出腹泻的症状[22]。新大陆骆驼科或美洲驼均没有弓形虫病感染的临床症状。怀孕的美洲驼注射弓形虫病的卵囊后不表现出临床症状，其后代也表现健康。尽管在母体已检测到抗体上升，但分娩出的幼畜在采食初乳前检测不到抗体[19]。通过对流产胎盘和美洲驼胎儿体内是否存在弓形虫DNA的检测结果表明，在美洲驼体内，弓形虫病不能通过胎盘传播。对50份胎儿DNA样品的检测中发现，有19份样品中发现有原虫性损伤，但未能检测到弓形虫的DNA。对这些样品进行PCR扩增时，有9份是新孢子虫阳性样品[34]。

弓形虫感染驼科动物的主要意义在于动物的肉质中含有组织囊，如果人食用未煮熟的肉，可引起弓形虫的感染[14,35]。

## 诊断

常用的诊断弓形虫的方法是检测中间宿主的抗体，几种检测方法可以联合使用（赛宾-费德曼Sabin-Feldman 染色试验、补体结合试验、间接血球凝集试验、改进的凝集试验、乳胶凝集试验和免疫印迹试验）。血清抗体检测呈阳性的数量一方面随检测方法的不同而变化，另一方面与检测系统中所使用的抗原有关。使用不同株的弓形虫抗原，用凝胶实验检测150份单峰驼血液样品，结果对比发现，阳性率在18%～30.7%[36]。

对于组织中是否存在弓形虫包囊，可以用组织病理学或生物分析法来确定，或给干净的猫饲喂可疑的肉来证实。在饲喂后9～11天，猫排泄出弓形虫的卵囊[14]，这些卵囊在形态上很难和哈氏哈芒球虫（*Hammondia hammondi*）区分开来。

## 防控

虽然驼科动物感染弓形虫没有明显的临床症状，但在肌肉中含有组织包囊，因此人食用含有组织包囊的肉便可感染。对骆驼弓形虫的感染，目前尚无有效的预防措施；唯一有效的途径是控制其感染源流浪猫和弓形虫的贮藏宿主啮齿动物的数量，便可有效地降低弓形虫的感染。

# 参考文献


[1] Abbas B., El Zubeir A.E.A. & Yasin T.T.M. (1987). – Survey for certain zoonotic diseases in camels in Sudan. *Rev. Elev. Méd.Vét. Pays Trop.*, 40, 231 – 233.

[2] Afzal M. & Sakkier M. (1994). – Survey of antibodies against various infectious disease agents in racing camels in AbuDhabi, United Arab Emirates. *Rev. sci. tech. Off. int. Epiz.*, 13 (3), 787 – 792.

[3] Berdyev A.S. (1972). – Present position of toxoplasmosis in Turkmenia [in Russian]. *Izv. Akad. Nauk Turkm. SSR. Ser. Biol. Nauk*, (6), 46 – 51.

[4] Bornstein S. & Musa B.E. (1987). – Prevalence of antibodies to some viral pathogens, *Brucella abortus* and *Toxoplasma gondii* in serum from camels (*Camelus dromedarius*) in Sudan. *J. Vet. B*, 34, 364 – 370.

[5] Chang K.H., Chavez A.V., Li O.E., Falcon N.P., Casas E.A. & Casas G.V. (2009). – Seroprevalencia de *Toxoplasma gondii* enllamas hembras de la Sierra Central del Peru. *Rev. Inv. Vet. Perú*, 20 (2), 306 – 311.

[6] Chavez-Velasquez A., Alvarez-Garcia G., Gomez-Bautista M., Casas-Astor E., Serrano-Martinez E. & Ortega-Mora L.M.(2005). –*Toxoplasma gondii* infection in adult llamas (*Lama glama*) and vicunˇ as (*Vicugna vicugna*) in the Peruvian Andeanregion. *Vet. Parasitol.*, 130, 93 – 97.

[7] Dubey J.P. (2011). – The history and life cycle of *Toxoplasma gondii. In Toxoplama gondii* – The model apicomplean: Perspectivesand methods (L.M. Weiss & K. Kim, eds). Elsevier, London, 1 – 19.

[8] Dubey J.P, Rickard L.G., Zimmerman G.L. & Mulrooney D.M. (1992). – Seroprevalence of *Toxoplasma gondii* in llamas (*Lamaglama*) in the northwest USA. *Vet. Parasitol.*, 44, 295 – 298.

[9] El Ridi A.M.S., Nada S.S.M., Aly A.S., Habeeb S.M. & Abdul Fattah M.M. (1990). – Serological studies on toxoplasmosisZagazig slaughterhouse. *J. Egypt. Soc. Parasitol.*, 20, 677 – 681.

[10] Elamin E.A., Elias S., Daugschis A & Rommel M. (1992). – Prevalence of *Toxoplasma gondii antibodies* in pastoral camels(*Camelus dromedarius*) in the Butana plains, mid eastern Sudan. *Vet. Parasitol.*, 43, 171 – 175.

[11] Eldin E.A.Z, El Khawad S.E., Kheir H.S.M. & Zain E.A. (1985). – A serological survey for *Toxoplasma* antibodies in cattle,sheep, goats and camels (*Camelus dromedarius*) in Sudan. *Rev. Elev. Méd. Vét. Pays Trop.*, 38, 247 – 249.

[12] Gomez F., Chavez A.V., Casas E.A., Serrano E.M. & Cardenas O. (2003). – Determinacion de la seroprevalencia detoxoplasmosis en alpacas y llamas en la estacion experimental Ina-Puno. *Rev. Inv. Vet. Peru*, 14 (1), 49 – 53.

[13] Gorman T., Arancibia J.P., Lorca M., Hird D. & Alcaino H. (1999). – Seroprevalence of *Toxoplasma gondii* infection in sheepand alpacas (*Llama pacos*) in Chile. *Prev. Vet. Med.*, 40, 143 – 149.

[14] Hilali M., Fatani A. & Al-Atiya S. (1995). – Isolation of *Toxoplasma, Isospora, Hammondia* and *Sarcocystis* from camel (*Camelusdromedarius*) meat in Saudi Arabia. *Vet. Parasitol.*, 58, 353 – 356.

[15] Hilali M., Romand S., Thulliez P., Kwok O.C.H. & Dubey J.P. (1998). – Prevalence of *Neospora caninum* and *Toxoplasmagondii* antibodies in sera from camels from Egypt. *Vet. Parasitol.*, 75, 269 – 271.

[16] Hussein M.F., Bakkar M.N., Basmaeil S.M. & Garelnabi A.R. (1988). – Prevalence of toxoplasmosis in Saudi Arabian camels(*Camelus dromedarius*). *Vet. Parasitol.*, 28, 175 – 178.

[17] Hutchison W.M, Dunachie J.F., Siim C. & Work K. (1969). – Life cycle of *Toxoplasma gondii. Br. Med. J.*, 4, 806.

[18] Ishag M.Y., Magzoub E. & Majid M. (2006). – Detection of *Toxoplasma gondii* tachyzoites in the milk of experimentallyinfected lactating she-camels. *J. Animal Vet. Adv.*, 5 (6), 456 – 458.

[19] Jarvinen J.A., Dubey J.P & Althouse G.C. (1999). – Clinical and serological evaluation of two llamas (*Lama glama*) infectedwith *Toxoplasma gondii* during gestation. *J. Parasitol.*, 85 (1), 142 – 144.

[20] Khalil K.M., Gadir A.E.A., Rahman M.M.A, Yassir O.M., Ahmed A.A. & Elrayah I.E. (2007). – Prevalence of *Toxoplasmagondii* antibodies in camels and their herders in three ecologically different areas in Sudan. *J. Camel Pract. Res.*, 14 (1), 11 – 13.

[21] Kozoed V.K., Blazek V. & Amin A. (1976). – Incidence of toxoplasmosis in domestic animals in Afghanistan. *Folia Parasitol.*,23, 273 – 275.

[22] Manal Y.I. & Maijd A.M. (2008). – Association of diarrhea with congenital toxoplasmosis in calf-camels (*Camelus dromedarius*).*Intern. J. Trop. Med.*, 3 (1), 10 – 11.

[23] Maronpot R.R. & Botros B.A.M. (1972). –*Toxoplasma* serologic survey in man and domestic animals in Egypt. *J. Egypt. Pub.Health Assoc.*, 47, 58 – 67.

[24] Michael S.A., El-Refai A.H. & Morsy T.A. (1977). – Incidence of *Toxoplasma* antibodies among camels in Egypt. *J. Egypt. Soc. Parasitol.*, 7, 129 – 132.

[25] More G., Pardini L., Basso W., Marin R., Bacigalupe D., Auad G., Venturini L. & Venturini M.C. (2008). – Seroprevalence of *Neospora caninum, Toxoplasma gondii* and *Sarcocystis* sp. in llamas (*Lama glama*) from Jujuy, Argentina. *Vet. Parasitol.*, 155,158 – 160.

[26] Okoh A.E.J., Agbonlahor D.E. & Momoh M. (1981). – Toxoplasmosis in Nigeria – a serological survey. *Trop. Anim. HealthProd.*, 13, 137 – 140.

[27] Patitucci A.N., Perez M.J., Barril G., Carcamo C.M. & Muno A. (2006). – Deteccion de anticuerpos sericos contra *Toxoplasmagondii* (Nicolle y Manceaux, 1909) enllamas (*Lama glama* Linneaus, 1758) y alpacas (*Lama pacos* Linneaus, 1758) de Chile*Arch. Med. Vet.*, 38 (2), 179 – 182.

[28] Ramirez J.R., Chavez A.V., Casas E.A. & Rosadio R.A. (2005). – Seroprevalencia de *Toxoplasma gondii* en alpacas decomunidades de la provincia de Canchis, Cusco. *Rev. Inv. Vet. Perú*, 6 (2), 169 – 174.

[29] Rifaat M.A., Morsy T.A., Sadek M.S.M., Khalid M.E., Azab M.E., Makled M.K., Safar E.H. & El-Din. O.M.N. (1977). – Incidenceof toxoplasmosis among farm animals in Suez Canal Governorates. *J. Egypt. Soc. Parasitol.*, 7, 135 – 340.

[30] Rifaat M.A., Morsy T.A., Sadek M.S.M., Khalid M.E., Azab M.E. & Safar E.H. (1978). – Prevalence of *Toxoplasma* antibodiesamong slaughtered animals in Lower Egypt. *J. Egypt. Soc. Parasitol.*, 8, 339 – 345.

[31] Sadrebazzaz A., Haddadzadeh H. & Shayan P. (2006). – Seroprevalence of *Neospora caninum* and *Toxoplasma gondii* in camels(*Camelus dromedarius*) in Mashrad, Iran. *Parasitol. Res.*, 98, 600 – 601.

[32] Saravia M.P., Chavez A.V., Casas E.A., Falcon N.P. & Pinto W.S. (2004). – Seroprevalencia de *Toxoplasma gondii* en llamasde una empresa pecuaria en Melgar, Puno. *Rev. Inv. Vet. Peru*, 15 (1), 49 – 55.

[33] Schuster R.K., Thomas K., Sivakumat S. & O'Donovan D. (2009). – The parasite fauna of stray domestic cats (*Felis catus*)in Dubai, United Arab Emirates. *Parasitol. Res.*, 105, 125 – 134.

[34] Serrano-Martinez E., Collantes-Fernandez E., Chavez-Velasquez A., Rodriguez-Bertos A., Casas-Astos E., Risco-Castillo V.,Rosadio-Alcantarra R. & Ortega-Mora L.M. (2007). – Evaluation of *Neospora caninum* and *Toxoplasma gondii* in alpaca (*Vicugnapacos*) and llama (*Lama glama*) aborted foetuses from Peru. *Vet. Parasitol.*, 150, 39 – 45.

[35] Shaapan R.M. & Fathia A.M.K. (2005). – Isolation of *Toxoplasma gondii* from camel meat in Egypt. *J. Egypt. Vet. Med. Assoc.*,65, 187 – 195.

[36] Shaapan R.M. & Khalil F. (2008). – Evaluation of different *Toxoplasma gondii* isolates as antigens used in the modifiedagglutination test for the detection of toxoplasmosis in camels and donkeys. *Am. Eurasian J. agric. environ. Sci.*, 3 (6), 837 – 841.

[37] Wernery U., Thomas R., Syriac G., Raghavan R. & Kletzka S. (2007). – Seroepidemiological studies for the detection ofantibodies against nine infectious disease in dairy dromedaries (Part I). *J. Camel Pract. Res.*, 14 (2), 85 – 90.

[38] Wolf D., Gauly M., Huanca W., Cardenas O., Bauer C. & Schares G. (2004). – Seroprevalence of *Neospora caninum* and *Toxoplasma gondii* in South American camelids. *In* South American camelids research (M. Gerken & C. Renieri, eds). Wageningen Academic Publishers, the Netherlands, 285 – 286.

（刘爱红译，刘光远、殷宏校）



## 5.1.9 新孢子虫病

犬新孢子虫（*Neospora caninum*）是一种引起动物感染的原虫。1984年首次在挪威的狗身上发现该病原，但直到1988年前，一直错误地认为是刚地弓形虫[2]。事实上，犬新孢子虫与之前了解的海氏哈芒球虫（*Hammondia heydorni*）非常相似（参见5.1.10 哈芒球虫病）。怀孕牛流产的一个主要原因是由犬新孢子虫引起的，因此研究犬新孢子虫已迫在眉睫。

## 形态学

犬新孢子虫的终末宿主是犬和其他犬科动物，他们排出未孢子化的卵囊，大小为(10～13)μm×(10～11)μm（图237）。孢子化卵囊含2个孢子囊，每个孢子囊含4个子孢子。速殖子和卵囊寄生于中间宿主和终末宿主体内，速殖子呈镰刀形，大小为（3～8）μm ×5 μm；组织包囊呈圆形到卵圆形，壁厚，大小为110μm，不能再分裂（图238）。在中枢神经系统和肌肉中可以观察到这些组织包囊，内含很多镰刀型的缓殖子，大小为(7～8)μm×2μm。

## 生活史

犬新孢子虫的生活史和刚地弓形虫病的生活史很相似，不同之处是犬为终末宿主。犬食入含有感染性包囊的中间宿主的组织或通过怀孕母畜的胎盘均能感染犬新孢子虫。在实验室，犬食入含有新孢子虫组织包囊的牛肉将感染犬新孢子虫，感染的潜伏期为5～10天，显露期为3～19天，在这一时期，有5 700~503 300个卵囊排出体外[4]。终末宿主通过食入孢子化卵囊而感染犬新孢子虫尚无充足的证据。

中间宿主草食动物通过食入孢子化卵囊而感染，牛食入约300个孢子化卵囊即可感染新孢子虫病。当母牛在怀孕期间，潜伏的感染被激活后，新孢子虫病可引起牛的垂直传播，且这种传播方式可以延续好几代[11]。

## 流行病学

新孢子虫病分布范围广，犬新孢子虫的宿主广泛，包括终末宿主犬科类动物、中间宿主牛和其他家畜以及一些野生动物。新孢子虫也可能是一种人兽共患寄生虫，但人是否感染新孢子虫目前尚未得到证实[3]。

犬排出新孢子虫卵囊的概率很低，因此，Schraes证实，在24 089只被检测的犬中，只有47只排出大小为9～14μm的卵囊。另外，用酶联免疫吸附试验（ELISA）检测了257份犬血清样品，检测结果显示，22%的样品呈犬新孢子虫抗体阳性[5]。这些数据表明，与猫感染弓形虫相比，犬排出卵囊持续时间很短。

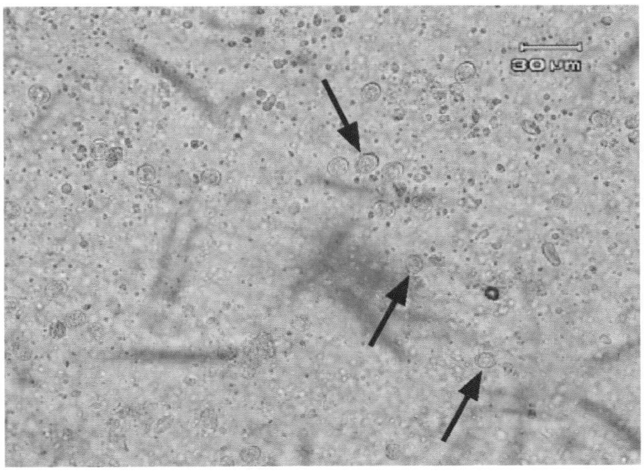


图237　犬排泄物中未孢子化的卵囊

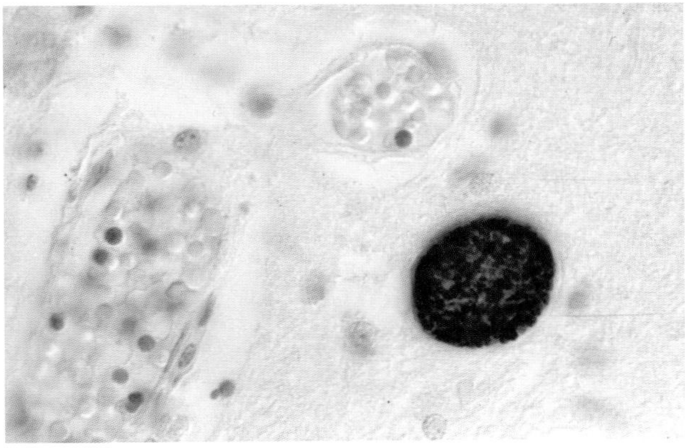

图238　犊牛脑部的新孢子虫孢囊的免疫染色照片（照片由德国Martin Perter 博士惠赠）

在埃及[6]和伊朗[7,9]，用凝集试验和间接荧光抗体（IFAT）两种方法检测单峰驼体内犬新孢子虫的感染情况，结果表明，抗体阳性率在3.2%～4.8%。Wernery等人证实[13]，在阿拉伯联合酋长国的奶驼场的单峰驼体内，犬新孢子虫的阳性率是13.7%。这些被检测的578峰母驼和541峰幼驼中的大多数都来自苏丹，令人值得注意的是，母驼和幼驼样品的血清阳性率没有明显差异，这些证据使得犬新孢子虫垂直传播的可能性得到有力支持。

在秘鲁，在流产的羊驼胎盘和美洲驼的胎儿体内首次发现新大陆骆驼的新孢子虫。对15个胎盘通过免疫组化法和聚合酶链反应检测证实，其中4个有由犬新孢子虫引起的脑组织损伤[12]。同样在秘鲁，用间接荧光抗体试验（IFAT）检测成年羊驼和美洲驼的新孢子虫，阳性率分别为35.9%和31.5%[1]。在阿根廷，美洲驼犬新孢子虫血清学阳性的为4.6%，而刚地弓形虫为30%[8]。然而，由于新孢子虫和刚地弓形虫病抗体有交叉反应而易出现假阳性，所以须认真分析间接荧光抗体的检测结果[14]。选用特异性好的免疫印迹试验，检测结果表明，在秘鲁的羊驼和美洲驼体内，血清学阳性率分别在0.6%和3.1%，但检测的驼马全为阴性。在德国的汉斯（Hesse），同样用免疫印迹实验检测羊驼和美洲驼，结果未能检测到犬新孢子虫，而弓形虫的阳性率则达到75%[14]。

### 临床症状和病理变化

新孢子虫的致病机理和临床症状在牛体内研究得比较详细，该病对成年动物的感染不明显。新孢子虫病在奶牛的主要表现是流产、死胎和不育。犬新孢子虫引起的胎盘、胎儿组织坏死和炎症反应是流产的主要原因。

新大陆骆驼的流产胎儿损伤主要在脑部，表现为非化脓性多病灶脑炎和脑膜脑炎。在心脏是局部和散在的化脓性心肌炎和钙化组织的病灶，长期的肝脏病变使得肝细胞坏死，在肺脏、肾脏和肾上腺可见非化脓性的浸润物。

### 诊断

在脑部可看到不能再分裂成小包囊的典型厚壁包囊和速殖子，用免疫组化方法可以鉴别这两种组织。检测新孢子虫病抗体常见的几种血清学方法是间接荧光抗体试验、酶联免疫吸附试验、凝集试验和免疫杂交技术。

### 防制

目前尚无有效的方法来预防新孢子虫病。通过消灭流浪的家养狗，尽量避免中间宿主与犬科动物的接触，从而降低中间宿主的感染机会。同时，为了防止其他寄生虫病的发生，不用新鲜的肉或动物内脏喂养牧羊犬，煮沸可以杀死寄生虫的卵囊。

## 参考文献

[1] Chavez-Velasquez A., Alverez-Garcia G., Collantez-Fernandez E., Casas-Astos E., Rosadio-Alcantara R., Serrano-Martinez E.& Ortega-Mora L.M. (2004). – First report of *Neospora caninum* in adult alpacas (*Vicugna pacos*) and llamas (*Lama glama*). *J. Parasitol.*, 90, 864 – 866.

[2] Dubey J.P., Carpenter J.L., Speer C.A., Tropper M.J. & Uggla A. (1988). – Newly recognized fatal protozoan disease of dogs. *J. Am. Vet. Med. Assoc.*, 192, 1 269 – 1 285.

[3] Dubey J.P., Schares G. & Ortega-Mora L.M. (2007). – Epidemiology and control of neosporosis and *Neospora caninum. Clin. Microbiol. Rev.*, 20 (2), 323 – 367.

[4] Gondim L.F.P., Gao L. & McAllister M.M. (2002). – Improved production of *Neospora caninum* oocysts, cyclical oral transmission between dogs and cattle, and in vitro isolation from oocysts. J. *Parasitol.*, 88, 1 159 – 1 163.

[5] Gozdzik K., Wrzesien R., Wielgosz-Ostolska A., Bien J, Kozak-Ljunggren M. & Cabaj W. (2011). – Prevalence of antibodies against Neospora caninum in dogs from urban areas in Central Poland. *Parasitol. Res.*, 108, 991 – 996.

[6] Hilali M., Romand S., Thulliez P., Kwok O.C.H. & Dubey J.P. (1998). – Prevalence of *Neospora caninum* and *Toxoplasma gondii* antibodies in sera from camels from Egypt. *Vet. Parasitol.*, 75, 269 – 271.

[7] Hosseininejad M., Pirali-Kheirabadi K. & Hosseini F. (2009). – Seroprevalence of *Neospora caninum* infection in camels (*Camelus dromedarius*) in Isfahan province, center of Iran. *Iranian J. Parasitol.*, 4 (4), 61 – 64.

[8] More G., Pardini L., Basso W., Marin R., Bacigalupe D., Auad G., Venturini L. & Venturini M.C. (2008). Seroprevalence of *Neospora caninum, Toxoplasma gondii and Sarcocystis sp. in llamas* (*Lama glama*) *from Jujuy, Argentina. Vet. Parasitol.*, 155, 158 – 160.

[9] Sadrebazzaz A., Haddadzadeh H. & Shayan P. (2006). – Seroplevalence of *Neospora caninum* and *Toxoplasma gondii* in camels (*Camelus dromedarius*) in Mashrad, Iran. *Parasitol. Res*., 98, 600 – 601.

[10] Schares G., Pantchew N., Barutzki D., Heydorn A.O., Bauer C. & Conraths F.J. (2005). – Oocysts of *Neospora caninum, Hammondia heydorni, Toxoplasma gondii* and *Hammondia hammondi* in faeces collected from dogs in Germany. *Int. J. Parasitol*., 35, 1 525 – 1 537.

[11] Schares G., Peters M. & Wurm R. (1998). – The efficiency of vertical transmission of *Neospora caninum* in dairy cattle analysed by serological techniques. *Vet. Parasitol*., 80, 87 – 98.

[12] Serrano-Martinez E., Collantes-Fernandez E., Rodriguez-Bertos A., Casas-Astos E., Alvarez-Garcia G., Chavez-Velasquez A. & Ortega-Mora L.M. (2004). – *Neospora* species associated abortion in alpacas (*Vicugna pacos*) and llamas (*Lama glama*). *Vet. Rec*., 155, 748 – 749.

[13] Wernery U., Thomas R., Raghavan R., Syriac G., Joseph S. & Georgy N. (2008). – Seroepidemiological studies for the detection of antibodies against 8 infectious diseases in dairy dromedaries of the United Arab Emirates using modern laboratory techniques – part II. *J. Camel Pract. Res*., 15, 139 – 145.

[14] Wolf D., Schares G., Cardenas O., Huanca W., Cordero A., Baerwald A., Conraths F.J., Gauli M., Zahner H. & Bauer C. (2005). – Detection of specific antibodies to *Neospora caninum* and *Toxoplasma gondii* in naturally infected alpacas (*Lama pacos*), llamas (*Lama glama*) and vicunas (*Lama vicugna*) from Peru and Germany. *Vet. Parasitol*., 130, 81 – 87.

（刘爱红译，刘光远、般宏校）

## 5.1.10 哈芒球虫病

在进行与肉孢子虫生活史相关的生物学分析研究中，用牛肉饲喂的犬排出没有孢子化的双芽等孢球虫卵囊[3]。将孢子化的卵囊感染牛后，成功地完成了该寄生虫完整生活史[4]。进一步研究表明，虫体在犬体内的发育过程中，仅存在于肠道上皮样细胞内。由于猫源刚地弓形虫（*Toxoplasma gondii*）曾被认为是等孢子球虫，并使用过双芽等孢球虫（*Isospora bigemina*）的名称，为与之区别，根据以下事实可以将此虫体命名为赫氏等孢球虫[8]。

（1）犬是这种寄生虫的唯一终末宿主。

（2）卵囊可以感染犬、羊、山羊和牛，不能感染猫，兔和小鼠。

（3）在组织切片处没找到组织内发育阶段。

（4）卵囊没有致病性。

（5）在犬体内产生的卵囊较少。

此后，赫氏等孢球虫被归类为新命名的种属，哈芒球虫(*Hammondia*）[1]。由于犬新孢子球虫与哈芒球虫的生活史和发育阶段的形态学完全相同，分子特征几乎相同，可以肯定地认为，前者是后者的次定同物异名关系，或者是同一个物种不同分离株[2]。

首次关于单峰驼哈芒球虫的报道是来自埃及。用骆驼肉饲喂犬，在7~11天的潜伏期后，持续5~7天排出哈芒球虫卵囊。同一时期，在苏丹进行了生物学测定实验，使用含有肉孢子球虫包囊的骆驼肉饲喂犬，在7~14天的潜伏期后可以在犬的粪便中发现小的、尚未孢子化的哈芒球虫卵囊。在实验后期，犬的粪便中发现了孢子化的肉孢子虫孢子囊。获自犬的未孢子化的哈芒球虫卵囊呈亚圆形，大小(8.3~11.8)μm × (10.7~14)μm[5,6]。在宿主体外完成孢子化过程（图239）。

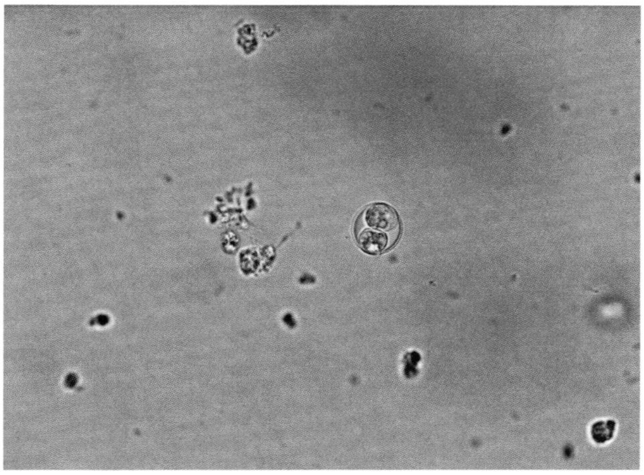

图239 犬粪便中发现的孢子化的弓形虫样卵囊。每个卵囊中含有两个子孢子，每个子孢子中分布着四个孢子

## 参考文献

[1] Dubey J.P. (1977). – *Toxoplasma, Hammondia, Besnoitia, Sarcocystis* and other cyst-forming coccidian of man and animals. *In* Parasitic protozoa, Vol 3 (J.P. Kreier, ed.). Academic Press, New York, 171 – 175.

[2] Heydorn A.O. & Mehlhorn H. (2002). – *Neospora caninum* is an invalid species name: an evaluation of facts and statements. *Parasitol. Res.*, 88, 175 – 184.

[3] Heydorn A.O. & Rommel M. (1972). – Beitrage zum Lebenszyklus der Sarcosporidien. II. Hund und Katze als Uebertrager der Sarcosporidien des Rindes. *Berl. Münch. Tierärztl. Wschr.*, 85, 121 – 123.

[4] Heydorn A.O. (1973). – Zum Lebenszyklus der kleinen Form von *Isospora bigemina* des Hundes. I. Rind und Hund als mogliche Zwischenwirte. *Berl. Münch. Tierärztl. Wschr.*, 86, 323 – 329.

[5] Hilali M., Fatani A. & Al-Atiya S. (1995). – Isolation of tissue cysts of *Toxoplasma, Isospora, Hammondia* and *Sarcocystis* from camel (*Camelus dromedarius*) meat in Saudi Arabia. *Vet. Parasitol.*, 58, 353 – 356.

[6] Hilali M., Nassar A.M. & El-Ghaysh A. (1992). – Camel (*Camelus dromedarius*) and sheep (*Ovis aries*) meat as source of dog infection with some coccidian parasites. *Vet. Parasitol.*, 43, 37 – 43.

[7] Nassar A.M., Hilali M. & Rommel M. (1983). – *Hammondia heydorni* infection in camels (*Camelus dromedarius*) and water buffaloes (*Bubalus bubalis*) in Egypt. *Z. Parasitenk.*, 69, 693 – 694.

[8] Tandros W. & Laarman J.J. (1976). Sarcocystis and related coccidian parasites: a brief general review, together with discussion on some biological aspects of their life cycles and a new proposal for their classification. *Acta Leiden*, 44, 1 107.

[9] Warrag M. & Hussein H. (1983). – The camel (*Camelus dromedarius*) as an intermediate host for *Hammondia heydorni*. *J. Parasitol.*, 69 (5), 926 – 929.

（刘军龙译，刘光远、殷宏校）

## 5.1.11 肉孢子虫病

肉孢子虫病是由肉孢子虫属（*Sarcocystis*）的各种能形成包囊的肉孢子虫引起的。肉孢子虫为二宿主的生活史，以草食动物为中间宿主，肉食动物为终末宿主。Mason[21]于1910年首次在埃及记载了骆驼的心肌层和食道发现了肉孢子虫，并把它命名为骆驼肉孢子虫（*S. cameli*）。该作者观察到薄壁和厚壁两种微囊，并认为其均属于骆驼肉孢子虫。在一些亚洲和非洲国家的单峰驼也报道过肉孢子虫，在美洲的骆驼中也可发现巨型和微小包囊的肉孢子虫。

### 形态学

根据Fatani等人[7]的研究，在单峰驼中薄壁的包囊大小为(141~400)μm × (70.5~188)μm，囊壁厚度在0.75~1μm。包囊腔被源自包囊壁的骨小梁分隔成若干部分。薄壁的肉孢子虫包囊可在各种肌肉组织包括食道中找到，而有厚壁的肉孢子虫包囊只存在于食道内。厚壁包囊大小为(170~194) μm × (118~188)μm，被1.5~2 μm的厚壁包裹；厚壁包囊的囊腔内没有分隔区。这两种类型的包囊都是充满了香蕉状的裂殖体[大小(14~17)μm × (3~5) μm]。

第一次饲喂感染试验中，用骆驼肉饲喂无寄生虫感染的幼犬，从犬排出物中可发现卵圆形的肉孢子虫的孢子囊，长度为10.7~14.3 μm，宽度在8.3~11.3 μm，每个卵囊含有四个子孢子和残体[10,11]。

随后通过实验感染犬，发现排出两种大小不同的卵囊，孢子囊大小在16.0μm × (9.9~11.5) μm的是一种新种，并命名为驼犬肉孢子虫（*S. camelocanis*）。另一类型孢子囊大小为(13.2~13.6) μm × (6.5~9.5)μm，命名为骆驼肉孢子虫(*S. ca meli*)[13]。

直到最近，才有学者对肉孢子虫的另外一种，类驼犬肉孢子虫（*S. camelicanis*）进行了详细的描述，其卵囊大小范围在(13.7~15.6) μm × (7.8~10.7)μm[1]。

由于在实验中既饲喂了骆驼肉也饲喂了骆驼食管道，所以不能弄清楚哪一个种产生薄壁包囊，哪一种形成厚壁包囊。虽然近期进行了超微结构和分子生物学特征的研究，但对于骆驼肉孢子虫（*S. cameli*）种的划分还存在疑问[24]。

Kuraev[16]报道了哈萨克斯坦骆驼的肉孢子虫病例，根据描述，组织期的孢子囊为细长椭圆形，长度为5~15μm，肉眼可见，两端为圆形。在饲喂实验中，犬排出的孢子囊大小为14 μm × 11μm。

在美洲的骆驼中已鉴定的3种肉孢子虫为原驼犬肉孢子虫（*S. tilopodi*）、羊驼肉孢子虫（*S. aucheniae*）和大羊驼驼犬肉孢子虫（*S. lamacanis*），其中只对羊驼肉孢子虫的生活史有详细的研究，其终末宿主是犬[25]。

## 生活史

骆驼科动物作为中间宿主，当摄食了感染性孢子囊污染的食物或水后即会被感染。子孢子侵入到肠壁内和血管内皮细胞并在此经历两个无性繁殖（孢内生殖）周期，第二代的裂殖子进入肌肉细胞并形成被虫样空泡包裹的包囊。虫体在宿主细胞内以多元内出芽方式进行多次繁殖，即可产生香蕉形裂殖子。随后包囊壁逐渐变厚，形成了具有种属特性的表面结构，隔膜将包囊分成若干室区，里面充满了感染性的缓殖子。

犬是骆驼科肉孢子虫的终末宿主，通过食用生的骆驼肉而感染。缓殖子从小肠释放出来后进入固有层细胞并形成配子母细胞。包囊的孢子化过程是在终末宿主的肠内完成的。薄壁的包囊通常在肠道破裂，孢子囊随着粪便排出体外，开始体外发育阶段。潜伏期可以持续两周，孢子囊的排出过程可持续2个月（表83）。

## 流行病学

感染途径为经口感染，犬在肉孢子虫的生活史中是一个必需的环节，骆驼通过摄取孢子囊污染的食物或饮水而被感染。肉孢子虫具有特异性中间宿主和终末宿主。

肉孢子虫病在埃塞俄比亚、埃及、索马里、苏丹、沙特阿拉伯、伊拉克、伊朗和阿富汗（表84）广泛流行，均在单峰驼中发现其肌肉包囊。此外在哈萨克斯坦也有双峰驼感染肉孢子虫的报道。Kuraev[16]检测了单峰驼和双峰驼共266头尸体，其中84份发现了肉孢子虫。在其报告中未说明骆驼的具体品种。Fukuyo等人报道在蒙古发现了4例野生双峰驼感染肉孢子虫的病例[8]。

在美洲驼中发现了3种肉孢子虫，其中，原驼犬肉孢子虫寄生在原驼，羊驼肉袍子虫寄生在美洲驼、羊驼、骆马中，大羊驼驼犬肉孢子虫寄生在美洲驼和羊驼中。肉孢子虫病在美洲的流行率很高，在秘鲁的羊驼中感染率达89.7%[6]，在阿根廷的美洲驼中高达96%[23]；根据Javir[14]等人的报道，在秘鲁羊驼因寄生虫致死的病例中，由肉孢子虫病引起的达29%。

## 临床症状和病理变化

肉孢子虫寄生在中间宿主身上时不表现主要临床症状，只有在高感染量的情况下方可引起急性和亚急性肉孢子虫病。由一代、二代的裂殖体引起的疾病可以在早期发现，被称为达尔梅尼病（Dalmeny disease，也称为牛类弓形虫病）[17]。美洲驼的急性症状表现为发热、沉郁、食欲减少、消瘦、共济失调、严重贫血（血球容积计<20%），出现淋巴细胞增多和流产，肌酸磷酸激酶、天冬氨酸转氨酶和山梨醇脱氢酶浓度升高。尸体剖检发现，在内脏浆膜的条纹肌和心肌有大量的出血点和

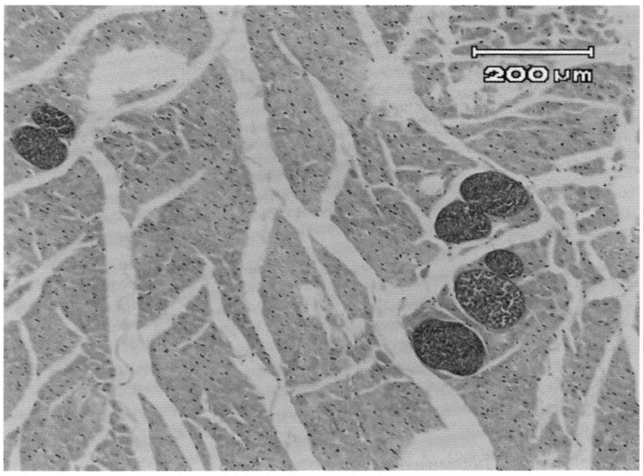


图240　肉孢子虫小囊观察，苏木精-伊红染色的组织切片

表83　骆驼科肉孢子虫的形态学特征

| 种类 | 包囊（μm） | 潜伏期（天） | 显露期（天） | 包囊 | 中间宿主 |
|---|---|---|---|---|---|
| 骆驼肉孢子虫<br>*S. cameli* | 13.2~13.6<br>6.5~9.0 | 9~13 | 55~57 | 微囊 | 单峰驼 |
| 驼犬肉孢子虫<br>*S. camelicanis* | 14.8×9.7 | 11 | 35 | 微囊 | 单峰驼 |
| 类驼犬肉孢子虫<br>*S. camelocanis* | 16.0×(9.9~11.5) | 9~13 | 37~45 | 微囊 | 单峰驼 |
| 羊驼肉孢子虫<br>*S. aucheniae* | 15.0×9.0 | 11 | 21 | 巨囊（不在中心） | 羊驼属，羊驼，骆马 |
| 原驼犬肉孢子虫<br>*S. tilopodi* | 未测到 | 6~16 | 19~61 | 巨囊 | 原驼 |
| 大羊驼犬肉孢子虫<br>*S. lamacanis* | 未测到 | 未测到 | 未测到 | 微囊 | 羊驼 |
| 某一种肉孢子虫<br>*Sarcocystis* sp | 14×11 | 8~10 | 20~25 | 巨囊 | 美洲驼 |

Kuraev[16]没有明确宿主的种类

表84　肉孢子虫包囊流行性及在不同的非洲和亚洲国家的单峰骆驼中的好发部位

| | 流行率（%） | | | | | | | |
|---|---|---|---|---|---|---|---|---|
| 国家 | 数量 | 分布率 | 食管 | 心肌 | 横膈膜 | 舌头 | 条纹肌 | 引用数 |
| 埃塞俄比亚 | 121 | 45.5 | 19.8 | 9.2 | 11.6 | n.i | 12.4 | 28 |
| 埃及 | 180 | 64.0 | n.i | n.i | n.i | n.i | n.i | 1 |
| 索马里 | 200 | 82.5 | n.i | n.i | n.i | n.i | n.i | 4 |
| 苏丹 | 100 | 81.0 | 52.0 | 29.0 | 36.0 | n.i | 42.0 | 12 |
| 沙特阿拉伯 | 103 | 88.4 | 72.6 | 71.8 | 79.6 | n.i | n.i | 7 |
| 沙特阿拉伯 | 624 | 64.0 | 58.0 | 34.0 | 63.0 | 43.0 | 43.0 | 2 |
| 约旦 | 110 | 21,8 | 18.2 | 0.9 | 7.3 | n.i | 4.5 | 19 |
| 伊拉克 | 33 | 91.6 | n.i | n.i | n.i | n.i | n.i | 18 |
| 伊朗 | 400 | 52.3 | 16.8 | 35.3 | 6.5 | 7.8 | 13.8 | 26 |
| 伊朗 | 250 | 83.6 | 58.8 | 48.0 | 41.6 | 28.0 | 46.8 | 27 |
| 阿富汗 | 101 | 66.33 | n.i | n.i | n.i | n.i | n.i | 15 |

n.i：未研究



瘀斑。组织切片观察血管、血管周围间隙和间质细胞渗透出现损伤，坏死主要发生在心脏、骨骼肌和肾脏；大羊驼犬肉孢子虫形成快速生长的微小孢子囊，从而引起心肌大面积出血和坏死；心肌呈深红色，心包和胸腔出现浆液出血性的流体。实验感染和未感染羊驼的心电图扫描参数没有差异[5]。

发生肉孢子虫病时，一般很少见到全身的肌肉发炎，旧大陆骆驼胴体也很容易通过卫生检验。完整的肉孢子虫肌包囊（图240）是良性的，位于肌纤维中，不引起宿主反应[3]。其他种类的肉孢子虫的包囊有溶血毒素和血凝集素，包囊死亡后释放麻蝇素，引起周围肌肉组织的变性，进一步导致炎症和钙化。

和旧大陆骆驼肉孢子虫相反，羊驼肉孢子虫和原驼犬肉孢子虫可在美洲骆驼肉中产生肉眼可见的巨囊，致使肉不再适合食用，在肉检中此类的骆驼肉检验不合格。有证据表明，食用生的或未煮熟的被肉孢子虫感染的羊驼肉，巨囊当中的毒素会导致食用者腹泻、发热和腹痛[20]。Mason和Orr[22]报道了进口到新西兰的雌性羊驼感染了肉孢子虫，但是状态表现良好。尸检显示动物具有肺水肿和心脏苍白并布满斑点，同时有很多长4mm鱼雷样病变遍布骨骼肌系统；组织学发现心脏显著损伤，肉芽肿的广泛存在表明了炎症性心肌坏死和囊肿化，钙化伴随大面积的间质纤维化。这些病变都是以变形的肉孢子囊为中心的。

### 诊断

在肉类中的肉孢子虫包囊可以根据包囊壁的典型结构进行鉴别，羊驼肉孢子虫和原驼犬肉孢子虫的巨型包囊可通过肉眼在尸体残骸或者肉检的时候发现，而其他骆驼肉孢子虫的微包囊只有在组织切片中偶尔发现。有针对性的科学研究表明，在制备的小量鲜肉样品压片中也可以看到微囊。缓殖子的观察通常利用碎肉样品的人工消化法进行，但单靠缓殖子的形态学观察并不能进行种类的鉴定。几例关于微囊在单峰驼寄生的组织偏好性研究表明，微囊更偏好于寄生在食道、心脏、隔膜和舌头部位的肌肉中。

在新大陆骆驼的流行病学研究中，血清学方法（间接荧光抗体试验和酶联免疫吸附试验）已用于肉孢子虫抗体的检测。

对犬的饲喂实验能够获得脱落的卵囊和孢子囊，可以用于种类的鉴定。然而，虫体发育阶段的形态学方面知识，特别是新大陆骆驼感染阶段，还不完全明了(表 83)。

### 疾病控制

避免犬因进食含有肉孢子虫的骆驼肉而被感染，可打断其生活史，肉孢子虫病就可控制。因此，死亡动物的尸体和没收的屠宰场的肉类必须进行适当的处理。蒸煮和低温冷冻能够杀死肉孢子虫的肌肉包囊。流浪犬应进行适宜的处理，进而阻断其对疾病的传播。

## 参考文献


[1] Abdel-Ghaffar F., Mehlhorn H., Bashtar A.R., Al-Rashed K., Sakran T. & El-Fayoumi H. (2009). – Life cycle of *Sarcocystis camelicanis* infecting the camel (*Camelus dromedarius*) and the dog (*Canis familiaris*), light and electron microscopic study. *Parasitol. Res.*, 106, 189 – 195.

[2] Al Goraishy S.A.R., Bashtar A.R., AL-Rasheid K.A.S. & Abdel-Ghaffar F.A. (2004). – Prevalence and ultra structure of *Sarcocystis* species infecting camels (*Camelus dromedarius*) slaughtered in Riyadh city, Saudi Arabia. *Saudi J. Biol. Sci.*, 11, 135 – 142.

[3] Bergmann V. & Kinder E. (1974). – Elektronenmikroskopische Befunde an der Parasit-Wirt- Grenze von *Sarcocystis miescheriana* in Muskelfasern des Schweins. *Monatsh. Veterinäed.*, 29, 956 – 958.

[4] Borrow H.A., Mohammed H.A. & Di Sacco B. (1989). – *Sarcocystis* in Somali camels. *Parasitologia*, 31 (2 3), 133 – 136.

[5] Bowler B. & Grandez, R. (2008). – Electrocardiographic parameters in alpacas infected with *Sarcocystis lamacanis. Intern. J. Appl. Res. Vet. Med.*, 6 (2), 87 – 92.

[6] Castro, E.C., Sam R.T., Lopez T.U., Gonzalez A.Z. & Silva M.I. (2004). – Evaluacion de la edad como factor de riesgo de seropositividad a *Sarcocystis* sp. en alpacas. *Rev. Inv. Vet. Peru*, 15 (1), 83 – 86.

[7] Fatani A., Hilali M., Al-Atia S., Al-Shami S. (1996). – Prevalence of *Sarcocystis* in camels (*Camelus dromedarius*) from Al-Ahsa, Saudi Arabia. *Vet. Parasitol.*, 62, 241 – 245.

[8] Fukuyo M., Battsetseg G. & Byambaa B. (2002). – Prevalence of *Sarcocystis* infection in meat-producing animals in Mongolia. *Southeast Asian J. Trop. Med. Public Health*, 33, 490 – 495.

[9] Gorman T.R., Alcaino H.A. & Munoz H. (1984). – *Sarcocystis* sp. in guanaco (*Lama guanicoe*) and effect of temperature on its viability. *Vet. Parasitol.*, 15, 95 – 101.

[10] Hilali M., Fatani A. & Al-Atiya S. (1995). – Isolation of *Toxoplasma, Isospora, Hammondia* and *Sarcocystis* from camel (*Camelus dromedarius*) meat in Saudi Arabia. *Vet. Parasitol.*, 58, 353 – 356.

[11] Hilali M., Nassar A.M. & El-Ghaysh A. (1992). – Camel (*Camelus dromedarius*) and sheep (*Ovis aries*) meat as source of

dog infection with some coccidian parasites. *Vet. Parasitol.*, 43, 37 – 43.

[12] Hussein H.S. & Warrag M. (1985). – Prevalence of *Sarcocystis* in food animals in the Sudan. *Trop. Animal Hlth Prod.*, 17, 100 – 101.

[13] Ishag M.Y., Majid A.M. & Magzoub A.M. (2006). – Isolation of a new *Sarcocystis* species from Sudanese camels (*Camelus dromedarius*). *Intern. J. Trop. Med.*, 1 (4), 167 – 169.

[14] Javir M.P., Zacarias C.C. & Leonico C.C. (2009). – Causas de mortalidad de alpacas en tres principales centros de produccion ubicados en puna seca y humeda del departamento de Puno. *REDVED*, 10, 1 – 13.

[15] Kirmse P. & Mohanbabu B. (1986). – *Sarcocystis* sp. in the one humped camel (*Camelus dromedarius*) from Afghanistan. *Br. Vet. J.*, 142, 73 – 74.

[16] Kuraev G.T. (1981). – Sarcocystosis of camels in Kazakhstan [in Russian]. *Veterinaria* (7), 81.

[17] La Perle K.M.D., Silveria F., Anderson D.E. & Blomme E.A.G. (2002). – Dalmeny disease in an alpaca (*Lama pacos*): sarcocystosis, eosinophilic myositis and abortion. *J. Comp. Pathol.*, 121 (3), 287 – 293.

[18] Latif B.M.A., Al-Delmi B.S., Mohammed S.M., Al-Bayati S. & Al-Amity A.M. (1999). – Prevalence of *Sarcocystis* spp. In meat-producing animals in Iraq. *Vet. Parasitol.*, 84, 85 – 90.

[19] Latif B.M.A. & Khamas W.A. (2007). – Light and ultrastructural morphology of sarcocystosis in one humped camel (*Camelus dromedarius*) in Northern Jordan. J. *Camel Pract. Res.*, 14 (1), 45 – 48.

[20] Leguia G. (1991). – The epidemiology and economic impact of llama parasites. *Parasitol. Today*, 7 (2), 54 – 56.

[21] Mason E.F. (1910). – *Sarcocystis* in the camel in Egypt. *J. Comp. Pathol. Therap.*, 23, 168 – 176.

[22] Mason P. & Orr M. (1993). – Sarcocystosis and hydatidosis in lamoids – disease we can do without. *Surveillance*, 20 (1), 14.

[23] More G., Pardini L., Basso W., Martin R., Bacigalupe D., Auad G., Venturini L. & Venturini M.C. (2008). – Seroprevalence of *Neospora caninum, Toxoplasma gondii* and *Sarcocystis* sp. in llamas (*Lama glama*) from Jujuy, Argentina. *Vet. Parasitol.*, 155, 158 – 160.

[24] Motamedi G.R., Dalimi A., Nouri A. & Aghaeipour K. (2011). – Ultrastructural and molecular characterization of *Sarcocystis* isolated from camel (*Camelus dromedarius*) in Iran. *Parasitol. Res.*, 108, 949 – 954.

[25] Schnieder T., Kaup F.-J., Drommer W., Thiel W. & Rommel M. (1984). – Zur Feinstruktur und Entwicklung von *Sarcocystis auchemiae* beim Lama. *Z. Parasitenkd.*, 70, 451 – 458.

[26] Shekarforoush S.S., Shakerian A. & Hasanpoor M.M. (2006). – Prevalence of *Sarcocystis* in slaughtered one-humped camels (*Camelus dromedarius*) in Iran. *Trop. Anim. Health Prod.*, 38, 301 – 303.

[27] Valinezhad A., Oryan A. & Ahmadi N. (2008). – *Sarcocystis* and its complications in camels (*Camelus dromedarius*) of Eastern Province of Iran. *Korean J. Parasitol.*, 46 (4), 229 – 234.

[28] Woldemeskel M. & Gumi B. (2001). – Prevalence of sarcocystis in one-humped camel (*Camelus dromedaries*) from Southern Ethiopia. *J. Vet. Med. B.*, 48, 223 – 226.

（刘军龙译，刘光远、殷宏校）

## 5.1.12 贝诺孢子虫病

目前已知的贝诺孢子虫属含有8个种，其中以贝氏贝诺孢子虫的危害性最大。贝诺孢子虫病也发生于驴、山羊、驯鹿以及兔、啮齿类动物和负鼠。在生活史已经明确的虫体中，猫是终末宿主，中间宿主因摄食了孢子化的卵囊而感染。虫体首先在不同的器官结缔组织中形成包囊，进而形成速殖子。据认为吸血昆虫可以在不同的中间宿主之间传播贝诺孢子虫。

目前有关骆驼的贝诺孢子虫病的知识较为缺乏，在原始文献中，只有两篇文章是关于单峰驼贝诺孢子虫病的研究。在印度，有关于单峰驼感染贝诺孢子虫样包囊的报道，与其他贝诺孢子虫不同的是，发现于肠的薄膜基层的包囊的大小为10~400μm，没有皮肤病灶。肠部的贝诺孢子虫病在伊朗也有报道，由大小不一、具有炎症与非炎症反应的包囊引起的机能障碍可以在小肠黏膜上观察到。在法尔斯省，5%的宰杀骆驼表现出这样的机能障碍。

### 参考文献


[1] Kharole M.U., Gupta S.K. & Singh J. (1981). – Note on besnoitiosis in a camel. *Indian J. Anim. Sci.*, 51, 802 – 804.

[2] Tafti A.K., Makeki M. & Oryan A. (2001). – Pathological study of intestines and mesenteric lymph nodes of camels (*Camelus dromedarius*) slaughtered in Iran. *J. Camel Pract. Res.*, 8 (2), 209 – 213.


（刘军龙译，刘光远、殷宏校）

## 5.1.13 泰勒虫病

泰勒虫病是由血液性梨形虫——泰勒虫引起的疾病，是寄生于骆驼的唯一一种蜱传性寄生虫病。在已有的文献中，骆驼泰勒虫（*Theileria camelensis*）仅仅为单峰驼的一种寄生虫。印度发现的单峰驼泰勒虫病的可信性遭到质疑，原因是自从这次报道之后就再也没有发现过该寄生虫病。

泰勒虫在哺乳动物宿主中仅以无性生殖的方式繁殖，存在于蜱唾液腺中的子孢子在宿主淋巴结的单核白细胞中形成大裂殖体，随后，裂殖子侵染红细胞；有性生殖则发生在媒介硬蜱的中肠之中。虫体的传播则发生在蜱的发育阶段，即幼蜱到若蜱阶段或者是若蜱到成蜱阶段。

在骆驼红细胞中的骆驼泰勒虫形态大部分呈现杆状，大小为(1.3~2.6)μm × (0.5~1)μm。圆环形的裂殖子直径为0.5~1.3μm，但很少能够观察到。

3个不同的研究报道表明，骆驼泰勒虫主要分布于埃及和利比亚，其中在埃及的流行率为30%和6.2%，在利比亚的流行率为6.5%。而另一个流行病学调查表明，在埃及，羊泰勒虫（*Theileria ovis*）在骆驼中的流行率为12.6%。但是这个报告是不可靠的，因为作者声称在血涂片检测中也发现了羊巴贝斯虫（*Babesia ovis*），其感染率为9.5%。骆驼泰勒虫在伊朗的骆驼中也有发现，但没有相关的流行病学数据公布。在感染泰勒虫的骆驼体获得的硬蜱中，嗜驼璃眼蜱（*Hyalomma dromedarii*）为优势种。

骆驼泰勒虫病被认为是无致病性的寄生虫病。然而，对28头自然感染的单峰驼的临床检测发现，患病动物表现出颈浅淋巴结肿大、消瘦、后肢虚弱、腹泻、食欲不振、被毛粗乱、黏膜苍白及大红细胞正常色素性贫血等症状。

最近，对于约旦的单峰驼血液的分子研究揭示了马巴贝斯虫（*B. caballi*）和马泰勒虫（*T. equi*）的存在，但是红细胞的显微镜检查并没有发现巴贝斯虫及泰勒虫。

依据免疫学实验方面的经验，作者认为单峰驼无法产生针对于环形泰勒虫（*T. annulata*）与马泰勒虫的抗体。

## 参考文献


[1] Al Saad K.M., Al Obaidi Q.T. & Al Obaidi W.A. (2006). – Clinical, haematological and biochemical study of theileriosis in one-humped Arabian camels (*Camelus dromedarius*). *Iraqi J. Vet. Sci.*, 20 (2), 211 – 218.

[2] El- Maghribi A.A. & Hosni M.M. (2009). – Detection of *Theileria* infection in dromedary camels. *Vet. Med. J. Giza*, 57 (1), 53 – 58.

[3] Mahran O.M. (2004). – Some studies on blood parasites in camels (*Camelus dromedarius*) at Shalatin City, Red Sea Governorate. *Assuit Vet. Med. J.*, 50 (102), 172 – 184.

[4] Mazyad S.A. & Khakaf S.A. (2002). – Studies on *Theileria* and *Babesia* infecting live and slaughtered animals in Al Arish and El Hasanah, North Sinai Governorate, Egypt. *J. Egypt. Soc. Parasitol.*, 32 (2), 601 – 610.

[5] Nassar A.M. (1992). – *Theileria* infection in camels (*Camelus dromedarius*) in Egypt. *Vet. Parasitol.*, 43, 147 – 149.

[6] Qablan M.A., Sloboda M., Jirku M., Obornik M., Dwairi S., Amr Z.A., Horin P., Lukes J. & Modry D. (2012). – Quest for the piroplasms in camels: identification of *Theileria equi* and *Babesia cabali* in Jordanian dromedaries by PCR. Vet. Parasitol., 186, 456 – 46.


（刘军龙译，刘光远、殷宏校）

5

寄生虫病

# 5.2 蠕虫病

## 5.2.1 吸虫病

### 肝片吸虫病

片形吸虫病是由片形属（*Fasciola*）吸虫（图241）引起的一种哺乳动物包括人类的重要疾病。片形吸虫病病原包括：肝片形吸虫（*F. hepatica*），为一种常见的体型较大的肝吸虫，见于温带和气候凉爽地区。大片吸虫（*F. gigantica*），见于气候温暖的地区。在热带时则分布于纬度较高的地区。第三种片形吸虫为属于片形科的大拟片形吸虫（*Fascioloides magna*），对骆驼有潜在的感染力，自然疫源地分布于北美洲，输入性病例则见于中欧。

### 形态学

肝片形吸虫成虫呈背腹扁平，月桂树叶形，体长3~5cm。活的虫体呈灰色至红棕色。口吸盘位于头部圆锥体顶端，腹吸盘位于躯体肩部膨大区。虫体前段覆盖小刺。大片吸虫体长可达7cm，其形态与肝片形吸虫相似。肝片形吸虫和大片吸虫的未胚化卵呈黄色，大小分别为(130~150)μm × (63~90)μm和(125~204)μm × (60~110)μm，内含一卵细胞，周围充满卵黄（图242）。

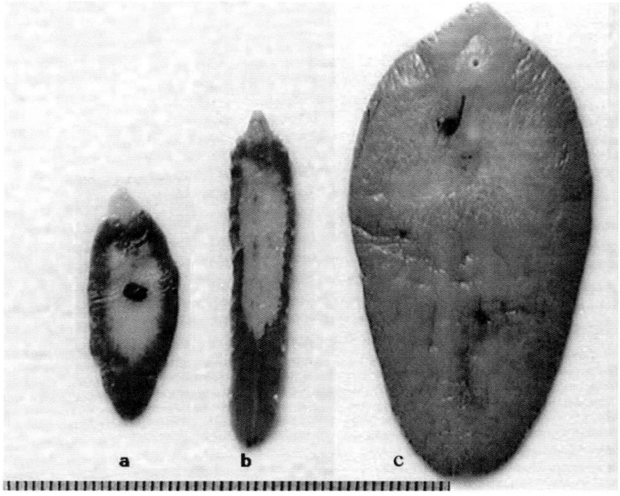


图241 片形科吸虫的成虫（a）肝片形吸虫（*F. hepatica*）；（b）大片吸虫（*F. gigantica*）；（c）大拟片形吸虫（*Fascioloides magna*）

### 生活史

片形吸虫和大拟片形吸虫的发育均需要两个中间宿主（图243）。成虫寄生于宿主肝脏的胆管，产生的卵随胆汁排泄至肠内容物中。虫卵随粪便排放到含蜗牛的水生或半水生环境中，进一步发育。在10℃以上条件下，卵细胞转变为第一期幼虫即毛蚴，毛蚴孵化，然后侵入蜗牛，移行至外套腔

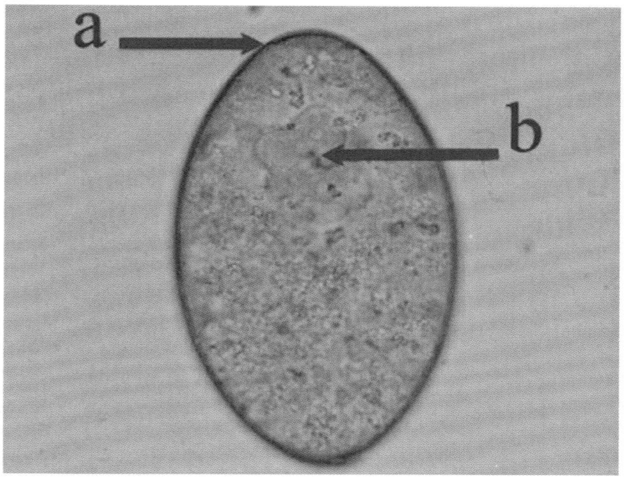


图242 片形吸虫虫卵 新鲜含多聚黄水晶（polylecithine）的虫卵有卵盖（a），充满卵黄，内含一卵细胞（b）

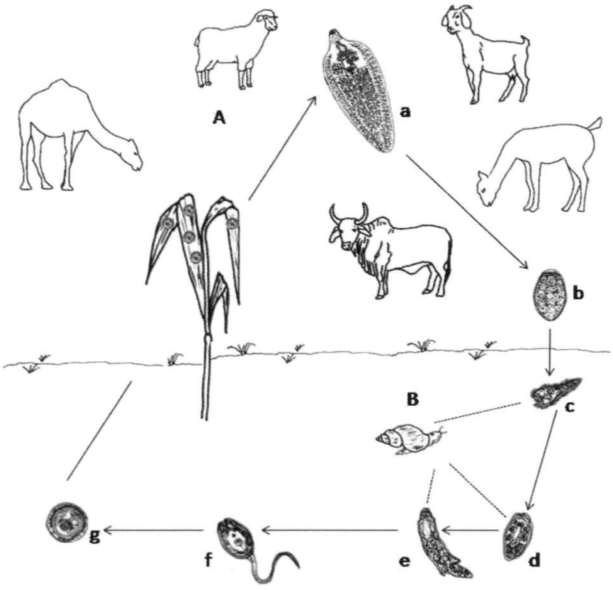


图243 肝片吸虫生活史（A）终末宿主——草食动物；（B）中间宿主——水蜗牛；（a）成虫；（b）虫卵；（c）毛蚴；（d）尾蚴；（e）雷蚴；（f）尾蚴；（g）囊蚴

变为无肠管的胞蚴。胞蚴发育产生第一代雷蚴（母雷蚴），沿着蜗牛肠管进入肝脏-胰腺。母雷蚴产生子代雷蚴或尾蚴，然后尾蚴离开中间宿主。肝片形吸虫的无性生殖能力是非常强大的，当一个毛蚴成功进入泥螺后可产生多达600个尾蚴。尾蚴能够游动，附着于植物上，变为囊蚴。早期的片形吸虫称为童虫，经胆汁激活后在终末宿主的空肠中离开囊蚴囊，穿透肠壁，经腹腔移行，穿过宿主内脏包膜最终进入肝脏。童虫在宿主体腔和肝实质内移行的时间可分别持续3周和5~8周。肝片形吸虫在肝脏的胆管中发育至成虫，在感染后2~4.5个月开始产卵。

## 流行病学

片形吸虫感染途径是通过食物经口感染。片形吸虫病呈全球性分布，宿主范围广泛，但是大多危害家畜，主要包括绵羊、牛和骆驼。感染出现发生于潮湿的草地，当动物在污染感染性囊蚴的水草区域放牧时遭受感染。在大片吸虫感染病例中，宿主也可能因为饮用池塘中的水而遭受感染。作者在非洲的经验表明，在干燥季节，水洞四周由淤泥围起来，缺少植被，而囊蚴漂浮在水面上。并非所有的童虫都会迁移到肝脏，它们有时可达到子宫，引起胎儿感染。片形吸虫属吸虫的感染率主要取决于中间宿主的存在。截口土窝螺（*Galba truncatula*）（图244）为肝片形吸虫的蜗牛宿主，分布于古北区北部、北美洲和南美高地。截口土窝螺是半水栖蜗牛，生活于泥土表面上或缓流小溪的河岸。它们也可能在脚印或凹槽中大量分布。

在古北区南部，大片吸虫利用另一种椎实螺——耳萝卜螺（*Radix auricularia*）作为中间宿主；另一种亲缘关系近的纳塔尔萝卜螺（*R. natalensis*）（图245）是非洲巨片形肝吸虫的主要中间宿主。这两种螺均需要永久性水源（图246）

耐热的小柱伪琥珀螺（*Pseudosuccinea columella*），原来仅分布于拉美国家，现已在南欧、南非澳大利亚也有分布，可作为肝片形吸虫的中间宿主[9]。实验感染也表明截口土窝螺对大片形吸虫囊蚴易感[3]。

在南美，新大陆骆驼的片形吸虫病由肝片形吸虫引起。由于羊驼和美洲驼在高纬度草原上放牧，所以肝片形吸虫的感染率相对较低[7]。Neyra等人发现秘鲁受检羊驼中只有7%的比例可查到片形吸虫卵[10]，而Issia等人报道阿根廷国家大羊驼中片形吸虫的感染率为0.5%，在阿根廷所有绵羊都受到了片形吸虫的感染[6]。尽管如此，来自巴塔哥尼亚穴居的40只大羊驼中有2只死亡，死亡率为5%，其死亡与片形吸虫感染相关[11]。来自阿根廷西北部半穴居的小羊驼片形吸虫的感染率变化范围为7.7%~25.7%[1]。但是，实验感染表明新大陆骆驼对肝片形吸虫的易感性很高[13]。单峰驼和双峰驼有时会感染肝片形吸虫，但是它们还是主要感染大片吸虫[4,12]。

近年来，人的片形吸虫病例已在安第斯山国

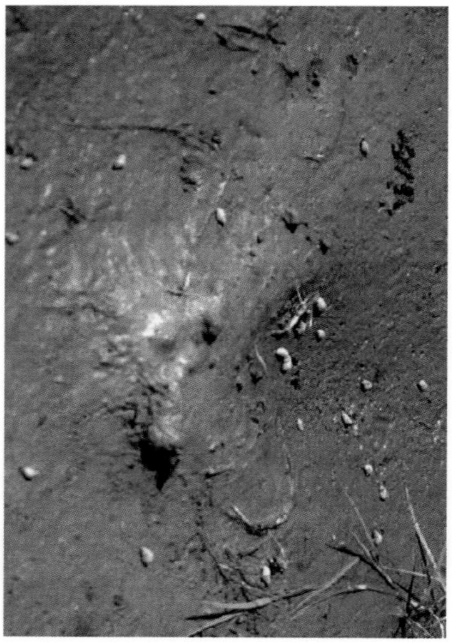

图244　截口土窝螺，是肝片形吸虫的主要中间宿主，体形小，半水栖斑蝶螺，可见于泥泞草原上，呈高密度分布

图245　纳塔尔萝卜螺（*Radix natalensis*）是非洲大片吸虫的主要中间宿主，它需要生活在永久性水源生境中

a 大片形吸虫在东非的生境。水体四周分布浓密植被，植被部分遭洪水淹没，当动物在此放牧时遭受感染

b 同一生境在干旱季节时的画面。现在，小水体仍然可见，泥泞的湖潭作为骆驼和其他家畜的水源。此时，动物因饮用漂浮在水体中囊蚴而遭受感染

图246

家（玻利维亚、秘鲁、智利和厄瓜多尔）报道，另外，人的片形吸虫病例也在埃及、伊朗和法国进行过报道[9]。

### 临床症状和病理学

肝片形吸虫主要引起慢性疾病，造成的经济损失包括增重减少，肉品检验中肝脏的废弃，产乳量和产毛量的减少，以及家畜繁殖力下降。

吸虫引起的肝炎和胆管炎导致亚临床和慢性临床症状，其病程和严重程度受宿主种类、年龄和感染强度等因素的影响。最初的症状包括食欲减退，腹部疼痛，体温升高等，在宿主受到感染后2~3周，童虫在体内移行时出现。Puente曾描述过羊驼感染肝片形吸虫后呈急性发作，羊驼在绵羊放牧过的草地上啃食牧草后遭受感染[8]。200只受感染的羊驼中有8只突然死亡，40只出现食欲不振、身体虚弱、黏膜苍白、呼吸困难和久卧不起等症状，另外50只羊驼遭受吸虫感染后表现程度不同的嗜睡和食欲废绝，而其余102只无明显症状。片形吸虫病后期，患病动物通常无症状。因此，Cafrune等人经对一群共15只自然感染的美洲驼（llama）进行了长达24个多月的检测和观察，结果表明这些感染动物并未出现明显的临床症状[2]。

在片形吸虫的急性移行期，感染动物因肝实质大范围遭受破坏和血管破裂，导致宿主腹腔出血。Puente在急性期死亡的羊驼体中观察到肝脏肿大、苍白和易碎，并有纤维蛋白沉积于肝脏表面[8]。肝实质出现许多血道，内充满大量未成熟虫体。病程后期，主要的病变可见于宿主胆管内及其周围（图247）。胆管炎、胆管周炎因带刺的童虫移行引起，也可能由于虫体分泌/排泄产物所致。Hamir和Smith曾描述感染羊驼出现重度胆管畸变[5]。病变集中于肝右叶，右叶萎缩、呈淡棕褐色，质地坚实，含有小的、隆起、白色至黄色、部分矿物化的圆形结节。

### 诊断

活体诊断基于发现淡黄色、长卵圆形虫卵，

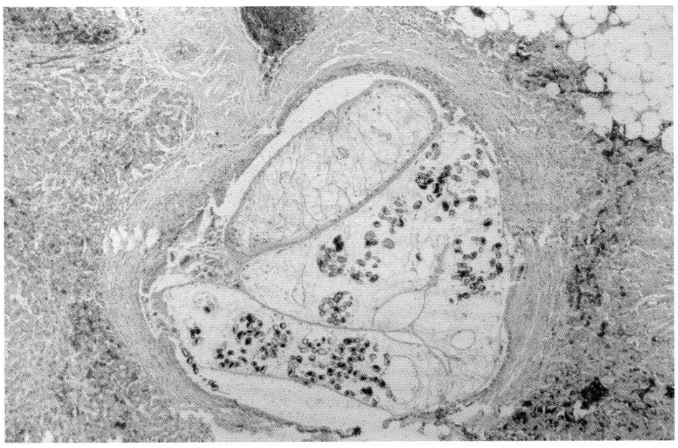

图247 双峰驼的肝片形吸虫。组织切片表明胆管增厚，管腔内充塞有一条吸虫成虫，虫体沿子宫方向被切开。照片由莱比锡大学兽医病理研究所的Weber惠赠

5 寄生虫病

其方法是利用粪沉淀法查卵。推荐的方法是在粪样沉淀物上滴加几滴甲基蓝，食物残渣着色而虫卵不着色。血清学检测是针对肝片形吸虫抗原的抗体，此方法用于人和新大陆骆驼[10]。死后可用于片形吸虫病的肉品检验或剖检。为此，必须将大的胆管切开后进行检查。

### 治疗与控制

多数抗吸虫病药物属于水杨酰苯胺（salicylanilide）、苯并咪唑（benzimidazole）和其他类型，可供选择使用。碘硫胺（rafoxanide）、氯氰碘硫胺（closantel）、羟氯硫胺（oxyclozanide）、硝碘酚腈（nitroxynil）、阿苯达唑（albendazole）和奈托比胺（netobimin）是治疗慢性肝片形吸虫病的有效化合物。这些药物大多数为口服剂，硝碘酚腈、克洛索隆（clorsulon）和氯氰碘硫胺也有注射剂型。此外，碘硫胺和氯氰碘硫胺除具有杀吸虫药的特性外，它们对苯并咪唑耐药的血矛线虫属（*Haemonchus*）虫株也有效。克洛索隆与伊维菌素（ivermectin）合用作为一种广谱驱虫药在市场有售。

然而，治疗肝片形吸虫病的药物首选是三氯苯咪唑（triclabendazole），它对肝片形吸虫的幼虫和成虫均有强的疗效，有可口服的片剂和混悬剂，也有与伊维菌素合剂注射剂型。

杀吸虫药特别是水杨酰苯胺和含有伊维菌素的合剂有相对长的半衰期，因此需要考虑长的休药期，以符合国家的相关法令。

肝片形吸虫可以通过草原的卫生改善使其减少。在草原上，应在潮湿的地方和水沟旁设置篱笆，在坚固的干燥地面上构筑适当的饮水池是最简单的方法之一，可以切断肝片形吸虫的传播链。尽管肝片形吸虫囊蚴抵抗力相对强，但是可以考虑将干燥的干草作为安全的食物来源。

## 参考文献

[1] Cafrune M.M., Aguirre D.H. & Freytes I. (2004). – Fasciolosis en vicunas (*Vicugna vicugna*) en semi-cautiverio de Molinos, Salta, Argentina, con notas de otros helmintos en este hospedador. *Vet. Arg.*, 21, 513 – 520.

[2] Cafrune M.M., Rebuffi G.E., Cabrera R.H. & Aguirre D.H. (1996). – *Fasciola hepatica* en llamas (*Lama glama*) de la Puna argentina. *Vet. Arg.* 13, 570 – 574.

[3] Dar Y.D., Rondelaud D. & Dreyfuss G. (2005). – Update of fasciolosis-transmitting snails in Egypt (review and comment). *J. Egypt. Soc. Parasitol.*, 35 (2), 477 – 490.

[4] El Bihari S. (1985). – Helminths of the camel: a review. *Br. Vet. J.*, 141, 315 – 326.

[5] Hamir A.N. & Smith B.B. (2002). – Severe biliary hyperplasia associated with liver fluke infection in an adult alpaca. *Vet. Pathol.*, 39, 592 – 594.

[6] Issia L., Pietrokovsky S., Sousa-Figueiredo J., Stothard J.R. & Wisnivesky-Colli C. (2009). – *Fasciola hepatica* infections in livestock flock, guanacos and coypus in two wildlife reserves in Argentina. *Vet. Parasitol.*, 165, 341 – 344.

[7] Leguia G. (1991). – The epidemiology and economic impact of llama parasites. *Parasitol. Today*, 7 (2), 54 – 56.

[8] Leguia Puente G. (1997). – Acute and sub acute fasciolosis of alpacas (*Lama pacos*) and treatment with triclabendazole. *Trop. Anim. Health Prod.*, 29, 31 – 32.

[9] Mas-Coma S., Bargues M.D. & Valero M.A. (2005). – Fascioliasis and other plant-borne trematode zoonoses. *Int. J. Parasitol.*, 35, 1 255 – 1 278.

[10] Neyra V., Chavarry E. & Espinoza J.R. (2002). – Cysteine proteinases Fas1 and Fas2 are diagnostic markers for *Fasciola hepatica* infection in alpacas (*Lama pacos*). *Vet. Parasitol.*, 105, 21 – 32.

[11] Olaechea F.V. & Abad M. (2005). – An outbreak of fascioliasis in semi-captive guanacos (*Lama guanicoe*) in Patagonia (Argentina). First report. *In* Proceedings of the 20th International Conference of the World Association for the Advancement

of Veterinary Parasitology, 16 – 20 October 2005, Christchurch, New Zealand, 169.

[12] Parsani H.R., Veer Singh V. & Momin R.R. (2008). – Common parasitic diseases of camel. *Vet. World*, 1 (10), 317 – 318.

[13] Timoteo O., Maco V Jr., Maco V., Neyra V., Yi P.J., Leguia G. & Espinoza J.R. (2005). – Characterization of the humoral immune response in alpacas (*Lama pacos*) experimentally infected with *Fasciola hepatica* against cysteine proteinases Fas1 and Fas2 and histopathological findings. *Vet. Immunol. Immunopathol.*, 106, 77 – 86.

（贾万忠译，殷宏校）

## 歧腔吸虫病（Dicrocoeliosis）

歧腔吸虫病是草食动物和其他哺乳动物包括人类的一种寄生虫病，分布广泛，由歧腔属（*Dicrocoelium*）吸虫的矛形的歧腔吸虫引起。主要包括3种：枝歧腔吸虫（*D. dendriticum*，图248），分布于欧洲、亚洲和北美洲；牛歧腔吸虫（*D. hospes*，图249）分布于西非、中非和东非；中华歧腔吸虫（*D. chinensis*，图250），分布于中国和日本，输入性病例已见于奥地利和意大利。

### 形态学

歧腔吸虫的成虫呈柳叶状，背腹扁平，体长取决于寄生宿主的种类，可达5~12mm。活的虫体呈半透明，体表光滑。躯体前部分布有两个口吸盘和生殖器官，呈淡粉红色，而躯体后部呈黑色，分布有子宫，子宫内充满棕色虫卵。白色卵黄腺分布于身体两侧。3种歧腔吸虫的睾丸外形不尽相同，它们位于腹吸盘后。歧腔吸虫胚化卵呈黄色，大小为(38~45)μm × (22~30)μm。通过肉眼从卵壳可以观察到虫卵内含一发育完全的毛蚴，拥有两个生殖球。

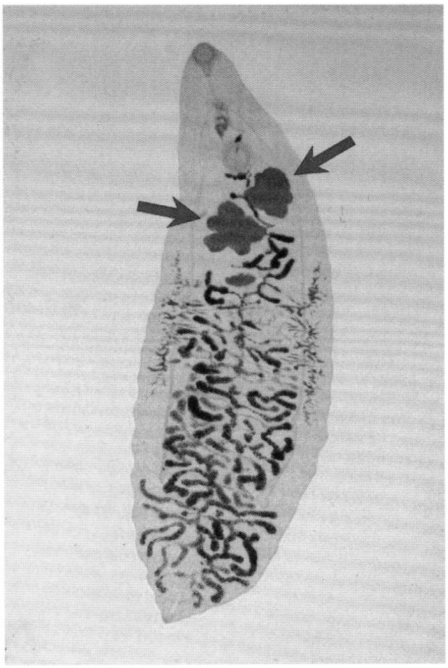

图248　枝歧腔吸虫 形如矛状，广泛分布于欧亚和北美。睾丸呈对角线位置排列

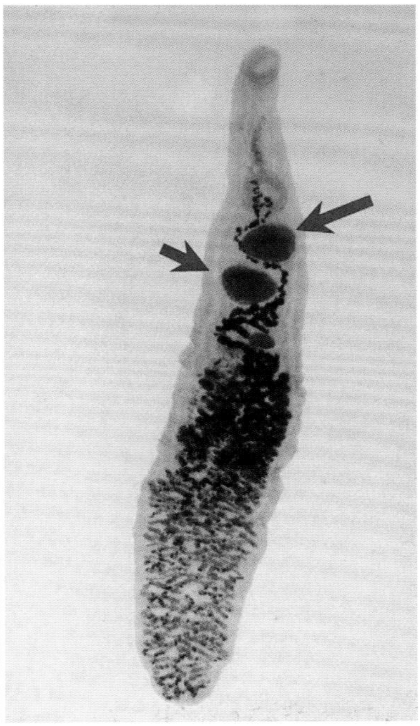

图249　牛歧腔吸虫，一非洲种，睾丸呈串联排列，这与枝歧腔吸虫不同

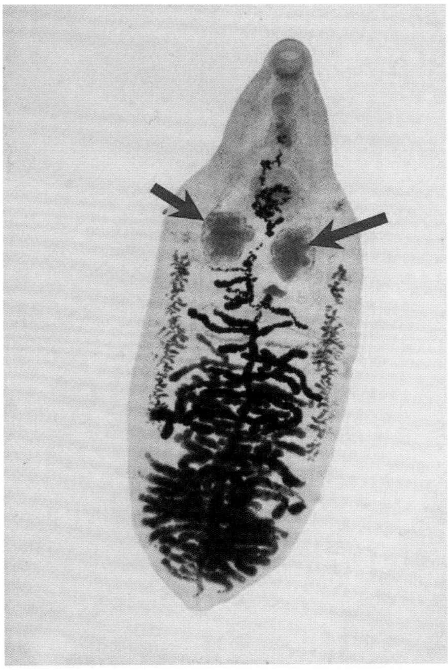

图250　中华歧腔吸虫 发于东亚，在19世纪末已通过输入性方式由梅花鹿传播至欧洲。由于睾丸呈平行排列，所以中华歧腔吸虫像是有“双肩结构”

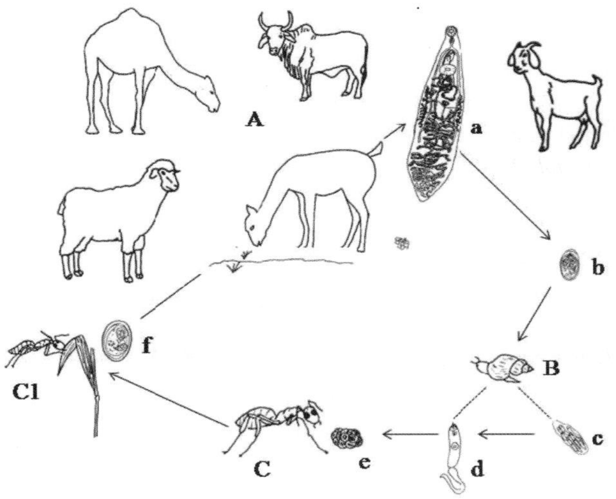


图251 歧腔吸虫生活史（A）终末宿主；（B）第一中间宿主——蜗牛；（C）第二中间宿主——蚂蚁；（C）受感染的蚂蚁通过痉挛的下颚粘附于水草上——（a）成虫；（b）虫卵；（c）胞蚴；（d）尾蚴；（e）黏液球；（f）囊蚴

图252 厚壳大蜗牛科（Helicellidae）蜗牛是歧腔吸虫主要的中间宿主。这些蜗牛生存不需要很潮湿条件，可在较干燥草原上存活

图253 受感染蜗牛产生和分泌含有数百只尾蚴的黏液球

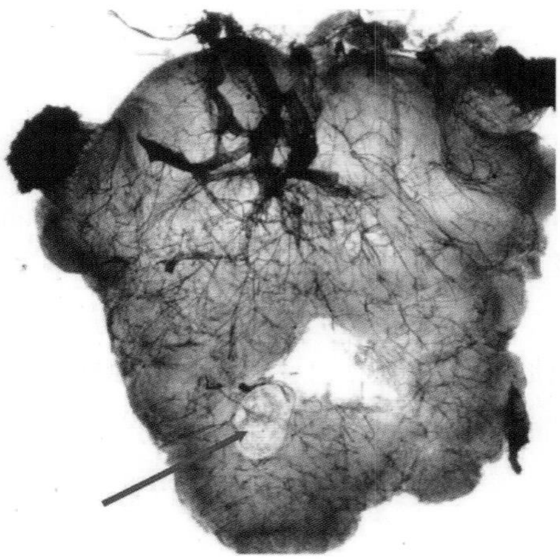

图254 在蚂蚁食管下神经节的“脑虫”（箭头所示）。单个虫体可改变蚂蚁的行为方式，它并不对终末宿主具有感染性

## 生活史

枝歧腔吸虫的发育需要3种陆生动物作为中间宿主（图251）。成虫产生胚化虫卵，通过胆汁进入宿主肠腔内容物中，然后随粪便排到体外。以腐烂植物包括反刍动物粪便为食的陆地旱生蜗牛（图252）食入虫卵，毛蚴进行一系列体形变化包括软体动物消化腺中两代胞蚴。最后，变为尾蚴。尾蚴通过黏液黏合在一起，然后一同从宿主被挤出体外（图253）。作为第二中间宿主的蚂蚁食入这些黏液球，释放出的尾蚴穿透蚂蚁的胃部。多数尾蚴停留在宿主腹腔，然后变为囊蚴包囊，然而单个虫体移行至食管下神经节，造成感染蚂蚁行为异常（图254）。与未受感染的蚂蚁相比，受感染蚂蚁爬向草和小型植被的上部，停留在牧草尖部（图255）。当受感染蚂蚁被终末宿主吞食后，吸虫在蚂蚁体内的生活史结束。囊蚴包囊在十二指肠中溶解，童虫沿胆管通过胆管系统从胆汁逆行至肝脏的胆管。

## 流行病学

歧腔吸虫感染途径是通过食物经口感染。歧腔吸虫病是牧场常见蠕虫病，呈全球性分布，宿主范围广泛，涉及众多哺乳动物，包括野生动物。在文献中记载有人感染矛形吸虫的病例。歧腔吸虫病

图255 受感染蚂蚁固定在草尖上。蚂蚁的这种异乎寻常行为是由“脑虫”引起的

在澳大利亚和南美洲尚未有报道。然而，新大陆骆驼歧腔吸虫病病例见于瑞典[2,3,10,11]、德国[5,6]、奥地利[1]和法国[4]，这些动物对歧腔吸虫特别易感。

在歧腔吸虫病专著中报道[8,9]，单峰驼和双峰驼可作为终末宿主，但是在近来的原始文献中未见报道。枝歧腔吸虫引起的歧腔吸虫病见于土库曼斯坦和土耳其的单峰驼，与哈萨克斯坦、乌兹别克斯坦和塔吉克斯坦的双峰驼。中华歧腔吸虫见于中国内蒙古自治区双峰驼中。虽然牛歧腔吸虫存在于西非、中非和东非等单峰驼种群中数量庞大的地区，但是到目前为止，尚未见有单峰驼感染各歧腔吸虫病例的报道。

隶属于槲果螺科（Cochlicopidae）、粒蛹螺科（Chrondrinidae）、艾纳螺科（Enidae）、琥珀蜗牛科（Zonitidae）、烟管螺科（Clausiliidae）、巴蜗牛科（Bradibaenidae）和厚壳大蜗牛科（Helicidae）的50多种不同蜗牛和隶属于蚁属（*Formica*）的15种蚂蚁分别作为歧腔吸虫第一和第二中间宿主。潜伏期长达7周。

### 临床症状和病理学

由于幼龄期的吸虫不能穿透肠道和肝脏被膜，所以临床症状通常不明显。感染动物在荷虫量大时可能出现贫血、水肿和消瘦。新大陆骆驼对歧腔吸虫反应强烈。据报道，美洲驼和羊驼在自然感染状态下全身状况迅速恶化，随后出现长卧不起、体温下降和不同程度的贫血。肝脏酶活性除$\gamma$-谷酰基转移酶外均表现为正常值[10]。歧腔吸虫在胆管中移行引起动物烦躁不安，导致分泌性细胞增殖，黏液增加。在病程后期，病理学变化以慢性、非化脓性胆管炎为主。新大陆骆驼在荷虫量大时会出现胆管纤维化、肝硬化等病变。值得提及的是，骆驼缺乏胆囊。感染吸虫的肝脏不宜作为人的食物进行消费，在肉品检验时必须按废弃物处理。

### 诊断

利用粪沉淀法检查粪样中的歧腔吸虫卵可以诊断歧腔吸虫病。虫卵大小为(36~46)μm × (20~30) μm，内含一完全发育的毛蚴。在肉品检疫或尸检中，胆管扩张应引起歧腔吸虫感染的怀疑，小的柳叶状吸虫可以通过肉眼看到，它们通常见于肝脏切面上。

### 治疗与控制

吡喹酮按每千克体重50mg剂量给药能大幅度降低受感染美洲驼排虫卵量[10]。阿苯达唑和奈托比胺在增加剂量至每千克体重15mg和20mg时对绵羊歧腔吸虫病的疗效良好[7]。



## 参考文献


[1] Burian E. (2010). – Endoparasitenstatus von Neuweltkameliden in Oberosterreich und im Burgenland. Thesis, Veterinary University Vienna, Austria.

[2] Gunsser I., Hanichen T. & Maierl J. (1999). – Leberegelbefall bei Neuweltkameliden, Parasitologie, Pathologie, Klinik und Therapie. *Tierärztl. Praxis*, 27, 187 – 192.

[3] Herzberg H. & Kohler L. (2006). – Prevalence and significance of gastrointestinal helminths and protozoa in South American Camelids in Switzerland. *Berl. Münch. Tierärztl. Wschr.*, 119, 291 – 294.

[4] Ollagnier C. (2007). – Recensement des parasites digestifs des petits camelides (Genre *Llama*) en France. Thesis, University Lyons, France.

[5] Rohbeck S. (2006). – Parasitosen des Verdauungstraktes und der Atemwege bei Neuweltkameliden: Untersuchungen zu ihrer Epidemiologie und Bekampfung in einer südhessischen Herde sowie zur Biologie von *Eimeria macusaniensis*. Thesis,Justus-Liebig-Universitaet Giessen, Germany.

[6] Schlögl C. (2010). – Erhebungen zum Vorkommen von Endo-sowie Ektoparasiten bei Neuweltkameliden. Thesis, Ludwig-Maximilians-University Munich, Germany.

[7] Schuster R. & Hiepe Th. (1993). – Bekämpfung der Dicrocoeliose beim Schaf. *Mh. Vet.-Med.*, 48, 653 – 657.

[8] Skrjabin K.I. & Evranova V.G. (1952). – Family Dicrocoeliidae Odhner, 1911. *In* Trematodes of animals and man [in Russian] (K.I. Skrjabin, ed.). Izd. Akademii Nauk SSSR, Moskva, 33 – 604.

[9] Tverdochlebov P.T. & Ajupov Ch.V. (1988). – Dicrocoeliosis in animals [in Russian]. Agropromizdat, Moscow.

[10] Wenker C., Hatt J.M., Herzberg H., Ossent P., Haenichen T., Brack A. & Isenbuegel E. (1998). – Dicrocoeliose bei Neuweltkameliden. *Tieraerztl. Praxis*, 26, 355 – 361.

[11] Zanolari P. & Meylan M. (2005). – Problematik des Leberegelbefalls bei Neuweltkameliden. *Lamas*, (2), 1 – 2.

（贾万忠译，殷宏校）

## 一些不太重要的吸虫感染

胰阔盘吸虫（*Eurytrema pancreaticum*，图256）隶属于歧腔科（Dicrocoeliidae）。该吸虫可抑制反刍动物胰腺的功能，也在单峰驼和双峰驼中分布[1,4,7,8,13]。流行区为亚洲大陆，包括中国北方地区到缅甸和印度，也包括哈萨克斯坦、日本、菲律宾、印度尼西亚和马来西亚。蜗牛和草猛分别为胰阔盘吸虫的第一和第二中间宿主[11,12]。胰阔盘吸虫导致的病理学变化在人体病例中进行了研究，病变包括胰腺管扩张、胰腺管上皮发生鳞状上皮化生以及淋巴滤泡形成。

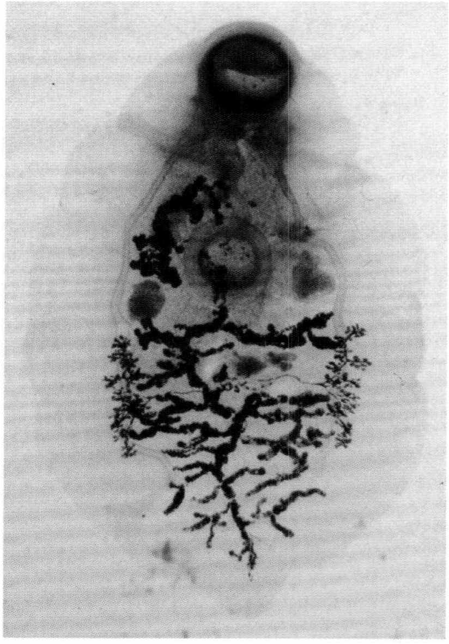

图256 胰阔盘吸虫，一种歧腔吸虫，分布于亚洲，可抑制草食动物胰腺的功能

有报道记载，四种分体科（Schistosomatidae，图257）流行于旧大陆骆驼。牛分体吸虫（*Schistosoma bovis*）和羊分体吸虫（*S. matthei*）分布于非洲单峰驼[3,10]，而印度分体吸虫（*S. indicum*）则分布于印度单峰驼[7]。土耳其斯坦东毕吸虫（*Orientobilharzia turcestanicum*）见于双峰驼和单峰驼，流行于前苏联、中国、蒙古、伊拉克

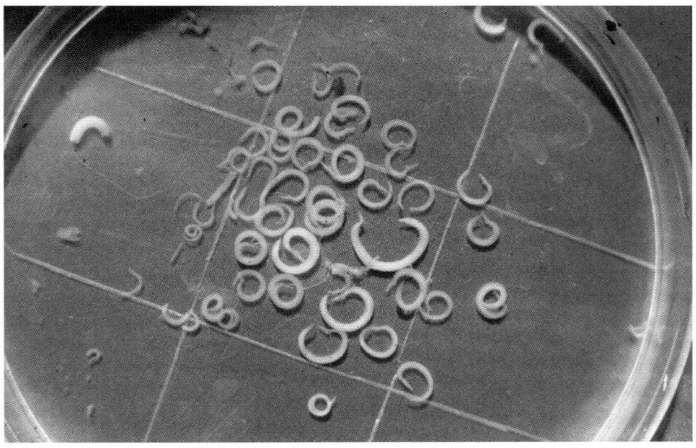

图257 羊血吸虫，一种能感染反刍动物和骆驼的血吸虫，主要分布于非洲

和伊朗[5,6,9]。成虫成对存在，分布于宿主肠系膜静脉，但是也能分布于其他静脉。纺锤样虫卵侵入小肠血管，进入内脏空腔。当粪便落入新鲜水体时虫卵开始孵化，虫卵内形成一发育完全的毛蚴。淡水蜗牛作为中间宿主，释放出叉尾蚴。终末宿主通过皮肤遭受感染，但是对于家畜而言，很可能是在受污染的池塘饮水时尾蚴通过口腔或食道进入血管而遭受感染的。未见有骆驼血吸虫感染时病理学变化的记录。其他宿主受吸虫感染后出现消瘦和腹泻症状。

分体吸虫卵存在于粪样中，用粪沉淀法可检出。为了防止毛蚴孵化在处理粪样时加入10%的福尔马林溶液。

吡喹酮按每千克体重50 mg的剂量给药可以作为治疗血吸虫病的选择药物。

隶属于同盘科（Paramphistomidae）的圆锥形或瘤胃吸虫偶尔见于骆驼。已有报道，鹿前后盘吸虫（*Paramphistomum cervi*）、荷包腹袋吸虫（*Gastrothylax crumnifer*）和长大卡妙吸虫（*Carmyerious spatiosus*）在巴基斯坦呈低感染流行状态[2]。前后盘吸虫是反刍兽动物典型的寄生虫之一，而骆驼的胃部结构似乎不适宜这类吸虫的感染。

## 参考文献


[1] Al-Khalidi N.W., Hassan M.W. & Al-Taee A.F. (1990). – Faecal incidence of *Fasciola* spp. and *Eurytrema pancreaticum* eggsin camels (*Camelus dromedarius*) in Iraq. *J. Vet. Parasitol.*, 4, 75 – 76.

[2] Anwar M. & Hayat C.S. (1999). – Gastrointestinal parasitic fauna of camel (*Camelus dromedarius*) slaughtered at Faisalabad abattoir. *Pakistan J. Biol. Sci.*, 2 (1), 209 – 210.

[3] Dakkak A. & Ouheli H. (1987). – Helminths and helminthoses of the dromedary. *In* Diseases of camels. *Rev. Sci. tech. Off. int. Epiz.*, 6 (2), 447 – 461.

[4] El Bihari (1985). – Helminths of the camel. A review. *Br. Vet. J.*, 141, 315 – 326.

[5] Erdenebileg O. (2001). – Camel diseases [in Mongolian]. Dalanzadgad, Ulaanbaatar, Mongolia.

[6] Farhati, Eldin de Pecoulas P., Rajguru-Kazemi M. & Bayssade-Dufour C. (1995). – Les schistosomes d'animaux d'Asie. *Med. Mal. Infect.*, 25, 107 – 100.

[7] Parasani H.R., Singh V. & Momin R.R. (2008). – Common parasites of camel. *Vet. World*, 1 (10), 317 – 318.

[8] Rewatkar S.G., Deshmukh S.S., Deshkar S.K., Maske D.K., Jumde P.D. & Bhangale G.M. (2009). Gastrointestinal helminths in migratory camel. *Vet. World*, 2 (7), 258.

[9] Skrjabin K.I. (1951). – Trematodes of animals and man. Volume 5. Suborder Schistosomata Skrjabin et Schulz, 1937 [in Russian] Izd. Akademii Nauk SSSR.

[10] Soliman K.N. (1956). – The occurrence of *Schistosoma bovis* (Sonsino, 1876) in the camel, *Camelus dromedarius*, in Egypt. *J. Egypt. Med. Assoc.*, 39, 171 – 181.

[11] Tang C. (1950). – The life history of *Eurytrema pancraticum* Janson, 1889. *J. Parasitol.*, 36, 559 – 573.

[12] Tang C.C., Mei T.L., Quiwen L.H. & Chang Q.Y. (1983). – Studies on the epidemiology of *Eurytrema pancraticum* in the

eastern Nei Mongol Autonomous region. *Acta Zool. Sin.*, (3), 85 – 93.

[13] Tang C.C. & Tang C.T. (1975). – Biology and epidemiology of *Eurytrema pancreaticum* (Janson, 1889) and *E. coelomaticum* (Giard et Billet, 1892). *J. Xiamen Univ. Nat. Sci.*, 2, 36 – 45.

（贾万忠译，殷宏校）

## 5.2.2 绦虫感染

### 裸头科绦虫病

裸头科绦虫病是由裸头科（Anoplocephalidae）绦虫引起的一种寄生虫病。本科绦虫能感染骆驼的有两个亚科——裸头亚科（Anoplocephalinae）和曲子宫亚科（Thysanosomatinae），由4个代表属组成，包括莫尼茨属（*Moniezia*）、无卵黄腺属（*Avitellina*）、斯泰勒属（*Stilesia*）和曲子宫属（*Thysanieza*）。

### 形态学

这些绦虫的特征是头节缺乏顶突，头节呈任何形式的甲胄型（图258）。莫尼茨绦虫体长可达数米（图259）。雌雄同体，每一节片含有两套雌性生殖腺，因此每一节片同时拥有两个生殖孔开口。数百个睾丸填充于卵巢之间的片段中部。通过位于成熟节片后部上的节间腺及其形状可以区别与鉴定莫尼茨绦虫（图260和图261）。孕卵节片长达2mm，宽15~25mm，含有多达20 000个的多形虫卵，虫卵内含有一六钩蚴，六钩蚴外层包裹一层梨形状胚膜。

中点无卵黄腺绦虫（*A. centripunctata*）体长可达1m。链体节片化程度低，由于节片之间的界线不清，因此链体上的节片不易看清楚（图262）。每一成熟节片的中部有一套雌性生殖器官，生殖孔在侧边开口，不规则地分布于两侧。约20个睾丸成两簇，横向分布，每一簇睾丸被纵向渗透压调节管道分隔开来。孕节呈圆形，含有40~50枚虫卵，虫卵包裹在一个子宫旁器中。

球点状斯泰勒绦虫（*Stilesia globipunctata*，球

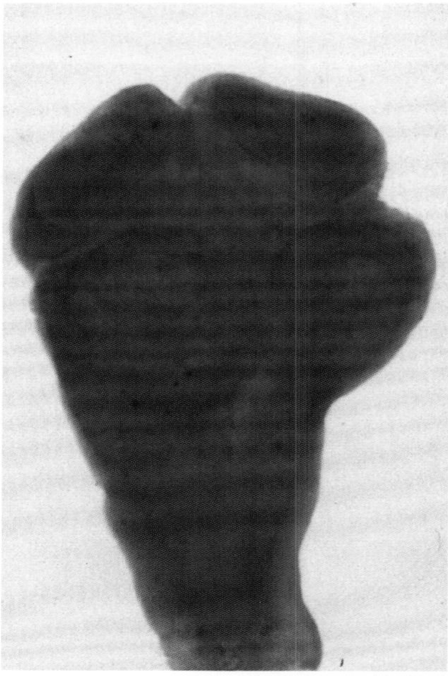

图258　条状斯泰勒绦虫（*Stilesia vittata*）的无甲胄头节

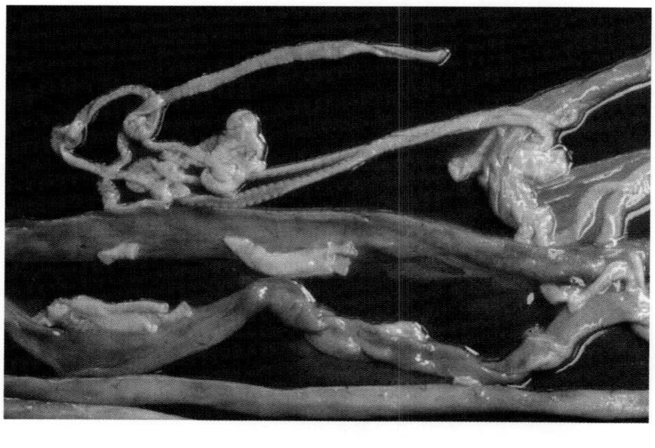

图259　幼驼小肠中的扩展莫尼茨绦虫（*M. expansa*）。链体的片段化非常清晰，肉眼可见

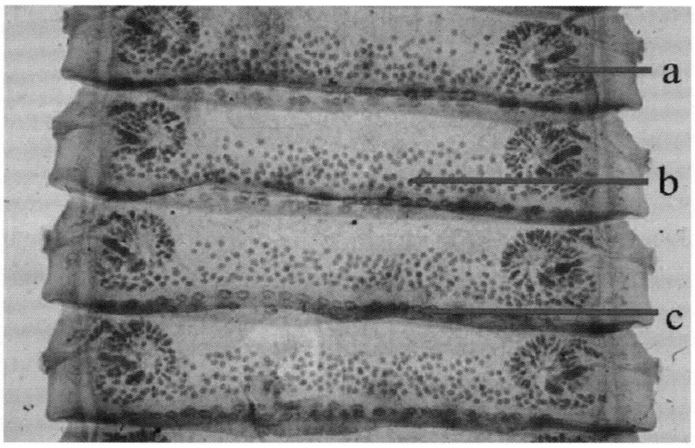


图260 扩展莫尼茨绦虫成熟节片。a. 莫尼茨绦虫节片含有两套雌性生殖腺；b. 100多个睾丸位于节片中部；c. 扩展莫尼茨绦虫玫瑰形的节间腺位于节片后缘

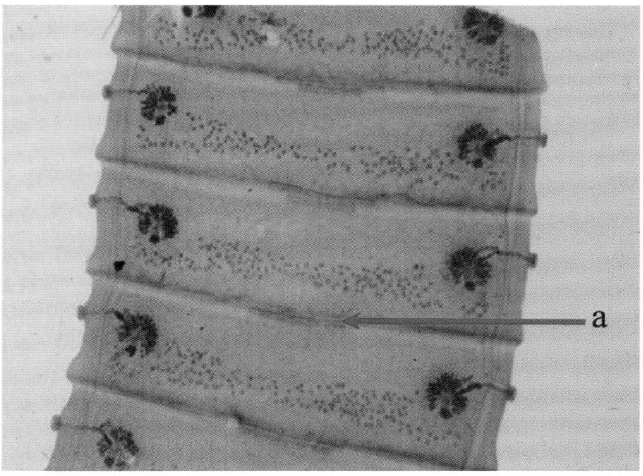


图261 贝氏莫尼茨绦虫（*M. benedeni*）的成熟节片，线状分布的节间腺（a）

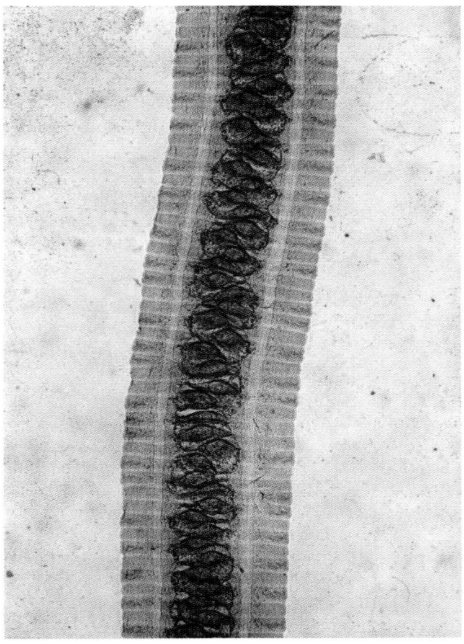

图262 中点无卵黄腺绦虫带状链体

图263 条状斯泰勒绦虫的链体较薄，可大量分布于骆驼十二指肠内

刺斯泰尔斯绦虫）和条状斯泰勒绦虫的链体长达60cm，但体宽仅有2mm（图263）。节片中只含一套雌性生殖器官，位于排泄管之间，但是稍微地偏向细孔一侧，生殖孔不规则分布。两组睾丸分布于两侧至排泄管，每一簇含睾丸4~7个。而条状斯泰勒绦虫每一簇的睾丸数目为5~10个。孕节含有2个子宫旁器，内充满子宫（图264）。

盖氏曲子宫绦虫（*Thysaniezia giardi*，同名为 *Helicometra ovilla*）体长数米，宽度长达1cm，乍看像莫尼茨绦虫。但是，与莫尼茨绦虫不同，每一节片只有一个生殖孔，呈不规则地交替分布，

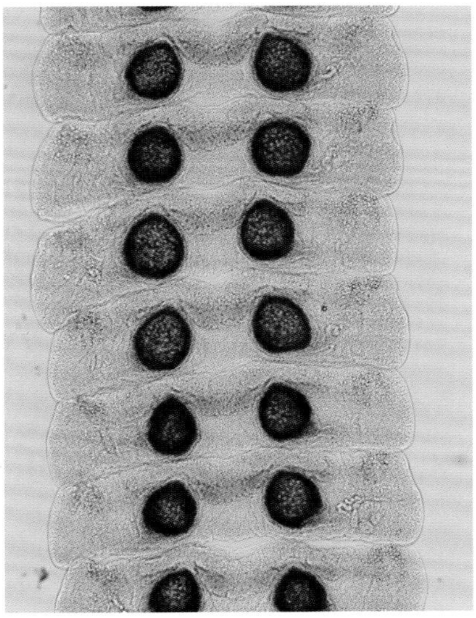

图264 条状斯泰勒绦虫孕节含有两个子宫旁器，子宫旁器中充满虫卵

110~150个睾丸成两簇，从渗透压调节管横向排列于节片两侧。卵巢和卵黄腺通过细孔相连，位于雌雄同体节片内。子宫初期呈管状，横向分布，发育后期被300多个子宫旁器所替代（图 265）。

反刍兽的另外两种裸头科绦虫——肝斯泰勒虫（*S. hepatica*）和放射穗体绦虫（*Thysanosoma actinoides*）分别见于非洲和南美洲反刍兽。这些绦虫可寄生于肝脏的胆管，但是尚未见在骆驼体的报道。

## 生活史

裸头科绦虫发育需要两个不同的寄生宿主，哺乳动物作为终末宿主，节肢动物作为中间宿主。扩展莫尼茨绦虫和贝氏莫尼茨绦虫的生活史研究的最为清楚（图266）。

经过5~7周的潜伏期，莫尼茨绦虫的孕节开始逐渐从链体末端脱落。卵袋状子宫破裂，虫卵与宿主小肠内容物混合，随粪便排出体外。甲螨或甲虫作为中间宿主（图267）。甲螨在永久性草场上分布密度高，已经在全世界发现，隶属于27个科共计127种甲螨可作为裸头科绦虫的中间宿主[8]。经过胃液消化的六钩蚴在甲螨体腔中移行，并发育为似囊尾蚴（图268）。当终末宿主在受污染的草原或

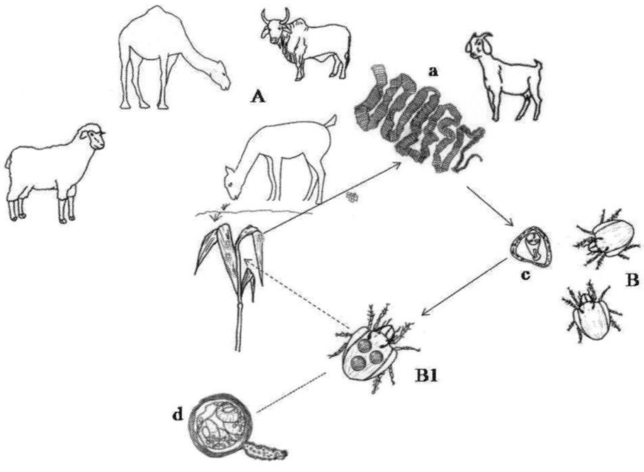


图266　莫尼茨绦虫生活史（A）终末宿主；（B）中间宿主——甲螨；（B1）受感染甲螨——（a）成虫；（b）莫尼茨绦虫虫卵；（c）似囊尾蚴

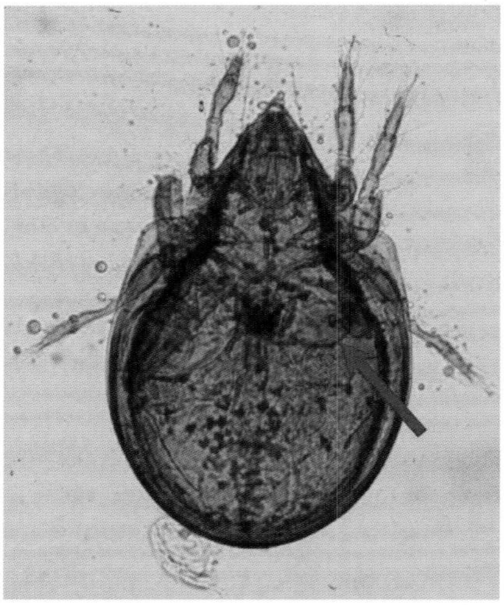

图267　棒菌甲螨（*Schelorbates latipes*）一种土壤螨，为莫尼茨绦虫的中间宿主。箭头指向位于甲螨体腔的一个似囊尾蚴

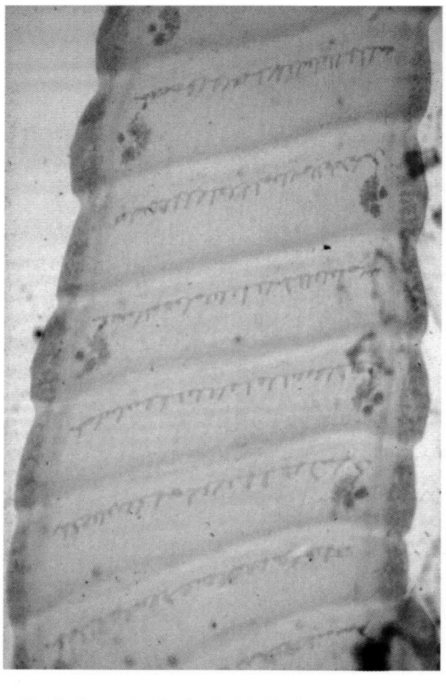

图265　盖氏曲子宫绦虫成熟节片。每一节片含一套生殖器官，管状子宫呈横向排列

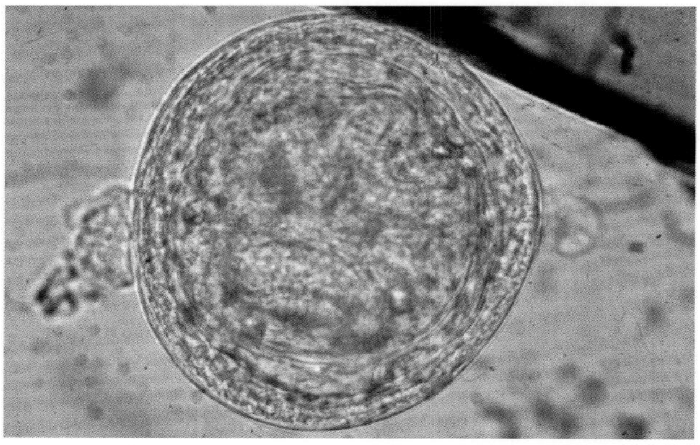

图268　莫尼茨绦虫的似囊尾蚴机械性地从甲螨体腔中溢出

者土壤上放牧时，或者食入新鲜牧草，苜蓿后遭受感染。

盖氏曲子宫绦虫也是将甲螨作为中间宿主[8]，但是对已发表的其他提及的绦虫的生活史资料应该谨慎对待。

## 流行病学

裸头科绦虫病的感染为食入性的经口感染。感染骆驼的裸头科绦虫种没有宿主特异性。小反刍兽和牛是主要宿主。野生反刍兽为保藏宿主。扩展莫尼茨绦虫和贝氏莫尼茨绦虫呈世界性分布，分布于非洲[3,9,13]和亚洲[2,5,18]的单峰驼，美洲[4,6,10,12,16]和欧洲[11,14,20]的新大陆骆驼。有研究资料进一步显示，斯克里亚宾莫尼茨绦虫（*M. skrjabini*）在蒙古双峰驼中也有分布[24]。

莫尼茨绦虫的中间宿主是营非寄生生活的甲螨。甲螨生活在地表或在土壤浅层，以真菌、藻类和其他有机材料为食，也包括蠕虫卵。甲螨能够钻入土壤1m之深，消夏或越冬。在适宜条件下，特别是在夜间或者清晨时分，这些厌旱性甲螨移行到植物。不同种类和密度的甲螨可在灌溉过的草坪、永久性草场或者水域旁边观察到。只有那些体长超过0.3mm、头部长有特殊口器的螨类可以作为中间宿主，这些螨类能够使莫尼茨绦虫卵的外层或者其他裸头科的绦虫的子宫旁器破裂。具有这些特征的螨属于翼甲螨属（*Achipteria*）、鲜甲螨属（*Cepheus*）、大翼甲螨属（*Galumna*）、丽甲螨属（*Liacarus*）、*Liebstadia*、垂盾甲螨属（*Scutovertex*）、菌甲螨属（*Scheloribates*）、毛甲螨属（*Trichoribates*）和合若甲螨属（*Zygoribatula*）等。在自然条件下，甲螨感染似囊尾蚴的比例不足1%，但是这些甲螨携带似囊尾蚴的数目较多，从而给终末宿主造成感染的威胁巨大[21]。受感染的甲螨可携带多达11个似囊尾蚴[22]。

鉴于甲螨的生命周期长达15~19个月，加之莫尼茨绦虫卵在5~8℃和足够潮湿的地方保存时可存活7月，所以草场污染后长达2年时间内仍具有感染性。但是，莫尼茨绦虫卵对热和干燥较为敏感，可在数天内死亡[23]。

肠道内寄生的斯泰勒属（*Stilesia*）分布于非洲[3,15]和亚洲[2,5,9,13,18]的单峰驼。中点无卵黄腺绦虫和肝斯泰勒绦虫只发现于苏丹单峰驼[1]。

盖氏曲子宫绦虫分布范围广泛，在全世界数种反刍兽中均有分布，但是在单峰驼中曲子宫属绦虫的感染仅见于一篇报道[7]。盖氏曲子宫绦虫作为新大陆骆驼的寄生虫只是在文献中提及[4,25]。

放射穗体绦虫（*Thysanosoma actinoides*）是一种能够定居在反刍兽肝脏中的胆管的绦虫，到目前为止，它只是在美国有报道，但是它也曾在沙特阿拉伯联合酋长国的一匹单峰驼上报道过[17]。

## 临床症状和病理学

反刍兽莫尼茨绦虫病和其他裸头科绦虫病的危害一直是一个有争议的话题。危害程度取决于荷虫量和宿主的年龄结构。到目前为止，还没有关于裸头科绦虫对骆驼影响的详细研究。然而，人们已经有一些有关动物裸头科绦虫病特别是小反刍兽莫尼茨绦虫病的经验[19]。由于肠道内寄生的绦虫会从肠道食糜中消耗机体的必要营养成分、维生素和微量元素，所以人们可观察到宿主感染后体重减轻。绦虫为了在严酷的小肠内生存，必须具备改变食糜pH值的能力。这会导致酶浓度或者酶活性的下降，这也可能导致通常在大肠内繁殖的细菌如厌氧菌在小肠滋生，这些细菌产生毒素，而产维生素微生物的生长则会受到抑制。幼龄动物特别是在第一个放牧季节容易受到感染。当宿主荷虫量大时，会导致宿主小肠内食糜停滞，直至便秘。

剖检时，感染动物荷虫量大时，小肠肿胀，通过肠壁可观察到淡黄色的绦虫虫体。莫尼茨绦虫、中点无卵黄腺绦虫和盖氏曲子宫绦虫通常位于空肠，而斯泰勒绦虫则寄生于十二指肠。

## 诊断

在莫尼茨绦虫和曲子宫绦虫感染的病例中，特别是潜伏期的开始阶段，当由15个孕节组成的节片从宿主排出时，可以在粪便表面观察到绦虫节片。仅在莫尼茨绦虫感染时粪便漂浮法检测呈阳性结果，因为在文中提及的其他绦虫感染时，裸头科绦虫的卵被包裹在子宫旁器中。扩展莫尼茨绦虫虫卵呈四面体，在显微镜观察时呈三棱锥体（图269），而贝氏莫尼茨绦虫虫卵呈立方体，看上去像投影的方块（图270）。

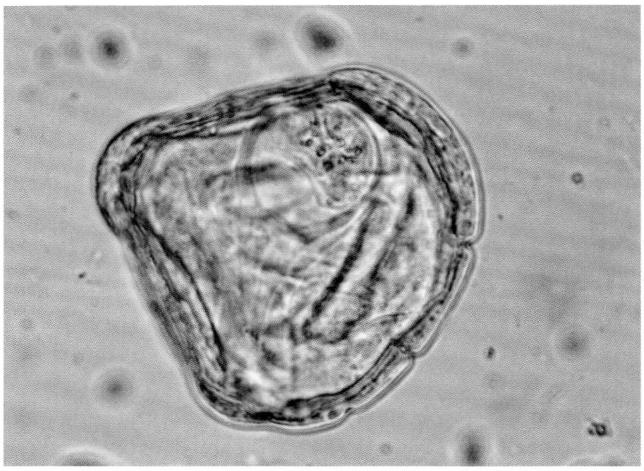

图269 莫尼茨绦虫虫卵，其投影呈三棱锥体。六钩蚴镶嵌在梨形胚膜中

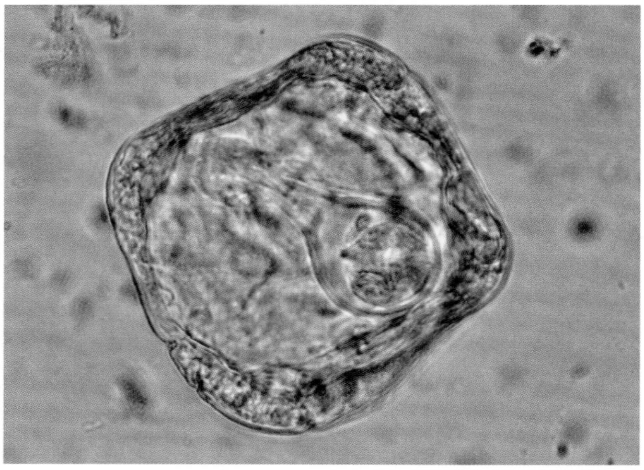

图270 贝氏莫尼茨绦虫虫卵，其投影呈方块

**治疗与控制**

有众多的市售的高效抗蠕虫药可用于裸头科绦虫病的治疗。多数抗虫药物为液体悬浮剂，但是也有一些抗虫药为片剂、玻璃水样或者糊剂。吡喹酮按每千克体重3.75~5mg剂量给药时可取得最佳抗虫效果。此剂量下，吡喹酮仅对肠道中的绦虫有效。在绦虫与胃肠道线虫或者肺丝虫混合感染的病例中，建议使用苯并咪唑类药物（benzimidazole），如苯硫咪唑（fenbendazole）和阿苯达唑（albendazole），用药剂量为每千克体重10~15mg。在世界一些地方，抗绦虫药如杀螺胺（niclosamid）、丁萘脒（bunamidin）、diclorphen、硫双二氯酚（bithionol）、sulphen和oksid仍在继续用于治疗裸头科绦虫病。

由于甲螨是恐干症性节肢动物，所以裸头科绦虫的感染可通过给家畜饲喂干草而得以控制。

## 参考文献

[1] Abdel Rahman M.B., Osman A.Y. & Hunter A.G. (2001). – Parasites of the one-humped camel (*Camelus dromedarius*) in the Sudan: a review. *Sudan J. Vet. Res.*, 17, 1 – 13.

[2] Anwar M. & Hayat C.S. (1999). – Gastrointestinal parasitic fauna of camel (*Camelus dromedarius*) slaughtered at Faisalabad abattoir. *Pakistan J. Biol. Sci.*, 2 (1), 209 – 210.

[3] Bekele T. (2002). – Epidemiological studies on gastrointestinal helminthes of dromedary (*Camelus dromedarius*) in semi-arid lands of eastern Ethiopia. *Vet. Parasitol.*, 105, 139 – 152.

[4] Beldomenico P.M., Uhart M., Bonco M.F., Marull C., Baldi R. & Peralta J.L. (2003). – Internal parasites of free ranging guanacos from Patagonia. *Vet. Parasitol.*, 118, 71 – 77.

[5] Borji H, Razmi Gh., Movassaghi A.R., Naghibi A.Gh. & Maleki M. (2010). – A study on gastrointestinal helminthes of camels in Mashhad abattoir, Iran. *Iranian J. Vet. Res. Shiraz Univers.*, 11 (2), 174 – 179.

[6] Chavez C., Guerrero C., Alva J. & Guerrero J. (1967). – El parasitismo gastrointestinal en alpaca. *Rev. Fac. Med. Vet. Lima*, 21, 9 – 19.

[7] Dakkak A. & Ouheli H. (1987). – Helminths and helminthoses of the dromedary. *In* Diseases of camels. *Rev. sci. tech.*

*Off. int. Epiz.*, 6 (2), 447 – 461.

[8] Denegri G. (1993). – Review of oribatid mites as intermediate hosts of tapeworms of the Anoplocephalidae. *Exp. Appl. Acarol.*, 17, 567 – 580.

[9] El Bihari S. (1985). – Helminths of the camel, a review. *Br. Vet. J.*, 141, 315 – 326.

[10] Fowler M.E. (1998). – Parasites. *In* Medicine and surgery of South American camelids (M.E. Fowler, ed.). Iowa State Press, Ames, IA, 231 – 269.

[11] Gareis A. (2008). – Feldstudie zum Vorkommen von Endoparasiten bei Neuweltkameliden in Ecuador. Thesis, Universitat Leipzig, Germany.

[12] Guerrero C. & Leguia G. (1987). – Enfermedades infecciosas y parasitarias de alpacas. *Rev. Camélidos Sudamericanos*, 4, 32 – 82.

[13] Haroun E.M., Mahmoud O.M., Magzoub M., Abdel Hamid Y. & Omer O.H. (1996). – The haematological effects of the gastrointestinal nematodes prevalent in camels (*Camelus dromedarius*) in central Saudi Arabia. *Vet. Res. Comm.*, 20, 25 – 264.

[14] Herzberg H. & Kohler L. (2006). – Prevalence and significance of gastrointestinal helminths and protozoa in South American Camelids in Switzerland. *Berl. Münch. Tierärztl. Wschr.*, 119, 291 – 294.

[15] Kendall S.B. (1974). – Some parasites of domestic animals in the Aswan Governorate Arab Republic of Egypt. *Trop. Anim. Health Prod.*, 6, 128 – 130.

[16] Leguia P. (1999). – Enfermedades parasitarias de camelidos sudamericanos. Editorial de Mar, Lima.

[17] Omer O.H. & Al-Sagair O. (2005). – The occurrence of *Thysanosoma actinoides* Diesing, 1834 (Cestoda: Anolocephalidae) in a Najdi camel in Saudi Arabia. *Vet. Parasitol.* 131, 165 – 167.

[18] Parasani H.R., Singh V. & Momin R.R. (2008). – Common parasites of camel. *Vet. World*, 1 (10), 317 – 318.

[19] Potemkina V.A. (1973). – Control of intestinal cestodoses of animals [in Russian]. Kolos, Moskva.

[20] Schlogl C. (2010). – Erhebungen zum Vorkommen von Endo- sowie Ektoparasiten bei Neuweltkameliden. Thesis, Ludwig- Maximilians-Universitat, Munich, Germany.

[21] Schuster R. (1988). – Untersuchungen zur Epizootiologie der Monieziose des Schafes in der DDR unter besonderer Berucksichtigung der Zwischenwirte. *Mh. Vet.-Med.*, 43, 233 – 235.

[22] Schuster R. (1995). Experimental studies on the influence of *Moniezia expansa* infection on oribatid mites. *J. Helminthol.*, 69, 177 – 179.

[23] Schuster R., Becker P. & Hiepe Th. (1989). – Untersuchungen zur Tenazitat von Eiern des Schafbandwurmes *Moniezia expansa. Wien. Tierärztl. Mschr.*, 75, 466 – 468.

[24] Sharhuu G. & Sharkhuu T. (2004). – The helminth fauna of wild and domestic ruminants in Mongolia, a review. *Europ. J. Wildl. Res.*, 50, 150 – 156.

[25] Spasskij A.A. (1951). – Essentials of cestodology. I. Anoplocephalata – Tapeworms of domestic and wild animals [in Russian]. Izd. Akademii Nauk SSSR, Moskva.

（贾万忠译，殷宏校）

## 棘球蚴病（echinococcosis）

棘球蚴病，又称包虫病（hydatidosis）或者囊性棘球蚴病（cystic echinococcosis），为草食和杂食哺乳动物的一种疾病，由细粒棘球绦虫（*Echinococcus granulosus*）的幼虫阶段引起。它也是一种极其危险的人兽共患病。棘球蚴为组织内寄生虫，主要寄生于肝和肺，有时也可以寄生于脾、子宫、肾、心脏和腹腔，偶尔见于肌肉、骨骼或眼睛。

在过去，棘球属绦虫（*Echinococcus*）共分为6个种，但是用于现代分子生物学技术对棘球属绦虫的分类研究仍在进行中。这6个种包括：细粒棘球绦虫（包括10个虫株或基因型）、多房棘球绦虫（*E. multilocularis*）、福氏棘球绦虫（*E. vogeli*）、少节棘球绦虫（*E. oligarthra*）、石渠棘球绦虫（*E. shiquicus*）和狮猫棘球绦虫（*E. felidis*）。然而，最新的研究结果表明，细粒棘球绦虫的某些虫株可以升格为独立种地位（译者注：最新的分类学修正结果认为棘球属绦虫至少包括9个种）。马株（G4）主要在马属动物流行，现修正为马棘球绦虫（*E. equinus*）；牛株（G5）重新命名为奥氏棘球绦虫（*E. ortleppi*）；之前的普通绵羊株（G1）与塔斯马尼亚绵羊株（G2）和水牛株（G3）一起组成细粒棘球绦虫狭义种（*E. granulosus* sensu stricto）；而细粒棘球绦虫广义种（*E. granulosus* sensu lato）包括骆驼株（G6）、猪株（G7）、两个鹿株（G8，G10）和波兰株（G9），这些虫株共同组成一个种，并命名为加拿大棘球绦虫（*E. canadensis*）。狮棘球绦虫与细粒棘球绦虫狭义种亲缘关系较近，是在非洲狮群中鉴定出的一个棘球绦虫种[16]。

### 形态学

细粒棘球绦虫体长6mm，由3~4个节片组成（图271）。头节直径为0.25~0.35 mm，其上分布4个吸盘和1个顶突，顶突上分布有2圈交替排列的带科绦虫小钩。成熟节片长大于宽，而孕卵节片则更长，节片内充满多达至1 500枚虫卵。

幼虫为单房囊，囊内充满液体（图 272和图273）。棘球蚴包囊，体形小至胡桃大至椰子，由双层膜组成，外层为致密的角质层，内膜为生发层（图274）。包囊外围被宿主结缔组织包裹。

过去，人们对角质层的特性了解甚少。近来，有研究表明，角质层由黏蛋白组成，富含半乳糖碳水化合物。它的功能是刺激宿主免疫系统产生非炎性应答反应[9,10]。

在内层生发膜上的育囊中产生原头蚴。棘球可育囊可产生数十万的原头蚴（棘球蚴砂，图275和图276）。

### 生活史

棘球属绦虫的生活史中需要更换宿主（图277）。成虫寄生于终末宿主——驯养和野生犬科动物（家犬、狼、狸猫/貉、丁狗、效狼、豺、狐狸和狮子）的小肠中；中间宿主主要是草食动物和杂食动物，当中间宿主通过摄入棘球绦虫卵而遭受感染。虫卵从终末宿主排出后即对任何潜在的中间

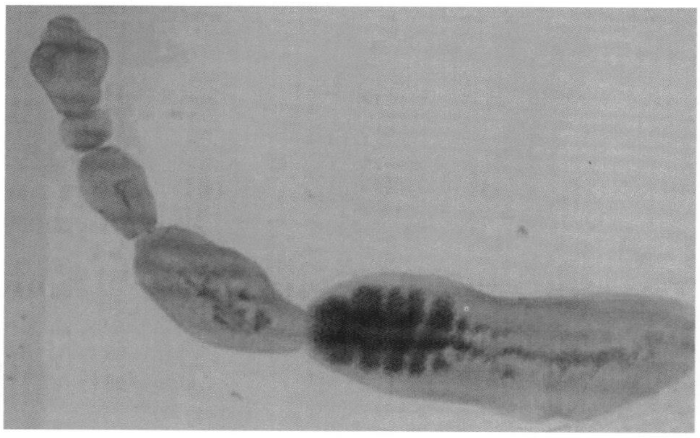

图271　细粒棘球绦虫成虫

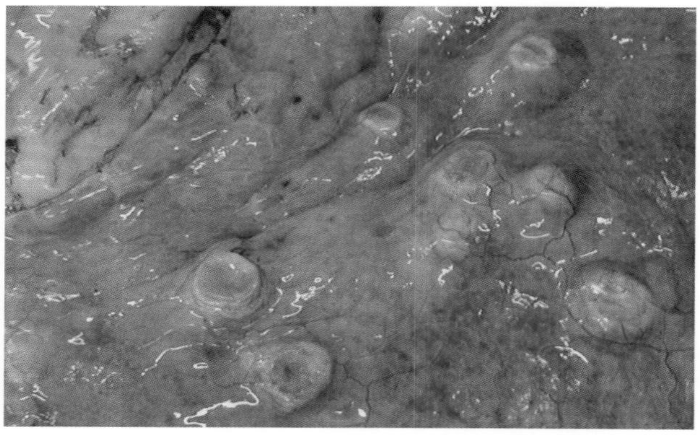

图272　骆驼肺部棘球蚴包囊

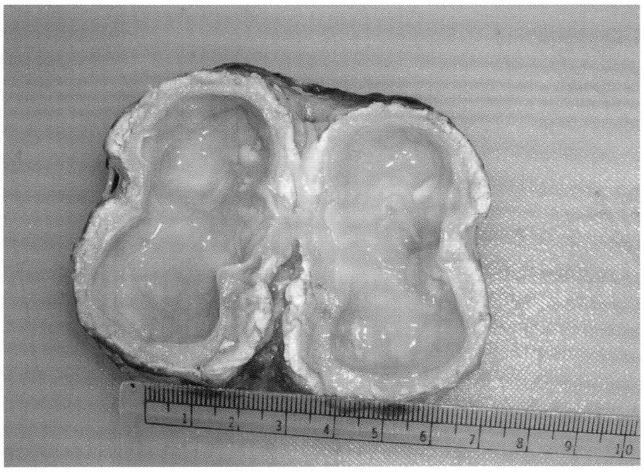


图273　分离出的棘球蚴包囊。生发层像一层胶状物

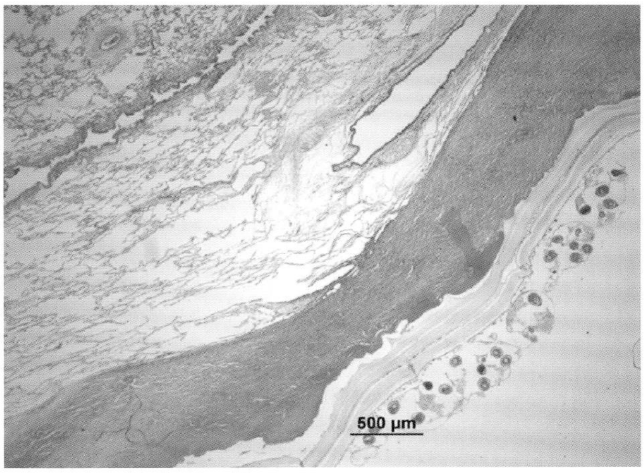


图274　棘球蚴包囊组织学结构。含有原头蚴的育囊最初是与生发层相连的。发育后期，这些育囊从生发层上脱落下来，以棘球蚴砂的形式漂浮在囊液中

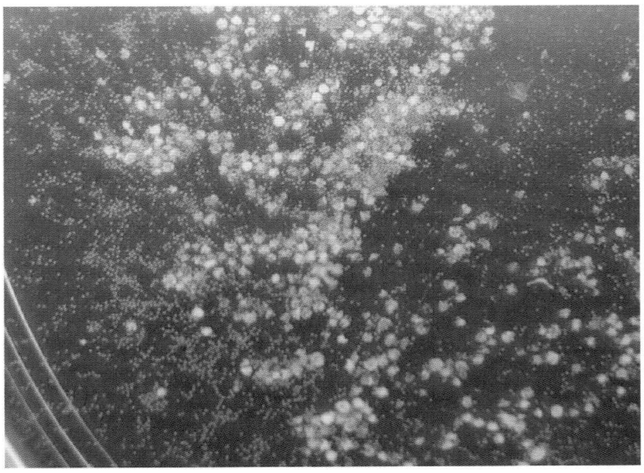

图275　棘球蚴砂，由育囊和原头蚴组成

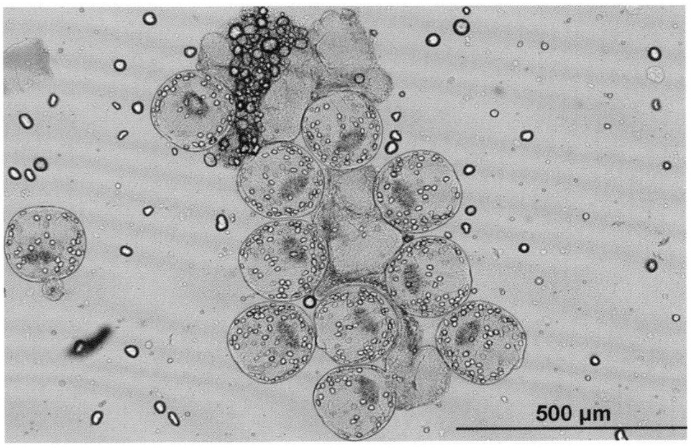


图276　细粒棘球蚴原头蚴

宿主有感染性。虫卵进入十二指肠后，虫卵外壳被宿主消化液中蛋白水解酶消化，胆汁刺激六钩蚴的孵化。然后幼虫穿过小肠黏膜，进入血管，移行至肝、肺和其他内脏器官，幼虫在定居器官中开始发育形成一个小包囊，然后逐渐发育为棘球蚴成熟包囊，大约需要数月。当犬科动物食入含可育囊的废弃脏器后，棘球绦虫将完成其整个生活史。当原头蚴在胆汁的刺激下头节外翻，然后利用吸盘和顶突上的小钩吸附在小肠黏膜上，原头蚴牢固固着于肠绒毛之间。

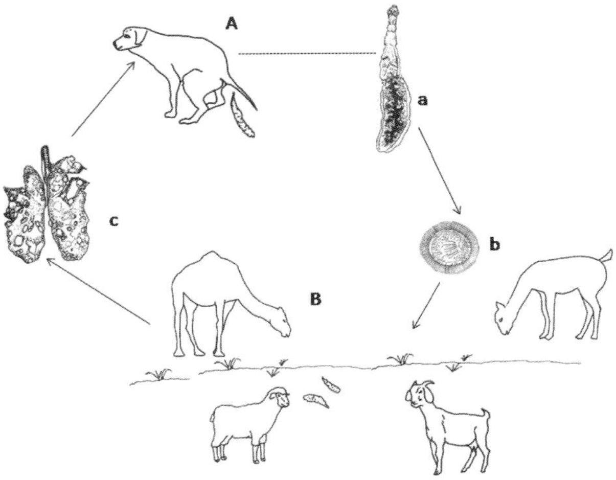


图277　细粒棘球绦虫生活史。（A）终末宿主；（B）中间宿主；（a）成虫；（b）虫卵；（c）肺棘球蚴

## 流行病学

棘球蚴的感染途径为经口感染，当中间宿主在虫卵污染的草原上放牧时即遭经口感染。有文献

报道称，单峰驼有时可经子宫内感染[15]。

棘球蚴病呈全球性分布，是地中海和黑海附近欧洲国家、北非和撒哈拉以南地区国家、中东、中亚和南美等的一种重要寄生虫病[4]。

细粒棘球绦虫骆驼株（G6）现定名为加拿大棘球绦虫，主要危害骆驼和山羊，分布于中东和北非以及东非。调查表明，加拿大棘球绦虫是肯尼亚图尔卡纳高度流行区人棘球蚴病的传染源[5]。除了人体病例外，加拿大棘球绦虫分布于肯尼亚的单峰驼、牛、绵羊、山羊和猪以及苏丹的单峰驼、牛和绵[11,21]。在毛利塔利亚，加拿大棘球绦虫可感染骆驼、牛和人[3]。在利比亚，所有从单峰驼中检查出来的棘球蚴均属于加拿大棘球绦虫的幼虫，从牛中检查出来的棘球蚴多数也属于加拿大棘球绦虫，而人体棘球蚴为细粒棘球绦虫普通绵羊株（G1）[1]。加拿大棘球绦虫也见于阿根廷、智利、中国、尼泊尔和伊朗等国家。然而，骆驼也常常感染广为分布的普通绵羊株（G1）[17,24]。

中东和北非阿拉伯国家棘球蚴病流行情况见表85。

有关新大陆骆驼棘球蚴病的资料不多。虽然近年来的调查结果表明，加拿大棘球绦虫在南美洲的人、绵羊、牛、猪和山羊中有分布，但是人们对该区域骆驼感染棘球蚴病的现状尚不清楚。一些旧文献记载表明，秘鲁的羊驼棘球蚴的感染率为19%，其中55%的棘球蚴为可育囊[18]。对来自智利的野生骆马（大羊驼）剖检表明，棘球蚴的感染率为12%，包囊为可育囊[7]。

终末宿主是通过在屠宰场摄食荷有棘球蚴包囊的废弃内脏器官或者捕食腐肉时遭受感染的。棘球绦虫在终末宿主的潜伏期为5~8周，具体的时间长短随虫种和虫株的不同而有差异。部分子宫内的虫卵在链体末段裂解时与宿主肠内容物混合。节片具有一定的爬行或者蠕动能力，这会帮助节片从粪便中排出。在节片爬行过程中，节片肌肉收缩，其结果是导致节片内子宫和虫卵在环境中不断散布。由于棘球绦虫虫卵外壳厚，所以它们具有较强的抵抗力，在外界环境中可保持感染性长达数天、数周，甚至数月之久，具体时间受外界环境的影响而有所不同。在犬体，成虫生存期约为6个月，只有个别报道，成虫期存活期长达2年。根据流行病学传播链，主要有3种传播方式在此进行讨论[3]。

（1）半野生循环赛型——豺和野犬作为终末宿主参与循环。后者因摄入荷有棘球蚴包囊的患病动物内脏时遭受感染，然后通过粪便向周围水域传播虫卵，水域周围家畜密集和活动频繁，从而加速家畜棘球蚴病的流行与发生。

（2）乡村循环型——因牧羊犬或者营地护卫犬而导致此循环型的发生，这些犬在白天拴在牧场或营地，夜间自由活动而且接近畜群。

表85　中东和北非阿拉伯国家棘球蚴病流行情况[22]

| 国家 | 犬（%）<br>剖检法 | 人 | 小反刍兽（%）<br>屠宰场资料 | 牛（%）<br>屠宰场资料 | 单峰驼（%）<br>屠宰场资料 |
|---|---|---|---|---|---|
| 土耳其 | 0.32~40 | (0.87~6.6)/10万（手术） | 11.2~50.7 | — | — |
| 伊朗 | 3.3~63.3 | 5.4%~5.9%（血清学） | 2~3 | 4 | 35.2 |
| 伊拉克 | 2.6~49.5 | 2/10万（手术）<br>7.7%（血清学） | 15 | 20.6 | — |
| 黎巴嫩 | — | — | — | 7 | 100 |
| 叙利亚 | — | — | 3 | — | 18.2 |
| 阿曼 | — | 0.3%（血清学） | — | — | 13（血清学） |
| 沙特阿拉伯 | — | — | 14.9~29.5 | 3.1 | 6 |
| 科威特 | — | 3.6/10万 | — | — | — |
| 阿尔及利亚 | — | (3.6~4.6)/10万 | — | — | — |
| 埃及 | — | 6.1%（血清学） | — | — | 18.9~31 |
| 利比亚 | 21.6~25.8 | 4.2/10万 | 17~33.4 | 1~13.9 | 1.4~40 |
| 摩洛哥 | 33 | (3.5~15.8)/10万 | 9.9 | 42 | — |
| 突尼斯 | — | 15/10万 | — | — | 10.1 |

（3）城市或郊区循环型——因流浪犬或者家庭和市场警卫犬所致，这些犬到处摄食动物废弃脏器，又同时与绵羊、山羊和骆驼等动物以及人类密切接触，这些动物在城区或饲养或被屠宰。

### 临床症状和病理学

棘球蚴病并不引起中间宿主的特异性临床症状，它的诊断依据剖检结果或者肉品检验时的结果。临床症状因棘球蚴包囊的感染强度、大小和寄生部位而异。生长期的棘球蚴将对宿主周围组织造成挤压，在荷囊量大时棘球蚴可影响宿主器官的功能。在骆驼，肺是棘球蚴寄生的最适宜器官[12]。作者对寄生于21只骆驼肺脏的116个包囊进行了检查，结果表明，93%的包囊含有棘球蚴砂，而所有53个肝脏包囊均为可育囊。与肺包囊相比，肝棘球蚴包囊囊壁较厚。肺作为棘球蚴包囊寄生的主要部位，导致参赛骆驼体能下降。肝棘球蚴可导致腹水和黄疸。当棘球蚴包囊偶尔寄生于心脏、肾、子宫、脑或眼部时也可出现其他各种临床症状。包囊破裂，会导致动物过敏性休克，进而引起动物死亡。育囊和原头蚴释放到腹腔或者胸腔，育囊和原头节继续发育，形成大的、游离而漂浮的外囊。棘球蚴囊外被覆宿主结缔组织纤维囊。非活动性包囊和陈旧性包囊会引起干酪化和钙化。

### 诊断

家畜棘球蚴病可在肉品检验或者剖检时做出诊断。棘球蚴可育囊为单房型包囊，淡黄色，大小差异很大，囊内充满囊液和棘球蚴砂。较新生的棘球蚴包囊中，可育囊与包囊的生发层相连，内含原头蚴。在这种情况下，从内膜取刮取物进行显微镜检查来进行诊断是可行的，这是一种值得推荐的诊断方法。外翻的原头节直径为150~200μm，小钩清晰可见。棘球蚴包囊并不总是位于薄壁组织器官的表面。为了排除假阴性结果在肉品检验中对肺脏进行触诊是必要的。棘球蚴包囊必须与肺和肝细菌性脓肿，如棒状杆菌（*Corynebacterium*）或红球菌（*Rhodococcus*）引起的脓肿相鉴别。

对于人棘球蚴病的活体诊断而言，影像学和多种以酶联免疫反应为基础的血清学方法可以利用，这些诊断技术会得到最为可靠的诊断结果[13]。虽然棘球绦虫不同种或者虫株之间宿主范围和形态学特征有所差异，但是其确切的分类地位的确定需要依靠分子生物学方法来实现[12]。

终末宿主棘球绦虫成虫感染的诊断可通过粪检法来进行。但是，棘球属绦虫虫卵与其他带科绦虫如泡状带绦虫（*Taenia hydatigena*）、羊带绦虫（*T. ovis*）和其他带属绦虫的卵不易区分。在这种情况下，过去往往使用槟榔碱泻下法进行治疗性驱虫来进行诊断。从宿主小肠排出的绦虫分布于粪便表面。这一传统方法正逐渐被更安全的特异性粪抗原检测法或者PCR（聚合酶链反应）法所替代[8,12]。

### 治疗与控制

目前，人们对家畜棘球蚴病的治疗还没有好的方法。所有的措施是针对切断传播循环链而设计的。在流行区，流浪犬可能成为潜在传染源，应该禁止给犬喂食带棘球蚴包囊的废弃内脏，死亡动物的尸体应焚烧或者深埋。比较好的做法是对废弃脏器进行煮沸，以免家犬遭受感染。对家犬进行有效的驱虫，驱虫药可选择吡喹酮异喹啉衍生物，剂量按每千克体重5.5mg给药。虽然异喹啉制剂对绦虫体壁有杀伤作用，但是绦虫卵仍具有感染性，因此，应对驱虫治疗后的家犬粪便进行适当处理。

盐酸萘衍生物（naphthalin derivate bunamidin hydrochloride）制剂效果不佳，已不再作为治疗犬棘球绦虫感染的药物。



## 参考文献

[1] Abushhewa M.H., Abushhiwa M.H., Nolan M.J., Jex A.R., Campbell, B.E., Jabbar A. & Gasser R.B. (2010). – Genetic classification of *Echinococcus granulosus* cysts from humans, cattle and camels in Libya using mutation scanning-based analysis of mitochondrial loci. *Mol. Cell Probes*, 24 (6), 346 – 351.

[2] Ahmadi N. & Dalimi A. (2006). – Characterization of *Echinococcus granulosus* isolates from human, sheep and camel

in Iran. *Inf. Genetics Evol.*, 6, 85 – 90.

[3] Bardonnet K., Piarroux Dia L., Schneegans F., Beurdeley A., Godot V. & Vuitton D.A. (2002). Combined eco-epidemiological and molecular biology approaches to assess *Echinococcus granulosus* transmission to humans in Mauritania, occurrence of the 'camel' strain and human cystic echinococcosis. *Trans. Roy. Soc. Trop. Med. Hyg.*, 96, 383 – 386.

[4] Cardona G.A. & Carmena D. (2013). – A review of the global prevalence, molecular epidemiology and economics of cystic echinococcosis in production animals. *Vet. Parasitol.*, 192, 10 – 32.

[5] Casulli A., Zeyhle E, Brunetti E., Pozio E., Meroni V., Genco F. & Filice C. (2010). – Molecular evidence of the camel strain (G6 genotype) of *Echinococcus granulosus* in humans from Turkana, Kenya. *Trans. Roy. Soc. Trop. Med. Hyg.*, 104, 29 – 32.

[6] Center for Food Security and Public Health (2011). – Echinococcosis. Available at: www.cfsph.iastate.edu/Factsheets/pdfs/echinococcosis.pdf (accessed 14 August 2013).

[7] Cunazza C. (1978). – Enfermedades y parasitos del guanaco. *In* El guanaco de Magallanes, Chile. Distribucion y biologia (J. Raedeke Kenneth & C. Cunazza, eds). CONAF, Santiago, Pub. Cientifica Nro. 4. Apendice 1.

[8] Deplazes P., Gottstein B., Eckert J., Jenkins D.J., Ewald D. & Jimenez-Palacios S. (1992). – Detection of *Echinococcus* coproantigens by enzyme-linked immunosorbent assay in dogs, dingoes and foxes. *Parasitol. Res.*, 78, 303 – 308.

[9] Diaz A., Casaravilla C., Allen J.E., Sim R.B. & Ferreira A.M. (2011). – Understanding the laminated layer of larval *Echinococcus* II: immunology. *Trends in Parasitol.*, 27 (6), 263 – 272.

[10] Diaz A., Casaravilla C., Irigoin F., Lin G., Previato J.O. & Feirra, F. (2011). – Understanding the laminated layer of larval *Echinococcus* I: Structure. *Trends in Parasitol.*, 27 (5), 204 – 212.

[11] Dinkel A., Njoroge E.M., Zimmermann A., Waelz M., Zeyhle E., Elamahdi I.E., Mackenstedt U. & Romig T. (2004). – A PCR system for the detection of species and genotypes of the *Echinococcus granulosus*-complex, with reference to the epidemiological situation in eastern Africa. *Int. J. Parasitol.*, 34, 645 – 653.

[12] Dinkel A., von Nickisch-Rosenegk M., Bilger B., Merli M., Lucius R. & Romig T. (1998). – Detection of *Echinococcus multilocularis* in the definitive host, coprodiagnosis by PCR as an alternative to necropsy. *J. Clin. Microbiol.*, 36 (7), 1 871 – 1 876.

[13] Eckert J. & Deplaces P. (2004). – Biological, epidemiological, and clinical aspects of echinococcosis, a zoonosis of increasing concern, *Clin. Microbiol. Rev.*, 17 (1), 107 – 135.

[14] Eckert J., Thompson R.C.A., Michael S.A., Kumaratilake L.M. & El-Sawah H.M. (1989). – *Echinococcus granulosus* of camel origin, development in dogs and parasite morphology. *Parasitol. Res.*, 75, 536 – 544.

[15] Elamin E.A., Ramadan R.O., Fatani A.E., Atiya S.A. & Abdin-Bey M.R. (2001). – Prenatal infection with a hydatid cyst in a camel (*Camelus dromedarius*). *Vet. Rec.*, 149, 59 – 60.

[16] Huttner M., Nakao M., Wassermann T., Siefert L., Boomker J.D.F., Dinkel A., Sako Y., Mackenstedt U., Romig T. & Ito A. (2008). – Genetic characterization and phylogenetic position of *Echinococcus felidis* Ortlepp, 1937 (Cestoda: Taeniidae) from the African lion. *Intern. J. Parasitol.*, 38, 861 – 868.

[17] Kamenetzky L., Gutierrez A.M., Canova S.G., Haag K.L., Guarnera E.A., Parra A., Garcia G.E. & Rosenzvit M.C. (2002). – Several strains of *Echinococcus granulosus* infect livestock and humans in Argentina. *Infect. Genetics Evol.*, 2, 129 – 136.

[18] Leguia G. (1991). – The epidemiology and economic impact of llama parasites. *Parasitol. Today*, 7, 54 – 56.

[19] Manterola C., Benavente F., Melo A., Vial M. & Roa J.C. (2008). – Description of *Echinococcus granulosus* genotypes in human hydatidosus in a region of southern Chile. *Parasitol. Intern.*, 57, 342 – 346.

[20] Nakao M., McManus D.P., Schanz P.M., Craig P.S. & Ito A. (2007). – A molecular phylogeny of the genus *Echinococcus* inferred from complete mitochondrial genomes. *Parasitology*, 134, 713 – 722.

[21] Omer R.A., Dinkel A., Romig T., Mackenstedt U., Elnahas A.A., Aradaib I.A., Ahmed M.E., Elmalik K.H. & Adam A. (2010). – A molecular survey of cystic echinococcosis in Sudan. *Vet. Parasitol.*, 169, 340 – 346.

[22] Sadjjadi S.M. (2006). – Present situation of echinococcosis in the Middle East and Arabic North Africa. *Parasitol. Internat.*, 55, S197 – S202.

[23] Schneider R., Gollackner B., Schindl M., Tucek G. & Auer H. (2010). – *Echinococcus canadensis* G7 (pig strain): an underestimated cause of cystic echinococcosis in Austria. *Am. J. Trop. Med. Hyg.*, 82 (5), 871 – 874.

[24] Shahnazi M., Hejazi H., Salehi M. & Andalib A.R. (2011). – Molecular characterisation of human and animal *Echinococcus granulosus* isolates in Isfahan, Iran. *Acta Trop.* 117 (1), 47 – 50.

[25] Soriano S.V., Pierangeli B., Pianciola L., Mazzeo M., Lazzarini L.E., Saiz M.S., Kossman A.V., Bergana H.F.J., Chartier K.& Basualdo J.A. (2010). – Molecular characterisation of *Echinococcus* isolates indicates goats as reservoir for *Echinococcus canadensis* G6 genotype in Neuquen, Patagonia Argentina. *Parasitol. Internat.*, 59, 626 – 628.

（贾万忠译，殷宏校）

## 囊尾蚴病和脑多头蚴病

囊尾蚴病是由带属绦虫（*Taenia*）的幼虫引起的一种寄生虫病。囊尾蚴（*Cysticercus*）是一包囊，含有一凹陷的原头蚴。在旧大陆骆驼中，有文献曾报道存在3种囊尾蚴。Pellegrini曾报道，最初的发现表明单峰驼囊尾蚴（*C. dromedarii*）分布于东非的厄立特里亚的单峰驼和牛的肌肉和内脏器官，如肝、脾、淋巴结和大脑等[9]。后来，单峰驼囊尾蚴也在埃及、索马里、苏丹和乍得发现[3]。除了牛和骆驼外，单峰驼囊尾蚴能够感染山羊和绵羊，但是主要的宿主是羚羊[7]。饲喂试验证明，单峰驼囊尾蚴是鬣狗囊尾蚴（*T. hyaenae*）的幼虫[9]。然而，它与克罗卡特囊尾蚴（*T. crocutae*）也有一些相似性，因此不能否定Pellegrini的实验材料中含有两个不同的种的结论[11]。单峰驼囊尾蚴包囊呈圆形或卵圆形，直径达1cm。球状原头蚴直径为0.5~1mm，有4个大吸盘。每个吸盘直径为280~350μm。顶突上有两圈钩，钩的数目为34~44个。较大的钩长度为187~212μm，较小的钩长度为112~137μm。当囊尾蚴定居在肌肉、舌和心脏时，囊尾蚴周围包裹一致密结构的囊。当寄生于肝脏时，薄的囊尾蚴能够轻易地从它们的结缔组织囊上脱落下来，而囊尾蚴于脑部寄生时它缺乏外层的囊[1]。近来有一报道称，埃及艾斯尤特自治省骆驼感染单峰驼囊尾蚴时，所有囊尾蚴均分布于心脏左心室，囊尾蚴有不同程度的变性，这表明骆驼并非是单峰驼囊尾蚴的主要中间宿主。

细颈囊尾蚴（*C. tenuicollis*）是泡状带绦虫（*T. hydatigena*）的幼虫，呈全球性分布。细颈囊尾蚴主要黏附于小型反刍兽和猪的肝脏实质、肠系膜和大网膜（图278）。Dakkak和Ouheli曾引用一些较旧的描述单峰驼细颈囊尾蚴感染的文献[2]。

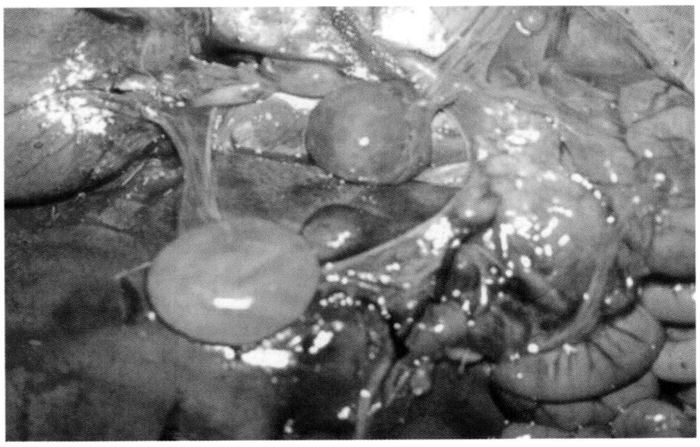

图278　细颈囊尾蚴。大部分细颈囊尾蚴附着于中间宿主的肠系膜上。个别情形下，细颈囊尾蚴寄生于草食动物的肝脏实质上

牛囊尾蚴（*C. bovis*）是人体带绦虫——牛带绦虫（*T. saginata*）的幼虫，在单峰驼肉品检验中也会碰到。Fahmy和El Affifi曾在埃及的一次肉品检验中检测到，单峰驼感染单峰驼囊尾蚴的同时携带有牛囊尾蚴样包囊，但是，未对其进行详细描述[5]。*T. hylicometra*绦虫的囊尾蚴与细颈囊尾蚴一起提及，在新大陆骆驼中有分布，但是它是一无效种[6]。*T. hylicometra*绦虫未曾在关于带科的相关专著中提及，因此它似乎是一无效种[1,8,11]。

脑多头蚴病是由隶属于多头属（*Multicpes*，译者注：大多数学者使用带属即*Taenia*作为本属的属名）的犬的绦虫的幼虫所致的一种寄生虫病。本病又称之为倒病、转圈病或者蹒跚病。脑多头蚴（*Coenurus cerebralis*）为多头绦虫（*M. multiceps*，译者注：大多数学者称其为多头带绦虫即*T. multiceps*）的幼虫，对畜牧业生产影响巨大。脑多头蚴主要寄生于小型反刍动物特别是绵羊的中枢神经系统。到目前为止，还没有报道称骆驼可感染脑多头蚴，但是教科书中已将单峰驼和双峰驼列为脑多头蚴的适宜中间宿主。脑多头蚴呈囊状，大小介于榛子和鸡蛋之间，囊内充满液体（图279）。囊上分布有多达数百个“芽”即原头蚴，成簇排列于囊的内生发层上。

多头带绦虫的生活史与细粒棘球绦虫相似。当中间宿主食入虫卵后，六钩蚴在小肠中孵化，然后随着血流移行至脑部。幼虫通过组织移行时会导致严重的中枢神经系统症状，有时对宿主是致命性的。感染后3个月，只有极少数幼虫能发育成为可育性脑多头蚴。

在过去，中枢脑多头蚴分布广泛，除美国、澳大利亚和新西兰外的许多养羊业发达国家均有脑多头蚴病的记载。中间宿主的明显临床症状、高效抗绦虫药和严格的肉品卫生检疫制度促进了许多地方对本病的根除。小型反刍兽中枢性脑多头蚴病仍在北非、中东和亚洲的国家存在[10]。

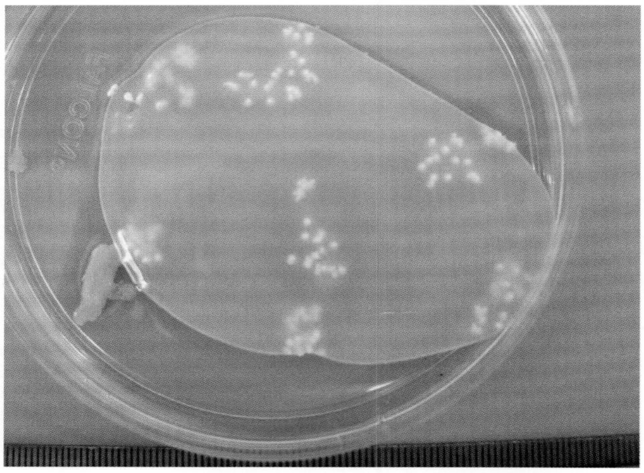

图279　单个分离的脑多头蚴。原头蚴成簇附着于生发层上

## 参考文献

[1] Abuladze K.I. (1964). – Essentials of cestodology. Vol. IV. Taeniata of animals and diseases caused by them [in Russian]. Izdatel'stvo Nauka, Moskva.

[2] Dakkak A. & Ouheli H. (1987). – Helminths and helminthoses of the dromedary. In Diseases of Camels. *Rev. sci. tech. Off. int. Epiz.*, 6 (2), 447 – 461.

[3] El Sergany, M.A., El Moussalami E. & El Nawawy F. (1971). – Histopathological studies on camel meat infested with *Cysticercus dromedarii. United Arab Republic Egyptian J. Vet. Sci.*, 7 (2), 191 – 200.

[4] Elbadri A.M., Hassan A.A., Zaghloli D.A., Elmataryi A.M.A. & Taher G.A. (2010). – Some studies on camel cysticercosis (*Cysticercus dromedarii*) in Assiut Governorate ICOPA XII, Abstract 1206.

[5] Fahmy M.A.M. & El Affifi A. (1964). – Cysticerci of the camel. *Zentralbl. Vet. Med.* B, 11, 147 – 150.

[6] Fowler M.E. (2010). – Medicine and surgery of South American camelids. Ames, Iowa State University Press, 254.

[7] Kutzer E. & Hinaidi H.K. (1968). – Beitrage zur Helminthenfauna Agyptens. 1. *Cysticercus cameli* Nomani, 1920. *Zentralbl. Vet. Med. B*, 15, 899 – 910.

[8] Loos-Frank B. (2000). – An update of Vester's (1969) 'Taxonomic revision of the genus *Taenia* Linnaeus (Cestoda)' in a table format. *System. Parasitol.*, 45, 155 – 183.

[9] Pellegrini D. (1947). – Il *Cysticercus dromedarii* Pelegrini, 1945, e lo stato larvale della *Taenia hyaenae* Baer, 1927. Identificacione morphologica esperimentale. *Boll. Soc. Ital. Med. Ig. Trop.* (*Sez. Eritrea*), 7 (5/6), 554 – 563.

[10] Sharma D.K. & Chauhan P.P.S. (2006). – Coenurosis status in Afro-Asian Region – a review. *Sm. Rum. Res.*, 64, 197 – 202.

[11] Vester A. (1969). – A taxonomic revision of the genus *Taenia* Linneus, 1758 s. str. *Onderstepoort J. Vet. Res.*, 36, 3 – 58.

（贾万忠译，殷宏校）

## 5.2.3 线虫感染

### 前言

根据寄生部位，骆驼线虫可分为4种类型。最大的一类为胃肠道线虫，这类线虫的代表科包括毛圆线虫科（Trichostrongylidae）、摩林线虫科（Molinidae）、钩口线虫科（Ancylostomidae）、夏伯特线虫科（Chabetriidae）、毛首线虫科（Trichuridae）、类圆线虫科（Strongyloididae）和胃线虫科（Habronematidae）。前四个科的种通常称为胃肠道圆线虫。网尾线虫科（Dictyocaulidae）和原圆线虫科（Protostrongylidae）寄生于肺，组成第二类。第三类线虫为组织寄生性，由盘尾丝虫科（Onchocercidae）组成。第四类线虫寄生于眼睑下，代表性科为吸吮线虫科（Thelaziidae）。

骆驼的线虫大多数为多宿主寄生，全球性分布，已适应寄生于其他宿主，主要为与骆驼有相同生存环境的反刍动物。在新大陆骆驼和旧大陆骆驼中描述过的线虫是捻转血矛线虫（*Haemonchus contortus*）和阴茎骆驼圆线虫（*Camelostrongylus mentulatus*），以及奥斯特属（*Ostertagia*）、古柏属（*Cooperia*）、细颈属（*Nematodirus*）、结节属（*Oesophagostomum*）、夏伯特属（*Chabertia*）和毛首属（*Trichuris*）的许多代表种。但是，也有一些线虫为宿主特异性种，例如羊驼纵纹线虫（*Graphinema auchenia*）秘鲁刺翼线虫（*Spiculopteragia peruvianus*）和查氏无峰驼线虫（*Lamanema chavesi*），到目前为止，这些线虫还在南美洲的新大陆骆驼中发现，或者毛里塔尼亚细颈线虫（*Nematodirus mauritanicus*）、嗜驼细颈线虫（*N. dromedarii*）、骆驼似细颈线虫（*Nematodirella cameli*）、扁盘尾丝虫（*Onchocerca fasciata*）和伊氏德拉伊丝虫（*Deraiophoronema evansi*）为旧大陆骆驼的特异性线虫，存在于这些骆驼的原始栖息地。其他宿主的其他线虫，如薄副鹿圆线虫（*Parelaphostrongylus tenuis*）、原线虫（*Protostrongylus* spp.）、贺氏细颈线虫（*Nematodirus helvetianus*）、马氏马歇尔线虫（*Mashallagia marshalli*）和加利福尼亚吸吮吸虫（*Thelazia californiensis*）只在自然分布区以外的新大陆骆驼中发现。骆驼的最常见线虫见表86。

除以上提到的线虫外，有时也会在骆驼中发现一些机会性线虫，如蛔虫（*Ascaris* spp.）和广圆线虫（*Angyostrongylus cantonensis*）分布于约旦的单峰驼[2]，蝇柔线虫（*Habronema muscae*）分布于美国的单峰驼[29]。

表86 骆驼的线虫

| 虫种（中文名） | 宿主 | 地理分布 | | | 参考文献 |
|---|---|---|---|---|---|
| | | 南美洲 | 亚洲 | 非洲 | |
| *Trichuris tenuis*（细毛首线虫） | G, L | √ | | | [6, 8] |
| *T. ovis*（羊毛首线虫） | D, L | √ | √ | | [3, 18, 19] |
| *T. globulosa*（球鞘毛首线虫） | D | | √ | √ | [11, 17, 23] |
| *T. skrijabin*（斯氏毛首线虫） | B | | √ | | [37] |
| *T. barbetonensis*（巴佰顿毛首线虫） | D | | √ | | [7] |
| *T. cameli*（驼毛首线虫） | D | | √ | | [11] |
| *Capillaria* sp.（毛细线虫） | A, L, G | √ | | | [9, 24, 34] |
| *Strongyloides papillosus*（乳头类圆线虫） | D | | √ | √ | [5, 27, 34] |
| *Trichostrogylus axei*（艾氏毛圆线虫） | A, L, G, V | √ | | | [9, 16,18, 24] |
| *T. probolurus*（突尾毛圆线虫） | B, D | | √ | √ | [3, 5, 11, 23, 30, 37] |
| *T. colubriformis*（蛇形毛圆线虫） | D, B, G | √ | √ | √ | [7, 9, 14, 18, 30] |
| *T. vitrinus*（透明毛圆线虫） | G | √ | | | [24] |
| *T. longispicularis*（长刺毛圆线虫） | A | √ | | | [15] |
| *Graphinema auchinae*（羊驼纵纹线虫） | A, L, V | √ | | | [9, 16] |
| *Teladorsagia circumcincta*（普通背带线虫） | A, V, D | √ | √ | | [3, 7, 9, 14] |

续表

| 虫种（中文名） | 宿主 | 地理分布 | | | 参考文献 |
|---|---|---|---|---|---|
| | | 南美洲 | 亚洲 | 非洲 | |
| *Ostertagia dahurica*（达乎尔奥斯特线虫） | B | | √ | | [37] |
| *O. ostertagi*（奥氏奥斯特线虫） | A, G, L, V | √ | | | [9, 14, 18, 24] |
| *O. lyrata*（竖琴奥特线虫） | A | √ | | | [9, 15] |
| *Marshalagia marshalli*（马氏马歇尔线虫） | C | | √ | | [7] |
| *Spiculopteragia peruvianus*（秘鲁刺翼线虫） | A | √ | | | [9, 15] |
| *Cooperia oncophora*（点状古柏线虫） | A, D, G, L, V | √ | √ | | [3, 7, 9, 14, 18] |
| *Camelostrongylus mentulatus*（阴茎骆驼圆线虫） | A, D, L, V | √ | √ | √ | [3, 7, 9, 11, 14, 16, 17, 23, 34] |
| *Haemonchus contortus*（捻转血矛线虫） | A, D, V | √ | √ | | [3, 14, 15] |
| *H. longistipes*（长柄血矛线虫） | D | | √ | √ | [6, 7, 11, 17, 23, 30] |
| *Impalaia tuberculata*（结节无囊线虫） | C | | | √ | [4] |
| *Nematodirus lamae*（羊驼细颈线虫） | A, L, V | √ | | | [9, 17, 24] |
| *N. battus*（巴氏细颈线虫） | G | √ | | | [24] |
| *N. oiratianus*（奥利春细颈线虫） | B, D | | √ | | [7, 37] |
| *N. spathinger*（匙形细颈线虫） | A, D, G, L, V | √ | √ | | [9, 14, 17, 18, 24, 34] |
| *N. filicollis*（线颈细颈线虫） | A, G | √ | | | [18, 24] |
| *N. mauritanicus*（毛里塔尼亚细颈线虫） | D | | √ | | [17] |
| *N. cameli*（驼细颈线虫） | D, B | | √ | | [7, 37] |
| *N. dromedarii*（单峰驼细颈线虫） | D | | √ | | [7] |
| *Lamanema chavesi*（查氏无峰驼线虫） | A, G, L, V | √ | | | [8, 9, 18, 19, 34] |
| *Oesophagostomum columbianum*（哥伦比亚结节线虫） | A, D | | | √ | [9, 29] |
| *O. venulosum*（微管结节线虫） | A, D | √ | √ | √ | [3, 9, 23, 30] |
| *Chabertia ovina*（绵羊夏伯特线虫） | B, D, V | √ | √ | | [3, 9, 37] |
| *Bunosomum trigonophorum*（三棱角仰口线虫） | A | √ | | | [25] |
| *Dictyocaulus filarial*（丝状尾网线虫） | A, L, G | √ | | | [9, 24] |
| *Skrjabinema ovis*（羊斯克里亚宾线虫） | G | √ | | | [24] |
| *Thelazia leesi*（李氏吸吮线虫） | D | | √ | | [32] |
| *Physocephalus dromedarii*（单峰驼膨首线虫）[a] | D | | √ | | [17, 28, 35, 38] |
| *P. cristatus*（有嵴膨首线虫）[b] | D | | √ | | [36] |
| *Parabronema skrjabini*（斯氏副柔线虫） | D | | √ | | [7, 17] |
| *Onchocerca gutturosa*（喉瘤盘尾丝虫） | D | | | √ | [20] |
| *O. armilata*（圈形盘尾丝虫） | D | | | √ | [20] |
| *O. fasciata*（福斯盘尾丝虫） | D | | √ | √ | [4, 10, 12, 13, 26 |
| *Deraiophoronema evansi*（埃氏双瓣线虫） | D | | √ | √ | [1, 7, 21, 22, 30, 31, 33, 39] |

A，Alpaca羊驼；B，Bactrian camel双峰驼；D，dromedary单峰驼；G，guanaco骆马；L，llama美洲驼；V，vicuna小羊驼；

a 为在旧文献记载中，这个种以*Physocephalus sexalatus* 或者*Ph. sexalatus dromedarii*亚种被提及。按照近来的研究[35]，它应该为独立种；

b 为在阿尔及利亚骆驼中的有嵴膨首线虫（*P. cristatus*），表面上描述为膨首线虫（*P. sexalatus*）的修饰化形式，或者后来将其升格为种。人们认为它是一疑问种

# 参考文献


[1] Ahmad S., Butt A.A., Muhammad G., Athar M. & Khan M.Z. (2004). – Haematobiochemical studies on the haemoparasitised camels. *Int. J. Agricult. Biol.*, 6 (2), 331 – 334.

[2] Al-Ani F.K., Sharrif L.A., Al-Rawashdeh O.F., Al Qudah K.M. & Al-Hammi Y. (1998). – Camel diseases in Jordan. *In* Proceedings of the 3rd Annual Meeting for Animal Production under Arid Conditions, Al Ain, UAE, 2–3 May 1998, 2, 77 – 92.

[3] Anwar M. & Hayat C.S. (1999). – Gastrointestinal parasitic fauna of camel (*Camelus dromedarius*) slaughtered at Faisalabad abattoir. *Pakistan J. Biol. Sci.*, 2 (1), 209 – 210.

[4] Banaja A.A. & Ghandour A.M. (1994). – A review of parasites of camels (*Camelus dromedarius*) in Saudi Arabia. *J. King Abdulaziz Univers. Sci.*, 6, 75 – 86.

[5] Bekele T. (2002). – Epidemiological studies on gastrointestinal helminths of dromedary (*Camelus dromedarius*) in semi-arid lands of eastern Ethiopia. *Vet. Parasitol.*, 105, 139 – 152.

[6] Beldomenico P.M., Uhart M., Bonco M.F., Marull C., Baldi R. & Peralta J.L. (2003). – Internal parasites of free ranging guanacos from Patagonia. *Vet. Parasitol.*, 118, 71 – 77.

[7] Borji H., Ramzi G.H., Movassaghi A.R., Naghibi A.G. & Maleki M. (2009). – A study on gastrointestinal helminths of camels in Mashhad abattoir. Iran. *Iranian J. Vet. Res. Shiraz Univers.*, 11 (2), 174 – 179.

[8] Carfune M.M., Aguirre D.H. & Rickard L.G. (2001). – First report of *Lamanema chavezi* (Trichostrongyloidea) in llamas (*Lama glama*) from Argentina. *Vet. Parasitol.*, 97, 165 – 168.

[9] Chavez C., Guerrero C., Alva, J. & Guerrero J. (1967). – El parasitismo gastrointestinal en alpaca. *Rev. Fac. Med. Vet. Lima*, 21, 9 – 19.

[10] Chema A.H., El-Bihari S., Ashour N.A. & Ali A.S. (1984). – Onchocerciasis in camels (*Camelus dromedarius*) in Saudi Arabia. *J. Helminthol.*, 58, 279 – 285.

[11] El Bihari S. (1985). – Helminths of the camel, a review. *Br. Vet. J.*, 141, 315 – 326.

[12] El-Massry A.A. & Derbala A.A. (2000). – Evidence of *Onchocerca fasciata* (Filaroidea: Onchocercidae) in camels (*Camelus dromedarius*), prevalence, nodular lesions appearance and parasite morphology. *Vet. Parasitol.*, 88, 305 – 312.

[13] Ghandour A.M., Al-Amoudi A.A. & Banaja A.A. (1991). – *Onchocerca fasciata* Railliet and Henry, 1910 and its nodule development in camels in Saudi Arabia. *Vet. Parasitol.*, 39, 67 – 77.

[14] Guerro C. (1960). – Helmintos en vicunas (*Vicugna vicugna*). *Rev. Fac. Med. Vet. Lima*, 15, 103 – 105.

[15] Guerro C.A. & Chavez C. (1967). – Helmintos comunicados por primera vez en alpacas (*Lama pacos*), con una descripcion de *Spiculopteragia peruvianus* n. sp. *Host. Bol. Chileno Parasit.*, 22, 147 – 150.

[16] Guerro C.A. & Rojas J.E. (1970). – *Graphinema aucheniae* n. gen. n. sp. (Nematoda) en auquenidos. *Bol. Chileno Parasit.*, 24, 134 – 136.

[17] Haroun E.M., Mahmoud O.M., Magzoub M., Abdel Hamid Y. & Omer O.H. (1996). – The haematological effects of the gastrointestinal nematodes prevalent in camels (*Camelus dromedarius*) in central Saudi Arabia. *Vet. Res. Comm.*, 20, 25 – 264.

[18] Hurtado E., Bustinza J. & Sanchez C. (1985). – Estudio parasitologico en llamas (*Lama glama*) del altiplano peruano. Resumenes 5. *In* Convencion Internacional sobre Camelidos Sudamericanos. 16 – 21 June 1985, Cusco, Peru.

[19] Hurtado E., Bustinza J. & Sanchez C. (1985). – Parasitismo gastrointestinal por examen de heces en guanacos (*Lama guanicoe*). Resumenes 5. *In* Convencion Internacional sobre Camelidos Sudamericanos, 16 – 21 June 1985, Cusco, Peru.

[20] Hussein H.S., El Mannan A.M.A. & El Sinnary K. (1988). – *Onchocerca armillata* Railliet and Henry, 1909 and *Onchocerca gutturosa* (Neumann, 1910) in camels (*Camelus dromedarius* L.) in the Sudan. *Vet Res. Commun.*, 12, 475 – 480.

[21] Karram M.H., Ibrahim H., Ali T.S.A., Manna A.M. & Abdel Ali T.S. (1991). – Clinical and haematological changes in camel infested with *Trypanosoma evansi* and microfilariae. *Assuit Vet. Med. J.*, 25, 118 – 128.

[22] Kataitseva T.Y. (1968). – The life cycle of *Dipetalonema evansi* Lewis, 1882 [in Russian]. *Dokl. AN SSSR*, 180, 1 262 –

1 264.

[23] Kendall S.B. (1974). – Some parasites of domestic animals in the Aswan Governorate Arab Republic of Egypt. *Trop. Anim. Health Prod.*, 6, 128 – 130.

[24] Larrieu E., Bigatti R., Lukovic R., Eddi C., Bonazzi E., Gomez E., Niez R. & Oporto N. (1982). – Contribucion al studio del parasitismo gastrointestinal en guanacos (*Lama guanicoe*) y llamas (*Lama glama*). *Gaceta Vet.*, 374, 958 – 960.

[25] Leguia G. (1991). – The epidemiology and economic impact of llama parasites. *Parasitol. Today*, 7, 54 – 56.

[26] Moghaddar N. & Zahedie A. (2006). – Prevalence and pathogenesis of *Onchocerca fasciata* infection in camels (*Camelus dromedarius*) in Iran. *J. Camel Pract. Res.*, 13, 31 – 35.

[27] Muhammad S.A., Farooq A.A., Akhtar M.S. & Hayat C.S. (2004). – Dipetalonemiasis in a dromedary camel and its treatment. *Pakistan Vet. J.*, 24 (4), 205 – 206.

[28] Mushkambarova M.G. & Dobrynin M.I. (1972). – On physocephalosis of one humped camels in Turkmenii [in Russian]. *Vestnik Akademii Nauk Turkmenskoj SSR, Seria Biologiceskie Nauki,* 4, 62 – 67.

[29] Myers D.A., Smyth C.D., Greiner E.C., Wiedner E., Abbott J., Marsella R. & Nunnery C. (2010). – Cutaneous periocular *Habronema* infection in a dromedary camel (*Camelus dromedarius*). *Vet. Dermatol.*, 21 (5), 527 – 530.

[30] Onyali I.O. & Onwuliri C.O.E. (1989). – Gastro-intestinal helminths of camels in Nigeria. *Trop. Animal Health Prod.*, 21, 24 – 246.

[31] Oryan A., Valinezhad A. & Bahrami S. (2008). – Prevalence and pathology of camel filariasis in Iran. *Parasitol. Res.*, 103, 1 125 – 1 131.

[32] Parasani H.R., Singh V. & Momin R.R. (2008). – Common parasites of camel. *Vet. World*, 1 (10), 317–318.

[33] Patak K.M.L., Singh Y. & Harsh D.L. (1998). – Prevalence of *Dipetalonema evansi* in camels of Rajasthan. *J. Camel Pract. Res.*, 5, 166 – 169.

[34] Rojas M., Lobato I. & Montalvo M. (1993). – Fauna parasitaria de camelidos sudamericanos y ovinos en pequenos rebanos mixtos familiars. *Rev. Rec Inv. IVITA*, 6 (1), 22 – 27.

[35] Schuster R.K., Wibbelt G. & Kinne J. (2013). – Light and scanning electron microscopy examination of *Physocephalus dromedarii* stat. nov. (Nematoda: Spirocercidae), an abomasal nematode of dromedaries (*Camelus dromedarius*). *J. Helminthol.* doi:10.1017/S0022149X13000497.

[36] Seurat L.-G. (1912). – Sur la presence, en Algerie, du *Spiroptera sexalata* Molin chez le dromadaire et chez l'ane. *Comptes Rendus Hebdomadaires des Séances et Mémoires de la Société de Biologie*, 1, 174 – 176.

[37] Sharhuu G. & Sharkhuu T. (2004). – The helminth fauna of wild and domestic ruminants in Mongolia – a review. *Eur. J. Wildl. Res.*, 50, 150 – 156.

[38] Skrjabin K.I., Sobolev A.A. & Ivashkin V.M. (1967). – Spirurata of animals and man and diseases caused by them. *In* Essentials of Nematology (K.I. Skrjabin, ed.). Vol. 9. Nauka, Moscow.

[39] Zariffarad M.R. & Hashemir F.R. (2000). – Study on tissue and blood protozoa of camels in southern Iran. *J. Camel Pract. Res.*, 7, 193 – 194.

（贯万忠译，般宏校）



## 寄生虫性胃肠道炎和结肠炎

骆驼的寄生虫性胃肠道炎的发生与寄生于皱胃和肠道中的许多种类的线虫相关。在骆驼胃肠道圆线目线虫中，隶属于血矛线虫属（*Haemonchus*）的皱胃线虫（abomasal worm）致病性最强。捻转血矛线虫（*H. contortus*）分布于新大陆骆驼和旧大陆骆驼，但是主要流行于单峰驼。其他皱胃线虫隶属于奥斯特科复合群（Ostertagiinae complex），包括奥斯特属（*Ostertagia*）、马歇尔属（*Marshallagia*）、背带线虫属（*Teladorsagia*）、短角鹿原线虫属（*Mazamastrongylus*又名*Spiculoptergia*，刺翼线虫属）和骆驼原线虫属（*Camelostrongylulus*）以及毛圆科（Trichostrongylinae）的羊驼纵纹线虫（*Graphinema auchniae*）、艾氏毛圆线虫（*Trichostrongylus axei*）和长刺毛圆线虫（*T. longispicularis*）。两种拟旋尾线虫（spiruroid）——斯克里亚宾副柔线虫（*Parabronema skrjabini*）和骆驼膨首线虫（*Physocephalus dromedarii*）也能够在旧大陆骆驼的皱胃中见到。小肠内可寄生下列各属的代表性虫种，这些属包括古柏线虫属（*Cooperia*）、细颈线虫属（*Nematodirus*）、无峰驼线虫（*Lamanema*）、无囊线虫属（*Impalaia*）、仰口线虫属（*Bunostomum*）、毛细线虫属（*Capillaria*）和类圆线虫属（*Strongyloides*）。而绵羊夏伯特线虫（*Chabertia ovina*）、结节线虫（*Oesophagostomum* spp.）和毛首鞭形线虫（*Trichuris* spp.）寄生于大肠，在感染严重时引起动物结肠炎。蛲虫样绵羊斯克里亚宾线虫（*Skrjabinema ovis*）寄生于直肠。

## 形态学

线虫是雌雄异体蠕虫，具有明显的二态性。雄虫有刺，作为辅助性交配器官。此外，对于所谓的有交合伞线虫，雄虫有一个交合伞作为交配器。但是，类圆线虫属（*Strongyloides*）是个例外，寄生代中的雌虫营孤雌生殖。

捻转血矛线虫和长柄血矛线虫（*H. longistipes*）两种线虫体长相近，其雄虫和雌虫体长分别为18~25mm和21~34mm。两者的性器官都有颈乳突（图280），鞘中有一微小的柳叶状结构。在雌虫体内，吸血寄生虫的卵巢沿红色的肠管螺旋转环绕，呈“理发店旋转立柱样”外观（图 281），阴门有副翼。雄虫交合伞背叶（dorsal lobe）不对称。两个种的丝网状交合刺长度方面有差别，捻转血矛线虫的交合刺长为488~544μm（图 282），而长柄血矛线虫的交合刺长为600~666μm（图 283）。新鲜虫卵内含16个卵裂球，捻转血矛线虫虫卵直径为70~80μm，而长柄血矛线虫的虫卵直径为60~70 μm。

奥氏特亚科（Ostertagiinae）线虫也称为媒介胃线虫，体形较短，长约10mm。只有马歇尔属（*Marshallagia*）体长是它的两倍。像血矛线虫属（*Haemonchus*），它们均有子宫颈乳突状结构。雌虫在子宫的远端有一特殊的括约肌结构，在多数种类中存在阴门副翼。阴茎骆驼圆线虫（*Camelostrongylus mentulatus*）的阴门副翼是缺失

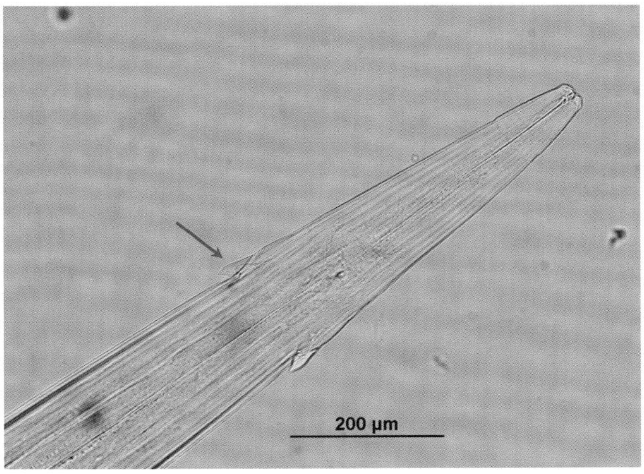


图280　长柄血矛线虫躯体后端具有子宫颈乳头状突起

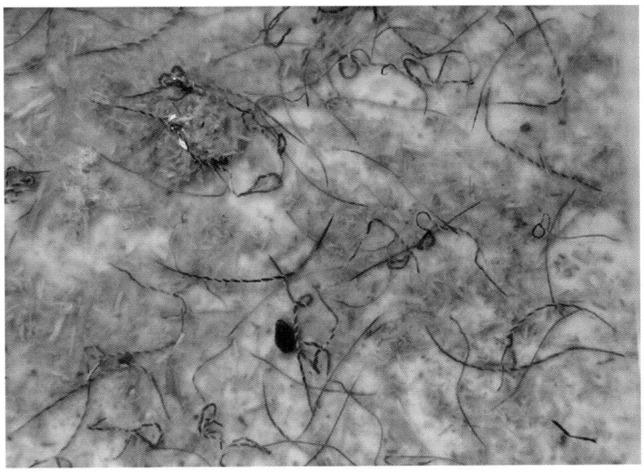

图281　单峰驼皱胃内容物，内含捻转血矛线虫。虫体因吸血获得营养物，而呈淡红色。而在雌虫的白色卵巢环绕于这些红色含血的肠管周围

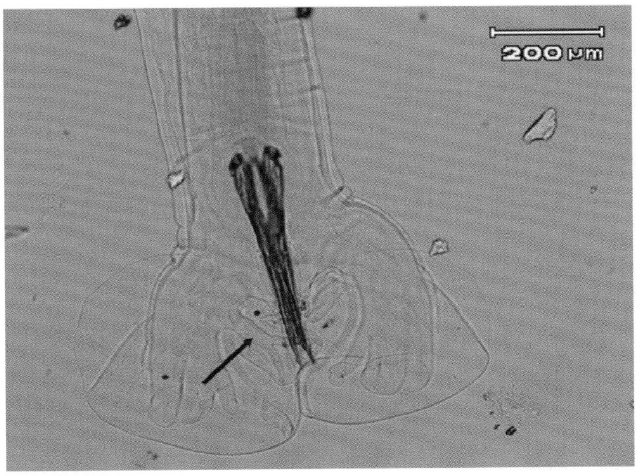


图282　捻转血矛线虫雄虫尾端，具有交合伞和交合刺。交合伞的不配对“Y”形背肋（箭头所示）是不对称的

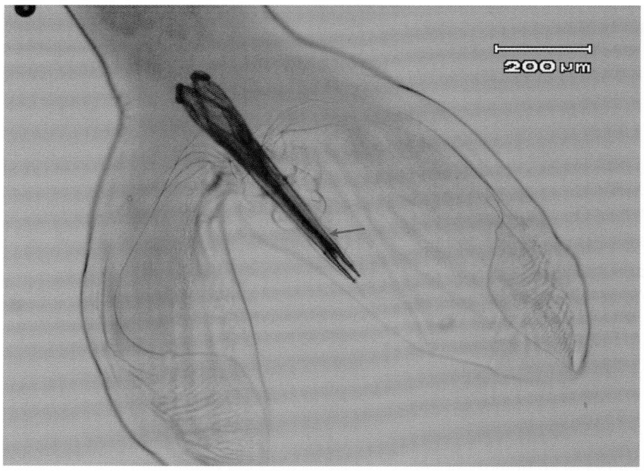


图283　长柄血矛线虫雄虫尾端，具有交合伞和交合刺。雄虫的交合刺长超过600 µm

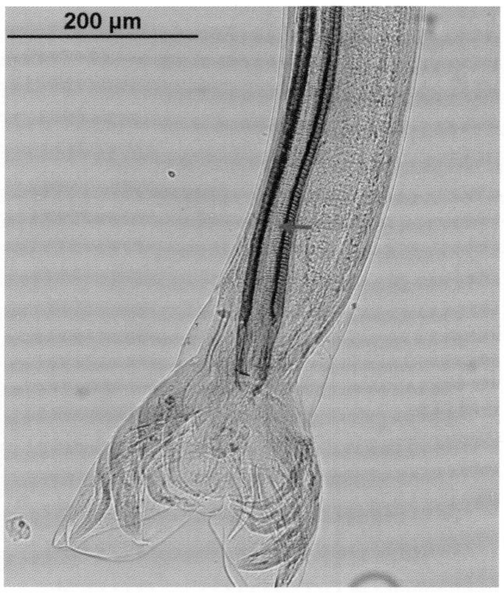


图284　阴茎骆驼圆线虫雄虫的尾端，具有交合伞和交合刺，长的交合刺呈对称排列，在它们的中部具有一海绵状的表面

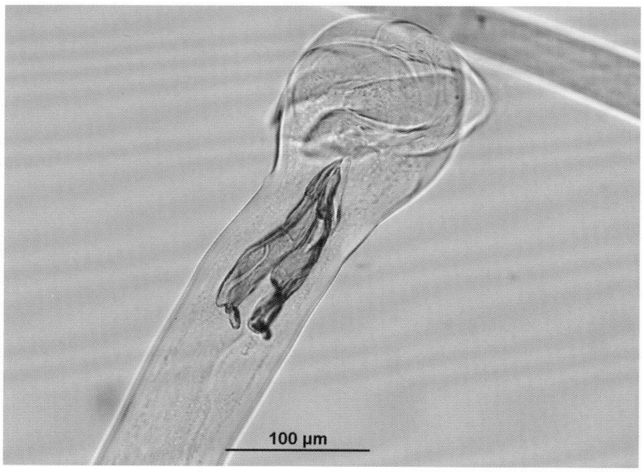


图285　蛇形毛圆线虫（*Trichostrongylus colubriformis*）雄虫远端，具有交合伞和交合刺。交合刺大小稍微不同

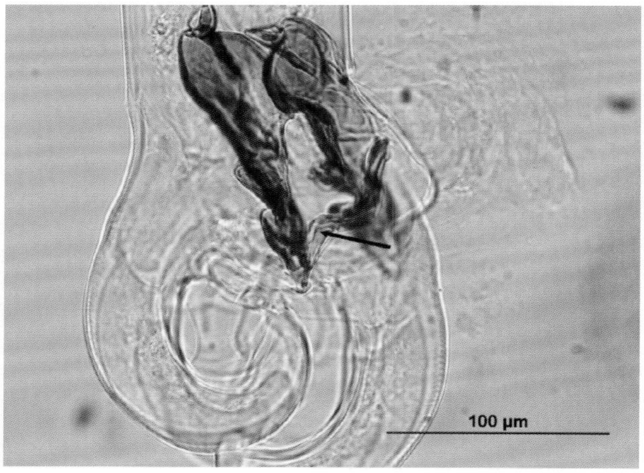


图286　突尾毛圆线虫（*Trichostrongylus probolurus*）雄虫远端，具有交合伞和交合刺。交合刺大小相同，弯曲程度比蛇形毛圆线虫大。在虫体的远端还可观察到引带结构（箭头所示）



的。雄虫的交合伞发育完善；交合刺对称排列，在远端发生种特异性融合（图284）。

毛圆线虫属（*Trichostrongylus*）包含体形最小的线虫种。雌虫和雄虫体长不超过10mm，缺乏子宫颈乳状突起。子宫远端具有括约肌，但是阴门副翼缺失。雄虫有短而特别弯曲的交合刺（图285和图286）。

中等体形的毛圆线虫如古柏属（*Cooperia*）和无囊线虫属（*Implaia*）的典型特征是具有一头囊（cephalic vesicle），但缺乏子宫颈和前交合伞乳头（prevursal papillae）以及副引器和引带（gubernaculum）（图287）。无囊线虫属的特征是更细长一些，有等长的交合刺和一个大的交合伞（图288）。

细颈线虫属（*Nematodirus*）是骆驼肠道内体形最长的毛圆线虫。雄虫和雌虫体长分别达15 mm和20mm。雌虫和雄虫都有头囊（图289）。雄虫的交合伞有两个发育完善的侧翼组成，但是其上只分布有一个小的背叶，由两个背肋支撑着。单峰驼细颈线虫（*N. dromedirus*）的交合刺特别长，可达5.5mm，这些刺在远端联合在一起，以柳叶形或小铲形结束（图290）。细颈线虫虫卵容易识别，虫

卵较大，椭圆形，内含8个卵裂球。

3个圆线虫——绵羊夏伯特线虫（*Chabertia ovina*）、微管食管口线虫（*Oesophagostomum venulosum*，微管结节线虫）和哥伦比亚结节线虫（*O. columbianum*）可在骆驼大肠中见到。它们呈白色，体形大小相似，雄虫长15mm，雌虫长20mm。而结节线虫有头囊，但是绵羊夏伯特线虫有非常明显的交合伞。

鞭虫属（*Trichuris*）鞭虫也寄生于宿主的大肠。这些线虫体长达8cm，可通过宽的后端然后快速变尖细的纤丝状前端结构鉴别（图291）。雄虫只有一根交合刺（图292）。

斯氏副柔线虫（*Parabronema skrjabini*）体长达45mm，虫体颜色呈特别的鲜红。雄虫体长只有18mm，后端呈螺丝状，两个交合刺不等长。雌雄两种虫体前部末端特征明显（图293）。单峰驼膨首线虫（*Physocephalus dromedarii*）也呈红色，但是体形较小，一直的咽和螺旋状增厚层为该虫种的特征（图294）。

## 生活史

骆驼的胃肠道线虫为直接发育型，但是斯克里亚宾副柔线虫和骆驼膨首线虫两种拟旋尾线虫是例外。前者是柔线科（Habronematidae）一代表种，为生物源性蠕虫（biohelminth），以蝇科的蝇为中间宿主，而骆驼膨首线虫是尾旋科（Spirocercidae）的成员，甲虫为其中间宿主。包裹在囊中的幼虫也存在于两栖类、爬行类、鸟类和

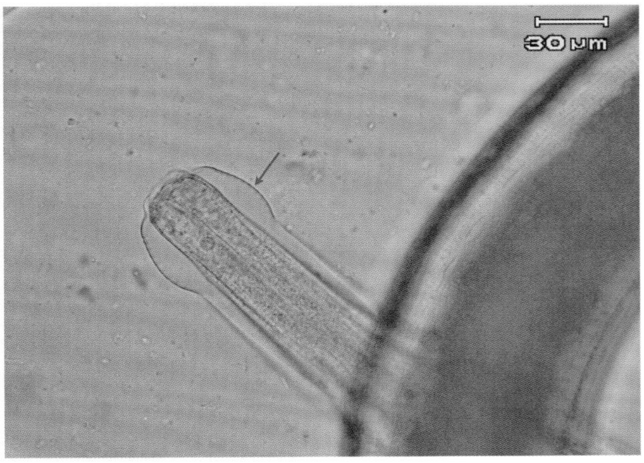


图287　无囊属线虫前端有一特别明显的头囊结构，是古柏亚科（Cooperiinae）线虫的特异性结构

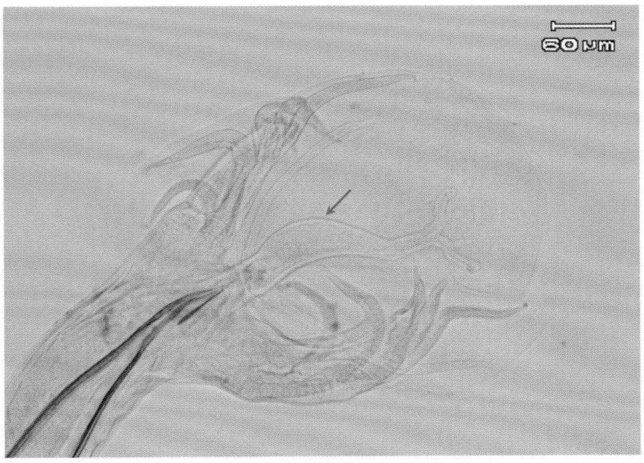


图288　无囊属线虫雄虫近尾端结构，具有交合伞和交合刺，交合伞的背肋很长（箭头所示）

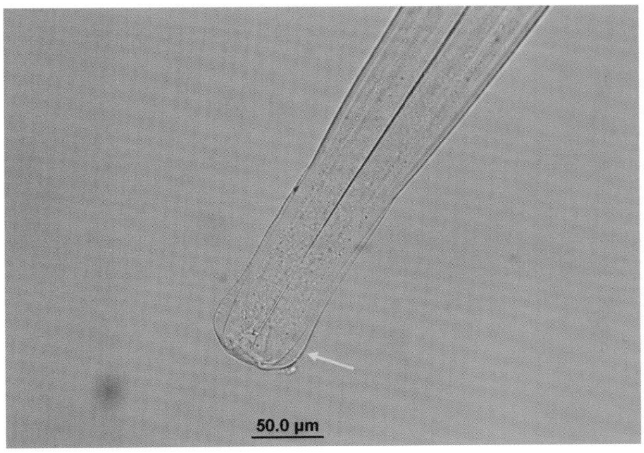


图289　单峰驼细颈线虫前端的小头囊

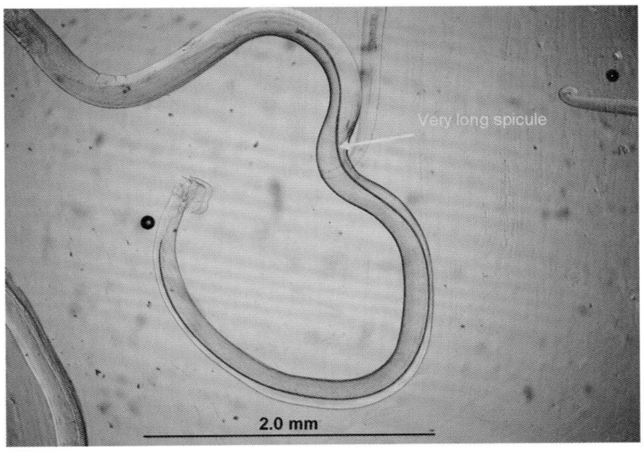


图290　单峰驼细颈线虫雄虫尾端，具有交合伞和极端长的交合刺，且交合刺融合于一起

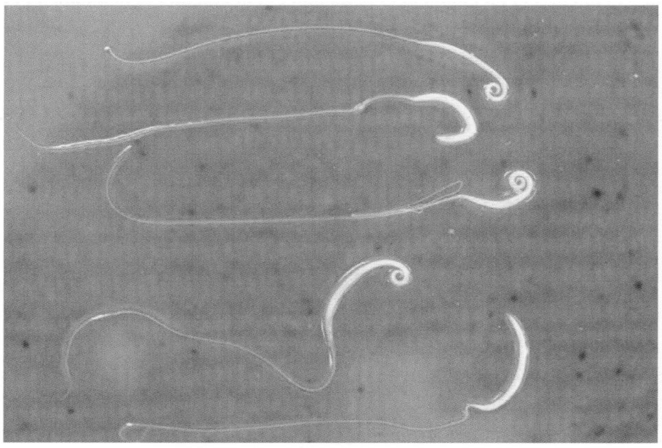

图291　球鞘毛首线虫（*Trichuris globulosa*），在鞭虫中，后端较宽，而前端较长且薄，雄虫通常有一卷着的尾端

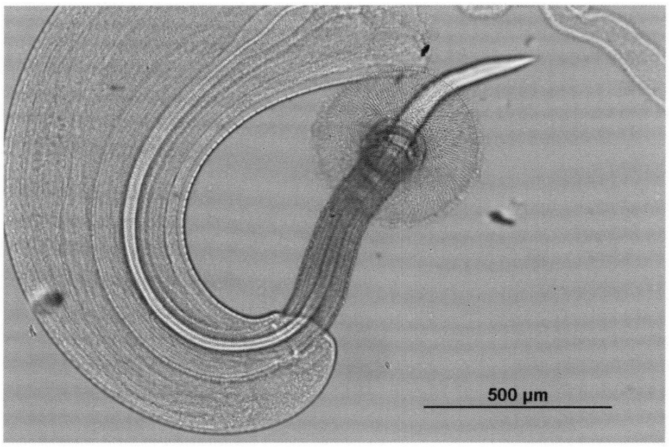


图292　球鞘毛首线虫雄虫尾端，在一薄片中分布有不配对的交合刺，尾端四周包裹着小的刺，以气球泡形式结束

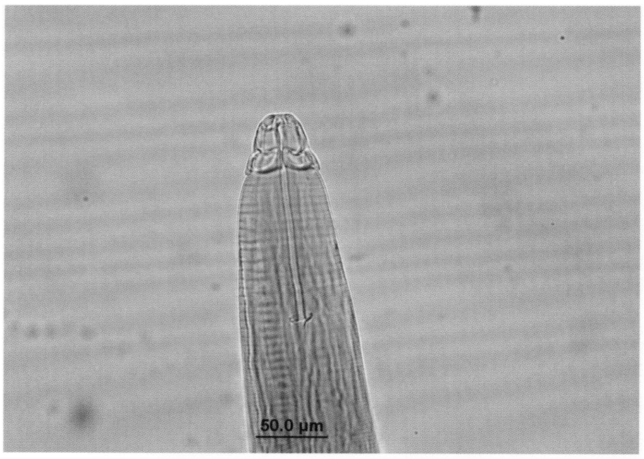


图293　斯氏副柔线虫前端。头囊盾牌样复合结构由侧伪上唇（pseudolabia）和“绶带”组成

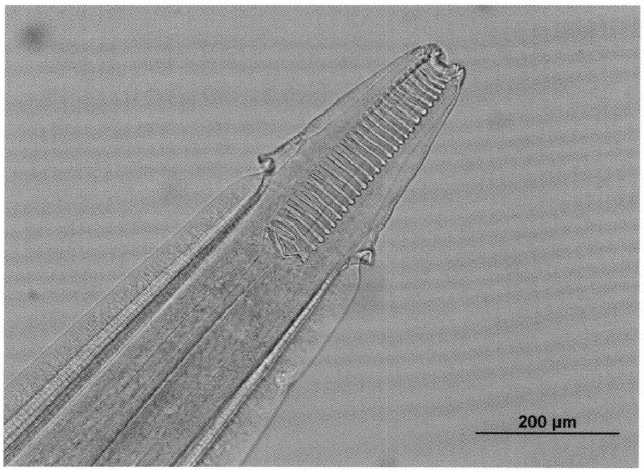


图294　单峰驼膨首线虫的前端，拥有螺旋状咽和膜状侧翼

蝙蝠。

在温暖和潮湿条件下，毛圆线虫和圆线虫在虫卵内的第一期幼虫发育很快。第一期幼虫离开卵壳，脱皮两次，变为丝状样、片状化的感染性第三期幼虫阶段。在夏季的适宜环境中，这两类发育过程持续5天，但是在较冷的环境中，它们的发育会延缓；如果温度低于0℃，则线虫会死亡，但是其他线虫对寒冷有较强的抵抗力。感染性幼虫可从粪便中移行，在潮湿的环境中，可以在植被上观察到。

细颈线虫属（*Nematodirus*）全部外生性幼虫的发育在卵壳内完成。毛圆线虫幼虫随着新鲜牧草被宿主吞食后，幼虫会失去卵壳，蜕皮两次，然后发育至成虫。潜伏期3~4周。

当毛首线虫属（*Trichuris*）和毛细线虫属（*Capillaria*）虫卵随着宿主粪便排出体外到环境中时卵内只含一个卵细胞，发育第一期幼虫阶段时即具有感染性。第一期幼虫只在宿主肠道内孵化。

仰口线虫属（*Bunostomum*）钩虫的第三期幼虫具备穿透宿主皮肤、通过血流移行至肺的能力。在肺部，幼虫脱皮变为第四期幼虫，然后重新进入消化道。经口摄入的幼虫可不经过在体内的移行即可发育。

最后，乳突类圆线虫（*Strongyloides papillosus*）具有寄生和营自由生活的双重能力。寄生生活的种群中包括全部的雌性虫体，这些雌性虫体通过孤雌生殖（parthenogenesis）产生含有胚胎的虫卵。孵化后，幼虫可能发育为营自由生活的雄虫和雌虫。然而，第三期幼虫被宿主吞食或者直接穿过宿主皮肤，通过静脉系统、肺和气管最终到小肠内寄生。骆驼幼仔在出生后吸食含有幼虫的初乳即可遭受感染。其他动物可出现胎盘感染，这种情形还未在骆驼上报道过。

## 流行病学

线虫感染途径最常见的是经口感染。而仰口线虫（*Bunostomum*）和乳头类圆线虫（*S. papillosus*）可通过皮肤直接侵入宿主。当终末宿主在放牧时摄入感染性幼虫而遭受感染。寄生虫性胃肠炎是由许多线虫感染引起的一种复合疾病。血矛线虫感染终末宿主的范围广泛。绵羊和单峰驼是捻转血矛线虫和长柄血矛线虫的主要宿主。在北非国家、印度和巴基斯坦，单峰驼的数据源于对粪样和屠宰场的调查结果，调查表明，血矛线虫是胃肠道蠕虫混合感染中优势虫种[4]。长柄血矛线虫是一种已对环境适应性非常强的线虫。这种寄生虫的生存策略是在干燥季节幼虫呈胁迫性发育，而在雨季虫体呈潜伏期感染状态。宿主与线虫的关系中值得注意的是单峰驼具有食粪癖，毛细线虫和毛首线虫会从宿主的这种行为中受益，它们感染性幼虫并不离开虫卵即可被宿主吞食。

在南美洲，新大陆骆驼的血矛线虫病常常被忽视，主要是由于骆驼通常在高纬度区域放牧，这些区域气候寒冷，不适宜于这类寄生虫的体外发育。在其他地区，当新大陆骆驼与其他小型反刍动物一起放牧时，血矛线虫病会变得重要。

在秘鲁，新大陆骆驼的线虫的发生具有明显的季节性，在雨季里，流行程度明显加重[6]。在这个季节，许多外界环境适合感染性幼虫的发育、生存和传播。细颈线虫属、无峰驼线虫（*Lamanema*）、毛首线虫属和毛细线虫属（*Capillaria*）的外源性发育阶段对外界环境的抵抗力强，这是由于整个幼虫的发育是在卵壳内完成的。无峰驼线虫的幼虫在0℃以下时仍能存活，在草场污染后长达2年的时间里仍然能够找到幼虫。近来的调查结果表明，查氏无峰驼线虫（*L. chavezi*）线虫用啮齿动物作为贮藏宿主[3,7]。关于人们对新大陆骆驼线虫暂时性低活力状态的认识还知之甚少。

## 临床症状和病理学

血矛线虫是嗜血性寄生虫。为了获得血液，这类寄生虫用位于口腔中的小叶结构刺穿宿主黏膜血管。血液学分析表明，嗜酸性细胞计数增加，红血球压积和血红蛋白容积降低。每一条线虫每天能够消耗多达0.05mL的血液。宿主贫血的程度取决于荷虫量。宿主只能通过代偿性红细胞生成补偿进入胃肠道的铁和蛋白质损失。急性血矛线虫病在骆驼并不常见。除了*C. mentulatus*外，其他皱胃线虫有时是旧大陆骆驼的寄生虫，但是它们在南美洲或自然疫源地外的新大陆骆驼中比较常见[10]。这些线虫的幼虫进入胃腺，破坏产生盐酸的胃壁细胞。结果导致宿主胃蛋白的消化功能紊乱，这是由于胃蛋白酶原活性下降。用秘鲁刺翼线虫（*Mazamastrongylus*（*Spiculopteragia*）*peruvianus*）的人工试验感染表明，感染羊驼食欲降低、体重下降和出现腹泻[6]。在英国，曾经报道过美洲驼的致死性病例[10]。尸体消瘦，黏膜苍白、主要病变发生在皱胃。在寄生部位，可见数千条线虫的存在，在黏膜表面分布许多小的出血点。皱胃内容物呈棕黑色，这是因为有变质血液进入胃内。在奥氏特线虫感染病例中，在剖检时可发现灰白色结节的形成。在小肠中导致肠炎的病变主要由毛细线虫、无峰驼线虫（*Lamanema*）、无囊线虫和古柏线虫引起。在这类病例中，在剖检时会发现宿主大量荷虫（图295）。查氏无峰驼线虫是相当独特的一种线虫，其第三期和第四期幼虫进行肠肝移行。对肠绒毛的严重损坏和肠黏膜的侵蚀导致绒毛萎缩，这些病变与幼虫在肠黏膜的寄生阶段相一致。幼虫移行引起卡他性和出血性肠炎以及黏膜区域性坏死[2]。新大陆骆驼在遭受大量的无峰驼线虫感染后，导致肉品卫生问题，这是因为幼虫移行导致肝脏出血和坏死。来自秘鲁和智利的羊驼和美洲驼由于一系列的病变而被拒绝上市的肝脏的比例分别达12%和18.6%[1,8]。

在大肠中寄生的线虫中，只有毛首线虫在荷虫量大的情况下可以引起感染宿主的结肠炎。幼虫穿过盲肠黏膜，引起机械性刺激和损伤，进而导致病灶卡他性炎症。但是，大多数情况下，骆驼的鞭虫感染引起的病变过程不明显。

## 诊断

骆驼的寄生性胃肠炎临床症状不特异。可出现厌食、体重减轻、全身虚弱和贫血等症状，但是这些症状在骆驼的其他疾病时也会出现。

体内诊断要依据粪样中检测查获虫卵。这些粪样必须在尽可能新鲜的状态下进行检查。可在骆驼中观察到以下5种类型的线虫卵。

（1）中等大小，卵壳薄，椭圆形到卵圆形虫卵，内含16个卵裂球。毛圆线虫的虫卵属于此类，如仰口线虫、古柏线虫和结节线虫，但是细颈线虫属和无峰驼线虫属线虫除外（图296）。

（2）体型大，卵壳薄，椭圆形虫卵，内含8个卵裂球，如细颈线虫和无峰驼线虫（*Lamanema*）线虫（图297）。

（3）中等大小，卵壳薄，椭圆形虫卵，内含一U型幼虫，如圆线虫（图298）。

（4）中等大小，卵壳厚，拥有两个极帽，如毛首线虫和毛细线虫（图299和图300）。

（5）中等大小，扁平的不对称虫卵，如斯克里亚宾线虫（*Skrjabinema*）。

鉴别第一组中的各个属需要一定的经验。通过人工培养使幼虫发育至第三期幼虫，然后根据一定数量虫体的特定大小、肠细胞的数目和长度等进行线虫属的准确鉴定是可能的。皱胃的旋尾线虫（*Parabronema skrjabini*）和骆驼泡首线虫（*Physocephalus dromedarii*）的虫卵用常规诊断方法不能从粪样中检测出来。

线虫成熟雄虫的辅助性生殖器官的特异性形态学特征可用于种水平的鉴定。

## 治疗与控制

有许多抗蠕虫药可用于寄生性胃肠炎的控制。

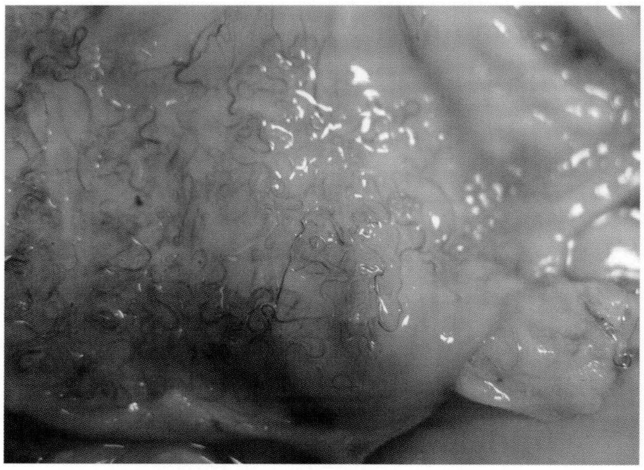

图295　在单峰驼小肠中的骆驼毛细线虫，成群感染

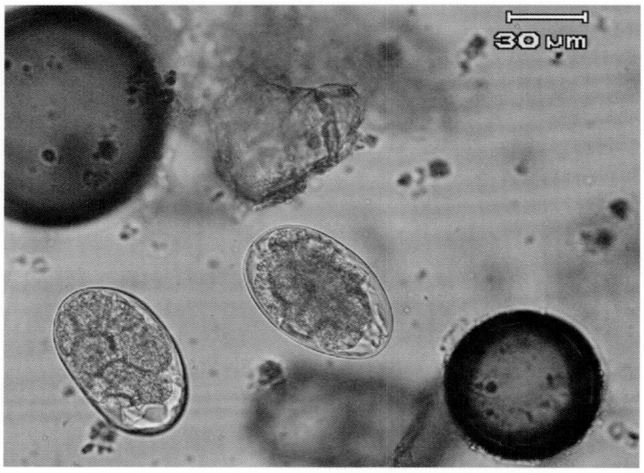


图296　血矛线虫卵。中等大小，内含16个卵裂球

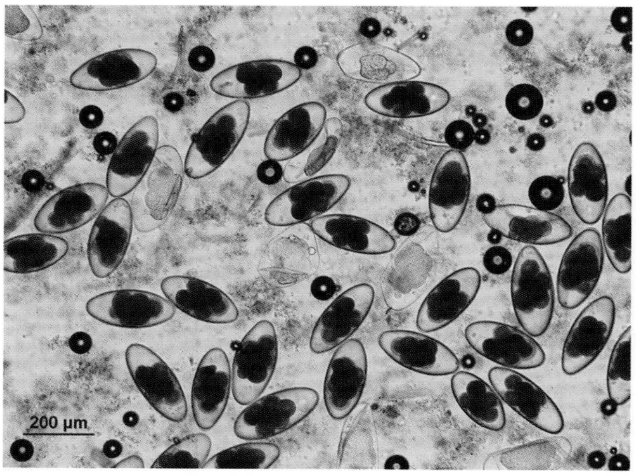


图297　细颈线虫卵，虫卵较大，内含8个卵裂球

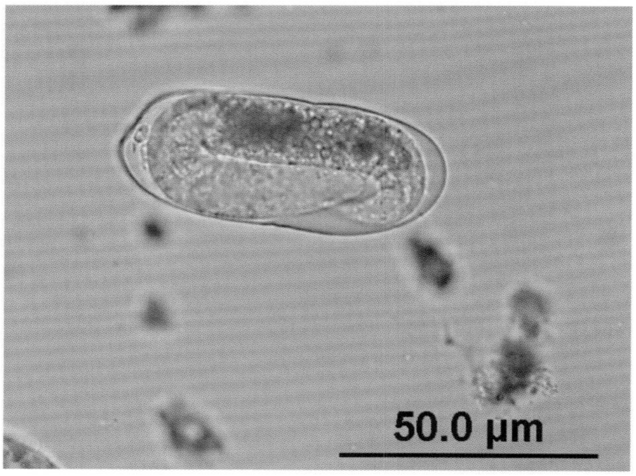


图298　卵胚生类圆线虫卵，薄壳的卵，内含一U型幼虫

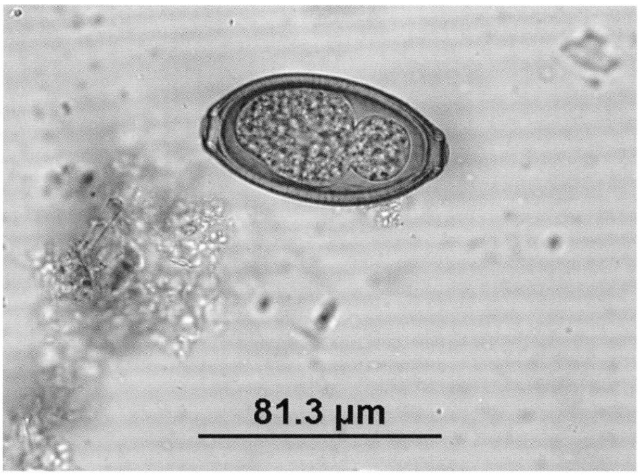


图299 厚壳的鞭虫卵，带有两个对称的极帽

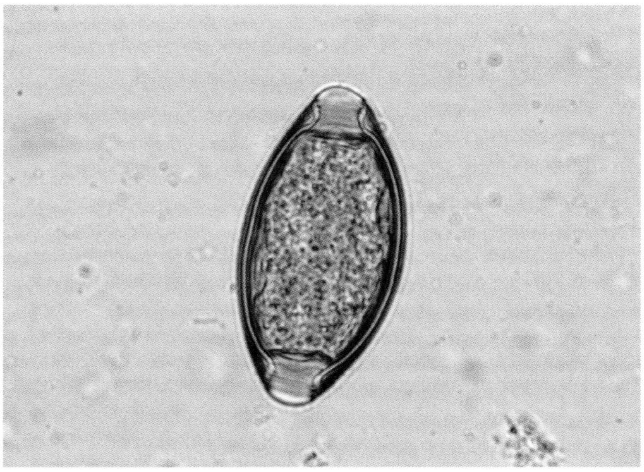

图300 毛细线虫卵。毛细线虫虫卵稍有不对称，带有可伸缩的极帽

推荐的抗线虫药归属于四种不同类型（见表87）。

大环内酯类也叫内外寄生虫兼杀药（endectocides），因为它们的作用方式，这类药物可直接作用于线虫和节肢动物。但是，这些针对骆驼疥螨的药物对线虫的作用是极其有限的，因为这类药物在动物体内的生物利用度较低。

其他类型的化合物主要是抗线虫，但是一些苯并咪唑替代物有杀绦虫甚至片形吸虫的特性。抗蠕虫药的大量使用导致抗药性毛圆线虫虫株的出现，特别是耐药捻转血矛线虫虫株。这一问题在来自美国的新大陆骆驼中发现，这些骆驼使用了两类抗蠕虫药[5]，但是目前还没有在旧大陆骆驼中报道过。

在表87中所列的适合于骆驼抗虫化合物液体口服剂中一些有时使用起来较麻烦。苯并咪唑以粉剂和颗粒剂型可供选择，它们可与浓缩饲料混合使用。其他抗蠕虫药以片剂或浇泼剂形式可供选择。注射剂也可选择使用。依普菌素透皮剂在单峰驼抗长柄血矛线虫感染试验中已经获得成功[9]。

表87 治疗胃肠道感染的蠕虫药

| 化合物类型 | 化合物 | 给药途径 |
|---|---|---|
| Imidazothiazoles咪唑噻唑类 | Levamisole左旋咪唑 | 口服，皮下注射，透皮（涂抹） |
| (Pro)Benzimidazoles苯并咪唑 | Thiabendazole噻苯咪唑 | 口服 |
| | Mebendazole甲苯达唑 | 口服 |
| | Fenbendazole芬苯达唑 | 口服 |
| | Oxfendazole奥芬达唑 | 口服 |
| | Oxibendazole奥苯达唑 | 口服 |
| | Albendazole阿苯达唑 | 口服 |
| | Febantel苯硫氨酯 | 口服 |
| | Netobimin奈托比胺 | 口服 |
| Tetrahydropyrimidines四氢尿嘧啶 | Morantel tartrate甲噻吩咪唑 | 口服 |
| | Pyrantel pamoate双羟萘酸噻嘧啶 | 口服 |
| Macrocyclic lactones大环内酯 | Ivermectin伊维菌素 | 口服，皮下注射，透皮 |
| | Doramectin多拉菌素 | 皮下注射 |
| | Eprinomectin依普菌素 | 涂抹 |
| | Moxidectin莫西菌素 | 口服，涂抹 |

## 参考文献

[1] Alva J., Guerrero C. & Nunez A. (1980). – Actividad antihelmintica del oxfendazole contra infecciones naturales de nematodes gastrointestinales de alpacas. *In* Resumenes 5. Convencion Internacional sobre Camelidos Sudamericanos, 16 – 21 June 1985, Cusco, Peru, 52.

[2] Cafrune M.M., Aguirre D.H. & Rickard L.G. (2001). – First report of *Lamanema chavezi* (Nematoda: Trichostrongyloidea) in llamas (*Lama glama*) from Argentina. *Vet. Parasitol.*, 97, 165 – 168.

[3] Cafrune M.M., Marin R.E., Rigalt F.A., Romero S.R. & Aguirre D.H. (2009). – *Lamanema chavezi* (Nematoda: Molineidae): epidemiological data of the infection in South American camelids of Northwest Argentina. *Vet. Parasitol.*, 166, 321 – 325.

[4] Chhabra M.B. & Gupta S.K. (2009). – Parasitic diseases of camels – an update. 2. Helminthoses. *J. Camel Pract. Res.*, 13 (2), 81 – 87.

[5] Gillespie R-A.M., Williamson L.H., Terrill T.H. & Kaplan R.M. (2010). – Efficacy of anthelmintics on South American camelid (llama and alpaca) farms in Georgia. *Vet. Parasitol.*, 172, 168 – 171.

[6] Leguia G. & Casas E. (1999). – Enfermedades parasitarias de camelidos sudamericanos. Editorial del Mar, Lima.

[7] Rickard L.G. & Hoberg E.P. (2000). – Reassignment of *Lamanema* from Nematodirinae to Molineinae (Nematoda: Trichostrongyloidea). *J. Parasitol.*, 86, 647 – 650.

[8] Rojas M., Lobato I. & Montalvo M. (1993). – Fauna parasitaria de camelidos sudamericanos y ovinos en pequenos rebanos mixtos familiares. *Rev. Rec Inv. IVITA.*, 6 (1), 22 – 27.

[9] Schuster R.K., Sivakumar S., Juhaz J., Nagy P. & Wernery U. (2009). – The efficacy of Eprinex (eprinomectin) on trichostrongylidae in dromedaries. *J. Camel Pract. Res.*, 16 (1), 93 – 96.

[10] Windsor R.S. (1997). – Type II ostertagiosis in llamas. *Vet. Rec.*, 141, 608.

（贾万忠译，般宏校）

## 网尾线虫病（dictyocaulosis）和原线虫病（protostrongylidosis）

骆驼肺线虫感染并不重要。这些肺线虫属于线虫的两个科：类毛样线虫超科（Trichostrongyloidea）的网尾线虫科（Dictyocaulidae）和后圆超科（Metastrongyloidea）的原圆线虫科（Protostrongylidae）。薄副鹿圆线虫（*Parelaphostrongylus tenuis*）是鹿类的肺外原圆线虫，它能够感染作为贮藏宿主的新大陆骆驼。薄副鹿圆线虫作为中枢神经系统紊乱的病因，在美国和加拿大落基山脉一带具有非常重要的地位。

### 形态学

肺线虫属（*Dictyocaulus*）呈白色，雄虫体长3~8cm，雌虫体长5~10cm。雄虫具有一小的交合伞。胎生网尾线虫（*D. viviparous*）和丝状网尾线虫（*D. filaria*）的交合刺分别长200~285μm和400~600μm。棕色的原圆线虫除带鞘囊尾线虫（*Cystocaulus ocreatus*）外，原圆线虫属（*Protostrongylus*）、缪勒线虫属（*Muellerius*）和新圆线虫属（*Neostrongylus*）其体形较小，寄生于肺组织小支气管或者小结节中。薄副鹿圆线虫体形纤细，结构精致，寄生于自然宿主——白尾鹿的颅腔中。

## 生活史

网尾线虫是土源性胚生线虫。成虫寄生于宿主肺支气管系统。第一期幼虫从虫卵孵化出后，逆行移行至气管，经宿主吞咽，随粪便排出体外。两次蜕皮后，在适宜的环境中5天内发育为感染性第三期幼虫，出现在草地上。被宿主摄食后，幼虫穿过肠黏膜，到达肠系膜淋巴结，并在此进行又一次蜕皮。第四期幼虫通过淋巴结和血流到达肺部。

原圆线虫是生物源性蠕虫，需要陆地螺作为中间宿主。随着粪便排出体外的幼虫通过接触感染后穿过螺的足在2~4周内发育为第三期幼虫。终末宿主通过摄食受感染的软体动物而遭受感染。幼虫利用与网尾线虫幼虫相类似的淋巴血管路径到达寄生部位——肺部。脑膜虫——薄副鹿圆线虫的生活史与原圆线虫其他种相似[14]。在白尾鹿中，成虫通常寄生于大脑和脊髓的脑膜之间，但是更常见于大脑的脑膜里。雌虫产下未胚化的虫卵，这些虫卵进而被运送到肺部，并在此胚化，孵化出第一期幼虫，侵入肺泡。当终末宿主摄入感染的腹足动物时而遭受感染。有人认为，第三期幼虫穿过肠壁，沿侧面脊髓神经进入脊椎管。幼虫在脊髓索中脱皮两次，进入大脑的静脉窦。薄副鹿圆线虫的整个生活史及该寄生虫和“驼鹿病”之间的相互关系可在实验条件下进行详细阐明[2,3]。

## 流行病学

肺线虫需要温和湿润的气候条件以及利于体外发育的永久性草场。绵羊肺丝虫——丝状网尾线虫被列为非洲和亚洲的单峰驼和双峰驼稀有寄生虫[8,10]。在伊朗中部的屠宰场调查中，发现单峰驼的丝状网尾线虫感染率为10%[19]。

比较陈旧的文献资料显示，来自哈萨克斯坦的双峰驼的骆驼网尾线虫（*D. cameli*）从形态学角度无法与胎生网尾线虫相区别。用从骆驼分离的肺线虫幼虫实验感染犊牛并不产生能完全发育的线虫[21]。但是，是否存在骆驼种特异性的骆驼网尾线虫仍有争议。

丝状尾网线虫也见于南美洲[6,11]和德国[20]新大陆骆驼。人们对骆驼感染原圆线虫的情况知之甚少。只有来自瑞典的一篇报道，认为新大陆骆驼粪样中存在缪勒线虫的幼虫[12]。

在北美洲，白尾鹿是薄副鹿圆线虫的自然终末宿主。这里，白尾鹿呈隐性感染，并不引起发病。虽然白尾鹿广泛分布于气候温和的南美洲，但是薄副鹿圆线虫仅限于落基山脉一带流行，这是因为西部大平原的气候对于螺和蛞蝓生存太过干燥。

至少有18种不同土壤螺已鉴定为薄副鹿圆线虫的中间宿主，感染率为0.1%~8.2%。实验感染表明，每个螺平均荷有8.5个感染性幼虫[15]。

据估计，42 000只羊驼中约有60%生活在北美洲，它们处于受脑膜虫感染的威胁中，在报道的死亡病例中有3%缘于薄副鹿圆线虫感染[4]。

## 临床症状和病理学

肺线虫病是一种严重的肺部疾病，其症状有咳嗽、呼吸困难和流鼻涕，对农场中的反刍兽是致命的。这类线虫的存在及其代谢产物可刺激杯状细胞的增加，纤毛细胞数减少。这会使得黏膜纤毛自动摆动功能丧失，呼吸道上皮细胞不能清除黏液，导致气管黏液增加。结果，内脏黏液滞留，阻塞深部气路，创造了厌氧菌生长繁殖的条件。但是，目前人们对于骆驼肺线虫病的过程尚未进行过研究。

有数篇文献报道新大陆骆驼感染副鹿圆线虫病的情况[1,5,7,16,18]。根据这些作者的观察，脑膜虫感染骆驼产生严重的神经性症状。机体针对感染，临床症状通常反映了病灶性、不对称脊髓病变，包括共济失调、身体坚硬、肌肉无力、运动幅度过大、躯体后部轻度瘫痪，头部歪斜，颈部成弓状，失明，体重逐渐减轻，精神抑郁，惊厥和死亡。感染后45~53天，临床症状逐渐显现，开始出现于后肢，逐渐发展到前肢。该病可能是急性的也可能是慢性的，数天内死亡或者发生病程长达数月至数年的共济失调。

## 诊断

肺线虫病的诊断要依据流行病学史和临床症状，并通过对新鲜粪便的检查来确诊。肺线虫的隐性感染可以通过贝尔曼-韦策尔技术（Berman-Wetzel technique）作出诊断，其依据是线虫的幼虫

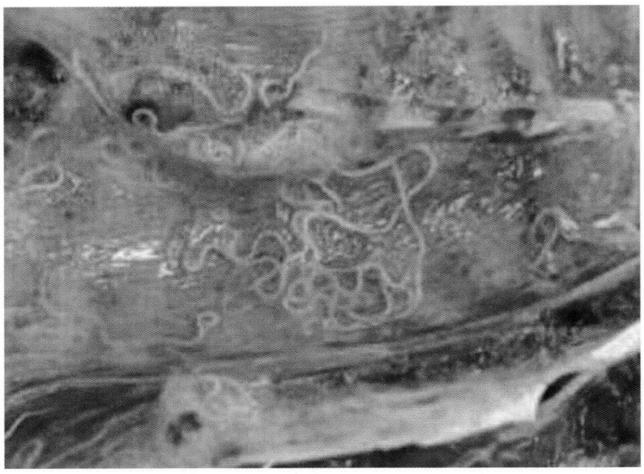

图301 寄生有胎生网尾线虫的肺脏。虫体寄生后刺激支气管分泌大量黏液

具有从新鲜粪堆中逃逸的趋势。网尾线虫不同种的第一期幼虫长度从300~550μm，虫体肠段前1/3处肠细胞上有黑色颗粒，后端呈圆形。原圆线虫的幼虫稍小，透明，无黑色颗粒，有特征性的种特异性后端。原圆线虫和缪勒线虫的第一期幼虫在尾部具有一个背刺，而原圆线虫则缺乏这一结构。

在尸检时，呼吸道也可出现其他病变。尾网线虫寄生于较大的支气管，而原圆线虫则在肺的实质中形成淡灰色和淡黄色至粉红色结节。

副鹿圆线虫病发生在北美洲地区，那里白尾鹿很常见。贮藏宿主在感染副鹿圆线虫后会呈现症状，而不是隐性感染，其最终的诊断结果要依据剖检结果做出。利用感染性薄副鹿圆线虫幼虫排泌/分泌抗原，人们开发出针对鹿科动物的血清学诊断方法[17]。

## 治疗与控制

治疗肺线虫感染的药物除噻嘧啶和甲噻吩嘧啶不再吸收外，其余与治疗胃肠道线虫的药物一样。

与网尾线虫病相关的区域，牛和绵羊可以用网尾线虫幼虫辐照疫苗进行免疫接种来进行控制。近年来，人们已经朝着开发基因工程重组抗原疫苗的方向发展[13]。

针对贮藏宿主脑膜线虫重组抗原疫苗的一些工作已在进行中。为了鉴定宿主-寄生虫相互关系中具有潜在重要意义的蛋白分子，发现薄副鹿圆线虫的天门冬氨酰蛋白酶抑制剂具有良好的免疫原性[9]。

## 参考文献

[1] Anderson D.E. (2007). – *Parelaphostrongylus tenuis* (meningeal worm) infection in llamas and alpacas. Ohio State University. Available at: http://goatconnection.com/articles/publish/article_ 126.shtml (accessed 15 August 2013).

[2] Anderson R.C. (1963). – The incidence, development, and experimental transmission of *Pneumostrongylus tenuis* Dougherty (Metastrongyloidea: Protostrongylidae) of the meninges of the white-tailed deer (*Odocoileus virginianus borealis*) in Ontario. *Can. J. Zool.*, 41, 775 – 792.

[3] Anderson R.C. (1964). – Neurologic disease in moose infected experimentally with *Pneumostrongylus tenuis* from white-tailed deer. *Pathol. Vet.*, 1, 289 – 322.

[4] Appleton J.A. & Duffy M.S. (2003). – Quest for a meningeal worm vaccine. *Alpacas Magazine*, 168 – 171.

[5] Baumgartner W. (1985). – Parelaphostrongylosis in llamas. *JAVMA*, 185 (11), 1 243 – 1 245.

[6] Beldomenico P.M., Uhart M., Bonco M.F., Marull C., Baldi R. & Peralta J.L. (2003). – Internal parasites of free ranging guanacos from Patagonia. *Vet. Parasitol.*, 118, 71 – 77.

[7] Brown T., Jordan H. & Demorest C. (1978). – Cerebrospinal parelaphostrongylosis in llamas. J. *Wildl. Dis.*, 14 (4), 441 – 444.

[8] Dakkak A. & Ouheli H. (1987). – Helminths and helminthoses of the dromedary. *In* Diseases of camels. *Rev. sci. tech. Off. int. Epiz.*, 6 (2), 447 – 461.

[9] Duffy M.S., MacAfee N., Burt M.D.B. & Appleton J.A. (2002). – An aspartyl protease inhibitor orthologue expressed by Parelaphostrongylus tenuis is immunogenic in an atypical host. *Clin. Diagn. Lab. Immun.*, 9 (4), 763 – 770.

[10] El Bihari S. (1985). – Helminths of the camel. A review. *Br. Vet. J.* 141, 315 – 326.

[11] Guerro C. & Leguia G. (1987). – Enfermedades parasitarias de alpacas. *Rev. Camelidos Sudamericanos*, 4, 32 – 82.

[12] Hertzberg H. & Kohler L. (2006). – Prevalence and significance of gastrointestinal helminths and protozoa in South American camelids in Switzerland. *Berl. Münch. Tierärztl. Wochenschr.*, 119, 291 – 294.

[13] Kohrmann A. (2002). – Immunisierungsversuche mit rekombinantem Major Sperm Protein von *Dictyocaulus viviparous*. Thesis, Tierarztliche Hochschule Hannover, Germany.

[14] Lankester M.W. (2001). – Extrapulmonary lungworms of cervids. *In* Parasitic diseases of wild mammals, 2nd Ed (W.M. Samuel, M.J. Pybus & A.A. Kocan, eds). Iowa State University Press, Ames, Iowa, 228 – 278.

[15] McCoy K.D. & Nudds T.D. (2000). – An examination of the manipulation hypothesis to explain prevalence of *Parelaphostrongylys tenuis* in gastropod intermediate host populations. *Can. J. Zool.*, 78, 294 – 299.

[16] Nagy D.W. (2004). – *Parelaphostrongylus tenuis* and other parasitic diseases of the ruminant nervous system. *Vet. Clin. North Am. Food Anim. Pract.*, 20, 393 – 412.

[17] Ogunremi O. (2010). – Serological diagnosis of *Parelaphostrongylus tenuis* in cervids of North America: a review. *In* Veterinary Parasitology (G. LaMann, ed.). Nova Science Publishers, New York, 223 – 239.

[18] Pugh D.G. (1995). – Clinical parelaphostrongylosis in llamas. Compendium on Continuing Education for the Practicing Veterinarian, 17, 600 – 606.

[19] Radfar M.H., Maimand A.E. & Sharify A. (2006). – A report on parasitic infections in camel (*Camelus dromedarius*) of Kerman slaughterhouse. *J. Fac. Vet. Med. Uni. Tehran*, 61 (2), 165 – 168.

[20] Rohbeck S. (2006). – Parasitosen des Verdauungstraktes und der Atemwege bei Neuweltkameliden: Untersuchungen zu ihrer Epidemiologie und Bekaempfung in einer suedhessischen Herde sowie zur Biologie von *Eimeria macusaniensis*. Thesis, Justus Liebig Universitat Giessen, Germany.

[21] Shumilina Z.V. (1953). – A study of Dictyocaulus disease in camels [in Russian]. *In* Papers on Helminthology dedicated to Academician K.I. Skrjabin on his 75 birthday (Z.G. Vasilkova, M.P. Gnedina, L.Ch. Gushanskaja, V.S. Ershov, A.M. Petrov, V.P. Podjanolskaja, K.M. Ryzikov, T.S. Skarbilovic & N.P. Shichobalova, eds). Izdatelstvo Akademii Nauk SSSR, Moscow, 793 – 800.

（贾万忠译，殷宏校）

## 盘尾丝虫病（Onchocercidosis）

盘尾丝虫病是旧大陆骆驼的一种线虫病，由盘尾丝虫科（Onchocercidae）线虫感染引起。扁盘尾丝虫（*Onchocerca fasciata*）和埃氏双瓣线虫（*Deraiophoronema/Dipetalonema evensi*）是骆驼特有的线虫，而喉瘤盘尾丝虫（*O. gutturosa*）和圈形盘尾丝虫（*O. armillata*）也寄生于骆驼，但是更多的时候出现于反刍兽的诊断中。

### 形态学

丝虫成虫细线状，颜色苍白，体长数厘米。盘尾属丝虫成虫很难测量其体长，这是因为它们会形成结节，但是来自血管中的埃氏双瓣线虫的雄虫和雌虫体长分别为9.5~10.5cm和17.5~19.5cm[6]。雄丝虫体形小于雌虫，在身体后端有两根不等的刺。雌虫含有后宫型子宫，生殖孔开口位于身体前半部。

### 生活史

生物源性埃氏双瓣线虫寄生于肺部和生殖系统的动脉，而盘尾丝虫在宿主的结缔组织中形成结节（图 302）。雌虫是胎生性的，产生无鞘的微丝蚴，喉瘤盘尾丝虫和圈形盘尾丝虫会侵害皮肤，而福斯盘尾丝虫和埃氏双瓣丝虫的微丝蚴分布于血流中。吸血昆虫作为媒介。埃氏双瓣丝虫通过蚊子（*Aedes*，伊蚊）传播，而福斯盘尾丝虫通过黑蝇（Simulidae，蚋科）作为媒介进行传播。关于骆驼的盘尾丝虫在媒介中的发育时间问题，人们仅对福斯盘尾丝虫做过些研究。微丝蚴发育时在媒介的飞行肌中进行两次蜕皮。在26~28℃环境下，第三期感染性幼虫形成和出现在鼻中需要9天时间[15]。从第10天起，可通过受感染蚊子的叮咬进行传播。

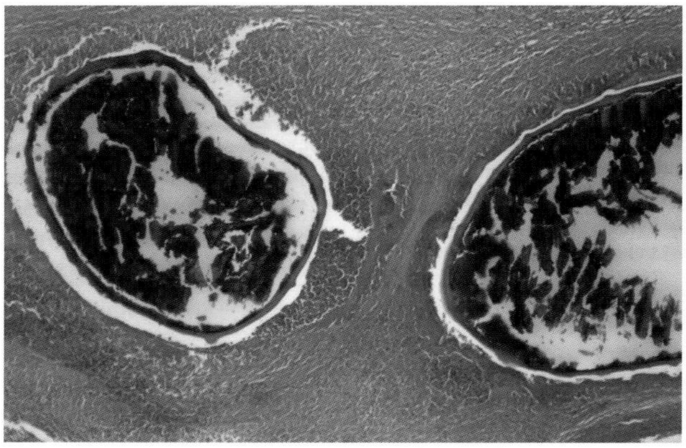

图302　在骆驼皮肤中钙化的盘尾丝虫形成皮下结节（苏木精-伊红染色）

## 流行病学

盘尾丝虫病是媒介传播性的寄生虫病。福斯盘尾丝虫在沙特阿拉伯[5,7,11]和伊朗[16]已有记载。El-Massry和Derbala发现从苏丹和索马里进口的骆驼中福斯盘尾丝虫的感染率为2.75%，但是在埃及本地骆驼中则无该寄生虫的感染[9]。在苏丹，骆驼同样也被诊断感染圈形盘尾丝虫（*O. armillata*）和喉瘤盘尾丝虫（*O. gutterosa*）[13]。在澳大利亚的骆驼中也发现感染有喉瘤盘尾丝虫[12]。其他骆驼特异性种——埃氏双瓣丝虫在埃及[14]、沙特阿拉伯[3]、印度[19]、巴基斯坦[2]、伊朗[5,17,20]和土库曼斯坦[15]均有报道。另外，澳大利亚有输入性病例[4]。盘尾丝虫微丝蚴在科威特的骆驼中被检测到，但是未对种进行鉴定[1]。

## 临床症状和病理学

只有埃氏双瓣丝虫感染能引起心功能不全、动脉硬化和寄生虫性睾丸炎，出现临床病症。Karram等曾描述过感染骆驼出现虚弱、黏膜苍白、食欲不振、体温升高、阴囊和睾丸肿胀等主要临床症状。小红细胞低色素性贫血与白细胞增多和嗜酸性粒细胞症相关。Oryan等描述过雄性骆驼感染埃氏双瓣丝虫形成的结节[18]。在睾丸结节区动脉感染埃氏双瓣丝虫时发生慢性炎症反应，包括动脉血管内皮和纤维结缔组织膜的增生和肥大性病变，内腔狭窄或闭塞。睾丸肥厚或萎缩，慢性睾丸炎，伴随淋巴细胞、嗜酸性粒细胞、巨噬细胞和成纤维细胞浸润。在肺部，出现肺泡壁坏死、肺不张、肺水肿和肺实质的纤维化以及伴随慢性间质性肺炎。

## 诊断

骆驼在丝虫轻度感染通常无明显症状。埃氏双瓣丝虫和圈形盘尾丝虫可出现于血流中（图303）。推荐的检测方法是在傍晚时分从身体表面血管采集血液，因为微丝蚴具有夜行性特性。为了检测扁盘尾丝虫的微丝蚴，可以使用Ference等人描述的方法[10]。为此，将结节性病灶样品放于生理盐水后在25℃温育24小时。微丝蚴从病灶逸出，然后用1000转／分转速离心10分钟。

埃氏双瓣丝虫微丝蚴大小的数据为267~295μm[8]到200~315μm[4]。骆驼的圈形盘尾丝虫的皮肤微丝蚴大小为148~246μm，只有来自牛的圈形盘尾丝虫微丝蚴大小的一半[13]。

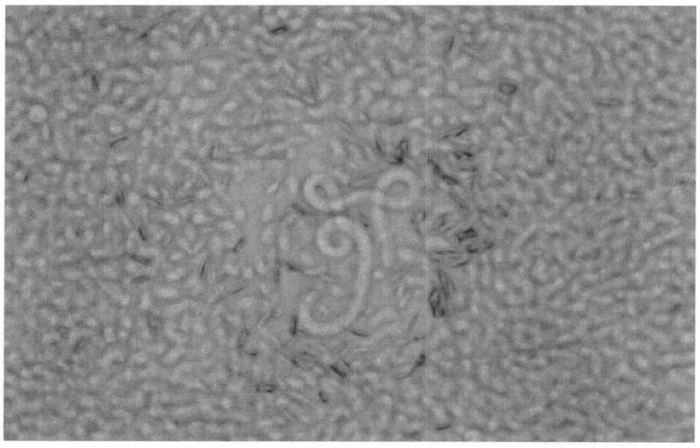

图303　在骆驼血液中的盘尾丝虫科的微丝蚴（很可能是埃氏双瓣丝虫）

## 治疗与控制

目前没有治疗丝虫成虫的有效药物。伊维菌素可杀灭循环系统中的微丝蚴[17]，阻止向其他骆驼进一步传播。抗生素（脱氧土霉素、利福平和土霉素）在人盘尾丝虫病治病中的应用已有描述。抗生素直接作用于沃尔巴克氏体属（*Wolbachia*）细菌的内共生，这些细菌能支持丝虫的代谢功能[19]，但是用抗生素治疗骆驼盘尾丝虫病尚无实际经验。

对黑蝇和蚊子的控制可分别降低宿主患盘尾丝虫病和双瓣线虫病的风险。

## 参考文献


[1] Abdul-Salam J. & Al-Taqui M. (1995). – Seasonal prevalence of *Onchocerca* like microfilaria in camels in Kuwait. *J. Egypt. Soc. Parasitol.*, 25 (1), 19 – 24.

[2] Ahmad S., Butt A.A., Muhammad G., Athar M. & Khan M.Z. (2004). – Haematobiochemical studies on the haemoparasitised camels. I*nt. J. Agricult. Biol.*, 6 (2), 331 – 334.

[3] Al-Khalifa M.S., Hussein H.S., Diab F.M. & Khalil G.M. (2009). – Blood parasites of livestock in certain regions in Saudi Arabia. *Saudi J. Biol. Sc.*, 16, 63 – 67.

[4] Anderson R.C. (2006). – Nematode parasites of vertebrates: Their development and transmission. 2nd Ed. Cabi Publishing, Wallingford, UK.

[5] Banaja A.A. & Ghandour A.M. (1994). – A review of parasites of camels (*Camelus dromedarius*) in Saudi Arabia. *J. King Abdulaziz Univers. Sci.*, 6, 75 – 86.

[6] Borji H. & Ramzi G.R. (2009). – Epidemiological study on haemoparasites of dromedary (*Camel dromedarius*) in Iran. *J. Camel Pract. Res.*, 16 (2), 217 – 219.

[7] Chema A.H., El-Bihari S., Ashour N.A. & Ali A.S. (1984). – Onchocerciasis in camels (*Camelus dromedarius*) in Saudi Arabia. *J. Helminthol.*, 58, 279 – 285.

[8] Elamin E.A., Mohammed G.A., Fadl M., Seham E., Saleem M.S. & Elbashir M.O.A. (1993). – An outbreak of cameline filariasis in the Sudan. *Br. Vet. J.*, 149, 195 – 200.

[9] El-Massry A. A, & Derbala A.A. (2000). – Evidence of *Onchocerca fasciata* (Filaroidea: Onchocercidae) in camels (*Camelus dromedarius*). 1. Prevalence, nodular lesions appearance and parasite morphology. *Vet. Parasitol.*, 88, 305 – 312.

[10] Ference S.A., Copemann D.B., Turk S.R. & Courney C.H. (1986). – *Onchocerca gutturosa* and *Onchocerca lienalis in* cattle, effect of age, sex and origin on prevalence of onchocercosis in subtropical and temperate regions of Florida and Georgia. *Am. J. Vet. Res.*, 47, 2 266 – 2 268.

[11] Ghandour A.M., Al-Amoudi A.A. & Banaja A.A. (1991). – *Onchocerca fasciata Railliet* and Henry, 1910 and its nodule development in camels in Saudi Arabia. *Vet. Parasitol.*, 39, 67 – 77.

[12] Holdsworth P.A. & Moorhouse D.E. (1985). – *Onchocerca gutturosa* in an Australian camel. *Austr. Vet. J.*, 63, 201 – 202.

[13] Hussein H.S., El Mannan A.M.A. & El Sinnary K. (1988). – *Onchocerca armillata* Railliet and Henry, 1909 and *Onchocerca gutturosa* (Neumann, 1910) in camels (*Camelus dromedarius* L.) in the Sudan. *Vet. Res. Commun.*, 12, 475 – 480.

[14] Karram M.H., Ibrahim H., Ali T.S.A., Manna A.M. & Abdel Ali T.S. (1991). – Clinical and haematological changes in camel infested with *Trypanosoma evansi* and microfilariae. *Assuit Vet. Med. J.*, 25, 118 – 128.

[15] Kataitseva T.Y. (1968). – The life cycle of *Dipetalonema evansi* Lewis, 1882 [in Russian]. *Dokl AN SSSR*, 180, 1 262 – 1 264.

[16] Moghaddar N. & Zahedie A. (2006). – Prevalence and pathogenesis of *Onchocerca fasciata* infection in camels (*Camelus dromedarius*) in Iran. *J. Camel Pract. Res.*, 13, 31 – 35.

[17] Muhammad S.A., Farooq A.A., Akhtar M.S. & Hayat C.S. (2004). – Dipetalonemiasis in a dromedary camel and its treatment. *Pakistan Vet. J.*, 24 (4), 205 – 206.

[18] Oryan A., Valinezhad A. & Bahrami S. (2008). – Prevalence and pathology of camel filariasis in Iran. *Parasitol. Res.*, 103, 1 125 – 1 131.

[19] Pathak K.M.L., Singh Y. & Harsh D.L. (1998). – Prevalence of *Dipetalonema evansi* in camels of Rajasthan. *J. Camel Pract. Res.*, 5, 166 – 169.

[20] Zariffarad M.R. & Hashemir F.R. (2000). – Study on tissue and blood protozoa of camels in southern Iran. *J. Camel Pract. Res.*, 7 (2), 193 – 194.


（贾万忠译，殷宏校）

## 5.3.1 蜱感染

蜱与螨共同构成蜱螨亚纲（Acari），隶属节肢动物门（Arthropoda）螯肢亚门（Chelicerata）。蜱具有以下特定器官：哈氏器（Haller' s organ），一种用来定位宿主的感觉器官；吉氏器（Gene' s organ），一种雌性生殖系统的附腺，可以分泌一种防水的蜡状物质覆盖到卵上，防止其脱水；口器上具有口下板。根据最新进展，世界蜱类区系包括3科896种，其中硬蜱科（Ixodidae）和软蜱科（Argasidae）分别包括702种和193种，单型科纳蜱科（Nuttalliellidae）仅包括纳马夸纳蜱（*Nuttalliella namaqua*）[20]。蜱不仅会吸食血液导致皮肤损伤，还是原虫、细菌和病毒的重要传播媒介。然而，蜱在骆驼中的媒介作用似乎没有在其他家畜中重要。

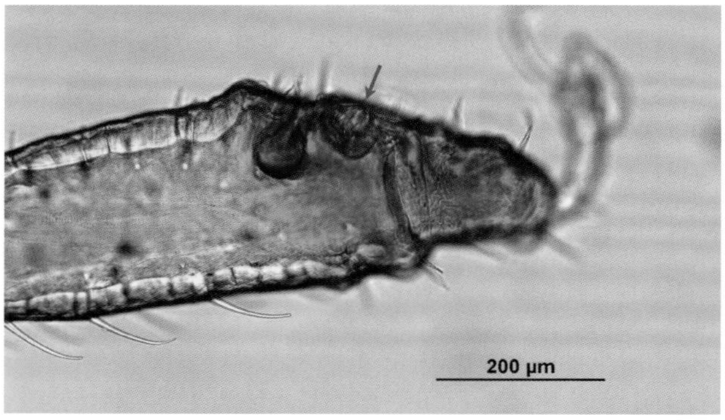

图304　哈氏器，蜱用来定位宿主的感觉器官，着生于第1对足的末节

图306　嗜驼璃眼蜱（*Hyalomma dromedarii*）的雄蜱和雌蜱。雌蜱和雄蜱在背板的大小上具有明显区别：雄蜱的背板覆盖整个背部，雌蜱仅覆盖一部分

### 形态

硬蜱的成蜱体型大小因种类、性别和饱血的程度而异，其范围在2~25mm。它们具有4对足，第1对足着生哈氏器（图304）。

口器在背腹面均可见，由须肢、螯肢和口下板组成（图305）。硬蜱具有性的二态现象（图306）。雄蜱的背板（盾板）覆盖整个身体，而雌蜱仅覆盖前体，后部具有弹性，能够摄取大量血液。有些种类饱食后的体重可增加到原来的100倍。大多数蜱类盾板的颜色呈单一的灰色、黑色、褐色或红色。花蜱属（*Amblyomma*）（图307和308）和革蜱属（*Dermacentor*）（图

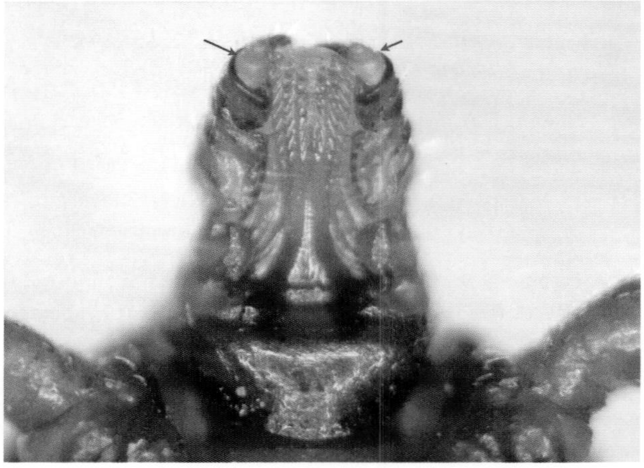
图305　硬蜱的口器（腹面观，由成对的须肢和螯肢及1个口下板组成）

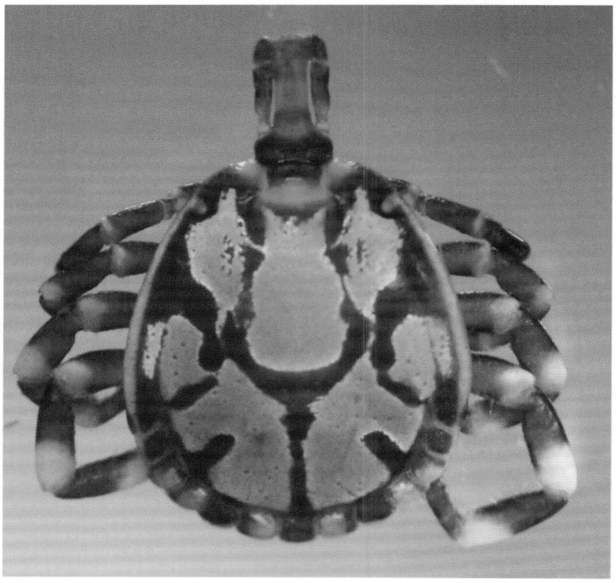
图307　宝石花蜱（*Amblyomma gemma*），一种美丽的具有华彩的非洲蜱

309）的种类，以及被称为斑马蜱的美丽扇头蜱（*Rhipicephalus pulchellus*）（图310）具有华丽的盾板。

硬蜱属级和种级鉴定可以根据其背部表面的特定结构，包括沟、刻点和缘垛（图311）；腹面结构包括气门板、肛侧板和肛下板（图312）；假头和口器的形状和大小；眼的有无及其形状。图305至图315是骆驼上发现的一些硬蜱的图片。

大多数未成熟期的硬蜱很难甚至不能从形态上进行鉴定。已经研发的含有有限种类的分子诊断工具可以对蜱卵进行鉴定[37,39,52]。幼蜱具有6条足（图316），无气门，而若蜱（图317）为8条足具有气门。幼蜱和若蜱阶段在身体前部具有1个小的盾板。

软蜱的假头和口器（除幼蜱阶段）仅从腹面可见。软蜱在所有阶段均没有盾板，并且性别很难区分。软蜱的若蜱在形态上跟成蜱非常相似，并且末期若蜱很难与刚蜕皮未吸血的成蜱区分。只有通过腹面的生殖孔来确定这些蜱处于成熟阶段。

## 生活史

所有硬蜱在胚后期都是专性暂时吸血的体外寄生虫，并且需要吸食血液才能促进发育。典型的

图308　彩饰花蜱（*Amblyomma variegatum*），与胞芽花蜱在华彩的类型上不同

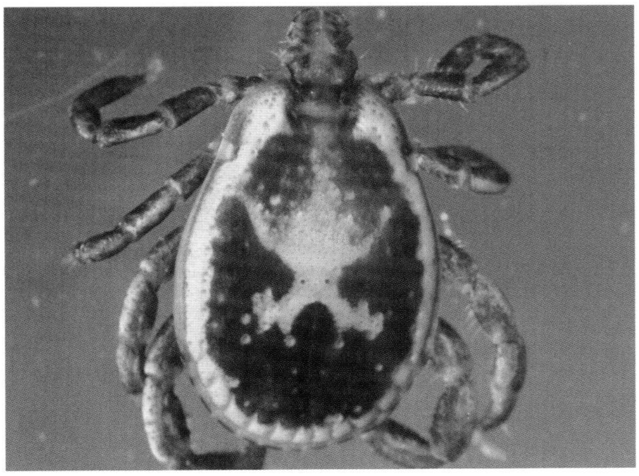

图310　美丽扇头蜱（*Rhipicephalus pulchellus*），东非在骆驼上最常见的一种蜱

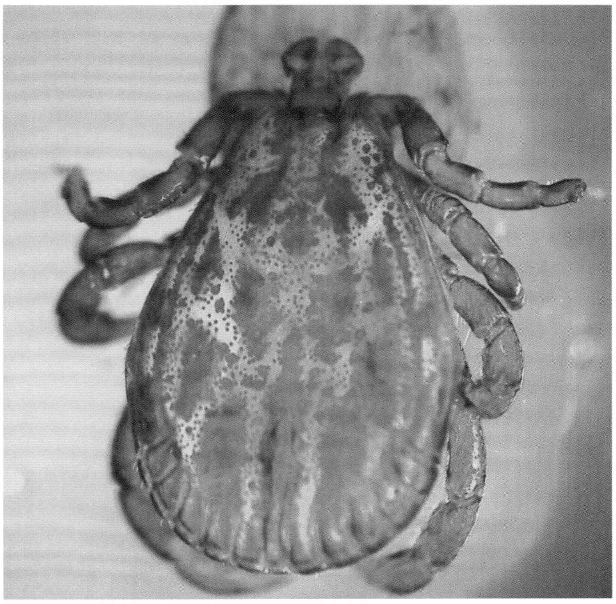

图309　草原革蜱（*Dermacentor nuttalli*），一种美丽的亚洲蜱

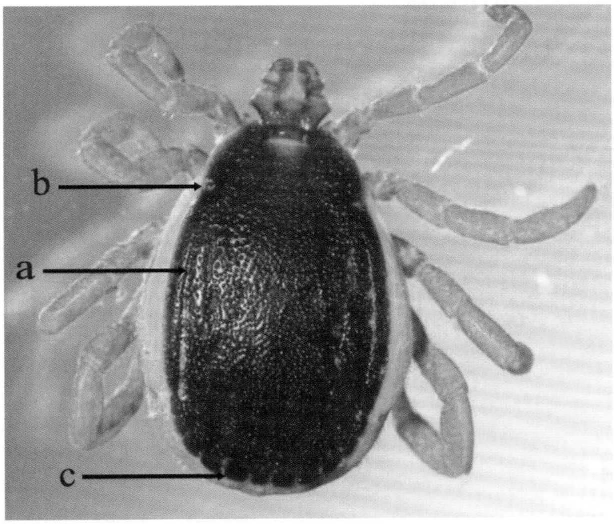


图311　埃氏扇头蜱（*Rhipicephalus evertsi*），盾板上具有粗糙的刻点和明显的侧沟（a），侧沟几乎达到眼（b）的部位。缘垛（c）位于盾板的尾部

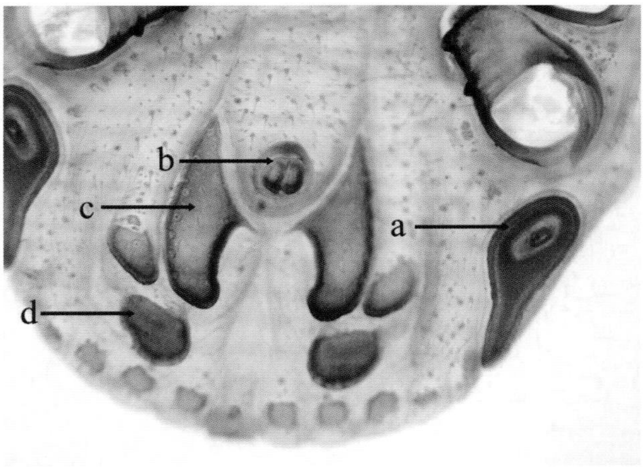


图312　嗜驼璃眼蜱（*Hyalomma dromedarii*），腹侧的气门板（a），肛门（b），肛侧板（c），肛下板（d）

图314　残缘璃眼蜱（*Hyalomma detritum*），具有一个光滑明亮的盾板，其上的后中沟和中侧沟明显

蜱的生活史包括1个非活动期（卵期）和3个可以活动的吸血时期（幼蜱、若蜱和成蜱），并且其生活史根据更换宿主和蜕皮的次数不同分为3种类型。

硬蜱的生活史可以根据它们在发育过程中寄生宿主的数目分为一宿主、二宿主或三宿主型。由于蜱不能长距离移动，只能在小范围内主动找寻宿主，因此寻找宿主是一件有风险的事情，尤其是对于栖息在恶劣环境如荒漠和半荒漠的物种。大多数骆驼上的硬蜱具有眼睛，嗜驼璃眼蜱远在10m开外就可以识别骆驼。硬蜱第1对足的跗节上具有哈氏器，可以识别潜在宿主皮肤上释放的氨气、丁酸和二氧化碳，以及酚和内酯。尽管他们能够

图313　无盾璃眼蜱（*Hyalomma impeltatum*），侧沟长而明显，盾板密布中等大小的刻点。注意足上的环带

图315　麻点璃眼蜱（*Hyalomma rufipes*），强壮，具有粗糙的密布刻点的盾板，足为红褐色具有明亮的灰色环带

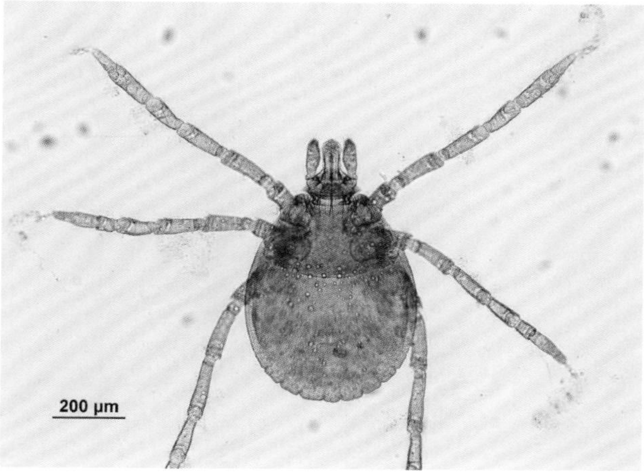


图316　璃眼蜱幼蜱。硬蜱在6条足的幼蜱阶段其背部盾板仅覆盖身体的前部

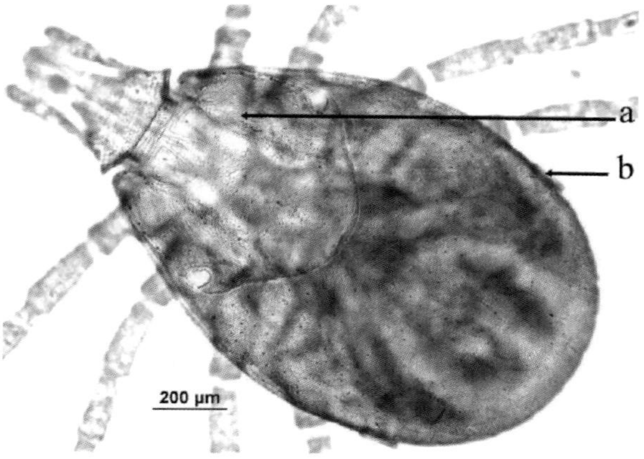


图317　璃眼蜱若蜱。若蜱阶段为4对足，同幼蜱一样，其背部盾板仅覆盖身体前部。幼蜱不具有气门板，而若蜱具有

在10 ~ 15m的范围内感受到宿主，但他们在转移到宿主之前的物理联系也是必须的。根据Waladde和Rice[49]的观点，蜱可以分为游猎型，即可以主动寻找和搜索宿主（如嗜驼璃眼蜱）；另一种为伏击型（如微小扇头蜱），他们会聚集到有利位置，做好当宿主靠近时附着到其身上的准备。一旦附着到宿主体上，蜱便会寻找合适的叮咬位点。这可能会在身体的任何部位，但特定的蜱会有其吸血位点的偏爱性，如耳朵、乳房、阴囊或肛门周围。蜱吸血有集群性，他们的唾液中包含麻醉物质，可以使他们的吸血点没有痛感。血管活性物质可以促进血管舒张，从而保障了蜱类口下板附近的集中吸血点的血液供应。蜱的吸血时间可长达2个星期，为了保护吸血位点，他们还通过口器分泌一种黏合剂到宿主皮肤上。一些种类的雄蜱不需吸血也不接触宿主就可以交配（硬蜱属），而有些雄蜱需要吸食大约饥饿时体重50%的少量血液后才能交配。这里，血液是促进精子成熟的必须条件。雌蜱吸血后体重可增加100倍甚至更多（图318）。雌蜱在吸血的过程中，通过分泌大量的低渗唾液将血液中过多的液体返回到宿主体内。因此，他们吸食血液的浓度是原来的2 ~ 3倍[26]。多数硬蜱的雌蜱在交配前不能完全饱血。受精后的雌蜱在宿主上吸血一次然后脱落，并伴随唾液腺的蜕化。

很多蜱在找到宿主之前就死亡了，他们的生存策略不仅表现在吸血次数的减少及一次性吸食大量血液，而且还会产生大量后代。多数硬蜱的交配发生在宿主身上（除一些前沟型硬蜱属的种类），雌蜱释放的性信息素在异性找寻中起重要作用。Varma[46]详细描述了蜱类复杂的交配过程，共包括几个阶段。蜱可以交配多次，只有最后一次获得的精囊用于受精。因此，与雌蜱交配就意味着终止吸血，这具有减少雌蜱防御的时间，减少干扰性竞争者及增加成为父亲的可能性。

牛蜱属包括4个重要种类：具环牛蜱（*Bo. annulatus*）、无色牛蜱（*Bo. decoloratus*）、嘉氏牛蜱（*Bo. geigyi*）和微小牛蜱（*Bo. microplus*）。最近牛蜱属成为扇头蜱属的亚属。

交配后的雌蜱饱血后离开宿主，在地面上藏身在隐蔽的地方来消化血液并产卵（图319），可以产200 ~ 15 000只卵[22]。

一宿主蜱（如微小扇头蜱）在幼蜱阶段寄生在宿主上，其幼蜱、若蜱和成蜱均吸食同一宿主的血液。二宿主蜱（如埃氏扇头蜱）幼蜱和若蜱阶段吸食同一宿主。饱血若蜱阶段脱离宿主，并蜕化为成蜱再侵袭第二宿主。三宿主蜱（如彩饰花蜱）每一个胚胎后阶段均在不同的宿主上吸血。根据周围的环境条件，一些种类可以表现为二宿主或三宿主类型（如小亚璃眼蜱或麻点璃眼蜱），或其整个生活史为一宿主、二宿主或三宿主型（如嗜驼璃眼蜱）。

嗜驼璃眼蜱是最重要的寄生在骆驼上的蜱，ElGhali和Hassan[15,16]在苏丹研究了嗜驼璃眼蜱的生活史和生态学特性，发现其表现为二宿主蜱，幼蜱-若蜱阶段经历16 ~ 27天，而成蜱需要吸血6 ~ 9天。在野外，18%身体扁平的成蜱可以在冬季存活

1个月，而夏季仅存活5%。单只未饱血的成蜱可以在冬季存活3个月。

饱血阶段的蜱通常在傍晚时分脱落。这样可以使蜱在变凉的夜间找到栖息地，以便他们消化血液并产卵。饱血雌蜱的产卵前期为10～15天。一只嗜驼璃眼蜱的雌蜱依其饱血程度的不同，其产卵量可高达7 000只。胚胎的发育依赖于温度，冬季要持续50天，夏季26～29天。卵的孵化率在阴暗处可达到82%～94%，而在阳光直射的条件下降低为56%～59%。幼蜱如果找不到宿主，则在一周内就会死亡。

北亚（俄罗斯、蒙古和中国）的草原革蜱栖息在完全不同的环境中。成蜱吸食家畜（包括双峰驼）的血液，他们在低于0℃的春季活动。经过5～17天饱血，但产卵要持续3个月。幼蜱和若蜱寄生在啮齿动物、野兔和喜鹊上，若蜱期可以存活6～7个月。在不同温度下，饱血草原革蜱的若蜱蜕化为成蜱需要11～73天。

多数软蜱具有多个宿主的生活史。在其生活史中，他们寄生在同种或不同种宿主的几个动物上，或在同一宿主个体上多次寄生。

一些种类6条足的幼蜱在他们蜕化为第一若蜱之前可能需要吸血，而其他一些种类的幼蜱，如萨氏钝缘蜱（*Ornithodoros savignyi*）不吸血。萨氏钝缘蜱的若蜱蜕皮4～5次，具有相似的栖息和摄食习性，每次蜕皮之间都需要休息[23]。成蜱不在宿主上交配，他们可以多次吸血。雌蜱吸血后产下一批卵。由于这些蜱一直栖息在隐蔽的栖息地，所以他们寻找宿主面临的风险就降低了。

图318　嗜驼璃眼蜱饱血成蜱

图319　一只嗜驼璃眼蜱雌蜱可以产卵达7 000只

梅氏残喙蜱（*Otobius megnini*）仅幼蜱和若蜱阶段寄生，而成蜱仅具有原始的口器，营自由生活。

软蜱的寿命可以长达几年。

## 流行病学

只有少数硬蜱将骆驼作为他们偏爱的宿主。这些包括嗜驼璃眼蜱、舒氏璃眼蜱（*Hy. schulzei*）、神秘璃眼蜱（*Hy. punt*）和黑缘璃眼蜱（*Hy. erythraeum*）（黑缘璃眼蜱被认为是天盾璃眼蜱的同物异名）。作者的经验证明嗜驼璃眼蜱很少出现在其他宿主上，如马、羚羊或瞪羚上。绵羊上从来没有发现过嗜驼璃眼蜱，尽管绵羊与感染嗜驼璃眼蜱很严重的骆驼距离很近。幼驼更容易感染嗜驼璃眼蜱。

舒氏璃眼蜱和印痕璃眼蜱（*Hy. impressum*）在形态上与嗜驼璃眼蜱相似，并常常混淆。很多其他种的蜱宿主范围广，包括骆驼。而有些种类偶尔寄生在骆驼上。一些蜱的分布范围内没有骆驼。具尾扇头蜱（*Rh. appendiculatus*）、新月扇头蜱（*Rh. lunulatus*）和埃氏扇头蜱（*Rh. evertsi*）主要分布在热带非洲和南非，而伊氏璃眼蜱（*Hy. isaaci*）主要分布在南亚和东南亚。

对于有些蜱，骆驼是偶然宿主。吉氏扇头蜱（*Rh. guilhoni*）更倾向于牛和绵羊。长基扇头蜱（*Rh. longicoxatus*）主要寄生在长颈鹿，血红扇头蜱（*Rh. sanguineus*）特异地-寄生在家犬上。由表88和89中可见，嗜驼璃眼蜱是非洲大部分地区和亚洲一些国家中在骆驼上最丰富的蜱，但在埃塞俄

表88　非洲国家蜱在骆驼上出现的概率（%）

| 种类 | 国家 | | | | |
|---|---|---|---|---|---|
| | 埃及[44] | 苏丹[8] | 埃塞俄比亚[53] | 肯尼亚[13] | 尼日利亚[28] |
| 嗜驼璃眼蜱 *Hyalomma dromedarii* | 96.0 | 89.0 | 5.9 | 3.6 | 47.0 |
| 无盾璃眼蜱 *Hy. impeltatum* | (+)[a] | 7.7 | 0.1 | | 18.5 |
| 边缘璃眼蜱 *Hy. marginatum* | (+)[a] | | 0.5 | | |
| 凹陷璃眼蜱 Hy. *excavatum* | | | 0.1 | | |
| 小亚璃眼蜱 *Hy. anatolicum* | | 3.3 | | | |
| 图兰璃眼蜱 *Hy. turanicum* | | 0.3 | 1.0 | 2.6 | 11.7 |
| 麻点璃眼蜱*Hy. rufipes* | | 0.3 | | 4.4 | 23.0 |
| 美丽扇头蜱 *Rhipicephalus pulchellus* | | | 85.2 | 72.4 | |
| 无色扇头蜱 *Rh. decoloratus* | | | 1.0 | | |
| 银边扇头蜱 *Rh. praetextatus* | | 0.3 | | | |
| 血红扇头蜱 *Rh. sanguineus* | | 0.1 | | | (+)[a] |
| 埃氏扇头蜱*Rh. evertsi* | | | 0.3 | | |
| 米萨姆扇头蜱 *Rh. muhsamae* | | | | 7.3 | |
| 具尾扇头蜱 *Rh. appendiculatus* | | | | 4.8 | |
| 变形扇头蜱*Rh. pravus* | | | | 0.4 | |
| 宝石花蜱 *Amblyomma gemma* | | | 4.0 | 2.2 | |
| 彩饰花蜱 *Am. variegatum* | | | 1.8 | 0.8 | (+)[a] |
| 丽表花蜱*Am. lepidum* | | | 0.1 | (+)[a] | |

[a]有，但数量少

比亚和肯尼亚，嗜驼璃眼蜱的地位被美丽扇头蜱（*Rh. pulchellus*）取代。嗜驼璃眼蜱的广泛分布说明该种对沙漠环境的适应。

国际文献上提到，仅在非洲的单峰驼上寄生有14种璃眼蜱、15种扇头蜱、3种花蜱和2种革蜱（表88）。嗜驼璃眼蜱、麻点璃眼蜱（*Hy. rufipes*）、截形璃眼蜱（*Hy. truncatum*）和美丽扇头蜱很常见。无盾璃眼蜱、印痕璃眼蜱、具尾扇头蜱、卡氏扇头蜱（*Rh. camicasi*）、宝石花蜱（*Am. gemma*）和彩饰花蜱是地区性常见种。萨氏钝缘蜱（*Ornithodorus savignyi*）偏好寄生在大型动物上[23, 50]，但由于他们饱血很快，通常在骆驼身上看不到。最近的文献中有少数病例的报道[42]。

其他种类的蜱要么很少或者偶尔有发现，或者仅在部分地区中度感染[13]。

蒙古的双峰驼寄生有亚洲璃眼蜱（*Hy. asiaticum*）、草原革蜱、达吉斯坦革蜱（*De. daghestanicus*）（一些作者认为达吉斯坦革蜱是银盾革蜱（*De. niveus*）的同物异名）和短小扇头蜱（*Rh. pumilio*）[9,19]（表89）。

表 89　亚洲国家蜱在骆驼上出现的频率（%）

| 种类 | 国家 | | | | | |
|---|---|---|---|---|---|---|
| | 沙特阿拉伯[1] | 也门[36] | 伊朗[40] | 阿富汗[25] | 印度[42] | 蒙古[1] |
| 嗜驼璃眼蜱 *Hyalomma dromedarii* | 96.0 | +++[a] | 93.0 | +[c] | 62.0 | |
| 边缘璃眼蜱 *Hy. marginatum* | | | 4.0 | +[c] | | |
| 伊氏璃眼蜱 *Hy. isaaci* | | | | | 0.3 | |
| 凹陷璃眼蜱 *Hy. excavatum* | | +[b] | | | | |
| 小亚璃眼蜱 *Hy. anatolicum* | | +[b] | 1.0 | +[c] | 37.0 | |
| 亚洲璃眼蜱 *Hy. asiaticum* | | | 1.0 | +[c] | | 64.9 |
| 舒氏璃眼蜱 *Hy. schulzei* | | | | +[c] | | |
| 盾糙璃眼蜱 *Hy. scupense* | | | | | 0.1 | |
| 花蜱未知种 | 4.0 | | | | | |
| 草原革蜱 *Dermacentor nuttalli* | | | | | | 16.7 |
| 达吉斯坦革蜱 *De. daghestanicus* | | | | | | 18.0 |
| 短小扇头蜱 *Rhipicephalas pumilio* | | | | | | 0.4 |
| 微小扇头蜱 *Rh. microplus* | | | | | 0.1 | |
| 萨氏钝缘蜱 *Ornithodorus savignyi* | | | | | (+)d | |

a最频繁；

b稍微频繁；

c在骆驼上发现；

d在骆驼栖息的地方发现

对新大陆骆驼上寄生的蜱知之甚少，这些知识主要基于病例报道。有一篇是关于小跗花蜱（*Am. parvitarsum*）和科氏血蜱（*Hae. juxtakochi*）在秘鲁羊驼和美洲驼寄生的研究简报[21, 46]。小跗花蜱是新大陆骆驼包括阿根廷、玻利维亚、智利、秘鲁的常见寄生虫，而科氏血蜱首次发现在羊驼上寄生。另一种花蜱，纽氏花蜱（*Am. neumanni*）在阿根廷的原驼上发现（Voltzit[47]研究了来自阿根廷一只原驼上寄生的蜱，并保存在美国国家蜱类收藏中心。根据Nava等[35]观点，Voltzit[47]的形态特征描述，以及所谓的纽氏花蜱的宿主种类，他们认为该蜱应为小跗花蜱）[47]。微小扇头蜱*Rh.* (*Boophilus*) *microplus*是阿根廷最常见的牛蜱，实验证明美洲驼也可以作为微小扇头蜱的宿主。尽管美洲驼生存的环境降水量低于300mm，且该地区没有微小扇头蜱的分布[2]。

其他蜱也会在起源地以外的地方发现。最近关于羊驼的重要外寄生虫的综述，提到了软蜱中的梅氏残喙蜱和硬蜱中的全环硬蜱（*Ixodes holocyclus*）和一些革蜱[7]。具刺耳蜱即梅氏残喙蜱最初分布在南北美洲，并导致动物耳炎[38]。该种蜱也可在耳朵深处发现，叮咬在密苏里州的羊驼的鼓膜上，导致脑脓肿[11]。随着农场动物和宠物在各大陆间的运输，梅氏残喙蜱被带到了非洲和亚洲的一些国家，并在欧洲和澳大利亚进口的动物中也有发现[27]。

安氏革蜱（*Dermacentor andersoni*）曾导致华盛顿两只美洲驼发生蜱瘫。移除蜱后，动物的症状改善[4]。在加利福尼亚报道了一个由太平洋硬蜱（*Ixodes pacifucus*）传播的埃立克体病[3]。

在澳大利亚的新大陆骆驼也有全环硬蜱导致蜱瘫的报道[12,24]。在欧洲，提到了蓖子硬蜱（*Ix. ricinus*）在德国南部和奥地利导致新大陆骆驼发生蜱瘫[41]。

## 危害

蜱可以导致以下3种危害.

（1）导致动物失血、贫血以及皮肤过敏。

（2）一些疾病的传播媒介。

（3）蜱中毒。

蜱做为外寄生虫其直接危害是导致动物血液流失，皮肤和皮革损伤。尽管蜱可以在骆驼上大量叮咬，但对宿主组织最主要的危害是其唾液对局部皮肤造成的破坏性影响。他们利用螯肢撕开宿主的表皮，并立即分泌唾液来软化和消化受损部位的组织。口下板会伴随着螯肢和唾液的活动慢慢地渗透到宿主的皮肤内。渗透通道附近组织的消化会导致毛细血管和淋巴管的破裂。叮咬处会导致出血和水肿，连接组织会闭合孔洞，导致皮革出现疤痕。

尽管骆驼的梨形虫病看起来危害不大，但骆驼上的蜱可以作为梨形虫或其他宿主动物其他媒介疾病的传播媒介，骆驼可以作为与这些疾病相关的蜱的贮存宿主。因此，骆驼上的璃眼蜱和革蜱被列为马梨形虫病的媒介[5,6]。骆驼上的蜱在环形泰勒虫病、东海岸热、牛的心水病传播中起重要作用[8,32]。传播克里米亚-刚果出血热的27种蜱中，彩饰花蜱、小亚璃眼蜱、边缘璃眼蜱和扇头蜱均在骆驼上发现[46]。此外，在内蒙古，草原革蜱和亚洲璃眼蜱是人的西伯利亚立克次体的传播媒介[30]。

蜱引起的骆驼蜱中毒，像蜱麻痹和蜱瘫，以及汗热病在其他宿主上更为常见。仅有1例在苏丹单峰驼上可能引起的蜱瘫病例[34]（图320）。在新大陆骆驼中，蜱瘫在美国[4]和澳大利亚[24]得到确诊。

## 蜱类鉴定

蜱通常要从宿主上移除再进行鉴定。通过指示宿主（牛、羊或狗）可以预测特定群落生境中蜱的密度。考虑到动物保护，布旗调查的方法取代了指示宿主。该方法是拖一条浅颜色的法兰绒布穿过可能有蜱等待宿主的植被。蜱可能会附着到布上，很容易被收集起来[10]。利用干冰做成的二氧化碳收集器。当干冰在底部有洞的桶内融化后，二氧化碳会将蜱吸引过来，并将他们粘到桶周围具有黏性的纸上。

基于形态学结构对蜱进行鉴定。Hoogstral[23]在1956年建立了蜱类属级水平的检索表，此后，其他作者[12,22,31,32,46]对其进行了少量修订。目前有一个非洲蜱类鉴定的图文手册[50]。

## 防控

氧化亚砷酸是第一种大规模在家畜上控制蜱的杀螨剂，但从没有在骆驼上用过。有很多合成的杀螨剂，包括氯化烃、有机磷酸酯、氨基甲酸盐、脒和拟除虫菊酯（如氯菊酯和氟氯苯氰菊酯，从过去到现在一直在使用）。尽管不主张在骆驼上使用这些药，但将阿维菌素制剂倒在骆驼身上也是可以使用的。

自1990年，在大型动物包括骆驼上控蜱的药物选择拟除虫菊酯中的氯氰菊酯。将1%的氯氰菊酯倒在骆驼上，沿着背部中线从前肩穿过驼峰一直到尾根部。剂量是每10kg体重1mL，但在感染严重的病例剂量要加倍[17]。

还有通过放有信息素和杀螨剂的设备来诱杀雄蜱。性信息素为ASP 2, 6-DCP和有机磷杀螨剂残杀威与聚氯乙烯树脂混合到一起，利用陶瓷模具塑造成许多小的球形物好像正在吸血的雌蜱。完成后，所谓的“蜱诱饵”（14mm长×4mm宽，重0.36g）在外形和信息素组成上酷似半饱血的雌

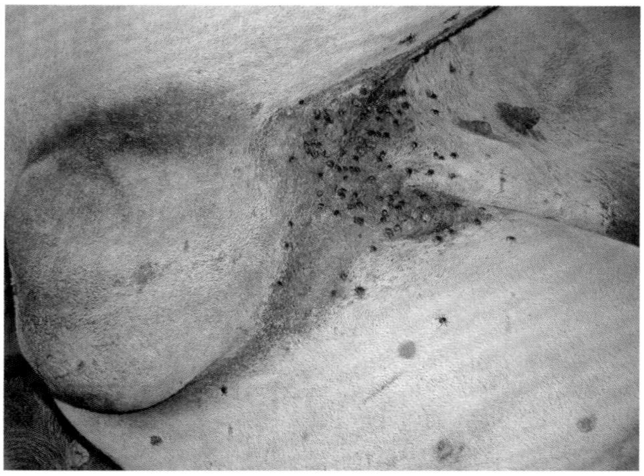

图320　一只寄生有大量蜱的公骆驼

蜱。这些球形物随后会粘到宿主的毛发上。由于会减少雄蜱的数量，大多数雌蜱不能交配而脱落死亡。“最后的招待”是将未释放的蜱信息素、乌嘌呤、黄嘌呤和血红素以及氯菊酯混合成油滴状应用到植被上，作为缓慢释放的制剂，在蜱到达宿主之前被杀死[43]。

除化学防控蜱外，人们自1981年开始尝试疫苗防控。1991年，第一个商业化的基于抗原Bm86疫苗Tick Guard 和Gavac投入使用。这些疫苗用于控制牛上的微小牛蜱，且直接与蜱的肠消化细胞中微绒毛的89-kDa糖蛋白反应。这些疫苗在其他蜱中也有作用，包括嗜驼璃眼蜱。与未免疫的对照宿主相比，被免疫的小牛中，饱血若蜱的数量下降95%，成蜱的体重下降72%，雌蜱的产卵量下降79%[48]。一些蜱的潜在候选免疫抗原也已经被识别[33]。

另外一种控制方法，珍珠鸡可以捕食蜱，随后在美洲对付达明硬蜱的实验获得成功[14]。

## 参考文献

[1] Abdally M.H. (2008). – Species of ticks on camels and their monthly population dynamics in Arar city, KSA. *Assuit Vet.Med. J.,* 54, 302 – 309.

[2] Aguirre D.H., Cafrune M.M. & Guglielmone A.A. (2000). – Experimental infestation of llamas (*Lama glama*) with *Boophilusmicroplus* (Acari: Ixodidae). *Exp. Appl. Acarol.*, 24, 661 – 664.

[3] Barlough J.E., Masdigan J.E., Turoff D.R., Clover J.R., Shelly S.M. & Dummler J.S. (1997). – An *Ehrlichia* strain from a llama(*Lama glama*) and llama associated ticks (*Ixodes pacificus*). *J. Clin. Microbiol.*, 35 (4), 1005 – 1007.

[4] Barrington G.M. & Parish S.M. (1995). – Tick paralysis in two llamas. *J. Am. Vet. Med. Assoc.*, 207 (4), 476 – 477.

[5] Battsetseg G., Batsukh Z., Igarashi I., Nagasawa H., Mikami T., Fujisaki K. (2001). – Detection of *Babesia caballi* and *Babesiaequi* in *Dermacentor nutalli* adult ticks. *Intern. J. Parasitol.*, 31, 384 – 386.

[6] Boldbaatar D., Xuan X, Battseteg B, Igarashi I., Battur B., Batsukh Z., Bayambaa B., and Fujisaki K. (2005). – Epidemiologicalstudy of equine piroplasmosis in Mongolia. *Vet. Parasitol.*, 127, 29 – 32.

[7] Bornstein S. (2010). – Important ectoparasites of alpaca (*Vicugna pacos*). *Acta Vet. Scandinavica,* 52, 1 – 6.

[8] Burridge M.J., Simmons L-A. & Allan S.A. (2000). – Introduction of potential heartwater vectors and other exotic ticks intoFlorida on imported Reptiles. *J. Parasitol.*, 86 (4), 700 – 704.

[9] Byambaa B., Dash M. & Tarasevich I.V. (1994). – Newly diagnosed rickettsial diseases [in Mongolian]. Esen erdene,Ulaanbaatar, Mongolia.

[10] Carrol J.F. & Schmidtmann E.T. (1992). – Tick sweep: modification of the tick drag-flag method for sampling nymphs ofthe Deer tick (Acari: Ixodidae). *J. Med. Entomol.*, 29 (2), 352 – 355.

[11] Chigerwe M., Middleton J.R., Pardo I., Johnson G.C. & Peters J. (2005). – Spinose ear ticks and brain abscession in analpaca (*Lama pacos*). *J. Camel Pract. Res.*, 12 (2), 145 – 147.

[12] Coles L. (2006). – Control of paralysis tick in the alpaca. *Alpacas Australia*, 51, 24 – 26.

[13] Dioli M. (2002). – Studies on hard ticks (Acari: Ixodida) infesting the one humped camel (*Camelus dromedarius*) in Kenyaand Southern Ethiopia. Thesis, Royal Veterinary College, London.

[14] Dufty D.C. (1992). – The effectiveness of helmethed guinea fowl in the control of the deer tick, the vector of Lyme disease. *The Wilson Bulletin*, 104 (2), 342 – 344.

[15] El Ghali A. & Hassan S.M. (2010). – Drop-off rhytmus and survival periods of *Hyalomma dromedarii* (Acari: Ixodidae) fedon camels (*Camelus dromedarius*) in the Sudan. *Vet. Parasitol.*, 170, 302 – 306.

[16] El Ghali A. & Hassan S.M. (2010). – Life cycle of the camel tick *Hyalomma dromedarii* (Acari: Ixodidae) under field conditionsin Northern Sudan. *Vet. Parasitol.*, 174, 305 – 312.

[17] El-Azy O.M. & Lucas S.F. (1996). – The sterilizing effect of pour-on flumethrin on the camel tick, *Hyalomma*

*dromedarii*(Acari: Ixodidae). *Vet. Parasitol.*, 61, 399 – 343.

[18] Elghali A. & Hassan S.M. (2009). – Ticks (Acari: Ixodidae) infesting camels (*Camelus dromedarius*) in Northern Sudan. *Onderstepoort J. Vet. Res.*, 76, 177 – 185.

[19] Erdenebileg O. (2001). – Camel diseases [in Mongolian]. Dalanzadgad, Ulaanbaatar, Mongolia.

[20] Guglielmone A.A., Robbins R.G., Apanaskevich D.A., Petney T.N., Estrada-Pena A., Horak I.G., Shao R. & Barker S. (2010). –The Argasidae, Ixodidae and Nutalliellidae (Acari: Ixodida) of the world: a list of valid species names. *Zootaxa*, 2528, 1 – 28.

[21] Guglielmone A.A., Romero J., Venzal J.M., Nava S., Mangold A.J. & Villavicencio J. (2005). – First record of *Haemaphysalis juxtakochi* Cooley, 1946 (Acari: Ixodidae) from Peru. *Syst. Appl. Acarol.*, 10, 33 – 35.

[22] Hiepe Th. & Ribbeck R. (1982). – Veterinarmedizinische Arachno-Entomologie. *In* Hiepe Th. (ed.) Lehrbuch der parasitologie.VEB Gustav-Fischer-Verlag, Jena.

[23] Hoogstral H. (1956). – African Ixoidea. I. Ticks of the Sudan (with special reference to Equatoria province and withpreliminary reviews of the genera *Boophilus, Margaropus*, and *Hyalomma*). Department of the Navy, Bureau of Medicine and Surgery, Washington DC.

[24] Jonsson N.N. & Rozmanec M. (2008). – Tick paralysis and hepatic lipidosis in a llama. *Aust. Vet. J.*, 75 (4), 250 – 253.

[25] Kaiser M.N. & Hoogstraal H. (1963). – The Hyalomma ticks (Ixodoidea, Ixodidae) of Afghanistan. *J. Parasitol.*, 49 (1),130 – 139.

[26] Kaufmann R.W. (2007). – Gluttony and sex in female ixodid ticks: how they compare to other blood-sucking arthropods? *J. Insect Physiol.*, 53, 264 – 273.

[27] Keirans J.E. & Pound J.M. (2003). – An annotated bibliography of the spinose ear tick, *Otobius megnini* (Duges, 1883) (Acari: Ixodida: Argasidae) 1883–2000. *Syst. Appl. Acarol. Special Publications*, 13, 1 – 68.

[28] Lawal M.D., Ameh I.G. & Ahmed A. (2007). – Some ectoparasites of *Camelus dromedarius* in Sokoto, Nigeria. *J. Entomol.*, 4 (2), 143 – 148.

[29] Leonovich S.A. (2004). – Phenol and lactone receptors in the distal sensilla of the Haller' s organ in *Ixodes ricinus* ticks and their possible role in host perception. *Exp. Appl. Acarol.*, 32(1–2), 89 – 102.

[30] Liu Q-H., Chen G-Y., Jin Y., Te M., Niu L-C., Dong S-P. & Walker D.H. (1995). – Evidence for high prevalence of spotted fever group rickettsial infections in diverse ecologic zones in Inner Mongolia. *Epidemiol. Infect.*, 115, 177 – 183.

[31] Mathisse J.G. & Colbo M.H. (1987). – The Ixodid ticks of Uganda. Entomolological Society of America, College Park, MD.

[32] Morel P.C. (1989). – Tick-borne diseases of livestock in Africa. *In* Manual of tropical veterinary parasitology (M.S. Fischer & R.R. Say, eds). Cambrian Printers, Aberystwyth, 299 – 460.

[33] Mulenga A., Sugimoto C. & Onuma M. (2000). – Issues in tick vaccine development: identification and characterisation of potential candidate vaccine antigens. *Microbes and Infection*, 2, 1 353 – 1 361.

[34] Musa M.T. & Osman O.M. (1990). – An outbreak of suspected tick paralysis in one-humped camels (*Camelus dromedarius*) in the Sudan. *Rev. Elev. Méd Vét. Pays Trop.*, 43 (4), 505 – 510.

[35] Nava S., Estrada-Peñ a A., Mangold A.J. & Guglielmone A.A. (2009). – Ecology of *Amblyomma neumanni* (Acari: Ixodidae). *Acta Trop.* 111, 226 – 236.

[36] Pergram G., Hoogstral H. & Wassef H.Y. (1982). – Ticks (Acari: Ixoidea) of the Yemen Arab Republic. I. Species infesting life stock. *Bull. Entomol. Res.*, 72, 215 – 227.

[37] Poucher K.L., Hutcheson H.J., Keirans J.E., Durden L.A. & Black W.C. (1999). – Molecular genetic key for the identification of 17 ixodid species in the United States (Acari: Ixodidae): a method's model. *J. Parasitol.* 85, 623 – 629.

[38] Rich G.B. (1957). – The ear tick, *Otobius megnini* (Duges) (Acarina: Argasidae), and its record in British Columbia. *Can. J. Comp. Med.*, 12, 415 – 418.

[39] Rumer L., Sheshukova O., Dautel H., Mantke O.D. & Niedrig D. (2011). – Differentiation of medically important Euro-Asian tick species *Ixodes ricinus, Ixodes persulcatus, Ixodes hexagonus* and *Dermacentor reticulatus* by polymerase chain reaction. *Vector Borne Zoonotic Dis.*, 11, 899 – 905.

[40] Salimabadi Y., Telmadarraiy Z., Vatandoost H., Chinikar S., Oshagi S., Moradi M., Mirabzadeh Ardakan E., Hekmat S. &

Nasiri A. (2010). – Hard ticks on domestic ruminants and their seasonal population dynamics in Yazd Province. *Iran. J. Arthropod-Borne Dis.*, 4 (1), 66 – 71.

[41] Schlogl C. (2010). – Erhebungen zum Vorkommen von Endo- sowie Ektoparasiten bei Neuweltkameliden. Thesis, Ludwig Maximilians-University, Munich.

[42] Singh S. & Chhabra M.B. (2009). – A note on ticks of camels in Haryana (India). *In* Selected research on camelid parasitology (T.K. Ghalot, T.K. & M.B. Chhabra, eds). Camel Publishing House, Bikaner, 221 – 222.

[43] Sonenshine D.E. (2006). – Tick pheromones and their use in tick control. *Annu. Rev. Entomol.*, 51, 557 – 580.

[44] van Straten M. and Jongejan F. (1993). – Ticks (Acari: Ixodidae) infesting the Arabian camel (*Camelus dromedarius*) in the Sinai, Egypt with a note on acaricidal efficacy of ivermectin. *Exp. Appl. Acarol.*, 17, 605 – 616.

[45] Thyron U. (1982). – Untersuchungen zur medikamentellen Bekampfung von *Dermacentor nutalli* beim Schaf unter den Bedingungen der Mongolischen Volksrepublik. Thesis, Humboldt-University, Berlin.

[46] Varma M.G.R. (1993). – Ticks and mites (Acari). *In* Medical insects and arachnids (R.P. Lane & R.W. Crosskey, eds). Chapman & Hall, London, 507 – 658.

[47] Voltzit A. (2007). – A review of neotropical *Amblyomma* species (Acari: Ixoxdidae). *Acarina* 15, 3 – 134.

[48] de Vos S., Zeinstra L., Taoufik O., Willadsen P. & Jongejan F. (2001). – Evidence for the utility of the Bm86 antigen from *Boophilus microplus* in vaccination against other tick species. *Exp. Appl. Acarol.*, 25, 245 – 261.

[49] Waladde S.M. & Rice M.J. (1982). – The sensory basis of tick feeding behavior. *In* Physiology of ticks (F.D. Obenchain & R. Galun, eds). Pergamon Press, Oxford, UK, 71 – 118.

[50] Walker A.R., Bouattour A., Camicas J.-L., Estrada-Penˇ a A., Horak I.G., Latif A.A., Pegram R.G. & Preston P.M. (2007). – Ticks of domestic animals in Africa: a guide to identification of species. Atalanta, Houten, The Netherlands.

[51] Yuval B., Deblinger R.D. & Spielman A. (1990). – Mating behavior of male deer ticks *Ixodes dammini* (Acari: Ixodidae). *J. Insect Behav.*, 3 (6), 765 – 771.

[52] Zahler M., Gothe, R. & Rinder H. (1995). – Diagnostic DNA amplification for individual tick eggs, larvae and nymphs. *Exp. Appl. Acarol.* 19, 731 – 736.

[53] Zeleke M. & Bekele T. (2004). – Species of ticks on camels and their seasonal population dynamics in Eastern Ethiopia.*Trop. Anim. Health Prod.*, 36, 225 – 231.

（陈泽译，殷宏校）

## 5.3.2 疥癣

疥癣是一种由寄生螨类引起的传染性皮肤疾病，可以导致严重的皮炎。疥螨是永久性寄生虫，它们栖息在皮肤内，以血液、淋巴液、皮肤碎屑或脂肪分泌物为食。骆驼上可以找到3种疥螨隶属2个科：疥螨科（Sarcoptidae）疥螨属（*Sarcoptes*）和痒螨科（Psoroptidae）痒螨属（*Psoroptes*）及足螨属（*Chorioptes*）。疥癣是所有种类骆驼最重要的皮肤疾病，是由疥螨（*Sarcoptes scabiei*）引起的。这些螨虫可以在100多种哺乳动物（包括人）体寄生，并且有很多在形态上很难区分的亚种或变种。不同的变种具有宿主特异性但特异性不高，它们可以从一个宿主传播到另一宿主。在过去，疥螨的种是根据分离于何种宿主而确立的，但基于rRNA基因的内部转录间隔2测定了来自4个国家9种宿主的23种疥螨，确定了疥螨是由单一的异源种类组成[42]。这一点与足螨和痒螨类似。足螨属包括德州足螨（*C.texanus*）和牛足螨（*C.bovis*）。从形态上看，这两种仅在雄螨的外侧末体刚毛的长度上有区别。骆驼上分离的螨虫通过分子鉴定发现属于牛足螨[14]。过去描述的所有不同种类的痒螨包括来自羊驼上的羊驼痒螨（*Psoroptes auchinae*）现在均认为是一个种[43]。然而，根据动物学分类规则，由于羊痒螨（*P. ovis*）具有分类优先权[40]，因此不再使用原先提议的马痒螨（*P. equi*）[43]。

## 形态学

疥螨（图321）形如海龟。雄虫和雌虫体长分别约250μm和500μm。假头宽度大于长度，足分为2组：前2对足的末端具有吸盘柄，着生在不分节的梗节或端跗节上，后2对足短且没有伸出身体边缘；在雌虫，末端具有长鬃；而雄虫仅第3对具有1个鬃，第4对具有吸盘柄，其结构与第1对相似。雌疥螨的背部表面覆盖鳞片。雄虫无交配吸盘和后叶。疥螨生活在皮肤上挖掘的隧道中。

足螨呈卵圆形。雄虫体长300～450μm，雌虫体长400～600μm。口锥钝，所有4对足均伸出身体边缘。足螨雄虫（图322）的所有足末端的爪垫着生在短的端跗节上（图323），而雌虫第3对足具有2个长鬃（图324）。通过交配吸盘和后叶很容易识别雄螨。

痒螨（图325）体型最大，雄虫体长500～600 μm，雌虫体长可达600～800μm。他们的口锥长而尖，所有的8条足均伸出身体边缘。梗节长，由3节组成，爪垫呈喇叭形（图326）。与足螨相似，痒螨

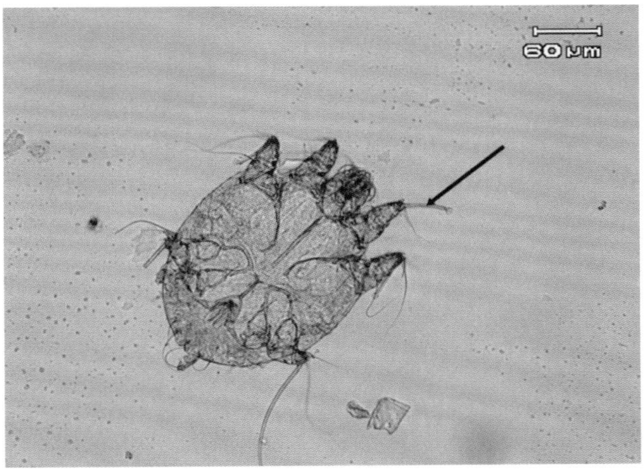


图321　疥螨是最小的疥癣螨。足短更容易适应隧道生活。雌虫和雄虫2对足的末端具有吸盘柄，并着生在长的不分节的梗节上（箭头所示）

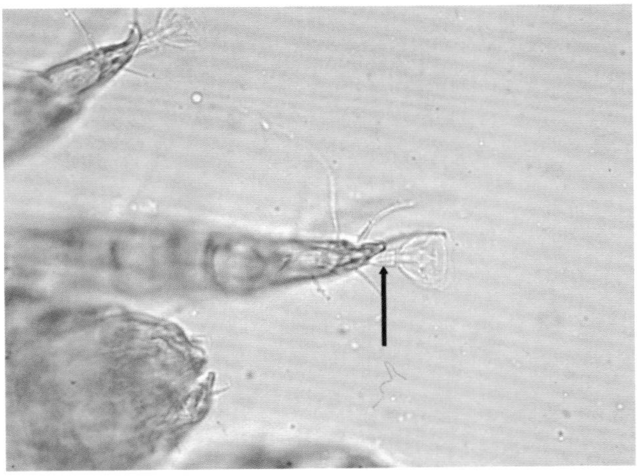

图322　雄性足螨具有交配吸盘（a）和后叶（b）（照片由位于德国Katharinenhof梅里亚公司的S. Rehbein博士提供）

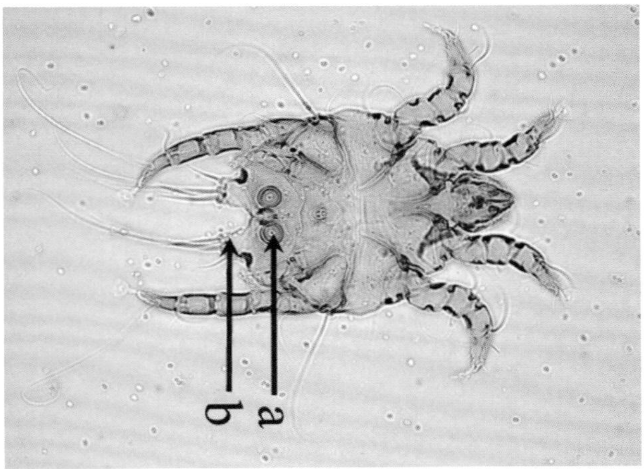


图323　足螨的端跗节（箭头）短且不分节（照片由吉森尤斯图斯-李比希大学寄生虫学院的Ch. Bauer博士提供）

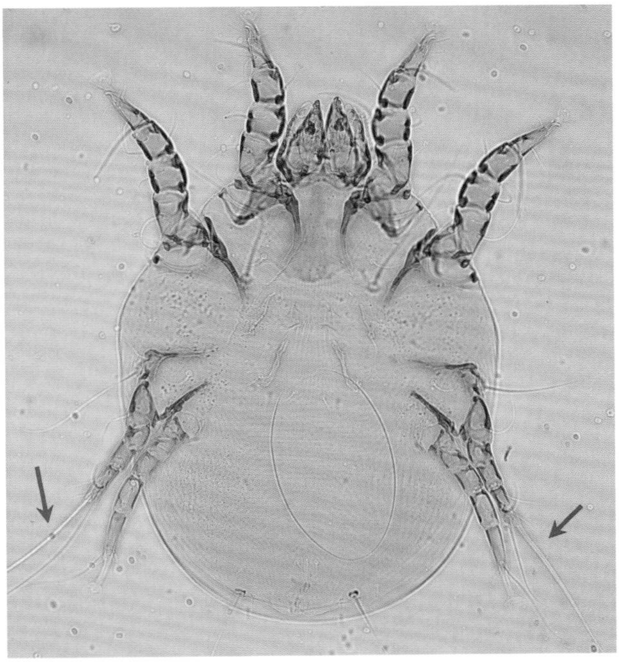

图324　足螨雌虫在第3对足上具有2个长鬃（照片由位于德国Katharinenhof梅里亚公司的S. Rehbein博士提供）

5 寄生虫病

的雌虫在第3对足上具有2个长鬃；雄虫可以通过交配吸盘和后叶来识别。

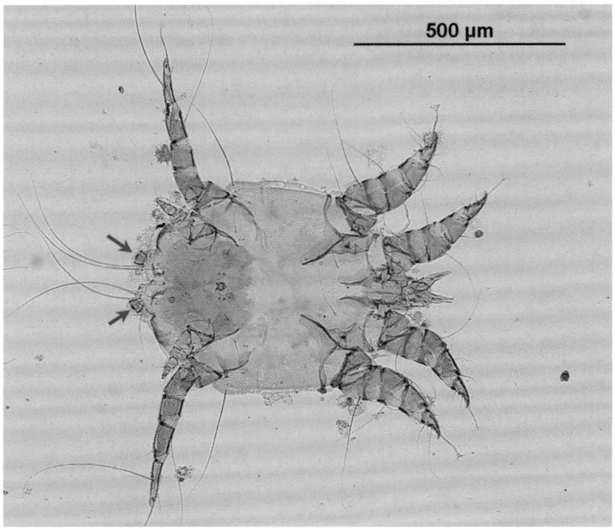


图325　痒螨的雄虫也具有交配吸盘

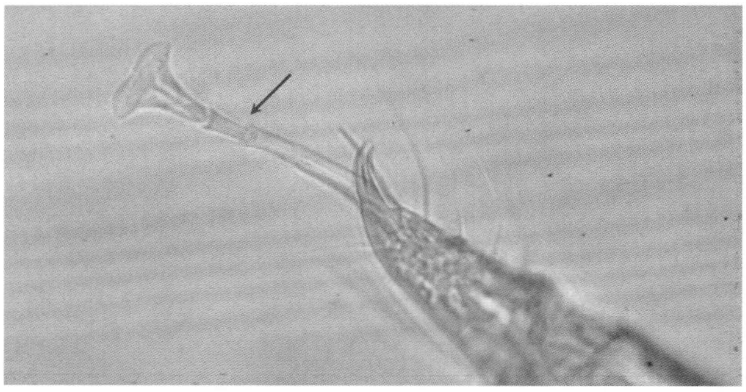

图326　痒螨的端跗节长且分节（照片由吉森尤斯图斯-李比希大学寄生虫学院的Ch. Bauer博士提供）

## 生活史

螨虫的生活史包括5个时期：卵、幼螨、第一若螨、第二若螨和成螨。刚受精的雌性疥螨会利用螯肢和第1对足的锐利边缘在宿主皮肤的表层挖深达25mm的洞（角质层和颗粒层），并在里面产卵。6条足的幼虫3～5天后孵化。幼螨、第1若螨和雄性第3若螨停留在隧道中，或迁移到毛囊中。而雌性第3若螨在皮肤表面与雄螨交配并蜕化为成螨。雌螨完成整个生活史需要21天，雄螨需要14天。雌性成螨的寿命可达4周。

足螨寄生在宿主的身体表面。2天后完成胚胎发育，整个生活史持续3周。雌螨孵化后3～7天开始产卵。成螨可以存活7～8天。

雄性痒螨的寿命为3～5周，雌性为4～6周。雌螨可产卵100只，在3周内完成整个生活史。

## 流行病学

疥癣是一种接触性疫病。所有螨虫的胚胎发育时期均可在皮肤表面发生，并可导致没有螨虫的宿主感染。宿主的种群密度越大，感染的风险越高。当宿主通过抓挠或在地面上翻滚导致螨虫脱落时会发生间接传染。根据不同的种类、温度和湿度，螨虫脱离宿主后可以存活数天甚至数周。梳毛器、毛毯、马鞍或其他器具可能会导致螨类传播。被螨虫感染的污染物在螨虫的传播过程中起了重要作用。

疥疮是旧大陆骆驼中最重要最常见的疥癣类型。尼日利亚[4]、厄立特里亚[37]、苏丹[6]、埃及[32]、阿拉伯联合酋长国[20]、沙特阿拉伯[25]、巴基斯坦[30]、印度[24,29]、阿曼[2]和前苏联的中亚各国[36]均有单峰驼疥癣的报道。疥癣也给澳大利亚的野生单峰驼[12]和蒙古的双峰驼[13]带来问题。

新大陆骆驼存在3种疥癣，大多数疥癣的报道均来自南美洲以外的国家。在南美洲，仅秘鲁有新大陆骆驼疥癣病的报道[3,41]。一篇关于美洲驼寄生虫经济影响的综述估计由于疥疮导致的经济损失可占外寄生虫引起的所有损失的95%[22]。作者认为疥螨在秘鲁羊驼和美洲驼的流行可分别达到40%和25%。痒螨引起的疥癣也存在，但问题不严重，因为仅仅导致皮肤表面损伤。有趣的是，在南美洲没有骆驼足螨病的报道[10]。

疥螨病在北美的美洲驼中很常见，其次为痒螨病[8]。足螨病在美国的美洲驼和羊驼中很少见[16]。

多数关于新大陆骆驼螨病的报道来自于英国，其中足螨看起来是新大陆骆驼中最常见的体外寄生虫[10]。作者调查了209只羊驼，认为足螨流行率为39.9%。在英国羊驼中也有暴发疥螨病的报道[38,39]。

在德国，足螨病也是最常见的。最近的研究表明，检查的254只新大陆骆驼中足螨的出现率为49%[33]。

关于英国的羊驼和美洲驼螨病现状的分析表明，畜群的大小及从秘鲁进口羊驼是导致畜群发生螨病的风险因子，而从智利和其他国家进口羊驼不作为风险因子[23]。

## 临床症状和病理变化

居住在皮肤的疥螨（图327）导致最严重的临床过程。螨虫的唾液可以溶解皮肤浅表层，宿主主要对其摄食和穴居行为做出反应。宿主通过生发层产生角蛋白来填充洞穴，这会导致角化过度。瘙痒是最主要的临床信号（图328）。

在与污染物接触的24天内，单峰驼出现了皮肤改变。单峰驼感染疥螨后，皮肤损伤的特点包括表皮脱落、红斑、破裂、流血、增厚和起皱、秃头、色素沉着过多[20]。鼻孔、嘴唇、眼眶、胸部、前后肩胛和股部、胸骨、大腿、尾部会阴、膝盖和踝关节经常被感染[11]。幼驼上的疥螨不治疗的话会导致死亡（图329）。

疥疮在单峰驼中还表现为免疫球蛋白G水平和嗜酸性粒细胞的增高[19]。

通过研究感染疥螨疥疮的单峰驼的血象和生化发现诸如红细胞、红细胞压积、血红蛋白、血清白蛋白、血清抗坏血酸等参数降低。这些变化在中度和严重感染的病例比温和感染或阴性对照更明显。然而在严重病例中，血清一氧化氮、丙二醛、蛋白质羰基增加[32]。

新大陆骆驼患疥螨疥疮的临床症状包括脱毛、结痂、皮肤增厚和严重瘙痒[39]（图330～图332）。皮肤的组织病理学检查可以发现多病灶角化过度、表皮里发现大量螨虫、多病灶到融合性的结痂、包括嗜酸性粒细胞在内的混合炎症细胞多病灶表皮浸润和迁移、多病灶到融合性色素断续，毛囊缺失及偶有脓肿[38]。疥疮是人畜共患病。人类感染该病是通过接触被感染的动物，而呈现伪疥螨病的临床症状（图 333）。然而，人类这种病仅限于自身，没有必要进行特定的杀螨剂处理。

痒螨和疥螨均生活在皮肤表面。痒螨寄生在宿主的皮肤表面，以皮肤细胞、细菌和淋巴的脂肪乳为食，从而导致宿主对螨虫粪便的超敏反应。组织病理学图片显示角膜下脓包、嗜酸性粒细胞脓包和由嗜酸性粒细胞、中性粒细胞、巨噬细胞和淋巴细胞构成的皮肤渗透。

新大陆骆驼上的痒螨主要涉及耳道病变导致外耳炎并伴随耳廓瘙痒、结痂和脱毛，并且外耳道不断有干皮脱落。足螨主要以有机物质为食，如毛发和皮屑。足螨病通常不严重，包括脱毛症、红斑、丘疹、结痂、渗出液和溃疡。病变在肢体远端、腹侧及尾根部[31]。在严重病例中，前额、耳朵、躯体、四肢和会阴均可涉及脱毛、鳞屑和

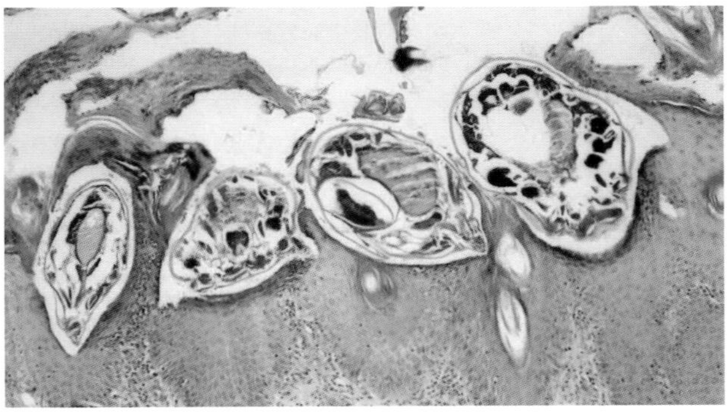

图327　疥螨在皮肤生发层的隧道中

图328　骆驼中的疥螨瘙痒症

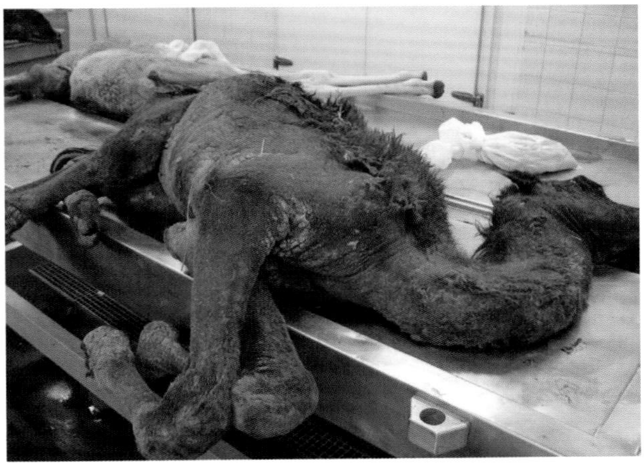

图329　幼驼上全身致死性疥疮

图330　羊驼头部的疥疮（照片由英国卢丁顿的兽医实验室联合局N. G. Vine博士提供）

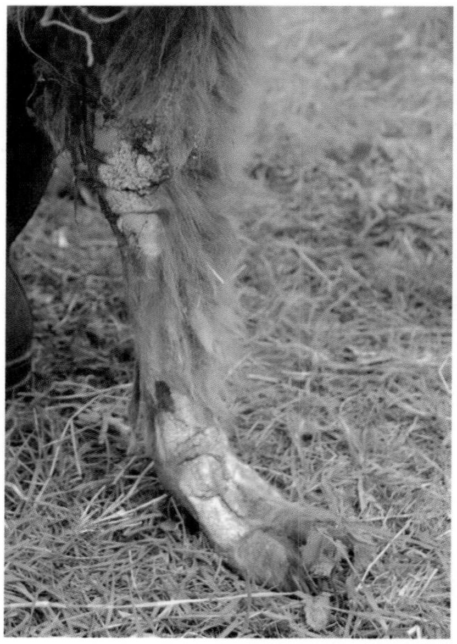

图331　羊驼腿上的疥疮（照片由英国卢丁顿的兽医实验室联合局N. G. Vine博士提供）

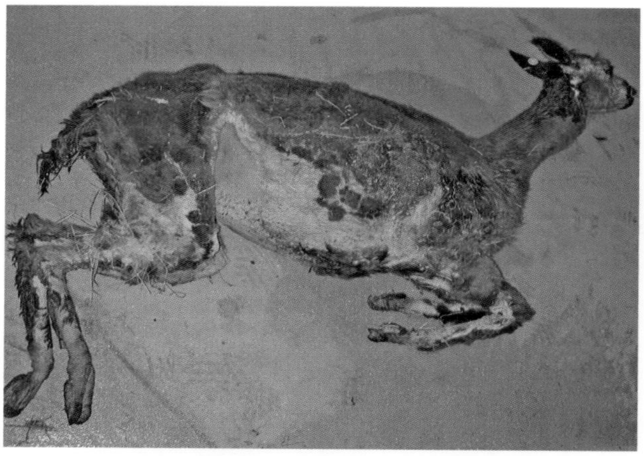

图332　羊驼致死性疥疮（照片由英国斯塔克罗斯兽医实验室的D. F. Twomey博士友情提供）

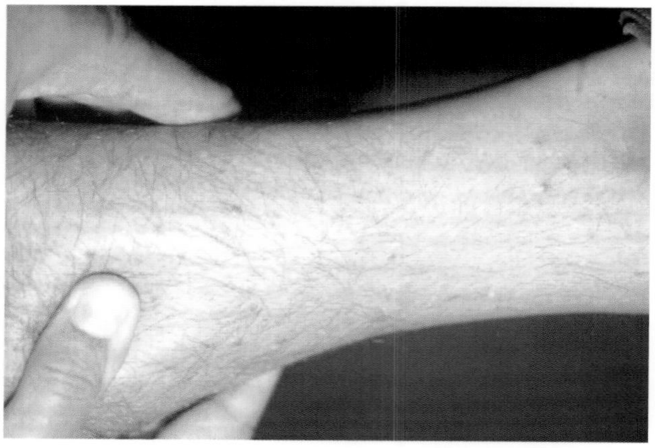

图333　人的伪疥螨病的临床照片。一个接触被疥螨感染的骆驼的管理人前臂上面的丘疹

结痂，及严重的红斑、糜烂、溃疡和苔藓样硬化[28]。

## 诊断

根据疥癣的临床症状和皮肤发病位置基本能确诊。癣、疹和羊痘疮必须区分开。必须在感染处和健康皮肤的边界取深层皮肤刮取物。教科书中推荐使用解剖刀片。然而作者的经验认为在职业安全性上，使用带有三角形钩的蹄刀更有效。

在实验室，样品需要加热到40℃，首先在解剖镜下验证是否有活螨。然后将皮肤刮取物用70℃的10%氢氧化钾处理30～60min来溶解碎屑和毛发，并在显微镜下检测螨虫。螨虫的大小、口器的形状、足上梗节的结构可以作为识别螨虫的依据特征。

对于疥螨感染，通过利用间接酶联免疫吸附试验识别抗体也可能检测出来[6]。现在，这些方法都商业化了，还需要进一步评估与骆驼抗体的结合。

## 防控

不同类型的疥疮必须遵循以下规则采用不同的治疗方法。

（1）由于所有的杀螨剂均没有杀卵效果，因此经过7 ~ 14天至少需要处理两次来根除活螨。

（2）群养的所有动物都必须处理，因为看起来健康的骆驼可以作为携带者。

（3）通过15%的水杨酸水浴移除更换的皮肤（如头皮、皮屑）可以提高处理效果。

（4）单只螨虫离开宿主后可以存活一段时间。因此，在防控措施中有必要清理器具、马鞍、毛毯及地面。

因为疥螨寄生在皮肤表层的隧道中，这就需要内吸模式的杀螨剂。过去处理骆驼疥螨的有机氯化学品，如林丹或六氯环已烷已被生态毒性小的有机磷、辛硫磷和敌百虫取代。这些化学试剂是节肢动物的神经毒素，会抑制乙酰胆碱酯酶，该酶是神经系统行使正确功能所必须的。有机磷可根据规划应用于清洗、喷雾和浇注。评估辛硫磷和六氯环已烷的功效表明利用水杨酸预先处理骆驼后，再用0.1%的辛硫磷乳剂可得到很好的结果[2]。氨基甲酸酯复合西维对节肢动物乙酰胆碱酯酶的作用具有与有机磷相似的功效。曾经有0.5%的喷雾处理单峰驼的疥螨病，但发现仅在第4次处理后就恢复了[35]。

在自然感染的骆驼中测试了双甲脒的效果（图334）。它可以降低螨虫的密度，改善动物的状况，但不能根除疥螨[34]。

合成的大环内脂具有完全不同的作用机理，正在取代有机磷和其他化学剂作为治疗疥疮的杀螨剂。阿维菌素类杀虫剂（伊维菌素、阿维菌素、多拉菌素、乙酰氨基阿维菌素）和米尔贝霉素（莫西菌素）直接作用于那些利用氨基丁酸作为神经递质的机体，包括节肢动物和线虫。这些药物结合并激活特定无脊椎动物类群的神经元和肌细胞的谷氨酸门控氯离子通道，导致这些细胞超极化并阻碍信号转导。阿维菌素类杀虫剂和米尔贝霉素不容易穿过哺乳动物的血脑屏障，因此在哺乳动物和鸟上应用是安全的。人们对在单峰驼上仅使用阿维菌素类杀虫剂[26]，使用阿维菌素类杀虫剂和米尔贝霉素[27]以及仅使用米尔贝霉素[5]，进行了一些合成的大环内脂的应用药物动力学研究。与山羊和奶牛相比，阿维菌素类杀虫剂在骆驼皮下注射（每千克体重0.2mg）后，乳汁和血清中的最大浓度很低（分别为1.79ng/mL和2.79ng/mL）。莫西菌素在乳汁和血清中的浓度很高（分别为28.75ng/mL和8.73ng/mL）。乙酰氨基阿维菌素是第二代阿维菌素类杀虫剂的一种外用制剂，主要针对产奶母牛防

图334　利用双甲脒喷雾法处理严重感染的骆驼

控外寄生虫和线虫，因此应用于产奶骆驼中也是安全的。骆驼乳汁中最大浓度为3.09 ng/mL。新大陆骆驼上利用伊维菌素[7,18]和牛上利用规定剂量的多拉菌素[17]也进行了相似的药物动力学研究得到相似的结果，这些说明新大陆骆驼上使用阿维菌素类杀虫剂的剂量应该增加。

伊维菌素的按推荐注射剂量每千克体重0.2mg治疗后，感染动物的状况进行改进，但仅通过一次注射不能根除骆驼上的疥疮。骆驼经过3周注射两次伊维菌素，治疗后3个月临床症状消失，但4周后又复发，甚至在没有感染的对照围栏里的骆驼也有发病，该围栏距离实验动物在10m开外[20]。或者动物饲养员在这次试验中患了伪疥螨病，或者鸟类再次传播了该病。疥疮只有重复使用伊维菌素才能得到根除。在一次实验中，患疥疮的骆驼注射了一次伊维菌素或多拉菌素。两组中，活螨均在处理后一周消失。多拉菌素处理的动物完全恢复，并且在8周观察的时间里没有螨虫，而伊维菌素处理的实验组在第3周又出现螨虫。作者认为多拉菌素具有长期功能的高效性归因于其分子的高脂溶性[1]。当剂量调到每千克体重0.2mg并持续2周后，伊维菌素和多拉菌素的功效相当[35]。

药用植物也被中医用于处理单峰驼上的疥疮。局部用药是洋葱、大蒜、柠檬和樟脑的混合物，或相思树树皮熬水进行水浴，在10天内可以使患疥疮的单峰驼上的临床症状和寄生虫痊愈[11]。应用于传统医学的其他复合物是天然原料油基泥浆与芸芥籽油或桐油混合，或亚麻阔叶达芙妮的新鲜切碎的叶子和嫩枝条与水或油混合，或烟草粉溶解在水中和刺山柑叶假木贼灰烬中与芝麻油混合[30]。

由于痒螨刺破宿主的皮肤，并以淋巴为食，新大陆骆驼的痒螨病可以通过内吸型和接触型杀螨剂处理。与此相反，足螨以表皮碎屑为食，不受内吸杀螨剂的影响。如前所述，与反刍动物和猪不同，阿维菌素类杀虫剂在骆驼上具有不同的药物动力学，因此，这类药物在新大陆骆驼上处理疥疮具有较低的药效。这在文献中有大量记述[15,21,39]。在一群羊驼中根除疥螨病需要注射伊维菌素多达12次[38]。其他推荐的药物，包括氟虫腈（0.25%，每千克体重3mL）[9]和双甲脒（药浴被感染区域，50 mL/10L水）[21]。当使用拟除虫菊酯类的产品时需要小心，因为该类药物尤其在幼龄羊驼使用时具有神经系统副作用的潜在风险。

## 参考文献

[1] Abdally M.H. (2010). – Acaricidal effect of Ivomec (ivermectin) and Dectomax (doramectin) on sarcoptic mange mites (*Sarcoptes sp.*) of Arabian camels (*Camelus dromedarius*) in Saudi Arabia. *J. Entomology*, 7 (2), 95 – 100.

[2] Abu-Samra M.T. (1999). – The efficacy of Sebacil E.C. 50%, Gamatox and Ivomec in the treatment of sarcoptic mange in camel (*Camelus dromedarius*). *J. Camel Pract. Res.*, 6 (1), 61 – 67.

[3] Alvardo J., Astrom G. & Heath G.B.S. (1966). – An investigation into remedies of sarna (sarcoptic mange) of alpacas in Peru. *Exp. Agric.*, 2 (4), 245 – 254.

[4] Basu A.K., Aliju A.L. & Mohammed A. (1995). – Prevalence of sarcoptic mange in camels (*Camelus dromedarius*) in Nigeria. *J. Camel Pract. Res.*, 2 (2), 141.

[5] Bengoumi M., Hidane K., Bendone-Ndong F., Van-Gool F. & Alvinerie M. (2007). – Pharmacokinetics of eprinomectin in plasma and milk in lactating camels (*Camelus dromedarius*). *Vet. Res. Commun.*, 31, 317 – 322.

[6] Bornstein S., Thebo P., Zakrisson M.T., Abu-Samra M.T. & Mohamed G.E. (1997). – Demonstration of serum antibody to *Sarcoptes scabiei* in naturally infected camels: a pilot study. *J. Camel Pract. Res.*, 4 (2), 183 – 185.

[7] Burkholder T.H., Jensen J., Chen H., Junkins K., Catfield J. & Boothe D. (2004). – Plasma evaluation in llamas (*Lama glama*) after standard subcutaneous dosing. *J. Zoo Wildl. Med.*, 35 (3), 395 – 396.

[8] Cheney J.M. & Allen G.T. (1989). – Parasitism in llamas. *Vet. Clin. North Am. Food Anim. Pract.*, 5 (1), 217 – 225.

[9] Curtis C.F., Chappel S.J. & Last R. (2001). – Concurrent sarcoptic and choroptic acariosis in a British llama (*Lama glama*). *Vet. Rec.*, 149, 208 – 209.

[10] D' Alterio G.L., Callaghan C., Just C., Manner-Smith A., Foster A.P. & Knowles G.T. (2005). – Prevalence of *Chorioptes* sp. mite infestation in alpaca (*Lama pacos*) in the South-West of England: implications for the skin. *Sm. Rum. Res.*, 57, 221 – 228.

[11] Dixit S.K., Tuteja F.C. & Sena D.S. (2009). – Sarcoptosis in dromedary camel – clinical observations and its therapeutic management. *Indian J. Animal Sci.*, 79 (3), 239 – 242.

[12] Edwards G.P., Zeng B., Saalfeld W.K., Vaarzon-Morel P. & McGregor M. (eds) (2008). – Managing the impacts of feral camels in Australia: a new way of doing business. DKCRC Report 47. Desert Knowledge Cooperative Research Centre, Alice Springs, Australia. Available at: www.desertknowledgecrc.com.au/publications/contractresearch.html (accessed 15 August 2013).

[13] Erdenebileg O. (2001). – Camel diseases [in Mongolian]. Dalanzadgad, Ulaanbataar, Mongolia.

[14] Essig A., Rinder H., Gothe R. & Zahler M. (1999). – Genetic differentiation of mites of the genus *Chorioptes* (Acari: Psoroptidae). *Exp. Appl. Acarol.*, 23, 309 – 318.

[15] Foster A., Jackson A. & D' Alterio G.L. (2007). – Skin diseases of South American camelids. *In Practice*, 29, 216 – 223.

[16] Fowler M.E. (2010). – Medicine and surgery of camelids, 3rd Ed. Blackwell Publishing Ltd, Oxford, UK, 243 – 248.

[17] Hunter R.P., Isaza R., Koch D.E., Dodd C.C. & Goatley M.A. (2004). – The pharmacokinetics of topical doramectin in llamas (*Lama glama*) and alpacas (*Lama pacos*). *J. Vet. Pharmacol. Therap.*, 27, 187 – 189.

[18] Jarvinen J.A., Miller J.A. & Oehler D.D. (2002). – Pharmacokinetics of ivermectin in lamas (*Lama glama*). *Vet. Rec.*, 150,344 – 346.

[19] Kataria A.K. & Kataria N. (2004). – Immunoradiometric assay of the serum IgG levels in dromedary camels. *J. Camel Pract. Res.*, 11 (1), 11 – 13.

[20] Kinne J. & Wernery U. (2003). – Experimental mange infection in camels (*Camelus dromedarius*). *J. Camel Pract. Res.*, 10 (1),1 – 8.

[21] Lau P., Hill P.B., Rybnicek J. & Steel L. (2007). – Sarcoptic mange in three alpacas treated successfully with mitraz. *Vet. Dermatol.*, 18 (4), 272 – 277.

[22] Leguia G. (1991). – The epidemiology and economic impact of llama parasites. *Parasitol. Today*, 7 (2), 54 – 56.

[23] Lusat J., Morgan E.R. & Wall R. (2009). – Mange in alpacas, llamas and goats in the UK: incidence and risk. *Vet. Parasitol.*,163, 179 – 184.

[24] Mathur M., Dadhich H. & Khare S. (1997). – Prevalence and histopathological observations of mange affected camel skin in different areas of Rajasthan. *J. Camel Pract. Res.*, 12 (1), 65 – 67.

[25] Mouchira M.M. & Khalid A.K. (2009). – Pathological studies on acariasis in dromedary (*Camelus dromedarius*) and llama (*Lama glama*) Camelidae. *Euro. J. Scient. Res., 38* (2), 159 – 171.

[26] Oukessou M., Badri M., Sutra J.F., Galtier P. & Alvinerie, (1996). – Pharmacokinetics of ivermectin in the camel (*Camelus dromedarius*). *Vet. Rec.*, 139, 425 – 426.

[27] Oukessou, M., Berrag, B. & Alvinerie M. (1999). – A comparative study of ivermectin and moxidectin in lactating camels (*Camelus dromedarius*). *Vet. Parasitol.*, 83, 151 – 159.

[28] Petrowski M. (1998). – Chorioptic mange in an alpaca herd. *In* Advances in veterinary dermatology, vol. 3 (K.W. Kwochka,T. Willemse, C. von Tscharner, *et al.*, eds). Butterworth-Heinemann, Oxford, 450 – 451.

[29] Premalatha N., Jayathangaraj M.G., Senthilkumar K., Sentilvel K., Vedagabady N. & Muralimanohar B. (2010). – Strategic treatment of scabies in captive camels (*Camelus dromedarius*). *Tamilnadu J. Vet. Animal Sci.*, 6 (4), 188 – 190.

[30] Raziq A., de Verdier K. & Younan M. (2010). – Ethnoveterinary treatments by dromedary camel herders in the Suleiman Mountainous Region in Pakistan: an observation and questionary study. *J. Ethnobiol. Ethnomed.*, 6, 1 – 12.

[31] Rosychuk R.A.W. (1989). – Lama dermatology. *Vet. Clin. North Am. Food Pract.*, 10, 228 – 239.

[32] Saleh M.A., Mahran O.M. & Al-Salahy M.B. (2011). – Circulating oxidative stress status in dromedary camels infested with sarcoptic mange. *Vet. Res. Commun.*, 35, 35 – 45.

[33] Schlogl C. (2010). – Erhebungen zum Vorkommen von Endo- sowie Ektoparasiten bei Neuweltkameliden. Thesis, Ludwig Maximilians-University Munich, Germany.

[34] Sing L., Kumar D. & Kataria A.K. (1996). – Efficacy of amitraz against *Sarcoptes scabiei* var. *cameli* infestation in camel (*Camelus dromedarius*). *J. Camel Pract. Res.*, 3 (1), 59 – 60.

[35] Singh I., Khurana R. & Khokhar R.S. (2007). – Comparative therapeutic effect of ivermectin, doramectin and carbaryl in camel mange. *In* Proceedings of the International Camel Conference 'Recent Trends in Camelids Research and Future forsaving Camels' (T.K. Gahlot, ed.). 16 – 17 February, Rajasthan, India, 218 – 220.

[36] Sopyev B., Divanov B. & Charyev C. (2005). – Diseases of camels, their preventive maintenances and treatment. *In* Desertification combat and food safety: the added value of camel producers (B. Faye & P. Esenov, eds). Ashkabad,Turkmenistan, 19 – 21 April 2004, IOS Press Amsterdam, 60 – 66.

[37] Teame G. (1997). – An assessment of the efficacy of deltamethrin with HCH for the treatment of sarcoptic mange in camels. *Trop. Anim. Hlt. Prod.*, 29, 333 – 334.

[38] Twomey D.F., Birch E.S. & Schock A. (2009). – Outbreak of sarcoptic mange in alpacas (*Vicugna pacos*) and control with repeated subcutaneous ivermectin injections. *Vet. Parasitol.*, 159, 186 – 191.

[39] Vine N.J., Keevill G. & Foster A.P. (2010). – Sarcoptic mange in alpacas. *Vet. Rec.*, 167, 946 – 947.

[40] Wall R. & Colbe K. (2006). – Taxonomic priority in Psoroptes mange mites: *P. ovis or P. equi. Exp. Appl. Acarol.*, 39, 159 – 162.

[41] Windsor R.H.S., Teran M. & Windsor R.S. (1992). – Effects of parasitic infestation on the productivity of alpacas (*Lama*

*pacos*). *Trop. Anim. Hlt. Prod.*, 24, 57 – 62.

[42] Zahler M., Essig A., Gothe R. & Rinder H. (1999). – Molecular analyses suggest monospecificity of the genus *Sarcoptes* (Acari: Sarcoptidae). *Int. J. Parasitol.*, 29 (5), 759 – 766.

[43] Zahler M., Hendrikx W.M.L., Essig A., Rinder H & Gothe R. (2000). – Species of the genus *Psoroptes* (Acari: Psoroptidae): a taxonomic consideration. *Exp. Appl. Acarol.*, 24, 213 – 225.

（陈泽译，殷宏校）

## 5.3.3 蠕形螨病

蠕形螨病是由前气门的蠕形螨引起的寄生性疾病。它们栖息在皮肤的毛囊和皮脂腺中（图335）。成螨体长200～300μm，香烟型，末体很短而尖，足粗短（图336）。受精后的雌虫进入毛囊并开始产卵。胚胎后生活史包括1个幼虫期和2个若虫期，随后为性二态的成虫并持续2～4周。

迁移的螨虫破坏毛囊基质和内毛根鞘，导致脱毛。随后，外毛根鞘闭合并形成小的结节，内有各发育期的螨虫、炎症产物和碎屑。

蠕形螨具有宿主特异性，通过母畜和幼畜间的接触传染。骆驼中蠕形螨病的研究很少。在伊朗，15%的单峰驼发现蠕形螨，主要寄生部位为眼睑[4]。它们也常发生在玻利维亚的羊驼和美洲驼上[6]。一篇关于羊驼重要外寄生虫的综述中以列表的形式提到了某种蠕形螨，但没有深入讨论[1]。南美洲以外，在新西兰的羊驼[3]、韩国的美洲驼[2]和德国的一只羊驼和一只美洲驼均有蠕形螨记载。蠕形螨在新大陆骆驼中没有引起明显的临床症状。表皮的毛囊炎和角化过度及真皮炎症细胞的渗润是主要的组织病理学变化。可用双甲脒（0.025%）药浴来消灭螨虫[2]。

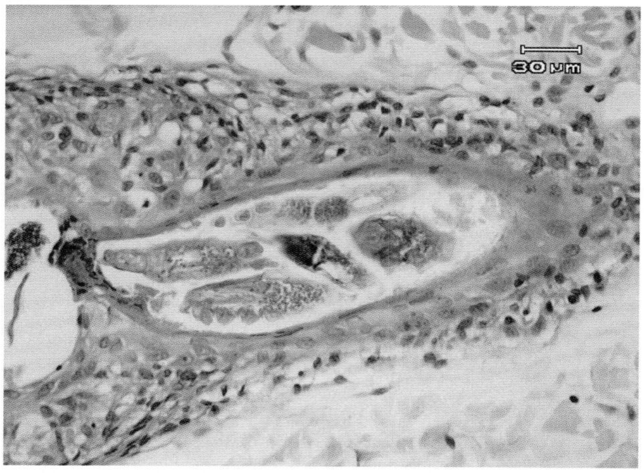


图335　毛囊中的蠕形螨，组织切片

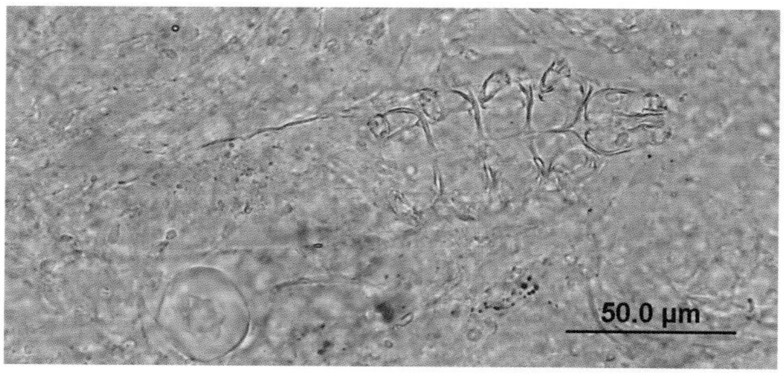


图336　皮肤中的蠕形螨

5 寄生虫病

### 参考文献

[1] Bornstein S. (2010). – Important ectoparasites of alpaca (*Vicugna pacos*). *Acta Vet. Scand.*, 52, 1 – 6.

[2] Eo K.-Y., Kwak D., Shin T., Yeo Y.G., Jung K.Y., Kwon S.C., Kim S. & Kwon O.D. (2010). – Skin lesions associated with *Demodex* sp. in a lama (*Lama peruana*). *J. Zoo. Wildl. Med.*, 41 (1), 178 – 180.

[3] Hill F.I., McKenna P.B. & Mirams C.H. (2008). –*Demodex* sp. infestation and suspected demodicosis of alpacas (*Vicugna pacos*) in New Zealand. *New Zealand Vet. J.*, 56 (3), 148 – 149.

[4] Rak H. & Rahgozar R. (1975). – Demodectic mange in the eye lid of domestic ruminants in Iran. *Bull. Soc. Pathol. Exot.*, 68,

591 – 593.

[5] Schlögl C. (2010). – Erhebungen zum Vorkommen von Endo- sowie Ektoparasiten bei Neuweltkameliden. Thesis, LudwigMaximilians-University, Munich.

[6] Squire F.A. (1972). – Entomological problems in Bolivia. *PANS*, 18, 249 – 268.

（陈泽译，殷宏校）

## 5.3.4 蝇蛆病

蝇蛆病（来源于希腊语myia，蝇）临床定义为由双翅目的幼虫引起的脊椎动物和人的一类疾病。幼虫阶段在宿主体表或体内，以宿主身体或消化道内容物内坏死的或正常的组织、分泌物、体液为食，完成其整个发育或至少部分的发育。蝇蛆病病原可分为专性或兼性寄生，兼性寄生的螺旋蝇蛆，根据其单独引起蝇蛆病、与其他种的蝇蛆共同引起蝇蛆病和在宿主濒死时引起蝇蛆病的特征，可划分为第一、第二和第三类寄生蝇[5]。

旧大陆骆驼(OWCs)伤口或创伤蝇蛆病的病原有麻蝇科（Sarcophagidae）和丽蝇科（Calliphoridae）的多个属的成员，可引起骆驼的表皮、真皮或皮肤蝇蛆病，如麻蝇科的黑须污蝇（*Wohlfahrtia magnifica*）、云朵污蝇（*W. nuba*）和酱麻蝇（*Sarcophaga dux*）和丽蝇科的蛆症金蝇（*Chrysomya bezziana*）、锥蝇（*Cochliomyia hominivorax*）和铜绿蝇（*Lucilia cuprina*）[23]。新大陆骆驼（NWCs）伤口蝇蛆病少见。近年来，在新西兰引进骆马和羊驼的风险分析文献中，提到几个螺旋蝇种，但在本土动物体，并未检测到皮肤蝇蛆病[5]。

与所有的有瓣蝇类一样，蝇蛆病的蝇为完全变态发育，受精的雌蝇将卵或幼虫产于宿主体，蝇蛆具有3个幼虫阶段（图337），接着是蛹期（图338）。成虫脱出蛹壳后，蛹上有一圆形孔洞。

图337 麻蝇的蝇蛆，三期幼虫

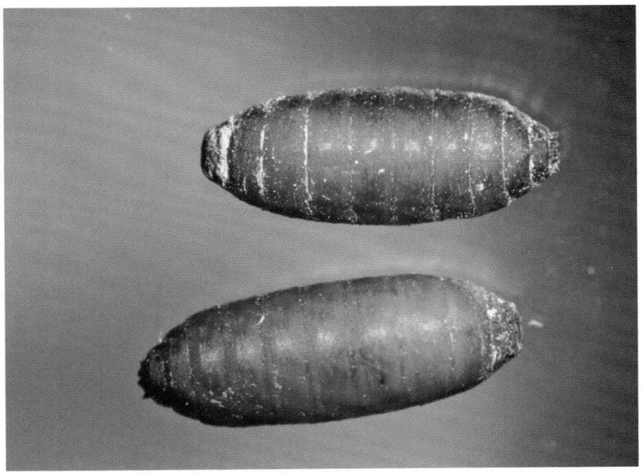

图338 麻蝇的蛹

### 麻蝇科蝇蛆引起的蝇蛆病

黑须污蝇为8~14mm长的白灰色蝇，拉长的黑色大斑点向背腹部中心聚集，周围分布较小的圆点。黑须污蝇是双峰驼和其他家畜最重要的伤口蝇蛆，在人医上也非常重要。该蝇主要分布于南欧地中海沿岸国家、俄罗斯南部、土耳其、伊朗、蒙古和中国[29]，在以色列、约旦、沙特阿拉伯[46]和西奈半岛[19]也有报道，但在旧热带界没有黑须污蝇[51]。在蒙古和中国，有人从骆驼伤口采集样

5 寄生虫病

品，对其形态和发育进行了研究[49].

胎生的雌蝇可产120~170个幼虫，黑须污蝇幼虫至少有一段时间以活组织为食，而在尸体上不发育，因此其为专性寄生虫，幼虫在5~7天内迅速成熟，对组织造成大范围的损伤。即使没有损伤，鼻、眼和生殖器官黏膜也可受到其侵袭[51]。蛹发育的适宜温度为18~22.5℃。

在蒙古共和国，有报道双峰驼的阴道蝇蛆病（图339）[37]。临床症状在6月初，产犊后4~6天出现，秋天恢复。

在戈壁滩的研究结果显示，老的雌骆驼泌尿生殖道蝇蛆病发病率较四岁的普遍，大部分发病动物在前一年就有蝇蛆病[45]。45群的712头被检查动物中，9.4% 有阴道蝇蛆病。6~7月份流行最严重，沙子覆盖区的流行(15%)远远高于砾石覆盖区(4.5%~9%)。

埃及单峰驼的污蝇蛆病也具有季节性，在3月产犊结束后出现[19]，5、6月发病升高，7月份下降，8月就检查不到伤口。雌性单峰驼表现阴蒂-阴道病变，雄性表现会阴蝇蛆病。污蝇将幼虫产于大量嗜驼璃眼蜱（*Hyalomma dromedarii*）感染引起的创伤部位。

云朵污蝇形态与黑须污蝇相似，但其成蝇较小，腹部具有黑色斑点（图340）。与黑须污蝇相比，该种适于在沙漠地区生存，也易于感染骆驼。云朵污蝇主要分布于非洲，从塞内加尔、摩洛哥一直到东非都有该污蝇的分布[51]，阿拉伯半岛[38]和印度[31]也有云朵污蝇。与黑须污蝇不同，云朵污蝇为兼性寄生蝇蛆病病原，在21℃下，云朵污蝇在体外完成其整个幼虫期的发育需35天，37℃下只需要14天，28℃最适于蛹的发育[4]。

迈泽麻蝇（Miser flesh fly），酱麻蝇（*Sarcophaga dux*）(同物异名. *Sarcophaga misera, Liosarcophaga misera, Parasarcophaga misera*)是另外一个麻蝇科致蝇蛆病的代表种，多地都报道其为骆驼蝇蛆病的病原[23]，成蝇7~12mm长，灰色胸部有3条黑色条纹，腹部有黑色划线(图 341)，触角呈深黑色，具有9或10条额鬃，这些形态特征可与其他属的蝇种相区分[42]。在古北区、东洋区和澳洲，酱麻蝇在法医鉴定方面非常重要，近来在印度[31]和泰国[42]也有酱麻蝇的报道，在沙特阿拉伯发现其可引起蝇蛆病。酱麻蝇在16~36℃下都可发育，最适发育温度为20~28℃。

### 丽蝇所致的皮肤蝇蛆病

旧大陆螺旋蝇蛆蛆症金蝇成蝇长8~12mm，体色呈金属绿或蓝色，腿呈深棕色或黑色（图342），蛆症金蝇和大头金蝇（*C. megacephala*）地理分布相同，根据眼的形状和下腋瓣的颜色与大头金蝇相区分。蛆症金蝇为专性寄生的伤口蝇蛆，其不同于其他属的蝇种，从不在腐烂物中发育。雌蝇在伤口边缘产150~500的一簇卵，有时也产在破损的软组织上，尤其是在血液污染的组织。18~24小时后孵化出幼虫，六七天后，幼虫发育完成，然后发育为蛹。蛹的发育与温度有很大关系，夏天需7~9天，较低温度下可达8周[52]。蛆症金蝇分布于印度和东洋区的其他地区、非洲和澳洲（但在澳大利亚和新西兰没有），在阿曼[9]、

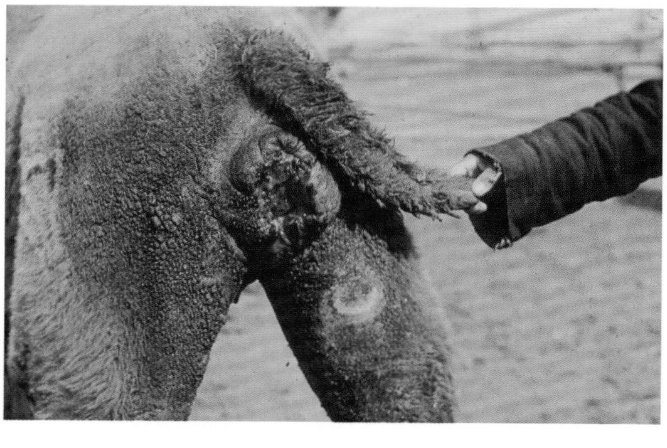

图339　黑须污蝇引起的蒙古双峰驼的阴道蝇蛆病(图片由柏林R. Ribbeck博士提供)

图340　云朵污蝇腹背侧特征性黑斑点

沙特阿拉伯[1,13]、阿拉伯联合酋长国[8,38,40]、巴林[26]、伊拉克[39]和伊朗[34]也有。蛆症金蝇是骆驼（图343）和其他家畜以及人常见的蝇蛆病病原。

线粒体DNA序列分析显示，蛆症金蝇有两个谱系，撒哈拉以南非洲谱系和亚洲海湾谱系，海湾和撒哈拉以南非洲之间的干旱的不毛之地可能是造成这一谱系分化的原因[20]。

其他金蝇种，如白头裸金蝇（*C. albiceps*）、绯颜裸金蝇（*C. rufifacies*）、赭金蝇（*C. varipes*），绿尾金蝇（*C. chloropyga*）和大头金蝇是第二类寄生蝇或兼性寄生蝇，其幼虫在没有第一类寄生蝇蛆（如铜绿蝇）的帮助，并不能攻击宿主[51]。

图341　酱麻蝇成蝇，其淡灰色胸背面有3条黑色条纹，腹部为方格状图案

图342　旧大陆螺旋蝇蛆蛆症金蝇成蝇具有白色下腋瓣，可与大头金蝇相区分

新大陆螺旋蝇蛆嗜人锥蝇原发于中美洲，分布于加勒比岛和南美洲的北部国家，直到智利北部和阿根廷。1988年由南美洲进口家畜时引入利比亚[10,11]，该病的此次暴发，致478家畜病例，包括骆驼和229个人病例。嗜人锥蝇成蝇长8~10mm，体色呈金属光泽的蓝墨绿色，眼黄红或棕色。胸部有3道横向的黑色条纹，头部有黑色刚毛。成蝇与旧大陆的螺旋蝇相似，但二者雌蝇的翅基片颜色不同。本属的另外一个种次生锥蝇（*C. macellaria*）为兼性第二类蝇蛆。嗜人锥蝇对热带气候非常适应，在实验室对其生活史的观察显示，幼虫在18~37℃下化蛹，化蛹的最适温度为20℃。在高温或低温情况下，都不能蛹化和羽化[15]。绿蝇中的铜绿蝇的成蝇体长6~9mm，呈绿色或铜绿色，可引起兼性原发性蝇蛆病（图344）。根据气温，在8小时到3天内孵化出一期幼虫，在几天内幼虫发育完全，在适宜温度下，一周内发育为蛹。在澳大利亚和南非，该蝇为羊的主要蝇蛆病病原，在东非也可引起单峰驼的蝇蛆病[23]。雌蝇在腐尸或腐烂的植物上产卵，但其也可被感染的伤口或褥疮或排泄物周围的毛发吸引。皮肤皱褶，特别是在被尿和粪便污染的会阴

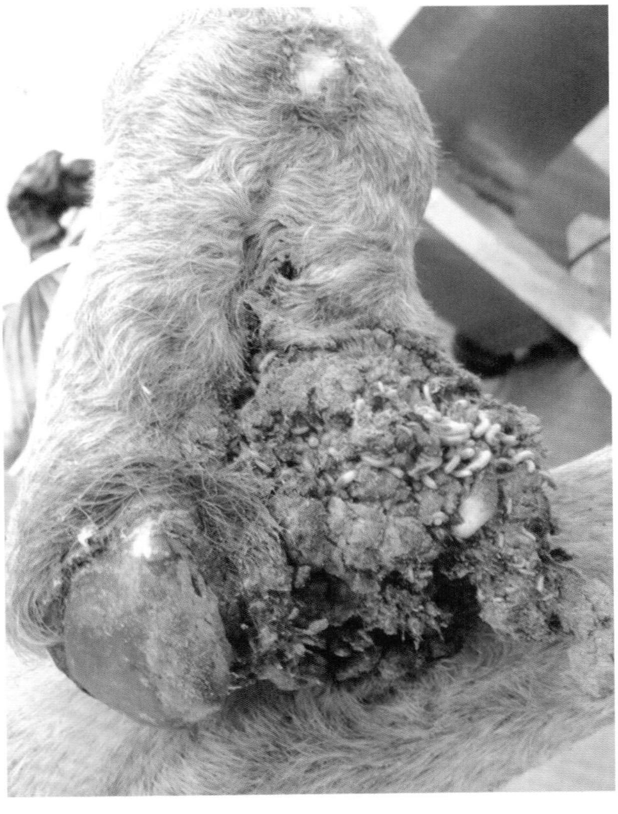

图343　蛆症金蝇引起的单峰驼下肢的皮肤病变

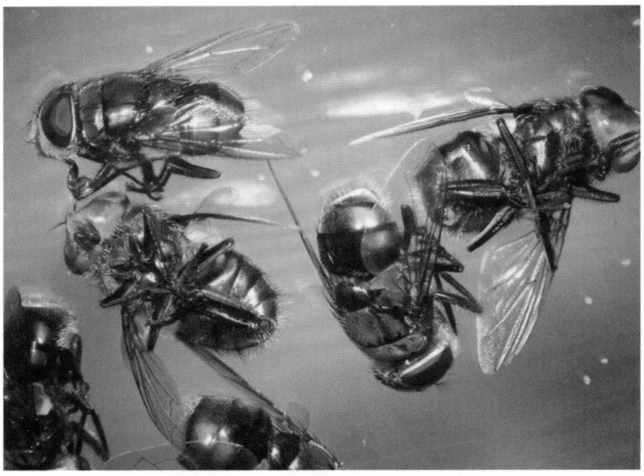

图344　绿蝇中的铜绿蝇致兼性蝇蛆病

部位，更加吸引成蝇，成为蝇蛆攻击的理想部位。

## 诊断

形态学鉴定时，应在伤口的深处采集蝇蛆，以免采集到兼性蝇蛆病病原[48]。幼虫的鉴定有一些关键部位[21,25,51]，其主要根据三龄幼虫的形态特征。蝇蛆表面特征（光滑或长毛）、后气门是否明显[隐藏在最后一节深的凹陷处（图345）或暴露在最后一节的后面]、后气门板的结构（图346）、气管干的可见度以及前气门的形状是主要的诊断特征。由于背气管干的颜色仅在活样本中可见，所以应在用80%酒精固定前进行检查。有些情况下，可以等幼虫化蛹并孵化出成蝇后再进行鉴定。

也可使用分子生物学方法[20,22,28,43]和表皮碳氢化合物分析做幼虫、成蝇和其地理起源的鉴定[8]。

虽然没有标准的血清学检测技术，但实验研究表明，也可以通过检测螺旋蝇蛆的抗体进行诊断[17,44]。

为了采集成蝇，可用风向网和气味诱捕的胶带[48]。在旧大陆澳大利亚螺旋蝇蛆的调查中，使用改良的桶网和最近研制的诱捕剂（Bezzilure, Department of Primary Industries, Yeeronpilli, Australia）进行成蝇的采集[36]。

## 防治

可用有机磷酸酯，如酚线磷、皮蝇磷和蝇毒磷处理感染的伤口，杀死蝇蛆。治疗后，蝇蛆离开创伤部位，落地死亡，然后清洗伤口，清除坏死组织。双氧水可使隐藏的幼虫爬出，然后再次使用杀虫剂，以免再次感染。疾病暴发时，可用0.25%的蝇毒磷水溶液、其他有机磷杀虫剂或除虫菊酯喷洒动物，以免感染。每千克体重0.2mg的大环内酯类、多拉菌素就可以完全保护牛在21天内不受感染，比伊维菌素效果更好[32]，而莫西菌素、依普菌素和多拉菌素的浇泼剂预防效果有限[47]。

在1988年，利比亚暴发螺旋蝇蛆病时，在新大陆的美国和墨西哥就在研究应用昆虫不育技术来控制螺旋蝇蛆病[27]。人工培养的蝇在出蛹前用铯-137短时照射，雄蝇将变为不育，但具有交配能力。

雌蝇只交配一次，当其与不育的雄蝇交配后，产的卵不能孵化。

新旧大陆的螺旋蝇蛆病被录入OIE的陆生动物卫生法典[48]，并对从有这些病原的国家进口家畜或动物园动物、隔离、运输和进口后检疫等做了详细的规范性规定。

## 狂蝇引起的鼻咽蝇蛆病

该类型蝇蛆病在旧大陆非常普遍，其由骆驼喉蝇（*Cephalopina titillator*）引起的，在南美洲外的新大陆，只有非常少的羊狂蝇（*Oestrus ovis*）和鹿蝇属（*Cephenemyia*）的病原被报道。

### 形态学

骆驼鼻蝇蛆病的病原骆驼喉蝇是单峰驼和双峰驼专一性的寄生虫，成蝇长8~10mm，胸棕灰色，眼棕色（图347）。

### 生活史

雌蝇口器退化，无法进食，在鼻孔产0.7mm长的幼虫，这些一期幼虫移行进入筛骨迷路和鼻腔鼻甲，在咽喉部位可发现大部分的二、三龄幼虫。发育成熟的幼虫（图348）从咽喉移行进入下鼻道，跌出鼻孔，落地蛹化（图349）[50]。在胃内也可见到三龄幼虫，说明有些幼虫经消化道离开宿主[41]。

### 流行病学

骆驼喉蝇为古北区的蝇种，其分布范围延伸至非洲东部和南部，甚至在澳大利亚也有分布[41].

骆驼鼻蝇感染率可高达70%以上，雄驼和雌驼的平均荷虫量分别为28和35个幼虫[7]，在苏丹的一

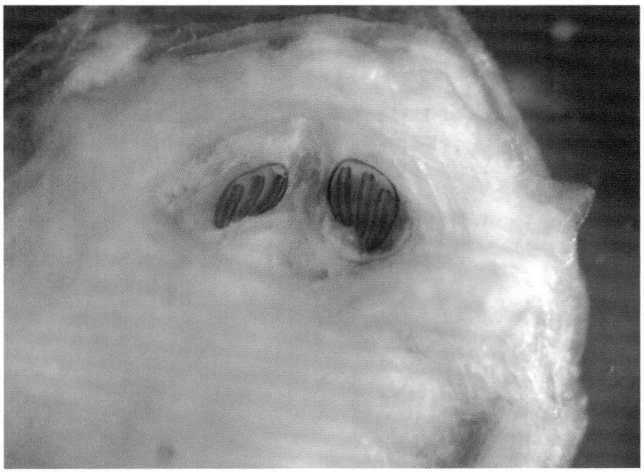

图345　麻蝇三龄幼虫，后气孔隐藏在深的凹陷内

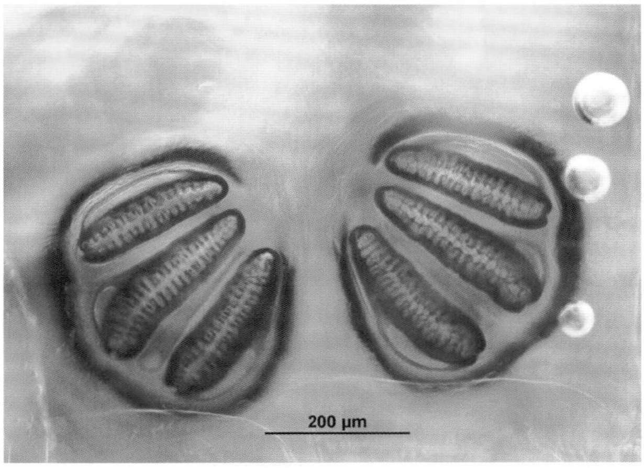


图346　蛆症金蝇三龄幼虫，后气孔被带有不清晰的扣状物的开口环围绕着

图347　骆驼喉蝇成蝇

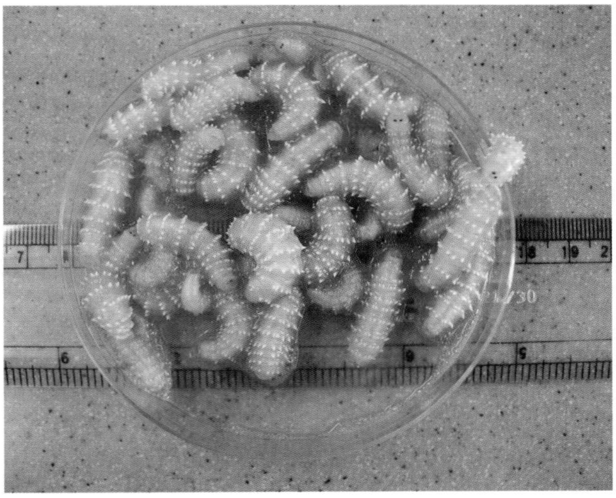

图348　骆驼喉蝇三龄幼虫，采自一头骆驼的鼻腔

头骆驼体发现的最大荷虫量为243只幼虫[33]，所有年龄段的骆驼都可被感染，幼龄驼（小于两岁）较老龄驼感染率低[2,14,35]，雄驼感染率稍高于雌驼[14]。在沙特阿拉伯一个屠宰场的研究显示，在温和气候条件下，90%的骆驼喉蝇一年可繁殖两代，流行的两个高峰分别在12月到1月和9月，然后迅速下降。在沙特阿拉伯的另外一个研究显示，除了4月和5月，骆驼喉蝇一期幼虫一年内其他月份都可以见到。也就是全年都可见到成蝇[2]。在中亚和东亚的大陆性气候条件下，所有幼虫在宿主体的发育可延续9~11个月[16]。

在骆驼和小反刍动物混养的地区，骆驼也可感染羊狂蝇引起的鼻蝇病，但这种偶然蝇蛆病的发病率很低，只能见到很少的一、二龄幼虫[2]。尚不清楚在旧大陆羊狂蝇是否能完成其整个生活史，

图349　骆驼喉蝇的蛹不显眼，地上很难发现

Fowler [18] 认为在新大陆有羊狂蝇，但没有找到最初的文献。有报道美洲驼感染鹿蝇属的蝇蛆（鹿的鼻蝇蛆），但只是在内窥镜内见到幼虫，而且治疗后消失[30]。

## 临床症状和病理变化

幼虫用身体前端的强有力的钩（图350）附着在鼻腔和咽喉内，使得宿主表现不安、组织损伤和呼吸障碍。感染动物食欲废绝，呼吸困难，喷鼻，打喷嚏，鼻孔内有幼虫排出。而且出现流产、不孕、奶量和体重下降[14]。骆驼喉蝇用两个强壮的口钩将其固定在黏膜上，当蝇蛆脱离后，附着过的部位有坚硬的微红色结节，结节顶端被切割成锯齿状，咽喉黏膜上也可见到深棕色或黑色的结节，表明早期被寄生过[24]。在自然感染的单峰驼，鼻咽部位的黏膜上皮会出现炎症反应、恶化、坏死等严重的组织变化和恢复的黏膜下腺体的囊性扩张等一系列的病理组织学变化，也会出现受人关注的脑膜炎扩散和甲状腺显著的胶性甲状腺肿，所有这些重症患畜还伴随肾脂沉积症。感染动物出现显著的甲状腺激素三碘甲状腺氨酸T3和甲状腺素T4的下降，而总脂蛋白胆固醇和低密度脂蛋白胆固醇升高[14]。

图350　骆驼喉蝇三龄幼虫前端有两强壮的口钩

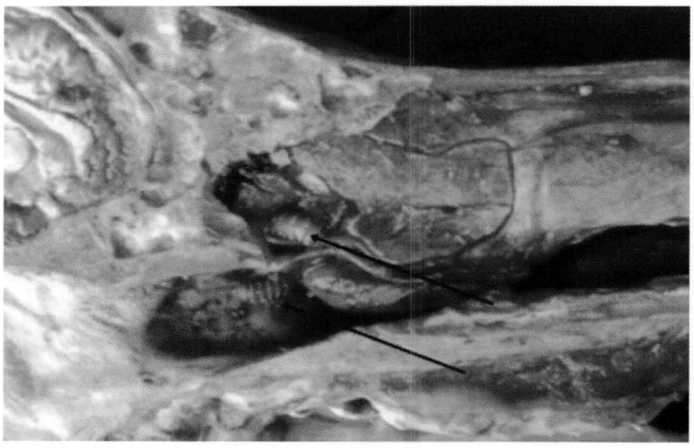

图351　鼻咽部的骆驼喉蝇三龄幼虫

## 诊断

鼻蝇蛆常在尸检或肉品检验时发现（图351）。由于这些寄生虫与宿主组织接触，可能骆驼会产生抗体，所以可用血清学方法进行检测。

## 防控

有机磷酸酯曾试验性的用于治疗骆驼喉蝇蛆病的治疗，在密闭的房间用敌敌畏 (Estrozol) 喷雾1小时，可产生100%的疗效，治疗后24小时后在奶中检测不到药物残留，5天后在肉中也检测不到，以每千克体重20 mg肌内注射三氯化硫磷sulfidofos (Etazid)，加在饮用水中（20L，0.03%）口服敌百虫metrifonat (Chlorofos) 或按每千克体重20mg在局部地区应用敌百虫metrifonat (Dioxafos)和倍硫磷（Sulfidofos-20）都可以获得很好的疗效，但需要较长的停药期[6]。

早期报道用蝇毒磷（coumaphos）、乙酰甲胺磷（trichlorphen）和异皮蝇磷（trichlormetaphos-3）鼻喷剂治疗鼻蝇蛆病[23]。在羊体，用硝碘酚腈杀吸虫剂flukicides nitroxinil（皮下，每千克体重10 mg）和雷复尼特 rafoxanide（口服，每千克体重7.5 mg）对羊狂蝇是有效的，鼻蝇蛆对大环内酯类药物敏感，在沙特阿拉伯，在4月和10月幼虫达到高峰时，治疗单峰驼蝇蛆病效果最好[2]，在大陆性气候的中亚和东亚，在9月或10月末，成蝇消失后治疗，效果最理想。

# 参考文献


[1] Abo-Shehada M.N. (2005). – Incidence of *Chrysomya bezziana* screw-worm myiasis in Saudi Arabia 1999/2000. *Vet Rec.*,156, 354 – 356.

[2] Alahmad A.M. (2002). – Seasonal prevalence of *Cephalopina titillator* larvae in camels in Riyadh Region, Saudi Arabia. *Arab Gulf J. Sci. Res.*, 20 (30), 161 – 164.

[3] Al-Mishnet F.A.M., Amoudi M.A. & Abou-Fannah S.S.M. (2001). – First record of *Sarcophaga* (*Liosarcophaga*) *dux* Thomson, 1868 (Diptera: Sarcophagidae) from Saudi Arabia. *Pakistan J. Zool.*, 33 (4), 313 – 315.

[4] Amoudi M.A. (1993). – Effect on temperature on the development stages of *Wohlfahrtia nuba* (Diptera: Sarcophagidae).*J. Egypt. Soc. Parasitol.*, 23 (3), 697 – 705.

[5] Anon. (2010). – Import risk analysis: llamas (*Lama glama*) and alpacas (*Vicugna pacos*) from specified countries. Draft for public consultation. MAF Biosecurity New Zealand, Wellington, New Zealand. Available at: www.biosecurity.govt.nz/files/biosec/consult/llamas-alpacas-specified-countries.pdf (accessed on 16 August 2013).

[6] Ashetova I.N. (1987). – Control of cephalopinosis in camels [in Russian]. *Veterinarija*, (5), 52 – 53.

[7] Bekele T. (2001). – Studies on *Cephalopina titillator*, the cause of 'Segale' in camels (*Camelus dromedarius*) in semi-arid areas of Somali State, Ethiopia. *Trop. Animal Health Prod.*, 33, 489 – 500.

[8] Brown W.V.R., Morton R., Lacey M.J., Spradbery J.P. & Mahon R.J. (1998). – Identification of the geographical source of adults of the Old World screw-worm fly *Chrysomya bezziana* Villeneuve (Diptera: Calliphoridae), by multivariate analysis of cuticular carbons. *Comp Bichem. Physiol.*, 119B, 391 – 399.

[9] Deeming J. (1996). – The Calliphoridae (Diptera: Cyclorrhapha) of Oman. *Fauna of Saudi Arabia*, 15, 264 – 279.

[10] El-Azazy O.M.E. (1989). – Wound myiasis caused by *Cochliomyia hominivorax* in Libya. *Vet. Rec.*, 124, 103.

[11] El-Azazy O.M.E. (1990). – Wound myiasis caused by *Cochliomyia hominivorax* in humans in Libya. *Transact. Royal Soc. Trop.Med. Hyg.*, 84 (5), 747 – 748.

[12] El-Azazy O.M.E. (1992). – Observations on the New World screwworm fly in Libya and the risk of its entrance into Egypt. *Vet. Parasitol.*, 42, 303 – 310.

[13] El-Azazy O.M.E. & El-Metenawy T.M. (2004). – Cutaneous myiasis in Saudi Arabia. *Vet. Rec.*, 154, 305 – 306.

[14] El-Rahman S.S.A. (2010). – Prevalence and pathology of nasal myiasis in camels slaughtered in El-Zawia Province – Western Libya: with reference to thyroid alteration and renal lipidosis. *Global Veterinaria*, 4 (2), 190 – 197.

[15] Elwaer O.R. & Elowni E.E. (1991). – Studies on the screwworm fly *Cochliomyia hominivorax* in Libya: effect on temperature on pupation and eclosion. *Vet. Parasitol.*, 77, 48 – 49.

[16] Erdenebileg O. (2001). – Camel diseases [in Mongolian]. Dalanzadgad, Ulaanbaatar, Mongolia.

[17] Figarola J.L., Berkebile D.R., Skoda S.R. & Foster J.E. (2001). – Identification of screwworms, *Cochliomyia hominivorax* (Coquerel) (Diptera: Calliphoridae), with a monoclonal antibody-based enzyme-linked immunosorbent assay (MAb-ELISA). *Vet. Parasitol.*, 102, 341 – 354.

[18] Fowler M.E. (2010). – Medicine and surgery of camelids, 3rd Ed. Wiley-Blackwell, Ames, Iowa, 239 – 241.

[19] Hadani A., Ben Yakov B. & Rosen S. (1989). – Myiasis caused by *Wohlfahrtia magnifica* (Schiner, 1862) in The Arabian camel (*Camelus dromedarius*) in the Peninsula of Sinai. *Rev. Elev Méd. Vét. Pays Trop.*, 42 (1), 33 – 38.

[20] Hall M.J.R., Edge W., Testa J., Adams Z.J.O. & Ready P.D. (2001). – Old World screwworm fly, *Chrysomya bezziana*, occurs as two geographical races. *Med. Vet. Entomol.*, 15, 393 – 402.

[21] Hall M.J.R. & Smith K.G.V. (1995). – Diptera causing myiasis in man. *In* Medical insects and arachnids (R.P. Lane &R.W. Crosskey, eds). Chapman and Hall, London, 429 – 469.

[22] Hall M.J.R., Testa J.M., Smith L., Adams Z.J.O., Khallaayoune K., Sotiraki S., Stefanakis A., Farkas R. & Ready P.D. (2009).– Molecular genetic analysis of populations of Wohlfahrt's wound myiasis fly, *Wohlfahrtia magnifica*, in outbreak population from Greece and Morocco. *Med. Vet. Entomol.*, 23 (Suppl. 1), 72 – 79.

[23] Higgins A.J. (1985). – The camel in health and disease. 4. Common ectoparasites of the camel and their control. *Br. Vet. J.*,141,

197.

[24] Hussein M.F., Elamin F.M., El-Taib N.T. & Basmaeil S.M. (1982). – The pathology of nasopharyngeal myiasis in Saudi Arabian camels (*Camelus dromedarius*). *Vet. Parasitol.*, 9, 253 – 260.

[25] Kettle D.S. (1984). – Medical and veterinary entomology. John Wiley & Sons, New York.

[26] Kloft W.J., Noll G.F. & Kloft E.S. (1981). – Introduction of *Chrysomya bezziana* Villeneuve (Diptera: Calliphoridae) into new geographic region by transit infestation. *Mitt. Dtsch. Gesellsch. Allg. Angew. Entomol.*, 3, 151 – 154.

[27] Kouba V. (2004). – History of the screwworm (*Cochliomyia hominivorax*) eradication in the Eastern Hemisphere. *Hist. Med. Vet.*, 29 (2), 43 – 53.

[28] Litjens P., Lessinger A.C. & De Azeredo-Espin A.M.L. (2001). – Characterization of the screwworm flies *Cochliomyia hominivorax* and *Cochliomyia macellaria* by PCR-RFLP of mitochondrial DNA. *Med. Vet. Entomol.*, 15, 183 – 188.

[29] Marinez I.R. & Leclercq M. (1994). – Data on distribution of screwworm fly *Wohlfahrtia magnifica* (Schiner) in Southwestern Europe (Diptera: Sarcophagidae). *Notes Fauniques de Gembloux*, (28), 53 – 60.

[30] Mattoon J.S., Gerros T.C., Parker J.E., Carter C.A. & Lamarche R.M. (1997). – Upper airway obstruction in a llama caused by aberrant nasopharyngeal bots. (*Cephenemyia* sp.). *Vet. Radiol. Ultrasound*, 38 (5), 384 – 386.

[31] Mitra B. & Sharma R.M. (2010). – Check list of Indian flesh flies (Insecta: Diptera: Sarcophagidae). Available at: http://zsi.gov.in/checklist/Indian_sarcophagidae.pdf.

[32] Moya-Bora G.E., Muniz R.A., Umehra O., Goncalves L.C.B., Silva D.S.F & McKenzie M.E. (1997). – Protective efficacy of doramectin and ivermectin against *Cochliomyia hominivorax*. *Vet. Parasitol.*, 72, 101 – 109.

[33] Musa M.T., Harrison M., Ibrahim A.M. & Taha T.O. (1989). – Observations on Sudanese camel nasal myiasis caused by the larvae of *Cephalopina titillator*. *Rev. Elev. Méd. Vét. Pays Trop.*, 42 (1), 27 – 31.

[34] Navidpour S., Hoghooghi, Rad N., Goodrazi H. & Pooladga A.R. (1996). Outbreak of *Chrysomya bezziana* in Khoozestan Province, Iran. *Vet. Rec.*, 139, 217.

[35] Oryan A., Valinezhad A. & Moraveji M. (2008). – Prevalence and pathology of camel nasal myiasis in eastern areas of Iran. *Trop. Biomed.*, 25 (1), 30 – 36.

[36] Rodrigues V.B. & Raphael B. (2008). – Review of the Old World Screw Worm fly trapping program conducted by AQIS inthe Torres Strait. Australian Government, Bureau of Rural Sciences, Canberra, Australia.

[37] Schuhmann H., Ribbeck R. & Beulig W. (1976). –*Wohlfahrtia magnifica* (Schiner, 1862) (Diptera: Sarcophagidae) als Ursacheeiner vaginalen Myiasis bei domestizierten zweihoeckrigen Kamelen in der Mongolischen Volksrepublik. *Arch. Exp. Vet.Med.*, 30, 799 – 806.

[38] Schuster R. & Deeming J.C. (2011). – Order Diptera, infraorder Muscomorpha. Cyclorrhaphous Diptera associated withvertebrates in the UAE. *In* Arthropod fauna of the UAE, vol. 4 (A. van Harten, ed.). Dar Al Ummah Printing, Abu Dhabi,UAE, 807 – 816.

[39] Siddig A., Al Jowari S., Al Izzi M., Hopkins J., Hall M.J.R. & Slingenbergh J. (2005). – Seasonality of Old World screwworm myiasis in the Mesopotamia valley in Iraq. *Med. Vet. Entomol.*, 19, 140 – 150.

[40] Spradbery J.P. & Kirk J. (1992). – Incidence of Old World Screwworm fly in the United Arab Emirates. *Vet. Rec.*, 130, 33.

[41] Spratt D.M. (1984). – The occurrence of *Cephalopina titillator* (Clark) (Diptera: Oestridae) in camels in Australia. *J. Austr. Entomol. Soc.*, 23, 229 – 230.

[42] Sukontason K., Bunchu N., Chaiwong T., Moophayak K. & Sukontason K.L. (2010). – Forensically important flesh fly species in Thailand morphology and development rate. *Parasitol. Res.*, 106, 1 055 – 1 064.

[43] Taylor D.B., Szalanski A.L. & Peterson R.D. (1996). – Identification of screwworm species by polymerase chain reaction restrictionfragment length polymorphism. *Med. Vet. Entomol.*, 10, 63 – 70.

[44] Thomas D.B. & Pruet J.H. (1992). – Kinetic development and decline of antiscrewworm (Diptera: Calliphoridae) antibodies in serum of infested sheep. *J. Med. Entomol.*, 29, 870 – 873.

[45] Valentin A., Baumann M.P.O., Schein E. & S. Bajanbileg S. (1997). – Genital myiasis (wohlfahrtiosis) in camel herds of Mongolia. *Vet. Parasitol.*, 73, 335 – 346.

[46] Verves J.G. (1986). – Sarcophagidae. *In* Catalogue of palaearctic Diptera, vol. 12 (A. Soos & L. Papp, eds).

Elsevier,Amsterdam, 58 – 193.

[47] Wardhaug K.G., Mahon R.J. & Bin Ahmad H. (2001). – Efficacy of macrocyclic lactones for the control of larvae of the Old World screw-worm fly (*Chrysomya bezziana*). *Aust. Vet. J.*, 79, 120 – 124.

[48] World Organisation for Animal Health (OIE) (2013). – Terrestrial Animal Health Code. OIE, Paris. Available at: www.oie.int/animal-health-in-the-world/oie-listed-diseases-2013/ (accessed 16 August 2013).

[49] Yasuda M. (1940). – Morphology of the larva of *Wohlfahrtia magnifica Schin.* found in a wound on a camel in Inner Mongolia. *J. Chosen Nat. Hist. Soc.*, 7 (29), 27 – 36.

[50] Zayed A.A. (1998). – Localisation and migration route of *Cephalopina titillator* (Diptera: Oestridae) larvae in the head of infested camels (*Camelus dromedarius*). *Vet. Parasitol.*, 80, 65 – 70.

[51] Zumpt F. (1965). – Myiasis in man and animals in the Old World. Butterworth, London.

（关贵全译，殷宏校）

## 5.3.5 虱病

虱病是由虱子寄生引起，感染驼科动物的4种吸血虱（虱目Anoplura）都属微胸虱属（*Microthoracius*），极长头微胸虱（*Microthoracius praelongiceps*）（一些作者认为极长头微胸虱（*M. praelongiceps*）是梅察微胸虱（*M. mazzai*）的同物异名[3]），梅察微胸虱（*M. mazzai*）和小微胸虱（*M. minor*）感染新大陆骆驼，而骆驼微胸虱（*M. cameli*）既可感染单峰驼，又可感染双峰驼。此外，新大陆骆驼也是咬虱/羽虱（丝角亚目Ischnocera）短头牛羽虱（*Bovicola breviceps*）的宿主。虱是宿主特异性的永久性寄生虫，其终身生活在宿主体，幼虫和成虫阶段都营寄生生活。

### 形态特征

微胸虱体短于2 mm，呈深棕色，背腹扁平，适于在宿主被毛下生活。腿上具爪，以便抓住宿主毛发，这是适应其生存条件的体现，因为虱离开宿主后将不能生存。微胸虱头成梭状，长度几乎与其腹一样长。吸血虱具有特殊口器，可刺入宿主皮肤。唇位于头前端，成短喙状，进食时向外伸出，用唇瓣齿抓住宿主皮肤，3根刺针刺入宿主血管吸血。

被称为咬虱的短头牛羽虱呈白色到棕褐色，体长可达4mm，头钝，比胸节宽，易与吸血虱区分。与吸血虱不同，咬虱不能刺入宿主皮肤，以皮屑、油脂和细菌为食。

### 生活史

虱是不完全变态发育昆虫，3个幼虫阶段在形态上与成虫相似，只是大小存在差异。为了确保其子代不离开宿主，雌虱将卵（“虮子”）用特殊的黏合剂黏附在宿主的毛发上，1~2周后，孵化出一期幼虫，然后在3周内蜕皮两次，发育为成虫，整个生活史可在4周内完成。

### 流行病学

虱可以水平接触或垂直接触传播，通常骆驼的吸血虱仅局限在很少的几个国家流行（表90），但过去推测骆驼微胸虱可能广泛分布。

最近的调查显示梅察微胸虱也可感染羊驼[3]，在智利梅察微胸虱和极长头微胸虱也在羊驼体发现[7]。在南美洲外，没有发现吸血虱感染新大陆的骆驼。咬虱短头牛羽虱在美洲驼体比羊驼体更常见，其最初只分布在智利[7]，然后随着动物的进口，被带入澳大利亚[12]、新西兰[9]和德国[8]，近来在英国的美洲驼群中也检测到短头牛羽虱[11]。

## 临床症状

吸血虱喜欢在动物的两侧、头、颈和肩胛部位寄生[1]，由于吸血虱刺穿宿主皮肤，所以感染的临床表现以皮肤擦伤为主要特征。吸血虱吸食血量较少，只吸食其饥饿体重30%的血量，原因是其可在几小时内消化掉这些血液，然后再次吸血。瘙痒症可使动物表现头、颈和肩胛部被毛不整和脱毛[3,5]，感染严重时动物会出现贫血。

从颈到尾的背中线是咬虱偏好寄生的部位，咬虱感染引起的被毛无光泽和不整的临床症状不如吸血虱明显，咬虱在宿主体爬行而引起动物表现瘙痒[6]。由于和其他种类的动物接触，血虱和咬虱可混合感染。冬季动物感染严重的情况下，对动物过冬造成较大困难。

## 诊断

瘙痒、被毛不整和脱毛是怀疑虱感染主要的临床症状。

将剪下的不同部位的毛分开，检查皮肤表面的虱子，同时查看毛上的“虮子”（通常在皮肤表面上5~10mm）[10]。

## 防治

每千克体重0.2mg的伊维菌素可以用于治疗新大陆骆驼的吸血虱[2]，但非口服给药并不能杀死咬虱。冬季虱感染较为严重，可用杀虫药喷雾（5%的西维因和5%的马拉松）或3%的倍硫磷浇泼剂浇泼[2]，每千克体重10mg的氯氰菊酯也可获得很好的效果[9,12]。只有将喷雾剂和浇泼剂用到皮肤表面，而不是被毛上，才有效。12~14天后再治疗一次，以杀灭初次治疗后从卵中孵化出的幼虱。

表90　骆驼科动物吸血虱的分布（根据 Durden 和 Musser 的描述[4]）

| 种 | 宿主 | 国家 |
|---|---|---|
| 骆驼微胸虱 | 单峰驼 | 阿尔及利亚，印度 |
| 梅察微胸虱 | 美洲驼 | 阿根廷 |
| 小微胸虱 | 美洲驼，驼马 | 阿根廷 |
| 极长头微胸虱 | 美洲驼，原驼，驼马 | 玻利维亚，秘鲁 |

## 参考文献

[1] Bornstein S. (2010). – Important ectoparasites of alpaca (*Vicugna pacos*). *Acta Vet. Scandinavica*, 52, 1–6.

[2] Cheney J.M. & Allen G.T. (1989). – Parasitism in llamas. *Vet. Clin. North Am. Food Anim. Pract.*, 5 (1), 217–225.

[3] Cicchino A.C., Cobenas M.E., Bulman G.M., Diaz J.C. & Laos A. (1998). – Identification of *Microthoracius mazzai* (Phthiraptera: Anoplura) as an economically important parasite of alpacas. *J. Med. Entomol.*, 35 (6), 922–930.

[4] Durden L.A. & Musser G.G. (1994). – The sucking lice (Insecta, Anoplura) of the world: a taxonomic checklist with records of mammalian hosts and geographical distributions. *AMNH Bulletin*, 218, 1–90.

[5] Foster A., Jackson A. & D'Alterio G.L. (2007). – Skin diseases of South American camelids. *In Practice*, 29, 216–223.

[6] Fowler M.E. (2010). – Parasites. *In* Medicine and surgery of South American camelids, 3rd Ed (M.E. Fowler, ed.). Wiley-Blackwell, Ames, IA, 231–269.

[7] Gonzalez-Acuna D., Cabezas I., Moreno L. & Castro D. (2007). – Nuevos registros de Phthiraptera (Artropoda: Insecta) en *Lama pacos* Linneus 1758, en Chile. *Arch. Med. Vet.*, 39 (1), 71–72.

[8] Mey E. & Gonzalez-Acuna D. (2007). – Uber einen Massenbefall von *Bovicola* (*Lepikentron*) *breviceps* (Rudow) (Insecta, Phithiraptera, Ischnocera, Bovicolidae) auf einem Alpaka *Vicugna vicugna* forma pacos in Thuringen (Deutschland), mit Anmerkungen zur Parthenogenese bei Tierlausen. *Rudolstäter Nat. Hist. Schr.*, 14, 71–82.

[9] Palma R.L., McKenna P.B. & Aitken P. (2006). – Confirmation of the occurrence of the chewing louse *Bovicola* (*Lepikentron*) *breviceps* (Insecta: Phthiraptera: Trichodectidae) on alpacas (*Lama pacos*) in New Zealand. *NZ Vet. J.*, 54 (5), 253–254.

[10] Scott D.W., Vogel J.W., Fleis R.I., Miller W.H. & Smith M.C. (2010). – Skin diseases in the alpaca (*Vicugna pacos*): a literature review and retrospective analysis of 68 cases (Cornell University 1997–2006). *Vet. Dermatol.*, 22, 2–16.

[11] Twomey D.F., Cooley W.A. & Wood R. (2010). – Confirmation of the chewing louse, *Bovicola breviceps*, in a British llama (*Lama glama*) herd. *Vet. Rec.*, 166, 790–791.

[12] Vaughan J.L. (2004). – Eradication of the camelid biting louse, *Bovicola breviceps*. *Aust. Vet. J.*, 82, 216–217.

（关贵全译，殷宏校）

## 5.3.6 咬蝇和扰蝇

骆驼周围的蝇子使动物很不安，咬蝇和扰蝇烦扰动物，并可使动物停止进食，从而降低动物生产性能[8]。咬蝇能传播传染病病原（炭疽、布鲁氏杆菌病和土拉菌病），也是锥虫的传播媒介。

全世界蝇科（Muscidae）中有4000种，大部分不是咬蝇，大约有50个种属于螫蝇亚科（Stomoxinae），它们以血液为食。被称为厩蝇的厩螫蝇（*Stomoxys calcitrans*）（图352）和角蝇（*Haematobia* spp）（在旧文献中，*Haematobia*有时称之为*Lyperosia*，其为*Haematobia*的次异名）（图353），为最常见的代表种。

厩蝇体型为小到中等大小，与家蝇相似，如细看，从上面可见细长坚硬的喙向前突出，口器由背面的上唇和腹面的下唇组成，中间有一中舌（图354），喙尖端的唇瓣叶进食时外翻，露出硬的齿。齿可以切割皮肤和固定喙部，刺入后，喙反复刺入抽出，以扩大血池的大小或定位于血管，喙插入皮肤后也可旋转[14]。角蝇的口器结构相似，但喙较短，下颚须较长。

蝇科的生活史由卵、3个幼虫阶段、蛹期和成蝇组成，根据气温高低，可在12天到6周内完成整个生活史。

厩蝇在宿主体停留时间较短，在空中交配，少数情况下，可将动物粪便作为其繁殖物。成蝇产卵于潮湿的各种腐烂有机物内，在室内，也可在草食动物粪便、尿和草的混合物内繁殖，可在室外腐烂的蔬菜、发酵的草屑或多腐殖质的土壤见到幼虫。角蝇大部分时间在宿主体，也在宿主体交配，其偏好在新鲜的奶牛粪便上繁殖[5]。

图 352 被称为厩蝇的厩螫蝇与家蝇很像，只有细看时才能发现其尖头上有一喙

图353 被称为角蝇的微小角蝇比厩螫蝇小，依据其长的下颚须很容易区分（箭头所指）

厩螫蝇采食其饥饿体重110%的血量，采食后48后消化[14]。由于厩螫蝇叮咬非常疼，动物竭力驱赶厩蝇。厩螫蝇离开宿主后，在叮咬部位可看到一小血滴。

图354　厩螯蝇口器由背面的上唇和腹面的下唇组成，中间有一中舌

图355　虻体型很大，叮咬非常疼

图356　麻虻属具有非常鲜亮的眼

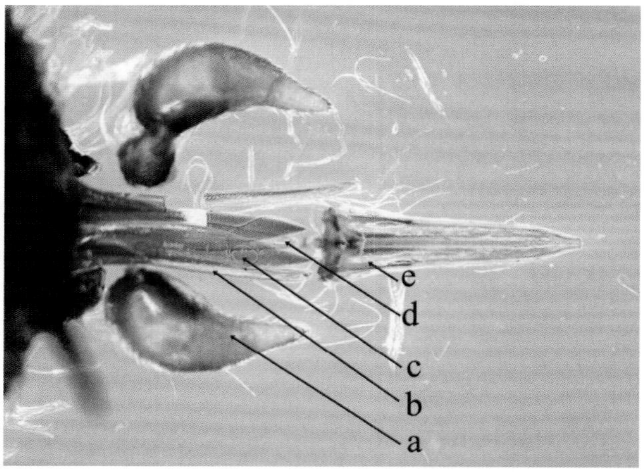


图357　虻的口器结构复杂，由下颚须(a)、上颚骨(b)、下颚骨(c)、上唇(d)和唇瓣(d)组成

在印度，厩螯蝇很喜欢在骆驼体吸食血液[8]，其也是几种细菌和病毒病的媒介，如炭疽、布鲁氏杆菌病、钩端螺旋体病和水泡性口炎。但其作为伊氏锥虫机械性传播媒介具有局限性，主要由于其喙裸露，没被覆盖。

全世界在虻科中发现4000个种，其在兽医领域是重要的机械性传播媒介，对于骆驼，其主要是伊氏锥虫的传播媒介。虻体型较大，长可达30mm，呈棕色、黑色或灰色（图355），头和喙较大，翅很特别，腋瓣大，腿上的大爪垫间的垫形爪间突为虻科的典型特征。一些种眼的颜色非常鲜亮[3]（图356），口器由下颚须、上唇、下颚骨、下咽和唇瓣组成（图357）。与厩蝇不同，虻的触角覆盖在喙上，在吸血过程中容易沾染上血液，使其更适于作为媒介传播疾病。

大部分虻在泥或潮湿的土壤中繁殖，繁殖地点有草场、林地、水池、小溪和其他地方（图358）。幼虫有6~13个发育期，但是大多数种的未成熟阶段尚不清楚。

仅雌虻吸血，雄虻可以鲜花的糖或蚜虫蜜露为食。虻并不是优秀的采食者，吸食血液可造成很大的伤口。通过观察、气味和体温发现宿主，虻可攻击宿主的任何部位，但偏好下腹部、腿和腹股沟区。叮咬时下颚刺穿皮肤，在唇瓣间吸上血液前，将含有抗凝成分的唾液注入损伤部位，血液通过上唇和咽部间的通道吸入。虻可一次吸入0.2mL的血液，每天需吸血3次。虻吸血时非常疼，宿主尽力驱赶虻吸血，所以，虻为了完成进食，很容易从一只动物到另一只动物移动，这样的移动对于疾病传播是至关重要的。

文献记载，很多非洲和亚洲虻都可作为伊氏锥虫机械性传播媒介，如虻属的带状虻（*Tabanus taeniola*），土灰虻（*T. amaenus*），双滴虻（*T. biguttanus*），牛虻（*T. bovinus*），二带原虻（*T. gratus*），白喙虻（*T. leucostomus*），细纹虻（*T. lineola*），黑齿虻（*T. mordax*），华虻（*T. mandarins*），微赤虻（*T. rubidus*），中华虻（*T. sinensis*），苏菲虻（*T. sufis*）和纹虻（*T. stiatus / striatus*）；麻虻属的小角麻虻（*Haematopota atellicorne*），冠麻虻（*H. coronata*），纤细麻虻（*H. tenuis*）和多毛麻虻（*H. lasiops*）；距虻属的马氏距虻（*Pangonia magnettii*）；宽胫虻属的阔宽胫虻（*Ancala latipes*）带宽胫虻（*A. fasciata*）和非洲宽胫虻（*A. africans*）；黄虻属的猎黄虻（*Atylotus agrestis*），昼黄虻（*A. diurnis*）和棕黄虻（*A. fuscipes*）；长喙虻属的马氏长喙虻（*Philoliche magretti*）和带长喙虻（*P. zonata*）；斑虻属的弯斑斑虻（*Chrysops streptobalia*）和斜带斑虻（*C. obliquefasciata*）[7,11,15,16,18,20]。

吸血性的采采蝇属于舌蝇属（*Glossina*），超过30个种，舌蝇周期性的传播非洲锥虫。与其他吸血性的昆虫相比，其种群密度较低，所以其作为叮咬害虫重要性较低。采采蝇长6.5~14mm，雌雄都吸血。成蝇休息时，翅封盖住腹部，很容易识别（图359），这使得从上面看，其形状像舌头（由希腊文 glossa得来，其意思为“舌头”）。翅的远端内侧细胞形同切肉刀（图360），触角芒上表面的毛具第二分支为其另一特征[10]，细长前突的口器由上、下唇，下咽部和两个下颚须组成（图361）。雌雄舌蝇平均每2~4天吸食其饥饿体重170%的血。

采采蝇为胎生，繁殖力相对较低，在其3个月的生命周期中雌蝇最多可产10个幼蝇。

采采蝇只在非洲撒哈拉沙漠以南地区分布，其他地方没有发现过，采采蝇共有3大种群，栖息在不同的群落生境。“福斯卡”种群分布于几内亚和尼日尔-刚果茂密的赤道林区，在真正的森林和树木繁茂草原的过渡带上；“须”种群的采采蝇具有相似的分布，但其在更北部也可见到，它们占据着西非

图358　湖、水渠和流速缓慢河流、泥泞的岸边是虻理想的繁殖地点

图359　吸血的舌蝇。从上看，合并的翅使得其看起来形状像舌头(图片由K. Seidl 和B. Baue博士提供、柏林自由大学寄生虫学和热带兽医学研究所)

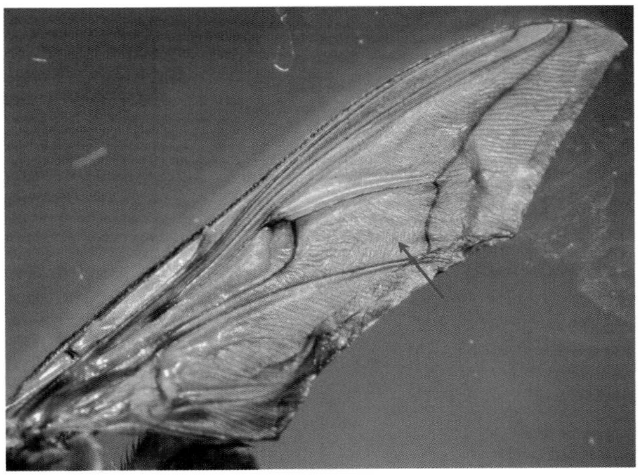

图360　舌蝇翅脉形同切肉刀

和中非的森林区域，河边森林贯穿于西非大草原、红树林沼泽和灌木植被区间；而“刺”种群主要分布于西非、中非和东非更加干旱的植被地带[9].

骆驼虱蝇（*Hypobosca camelina*）为具翅虱，属蛹生目（Pupipara），体结实，红褐色、布有淡黄色斑点，腿棕色多毛，体长可达1cm。虽然其善于飞翔，但喜欢停留在宿主体，具成对的背离爪，其可用跗爪抓在骆驼的毛上。骆驼虱蝇是吸血性昆虫，吸食渗出血液或直接从血管中吸血，喙由上下唇和下咽部组成，为唾液腺的通道。

骆驼虱蝇在单峰驼分布的非洲、阿拉伯半岛和印度的很多地方广泛分布，雌蝇在干燥的土壤中产幼虫，一次只产一个幼虫，并将幼虫掩盖[8]。另一个在骆驼体可见的虱蝇是变异型虱蝇（*Hypobosca variegata*）[6]。骆驼体的虱蝇主要在会阴和四肢[13]，由于虱蝇传播粪型锥虫，所以其可能也是伊氏锥虫的传播媒介，但随后的调查并不支持这一观点[17]。

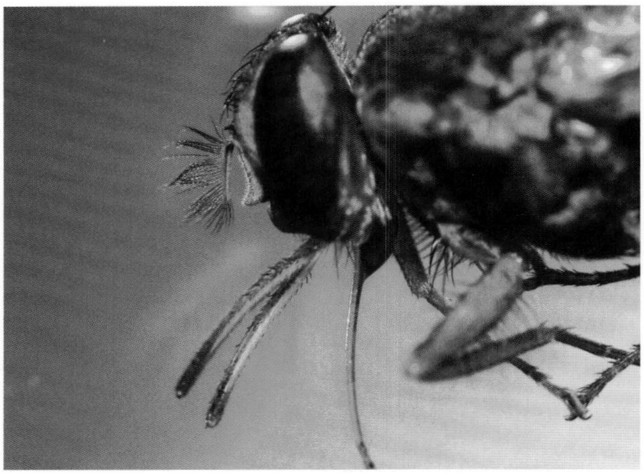

图361　采采蝇头和口器

蚊、蠓、蚋和白蛉都属于双翅目长角亚目（Nematocera）的小型吸血昆虫，对其是否寄生于骆驼知之甚少，但蚊、蠓和蚋在骆驼丝虫的传播过程中扮演着重要角色[1]，同时在白蛉吸食的血液中检测到了骆驼的血液[21]。这些昆虫繁殖区不相同，蚋（蚋科，Simuliidae）在流水（河流、小溪）中产卵[4]。

蚊（蚊科）幼虫发育也需要水，但其可在永久或临时的静水、流水，或者自然、人工的水容器中繁殖[19]。而蠓（蠓科Ceratopogonidae）的繁殖发育很少依赖于水，他们的栖息地范围从潮湿的堆肥或落叶到池塘边的泥土，甚至在混有动物粪便的泥土中都可以繁殖[2]。白蛉（白蛉亚科Phlebotominae）的发育也很少依赖于水，它们对繁殖地的唯一要求就是潮湿，并具有机物，白蛉的繁殖地总是在相对凉爽、潮湿的陆地[12]。

长角亚目的昆虫仅雌虫吸血，大部分蚊子和白蛉在黄昏和晚上叮咬吸血，而蚋和蠓在白天吸食血液。

## 参考文献

[1] Anderson R.C. (2006). – Nematode parasites of vertebrates. Their development and transmission. 2nd Ed. CPI Antony Rowe, Eastbourne. 650 pp.

[2] Boorman J. (1995). – Biting midges (Ceratopogonidae). *In* Medical insects and arachnids. Chapman and Hall (R.P. Lane & R.W. Crosskey, eds). London, 228 – 309.

[3] Chainey J.E. (1995). – Horse-flies, deer flies and clegs (Tabanidae). *In* Medical insects and arachnids. Chapman & Hall (R.P. Lane & R.W. Crosskey, eds). London, 310 – 332.

[4] Crosskey R.W. (1995). – Blackflies (Simuliidae). *In* Medical insects and arachnids. Chapman & Hall (R.P. Lane & R.W. Crosskey, eds). London, 241 – 287.

[5] Crosskey R.W. (1995). – Stable flies and horn-flies (blood sucking Muscidae). *In* Medical insects and arachnids. Chapman & Hall (R.P. Lane & R.W. Crosskey, eds). London, 389 – 402.

[6] Dia M.L., Elsen P., Cuisance D., Diop C., Thiam A. & Chollet J.Y. (1998). – Abundance and seasonal variations of tabanids in southern Trarza (Mauritania). *In* Tropical veterinary medicine: molecular epidemiology, hemoparasites and their vectors and genetic topics (F. Jongejan, W. Goff & E. Camus, eds). *Ann. New York Acad. Sci.*, 849, 456 – 460.

[7] Dirie M.F., Wallbanks K.R., Aden A.A., Bornstein S. & Ibrahim M.D. (1989). – Camel trypanosmiasis and its vectors in

Somalia. *Vet. Parasitol.*, 32, 285 – 291.

[8] Higgins A.J. (1985). – Common ectoparasites and their control. *Br. Vet. J.*, 141 (2), 197 – 216.

[9] Itard J. (1989). – African animal trypanosomosis. *In* Manual of tropical veterinary parasitology (M. Shah-Fischer & R.R. Say, eds). CAB International, Wallingford, 177 – 292.

[10] Jordan A.M. (1995). – Tsetse-flies (Glossinidae). *In* Medical insects and arachnids. Chapman & Hall (R.P. Lane & R.W. Crosskey, eds). London, 333 – 388.

[11] Kigaye M.K. & Jiffar T. (1991). – A survey of ectoparasites of cattle in Harar and Dire Dawa districts Hararghe administrative region of Ethiopia. *Bull. Anim. Health Prod. Afr.*, 39, 15 – 24.

[12] Lane R.P. (1995). – Sandflies (Phlebotominae). *In* Medical insects and arachnids. Chapman & Hall (R.P. Lane & R.W. Crosskey, eds). London, 78 – 119.

[13] Lawal M.D., Ahme I.G. & Ahmed A. (2007). – Some ectoparasites of *Camelus dromedarius* in Sokoto, Nigeria. *J. Entomol.*, 4, 143 – 148.

[14] Lehane M.J. (1991). – Blood sucking insects. Harper Collins Academic, London. 288 pp.

[15] Lun Z-R., Fang Y., Wang C-J. & Brun R. (1993). – Trypanosomiasis of domestic animals in China. *Parasitol. Today*, 9 (2), 41 – 45.

[16] Ouhelli H. & Dakkak A. (1987). – Protozoal diseases of dromedaries. *In* Diseases of camels. *Rev. sci. tech. Off. int. Epiz.*,6 (2), 417 – 422.

[17] Ovieke F.A. & Reid G. (2003). – The mechanical transmission of *Trypanosoma evansi* by *Haematobia minuta* (Diptera: Muscidae) and *Hippobosca camelina* (Diptera: Hippoboscidae) from an infected camel to a mouse and the survival of trypanosomes in fly mouthparts and gut (a preliminary record). *Folia Vet.*, 47 (1), 38 – 41.

[18] Rahman A.H.A. (2005). – Observation on the trypanosomosis problem outside tsetse belts of Sudan. *Rev. sci. tech.Off. int. Epiz.*, 24 (3), 962 – 972.

[19] Service M.W. (1995). – Mosquitoes (Culicidae). *In* Medical insects and arachnids (R.P. Lane & R.W. Crosskey, eds). Chapman & Hall, London, 120 – 240.

[20] Sinshaw A., Abebe G., Desquesnes M. & Yoni W. (2006). – Biting flies and *Trypanosoma vivax* infection in three highland districts bordering the lake Tana, Ethiopia. *Vet. Parasitol.*, 142, 35 – 46.

[21] Teshome G-M., Meshesha B., Nega B., Asrat H. & Talemtsehay M. (2010). – Further studies on the phlebotomine sandflies of the kala-azar endemic lowlands of Humera-Metema (north-west Ethiopia) with observations on their natural blood meal sources. *Parasite Vector*, 3, 6.

（关贵全译，殷宏校）

## 5.3.7 蠕形蚤病

蠕形蚤病是由蠕形蚤科（Vermipsyllidae）的跳蚤引起的家养、野生动物以及人的寄生虫病。与其他蚤不同，蠕形蚤是稳定的寄生性蚤。在蒙古双峰驼和反刍动物体发现两个种，花蠕形蚤（*Vermipsylla alacurt*）和羊长喙蚤（*Dorcardia ioffi*）[1,2]。其饥饿体长2.5~3.5mm，饱血的羊长喙蚤可达16mm，看上去像条纹样蠕虫，因此本地人称之为“花虫（alacurt）”。成虫前的发育在春夏季土壤表面，成虫的感染在秋季，在宿主体过冬。由于疾病在冬季发生，所以常用粉末杀虫剂进行治疗。

## 参考文献

[1] Ribbeck R., Splisteser H., Rauch H. & Hiepe Th. (1979). – Probleme der Ektoparasitenbekampfung in der Mongolischen Volksrepublik. *Angew. Parasitol.*, 20, 221 – 229.

[2] Zedev B. (1976). – Untersuchung uber Biologie, Vorkommen und Verbreitung von *Vermipsylla* spp. (Siphonaptera, Vermipsyllidae) bei Nutz- und Wildtieren in der Mongolischen Volsrepublik. *Mh. Vet.-Med.*, 31, 788 – 791.

（关贵全译，殷宏校）

## 5.3.8 舌形虫病

舌形虫病是由舌形虫科（Linguatulidae）的“鼻蠕虫”，锯齿状舌形虫（*Linguatula serrata*）幼虫和成虫阶段寄生引起的疾病，该病原属分类上介于节肢动物和环节动物之间的舌形动物门(Pentastoma)。

### 形态特征

无色透明的成虫体型细长、舌形双侧扁平，具有90条条纹。雄虫18~20mm长，妊娠的雌虫可达120mm长。前腹侧面每面具两条带几丁质钩的狭缝，中间是U形的口。淡红色并被黏液覆盖的卵胎生卵长90μm，内含一发育完全的胚胎，胚胎具有和成虫一样的形态特征[5]。

### 生活史

犬科动物是其主要的终末宿主，草食哺乳动物为中间宿主。人也是其终末宿主，当吃生的反刍动物器官做的菜肴而被感染。感染后6个月，雌舌形虫开始产卵，打喷嚏时卵被排出，或吞咽经消化道随粪便排出终末宿主体。中间宿主摄入污染的食物而感染，早期幼虫刺穿肠壁，随血流移行至肺、肝、肾和淋巴结，然后发育为二龄幼虫，并蜕皮8次。最后的幼虫或终末幼虫阶段（若虫）长4~5mm，终末幼虫经食道和咽喉移行出胃，进入鼻和鼻旁窦，进行最后一次蜕皮，发育为成虫。

### 流行病学

过去，欧洲许多国家有犬的舌形虫病[5]，随着肉品检测的加强，现在舌形虫病流行范围缩小，仅局限在南部国家。在埃及的研究显示，360只犬中，30只感染1~3条舌形虫成虫[7]。苏丹，50%的流浪狗感染[12]，伊朗的感染率高达62%，3~4岁的犬感染率最高，为92%[8]。

近来报道了许多中间宿主感染舌形虫幼虫的报道，如伊朗的黄牛、水牛甚至猫[1,10,11,13]，苏丹[12]和印度[9]的小反刍动物。

伊朗的双峰驼也有本病的报道[2,3]，没有一头小于4岁的单峰驼感染，而4~8岁的仅发现在肠系膜淋巴结感染，老单峰驼发现在肝脏和肺脏有若虫感染。

舌形虫病是一种人畜共患病。

### 临床症状和病理变化

舌形虫成虫摄食分泌物和淋巴液，其几丁质钩刺激鼻黏膜，从而引起浆性和脓性流涕和喷嚏，随病程发展，犬的嗅觉也受到影响。人的舌形虫病被称之为马拉拉（Marrara，马拉拉是一道用反刍动物和骆驼的生的肝、肺、气管和瘤胃做成的苏丹大众菜）或哈尔宗（Halzoun，哈尔宗是在黎巴嫩舌形虫若虫移行引起的口咽感染的地方名）综合征——一种由若虫引起的上呼吸道和颊鼻咽黏膜的过敏反应。虽然宿主摄入的幼虫会在淋巴结和腹腔造成机械性损伤，但大型的中间宿主体不表现临床症状。大量的感染会引起啮齿动物和小型反刍动物死亡[6]。

### 诊断

肉眼可见薄壁组织器官表面的结节，流行病

学调查时，通常将肠淋巴结切片，置于生理盐水，38℃ 5~6小时，活的若虫会钻出组织。另外，可研碎淋巴结、肺和肝组织，然后用人工消化液消化。若虫长4~6mm、宽1mm、灰白色、细长、舌形，具明显的外节，前端口腔边具两对镰形双生钩[9]。

感染终末宿主后表现流涕，显微镜镜检鼻液，可查到舌形虫虫卵。剖检可在紧挨软腭边的黏液包埋的鼻咽部可见幼龄舌形虫，雌虫喜欢寄生在紧挨耳咽管的鼻甲骨中段或下段[5]。

### 防控

无特殊的控制骆驼舌形虫病的治疗方法，常用伊维菌素治疗被称为驯鹿窦蠕虫的北极舌形虫（*L. arctica*）成虫感染，与未治疗动物相比，治疗组感染率和感染密度下降[4]。舌形虫病关系到肉制品卫生的问题，防控的关键是阻止用感染的器官喂狗。

## 参考文献


[1] Esmaeilzadeh S., Mohammadian B. & Rezaei A. (2008). –*Linguatula serrata* in a cat. *Iran. J. Vet. Res., Shiraz Univers.*, 9 (4),387 – 389.

[2] Haddadzadeh H.R., Athari S.S., Abedini R., Khazraiinia S., Khazraiinia P., Nabian S. & Haji-Mohamadi B. (2010). – One humped camel (*Camelus dromedarius*) infestation with *Linguatula serrata* in Tabriz, Iran. *Iranian J. Arthropod-Borne Dis.*,4 (1), 54 – 59.

[3] Haddadzadeh H., Athari S.S. & Hajimohammadi B. (2009). – The first record of *Linguatula serrata* infection of two-humped camel in Iran. *Iranian J. Parasitol.*, 4 (1), 59 – 61.

[4] Haugerund R., Nilssen A.C. & Rognmo A. (1993). – On the efficacy of ivermectin against the reindeer sinus worm *Linguatula arctica* (Pentastomida), with a review on ivermectin treatment in reindeer. *Rangifer*, 13 (3), 157 – 162.

[5] Heymons R. (1942). – Der Nasenwurm des Hundes (*Linguatula serrata*, Froehlich), seine Wirte und Beziehungen zur europäischen Tierwelt, seine Herkunft und praktische Bedeutung auf Grund unserer bisherigen Kenntnisse. *Z. f. Parasitenk.*,12, 607 – 638.

[6] Hiepe Th. & Ribbeck R. (1982). – Veterinarmedizinische arachno-entomologie. *In* Lehrbuch der parasitologie (Th. Hiepe,ed.). VEB Gustav-Fischer-Verlag, Jena.

[7] Khalil M. (1970). – Incidence of *Linguatula serrata* infection in Cairo mongrel dogs. *J. Parasitol.*, 56 (3), 485.

[8] Meshgi B. & Asgarian O. (2003). – Prevalence of *Linguatula serrata* infestation in Share Kord, Iran. *J. Med. Vet. B.*, 50, 466 – 467.

[9] Ravindran R., Lakshmanan B., Ravishankar C. & Subramanian H. (2008). – Prevalence of *Linguatula serrata* in domestic animals in south India. *Southeast Asian J. Trop. Med. Public Health*, 39 (5), 808 – 812.

[10] Tajik H. & Jalali S.S. (2010). –*Linguatula serrata* prevalence and morphometrical features: an abattoir survey on water buffaloes in Iran. *It. J. anim. Sci.*, 9 (3), 348 – 351.

[11] Tajik H., Tavassoli M., Dalirnaghadeh B. & Danehloipour M. (2006). – Mesenteric lymph nodes infection with *Linguatula serrata* nymphs in cattle. *Iranian J. Vet. Res., Univers. Shiraz*, 7 (4), 82 – 85.

[12] Yagi H., El Bahari S., Mohammed H.A., Ahmed E.R.S., Mustafa B., Mahmoud A., Saad M.B.A., Sulaiman S.M. & El Hassan A.M. (1996). – The Marrara syndrome: a hypersensitivity reaction of the upper respiratory tract and buccopharyngeal mucosato nymphs of *Linguatula serrata. Acta Trop.*, 62, 127 – 134.

[13] Youssefi M.R. & Moalem S.H.H. (2010). – Prevalence of *Linguatula serrata* nymphs in cattle in Babol slaughterhouse, North of Iran 2010. *World J. Zool.*, 5 (3), 197 – 199.


（关贵全译，殷宏校）

# 附录

本书附录中列出了有关骆驼的最重要的书籍、一些重要会议的论文集和手册。另外还附上了世界动物卫生组织（OIE）骆驼疫病特别小组第二次会议报告（也可见于www.oie.int)。

## 所有图表中使用的缩略词

Ab-ELISA, 抗体酶联免疫吸附试验antibody enzyme-linked immunosorbent assay

AHS, 非洲马瘟African horse sickness

BHV, 牛疱疹病毒bovine herpes virus

BT, 蓝舌病bluetongue

BVD, 牛病毒性腹泻bovine viral diarrhoea

CATT, 锥虫卡片凝集试验card agglutination trypanosoma test

CCHF, 克里米亚刚果出血热Crimean–Congo haemorrhagic fever

c-ELISA, 抗体酶联免疫吸附试验competitive enzyme-linked immunosorbent assay

CF, 补体结合试验complement fixation

CIRAD, 法国农业研究国际合作中心Centre de CoopérationInternationale pour la RechercheAgronomiqueenDéveloppement

CVRL, 中央兽医研究实验室（迪拜，阿联酋）Central Veterinary Research Laboratory (Dubai, UAE)

EHV, 马疱疹病毒equine herpes virus

FAT, 荧光抗体试验fluorescent antibody test

FMD,口蹄疫 foot and mouth disease

HI, 血凝抑制haemagglutination inhibition

IBR,传染性鼻气管炎 infectious bovine rhinotrachitis

IHC,免疫组织化学 immunohistochemistry

MAT,微量凝集试验 microscopic agglutination test

NSP ELISA,非结构蛋白酶联免疫吸附试验 non-structural protein enzyme-linked immunosorbentassay

OIE, 世界动物卫生组织World Organisation for Animal Health

OIE Terrestrial Manual, 《OIE陆生动物诊断试验和疫苗手册》OIE Manual of Diagnostic Tests andVaccines for Terrestrial Animals

PCR, 聚合酶链式反应polymerase chain reaction

PPR, 小反刍兽疫peste des petits ruminants

RBT,玫瑰花环试验 Rose–Bengal test

RVF,裂谷热 Rift Valley fever

SAT, 血清凝集试验sero-agglutination test

TEM, 透射电镜transmission electron microscopy

VNT, 病毒中和试验virus neutralisation test

# 附录1

## 有关骆驼疾病的重要书籍名录

[1] Allen W.R., Higgins A.J., Mayhew I.G., Snow D.H. & Wade J.F. (1992). – Proceedings of the 1st International Camel Conference, 2 ~ 6 February. R. & W. Publications, Newmarket, UK.

[2] Breulmann M., Boer B., Wernery U., Wernery R., El Shaer H., Gallagher D., Peacock J., Chaudhary S.A., Brown G. & Norton J. (2007). – The camel from traditional to modern times. Unesco Doha Brochure, Qatar.

[3] Emmerich J.U., Ganter M. & Wittek T.H. (2013). – Dosierungsvorschläge für Arzneimittel bei kleinen Wiederkäuern und Neuweltkameliden. Schattauer, Leipzig, Hannover and Vienna.

[4] Duncanson G.R. (2012). – Veterinary treatment of llamas and alpacas. Cabi, Wallingford, UK.

[5] Curasson G. (1947). – Le chameau et ses maladies. Vigot Frères, Éditeurs, Paris.

[6] Erdenebileg O. (2001). – Camel diseases [in Mongolian]. Dalanzadgad, Mongolyn Mal Ėmnėlėgiin Kholboo.

[7] Erwin R. (2010). – The camel. Reaktion Books, London.

[8] Faye B. (1997). – Guide de l'élevage du dromadaire. 1st Ed. Sanofi, France.

[9] Fowler M.E. (2010). – Medicine and surgery of camelids. 3rd Ed. Wiley-Blackwell, Oxford, UK.

[10] Gahlot T.K. (2000). – Selected topics on Camelids. Camel Publishing House, Bikaner, India. Available at: www.camelsandcamelids.com.

[11] Gahlot T.K. (2007). – Proceedings of the International Camel Conference, College of Veterinary and Animal Science. Rajasthan Agricultural University, Sankhla Printers, Bikaner, India.

[12] Gahlot T.K. & Chhabra M.B. (2009). – Selected research on camelid parasitology. Camel Publishing House. Available at: www.camelsandcamelids.com.

[13] Gahlot T.K., Tibary A., Wernery U. & Zhao X.X. (2002). – Selected bibliography on camelids 1991–2000. Camel Publishing House, Bikaner, India. Available at: www.camelsandcamelids.com.

[14] Gauly M., Vaughan I. & Cebra Ch. (2011). – Neuweltkameliden. 3rd Ed. Enke Verlag, Stuttgart, Germany.

[15] Higgins A. (1986). – The camel in health and disease. Baillière Tindall, London.

[16] Jensen I.M. (2006). – Camelid drug formulary. 1st Ed. Game Ranch Health, Texas. Available at: www.gameranchhealth.com.

[17] Journal of Camel Practice and Research. Biannual journal published by Camel Publishing House, Bikaner, India.

[18] Juhasz J., Skidmore J.A. & Nagy P. (2012). – Proceedings of the ICAR 2012 Satellite meeting on Camelid reproduction, 3 ~ 5August, Vancouver, Canada.

[19] Knoll E.-M. & Burger P. (2012). – Camels in Asia and North Africa. Eds Verlags (Österreichische Akademie der Wissenschaften), Austrian Academy of Sciences Press.

[20] Köhler-Rollefsen I., Mundy P. & Mathias E. (2001). – A field manual of camel diseases. ITDG Publishing, London.

[21] Manefield G.W. & Tinson A.H. (1996). – Camels. A compendium. University of Sydney. Postgraduate Foundation inVeterinary Science, National Library, Australia.

[22] Megersa B. (2010). – An epidemiological study of major camel diseases in the Borana lowland, Southern Ethiopia. Drylands Coordination Group (DCG) Report No. 58, Oslo, Norway.

[23] Moallin A.S.M. (2009). – Observation on diseases of the dromedary in Central Somalia. Available from the author by emailing amkutub@hotmail.com.

[24] Proceedings of International Camel Health Conferences, USA. College of Veterinary Medicine, Columbus Ohio State University, Columbus, Ohio, USA.

[25] Saltin B. & Rose R.J. (1994). – The racing camel (Camelus dromedarius). Acta Physiol. Scand. Supplement 617.

[26] Schulz U. (2008). – El camello en Lanzarote. Association for Rural Development of Lanzarote (ADERLAN), Spain.

[27] Schwartz H.J. & Dioli M. (1992). – The one-humped camel (*Camelus dromedarius*) in Eastern Africa: a pictorial guide to diseases, healthcare and management, Verlag Josef Margraf, Weikersheim, Germany.

[28] Tibary A. & Anouassi A. (1997). – Theriogenology in Camelidae. Actes Editions, Institut Agronomique et Vétérinaire Hassan II, Rabat-Instituts, Morocco.

[29] Wernery U., Fowler M.E. & Wernery R. (1999). – Color atlas of camelid haematology. Blackwell Wissenschafts, Berlin and Vienna.

[30] Wernery U. & Kaaden O.-R. (2002). – Infectious diseases in camelids. Blackwell Science, Berlin and Vienna.

[31] Wilson R.T. (1989). – Ecophysiology of the camelidae and desert ruminants. Springer Verlag, Berlin and Heidelberg, Germany.

[32] Yagil R. (1985). – The desert camel. Comparative physiological adaptation. Karger, Basel.

## 附录2

骆驼有关的传染病：OIE骆驼疫病特别小组第二次会议报告2010年5月3－5日，巴黎

（1）骆驼病毒病。

（2）骆驼细菌病。

（3）骆驼寄生虫病和真菌病。

# （1）骆驼病毒病

## 单峰驼

第一类：已知的主要疫病。
第二类：虽没有报告疫病但携带潜在病原者。
第三类：次要疫病。

附表Ⅰ 单峰驼病毒病

| 疫病 | 病原鉴定 | 血清学检测 | 推荐技术 | |
|---|---|---|---|---|
| | | | 诊断 | 防控 |
| 第一类 | | | | |
| 骆驼痘[a] | 2008年版《OIE陆生动物诊断试验和疫苗手册》，2.9.2节，1177页：TEM，病毒分离，IHC和PCR | ELISA和VNT[a] | ELISA试剂盒已建立，但需要验证 | 疫苗免疫 |
| 接触传染性脓疱 | TEM，IHC 和 PCR | 无 | 病毒分离是必须的 | 正在研发疫苗 |
| 乳头状瘤病 | TEM，PCR 和 IHC | 无 | | 自家疫苗 |
| 狂犬病[a] | 2008年版《OIE陆生动物诊断试验和疫苗手册》，2.1.13节，304页：FAT 和 IHC | VNT[a]<br>正在研发血清型方法 | | 使用牛的剂量免疫骆驼，但需要进一步调研 |
| 裂谷热[a] | 2008年版《OIE陆生动物诊断试验和疫苗手册》，2.1.14节，323页：培养，AGID，PCR和组织病理学 | c-ELISA和VNT[a] | ①ELISA方法需要更多样品验证<br>②应研究易感性和病毒血症的持续时间 | 调研疫苗免疫 |
| 第二类 | | | | |
| 非洲马瘟 | 2008年版《OIE陆生动物诊断试验和疫苗手册》，2.5.14节，823 页：病毒分离，PCR，ELISA 和 VN | 无 | ①研究不同毒力株和血清型的敏感性<br>②应研究病毒血症的持续时间<br>③研发ELISA | |
| 蓝舌病 | 2008年版《OIE陆生动物诊断试验和疫苗手册》，2.1.3节，158 页：病毒分离，免疫学方法和PCR | c-ELISA | 应研究强毒株的易感性、血清型和带毒状态 | ①调研疫苗免疫<br>②在贸易上采取跟牛同样的措施 |
| 牛病毒性腹泻[a] | 2008年版《OIE陆生动物诊断试验和疫苗手册》，2.4.8节，698页：病毒分离，PCR，IHC 和 ELISA | c-ELISA 和VNT[a] | ①检测奶样的血清学方法需要验证<br>②需要病毒分离方法 | 正在调研易感性 |
| 小反刍兽疫 | 2008年版《OIE陆生动物诊断试验和疫苗手册》，2.7.11节，1036页：病毒分离，AGID和 PCR | 无 | ①应对c-ELISA 验证 | 正在调研强毒株的易感性 |

续表

| 疫病 | 病原鉴定 | 血清学检测 | 推荐技术 | |
|---|---|---|---|---|
| | | | 诊断 | 防控 |
| 第三类 | | | | |
| 克里米亚-刚果出血热[b] | 病毒分离[b]和 PCR | 无 | ①应对反刍动物用的c-ELISA进行验证<br>②开展血清学监测 | |
| 疱疹病毒感染[a] | 2008年版《OIE陆生动物诊断试验和疫苗手册》，2.5.9 节（EHV），894页；2.4.13节（IBR），752页：PCR，病毒分离和免疫荧光 | VNT[a] | ①血清学方法需要验证<br>②应调研对EHV 4 和BHV 1易感性 | 正在调研使用马的免疫程序的效果 |
| 西尼罗河热 | 2008年版《OIE陆生动物诊断试验和疫苗手册》，2.1.20节，377页：PCR和病毒分离 | c-ELISA | 应调研对两个毒株的易感性 | |

[a]需要在生物安全3级实验室进行

[b]需要在生物安全4级实验室进行

**双峰驼**

第一类：已知的主要疫病。
第二类：虽没有报告疫病但携带潜在病原者。
第三类：次要疫病。

附表Ⅱ　双峰驼病毒病

| 疫病 | 病原鉴定 | 血清学检测 | 推荐技术 | |
|---|---|---|---|---|
| | | | 诊断 | 防控 |
| 第一类 | | | | |
| 骆驼痘[a] | 2008年版《OIE陆生动物诊断试验和疫苗手册》，2.9.2节，1177页：TEM，病毒分离，IHC和PCR | ELISA和VNT[a] | ELISA试剂盒已建立，但需要验证 | 疫苗免疫，但免疫程序需要调研 |
| 接触传染性脓疱 | TEM和 IHC | 无 | 病毒分离 | 正在研发疫苗 |
| 口蹄疫 | 2008年版《OIE陆生动物诊断试验和疫苗手册》，2.1.5节，190页：PCR和病毒分离 | NSP c-ELISA | ①使用NSP c-ELISA两次检查<br>②应进行更多的调研 | 疫苗免疫，但免疫程序需要调研 |
| 甲型流感病毒感染 | 病毒分离，PCR 和ELISA | HI | 应调研对不同血清型的易感性 | 正在调研使用马的免疫程序的效果 |

续表

| 疫病 | 病原鉴定 | 血清学检测 | 推荐技术 | |
|---|---|---|---|---|
| | | | 诊断 | 防控 |
| 狂犬病[a] | 2008年版《OIE陆生动物诊断试验和疫苗手册》，2.1.13节，304页：FAT 和 IHC | VNT[a]<br>正在研发血清学方法 | | 使用牛的剂量免疫骆驼，但需要进一步调研 |
| 第二类 | | | | |
| 牛病毒性腹泻[a] | 2008年版《OIE陆生动物诊断试验和疫苗手册》，2.4.8节，698页：病毒分离，PCR，IHC 和 ELISA | VNT[a] | ①血清学方法需要验证<br>②调研易感性 | |
| 第三类 | | | | |
| 蓝舌病 | 2008年版《OIE陆生动物诊断试验和疫苗手册》，2.1.3节，158 页：病毒分离，免疫学方法PCR | c-ELISA | 应研究强毒株的易感性、血清型和带毒状态 | ①调研疫苗免疫<br>②在贸易上采取跟牛同样的措施 |
| 克里米亚-刚果出血热[b] | 病毒分离[b]和 PCR | 无 | ①应对反刍动物用的c-ELISA进行验证<br>②开展血清学监测 | |
| 疱疹病毒感染[a] | 2008年版《OIE陆生动物诊断试验和疫苗手册》，2.5.9 节（EHV），894页；2.4.13节（IBR），752页：PCR，病毒分离和免疫荧光 | VNT[a] | ①血清学方法需要验证<br>②应调研对EHV 4 和BHV 1易感性 | 正在调研使用马的免疫程序的效果 |

[a]需要在生物安全3级实验室进行

[b]需要在生物安全4级实验室进行

## 新大陆骆驼

第一类：已知的主要疫病。

第二类：虽没有报告疫病但携带潜在病原者。

第三类：次要疫病。

附表Ⅲ 新大陆骆驼病毒病

| 疫病 | 病原鉴定 | 血清学检测 | 推荐技术 | |
|---|---|---|---|---|
| | | | 诊断 | 防控 |
| 第一类 | | | | |
| 牛病毒性腹泻[a] | 2008年版《OIE陆生动物诊断试验和疫苗手册》，2.4.8节，698页：病毒分离，PCR，IHC 和 ELISA | c-ELISA和VNT[a] | 血清学方法需要验证 | 调研使用牛的免疫程序的效果 |

续表

| 疫病 | 病原鉴定 | 血清学检测 | 推荐技术 | |
|---|---|---|---|---|
| | | | 诊断 | 防控 |
| 蓝舌病 | 2008年版《OIE陆生动物诊断试验和疫苗手册》，2.1.3节，158 页：病毒分离，免疫学方法PCR | c-ELISA | 应研究不同强毒株和血清型的易感性 | ①使用绵羊的程序进行免疫<br>②在贸易上采取跟牛同样的措施 |
| 疱疹病毒感染[a] | 2008年版《OIE陆生动物诊断试验和疫苗手册》，2.5.9 节（EHV），894页；2.4.13节（IBR），752页：PCR，病毒分离和免疫荧光 | VNT[a] | 血清学方法需要验证应调研对EHV 4 和BHV 1易感性 | 调研疫苗免疫 |
| 第二类 | | | | |
| 接触传染性脓疱 | TEM和 IHC | 无 | 病毒分离 | 正在研发疫苗 |
| 第三类 | | | | |
| 骆驼痘[a] | 2008年版《OIE陆生动物诊断试验和疫苗手册》，2.9.2节，1177页：TEM，IHC和PCR | ELISA和VNT[a] | ELISA试剂盒已建立，但需要验证 | |
| 马脑脊髓炎 | 2008年版《OIE陆生动物诊断试验和疫苗手册》，2.5.5节，858页和2.5.14节，931页: PCR和病毒分离 | 无 | ①调研易感性<br>②对已有的血清型方法验证 | |
| 狂犬病[a] | 2008年版《OIE陆生动物诊断试验和疫苗手册》，2.1.13节，304页：FAT 和 IHC | VNT[a] | | 应调研疫苗免疫程序 |
| 西尼罗河热和其他黄病毒[a] | 2008年版《OIE陆生动物诊断试验和疫苗手册》，2.1.20节，377页：PCR和病毒分离 | c-ELISA和VNT[a] | 应调研易感性易感性 | |

[a]需要在生物安全3级实验室进行

## （2）骆驼细菌病

单峰驼

第一类：已知的主要疫病。

第二类：虽没有报告疫病但携带潜在病原者。

第三类：次要疫病。

附表Ⅳ　单峰驼细菌病

| 疫病 | 病原鉴定 | 血清学检测 | 推荐技术 | |
|---|---|---|---|---|
| | | | 诊断 | 防控 |
| 第一类 | | | | |
| 炭疽[a] | 2008年版《OIE陆生动物诊断试验和疫苗手册》，2.1.1节，135页：免疫荧光，PCR，炭疽杆菌的培养和鉴定 | 无 | 无 | ①在流行区免疫<br>②需要进行疫苗的田间中试 |
| 布氏杆菌病[a]（马尔他布氏杆菌病） | 2008年版《OIE陆生动物诊断试验和疫苗手册》，2.4.31节，624页：染色，培养和 PCR | CF，RBT，SAT，c-ELISA[a] | CF，RBT，SAT 和 c-ELISA需要验证 | ①疫苗免疫<br>②需要对疫苗免程序疫进行验证 |
| 梭菌感染 | 病原分离和定型，检测毒素 | 已有毒素分型（产气荚膜）的ELISA和PCR | 正在研发多重PCR | 调研疫苗免疫 |
| 大肠杆菌病 | 2008年版《OIE陆生动物诊断试验和疫苗手册》，2.9.11节，1294页：培养，免疫学方法和PCR | 无 | ①应对大多数致病性的血清型进行鉴定<br>②应研发血清型检测方法 | 研发疫苗 |
| 嗜皮菌病（刚果嗜皮菌） | 2008年版《OIE陆生动物诊断试验和疫苗手册》，2.4.10节，725页：培养，免疫学方法和PCR | 无 | ① 应对大多数致病性的生物型进行鉴定<br>②应研发血清学方法 | 研发疫苗 |
| 出血性败血症（多杀性巴氏杆菌和溶血性曼氏杆菌） | 2008年版《OIE陆生动物诊断试验和疫苗手册》，2.4.12节，739页：培养和PCR | 无 | 易感性和病原学已有的数据有矛盾，应进行调研 | 应对疫苗免程序疫进行调研 |
| 副结核病（约内氏病） | 2008年版《OIE陆生动物诊断试验和疫苗手册》，2.1.11节，276页：培养和PCR | 无 | 应验证血清学检测方法 | 在完成方法验证后对阳性动物扑杀 |
| 脓肿性疾病（干酪性淋巴腺炎） | 细菌分离和定型 | 无 | 应研发假结核棒状杆菌（*Corynebacterium pseudotuberculosis*）和金黄色葡萄球菌（*Staphylococcus aureus*）的血清型检测方法 | 研发疫苗 |

续表

| 疫病 | 病原鉴定 | 血清学检测 | 推荐技术 | |
|---|---|---|---|---|
| | | | 诊断 | 防控 |
| 沙门氏菌病 | 2008年版《OIE陆生动物诊断试验和疫苗手册》，2.9.9节，1267页：培养和PCR | | ①应对大多数致病性的血清型进行鉴定并明确易感性<br>②应研发血清学检测方法 | 研发疫苗 |
| 第二类 | | | | |
| 钩端螺旋体病 | 2008年版《OIE陆生动物诊断试验和疫苗手册》，2.1.9节，251页：PCR | MAT | ①应对大多数流行的血清型进行鉴定<br>②明确易感性 | 研发疫苗 |
| Q热 | 2008年版《OIE陆生动物诊断试验和疫苗手册》，2.1.12节，292页：病原分离，染色和PCR | CF | ①应调研易感性<br>②应验证血清学检测方法 | 研发疫苗 |
| 结核病[a] | 2008年版《OIE陆生动物诊断试验和疫苗手册》，2.4.7节，683页：直接鉴定，培养和PCR | RT-test[a] | 应开展血清学监测；应对皮内变态实验进行调研 | 在完成方法验证后对阳性动物扑杀 |
| 第三类 | | | | |
| 衣原体病 | 病原分离与鉴定 | c-ELISA | 验证血清学检测方法 | |
| 鼻疽[a] | 2008年版《OIE陆生动物诊断试验和疫苗手册》，2.5.11节，919页：培养和PCR | CF[a] | 验证血清学检测方法 | 对阳性动物扑杀 |
| 鼠疫(鼠疫耶尔森氏菌) | 分离病原 | 无 | 研发血清学方法 | 扑杀感染动物 |

[a]需要在生物安全3级实验室进行

## 双峰驼

第一类：已知的主要疫病。

第二类：虽没有报告疫病但携带潜在病原者。

第三类：次要疫病。

附表Ⅴ 双峰驼细菌病

| 疫病 | 病原鉴定 | 血清学检测 | 推荐技术 | |
|---|---|---|---|---|
| | | | 诊断 | 防控 |
| 第一类 | | | | |
| 炭疽[a] | 2008年版《OIE陆生动物诊断试验和疫苗手册》，2.1.1节，135页：免疫荧光，PCR，炭疽杆菌的培养和鉴定 | 无 | 无 | ①在流行区免疫<br>②需要进行疫苗的田间中试 |

续表

| 疫病 | 病原鉴定 | 血清学检测 | 推荐技术 | |
|---|---|---|---|---|
| | | | 诊断 | 防控 |
| 布氏杆菌病[a]（流产和马尔他布氏杆菌病） | 2008年版《OIE陆生动物诊断试验和疫苗手册》，2.4.31节，624页：染色，培养和 PCR | CF，RBT，SAT，c-ELISA[a] | 检测流产布鲁菌和马尔他布鲁菌的CF，RBT，SAT 和c-ELISA需要验证；其他方法也需要验证 | ①根据细菌的种类（流产布鲁菌或马尔他布鲁菌）进行疫苗免疫<br>②需要对疫苗免程序疫进行验证 |
| 梭菌感染 | 病原分离和定型，检测毒素 | 已有毒素分型（产气荚膜）的ELISA和PCR | 正在研发多重PCR | 调研疫苗免疫 |
| 大肠杆菌病 | 2008年版《OIE陆生动物诊断试验和疫苗手册》，2.9.11节，1294页：培养，免疫学方法和PCR | 无 | ①应对大多数致病性的血清型进行鉴定<br>②应研发血清型检测方法 | 研发疫苗 |
| 副结核病（约内氏病） | 2008年版《OIE陆生动物诊断试验和疫苗手册》，2.1.11节，276页：培养和PCR | 无 | 应验证血清学检测方法 | 在完成方法验证后对阳性动物扑杀 |
| 鼠疫(鼠疫耶尔森氏菌) | 分离病原 | 无 | 研发血清学方法 | ①扑杀感染动物<br>②控制传播媒介 |
| 脓肿性疾病（干酪性淋巴腺炎） | 细菌分离和定型 | 无 | 应研发假结核棒状杆菌（*Corynebacterium pseudotuberculosis*）和金黄色葡萄球菌（*Staphylococcus aureus*）的血清型检测方法 | 研发疫苗 |
| 沙门氏菌病 | 2008年版《OIE陆生动物诊断试验和疫苗手册》，2.9.9节，1267页：培养和PCR | | ①应对大多数致病性的血清型进行鉴定并明确易感性<br>②应研发血清学检测方法 | 研发疫苗 |
| 结核病[a] | 2008年版《OIE陆生动物诊断试验和疫苗手册》，2.4.7节，683页：直接鉴定，培养和PCR | RT-test[a] | ①应开展血清学监测<br>②应对皮内变态实验进行调研 | 在完成方法验证后对阳性动物扑杀 |
| 第二类 | | | | |
| 钩端螺旋体病 | 2008年版《OIE陆生动物诊断试验和疫苗手册》，2.1.9节，251页：PCR | MAT | ①应对大多数流行的血清型进行鉴定<br>②应调研易感性 | 研发疫苗 |
| Q热 | 2008年版《OIE陆生动物诊断试验和疫苗手册》，2.1.12节，292页：染色，病原分离和PCR | CF | ①应调研易感性<br>②验证血清学检测方法 | 研发疫苗 |
| 第三类 | | | | |
| 鼻疽[a] | 2008年版《OIE陆生动物诊断试验和疫苗手册》，2.5.11节，919页 | CF[a] | 验证血清学检测方法 | 对阳性动物扑杀 |
| 衣原体病 | 病原分离与鉴定 | c-ELISA | 验证血清学检测方法 | |

续表

| 疫病 | 病原鉴定 | 血清学检测 | 推荐技术 | |
|---|---|---|---|---|
| | | | 诊断 | 防控 |
| 出血性败血症（多杀性巴氏杆菌和溶血性曼氏杆菌） | 2008年版《OIE陆生动物诊断试验和疫苗手册》，2.4.12节，739页：培养和PCR | 无 | 易感性和病原学已有的数据有矛盾，应进行调研 | 应对疫苗免程序疫进行调研 |

[a]需要在生物安全3级实验室进行

## 新大陆骆驼

第一类：已知的主要疫病。

第二类：虽没有报告疫病但携带潜在病原者。

第三类：次要疫病。

附表Ⅵ　新大陆骆驼细菌病

| 疫病 | 病原鉴定 | 血清学检测 | 推荐技术 | |
|---|---|---|---|---|
| | | | 诊断 | 防控 |
| 第一类 | | | | |
| 炭疽[a] | 2008年版《OIE陆生动物诊断试验和疫苗手册》，2.1.1节，135页：免疫荧光，PCR，炭疽杆菌的培养和鉴定 | 无 | 无 | ①在流行区免疫<br>②需要进行疫苗的田间中试 |
| 布氏杆菌病[a]（流产和马尔他布氏杆菌病） | 2008年版《OIE陆生动物诊断试验和疫苗手册》，2.4.31节，624页：染色，培养和 PCR | CF，RBT，SAT，c-ELISA[a] | 检测流产布鲁菌和马尔他布鲁菌的CF，RBT，SAT 和 c-ELISA需要验证；其他方法也需要验证 | ①根据细菌的种类（流产布鲁菌或马尔他布鲁菌）进行疫苗免疫<br>②需要对疫苗免程序疫进行验证 |
| 大肠杆菌病 | 2008年版《OIE陆生动物诊断试验和疫苗手册》，2.9.11节，1294页：培养和PCR | 无 | ①应对大多数致病性的生物型进行鉴定<br>②应研发血清学方法 | 研发疫苗 |
| 肠毒血症 | 分离和鉴定细菌 | 已有鉴定毒素的ELISA和PCR | 应研发多重PCR | 应根据现有的类毒素疫苗确定免疫程序 |
| 钩端螺旋体病 | 2008年版《OIE陆生动物诊断试验和疫苗手册》，2.1.9节，251页：PCR | MAT | 应对大多数流行的生物型进行鉴定，明确易感性 | 研发疫苗 |
| 沙门氏菌病 | 2008年版《OIE陆生动物诊断试验和疫苗手册》，2.9.9节，1267页：培养，免疫学方法和PCR | 无 | ①应研发血清学检测方法<br>②应对最常见的血清型进行鉴定 | 研发疫苗 |
| 结核病[a] | 2008年版《OIE陆生动物诊断试验和疫苗手册》，2.4.7节，683页：直接鉴定，培养和PCR | RT-test[a] | 结核菌素试验无效，应研发血清学检测方法 | 在完成方法验证后对阳性动物扑杀 |

续表

| 疫病 | 病原鉴定 | 血清学检测 | 推荐技术 | |
|---|---|---|---|---|
| | | | 诊断 | 防控 |
| 第二类 | | | | |
| 副结核病（约内氏病） | 2008年版《OIE陆生动物诊断试验和疫苗手册》，2.1.11节，276页：培养和PCR | 无 | 应验证血清学检测方法 | 在完成方法验证后对阳性动物扑杀 |
| 脓肿性疾病（干酪性淋巴腺炎） | 细菌分离和定型 | 无 | 应研发假结核棒状杆菌（*Corynebacteriumpseudotuberculosis*）和金黄色葡萄球菌（*Staphylococcus aureus*）的血清型检测方法 | 研发疫苗 |
| Q热 | 2008年版《OIE陆生动物诊断试验和疫苗手册》，2.1.12节，292页：染色，病原分离和PCR | CF | ①应调研易感性<br>②应验证血清学检测方法 | 研发疫苗 |
| 第三类 | | | | |
| 放线杆菌病 | 病原分离与鉴定 | 无 | 验证血清学检测方法 | |
| 巴氏杆菌病(出血性败血症) | 2008年版《OIE陆生动物诊断试验和疫苗手册》，2.4.12节，739页：病原培养和PCR | 无 | 应调研易感性 | |

[a]需要在生物安全3级实验室进行

# （3）骆驼寄生虫病和真菌病

**单峰驼**

第一类：已知的主要疫病。

第二类：次要疫病。

附表Ⅶ 羊峰驼寄生虫病

| 疫病 | 病原鉴定 | 血清学检测 | 推荐技术 | |
|---|---|---|---|---|
| | | | 诊断 | 防控 |
| 第一类 | | | | |
| 狂蝇感染 | 直接进行病原鉴定 | 无 | 进行病原的形态学鉴定 | 研发新的治疗方法 |
| 球虫病 | 直接进行病原鉴定：幼龄骆驼感染艾美尔球虫、等孢子虫和隐孢子虫 | 无 | ①进行病原的形态学鉴定<br>②研发PCR | 研发新的治疗方法和疫苗 |
| 胃肠道寄生虫病 | 对毛园线虫和血矛线虫等直接进行病原鉴定 | 无 | 进行病原的形态学鉴定 | 应调研治疗程序和抗药性 |
| 包虫病<br>棘球蚴病 | 2008年版《OIE陆生动物诊断试验和疫苗手册》，2.1.4节，175页：直接病原鉴定和PCR | ELISA | 可使用抗骆驼的二抗标记的ELISA | ①治疗感染狗<br>②研发疫苗 |
| 螨（疥螨） | 2008年版《OIE陆生动物诊断试验和疫苗手册》，2.9.8节，1255页：直接病原鉴定 | c-ELISA | 为了与其他皮肤病（痒螨、嗜皮菌病）鉴别诊断，应进行病原的形态学鉴定 | 应进行检疫并治疗<br>研发疫苗 |
| 嗜皮菌病 | 直接进行病原鉴定 | 无 | 进行病原的形态学鉴定 | 有疫苗（最初是牛用疫苗），但应验证免疫程序 |
| 蜱感染 | 直接进行病原鉴定 | 无 | 进行病原的形态学鉴定 | 研发新的治疗方法和疫苗 |
| 锥虫病 | 2008年版《OIE陆生动物诊断试验和疫苗手册》，2.4.8节，352页：PCR | CATT和间接ELISA（两种方法都没有商品化产品） | ①间接ELISA可使用抗骆驼的二抗标记<br>②PCR | 为了贸易，应开展系统性控制<br>治疗阳性动物<br>调研抗药性 |
| 第二类 | | | | |
| 蝇蛆病和其他蝇类感染 | 直接进行病原鉴定 | 无 | 进行病原的形态学鉴定 | 伊维菌素 |
| 新孢子虫病 | 直接进行病原鉴定，PCR | ELISA | 调研易感性<br>使用ELISA进行诊断<br>研发PCR | 研发疫苗 |
| 弓形虫病 | 2008年版《OIE陆生动物诊断试验和疫苗手册》，2.9.10节，1284页：分离，组织切片，PCR，包囊检测 | SAT ELISA | 调研易感性 | |

## 双峰驼

第一类：已知的主要疫病。
第二类：次要疫病。

附表Ⅷ 双峰驼寄生虫病

| 疫病 | 病原鉴定 | 血清学检测 | 推荐技术 | |
|---|---|---|---|---|
| | | | 诊断 | 防控 |
| 第一类 | | | | |
| 狂蝇感染 | 直接进行病原鉴定 | 无 | 进行病原的形态学鉴定 | 研发新的治疗方法 |
| 球虫病 | 直接进行病原鉴定：幼龄骆驼感染艾美尔球虫、等孢子虫和隐孢子虫 | 无 | ①进行病原的形态学鉴定<br>②研发PCR | 研发新的治疗方法和疫苗 |
| 胃肠道寄生虫病 | 对毛园线虫和血矛线虫等直接进行病原鉴定 | 无 | 进行病原的形态学鉴定 | 应调研治疗程序和抗药性 |
| 包虫病<br>棘球蚴病 | 2008年版《OIE陆生动物诊断试验和疫苗手册》，2.1.4节，175页：直接病原鉴定和PCR | ELISA | 可使用抗骆驼的二抗标记的ELISA | ①治疗感染狗<br>②研发疫苗 |
| 螨（疥螨） | 2008年版《OIE陆生动物诊断试验和疫苗手册》，2.9.8节，1255页：直接病原鉴定 | c-ELISA | 为了与其他皮肤病（痒螨、嗜皮菌病）鉴别诊断，应进行病原的形态学鉴定 | 应进行检疫并治疗<br>研发疫苗 |
| 嗜皮菌病 | 直接进行病原鉴定 | 无 | 进行病原的形态学鉴定 | 有疫苗（最初是牛用疫苗），但应验证免疫程序 |
| 蜱感染 | 直接进行病原鉴定 | 无 | 进行病原的形态学鉴定 | 研发新的治疗方法和疫苗 |
| 锥虫病 | 2008年版《OIE陆生动物诊断试验和疫苗手册》，2.4.8节，352页：PCR | CATT和间接ELISA（两种方法都没有商品化产品） | ①间接ELISA可使用抗骆驼的二抗标记<br>②PCR | 为了贸易，应开展系统性控制<br>治疗阳性动物<br>调研抗药性 |
| 第二类 | | | | |
| 蝇蛆病和其他蝇类感染 | 直接进行病原鉴定 | 无 | 进行病原的形态学鉴定 | 伊维菌素 |
| 新孢子虫病 | 直接进行病原鉴定，PCR | ELISA | 调研易感性<br>使用ELISA进行诊断<br>研发PCR | 研发疫苗 |
| 弓形虫病 | 2008年版《OIE陆生动物诊断试验和疫苗手册》，2.9.10节，1284页：分离，组织切片，PCR，包囊检测 | SAT ELISA | 调研易感性 | |

## 新大陆骆驼

第一类：已知的主要疫病。
第二类：次要疫病。

附表Ⅸ　新大陆骆驼寄生虫病

| 疫病 | 病原鉴定 | 血清学检测 | 推荐技术 | |
|---|---|---|---|---|
| | | | 诊断 | 防控 |
| 第一类 | | | | |
| 球虫病 | 直接进行病原鉴定：幼龄骆驼感染艾美尔球虫、等孢子虫和隐孢子虫 | 无 | ①进行病原的形态学鉴定<br>②研发PCR | 研发新的治疗方法和疫苗 |
| 包虫病<br>棘球蚴病 | 2008年版《OIE陆生动物诊断试验和疫苗手册》，2.1.4节，175页：直接病原鉴定，粪抗原检测和PCR | ELISA | 可使用抗骆驼的二抗标记的ELISA | ①治疗感染狗<br>②研发疫苗 |
| 螨（疥螨） | 2008年版《OIE陆生动物诊断试验和疫苗手册》，2.9.8节，1255页：直接病原鉴定 | c-ELISA | 为了与其他皮肤病（痒螨、嗜皮菌病）鉴别诊断，应进行病原的形态学鉴定 | ①应进行检疫并治疗<br>②研发疫苗 |
| 新孢子虫病 | PCR，IF | ELISA | 调研易感性 | 对已有的疫苗进行评价 |
| 肉孢子虫病 | 直接进行病原鉴定 | ELISA | 需要对ELISA验证 | 研发疫苗 |
| 吸虫病 | 直接进行病原鉴定 | 只有检测肝片吸虫的ELISA | 尸体剖检时应对大型和小型的吸虫都进行鉴定 | 治疗。应验证治疗程序 |
| 第二类 | | | | |
| 球孢子菌病 | 直接进行病原鉴定（尸体解剖） | CF和AGID | | 可治疗 |
| 嗜皮菌病 | 直接进行病原鉴定 | 无 | 进行病原的形态学鉴定 | 有疫苗（最初是牛用疫苗），但应验证免疫程序 |

（殷宏译，储岳峰校）

# 索引

**注：表格位置的页码为黑体，插图的页码为斜体，vs.表示比较或对比。缩写词列于本书的开始部分.**


**A**

abomasum haemorrhage, endotoxicosis, 第三胃出血，内毒素中毒, *40*

abortion 流产

bovine viral diarrhoea, 牛病毒性腹泻，270

rate of, 比例，152

Rift Valley Fever, 裂谷热，276

trypanosomosis, 锥虫病，361

abscesses，脓肿

caseous lymphadenitis, 干酪样淋巴结炎，170, *170*

lymph nodes, 淋巴结，166

*Rhodococcus equi* infection, 马红球菌感染，*49*, 50

*Staphylococcus aureus* dermatitis, 金黄色葡萄球菌皮炎，177, *177*

abscess pneumonia, 脓肿性肺炎，**128**

acaricides, mange treatment, 杀螨药物，螨病治疗，*451*

*Achipteria,* 翼甲螨, 409

*Achromobacter pyogenes* infection, 化脓无色杆菌感染，208

actinobacillosis, NWC, 放线菌病，新大陆骆驼，**486**

*Actinomyces,* 放线菌

lungs, 肺脏，**131**

nasal swabs, 鼻试子，132

*Actinomyces lignieresii* infection, 林氏放线菌感染，208

*Actinomyces pyogenes* infection 化脓放线菌感染

caseous lymphadenitis, 干酪性淋巴腺炎，**168**

infective reproductive losses, 传染性繁殖障碍，153

integument,皮肤，**178**

*Actinomyces viscosus* infection, 林氏放线杆菌感染，208

active metritis, chronic, 慢性活动性子宫炎，**158**

acute suppurative endometritis, 急性化脓性子宫内膜炎，**158**

acute suppurative metritis, 急性化脓性子宫炎，**158**

acyclovir, equine herpesvirus-1 infection treatment, 阿昔洛韦，马疱疹病毒1型感染治疗，302

*Adenoviridae,* 腺病毒，312

adenovirus, literature review,腺病毒，文献综述，**221**

*Aedes,* 伊蚊，275

*Aegyptianella pullorum,* 鸡埃及小体，**76**

aerobic bacteria, 需氧菌

integument, 皮肤，**178**

nasal swabs, 鼻试子，**133**

African horse sickness (AHS), 非洲马瘟，**280**, 317–321

aetiology,病原学，*317*

clinical signs,临床症状 318

control, 控制, 318-319

diagnosis, 诊断，318

Dromedary, 单峰驼，**478**

epidemiology, 流行病学，317-318

geographical distribution, 地理分布，317

literature review, 文献综述，**216**

pathology, 病理学，318

transmission,传播，317-318

treatment, 防控, 318-319

agar gel immunodiffusion test (AGID)，琼脂免疫扩散试验

African horse sickness diagnosis, 非洲马瘟诊断，318

aspergillosis, 曲霉病，337

caseous lymphadenitis, 干酪性淋巴腺炎，172

paratuberculosis diagnosis,副结核病的诊断，107-109

peste des petits ruminants diagnosis, 小反刍兽疫诊断，263-264

retrovirus infections, 逆转录病毒感染，**321**，

agglutination tests，凝集试验

modified, toxoplasmosis diagnosis,改进的弓形虫诊断方法，384

neosporosis diagnosis, 新孢子虫病的诊断，388

AGID *see* agar gel immunodiffusion test (AGID)，琼脂免疫扩散试验（AGID）

AHS *see* African horse sickness (AHS)，非洲马瘟(AHS)

Akabane disease, 赤羽病，**280, 324**

Akabane virus, 赤羽病毒，324

literature review, 文献综述，**229**

albendazole，阿苯达唑

anoplocephalidosis treatment, 裸头科绦虫病治疗，410
dicrocoeliosis treatment, 歧腔吸虫病的治疗，403
fasciolosis treatment, 肝片吸虫病的治疗，400
parasitic gastroenteritis treatment,寄生虫性胃肠炎治疗，**429**
*Allescheria* infection, mycotic dermatitis,霉样真菌感染，霉菌性皮炎，**332**
alpaca *(Viguna pacos),* 羊驼,*5*
biology, 生物学，**8–9**
distribution, 分布，7
domestication, 驯化，12
estimated population,估计的数量，**6**
evolution,进化，4
*Alphavirus,* 甲病毒，280
alternariosis, cutaneous, 皮肤链格孢病，**345**, 346
*Amblyomma,*花蜱，437
frequency of infection, 感染频度，**442**
*Amblyomma gemma,* 宝石花蜱，*436*
*Amblyomma variegatum,* 彩饰花蜱，*437*
amitraz, mange treatment, 双甲脒，螨的治疗，451
amoxicillin，阿莫西林，**97**
leptospirosis treatment, 钩端螺旋体病的治疗，*72*
salmonellosis treatment, 沙门氏菌病的治疗，**97**
amoxicillin–clavulanate, salmonellosis treatment, 阿莫西林，沙门氏菌病的治疗，**97**
amphotericin B, coccidiomycosis treatment, 两性霉素B，球孢子菌病的治疗，342
ampicillin，氨苄青霉素，**97**
infectious mastitis treatment, 传染性乳房炎的治疗，193
salmonellosis treatment, 沙门氏菌病的治疗，97
anaerobic infections *see Clostridium* infections, 厌氧菌感染参见 梭菌感染
*Anaplasma bovis,* 牛无浆体，**76**
*Anaplasma canis,* 犬无浆体，**76**
*Anaplasma centrale,* 中央无浆体，**76**
*Anaplasma haemolamae* infection, 嗜血支原体感染，**81**–82
*Anaplasma marginale* infection, 边缘无浆体感染，75, *77*
diagnosis, 诊断，**80–81**
OWC, 旧大陆骆驼，75–**76**
*see also* anaplasmosis 参见 无浆体病
*Anaplasma ondrii,* 翁迪里无浆体，*76*
*Anaplasma ovina,* 羊无浆体,*76*
*Anaplasma ovis,* 绵羊无浆体，*76*
*Anaplasma (Ehrlichia) phagocytophilum,* 嗜粒细胞无浆体，*76*
*Anaplasma phagocytophilum* infection, 嗜粒细胞无浆体感染，76, *77*
diagnosis, 诊断，**81**
*see also* anaplasmosis 参见 无浆体病
*Anaplasma (Cowdria) ruminantium,* 反刍兽无浆体，**76**
anaplasmosis, 无浆体病，75–84
canine, 犬，**81**
control, 控制, 83
diagnosis, 诊断，**80**–82, **81**
equine,马，**81**
literature studies,文献综述，80
ovine,羊，**81**
treatment, 治疗，82
*Anopheles,* Rift Valley Fever vectors, 按蚊，裂谷热媒介，275
Anoplocephalidosis，裸头科绦虫病, 406–412
clinical signs,临床症状，409
control, 控制, *410*
diagnosis, 诊断，409
epidemiology, 流行病学，408
infection route,感染途径 408–409
intermediate hosts,中间宿主，409, *409*
morphology, 形态学，406–408
parasite life cycles,寄生虫生活史，409, *409*
pathology,病理学，409
treatment, 治疗，410
anthrax, 炭疽，34–37
aetiology,病原学，34
Bactrian camel, 双峰驼，**476**
clinical signs, 临床症状，34–*36*, *36*
control, 控制,35-36
diagnosis, 诊断，35, *36*
Dromedary, 单峰驼，**475**
epidemiology, 流行病学,34-35
NWC, 新大陆骆驼，**485**
pathology, 病理学，35
prevention,预防，35–36
zoonotic implications,公共卫生意义，34
antibiotics，抗生素
anthrax treatment, 炭疽治疗，35-36
broad-spectrum, 广谱，133
infective reproductive loss treatment,传染性繁殖障碍的治疗，157
onchocercidosis treatment，盘尾丝虫病的治疗, 434
tetanus treatment, 破伤风治疗，202
antibodies *see* immunoglobulins 抗体 参见 免疫球蛋白
antibody detection assay，抗体检测方法，121
brucellosis diagnosis,布鲁氏菌病的诊断，142
mycoplasmosis diagnosis, 支原体病的诊断，***80***
tuberculosis diagnosis,结核病的诊断，121
anticoccidials, *Eimeria* coccidiosis treatment, 抗球虫药物，艾美尔球虫病的治疗，376

antigen-detection enzyme-linked immunosorbent assays, 抗原检测酶联免疫吸附试验
trypanosomosis diagnosis, 锥虫病诊断, 362
antimicrobials, 抗细菌药物
colibacillosis treatment, 大肠杆菌的治疗, 104
salmonellosis treatment, 沙门氏菌病的治疗, 97, **97**
antiserum administration 抗血清疗法
endotoxicosis treatment, 内毒素中毒治疗, 45
polyvalent, botulism treatment, 多价, 肉毒梭菌中毒治疗, 33
anti-tetanus serum, tetanus treatment, 抗破伤风血清, 破伤风治疗, 202
antitoxin, tetanus, 抗毒素, 破伤风, 202
*Aphthovirus* infections *see* foot and mouth disease (FMD), 口蹄疫病毒属 参见 口蹄疫
arbovirus infections, 虫媒病毒感染, 324-327, **324**
arboviruses, 虫媒病毒, 280
*Arcanobacterium haemolyticum,* nasal swabs, 溶血隐秘杆菌, 鼻试子, **133**
arsenious oxide, tick infestation control, 氧化亚砷酸, 蜱的防控, 443
aspergillosis, 曲霉病, 337–338
aetiology, 病原学, 337
clinical findings, 临床症状, 337–338
diagnosis, 诊断, 337
epidemiology, 流行病学, 337
lesions, 病灶, 337–338, *338*
prevention, 预防, 338
treatment, 治疗, 338
*Aspergillus,* 曲霉, 337
nasal swabs, 鼻试子, **133**
*Aspergillus fumigatus* infection, 烟曲霉感染
endotoxicosis, 内毒素中毒, 43–45
rumenitis, 瘤胃炎, 338
*Aspergillus niger* infection, pyogranulomatous pneumonia, 黑曲霉感染, 脓肺炎, 338
aspiration pneumonia, 吸入性肺炎, ***133***
ATP test, infectious mastitis, 三磷酸腺苷(ATP)试验, 传染性乳房炎, 190
attempted yawning, rabies, 试图不断地打哈欠, 狂犬病, *226*, 226–227
Aujeszky disease, 伪狂犬病, **325**, 326
Australia, 澳大利亚, 3
avermectins, mange treatment, 阿维菌素, 螨的治疗, 451
avian aegyptianellosis, diagnosis, 禽埃及小体病的诊断, **81**
*Avitellina,* 无卵黄腺绦虫, 406
infection *see* anoplocephalidosis 感染 参见 裸头科绦虫病
*Avitellina centripunctata,* 中点无卵黄腺绦虫, 406, *407*

**B**
bacillary haemoglobinuria, *Clostridium* infections, 细菌性血红素尿, 梭菌感染, 24
*Bacillus* infections, 杆菌感染, *23*
lungs, 肺脏, **131**
*Bacillus cereus* infection *see* endotoxicosis, 蜡样芽孢杆菌感染 参见 内毒素中毒
*Bacillus cereus* intoxication, 蜡样芽孢杆菌毒素中毒, 38
Bacteria 细菌
aerobic *see* aerobic bacteria, 需氧 参见 需氧菌
endotoxicosis, 内毒素中毒, *41*
infective reproductive losses, 传染性繁殖障碍, **154**, 155, **156**, ***157***
lungs, 肺脏, **131**
bacterial diseases, 细菌病, 19–210, 475–486
Bactrian camel, 双峰驼, 476–485
digestive system, 消化系统, 91–111
Dromedary, 单峰驼, 475–476
endotoxicosis, 内毒素中毒, *40*
integument, 皮肤, 166–185
minor disease, 其他疫病, 208–209
nervous system, 神经系统, 200–210
NWC, 新大陆骆驼, **485–486**
respiratory system, 呼吸系统, *113*–136
systemic diseases, 全身性疾病, 21–89
udder, 乳房, 186–199
urogenital system, 泌尿生殖系统, 137–165
vaccinations, 疫苗接种, **353**
*see also specific infections* 参见具体疫病
*Bacteroides fragilis* infection, 脆弱拟杆菌感染, 208
Bactrian camel *(Camelus bactrianus)*, 双峰驼, ***6***, *7*
bacterial disease, 476–485
distribution, 分布, 6, *7*
domestication, 驯化, 7
embryology, 胚胎, *4*
evolution, 进化, ***4***
fungal disease, 真菌病, **488**
historical aspects, 历史情况, 10–11, *11*
parasitic disease, 寄生虫病, **488**
population, 种群数量, **8**
viral disease, 病毒病, **473**
balanoposthitis, infectious pustular, 传染性龟头包皮炎, **301**
Banzai virus, Banzai病毒, **325**
literature review, 文献综述, **223**
BCT (buffy coat technique), trypanosomosis diagnosis, 血沉棕黄层技术, 锥虫病的诊断, 361
BD *see* Borna disease (BD), BD 参见 博尔纳病
benzalkonium chloride (Zephiran), 氯化苯甲氢铵, 109–110
benzimidazoles, 苯并咪唑类药物

anoplocephalidosis treatment, 裸头科绦虫病, 410
fasciolosis treatment, 肝片吸虫病的治疗, 400
parasitic gastroenteritis treatment, 寄生虫性胃肠炎治疗, **429**
besnoitiosis, 贝诺孢子虫病, 395
*Betaretrovirus*, β 逆转录病毒, 321
biting flies, 咬蝇, 465–469
black disease (infectious necrotising hepatitis), *Clostridium* infections, 羊黑疫(传染性坏死性肝炎), 梭菌感染, 24
blackleg, *Clostridium* infections,黑脚病, 梭菌感染, 22
blood smears 血涂片
mycoplasmosis diagnosis, 支原体病的诊断, 79
trypanosomosis diagnosis, 锥虫病的诊断, 361
bluetongue (BT),蓝舌病, 279–286
aetiology,病原学, 279–280
Bactrian camel, 双峰驼, **473**
clinical signs,临床症状, 283
control, 控制,283
diagnosis, 诊断, 281, 283
differential diagnosis,鉴别诊断, 282
Dromedary, 单峰驼, **478**
epidemiology, 流行病学, ***281***–283
experimental infections, 实验感染, 281–282
literature review, 文献综述, **216**
NWC, 新大陆骆驼, **474**
pathology, 病理学, 282
prevention,预防, 283
serology,血清学, **281**
transmission, 传播, 279
Borna disease (BD), 博尔纳病, 231–234
aetiology,病原学, 231
clinical signs,临床症状, 231, *231*
diagnosis, 诊断, 231
epidemiology, 流行病学, *230*
literature review, 文献综述, **217**
pathology,病理学, 231
prevention,预防, 231
treatment, 治疗, 231
botulism, 肉毒梭菌中毒, 32–34
aetiology, 病原学, 32
clinical signs, 临床症状, 32
diagnosis, 诊断, *33*
epidemiology, 流行病学, **32**–*33*
prevention, 预防, 33
treatment, 治疗, 33
bovine herpesvirus type 1, 牛疱疹病毒1型, 302–303, **303**
distribution, 分布, 300
prevention, 预防, **301**
serology, 血清学, 302-303
transmission, 传播, 300
bovine respiratory syncytial virus (BRSV), 牛呼吸道合胞体病毒, 312
bovine viral diarrhoea (BVD), 牛病毒性腹泻, **266**–274
abortion, 流产, 270
aetiology, 病原学, 267, 269
Bactrian camel, 双峰驼, 473
clinical signs, 临床症状, 269–271
diagnosis, 诊断, 270
Dromedary, 单峰驼, **478**
economic losses, 经济损失, 271
epidemiology, 流行病学, **269–271**
experimental infections,实验感染, 270
literature review,文献综述, **217**
literature reviews, 文献综述, **267–268**
NWC, 新大陆骆驼, **474**
pathology, 病理学, **269–271**
prevention,预防, 268
serological studies, 血清学研究, 270
serology,血清学, 270
susceptibility,易感性, **269**
treatment, 防控, 271
viral co-infections, 病毒混合感染, **270**
bovine viral diarrhoea virus (BVDV), 牛病毒性腹泻, 266
genotypes, 基因型, 266
Bowman's capsule, endotoxicosis, 鲍氏囊, 内毒素中毒, 41, *41*
broad-spectrum antibiotics, pneumonia therapy, 广谱抗生素, 肺炎的治疗, 133
bronchopneumonia, chronic, 慢性支气管肺炎, 131
BRSV (bovine respiratory syncytial virus), 牛呼吸道合胞体病毒, 312
*Brucella*, 布鲁氏菌, **137**
*Brucella abortus*, 流产布鲁氏菌, **137**
*Brucella canis*, 犬布鲁氏菌, **137**
*Brucella melitensis*, 马尔他布鲁氏菌, **137**
*Brucella neotomae*, 沙林鼠布鲁氏菌, **137**
*Brucella ovis*, 绵羊布鲁氏菌, **137**
*Brucella suis*, 猪布鲁氏菌, **137**
brucellosis, 布鲁氏菌, 137–151
aetiology, 病原学, 137
Bactrian camel, 双峰驼, **476**
breeding stock, 育种群, 141
*Brucella* isolated species, 分离到的布鲁氏菌, **139**
clinical signs, 临床症状, 137–139, 142–143
control, 控制, 145
diagnosis, 诊断, 137, 139, 143–145, **144**, *144*
Dromedary, 单峰驼, **475**
epidemiology, 流行病学, 137–139, 141–142

infectious routes, 感染途径，138
literature summary, 文献摘要，**138–141**
NWC, 新大陆骆驼，**485**
pathology,病理学，142
treatment, 治疗，145
zoonotic implications, 公共卫生意义，137–138, **138**
BT *see* bluetongue (BT), BT, 参见 蓝舌病
bubonic plague, 腺鼠疫，65
buffy coat technique (BCT), trypanosomosis diagnosis, 血沉棕黄层技术，锥虫病的诊断，361
*Bunostomum trigonophorum,* 416
*Bunyaviridae,* 布尼亚病毒科，**280**
literature review, 文献综述，**224**
Bunyavirus, 布尼亚病毒，**280**
*Burkholderia mallei* infection *see* glanders，鼻疽伯氏菌感染 参见 鼻疽
*Burkholderia pseudomallei* infection *see* melioidosis，类鼻疽伯氏菌感染 参见 类鼻疽
BVD *see* bovine viral diarrhoea (BVD)，牛病毒性腹泻(BVD)
BVDV *see* bovine viral diarrhoea virus (BVDV)，牛病毒性腹泻病毒(BVDV)

**C**

California mastitis test (CMT)，乳房炎加利福尼亚体细胞检测法
infectious mastitis, 传染性乳房炎，190
udder, 乳房，186
*Camelostrongylus mentulatus,* 阴茎骆驼圆线虫，**420**
infection *see* parasitic gastroenteritis/colitis，感染 参见 寄生虫性胃肠炎/结肠炎
morphology,形态学，424，*424*
camelpox, 骆驼痘，235–*245*
aetiology,病原学，235
Bactrian camel, 双峰驼，**473**
clinical signs,临床症状，237–239, ***238***, *239*
control, 控制，239–240
diagnosis, 诊断，237
Dromedary, 单峰驼，**478**
epidemiology, 流行病学，235–*236*
experimental infection,实验感染，237, ***238***
literature reviews,文献综述，*236*
morbidity,发病率，237, **238**
mortality, 死亡率，237, **238**
NWC, 新大陆骆驼，474
pathology,病理学，237–239
serology, 血清学，235–237
treatment, 治疗，239–240
vectors,媒介，235
zoonotic implications,公共卫生意义，235
camel tick *see Hyalomma dromedarii*，骆驼蜱 参见 嗜驼璃眼蜱
*Camelus bactrianus see* Bactrian camel *(Camelus bactrianus)*，双峰驼
*Camelus dromedarius see* dromedary *(Camelus dromedarius)*，单峰驼
*Campylobacter fetus* infection, infective reproductive losses, 胎儿弯曲菌，153
*Candida albicans* infection, mycotic dermatitis, 白色念珠菌感染，霉菌性皮炎，**332**
Candidosis，念珠菌病，339–341
aetiology, 病原学，339
clinical findings, 临床症状，339, *340*, 341, *341*
diagnosis, 诊断，341
epidemiology, 流行病学，339
lesions, 病灶，339, *340, 341*
prevention, 预防，341
treatment, 治疗，341
canine anaplasmosis,犬无浆体病，**81**
*Capillaria,* 毛细线虫，**419**
life cycle, 生活史，428
captan, mycotic dermatitis treatment, 克菌丹，霉菌性皮炎治疗，333
card agglutination test for trypanosomosis (CATT), 锥虫病卡片凝集试验，361–363
*Carmyerious spatiosus,* 长大卡妙吸虫，405
caseous lymphadenitis (CLA), 干酪性淋巴腺炎，166–176
aetiology, 病原学，166
Bactrian camel, 双峰驼，**485**
causative agents, 病原，168
clinical signs, 临床症状，169, *170*
control, 控制，170–171
diagnosis, 诊断，172
Dromedary, 单峰驼，**475**
epidemiology, 流行病学，167–169
experimental infections, 实验感染，168–169
incidence,发病，167
infection route, 感染途径，167
lymphadenitis, 淋巴腺炎，168
NWC, 新大陆骆驼，**486**
oral cavity mucous membranes, 口腔黏膜，168
pathology, 病理学，169
serology,血清学，172
treatment, 治疗，170–171
caseous necrosis, *Rhodococcus equi* infection, 马红球菌感染，*49*
catarrhal endometritis, 卡他性子宫内膜炎，**157**
chronic, 慢性，**158**

catarrhal pneumonia, 卡他性肺炎, **128**
CATT (card agglutination test for trypanosomosis), 锥虫病卡片凝集试验, 361 - 363
CCHF *see* Crimean–Congo haemorrhagic fever (CCHF), 克里米亚-刚果出血热(CCHF)
ceftiofur, salmonellosis treatment, 头孢噻呋, 沙门氏菌病的治疗, **97**
cELISA *see* competitive enzyme-linked immunosorbent assay (cELISA), 参见 竞争性ELISA
cell culture bioassays, aspergillosis, 细胞培养生物分析法, 曲霉病, 337
Central European tick-borne encephalitis, 中欧蜱媒脑炎, **280**
Central Veterinary Research Laboratory (CVRL), 中央兽医研究实验室, 187
*Cephalopina* infestation, 狂蝇感染
  Bactrian camel, 双峰驼, **488**
  Dromedary, 单峰驼, **487**
*Cephalopina titllator,* 骆驼喉蝇, *459, 460*
cephalosporins, caseous lymphadenitis treatment, 头孢菌素, 干酪性淋巴腺炎治疗, 171
*Cepheus,* 鲜甲螨, 409
Ceratopogonidae, 蠓科, 468
cestode infections, 绦虫感染, 406–418
  *see also specific diseases/disorders* 参见具体的疫病
CFT *see* complement fixation test (CFT), 补体结合试验
*Chabertia ovina,* 绵羊夏伯特线虫, 420
  infection *see* parasitic gastroenteritis/colitis, 感染 参见 寄生虫性胃肠炎/结肠炎
  morphology, 形态学, 423–426
Chlamydiaceae, 衣原体科, **163**
chlamydiosis, 衣原体病, 163–165
  Bactrian camel, 双峰驼, **485**
  Dromedary, 单峰驼, **476**
  keratoconjunctivitis, 角膜结膜炎, *164*
chloramphenicol, 氯霉素
  anaplasmosis/mycoplasmosis treatment, 无浆体病/支原体病的治疗, 82
  chlamydiosis treatment, 衣原体病治疗, 164
cholangitis, fasciolosis, 胆管炎, 肝片吸虫病, 399
*Chorioptes*, 足螨
  infection *see* mange, 感染 参见 螨
  life cycle, 生活史, 448
  morphology, 形态学, 446, *447*
chronic active metritis, 慢性活动性子宫炎, **158**
chronic bronchopneumonia, 慢性支气管肺炎, 131
chronic catarrhal endometritis, 慢性卡他性子宫内膜炎, 157
chronic enteritis, 慢性肠炎, 96
chronic haematuria, leptospirosis, 血尿症, 钩端螺旋体病, *70,* 69–70
chronic non-suppurative metritis, 慢性非化脓性子宫炎, **158**
chronic non-suppurative pneumonia, 慢性非脓性肺炎, **129**
chronic proliferative pneumonia, 慢性增生性肺炎, **130**
*Chrysomya albiceps,* 白头裸金蝇, *457*
*Chrysomya bezziana,* 蛆症金蝇, *456, 456–457, 457*
*Chrysomya chloropyga,* 绿尾金蝇, 457
*Chrysomya megacephala,* 大头金蝇, 457
*Chrysomya rufifacies,* 绯颜裸金蝇, 457
*Chrysomya variceps,* 赭金蝇, 457
*Chrysosporium* infection, mycotic dermatitis, 金孢子菌感染, 霉菌性皮炎, **332**
CLA *see* caseous lymphadenitis (CLA), 干酪性淋巴腺炎 (CLA)
classification of camelids, 骆驼的分类, 5
clorsulon, fasciolosis treatment, 克洛索隆, 肝片吸虫病的治疗, 400
closantel, fasciolosis treatment, 氯氰碘硫胺, 肝片吸虫病的治疗, 400
*Clostridium* infections, 梭菌性疾病, **21**, 21–32, *23*
  aetiology, 病原学, 22
  bacillary haemoglobinuria, 细菌性血红素尿, 24
  Bactrian camel, 双峰驼, 476
  blackleg, 黑脚病, 22
  clinical signs, 临床症状, 22
  control, 控制, 28–30
  Dromedary, 单峰驼, 475
  endotoxaemia *see Clostridium* enterotoxaemia 肠毒血症 参见 梭菌肠毒血症
  epidemiology, 流行病学, 22
  gas oedema complex, 气性水肿综合征, 22
  ictero haemoglobinuria, 黄疸血红素尿, 24
  infectious necrotising hepatitis (black disease), 传染性坏死性肝炎(羊黑疫), 24
  malignant oedema, 恶性水肿, 22
  treatment, 治疗, 27–29
  zoonotic implications, 公共卫生意义, **21**
*Clostridium botulinum,* 肉毒梭菌, **21**
  infection *see* botulism, 感染 参见 肉毒梭菌中毒
  toxins, 毒素, **33**
*Clostridium chauvoei,* 气肿疽梭菌, **21**
  blackleg, 黑脚病, 22
*Clostridium* enterotoxaemia, 梭菌肠性肠毒血症, 24–27
  clinical features, 临床特征, *27*
  control, 控制, 30–31
  diagnosis, 诊断, *25*, 26–29
  kidneys, 肾脏, 25, *25*

lung haemorrhages, 肺脏出血, *25*
motality, 死亡率, 30
nutrition,营养, **26**
NWCs, 新大陆骆驼, 24
OWCs, 旧大陆骆驼, 24–*27*
petechial haemorrhages, 瘀斑出血, *25*
serology, 血清学, 26
treatment, 治疗, 28–30
trypanosomosis, 锥虫病, 25
*Clostridium equi* infection, caseous lymphadenitis, 马棒状杆菌, 干酪性淋巴腺炎, 167
*Clostridium haemolyticus,* 溶血梭菌,**21**
*Clostridium novyi,* 诺维氏芽孢梭菌,**21**
*Clostridium perfringens,* 产气荚膜梭菌,**21**
enterotoxaemia, 肠毒血症, 24
*Clostridium psudotuberculosis,* integument, 假结核棒状杆菌, 皮肤, **178**
*Clostridium renale* infection, caseous lymphadenitis, 牛肾盂炎棒杆菌, 干酪性淋巴腺炎, **167**
*Clostridium septicum,* 腐败梭菌,**21**
*Clostridium sordelli,* 索氏梭菌, **21**
*Clostridium tetani,* 破伤风梭菌,**21**
infection *see* tetanus 感染 参见 破伤风
*Clostridium ulcerans* infection, caseous lymphadenitis, 溃疡棒状杆菌, 干酪性淋巴腺炎, 168
cloxacillin, infectious mastitis treatment, 邻氯青霉素, 传染性乳房炎的治疗, 193
CMT *see* California mastitis test (CMT), 乳房炎加利福尼亚体细胞检测法
coagulase-negative staphylococci, 凝固酶阴性葡萄球菌, **133**
coccidiomycosis, 球孢子菌病, *342,* 342
NWC, 新大陆骆驼, **489**
Coccidiosis, 球虫病
Bactrian camel, 双峰驼, **488**
Dromedary, 单峰驼 , **487**
*Eimeria* infection *see Eimeria* coccidiosis, 艾美尔球虫感染 参见 艾美尔球虫病
NWC, 新大陆骆驼, **489**
*Cochliomyia hominivorax,* 嗜人锥蝇, *457–458*
coenurosis, 脑多头蚴病, 417–418
colibacillosis, 大肠杆菌病, 100–105
aetiology, 病原学, 100–101
Bactrian camel, 双峰驼, **476**
clinical signs,临床症状, 100, 101–102, *102, 103*
control, 控制, 103–104
diagnosis, 诊断, 103
Dromedary, 单峰驼, **475**
epidemiology, 流行病学, 101–*102*
NWC, 新大陆骆驼, **486**
pathology,病理学, 102
treatment, 治疗, 103–104
zoonotic implications, 公共卫生意义, 101
colisepticaemia, candidosis, 大肠杆菌性败血症, 念珠菌病, 339
colistin, salmonellosis treatment, 黏菌素, 沙门氏菌病的治疗, **97**
colitis, parasitic *see* parasitic gastroenteritis/colitis, 寄生性结肠炎 参见 寄生性结胃肠炎/结肠炎
colon mucosa, paratuberculosis, 结肠黏膜, 副结核病, *108*
colostrum, colibacillosis treatment, 初乳, 大肠杆菌病的治疗, 103–104
competitive enzyme-linked immunosorbent assay (cELISA), 竞争性酶联免疫吸附试验
*Anaplasma phagocytophilum* infection, 嗜粒细胞无浆体, 76
brucellosis diagnosis, 布鲁氏菌病诊断, 144
glanders diagnosis, 鼻疽诊断, 52-53
Q fever diagnosis, Q热的诊断, 86
rinderpest diagnosis, 牛瘟诊断, 316
West Nile virus diagnosis, 西尼罗河病毒, 287
complement fixation test (CFT), 补体结合试验
African horse sickness diagnosis, 非洲马瘟诊断, 319
brucellosis diagnosis, 布鲁氏菌病诊断, 142–143
glanders diagnosis,鼻疽诊断, 52
melioidosis diagnosis, 类鼻疽诊断, 55–56
toxoplasmosis diagnosis,弓形虫诊断, 384
Congo haemorrhagic virus, 文献综述, **224**
contagious erythema, 接触传染性脓疱, 246–251
aetiology,病原学, *246*
Bactrian camel, 双峰驼, 473
clinical signs, 临床症状, 247, *247–248,*
control, 控制, 248-249
diagnosis, 诊断, 248
differential diagnosis, 鉴别诊断, 248
Dromedary, 单峰驼, **478**
epidemiology, 流行病学, 246–248
morbidity, 发病率, **247**, 248
mortality, 死亡率, **247**
NWC, 新大陆骆驼, **474**
pathology, 病理学, 248
transmission, 传播, 246
treatment, 治疗, 248
zoonotic implications, 公共卫生意义, 245, *245*
contagious pustular dermatitis *see* contagious erythema, 传染性化脓性皮炎 参见 接触传染性脓疱
*Cooperia*, 古柏线虫
infection *see* parasitic gastroenteritis/colitis, 感染

参见 寄生虫性胃肠炎/结肠炎，
morphology,形态学，423
*Cooperia oncophora,* 点状古柏线虫，**420**
coronavirus, literature review, 冠状病毒，文献综述，**220**
coronavirus neonatal diarrhoea, 冠状病毒新生畜腹泻，289, 291
diagnosis, 诊断，290
epidemiology, 流行病学，288
*Corynebacterium* infection，棒状杆菌感染
infectious mastitis, 传染性乳房炎，188
lungs, 肺脏，**131**
*Corynebacterium pseudotuberculosis* infection *see* caseous lymphadenitis (CLA)，假结核棒状杆菌感染 参见 干酪性淋巴腺炎
*Corynebacterium pyogenes* infection, infectious mastitis, 化脓棒状杆菌，传染性乳房炎，192
counter electrophoresis, peste des petits ruminants diagnosis, 对流免疫电泳，小反刍兽疫诊断， 263
*Coxiella burnetii* infection *see* Q fever，贝氏柯克斯体 参见 Q热
Crimean–Congo haemorrhagic fever (CCHF) 克里米亚-刚果出血热, 324, 324-325
Bactrian camel, 双峰驼， **479**
Dromedary, 单峰驼，**478**
tick infestations, 蜱感染，436
crossbreeding, 杂交，10, *10*
cryptococcosis, 隐球菌病，**344**
*Cryptococcus neoformans* infection, mycotic dermatitis, 新型隐球菌感染，霉菌性皮炎，331
cryptosporidiosis, 隐孢子虫病，377–380
clinical signs,临床症状，*378,* 378–379
diagnosis, 诊断， 379
epidemiology, 流行病学，378
life cycle,生活史，378
morphology, 形态学，377-378
pathology,病理学，378–379
treatment, 治疗，379
*Cryptosporidium* infection, neonatal diarrhoea, 隐孢子虫感染，新生畜腹泻，289
*Culex,* Rift Valley Fever vectors, 库蚊，裂谷热媒介，274
*Culicoides* 库蠓
African horse sickness transmission, 非洲马瘟传播，317
bluetongue vector,蓝舌病媒介，278
culture 培养
brucellosis diagnosis, 布鲁氏菌病的诊断，143
candidosis diagnosis, 念珠菌病的诊断，*341, 341*
*Clostridium* enterotoxaemia diagnosis, 梭菌肠毒血症的诊断，26，*26*
dermatophilosis diagnosis, 嗜皮菌病的诊断，*180, 182–183*
infective reproductive losses,传染性繁殖障碍，151
paratuberculosis diagnosis, 副结核病的诊断，107-110
*Rhodococcus equi* infection diagnosis, 马红球菌感染诊断，*49*
tuberculosis diagnosis,结核病诊断，121
cutaneous alternariosis, 皮肤链格孢病，**346**
CVRL (Central Veterinary Research Laboratory), 中央兽医研究实验室，187
cypermetrin, pediculosis treatment, 氯氰菊酯，虱病治疗，464
cysticercosis, 囊尾蚴病，417–418
*Cysticercus bovis,* 牛囊尾蚴，417
*Cysticercus tenuicollis,* 细颈囊尾蚴，*417–418, 417*

**D**

Dalmeny disease, 达尔梅尼病，391–392
*Deltaretrovirus,* δ逆转录病毒，320
*Demodex* infection, 蠕形螨感染，454–455, *454*
demodicosis, 蠕形螨病，454–455, *454*
denaturing gradient gel electrophoresis (DGGE), 变性梯度凝胶电泳
mycoplasmosis diagnosis, 支原体病的诊断，80
dengue virus, 登革热病毒，**325**
literature review, 文献综述，**223**
dental infections,牙齿感染，208-209
*Deraiophoronema evansi,* 埃氏双瓣线虫，**420**
*Dermacentor,* 革蜱，*437*
*Dermacentor,* frequency of infection, 革蜱，感染频度，**442**
*Dermacentor andersoni,* epidemiology, 安氏革蜱，流行病学，442
*Dermacentor nuttalli,* 草原革蜱，*437*
life cycle, 生活史，437-440
dermatitis, mycotic *see* mycotic dermatitis，霉菌性皮炎
dermatophilosis, 嗜皮菌病，179–185
aetiology, 病原学 *180*
clinical signs, 临床症状，181–182, *182*
control, 控制, 183
culture, 培养，*180*
diagnosis, 诊断， *182–183*
Dromedary, 单峰驼，**475**
epidemiology, 流行病学，*180*
pathology, 病理学，181–182
prevalence, 流行，179
treatment, 治疗, 183
vector, 媒介，180
zoonotic implications, 公共卫生意义，179

*Dermatophilus chelonae see* dermatophilosis，海龟嗜皮菌 参见 癣菌病
*Dermatophilus congolensis*，刚果嗜皮菌
infection *see* dermatophilosis，感染 参见 癣菌病
integument，皮肤，**178**
dermatophytosis *see* ringworm (dermatophytosis)，癣菌病
DGGE (denaturing gradient gel electrophoresis), mycoplasmosis diagnosis，变性梯度凝胶电泳，支原体病的诊断，78
Dhori virus，托里病毒，325
literature review，文献综述，**224**
diarrhoea, neonatal *see* neonatal diarrhoea，新生畜腹泻
DIC (disseminated intravascular coagulation), endotoxicosis,弥散性血管内凝血，内毒素中毒，37
dicrocoeliosis，歧腔吸虫病，401–406
clinical signs，临床症状，403
control，控制，403
epidemiology，流行病学，402–403
intermediate hosts，中间宿主，*402*
pathology，病理学，403
treatment，治疗，403
*Dicrocoelium*，歧腔吸虫
infection *see* dicrocoeliosis 感染 参见 歧腔吸虫病
life cycle，生活史，402, *402*
morphology，形态学，401, *401*
*Dicrocoelium chinesis,* 中华歧腔吸虫，*401*
*Dicrocoelium dendriticum,* 枝歧腔吸虫，,*401*
*Dicrocoelium hospes,* 牛歧腔吸虫，*401*
dictyocaulosis，网尾线虫病，430–433
clinical signs，临床症状，431
control，控制，432
diagnosis，诊断，431
epidemiology，流行病学，431
parasite life cycle，寄生虫生活史，431
parasite morphology，寄生虫形态学，430-431
pathology，病理学，431
treatment，治疗，432
*Dictyocaulus filaria,* 丝状尾网线虫，**420**
diffuse proliferative enteritis, paratuberculosis，弥漫性增生性肠炎，副结核病，*109*
digestive system, bacterial disease，消化系统，细菌病，91–112
dimethyl sulfoxide (DMSO), endotoxicosis treatment，二甲基亚砜，内毒素中毒治疗，45
diminazene aceturate, trypanosomosis treatment，重氮氨苯脒乙酰甘氨酸盐，锥虫病治疗，**363**
diphtheroids, nasal swabs，类白喉菌 鼻试子，133
direct immunofluorescence assay，直接荧光试验
cryptosporidiosis diagnosis，隐孢子虫病诊断，379
giardiosis diagnosis，贾地鞭毛虫病诊断，368
Directogen FLU-A test, influenza diagnosis，检测马流感的FLU-A检测试剂盒，306，*306*
disinfection, infectious mastitis treatment，消毒，传染性乳房炎的治疗，192，*192*
disseminated intravascular coagulation (DIC), endotoxicosis,弥散性血管内凝血，内毒素中毒，43
DMSO (dimethyl sulfoxide), endotoxicosis treatment，二甲基亚砜，内毒素中毒治疗，45
domestication，驯化，5–6, 11–12
doramectin，多拉菌素
mange treatment，螨的治疗，451
parasitic gastroenteritis treatment，寄生虫性胃肠炎治疗，**429**
*Dorcardia ioffi,* 羊长喙蚤，469
Dromedary *(Camelus dromedarius),* 单峰驼，*6, 7*
bacterial diseases,细菌病，**482–486**
bacterial integument disease,细菌性皮肤病，**181**
distribution,分布，6, *7*
domestication，驯化，*11*
embryology，胚胎学，*4*
evolution，进化，4
fertility rate，生育率，152
fungal disease,真菌病，487
historical aspects，历史情况，*12*
parasitic disease，寄生虫病，**487**
physiological characteristics，生理特征，**15**
population，种群数量，**8**
uses of，使用，13
viral disease，病毒病，**478**

**E**

ear tip necrosis, salmonellosis，耳尖坏死 沙门氏菌病，93, **93**
Eastern equine encephalitis，东部马脑炎，287
literature review，文献综述，**217**
EAV *see* equine arteritis virus (EAV)，马动脉炎病毒
EBL *see* enzootic bovine leukosis (EBL)，地方流行性牛白血病
ecchymosis, endotoxicosis，皮下血肿，内毒素中毒，*39*
echinococcosis，棘球蚴
Bactrian camel，双峰驼，**488**
Dromedary，单峰驼，**487**
NWC，新大陆骆驼，**489**
*Echinococcus* infection *see* echinococcosis; hydatidosis，棘球蚴感染 参见 棘球蚴病，包虫病
ecthyma contagiosum, literature review，接触传染性脓疱，文献综述，**219**
eczema, exudative，渗出性湿疹，176

*Ehrlichia bovis see Anaplasma bovis*, 牛埃里克体 参见 牛无浆体
EHV-1 (equine herpesvirus-1), 马疱疹病毒1型，301-302
*Eimeria* 艾美尔球虫
life cycle,生活史，372–374, ***373, 373***
morphology,形态学，371, *372*
oocysts, 卵囊，*373*
*Eimeria alpacae,* 羊驼艾美尔球虫，372, *372*,***373*** ，374
*Eimeria bactriani,* 双峰驼艾美尔球虫，**371**
*Eimeria cameli,* 骆驼艾美尔球虫，**371**, *372, 374*
*Eimeria* coccidiosis, 艾美尔球虫病，371–377
clinical signs,临床症状，*375, 375*
control, 控制, 376
diagnosis, 诊断， 372, 375
epidemiology, 流行病学，374–375
morphology, 形态学，371, *372*
oral infection, 经口感染，374–375
pathology, 病理学，375
treatment, 治疗, 376
*Eimeria dromedarii,* 单峰驼艾美尔球虫，**371**, *374*
*Eimeria ivitaensis,* 伊维塔艾美尔球虫，**371**, *372, 374*
*Eimeria lamae,* 无峰驼艾美尔球虫，**371**, *372, 374*
*Eimeria macusaniensis,* 曼库塞尼艾美尔球虫，**371**, *372, 374*
*Eimeria pellerdyi,* 派氏艾美尔球虫，**371**
*Eimeria peruviana,* 秘鲁艾美尔球虫，**371**
*Eimeria punoensis,* 普诺艾美尔球虫，**371**, *372, 374*
*Eimeria rajasthani,* 拉氏艾美尔球虫，**371**, *374*
ELISA *see* enzyme-linked immunosorbent assay (ELISA), ELISA，参见 酶联免疫吸附试验
embryology, 胚胎学，4, *4*
EMCV (encephalomyocarditis virus), 脑心肌炎病毒，258
encephalitis, Western equine, 西方马脑炎马，**279**
encephalomyelitis，马脑炎
equine *see* equine encephalomyelitis，马的 参见 马脑炎
literature review, 文献综述，**223**
encephalomyocarditis virus (EMCV), 脑心肌炎病毒，258
endometrial granulomas, infective reproductive losses, 子宫内膜肉芽肿，传染性繁殖障碍，153,*153*
Endometritis，子宫内膜炎
with abscessation, 脓肿，**158**
acute suppurative, 急性化脓性，**158**
catarrhal, 卡他性, **157**
chronic catarrhal, 慢性卡他性，**158**
haemorrhagic, 出血性，**158**
infective reproductive losses, 传染性繁殖障碍，**153**
necrotic, 坏死，**158**
treatment, 治疗, **157–160**
endotoxaemia *see* endotoxicosis，肠毒血症 参见 内毒素中毒
endotoxicosis, 内毒素中毒，37–47
aetiology, 病原学，37
aspergillosis, 曲霉病，337
*Aspergillus fumigatus,* 烟曲霉菌，44–45
clinical pathology,临床病理学，*41,* 41–44
clinical signs,临床症状，38–42, *40, 41, 42*
compartment 1 properties, 第一胃特征，41–42, **42**
control, 控制, 46–47
disseminated intravascular coagulation, 弥散性血管内凝血，**42**
mortality, 死亡率，*39*
pathology, 病理变化，38–42,*43*
serology,血清学，**39**, **42**，42–43
treatment, 治疗, 45–46
Endovac-Bovi, endotoxicosis treatment, 内毒素疫苗，内毒素治疗，46
Enteritis，肠炎
chronic, 肠炎慢性，96
diffuse proliferative, 弥漫性增生性肠炎，*109*
haemorrhagic, 出血，*375*
*Enterobacter*，肠杆菌
lungs, 肺脏，**131**
nasal swabs, 鼻试子，132
enteropathogenic *Escherichia coli* (EPEC), 肠致病性大肠杆菌，100
enterotoxaemia，肠毒血症
*Escherichia coli* infections, 大肠杆菌感染，102
NWC, 新大陆骆驼, **485**
Enterotoxaemia Antigen Test, 肠毒血症抗原检测，28
enterotoxaemia complex *see Clostridium* enterotoxaemia，肠毒血症综合征 参见 梭菌肠性肠毒血症
enterotoxigenic *Escherichia coli* (ETEC), 肠产毒素性大肠杆菌，100
neonatal diarrhoea, 新生畜腹泻，289
enzootic bovine leukosis (EBL), 地方性牛白血病，321–323
clinical features, 临床症状，*321*
epidemiology, 流行病学，320
pathogenesis, 致病机理，322
serology, 血清学，322
enzyme-linked immunosorbent assay (ELISA)，酶联免疫吸附试验
African horse sickness diagnosis, 非洲马瘟诊断，318
bluetongue diagnosis,蓝舌病的诊断，282
botulism diagnosis, 肉毒梭菌中毒诊断，33
BRSV, 牛呼吸道合胞体病毒，310
brucellosis diagnosis, 布鲁氏菌病，137
camelpox diagnosis, 骆驼痘诊断，236

caseous lymphadenitis, 干酪性淋巴腺炎，172
*Clostridium* enterotoxaemia diagnosis, 梭菌肠毒血症的诊断，28, *29*
competitive *see* competitive enzyme-linked immunosorbent assay (cELISA) 竞争性 参见 竞争性酶联免疫吸附试验
contagious erythema diagnosis, 接触传染性脓疱，247
dermatophilosis diagnosis, 嗜皮菌病的诊断，*182*
equine herpesvirus-1 infection diagnosis, 马疱疹病毒1型感染诊断，301
influenza diagnosis,流感诊断，306
melioidosis diagnosis, 类鼻疽诊断，55–56
neosporosis diagnosis, 新孢子虫病的诊断，388
parainfluenza virus 3 diagnosis, 3型副流感病毒诊断，310-311
paratuberculosis diagnosis, 副结核病的诊断，107-110
passive immunoglobulin transfer, 免疫球蛋白的被动转移 295–296
peste des petits ruminants diagnosis, 小反刍兽疫诊断，263-264, *264*
Q fever diagnosis, Q热的诊断，86-87
rabies diagnosis, 狂犬病诊断，228
Rift Valley Fever diagnosis, 裂谷热诊断，275, 276
rotavirus neonatal diarrhoea diagnosis, 轮状病毒新生畜腹泻，290，*291*
trypanosomosis diagnosis, 锥虫病的诊断，361
tuberculosis diagnosis, 结核病诊断，118
EPEC (enteropathogenic *Escherichia coli*), 肠致病性大肠杆菌，100
ephemeral fever, 流行热，**279**
epizootic haemorrhagic diseases, 流行性出血热，**279**
eprinomectin, parasitic gastroenteritis treatment, 依普菌素，寄生虫性胃肠炎治疗，**429**
equine anaplasmosis diagnosis,马无浆体病的诊断，**81**
equine arteritis virus (EAV), 马动脉炎炎病毒,**324**，**325**
literature review, 文献综述，**223**
equine encephalomyelitis,马脑炎，**279**
NWC, 新大陆骆驼，**480**
equine herpesvirus-1 (EHV-1), 马疱疹病毒1型，300-301
equine piroplasmosis, tick infestations,马梨形虫，蜱侵袭，443
equine rhinitis A virus，马鼻炎病毒甲型，155
infective reproductive losses, 传染性繁殖障碍，155
literature review, 文献综述，**219**
erythema, contagious *see* contagious erythema，接触传染性脓疱，478
*Escherichia* infections, lungs, 埃希氏菌感染，肺脏，**130**
*Escherichia coli,* integument, 大肠埃希氏菌，**178**
*Escherichia coli* infection，大肠埃希氏菌感染，353
caseous lymphadenitis, 干酪性淋巴腺炎，168
enterotoxaemia, 肠毒血症，102
infective reproductive losses, 传染性繁殖障碍，151
nasal swabs, 鼻试子，**133**
*see also* colibacillosis 参见 大肠杆菌病
ETEC *see* enterotoxigenic *Escherichia coli* (ETEC)，产肠毒素大肠杆菌，101
*Eurytrema pancreatum,* 胰阔盘吸虫，404, *404*
evolution, 进化，4–5
extinction threats, 灭绝危险，3
exudative eczema, *Staphylococcus aureus* dermatitis, 渗出性湿疹，金黄色葡萄球菌皮炎，176
eye diseases,眼病，209

## F

faecal isolation, salmonellosis diagnosis, 粪便中细菌分离，沙门氏菌病的诊断，96
faecal sampling/smears，粪样品/涂片
cryptosporidiosis diagnosis, 隐孢子虫病诊断，379, *379*
*Eimeria* coccidiosis diagnosis, 艾美尔球虫病的诊断，375
hydatidosis diagnosis, 包虫病的诊断，415
isosporosis diagnosis, 等孢球虫病，380
parasitic gastroenteritis diagnosis,寄生虫性胃肠炎的诊断，427
salmonellosis diagnosis, 沙门氏菌病的诊断，*96*
*Fasciola,* 片形吸虫，397
infection *see* fasciolosis，感染 参见 肝片吸虫病
intermediate host, 中间宿主，*398*
life cycle, 生活史，397–398, *398*
morphology, 形态学，397，*397*
*Fascioloides magna*，大拟片形吸虫，397
infection *see* fasciolosis, 感染 参见 肝片吸虫病
life cycle, 生活史，397–398
morphology,形态学，397, *397*
fasciolosis,肝片吸虫病，397–401
clinical signs, 临床症状，*399, 399*
control, 控制, 400
diagnosis, 诊断，399–400
epidemiology, 流行病学，*398–399*
intermediate hosts,中间宿主，*398, 398–399*
pathology,病理学，*399*
treatment, 治疗，400
FAT (fluorescence antibody test), parainfluenza virus 3，荧光抗体试验，3型副流感病毒，310-311
diagnosis, 诊断，310-311
fatal ingestion, 食入致命杂物，*211, 211*
fatty liver degeneration, endotoxicosis, 脂肪变性，内毒素中毒，*41*

febantel, parasitic gastroenteritis treatment, 芬苯达唑，寄生虫性胃肠炎治疗，**429**
feeding experiments, sarcocystosis diagnosis, 饲喂实验，肉孢子虫病的诊断，393
feeds, mycotoxin testing,饲料，真菌毒素检测，344, **344**
feline infectious anaemia, diagnosis, 猫传染性贫血，**81**
fenbendazole，苯硫咪唑，406
anoplocephalidosis treatment, 裸头科绦虫病，406
giardosis treatment, 贾地鞭毛虫病治疗，368
fertility rate, 生育率，152
fetal losses, infective reproductive losses, 胎儿夭折，传染性繁殖障碍，**160**
fever 热
ephemeral, 暂时( 流行 )，**279**
Rift Valley Fever, 裂谷热，275
tick-borne, 蜱传播，**81**
undulant, 波状热，137–138
*Flaviviridae,* 黄病毒科，**279**
*Flavivirus,* 黄病毒，**279**
literature review, 文献综述，**224**
florphenicol, salmonellosis treatment, 氟苯尼考，沙门氏菌病的治疗，**97**
fluid replacement therapy，输液补液疗法，45
endotoxicosis treatment,内毒素中毒治疗，45
neonatal diarrhoea treatment, 新生畜腹泻的防控，295
flumethrin, tick infestation control, 氟氯苯氰菊酯，蜱的防控，443
fluorescence antibody test (FAT), parainfluenza virus 3diagnosis, 荧光抗体试验，3型副流感病毒诊断，310-311
fluoroquinolones, salmonellosis treatment, 氟喹诺酮类，沙门氏菌病的治疗，**97**
FMD *see* foot and mouth disease (FMD)，FMD 参见 口蹄疫
FMDV (foot and mouth disease virus)，口蹄疫，**219–220**
focal interstitial pneumonia, paratuberculosis, 间质性肺炎，副结核病，107
foot and mouth disease (FMD),口蹄疫，253-258
aetiology,病原学，255
Bactrian camel, 双峰驼，254-256, **256**, *257,* 258，**479**
diagnosis, 诊断，257
Dromedary, 单峰驼，255-258 *257,* 258
epidemiology, 流行病学，253
experimental infections, 实验感染，253-254
NMC,新大陆骆驼，253
prevention,预防，257
serology, 血清学，254, 254-257
treatment, 防控，257
foot and mouth disease virus (FMDV)，口蹄疫，**219–220**
foot lesions, 蹄感染，209
foreign bodies, rabies, 异物，狂犬病，227, *227*
formosulfathiazole, *Eimeria* coccidiosis treatment, 甲醛磺胺噻唑，艾美尔球虫病的治疗，376
fungal disease, 真菌病，329-349，**487–489**
Bactrian camel, 双峰驼，**488**
Dromedary, 单峰驼，**487**
NWC, 新大陆骆驼，**489**
opportunistic infections, 条件性感染，319
vaccinations, 疫苗接种，**354**
*see also specific diseases/disorders*，参见具体的疫病
*Fusobacterium necrophorum* infection, 坏死梭杆菌感染，208–209

**G**

*Galumna,* anoplocephalidosis intermediate hosts, 大翼甲螨，裸头科绦虫的中间宿主，409
gas oedema complex, *Clostridium* infections, 气性水肿综合征，梭菌感染，22
gastroenteritis, parasitic *see* parasitic gastroenteritis/colitis，感染 参见 寄生虫性胃肠炎/结肠炎
gastrointestinal parasitosis，胃肠炎，488
Bactrian camel, 双峰驼，**488**
Dromedary, 单峰驼，**487**
*Gastrothylax crumnifer,* 荷包腹袋吸虫，405
genetics, 遗传学，10
*Geodermatophilus obscurus see* dermatophilosis，阴暗地嗜皮菌参见 嗜皮菌病
giardiosis, 贾地鞭毛虫病，367–369
diagnosis, 诊断，367, *368, 368*
life cycle,生活史，*368, 368*
Giemsa staining, candidosis diagnosis, 姬姆萨染色，念珠菌病的诊断，*341*
glanders,鼻疽 51–53
aetiology,病原学，51
Bactrian camel, 双峰驼，**483**
clinical signs, 临床症状，51, *52*
diagnosis, 诊断，52–53
Dromedary, 单峰驼，**478**
epidemiology, 流行病学，*52*
pathology,病理变化，*52*
*Glossina,* 舌蝇，467, *467*
Gram-negative bacterial infections,革兰氏阴性菌感染，208
Gram staining, candidosis diagnosis, 姬姆萨染色，念珠菌病的诊断，341
Granulomas，，393
aspergillosis, 曲霉病，337
coccidiomycosis, 球孢子菌病，342
endometrial, 子宫内膜，153, *153*
granulomatous pneumonia, 肉芽肿性肺炎，**128**

*Graphinema auchinae,* 羊驼纵纹线虫, **420**
griseofulvin, mycotic dermatitis treatment, 灰黄霉素，霉菌性皮炎治疗, 333
guanaco *(Lama guanaco),* 原驼, *5*
biology, 生物学, 8-9
distribution,分布, *7*
estimated population,估计的数量, *6*
evolution, 演变, 4

**H**

haemagglutination inhibition tests (HIT), 血凝抑制试验
African horse sickness diagnosis, 非洲马瘟诊断, 318
caseous lymphadenitis, 干酪性淋巴腺炎, 172
haemagglutination tests, caseous lymphadenitis, 血凝试验，干酪性淋巴腺炎, 172
*Haematobia,* 角蝇, *465*
*Haematobia minuta,* 微小角蝇, *465*
*Haematopota,* 麻虻, *466*
haematuria, leptospirosis, 血尿症，钩端螺旋体病, *70,* 68-74
*Haemobartonella* infections *see* mycoplasmosis, 血巴尔通体参见 支原体病
haemolytic diplococci, nasal swabs, 溶血性双球菌，鼻试子, **133**
$\alpha$-haemolytic streptococci, nasal swabs, α –溶血性链球菌，鼻试子, **133**
$\beta$-haemolytic streptococci, nasal swabs, β –溶血性链球菌，鼻试子, **133**
haemonchosis, parasitic gastroenteritis/colitis,血矛线虫病，寄生虫性胃肠炎/结肠炎, 427
*Haemonchus* 血矛线虫, 420
infection *see* parasitic gastroenteritis/colitis, 感染参见 寄生虫性胃肠炎/结肠炎,
life cycle, 生活史, *425*
morphology, 形态学, 423–426
*Haemonchus contortus,* 捻转血矛线虫, **420**
morphology, 形态学, *423–425, 423*
*Haemonchus longistipes,* 长柄血矛线虫, **420**
morphology, 形态学, *423,* 423–424, *424*
*Haemophilus* infection, lungs, 嗜血杆菌感染, **131**
haemorrhages, endotoxicosis, 出血，内毒素中毒, *40,* 41, *41,* **42**
haemorrhagic diathesis, 出血性素质, 37
haemorrhagic diseases, 出血症, 38
epizootic, 流行性, **279**
haemorrhagic endometritis, 出血性子宫内膜炎, **158**
haemorrhagic enteritis 出血性肠炎
*Eimeria* coccidiosis, 艾美尔球虫病, *375*
salmonellosis, 沙门氏菌病, 91, *91*
haemorrhagic pneumonia, 出血性肺炎, **129**
haemorrhagic septicaemia, 出血性败血症, 59-64
Bactrian camel, 双峰驼, **483**
clinical course,临床病程, 62–63
control, 控制, 61
diagnosis, 诊断, 61
Dromedary, 单峰驼, **478**
serological studies, 血清学研究, 61
treatment, 治疗, 61
Haller's organ, 哈氏器, 436, *436*
Hammondiosis, 哈芒球虫病, 389–390
heartwater, diagnosis, 心水病的诊断, **81**
*Helicobacter pylori* infection, 幽门螺旋杆菌感染, 208
helminthoses, 蠕虫病, 397–430
cestode infections, 绦虫感染, 406–418
trematode infections, 吸虫感染, 397–405
*see also specific diseases/disorders*, 参见 具体的疫病
hepatitis, interstitial, 间质性肝炎, *109*
herpesviruses, 疱疹病毒, 300-304
*see also specific herpesviruses* 参见 具体的疱疹病毒
herpesvirus infections, 疱疹病毒感染, 479
Bactrian camel, 双峰驼, **479**
Dromedary, 单峰驼, **478**
literature review, 文献综述, **218**
NWC, 新大陆骆驼, **480**
$\gamma$ -hexachlorocyclohexane ($\gamma$ -HCH), mange treatment, 六氯环己烷，螨的治疗, 451
histology 历史
infective reproductive loss diagnosis, 传染性繁殖障碍的诊断, 156
listeriosis, 李斯特菌病 *205*
histopathology, enzootic bovine leukosis, 病理学变化，地方性牛白血病, 322
*Histoplasma farciminosum* infection, caseouslymphadenitis, 皮疽组织胞浆菌感染，干酪性淋巴腺炎, 168
histoplasmosis, 组织胞浆菌病, **344**
historical aspects,历史回故, 3–4
HIT *see* haemagglutination inhibition tests (HIT), 血凝抑制试验, 276
homidium bromide, trypanosomosis treatment, 溴化乙锭，锥虫病治疗, **363**
homidium chloride, trypanosomosis treatment, 胡米氯铵，锥虫病治疗, **363**
*Hyalomma*, 璃眼蜱, 357
dermatophilosis vector, 嗜皮菌病媒介, *180*
frequency of infection, 感染频度, **442**
larva, 幼蜱, *439*,
nymph,若蜱 *439*
plague vectors, 鼠疫媒介, 65

*Hyalomma detritum*, 残缘璃眼蜱, *438*
*Hyalomma dromedarii* (camel tick), 嗜驼璃眼蜱(骆驼蜱), *436, 438*
camelpox vectors, 骆驼痘媒介, 234
epidemiology, 流行病学, 440
life cycle,生活史, 437
*Hyalomma erythraeum*,黑缘璃眼蜱, 440
*Hyalomma impeltatum*, 无盾璃眼蜱, 438
*Hyalomma punt*,神秘璃眼蜱, 440
*Hyalomma rufipes*, 麻点璃眼蜱, 438
*Hyalomma schulzei*,舒氏璃眼蜱, 440
hydatidosis, 包虫病, 412–417
Bactrian camel, 双峰驼, **488**
clinical signs,临床症状, 415
control, 控制, 415
diagnosis, 诊断, 415
Dromedary, 单峰驼, **487**
epidemiology, 流行病学, 413–415
life cycle, 生活史, *412–413 413*
NWC, 新大陆骆驼, **489**
parasite morphology,寄生虫形态学, 412*, 412, 413*
pathology,病理学, 415
prevalence, 流行, 413–415, **414**
treatment, 治疗, 415
hydrometritis, **15**
hygiene, infectious mastitis treatment, 卫生质量,传染性乳房炎的治疗, 189
*Hypobosca camelina*, 骆驼虱蝇, *468*
hypo-pericardium, *Staphylococcus aureus* dermatitis, 心包积液, 金黄色葡萄球菌皮炎, *177*

**I**

ictero haemoglobinuria, *Clostridium* infections, 黄疸血红素尿, 梭菌感染, 24
IFA *see* indirect fluorescent antibody test (IFA),间接荧光抗体试验
IFAT (immunofluorescence antibody test), equine, 马的间接荧光抗体试验
herpesvirus-1 infection diagnosis, 马疱疹病毒1型感染诊断, 301
imidazothiazoles, parasitic gastroenteritis treatment, 咪唑噻唑类,寄生虫性胃肠炎治疗, 429
immunofluorescence, 免疫荧光, 361
aspergillosis, 曲霉病, 337
contagious erythema diagnosis, 接触传染性脓疱, 247
direct assay *see* direct immunofluorescence assay, 直接 参见 直接接荧光抗体试验
rabies diagnosis,狂犬病诊断, 228
immunoglobulins 免疫球蛋白, 295
colostrum, 初乳, 293–294*,* **293**
neonatal diarrhoea treatment, 新生畜腹泻的防控, 295
passive acquisition, 被动获得, 292-295, *292*
structure, 结构, **293**
subclasses,亚群, **293**, 294
immunohistochemistry, 免疫组织化学, 205
camelpox diagnosis, 骆驼痘诊断, 236
enzootic bovine leukosis, 地方性牛白血病, 320, *322*
peste des petits ruminants diagnosis, 小反刍兽疫诊断, 263
immunohistology, listeriosis,免疫组织化学, 李斯特菌病, *205*
immunology, *Rhodococcus equi* infection diagnosis, 马红球菌感染,47-50
immunotherapy, passive,被动免疫疗法, 287
*Impalaia*, 无囊线虫,420
infection *see* parasitic gastroenteritis/colitis, 感染 参见 寄生虫性胃肠炎/结肠炎
morphology, 形态学, 423-426,
*Impalaia tuberculata*, 结节无囊线虫,**420**
indirect fluorescent antibody test (IFA) ,间接荧光抗体试验, 247
African horse sickness diagnosis, 非洲马瘟诊断, 318
neosporosis diagnosis, 新孢子虫病的诊断, 388
indirect haemagglutination test, toxoplasmosis diagnosis, 间接血球凝集试验, 弓形虫病诊断, 384
indirect immunofluorescence assay, contagious erythema diagnosis,间接荧光抗体试验,接触传染性脓疱诊断, 247
indirect serological tests, brucellosis diagnosis, 间接血清学试验,布鲁氏菌病诊断, 143, **144**
infectious bovine rhinotracheitis, 传染性鼻气管炎, **301**
infectious mastitis, 传染性乳房炎, 188–199
aetiology, 病原学, 188–192
ATP test, 三磷酸腺苷(ATP)试验 190
California mastitis test, 乳房炎加利福尼亚体细胞检测法, 190
causative organisms, 病原菌, 191–192
clinical signs, 临床症状, *130*
diagnosis, 诊断, **190**
literature review, 文献综述, **190**
pathology, 病理学, 192
somatic cell count,体细胞计数, 190
treatment, 治疗, 192-193
infectious necrotising hepatitis (black disease), *Clostridium* infections, 传染性坏死性肝炎(羊黑疫), 梭菌感染, 24
infectious pododermatitis, 传染性蹄皮炎, *209*
infectious pustular balanoposthitis, 传染性龟头包皮炎, **301**

infectious pustular vulvovaginitis, 传染性脓疱样外阴道炎, **301**
infective reproductive losses, 传染性繁殖障碍, 151–163
abortion *see* abortion 参见 流产
*Actinomyces pyogenes,* 化脓放线菌,153
*Campylobacter fetus,*胎儿弯曲菌,153, *153*
control, 防控,157
culture,培养, 155
diagnosis, 诊断, 156–157
endometrial granulomas, 子宫内膜肉芽肿,153, *153*
endometritis, 子宫内膜炎, **154**
epidemiology, 流行病学, 153-156
equine rhinitis A virus,甲型马鼻炎病毒, 155
*Escherichia coli,* 大肠杆菌,154
fetal losses,胎儿夭折, **160**
male infertility,公畜不育, **160**
*Mycoplasma argini,* 精氨酸支原体,155
neonatal losses, 新生儿夭折, **160**
pathology, 病理学, *153*-**156**
picornaviruses, 小RNA病毒, 155
*Pseudomonas aeruginosa,*铜绿假单胞菌, 153, 154
scientific reports,科技报告, 150-151
*Streptococcus zooepidemicus,* 兽疫链球菌 **154**
treatment, 治疗, **157–160**
*Trichomonas fetus,* 胎儿三毛滴虫,153
*Ureaplasma diversum,* 差异脲原体,155
influenza, 流感, 305–307
aetiology, 病原学, 305
control, 控制,307
diagnosis, 诊断, 302, *306*
epidemics,流行, 305
epidemiology, 流行病学, 305–306
experimental infection,实验感染, 305-307
literature review, 文献综述, **219**
treatment, 控制, 308
influenza A infection, Bactrian camel, 甲型流感病毒感染, 双峰驼, **473**
infundibuloriosis,纤毛虫病, *370,* 370–371
immunofluorescence antibody test (IFAT), equine, 马的接触传染性脓疱
herpesvirus-1 infection diagnosis, 马疱疹病毒1型感染诊断, 300
insecticides, trypanosomosis control,杀昆虫剂, 锥虫病的控制, 363
integumental bacterial disease, 皮肤细菌病, 166–185
disease causes, 病因, **167**
Dromedary, 单峰驼, **178**
species isolation, 分离的菌种, 177-**178**
interferon gamma assay, tuberculosis diagnosis, γ -干扰素检测方法, 结核病诊断, 120, *122*
interstitial hepatitis, paratuberculosis, 间质性肝炎, 副结核病, *109*
interstitial pneumonia, 间质性肺炎, **129**
intradermal testing, paratuberculosis diagnosis,皮内试验, 副结核病的诊断, 107
intranuclear inclusion bodies, Borna disease, 核内包涵体, 博尔纳病, 228, *230*
isometamidium chloride, trypanosomosis treatment,氮氨菲啶氯化物, 锥虫病治疗, 363, **363**
*Isospora bigemina*, 双芽等孢球虫
infection, 感染, 387–388
morphology, 形态学, *387*
*Isospora orlovi*, 奥氏等孢球虫
infection *see* isosporosis, 感染, 参见等孢球虫病
life cycle, 生活史, 381
morphology, 形态学, *381*
isosporosis, 等孢球虫, 380-383
clinical signs, 临床症状, *381*
diagnosis,诊断, 382, *382*
epidemiology, 流行病学, 381
life cycle,生活史, 381
morphology, 形态学, 381
pathology, 病理学, 381
treatment, 治疗, 382
ivermectin, 伊维菌素
mange treatment, 螨的治疗, 450
parasitic gastroenteritis treatment,寄生虫性胃肠炎治疗, **428**
pediculosis treatment,虱病治疗, 464
ixodid ticks, life cycle,硬蜱, 生活史, 437–440

**J**

Japanese encephalitis, 日本脑炎, **286**
jaw, osteomyelitis, 下颌骨骨髓炎, *209*
JLIDS (juvenile llama immunodeficiency syndrome),青年美洲驼免疫缺陷综合征, 295
Johne's disease *see* paratuberculosis (Johne's disease), 副结核病, 105–112
juvenile llama immunodeficiency syndrome (JLIDS), 青年美洲驼免疫缺陷综合征, 295

**K**

Kadam virus,卡达姆山病毒, 324, **324**
literature review, 文献综述, 221
keratoconjunctivitis, 角膜结膜炎, *163*
ketoconazole, candidosis treatment,酮康唑, 念珠菌病的治疗, 340
*Klebsiella infection*, lungs,克雷伯氏菌感染, 肺脏, **129**

*Klebsiella pneumoniae*, 肺炎克雷伯氏菌，nasal swabs, 鼻试子，131-132

**L**

lactic acid syndrome, 乳酸中毒，38
lactones, macrocyclic, 大环内酯，**429**
*Lama glama see* llama *(Lama glama)*，美洲驼，5
*Lama guanaco see* guanaco *(Lama guanaco)*，原驼，5
*Lamanema chavesi*, 查氏无峰驼线虫，**420**
LAT *see* latex agglutination test (LAT)，乳胶凝集试验，384
latex agglutination test (LAT)，乳胶凝集试验,384
  toxoplasmosis diagnosis,弓形虫诊断，384
  trypanosomosis diagnosis, 锥虫病的诊断，361
*Lentivirus*,慢病毒，320
leptospirosis,钩端螺旋体病，68–74
  aetiology, 病原学，68–69
  clinical signs, 临床症状，*69*，69–*70*
  diagnosis, 诊断，*70-72*
  Dromedary, 单峰驼，478
  epidemiology, 流行病学，69–*70*
  NWC, 新大陆骆驼, **480**
  prevalence,流行，**71**
  serotypes, 血清型，**71**
  treatment, 治疗, 72
  zoonotic implications,公共卫生意义，68
leucocytes, endotoxicosis,白细胞，内毒素中毒，41, *41*
leukaemia, lymphoblastic *see* lymphoblastic leukaemia，淋巴母细胞白血病
levamisole, parasitic gastroenteritis treatment,左旋咪唑，寄生虫性胃肠炎治疗，*428*
*Liacarus*, 丽甲螨，409
Lice, 虱
  distribution,分布，463
  *see also individual species*，参见 具体的种
*Liebstadia*, anoplocephalidosis intermediate hosts, *Liebstadia*,裸头科绦虫的中间宿主,*409*
lindane, mange treatment, 林丹，螨的治疗，450
*Linguatula serrata*,锯齿状舌形虫，470-471
linguatulosis,舌形虫病，470-471
*Listeria monocytogenes* infection *see* listeriosis，单核细胞增多李斯特菌 参见 李斯特菌病
listeriosis, 李斯特菌病，204-207
  aetiology, 病原学，204
  clinical signs,临床症状，*205*, 205-206
  control, 控制, 206
  diagnosis, 诊断, 206
  epidemiology,流行病学，*204*
  pathology, 病理学，205-206
  treatment, 治疗, 206
  zoonotic implications, 公共卫生意义，*204*
liver enlargement, paratuberculosis, 肝脏肿大，副结核病，*108*
llama *(Lama glama)*,美洲驼，*5*
  distribution,分布，*7*
  domestication, 驯化, 12
  estimated population, 估计的数量，**6**
lockjaw *see* tetanus，锁喉症 参见 破伤风
Loewenstein–Jensen media, tuberculosis diagnosis, L-J培养基，结核病的诊断，121, *122*
louping ill, 跳跃病，**279**, 287
  literature review, 文献综述，**215**
lungs，肺脏，131
  anatomy,解剖学，113
  bacterial flora,细菌菌群，**131**
  hydatidosis, 包虫病，412-417
  paratuberculosis, 副结核病，105
  pleural curtain,胸膜屏障，113, *113*
lungworm infections, 肺线虫感染，430-433
  *see also* dictyocaulosis; protostrongylidosis，参见网尾线虫病；原线虫病，430
lymphadenitis, caseous lymphadenitis,淋巴腺炎，干酪性淋巴腺炎，168
lymph nodes，淋巴结，166
  abscesses, 脓肿，163
  endotoxicosis, 内毒素中毒，39–40, *40*
lymphoblastic leukaemia,淋巴母细胞白血病，*321*
  aetiology,病原学，**320**
lymphocyte proliferation assay, tuberculosis diagnosis淋巴细胞增殖试验,结核病诊断，120
lymphosarcoma,淋巴肉瘤，322

**M**

macrocyclic lactones, parasitic gastroenteritis treatment, 大环内酯，寄生虫性胃肠炎治疗，*429, 428*
MAECT(mini-anion exchange centrifugation technique), trypanosomosis diagnosis,小型阴离子交换离心技术，锥虫病的诊断, 363
*Malassezia* infection, mycotic dermatitis,马拉色菌感染，霉菌性皮炎，**331**
male infertility, infective reproductive losses, 公畜不育，传染性繁殖障碍，**160**
malignant catarrhal fever, literature review,恶性卡他热，文献综述，**219**
malignant oedema, *Clostridium* infections, 恶性水肿, 梭菌感染, 22
mallein tests, melioidosis diagnosis,鼻疽菌素试验，类鼻疽诊断, 56
Malta fever,马耳他热，137-138

mandibular osteomyelitis, 下颌骨骨髓炎, *209*
mange,螨, 446–454
  Bactrian camel, 双峰驼, **479**
  clinical signs, 临床症状, 449–450
  control, 控制, *450*-452
  diagnosis, 诊断, 450
  Dromedary, 单峰驼, **478**
  epidemiology, 流行病学, 448
  NWC, 新大陆骆驼,**485**
  parasitic life cycle, 寄生虫生活史史, 448
  pathology, 病理学, 449-450
  transmission,传播, 448
  treatment, 治疗, 450-452
*Mannheimia haemolytica* infection *see* pasteurellosis, 溶血性曼氏杆菌参见巴氏杆菌病
*Mansonia,* 曼蚊, 274
MAPIA (multiantigen print immunoassay), tuberculosis diagnosis, 多抗原线条免疫分析法, 结核病诊断, 119–120
*Marshalagia marshalli,* 马氏马歇尔线虫, **420**
mastitis, infectious *see* infectious mastitis, 传染性乳房炎
MAT (microscopic agglutination test), leptospirosis diagnosis, 微量凝集试验, 螺旋体病诊断, 72, *72*
meat source,肉品来源, 4
mebendazole, parasitic gastroenteritis treatment, 甲苯达唑, 寄生虫性胃肠炎治疗, **428**
medicinal plants, mange treatment, 药用植物, 螨的治疗, 451
melarsamin, trypanosomosis treatment,美拉索明, 锥虫病的治疗, **364**
melarsopol, trypanosomosis treatment, 美拉胂醇, 锥虫病的治疗, **364**
melioidosis, 类鼻疽, 54–57
  aetiology, 病原学, 54
  clinical signs, 临床症状, 54,*55*
  control, 控制,56
  diagnosis, 诊断, 55-56
  epidemiology, 流行病学, 54
  pathology, 病理学, 54-55
  treatment, 治疗, 56
MERS-CoV (Middle East respiratory coronavirus),中东呼吸综合征冠状病毒, 325
metrifonat, nasopharyngeal myiasis treatment, 敌百虫, 鼻咽蝇蛆病的治疗, 460
metritis, 子宫炎
  acute suppurative, 急性化脓性, **157**
  bacterial isolation, 细菌分离, **157-158**
  chronic active, 慢性活动性, *157*
  chronic non-suppurative, 慢性非化脓性, **158**
mHCT (microhaematocrit centrifugation technique),trypanosomosis diagnosis, 微血细胞比容离心技术, 锥虫病的诊断, 361
miconazole, candidosis treatment, 咪康唑, 念珠菌病的治疗, 341
microhaematocrit centrifugation technique (mHCT),trypanosomosis diagnosis, 微血细胞比容离心技术, 锥虫病的诊断, 361
micro-organism isolation, brucellosis diagnosis,病原分离, 布鲁氏菌病的诊断, 141
microscopic agglutination test (MAT), leptospirosis diagnosis微量凝集试验, 钩端螺旋体病诊断, 72, *72*
microscopic lesions, rabies,显微病灶, 狂犬病, 227
microscopy, 显微镜
  anaplasmosis/mycoplasmosis diagnosis,无浆体病/支原体病的诊断, 80
  contagious erythema, 接触传染性脓疱, 247-248
*Microsporum*, 小孢子菌
  culture,培养, 333
  *Trichophyton vs.,* 毛癣菌, *331*
*Microsporum canis,*犬小孢子菌, *335*
  mycotic dermatitis,霉菌性皮炎, **331**
*Microsporum gypseum,*石膏样小孢子菌, *334*
  mycotic dermatitis, 霉菌性皮炎, **331**
*Microsporum nanum* infection, mycotic dermatitis,猪小孢子菌感染, 霉菌性皮炎, 332
*Microthoracius* infection,微胸虱感染, 463-465
*Microthoracius cameli,*骆驼微胸虱, **463**
*Microthoracius mazzai,* 梅察微胸虱,**463**
*Microthoracius minor,*小微胸虱, **463**
*Microthoracius praelongiceps,* 极长头微胸虱, **463**
Middle East respiratory coronavirus (MERS-CoV), 中东呼吸综合征冠状病毒, 325
midges,蠓, 463
  *see also individual species*参加具体的种
milbemycins, mange treatment,米尔贝霉素, 螨的治疗, 451
milk/milking, 产奶, 4, *183, 183, 185*
  hygienic quality,卫生质量, *189*
  nutrition content, 营养成分, *186,187*
  pasteurization, 巴氏消毒, 187
milk ring test, brucellosis diagnosis, 乳环试验, 布鲁氏菌病诊断, 143, *143*
mini-anion exchange centrifugation technique (MAECT),trypanosomosis diagnosis,小型阴离子交换离心技术, 锥虫病的诊断, 362
MLV (modified live virus) vaccine, rabies, 改进的狂犬病灭活疫苗, *230*
modified agglutination test, toxoplasmosis diagnosis,改

进的凝集试验，弓形虫诊断，384
modified live virus (MLV) vaccine, rabies, 改进的狂犬病灭活疫苗，230
Mongolia,蒙古国，3
*Moniezia,* 莫尼茨绦虫，*406*
infection *see* anoplocephalidosis，感染 参见 裸头科绦虫病.
life cycle, 生活史，*408*
*Moniezia benedeni,*贝氏莫尼茨绦虫，*407,410*
infection *see* anoplocephalidosis，感染，参加裸头科绦虫病
*Moniezia expansa,*扩展莫尼茨绦虫，*407,409*
infection *see* anoplocephalidosis，感染 参见 裸头科绦虫病
moniliasis *see* candidosis，参见念珠菌病
morantel tartrate, parasitic gastroenteritis treatment,甲噻吩咪唑，寄生虫性胃肠炎治疗，**428**
*Moraxella* infections, 摩拉克氏菌感染，209
mosquitoes, 蚊子，463
Rift Valley Fever vectors, 裂谷热媒介，274
West Nile virus transmission,西尼罗河病毒传播，286
mouse test, botulism diagnosis, 老鼠试验，肉毒中毒诊断，33–34, **34**
moxidectin，莫西菌素
mange treatment, 螨的治疗，450
parasitic gastroenteritis treatment, 寄生虫性胃肠炎治疗，**428**
mucopurulent discharge, glanders,黏脓性的分泌物，*52*
mucormycosis,毛霉菌病，343, *343*
mucous membranes, oral cavity, 口腔黏膜，168
multiantigen print immunoassay (MAPIA), tuberculosis，多抗原线条免疫分析法，结核病
diagnosis,诊断，119–120
Muscidae, 蝇科，459-460
*Mycella,* mycotic dermatitis,无孢菌，霉菌性皮炎，**332**
*Mycobacterium* species,分枝杆菌属的细菌，**117**
*Mycobacterium* infection, lungs,分枝杆菌感染，肺脏，**131**
*Mycobacterium aquae,*水分枝杆菌，**117**
*Mycobacterium aquae* var. *ureolyticum,* 水分枝杆菌解脲亚种，**117**
*Mycobacterium avium* subsp. *paratuberculosis* (MAP)，禽分枝杆菌副结核亚种
infection *see* paratuberculosis (Johne's disease)，感染 参见 副结核病
*Mycobacterium bovis* infection, 牛分枝杆菌感染，114, 117
*Mycobacterium caprae,* 山羊分枝杆菌，**117**
*Mycobacterium kansasii,* 堪萨斯分枝杆菌，**117**
*Mycobacterium microti,* 田鼠分枝杆菌,**117**
*Mycobacterium paratuberculosis* infection, 副结核分枝杆菌感染，114
*Mycobacterium pinnipedii,*鳍脚亚目动物分枝杆菌，**117**
*Mycobacterium smegmatis,*耻垢分枝杆菌,**117**
*Mycobacterium tuberculosis* infection *see* tuberculosis，结核分枝杆菌感染，参见结核病
*Mycoplasma argini,* infective reproductive losses,精氨酸支原体，传染性繁殖障碍，155
mycoplasmosis, 支原体病，77-84
aetiology, 病原学，77-78
control, 控制,82
diagnosis, 诊断，77–78, 80–82, **81**
infective organisms, 病原，**78**
literature studies,文献综述，**80**
microbiology,微生物，*79*
porcine,猪 **81**
treatment, 治疗，82
mycotic dermatitis, 霉菌性皮炎，331-336
aetiology,病原学，331
causative organisms,致病菌，331,**333**
clinical signs, 临床症状，331,333
diagnosis, 诊断，333-336
epidemiology, 流行病学，331-333
histopathology,组织病理学，333
pathology, 病理学，335
prevention,预防 333-336
treatment, 治疗，333-336
*see also* ringworm (dermatophytosis)，参见癣菌病
myiasis, 蝇蛆病，455-465
Bactrian camel,双峰驼，**488**
causative agents,病原，455-460
control, 控制,456-460
diagnosis, 诊断，456
Dromedary, 单峰驼，**487**
nasopharyngeal, 鼻咽，458-460
Sarcophagidae flies, 麻蝇，455-456
treatment, 治疗，456-460
*see also individual species*，参见具体的种
myositis, sarcocystosis, 肌肉炎，肉孢子虫病，**392**

**N**

Nairobi sheep disease, 内罗毕羊病，280
*Nairovirus,* 内罗病毒，280
nasal swabs, pneumonia, 鼻试子，肺炎，**133**
nasopharyngeal myiasis,鼻咽蝇蛆病，458-460
necrotic endometritis, 坏死性子宫内膜炎，**158**
necrotic pneumonia,坏死性肺炎，**129**
necrotising hepatitis, salmonellosis, 坏死肝炎，沙门氏菌病，96

Negri bodies, rabies diagnosis,内基体，狂犬，病诊断，228-229,*228*
*Neisseria,* nasal swabs,奈瑟氏菌，鼻试子，**133**
Nematocera, 长角亚目，*468*
nematode infections,线虫感染，419-420, 419-430
*Nematodirus*，细颈线虫，420
infection *see* parasitic gastroenteritis/colitis，感染参见寄生虫性胃肠炎/结肠炎
life cycle, 生活史，428
morphology,形态学，425
*Nematodirus battus,*巴氏细颈线虫,**420**
*Nematodirus cameli,* 驼细颈线虫,**420**
*Nematodirus dromedarii,* 单峰驼毛细线虫，**420**,*428*
morphology,形态学，*425,426*
*Nematodirus filicollis,*线颈细颈线虫，**420**
*Nematodirus lamae,* 羊驼细颈线虫,**420**
*Nematodirus mauritanicus,* 毛里塔尼亚细颈线虫，**420**
*Nematodirus oriatinus,* 奥利春细颈线虫，**420**
*Nematodirus spathinger,* 匙形细颈线虫,**420**
neomycin, infectious mastitis treatment,新霉素，传染性乳房炎的治疗，*193*
neonatal diarrhoea, 新生畜腹泻，289-300
causative agents, 病原，289
coronavirus *see* coronavirus neonatal diarrhea，冠状病毒 参见 冠状病毒新生畜腹泻
diagnosis, 诊断， 291-296
epidemiology, 流行病学，289-296
literature review, 文献综述，**219**
treatment, 防控， 296
neonatal losses, infective reproductive losses, 新生儿夭折，传染性繁殖障碍，**160**
*Neorickettsia helminoeca,* 蠕虫新立克次氏体， **76**
*Neorickettsia risticii,* 立氏新立克次氏体， **76**
*Neospora caninum*，犬新孢子虫，386
life cycle, 生活史，387
morphology, 形态学，**387**,*387*
neosporosis, 新孢子虫病，386-389
Bactrian camel, 双峰驼，**488**
clinical signs,临床症状，388
control, 控制,386
diagnosis, 诊断， 388
Dromedary, 单峰驼，**487**
epidemiology, 流行病学，387-388
intermediate hosts,中间宿主， 387
NWC, 新大陆骆驼，**489**
pathology, 病理学，388
nervous system, bacterial disease, 神经系统，细菌病，200-210
netobimin，奈托比胺
dicrocoeliosis treatment,歧腔吸虫病的治疗，403
fasciolosis treatment, 肝片吸虫病的治疗，400
parasitic gastroenteritis treatment, 寄生虫性胃肠炎治疗，**428**
nitroxynil, fasciolosis treatment,硝碘酚腈，肝片吸虫病的治疗，400
non-fatty acids, milk, 非饱和脂肪酸，186
non-pathogenic viral disease,无致病性的病毒病， 311-322
non-suppurative metritis, chronic, 慢性非化脓性子宫炎，**158**
non-suppurative pneumonia, chronic,慢性非化脓性肺炎，**129**
nuisance flies,扰蝇，459-464
nystatin, candidosis treatment,制霉菌素，念珠菌病的治疗， 341

**O**

oedema, malignant, 恶性水肿, 22
*Oesophagostomum columbianum,* 哥伦比亚结节线虫，**420**
infection *see* parasitic gastroenteritis/colitis，感染参见 寄生虫性胃肠炎/结肠炎
morphology, 形态学，423-426
*Oesophagostomum venulosum,*微管结节线虫，**420**
infection *see* parasitic gastroenteritis/colitis，感染参见 寄生虫性胃肠炎/结肠炎
morphology, 形态学，423-426
Ogawa media, tuberculosis diagnosis, 小川培养基，结核病的诊断，121
omeprazole, endotoxicosis treatment, 奥美拉唑，内毒素中毒治疗，46
*Onchocerca armillata,* 圈形盘尾丝虫,**420**
*Onchocerca fasciata,* 福斯盘尾丝虫，**420**
*Onchocerca gutturosa,*喉瘤盘尾丝虫,**420**
Onchocercidae, 盘尾丝虫科，433,*433*
onchocercidosis, 盘尾丝虫病，433-435
clinical signs,临床症状，429
control, 控制,429
diagnosis, 诊断， 429
epidemiology, 流行病学，429
parasite life cycle,寄生虫生活史， 433-434
parasite morphology,寄生虫形态学， 433,*434,434*
treatment, 治疗, *434*
oral cavity mucous membranes, caseous lymphadenitis, 口腔黏膜，干酪性淋巴腺炎， 168
*Orbivirus,* 环状病毒，**280**
ORF *see* contagious erythema，羊口疮，接触传染性脓疱,245

*Ornithodoros,* plague vectors, 钝缘蜱，鼠疫媒介，65
*Ornithodorus savignyi,* frequency of infection, 萨氏钝缘蜱，感染频度，442
*Orthopoxvirus cameli see* camelpox，骆驼正痘病毒 参见骆驼痘
osteomyelitis, mandibular,下颌骨骨髓炎，*209*
*Osteragia dahurica,* 达乎尔奥斯特线虫，**420**
*Osteragia lyrata,* 竖琴奥特线虫, **420**
*Osteragia ostertagi,* 奥氏奥斯特线虫，**420**
otitis, candidosis,中耳炎，念珠菌病，341
ovine anaplasmosis, diagnosis,羊无浆体病的诊断，**81**
oxfendazole, parasitic gastroenteritis treatment,奥芬达唑，寄生虫性胃肠炎治疗，**429**
oxibendazole, parasitic gastroenteritis treatment,奥苯达唑，寄生虫性胃肠炎治疗，**428**
oxyclozamide, fasciolosis treatment,羟氯硫胺，肝片吸虫病的治疗，400
oxytetracycline, brucellosis treatment,四环素，布鲁氏菌病的治疗，145

**P**

papillomatosis, 乳头状瘤，251-253
  aetiology, 病原学，251
  clinical signs,临床症状，251-252,*252*
  control, 控制, 253
  diagnosis, 诊断、*252*, **253**
  Dromedary, 单峰驼、478
  epidemiology, 流行病学，251-252
  literature review,文献综述，**221**
  pathology,病理学，252-253
  transmission,传播，252
  treatment, 治疗,253
*Parabronema skrjabini,*斯氏副柔线虫, **420**
  infection *see* parasitic gastroenteritis/colitis，感染 参见 寄生虫性胃肠炎/结肠炎
  morphology,形态学，**426**,*426*
parainfluenza virus, literature review, 副流感，文献综述，**221**
parainfluenza virus 3 (PI-3), 副流感病毒，311–312
paralytic syndrome (dumb form), rabies,瘫痪症状，狂犬病，227
*Paramphistomium cervi,* 鹿前后盘吸虫，*405*
*Parapoxvirus ovis see* contagious erythema，羊副痘病毒，接触传染性脓疱，419
parasitic disease, 寄生虫病，355–471, **487-489**
  active host finding,主动寻找宿主，*357*
  arthropod infections, 节肢动物感染，436-471
  Bactrian camel,双峰驼, **488**
  Dromedary,单峰驼，**487**
  helminthoses *see* helminthoses，蠕虫病，397–430
  locations, 寄生部位，357, *357*
  NWC, 新大陆骆驼, 489
  protozoal infections, 原虫感染，358-395
parasitic gastroenteritis/colitis, 寄生虫性胃肠炎治疗，422-428
  clinical signs,临床症状，427
  control, 控制,428-429
  diagnosis, 诊断，427-428
  epidemiology, 流行病学，427
  parasite life cycle,寄生虫生活史，425
  pathology,病理学，427
  treatment, 治疗,428-429, **429**
paratuberculosis (Johne's disease), 副结核病，105–112
  aetiology, 病原学，106
  Bactrian camel, 双峰驼, **476**
  clinical signs,临床症状、106, *107,108*
  control, 控制, 110–111
  diagnosis,诊断，107-110
  Dromedary, 单峰驼，**475**
  epidemiology,流行病学，106
  histopathology,组织病理学，108
  NWC, 新大陆骆驼, **486**
  pathology, 病理学、107,109
  treatment, 治疗, 110–111
  zoonotic implications, 公共卫生意义，106
PAS *see* periodic acid–Schiff (PAS) stains，PAS 参见过碘酸希夫反应染色
passive immunotherapy, West Nile virus,被动免疫疗法，西尼罗河病毒，286
*Pasteurella,* lungs, 巴氏杆菌，肺脏，**131**
*Pasteurella multocida* infection,多杀性巴氏杆菌感染,**57,** 57-58
*Pasteurella pneumotropica* infection, 嗜肺巴氏杆菌感染，**58**
  *see also* pasteurellosis，参见 巴氏杆菌病
pasteurellosis, 巴氏杆菌病，57-64
  aetiology, 病原学，**58**
  clinical signs,临床症状，58-59
  epidemiology, 流行病学，58-59
  mortality,死亡率，**61**
  NWC, 新大陆骆驼, **480**
  occurrence, 发病情况，**62**
  predisposing factors, 诱因，57
pasteurization, milk, 巴氏消毒奶，187
pasture hygiene, fasciolosis control,草场卫生，肝片吸虫的控制，*400*
pathogenic viral disease,致病性病毒病，226-310
PCR *see* polymerase chain reaction (PCR)，PCR 参见 聚合

酶链式反应
pediculosis, 虱病，463-464
penicillin，青霉素
caseous lymphadenitis treatment,干酪性淋巴腺炎治疗，171
listeriosis treatment,李斯特菌病的治疗，206
tetanus,破伤风，202
*Penicillium vinaceum* infection, mycotic dermatitis,葡萄酒色青霉感染，霉菌性皮炎，**332**
peptidoglycan recognition protein (PGRP), 肽聚糖识别蛋白，188
pericarditis, *Staphylococcus aureus* dermatitis, 心包炎，金黄色葡萄球菌皮炎，*177*
pericholangitis, fasciolosis, 胆管周炎，肝片吸虫病，*399*
periodic acid–Schiff (PAS) stains，过碘酸希夫反应染色
isosporosis diagnosis,等孢球虫病的诊断，*381*
mycotic dermatitis,霉菌性皮炎，331
trypanosomosis,锥虫病，362, *362*
peste des petits ruminants (PPR),小反刍兽疫，261-265
aetiology, 病原学，261
clinical signs,临床症状，263
control, 防控, 264
diagnosis, 诊断，263-264
Dromedary, 单峰驼，**478**
epidemiology, 流行病学，262-263
geographical distribution, 地理分布，*262*
literature review, 文献综述，**221**
mortality, 死亡率，263
pathology,病理学，264
rinderpest *vs.*,牛瘟，261
treatment, 防控，264
petechiae, endotoxicosis,出血点，内毒素中毒，*40*
Peyer's patches, paratuberculosis, 派尔结节，副结核病，*107*
pharyngeal plague, 咽型鼠疫，**65**
pheromone acaricide-containing devices, tick infestationcontrol, 信息素和杀螨剂的设备，蜱控制，443–444
Phlebotominae, 白蛉亚科，468
*Phlebovirus*,白蛉病毒，**280**
*see also* Rift Valley Fever (RVF)，参见裂谷热
phoxim, mange treatment, 辛硫磷，螨的治疗，**451**
phycomycosis,藻菌病，**344**, *345*
*Physocephalus cristatus*, 有嵴膨首线虫，**420**
*Physocephalus dromedarii*, 单峰驼膨首线虫，**420**
infection *see* parasitic gastroenteritis/colitis，感染参见寄生虫性胃肠炎/结肠炎
morphology, 形态学，426,*426*
PI-3 (parainfluenza virus 3),3型副流感病毒，310-311
picornaviruses, 小RNA病毒，258-259
infective reproductive losses,传染性繁殖障碍，155
piroplasmosis梨形虫病
equine, 马，443
tick infestations, 蜱感染，443
plague, 鼠疫，65–68
aetiology, 病原学，65
Bactrian camel, 双峰驼, **476**
clinical signs,临床症状，65–66
control, 控制,66
Dromedary, 单峰驼，**476**
epidemiology, 流行病学，65–66
meat consumption,肉品消费，66
pharyngeal,喉，65
treatment, 治疗, 66
vectors, 媒介，65
plants, medicinal, 药用植物，451
plastic ingestion, 食入塑料，211，*211*
pleural curtain, lungs,胸膜屏障，肺脏，*113*, *113*
*Pneumococcus*, lungs,肺炎球菌，肺脏，**131**
pneumonia, 肺炎，128-133
aetiology, 病原学，130
catarrhal,卡他性，**128**
chronic non-suppurative, 慢性非脓性，**129**
clinical signs, 临床症状，130,130-133
epidemiology, 流行病学，130,130-133
granulomatous, 肉芽肿性，**128**
haemorrhagic, 出血性，**128**
incidence, 发病，130
interstitial, 间质性，**128**
nasal swabs, 鼻试子，**132**
necrotic, 坏死性，**129**
pathology,病理学，**130**,130-133
pyogranulomatous, 脓性肉芽肿，338
*Rhodococcus equi* infection, 马红球菌感染，49, *49*, 49–50
*Streptococcus*,链球菌，131
suppurative,化脓性，**129**
therapy, 治疗，133
pododermatitis,蹄皮炎，**209**
infectious,传染性，*209*
polymerase chain reaction (PCR)，聚合酶链式反应
bluetongue diagnosis, 蓝舌病的诊断，280,282
brucellosis diagnosis,布鲁氏菌病诊断，136, 143-144
camelpox diagnosis,骆驼痘诊断，236
*Clostridium* enterotoxaemia diagnosis, 梭菌肠毒血症的诊断，29, *29*
colibacillosis diagnosis,大肠杆菌病的诊断，*103*
contagious erythema diagnosis,接触传染性脓疱，247

equine herpesvirus-1 infection diagnosis,马疱疹病毒1型感染诊断,301
leptospirosis diagnosis, 钩端螺旋体病的诊断,72
melioidosis diagnosis, 类鼻疽诊断,56
mycoplasmosis diagnosis, 支原体,78
peste des petits ruminants,小反刍兽疫,263-264
peste des petits ruminants diagnosis, 小反刍兽疫诊断,261
Q fever diagnosis, Q热的诊断,84, 85,86
rabies diagnosis, 狂犬病诊断,228
retrovirus infection diagnosis,逆转录病毒感染诊断,319
reverse transcriptase *see* reverse-transcriptase polymerase chain reaction (RT-PCR) 反转录-PCR
*Rhodococcus equi* infection diagnosis, 马红球菌感染诊断,49
salmonellosis diagnosis, 沙门氏菌病的诊断, 96–97
trypanosomosis diagnosis, 锥虫病的诊断, 361-362
tuberculosis diagnosis, 结核病诊断,120
polymyxin B, endotoxicosis treatment, 多黏菌素B,内毒素中毒治疗, 45
polyvalent antiserum, botulism treatment, 多价抗血清,肉毒梭菌中毒的治疗,33
porcine mycoplasmosis, diagnosis, 猪支原体病的诊断,**81**
potentiated sulphonamides, salmonellosis treatment, 磺胺增效类药物,沙门氏菌病的治疗,**97**
Potiskum virus, 波蒂斯库姆病毒, **325**
literature review, 文献综述,**223**
Potomac horse fever diagnosis,波托马克马热的诊断, **81**
PPR *see* peste des petits ruminants (PPR) , PPR 参见 小反刍兽疫
Praziquantel,吡喹酮
anoplocephalidosis treatment, 裸头科绦虫病的治疗,410
dicrocoeliosis treatment, 歧腔吸虫病的治疗,403
echinococcosis treatment, 棘球蚴病的治疗,415
probenzimidazoles, parasitic gastroenteritis treatment, 苯并咪唑,寄生虫性胃肠炎治疗, 429
proliferative enteritis, diffuse, 弥漫性增生性肠炎, *109*
proliferative pneumonia, chronic,慢性增生性肺炎, **129**
propionyl promazine, tetanus, 丙酰丙嗪,破伤风, *202*
*Proteus* 变形杆菌
integument, 皮肤,**178**
lungs, 肺脏,**131**
protostrongylidosis, 原线虫病,430-433
clinical signs, 临床症状,431
control, 控制,432
diagnosis, 诊断, 432
epidemiology, 流行病学,431
parasite life cycle, 寄生虫生活史,431
parasite morphology, 寄生虫形态学,430-431
pathology,病理学, 431
treatment, 治疗, 432
protozoal parasitic disease, 原虫病,358–395
*see also specific diseases/disorders,*参见具体的疾病
*Pseudoarachiniotus* infection, mycotic dermatitis, 假蛛网霉感染,霉菌性皮炎, **332**
*Pseudomonas*, 假单孢菌
integument, 皮肤,**178**
lungs, 肺脏,**131**
*Pseudomonas aeruginosa,* infective reproductive losses, 铜绿假单胞菌,传染性繁殖障碍, 153,154
*Pseudorotium* infection, mycotic dermatitis, *Pseudorotium*感染,霉菌性皮炎, **332**
pseudotuberculosis *see* caseous lymphadenitis (CLA),假结核, 参见干酪性淋巴腺炎
*Psoroptes*,痒螨
infection *see* mange, 感染 参见 螨
morphology, 形态学,446–447, *447*
pustular balanoposthitis, infectious, 传染性龟头包皮炎, **301**
pustular vulvovaginitis, infectious, 传染性脓疱样外阴道炎,**301**
pyogenic affections, *Staphylococcus aureus* dermatitis, 化脓,金黄色葡萄球菌皮炎,176–178
pyogenic disease *see* caseous lymphadenitis (CLA),化脓性疾病 参见 干酪羊淋巴腺炎
pyogranulomatous pneumonia, *Aspergillus niger*,脓性肉芽肿肺炎,黑曲霉, *338*
pyometra, 宫腔积脓,**158**
pyrantel pamoate, parasitic gastroenteritis treatment, 双羟萘酸噻嘧啶,寄生虫性胃肠炎治疗, *429*
*Pythium insidiosum,* mycotic dermatitis, 诡谲腐霉, 霉菌性皮炎, **332**

**Q**

Q fever, Q热, 84–90
aetiology,病原学,84
Bactrian camel, 双峰驼, **485**
clinical signs, 临床症状, 86
control, 控制,87
diagnosis, 诊断,84–85, 86, **87**
Dromedary, 单峰驼,**476**
epidemiology, 流行病学,84–86
NWC, 新大陆骆驼, **486**
serology,血清学,85
seroprevalence studies,血清学调查, 84
treatment, 治疗,87

zoonotic implications,公共卫生意义，84
Quaranfil virus, 夸兰菲尔病毒，324
literature review, 文献综述，**224**
quinapyramine methyl sulphate, 喹匹拉明，363, **363**

**R**

rabies, 狂犬病，226–230
aetiology,病原学，226
Bactrian camel, 双峰驼，**473**
clinical signs,临床症状，227–228
control, 控制，229–230
diagnosis, 诊断，228–229
Dromedary, 单峰驼，478
epidemiology, 流行病学，226–227
incubation period, 潜伏期，226
literature review, 文献综述，222
NWC, 新大陆骆驼，**474**
pathology,病理学，227–228
susceptibility,易感性，226
transmission, 传播，226
treatment, 治疗，229–230
rabies-like retrovirus, 狂犬样反转录病毒，**325**
rabies-like virus, 狂犬样病毒，**325**
racing, 比赛，13–14, *14*
radial diffusion, passive immunoglobulin transfer, 单向放射免疫扩散检测法，**294** *294*
rafoxanide, fasciolosis treatment, 碘硫胺，肝片吸虫病的治疗，400
raging fury, rabies, 狂躁，狂犬病，*227*
rapid test (RT), tuberculosis diagnosis,快速检测方法，结核病诊断，119–120
RBPT (rose bengal plate test), brucellosis diagnosis, 虎红平板试验，布鲁氏菌病诊断，143
renal calcifications, leptospirosis,肾脏钙化，钩端螺旋体病，*70*
*Reoviridae,* 呼肠孤病毒，**279**
reovirus, 呼肠孤病毒，**324**
literature review, 文献综述，**223**
reproductive losses, infectious *see* infective reproductive losses，传染性繁殖障碍
respiratory syncytial virus, 呼吸道合胞体病毒，**221**
respiratory system, bacterial disease, 呼吸系统细菌病，113–136
respiratory viruses,呼吸道病毒，310–313
literature review, 文献综述，**221**
*see also specific viruses*，参见具体病毒
retrovirus infections, 逆转录病毒感染，320–323
reverse-transcriptase polymerase chain reaction (RT-PCR)，反转录PCR
African horse sickness diagnosis, 非洲马瘟诊断，*320*
BRSV, 牛呼吸道合胞体病毒，311
influenza diagnosis, 流感诊断，306
parainfluenza virus 3 diagnosis, 3型副流感病毒诊断，310–311
*Rhabdovirus,* 弹状病毒，**280**
*Rhicephalus evertsi,* 埃氏扇头蜱，*437*
*Rhicephalus* infection, 扇头蜱感染，**442**
*Rhicephalus pulchellus,*美丽扇头蜱，437, *437*
*Rhodococcus equi,* nasal swabs, 马红球菌 鼻试子，**132**
*Rhodococcus equi* infection, 马红球菌感染，47–51
aetiology, 病原学，47
clinical signs,临床症状，48–50, *51,* 52
control, 控制，50–51
diagnosis, 诊断，*47,* 49
epidemiology, 流行病学，48
treatment, 治疗，49–50
rickettsial diseases, 立克次体病，74
*see also* anaplasmosis; mycoplasmosis，参见 无浆体病和支原体病
Rift Valley Fever (RVF), 裂谷热，274–278
aetiology,病原学，274
climate effects, 气候影响，274–275
clinical signs,临床症状，275
control, 控制，276–277
diagnosis, 诊断，275, *276*
Dromedary, 单峰驼，**478**
epidemics, 流行病，273–276
epidemiology, 流行病学，274–275
literature review, 文献综述，**222**
treatment, 控制，276–277
vector, 媒介，274
zoonotic implications, 公共卫生意义，274
rinderpest, 牛瘟，313–316
aetiology, 病原学，313
animal reservoirs,储藏病原动物，315
clinical presentation, 临床症状，315
clinical signs,临床症状，313–315
diagnosis, 诊断，317
epidemiology, 流行病学，315
experimental infections, 实验感染，316–317
literature review, 文献综述，221-223
pathology, 病理学，314–316
peste des petits ruminants *vs.,* 小反刍兽疫，261
prevention,预防，317
treatment, 控制，317
ringworm (dermatophytosis), 癣菌病，331, 331–333, *335*
Bactrian camel, 双峰驼，**488**
Dromedary, 单峰驼，**487**

incidence, 发病,332
NWC, 新大陆骆驼, **489**
*see also* mycotic dermatitis,参见 霉菌性皮炎
RNA viruses, RNA病毒, **279**
rose bengal plate test (RBPT), brucellosis diagnosis, 虎红平板试验,布鲁氏菌病的诊断,143
rotavirus, literature review, 轮状病毒,文献综述,220
rotavirus neonatal diarrhoea, 轮状病毒新生畜腹泻, 288, 289, 290
diagnosis, 诊断, 290, *291*
epidemiology, 流行病学,288
seasonality, 季节性,*289*
RT (rapid test), tuberculosis diagnosis, 快速检测方法, 结核病诊断, 119–120
RT-PCR *see* reverse-transcriptase polymerase chain reaction (RT-PCR), 反转录PCR
rumenitis, *Aspergillus fumigatus,* 瘤胃炎,烟曲霉, 338
ruminants, in comparison, 反刍动物,比较, 8-**9**
RVF *see* Rift Valley Fever (RVF),RVF 参见 裂谷热

**S**
Sabin–Feldman test, toxoplasmosis diagnosis, 赛宾–费德曼染色试验,弓形虫诊断, 384
salicylanides, fasciolosis treatment, 水杨酰苯胺,肝片吸虫病的治疗,400
*Salmonella bongori,* 邦戈尔沙门氏菌, 91
*Salmonella enterica,* 肠道沙门氏菌, 91
*Salmonella* infection, neonatal diarrhoea, 沙门氏菌感染, 新生畜腹泻, 288
*Salmonella subterranea,* 地下沙门氏菌,91
salmonellosis, 沙门氏菌病,91–100
aetiology,病原学, 91
Bactrian camel, 双峰驼, **485**
clinical signs, 临床症状,91, **93**, 93–96
control, 控制,97–98
diagnosis, 诊断, *93*, 96–97
Dromedary, 单峰驼,**475**
epidemiology, 流行病学,91, 93–96
literature, 文献综述,**92**
mortality, 死亡率,93
NWC, 新大陆骆驼, **485**
pathology,病理学, *96*
prevention, 预防,96–97
serotypes,血清型, 93, **94**, 95
serovars, 血清型, **95**
treatment, 治疗, 96–97
zoonotic implications, 公共卫生意义,91, 93–94
salmon poisoning, diagnosis, 鲑鱼中毒, **81**
sandflies, 白蛉,463
*Sarcocystis,* 肉孢子虫,**390**
life cycle, 生活史, 391
morphology, 形态学,391, *391*
*Sarcocystis aucheriae,* 羊驼肉孢子虫, **391**, 392
*Sarcocystis cameli,* 骆驼肉孢子虫,**391**
*Sarcocystis camelicanis,* 驼犬肉孢子虫,**391**
*Sarcocystis lamacanis,* 大羊驼犬肉孢子虫,**391**
*Sarcocystis tilopodi,* 原驼犬肉孢子,391, 392
sarcocystosis, 肉孢子虫病,390–394
clinical signs, 临床症状,391–392
control, 控制, 393
diagnosis, 诊断,393
epidemiology, 流行病学,**391**
NWC, 新大陆骆驼, **489**
pathology, 病理学,391–392
prevalence, 流行,**392**
*Sarcophaga,* 麻蝇,*455*
*Sarcophaga dux,* 酱麻蝇,456, *456*
Sarcophagidae flies, myiasis, 麻蝇科的蝇类, 蝇蛆病, 455–456
*Sarcoptes*,疥螨
life cycle, 生活史,*448*
morphology, 形态学,446, *447*
*Sarcoptes* infection *see* mange,疥螨感染 参见 螨
SBV *see* Schmallenberg virus (SBV), SBV 参见 施马伦贝格病毒
scabby mouth *see* contagious erythema,疮痂嘴,参见接触传染性脓疱
SCC *see* somatic cell count (SCC),参见体细胞计数
*Scheloribates,* 菌甲螨,*408*
anoplocephalidosis intermediate hosts, 裸头科绦虫的中间宿主,409
*Schistosoma bovis,* 牛分体吸虫,404–405
*Schistosoma matthei,* 羊分体吸虫,*404,* 404–405
schistosomosis,分体吸虫病, 404–405
Schmallenberg virus (SBV), 施马伦贝格病毒,324
literature review, 文献综述,**223**
*Scutovertex,*垂盾甲螨,409
sedimentation method, fasciolosis diagnosis, 沉淀法,肝片吸虫病的诊断,399–400
septicaemia, haemorrhagic *see* haemorrhagic septicaemia,参见 出血性败血症
serological studies,血清学研究
bovine viral diarrhoea,牛病毒性腹泻, 265
coccidiomycosis, 球孢子菌病, 342
haemorrhagic septicaemia,出血性败血症, 63
Rift Valley Fever, 裂谷热,273
Serology, 血清学
*Adenoviridae,* 腺病毒,311

bluetongue, 蓝舌病, **281**
brucellosis,布鲁氏菌病, 138
camelpox, 骆驼痘, 234–244
endotoxicosis, 内毒素中毒, 42
fasciolosis, 肝片吸虫病, 400
leptospirosis, 钩端螺旋体, 70
mange, 螨, 446
melioidosis, 类鼻疽, 55–56
neosporosis, 新孢子虫病, 386
peste des petits ruminants, 小反刍兽疫, 261
rabies, 狂犬病, 228, *228*
retrovirus infections, 逆转录病毒感染, 320
*Rhodococcus equi* infection, 马红球菌感染, 49-50
sarcocystosis, 肉孢子虫病, 390
toxoplasmosis, 弓形虫病, 383–384
seroprevalence studies, Q fever,血清学调查, Q热, 84
serotypes, leptospirosis,血清型, 钩端螺旋体病, **71**
shiga toxin-producing *Escherichia coli* (STEC), 产志贺氏毒素大肠杆菌, 100
*Shigella,* caseous lymphadenitis, 志贺氏菌, 干酪性淋巴腺炎, 167
SICTT (single intradermal comparative tuberculin test), 单次结核菌素皮内比较试验, 119
silent fury, rabies, 沉静型, 狂犬病, *227*
single intradermal comparative tuberculin test (SICTT), 单次结核菌素皮内比较试验, 119
single intradermal tuberculin test (SITT), 单次结核菌素皮内比较试验, 119
SITT (single intradermal tuberculin test), 单次结核菌素皮内比较试验, 119
skin diseases *see* integumental bacterial disease, 皮肤病, 参见 皮肤细菌病
skin necrosis, dermatophilosis,皮肤坏死, 嗜皮菌病, *182*
*Skrabinema ovis,* 羊斯克里亚宾线虫, **420**
smears, infective reproductive loss diagnosis,抹片, 传染性繁殖障碍的诊断, 157–158
somatic cell count (SCC) ,体细胞计数
infectious mastitis, 传染性乳房炎, *191*
udder, 乳房, *186*
*Spiculopteragia peruvianus,* 秘鲁刺翼线虫, **420**
splenomegaly, paratuberculosis, 脾肿大, 副结核病, *108*
*Sporothrix schenckii* infection, mycotic dermatitis, 申克孢子丝菌感染, 霉菌性皮炎, 332
staphylococci, coagulase-negative, 凝固酶阴性葡萄球菌, **133**
*Staphylococcus*, 葡萄球菌
infectious mastitis, 传染性乳房炎, 191–192
integument, 皮肤, **178**
lungs, 肺脏, **131**
*Staphylococcus aureus*, 金黄色葡萄球菌
caseous lymphadenitis, 干酪性淋巴腺炎, 168
infectious mastitis, 传染性乳房炎, 190
integument, 皮肤, **178**
nasal swabs, 鼻试子, **132**
*Staphylococcus aureus* dermatitis, 金黄色葡萄球菌皮炎, 176–179
clinical signs, 临床症状, *177,* 177–178
control, 控制,178–179
epidemiology, 流行病学, 176–177
pathology, 病理学, 176–178
treatment, 治疗, 178–179
STEC (shiga toxin-producing *Escherichia coli*), 产志贺氏毒素大肠埃希菌, 100
sterile insect release, myiasis control, 昆虫不育技术, 蝇蛆病控制, 456–460
stiff-neck syndrome, tetanus, 硬颈综合征, 破伤风, 200–201
*Stilesia,* 斯泰勒绦虫, 406, *406*
infection *see* anoplocephalidosis, 感染 参见 裸头科绦虫病
*Stilesia globipunctata,* 球点状斯泰勒绦虫, 406
infection *see* anoplocephalidosis, 感染 参见 裸头科绦虫病
*Stilesia hepatica,* 肝斯泰勒虫, 407
*Stilesia vittata,* 条状斯泰勒绦虫, *406*, 407
*Stomoxys calcitrans,* 厩螫蝇, 465, *465*
bites, 叮咬, 465
Strauss reaction, glanders, 斯特劳斯反应, 鼻疽, *53*
*Streptococcus*, 链球菌
infectious mastitis, 传染性乳房炎, 191–192
integument, 皮肤, **178**
lungs, 肺脏, **131**
pneumonia,肺炎, 131
*Streptococcus agalactiae,* infectious mastitis, 无乳链球菌, 传染性乳房炎, 191, *191*
*Streptococcus equi* subsp. *zooepidemicus,*马链球菌兽疫亚种, 208
*Streptococcus zooepidemicus,* infective reproductive losses, 兽疫链球菌, 传染性繁殖障碍, 153
*Streptomyces,* nasal swabs, 链霉菌, 鼻试子, **133**
Streptomycin, 链霉素
brucellosis treatment, 布鲁氏菌病, 145
dermatophilosis treatment, 嗜皮菌病的治疗, 183
leptospirosis treatment, 钩端螺旋体病的治疗, 72
plague treatment, 鼠疫治疗, 66
*Strongyloides,* life cycle, 类圆线虫, 生活史, *428*
*Strongyloides papillosus,* 乳头类圆线虫, **419**
subacute suppurative endometritis, 亚急性化脓性子宫内

膜炎，**157**
subendocardial haemorrhage, endotoxicosis, 心内膜下出血，*39*
sulfadimidine, *Eimeria* coccidiosis treatment, 磺胺二甲嘧啶，艾美尔球虫病的治疗，376
sulfidofos, nasopharyngeal myiasis treatment, 倍硫磷，鼻咽蝇蛆病治疗，460
sulphonamides, potentiated, 磺胺增效类药物，**97**
sulphur in mustard oil, candidosis treatment, 芥子油，念珠菌病的治疗，341
suppurative endometritis, acute, 急性化脓性子宫内膜炎，158
suppurative metritis, acute, 急性化脓性子宫炎，**158**
suppurative pneumonia, 化脓性肺炎，**129**
suramin, trypanosomosis treatment, 苏拉明，363，**364**

**T**

Tabanidae, 虻科，466–467
*Tabanus,* 虻，*461*
tapeworms *see* anoplocephalidosis, cysticercosis, hydatidosis，绦虫 参见 裸头科绦虫病，囊虫病，包虫病
tetanus, 破伤风，200–203
    aetiology, 病原学，200–201
    clinical signs,临床症状，201–202, *202*
    control, 控制,202
    diagnosis, 诊断，202, *202*
    epidemiology, 流行病学，200–201
    infective route, 感染途径，200
    prevention, 预防，202
    stiff-neck syndrome, 硬颈综合征，200–201
    wry-neck syndrome, "歪脖"综合征，200, *201*
tetracyclines，四环素
    *Anaplasma phagocytophilum* infection, 嗜粒细胞无浆体感染，76
    anaplasmosis/mycoplasmosis treatment, 无浆体病/支原体病的治疗，82
    caseous lymphadenitis treatment, 干酪性淋巴腺炎治疗，171
    chlamydiosis treatment, 衣原体病治疗，164
    plague treatment,鼠疫治疗，66
tetrahydropyrimidines, parasitic gastroenteritis treatment, 四氢尿嘧啶，寄生虫性胃肠炎治疗，*429*
tetrazolium cleavage test, *Clostridium* enterotoxaemia diagnosis，噻唑蓝分解实验( MTT )，梭菌肠毒血症诊断，28, *29*
theileriosis, 泰勒虫病，395–396
thiabendazole, parasitic gastroenteritis treatment, 噻苯咪唑，寄生虫性胃肠炎治疗，*429*
Thogoto virus, 索戈托病毒，324
    literature review, 文献综述，**224**
*Thysaniezia,* 曲子宫绦虫，406
*Thysaniezia giardi,* 盖氏曲子宫绦虫，408, *408,* 409
*Thysanosoma actinoides,* 放射穗体绦虫，408–409
tick-borne fever, diagnosis,蜱传热的诊断，**81**
tick infestations, 蜱侵袭，436–445
    adverse effects, 副作用，443
    Bactrian camel, 双峰驼，**488**
    control, 控制，443–444
    Dromedary, 单峰驼，**487**
    epidemiology, 流行病学，440
    frequency of infection,感染频度，**442**
    identification, 鉴定，443
    life cycle,生活史，438–440
    morphology,形态学，436–438
    zoonotic implications,公共卫生意义，443
    *see also individual species; specific infections*，参见具体的种，具体的病
*Togaviridae,* 披膜病毒，**280**
    literature review, 文献综述，**224**
toltrazuril, *Eimeria* coccidiosis treatment, 托曲珠利，艾美尔球虫病的治疗，376
*Toxoplasma*，弓形虫
    life cycle,生活史，383-384
    morphology,形态学，383，*383*
toxoplasmosis, 弓形虫病，383–386
    Bactrian camel, 双峰驼，**488**
    clinical signs,临床症状，384
    control, 控制，384
    diagnosis, 诊断，384
    Dromedary, 单峰驼，487
    epidemiology, 流行病学，384
    serology, 血清学，383–384
    treatment, 治疗，384
tracheal ulcers, endotoxicosis, 气管溃疡，内毒素中毒,*39*
trematode infections, 吸虫感染，397–405
    *see also specific diseases/disorders*，参见具体的疾病
trematodosis, NWC, 吸虫病，新大陆骆驼，**489**
trichlorfon, mange treatment,敌百虫，螨的治疗，*451*
*Trichomonas fetus,* infective reproductive losses, 胎儿三毛滴虫，传染性繁殖障碍，*153*
*Trichophyton*，毛癣菌，331
    culture,培养，333, *334*
    *Microsporum vs.,* 小孢子菌，*331*
*Trichophyton dankaliense* infection, 丹卡利毛癣菌感染，**332**
*Trichophyton mentagrophytes,* 须癣毛癣菌，*334*
    mycotic dermatitis, 霉菌性皮炎，**331**

*Trichophyton sarkisovii* infection, 萨氏毛癣菌感染, **332**
*Trichophyton schoenleinii,* 许兰毛癣菌, *335*
mycotic dermatitis, 霉菌性皮炎, **331**
*Trichophyton tonurans* infection, 脱毛毛癣菌, **332**
*Trichophyton verrucosum,* 疣状毛癣菌, *335*
culture, 培养, *334*
mycotic dermatitis, 霉菌性皮炎, **331**
*Trichoribates,* 毛甲螨, 409
*Trichostrongylus*, 毛圆线虫, 419
infection *see* parasitic gastroenteritis/colitis, 感染 参见 寄生虫性胃肠炎/结肠炎,
morphology, 形态学, 424
*Trichostrongylus axei,* 艾氏毛圆线虫, **419**
*Trichostrongylus colubriformis,* 蛇形毛圆线虫, **419**, *424*
*Trichostrongylus longispiculartis,*长刺毛圆线虫, **419**
*Trichostrongylus probulrus,* 突尾毛圆线虫, **419**, *424*
*Trichostrongylus vitrinus,* 透明毛圆线虫, **419**
*Trichuris*, 毛首线虫
infection *see* parasitic gastroenteritis/colitis, 感染 参见 寄生虫性胃肠炎/结肠炎
life cycle, 生活史 *428*
morphology, 形态学, *426*
*Trichuris barbetonensis,* 巴佰顿毛首线虫, **419**
*Trichuris cameli,* 驼毛首线虫 **419**
*Trichuris globulosa,* 球鞘毛首线虫, **419**
morphology, 形态学, *426*
*Trichuris ovis,* 羊毛首线虫, **419**
*Trichuris skjabini,* 斯氏毛首线虫, **419**
*Trichuris tenuis,*细毛首线虫, **419**
triclabendazole, fasciolosis treatment, 三氯苯咪唑, 肝片吸虫病的治疗, 400
tritrichomonosis, 三毛滴虫病, 369, *369*
trypanocides, 杀锥虫药, **363**
*Trypanosoma brucei,* 布氏锥虫, *358*
*Trypanosoma congolense,* 刚果锥虫, *358*
*Trypanosoma evansi,* 伊氏锥虫, 358, *358*
historical aspects,历史回顾, 360
infection *see* trypanosomosis 感染 参见 锥虫病
*Trypanosoma simiae,* 猴锥虫, *358*
*Trypanosoma vivax,* 活跃锥虫, *358*
trypanosomosis, 锥虫病, 358–367
Bactrian camel, 双峰驼, **488**
clinical signs,临床症状, 361, *362*
*Clostridium* enterotoxaemia,梭菌性肠毒血症, *25*
control, 控制, 363–364
diagnosis, 诊断, 362, *362,* 362–363
Dromedary, 单峰驼, **487**
epidemiology, 流行病学, 358–361
morphology,形态学, *358,* 358–359
pathology, 病理学, 361,
prevalence, 流行, **359**
therapy, 治疗方法, **363**, 363–364
transmission, 传播, 359, 360
vectors, 媒介, 360
tsetse flies, 采采蝇, 467, *467*
TST (tuberculin skin test),结核菌素皮内比较试验, 118, *118*
tuberculin skin test (TST), 结核菌素皮内比较试验, 118, *118*
tuberculosis,结核病, 113–127
aetiology, 病原学, 114
ante mortem diagnosis, 生前诊断, *118,* 118–121
Bactrian camel, 双峰驼, **485**
clinical signs, 临床症状, 118, *122*
control, 控制,119, *121*, 123
Dromedary, 单峰驼, **476**
economic losses, 经济损失, 117
epidemiology, 流行病学, 114, 117–118
incidence, 发病, *113*
literature summary, 文献综述, **115–117**
NWC, 新大陆骆驼, **485**
post mortem diagnosis,死后诊断, *121*
transmission mode, 传播方式, 114
treatment, 治疗, *121*, 123
zoonotic implications, 公共卫生意义, 113–114

**U**

Udder, 乳房, 186–199
anatomy, 解剖学, 188, *188*
bacterial disease, 细菌病, 186–199
Californian mastitis test, 乳房炎加利福尼亚体细胞检测法, 186
publications, 出版, 186
somatic cell content,体细胞计数, 186
Uganda virus, 乌干达病毒, **325**
literature review, 文献综述, **223**
ulcers, endotoxicosis, 溃疡, 内毒素中毒, *39*
undulant fever, 波状热, 137–138
*Ureaplasma diversum,* infective reproductive losses, 差异脲原体, 传染性繁殖障碍, 155
urogenital system, bacterial disease, 泌尿生殖系统, 细菌病, 137–165
use of camelids, 骆驼的利用, 7–8

**V**

vaccinations, 疫苗接种, 352, **352, 353**
African horse sickness, 非洲马瘟, 320
anthrax, 炭疽, 37, **353**
bacterial disease, 细菌病, **353**

bluetongue, 蓝舌病, 284
bovine herpesvirus type 1, 牛疱疹病毒1型, 302
bovine viral diarrhoea, 牛病毒性腹泻/黏膜病, **354**
brucellosis, 布鲁氏菌病, 144–145
camelpox, 骆驼痘, 234–244, **354**
caseous lymphadenitis, 干酪性淋巴腺炎, 171
*Clostridium chauvoei,* 气肿疽梭菌, **353**
*Clostridium* enterotoxaemia, 梭菌肠毒血症, 30, *30*
*Clostridium novyi,* 诺维氏芽孢梭菌, **353**
*Clostridium perfringens,* 产气荚膜梭菌, **353**
*Clostridium septicum*, 腐败梭菌, **353**
coccidiomycosis, 球孢子菌病, 342
colibacillosis, 大肠杆菌病, 104
dermatophytoses, 皮肤真菌病, **354**
ecthyma contagiosum, 接触传染性脓疱, **354**
enterotoxaemia, 内毒素血症, **353**
equine herpesvirus-1, 马疱疹病毒1型, 302, **354**
*Escherichia coli,* 大肠杆菌, **353**
foot and mouth disease, 口蹄疫, 251
fungal disease, 真菌病, **354**
haemorrhagic septicaemia, 出血性败血症, 61
leptospirosis, 钩端螺旋体病, **353**
mycotic dermatitis, 霉菌性皮炎, 331
neonatal diarrhoea, 新生畜腹泻, 296
neonatal viral diarrhoea, 新生畜腹泻, **354**
*Orthopoxvirus,* 正痘病毒, **354**
papillomatosis, 乳头状瘤, **354**
papovavirus, 乳头状瘤病毒, **354**
parapoxvirus, 副痘病毒, **354**
paratuberculosis, 副结核病, 110–111
peste des petits ruminants, 小反刍兽疫, 265
plague, 鼠疫, 66
rabies, 狂犬病, 227–230, **354**
respiratory viruses, 呼吸道病毒, 310-313
Rift Valley Fever, 裂谷热, 273
rinderpest, 牛瘟, 313
salmonellosis, 沙门氏菌病, 97–98, **353**
*Staphylococcus aureus* dermatitis, 金黄色葡萄球菌皮炎, 178
tetanus, 破伤风, 200–201, **353**
tick infestation control, 蜱的控制, 443
tuberculosis, 结核病, 117
viral disease, 病毒病, **354**
West Nile virus, 西尼罗河病毒, 287

vector control, trypanosomosis, 媒介控制, 锥虫病, 364
Venezuelan equine encephalitis, 委内瑞拉马脑炎, **279**
Vermipsyllidosis, 蠕形蚤病, 469
*Vermipsylla alacurt,* 花蠕形蚤, 469
vesicular stomatitis (VS), 水泡性口炎, 259, **279**
*Vicugna vicugna see* vicuna *(Vicugna vicugna)*, 驼马
vicuna *(Vicugna vicugna),* 驼马, *5*
biology, 生物学, 8-9
distribution, 分布, *7*
estimated population, 估计的数量, **6**
evolution, 进化, 4
*Viguna pacos see* alpaca *(Viguna pacos)*, 参见羊驼
viral disease, 病毒病, 213–322, **216–224**, **479–475**
Bactrian camel, 双峰驼, **473**
Dromedary, 单峰驼, 478
non-pathogenic, 无致病性, 310–322
NWC, 新大陆骆驼, **474**
pathogenic, 致病性, 226–311
vaccinations, 疫苗接种, **354**
*see also specific diseases/disorders* 参见具体的疫病
viral isolation, rabies diagnosis, 病毒分离, 狂犬病诊断, 228
visual inspection, sarcocystosis diagnosis, 视检, 肉孢子虫病的诊断, 393
VS (vesicular stomatitis), 疱疹性口炎, 260, **279**
vulvovaginitis, infectious pustular, 传染性脓疱样外阴道炎, **301**

**W**

Wanowrie virus, Wanowrie 病毒, 325
literature review, 文献综述, **224**
warts *see* papillomatosis, 疣 参见 乳头状瘤
Wesselbron fever, 韦瑟尔斯布朗热, **281, 325**
Wesselbron virus, 韦瑟尔斯布朗病毒, **224**
Western blotting
contagious erythema diagnosis, 接触传染性脓疱的诊断, 247
toxoplasmosis diagnosis, 弓形虫诊断, 384
Western equine encephalitis, 西方马脑炎, **279**
West Nile fever, 西尼罗河热, **279**
Dromedary, 单峰驼, 478
NWC, 新大陆骆驼, **474**
West Nile virus, 西尼罗河热病毒, **324**
encephalitis, 脑炎, 287–288
literature review, 文献综述, 223
WNS (wry-neck syndrome), tetanus, "歪脖"综合征, 破伤风, 200, *201*
*Wohlfahrtia magnifica,* 黑须污蝇, 455, *456*
*Wohlfahrtia nuba,* 云朵污蝇, 456, *456*
wool rot, dermatophilosis, 毛腐病, 嗜皮菌病, 181
wool source, 产毛, 4
wound healing, 创伤愈合, 166
wry-neck syndrome (WNS), tetanus, "歪脖"综合征, 破伤风, 200, *201*

**Y**

yellow fever, 黄热病,325
    literature review, 文献综述,223
*Yersinia pestis* infection *see* plague,鼠疫耶尔森氏菌，参见鼠疫
*Yersinia pseudotuberculosis* infection, 假结核耶尔森菌，208

**Z**

Zephiran (benzalkonium chloride), 氯化苯甲氢铵，109–110
zygomycosis, 结合菌病,**344**
*Zygoribatula,* 若甲螨,409

# 索引

注：表格位置的页码为黑体，插图的页码为斜体，vs.表示比较或对比。缩写词列于本书的开始部分（xi–xii页）.


**a**

“歪脖”综合征，wry–neck syndrome，200，*201*

“歪脖”综合征，破伤风，WNS (wry–neck syndrome)，tetanus，200，*201*

3型副流感病毒诊断，parainfluenza virus 3 diagnosis，310–311

L–J培养基，结核病的诊断，Loewenstein–Jensen media，tuberculosis diagnosis，**121**，*122*

Q热，Q fever，84–90

Q热的诊断，Q fever diagnosis，86–87

RNA病毒，RNA viruses，**279**

α–溶血性链球菌，鼻试子，α–haemolytic streptococci，nasal swabs，**133**

β–溶血性链球菌，鼻试子，β–haemolytic streptococci，nasal swabs，**133**

*β*逆转录病毒，*Betaretrovirus*，321

γ–干扰素检测方法，结核病诊断，interferon gamma assay，tuberculosis diagnosis，120，*122*

*δ*逆转录病毒，*Deltaretrovirus*，320

阿莫西林，沙门氏菌病的治疗，amoxicillin–clavulanate，salmonellosis treatment，**97**

阿维菌素，螨的治疗，451

阿昔洛韦，马疱疹病毒1型感染治疗，acyclovir，equine herpesvirus–1 infection treatment，**302**

埃氏扇头蜱，*Rhicephalus evertsi*，*437*

埃氏双瓣线虫，*Deraiophoronema evansi*，**420**

埃希氏菌感染，肺脏，*Escherichia* infections，lungs，**130**

艾美尔球虫，*Eimeria*，371

艾美尔球虫病，*Eimeria* coccidiosis，371–377

艾美尔球虫病的诊断，*Eimeria* coccidiosis diagnosis，375

艾美尔球虫感染 参见 艾美尔球虫病，*Eimeria* infection *see Eimeria* coccidiosis，371

艾氏毛圆线虫，*Trichostrongylus axei*，**419**

安氏革蜱，流行病学，*Dermacentor andersoni*，epidemiology，442

按蚊，裂谷热媒介，*Anopheles*，Rift Valley Fever vectors，275

奥芬达唑，寄生虫性胃肠炎治疗，oxfendazole，parasitic gastroenteritis treatment，**428**，**429**

奥利春细颈线虫，*Nematodirus oriatinus*，**420**

奥美拉唑，内毒素中毒治疗，omeprazole，endotoxicosis treatment，46

奥氏奥斯特线虫，*Osteragia ostertagi*，**420**

奥氏等孢球虫，*Isospora orlovi*，380，381，*381*

澳大利亚，Australia，3

**b**

巴佰顿毛首线虫，*Trichuris barbetonensis*，**419**

巴氏杆菌病病，pasteurellosis，57–64

巴氏细颈线虫，*Nematodirus battus*，**420**

巴氏消毒，pasteurization，187

巴氏消毒奶，pasteurization milk，187

巴氏杆菌，肺脏，*Pasteurella*，lungs，**131**

白蛉，sandflies，463

白蛉亚科，Phlebotominae，468

白色念珠菌感染，霉菌性皮炎，*Candida albicans* infection，mycotic dermatitis，**332**

白头裸金蝇，*Chrysomya albiceps*，*457*

白细胞，内毒素中毒，leucocytes，endotoxicosis，**41**，*41*

邦戈尔沙门氏菌，*Salmonella bongori*，91

包虫病，hydatidosis，**412**，**417**

包虫病的诊断，hydatidosis diagnosis，415

宝石花蜱，*Amblyomma gemma*，*436*

鲍氏囊，内毒素中毒，Bowman's capsule，endotoxicosis，41，*41*

贝诺孢子虫病，besnoitiosis，395

贝氏柯克斯体 参见 Q热*Coxiella burnetii* infection *see* Q fever，84

贝氏莫尼茨绦虫，*Moniezia benedeni*，*407*，*410*

倍硫磷，鼻咽蝇蛆病治疗，sulfidofos，nasopharyngeal myiasis treatment，460

被动获得，passive acquisition，292–295，*292*

被动免疫疗法，immunotherapy，passive，287

被动免疫疗法，西尼罗河病毒，passive immunotherapy，West Nile virus，286

苯并咪唑，寄生虫性胃肠炎治疗，probenzimidazoles，parasitic gastroenteritis treatment，429

苯并咪唑类药物，benzimidazoles，400，410

苯硫咪唑，fenbendazole，368，406

鼻疽，glanders，51–53

鼻疽伯氏菌感染 参见鼻疽，*Burkholderia mallei* infection *see* glanders，51

鼻疽菌素试验，类鼻疽诊断，mallein tests，melioidosis diagnosis，56

鼻疽诊断，glanders diagnosis，52–53

鼻试子，nasal swabs，132，133

鼻试子，肺炎，nasal swabs，pneumonia，**133**

鼻咽，nasopharyngeal，458–460
鼻咽蝇蛆病，nasopharyngeal myiasis，458–460
比赛，racing，13–14，*14*
吡喹酮，Praziquantel，403，410，415
边缘无浆体感染，*Anaplasma marginale* infection，75，*77*
变形杆菌，*Proteus*，131，178
变性梯度凝胶电泳，支原体病的诊断，DGGE (denaturing gradient gel electrophoresis)，mycoplasmosis diagnosis，78
丙酰丙嗪，破伤风，propionyl promazine，tetanus，*202*
病毒病，viral disease，213–322，**216–224**，**479**–473
病毒分离，狂犬病诊断，viral isolation，rabies diagnosis，228
病毒混合感染，viral co–infections，**270**
病理学变化，地方性牛白血病，histopathology，enzootic bovine leukosis，322
病原分离，布鲁氏菌病的诊断，micro–organism isolation，brucellosis diagnosis，141
波蒂斯库姆病毒，Potiskum virus，**325**
波托马克马热的诊断，Potomac horse fever diagnosis，**81**
博尔纳病，Borna disease (BD)，231–234
补体结合试验，complement fixation test (CFT)，52，55–56，142–143，319，384
布鲁氏菌，brucellosis，137–151
布尼亚病毒科，*Bunyaviridae*，**280**
布尼亚病毒，Bunyavirus，**280**
布氏锥虫，*Trypanosoma brucei*，*358*

**c**

采采蝇，tsetse flies，467，*467*
彩饰花蜱，*Amblyomma variegatum*，*437*
残缘璃眼蜱，*Hyalomma detritum*，*438*
草场卫生，肝片吸虫的控制，pasture hygiene，fasciolosis control，*400*
草原革蜱，*Dermacentor nuttalli*，*437*
查氏无峰驼线虫，*Lamanema chavesi*，**420**
差异脲原体，传染性繁殖障碍，*Ureaplasma diversum*，infective reproductive losses，155
差异脲原体，*Ureaplasma diversum*，155
产毛，wool source，4
产奶，milk/milking，4，*183*，*183*，*185*
产气荚膜梭菌，*Clostridium perfringens*，**353**
产志贺氏毒素大肠杆菌，shiga toxin–producing *Escherichia coli* (STEC)，100
肠产毒素性大肠杆菌，enterotoxigenic *Escherichia coli* (ETEC)，100
肠道沙门氏菌，*Salmonella enterica*，91
肠毒血症，enterotoxaemia，24，102
肠毒血症抗原检测，Enterotoxaemia Antigen Test，28
肠毒血症综合征 参见 梭菌肠性肠毒血症 enterotoxaemia complex *see Clostridium* enterotoxaemia，24
肠杆菌，*Enterobacter*，*131*，132
肠炎慢性，chronic，96
肠致病性大肠杆菌，EPEC (enteropathogenic *Escherichia coli*)，100
沉淀法，肝片吸虫病的诊断，sedimentation method，fasciolosis diagnosis，399–400
沉静型，狂犬病，silent fury，rabies，*227*
耻垢分枝杆菌，*Mycobacterium smegmatis*，**117**
赤羽病，Akabane disease，**280**，**324**
赤羽病毒，Akabane virus，324
虫媒病毒，arboviruses，280
虫媒病毒感染，arbovirus infections，324–327，**324**
出血，内毒素中毒，haemorrhages，endotoxicosis，*40*，41，*41*，**42**
出血点，内毒素中毒，petechiae，endotoxicosis，*40*
出血性败血症，haemorrhagic septicaemia，59–63，61
出血性肠炎，haemorrhagic enteritis，*91*，91，375
出血性肺炎，haemorrhagic pneumonia，**129**
出血性素质，haemorrhagic diathesis，37
出血性子宫内膜炎，haemorrhagic endometritis，**158**
出血症，haemorrhagic diseases，38
初乳，colostrum，293–294，**293**
初乳，大肠杆菌病的治疗，colostrum，colibacillosis treatment，103–104
储藏病原动物，animal reservoirs，315
传播，transmission，226，246，279，300，**317–318**，359，360，252，448
传播方式，transmission mode，114
传染性，infectious，*209*
传染性乳房炎，infectious mastitis，190
传染性鼻气管炎，infectious bovine rhinotracheitis，**301**
传染性繁殖障碍，infective reproductive losses，151–163，**153**，**154**，155，**156**，***157***
传染性繁殖障碍的诊断，infective reproductive loss diagnosis，156
传染性繁殖障碍的治疗，infective reproductive loss treatment，157
传染性龟头包皮炎，pustular balanoposthitis，infectious，**301**
传染性坏死性肝炎（羊黑疫），梭菌感染，infectious necrotising hepatitis (black disease)，*Clostridium* infections，24
传染性脓疱样外阴道炎，infectious pustular vulvovaginitis，**301**
传染性乳房炎，infectious mastitis，188–199，190，*191*–192
传染性乳房炎的治疗，infectious mastitis treatment，193
传染性蹄皮炎，infectious pododermatitis，*209*
疮痂嘴，接触传染性脓疱，scabby mouth *see* contagious erythema，245
创伤愈合，wound healing，166
垂盾甲螨，*Scutovertex*，409
脆弱拟杆菌感染，*Bacteroides fragilis* infection，208

**d**

达尔梅尼病，Dalmeny disease，391–392
达乎尔奥斯特线虫，*Osteragia dahurica*，**420**
大肠杆菌病，colibacillosis，100–105
大肠杆菌病的诊断，colibacillosis diagnosis，*103*
大肠杆菌的治疗，colibacillosis treatment，104
大肠杆菌感染，*Escherichia coli* infections，102
大肠杆菌性败血症，念珠菌病，colisepticaemia，candidosis，339
大环内酯，lactones，macrocyclic，**429**
大环内酯，寄生虫性胃肠炎治疗，macrocyclic lactones，parasitic gastroenteritis treatment，429
大拟片形吸虫，*Fascioloides magna*，397，***397–398***
大头金蝇，*Chrysomya megacephala*，457
大羊驼犬肉孢子虫，*Sarcocystis lamacanis*，**391**

大翼甲螨，裸头科绦虫的中间宿主，*Galumna*，anoplocephalidosis intermediate hosts，409
丹卡利毛癣菌感染，*Trichophyton dankaliense* infection，**332**
单次结核菌素皮内比较试验，SICTT (single intradermal comparative tuberculin test)，119
单次结核菌素皮内比较试验，single intradermal comparative tuberculin test (SICTT)，119
单次结核菌素皮内比较试验，single intradermal tuberculin test (SITT)，119
单次结核菌素皮内比较试验，SITT (single intradermal tuberculin test)，119
单峰驼，Dromedary *(Camelus dromedarius)*，*6*，*7*，255–258 *257*，258，**475**，**478**，**487**，**487**
单峰驼艾美尔球虫，*Eimeria dromedarii*，**371**，*374*
单峰驼毛细线虫，*Nematodirus dromedarii*，**420**，*428*
单峰驼膨首线虫，*Physocephalus dromedarii*，**420**
单向放射免疫扩散检测法，radial diffusion，passive immunoglobulin transfer，**294** *294*
胆管炎，肝片吸虫病，cholangitis，fasciolosis，399
胆管周炎，肝片吸虫病，pericholangitis，fasciolosis，*399*
弹状病毒，*Rhabdovirus*，**280**
氮氨菲啶氯化物，锥虫病治疗，isometamidium chloride，trypanosomosis treatment，363，**363**
登革病毒，dengue virus，**325**
等孢球虫，isosporosis，380–383
等孢球虫病，isosporosis diagnosis，380
等孢球虫病的诊断，isosporosis diagnosis，*381*
敌百虫，鼻咽蝇蛆病的治疗，metrifonat，nasopharyngeal myiasis treatment，460
敌百虫，螨的治疗，trichlorfon，mange treatment，*451*
地方流行性牛白血病，EBL *see* enzootic bovine leukosis (EBL)
地方性牛白血病，enzootic bovine leukosis (EBL)，321–323，*322*
地下沙门氏菌，*Salmonella subterranea*，91
第三胃出血，内毒素中毒，abomasum haemorrhage，endotoxicosis，*40*
第一胃特征，compartment 1 properties，41–42，**42**
点状古柏线虫，*Cooperia oncophora*，**420**
碘硫胺，肝片吸虫病的治疗，rafoxanide，fasciolosis treatment，400
东部马脑炎，Eastern equine encephalitis，287
毒素，toxins，**33**
钝缘蜱，鼠疫媒介，*Ornithodoros*，plague vectors，65
多价抗血清，肉毒梭菌中毒的治疗，polyvalent antiserum，botulism treatment，33
多抗原线条免疫分析法，结核病，multiantigen print immunoassay (MAPIA)，tuberculosis，119–120
多拉菌素，doramectin，451，429
多黏菌素B，内毒素中毒治疗，polymyxin B，endotoxicosis treatment，45
多杀性巴氏杆菌感染，*Pasteurella multocida* infection，**57**，57–58
对流免疫电泳，小反刍兽疫诊断，counter electrophoresis，peste des petits ruminants，diagnosis，263

**e**

恶性卡他热，文献综述，malignant catarrhal fever，literature review，**219**
恶性水肿，oedema，malignant，22
恶性水肿，梭菌感染，malignant oedema，*Clostridium* infections，22
耳尖坏死沙门氏菌病，ear tip necrosis，salmonellosis，93，**93**
二甲基亚砜，内毒素中毒治疗，dimethyl sulfoxide (DMSO)，endotoxicosis treatment，45

**f**

反刍动物，比较，ruminants，in comparison，8–**9**
反刍兽无浆体，*Anaplasma (Cowdria) ruminantium*，**76**
反转录PCR，reverse–transcriptase polymerase chain reaction (RT–PCR)，306，311，320
放射穗体绦虫，*Thysanosoma actinoides*，408–409
放线菌，*Actinomyces*，131，132
放线菌病，新大陆骆驼，actinobacillosis，NWC，**486**
非饱和脂肪酸，non–fatty acids，milk，186
非洲马瘟，African horse sickness (AHS)，**280**，**317–321**
绯颜裸金蝇，*Chrysomya rufifacies*，457
肺线虫感染，lungworm infections，430–433
肺炎克雷伯氏菌，*Klebsiella pneumoniae*，鼻试子，nasal swabs，131–132
肺炎球菌，肺脏，*Pneumococcus*，lungs，**131**
肺脏出血，lung haemorrhages，*25*
分体吸虫病，schistosomosis，404–405
分枝杆菌感染，肺脏，*Mycobacterium* infection，lungs，**131**
分枝杆菌属的细菌，*Mycobacterium* species，**117**
芬苯达唑，寄生虫性胃肠炎治疗，febantel，parasitic gastroenteritis treatment，**429**
粪便中细菌分离，沙门氏菌病的诊断，faecal isolation，salmonellosis diagnosis，96
粪样品/涂片，faecal sampling/smears，96，399，380，397，415
氟苯尼考，沙门氏菌病的治疗，florphenicol，salmonellosis treatment，**97**
氟喹诺酮类，沙门氏菌病的治疗，fluoroquinolones，salmonellosis treatment，**97**
氟氯苯氰菊酯，蜱的防控，flumethrin，tick infestation control，443
福斯盘尾丝虫，*Onchocerca fasciata*，**420**
腐败梭菌，*Clostridium septicum*，**21**，**353**
副痘病毒，parapoxvirus，**354**
副结核病，paratuberculosis (Johne's disease)，105–112
副结核分枝杆菌感染，*Mycobacterium paratuberculosis* infection，114
副流感，文献综述，parainfluenza virus，literature review，**221**
副流感病毒，parainfluenza virus 3 (PI–3)，311–312

**g**

改进的弓形虫诊断方法，modified，toxoplasmosis diagnosis，384
改进的狂犬病灭活疫苗，MLV (modified live virus) vaccine，rabies，*230*
改进的凝集试验，弓形虫诊断，modified agglutination test，toxoplasmosis diagnosis，384
盖氏曲子宫绦虫，*Thysaniezia giardi*，408，*408*，409
肝片吸虫病，fasciolosis，397–401
肝片吸虫病的治疗，fasciolosis treatment，400
肝斯泰勒虫，*Stilesia hepatica*，407

肝脏肿大，副结核病，liver enlargement，paratuberculosis，*108*
杆菌感染，*Bacillus* infections，*23*
干酪性淋巴腺炎，caseous lymphadenitis (CLA)，166–176
干酪性淋巴腺炎治疗，caseous lymphadenitis treatment，171
刚果嗜皮菌，*Dermatophilus congolensis*，*178*
刚果锥虫，*Trypanosoma congolense*，*358*
哥伦比亚结节线虫，*Oesophagostomum columbianum*，**420**
革兰氏阴性菌感染，Gram–negative bacterial infections，208
革蜱，*Dermacentor*，*437*
革蜱，感染频度，*Dermacentor*，frequency of infection，**442**
弓形虫病，toxoplasmosis，383–386
弓形虫病诊断，toxoplasmosis diagnosis，384
公畜不育，传染性繁殖障碍，male infertility，infective reproductive losses，**160**
宫腔积脓，pyometra，**158**
钩端螺旋体，leptospirosis，70
钩端螺旋体病，leptospirosis，**68–74**
钩端螺旋体病的诊断，leptospirosis diagnosis，72
钩端螺旋体病的治疗，leptospirosis treatment，72
古柏线虫，*Cooperia*，*423*
冠状病毒，文献综述，coronavirus，literature review，**220**
冠状病毒新生畜腹泻，coronavirus neonatal diarrhoea，289，291
广谱抗生素，肺炎的治疗，broad–spectrum antibiotics，pneumonia therapy，133
鲑鱼中毒，salmon poisoning，diagnosis，**81**
诡谲腐霉，霉菌性皮炎，*Pythium insidiosum*，mycotic dermatitis，**332**

**h**

哈芒球虫病，Hammondiosis，389–390
哈氏器，Haller's organ，436，*436*
荷包腹袋吸虫，*Gastrothylax crumnifer*，405
核内包涵体，博尔纳病，intranuclear inclusion bodies，Borna disease，228，*230*
黑脚病，梭菌感染，blackleg，*Clostridium* infections，22
黑曲霉感染，脓肺炎，*Aspergillus niger* infection，pyogranulomatous pneumonia，338
黑须污蝇，*Wohlfahrtia magnifica*，455，*456*
黑缘璃眼蜱，*Hyalomma erythraeum*，440
喉，pharyngeal，65
喉瘤盘尾丝虫，*Onchocerca gutturosa*，**420**
猴锥虫，*Trypanosoma simiae*，*358*
呼肠孤病毒，*Reoviridae*，**279**
呼肠孤病毒，reovirus，**324**
呼吸道病毒，respiratory viruses，310–313
呼吸道合胞体病毒，respiratory syncytial virus，**221**
呼吸系统，respiratory system，*113*–136
呼吸系统细菌病，respiratory system，bacterial disease，113–136
胡米氯铵，锥虫病治疗，homidium chloride，trypanosomosis treatment，**363**
虎红平板试验，布鲁氏菌病诊断，RBPT (rose bengal plate test)，brucellosis diagnosis，143
花蜱，*Amblyomma*，437
花蠕形蚤，*Vermipsylla alacurt*，469
化脓，金黄色葡萄球菌皮炎，pyogenic affections，*Staphylococcus aureus* dermatitis，176–178
化脓棒状杆菌，传染性乳房炎，*Corynebacterium pyogenes* infection，infectious mastitis，192
化脓放线菌，*Actinomyces pyogenes*，153
化脓无色杆菌感染，*Achromobacter pyogenes* **infection**，**208**
化脓性肺炎，suppurative pneumonia，**129**
坏死肝炎，沙门氏菌病，necrotising hepatitis，salmonellosis，96
坏死梭杆菌感染，*Fusobacterium necrophorum* infection，208–209
坏死性肺炎，necrotic pneumonia，**129**
坏死性子宫内膜炎，necrotic endometritis，**158**
环状病毒，*Orbivirus*，**280**
黄病毒，*Flavivirus*，**279**
黄病毒科，*Flaviviridae*，**279**
黄疸血红素尿，梭菌感染，ictero haemoglobinuria，*Clostridium* infections，24
黄热病，yellow fever，325
磺胺二甲嘧啶，艾美尔球虫病的治疗，sulfadimidine，*Eimeria* coccidiosis treatment，376
磺胺增效类药物，沙门氏菌病的治疗，potentiated sulphonamides，salmonellosis treatment，**97**
灰黄霉素，霉菌性皮炎治疗，griseofulvin，mycotic dermatitis treatment，333
活跃锥虫，*Trypanosoma vivax*，*358*

**j**

肌肉炎，肉孢子虫病，myositis，sarcocystosis，**392**
鸡埃及小体，*Aegyptianella pullorum*，**76**
姬姆萨染色，念珠菌病的诊断，Giemsa staining，candidosis diagnosis，*341*
极长头微胸虱，*Microthoracius praelongiceps*，**463**
急性化脓性子宫内膜炎，suppurative endometritis，acute，158
急性化脓性子宫炎，acute suppurative metritis，**158**
棘球蚴，echinococcosis，412–417
棘球蚴病的治疗，echinococcosis treatment，415
寄生虫病，parasitic disease，355–471，**487–489**
寄生虫性胃肠炎的诊断，parasitic gastroenteritis diagnosis，427
寄生虫性胃肠炎治疗，parasitic gastroenteritis/colitis，422–428
寄生性结肠炎 参见 寄生性结胃肠炎/结肠炎colitis，parasitic *see* parasitic gastroenteritis/colitis
甲苯达唑，寄生虫性胃肠炎治疗，mebendazole，parasitic gastroenteritis treatment，**428**
甲病毒，*Alphavirus*，280
甲醛磺胺噻唑，艾美尔球虫病的治疗，formosulfathiazole，*Eimeria* coccidiosis treatment，376
甲噻吩咪唑，寄生虫性胃肠炎治疗，morantel tartrate，parasitic gastroenteritis treatment，**428**
甲型流感病毒感染，双峰驼，influenza A infection，Bactrian camel，**473**
甲型马鼻炎病毒，equine rhinitis A virus，155
贾地鞭毛虫病，giardiosis，367–369
贾地鞭毛虫病诊断，giardiosis diagnosis，368
贾地鞭毛虫病治疗，giardiosis diagnosis，368
假单孢菌，*Pseudomonas*，131，178
假结核参见 干酪性淋巴腺炎，pseudotuberculosis *see* caseous lymphadenitis (CLA)，
假结核棒状杆菌，皮肤，*Clostridium psudotuberculosis*，integu-

ment, **178**
假结核棒状杆菌感染 参见 干酪性淋巴腺炎 *Corynebacterium pseudotuberculosis* infection *see* caseous lymphadenitis (CLA)
假结核耶尔森菌, *Yersinia pseudotuberculosis* infection, 208
假蛛网霉感染, 霉菌性皮炎, *Pseudoarachiniotus* infection, mycotic dermatitis, **332**
间接血清学试验, 布鲁氏菌病诊断, indirect serological tests, brucellosis diagnosis, 143, **144**
间接血球凝集试验, 弓形虫病诊断indirect haemagglutination test, toxoplasmosis diagnosis, 384
间质性肺炎, interstitial pneumonia, **129**
间质性肺炎, 副结核病, focal interstitial pneumonia, paratuberculosis, 107
间质性肝炎, hepatitis, interstitial, *109*
间质性肝炎, 副结核病, interstitial hepatitis, paratuberculosis, *109*
检测马流感的FLU–A检测试剂盒, Directogen FLU–A test, influenza diagnosis, 306, *306*
酱麻蝇, *Sarcophaga dux*, 456, *456*
角膜结膜炎, **keratoconjunctivitis**, *163*
角蝇, *Haematobia*, *465*
接触传染性脓疱, contagious erythema, 246–251
接触传染性脓疱的诊断, contagious erythema diagnosis, 247
节肢动物感染, arthropod infections, 436–471
结肠黏膜, 副结核病, colon mucosa, paratuberculosis, *108*
结合菌病, zygomycosis, **344**
结核病, tuberculosis, 113–127, 117
结核病的诊断, tuberculosis diagnosis, 121
结核菌素皮内比较试验, TST (tuberculin skin test), 118, *118*
结节无囊线虫, *Impalaia tuberculata*, **420**
芥子油, 念珠菌病的治疗, sulphur in mustard oil, candidosis treatment, 341
疥螨, *Sarcoptes*, 446–448
金孢子菌感染, 霉菌性皮炎, *Chrysosporium* infection, mycotic dermatitis, **332**
金黄色葡萄球菌, *Staphylococcus aureus*, 176–179
金黄色葡萄球菌皮炎, *Staphylococcus aureus* dermatitis, 176–179
精氨酸支原体, *Mycoplasma argini*, 155
精氨酸支原体, 传染性繁殖障碍, *Mycoplasma argini*, infective reproductive losses, 155
厩螫蝇, *Stomoxys calcitrans*, 465, *465*
锯齿状舌形虫, *Linguatula serrata*, **470–471**
菌甲螨, *Scheloribates*, *408*

**k**

卡达姆山病毒, Kadam virus, **324**, **324**
卡他性肺炎, catarrhal pneumonia, **128**
卡他性子宫内膜炎, catarrhal endometritis, **157**
堪萨斯分枝杆菌, *Mycobacterium kansasii*, **117**
抗破伤风血清, 破伤风治疗, anti–tetanus serum, tetanus treatment, 202
抗球虫药物, 艾美尔球虫病的治疗, anticoccidials, *Eimeria* coccidiosis treatment, 376
克菌丹, 霉菌性皮炎治疗, captan, mycotic dermatitis treatment, 333
克雷伯氏菌感染, 肺脏, *Klebsiella* infection, lungs, **129**
克里米亚–刚果出血热(CCHF), CCHF *see* Crimean–Congo haemorrhagic fever (CCHF), **324**, 324–325
克里米亚–刚果出血热, Crimean–Congo haemorrhagic fever (CCHF), **324**, 324–325
克洛索隆, 肝片吸虫病的治疗, clorsulon, fasciolosis treatment, 400
口腔黏膜, mucous membranes, oral cavity, 168
口腔黏膜, 干酪性淋巴腺炎, oral cavity mucous membranes, caseous lymphadenitis, 168
口蹄疫, foot and mouth disease virus (FMDV), **219–220**
库蠓, *Culicoides* , *278*, *317*
库蚊, 裂谷热媒介, *Culex*, Rift Valley Fever vectors, 274
夸兰菲尔病毒, Quaranfil virus, 324
快速检测方法, 结核病诊断, rapid test (RT), tuberculosis diagnosis, 119–120
狂犬病, rabies, 226–230
狂犬样病毒, rabies–like virus, **325**
狂犬样反转录病毒, rabies–like retrovirus, **325**
狂蝇感染, *Cephalopina* infestation, 487, 488
狂躁, 狂犬病, raging fury, rabies, *227*
喹匹拉明, quinapyramine methyl sulphate, 363, **363**
溃疡, 内毒素中毒, ulcers, endotoxicosis, *39*
溃疡棒状杆菌, 干酪性淋巴腺炎, *Clostridium ulcerans* infection, caseous lymphadenitis, 168
昆虫不育技术, 蝇蛆病控制, sterile insect release, myiasis control, 456–460
扩展莫尼茨绦虫, *Moniezia expansa*, *407*, *409*

**l**

拉氏艾美尔球虫, *Eimeria rajasthani*, **371**, *374*
蜡样芽孢杆菌毒素中毒, *Bacillus cereus* intoxication, 38
蓝舌病, bluetongue (BT), 279–286
蓝舌病的诊断, bluetongue diagnosis, 280, 282
蓝舌病媒介, bluetongue vector, 278
老鼠试验, 肉毒中毒诊断, mouse test, botulism diagnosis, 33–34, **34**
类白喉菌鼻试子, diphtheroids, nasal swabs, 133
类鼻疽, melioidosis, 54–57
类鼻疽伯氏菌感染 参见 类鼻疽, *Burkholderia pseudomallei* infection *see* melioidosis, 54
类鼻疽诊断, melioidosis diagnosis, 55–56
类圆线虫, 生活史, *Strongyloides*, life cycle, *428*
梨形虫病, piroplasmosis, 443
李斯特菌病, listeriosis, **204–207**
李斯特菌病的治疗, listeriosis treatment, 206
立克次体病, rickettsial diseases, 74
立氏新立克次氏体, *Neorickettsia risticii*, **76**
丽甲螨, *Liacarus*, **409**
链霉菌, 鼻试子, *Streptomyces*, nasal swabs, **133**
链霉素, Streptomycin, **66**, **72**, **145**, **182**
链球菌, *Streptococcus*, **131**
两性霉素B, 球孢子菌病的治疗, amphotericin B, coccidiomycosis treatment, 342
裂谷热, Rift Valley Fever (RVF), 274–278
裂谷热媒介, Rift Valley Fever vectors, 274
裂谷热诊断, Rift Valley Fever diagnosis, 275, 276
邻氯青霉素, 传染性乳房炎的治疗, cloxacillin, infectious mas-

titis treatment，193
林丹，螨的治疗，lindane，mange treatment，**450**
淋巴母细胞白血病，lymphoblastic leukaemia，*321*
淋巴肉瘤，lymphosarcoma，322
淋巴细胞增殖试验，结核病诊断，lymphocyte proliferation assay，tuberculosis diagnosis120
淋巴腺炎，干酪性淋巴腺炎，lymphadenitis，caseous lymphadenitis，168
蛉病毒，*Phlebovirus*，**280**
流产布鲁氏菌，*Brucella abortus*，**137**
流感，influenza，305–307
流感诊断，influenza diagnosis，306
流行热，ephemeral fever，**279**
流行性出血热，epizootic haemorrhagic diseases，**279**
瘤胃炎，rumenitis，338
瘤胃炎，烟曲霉，rumenitis，*Aspergillus fumigatus*，338
六氯环己烷，螨的治疗，γ –hexachlorocyclohexane (γ –HCH)，mange treatment，451
鹿前后盘吸虫，*Paramphistomium cervi*，*405*
卵囊，oocysts，*373*
轮状病毒，文献综述，rotavirus，literature review，220
轮状病毒新生畜腹泻，rotavirus neonatal diarrhoea diagnosis，290，*291*
裸头科绦虫病，Anoplocephalidosis，406–412
裸头科绦虫病的治疗，anoplocephalidosis treatment，410
裸头科绦虫病的中间宿主，anoplocephalidosis intermediate hosts, 409
骆驼艾美尔球虫，*Eimeria cameli*，**371**，*372*，*374*
骆驼的分类，classification of camelids，**5**
骆驼的利用，use of camelids，7–8
骆驼痘诊断，camelpox diagnosis，236
骆驼喉蝇，*Cephalopina titllator*，*459*，*460*
骆驼蜱 参见 嗜驼璃眼蜱，camel tick *see Hyalomma dromedarii*，357
骆驼肉孢子虫，*Sarcocystis cameli*，**391**
骆驼虱蝇，*Hypobosca camelina*，*468*
骆驼微胸虱，*Microthoracius cameli*，**463**
骆驼正痘病毒 参见 骆驼痘，*Orthopoxvirus cameli see* camelpox
绿尾金蝇，*Chrysomya chloropyga*，457
氯化苯甲氢铵，benzalkonium chloride (Zephiran)，109–110
氯化苯甲氢铵，Zephiran (benzalkonium chloride)，109–110
氯霉素，chloramphenicol，82，164
氯氰碘硫胺，肝片吸虫病的治疗，closantel，fasciolosis treatment，400
氯氰菊酯，虱病治疗，cypermetrin，pediculosis treatment，464

**m**

麻点璃眼蜱，*Hyalomma rufipes*，438
麻虻，*Haematopota*，*466*
麻蝇，Sarcophagidae flies，455–456
麻蝇科的蝇类，蝇蛆病，Sarcophagidae flies，myiasis，455–456
马棒状杆菌，干酪性淋巴腺炎，*Clostridium equi* infection，caseous lymphadenitis，167
马鼻炎病毒甲型，equine rhinitis A virus，155
马动脉炎炎病毒，equine arteritis virus (EAV)，**324**，**325**
马尔他布鲁氏菌，*Brucella melitensis*，**137**
马耳他热，Malta fever，137–138
马红球菌 鼻试子，*Rhodococcus equi*，nasal swabs，**132**
马红球菌感染，*Rhodococcus equi* infection，49–50
马红球菌感染诊断，*Rhodococcus equi* infection diagnosis，*49*
马拉色菌感染，霉菌性皮炎，*Malassezia* infection，mycotic dermatitis，**331**
马梨形虫，equine piroplasmosis，443
马链球菌兽疫亚种，*Streptococcus equi* subsp. *zooepidemicus*，208
马脑炎，encephalomyelitis，**279**
马脑炎，equine encephalomyelitis，**279**
马疱疹病毒1型，EHV–1 (equine herpesvirus–1)，301–302
马疱疹病毒1型，equine herpesvirus–1 (EHV–1)，300–301
马疱疹病毒1型感染诊断，equine herpesvirus–1 infection diagnosis，301
马氏马歇尔线虫，*Marshalagia marshalli*，**420**
马无浆体病的诊断，equine anaplasmosis diagnosis，**81**
螨，mange，446–454
螨的治疗，mange treatment，450
曼库塞尼艾美尔球虫，*Eimeria macusaniensis*，**371**，*372*，*374*
曼蚊，*Mansonia*，274
慢病毒，*Lentivirus*，**320**
慢性肠炎，chronic enteritis，96
慢性非化脓性肺炎，non–suppurative pneumonia，chronic，**129**
慢性非化脓性子宫炎，chronic non–suppurative metritis，**158**
慢性非脓性肺炎，chronic non–suppurative pneumonia，**129**
慢性活动性子宫炎，chronic active metritis，**158**
慢性卡他性子宫内膜炎，chronic catarrhal endometritis，157
慢性增生性肺炎，proliferative pneumonia，chronic，**129**
慢性支气管肺炎，bronchopneumonia，chronic，131
猫传染性贫血，feline infectious anaemia，diagnosis，**81**
毛腐病，嗜皮菌病，wool rot，dermatophilosis，181
毛甲螨，*Trichoribates*，409
毛里塔尼亚细颈线虫，*Nematodirus mauritanicus*，**420**
毛霉菌病，mucormycosis，343，*343*
毛细线虫，*Capillaria*，**419**
毛癣菌，*Trichophyton vs.*，*331*
毛圆线虫，*Trichostrongylus*，424
梅察微胸虱，*Microthoracius mazzai*，**463**
媒介控制，锥虫病，vector control，trypanosomosis，364
酶联免疫吸附试验，enzyme–linked immunosorbent assay (ELISA),28–29,33,55–56,86–87,118,137,172,182,228,236,247,275,276,290,291,306,310,388
霉菌性皮炎，mycotic dermatitis，331–336
霉样真菌感染，霉菌性皮炎，*Allescheria* infection，mycotic dermatitis，**332**
美拉胂醇，锥虫病的治疗，melarsopol，trypanosomosis treatment，**364**
美丽扇头蜱，*Rhicephalus pulchellus*，437，*437*
美洲驼，llama *(Lama glama)*，*5*
美洲驼，*Lama glama see* llama *(Lama glama)*，*5*
虻，*Tabanus*，*461*
虻科，Tabanidae，466–467
蒙古国，Mongolia，3
蠓，midges，463
蠓科，Ceratopogonidae，468
咪康唑，念珠菌病的治疗，miconazole，candidosis treatment，

341
咪唑噻唑类，寄生虫性胃肠炎治疗，imidazothiazoles，parasitic gastroenteritis treatment，**429**
弥漫性增生性肠炎，副结核病，diffuse proliferative enteritis，paratuberculosis，*109*
弥散性血管内凝血，内毒素中毒，DIC (disseminated intravascular coagulation)，endotoxicosis，37，42
米尔贝霉素，螨的治疗，milbemycins，mange treatment，451
泌尿生殖系统，urogenital system，137–165
泌尿生殖系统，细菌病，urogenital system，bacterial disease，137–165
秘鲁艾美尔球虫，*Eimeria peruviana*，**371**
秘鲁刺翼线虫，*Spiculopteragia peruvianus*，**420**
绵羊布鲁氏菌，*Brucella ovis*，**137**
绵羊无浆体，*Anaplasma ovis*，*76*
绵羊夏伯特线虫，*Chabertia ovina*，**420**
免疫球蛋白的被动转移，passive immunoglobulin transfer，295–296
免疫组织化学，李斯特菌病，immunohistology，listeriosis，*205*
灭绝危险，extinction threats，3
摩拉克氏菌感染，*Moraxella* infections，209
抹片，传染性繁殖障碍的诊断，smears，infective reproductive loss diagnosis，157–158
莫尼茨绦虫，*Moniezia*，*406*
莫西菌素，moxidectin，428,450

**n**

奈瑟氏菌，鼻试子，*Neisseria*，nasal swabs，**133**
奈托比胺，netobimin，400,403,428
囊尾蚴病，cysticercosis，417–418
脑多头蚴病，coenurosis，417–418
脑心肌炎病毒，encephalomyocarditis virus (EMCV)，258
脑炎，encephalitis，287–288
内毒素血症，enterotoxaemia，**353**
内毒素疫苗Endovac–Bovi ，内毒素治疗，Endovac–Bovi，endotoxicosis treatment，46
内毒素中毒，endotoxicosis，37–47
内毒素中毒治疗，endotoxicosis treatment，45
内基体，狂犬病诊断，Negri bodies，rabies diagnosis，228–229，*228*
内罗毕羊病，Nairobi sheep disease，280
内罗病毒，*Nairovirus*，280
逆转录病毒感染，retrovirus infections，320–323
黏菌素，沙门氏菌病的治疗，colistin，salmonellosis treatment，**97**
黏脓性的分泌物，mucopurulent discharge，glanders，*52*
捻转血矛线虫，*Haemonchus contortus*，**420**
念珠菌病，Candidosis，339–341
念珠菌病的诊断，candidosis diagnosis，*341*，*341*
凝固酶阴性葡萄球菌，staphylococci，coagulase–negative，**133**
凝固酶阴性葡萄球菌，coagulase–negative staphylococci，**133**
牛病毒性腹泻，bovine viral diarrhoea (BVD)，**266**–274
牛病毒性腹泻/黏膜病，bovine viral diarrhoea，**354**
牛分体吸虫，*Schistosoma bovis*，404–405
牛分枝杆菌感染，*Mycobacterium bovis* infection，114，117
牛呼吸道合胞体病毒，bovine respiratory syncytial virus (BRSV)，312
牛囊尾蚴，*Cysticercus bovis*，417
牛疱疹病毒1型，bovine herpesvirus type 1，302–303，**303**
牛歧腔吸虫，*Dicrocoelium hospes*，*401*
牛肾盂炎棒杆菌，干酪性淋巴腺炎，*Clostridium renale* infection，caseous lymphadenitis，**167**
牛瘟，rinderpest，313–316
牛瘟诊断，rinderpest diagnosis，316
牛无浆体，*Anaplasma bovis*，**76**
脓性肉芽肿肺炎，黑曲霉，pyogranulomatous pneumonia，*Aspergillus niger*，*338*
脓肿，abscesses，163
脓肿性肺炎，abscess pneumonia，**128**
诺维氏芽孢梭菌，*Clostridium novyi*，**21**，**353**

**p**

派氏艾美尔球虫，*Eimeria pellerdyi*，**371**
派尔结节，副结核病，Peyer's patches，paratuberculosis，*107*
盘尾丝虫病，onchocercidosis，433–435
盘尾丝虫病的治疗，onchocercidosis treatment，434
盘尾丝虫科，Onchocercidae，433，*433*
疱疹病毒，herpesviruses，300–304
疱疹性口炎，VS (vesicular stomatitis)，260，**279**
胚胎学，embryology，4，*4*
披膜病毒，*Togaviridae*，**280**
皮肤坏死，嗜皮菌病，skin necrosis，dermatophilosis，*182*
皮肤链格孢病，alternariosis，cutaneous，**345**，346
皮肤细菌病，integumental bacterial disease，166–185
皮肤真菌病，dermatophytoses，**354**
皮疽组织胞浆菌感染，干酪性淋巴腺炎，*Histoplasma farciminosum* infection，caseous lymphadenitis，168
皮内试验，副结核病的诊断，intradermal testing，paratuberculosis diagnosis，107
皮下血肿，内毒素中毒，ecchymosis，endotoxicosis，*39*
脾肿大，副结核病，splenomegaly，paratuberculosis，*108*
蜱传热的诊断，tick–borne fever，diagnosis，**81**
蜱的控制，tick infestation control，443
蜱感染，tick infestations，436–445
片形吸虫，*Fasciola*，397
破伤风，tetanus，200–201，**353**
破伤风梭菌，*Clostridium tetani*，**21**
破伤风治疗，tetanus treatment，202
葡萄酒色青霉感染，霉菌性皮炎，*Penicillium vinaceum* infection，mycotic dermatitis，**332**
普诺艾美尔球虫，*Eimeria punoensis*，**371**，*372*，*374*

**q**

歧腔吸虫*Dicrocoelium*，*401*
歧腔吸虫病，dicrocoeliosis，401–406
歧腔吸虫病的治疗，dicrocoeliosis treatment，403
鳍脚亚目动物分枝杆菌，*Mycobacterium pinnipedii*，**117**
气管溃疡，内毒素中毒，tracheal ulcers，endotoxicosis，*39*
气候影响，climate effects，274–275
气性水肿综合征，梭菌感染，gas oedema complex，*Clostridium* infections，22
气肿疽梭菌，*Clostridium chauvoei*，**21**，**353**
禽埃及小体病的诊断，avian aegyptianellosis，diagnosis，**81**
禽分枝杆菌副结核亚种，*Mycobacterium avium* subsp. *paratu-*

*berculosis* (MAP), 105
羟氯硫胺，肝片吸虫病的治疗，oxyclozamide，fasciolosis treatment，400
青霉素，penicillin，71,202,206
青年美洲驼免疫缺陷综合征，juvenile llama immunodeficiency syndrome (JLIDS)，295
琼脂免疫扩散试验，agar gel immunodiffusion test (AGID)，107–109,172,263–264,318,321,337
球孢子菌病，coccidiomycosis，*342*，342
球虫病，Coccidiosis，371–377
球点状斯泰勒绦虫，*Stilesia globipunctata*，406
球鞘毛首线虫，*Trichuris globulosa*，**419**
蛆症金蝇，*Chrysomya bezziana*，*456*，*456–457*，*457*
曲霉，*Aspergillus*，337
曲霉病，aspergillosis，337–338
曲子宫绦虫，*Thysaniezia*，406
圈形盘尾丝虫，*Onchocerca armillata*，**420**
全身性疾病，systemic diseases，21–89
犬，canine，**81**
犬布鲁氏菌，*Brucella canis*，**137**
犬无浆体，*Anaplasma canis*，**76**
犬无浆体病，canine anaplasmosis，**81**
犬小孢子菌，*Microsporum canis*，*335*
犬新孢子虫，*Neospora caninum*，*386*

**r**

扰蝇，nuisance flies，459–464
日本脑炎，Japanese encephalitis，**286**
溶血梭菌，*Clostridium haemolyticus*，**21**
溶血性双球菌，鼻试子，haemolytic diplococci，nasal swabs，**133**
溶血隐秘杆菌，鼻试子，*Arcanobacterium haemolyticum*，nasal swabs，**133**
肉孢子虫病，sarcocystosis，390–394
肉毒梭菌，*Clostridium botulinum*，**21**
肉毒梭菌中毒，botulism，32–34
肉毒梭菌中毒诊断，botulism diagnosis，33
肉品来源，meat source，4
肉品消费，meat consumption，66
肉芽肿，Granulomas，393
肉芽肿性肺炎，granulomatous pneumonia，**128**
蠕虫病，helminthoses，397–430
蠕虫新立克次氏体，*Neorickettsia helminoeca*，**76**
蠕形螨病，demodicosis，454–455，*454*
蠕形螨感染，*Demodex* infection，454–455，*454*
蠕形蚤病，Vermipsyllidosis，469
乳房，udder，186–199
乳房炎加利福尼亚体细胞检测法，California mastitis test (CMT)，190
乳环试验，布鲁氏菌病诊断，milk ring test，brucellosis diagnosis，143
乳胶凝集试验，latex agglutination test (LAT)，384
乳酸中毒，lactic acid syndrome，**38**
乳头类圆线虫，*Strongyloides papillosus*，**419**
乳头状瘤，papillomatosis，251–253
乳头状瘤病毒，papovavirus，**354**
若甲螨，*Zygoribatula*，409

**s**

萨氏钝缘蜱，感染频度，*Ornithodorus savignyi*，frequency of infection，**442**
萨氏毛癣菌感染，*Trichophyton sarkisovii* infection，**332**
噻苯咪唑，寄生虫性胃肠炎治疗，thiabendazole，parasitic gastroenteritis treatment，*429*
噻唑蓝分解实验( MTT )，梭菌肠毒血症，tetrazolium cleavage test，*Clostridium* enterotoxaemia，28,29
赛宾–费德曼染色试验，弓形虫诊断，Sabin–Feldman test，toxoplasmosis diagnosis，384
三磷酸腺苷( ATP )试验，传染性乳房炎，ATP test，infectious mastitis，190
三氯苯咪唑，肝片吸虫病的治疗，triclabendazole，fasciolosis treatment，400
三毛滴虫病，tritrichomonosis，369，*369*
杀昆虫剂，锥虫病的控制，insecticides，trypanosomosis control，363
杀螨药物，螨病治疗，acaricides，mange treatment，*451*
杀锥虫药，trypanocides，**363**
沙林鼠布鲁氏菌，*Brucella neotomae*，**137**
沙门氏菌病，salmonellosis，91–100
沙门氏菌病的诊断，salmonellosis diagnosis，96–97
沙门氏菌病的治疗，salmonellosis treatment，97,**97**
沙门氏菌感染，新生畜腹泻，*Salmonella* infection，neonatal diarrhoea，288
山羊分枝杆菌，*Mycobacterium caprae*，**117**
扇头蜱感染，*Rhicephalus* infection，**442**
舌形虫病，linguatulosis，**470–471**
舌蝇，*Glossina*，467，*467*
蛇形毛圆线虫，*Trichostrongylus colubriformis*，**419**，*424*
申克孢子丝菌感染，霉菌性皮炎，*Sporothrix schenckii* infection，mycotic dermatitis，**332**
神经系统，细菌病，nervous system，bacterial disease，200–210
神秘璃眼蜱，*Hyalomma punt*，440
肾脏钙化，钩端螺旋体病，renal calcifications，leptospirosis，*70*
渗出性湿疹，金黄色葡萄球菌皮炎，exudative eczema，*Staphylococcus aureus* dermatitis，176
生育率，fertility rate，152
虱，**Lice**，463
虱病，pediculosis，463–464
虱病治疗，pediculosis treatment，464
施马伦贝格病毒，Schmallenberg virus (SBV)，324
石膏样小孢子菌，*Microsporum gypseum*，*334*
食入塑料，plastic ingestion，211，*211*
食入致命杂物，fatal ingestion，211，*211*
匙形细颈线虫，*Nematodirus spathinger*，**420**
试图不断地打哈欠，狂犬病，attempted yawning，rabies，*226*，226–227
视检，肉孢子虫39病的诊断，visual inspection，sarcocystosis diagnosis，393
嗜肺巴氏杆菌感染，*Pasteurella pneumotropica* infection，**58**
嗜粒细胞无浆体，*Anaplasma (Ehrlichia) phagocytophilum*，*76*
嗜粒细胞无浆体感染，*Anaplasma phagocytophilum* infection，76，*77*
嗜皮菌病，dermatophilosis，179–185
嗜皮菌病的诊断，dermatophilosis diagnosis，*180*，*182–183*
嗜皮菌病的治疗，dermatophilosis treatment，183

嗜皮菌病媒介，dermatophilosis vector，*180*
嗜人锥蝇，*Cochliomyia hominivorax*，*457–458*
嗜驼璃眼蜱（骆驼蜱），*Hyalomma dromedarii* (camel tick)，*436*，*438*
嗜血杆菌感染，*Haemophilus* infection，lungs，**131**
嗜血支原体感染，*Anaplasma haemolamae* infection，**81**–82
兽疫链球菌，传染性繁殖障碍，*Streptococcus zooepidemicus*，infective reproductive losses，153
舒氏璃眼蜱，*Hyalomma schulzei*，440
鼠疫，plague，65–68
鼠疫媒介，plague vectors，65
鼠疫耶尔森氏菌 参见鼠疫，*Yersinia pestis* infection *see* plague
鼠疫治疗，plague treatment，66
竖琴奥特线虫，*Osteragia lyrata*，**420**
双峰驼，Bactrian camel，254–257，258，**479**，**473**，**476–485**，**488**
双峰驼，*Camelus bactrianus*，**6**，*7*，**479**，**483**，**488**
双峰驼艾美尔球虫，*Eimeria bactriani*，**371**
双甲脒，螨的治疗，amitraz，mange treatment，451
双羟萘酸噻嘧啶，寄生虫性胃肠炎治疗，pyrantel pamoate，parasitic gastroenteritis treatment，*429*
双芽等孢球虫，*Isospora bigemina*，387–388
水分枝杆菌，*Mycobacterium aquae*，**117**
水分枝杆菌解脲亚种，*Mycobacterium aquae* var. *ureolyticum*，**117**
水泡性口炎，vesicular stomatitis (VS)，259，**279**
水杨酰苯胺，肝片吸虫病的治疗，salicylanides，fasciolosis treatment，400
丝状尾网线虫，*Dictyocaulus filaria*，**420**
斯氏副柔线虫，*Parabronema skrjabini*，**420**
斯氏毛首线虫，*Trichuris skjabini*，**419**
斯泰勒绦虫，*Stilesia*，406，*406*
斯特劳斯反应，鼻疽，Strauss reaction，glanders，*53*
四环素，tetracyclines，66,76,82,164,171
四氢尿嘧啶，寄生虫性胃肠炎治疗，tetrahydropyrimidines，parasitic gastroenteritis treatment，*429*
四环素，布鲁氏菌病的治疗，oxytetracycline，brucellosis treatment，145
饲料，真菌毒素检测，feeds，mycotoxin testing，344，**344**
饲喂实验，肉孢子虫病的诊断，feeding experiments，sarcocystosis diagnosis，393
苏拉明，suramin，trypanosomosis treatment，363，**364**
梭菌肠毒血症，*Clostridium* enterotoxaemia，30，*30*
梭菌肠毒血症的诊断，*Clostridium* enterotoxaemia diagnosis，28，*29*
梭菌性疾病，*Clostridium* infections，**21**，21–32，*23*
索戈托病毒，Thogoto virus，324
索氏梭菌，*Clostridium sordelli*，**21**

**t**

胎儿三毛滴虫，传染性繁殖障碍，*Trichomonas fetus*，infective reproductive losses，*153*
胎儿弯曲菌，*Campylobacter fetus* infection，infective reproductive losses，153
胎儿夭折，传染性繁殖障碍，fetal losses，infective reproductive losses，**160**
肽聚糖识别蛋白，peptidoglycan recognition protein (PGRP)，188
泰勒虫病，theileriosis，395–396
瘫痪症状，狂犬病，paralytic syndrome (dumb form)，rabies，227
炭疽治疗，anthrax treatment，35–36
绦虫感染，cestode infections，406–418
蹄感染，foot lesions，209
蹄皮炎，pododermatitis，**209**
体细胞计数，somatic cell content，186
田鼠分枝杆菌，*Mycobacterium microti*，**117**
条状斯泰勒绦虫，*Stilesia vittata*，*406*，*407*
跳跃病，louping ill，**279**，**287**
铜绿假单胞菌，*Pseudomonas aeruginosa*，153，154
铜绿假单胞菌，传染性繁殖障碍，*Pseudomonas aeruginosa*，infective reproductive losses，153，154
酮康唑，念珠菌病的治疗，ketoconazole，candidosis treatment，**340**
头孢菌素，干酪性淋巴腺炎治疗，cephalosporins，caseous lymphadenitis treatment，171
头孢噻呋，沙门氏菌病的治疗，ceftiofur，salmonellosis treatment，**97**
突尾毛圆线虫，*Trichostrongylus probulrus*，**419**，*424*
托里病毒，Dhori virus，325
托曲珠利，艾美尔球虫病的治疗，toltrazuril，*Eimeria* coccidiosis treatment，376
脱毛毛癣菌，*Trichophyton tonurans* infection，**332**
驼马，vicuna (*Vicugna vicugna*)，*5*
驼毛首线虫，*Trichuris cameli*，**419**
驼犬肉孢子虫，*Sarcocystis camelicanis*，**391**
驼细颈线虫，*Nematodirus cameli*，**420**

**w**

网尾线虫病，dictyocaulosis，430–433
微管结节线虫，*Oesophagostomum venulosum*，**420**
微量凝集试验，钩端螺旋体病microscopic agglutination test (MAT)，leptospirosis diagnosis，466
微生物，microbiology，*79*
微小角蝇，*Haematobia minuta*，*465*
微胸虱感染，*Microthoracius* infection，463–465
微血细胞比容离心技术，锥虫病的诊断，mHCT (microhaematocrit centrifugation technique)，trypanosomosis diagnosis，361
韦瑟尔斯布朗病毒，Wesselbron virus，**224**
韦瑟尔斯布朗热，Wesselbron fever，**281**，**325**
伪狂犬病，Aujeszky disease，**325**，326
委内瑞拉马脑炎，Venezuelan equine encephalitis，**279**
卫生质量，传染性乳房炎的治疗，hygiene，infectious mastitis treatment，189
胃肠炎，gastrointestinal parasitosis，429
蚊子，mosquitoes，463
翁迪里无浆体，*Anaplasma ondrii*，*76*
乌干达病毒，Uganda virus，**325**
无孢菌，霉菌性皮炎，*Mycella*，mycotic dermatitis，**332**
无盾璃眼蜱，*Hyalomma impeltatum*，438
无峰驼艾美尔球虫，*Eimeria lamae*，**371**，*372*，*374*
无浆体/支原体的治疗，anaplasmosis/mycoplasmosis treatment，82
无浆体病，anaplasmosis，75–84
无浆体病/支原体病的诊断，anaplasmosis/mycoplasmosis di-

agnosis 80
无卵黄腺绦虫，*Avitellina*，406
无囊线虫，*Impalaia*，423–426
无乳链球菌，传染性乳房炎，*Streptococcus agalactiae*，infectious mastitis，191，*191*
无致病性的病毒病，non–pathogenic viral disease，311–322

**X**

西方马脑炎，Western equine encephalitis，**279**
西方马脑炎马，encephalitis，Western equine，**279**
西尼罗河热，West Nile fever，**279**
西尼罗河热病毒，West Nile virus，**324**
吸虫感染，trematode infections，397–405
吸入性肺炎，aspiration pneumonia，***133***
细胞培养生物分析法，曲霉病，cell culture bioassays，aspergillosis，337
细颈囊尾蚴，*Cysticercus tenuicollis*，*417–418*，*417*
细颈线虫，*Nematodirus*，420–428
细菌病，bacterial diseases，19–210，475–486
细菌分离，bacterial isolation，**157–158**
细菌菌群，bacterial flora，**131**
细菌性皮肤病，bacterial integument disease，**181**
细菌性血红素尿，bacillary haemoglobinuria，24
细菌性血红素尿，梭菌感染，bacillary haemoglobinuria，*Clostridium* infections，24
细毛首线虫，*Trichuris tenuis*，**419**
下颌骨骨髓炎，jaw，osteomyelitis，*209*
下颌骨骨髓炎，mandibular osteomyelitis，*209*
纤毛虫病，infundibuloriosis，*370*，370–371
鲜甲螨，*Cepheus*，409
显微病灶，狂犬病，microscopic lesions，rabies，227
线虫感染，nematode infections，419–420，419–430
线颈细颈线虫，*Nematodirus filicollis*，**420**
腺病毒，*Adenoviridae*，**311**，**312**
腺病毒，文献综述，adenovirus, literature review，**221**
腺鼠疫，bubonic plague，65
消毒，传染性乳房炎的治疗，disinfection，infectious mastitis treatment，192，*192*
消化系统，细菌病，digestive system，bacterial disease，91–112
硝碘酚腈，肝片吸虫病的治疗，nitroxynil，fasciolosis treatment，400
小RNA病毒，picornaviruses，258–259
小孢子菌，*Microsporum vs.*，*331*
小川培养基，结核病的诊断，Ogawa media，tuberculosis diagnosis，121
小反刍兽疫，peste des petits ruminants (PPR)，261–265
小微胸虱，*Microthoracius minor*，**463**
小型阴离子交换离心技术，锥虫病的诊断，mini–anion exchange centrifugation technique (MAECT)，trypanosomosis diagnosis，362
心包积液，金黄色葡萄球菌皮炎，hypo–pericardium，*Staphylococcus aureus* dermatitis，*177*
心内膜下出血，subendocardial haemorrhage，endotoxicosis，*39*
心水病的诊断，heartwater，diagnosis，**81**
辛硫磷，螨的治疗，phoxim，mange treatment，**451**
新孢子虫病，neosporosis，386–389
新孢子虫病的诊断，neosporosis diagnosis，388
新大陆骆驼，NWC，24，253，474，480，485–486，489，**485**，**489**
新霉素，传染性乳房炎的治疗，neomycin，infectious mastitis treatment，*193*
新生畜腹泻，neonatal diarrhoea，289–300
新生畜腹泻的治疗，neonatal diarrhoea treatment，295
新生儿夭折，neonatal losses，**160**
新生儿夭折，传染性繁殖障碍，neonatal losses，infective reproductive losses，**160**
新型隐球菌感染，霉菌性皮炎，*Cryptococcus neoformans* infection，mycotic dermatitis，331
信息素和杀螨剂的设备，蜱控制，pheromone acaricide–containing devices，tick infestationcontrol，443–444
型副流感病毒，PI–3 (parainfluenza virus 3)，3310–311
胸膜屏障，pleural curtain，**113**，*113*
胸膜屏障，肺脏，pleural curtain，lungs，113，*113*
溴化乙锭，锥虫病治疗，homidium bromide，trypanosomosis treatment，**363**
须癣毛癣菌，*Trichophyton mentagrophytes*，*334*
许兰毛癣菌，*Trichophyton schoenleinii*，*335*
癣菌病，ringworm (dermatophytosis)，331，331–333，*335*
血沉棕黄层技术，锥虫病的诊断，BCT (buffy coat technique)，trypanosomosis diagnosis，361
血矛线虫，*Haemonchus*，*427*
血矛线虫病，寄生虫性胃肠炎/结肠炎，haemonchosis，parasitic gastroenteritis/colitis，427
血尿症，钩端螺旋体病，haematuria，leptospirosis，*70*，68–74
血凝试验，干酪性淋巴腺炎，haemagglutination tests，caseous lymphadenitis，172
驯化，domestication，5–6，11–12
牙齿感染，dental infections，208–209
亚急性化脓性子宫内膜炎，subacute suppurative endometritis，**157**
咽型鼠疫，pharyngeal plague，**65**
烟曲霉菌，*Aspergillus fumigatus*，44–45
眼病，eye diseases，209
羊分体吸虫，*Schistosoma matthei*，*404*，404–405
羊副痘病毒，接触传染性脓疱，*Parapoxvirus ovis see* contagious erythema，419
羊黑疫(传染性坏死性肝炎)，梭菌感染，black disease (infectious necrotising hepatitis)，*Clostridium* infections，24
羊口疮，参见 接触传染性脓疱，ORF *see* contagious erythema，
羊毛首线虫，*Trichuris ovis*，**419**
羊无浆体病的诊断，ovine anaplasmosis，diagnosis，**81**
羊斯克里亚宾线虫，*Skrabinema ovis*，**420**
羊驼，alpaca *(Viguna pacos)*，*5*
羊驼，*Viguna pacos*，5
羊驼艾美尔球虫，*Eimeria alpacae*，372，*372*，***373***，374
羊驼肉孢子虫，*Sarcocystis aucheriae*，**391**，392
羊驼细颈线虫，*Nematodirus lamae*，**420**
羊驼纵纹线虫，*Graphinema auchinae*，**420**
羊无浆体，*Anaplasma ovina*，*76*
羊长喙蚤，*Dorcardia ioffi*，469
氧化亚砷酸，蜱的防控，arsenious oxide，tick infestation control，443
痒螨，*Psoroptes*，446–447
咬蝇，biting flies，465–469

药用植物，plants，medicinal，451
药用植物，螨的治疗，medicinal plants，mange treatment，451
伊氏锥虫，*Trypanosoma evansi*，358，*358*
伊维菌素，ivermectin,428,450,464
伊维塔艾美尔球虫，*Eimeria ivitaensis*，**371**，*372*，*374*
伊蚊，*Aedes*，**275**
衣原体病，chlamydiosis，163–165
衣原体病治疗，chlamydiosis treatment，164
衣原体科，Chlamydiaceae，**163**
依普菌素，寄生虫性胃肠炎治疗，eprinomectin，parasitic gastroenteritis treatment，**429**
胰阔盘吸虫，*Eurytrema pancreatum*，404，*404*
遗传学，genetics，10
异物，狂犬病，foreign bodies，rabies，227，*227*
疫苗接种，vaccinations，352–**354**
翼甲螨，*Achipteria*，**409**
阴茎骆驼圆线虫，*Camelostrongylus mentulatus*，**420**
隐孢子虫病，cryptosporidiosis，377–380
隐孢子虫感染，新生畜腹泻，*Cryptosporidium* infection，neonatal diarrhoea，289
隐球菌病，cryptococcosis，**344**
荧光抗体试验，3型副流感病毒，FAT (fluorescence antibody test)，parainfluenza virus 3，310-311
荧光抗体试验，3型副流感病毒，fluorescence antibody test (FAT)，parainfluenza virus 3，310-311
蝇科，Muscidae，459–460
蝇蛆病，myiasis，455–465
硬颈综合征，破伤风，stiff–neck syndrome，tetanus，200–201
硬蜱，生活史，ixodid ticks，life cycle，437–440
幽门螺旋杆菌感染，*Helicobacter pylori* infection，208
疣状毛癣菌，*Trichophyton verrucosum*，*335*
有嵴膨首线虫，*Physocephalus cristatus*，**420**
幼蜱，larva，*439*，
瘀斑出血，petechial haemorrhages，*25*
育种群，breeding stock，141
原虫病，protozoal parasitic disease，358–395
原虫感染，protozoal infections，358–395
原驼，guanaco *(Lama guanaco)*，*5*
原驼犬肉孢子，*Sarcocystis tilopodi*，**391**，392
原线虫病，protostrongylidosis，430–433
云朵污蝇，*Wohlfahrtia nuba*，456，*456*

**z**

杂交，crossbreeding，10，*10*
藻菌病，phycomycosis，**344**，*345*
长柄血矛线虫，*Haemonchus longistipes*，**420**
长刺毛圆线虫，*Trichostrongylus longispiculartis*，**419**
长大卡妙吸虫，*Carmyerious spatiosus*，405
长角亚目，Nematocera，*468*
赭金蝇，*Chrysomya variceps*，457
真菌病，fungal disease，329–349，**487–489**
正痘病毒，*Orthopoxvirus*，**354**
支原体，mycoplasmosis diagnosis，78
支原体病，mycoplasmosis，77–84
支原体病的诊断，mycoplasmosis diagnosis，***80***
枝歧腔吸虫，*Dicrocoelium dendriticum*，*401*
脂肪变性，内毒素中毒，fatty liver degeneration，endotoxicosis，*41*
志贺氏菌，干酪性淋巴腺炎，*Shigella*，caseous lymphadenitis，167
制霉菌素，念珠菌病的治疗，nystatin，candidosis treatment，341
致病性病毒病，pathogenic viral disease，226–310
中点无卵黄腺绦虫，*Avitellina centripunctata*，406，*407*
中东呼吸综合征冠状病毒，Middle East respiratory coronavirus (MERS–CoV)，325
中耳炎，念珠菌病，otitis，candidosis，341
中华歧腔吸虫，*Dicrocoelium chinesis*，387，*398*，*398–399*，*401*，*402*，409，*409*
中欧蜱媒脑炎，Central European tick–borne encephalitis，**280**
中央兽医研究实验室，Central Veterinary Research Laboratory (CVRL)，187
中央无浆体，*Anaplasma centrale*，**76**
重氮氨苯脒乙酰甘氨酸盐，锥虫病治疗，diminazene aceturate，trypanosomosis treatment，**363**
猪布鲁氏菌，*Brucella suis*，**137**
猪小孢子菌感染，霉菌性皮炎，*Microsporum nanum* infection，mycotic dermatitis，**332**
猪支原体病的诊断，porcine mycoplasmosis，diagnosis，**81**
主动寻找宿主，active host finding，*357*
锥虫病，trypanosomosis，358–367
锥虫病的诊断，trypanosomosis diagnosis，**361**，361，361–362
锥虫病卡片凝集试验，card agglutination test for trypanosomosis (CATT)，361–363
子宫内膜，endometrial，153，*153*
子宫内膜肉芽肿，endometrial granulomas，153，*153*
子宫内膜肉芽肿，传染性繁殖障碍，endometrial granulomas，infective reproductive losses，153，*153*
子宫内膜炎，endometritis，**154**
足螨，*Chorioptes*，447
组织胞浆菌病，histoplasmosis，**344**
左旋咪唑，寄生虫性胃肠炎治疗，levamisole，parasitic gastroenteritis treatment，*428*